Practical Functional Anatomy for Spinal Cord Injury Rehabilitation

Hyun-Yoon Ko

Practical Functional Anatomy for Spinal Cord Injury Rehabilitation

A Guide for Physiatrists and Rehabilitation Specialists

Springer

Hyun-Yoon Ko
Parkside Rehabilitation Hospital
Busan, Korea (Republic of)

ISBN 978-981-95-6456-9 ISBN 978-981-95-6457-6 (eBook)
https://doi.org/10.1007/978-981-95-6457-6

This Springer imprint is published by the registered company Springer Nature Singapore Pte Ltd.
The registered company address is: 152 Beach Road, #21-01/04 Gateway East, Singapore 189721, Singapore

To my beloved parents, whose boundless love and unwavering faith have illuminated my path with a light that never dims.

To my distinguished mentors, and with deepest gratitude to Professor John Ditunno (1932–2024), who guided me into the profound world of spinal cord medicine. Your wisdom continues to be the compass by which I navigate.

To my patients, who have entrusted me with their most vulnerable moments and become my greatest teachers. Your stories and strength have transformed the way I understand both medicine and humanity.

And to my beloved Insun, whose love transcends the ordinary bounds of partnership. Your grace, patience, and understanding have been the quiet force behind every word written here. This work, like my life, is infinitely richer for having you by my side.

Preface

The relationship between anatomical structure and function forms the cornerstone of rehabilitation medicine. This connection takes on special significance in spinal cord medicine, where understanding extends far beyond the musculoskeletal system to include the body's complete networks of function. Unlike medical specialties focusing on specific organ systems, spinal cord medicine requires comprehensive knowledge of human anatomy—from kinesiology and neuroanatomy to the complex workings of the autonomic nervous system and the management of bladder, bowel, sexual, and respiratory functions.

Throughout years of specializing in spinal cord medicine and teaching medical students and residents, a consistent observation has emerged: a solid foundation in functional anatomy and neuroanatomy of relevant organ systems is essential for developing strong clinical capabilities. This observation, combined with extensive clinical experience, has guided the organization and content of this book. The anatomical concepts presented here have been carefully selected based on their direct relevance to understanding spinal cord injury and providing effective clinical care.

Practical Functional Anatomy for Spinal Cord Injury Rehabilitation: A Guide for Physiatrists and Rehabilitation Specialists emerges from decades of clinical experience and numerous interactions with colleagues, students, residents, and most importantly, patients. While rehabilitation medicine is fundamentally rooted in functional anatomy, there is a clear need for resources that connect anatomical knowledge with clinical application in spinal cord medicine. The book contains 13 carefully structured chapters, each examining essential anatomical systems and their clinical roles in spinal cord injury rehabilitation. Beginning with a comprehensive review of spinal cord neuroanatomy, the text progresses through crucial topics including motor unit function, autonomic nervous system organization, and the anatomical foundations of various body systems.

The content spans from the examination of skeletal muscle, skin, and fascia to the study of vascular anatomy and respiratory systems, providing rehabilitation specialists with the depth of anatomical knowledge needed for excellence in clinical practice. Special attention has been given to the functional anatomy of the extremities, urinary tract, and gastrointestinal system, as well as the neuroanatomical basis of essential functions like swallowing and spasticity management. Each chapter has been designed not only to convey theoretical knowledge but also to create meaningful connections between anatomical understanding and practical clinical application—a bridge that has proven particularly valuable in teaching medical students and residents.

As I write this preface, I reflect on my mentor, Professor John Ditunno (1932–2024), who passed away this year. During my year-long fellowship with my wife at Thomas Jefferson University College of Medicine in Philadelphia, his extraordinary generosity and intellectual guidance shaped both my professional and personal life. Professor Ditunno opened my eyes to new perspectives in spinal cord medicine and remained a mentor and friend throughout the years. His wisdom, humanity, and dedication to spinal cord medicine will stay with me always, and I extend my deepest condolences to his family, who showed such warmth during our time in Philadelphia.

The knowledge presented in these pages represents a synthesis of academic foundations with practical clinical application, merging current evidence-based practices with fundamental anatomical principles. Each chapter reflects not only established medical knowledge but also insights gained through years of clinical practice and teaching experience, making the content particularly relevant for those working in rehabilitation settings.

This book is offered as a comprehensive resource for all professionals in spinal cord injury rehabilitation, from experienced practitioners to those beginning their journey in this field. I hope these pages will prove especially valuable in clinical settings, supporting healthcare professionals in their daily practice and contributing to improved patient care outcomes. By providing a deeper understanding of functional anatomy, this work seeks to enhance the quality of care delivered to patients with spinal cord injuries, ultimately leading to better outcomes and improved quality of life for those we serve.

Deep gratitude is extended to the many colleagues, students, and most importantly, patients who have contributed to this understanding of spinal cord medicine over the years. Their insights, challenges, and successes have helped shape the practical approach presented here. Special appreciation goes to the medical students and residents whose thoughtful questions have helped refine these teaching concepts, and to the authors and publishers who have generously granted permission to reproduce important figures that enhance the educational value of this work.

Busan, South Korea
December 2024

Hyun-Yoon Ko

Competing Interests The author has no competing interests to declare that are relevant to the content of this manuscript.

Contents

1 Functional Neuroanatomy of the Spinal Cord: A Comprehensive Review of the Structure, Organization, and Clinical Implications

The spinal cord exhibits a distinct anatomical differentiation consisting of three meningeal layers: the dura mater, a strong fibrous outer layer; the arachnoid membrane, involving trabeculae that traverse the subarachnoid space; and the pia mater, a delicate vascular membrane closely attached to the spinal cord's neural tissue (Standring 2020). The exterior topography displays a ventral median fissure, approximately 3 mm deep, and a dorsal median sulcus of lesser depth. The midline depressions divide the spinal cord into symmetrical halves, each displaying anterolateral and posterolateral sulci that indicate the emerging locations of nerve rootlets (Vanderah and Gould 2016).

The structural stability of the spinal cord relies on specialized ligamentous components derived from the pia mater. The paired denticulate ligaments, consisting of 20 or 21 triangular extensions on each side, project laterally through the subarachnoid space. These ligaments originate from the pia mater between successive spinal nerve roots and insert into the dura mater, forming a suspensory system that extends from the foramen magnum to the conus medullaris at the inferior aspect of the first lumbar vertebra (L1) (Bican et al. 2013; Watson et al. 2009). The septum posticum, a thin membranous extension of the pia mater, spans from the dorsal surface of the spinal cord to the arachnoid membrane, providing additional structural support (Standring 2020).

The filum terminale is a specialized fibrous extension consisting of two separate segments. The filum terminale internum, approximately 8 cm in length, contains glial cells, ependymal remnants, and small blood vessels, extending from the conus medullaris to its dural penetration at the second sacral vertebra (S2). The filum terminale externum, roughly 8–10 cm long, continues as a purely connective tissue structure to its coccygeal attachment (Vanderah and Gould 2016).

The internal architecture exhibits precise laminar organization. The white matter contains myelinated and unmyelinated axons arranged in specific tracts: the ventral funiculus between the ventral median fissure and anterolateral sulcus, the lateral funiculus between the anterolateral and posterolateral sulci, and the dorsal funiculus between the posterolateral and dorsal median sulci (Kierszenbaum and Tres 2016). Each funiculus contains multiple ascending and descending pathways with specific functional roles in sensory perception and motor control. The white matter volume demonstrates a craniocaudal reduction pattern, reflecting the progressive decrease in ascending and descending fiber numbers (Standring 2020).

The gray matter, organized according to Rexed laminae (I–X), contains neuronal cell bodies arranged in functional columns. The ventral horns (laminae VII–IX) contain alpha (α)- and γ-motoneurons, while the dorsal horns

H.-Y. Ko, *Practical Functional Anatomy for Spinal Cord Injury Rehabilitation*,
https://doi.org/10.1007/978-981-95-6457-6_1

(laminae I–VI) process sensory information. The lateral horn (intermediolateral cell column), present from T1 to L2/L3 segments, contains preganglionic sympathetic neurons, while parasympathetic preganglionic neurons reside in the lateral gray matter of S2–S4 segments (Watson et al. 2009). The diameter of the central canal of the human spinal cord varies depending on the age, location within the spinal cord, and individual differences. Generally, it ranges from 0.1 to 0.2 mm in diameter during early life. The central canal maintains continuity with the fourth ventricle and contains cerebrospinal fluid (CSF), surrounded by a specialized region known as the substantia gelatinosa centralis (lamina X) (Vanderah and Gould 2016).

The spinal cord's vascular supply demonstrates a complex pattern. The anterior spinal artery supplies the ventral two-thirds of the cord, while paired dorsal spinal arteries supply the dorsal third. These vessels receive reinforcement through radicular arteries, with the artery of Adamkiewicz (great ventral radicular artery) providing significant supplementary flow to the thoracolumbar region.

For reference, in neuroanatomical terms of orientation, the expression of orientation can become confusing in upright vertebrates like humans. In most cases, such as with the spinal cord, "anterior" and "ventral" point in the same direction, allowing terms like "ventral or anterior" and "dorsal or posterior" to be used interchangeably. Additionally, directions can be expressed as "superior (rostral or cranial)" and "inferior (caudal)." However, in the case of the brain, the orientation is expressed differently: "anterior or rostral," "posterior or caudal," "superior or dorsal," and "inferior or ventral."

1.1 Overview of the Anatomy of the Spinal Cord

The spinal cord, a vital component of the central nervous system, extends from the cervicomedullary junction to the conus medullaris and is responsible for motor, sensory, and autonomic innervations of the trunk, extremities, neck, and internal organs (Standring 2020). Despite its relative structural and functional simplicity compared to the brain and its comprising only about 2% of the human nervous system, spinal cord injuries can lead to significant neurological deficits due to the dense and compact arrangement of its structures (Cho 2015). The topography and cytoarchitecture of the human spinal cord are well known, but the functional implications of well-described structures are difficult to explain.

Anatomically, the spinal cord is part of the central nervous system, extending from the cervicomedullary junction to the tip of the conus medullaris. The rostral limit of the conus medullaris lacks a distinct anatomical reference point. The region of the spinal cord immediately rostral to the conus medullaris is referred to as the epiconus. Usually, this region includes spinal cord segments ranging from L4 to S2 (Baehr and Frotscher 2012; Toribatake et al. 1997) (Fig. 1.1). Clinically, the spinal cord also includes the cauda equina, a bundle of lumbar and sacral nerve roots below the conus medullaris in the spinal canal (Barson 1970). The spinal cord's length ranges from 41 to 43 cm in females to 45 cm in males, with a width of approximately 1.27 cm in the cervical and lumbar regions, 64 mm in the thoracic region, and an average weight of about 30 g (Bican et al. 2013). Although there is no significant correlation between spinal cord length and height, taller individuals tend to have a higher conus medullaris.

The spinal cord exhibits enlarged regions in the cervical and lumbar areas, which contain neurons responsible for innervating the upper and lower limbs, respectively (Barson 1970). Extending from the foramen magnum to the sacral hiatus at the level of the fourth sacral vertebra, the spinal canal houses a variety of neural and nonneural structures. During embryonic development, up to 14 weeks postconception, the spinal cord spans the entire length of the spinal canal, with spinal nerves exiting the vertebral column via corresponding intervertebral foramina. As the vertebral column undergoes subsequent growth and lengthening, the spinal cord exhibits relatively slower growth, a process termed ascensus medullae.

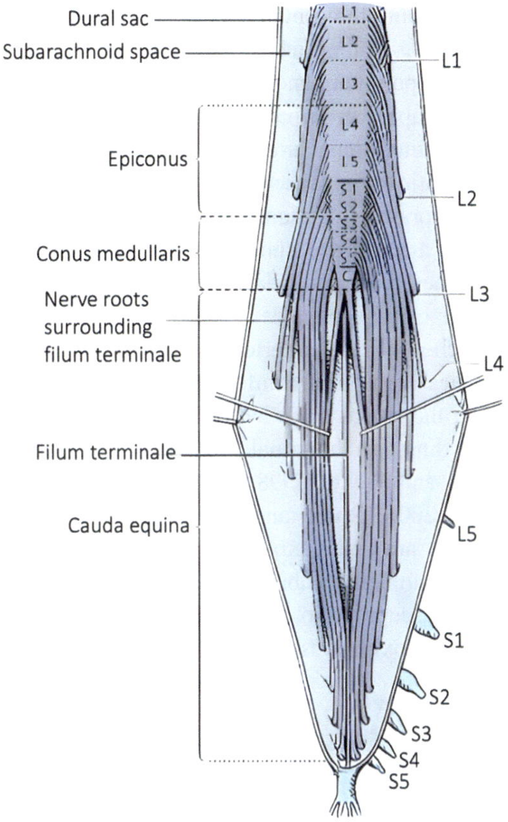

Fig. 1.1 Diagram illustrating the anatomical regions of the epiconus, conus medullaris, and cauda equina, depicting their structure and spatial relationships. (Adapted with permission from Baehr and Frotscher (2012))

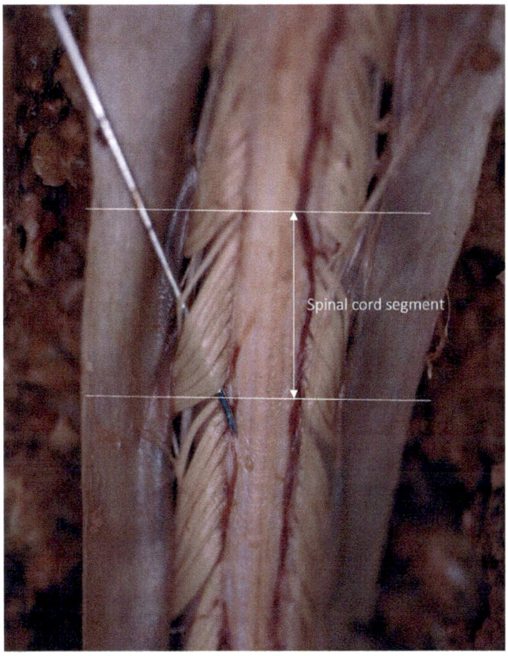

Fig. 1.2 Spinal cord segment. A spinal cord segment is refers to the portion of the spinal cord that gives rise to a single pair of spinal nerves. It consists of gray and white matter associated with the rootlets (filaments) of one pair of dorsal and ventral roots, which converge to form a spinal nerve

In adults, the spinal cord extends from the foramen magnum to the inferior aspect of the first lumbar vertebra or the interspace between the first and second lumbar vertebrae, terminating as the conus medullaris (Barson 1970). The orientation of the spinal rootlets varies along the length of the spinal cord, with cervical and upper thoracic rootlets positioned perpendicularly to the cord, while lower thoracic, lumbar, and sacral rootlets assume an increasingly oblique orientation. This variation is attributed to the differential growth rates between the spinal cord and vertebral column. Consequently, a discrepancy exists between the spinal cord segments and their corresponding vertebral levels, particularly in regions caudal to the upper cervical segments. This mismatch becomes progressively more evident in the more caudal segments of the spinal cord, as the vertebral column experiences a faster growth rate compared to the spinal cord during development (Müller and O'Rahilly 2014).

A spinal cord segment is defined as the portion of the spinal cord spanning between the attachment sites of the ventral and dorsal root filaments (rootlets) for each nerve root (Fig. 1.2). The ventral rootlets, often numbering no more than eight, converge to form a single ventral root, which contains the axons of motor and, in some locations, sympathetic neurons. The dorsal rootlets, usually numbering no more than eight, unite and enter the small enlargement known as the dorsal root ganglion, which contains the unipolar nerve cell bodies. There are 31 segments in total, extending from the first cervical (C1) to the fifth sacral (S5) segment, each with distinct quantitative characteristics. Although each spinal cord segment weighs approximately 1 gram, their lengths vary across different regions: around

25 mm in the mid-thoracic region, approximately 12 mm in the mid-cervical region, and about 10 mm in the mid-lumbar region (Fig. 1.3). Segment T6 is the longest of all segments, with an average length of 22.4 mm. Within the cervical and lumbar regions, the longest segments are C5 and L1, respectively, both averaging 15.5 mm in length.

The anterodorsal diameter of each segment is relatively consistent, ranging from 6 to 8 mm. However, the lateral diameters vary, generally decreasing as the segment length increases. Notably, the cervical and lumbar enlargements exhibit greater diameters compared to other spinal cord regions. These enlargements are quantitatively characterized primarily by their increased lateral diameters (Ko et al. 2004).

The spinal cord's gray matter, located centrally, comprises the sensory dorsal and motor ventral horns, which contain nerve cell bodies. Surrounding the gray matter is the white matter, consisting of long motor and sensory tracts distributed peripherally. The spinal cord receives its blood supply from one ventral and two dorsal longitudinal spinal arteries (Turnbull et al. 1966). Originating from the vertebral artery at the level of the foramen magnum, the ventral and posterior spinal arteries are supplemented by additional branches from radicular arteries that enter the spinal canal alongside the nerve roots. These branches of the anterior and posterior spinal arteries form a circumferential network encircling the spinal cord, with penetrating branches extending into the spinal cord from the periphery (Cole and Weller 1998; Etz et al. 2011; Novy et al. 2006; Rubin and Rabinstein 2013). The arteries are fully exposed to the cerebrospinal fluid within the subarachnoid space, rendering them susceptible to damage from tumors, infections, or toxins present in this space.

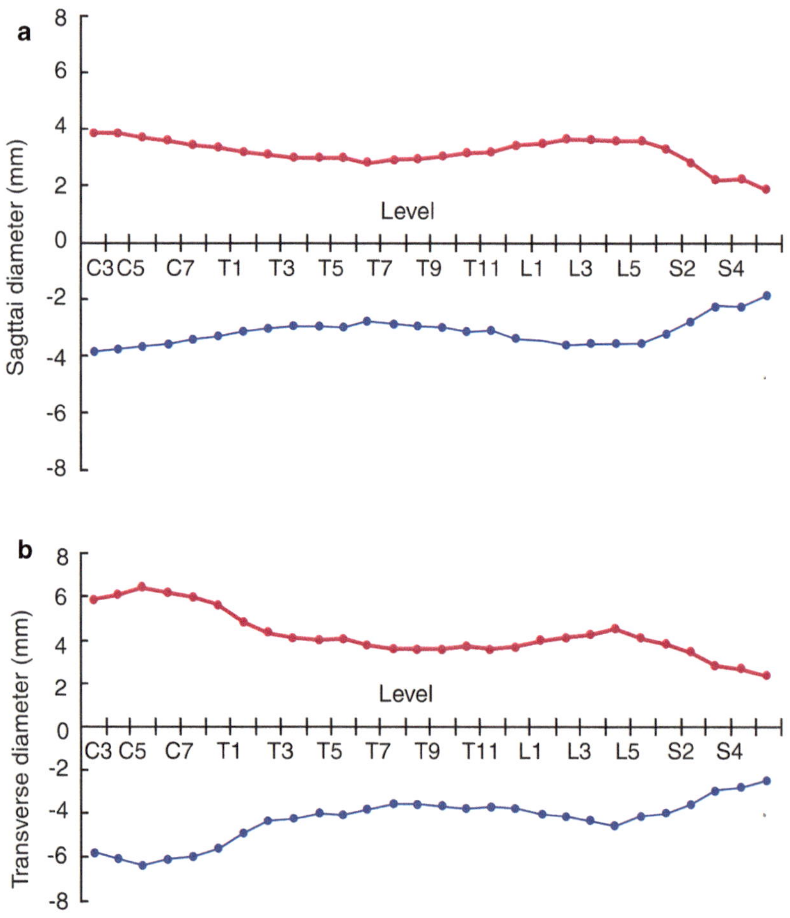

Fig. 1.3 (**a**) Sagittal diameter and (**b**) lateral diameter of each spinal cord segment, illustrating the variations in dimensions along its length. (Adapted with permission from Ko et al. (2004))

The primary descending motor pathway, located in the lateral part of the spinal cord, projects into the ventral horn, where it synapses with segmental motor nuclei. Lower motor neurons, both α- and γ-types, extend from these motor nuclei via the ventral roots to innervate extrafusal and intrafusal muscle fibers (Norenberg et al. 2004). Clinical examination of motor function in the spinal cord predominantly focuses on the pyramidal tract. Other motor pathways, associated with the extrapyramidal system, are involved in more proximal, automatic movements. The complexity of the relationships between descending tracts and motor function is not typically of clinical or physiological importance in spinal cord pathology (Norenberg et al. 2004).

Sensory afferent neurons, with cell bodies located in the dorsal (posterior) root ganglia, send their proximal axons into the spinal cord via the dorsal roots. Large myelinated sensory fibers, which convey low-threshold cutaneous mechanoreceptor, muscle, and joint information, traverse the dorsal horn and ascend ipsilaterally in the dorsal columns to synapse in the dorsal column nuclei (Bican et al. 2013). Conversely, small myelinated and unmyelinated fibers, which transmit sensory information related to temperature, pain, and muscle fatigue, synapse in the dorsal horn laminae before ascending contralaterally via the spinothalamic tracts. These fibers cross at or near the segmental level in the ventral part of the spinal cord near the central canal. This anatomical arrangement makes them vulnerable to pressure effects from a syrinx, potentially leading to classical dissociated sensory loss (Bican et al. 2013; Kim et al. 2015).

1.2 Developmental Segmentation of the Spinal Cord

In developmental and embryological contexts, metamerism denotes the segmentation of the body into repeated units called metameres or segments, including dermatomes, myotomes, and sclerotomes. The spinal cord is basically a segmental structure, with each of the 31 pairs of spinal nerves serving a metamere or body segment that originates from an embryonic somite (Standring 2020).

The embryological development initiates approximately 25 days postconception when somites emerge as paired structures longitudinally on either side of the neural groove. This process begins with four occipital somite pairs, advancing in a rostrocaudal sequence. During the subsequent 10-day period, additional segments form systematically, establishing eight cervical, 12 thoracic, five lumbar, five sacral, and three to six coccygeal segments (Sadler 2012). The precise timing and molecular signaling that control this segmentation involve multiple morphogens and transcription factors, including Hox genes, which determine segment-specific identities (Schoenwolf et al. 2015).

Each somite undergoes differentiation into two specific components. The ventromedial sclerotome develops into vertebral elements, while the dorsolateral dermomyotome generates both muscular components (myotome) and cutaneous tissues (dermatome) (Fig. 1.4). This developmental pattern establishes the foundation for sensory and motor organization in the mature spinal cord (Albright et al. 2008). The dermatomal arrangement, representing cutaneous areas innervated by specific dorsal root ganglia, demonstrates both precision and adaptability in sensory distribution (Lee et al. 2008; Watson et al. 2010).

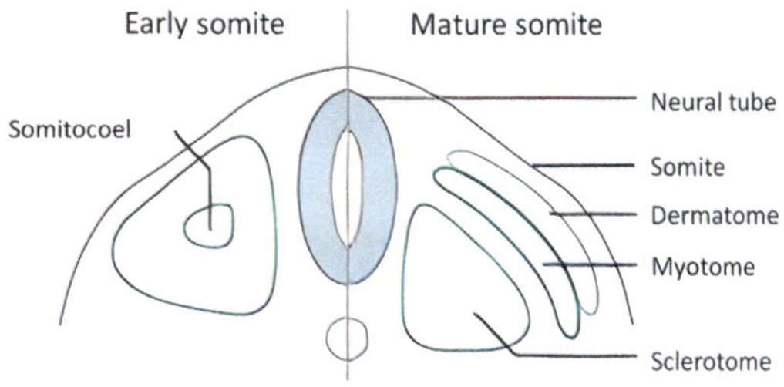

Fig. 1.4 Condensation of cells around the somitocoel differentiates into dermamyotomes and sclerotomes. The dermamyotome subsequently divides into dermatome and myotome, with myotomal cells migrating into the limb bud. (Adapted with permission from Webster and de Wreede (2016))

Clinical assessment of dermatomes reveals variable degrees of overlap between adjacent segments. This overlap varies by anatomical region: thoracic dermatomes exhibit relatively clear boundaries, while cervical and lumbar regions show more extensive overlap (Fig. 1.5). This arrangement provides redundancy in sensory coverage, enhancing the resilience of the sensory system. Biomechanical studies have shown that this segmental organization also influences the mechanical properties and function of spinal structures (Jaumard et al. 2011).

Motor innervation patterns follow segmental organization but exhibit more complex arrangements than sensory distribution (Heimer 1983). The relationship between muscle location and innervation reflects embryological migration patterns of myotomes. Most muscles maintain innervation from spinal segments corresponding to their developmental origin, even after migration to distant locations (Standring 2020). This principle aids in understanding clinical presentations of spinal cord pathologies affecting motor function.

The developmental basis of spinal cord segmentation has direct implications for clinical practice in neurology and rehabilitation medicine. The segmental organization extends to spinal cord microstructure and vascular supply patterns, influencing both normal function and pathological conditions.

1.2.1 Dermatomal Development

Initial limb buds project perpendicularly from the trunk, exhibiting distinct ventral and dorsal surfaces, alongside cephalic (preaxial) and caudal (postaxial) borders (Palastanga and Soames 2012; Standring 2020). The origin of these patterns is thought to derive from an early segmental arrangement in limb buds, mirroring the trunk's segmental configuration, which later reorganizes to match the adult form.

During embryogenesis, dermatomes initially display an orderly arrangement. However, the limb outgrowth along the ventral axial line distorts this pattern, positioning the C4 dermatome above T2 at the sternal angle level. Subsequently, the C5 dermatome is found adjacent to the T1 dermatome (Vanderah and Gould 2016) (Fig. 1.6). In vertebrate development, limb rotation is a key feature, with the upper and lower limbs rotating in opposite directions, the upper limb laterally and the lower limb medially (Fig. 1.7). Cervical dermatomes maintain an orderly distribution in the upper extremity with minimal rotation. This results in the anatomical positioning of the thumb as the lateral digit of the hand and the great toe as the medial digit of the foot, which is important for understanding the functional orientation of the limbs (Moore et al. 2018).

Initially, each limb bud extends perpendicularly to the body surface, featuring ventral and dorsal surfaces, and cephalic (preaxial) and caudal (postaxial) borders (Sadler 2012). As the limb elongates and differentiates, it folds ventrally, resulting in the ventral surface becoming medial. During this process, the convexities of the elbow and knee are directed laterally (Moore et al. 2018) (Fig. 1.7). Subsequently, the upper and lower limbs undergo opposite rotations. This rotational process results in the knee's convexity toward the cranial end of the body, contributing to the final anatomical orientation of the limbs (Schoenwolf et al. 2015). By the end of the eighth week of embryonic development, the medial rotation of the lower limb reverses the preaxial and postaxial borders, forming a spiral dermatome arrangement (Fig. 1.6). Consequently, the great toe and tibia are positioned medially, while the little toe and fibula are positioned laterally. Hence, the tibial border corresponds to the original preaxial border, and the fibular border corresponds to the postaxial border of the lower limb. The great toe receives innervation from a more rostral dermatome (L4) compared to the little toe (S1). The lower extremity extends from the trunk, with the most caudal dermatomes innervating the perineum rather than the foot (Felten et al. 2016).

Dermatomes, defined as segments of sensory innervation in the skin, correlate with the sensory fibers of specific spinal nerves. Dermatome maps have significant clinical utility, aiding in the

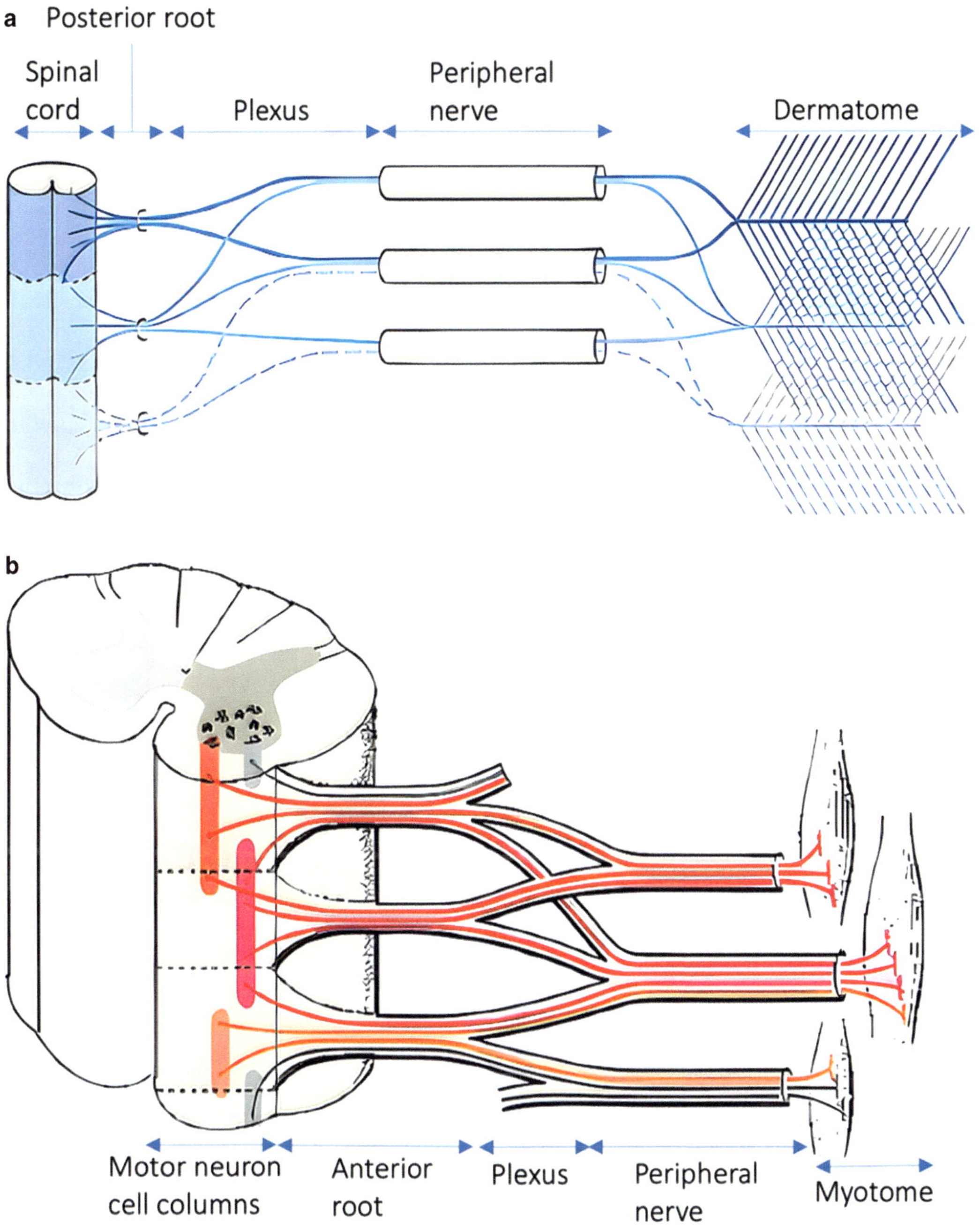

Fig. 1.5 Redistribution of afferent and efferent nerve fibers to the dermatome and myotome. (**a**) Sensory fibers from a single spinal cord segment are reorganized to innervate specific segmental regions of the skin (dermatome). (**b**) Each muscle is innervated by a single peripheral nerve derived from multiple anterior roots or spinal cord segments (myotome). (Adapted with permission from Baehr and Frotscher (2012))

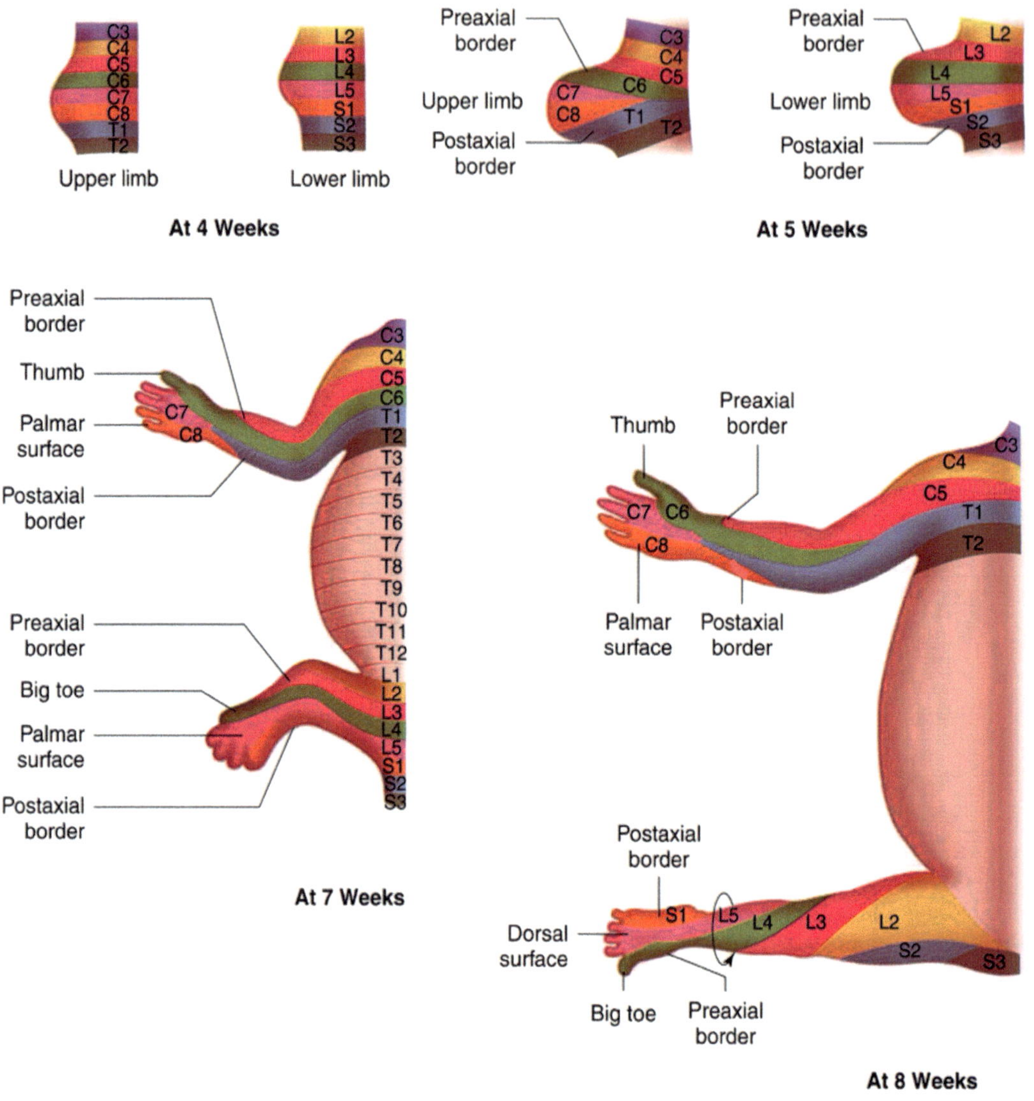

Fig. 1.6 Ventral dermatome pattern changes during limb development with rotation. Lower limb rotation reverses the preaxial and postaxial borders, resulting in a spiral dermatome arrangement. Spinal nerve segments on the ventral surface of the lower extremity extend medially and inferiorly, with the great toe innervated by a more rostral dermatome (L4) compared to the little toe (S1). The lower extremity develops as an extension of the trunk, with the most caudal dermatomes (sacral and coccygeal) innervating the perineum rather than the foot. Cervical dermatomes exhibit minimal rotation, maintaining a relatively orderly distribution in the upper extremity. (Adapted with permission from Felten et al. (2016))

diagnosis of neuropathological conditions affecting spinal nerves or the spinal cord (Heimer 1983). It is essential to consider the substantial overlap among adjacent dermatomes, which affects the clinical assessment of sensory deficits (Netter 2019).

1.2.2 Myotome Development in Limb Formation

The ventral rami entering the developing limb bud traverse ventrally to the myotomes, subsequently bifurcating into dorsal and ventral divisions.

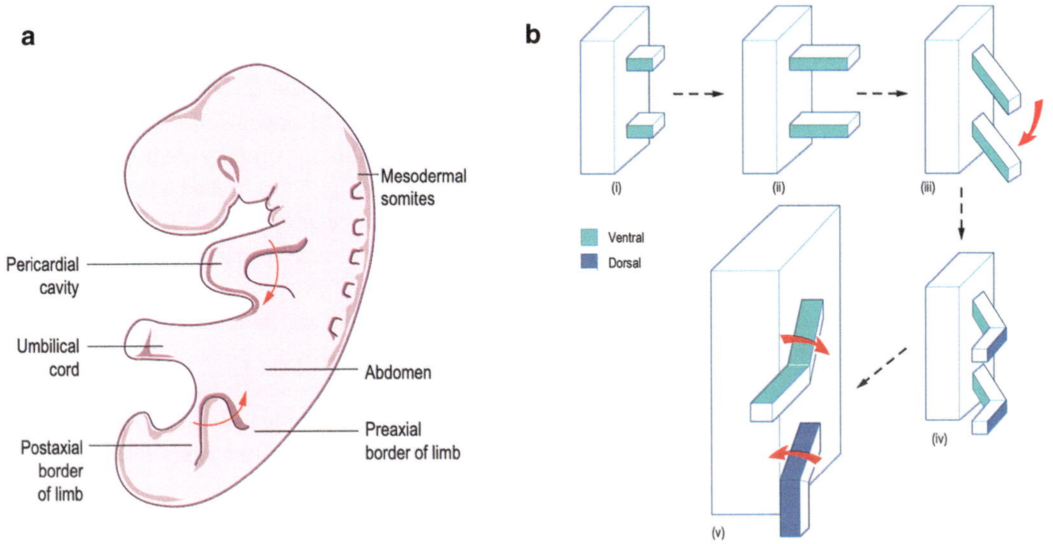

Fig. 1.7 (**a**) Limb buds indicating the direction of their future rotation and (**b**) the resulting rotation of the limbs. (Adapted and modified with permission from Palastanga and Soames (2012))

These divisions merge with corresponding segments from the adjacent ventral rami, forming the nerves that innervate the dorsal (extensors) and ventral (flexors) aspects of the limb bud (Sadler 2012; Schoenwolf et al. 2015). Prior to the rotational events that establish the adult limb configuration, these nerves supply the muscle sheets and overlying skin on both surfaces of the limbs before they rotate into their adult positions. Throughout the folding, rotation, and migration of muscle masses, the nerve–muscle connections are preserved (O'Rahilly and Müller 2001).

During the development of the lower limb, rotational changes result in the extensor and flexor surfaces being oriented ventrally and dorsally, respectively. This reorientation affects the innervation patterns. The muscles on the ventral aspect of the thigh and leg are innervated by nerves originating from the dorsal division of the lumbar and lumbosacral plexuses, specifically the femoral and common peroneal nerves. In contrast, the muscles on the dorsal aspect of the thigh and leg, as well as those in the sole of the foot, are innervated by the tibial nerve, which arises from the ventral division of the lumbosacral plexus (Drake et al. 2020; Netter 2019; Standring 2020).

1.2.2.1 Myotome Formation and Innervation Patterns of the Upper Extremity

The development of the upper limb begins with the development of the limb bud, containing primitive muscle tissue. During the process of growing, the periosteum of the humerus forms intermuscular septa that divide the arm into ventral and dorsal compartments, which resemble the arrangement found in adults. Similar to the lower limb, specific muscles in the ventral compartment separate into an adductor group. However, this muscular mass has undergone phylogenetic regression in the upper limb, with only the coracobrachialis remaining. The powerful adduction of the upper limb is facilitated by significant muscles, such as the latissimus dorsi dorsally and pectoralis major ventrally, which have migrated into the limb. In the forearm, the radius (preaxial bone) and ulna (postaxial bone) are connected by an interosseous membrane and intermuscular septa attaching to the investing fascia. These septa form compartments that contain muscles with similar or related functions (Cole et al. 2009; Palastanga and Soames 2012).

The segmented innervation of the upper limb muscles mirrors the developmental compartmentalization of the brachial plexus. Preaxial ventral compartment muscles, located in the same anatomical plane, receive motor innervation from the lateral and medial cords, which originate from the ventral divisions of the brachial plexus nerve trunks. In contrast, the postaxial dorsal compartment muscles are innervated by the dorsal cord, derived from the dorsal divisions (Moore et al. 2018).

These pre- and postaxial divisions are also evident in the nerves; the median, musculocutaneous, and ulnar nerves provide preaxial control, while the radial nerve innervates the postaxial musculature below the shoulder girdle. Within the pectoral girdle, the embryologically distinct clavicle and coracoid process undergo intramembranous ossification as ventral preaxial elements, giving rise to preaxially innervated musculature. The postaxial scapula undergoes endochondral ossification, with its muscles receiving postaxial innervation. Consequently, the segmented origins of the brachial plexus nerves, reflecting the primitive serial morphology of the embryo, continue postnatally through the distinct pre- and postaxial innervations of the upper limb muscles and skeletal elements (Schoenwolf et al. 2015).

As compared to the sensory innervation pattern in the upper extremity, the motor innervation pattern is organized in a sequential manner: The C5 nerve allows shoulder movements, C6 influences the elbow and partially the forearm, C7 controls wrist movements, and C8, along with C7, innervates the fingers and thumb. This distribution extends from the C5 nerve proximally to the T1 nerve distally, including the intrinsic movements of the hand. Unlike the lower limb, the upper boundary of the upper limb is not as distinctly demarcated. Although muscle attachments extend to the head, neck, and thorax, the upper limit of the upper limb can be pragmatically defined as the superior surface of the clavicle ventrally and the superior border of the scapula dorsally. This description offers a concise understanding of the nerve distribution and upper limit demarcation in the upper limb, as detailed in the context of anatomy and human movement (Drake et al. 2020; Palastanga and Soames 2012).

1.2.2.2 Myotome Formation and Innervation Patterns of the Lower Extremity

The developmental organization of motor innervation in the lower extremity demonstrates systematic patterns that establish functional neuromuscular relationships. While the segmental arrangement may appear less defined than in the trunk, there exists a functional correlation between deep muscle innervation and cutaneous sensory supply. Muscles positioned beneath specific dermatomes typically receive motor innervation from the same spinal segments that provide sensory input to the overlying skin (Heimer 1983; Standring 2020).

The embryological development of lower extremity muscles progresses through multiple stages. Myogenic precursor cells migrate from somites into the developing limb bud, where they undergo proliferation and differentiation. These cells form muscle masses that receive innervation from specific spinal segments through the lumbosacral plexus (Sadler 2012). The distribution of motor neurons follows patterns determined by molecular signals and morphogens that guide axonal growth and synapse formation.

The anatomical arrangement of lower limb muscles exhibits unique characteristics. Many muscles cross multiple joints, creating a biomechanical system capable of producing complex movements. For example, the rectus femoris spans both hip and knee joints, while the gastrocnemius influences both knee and ankle function. This multi-joint arrangement necessitates sophisticated neural control mechanisms for coordinated movement (Schoenwolf et al. 2015). The motor control system includes primary movers, synergists, and stabilizers, all receiving specific innervation patterns. Synergistic muscles often share spinal segmental innervation, reflecting their functional relationships. Stabilizer muscles, particularly those controlling postural adjustments, demonstrate continuous activation patterns regulated by spinal circuits and supraspinal inputs (Watson et al. 2010).

Clinical assessment of lower extremity muscle function requires an understanding of these complex relationships. Neurological examination must consider both the segmental innervation and the functional interdependence of muscle groups. For instance, weakness in proximal muscles may affect the function of more distal segments through biomechanical chain reactions (Albright et al. 2008). The evaluation of muscle strength and coordination must account for these interactions to provide accurate clinical assessments (Palastanga and Soames 2012; Sisto et al. 2008).

1.3 Composition of Spinal Cord White Matter

Many axons within the nervous system are covered by a myelin sheath, giving them a white, shiny appearance. This feature leads to the term "white matter," which mainly consists of these myelinated axon bundles. In contrast, "gray matter" gets its name from its gray color due to the abundance of neuronal cell bodies and a dense network of unmyelinated nerve fibers.

In the brain, gray matter is found predominantly in two main areas: the cerebral cortex, forming the brain's outer layer, and several large subcortical nuclei, such as the caudate nucleus, thalamus, and lentiform nucleus. Conversely, in the spinal cord, gray matter is centrally located and forms a distinct H-shaped pattern in cross-sectional views. This central gray matter is surrounded by the white matter, which includes extensive ascending and descending neuronal pathways (Heimer 1983).

The white matter, positioned on the spinal cord's periphery, encompasses the gray matter and includes ascending and descending tracts. These tracts are composed of nerve fibers originating from the dorsal root ganglia, spinal gray matter neurons, and higher-level neurons in the brain. Additionally, the white matter contains short-range intrasegmental fibers connecting neurons within the same spinal segment and intersegmental fibers that ascend or descend a few segments before entering the gray matter (Critchley and Eisen 1997). Besides nerve fibers, the white matter also contains neuroglial cells and blood vessels. The glial cells present include oligodendrocytes, astrocytes, and microglia.

The development of ascending neurons in humans begins at around 10 weeks of gestation (Clowry et al. 2005). By 13 weeks, the lateral corticospinal tract reaches the caudal medulla oblongata, and the pyramidal decussation is completed 2 weeks later (ten Donkelaar et al. 2004; Armand 1982). The lateral corticospinal tract extends to the cervical spinal cord between 14 and 16 weeks of gestation. The corticospinal tract subsequently invades the lower thoracic spinal cord by 17 weeks and the lumbosacral spinal cord by 27 weeks (Armand 1982) (Fig. 1.8). The maturation of the spinal motor center is closely related to neuronal activity, following an activity-dependent pattern.

The volume of white matter in the spinal cord increases in the higher segments, with the cervical cord region exhibiting the most substantial development. This volume diminishes progressively at the lower spinal levels (Breig and el-Nadi 1966; Breig et al. 1966; Holmes et al. 1996; Kameyama et al. 1996) (Fig. 1.9). The cervical spinal cord segments have a higher concentration of white matter due to the convergence of descending neurons from the brain and ascending neurons passing through this region to the brain. Consequently, the sacral spinal cord, situated at the lowest level, contains the least amount of white matter since it lacks most ascending and descending fibers. Conversely, the cervical and lumbar enlargements show increased gray matter to accommodate the neurons of the upper and lower limbs, respectively.

In the spinal cord, the ventral sulcus is a significant landmark within the white matter. This sulcus does not penetrate as deeply as the gray matter, which permits nerve fibers to cross from one side of the spinal cord to the other through the ventral white commissure, situated ventrally to the central canal. The ventral commissure is much more massive than the dorsal commissure, which is composed only of a small group of myelinated fibers (Parent 1996). Contrarily, on the dorsal aspect of the spinal cord, there is no true sulcus. Instead, the dorsal median septum extends centrally and aligns with the depth of the gray matter. This anatomical arrangement results in the absence of a dorsal white commissure, as fibers do not cross at this level.

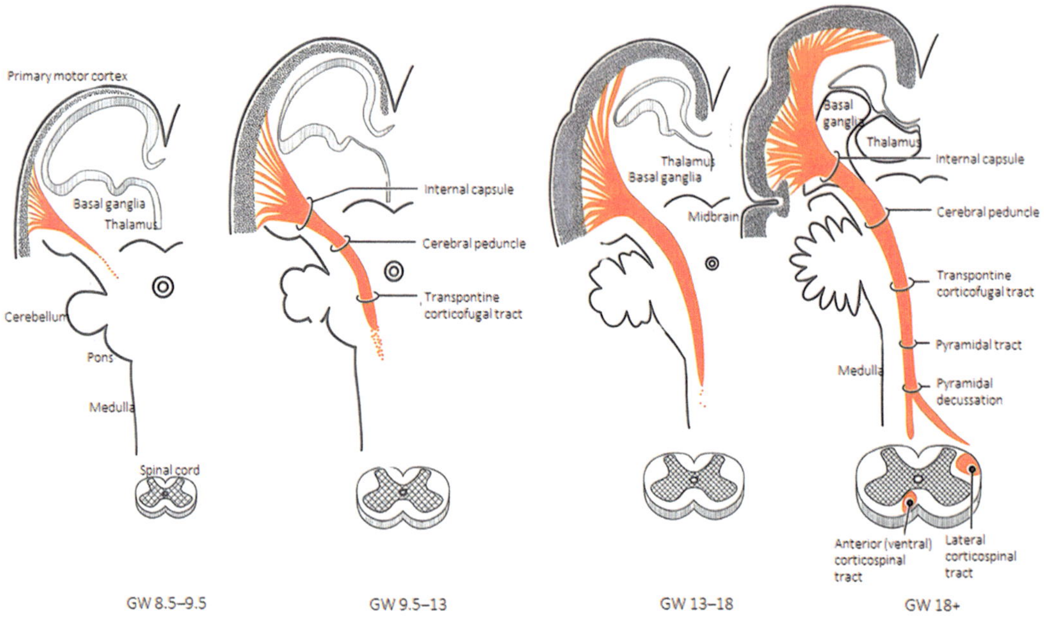

Fig. 1.8 Diagram illustrating the development and lengthening of the human corticospinal tract. Between gestational weeks (GW) 8.5 and GW 9.5, the earliest descending corticofugal fibers penetrate the basal ganglia, forming the internal capsule. The corticospinal tract volume increases significantly between GW 13 and GW 18. At about GW 18, or shortly thereafter, the corticospinal tract fibers within the pyramids split into two descending components: the larger contralaterally projecting lateral corticospinal tract and the smaller ipsilaterally projecting ventral corticospinal tract. (Adapted with permission from Altman and Bayer (2001))

Fig. 1.9 Cross-sectional mean area and white and gray matter areas of each segment of the spinal cord. (Adapted with permission from Kameyama et al. (1996))

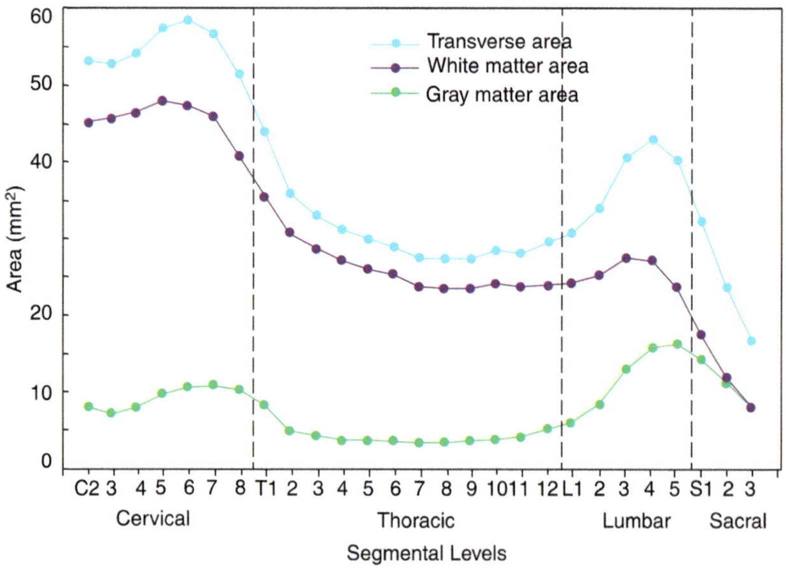

The white matter in the spinal cord is segmented into ventral, dorsal, and lateral funiculi by the gray matter's ventral and dorsal horns and by the ventral and dorsal sulci (Nolte and Angevine 2013) (Fig. 1.10). These funiculi house ascending and descending tracts (or fasciculi), which are categorized into functional groups. Cho (2015) provides a detailed overview of the primary ascending and descending tracts (Fig. 1.11; Table 1.1). The ascending pathways include (1) the dorsal column–medial lemniscal system, which conveys sensory information on vibration and proprioception via the fasciculus cuneatus and gracilis; (2) the anterolateral system, which transmits nociceptive, thermoreceptive, and tactile information through the ventral and lateral spinothalamic tracts and the spinoreticular tract; and (3) the cerebellar input system, which conveys proprioceptive information from the limbs through the ventral and dorsal spinocerebellar tracts (Bican et al. 2013; Patestas and Gartner 2006).

Moreover, the spinal cord comprises association pathways that connect various segments within the cord. Known as the fasciculi proprii, these pathways are present in all three funiculi and are typically found adjacent to the gray matter. Lissauer's tract, situated between the dorsal horn and the spinal cord surface, primarily serves as a propriospinal pathway (Heimer 1983).

The descending pathways are categorized into two main systems: the lateral motor system and the medial motor system (Patestas and Gartner 2006) (Fig. 1.11). The lateral motor system, comprising the lateral corticospinal tract and rubrospinal tract, governs the movement of

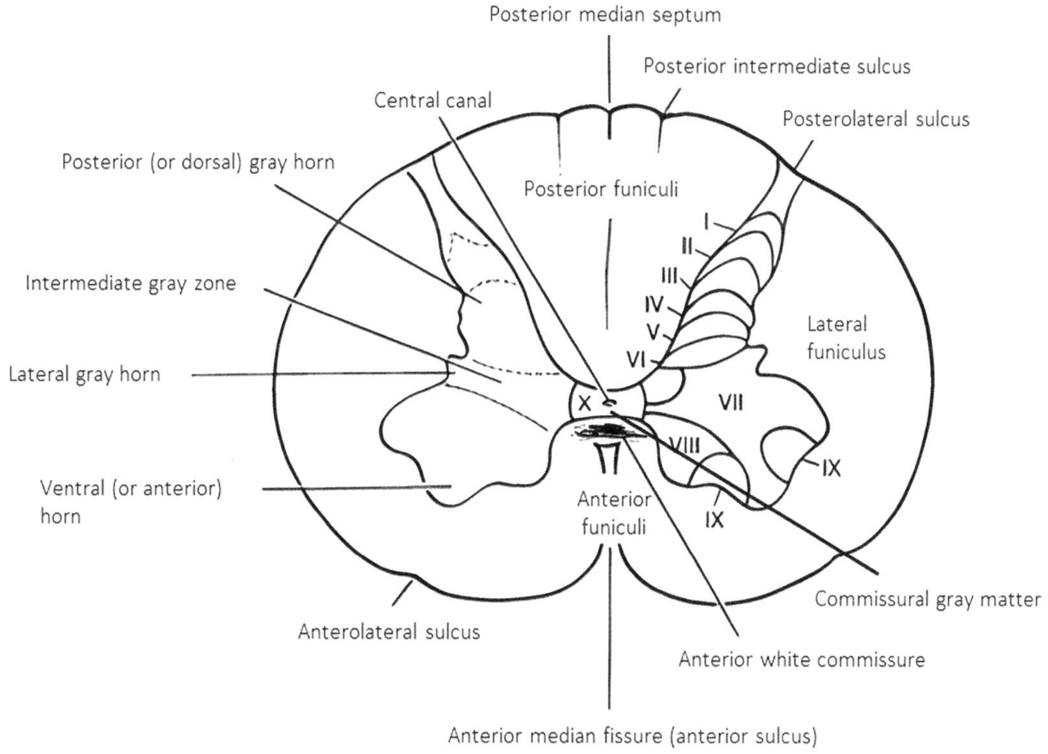

Fig. 1.10 Subdivisions of the white matter and gray matter in the spinal cord. The white matter is organized into three funiculi, while the gray matter is divided into two horns (ventral and dorsal horns) and an intermediate zone. Rexed laminae within the gray matter are depicted on the right. (Adapted with permission from Strominger et al. (2012))

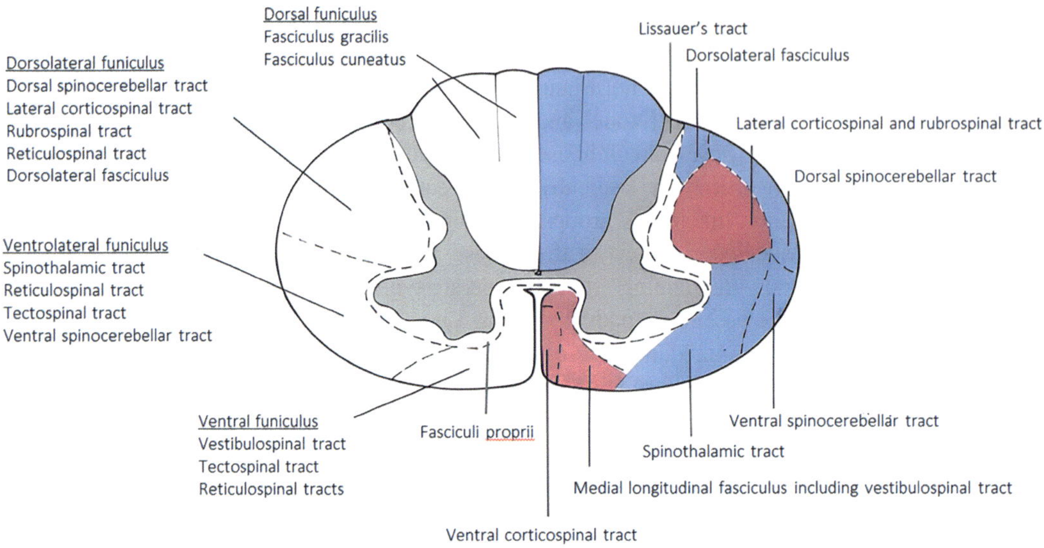

Fig. 1.11 Ascending and descending tracts of the spinal cord, categorized by their corresponding to functional subgroups within the white matter

contralateral limbs. Conversely, the medial motor system, which includes the ventral corticospinal tract, the medial and lateral vestibulospinal tracts, the reticulospinal tract, and the tectospinal tract, is responsible for controlling bilateral trunk muscles, head and neck positioning, balance, and other posture- and gait-related functions (Patestas and Gartner 2006).

In embryonic development, the earliest nerve fibers forming tracts tend to be covered by subsequent fibers of similar origin, resulting in a laminated structure within individual fiber tracts (Blumenfeld 2010). Ascending fibers from lower segments overlap the preexisting fibers at higher levels, creating a distinct laminar arrangement in the cervical spinal cord. In this arrangement, axons originating from the cervical region are situated deeper than those from the thoracic, lumbar, and sacral regions, with the latter occupying the most superficial position (Fig. 1.12). This lamination pattern is less commonly observed in tracts consisting of short ascending and descending fibers, such as the fasciculi proprius. Long descending tracts in the spinal cord also exhibit lamination, with fibers growing from the brain toward the spinal cord surface. The earliest descending fibers are embedded in the white matter by fibers of later origin. Within the larger lateral corticospinal tracts, fibers terminating at the cervical cord levels are located most medially, while those projecting to the lumbosacral levels are positioned most laterally (Armand 1982; Moore et al. 2018).

1.3.1 Fasciculi Proprii and Interneurons

The white matter surrounding the gray matter in the spinal cord predominantly contains propriospinal fibers, referred to as fasciculi proprii (Fig. 1.13). The fasciculus proprius is a region of white matter situated immediately adjacent to the gray matter. The structure is composed of both descending and ascending axons of propriospinal neurons. These intersegmental fibers contribute to facilitate the connectivity between various spinal segments. The fasciculi proprii, which are association pathways within the spinal cord, enable intersegmental communication. These pathways are found in the ventral and lateral funiculi, typically adjacent to the gray matter (Heimer 1983).

The gray matter of the spinal cord exhibits a complex cellular composition, containing neurons, neuroglia, blood vessels, and numerous

Table 1.1 Summary of the primary ascending (a) and descending (b) tracts of the spinal cord, including their functional roles

Tract	Origin	Termination	Crossed/uncrossed	Main function
(a) *Ascending tracts*				
Fasciculus gracilis	Dorsal root ganglia below T6	Nucleus gracilis (medulla)	U	Proprioception and vibration
Fasciculus cuneatus	Dorsal root ganglia above T6	Nucleus cuneatus (medulla)	U	Proprioception and vibration
Posterior spinocerebellar	Clarke's nucleus	Cerebellum	U	Proprioceptive, pressure and touch input to the cerebellum
Anterior spinocerebellar	Anterior horn especially lumbosacral	Cerebellum	C	Proprioceptive, pressure and and touch input to the cerebellum
Lateral spinothalamic	Rexed laminae I, III, IV, and V	Thalamus (ventral posterior nucleus)	C	Pain and temperature
Anterior spinothalamic	Rexed laminae I, III, IV, and V	Thalamus (ventral posterior nucleus)	C	Light touch and pressure
(b) *Descending tracts*				
Pyramidal system				
Lateral corticospinal	Layer V neurons in the contralateral motor cortex (major)	Medial and lateral motor nuclei at all levels	C (85–90%)	Skilled movements mediated by distal limb muscles
Anterior corticospinal	Layer V neurons in contralateral motor cortex	Medial and lateral motor nuclei in cervical and upper thoracic levels	U	Descending motor input to motor neurons that innervate neck musculature
Extrapyramidal system				
Rubrospinal	Contralateral Red nucleus (midbrain)	Spinal neurons primarily at cervical levels	C	Posture and locomotion primarily flexor activities (small tract in man)
Reticulospinal	Bilateral reticular formation (pons and medulla)	Spinal neurons for trunk and proximal limb musculature	C and U	Posture and locomotion, respiration, modulation of pain, vasomotor tone
Tectospinal	Superior and inferior colliculi	Motor neurons for neck muscles	C	Reflexic head movements toward visual and auditory stimuli
Vestibulospinal	Ipsilateral vestibular nuclei (medulla)	Spinal neurons for trunk musculature	U	Posture and locomotion primarily extensor activities
Central autonomic tract				
Sympathetic	Hypothalamus and brain stem nuclei	Preganglionic sympathetic neurons in intermediolateral nucleus (T1–L2)	C and U	Descending sympathetic outflow to entire body
Parasympathetic	Hypothalamus and brain stem nuclei	Preganglionic parasympathetic neurons in the intermediolateral nucleus (S2–S4)	C and U	Parasympathetic supply to the distal colon, rectum, bladder, and sexual organs

Adapted with permission from Critchley and Eisen (1997)

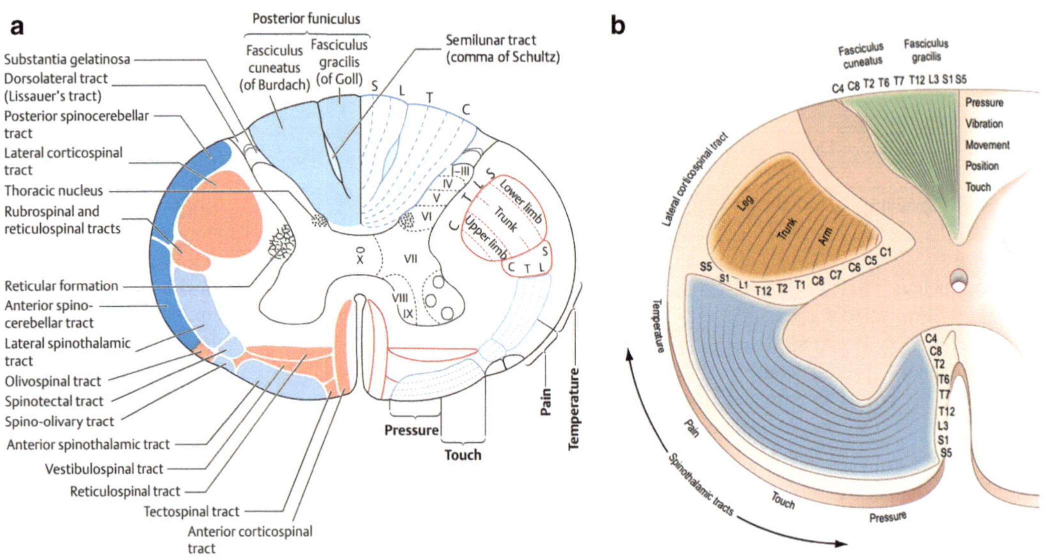

Fig. 1.12 Somatotopic organization of spinal cord tracts. (**a**) Rexed laminae in the gray matter are labeled with Roman numerals. (**b**) The ascending and descending tracts, except for the dorsal funiculus, exhibit a similar somatotopic organization. (Adapted with permission from Baehr and Frotscher (2012) and Standring (2020))

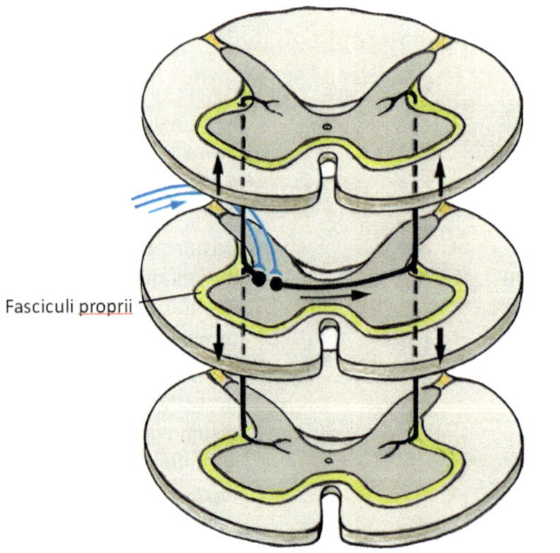

Fig. 1.13 Fasciculi proprii. The fasciculi proprii are composed of ascending and descending fibers that form association pathways, connecting different spinal cord segments. These fibers are located immediately adjacent to the gray matter. (Adapted with permission from Baehr and Frotscher (2012))

nerve fibers—both myelinated and unmyelinated axons. Within this structure, interneurons represent the predominant neuronal population, similar to other regions of the central nervous system (Watson et al. 2009). These cells, also known as association or intercalated neurons, perform integrative functions that coordinate various neural circuits (Fig. 1.14).

Spinal cord interneurons display distinct anatomical patterns based on their axonal projections. Three main categories have been identified: intrasegmental interneurons with short-to-medium-length axons confined within the same segment; intersegmental interneurons that extend into adjacent segments; and commissural interneurons that cross the spinal cord midline (Critchley and Eisen 1997; Flynn et al. 2011). The term "propriospinal neurons" specifically refers to neurons connecting different spinal cord segments. These interneuronal networks form components of various reflex circuits and intrinsic neural pathways. They function as intermediaries between multiple neural systems, including long descending pathways and spinal motor-sensory mechanisms, as well as between peripheral afferent fibers and neurons forming ascending pathways (Heimer 1983). This arrangement allows for complex information processing and integration within the spinal cord.

The nucleus proprius, a cell column situated in the head of the dorsal horn of the gray matter (Rexed laminae III and IV), contains nerve cells

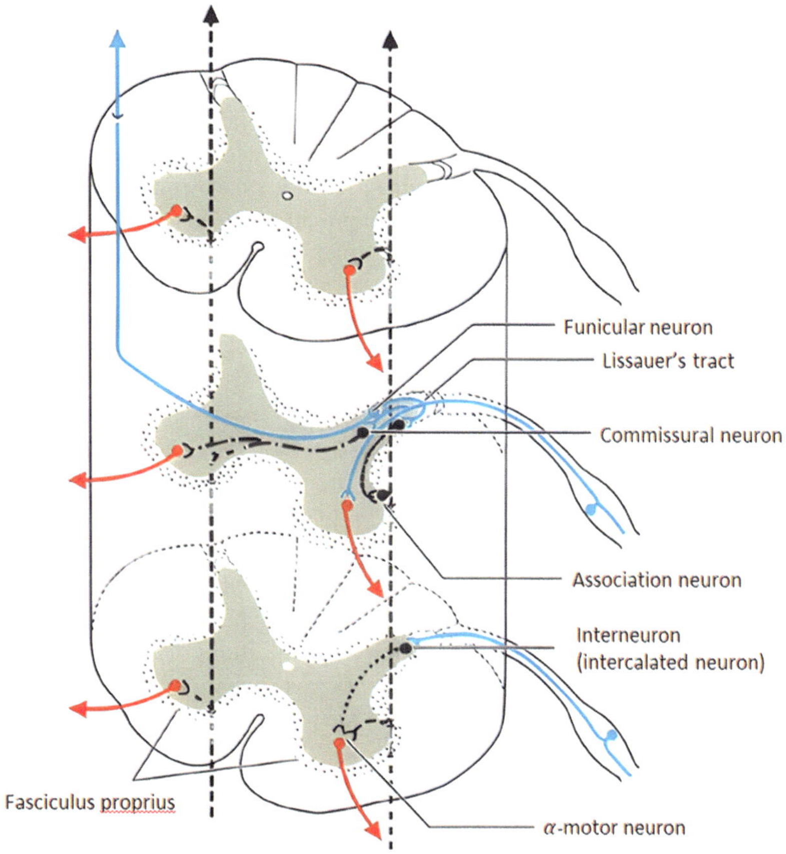

Fig. 1.14 Components of the intrinsic neurons and their connections. Interneurons, including those with T-shaped axonal branching, transmit afferent signals to both higher and lower spinal cord segments through crossed and uncrossed pathways. These interneurons are also referred to as intercalated or internuncial neurons. (Adapted with permission from Baehr and Frotscher (2012))

of various sizes with diverse anatomical connections and functional properties. Some of these neurons contribute to the long ascending spinothalamic pathways that project to the thalamus, while others at the base of the dorsal horn give rise to spinocerebellar pathways. Additionally, some neurons are part of the propriospinal system, and numerous small neurons function as interneurons in spinal reflexes or as targets for descending brain pathways (Heimer 1983; Standring 2020).

1.3.2 Ascending Tracts and Sensory Information Transmission

Ascending tracts in the spinal cord are responsible for conveying sensory information to higher relay nuclei within the main sensory pathway, such as the thalamus, or providing proprioceptive sensory input to the cerebellum. The sensory fibers that are transmitting the action potential from the peripheral receptors are referred to as first-order neurons. These neurons are the first step in a network of neurons that eventually transmit to a higher center, such as the cerebral cortex for the conscious awareness of the stimulus or the cerebellar cortex for the control of motor patterns (Cramer and Darby 2014). Typically, these tracts consist of second-order sensory afferents originating from neurons in the gray matter, primarily within the dorsal horn. When sensory information is consciously perceived, the second-order neuron decussates and then connects with a third-order neuron in the thalamus. An exception to this arrangement is found in the dorsal columns, where the fasciculus cuneatus and fasciculus gracilis are composed of primary sensory afferents from the dorsal root ganglia. These primary afferents enter the spinal cord and ascend directly to the medulla without synapsing in the dorsal horn (Critchley and Eisen 1997).

Typically, an ascending tract can be categorized into one of three groups according to the three distinct roles given by the somatosensory system. One group consists of tracts that provide information on the type, location, and stimulation intensity. These tracts are classified as direct or discriminatory. Another category includes tracts that provide information related to the activation of reflexes that are potentially associated with arousal, affective, and motivational responses to a stimulus. The third category of tracts consists of those that provide information related to the unconscious monitoring and control of motor activity, such as maintaining posture and executing movements (Benarroch et al. 1999; Kandel et al. 2013).

The details of the specific type of information conveyed by the ascending tract, the site of crossing of the tract, and the termination site are presented in Table 1.1.

1.3.2.1 Anterolateral System

Small-diameter A-delta (Aδ) fibers and unmyelinated C fibers, which convey nociceptive and thermoreceptive signals, penetrate the spinal cord at the dorsal horns. Upon entering the spinal cord, some primary sensory fibers bifurcate, allowing them to ascend and descend for one or two segments within the Lissauer's tract (fasciculus) before they penetrate the gray matter. Positioned between the posterolateral surface of the spinal cord and the dorsal horn's tip, Lissauer's tract is proximal to the substantia gelatinosa. This tract appears lighter in myelin-stained preparations compared to other white matter areas because it contains finely myelinated and unmyelinated fibers associated with the substantia gelatinosa (Vanderah and Gould 2016). Functionally, Lissauer's tract serves mainly as a propriospinal pathway, situated between the dorsal horn and the spinal cord's external surface (Heimer 1983).

After traveling through Lissauer's tract, these first-order fibers synapse in the dorsal horn. The second-order neurons, primarily located in laminae I and V of the gray matter's dorsal horn, receive these signals. Following synapse formation, the second-order fibers cross the ventral commissure, positioned ventral to the central canal, and, subsequently, ascend within the contralateral spinothalamic tract located in the anterolateral white matter (Afifi and Bergman 2005; Bican et al. 2013).

The ventral spinothalamic tract's axons traverse the ventral funiculus of the spinal cord, whereas the lateral spinothalamic tract's axons navigate both the ventral and lateral funiculi. Collectively known as the anterolateral system, these tracts are responsible for different sensory modalities: The ventral spinothalamic tract transmits light (crude) touch and pressure sensations, while the lateral spinothalamic tract conveys pain, temperature, tickle, itch, and sexual sensations. The spinothalamic tract comprises fibers that decussate in the ventral white commissure, which is a true commissure facilitating communication between the spinal cord's right and left halves. It should not be confused with the ventral commissure located in the telencephalon, connecting the temporal lobes and olfactory nuclei. Similarly, the dorsal white commissure should not be mistaken for the dorsal commissure, located in the diencephalon (Schuenke et al. 2020).

Regarding the ventral spinothalamic tract, the axons of the first-order neurons initially bifurcate, forming a T-shaped pattern (Fig. 1.15). These axons descend 1–2 segments and ascend 2–15 segments within the spinal cord before synapsing with second-order neurons in the dorsal horn. The second-order neuron axons then cross via the ventral white commissure and ascend within the contralateral ventral funiculus.

For the lateral spinothalamic tract, the axons of the first-order neurons immediately synapse with second-order neurons upon entering the gray matter at the same level of spinal cord entry. The axons of these second-order neurons cross the midline in the ventral white commissure and ascend within the contralateral lateral funiculus. Understanding these pathways is essential, par-

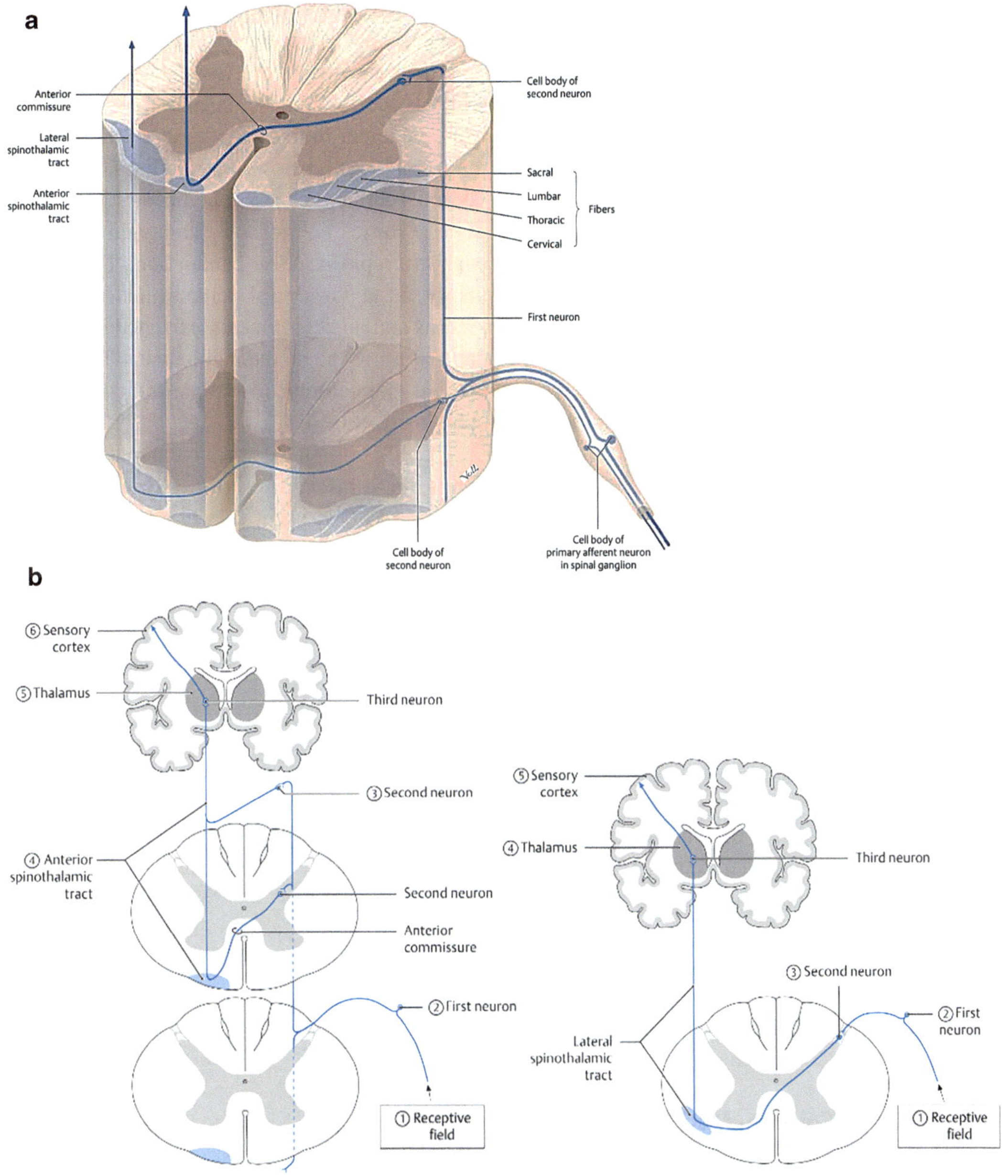

Fig. 1.15 (**a**) Pathway of the anterior and lateral spinothalamic tracts in a transverse spinal cord section. (**b**) Anterior and lateral spinothalamic tracts and their central connections. (Adapted from Schuenke et al. (2020))

ticularly in the clinical assessment of patients with Brown-Séquard syndrome (Schuenke et al. 2020).

Lesions affecting the spinothalamic tracts result in diminished or complete loss of sensation to various stimuli, including pain, temperature, and crude touch stimuli. Given the anatomical proximity of the ventral and lateral spinothalamic tracts, isolated lesions involving only one tract are exceedingly rare. Such lesions typically produce contralateral sensory deficits relative to the lesion site, corresponding to the same side as the dorsal root ganglion, which houses the first-order neuron cell bodies. Consequently, a lesion involving the left first-order neuron (part of the peripheral nervous system) or the left second-order neuron (part of the central nervous system) prior to decussation will cause a deficit on the left side of the body. Conversely, damage to the right third- or fourth-order neurons (both within the central nervous system) will result in sensory deficits on the left side (Schuenke et al. 2020).

Both the ventral and lateral spinothalamic tracts, which are crossed pathways, originate from neurons in the principal sensory nucleus (laminae III and IV) and from some neurons in laminae I and V of the dorsal horn. These tracts ascend ipsilaterally for one or two segments before crossing to the contralateral side. Cervical fibers are located medially within the tract, while sacral fibers are situated laterally (Afifi and Bergman 2005; Mai and Paxinos 2011). The lateral spinothalamic tract is responsible for transmitting pain and temperature sensations, along with tickling, itching, sexual sensations, and muscle fatigue. The ventral spinothalamic tract conveys light touch and pressure.

Dorsal root fibers can ascend several segments in the dorsal columns (Lissauer's tract) before synapsing in the dorsal horn. Despite a lesion in the spinothalamic system on one side, touch and pressure sensations are often retained bilaterally. Small or early lesions of the spinal commissures, such as in syringomyelia, can abolish pain and temperature sensations on both sides without affecting tactile perception, as pain and temperature pathways are located more dorsally compared to those for touch sensation.

1.3.2.2 Dorsal Column–Medial Lemniscal System

In the cervical and upper thoracic regions, the dorsal intermediate sulci separate into the fasciculi gracilis and cuneatus. The fasciculus gracilis ("slender fasciculus") and fasciculus cuneatus ("wedge-shaped fasciculus") represent two significant ascending tracts within the dorsal funiculus. These tracts are responsible for carrying fibers related to position sense (conscious proprioception) and fine cutaneous sensation, which includes touch, vibration, fine pressure sense, and two-point discrimination, to the telencephalon (Watson and Harrison 2012). The fasciculus gracilis transmits fibers from the lower limbs, whereas the fasciculus cuneatus transmits fibers from the upper limbs, and, therefore, is absent in the spinal cord below the T3 level. The first neuron's cell bodies reside in the dorsal root ganglion. These fibers are heavily myelinated, enabling rapid impulse conduction. They travel uncrossed to the dorsal column nuclei (nucleus gracilis and nucleus cuneatus) located in the caudal portion of the medulla oblongata. The fasciculi are somatotopically organized (Schuenke et al. 2020) (Fig. 1.16).

The topographical arrangement of fibers within these fasciculi follows a specific pattern, with sacral fibers occupying the innermost position, succeeded by lumbar, thoracic, and cervical fibers (Smith and Deacon 1984). Furthermore,

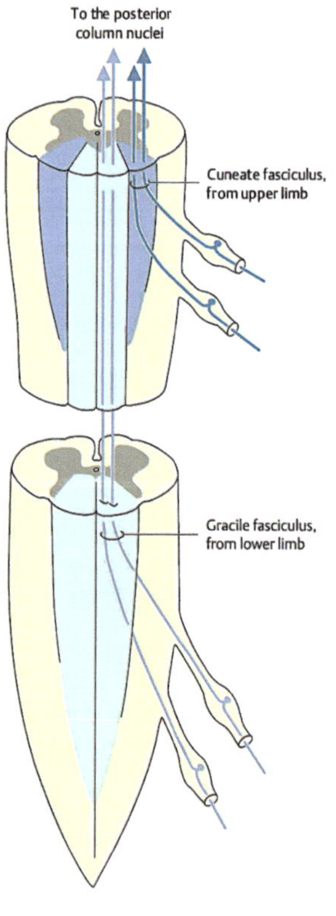

Fig. 1.16 Posterior funiculus, illustrating the posterior column, including the fasciculus gracilis and fasciculus cuneatus. (Adapted with permission from Baehr and Frotscher (2012))

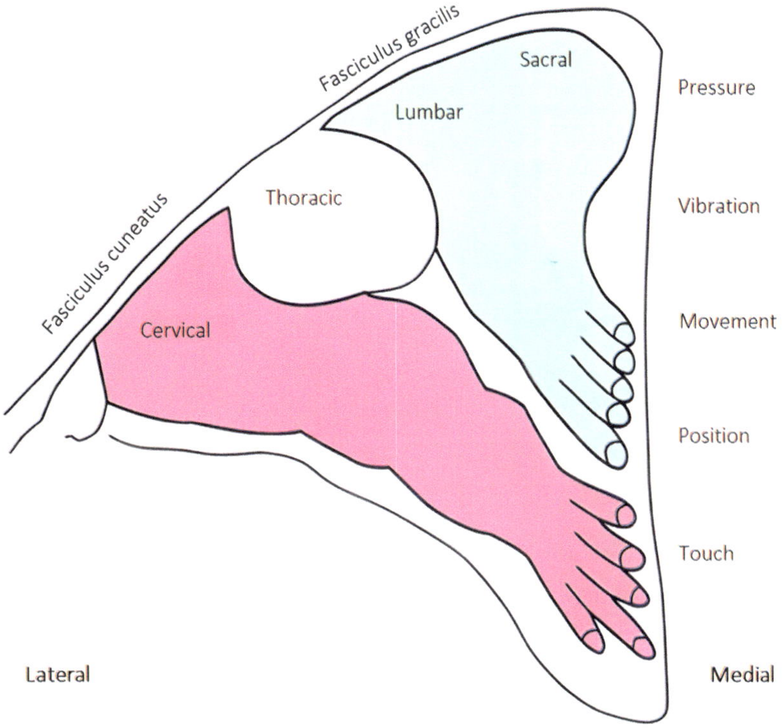

Fig. 1.17 Schematic representation of the somatotopic organization of the spinal cord levels and sensory modalities within the posterior funiculus

the arrangement of sensory modalities from the ventral to dorsal regions demonstrates a sequence of touch, position, movement, vibration, and pressure (Standring 2020) (Fig. 1.17). Notably, these fibers maintain their ipsilateral course throughout the spinal cord, without decussation or synaptic interruption. The fasciculi gracilis and cuneatus project to their respective nuclei, the nucleus gracilis and nucleus cuneatus, situated in the caudal medulla's dorsal region (Parent 1996; Purves et al. 2018).

The dorsal column refers to the first-order fibers that are situated on the same side of the spinal cord in the dorsal white column. Second-order sensory afferent neurons arise from these nuclei, and their axons decussate, forming the internal arcuate fibers. These fibers ascend through the medial lemniscus and pass through the thalamic ventral posterolateral nucleus (VPL) before ending in the postcentral gyrus of the cerebral cortex (Brodmann areas 3, 1, and 2) via the internal capsule (Purves et al. 2018) (Fig. 1.18). This intricate sensory pathway ensures the precise transmission of diverse sensory modalities from the peripheral receptors to the higher cortical centers, facilitating the perception and integration of sensory information crucial for motor control and overall sensory experience (Abraira et al. 2017; Purves et al. 2018).

1.3.2.3 Cerebellar Input System

The spinocerebellar tracts, situated within the lateral funiculus of the spinal cord, differ from the previously discussed ascending tracts by directing information not to the cerebral cortex but to the cerebellum, bypassing the thalamus. Consequently, the sensory input transmitted through these tracts remains subconscious, contributing to the automatic coordination of motor functions such as running or cycling (unconscious proprioception). The ventral and dorsal spinocerebellar tracts maintain the same somatotopic organization from ventral to dorsal region (Schuenke et al. 2020).

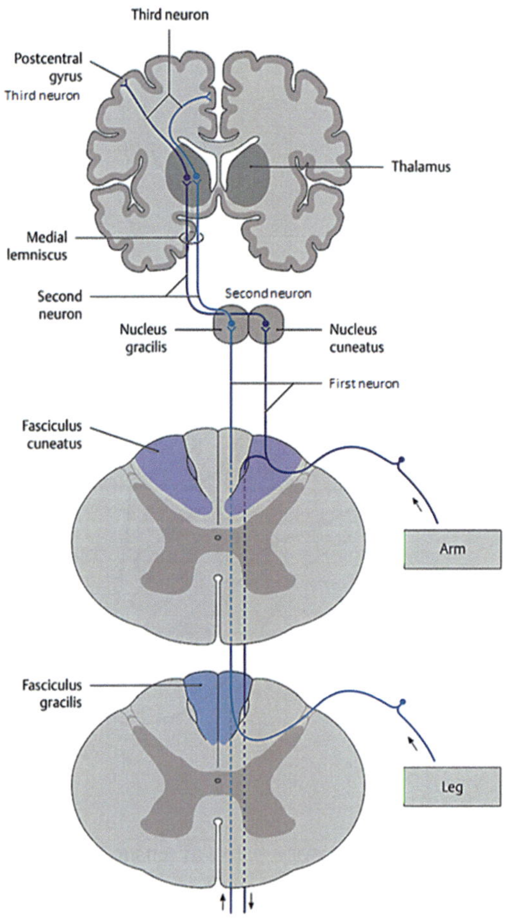

Fig. 1.18 Pathways of the fasciculus cuneatus and fasciculus gracilis, including their central connections. (Adapted from Schuenke et al. (2020))

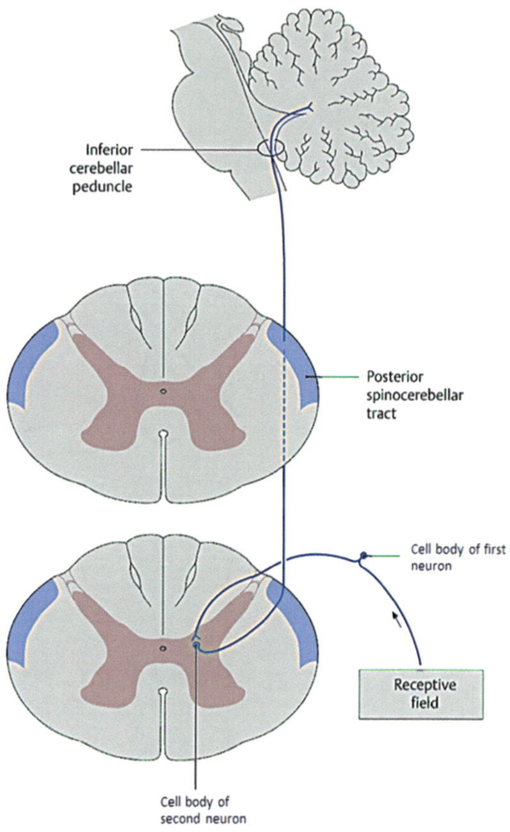

Fig. 1.19 Pathway of the posterior spinocerebellar tract and its central connections. (Adapted from Schuenke et al. (2020))

These tracts serve as the primary afferent pathways to the cerebellum, delivering feedback on limb movement and spinal interneuron activity during locomotion. The dorsal spinocerebellar tract, an uncrossed pathway comprising second-order sensory afferents, originates from Clarke's nucleus at levels T1(C8)–L2(L3) (lamina VII) and is positioned at the dorsal periphery of the lateral funiculi (Afifi and Bergman 2005; Mai and Paxinos 2011). The lower thoracic and upper lumbar segments exhibit the best development of Clarke's nucleus (Nolte and Angevine 2013). The primary role of the dorsal spinocerebellar tract is to transmit proprioceptive and cutaneous touch and pressure information from the trunk and lower limbs (Nolte and Angevine 2013; Standring 2020). This tract extends into the brain stem and enters the cerebellum via the ipsilateral inferior cerebellar peduncle (Fig. 1.19).

In contrast, the ventral spinocerebellar tracts, which are smaller and less prominent, are also located at the periphery of the lateral funiculi. These tracts consist of crossed fibers arising from the ventral horn, ascending through the restiform bodies to reach the cerebellar vermis. They ascend from the lower spinal cord regions, notably from neurons within the lumbar enlargement, and occupy a more superficial position in the lateral funiculi of the cervical spinal cord segments. Most of the tract fibers decussate in the ventral white commissure (Parent 1996) (Fig. 1.20).

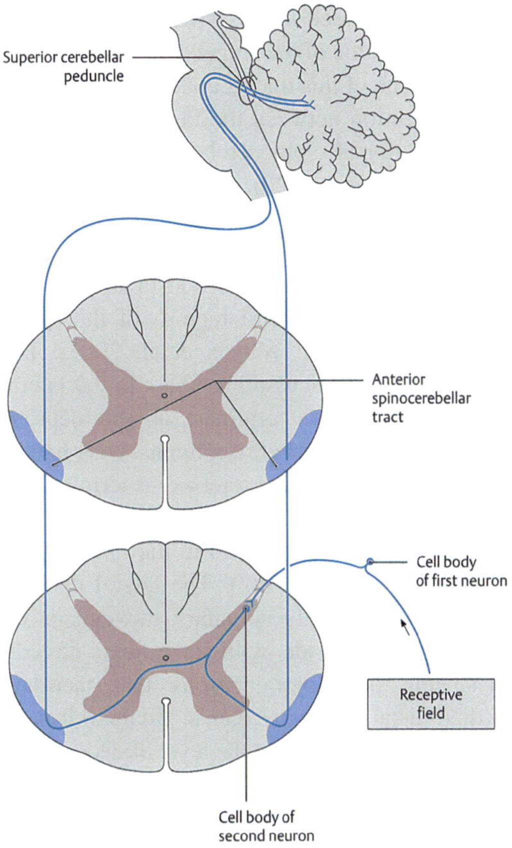

Fig. 1.20 Pathway of the anterior spinocerebellar tract and its central connections. (Adapted from Schuenke et al. (2020))

1.3.2.4 Supplementary Ascending Tracts

The spinoreticular tract ascends ipsilaterally in the anterolateral spinal cord without decussation. It terminates in the reticular formation of the brain stem. From there, projections extend to the intralaminar thalamic nuclei and subsequently to the limbic system (Nieuwenhuys et al. 2008). Several other tracts, composed of second-order neurons originating in the spinal gray columns, have been identified. These include the spino-olivary tracts, which terminate in the inferior olivary nuclei; the spinopontine tracts, ending in the pontine nuclei; and the spinovestibular tracts, which reach the lateral vestibular nuclei (Purves et al. 2018). While the anatomical locations of these fibers are well established, their precise functions in the human spinal cord remain incompletely understood.

1.3.3 Descending Tracts and Motor Information Transmission

1.3.3.1 Corticospinal Tracts

The descending motor pathways are categorized into two main systems: the lateral and medial motor systems. Furthermore, these pathways can be further classified as pyramidal and extrapyramidal motor tracts, along with the central autonomic tract (Cho and Bhattacharyya 2018). The pyramidal system includes the lateral and ventral corticospinal tracts as well as the corticobulbar tract. Although the term "extrapyramidal" is somewhat imprecise, it remains prevalent in medical literature.

The fundamental organization of these systems involves upper motor neurons projecting to lower motor neurons situated in the spinal cord and brain stem. Traditionally, medical education depicts the motor pathway as a monosynaptic circuit: originating in the motor cortex, traversing the descending corticospinal tract, and synapsing with motor neurons in the ventral horn of the spinal cord. However, this model is overly simplistic (Lemon 2008). The corticospinal tract extensively projects to all layers of the spinal cord, including the dorsal horn, affecting nociceptive, somatosensory, autonomic, and motor functions. This influence is exerted more through spinal interneurons rather than direct connections to motor neurons (Cho and Bhattacharyya 2018; Haines and Mihailoff 2018; Lemon 2008).

The corticospinal tract is the most predominant and functionally significant descending pathway in the human spinal cord, integral to the pyramidal system. These tracts traverse the ventral medulla oblongata's pyramids (Coppola 1973; Dumitru and Lang 1986). Approximately 40% of the corticospinal tract fibers arise from the primary motor cortex in the precentral gyrus (Brodmann area 4), while the remaining fibers originate from other cortical areas in the frontal and parietal lobes. This tract plays an important role in facilitating precise and dexterous movements of the digits and joints (Armand 1982; Moore et al. 2018).

The spatial organization of the corticospinal tract fibers within the spinal cord follows a medial-to-lateral pattern, with cervical fibers positioned most medially and sacral fibers positioned most laterally (Armand 1982; Moore et al. 2018). At the caudal medulla oblongata, approximately 85–90% of the corticospinal tract fibers decussate, while the remaining fibers continue ipsilaterally into the spinal cord (Armand 1982) (Fig. 1.21). The lateral corticospinal tract, formed by crossed fibers, descends through the lateral funiculus of the spinal cord white matter. The position of this tract is anatomically defined by its location between the fasciculus proprius and the dorsal spinocerebellar tract (Netter 2019; Standring 2020). The existence of the dorsal spinocerebellar tract decreases below the L2 or L3 segment of the spinal cord. If this tract is not present, the lateral corticospinal tract may extend to the peripheral regions of the spinal cord (Parent 1996; Watson et al. 2009). The majority of the decussating fibers in the lateral corticospinal tracts terminate on interneurons located in the ventral gray columns. Over half of the lateral corticospinal tract axons terminate in the cervical and upper thoracic spinal cord, while only a quarter extend to the lumbosacral segments (Armand 1982). The lateral corticospinal tract controls voluntary movements on the contralateral side, with some fibers directly synapsing on motor neurons that facilitate skilled distal limb movements. Bilateral lesions of the pyramidal tract at the level of decussation result in a loss of skilled hand movements but preserve gross locomotor function (Critchley and Eisen 1997; Lawrence and Kuypers 1968). The size of these tracts exhibits individual and bilateral variability.

The uncrossed fibers are further divided, with 5% joining the lateral corticospinal tract and 15% forming the ventral corticospinal tract in the ventral column of the white matter. The ventral corticospinal tract fibers synapse on motor neurons or interneurons in the medial aspects of the ventral horn or intermediate gray matter, influencing the activity of motor neurons innervating axial musculature. Although some fibers cross in the ventral white commissure before forming synapses, others remain uncrossed. The majority of the ventral corticospinal tract fibers terminate in the cervical and thoracic segments, potentially contributing to the control of neck and shoulder muscles. However, lesions affecting this tract rarely result in significant weakness, possibly due to the bilateral distribution of fibers from the contralateral tract (Vanderah and Gould 2016).

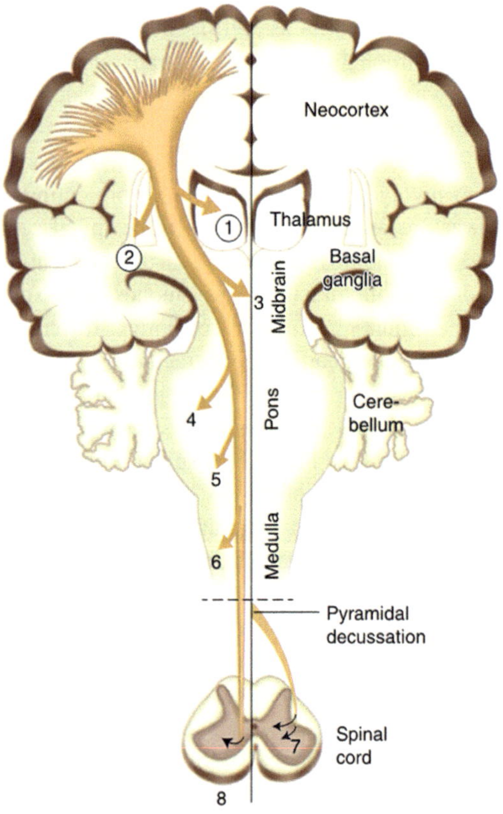

Fig. 1.21 Origin and trajectory of fibers of the descending pyramidal tract in the human brain and spinal cord. Numbers indicate the following components of the pyramidal tract: (1) corticothalamic; (2) corticostriatal; (3) corticotectal; (4) corticopontocerebellar; (5) corticoolivary; (6) corticocuneate and corticogracile. Beyond the pyramidal decussation: (7) the crossed lateral corticospinal tract; (8) the uncrossed ventral corticospinal tract. (Adapted with permission from Altman and Bayer (2001))

1.3.3.2 Supplementary Descending Tracts

The spinal cord receives axonal projections from various nuclei situated in the brain stem, collectively forming the extrapyramidal system, which is evolutionarily older than the pyramidal tracts. The primary origins of the spinal components of this system include neurons located in the vestibular nuclei, red nuclei, superior colliculi, and the reticular formation of the brain stem (Haines and Mihailoff 2018). Five distinct tracts of this system are identified bilaterally in the human spinal cord: the medial longitudinal fasciculus, tectospinal, vestibulospinal, reticulospinal, and rubrospinal tracts (Watson et al. 2009).

The tectospinal tract originates in the midbrain and contributes to the coordination of head and eye movements. Originating from the lateral and medial vestibular nuclei, the vestibulospinal tract is involved in maintaining head and neck posture and balance. The reticulospinal tract, derived from the reticular formation of the brain stem, contributes to automatic postural control and locomotor movements by modulating spinal reflexes. The rubrospinal tract, arising from the midbrain red nucleus, crosses within the brain stem and is believed to influence muscle tone in the distal extremities (ten Donkelaar 1988).

In clinical observations, differences in postures due to lesions in the brain stem provide insights into the functional significance of these tracts. A decorticate posture, characterized by flexion of the upper extremities, indicates lesions above the red nucleus. Conversely, a decerebrate posture, marked by extension of the upper extremities, corresponds to lesions below the red nucleus (Posner et al. 2019). It is hypothesized that the rubrospinal projections exert the most significant influence on the flexor motor neurons of the upper extremities (Mtui et al. 2016).

1.4 Composition of Spinal Cord Gray Matter

The spinal cord cross section reveals a distinctive H-shaped (butterfly-shaped) central gray matter structure. This structure consists of symmetrical, crescent-shaped masses connected by a commissural bridge of tissue that includes the central canal (Standring 2020). The gray matter is predominantly composed of neuronal cell bodies, neuroglia, and vasculature. Additionally, it contains a complex network of nerve fibers, including both myelinated and unmyelinated axons (Heimer 1983). The multipolar neurons constituting the gray matter can be categorized into four principal groups: motor neurons, the axons of which leave the spinal cord and innervate the effector tissues, including skeletal, smooth, and cardiac muscles, as well as glands; tract neurons, the axons of which ascend through the white matter to higher neural centers; interneurons, which have short processes and facilitate local circuit connections within the spinal cord; and propriospinal neurons, the axons of which provide communication between different segments of the spinal cord (Watson et al. 2009). Table 1.2 provides a summary of the many types of neuron cell bodies found in each lamina, the primary afferent fiber types that terminate in each lamina, and the specific laminae where descending (motor) fibers from higher centers synapse.

Interneurons, also known as association or intercalated neurons, constitute the majority of neurons within the spinal cord's gray matter and throughout the central nervous system. These cells play a significant role in integrative processes. Spinal interneurons can be categorized based on their axonal projections: intrasegmental interneurons with short axons confined to the same segment, intersegmental interneurons projecting to adjacent segments, and commissural interneurons extending to the contralateral side of the spinal cord. Neurons connecting different spinal segments are alternatively termed propriospinal neurons (Parent 1996). These interneurons are integral components of various reflex arcs and intrinsic neural circuits. They also mediate interactions between long descending pathways and spinal motor and sensory mechanisms, as well as between peripheral afferent fibers and spinal neurons that give rise to long ascending pathways (Kandel et al. 2013; Purves et al. 2018).

Propriospinal neurons are classified into three types based on the length of their axons: long

Table 1.2 Summary of inputs and neuronal organization within the Rexed laminae

Types of neurons present				Primary afferent fiber type	Termination of descending motor fibers
Rexed lamina	Interneuron	Propriospinal	Motor		
I	+			A-δ, (C) Nociceptive and temperature	
II	+			C, (A-δ) Nociceptive and mechanoreceptive	
III	+			A-β Mechanoreceptive	
IV	+			A-β Mechanoreceptive	(+)
V	+	+		A-δ, III, IV Nociceptive, thermal, and visceral	+
VI	+	+		I, II Proprioceptive	+
VII	+	+	+	I, II Proprioceptive and autonomic	+
VIII	+	+		(I, II) Descending motor	+
IX	+	+	+	(I, II) Motor neurons	(+)
X	+			A-δ Visceral and nociceptive	

Entries in parentheses indicate minor contributions

propriospinal neurons that span the entire length of the spinal cord, mostly within the ventral funiculus and the ventral part of the lateral funiculus, which have bilateral descending axonal projections, with their ascending axons predominantly crossing to the contralateral side (Standring 2020); intermediate propriospinal neurons, occupying a middle ground, primarily projecting ipsilaterally for more than eight segments but not spanning the entire cord length; and short propriospinal neurons, which, in contrast, project ipsilaterally for six to eight segments within the lateral funiculus.

The majority of propriospinal fibers are situated in close proximity to the gray matter within the fasciculus proprius, though some traverse more lateral regions of the funiculi (Flynn et al. 2011). Medial propriospinal neurons are characterized by long axons with extensive branching patterns. Some of these neurons extend the full length of the spinal cord, facilitating coordination between neck and pelvic axial musculature for postural adjustments. Conversely, laterally positioned propriospinal fibers influence neurons innervating more distal muscles, communicate across fewer segments, and exhibit less extensive branching (Conta and Stelzner 2004; Standring 2020).

The interconnectivity of propriospinal neurons with other spinal cord segments contributes not only to the coordination of somatic motor activities but also to autonomic functions. These include the regulation of sweat glands, vascular smooth muscle, and visceral organs such as the bladder and bowel (Standring 2020).

The spinal cord's gray matter is structurally divided into ventral and dorsal columns (or horns) by an imaginary transverse line intersecting the central canal. A slender band of gray matter, known as the gray commissure, traverses the midline, uniting the left and right halves of the gray matter. The central canal, lined with ependyma, is situated at the center of this commissure. The intermediolateral (or lateral) horn, containing neurons of the autonomic nervous system, is present exclusively in the thoracic and upper lumbar spinal cord segments. This intermediate horn, spanning from T1 to L2, receives visceral afferent signals and houses cell bodies of preganglionic, visceral efferent neurons, whose axons exit via the ventral roots (Afifi and Bergman 2005; Clifton et al. 1976).

The dorsal gray columns or horns function as primary receptive zones for afferent impulses from the dorsal roots. The dorsal column or horn extends posterolaterally and is separated from the posterolateral sulcus by Lissauer's tract, a thin but significant layer of white matter. Conversely, the ventral gray columns or horns contain nerve cells whose axons emerge from the cord in ventral roots, innervating skeletal muscles. These ventral gray columns are most prominent in the cervical and lumbar enlargements, where they house motor neurons for the extremities' musculature. Notably, the thoracic lateral horn encompasses preganglionic sympathetic neurons (Fig. 1.22).

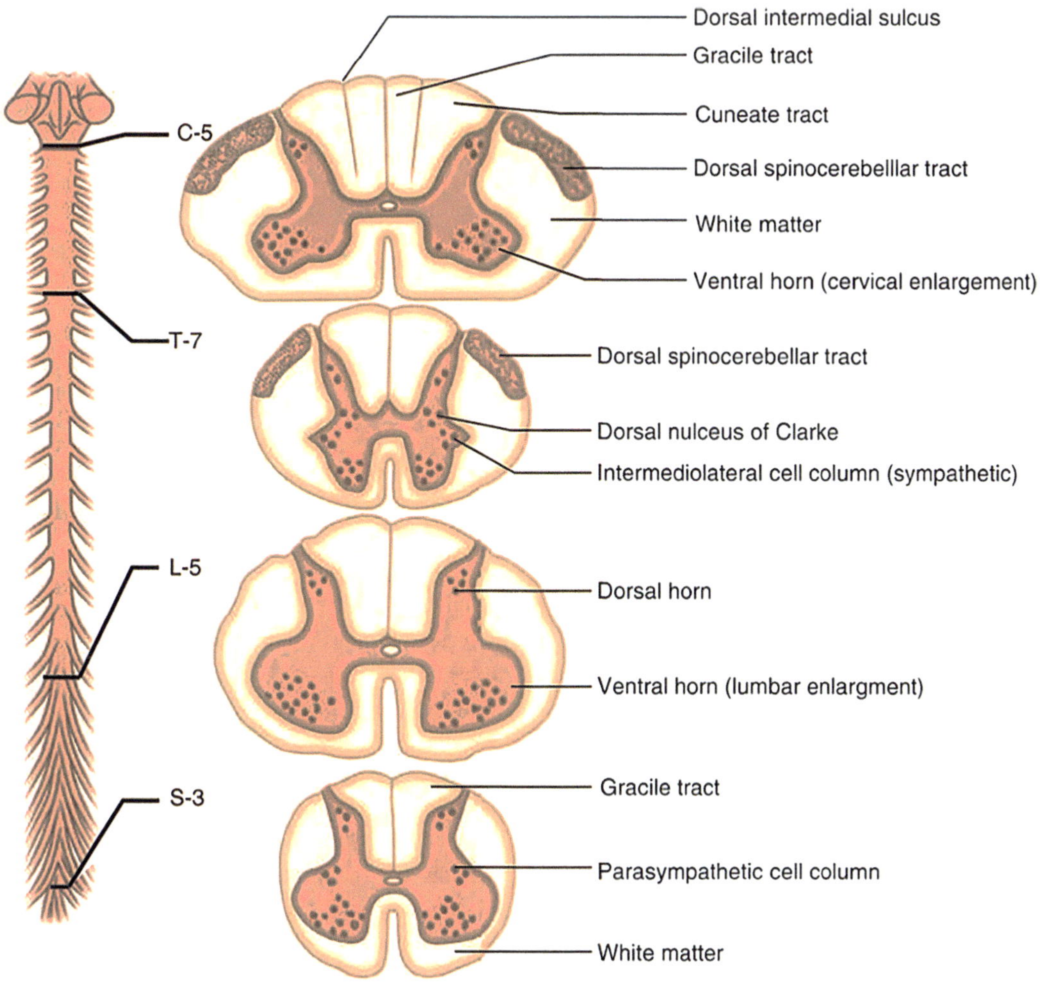

Fig. 1.22 Variations in spinal cord segments at different levels. The intermediate horn in the thoracic and upper lumbar regions receives visceral afferent input and contains the cell bodies of preganglionic visceral efferent neurons, whose axons exit via the ventral roots

The spinal cord's gray matter plays a fundamental role in integrating multiple peripheral and central inputs. In the cervical region, the gray matter intermingles with white matter in the concavity of either gray mass, forming the spinal reticular formation. The central canal, a continuation of the fourth ventricle, is lined with ciliated columnar epithelium and contains cerebrospinal fluid.

1.4.1 Architectural Lamination of the Gray Matter

The gray matter of the spinal cord exhibits a complex laminar organization, comprising ten distinct zones or laminae, labeled I–X in a dorsal-to-ventral sequence. These divisions, known as Rexed laminae, were established to aid in the study of the cytoarchitecture of neurons—specifically their size, density, and morphology—based on a fundamental anatomical nomenclature developed for the spinal cord gray matter of cats (Afifi and Bergman 2005; Blumenfeld 2010; Rexed 1952) (Fig. 1.23). The organization of spinal cord laminae displays a complex cytoarchitecture, where each lamina consists of a wide range of neuronal types and synaptic connections. Each lamina contains at least one of the four principal neuronal categories: motor neurons, tract neurons, interneurons, and propriospinal neurons (Rexed 1952). The diverse cellular composition allows for complex information processing and integration within each laminar structure. In addition to their own neurons, spinal cord laminae function as termination sites for various neural projections. These include primary afferent fibers, which transmit sensory information from the periphery, as well as descending tract fibers originating from supraspinal regions (Todd 2010). Furthermore, propriospinal neurons and interneurons from adjacent laminae also establish synaptic connections within each lamina, contributing to the formation of local neural circuits (Flynn et al. 2011).

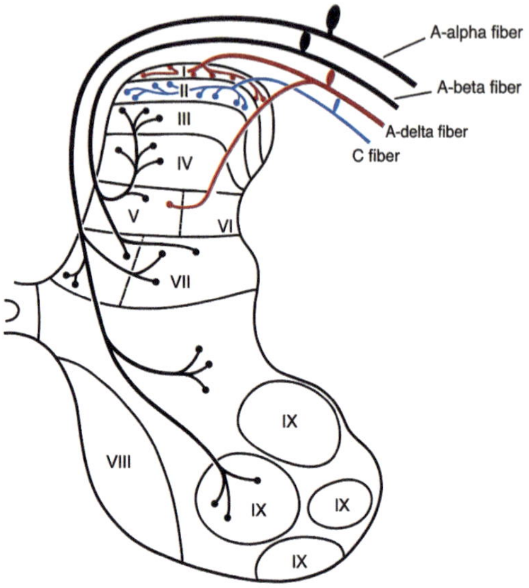

Fig. 1.23 Organization of the spinal cord gray matter. Heavily myelinated A-α fibers from neuromuscular spindles and Golgi tendon organs terminate in laminae VI, VII, and IX. A-β fibers from cutaneous mechanoreceptors terminate in laminae III-VI. A-δ and C fibers from nociceptors terminate in laminae I–V. (Adapted with permission from Strominger et al. (2012))

The dorsal horn comprises laminae I–VI, with laminae VII–X positioned at the base of the ventral horn and the central region of the dorsal horn. Not all laminae are clearly distinguishable from one another. Laminae I–VIII span the entire spinal cord length, though their size varies significantly between segments.

Lamina I, the marginal zone, forms a thin layer covering the substantia gelatinosa. Lamina II is synonymous with the substantia gelatinosa, although this structure often encompasses both laminae II and III (Heimer 1983). Laminae III–VI constitute the body of the dorsal horn. Lamina VII corresponds approximately to the intermediate gray matter, including Clarke's nucleus, and extends into the ventral horn. Lamina VIII includes interneuronal zones of the ventral horn, while lamina IX contains motor neuron clusters within the ventral horn. Lamina X encircles the central canal (Vanderah and Gould 2016). Lamina IX is not a true lamina in the traditional sense but is rather composed of interspersed motor neuron pools within laminae VII and VIII and along the edge of lamina X. It houses large α-motoneurons, which innervate striated musculature, and smaller γ-motoneurons, which innervate muscle spindles. These neurons are surrounded by specific interneurons, notably Renshaw cells, which induce recurrent inhibition of motor neurons, thereby providing an essential negative feedback mechanism for the motor neuron activity. Lamina VII occupies the intermediate gray zone and contains Clarke's nucleus. Lamina X, the central gray matter of the spinal cord, extends along its entire length, encircling the central canal region (Table 1.3).

1.4.1.1 Dorsal Gray Matter (Laminae I–IV)

The dorsal gray horn of the spinal cord serves as the entry point for afferent sensory information into the central nervous system. Within this region, laminae I–IV are situated. Within the dorsal and intermediate horns, two anatomically distinct nuclei are present: the substantia gelatinosa (lamina II) and the nucleus dorsalis, also known as Clarke's nucleus (lamina VII) (Zeman and Innes 1963).

Laminae I–IV serve as the main sites where cutaneous afferent fibers and their collateral branches terminate. Laminae I and II, collectively referred to as the superficial dorsal horn, play a significant role in the processing of nociception. Almost all of the A-delta- and C-fibers end their course at this location. The tip of the dorsal horn is formed by Rexed lamina I. Lamina I is alternatively referred to as the marginal zone of Waldeyer. The remaining gray matter constitutes the nucleus proprius (laminae III and IV) of the dorsal gray column. The nucleus proprius neurons exhibit diverse anatomical connections and functional properties. Some neurons in this region project to the thalamus as part of the spinothalamic system, while others, located at the base of the dorsal horn, give rise to spinocerebellar pathways. Additionally, certain neurons in this area are involved in the propriospinal system or function as interneurons in spinal reflexes. These cells also serve as targets for descending pathways originating in the brain (Heimer 1983).

Lamina II, or the substantia gelatinosa, is characterized by the highest neuronal density within the spinal gray matter. This distinctive area caps the dorsal horns of the spinal cord gray matter. In myelin-stained preparations, this region appears distinctly paler than the surrounding gray matter due to its composition of predominantly finely myelinated and unmyelinated sensory fibers. These fibers play a significant role in the transmission of nociceptive and thermal sensations (Vanderah and Gould 2016) (Fig. 1.24).

The neurons of lamina II consist mostly of interneurons, which can be either excitatory or inhibitory. These interneurons have dendrites that branch out inside the lamina and also project into neighboring laminae. The substantia gelatinosa is present throughout all spinal cord segments and contains numerous small cells (Golgi type II) that function as interneurons within the pain pathway. These interneurons form synapses with larger neurons in the adjacent nucleus proprius, which, in turn, give rise to axons that contribute to the lateral spinothalamic tract (Bican et al. 2013). Laminae I, II, and V receive input from noxious stimuli, whereas laminae III and IV process light touch and proprioceptive input. Laminae III and IV are commonly referred to as the nucleus proprius. They include interneurons and some tract neurons (Bican et al. 2013).

Anatomical studies in felines and primates have revealed specific somatotopic patterns in the cervical spinal cord dorsal horn gray matter (Altman and Bayer 2001; Brown and Fuchs 1975; Nyberg and Blomqvist 1985; Whitsel et al. 1969; Whitsel et al. 1970). Research has demonstrated a consistent mediolateral organization extending from the central canal, following a palm–finger–hand sequence and continuing to the forearm dorsum (Altman and Bayer 2001). This anatomical arrangement explains the progressive development of sensory symptoms in syringomyelia. When a syrinx develops in the C7

Table 1.3 Key subdivisions of spinal cord gray matter (Rexed laminae), including their regions, segmental levels, and associated functions

Region and nucleus		Levels	Rexed lamina	Fibers	Main function
Dorsal gray column	Marginal zone	All	I	A-δ	Sensation of temperature and fast pain Modulate nociceptive sensory input Some spinothalamic tract cells
	Substantia gelatinosa	All	II	C	Sensation of slow pain. Modulate nociceptive sensory input Modulate transmission of pain and temperature information
	Nucleus proprius	All	III	A-β	Mechanoreceptors for touch and proprioception
	Nucleus proprius	All	IV	A-β	Mechanoreceptors for touch and proprioception
	Nucleus dorsalis	All	V	A-δ, C	Receives information on pain sensation and movement
	Base of dorsal horn	All	VI	Ia, A-α Ib, A-α, Golgi	Spinal reflexes, integration of somatic motor function
Intermediate gray column	Clarke's nucleus	C8–L3	VII	Nucleus dorsalis	Preganglionic parasympathetic neurons Origin of posterior spinocerebellar tract
	Intermediolateral nucleus	T1–L2	VII	Intermediolateral nucleus	Origin of preganglionic sympathetic fibers
	Intermediolateral nucleus (sacral parasympathetic nucleus)	S2–S4	VII	Preganglionic sacral autonomic nucleus	Origin of preganglionic parasympathetic fibers
Ventral gray column	Commissural nucleus	All	VIII	Descending tracts	Motor neurons Modulate muscular tone and movement
	Medial motor nuclei	All	IX	Somatic α- and δ-motor neurons	Motor neurons innervating trunk musculature
	Lateral motor nuclei	C5–T1 and L2–S3	IX	Somatic α- and δ-motor neurons	Innervation of extrafusal and intrafusal fibers Motor neurons innervating upper and lower limb musculature
Perimeter of the central canal	Grisea centralis	All	X	Anterior commissure tracts	Decussation of axons

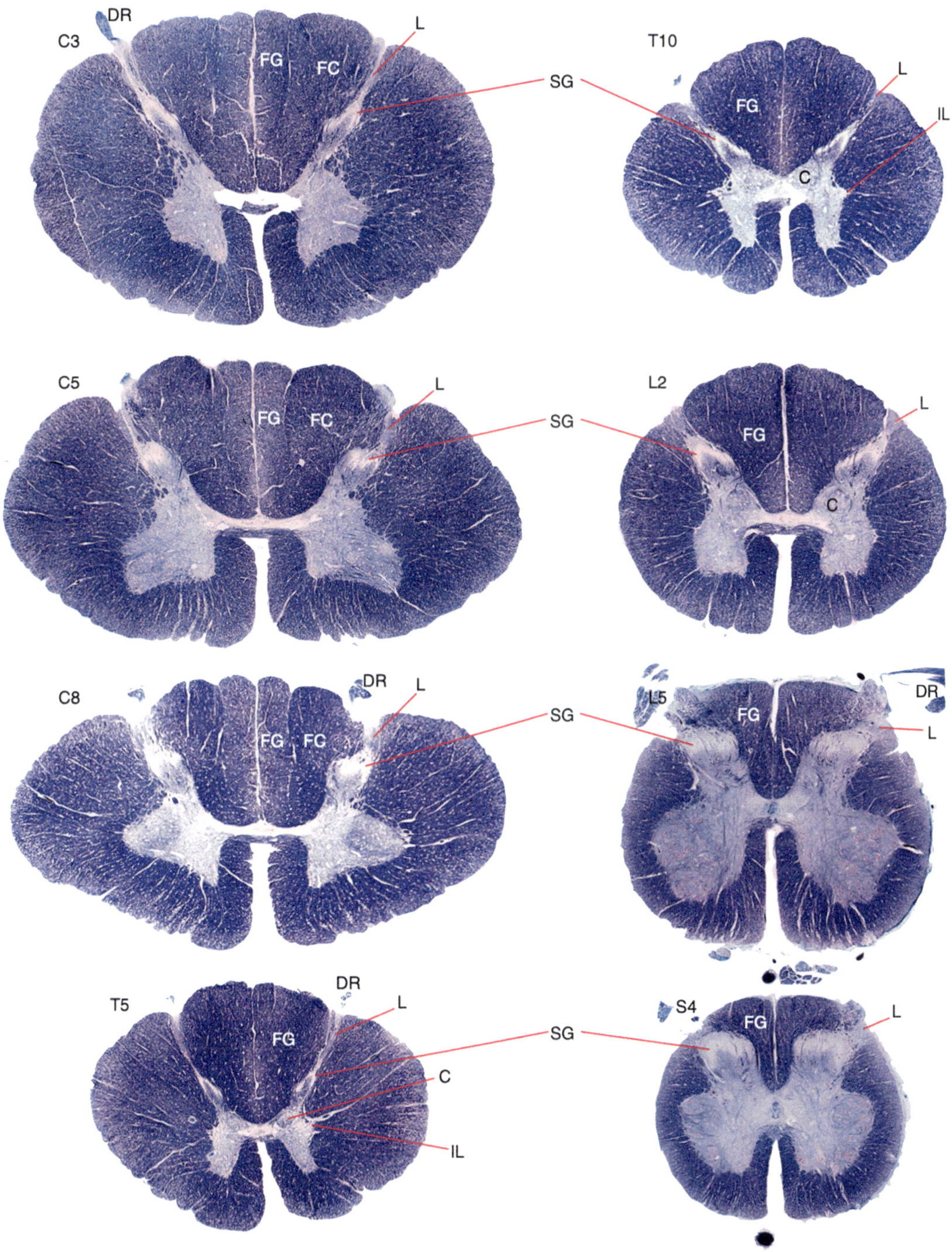

Fig. 1.24 Cross-sectional views of the spinal cord at different levels. The ventral horns display prominent lateral extensions at C5, C8, and L5. Abbreviations: C, Clarke's nucleus; DR, dorsal root; FC, fasciculus cuneatus; FG, fasciculus gracilis; IL, intermediolateral cell column; L, Lissauer's tract; SG, substantia gelatinosa. (Adapted with permission from Vanderah and Gould (2016))

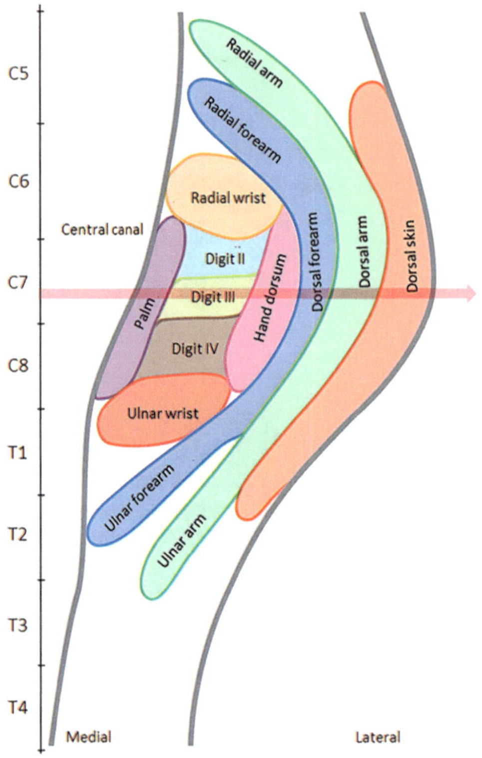

Fig. 1.25 Somatotopic arrangement of cutaneous afferents within laminae II–III of the cervical and upper thoracic spinal cord in the cat. Sensory fibers originating from the dorsal skin, arm, forearm, and hand are positioned laterally to medially, while digits II, III, and IV are represented in a rostral-to-caudal sequence. (Adapted with permission from Altman and Bayer (2001))

segment (indicated by a light red arrow in Fig. 1.25), patients initially experience sensory alterations—particularly paresthesia and pain—in the palm and middle finger regions. As the syrinx expands, these symptoms progress laterally, affecting the hand dorsum and extending to the forearm (Ko 2023; Standring 2020).

1.4.1.2 Intermediate Gray Matter (Laminae V–VIII and Autonomic Nuclei)

The intermediate gray matter, situated between the ventral and dorsal horns of the spinal cord gray matter, is characterized by a diverse cellular composition with small-to-medium-sized neurons. This region includes laminae V–VIII and autonomic nuclei, playing a significant role in various spinal cord functions (Vanderah and Gould 2016). This region prominently displays the intermediolateral and intermediomedial cell columns, which extend from T1 to L2 and perform an important part in the formation of the lateral horn. These columns include preganglionic visceral motor neurons that are essential for the sympathetic division of the autonomic nervous system. At the levels of S2–S4, there are similar cell columns that function in the parasympathetic division (Heimer 1983).

Lamina VII forms most of the intermediate region of the gray matter and also extends into the ventral horn. The intermediate zone is primarily observed in the thoracic and upper lumbar segments and contains preganglionic autonomic neurons for sympathetic function. These neurons receive input from hypothalamic and brain stem nuclei via the descending central autonomic tract. Lamina VII of the gray matter in these segments includes cells projecting preganglionic sympathetic fibers and the nucleus dorsalis (Clarke's nucleus), which forms the dorsal spinocerebellar tract (Rexed and Brodal 1951). Clarke's nucleus occupies a medial area near the dorsal horn base in the T1–L2 segments, reaching its maximum size in the lower thoracic and upper lumbar regions. This nucleus comprises large nerve cells receiving collateral fibers from the medial divisions of dorsal roots. It serves as a key relay station for proprioceptive information to the cerebellum and, potentially, to the thalamus, thereby integrating it into the sensory processing system of the dorsal horn (Vanderah and Gould 2016).

Axons from small nerve cells in the intermediolateral column exit the spinal cord via ventral roots and synapse with postganglionic sympathetic neurons in sympathetic chain ganglia. Additional visceral efferent neurons are located in the S2–S4 segments, forming a parasympathetic efferent column lateral to the central gray matter and central canal in the intermediate zone. These neurons give rise to pelvic splanchnic nerves, providing parasympathetic input to the distal colon, rectum, bladder, and sexual organs.

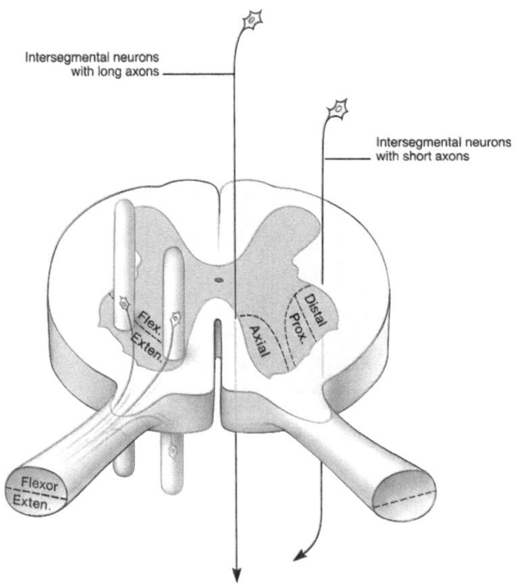

Fig. 1.26 The α- and γ-motoneurons in lamina IX of the ventral horn are organized based on functional roles. Medially located motor neurons innervate axial muscles, while those positioned laterally supply proximal and distal limb muscles. Motor neurons situated dorsally are associated with flexor muscles, and those ventrally located innervate extensor muscles. Medial motor neurons are interconnected by long-axoned intersegmental neurons, whereas lateral motor neurons are linked by intersegmental neurons with shorter axons. The cylindrical structure spanning multiple spinal segments in lamina IX illustrates the distribution of lower motor neurons innervating specific or related muscles. (Adapted with permission from Vanderah and Gould (2016))

Their fibers synapse with postganglionic parasympathetic neurons within the walls of these target organs.

1.4.1.3 Ventral Gray Matter (Lamina IX)

In histological cross sections of spinal cord preparations stained to reveal the cellular structure, several groups of large somatic motor neurons are observed within the ventral horn. These neuronal clusters are particularly notable in the spinal cord enlargements, where lateral and medial cell groups are evident. These clusters are part of longitudinally organized neuronal columns that vary in length (Watson et al. 2012) (Fig. 1.26). The medial cell column, which extends throughout the spinal cord, provides innervation to the trunk and neck musculature. It supplies the axons that provide nerve supply to the epaxial muscle groups (also known as the "true" or "deep" back muscles) and hypaxial muscle groups (including the prevertebral muscles, intercostal muscles, and ventral abdominal wall muscles) (Fig. 1.27). Furthermore, the diaphragm undergoes development as an axial muscle, resulting in the proximity of the phrenic nucleus to other motor neurons that innervate the axial musculature. In contrast, the lateral columns, responsible for supplying the limb muscles, are localized exclusively to the cervical and lumbar enlargements (Watson et al. 2009). Within these lateral motor neuron groups, a somatotopic organization is evident: Neurons that innervate distal limb muscles are situated dorsally relative to those that supply the proximal limb muscles (Heimer 1983; Romanes 1951).

The ventral horns of the spinal cord contain two primary types of motor neurons: α-motoneurons and γ-motoneurons. Alpha-motoneurons, which are larger in size, innervate the extrafusal fibers of skeletal muscles. Interspersed among the large motor neurons are smaller neurons called γ-motoneurons; they send their axons to intrafusal muscle fibers and are, therefore, also referred to as fusimotor neurons. These intrafusal fibers are connected to specialized receptors within muscles, known as muscle spindles (Heimer 1983; Vanderah and Gould 2016). A third neuronal type, intrasegmental neurons or interneurons, characterized by short axons, is predominantly found in the medial region of the ventral gray matter.

The arrangement of neurons in the ventral gray column (lamina IX) is somatotopically organized. Neurons located medially innervate axial and proximal limb muscles, whereas those situated laterally innervate distal limb muscles. Furthermore, neurons positioned more ventrally innervate extensor muscles, while those located more laterally innervate flexor muscles (Fig. 1.28). Additionally, a specific cluster of ventrolateral neurons located in the S2 and S3

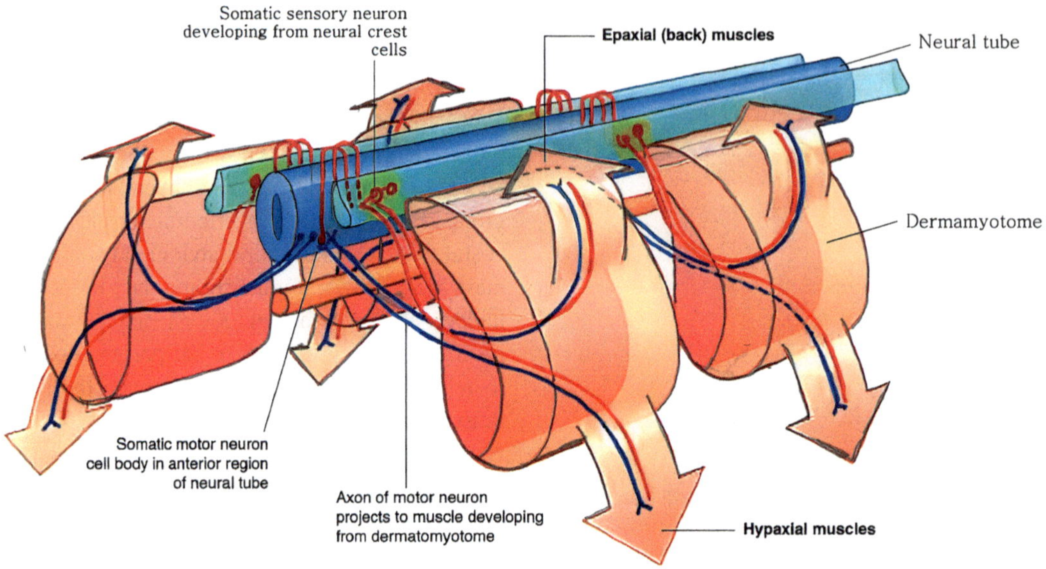

Fig. 1.27 Somatic sensory and motor neurons. The medial cell column innervates both epaxial and hypaxial muscle groups. Motor nerves are depicted with blue lines, and sensory nerves are represented with red lines. (Adapted with permission from Drake et al. (2020))

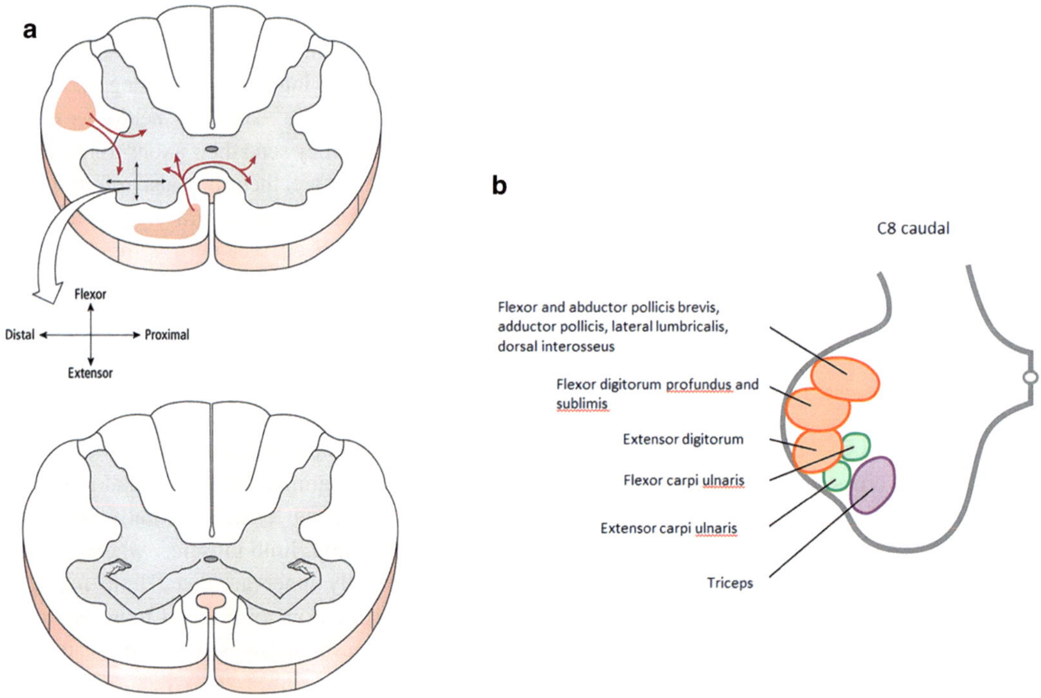

Fig. 1.28 Somatotopic organization of neurons in the ventral gray column. (**a**) Medial neurons innervate axial and proximal limb muscles, while lateral neurons supply distal limb muscles. Ventral neurons are associated with extensor muscles, and dorsal neurons are linked to flexor muscles. (**b**) Diagram depicting the distribution of motor neurons for individual muscles in the caudal portion of the C8 segment

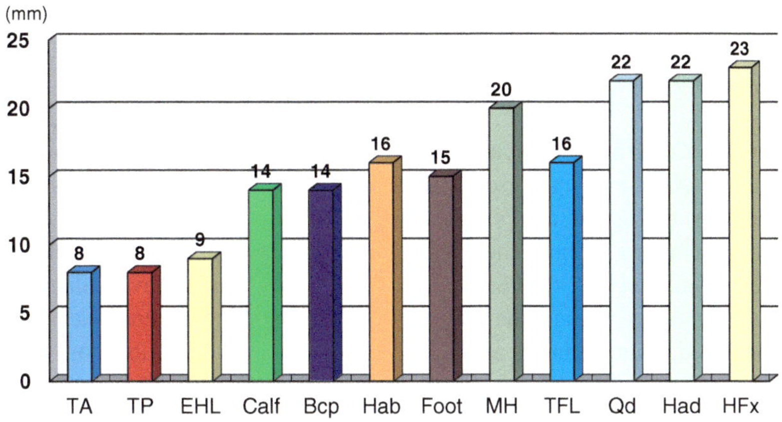

Fig. 1.29 Length of the motor neuronal columns associated with leg muscles. (Modified from Sharrard (1955))

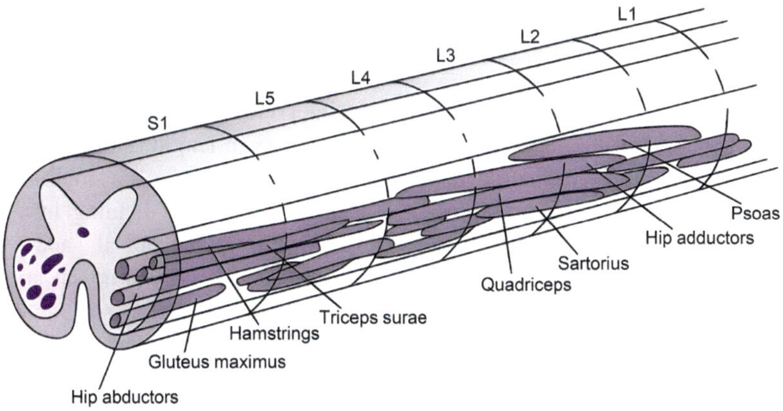

Fig. 1.30 Approximate location in the transverse plane and longitudinal extent of the ventral horn cell groups innervating muscles in the lumbosacral segments of the human spinal cord. The S1 cluster includes motor neurons for the triceps surae, hamstrings, and gluteal muscles, while the L2–3 cluster corresponds to the quadriceps, hip flexors, and adductors. This figure is based on clinicopathological studies of poliomyelitis. (Adapted from Hsu et al. (2008) with modifications from Sharrard (1955))

segments forms Onuf nucleus. These neurons innervate the perineal musculature (Standring 2020).

The anatomical characteristics of motor neuron groups significantly influence the severity of motor paralysis and the potential for functional recovery following partial damage to a spinal cord segment. Differences in the size and length of the motor neuron columns for various muscles can impact the extent of injury and recovery. Muscles with smaller or shorter motor neuron columns are less likely to sustain damage, but the injury can be more severe if they are affected (Elliott 1942; Sharrard 1955) (Figs. 1.29 and 1.30). This neuroanatomical organization has significant implications for understanding the motor function, the patterns of paralysis observed in spinal cord injuries, and the potential for recovery. The somatotopic arrangement and the varying sizes of motor neuron columns provide a foundation for predicting and interpreting the clinical manifestations of spinal cord lesions at different levels and extents.

1.4.1.4 Central Canal (Lamina X)

Lamina X is a unique group of neurons that encircles the spinal cord's central canal. This lamina consists of both the dorsal and ventral gray commissures, forming a structural and functional unit around the central canal. It is composed of interneurons and some decussating axons. The central canal itself is a continuous lumen that extends throughout the entire length of the spinal cord, lined with specialized ependymal cells that play a role in cerebrospinal fluid dynamics and neural stem cell function (Saker et al. 2016).

Anatomically, the central canal originates from the fourth ventricle in the medulla oblongata and descends caudally, terminating in the upper portion of the filum terminale. This structure maintains communication between the ventricular system of the brain and the subarachnoid space surrounding the spinal cord, facilitating the circulation of cerebrospinal fluid (Yasui et al. 1999). A notable anatomical feature is the presence of a small expansion of the central canal, termed the terminal ventricle, located within the conus medullaris. This structure is typically present during the early developmental stages but undergoes obliteration around the fourth decade of life. The gradual disappearance of the terminal ventricle may have implications for cerebrospinal fluid dynamics and, potentially, for the pathophysiology of certain spinal cord disorders (Milhorat et al. 1994).

1.5 Spinal Roots and Nerves

The spinal cord is associated with 31 pairs of spinal nerves, each corresponding to a specific spinal cord segment. Every spinal nerve is composed of a motor root and a sensory root. These roots emerge from the spinal cord at distinct locations: the motor root originates from the anterolateral surface, while the sensory root arises from the posterolateral surface of the spinal cord at each segmental level. It is noteworthy that the first cervical nerve and the coccygeal nerve lack a dorsal root, resulting in no dermatomal representation. The formation of each root involves the convergence of multiple small rootlets, also known as filaments. Typically, the motor root comprises four to seven filaments that exit the cord at the anterolateral sulcus. In contrast, the sensory root, entering the cord at the posterolateral sulcus, consists of four to ten filaments.

The organization of spinal nerves follows a complex yet systematic arrangement that begins at their formation and continues through their peripheral distribution. Upon the union of dorsal and ventral roots, spinal nerves emerge through the intervertebral foramina, measuring only a few millimeters in length (Drake et al. 2020). These nerves initially give off a small recurrent meningeal branch before dividing into their primary components: the dorsal and ventral rami (Standring 2020). Both dorsal and ventral rami carry motor, sensory, and autonomic fibers, though their distribution patterns differ significantly. The dorsal ramus, which is relatively smaller, divides into medial and lateral branches. The medial branch primarily transmits sensory information from the skin, while the lateral branch provides motor innervation to the paraspinal muscles at the corresponding segmental level (Bican et al. 2013). This arrangement maintains precise segmental control of the axial musculature and overlying skin.

The ventral ramus, being considerably larger, demonstrates more extensive distribution patterns. At thoracic levels, these rami form intercostal nerves, while in the cervical, lumbar, and sacral regions, they contribute to the formation of complex neural networks known as the brachial and lumbosacral plexuses (Moore et al. 2018). This plexus formation allows for more sophisticated control of limb movements and sensory processing.

The autonomic connections of spinal nerves exhibit distinct patterns along the neuraxis. Each ventral primary ramus receives postganglionic sympathetic fibers through a gray ramus communicans from its corresponding sympathetic chain ganglion (Mtui et al. 2016). These gray rami communicantes contain unmyelinated fibers that distribute sympathetic innervation to peripheral tissues. The thoracolumbar region (T1 through L2/3) displays additional autonomic complexity. In these segments, the ventral primary rami also provide white rami communicantes to their respective sympathetic chain ganglia. Unlike their gray rami communicantes, these white rami communicantes contain myelinated preganglionic sympathetic fibers, establishing a bidirectional autonomic communication system. The sacral segments (S1–S3) present a unique autonomic arrangement. Their ventral primary rami contribute preganglionic parasympathetic fibers that bypass the sympathetic chain, traveling directly to their target organs. This direct parasympathetic pathway represents a distinct organizational principle from the thoracolumbar sympathetic system, reflecting the functional specialization of sacral autonomic control.

The sensory root contains the central processes of bipolar sensory neurons, which are situated in the dorsal root ganglion. An interesting aspect of spinal nerve anatomy is that some primary sensory afferents, particularly those of visceral origin, enter the spinal cord through the motor root but, ultimately, terminate in the dorsal horn. Another notable feature is the presence of a substantial number of unmyelinated axons in all motor roots, constituting 27% of the total axonal population in these roots. While the precise function of these unmyelinated axons in the motor roots remains unclear, evidence suggests that a significant proportion may serve sensory functions (Coggeshall et al. 1974).

The ventral roots function as the motor output pathway. These roots contain large-diameter α-motoneuron axons, which innervate the extrafusal striated muscle fibers, and γ-motoneuron axons, which innervate the intrafusal striated muscle spindles, along with preganglionic autonomic fibers. In contrast, the dorsal roots consist of various afferent fiber types, organized somatotopically with the largest-diameter fibers positioned medially relative to smaller ones (Fig. 1.31). The largest afferent fibers, classified as Ia and Ib, are heavily myelinated and function as the afferent component of muscle stretch reflexes, transmitting information from muscle spindles and Golgi tendon organs. A-β fibers, of medium size, convey impulses from mechanoreceptors located in the skin and joints. Noxious and thermal sensations are transmitted by small, thinly myelinated Aδ fibers and unmyelinated C fibers (Bican et al. 2013) (Table 1.4).

The spinal nerve roots are enveloped by a thin layer of arachnoid cells, which encapsulate the endothelial compartment of the nerve. This com-

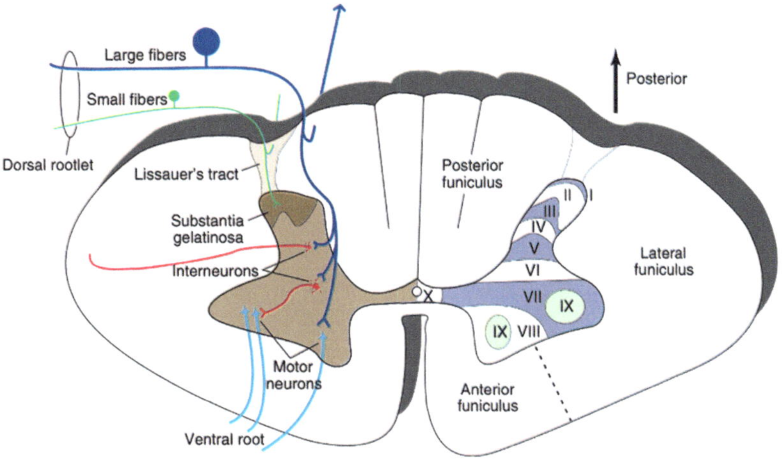

Fig. 1.31 Somatotopically organization of the dorsal roots: larger-diameter fibers are positioned medially relative to smaller-diameter fibers. (Adapted with permission from Nolte and Angevine (2013))

Table 1.4 Classification of afferent and efferent nerve fibers by type and function

Fiber type	Myelin sheath	Conduction velocity (m/s)	Fiber diameter (μm)	Function
Afferent				
Ia (Aα)	Yes	70–120	12–22	Primary sensory from muscle spindles
Ib (Aα)	Yes	70–120	10–20	Sensory fibers of Golgi tendon organs
II (Aβ)	Yes	30–70	6–12	Secondary sensory terminals of muscle spindles
III (Aδ)	Thin	10--30	1–6	Nociceptive, touch, pressure
IV (C)	No	0.5–2	>1.5	Nociceptive in muscle, temperature (?)
Efferent				
Aα	Yes	70–120	12–22	Motor to skeletal muscle fibers (extrafusal)
Aγ	Yes	10–45	2–10	Fusimotor to intrafusal fibers of muscle spindles
Sympathetic	Thin	3–15	3	Preganglionic to soma in ganglion
Sympathetic	No	0.5–2	>1.5	Postganglionic to glands and smooth and cardiac muscles

Adapted with permission from Patestas and Gartner (2006)

partment contains myelinated and unmyelinated nerve fibers, accompanied by Schwann cells and endoneurial blood vessels. The dura mater extends to form dural sleeves within each intervertebral foramen, eventually merging with the epineurium of the spinal nerves. Pia mater surrounds the roots as they traverse the subarachnoid space, terminating at their respective dural sleeves, which emerge from the ventrolateral aspect of the dural sac. Notably, nerve roots lack epineurial covering within the subarachnoid space (Fig. 1.32). This anatomical configuration renders nerve roots in the subarachnoid space more vulnerable to injury and increased sensitivity compared to spinal nerves (Nowicki et al. 1996).

The cervical spinal nerves exhibit a distinctive spatial arrangement: The first seven cervical spinal nerves exit superior to their corresponding cervical vertebrae. The C8 nerve emerges between the seventh cervical and first thoracic vertebrae. All subsequent spinal nerves caudal to this level exit inferior to their respective vertebrae. In the upper cervical region, nerve roots exit the spinal canal via the intervertebral foramina at approximately the same level they depart from the spinal cord. However, as the spinal cord terminates at the L1 or L2 vertebral level, the segmental level of the cord is positioned superior to the intervertebral foramen at most levels. Consequently, at progressively caudal levels, the roots follow a more oblique and downward trajectory within the subarachnoid space from the spinal cord to the intervertebral foramen. For instance, the lower sacral roots originate from the cord at L1 and extend several centimeters into the sacral region before exiting the spinal canal (Cramer and Darby 2014).

When assessing affected nerve roots, it is imperative to consider the level of dissociation. In the supine position, the nerve roots in the lower lumbar spine assume a U-shaped configuration within the dorsal aspect of the dural sac. The sacral roots are positioned dorsally and medially, while the lumbar roots are situated more laterally and ventrally (Bogduk 2005).

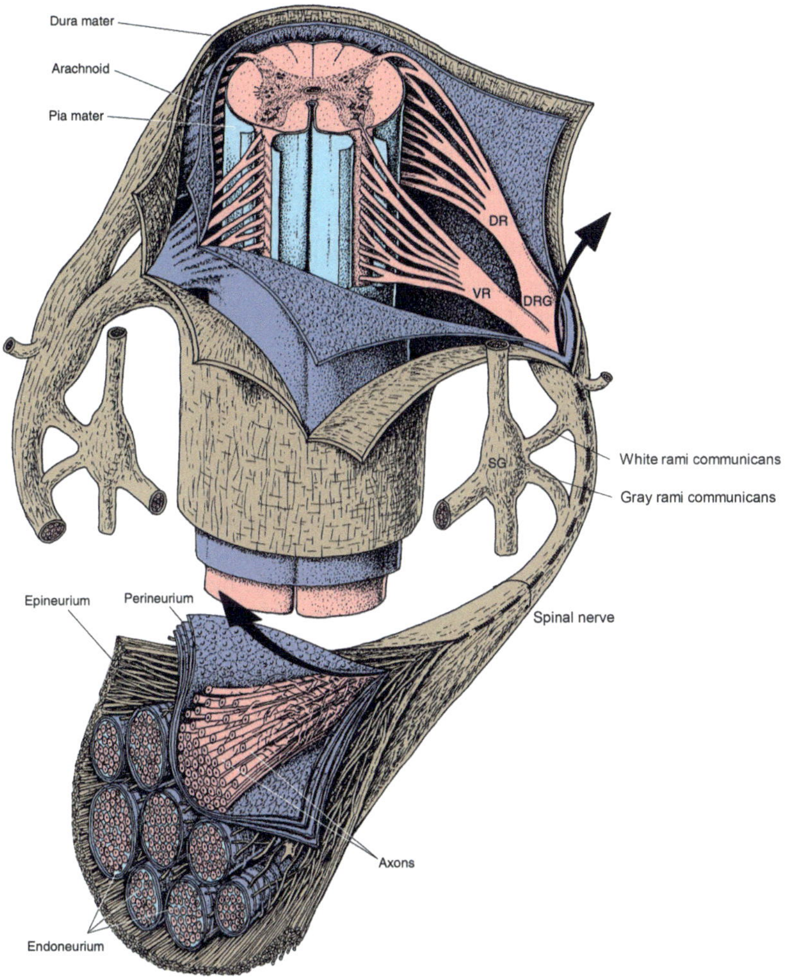

Fig. 1.32 Continuity of the spinal meninges with the sheaths of peripheral nerves. The arachnoid membrane continues as the perineurium of peripheral nerves, and the dura mater transitions into the epineurium. An *arrow* indicates the connection between the spinal subarachnoid space and the extracellular space within nerve fascicles, emerging from the cut end of the nerve and near a dorsal root ganglion (DRG). For clarity, the pia mater is reflected from the ventral rootlet exit zone. Abbreviations: DR, dorsal root; SC, sympathetic ganglion; VR, ventral root. (Adapted with permission from Vanderah and Gould (2016))

1.6 Meninges, Subarachnoid Space, and Cerebrospinal Fluid

The spinal cord is enveloped by three protective membranes: the dura mater, arachnoid, and pia mater, listed from outermost to innermost. These membranes, collectively known as meninges, originate from the condensation of embryonic mesenchyme surrounding the neural tube. The arachnoid and pia mater are jointly referred to as leptomeninges, derived from the Greek word "lepto," meaning thin (Bican et al. 2013). These protective layers extend continuously from the brain to the spinal cord.

1.6.1 Dura Mater

The dura mater, identifiable by 8 weeks of gestation, is a robust membrane composed of fibrous connective tissue. Its inner surface features a layer of simple squamous epithelium, while the outer surface blends with epidural areolar and adipose tissues, separating it from the spinal canal walls. At the foramen magnum, the dura mater's outer layer functions as the cranial periosteum. The dural sac attaches to the skull at the foramen magnum and terminates caudally at the inferior border of the second sacral vertebra. However, its fibrous wall extends to the coccyx, merging with the periosteum on the dorsal surface of this bone to form the coccygeal ligament, which anchors the spinal cord within the spinal canal.

In contrast to cranial anatomy, the spinal dura mater is separated from the vertebral periosteum by the epidural space, which contains fatty and loose connective tissue, as well as the internal vertebral epidural venous plexus (Batson's venous plexus). This plexus is considered significant in the dissemination of infection or metastatic cancer (Bican et al. 2013). The adipose tissue in the epidural space is more abundant dorsally than ventrally. The nerve roots from the spinal cord or cauda equina, which pass through the intervertebral foramen, are ensheathed by an extension of the dura mater known as the dural sleeve. This dural layer forms a significant barrier, largely preventing the invasion of metastatic tumors from the bone or the epidural space into the spinal cord. It also delays the spread of infection from the extradural compartment into the subarachnoid space (Parent 1996).

1.6.2 Arachnoid Membrane

The spinal arachnoid membrane is a delicate, avascular connective tissue layer composed of squamous cells. This thin membrane adheres to the inner surface of the dura mater, terminating as a sac at the level of the second sacral vertebra. A notable anatomical distinction between the spinal and cerebral meninges is the minimal space separating the dura mater from the arachnoid membrane in the spinal region. This anatomical feature results in a lower incidence of subdural hemorrhage in the spinal cord compared to the brain (Fig. 1.33).

The arachnoid membrane's primary attachments to surrounding structures occur at the denticulate ligaments and spinal nerve roots. The subarachnoid area, located between the arachnoid and the pia mater, is filled with cerebrospinal fluid (CSF). This fluid-filled space provides a hydraulic buffer for the spinal cord and maintains a continuous connection with a substantial cistern located superior to the foramen magnum. In the caudal region, the subarachnoid space exhibits an enlargement between the second lumbar and second sacral vertebrae. This expanded area, known as the lumbar cistern, surrounds the cauda equina and filum terminale. The lumbar cistern's anatomical position and increased CSF volume make it a preferred site for various clinical procedures, such as lumbar punctures and the administration of spinal anesthesia (Boon et al. 2004).

Fig. 1.33 Meningeal layers of the spinal cord. Unlike the brain, the space between the dura mater and the arachnoid in the spinal cord is narrow

The arachnoid membrane, along with the pia mater, constitutes the leptomeninges. These structures play a significant role in the protection and support of the spinal cord, contributing to its overall functionality and resilience against mechanical stresses (Greenberg 2016; Vandenabeele et al. 1996).

1.6.3 Pia Mater

The spinal cord pia mater is a thin, membranous structure comprising two distinct layers. The external layer consists of connective tissue, while the internal layer incorporates terminal processes of neuroglia cells, forming the pia–glial membrane. Longitudinal arrangement characterizes the outer fibers of the pia mater. As it progresses distally toward the conus medullaris, the pia mater tapers and extends into the filum terminale (Barson 1970).

The filum terminale internum, extending caudally from the conus medullaris, is primarily composed of connective tissue, glia, and leptomeninges. Its diameter typically does not exceed 2 mm. A branch of the anterior spinal artery runs within it, causing enhancement following contrast agent administration. This structure penetrates the caudal end of the dural sac, securing the spinal cord's caudal end to the caudal end of the spinal dural sheath (Saker et al. 2017). The extradural continuation, termed filum terminale externum, attaches to the dorsum of the first or second coccygeal segment. This structure anchors the caudal end of the dural sheath to the caudal end of the spinal canal (Standring 2020).

On each lateral aspect of the cord, midway between the dorsal and ventral spinal nerve roots, the spinal pia mater forms two fibrous bands known as denticulate ligaments. These scalloped bands project through the arachnoid and attach to the dura mater via a series of pointed processes, typically numbering 21 on each side. The denticulate ligaments play a key role in suspending the spinal cord within its fluid-filled environment (Yasui et al. 1999).

1.6.4 Anatomical Spaces Between the Spinal Meninges

The protective covering of the spinal cord comprises three potential anatomical spaces: the epidural, subdural, and subarachnoid spaces. The outermost space, the epidural space, is situated external to the dura mater. This space contains predominantly loose connective tissue, adipose deposits, and the internal vertebral venous plexus. The epidural space exhibits anatomical variations, becoming narrower at the foramen magnum superiorly and the sacral hiatus inferiorly. At these locations, the spinal dura mater fuses with the endosteal layer of the cranial dura. The cervical and thoracic regions of the epidural space are characterized by the presence of a substantial basivertebral vein (Groen et al. 1997). Contrary to traditional anatomical descriptions, the subdural space does not constitute a true anatomical space under physiological conditions (Haines et al. 1993).

The subarachnoid space, situated between the arachnoid and pia mater, extends from the cranial cavity to the level of the second sacral vertebra. This space contains the cerebrospinal fluid (CSF) and a loose network of collagen fibers and fibroblasts, known as arachnoid trabeculae. The subarachnoid space also encompasses the blood vessels supplying the spinal cord surface, exposing them to the cerebrospinal fluid. This anatomical arrangement is clinically significant, as it allows for direct contact between the cerebrospinal fluid and both spinal nerve roots and vascular structures. Consequently, any substance introduced into the subarachnoid space, whether therapeutic agents or pathogenic entities, can potentially affect both neural and vascular components (Bican et al. 2013).

The cerebrospinal fluid within the subarachnoid space provides buoyancy and mechanical protection to the spinal cord. Additional stabilization is achieved through the denticulate ligaments and the dorsal and posterolateral arachnoid ligaments. Under normal circumstances, the cerebrospinal fluid circulates freely between the cranial and spinal subarachnoid spaces. Disruption of this circulation can result in spinal cord disorders, such as syringomyelia (Brodbelt and Stoodley 2007). Dorsal and ventral nerve roots originating from the spinal cord occupy a significant portion of the subarachnoid space.

1.7 Vascular Anatomy of the Spinal Cord

The vascular network of the spinal cord undergoes remarkable changes during human development. A notable characteristic of the neonatal spinal cord is its proportionally higher blood supply relative to mature individuals, reflecting the metabolic demands of early development (Etz et al. 2011).

The developmental process of the spinal cord involves substantial volumetric changes. From an initial volume of approximately 6 mL in neonates, the spinal cord expands dramatically, reaching a volume more than 13 times larger in adults (Haines and Mihailoff 2018). This volumetric increase represents a significant adaptation to support the growing neurological functions and the increasing body size. However, the growth pattern of the spinal cord's blood vessels does not match this dramatic volume increase. The vascular network experiences a relatively modest expansion, with vessel diameters increasing by less than 50% from birth to adulthood. The artery of Adamkiewicz, which functions as a major blood supply to the spinal cord, achieves a maximum diameter of only 1.2 mm in adults. Other spinal cord vessels maintain even smaller diameters, ranging from 0.1 to 0.8 mm (Melissano et al. 2010; Melissano et al. 2015). This disproportionate relationship between spinal cord growth and vascular expansion creates a unique physiological condition. The limited growth of blood vessel diameters relative to the substantial increase in spinal cord volume may affect blood flow dynamics and could increase susceptibility to ischemic conditions in specific regions of the spinal cord.

1.7.1 Arterial Supply of the Spinal Cord

The spinal cord receives its arterial blood supply primarily from two sources: the vertebral arteries and the radicular arteries, which originate from the thoracic and abdominal aorta (Santillan et al. 2012). The vertebral artery typically emerges from the subclavian artery, although it may occasionally arise from the innominate artery or directly from the aorta. In more than 80% of cases, the vertebral artery originates at the superior dorsal aspect of the first segment of the subclavian artery, representing its most proximal and largest branch (Meila et al. 2012). The most frequent variation in origin relates to a more proximal location. On the left side, the vertebral artery emerges from the aortic arch between the left common carotid and left subclavian arteries in approximately 2.4–5.8% of instances (Komiyama et al. 2001). In such cases, the vertebral artery enters the foramen of the transverse process of the fifth cervical vertebra rather than the sixth cervical vertebra as is typical (Bruneau et al. 2006).

The vertebral artery's course is divided into four segments (Fig. 1.34):

- V1 segment: From origin to entry into the first transverse foramen (C6 level in 87.5% of cases) (Daseler and Anson 1959).
- V2 segment: Traverses the transverse foramina from C6 to C2.
- V3 segment: From C2 to dural penetration.
- V4 segment: Intradural portion, terminating at the vertebrobasilar junction.

The V2 segment, being the most susceptible to injury during common cervical spine procedures, requires careful radiographic evaluation prior to surgery due to potential tortuosity (Williams et al. 2017).

The vertebral arteries give rise to one anterior spinal artery and two posterior spinal arteries. Near the point where the vertebral arteries con-

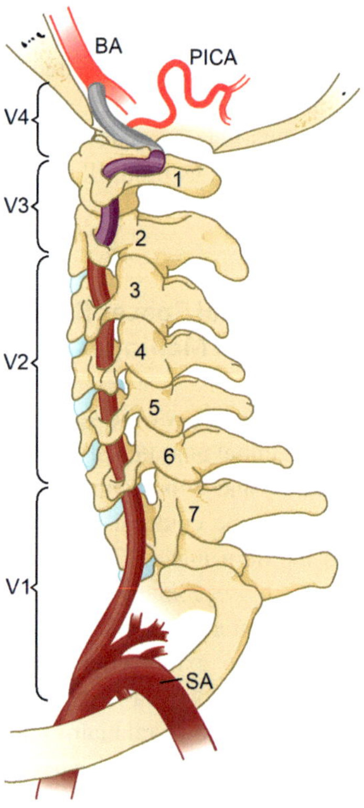

Fig. 1.34 Anatomy and course of the vertebral artery, divided into four segments. Abbreviations: BA, basilar artery; PICA, posterior inferior cerebellar artery; SA, subclavian artery. (Adapted with permission from Williams et al. (2017))

verge to form the basilar artery, each artery gives off a branch directed caudally and medially. These branches unite at the midline, typically fusing about 2 cm from their origin to form the anterior spinal artery. This vessel extends the entire length of the spinal cord, closely following the ventral median fissure, with its diameter varying according to the spinal segment. The two vertebral arteries merge to form a single midline artery, running along the ventral median fissure, and it is considered the longest artery in the body (Santillan et al. 2012). Dorsal to this structure, two posterior spinal arteries originate from either the vertebral or the dorsal inferior cerebellar arteries, following a course along the dorsal rootlet attachments on both sides (Vanderah and Gould 2016). These three principal arteries—one ventral and two dorsal—provide direct blood supply to the spinal cord. They receive supplementary input from the subclavian artery via the vertebral artery, thyrocervical trunk, and costocervical trunk along the vertebral column (Thron 2016) (Fig. 1.35).

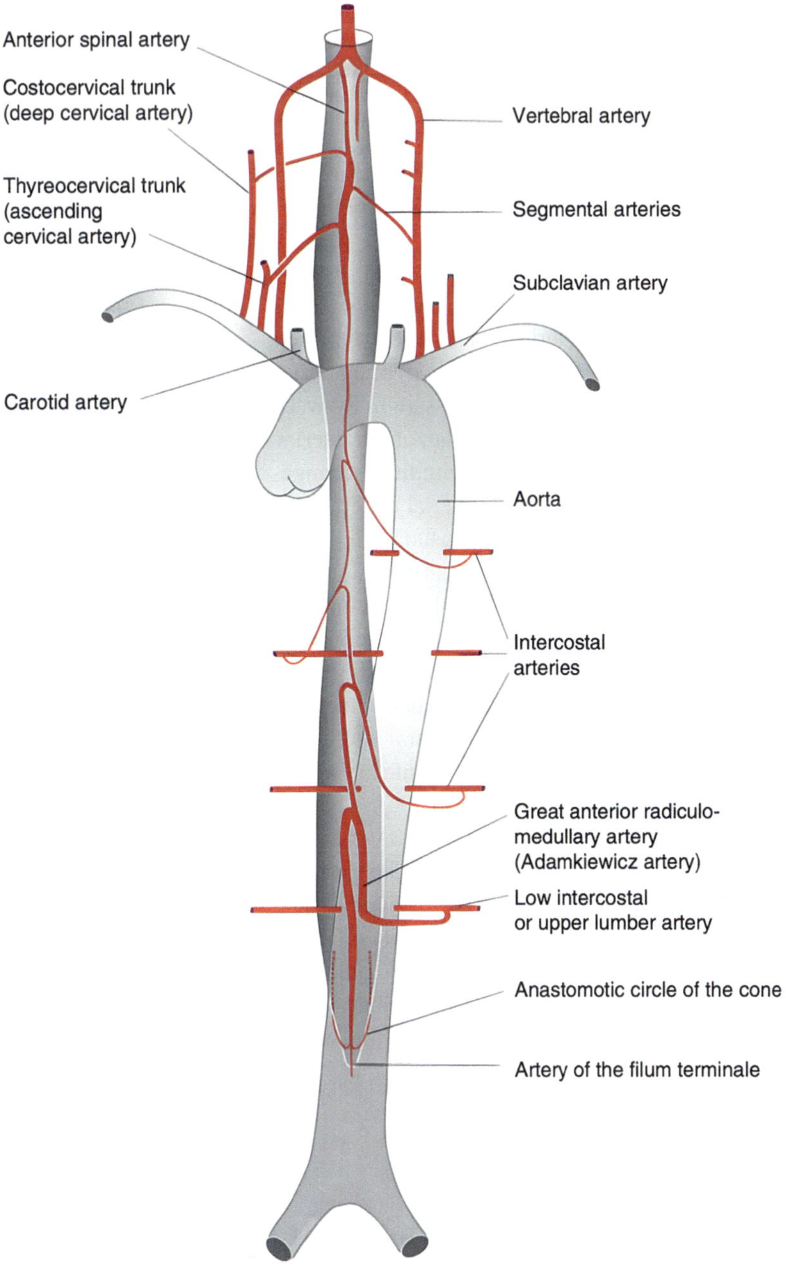

Fig. 1.35 Main arterial source supplying the anterior spinal artery. (Adapted with permission from Thron (2016))

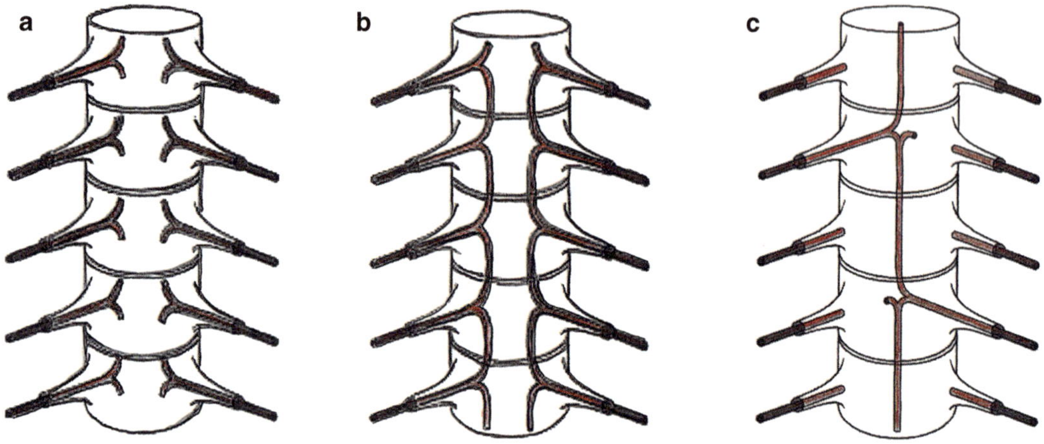

Fig. 1.36 Embryonic desegmentation of the anterior spinal artery from 31 bilateral segmental feeders. (**a**) Metameric stage: Each segmental artery follows its corresponding spinal nerve to supply a specific myelomeric segment of the neural tube. (**b**) Longitudinal anastomosis: Each segmental artery anastomoses with adjacent arteries to form a paired ventral longitudinal neural arterial axis. (**c**) Fusion and desegmentation: The ventral arteries fuse at the midline to form a single ventral spinal artery. Most radicular arterial sources regress, leaving a few to supply blood flow to the ventral spinal system, while others remain unimetameric to supply the nerve, dura, and vertebral body. (Adapted with permission from Lasjaunias et al. (2001))

The primary arterial supply to the spinal cord is variable, with the anterior spinal artery typically arising from two branches of the vertebral arteries at the level of the pyramidal decussation. Between the sixth week and the fourth month of intrauterine development, progressive longitudinal changes occur, resulting in the mature arterial pattern. This process develops in a craniocaudal direction. During this period, the two ventral neural arterial channels of the spinal cord fuse along the midline to form a single ventral longitudinal arterial axis, the anterior (ventral) spinal artery (Fig. 1.36a, b). Concurrently, desegmentation occurs, characterized by the regression of most ventral and dorsal radicular arterial sources that contribute to the ventral axis and the pial network.

The embryological development of the anterior spinal artery involves contributions from 31 bilateral segmental arteries. By the end of the fourth gestational month, the majority of these segmental arteries regress and obliterate (Fig. 1.36c). At the thoracic and lumbar levels, segmental arteries originating from the dorsal surface of the aorta are referred to as intercostal and lumbar arteries, respectively. In adults, only 2–14 (average 6) of these segmental arteries persist (Melissano et al. 2010; Melissano et al. 2015):

- Cervical level: 2–3 (left = right).
- Thoracic level: 2–3 (left > right).
- Lumbosacral level: 0–1 (left > right).

The segmental arteries of the vertebral column give rise to medullary feeder arteries, which traverse the intervertebral foramina alongside spinal nerves. In addition to these feeders, radicular arteries, which nourish the spinal nerve roots, also develop from the segmental arteries and the longitudinal arterial trunks (Windle 1980). While the dorsal medullary feeder arteries are more numerous, they are smaller in size compared to the ventral feeder arteries. The ventral medullary feeder arteries vary in number, ranging from 2 to 17, and their anatomical locations are also variable (Santillan et al. 2012) (Fig. 1.37). In the tho-

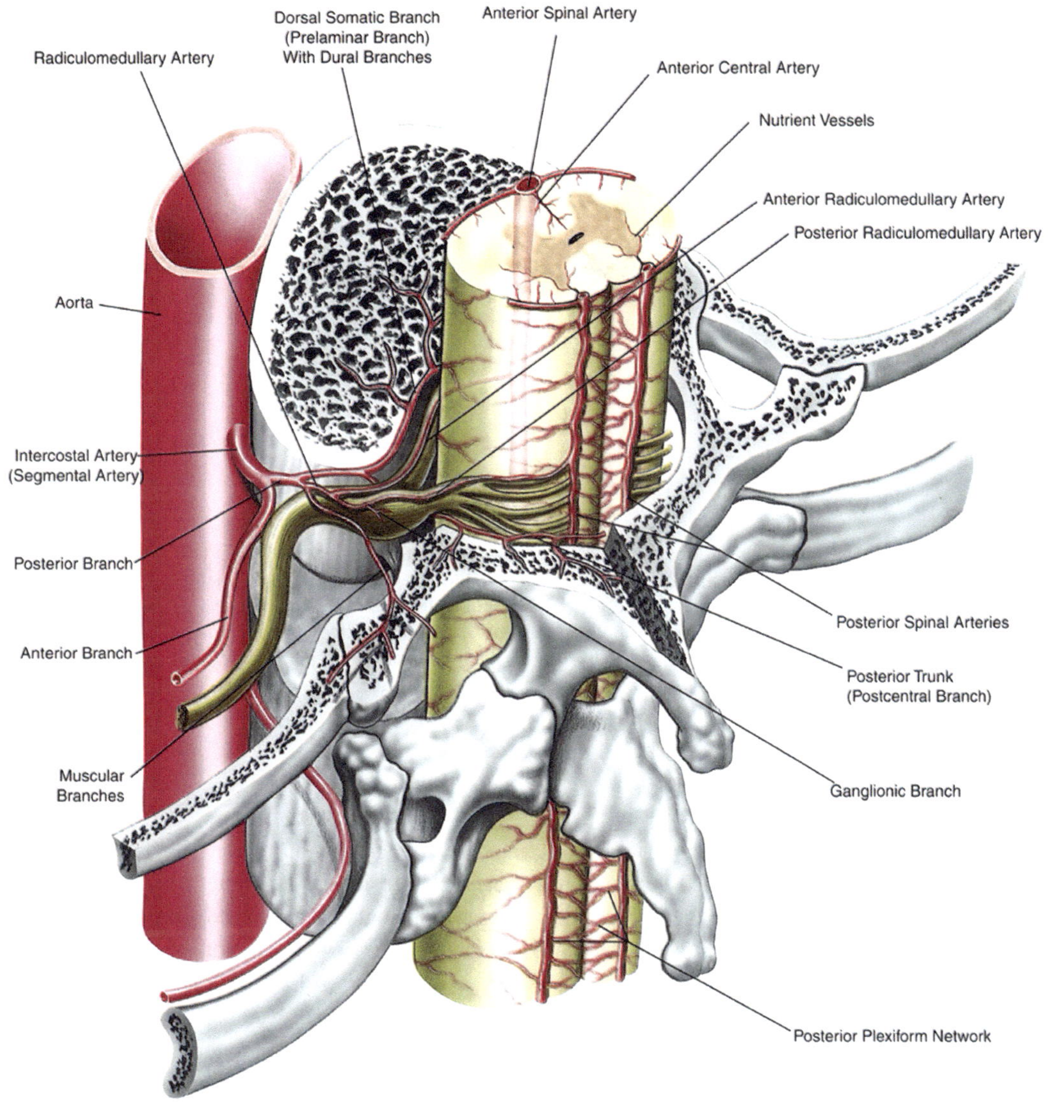

Fig. 1.37 Schematic representation of the spinal cord arterial system including segmental arteries (intercostal and lumbar arteries), radiculomedullary arteries, and spinal arteries. (Adapted from Uflacker and Guimaraes (2021))

racic region, these feeder arteries are fewer and smaller compared to those at higher and lower levels of the spinal cord. It is important to differentiate between medullary feeders and radicular arteries, as they serve distinct functions in spinal cord vasculature.

Blood supply to the spinal cord parenchyma is provided by small branches of the ventral longitudinal arterial trunk, specifically the anterior spinal artery, known as sulcal arteries. These arteries penetrate the cord via the ventral sulcus (Fig. 1.38). Sulcal arteries course dorsally through the ventral median sulcus and then laterally branch out to supply each half of the cord at each level (Martirosyan et al. 2011). The density of sulcal arteries ranges from 2 to 12 per centime-

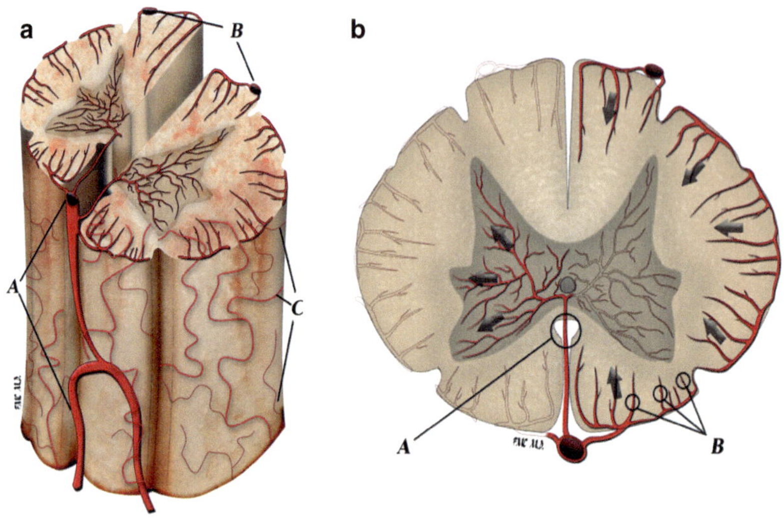

Fig. 1.38 (**a**) The spinal cord is covered by a net-like anastomosing vascular system, including the anterior spinal artery (A) and the two posterior spinal arteries (B). The pial network (C) and intramedullary anastomoses primarily facilitate slower circulatory adjustments. (**b**) The intrinsic arterial system of the spinal cord comprises a central (centrifugal) system and a peripheral system. The central system includes sulcal arteries originating from the anterior arterial tract and passing through the anterior median fissures (A). The peripheral system consists of numerous small arteries, referred to as "rami perforantes" (B). (Adapted with permission from Melissano et al. (2010), open access)

ter along the ventral longitudinal arterial trunks, with a higher concentration at the cervical and lumbosacral enlargements due to the larger spinal gray matter in these regions compared to the thoracic cord (Bosmia et al. 2015a, 2015b; Griepp et al. 2012).

The spinal column receives blood supply from radicular branches of segmental arteries, which accompany ventral nerve roots through intervertebral foramina. The distribution of these radicular branches is not uniform along the spinal cord. Cervical and lumbar enlargements are characterized by a higher density of radicular branches, while the thoracic cord receives fewer radicular arteries. This pattern likely reflects the reduced metabolic demands of the thoracic region, which contains less gray matter.

Consequently, the anterior spinal artery may be discontinuous in the thoracic region. Additionally, there is a significant alteration in the arterial supply in the upper thoracic region where the subclavian artery feeds the radicular arteries, transitioning to direct aortic supply. This area represents a watershed zone, particularly susceptible to systemic hypotension or occlusion of a single vessel. The mid-thoracic region (T4–T8) serves as a watershed area, positioned between the more robust anterior spinal artery in the rostral region and the caudal region, which receives blood from the larger radicular artery of Adamkiewicz (Murthy et al. 2010).

The artery of Adamkiewicz, a large radicular artery, plays a significant role in supplying the anterior spinal artery from the lower thoracic to upper lumbar regions. Originating from the abdominal aorta at the T9 level, it traverses the intervertebral foramen before anastomosing with the anterior spinal artery (Nijenhuis et al. 2007) (Fig. 1.39). This artery exhibits variability in its location, more frequently found on the left side and potentially arising anywhere between the T7 and L4 segments. In 30% of individuals, it

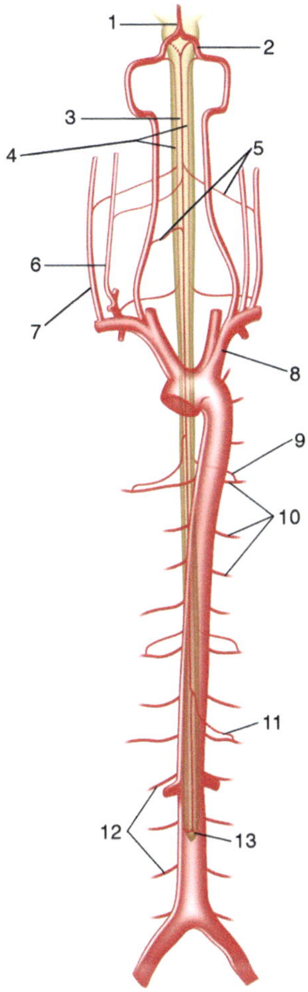

Fig. 1.39 Overview of main arterial system of the spinal cord, including the artery of Adamkiewicz and segmental arteries (intercostal and lumbar arteries). Key structures: (1) Basilar artery; (2) vertebral artery; (3) anterior spinal artery; (4) posterior spinal arteries; (5) anterior radiculomedullary artery; (6) ascending cervical artery; (7) deep cervical artery; (8) subclavian artery; (9) posterior radiculomedullary artery; (10) segmental arteries (posterior intercostal arteries); (11) great anterior radiculomedullary artery or artery of Adamkiewicz; (12) segmental arteries (lumbar arteries); (13) rami cruciantes. (Adapted from Santillan et al. (2012))

accompanies the left T10 ventral root, with the remainder associated with roots from T9 to L1 (Fig. 1.40). The artery of Adamkiewicz typically displays a cranial orientation. Furthermore, the Adamkiewicz artery transverses the intervertebral foramen in a ventral and superior direction (Murthy et al. 2010).

The anterior spinal artery supplies the ventral two-thirds of the spinal cord, including the nucleus dorsalis (Clarke's column) and the corticospinal tracts. In contrast, the dorsal one-third of the spinal cord, including the dorsal funiculi and dorsal gray matter, receives its blood supply from pia-penetrating arteriolar branches originating from the dorsolateral longitudinal arterial trunks, also known as the posterior spinal arteries (Zhang et al. 2013). A notable characteristic of the spinal cord vasculature is the lack of anastomoses between arteries entering the spinal cord, except at the conus medullaris. The conus medullaris features an anastomotic ring that incorporates segmental branches from lumbar and iliolumbar arteries, establishing a connection between the anterior and posterior arterial systems.

The posterior spinal arteries typically originate either directly from the vertebral artery in the cervical region or from the dorsal inferior cerebellar artery on the dorsolateral surface of the medulla oblongata. Post-origin, each artery bifurcates into two anastomotic tracts. Two dorsal spinal arteries, arising from the dorsal radicular branch arteries, form a small vessel plexus on the dorsal surface of the cord. These posterior spinal arteries also receive blood from various segmental arteries, including the ascending cervical, profound cervical, dorsal intercostal, and lumbar arteries. Circumflex vessels provide additional connections between the ventral and dorsal systems. While the posterior spinal arteries are less frequently affected by pathological processes compared to the ventral system, they play a significant role in supplying approximately the dorsal one-third of the cord, particularly the dorsal columns. This supply is mediated through the pial arterial plexus, which penetrates the cord's surface.

Within the spinal cord parenchyma, a complex capillary network is formed, with notably greater complexity observed in the gray matter compared

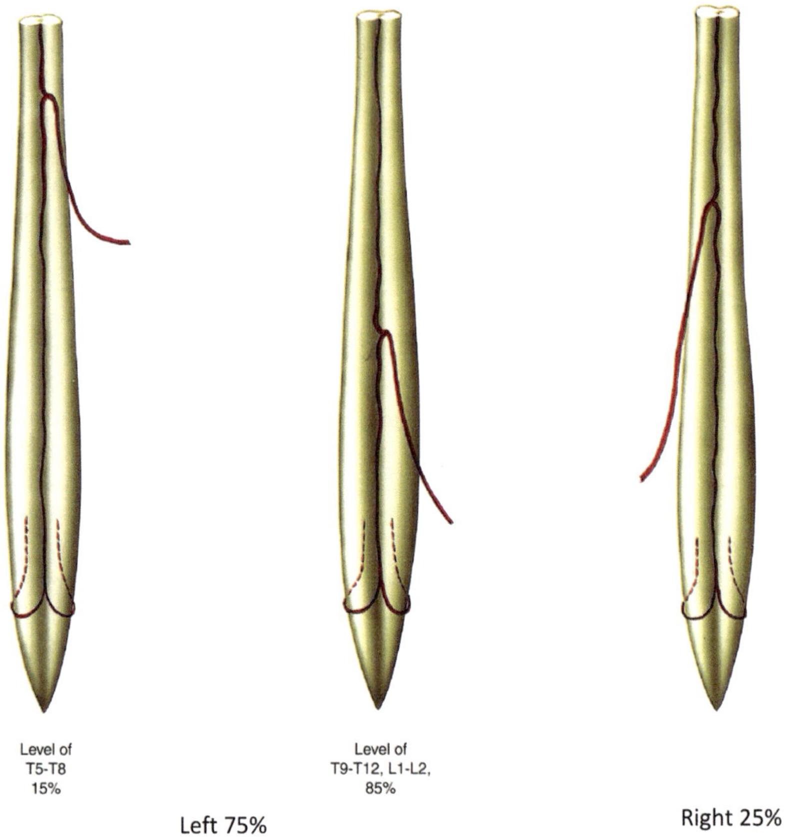

Fig. 1.40 Variability in the levels of origin of the artery of Adamkiewicz. (Adapted from Uflacker and Guimaraes (2021))

to the white matter (Scharrer 1944). Capillary density varies across different regions of the cord. For instance, the corticospinal tract exhibits approximately twice the capillary density of the fasciculus cuneatus (Zeman and Innes 1963). This differential vascular density likely reflects the varying metabolic demands of different spinal cord regions.

1.7.2 Venous Drainage of the Spinal Cord

The spinal cord's venous drainage demonstrates a complex organization with distinct anatomical patterns and regional variations. The system comprises six longitudinal, plexiform venous channels: an anterior (ventral) midline channel, a posterior (dorsal) midline channel, and two pairs of lateral channels that run along the attachment points of dorsal and ventral rootlets. These channels direct blood flow into radicular veins, which subsequently drain into the epidural venous plexus (Vanderah and Gould 2016).

The venous drainage system is divided into intrinsic and extrinsic components. The intrinsic system includes the outflow from the vertebral body through the basilar vein, while the extrinsic system encompasses all other venous structures involved in spinal cord drainage (Groen et al.

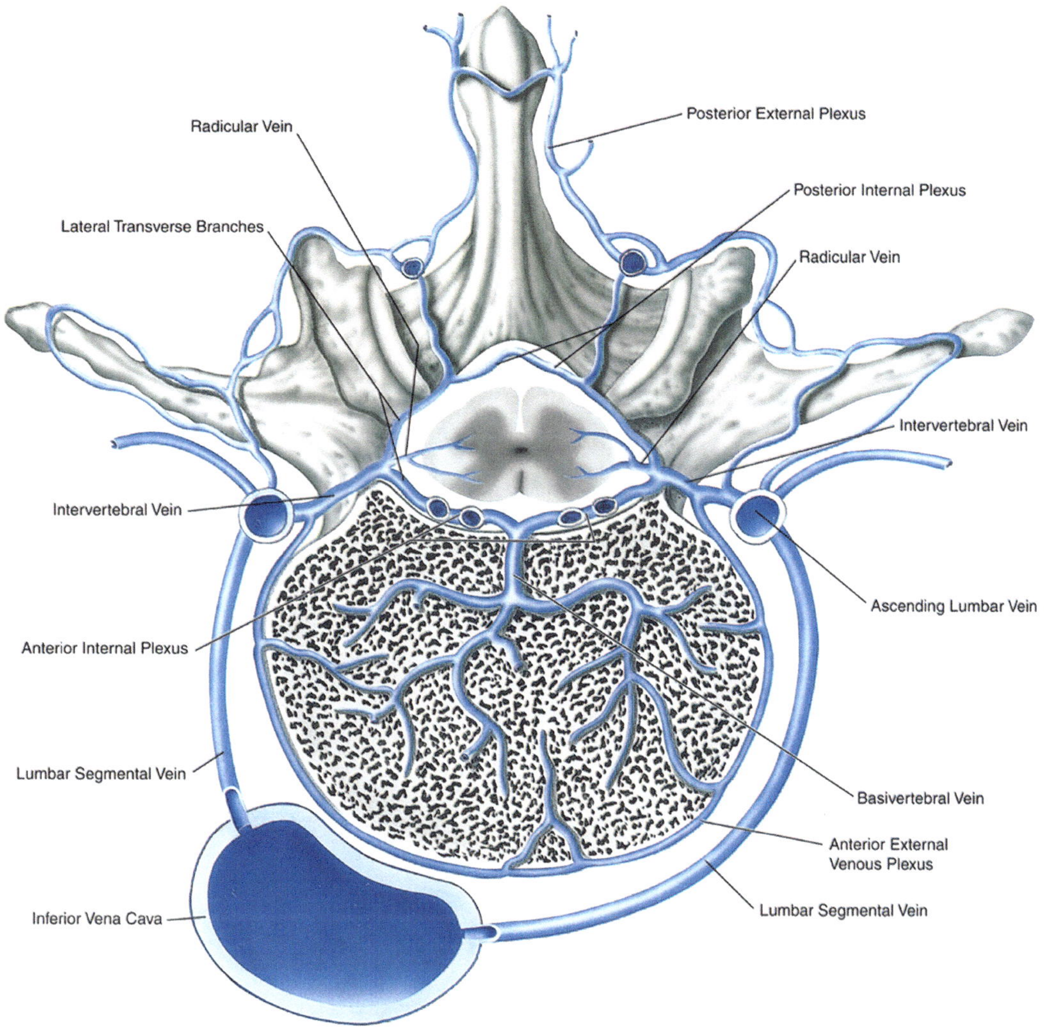

Fig. 1.41 Axial view of the vertebrae illustrating the close relationship between the anterior internal venous plexus and the posterior aspect of the vertebrae. Four vertical channels are visible in cross section, interconnected by lateral transverse veins. (Adapted from Uflacker and Guimaraes (2021))

1997) (Fig. 1.41). This dual system allows for efficient blood removal from both the deep and superficial regions of the spinal cord. The longitudinal organization of spinal cord veins parallels the arterial system, with ventral and dorsal spinal veins running along the length of the cord. However, these veins show notable caliber variations throughout their course. The ventral spinal veins typically accompany the anterior spinal artery, maintaining a close anatomical relationship (Santillan et al. 2012).

A distinctive feature of the spinal cord's vascular organization is the differential distribution density between arterial and venous networks. While the arterial network shows greater concentration on the ventral aspect, the venous network exhibits higher density on the dorsal portion of the cord (Santillan et al. 2012) (Fig. 1.42). In the

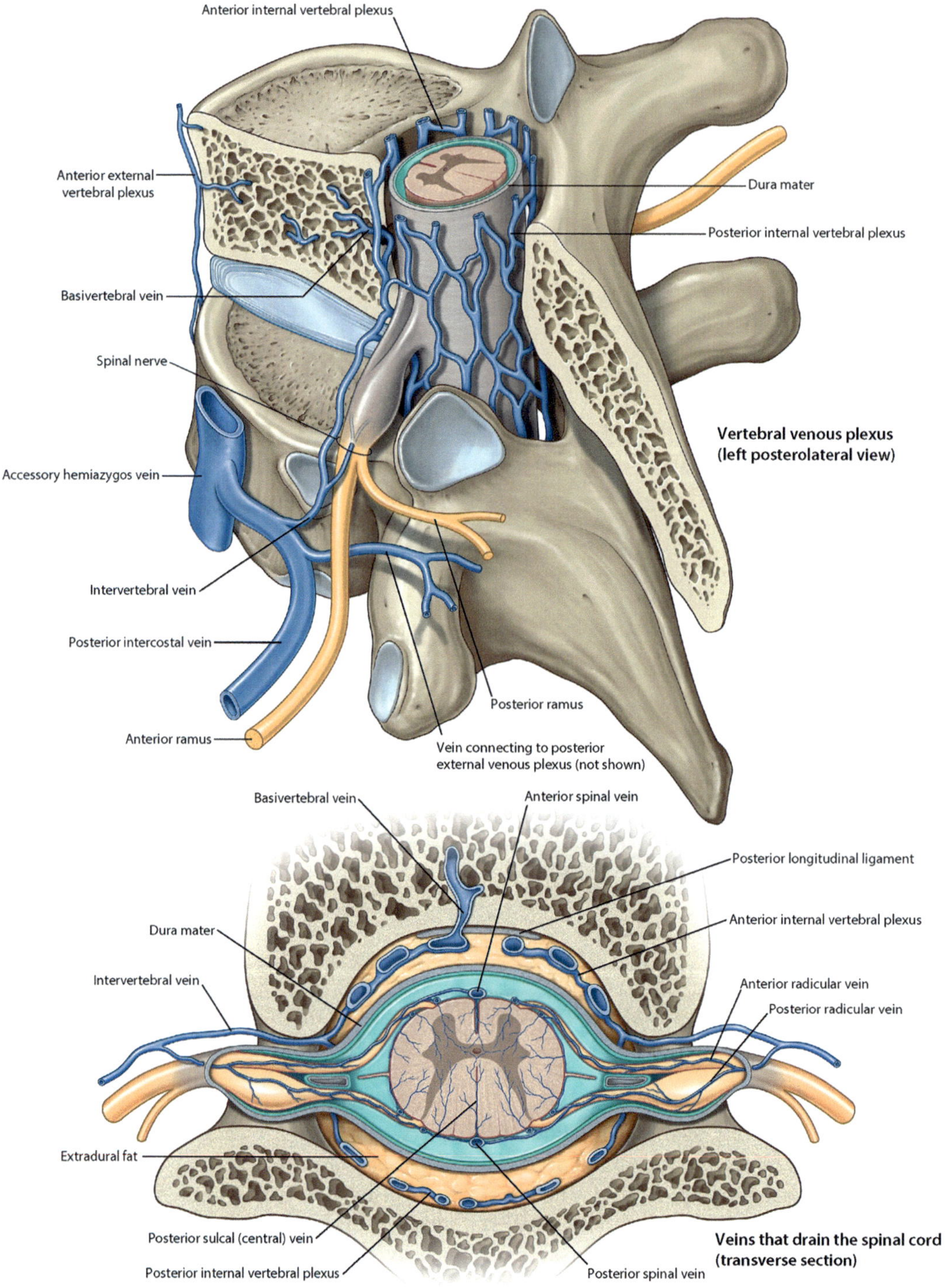

Fig. 1.42 Venous drainage of the spinal cord. The venous network is denser on the dorsal surface of the spinal cord, whereas the arterial network is more concentrated on the ventral aspect. (Adapted with permission from Drake et al. (2020))

dorsal region, prominent veins follow independent paths from the spinal arteries, draining into dorsal radicular veins that connect with every second or third nerve root.

The lumbar enlargement region displays a particularly well-developed venous plexus, reflecting the increased metabolic requirements of this area. Both ventral and dorsal venous plexuses ultimately channel blood into segmental veins, which then drain into the ascending lumbar veins, the azygos system, and the pelvic venous plexuses, forming an integrated drainage network.

References

Abraira VE, Kuehn ED, Chirila AM, Springel MW, Toliver AA, Zimmerman AL, et al. The cellular and synaptic architecture of the mechanosensory dorsal horn. Cell. 2017;168(1–2):295–310.e19.

Afifi AK, Bergman RA. Functional neuroanatomy: text and atlas. 2nd ed. New York: McGraw-Hill; 2005.

Albright AL, Pollack IF, Adelson PD. Principles and practice of pediatric neurosurgery. 2nd ed. New York: Thieme; 2008.

Altman J, Bayer SA. Development of the human spinal cord: an interpretation based on experimental studies. 1st ed. New York: Oxford University Press; 2001.

Armand J. The origin, course and terminations of corticospinal fibers in various mammals. Prog Brain Res. 1982;57:329–60.

Baehr M, Frotscher M. Duus' topical diagnosis in neurology. Anatomy, physiology, signs, symptoms. 5th ed. New York: Thieme; 2012.

Barson AJ. The vertebral level of termination of the spinal cord during normal and abnormal development. J Anat. 1970;106(Pt 3):489–97.

Benarroch EE, Westmoreland BF, Daube JR, Reagan TJ, Sandok BA. Medical neurosciences: an approach to anatomy, pathology, and physiology by systems and levels. 4th ed. Philadelphia: Lippincott Williams & Wilkins; 1999.

Bican O, Minagar A, Pruitt AA. The spinal cord: a review of functional neuroanatomy. Neurol Clin. 2013;31(1):1–18.

Blumenfeld H. Neuroanatomy through clinical cases. 2nd ed. Sunderland: Sinauer Associates; 2010.

Bogduk N. Clinical anatomy of the lumbar spine and sacrum. 4th ed. Edinburgh: Churchill Livingstone; 2005.

Boon JM, Abrahams PH, Meiring JH, Welch T. Lumbar puncture: anatomical review of a clinical skill. Clin Anat. 2004;17(7):544–53.

Bosmia AN, Hogan E, Loukas M, Tubbs RS, Cohen-Gadol AA. Blood supply to the human spinal cord: part I. Anatomy and hemodynamics. Clin Anat. 2015a;28(1):52–64.

Bosmia AN, Tubbs RS, Hogan E, Bohnstedt BN, Denardo AJ, Loukas M, et al. Blood supply to the human spinal cord: part II. Imaging and pathology. Clin Anat. 2015b;28(1):65–74.

Breig A, el-Nadi AF. Biomechanics of the cervical spinal cord. Relief of contact pressure on and overstretching of the spinal cord. Acta Radiol Diagn (Stockh). 1966;4(6):602–24.

Breig A, Turnbull I, Hassler O. Effects of mechanical stresses on the spinal cord in cervical spondylosis. A study on fresh cadaver material. J Neurosurg. 1966;25(1):45–56.

Brodbelt AR, Stoodley MA. CSF pathways: a review. Br J Neurosurg. 2007;21(5):510–20.

Brown PB, Fuchs JL. Somatotopic representation of hindlimb skin in cat dorsal horn. J Neurophysiol. 1975;38(1):1–9.

Bruneau M, Cornelius JF, Marneffe V, Triffaux M, George B. Anatomical variations of the V2 segment of the vertebral artery. Neurosurgery. 2006;59(1 Suppl 1):ONS20–4.

Cho TA. Spinal cord functional anatomy. Continuum (Minneap Minn). 2015;21(1 Spinal Cord Disorders):12–35.

Cho TA, Bhattacharyya S. Approach to myelopathy. Continuum (Minneap Minn). 2018;24(2, Spinal Cord Disorders):386–406.

Clifton GL, Coggeshall RE, Vance WH, Willis WD. Receptive fields of unmyelinated ventral root afferent fibers in the cat. J Physiol. 1976;256(3):573–600.

Clowry GJ, Moss JA, Clough RL. An immunohistochemical study of the development of sensorimotor components of the early fetal human spinal cord. J Anat. 2005;207(4):313–24.

Coggeshall RE, Coulter JD, Willis WD Jr. Unmyelinated axons in the ventral roots of the cat lumbosacral enlargement. J Comp Neurol. 1974;153(1):39–58.

Cole J, Weller RO. Introduction to the clinical presentations of spinal cord disease from a pathophysiological perspective. In: Engler GL, Cole J, Merton WL, editors. Spinal cord diseases: diagnosis and treatment. New York: Marcel Dekker, Inc; 1998. p. 1–13.

Cole P, Kaufman Y, Hatef DA, Hollier LH Jr. Embryology of the hand and upper extremity. J Craniofac Surg. 2009;20(4):992–5.

Conta AC, Stelzner DJ. Differential vulnerability of propriospinal tract neurons to spinal cord contusion injury. J Comp Neurol. 2004;479(4):347–59.

Coppola AR. "Cruciate paralysis": a complication of surgery. South Med J. 1973;66:684.

Cramer GD, Darby SA. Clinical anatomy of the spine, spinal cord, and ANS. 3rd ed. St. Louis: Elsevier Health Sciences; 2014.

Critchley E, Eisen A. Spinal cord disease: basic science, diagnosis and management. London: Springer; 1997.

Daseler EH, Anson BJ. Surgical anatomy of the subclavian artery and its branches. Surg Gynecol Obstet. 1959;108(2):149–74.

Drake RL, Vogl AW, Mitchell AWM, et al. Gray's anatomy for students. 4th ed. Philadelphia: Elsevier; Churchill Livingstone; 2020.

Dumitru D, Lang JE. Cruciate paralysis. Case report. J Neurosurg. 1986;65(1):108–10.

Elliott HC. Studies on the motor cells of the spinal cord. I. Distribution in the normal human cord. Am J Anat. 1942;70(1):95–117.

Etz CD, Kari FA, Mueller CS, Silovitz D, Brenner RM, Lin H-M, et al. The collateral network concept: a reassessment of the anatomy of spinal cord perfusion. J Thorac Cardiovasc Surg. 2011;141(4):1020–8.

Felten DL, O'Banion MK, Maida MS. Netter's atlas of neuroscience. 3rd ed. Philadelphia: Elsevier; 2016.

Flynn JR, Graham BA, Galea MP, Callister RJ. The role of propriospinal interneurons in recovery from spinal cord injury. Neuropharmacology. 2011;60(5):809–22.

Greenberg MS. Handbook of neurosurgery. 8th ed. New York: Thieme; 2016.

Griepp EB, Di Luozzo G, Schray D, Stefanovic A, Geisbüsch S, Griepp RB. The anatomy of the spinal cord collateral circulation. Ann Cardiothorac Surg. 2012;1(3):350–7.

Groen RJ, Groenewegen HJ, van Alphen HA, Hoogland PV. Morphology of the human internal vertebral venous plexus: a cadaver study after intravenous Araldite CY 221 injection. Anat Rec. 1997;249(2):285–94.

Haines DE, Mihailoff GA. Fundamental neuroscience for basic and clinical applications. 5th ed. Philadelphia: Elsevier; 2018.

Haines DE, Harkey HL, al-Mefty O. The "subdural" space: a new look at an outdated concept. Neurosurgery. 1993;32(1):111–20.

Heimer L. The human brain and spinal cord: functional neuroanatomy and dissection guide. New York: Springer-Verlag; 1983.

Holmes A, Han ZH, Dang GT, Chen ZQ, Wang ZG, Fang J. Changes in cervical canal spinal volume during in vitro flexion-extension. Spine (Phila Pa 1976). 1996;21(11):1313–9.

Hsu JD, Michael JW, Fisk JR. AAOS atlas of orthoses and assistive devices. 4th ed. Philadelphia: Mosby; 2008.

Jaumard NV, Welch WC, Winkelstein BA. Spinal facet joint biomechanics and mechanotransduction in normal, injury and degenerative conditions. J Biomech Eng. 2011;133(7):071010.

Kameyama T, Hashizume Y, Sobue G. Morphologic features of the normal human cadaveric spinal cord. Spine (Phila Pa 1976). 1996;21(11):1285–90.

Kandel ER, Schwartz JH, Jessell TM, Siegelbaum SA, Hudspeth AJ. Principles of neural science. 5th ed. New York: McGraw-Hill; 2013.

Kierszenbaum AL, Tres L. Histology and cell biology: an introduction to pathology. 4th ed. Philadelphia: Elsevier; 2016.

Kim HJ, Tetreault LA, Massicotte EM, Arnold PM, Skelly AC, Brodt ED, et al. Differential diagnosis for cervical spondylotic myelopathy: literature review. Spine (Phila Pa 1976). 2015;40(1):S78–88.

Ko HY, Park JH, Shin YB, Baek SY. Gross quantitative measurements of spinal cord segments in human. Spinal Cord. 2004;42(1):35–40.

Komiyama M, Morikawa T, Nakajima H, Nishikawa M, Yasui T. High incidence of arterial dissection associated with left vertebral artery of aortic origin. Neurol Med Chir (Tokyo). 2001;41(1):8–11.

Lasjaunias P, Berenstein A, ter Brigge KG. Surgical neuroangiography. I clinical vascular anatomy and variation. 2nd ed. New York: Springer; 2001.

Lawrence DG, Kuypers HG. The functional organization of the motor system in the monkey. I. The effects of bilateral pyramidal lesions. Brain. 1968;91(1):1–14.

Lee MW, McPhee RW, Stringer MD. An evidence-based approach to human dermatomes. Clin Anat. 2008;21(5):363–73.

Lemon RN. Descending pathways in motor control. Annu Rev Neurosci. 2008;31:195–218.

Mai JK, Paxinos G, editors. The human nervous system. 3rd ed. San Diego: Academic; 2011.

Martirosyan NL, Feuerstein JS, Theodore N, Cavalcanti DD, Spetzler RF, Preul MC. Blood supply and vascular reactivity of the spinal cord under normal and pathological conditions. J Neurosurg Spine. 2011;15(3):238–51.

Meila D, Tysiac M, Petersen M, Theisen O, Wetter A, Mangold A, et al. Origin and course of the extracranial vertebral artery: CTA findings and embryologic considerations. Clin Neuroradiol. 2012;22(3):327–33.

Melissano G, Civilini E, Bertoglio L, Calliari F, Campos Moraes Amato A, Chiesa R. Angio-CT imaging of the spinal cord vascularisation: a pictorial essay. Eur J Vasc Endovasc Surg. 2010;39(4):436–40.

Melissano G, Bertoglio L, Rinaldi E, Civilini E, Tshomba Y, Kahlberg A, et al. An anatomical review of spinal cord blood supply. J Cardiovasc Surg. 2015;56(5):699–706.

Milhorat TH, Kotzen RM, Anzil AP. Stenosis of central canal of spinal cord in man: incidence and pathological findings in 232 autopsy cases. J Neurosurg. 1994;80(4):716–22.

Moore KL, Dalley AF, Agur AMR. Clinically oriented anatomy. 8th ed. Philadelphia: Wolters Kluwer; 2018.

Mtui E, Gruener G, Dockery P. Fitzgerald's clinical neuroanatomy and neuroscience. 7th ed. Philadelphia: Elsevier; 2016.

Müller F, O'Rahilly R. The early development of the nervous system in staged insectivore and primate embryos. J Comp Neurol. 2014;527(7):1368–84.

Murthy NS, Maus TP, Behrns CL. Intraforaminal location of the great anterior radiculomedullary artery (artery of Adamkiewicz): a retrospective review. Pain Med. 2010;11(12):1756–64.

Netter FH. Atlas of human anatomy. 7th ed. Philadelphia: Elsevier; 2019.

Nieuwenhuys R, Voogd J, van Huijzen C. The human central nervous system. 4th ed. Berlin: Springer; 2008.

Nijenhuis RJ, Jacobs MJ, Schurink GW, Kessels AG, van Engelshoven JM, Backes WH. Magnetic resonance angiography and neuromonitoring to assess spinal cord blood supply in thoracic and thoracoabdominal aortic aneurysm surgery. J Vasc Surg. 2007;45(1):71–7.

Nolte J, Angevine JB. The human brain in photographs and diagrams. 4th ed. Philadelphia: Elsevier; 2013.

Norenberg MD, Smith J, Marcillo A. The pathology of human spinal cord injury: defining the problems. J Neurotrauma. 2004;21(4):429–40.

Novy J, Carruzzo A, Maeder P, Bogousslavsky J. Spinal cord ischemia: clinical and imaging patterns, pathogenesis, and outcomes in 27 patients. Arch Neurol. 2006;63(8):1113–20.

Nowicki BH, Haughton VM, Schmidt TA, Lim TH, An HS, Riley LH 3rd. Occult lumbar lateral spinal stenosis in neural foramina subjected to physiologic loading. AJNR Am J Neuroradiol. 1996;17(9):1605–14.

Nyberg G, Blomqvist A. The somatotopic organization of forelimb cutaneous nerves in the brachial dorsal horn: an anatomical study in the cat. J Comp Neurol. 1985;242(1):28–39.

O'Rahilly R, Müller F. Human embryology and teratology. 3rd ed. New York: Wiley-Liss; 2001.

Palastanga N, Soames R. Anatomy and human movement: structure and function. 6th ed. Philadelphia: Elsevier; 2012.

Parent A. Carpenter's human neuroanatomy. 9th ed. Baltimore: Williams & Wilkins; 1996.

Patestas MA, Gartner LP. A textbook of neuroanatomy. Oxford: Blackwell Publishing; 2006.

Posner JB, Saper CB, Schiff ND, Claassen J. Plum and Posner's diagnosis of stupor and coma. 5th ed. Oxford: Oxford University Press; 2019.

Purves D, Augustine GJ, Fitzpatrick D, Hall WC, LaMantia AS, White LE. Neuroscience. 6th ed. Oxford: Oxford University Press; 2018.

Rexed B. The cytoarchitectonic organization of the spinal cord in the cat. J Comp Neurol. 1952;96(3):414–95.

Rexed B, Brodal A. The nucleus cervicalis lateralis: a spinocerebellar relay nucleus. J Neurophysiol. 1951;14(5):399–407.

Romanes GJ. The motor cell columns of the lumbosacral spinal cord of the cat. J Comp Neurol. 1951;94(2):313–63.

Rubin MN, Rabinstein AA. Vascular diseases of the spinal cord. Neurol Clin. 2013;31(1):153–81.

Sadler TW. Langman's medical embryology. 12th ed. Philadelphia: Wolters Kluwer; 2012.

Saker E, Henry BM, Tomaszewski KA, Loukas M, Iwanaga J, Oskouian RJ, et al. The human central canal of the spinal cord: a comprehensive review of its anatomy, embryology, molecular development, variants, and pathology. Cureus. 2016;8(12):e927.

Saker E, Henry BM, Tomaszewski KA, Loukas M, Iwanaga J, Oskouian RJ, et al. The filum terminale internum and externum: a comprehensive review. J Clin Neurosci. 2017;40:6–13.

Santillan A, Nacarino V, Greenberg E, Riina HA, Gobin YP, Patsalides A. Vascular anatomy of the spinal cord. J Neurointerv Surg. 2012;4(1):67–74.

Scharrer E. The blood vessels of the nervous tissue. Q Rev Biol. 1944;19(4):308–18.

Schoenwolf GC, Bleyl SB, Brauer PR, Francis-West PH. Larsen's human embryology. 5th ed. Philadelphia: Churchill Livingstone; 2015.

Schuenke M, Schulte E, Schumacher U. Thieme atlas of anatomy: head, neck, and neuroanatomy. 3rd ed. New York: Thieme; 2020.

Sharrard WJ. The distribution of the permanent paralysis in the lower limb in poliomyelitis: a clinical and pathological study. J Bone Joint Surg Br. 1955;37-B(4):540–58.

Sisto SA, Druin E, Macht Sliwinski M. Spinal cord injuries: management and rehabilitation. St. Louis: Mosby Elsevier; 2008.

Smith MC, Deacon P. Topographical anatomy of the posterior columns of the spinal cord in man: the long ascending fibres. Brain. 1984;107(3):671–98.

Standring S. Gray's anatomy: the anatomical basis of clinical practice. 42nd ed. Edinburgh: Churchill Livingstone; 2020.

Strominger NL, Demarest RJ, Laemle LB. Noback's human nervous system: structure and function. 7th ed. New York: Humana Press; 2012.

ten Donkelaar HJ. Evolution of the red nucleus and rubrospinal tract. Behav Brain Res. 1988;28(1–2):9–20.

ten Donkelaar HJ, Lammens M, Wesseling P, Hori A, Keyser A, Rotteveel J. Development and malformations of the human pyramidal tract. J Neurol. 2004;251(12):1429–42.

Thron AK. Vascular anatomy of the spinal cord: radioanatomy as the key to diagnosis and treatment. 2nd ed. Cham: Springer; 2016.

Todd AJ. Neuronal circuitry for pain processing in the dorsal horn. Nat Rev Neurosci. 2010;11(12):823–36.

Toribatake Y, Baba H, Kawahara N, Mizuno K, Tomita K. The epiconus syndrome presenting with radicular-type neurological features. Spinal Cord. 1997;35(3):163–70.

Turnbull IM, Brieg A, Hassler O. Blood supply of cervical spinal cord in man. A microangiographic cadaver study. J Neurosurg. 1966;24(6):951–65.

Uflacker A, Guimaraes M. Uflacker's atlas of vascular anatomy: an angiographic approach. 3rd ed. Philadelphia: Wolters Kluwer; 2021.
Vandenabeele F, Creemers J, Lambrichts I. Ultrastructure of the human spinal arachnoid mater and dura mater. J Anat. 1996;189(Pt 2):417–30.
Vanderah TW, Gould DJ. Nolte's the human brain: an introduction to its functional anatomy. 7th ed. Philadelphia: Elsevier; 2016.
Watson C, Harrison M. The location of the major ascending and descending spinal cord tracts in all spinal cord segments in the mouse: actual and extrapolated. Anat Rec. 2012;295(10):1692–7.
Watson C, Paxinos G, Kayalioglu G. The spinal cord: a Christopher and Dana Reeve Foundation text and atlas. New York: Academic; 2009.
Watson C, Kirkcaldie MTK, Paxinos G. The brain: an introduction to functional neuroanatomy. London: Academic; 2010.
Watson C, Paxinos G, Puelles L. The mouse nervous system. London: Academic; 2012.
Webster S, de Wreede R. Embryology at a glance. 2nd ed. West Sussex: Wiley Blackwell; 2016.
Whitsel BL, Petrucelli LM, Sapiro G. Modality representation in the lumbar and cervical fasciculus gracilis of squirrel monkeys. Brain Res. 1969;15(1):67–78.
Whitsel BL, Petrucelli LM, Sapiro G, Ha H. Fiber sorting in the fasciculus gracilis of squirrel monkeys. Exp Neurol. 1970;29:227–42.
Williams KA, Rauschning W, Prasad S. Applied anatomy of the cervical spine. In: Steinmetz MP, Benzel EC, editors. Benzel's spine surgery. 4th ed. Philadelphia: Elsevier; 2017.
Windle WF. The spinal cord and its reaction to traumatic injury. New York: Marcel Dekker; 1980.
Yasui K, Hashizume Y, Yoshida M, Kameyama T, Sobue G. Age-related morphologic changes of the central canal of the human spinal cord. Acta Neuropathol. 1999;97(3):253–9.
Zeman W, Innes JR. Craigie's neuroanatomy of the rat. New York: Academic; 1963.
Zhang S, Wadhwa R, Haydel J, Toms J, Johnson K, Guthikonda B. Spine and spinal cord trauma: diagnosis and management. Neurol Clin. 2013;31(1):183–206.

Recommended Additional Reading

Byrne TN, Benzel EC, Waxman SG. Diseases of the spine and spinal cord. Oxford: Oxford University Press; 2000.
Jacobson S, Marcus EM, Pugsley S. Neuroanatomy for the neuroscientist. 3rd ed. Cham: Springer; 2018.
Kaiser MG, Haid RW, Shaffrey CI, Fehlings MG. Degenerative cervical myelopathy and radiculopathy: treatment approaches and options. Cham: Springer; 2019.
Kirshblum S, Lin VM. Spinal cord medicine. 3rd ed. New York: Demos Medical; 2019.
Ko HY. Management and rehabilitation of spinal cord injuries. 2nd ed. Singapore: Springer; 2022.
Ko HY. A practical guide to care of spinal cord injuries: clinical questions and answers. Singapore: Springer; 2023.
König A, Spetzger U. Degenerative diseases of the cervical spine: therapeutic management in the subaxial section. Cham: Springer; 2017.
Mancall EL, Brock DG. Gray's clinical neuroanatomy: the anatomical basis for clinical neuroscience. Philadelphia: Elsevier; 2011.
ten Donkelaar HJ. Clinical neuroanatomy: brain circuitry and its disorders. New York: Springer; 2011.
Zholudeva LV, Lane MA. Spinal interneurons: plasticity after spinal cord injury. London: Academic; 2023.

2 Anatomy and Function of Motor Units and Peripheral Nerves

The motor unit constitutes the fundamental functional unit of neuromuscular control and movement production (Kandel et al. 2013). Each motor unit comprises a specific anatomical sequence: the ventral horn cell (α-motoneuron), its axonal projection, ventral roots, spinal nerves, neural plexus, peripheral nerves, neuromuscular junctions (motor endplates), and the muscle fibers receiving innervation (Fig. 2.1). The perikaryon of the ventral horn cell contains the nucleus and the cellular organization essential for protein synthesis and metabolic functions. This cellular structure supports the preservation of neural function and the formation of specific cellular projections—axons and dendrites (Purves et al. 2018).

At the molecular level, the peripheral nervous system exhibits unique characteristics in myelin formation and maintenance. Unlike the central nervous system myelin produced by oligodendrocytes, peripheral myelin is generated by Schwann cells. This structural difference contributes to the peripheral nervous system's enhanced regenerative capacity following injury. The system encompasses multiple specialized components: spinal nerve roots emerging from the spinal cord, dorsal root ganglia containing primary sensory neuron cell bodies, peripheral nerves, and both motor and autonomic nerve terminals (Watson et al. 2009).

The spinal cord gives rise to 31 pairs of spinal nerves: 8 cervical, 12 thoracic, 5 lumbar, 5 sacral,

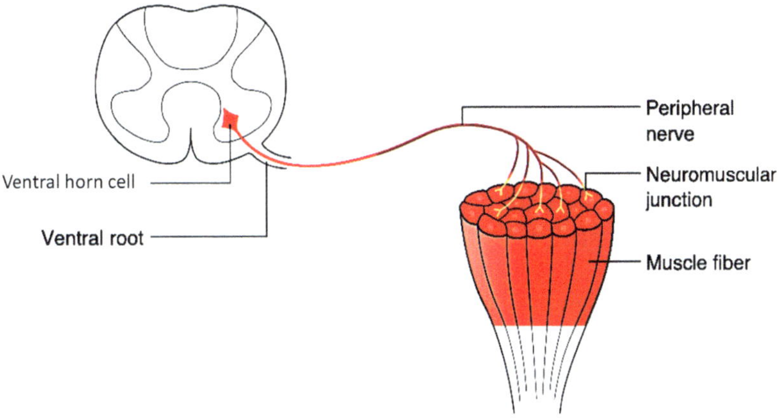

Fig. 2.1 The motor unit. An α-motoneuron located in the ventral horn extends an axon that branches and arborizes in the periphery, innervating scattered muscle fibers within a muscle fascicule. (Adapted from Campbell (2013))

H.-Y. Ko, *Practical Functional Anatomy for Spinal Cord Injury Rehabilitation*,
https://doi.org/10.1007/978-981-95-6457-6_2

and 1 coccygeal (Netter 2019). Each spinal nerve originates from the convergence of dorsal and ventral roots within the intervertebral foramen. The dorsal root carries sensory information through pseudounipolar neurons, whose cell bodies reside in the dorsal root ganglion, while the ventral root contains axons from α-motoneurons and preganglionic autonomic neurons (Kiernan and Rajakumar 2014).

Following their formation, spinal nerves divide into dorsal and ventral rami, each serving specific anatomical territories. The dorsal rami maintain relatively simple segmental patterns, innervating the paraspinal muscles, facet joints, and overlying skin through medial and lateral branches. The ventral rami demonstrate more complex organizational patterns, particularly in regions forming neural plexuses. Thoracic ventral rami preserve segmental organization, forming intercostal and subcostal nerves. In the cervical, brachial, and lumbosacral regions, adjacent ventral rami reorganize into neural plexuses through complex anastomotic patterns.

The cervical plexus, composed of the ventral rami of C1–C4, supplies sensory innervation to the neck and upper thorax, while also providing motor input to the deep neck muscles. The brachial plexus, originating from the ventral rami of C5–T1, undergoes a systematic reconstruction through roots, trunks, divisions, cords, and terminal branches to innervate the upper extremity. The lumbosacral plexus, consisting of the L1–S4 ventral rami, innervates the lower extremity and pelvic structures via a similar anatomical rearrangement (Standring 2020).

2.1 Ventral (Anterior) Horn Motor Neurons in the Spinal Cord

The ventral horn motor neurons, located in the Rexed lamina IX of the spinal cord, comprise several neuronal populations essential for motor control. Among these populations are α-motoneurons (30–70 μm in diameter) and smaller γ-motoneurons (<37 μm in diameter) (Vanderah and Gould 2016). The α-motoneurons provide direct innervation to skeletal muscle fibers, while γ-motoneurons, also known as fusimotor neurons, specifically innervate intrafusal fibers within muscle spindles (Kandel et al. 2013). The morphological characteristics of α-motoneurons include multiple dendrites (three to 20 per cell), which extend into laminae VII and VIII. These dendrites can span adjacent spinal segments, facilitating the integration of motor control signals. Their axons emerge from the axon hillock with an initial unmyelinated segment of approximately 3.5 μm diameter, which increases as they enter the white matter.

The organization of these motor neurons follows a precise somatotopic arrangement within lamina IX (Fig. 2.2). Neurons positioned medially control the axial and proximal limb musculature, while lateral neurons govern the distal limb muscles. Additionally, neurons in ventral positions predominantly innervate the extensor muscles, whereas those located more laterally control the flexor muscles (Craw 1928; Watson and Harrison 2012) (Fig. 2.3).

The distribution and density of motor neurons vary significantly across spinal segments. The cervical and lumbar enlargements contain the highest concentration of motor neurons, corresponding to the innervation of the upper and lower limbs. Segments L1–S5 contain approximately 52,000–62,000 motor neurons, with L5 typically housing around 10,000 neurons. The S1 segment exhibits the highest density of motor neurons per cross section (Standring 2020). Age-related changes impact these populations, with an estimated loss of 200 lamina IX neurons per decade after age 60 (Watson and Harrison 2012).

An important component of motor control involves Renshaw cells, specialized interneurons receiving collateral input from α-motoneurons. These cells participate in a negative feedback loop that regulates motor neuron firing rates through recurrent inhibition (Fig. 2.4). When motor neuron firing increases, Renshaw cells enhance their inhibitory influence, stabilizing the output. They also form inhibitory connections with Ia inhibitory interneurons affecting antagonistic motor neurons, creating a coordinated control system (Burke et al. 1970).

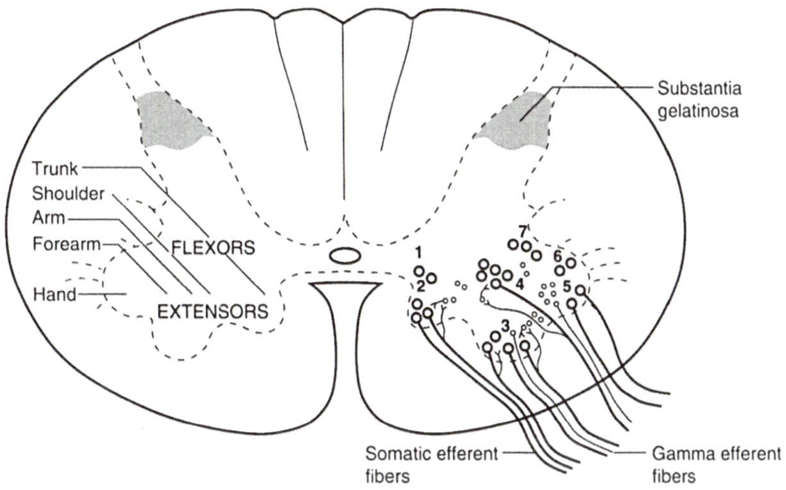

Fig. 2.2 Motor nuclei in ventral gray horn of a lower cervical segment of spinal cord. On the left is shown the general location of ventral horn cells that send motor axons to specific muscle groups of the upper extremity. Motor nuclei indicated on the right are as follows: (1) posteromedial, (2) anteromedial, (3) anterior, (4) central, (5) anterolateral, (6) posterolateral, (7) retroposterolateral. Smaller ventral horn cells send axons (γ-efferents) to supply small muscle fibers of neuromuscular spindle. (Adapted from Parent (1996))

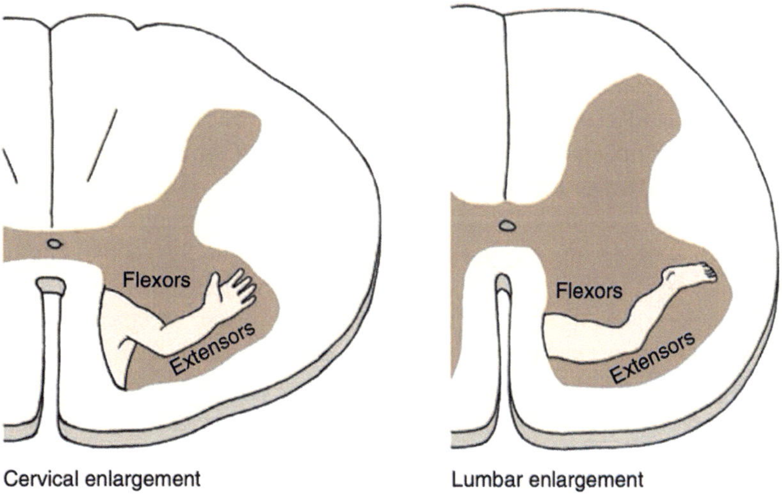

Fig. 2.3 Somatotopic arrangement of motor neurons in laminae VIII and IX and the location of the neurons innervating the flexors and extensors and proximal and distal muscles. (Adapted from Siegel and Sapru (2011))

The susceptibility of motor neuron columns to injury varies based on their size and length (see Figs. 1.29 and 1.30). Motor neuron pools controlling smaller muscles, while generally less vulnerable to initial damage due to their compact organization, may experience more severe functional deficits when injured (Elliott 1942; Sharrard 1955). This anatomical feature has significant implications for recovery patterns following spinal cord injury.

Though various classification schemes exist for ventral horn motor neurons, the current convention recognizes two primary groups: medial and lateral, each with distinct subdivisions. This simplified organization reflects the most anatomically and functionally evident groupings.

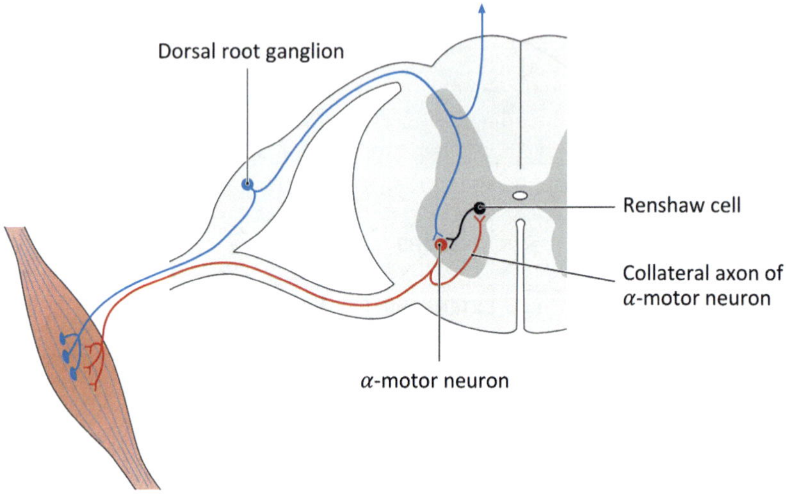

Fig. 2.4 Effect of the Renshaw cell on α-motoneuron. (Adapted from Schuenke et al. (2016))

2.1.1 Medial Motor Nuclear Group

The spinal cord contains a medial motor nuclear complex with dorsomedial and ventromedial subdivisions, representing a fundamental organization pattern of motor control (Parent 1996). The ventromedial division extends throughout the spinal cord's length, showing marked cellular density in segments C1, C2, C4, T1, T2, L3, L4, S2, and S3. This cellular arrangement continues rostrally into the medulla as the hypoglossal nucleus, forming an anatomical and functional continuum (Watson et al. 2009).

The dorsomedial subdivision exhibits smaller dimensions but maintains clear anatomical boundaries within cervical and lumbar enlargements. This subdivision may be anatomically indistinct or absent in sacral segments. Together, these medial motor neurons form synaptic connections with the axial musculature, including both superficial and deep muscles attached to the vertebral column (Kandel et al. 2013).

2.1.2 Lateral Motor Nuclear Group

The lateral motor nuclear group manages peripheral musculature through a systematic anatomical arrangement (Watson et al. 2009). In thoracic segments, this nuclear group remains compact and unified, primarily controlling intercostal and ventrolateral trunk musculature. The complex expands significantly in cervical and lumbar enlargements, forming multiple subgroups that control limb movements (Standring 2020) (Fig. 2.3).

The nuclear organization follows a mediolateral topography corresponding to proximal–distal muscle innervation patterns. Moving laterally from the ventral horn's medial border, motor neurons sequentially innervate the trunk muscles, girdle muscles (shoulder/pelvis), proximal limb muscles (upper arm/thigh), and distal limb muscles (forearm/leg). The retrodorsolateral subdivision specifically controls the hand and foot musculature (Kandel et al. 2013).

Ventrolateral cell columns provide innervation to the dorsal divisions of ventral primary rami, while dorsomedially positioned neurons supply the ventral divisions. Spinal border cells, which are multipolar neurons situated along the ventral gray matter margins in thoracic through sacral segments, form the nucleus pericornualis anterior and contribute to anterior spinocerebellar pathways (Drake et al. 2020).

The organization of motor neurons follows specific patterns related to muscle function. Flexor-associated neurons occupy the dorsolateral regions, while extensor motor neurons lie ventrally (Fig. 2.3). Between the medial and lateral columns in segments C3–C5, ventrally positioned neurons give rise to the phrenic nerve, with dendrites showing characteristic longitudinal arrangements.

The nucleus of Onuf, located in the ventral sacral cord among unmyelinated fibers, contributes to the pudendal nerve (S2–S4). This specialized nuclear group controls the pelvic floor musculature, including external urethral and anal sphincters. It also innervates the ischiocavernosus and bulbospongiosus muscles, demonstrating sexual dimorphism, with greater neuronal density in males (Watson and Harrison 2022).

2.2 Spinal Root Anatomy

At each segmental level of the spinal cord, the ventral and dorsal roots originate from the anterolateral and posterolateral surfaces, respectively. Notably, the first cervical nerve and coccygeal nerve lack dorsal roots and, thus, do not exhibit dermatomal distribution (Netter 2019). Each spinal root comprises the convergence of three or more small rootlets, or filaments. Typically, the ventral root is made up of four to seven filaments that emerge from the anterolateral sulcus, whereas the dorsal root, consisting of four to ten filaments, enters at the posterolateral sulcus (Standring 2020).

The ventral root has motor neuron axons at all levels. In addition, the roots at levels T1–L2 (sympathetic) and S2–S4 (parasympathetic) also contain preganglionic autonomic fibers originating from the intermediolateral nucleus, whereas the dorsal root contains the central processes of bipolar sensory neurons, which are located in the dorsal root ganglion. Certain primary sensory afferents, particularly those originating from visceral structures, enter the spinal cord through the ventral root but terminate in the dorsal horn (Purves et al. 2018).

Interestingly, all ventral roots contain a substantial proportion of unmyelinated axons, comprising approximately 27% of the total ventral root axon population. The exact role of these unmyelinated axons remains unclear, but there is a strong possibility that many of them serve sensory functions (Coggeshall et al. 1975).

The ventral roots of the spinal cord function as motor output pathways. These roots consist of large-diameter α-motoneuron axons that innervate extrafusal striated muscle fibers, as well as γ-motoneuron axons that supply intrafusal striated muscle spindles. Additionally, preganglionic autonomic fibers are present within these ventral roots (Standring 2020).

The dorsal roots contain various types of afferent fibers, arranged somatotopically with the largest-diameter fibers positioned medially to the smaller ones (Fig. 2.5). The largest afferent fibers (Ia and Ib) are heavily myelinated and conduct the afferent limb of muscle stretch reflexes while transmitting information from muscle spindles and Golgi tendon organs. Medium-sized fibers (A-β) transmit impulses from the mechanoreceptors in the skin and joints. Small, thinly myelinated A-δ and unmyelinated C fibers convey noxious and thermal sensations (Kandel et al. 2013).

After the dorsal and ventral roots combine to create spinal nerves, they divide into a smaller primary dorsal ramus and a larger primary ventral ramus. Both rami consist of motor, sensory, and autonomic fibers. The dorsal ramus is composed of a medial sensory branch and a lateral motor branch, supplying the skin and the paraspinal muscle at the segmental level, respectively. The ventral ramus, being significantly larger, contributes to the brachial and lumbosacral plexuses, as well as to segmental branches such as intercostal nerves (Bican et al. 2013).

A thin layer of arachnoid cells envelops the nerve roots, enclosing the endothelial compartment of the nerve. This compartment includes myelinated and unmyelinated nerve fibers accompanied by Schwann cells and endoneurial blood vessels. The dura mater forms dural sleeves at each intervertebral foramen, merging with the epineurium of each spinal nerve. Pia mater surrounds the roots as they traverse the subarachnoid space until they reach their respective dural sleeve, which extends from the anterolateral surface of the dural sac (Cramer and Darby 2014). It is noteworthy that nerve roots lack dural covering and epineurium in the subarachnoid space. This anatomical feature makes nerve roots in the sub-

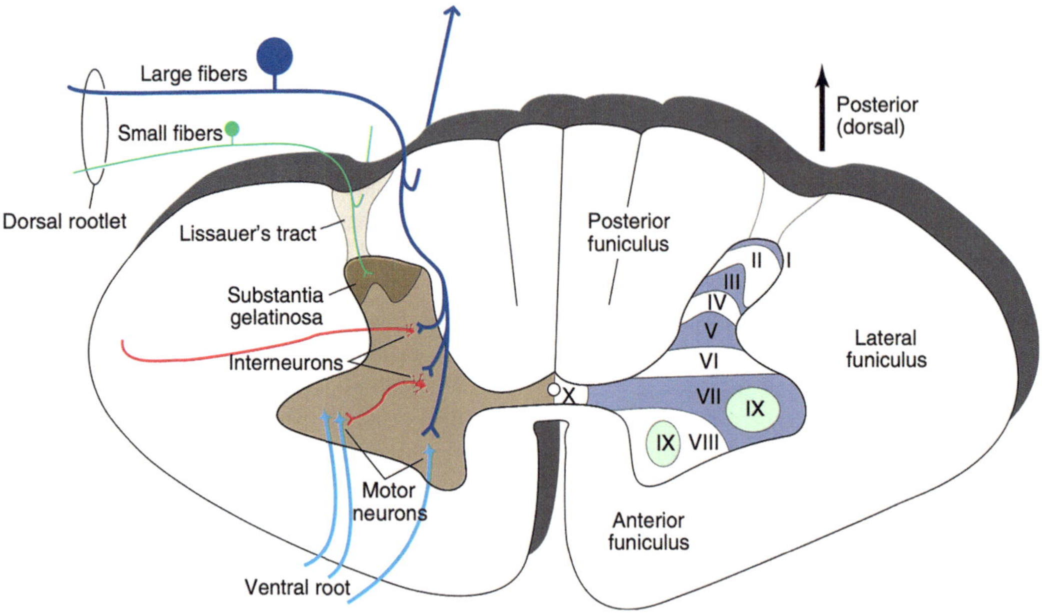

Fig. 2.5 Schematic representation of the spinal cord at the C8 segment. The dorsal roots are arranged somatotopically, with the largest-diameter fibers located medially relative to smaller-diameter fibers. (Adapted with permission from Vanderah and Gould (2016))

arachnoid space more sensitive and susceptible to injury compared to spinal nerves (Watson et al. 2009) (see Fig. 1.32).

2.3 Functional and Clinical Significance of Intervertebral Foraminal Anatomy

The intervertebral foramen is an anatomically complex osseous passage that requires a detailed understanding due to its significant role in spinal nerve protection and function (Bogduk 1995). Each foramen is bounded superiorly and inferiorly by pedicles of adjacent vertebrae, ventrally by the posterolateral aspect of the vertebral bodies and intervertebral disc, and dorsally by the articular processes and ligamentum flavum (Cramer and Darby 2014).

Within this three-dimensional space, multiple structures coexist in close proximity: the spinal nerve with its dorsal and ventral roots, the dorsal root ganglion, radicular vessels, and the surrounding adipose tissue. The anatomical arrangement of these components creates a dynamic environment that responds to spinal movements and postural changes (Standring 2020). This mobility can affect the cross-sectional area of the foramen, potentially influencing neural function during various neck positions and spinal movements.

Understanding foraminal anatomy becomes particularly relevant in conditions affecting the proximal motor unit. The location of the dorsal root ganglion within the foramen makes it susceptible to mechanical compression, while the spinal nerve's course through this space can be affected by pathological changes in the surrounding structures (Jaumard et al. 2011). In cases of traumatic injuries, including spinal root or nerve avulsion, the anatomical constraints of the foramen can influence both injury patterns.

2.3.1 Cervical Intervertebral Foramen

The cervical intervertebral foramina are formed by the superior and inferior vertebral notches of adjacent cervical vertebrae. These openings exhibit an oblique ventral orientation, with angles of approximately 45° from the midsagittal plane in the upper and middle cervical regions and 55° in the lower cervical regions. They also slope inferiorly at a 10° angle relative to a horizontal plane aligned with the superior vertebral end plates. Unique to the cervical spine, the uncinate processes form the ventral boundaries of these foramina, creating an almost oval shape. The vertical dimension of these foramina progressively increases from the C3–4 to the C7–T1 levels (Ebraheim et al. 1996).

Intervertebral foraminal dimensions are dynamic and influenced by cervical spine movements. Flexion from a neutral position enlarges the vertical diameter, whereas extension diminishes it (Yoo et al. 1992; Muhle et al. 2001). Therapeutic traction has been shown to augment both the area and height of these foramina, with increases of 18.9% and 10.4%, respectively, observed under 15 kg of traction weight (Liu et al. 2008). Extension elevates internal foraminal pressure, while shoulder abduction reduces it (Farmer and Wisneski 1994). This pressure reduction during upper extremity abduction corresponds with a positive Bakody's sign, characterized by alleviation of distal upper limb pain through arm abduction. Furthermore, ipsilateral axial rotation at 20° and 40° can narrow the foramina by up to 15% and 23%, respectively, while contralateral rotation at these angles expands them by up to 9% and 20% (Muhle et al. 2001).

The cross-sectional area of a typical cervical intervertebral foramen includes the spinal nerve, with its dorsal and ventral roots and dural sleeve, comprising 35–50% of the total foraminal area (Sunderland 1974). In a neutral cervical position, these roots are situated in the inferior portion of the foramen, at or below the disc level (Pech et al. 1985). The superior part of the foramen contains epidural fat and blood vessels. The dorsal root, positioned dorsally and slightly superior to the ventral root, contacts the superior articular process. The dorsal root ganglion resides near a small notch on the ventral aspect of this process in the cervical region. The ventral root aligns with the uncinate process. The adipose tissue between the dorsal and ventral roots functions as a cushioning element within the intervertebral foramen (Pech et al. 1985).

2.3.2 Thoracic Intervertebral Foramen

The thoracic intervertebral foramina show unique anatomical features compared to those in the cervical region. The cervical intervertebral foramina are oriented obliquely in an anterolateral direction, whereas the thoracic intervertebral foramina project directly laterally, very similar to the pattern seen in the lumbar region (Sunderland 1974). The first ten thoracic intervertebral foramina (T1–T10) are uniquely associated with their corresponding ribs, whereas the 11th and 12th ribs lack direct connections to the intervertebral foramina.

The proportion of the intervertebral foramen occupied by the spinal nerve varies across different spinal regions. In the thoracic region, only one-twelfth of the foramen's cross-sectional area is filled by the spinal nerve. In contrast, the cervical region shows a much higher nerve-to-foramen ratio, with up to half of the intervertebral foramen's space occupied by the spinal nerve and its accompanying dural root sleeve (Sunderland 1974). The lumbar spine exhibits the spinal nerve occupying approximately one-third of the area of the foramen. These anatomical differences may explain the lower frequency of radiculopathy due to intervertebral disc protrusion in the thoracic region compared to the cervical and lumbar areas. The reduced nerve-to-foramen ratio in the thoracic spine might decrease the probability of nerve compression by herniated disc material, as the nerve occupies a relatively smaller portion of the available foramen space.

2.3.3 Lumbar Intervertebral Foramen

The lumbar intervertebral foramina exhibit several distinctive features, primarily due to the unique features of lumbar and sacral spinal nerves. The spinal cord typically terminates at the L1–L2 level, necessitating the descent of lumbar and sacral dorsal and ventral roots within the subarachnoid space of the lumbar vertebral canal, a region referred to as the lumbar cistern (Bogduk 2005).

The dorsal and ventral rootlets or roots exit the lumbar cistern by entering a dura mater sleeve. This typically occurs slightly below the intervertebral disc level above the eventual intervertebral foramen. Upon leaving the lumbar cistern, these roots traverse obliquely downward and laterally, maintaining a substantial dural covering, the dural root sleeve. This sleeve encloses the neural elements, along with the radicular arteries and veins, until they exit the intervertebral foramen. The largest lumbar dural root sleeve is at L5 (5.6 mm in diameter), while the smallest is at L1 (3.5 mm in diameter) (Torun et al. 2006). As the sleeve reaches the intervertebral foramen's lateral edge, it merges with the spinal nerve's epineurium.

The L5 nerve root canal's anatomy is notably unique, rendering the L5 roots and nerve susceptible to compression at various locations. This canal is longer and oriented more obliquely ventrally compared to other lumbar nerve root canals. Additionally, the L5 vertebra's lateral recess is the deepest and often the narrowest from vental to dorsal within the spinal canal. This narrow recess can sometimes compress the L5 nerve (Rauschning 1987). After passing through the lateral recess, the L5 roots and their dural sleeve continue along the L5 nerve root canal, wrapping around the dorsal and lateral aspects of the L5 vertebral body. The roots proceed around the posterolateral aspect of the L5 intervertebral disc and unite to form the spinal nerve. The L5 nerve, the largest of the lumbar nerves, exits through the smallest intervertebral foramen in the lumbar spine, making it particularly prone to compression within this foramen (Cramer and Darby 2014; Olsewski et al. 1990).

2.4 Spinal Nerves

The human nervous system comprises 31 pairs of spinal nerves, each corresponding to a specific spinal cord segment. Every spinal nerve is composed of a sensory root and a motor root, which combine to form a mixed nerve carrying both afferent and efferent signals. In the cervical region, the first seven spinal nerves emerge superior to their corresponding cervical vertebrae. The C8 nerve, however, exits between the seventh cervical and first thoracic vertebrae. From this point caudally, all subsequent spinal nerves emerge inferior to their respective vertebrae (Cramer and Darby 2014; Kandel et al. 2013; Standring 2020). In the upper cervical region, nerve roots exit the spinal canal through the intervertebral foramina at approximately the same level as their origin from the spinal cord. This arrangement changes progressively in the caudal direction (Watson et al. 2009).

It is important to note that the spinal cord terminates at the level of the L1 or L2 vertebra in adults. Consequently, at lower levels, the segmental level of the cord is situated superior to the corresponding intervertebral foramen. This anatomical configuration results in an increasingly oblique and descending course of the nerve roots within the subarachnoid space as they travel from the cord to their respective foramina. This arrangement is particularly pronounced in the lower spinal segments. For instance, the lower sacral roots originate from the spinal cord at the L1 level and traverse several centimeters through the subarachnoid space before exiting the spinal canal in the sacral region. This extended intrathecal course of the lower spinal nerves forms the cauda equina (Netter 2019).

2.5 Organization of the Cervical, Brachial, and Lumbosacral Plexuses

The comprehension of nerve plexus branch origins and their corresponding innervation patterns for both cutaneous regions and muscles is fundamental for the precise localization of plexus lesions. Plexuses function similarly to fascicular nerve fiber branching, providing protection against complete denervation when only a portion of the plexus is affected. Typically, nerves originating from a single spinal cord segment supply multiple muscles. Conversely, most muscles receive innervation from multiple spinal cord segments. This distribution pattern confers resilience to muscular function in the event of nerve injuries (Wineski 2019).

Furthermore, there is considerable overlap in the cutaneous areas innervated by adjacent spinal cord segments. This arrangement minimizes the risk of complete sensory loss in a specific region due to localized nerve damage. The term "dermatome" describes the cutaneous area innervated by a single spinal cord segment, while "myotome" refers to the group of muscles primarily supplied by a single spinal cord segment (Moore et al. 2018; Wineski 2019).

Plexuses redistribute nerve fibers from spinal segments to peripheral nerves, facilitating the innervation of specific muscle groups and cutaneous areas. This reorganization is particularly significant in the context of spinal cord injury rehabilitation, as it influences the pattern of functional loss and potential recovery. The brachial and lumbosacral plexuses exemplify this complex organization. In the upper limb, the brachial plexus forms from the ventral rami of the C5–T1 spinal nerves, with occasional contributions from C4 or T2. This plexus gives rise to major nerves such as the musculocutaneous, median, ulnar, and radial nerves, each innervating specific muscle groups and cutaneous areas of the upper limb. Similarly, the lumbosacral plexus, formed by the ventral rami of the L1–S4 spinal nerves, supplies the lower limb. It gives rise to nerves such as the femoral, obturator, and sciatic nerves, which innervate the muscles and skin of the lower limb (Drake et al. 2020).

2.5.1 Anatomical and Functional Aspects of the Cervical Plexus

The cervical plexus is composed of the ventral rami of C1 through C4 spinal nerves (Fig. 2.6). This plexus provides cutaneous innervation to the dorsolateral regions of the head, neck, and shoulder and controls the movement of several muscle groups (Standring 2020). This extensive cutaneous innervation plays an important role in proprioception and sensory feedback from these regions, which helps in coordinated movement and spatial awareness (Netter 2019).

Motor fibers originating from the cervical plexus traverse to several key muscle groups:

- Deep cervical muscles: These muscles, including the longus colli, longus capitis, and rectus capitis ventral and lateralis, are essential for the fine control of head and neck movements (Cleland and Koppenhaver 2011).
- Hyoid muscles: The infrahyoid muscles, such as the sternohyoid, sternothyroid, and omohyoid, receive motor innervation from the cervical plexus. These muscles play a role in laryngeal elevation during swallowing and speech (Standring 2020).
- Diaphragm: The phrenic nerve, a major branch of the cervical plexus (primarily C3–C5), innervates the diaphragm, the principal muscle of respiration.
- Sternocleidomastoid and trapezius muscles: While these muscles receive primary innervation from the spinal accessory nerve (CN XI), they also receive contributions from the cervical plexus, enhancing their complex functions in head and neck movements (Kierner et al. 2000).

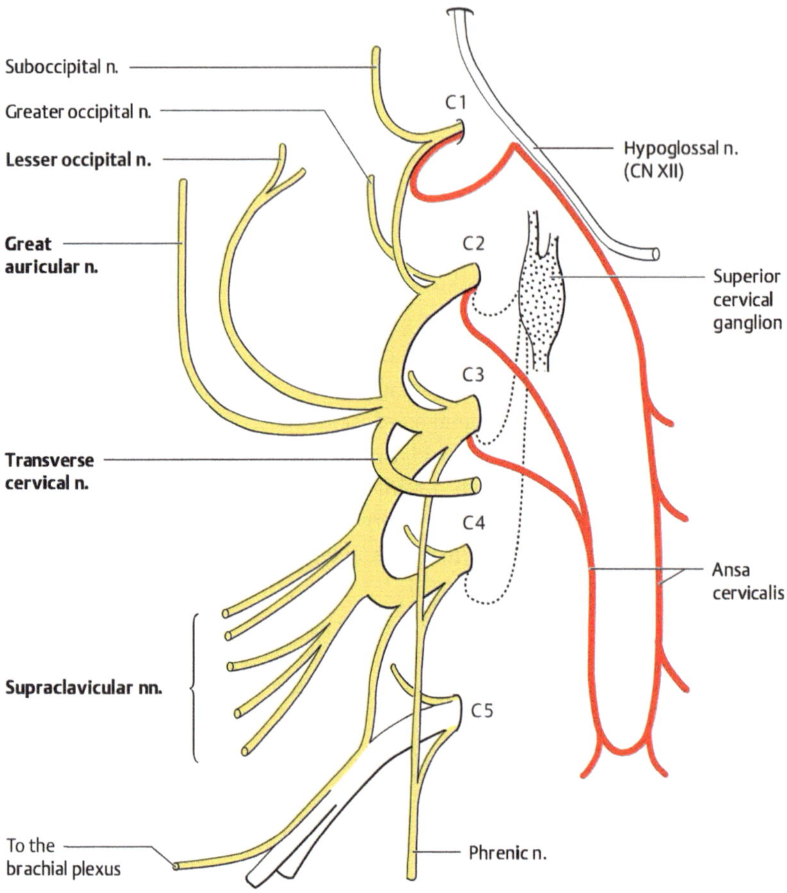

Fig. 2.6 Schematic illustration of the cervical plexus. (Adapted with permission from Baehr and Frotscher (2012))

2.5.2 Anatomical and Functional Aspects of the Brachial Plexus

The brachial plexus originates from the ventral rami of the C5 through T1 spinal nerves, sometimes including contributions from C4 and T2. This intricate neural structure is responsible for innervating the upper extremity. Following the convergence of the ventral rami within the plexus, numerous branches emerge, including five primary terminal branches: the axillary, musculocutaneous, radial, ulnar, and median nerves (Fig. 2.7).

The axillary nerve (C5–C6) provides cutaneous innervation to the deltoid region and motor supply to the deltoid and teres minor muscles. The musculocutaneous nerve (C5–C7) facilitates sensory function in the ventrolateral and dorsolateral forearm aspects while innervating the arm flexors, including the biceps brachii. The radial nerve (C5–C8, potentially T1) exhibits an extensive distribution pattern in both the arm and the forearm. Its cutaneous branches innervate the dorsal arm and forearm, while the superficial radial nerve supplies sensation to the lateral dorsum of the hand and the first three and a half digits, excluding the nails. Additionally, the radial nerve innervates the upper extremity extensor muscles.

The ulnar nerve (C8–T1, occasionally C7) traverses the arm to innervate structures in the forearm and hand. It provides motor supply to one and a half forearm flexors (ulnar side) and intrinsic hand muscles, including the hypothenar, all interossei, and some thenar muscles. Its cutaneous distribution is limited to the hand, covering the ulnar half and including the fifth digit and the medial half of the fourth digit. The median nerve

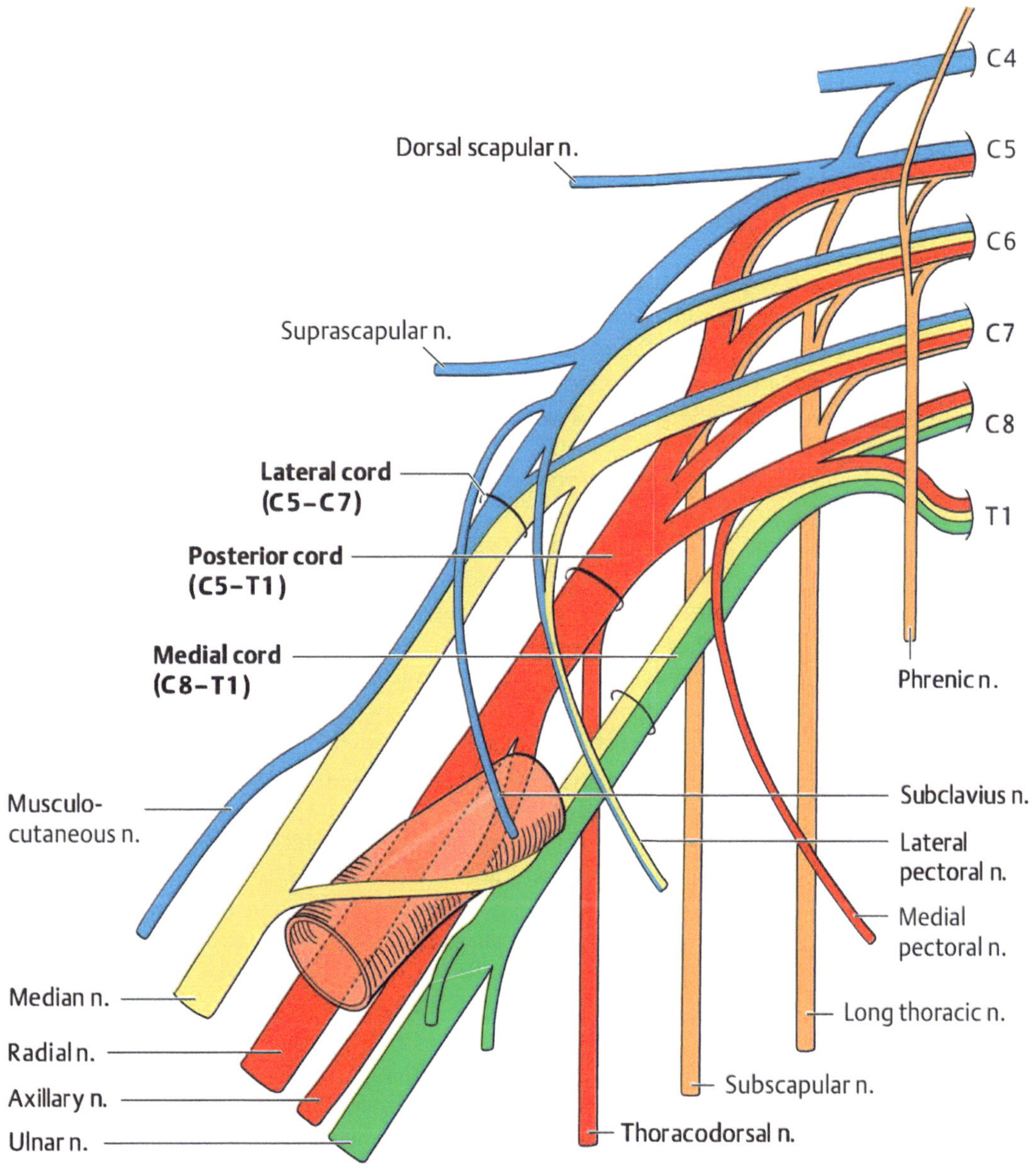

Fig. 2.7 Schematic illustration of the brachial plexus. (Adapted with permission from Baehr and Frotscher (2012))

(C6–C8 and T1 and, sometimes, C5) supplies motor fibers to the forearm flexors (excluding those with ulnar innervation) and certain intrinsic hand muscles, primarily the thenar group. Its sensory distribution, like the ulnar nerve, is limited to the hand, covering the lateral palmar surface and the first three and a half digits, including the nails (Standring 2020).

The brachial plexus is vulnerable to injury due to its relatively superficial location, proximity to osseous structures, and the extensive mobility of the cervical and shoulder girdle regions. The plexus formation begins with the union of the C5 and C6 spinal nerves at the lateral border of the scalenus medius muscle, forming the upper trunk. The C8 and T1 spinal nerves converge dorsally to the scalenus ventral muscle, forming the lower trunk, while the C7 alone forms the middle trunk. These three trunks, each bifurcating into ventral and dorsal divisions, course inferolaterally, positioned above or behind the clavicle.

The dorsal divisions of all trunks combine to form the dorsal cord, initially above and then behind the axillary artery. The ventral divisions of the upper and middle trunks merge to create the lateral cord, situated lateral to the axillary artery. The ventral division of the lower trunk courses dorsally and then medially to the axillary artery, forming the medial cord (Fig. 2.7).

It is noteworthy that the dorsal (posterior) cord primarily innervates the extensor muscles of the upper extremity, while the lateral and medial cords predominantly supply the flexor muscles. However, the innervation pattern is more intricate, involving multiple branches that supply various upper limb muscles. For instance, the radial nerve, originating from the dorsal cord, innervates most extensor muscles in the arm and forearm. The median and ulnar nerves, arising from the lateral and medial cords, respectively, are primarily responsible for innervating the flexor muscles of the forearm and hand (Moore et al. 2018; Standring 2020).

2.5.3 Anatomical and Functional Aspects of the Lumbosacral Plexus

The lumbosacral plexus, a complex network of nerves, plays a fundamental role in lower limb innervation and pelvic function. This intricate system comprises two distinct interconnected components: the lumbar and sacral plexuses (Moore et al. 2018) (Fig. 2.8).

2.5.3.1 Lumbar Plexus

The lumbar plexus originates from the ventral rami of spinal nerves L1 through L4. These neural elements converge within the psoas major muscle, forming the plexus. The obturator nerve, a key component, stems from the ventral divisions of L2–L4 nerves (Drake et al. 2020). Multiple branches emerge from this plexus, including the following:

- Muscular branches to psoas and iliacus.
- Lateral cutaneous nerve of the thigh.
- Iliohypogastric nerve.
- Ilioinguinal nerve.
- Genitofemoral nerve.

2.5.3.2 Sacral Plexus

The sacral plexus is primarily derived from the lumbosacral trunk, including fibers from the L4 and L5 ventral rami and the ventral rami of S1–S4. These ventral rami bifurcate into dorsal and ventral branches, similar to the lumbar plexus. The dorsal branches of the lumbar trunk (L4–L5) and the S1–S2 ventral rami converge to form the lateral component of the sciatic nerve, which subsequently gives rise to the common peroneal nerve. Conversely, the ventral branches from these segmental levels coalesce to create the medial trunk of the sciatic nerve, ultimately forming the dorsal tibial nerve (Netter 2019).

In addition to the primary branches of the sciatic nerve, the sacral plexus also produces several other important nerves. The following components are included:

- Superior and inferior gluteal nerves (motor innervation to the gluteal muscles).
- Pudendal nerve.
- Dorsal cutaneous nerve of the thigh (sensory innervation to the dorsal thigh).
- Various smaller nerves supplying the pelvic and hip regions.

The lumbar plexus is situated dorsal to or within the fasciculi of the psoas major muscle. In contrast, the sacral plexus lies within the pelvis, ventral to the piriformis muscle and the sacrum (Standring 2020; Wineski 2019).

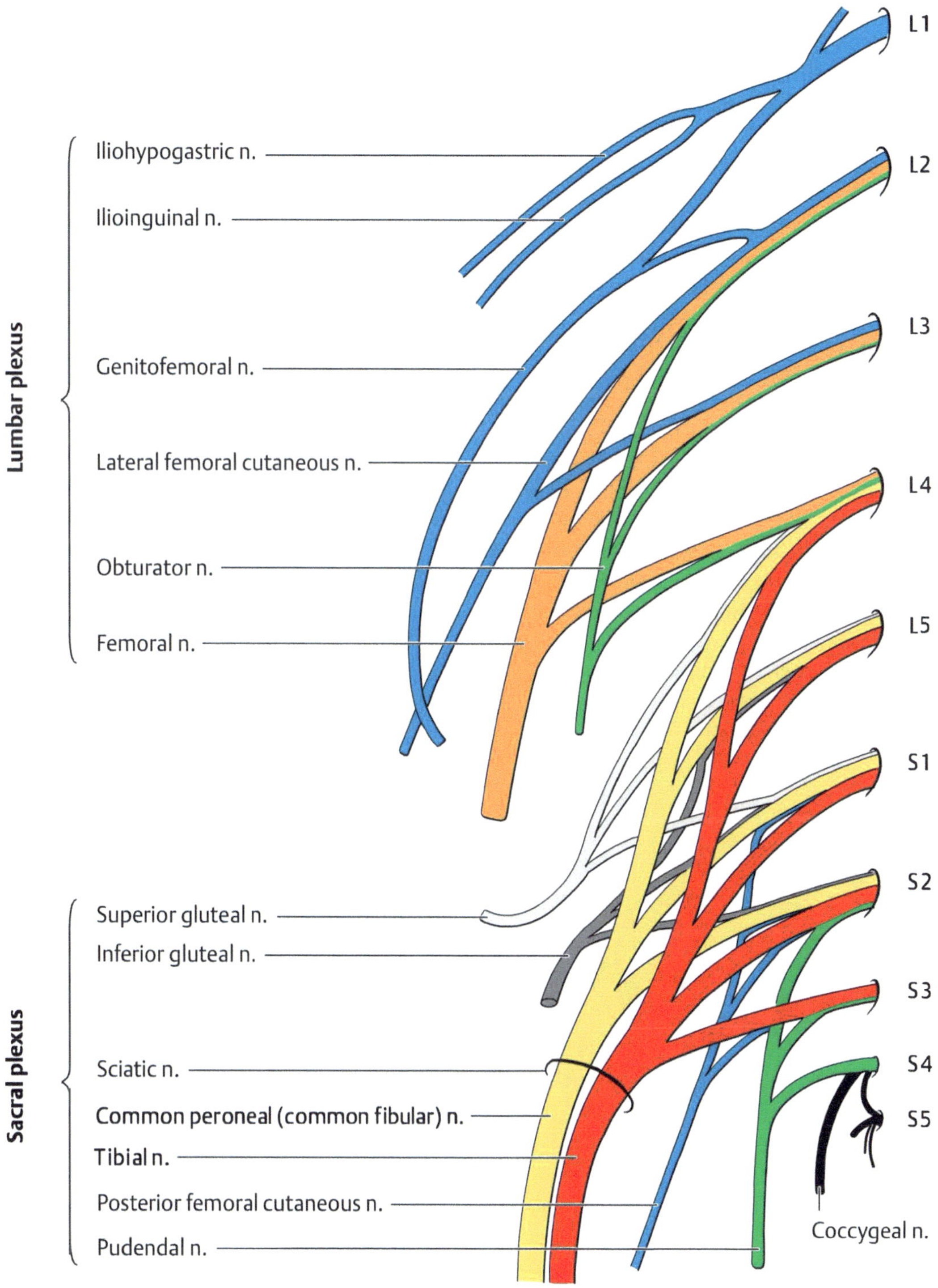

Fig. 2.8 Schematic illustration of the lumbosacral plexus. (Adapted with permission from Baehr and Frotscher (2012))

2.6 Peripheral Nerve

2.6.1 Peripheral Nerve Structures and Composition

Peripheral nerves exhibit a complex hierarchical structure, composed of multiple layers encasing individual and grouped axons (Sunderland 1945). Myelinated axons are individually enclosed by the endoneurium, a tubular fibrous sheath. These axons are aggregated and surrounded by the perineurium, a larger fibrous covering. Unmyelinated axons, which lack individual endoneurial coverings, form discrete bundles running parallel to myelinated axons within the perineurial enclosure. A "nerve fascicle" is defined as a group of axons contained within a single perineurial sheath (Fig. 2.9). Axons within a fascicle typically maintain their grouping throughout the nerve's length, extending to their peripheral termination points (Stewart 2003). The epineurium, an external fibrous tissue layer, binds multiple fascicles together, constituting the outermost visible layer of the nerve at the macroscopic level (Topp and Boyd 2006).

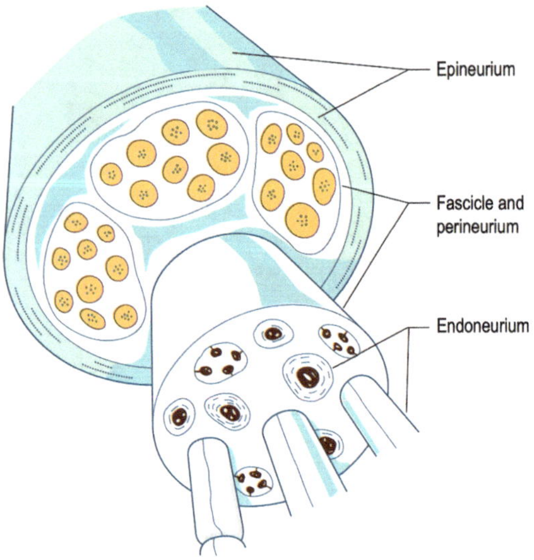

Fig. 2.9 Structural components of a peripheral nerve, including the axon, endoneurium, fascicle with perineurium, and epineurium. (Adapted with permission from Palastanga and Soames (2012))

Ventral and dorsal roots in the subarachnoid space consist of axons enveloped by the endoneurium, an extension of the spinal cord's pia mater. Notably, these roots lack both perineurium and epineurium, making them more vulnerable to traumatic and toxic damage (Haller and Low 1971). As the spinal root extends toward the intervertebral foramen, a structural transition occurs: the arachnoid membrane differentiates into perineurium, while the dura mater develops into epineurium, thus forming the peripheral nerve structure. This metamorphosis represents a significant architectural change as nerve components emerge from the spinal column (see Fig. 1.32).

The multilayered organization of peripheral nerves provides protection and support for the axons they contain, facilitating efficient signal transmission while offering resilience against mechanical stresses (Lundborg 1988; Mackinnon 2015).

2.6.2 Myelination of Peripheral Nerve

The peripheral nervous system comprises two primary categories of nerve fibers: myelinated and unmyelinated. Myelinated fibers consist of axons encased in multiple concentric layers of myelin, a lipid-rich substance produced by Schwann cells. Unmyelinated axons are surrounded either individually or in groups by the cytoplasm of adjacent Schwann cells and are typically smaller in diameter compared to myelinated axons. This structural distinction is fundamental to the functional efficiency of nerve impulse transmission within the peripheral nervous system.

In the peripheral nervous system, myelination involves the envelopment of individual axons by myelin sheaths, which are generated by Schwann cells. These sheaths serve dual purposes: They provide axonal insulation and markedly enhance the speed of electrical impulse conduction (Sherman and Brophy 2005). The process begins with the Schwann cell wrapping its membrane around the axon in a spiral manner, during which its cytoplasm is squeezed out, forming the myelin sheath. Additionally, the axon is surrounded peripherally by the Schwann cell's cytoplasm, and the outermost membrane forms the neurolemma, providing an extra protective layer (Nave and Werner 2014). This structural arrangement contributes to the overall integrity and functionality of myelinated nerve fibers in the peripheral nervous system.

The distinction between myelinated and unmyelinated fibers has significant implications for nerve conduction velocity and the efficiency of signal transmission. Myelinated fibers, with their insulating sheaths, facilitate saltatory conduction, allowing for the rapid propagation of action potentials along the axon. This mechanism is particularly important for functions requiring swift neural responses, such as motor control and certain sensory processes (Mackinnon 2015; Seidl 2014).

In the peripheral nervous system, individual axons are typically myelinated by multiple Schwann cells, with each Schwann cell interacting with only one axon. The junctions where the myelin sheaths from adjacent Schwann cells meet are termed nodes of Ranvier. A nerve fiber in the peripheral nervous system is thus composed of an axon, its myelin sheath, and the surrounding Schwann cells (Salzer 2015).

Numerous axons in the peripheral nervous system remain unmyelinated. These axons lack a tightly wound myelin sheath and, instead, reside within invaginations of Schwann cell membranes. This arrangement allows a single Schwann cell to interact with multiple axons. While unmyelinated axons have less physical protection than myelinated ones, their primary functional difference lies in impulse conduction velocities: Unmyelinated axons conduct impulses at significantly lower speeds compared to myelinated axons (Debanne et al. 2011).

Central nervous system myelination is predominantly carried out by oligodendrocytes, a specialized glial cell type. Despite the similarities in the basic myelination mechanisms of the central nervous system and the peripheral nervous system, a key difference exists: One oligodendrocyte can myelinate multiple axons, unlike Schwann cells in the peripheral nervous system, which myelinate only a single axon each. Oligodendrocytes, like Schwann cells, envelop axons with their cell membranes to form myelin sheaths. Meanwhile, unmyelinated axons in the central nervous system are found within the cytoplasmic extensions of oligodendrocytes, which provide structural support without forming myelin sheaths (Simons and Nave 2015) (Fig. 2.10). Table 2.1 illustrates the distinctions between the peripheral nervous system and the central nervous system.

2.6.3 Classification and Properties of Mammalian Nerve Fibers

Mammalian nerve fibers are classified into three main types: A, B, and C, each characterized by distinct physiological properties, conduction velocities, and diameters. An alternative classification system employs Roman numerals: Groups

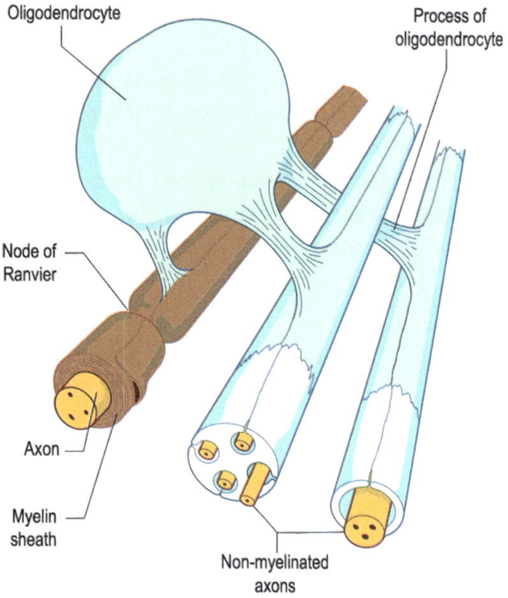

Fig. 2.10 Relationship between an oligodendrocyte and the several myelinated and unmyelinated axons it ensheathes. (Adapted with permission from Palastanga and Soames (2012))

Table 2.1 Comparative overview of myelination in the peripheral nervous system (PNS) and central nervous system (CNS)

Aspect	PNS myelination	CNS myelination
Myelinating cells	Schwann cells	Oligodendrocytes
Process	Each Schwann cell myelinates a single axon	One oligodendrocyte myelinates multiple axons simultaneously
Neurolemma	Presence of neurolemma, aiding in nerve regeneration	No neurolemma, limiting regeneration capacity
Nodes of Ranvier	Regularly spaced	Differently spaced and organized
Regeneration	Higher capacity for regeneration due to neurolemma	Limited regeneration capacity
Response to injury	Schwann cells can dedifferentiate and promote regeneration	Limited regenerative response from oligodendrocytes

I and II correspond to Aα fibers, group III to Aδ fibers, and group IV to C fibers, providing a complementary classification system (Ropper et al. 2014) (Table 2.2).

A fibers, characterized by their large size and myelination, are responsible for the rapid conduction of motor or sensory impulses (Bear et al. 2020). The diameter of these fibers significantly influences their conductivity; internal longitudinal resistance is inversely related to diameter, allowing larger fibers to conduct impulses at higher velocities. This group is further subdivided into α, β, γ, and δ types, based on their conduction rates, with α fibers exhibiting the fastest conduction and δ fibers exhibiting the slowest. A fibers demonstrate increased susceptibility to mechanical pressure and oxygen deprivation, a factor of clinical significance (Kandel et al. 2013; Ropper et al. 2014; Purves et al. 2018).

B fibers consist of smaller, myelinated efferent preganglionic fibers that are specifically associated with the sympathetic system. These fibers have a slower conduction rate compared to A fibers and provide an important function in the autonomic nervous system (Hall 2016).

C fibers represent the smallest fibers in this classification. They are unmyelinated and include postganglionic sympathetic fibers as well as somatic efferent fibers (Hallett et al. 2015). C fibers primarily transmit sensory information, with a particular role in conveying deep pain sensations (Wineski 2019).

The physiological properties of these fiber types have significant implications for neurological function and pathology. For instance, the differential vulnerability of A fibers to mechanical pressure can explain certain patterns of sensory and motor deficits in compressive neuropathies (Caliandro et al. 2016). Similarly, the slower conduction of C fibers contributes to the delayed perception of certain pain sensations (Woolf and Ma 2007).

Table 2.2 Classification and characteristics of mammalian nerve fiber types

Fiber type		Function	Afferent fibers (from skeletal muscle and articular receptors)	Efferent fibers	Myelination	Fiber diameter (μm)	Conduction velocity (m/s)
A	α	Proprioception, somatic motor	Type Ia (from mechanoreceptors, muscle spindles) Type Ib (from sensory fibers of Golgi tendon organs)	α-motoneurons innervating extrafusal muscle fibers	Heavily myelinated	12–22	70–120
	β	Touch, pressure	Type II (from mechanoreceptors, secondary afferents of muscle spindle)		Myelinated	6–12	30–70
	γ	Motor to muscle spindles		γ-motoneurons innervating intrafusal fibers of muscle spindles	Myelinated	2–10	15–30
	δ	Pain, temperature, touch	Type III (from mechanoreceptors and nociceptors)		Thinly myelinated	1–6	12–30
B		Preganglionic autonomic		Thinly myelinated preganglionic autonomic fibers	Thinly myelinated	<3	3–15
C	Dorsal root	Pain, reflex responses	Type IV (from nociceptors)		Unmyelinated	0.4–1.2	0.5–2
	Sympathetic	Postganglionic sympathetics	Type IV	Unmyelinated postganglionic autonomic fibers	Unmyelinated	0.3–1.3	0.7–2.3

2.7 Neuromuscular Junction Architecture and Physiology

Neuromuscular junctions are critical structures that serve as interfaces between motor neurons and muscle cells, facilitating muscular contractions. These synapses share structural and functional similarities with neuronal synapses in the nervous system. There are two types of motor neurons involved: α-motoneurons, which innervate extrafusal muscle fibers responsible for force production, and γ-motoneurons, which innervate intrafusal fibers that regulate muscle spindle sensitivity. Both types form specialized nerve endings at the neuromuscular junctions (Kandel et al. 2013).

As a motor axon approaches its target muscle fiber, it undergoes a transformation. The myelin sheath is shed, and the axon terminal expands into a flattened structure adjacent to the muscle cell membrane. This nerve expansion is encapsulated by a Schwann cell, which provides insulation for the neuromuscular junction from the surrounding tissue (Purves et al. 2018).

The muscle cell membrane at the point of contact with the nerve cell undergoes specific modifications. It forms an elevated region with accumulated sarcoplasm and organelles, creating a distinct protrusion beneath the nerve terminal. The portion of the muscle cell membrane in direct contact with the nerve cell is termed the postjunctional membrane, while the corresponding part of the nerve cell membrane is known as the prejunctional membrane (Standring 2020).

In this arrangement, the postjunctional membrane forms invaginations that accommodate corresponding projections of the prejunctional membrane. This configuration creates a synaptic cleft approximately 20–50 nm wide, which contains a specialized extracellular matrix that supports neurotransmission. The postjunctional membrane also features secondary synaptic clefts, which increase its surface area and enhance its receptivity to neurotransmitters (Sanes and Lichtman 1999).

Acetylcholine, the primary neurotransmitter at the neuromuscular junctions, is stored in synaptic vesicles within the terminal expansion of the motor neuron. When an action potential reaches the nerve terminal, it triggers the release of acetylcholine into the synaptic cleft. The neurotransmitter then diffuses across the cleft to interact with receptors on the postjunctional membrane, initiating an action potential in the muscle membrane. This electrical signal propagates into the muscle cell, ultimately resulting in the contraction of muscle fibers (Hall 2016).

2.8 Skeletal Muscles

Skeletal muscle constitutes more than one-third of the total human body mass and is composed of non-branching striated muscle fibers. These fibers are enveloped by the areolar tissue, a variant of loose connective tissue. The morphological diversity of skeletal muscles is significant, encompassing flat, sheet-like formations; short and thick structures; and long, slender configurations. The length of a muscle, excluding its tendinous components, is directly related to the required contraction distance (Lieber and Fridén 2000).

Scientific investigations have established that muscle fibers possess the capability to contract to approximately half their original length. This contraction capacity is significantly influenced by the specific arrangement of fibers within a given muscle. Notwithstanding the variability in fiber arrangement, all muscle movements are facilitated through contraction, which involves the shortening of muscle fibers. This contractile process generates a pulling force across joints, resulting in alterations of the positions of adjacent bones, thereby enabling movement.

The structural organization of skeletal muscles is hierarchical. Individual muscle fibers, also known as myofibers, are multinucleated cells formed by the fusion of myoblasts during development. These fibers are bundled into fascicles, which are then grouped to form the complete muscle. Each level of this organization is surrounded by connective tissue: endomysium around individual fibers, perimysium around fascicles, and epimysium encasing the entire muscle (Gillies and Lieber 2011).

At the molecular level, the contractile apparatus within muscle fibers consists of myofibrils composed of thick and thin filaments. The thick filaments are primarily made of myosin, while the thin filaments consist mainly of actin. The sliding filament theory explains muscle contraction as the interaction between these filaments, facilitated by cross-bridge cycling and adenosine triphosphate (ATP) hydrolysis (Geeves and Holmes 1999).

The functional properties of skeletal muscles are determined by their fiber type composition. Broadly, muscle fibers are categorized into slow-twitch (type I) and fast-twitch (type II) fibers. Type I fibers are characterized by high oxidative capacity and fatigue resistance, making them suitable for endurance activities. Type II fibers, further subdivided into IIa and IIx in humans, have a higher glycolytic capacity and are adapted for rapid, powerful contractions (Schiaffino and Reggiani 2011).

Spinal cord injuries can lead to alterations in muscle innervation, resulting in atrophy and changes in fiber type composition. These changes can significantly impact muscle function and rehabilitation strategies. For instance, the loss of neural input can lead to a shift from slow-twitch to fast-twitch fiber predominance, affecting muscle endurance and metabolic properties (Biering-Sørensen et al. 2009; Shields 2002).

2.8.1 Anatomy of Skeletal Muscles

2.8.1.1 Fiber Arrangement of Skeletal Muscles

The arrangement of muscle fibers is a key determinant of skeletal muscle function. Muscles exhibit two primary fiber arrangements: parallel and pennate. Parallel fibers are oriented in alignment with the muscle's line of pull, manifesting in two main forms. The fusiform structure, exemplified by the biceps brachii, allows for substantial shortening during contraction, albeit with reduced power output. Alternatively, a flat, sheet-like arrangement is observed in muscles such as the external oblique of the abdomen (Fig. 2.11).

Pennate muscles, characterized by fibers positioned obliquely to the line of pull, offer enhanced contractile force at the expense of shortening capacity. This increased power generation is attributed to the higher density of fibers within a

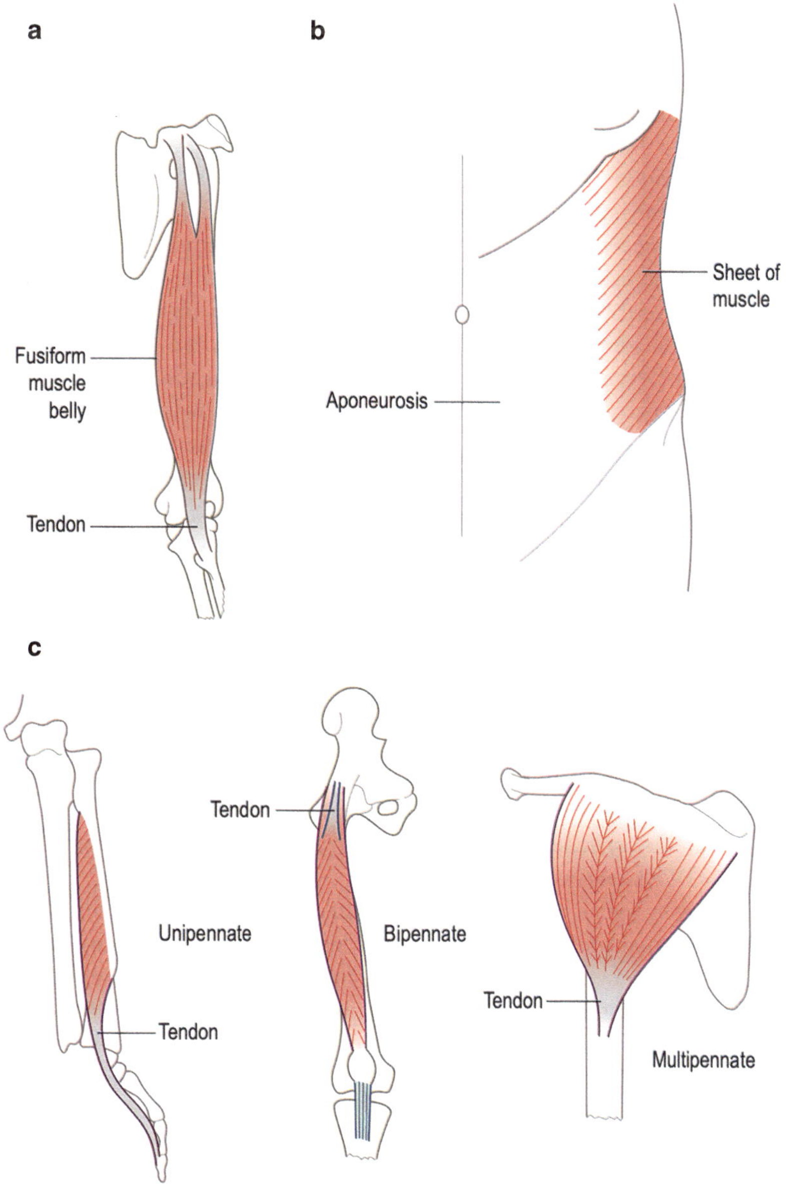

Fig. 2.11 Fiber arrangements in muscles: (**a**) fusiform, (**b**) sheet, (**c**) pennate. (Adapted with permission from Palastanga and Soames (2012))

given area. Pennate muscles are further categorized into unipennate, bipennate, and multipennate subtypes. Unipennate muscles, like the flexor pollicis longus, feature fibers attaching to one side of the tendon. Bipennate muscles, such as the rectus femoris, possess a central septum, with fibers attaching bilaterally and to a continuous central tendon. Multipennate muscles, exemplified by the deltoid, comprise several intermediate septa, each exhibiting a bipennate arrangement, thus augmenting the muscle's overall power and efficiency in force production.

The functional capabilities of muscle fibers are substantially influenced by the structural variety of their arrangements. Parallel-fibered muscles, especially those having a fusiform shape, are highly effective in tasks that involve a wide range of motion. Due to their ability to contract, they are highly suitable for activities that require substantial movement, such as the bending of the elbow by the biceps brachii (Lieber and Fridén 2000). Nevertheless, the increased range of motion is accompanied by a decrease in force production when compared to pennate arrangements.

Pennate muscles are designed to generate force efficiently due to their oblique arrangement of muscle fibers. The fiber's angular configuration enables a higher concentration of contractile units to be efficiently accommodated within a specific muscle volume, leading to increased force production. This architectural characteristic is especially beneficial in muscles that are responsible for supporting weight or generating strong, limited-range movements. An example of a muscle that is important for tasks such as standing up from a chair or going up the stairs is the quadriceps femoris, specifically the bipennate rectus femoris (Wickiewicz et al. 1983).

Following spinal cord injury, changes in muscle innervation and usage patterns can lead to alterations in muscle architecture. These changes may include shifts in fiber type composition, reductions in muscle cross-sectional area, and modifications in pennation angle (Castro et al. 1999). For example, in muscles with parallel fiber arrangement, rehabilitation protocols might focus on maintaining or improving the range of motion to capitalize on their capacity for extensive shortening. Conversely, for pennate muscles, interventions may prioritize strength training to preserve or enhance their force-generating capabilities (Gorgey and Dudley 2007).

2.8.1.2 Attachment and Actions of Skeletal Muscles

Skeletal muscles connect to bones and other tissues primarily through connective tissue components. Typically, the perimysium and epimysium of muscles frequently integrate directly with the periosteum or joint capsule. The attachment region of connective tissue manifests as a flattened or depressed area on the bone without leaving a clear mark. Commonly, the muscle's connective tissue elements converge to form a tendon, composed of collagen fiber bundles. It is noteworthy that muscle fibers do not directly continue into tendon fibers. Tendons, characterized by their high tensile strength, can take various forms, such as round cords, flattened bands, or thin sheets known as aponeuroses. Tendon attachments to the bone typically create a smooth impression on the bone, except in cases involving both fleshy and tendinous fibers or extensive aponeurosis attachments, which may roughen the bone surface (Moore et al. 2018).

Muscles attach at two distinct sites on different bones, traditionally labeled as "origin" and "insertion," based on the primary action and relative movement, which is less mobile or more mobile, respectively. However, this labeling can be misleading since muscles can generate movement from either attachment point. Therefore, the term "attachment" is now favored. Muscle contraction, initiated by stimulation, draws the attachment points closer together (Lieber and Fridén 2000).

Muscle contractions are categorized based on changes in length and tension. If a muscle contracts and changes length while maintaining relatively constant tension, it is classified as "isotonic." Isotonic contractions are further subdivided into "concentric" (muscle shortening) and "eccentric" (muscle lengthening). Eccentric contractions occur during controlled movement against an applied force. In contrast, "isometric" contractions occur when the muscle length remains unchanged, often due to external resistance, resulting in increasing tension (Frontera et al. 2019).

In the analysis of musculoskeletal dynamics, specific terminologies are employed to categorize muscles based on their roles in movement execution. The term "prime mover" designates muscles or muscle groups that initiate and drive particular movements (Neumann 2017). These muscles are primarily responsible for generating the force necessary to perform a specific action. Conversely, "antagonists" are defined as muscles that exert an opposing force to the action of the prime movers (Palastanga and Soames 2012). These muscles play a significant role in movement control and precision by providing counterbalance to the prime movers' action. Muscles that work in opposition are arranged in sets known as antagonistic pairs. A prime illustration of this is the relationship between the biceps brachii and triceps brachii, which oppose each other in elbow flexion and extension.

Anatomical literature typically characterizes primary acting muscles based on their fundamental role: They are situated on the same aspect of a bone, influence the same joint, and facilitate movement in a consistent direction. In contrast, their counterparts affect the same joint but generate motion in the opposite direction. In some cases, muscle pairs may act as agonists in one type of motion and antagonists in another. For example, the flexor carpi ulnaris and flexor carpi radialis work together as agonists during wrist flexion yet oppose each other during abduction and adduction. Similarly, the ventral and dorsal portions of the deltoid act as agonists in shoulder abduction but function as antagonists in flexion and extension (Neumann 2017; Standring 2020).

"Synergists," on the other hand, are muscles that work in conjunction with the prime movers to stabilize joints and prevent unwanted movements during the execution of a specific action (Harvey 2008; Nordin and Frankel 2001). Furthermore, this classification system aids in the assessment of muscle function and the development of therapeutic interventions. For example, in cases of muscle imbalance or weakness following spinal cord injury, strengthening exercises may be designed to target specific prime movers while also addressing the role of antagonists and synergists to achieve optimal functional outcomes (Sisto et al. 2009).

It is important to acknowledge that the functions of muscles might vary based on the particular movement or motion being executed. A muscle that functions as the main force generator in one movement may serve as an assist or opponent in another, demonstrating its complex and adaptable nature of the musculoskeletal system (Lieber 2002).

2.8.1.3 Sensory Mechanisms in Skeletal Muscles

Skeletal muscles incorporate diverse sensory receptors that are fundamental for nociception and the regulation of muscular activity, tension, and dimension. Among these are free nerve endings, Golgi tendon organs, and muscle spindles. Free nerve endings are primarily responsible for nociception and are unevenly distributed within the muscle tissue, with a higher density at the myotendinous junction. Golgi tendon organs are more complex structures located within the tendinous tissue. These structures consist of axonal terminal branches intertwined with collagenous fibers and are enclosed within a fibrous capsule (Fig. 2.12). The compression and elongation of collagen fibers during muscular contraction activate these nerve terminals. This structure enables Golgi tendon organs to detect both the intensity of muscle contraction and the force exerted by the muscle.

Muscle spindles are specialized sensory receptors situated within the muscle belly. They consist of two types of intrafusal muscle fibers, encased in a fibrous capsule, distinct from the extrafusal fibers responsible for movement. The intrafusal fibers are classified into nuclear bag fibers and nuclear chain fibers based on the arrangement of their nuclei. Group Ia neurons form spiral endings around the mid-region of both types of intrafusal fibers, whereas group II neurons form spirals on nuclear chain fibers and spray-like endings on nuclear bag fibers. These structures play a pivotal role in providing

Fig. 2.12 Innervation of skeletal muscle, illustrating afferent and efferent pathways

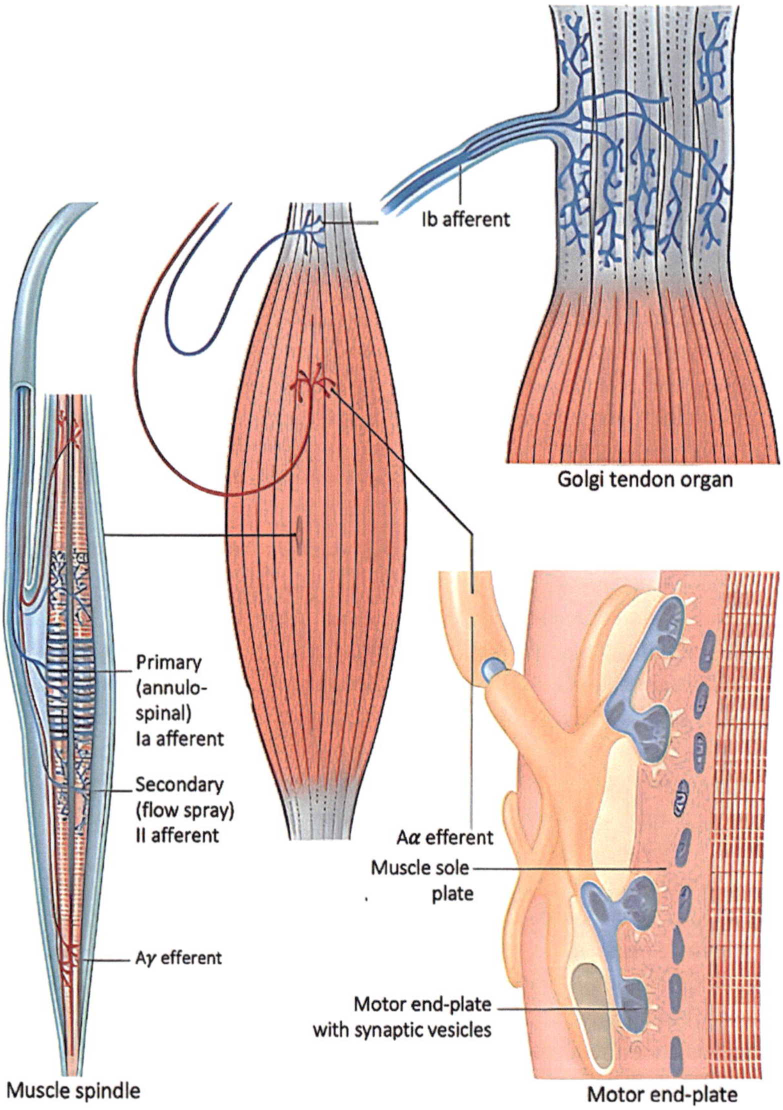

feedback on muscle stretch and proprioception (Proske and Gandevia 2012).

Alterations in muscle length, either through stretching or contraction, affect the tension of intrafusal fibers, subsequently influencing the activity of both Ia and II sensory neurons. This neuronal activity is essential for relaying information to the central nervous system concerning muscle length, the rate of change in length, and, indirectly, the position of joints. Various subtypes of nuclear bag and chain fibers are specialized to detect specific aspects of muscle length alteration and its velocity. Additionally, muscle spindles receive motor innervation from γ-motoneurons, which connect to the ends of intrafusal fibers adjacent to their central sensory regions. Activation of these neurons results in the contraction of peripheral intrafusal fiber segments, effectively stretching the spindle's central sensory zone and modulating its sensitivity. This process synchronizes the spindle's sensory capabilities with overall muscle length changes during contraction or relaxation states.

In the absence of γ-motoneuron stimulation, muscle spindles would exhibit limited responsiveness, primarily reacting to muscle stretches at maximal extension. However, γ-motoneuron–mediated contraction of intrafusal fibers sustains tension in the spindle's central sensory region, facilitating continuous monitoring of muscle length throughout the full range of motion (Palastanga and Soames 2012). The intricate interplay between intrafusal and extrafusal fibers, mediated by γ-motoneurons, allows for precise proprioceptive feedback.

2.8.2 Classification of Muscle Fiber Types

Skeletal muscles comprise various fiber types, primarily categorized into slow-twitch (type I) and fast-twitch (type II) fibers. The fast-twitch category is further subdivided into types IIa and IIb. The study of muscle fiber types is fundamental to understanding muscle metabolism, energy utilization, and mechanical responses (Schiaffino and Reggiani 2011).

2.8.2.1 Characteristics of Slow-Twitch Fibers

Type I fibers, also known as slow-twitch fibers, are characterized by their oxidative metabolism. These fibers appear red due to their high myoglobin content. Slow-twitch fibers exhibit prolonged contraction times and are well-suited for sustained, low-intensity activities. Athletes specializing in endurance sports typically possess a higher proportion of these fibers (Zierath and Hawley 2004).

2.8.2.2 Characteristics of Fast-Twitch and Intermediate Fiber Types

Type II fibers, or fast-twitch fibers, are further classified into two subtypes: type IIa (oxidative–glycolytic) and type IIb (glycolytic). Type IIa fibers, also referred to as intermediate fast-twitch fibers, are red in appearance. These fibers can maintain activity for extended periods while also being capable of producing brief bursts of force before fatigue onset. In contrast, type IIb fibers, which are white in color, are responsible for rapid force generation but fatigue quickly (Bottinelli and Reggiani 2000).

Most muscles contain a mixture of these fiber types. For instance, the vastus lateralis muscle typically exhibits an approximately equal distribution of fast-twitch and slow-twitch fibers (Staron et al. 2000). The specific composition of fiber types within a muscle influences its training adaptations and functional capabilities. This information is particularly relevant in developing targeted rehabilitation strategies for individuals with spinal cord injuries (Biering-Sørensen et al. 2009).

The distribution of fiber types varies among different muscle groups and can be influenced by factors such as genetics and training. For example, muscles heavily utilized in sprinting and jumping activities, such as the gastrocnemius, often display a higher concentration of fast-twitch fibers. Conversely, muscles primarily engaged in endurance activities typically contain a greater proportion of slow-twitch fibers (Trappe et al. 2015).

2.8.3 Physiology of Skeletal Muscle Contraction

Understanding the physiology of skeletal muscle contraction requires examining its complex structure. Muscles are made up of fascicles, each containing multiple muscle fibers. These fibers are further organized into motor units, each controlled by a single axon. It is important to note that muscle fibers within a motor unit are dispersed rather than adjacent, which allows for a more even distribution of contractile force across the muscle. Interestingly, most muscle fibers do not extend the full length of the muscle (Lieber and Fridén 2000).

At the neuromuscular junction, where the terminal axon meets the muscle fiber, several structural changes occur. The axon loses its myelin sheath and expands to form a notable terminal within a groove on the muscle fiber. This terminal area, known as the presynaptic portion of the neuromuscular junction, is abundant in synaptic vesicles filled with the neurotransmitter acetylcholine. These vesicles are densely situated near the presynaptic membrane, ready for quick release (Sanes and Lichtman 1999). The presynaptic region is separated from the postsynaptic membrane by a synaptic cleft approximately 50 nm wide. The postsynaptic membrane contains extensive junctional folds that significantly expand its surface area. This architecture accommodates a high density of acetylcholine receptors and acetylcholinesterase, the enzyme responsible for regulating neurotransmitter activity (Fagerlund and Eriksson 2009; Kandel et al. 2013) (Fig. 2.13).

The process of muscle contraction initiation begins with a depolarization wave arriving at the nerve terminal. This event opens voltage-sensitive channels in the synaptic membrane, allowing calcium ions to enter the nerve terminal following their concentration gradient. The influx of calcium ions prompts synaptic vesicles to move toward the presynaptic membrane and release acetylcholine into the synaptic cleft (Katz and Miledi 1968).

The subsequent binding of acetylcholine to receptors on the postsynaptic membrane is essential for neuromuscular transmission. This interaction opens sodium and potassium ion channels, resulting in a localized depolarization known as the end plate potential. If this potential reaches the necessary activation threshold, it triggers a depolarization wave that propagates along the muscle membrane (Fatt and Katz 1951).

Acetylcholine's effect is transient, typically ending within milliseconds after its release. This

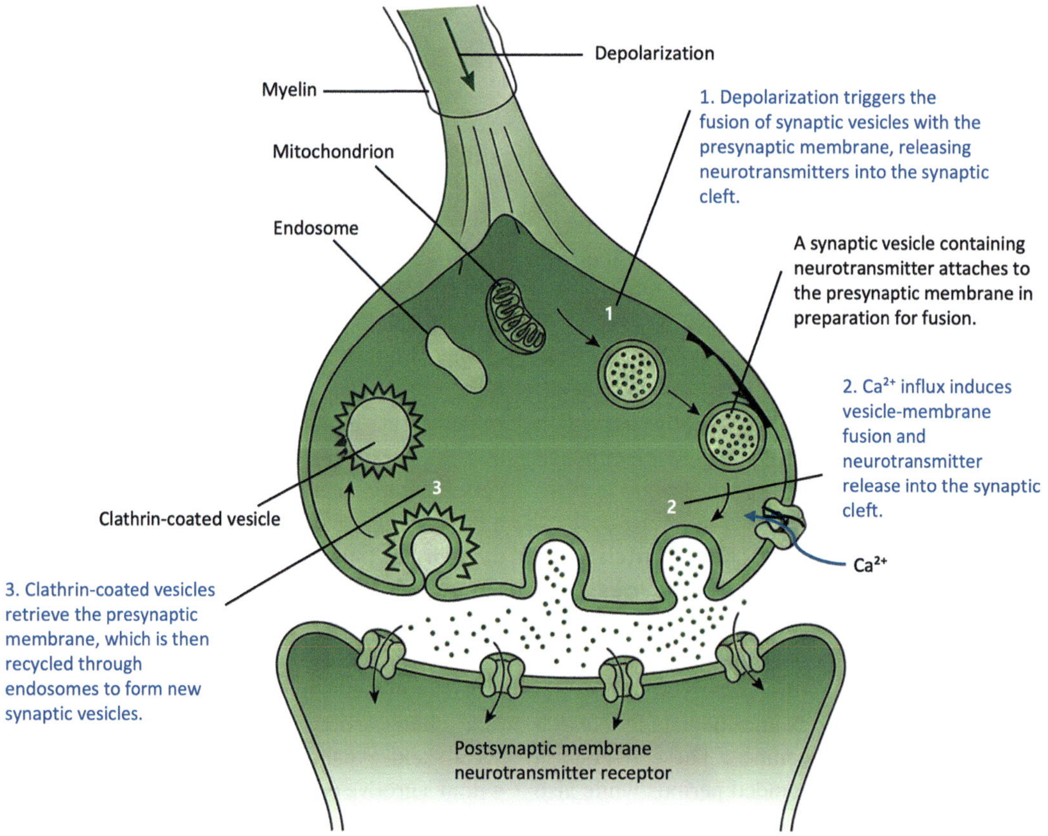

Fig. 2.13 Neuromuscular junction: illustrating depolarization, calcium influx, presynaptic neurotransmitter release, and interaction with the postsynaptic membrane. (Adapted with permission from Patestas and Gartner (2006))

rapid cessation is due to the action of the enzyme cholinesterase, which breaks down acetylcholine into choline and acetate. To maintain effective neuromuscular transmission, an excess of acetylcholine is released, a phenomenon termed the "safety factor" (Wood and Slater 2001).

The muscle membrane's depolarization initiates impulse propagation through the transverse tubule system (T system), an extension of the sarcolemma that penetrates the muscle fiber near contractile elements. At the intersection of the T system and sarcoplasmic reticulum, this electrical signal induces calcium ion release, a process known as excitation–contraction coupling. The sarcoplasmic reticulum, a network of cisternae surrounding myofibrils, functions as the primary calcium reservoir in muscle fibers. It contains high concentrations of calsequestrin, a calcium-binding protein that allows for a 40-fold increase in calcium storage capacity. The calcium pump plays a key role in reducing calcium levels in the myofibrillar cytosol during muscle relaxation. Muscle fibers consist of 100–1000 myofibrils, each encircled by the sarcoplasmic reticulum and immersed in electrolytes and protein enzymes. Individual myofibrils are composed of approximately 1500 thick myosin filaments and 2000 thin actin filaments, which interlock. The Z discs, containing filamentous proteins, serve as attachment points for actin filaments. The region between two Z discs constitutes the sarcomere, which is the functional unit of the myofibril. Under polarized light microscopy, the sarcomere's structure reveals isotropic I light bands, containing actin filaments, and anisotropic A dark bands, comprising myosin and overlapping actin filaments (Gordon et al. 2000) (Fig. 2.14).

The neuromuscular junction facilitates the transmission of neural signals to muscle fibers. When an action potential reaches the presynaptic terminal, it triggers the release of acetylcholine into the synaptic cleft. This neurotransmitter binds to receptors on the postsynaptic membrane, initiating a local depolarization that spreads along the muscle fiber (Kanning et al. 2010).

The contractile process involves the sliding filament theory, where myosin heads interact

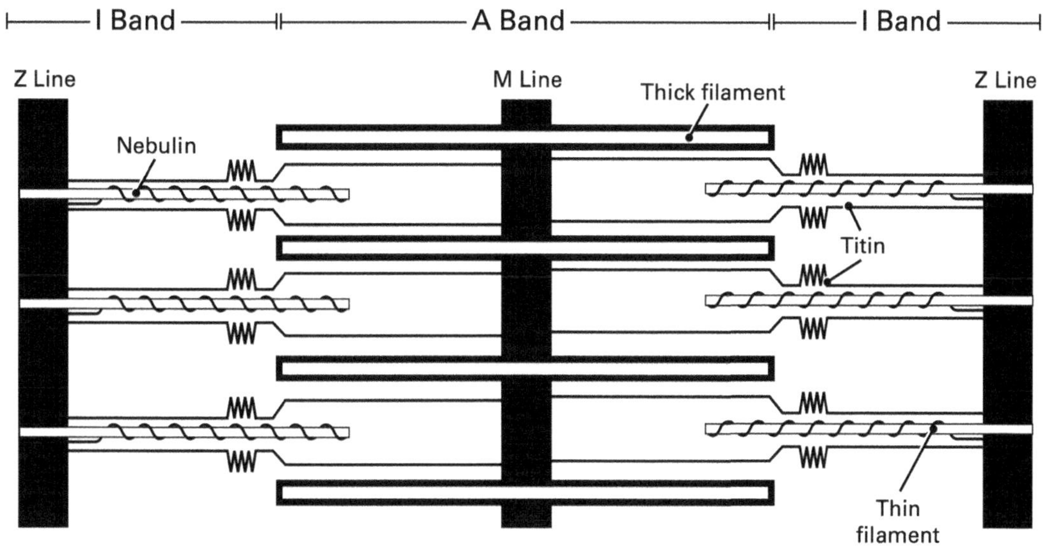

Fig. 2.14 Diagram of the sarcomere. Myosin thick filaments are anchored to the M line, while actin thin filaments are anchored to the Z line. The interdigitation of these filaments is organized by the giant protein titin. Nebulin spans the length of the thin filament, regulating its dimensions. Muscle contraction occurs through the interaction of myosin and actin, which increases the overlap between thick and thin filaments. (Adapted with permission from Brushart (2011))

with actin filaments in the presence of calcium ions and ATP. This interaction results in the shortening of the sarcomere and, consequently, muscle contraction. The troponin–tropomyosin complex regulates this process by controlling the availability of myosin-binding sites on actin filaments (Gordon et al. 2000).

2.8.4 Muscle–Joint Movement Interaction in Kinesiology

Degrees of freedom in kinesiology refer to the number of independent angular motions possible at a joint. A joint can exhibit up to three degrees of freedom, aligned with the three cardinal planes: sagittal, frontal, and transverse. Each rotational motion in these planes comprises six degrees of freedom. Joint movements along these degrees of freedom pivot around the fulcrum of angular motion, termed the axis of rotation, which is located within or very close to the joint (Levangie and Norkin 2005). Body movement, whether translational or rotational, can occur through active movement involving muscle contraction or passive movement without active muscle involvement (Neumann 2017).

Joint movement can be categorized into two kinematic patterns (Fig. 2.15):

- Movement of the distal segment relative to a fixed proximal segment.
- Movement of the proximal segment relative to a fixed distal segment.

For instance, knee flexion represents the relative movement between the thigh and lower leg, exemplifying both proximal-on-distal and distal-on-proximal segment kinematics (Karandikar and Vargas 2011). In the upper extremity, during typical activities, the proximal segment is usually stabilized by muscles, ligamentous structures, gravity, or its inertia, while the distal segment rotates more freely, demonstrating distal-on-proximal segment kinematics. However, the upper extremities can also perform proximal-on-distal segment kinematics, such as during push-ups or seat transfers (Kisner and Colby 2007).

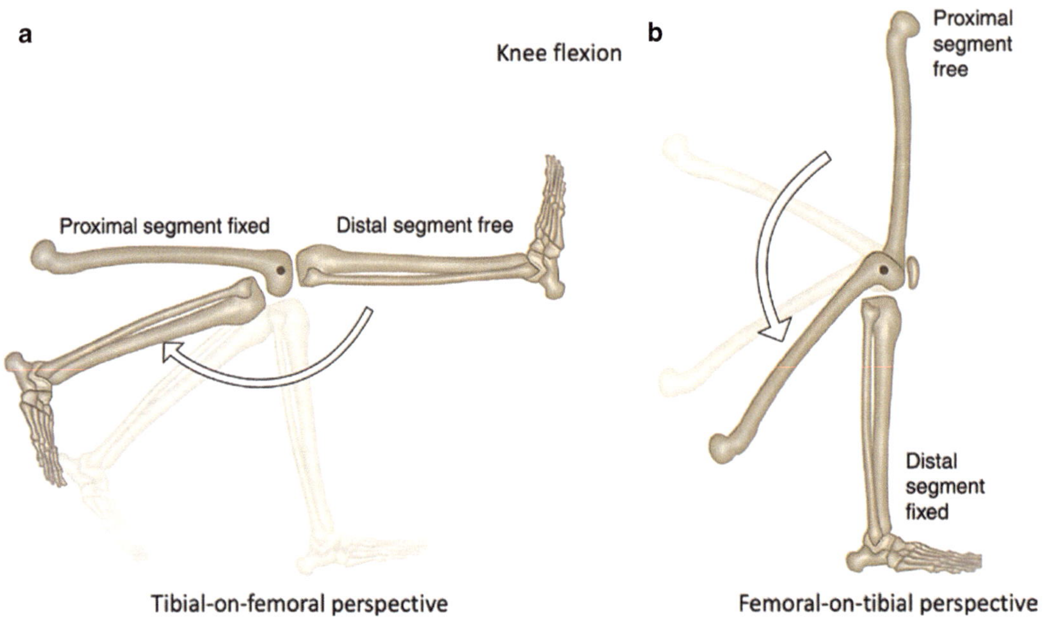

Fig. 2.15 Sagittal plane motion at the knee illustrating kinematics of (**a**) distal-on-proximal segment movement and (**b**) proximal-on-distal segment movement. (Adapted with permission from Ko (2022))

A kinetic chain refers to a series of articulated, segmented links, such as the interconnected pelvis, thigh, lower leg, and foot of the lower extremity. The terms "open kinetic chain" and "closed kinetic chain" are defined by the fixation status of the distal end of an extremity:

- Open kinematic chain: The distal segment of a kinematic chain is not fixed to an immovable object, allowing free movement of the distal segment.
- Closed kinematic chain: The distal segment of the kinematic chain is fixed to an immovable object, allowing free movement of the proximal segment (McGinnis 2013).

The application of closed kinetic chain kinematics in individuals with spinal cord injuries, particularly in those with lesions as high as C5 or C6, can significantly enhance their ability to perform pressure relief maneuvers, execute transfers, and propel wheelchairs. High tetraplegics face significant challenges in achieving independence in daily activities due to the lack of active antigravity strength for elbow extension (Sisto et al. 2009). For these individuals, the ventral deltoid, biceps, and brachialis muscles, innervated by the C5 and/or C6 spinal segments, are utilized within a closed kinetic chain to transform elbow and shoulder flexion movements into elbow extension. This is achieved when the hand is fixed on the wheelchair hand rim through friction or on the floor (Harvey 2008; Karandikar and Vargas 2011; Little et al. 2000) (Fig. 2.16).

It is important to note that this mechanism does not involve active extension of the triceps muscle. Instead, it represents a substitution strategy employing muscles with intact innervation to produce elbow extension within a closed-chain system. This biomechanical adaptation allows individuals with high cervical spinal cord injuries to perform functional tasks that would otherwise be impossible due to the lack of active triceps function (Mulcahey et al. 2007; Nene et al. 1996).

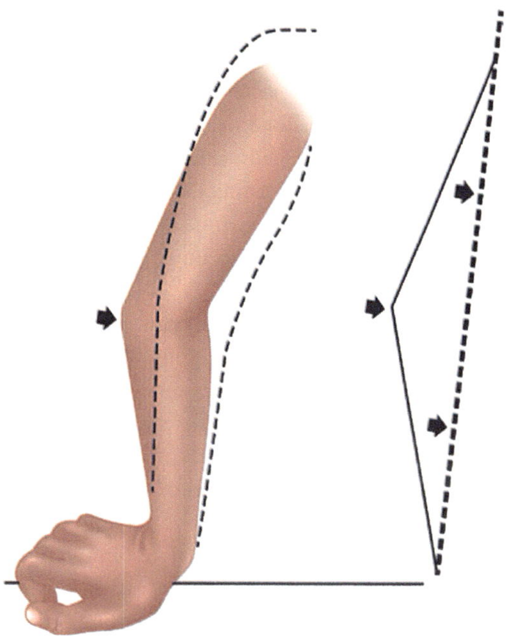

Fig. 2.16 Substitution for elbow extension using closed kinetic chain kinematics. Humeral adduction with forearm supination facilitates elbow joint extension. (Adapted with permission from Ko (2022))

2.8.4.1 One-Joint and Two-Joint Muscles

The anatomical terms, the origin and insertion sites of a muscle, are determined by the relative mobility of these points. Generally, the origin is less mobile compared to the insertion site. The functions and movements accomplished by a muscle can vary depending on which site is stabilized, affecting joint movement between the origin and insertion (Levangie and Norkin 2005; Neumann 2017).

Muscle actions can either move a segment at one end of its attachment or influence two segments at both ends. Most muscles span only one joint, with their primary action focused on that joint. For instance, the soleus, primarily an ankle plantar flexor, can indirectly force knee extension when the foot is grounded despite not crossing the knee joint (Escamilla et al. 1998).

Two-joint muscles present a unique case, crossing two joints and generating multiple, often opposing movements. The rectus femoris exemplifies this, producing both hip flexion and knee extension. The action of the two-joint muscle is influenced by body position and its interaction with external objects, such as the ground. The rectus femoris, for instance, primarily contributes to knee extension due to the hip joint's positioning. This positioning causes the muscle's force to act near the hip, limiting its efficacy in producing hip flexion (Hamill et al. 2015; Herzog et al. 1998).

Functional limitations can occur in two-joint muscles at specific joint positions. When a two-joint muscle is stretched to its maximum length, it experiences passive insufficiency, where the antagonistic muscle cannot be further elongated, hence limiting the full range of motion. Tight hamstrings can cause passive insufficiency, which prevents full knee extension. On the other hand, active insufficiency occurs when a two-joint muscle is shortened to the extent that it can no longer provide its maximal amount of force. For example, when the wrist is flexed, the finger flexors are unable to provide the maximum grip force due to their shortened state (Hamill et al. 2015; Gajdosik et al. 1993).

2.8.4.2 Functional Mechanics of the Levers of Human Movement

The comprehension of lever mechanics is fundamental for analyzing human movement patterns. The human body utilizes three distinct classes of levers to facilitate motion, each characterized by specific configurations of the fulcrum (axis of rotation), effort arm (force application point), and resistance arm (load location) (Fig. 2.17).

First-class levers in the human body are structured with the fulcrum situated between the effort and resistance arms. This configuration is primarily utilized for maintaining equilibrium and altering force direction. An example is standing on one leg, where the hip joint acts as the fulcrum. The load is represented by the body weight on the opposite side of the hip, while the gluteus medius and minimus muscles contract to generate force on the same side. This configuration generally does not provide a mechanical benefit in terms of increasing force.

Second-class levers position the fulcrum at one extremity, with the effort applied at the opposite end and the load centrally located. This arrangement offers a mechanical advantage, permitting the movement of large loads with less force at a reduced speed. The act of plantar flexion exemplifies this lever class. In this motion, the metatarsal heads serve as the fulcrum, the body mass transmitted through the tibia constitutes the load, and the gastrocnemius–soleus complex provides the effort. The resistance arm spans from the tibia to the metatarsal heads, while the effort arm extends from the calcaneal insertion of the triceps surae to the metatarsal heads (Neumann 2017).

A third-class lever is the most prevalent lever type in the human body. In this configuration, the fulcrum is located at one end, the load is located at the opposite end, and the force is applied in between. Although it operates at a

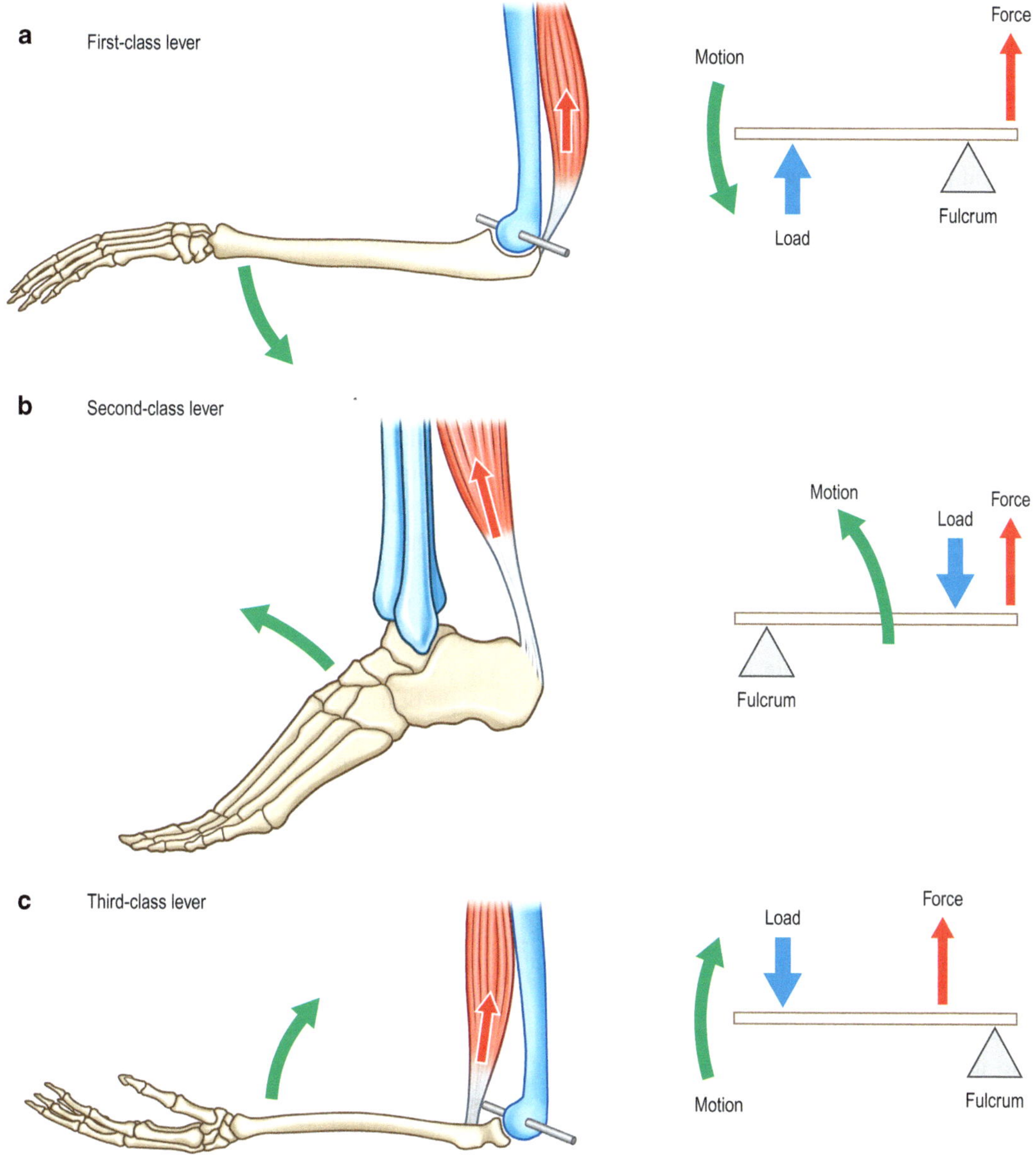

Fig. 2.17 Classes of levers. (**a**) First-class lever: forces are applied on either side of the fulcrum. (**b**) Second-class lever: the muscle or internal force has greater leverage than the external force due to its position farther from the fulcrum (pivot), both being on the same side of the fulcrum. (**c**) Third-class lever: the muscle or internal force is positioned closer to the axis of rotation than the external force. (Adapted with permission from Ko (2022))

mechanical disadvantage, moving less weight, it allows for greater speed. An example of this is the biceps brachii muscle acting across the elbow joint. The elbow joint serves as the fulcrum, with the forearm and hand representing the load, while the contraction of the biceps provides the force. Here, the load arm is the distance from the elbow to the forearm and the hand's center of mass, and the force arm is the distance from the elbow joint to the biceps attachment (Palastanga and Soames 2012).

2.9 Muscle Weakness

Muscle weakness is defined as the inability of a rested muscle to exert its maximum force during testing. This evaluation reflects multiple physiological processes, including cerebral input, neuromuscular coordination, motor unit recruitment, peripheral nerve function, neuromuscular transmission, and the muscle's cross-sectional area. Any disruption within this pathway can lead to a weakened muscle contraction (Frontera and Ochala 2015).

Several methods exist for measuring muscle strength, with manual muscle testing being the most commonly employed by clinicians. In this technique, the patient is asked to maintain a brief isometric contraction against resistance provided by the examiner. Muscle weakness should be apparent at the onset of the contraction, making factors such as substrate availability or excitation–contraction coupling less relevant in static strength assessments. The muscle's ability to resist is directly proportional to the number of activated motor units, their firing rate and pattern, and the cross-sectional area of the contracting muscle fibers. Reduced muscle force is predominantly due to the loss of functional skeletal muscle fibers, fiber atrophy, significant neuromuscular junction blockade, or impairments in motor drive or peripheral nerve function (Aagaard et al. 2001).

Physiatrists often recognize that manual muscle testing has limited sensitivity in detecting mild motor weakness. Historical studies, particularly those involving polio patients, have shown that despite a 50% loss of muscle fibers, patients often presented with normal muscle grades during manual muscle testing (Beasley 1961; Sharrard 1955).

2.10 Muscle Atrophy

Experienced physiatrists often observe a discrepancy between manual muscle testing results and functional performance in patients following extended hospitalization. While strength assessments may indicate normal or near-normal levels, significant functional limitations frequently emerge during initial mobilization. This discrepancy arises from physiological alterations in skeletal muscle and other organ systems induced by immobilization (Dirks et al. 2018). Immobilization is not the only cause of muscle disuse atrophy; even when immobilized, a muscle with intact innervation is not entirely disused but rather used less. Thus, the physiological definition of disuse involves altered neuronal discharge, not just reduced activity or immobilization (Booth and Kelso 1973).

Consider spinal cord injury as an example of true disuse atrophy, which results from upper motor neuron innervation loss. Animal studies involving spinal cord transection demonstrate a notable shift from type 1 (slow) to type 2 (fast) muscle fibers, accompanied by alterations in muscle properties, including diminished endurance. This fiber transition is caused by changes in weight-bearing, muscle length alterations, and a substantial reduction in neuronal input. Individuals with chronic paraplegia experience muscle atrophy, and a decrease in type I fibers is observed, whereas there is a simultaneous increased proportion of type 2B fibers. The altered aerobic capacity of muscles following spinal cord injury has a substantial impact on how disused muscles respond to exercise (Lotta et al. 1991).

Muscle denervation atrophy shows the importance of the nerve–muscle relationship in disuse atrophy. This type of atrophy is severe, affecting both fast and slow fibers, resulting in reduced fiber diameter and muscle strength. Additionally, it causes structural changes in the muscle, such as changes in the proportions of types of muscle fibers, which could potentially affect exercise performance in the future. Investigating the potential of retraining muscles to regain their premorbid state before illness or injury is important (Gordon et al. 2011; Lieber et al. 1986).

Limb immobilization can induce a shift from type 1 to type 2 fibers, although the effects are less pronounced than those observed with spinal transection. Debate persists regarding whether these histological changes result more from selective type 1 atrophy or actual fiber type conversion. The most significant reduction in strength occurs during the first week of immobilization, with a gradual decrease thereafter. Physiological extensors experience more pronounced atrophy than flexors during immobilization, potentially due to their postural role and higher percentage of type 1 fibers (Berg et al. 1997).

Animal studies have revealed an inverse relationship between muscle strength and the extent of muscle atrophy during limb immobilization. When immobilized in a shortened state, segmental necrosis occurs at the fiber ends, with sarcomeres being removed both in series and in parallel, leading to a reduced muscle cross-sectional area. Conversely, immobilization in a lengthened state results in the addition of sarcomeres (Lotta et al. 1991; Williams and Goldspink 1973).

References

Aagaard P, Andersen JL, Dyhre-Poulsen P, Leffers AM, Wagner A, Magnusson SP, et al. A mechanism for increased contractile strength of human pennate muscle in response to strength training: changes in muscle architecture. J Physiol. 2001;534(2):613–23.

Baehr M, Frotscher M. Duus' topical diagnosis in neurology. Anatomy, physiology, signs, symptoms. 5th ed. New York: Thieme; 2012.

Bear MF, Connors BW, Paradiso MA. Neuroscience: exploring the brain. 4th ed. Burlington: Jones & Bartlett Learning; 2020.

Beasley WC. Quantitative muscle testing: principles and applications to research and clinical services. Arch Phys Med Rehabil. 1961;42:398–425.

Berg HE, Larsson L, Tesch PA. Lower limb skeletal muscle function after 6 wk of bed rest. J Appl Physiol. 1997;82(1):182–8.

Bican O, Minagar A, Pruitt AA. The spinal cord: a review of functional neuroanatomy. Neurol Clin. 2013;31(1):1–18.

Biering-Sørensen B, Kristensen IB, Kjaer M, Biering-Sørensen F. Muscle after spinal cord injury. Muscle Nerve. 2009;40(4):499–519.

Bogduk N. The anatomical basis for spinal pain syndromes. J Manip Physiol Ther. 1995;18(9):603–5.

Bogduk N. Clinical anatomy of the lumbar spine and sacrum. 4th ed. Edinburgh: Elsevier Churchill Livingstone; 2005.

Booth FW, Kelso JR. Effect of hind-limb immobilization on contractile and histochemical properties of skeletal muscle. Pflugers Arch. 1973;342(3):231–8.

Bottinelli R, Reggiani C. Human skeletal muscle fibres: molecular and functional diversity. Prog Biophys Mol Biol. 2000;73(2-4):195–262.

Brushart TM. Nerve repair. New York: Oxford University Press; 2011.

Burke RE, Jankowska E, ten Bruggencate G. A comparison of peripheral and rubrospinal synaptic input to slow and fast twitch motor units of triceps surae. J Physiol. 1970;207(3):709–32.

Caliandro P, La Torre G, Padua R, Giannini F, Padua L. Treatment for ulnar neuropathy at the elbow. Cochrane Database Syst Rev. 2016;11(11):CD006839.

Campbell WW. Dejong's the neurologic examination. 7th ed. Philadelphia: Wolters Kluwer; 2013.

Castro MJ, Apple DF Jr, Hillegass EA, Dudley GA. Influence of complete spinal cord injury on skeletal muscle cross-sectional area within the first 6 months of injury. Eur J Appl Physiol Occup Physiol. 1999;80(4):373–8.

Cleland J, Koppenhaver S. Netter's orthopaedic clinical examination: An evidence-based approach. 2nd ed. Philadelphia: Elsevier; 2011.

Coggeshall RE, Applebaum ML, Fazen M, Stubbs TB 3rd, Sykes MT. Unmyelinated axons in human ventral roots, a possible explanation for the failure of dorsal rhizotomy to relieve pain. Brain. 1975;98(1):157–66.

Cramer GD, Darby SA. Clinical anatomy of the spine, spinal cord, and ANS. 3rd ed. St. Louis: Elsevier Health Sciences; 2014.

Craw CH. The distribution of the nerve cells in the ventral columns of the spinal cord. J Comp Neurol. 1928;45:283–99.

Debanne D, Campanac E, Bialowas A, Carlier E, Alcaraz G. Axon physiology. Physiol Rev. 2011;91(2):555–602.

Dirks ML, Wall BT, van Loon LJC. Interventional strategies to combat muscle disuse atrophy in humans: focus on neuromuscular electrical stimulation and dietary protein. J Appl Physiol. 2018;125(3):850–61.

Drake RL, Vogl AW, Mitchell AWM. Gray's anatomy for students. 4th ed. Philadelphia: Elsevier; 2020.

Ebraheim NA, An HS, Xu R, Ahmad M, Yeasting RA. The quantitative anatomy of the cervical nerve root groove and the intervertebral foramen. Spine. 1996;21(14):1619–23.

Elliott HC. Studies on the motor cells of the spinal cord: I. Distribution in the normal human cord. Am J Anat. 1942;70(1):95–117.

Escamilla RF, Fleisig GS, Zheng N, Barrentine SW, Wilk KE, Andrews JR. Biomechanics of the knee during closed kinetic chain and open kinetic chain exercises. Med Sci Sports Exerc. 1998;30(4):556–69.

Fagerlund MJ, Eriksson LI. Current concepts in neuromuscular transmission. Br J Anaesth. 2009;103(1):108–14.

Farmer JC, Wisneski RJ. Cervical spine nerve root compression. An analysis of neuroforaminal pressures with varying head and arm positions. Spine. 1994;19(16):1850–5.
Fatt P, Katz B. An analysis of the end-plate potential recorded with an intra-cellular electrode. J Physiol. 1951;115(3):320–70.
Frontera WR, Ochala J. Skeletal muscle: a brief review of structure and function. Calcif Tissue Int. 2015;96(3):183–95.
Frontera WR, DeLisa JA, Basford JR, Bockenek WL, Chae J, Robinson LR. DeLisa's physical medicine and rehabilitation: principles and practice. 6th ed. Philadelphia: Wolters Kluwer; 2019.
Gajdosik RL, Rieck MA, Sullivan DK, Wightman SE. Comparison of four clinical tests for assessing hamstring muscle length. J Orthop Sports Phys Ther. 1993;18(5):614–8.
Geeves MA, Holmes KC. Structural mechanism of muscle contraction. Annu Rev Biochem. 1999;68:687–728.
Gillies AR, Lieber RL. Structure and function of the skeletal muscle extracellular matrix. Muscle Nerve. 2011;44(3):318–31.
Gordon AM, Homsher E, Regnier M. Regulation of contraction in striated muscle. Physiol Rev. 2000;80(2):853–924.
Gordon T, Tyreman N, Raji MA. The basis for diminished functional recovery after delayed peripheral nerve repair. J Neurosci. 2011;31(14):5325–34.
Gorgey AS, Dudley GA. Skeletal muscle atrophy and increased intramuscular fat after incomplete spinal cord injury. Spinal Cord. 2007;45(4):304–9.
Hall JE. Guyton and Hall textbook of medical physiology. 13th ed. Philadelphia: Elsevier; 2016.
Haller FR, Low FN. The fine structure of the peripheral nerve root sheath in the subarachnoid space in the rat and other laboratory animals. Am J Anat. 1971;131(1):1–19.
Hallett M, Albanese A, Dressler D, et al. Practical guidance for CD management involving treatment of botulinum toxin: a consensus statement. J Neurol. 2015;262(10):2201–13.
Hamill J, Knutzen KM, Derrick TR. Biomechanical basis of human movement. 4th ed. Philadelphia: Wolters Kluwer Health; 2015.
Harvey LA. Management of spinal cord injuries: a guide for physiotherapists. Edinburgh: Butterworth-Heinemann; 2008.
Herzog W, Leonard TR, Wu JZ. Force depression following skeletal muscle shortening is long lasting. J Biomech. 1998;33(12):1539–48.
Jaumard NV, Welch WC, Winkelstein BA. Spinal facet joint biomechanics and mechanotransduction in normal, injury and degenerative conditions. J Biomech Eng. 2011;133(7):071010.
Kandel ER, Schwartz JH, Jessell TM, Siegelbaum SA, Hudspeth AJ. Principles of neural science. 5th ed. New York: McGraw-Hill; 2013.
Kanning KC, Kaplan A, Henderson CE. Motor neuron diversity in development and disease. Annu Rev Neurosci. 2010;33:409–40.
Karandikar N, Vargas OO. Kinetic chains: a review of the concept and its clinical applications. PM&R. 2011;3(8):739–45.
Katz B, Miledi R. The role of calcium in neuromuscular facilitation. J Physiol. 1968;195(2):481–92.
Kiernan JA, Rajakumar N. Barr's the human nervous system: an anatomical viewpoint. 10th ed. Philadelphia: Wolters Kluwer; 2014.
Kierner AC, Zelenka I, Heller S, Burian M. Surgical anatomy of the spinal accessory nerve and the trapezius branches of the cervical plexus. Arch Surg. 2000;135(12):1428–31.
Kisner C, Colby LA. Therapeutic exercise: foundations and techniques. 5th ed. Philadelphia: FA Davis; 2007.
Ko HY. Management and rehabilitation of spinal cord injuries. 2nd ed. Singapore: Springer; 2022.
Levangie PK, Norkin CC. Joint structure and function: a comprehensive analysis. 4th ed. Philadelphia: FA Davis; 2005.
Lieber RL. Skeletal muscle structure, function, and plasticity. 2nd ed. Philadelphia: Lippincott Williams & Wilkins; 2002.
Lieber RL, Fridén J. Functional and clinical significance of skeletal muscle architecture. Muscle Nerve. 2000;23(11):1647–66.
Lieber RL, Johansson CB, Vahlsing HL, Hargens AR, Feringa ER. Long-term effects of spinal cord transection on fast and slow rat skeletal muscle. I. Contractile properties. Exp Neurol. 1986;91(3):423–34.
Little JW, Goldstein B, Hammond MC. Spinal cord injury rehabilitation. In: Belandres PV, Dillingham TR, editors. Rehabilitation of the injured combatant. Washington, DC: Department of the Army; 2000.
Liu J, Ebraheim NA, Sanford CG Jr, Patil V, Elsamaloty H, Treuhaft K, et al. Quantitative changes in the cervical neural foramen resulting from axial traction: in vivo imaging study. Spine J. 2008;8(4):619–23.
Lotta S, Scelsi R, Alfonsi E, Saitta A, Nicolotti D, Epifani P, et al. Morphometric and neurophysiological analysis of skeletal muscle in paraplegic patients with traumatic cord lesion. Paraplegia. 1991;29(4):247–52.
Lundborg G. Intraneural microcirculation. Orthop Clin North Am. 1988;19(1):1–12.
Mackinnon SE. Nerve surgery. New York: Thieme; 2015.
McGinnis PM. Biomechanics of sport and exercise. 3rd ed. Champaign: Human Kinetics; 2013.
Moore KL, Dalley AF, Agur AMR, editors. Clinically oriented anatomy. 8th ed. Wolters Kluwer: Philadelphia; 2018.
Muhle C, Resnick D, Ahn JM, Südmeyer M, Heller M. In vivo changes in the neuroforaminal size at flexion-extension and axial rotation of the cervical spine in healthy persons examined using kinematic magnetic resonance imaging. Spine. 2001;26(13):E287–93.

Mulcahey MJ, Hutchinson D, Kozin S. Assessment of upper limb in tetraplegia: considerations in evaluation and outcomes research. J Rehabil Res Dev. 2007;44(1):91–102.

Nave KA, Werner HB. Myelination of the nervous system: mechanisms and functions. Annu Rev Cell Dev Biol. 2014;30:503–33.

Nene AV, Hermens HJ, Zilvold G. Paraplegic locomotion: a review. Spinal Cord. 1996;34(9):507–24.

Netter FH. Atlas of human anatomy. 7th ed. Philadelphia: Elsevier; 2019.

Neumann DA. Kinesiology of the musculoskeletal system: foundations for rehabilitation. 3rd ed. St. Louis: Mosby; 2017.

Nordin M, Frankel VH. Basic biomechanics of the musculoskeletal system. 4th ed. Philadelphia: Lippincott Williams & Wilkins; 2001.

Olsewski JM, Simmons EH, Kallen FC, Pope MH, Heiple KG. Morphometry of lumbar spine: anatomical perspectives related to transpedicular fixation. J Bone Joint Surg. 1990;72(4):541–9.

Palastanga N, Soames R. Anatomy and human movement: structure and function. 6th ed. Philadelphia: Elsevier Health Sciences; 2012.

Parent A. Carpenter's human neuroanatomy. 9th ed. Philadelphia: Lippincott Williams & Wilkins; 1996.

Patestas MA, Gartner LP. A textbook of neuroanatomy. Oxford: Blackwell Publishing; 2006.

Pech P, Daniels DL, Williams AL, Haughton VM. The cervical neural foramina: correlation of microtomy and CT anatomy. Radiology. 1985;155(1):143–6.

Proske U, Gandevia SC. The proprioceptive senses: their roles in signaling body shape, body position and movement, and muscle force. Physiol Rev. 2012;92(4):1651–97.

Purves D, Augustine GJ, Fitzpatrick D, Hall WC, LaMantia AS, White LE. Neuroscience. 6th ed. New York: Oxford University Press; 2018.

Rauschning W. Normal and pathoanatomy of the lumbar root canals. Spine. 1987;12(10):1008–19.

Ropper AH, Samuels MA, Klein JP. Adams and Victor's principles of neurology. 10th ed. New York: McGraw-Hill Education; 2014.

Salzer JL. Schwann cell myelination. Cold Spring Harb Perspect Biol. 2015;7(8):a020529.

Sanes JR, Lichtman JW. Development of the vertebrate neuromuscular junction. Annu Rev Neurosci. 1999;22:389–442.

Schiaffino S, Reggiani C. Fiber types in mammalian skeletal muscles. Physiol Rev. 2011;91(4):1447–531.

Schuenke M, Schulte E, Schumacher U. Thieme atlas of anatomy. Head, neck, and neuroanatomy. 2nd ed. New York: Thieme; 2016.

Seidl AH. Regulation of conduction time along axons. Neuroscience. 2014;276:126–34.

Sharrard WJ. The distribution of the permanent paralysis in the lower limb in poliomyelitis: a clinical and pathological study. J Bone Joint Surg Br. 1955;37-B(4):540–58.

Sherman DL, Brophy PJ. Mechanisms of axon ensheathment and myelin growth. Nat Rev Neurosci. 2005;6(9):683–90.

Shields RK. Muscular, skeletal, and neural adaptations following spinal cord injury. J Orthop Sports Phys Ther. 2002;32(2):65–74.

Siegel A, Sapru HN. Essential neuroscience. 2nd ed. Philadelphia: Wolters Kluwer; 2011.

Simons M, Nave KA. Oligodendrocytes: myelination and axonal support. Cold Spring Harb Perspect Biol. 2015;7(5):a020479.

Sisto SA, Druin E, Sliwinski MM. Spinal cord injuries: management and rehabilitation. St. Louis: Mosby; 2009.

Standring S. Gray's anatomy: the anatomical basis of clinical practice. 42nd ed. Edinburgh: Churchill Livingstone; 2020.

Staron RS, Hagerman FC, Hikida RS, Murray TF, Hostler DP, Crill MT, et al. Fiber type composition of the vastus lateralis muscle of young men and women. J Histochem Cytochem. 2000;48(5):623–9.

Stewart JD. Peripheral nerve fascicles: anatomy and clinical relevance. Muscle Nerve. 2003;28(5):525–41.

Sunderland S. The intraneural topography of the radial, median and ulnar nerves. Brain. 1945;68:243–99.

Sunderland S. Meningeal-neural relations in the intervertebral foramen. J Neurosurg. 1974;40(6):756–63.

Topp KS, Boyd BS. Structure and biomechanics of peripheral nerves: nerve responses to physical stresses and implications for physical therapist practice. Phys Ther. 2006;86(1):92–109.

Torun F, Dolgun H, Tuna H, Attar A, Uz A, Erdem A. Morphometric analysis of the roots and neural foramina of the lumbar vertebrae. Surg Neurol. 2006;66(2):148–51.

Trappe S, Luden N, Minchev K, Raue U, Jemiolo B, Trappe TA. Skeletal muscle signature of a champion sprint runner. J Appl Physiol (1985). 2015;118(12):1460–6.

Vanderah TW, Gould DJ. Nolte's the human brain: an introduction to its functional anatomy. 7th ed. Philadelphia: Elsevier; 2016.

Watson C, Harrison M. The location of the major ascending and descending spinal cord tracts in all spinal cord segments in the mouse: actual and extrapolated. Anat Rec. 2012;295(10):1692–7.

Watson C, Harrison M. Spinal cord organization. Neurosci Bull. 2022;38(1):17–35.

Watson C, Paxinos G, Kayalioglu G. The spinal cord: a Christopher and Dana Reeve Foundation text and atlas. Amsterdam: Academic Press; 2009.

Wickiewicz TL, Roy RR, Powell PL, Edgerton VR. Muscle architecture of the human lower limb. Clin Orthop Relat Res. 1983;179:275–83.

Williams PE, Goldspink G. The effect of immobilization on the longitudinal growth of striated muscle fibres. J Anat. 1973;116(Pt 1):45–55.

Wineski LE. Snell's clinical anatomy by regions. 10th ed. Philadelphia: Wolters Kluwer; 2019.

Wood SJ, Slater CR. Safety factor at the neuromuscular junction. Prog Neurobiol. 2001;64(4):393–429.
Woolf CJ, Ma Q. Nociceptors—noxious stimulus detectors. Neuron. 2007;55(3):353–64.
Yoo JU, Zou D, Edwards WT, Bayley J, Yuan HA. Effect of cervical spine motion on the neuroforaminal dimensions of human cervical spine. Spine. 1992;17(10):1131–6.
Zierath JR, Hawley JA. Skeletal muscle fiber type: influence on contractile and metabolic properties. PLoS Biol. 2004;2(10):e348.

Recommended Additional Reading

Clemente CD. Anatomy: a regional atlas of the human body. 6th ed. Philadelphia: Wolters Kluwer; 2011.
Downey JA, Myers SJ, Gonzalez EG, Lieberman JS. The physiological basis of rehabilitation medicine. Boston: Butterworth-Heinemann; 1994.
Hanna AS. Anatomy and exposures of spinal nerves. New York: Springer; 2015.
ten Donkelaar HJ. Clinical neuroanatomy: brain circuity and its disorders. 2nd ed. Cham: Springer; 2020.

3 Structure and Functional Organization of the Autonomic Nervous System

The autonomic nervous system (ANS) serves as a motor system that controls the cardiac muscle, smooth muscle, and glandular secretory cells. Unlike the direct innervation pattern seen in somatic motor systems, the ANS employs a characteristic two-neuron arrangement (Brodal 2010). This system consists of preganglionic (or presynaptic) neurons, whose cell bodies reside within the central nervous system (CNS), forming synapses with postganglionic (or postsynaptic) neurons located in peripheral ganglia. These postganglionic neurons then extend to innervate various target organs and tissues. The preganglionic neurons produce myelinated axons, while postganglionic axons are unmyelinated (Kiernan and Rajakumar 2014; Patestas and Gartner 2006).

The ANS maintains homeostatic balance through three main divisions: sympathetic, parasympathetic, and enteric systems. While the enteric nervous system exists within the gastrointestinal tract wall and can operate independently, it receives modulatory input from both sympathetic and parasympathetic divisions (Costa et al. 2015).

Sympathetic preganglionic neurons emerge from the thoracic and upper lumbar spinal cord segments. Their axons exit through anterior roots and travel via anterior rami before entering the paravertebral sympathetic chain ganglia. Many of these fibers form synapses in nearby ganglia, with their postganglionic counterparts rejoining anterior rami to course alongside somatic nerves through various plexuses. Some postganglionic sympathetic fibers follow blood vessels to reach their destinations (Drake et al. 2020) (Fig. 3.1). In the skin, these fibers provide innervation to sweat glands and blood vessels, following patterns similar to somatic sensory nerve distribution (Barker and Barasi 1999; Hall 2015).

The parasympathetic division originates from both cranial and sacral regions. Cranial nerves III, VII, and IX provide parasympathetic innervation to structures such as the pupil, ciliary muscles, and various glands. The vagus nerve (cranial nerve X) delivers extensive parasympathetic input to thoracic and abdominal organs (Goyal and Hirano 1996; Powley et al. 2016). The sacral component, comprising pelvic splanchnic nerves from S2 to S4, innervates genitourinary organs and the distal gastrointestinal tract. Parasympathetic postganglionic neurons typically reside near their target organs (Purves et al. 2018; Standring 2020).

The ANS coordinates visceral functions by integrating neural components from both the CNS and the peripheral nervous system (PNS). This integration enables precise regulation of the internal environment through the control of glands, smooth muscles, and cardiac muscles. The visceral efferent pathways incorporate peripheral synapses, requiring at least two neurons—preganglionic and postganglionic—to connect the CNS with target structures (Watson et al. 2010) (Fig. 3.2). Preganglionic neurons,

H.-Y. Ko, *Practical Functional Anatomy for Spinal Cord Injury Rehabilitation*,
https://doi.org/10.1007/978-981-95-6457-6_3

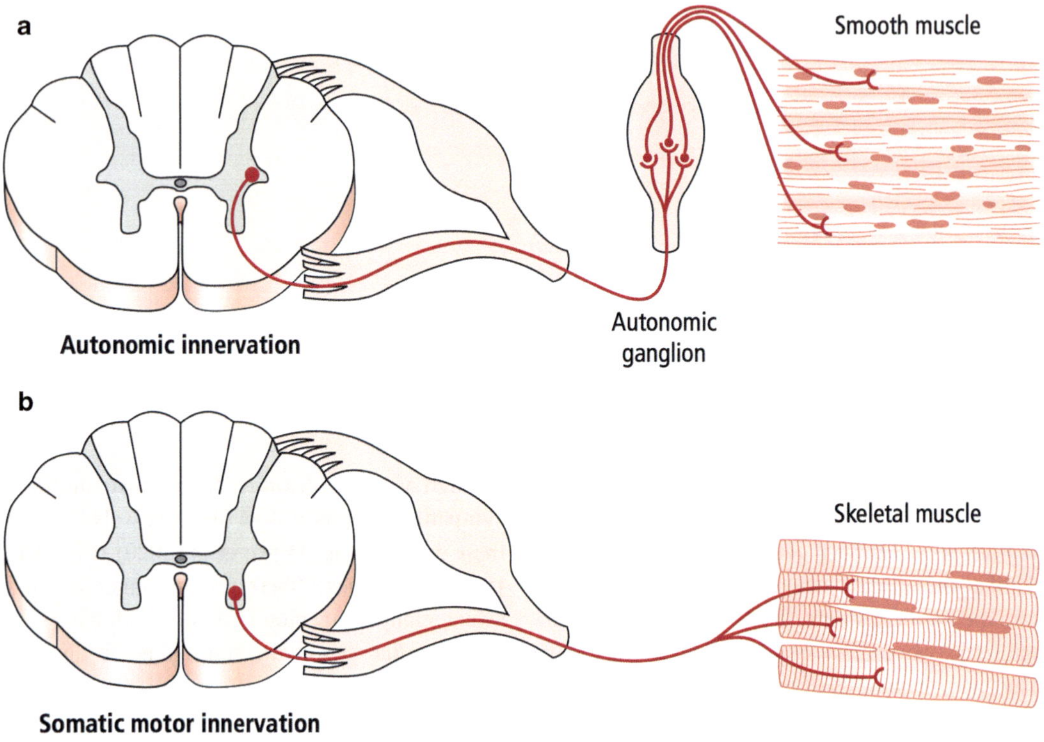

Fig. 3.1 Diagram illustrating the differences between autonomic innervation (**a**) and somatic motor innervation (**b**). The autonomic system involves two neurons, whereas the somatic motor system utilizes a single motor neuron. (Adapted with permission from Patestas and Gartner (2006))

situated in the brainstem visceral efferent nuclei and spinal cord lateral gray columns, extend their myelinated axons through specific cranial and spinal nerves. These neurons typically synapse with multiple postganglionic neurons, allowing for widespread autonomic control.

Historically, the sympathetic and parasympathetic systems were viewed as antagonistic in function, with sympathetic responses being more generalized and parasympathetic responses more localized. However, the modern understanding posits that these systems function synergistically to regulate visceral and homeostatic processes (Paxinos 1990). While sympathetic activation can manifest as a generalized reaction, such as in fight-or-flight situations, it can also occur in a more localized manner. Sympathetic stimulation typically results in vasoconstriction, increased cardiac output, and elevated blood pressure, preparing the body for increased activity. In contrast, parasympathetic activity slows the heart rate and enhances glandular secretions and peristalsis, contributing to energy conservation. Although the sympathetic, parasympathetic, and enteric systems differ in their structures, they operate as a coordinated network. Most structures innervated by the ANS receive input from both sympathetic and parasympathetic systems, while the enteric system operates independently within the gastrointestinal tract.

Autonomic responses are not exclusively triggered by visceral afferent reflexes. Somatic stimuli, such as those arising from the special senses or the skin, can initiate autonomic responses. For instance, increases in blood pressure or pupil dilation may be driven by somatic receptor activation. Peripheral autonomic activities are integrated at higher centers within the brainstem and cerebrum, involving brainstem nuclei, the thalamus, hypothalamus, and limbic structures, alongside the ascending and descending pathways connecting these regions.

In terms of neurotransmission, both sympathetic and parasympathetic preganglionic neurons are cholinergic. Parasympathetic postganglionic

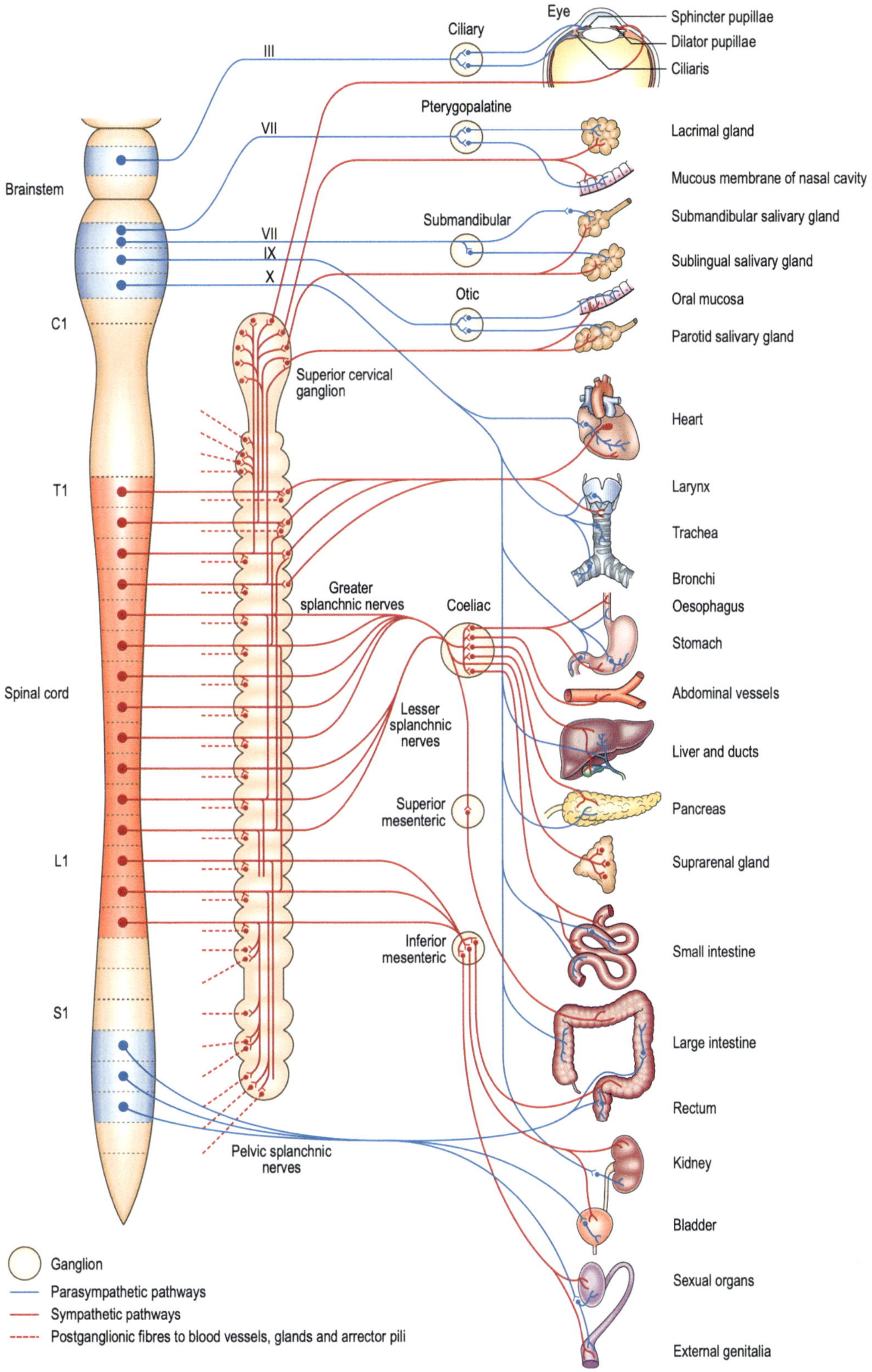

Fig. 3.2 Efferent pathways of the autonomic nervous system. (Adapted with permission from Standring (2020))

neurons also release acetylcholine, while sympathetic postganglionic neurons primarily release norepinephrine. However, the discovery of neurons that utilize alternative neurotransmitters, such as adenosine 5′-triphosphate (ATP) and various peptides, has expanded the understanding of autonomic signaling. Key co-transmitters in sympathetic neurons include ATP and neuropeptide Y, while parasympathetic neurons often release vasoactive intestinal polypeptide (VIP). In the enteric system, neurotransmitters such as ATP, VIP, and substance P play significant roles (Burnstock 2017).

Typical anatomical and functional differences were summarized in Table 3.1 before commencing the detailed description of the anatomy and function of the autonomic nervous system.

Table 3.1 Key distinguishing features of the sympathetic and parasympathetic nervous systems

Sympathetic	Parasympathetic
Originates from the thoracic and lumbar regions of the spinal cord (T1–L2)	Originates from the brainstem (cranial nerve III, VII, IX, X) and sacral region of the spinal cord (S2–S4)
Ganglia are located in the paravertebral sympathetic ganglion chain or collateral ganglia	Terminal ganglia are located near or embedded within the target tissue
Short cholinergic preganglionic fibers and long adrenergic postganglionic fibers	Long cholinergic preganglionic fibers and short cholinergic postganglionic fibers
The ratio of preganglionic fibers to postganglionic fibers is 1:20	The ratio of preganglionic fibers to postganglionic fibers is 1:3
Divergence facilitates the coordination of neuronal activity across multiple levels of spinal cord	Exhibits limited divergence
Activity frequently results in mass discharge of the entire system	Activity is typically directed to discrete organs
The primary neurotransmitter of postganglionic neurons is norepinephrine	The primary neurotransmitter of postganglionic neurons is acetylcholine
Predominates during emergency “fight-or flight” responses and periods of exercise	Predominates during quiet, resting conditions

Adapted with permission from McCorry (2007)

3.1 Anatomy and Function of the Autonomic Nervous System

The ANS exhibits specific anatomical organization. Sympathetic preganglionic neurons reside in the thoracic and upper lumbar spinal cord (T1–L2 or L3), while parasympathetic preganglionic neurons originate from the brainstem nuclei and sacral spinal cord segments S2–S4 (Standring 2020). The postganglionic neurons demonstrate differential anatomical patterns—sympathetic neurons aggregate near the spinal cord in chain or collateral ganglia, while parasympathetic neurons lie adjacent to their target organs (Benarroch 1993).

The ANS operates via three main pathways: afferent input, central integration, and efferent output. At ganglionic synapses, both divisions utilize acetylcholine, but their postganglionic neurotransmission differs. Parasympathetic postganglionic fibers release acetylcholine acting on muscarinic receptors, while sympathetic fibers primarily employ norepinephrine (Gibbins 2013; Valenza et al. 2019).

The central autonomic network (CAN) processes multiple inputs and generates coordinated responses. Located in the forebrain and brainstem regions, it integrates somatosensory and visceral information while modulating sleep cycles and emotional responses (Benarroch 1993) (Table 3.2). The hypothalamus maintains homeostasis through the regulation of autonomic, motor, and endocrine functions (Jänig 2006).

The CAN includes the insular cortex, anterior cingulate cortex, amygdala, hypothalamus, periaqueductal gray, and nucleus tractus solitarius. This network processes information hierarchically to regulate basic physiological functions and behavioral responses. Fast neurotransmission occurs via glutamate (AMPA/NMDA receptors) and GABA (GABA-A/B receptors), modulated by slower-acting acetylcholine, monoamines, neuropeptides, and nitric oxide (Benarroch 1993).

The peripheral components include sympathetic and parasympathetic divisions, along with the enteric nervous system. The enteric nervous

Table 3.2 The central autonomic network (CAN)

Area	Function
Insular cortex	Primary viscerosensory (interoceptive) cortex receiving pain and temperature information
Anterior cingulate cortex	Regulation of affective behavior, modulates bodily arousal
Amygdala	Emotional significance of sensory input, conditioned fear response
Hypothalamus	Homeostasis and adaptation: thermoregulation, osmoregulation, food intake, stress response, reproduction, sleep-wake cycle, feeding, reward response
Periaqueductal gray	Integration of autonomic, somatic and antinociceptive responses to stressors (e.g., pain)
Parabrachial nucleus	Relay center for converging visceral, nociceptive and thermoregulatory stimuli to the hypothalamus
Nucleus of the solitary tract	First relay station of taste and visceral afferents, central relay station for all medullary reflexes (cardiovascular, respiratory, and gastrointestinal)
Ventrolateral medulla	Vasomotor tone, cardiac function, respiration
Medullary raphe	Thermoregulation and respiratory chemosensitivity

Adapted with permission from Struhal et al. (2017)

system resides in the submucosal (Meissner) and myenteric (Auerbach) plexuses, regulating gastrointestinal functions (Porges 2007).

Sympathetic preganglionic neurons in the intermediolateral nucleus (T1–L3) project to the norepinephrine-releasing ganglia, except for cholinergic sweat gland innervation. Parasympathetic control originates from the brainstem nuclei and sacral segments, with vagal regulation of cardiorespiratory and gastrointestinal functions. Sacral parasympathetic neurons regulate micturition, defecation, and sexual functions (Gibbins 2013).

3.1.1 Central Autonomic Nervous System

The central autonomic network (CAN) regulates autonomic output through interconnected neural structures distributed across multiple levels of the neuraxis. This network coordinates preganglionic sympathetic and parasympathetic neuronal activity to modulate visceral functions, including cardiovascular control, gastrointestinal processes, and thermoregulation. Furthermore, the CAN adapts physiological responses to internal and external environmental changes (Benarroch 1993; Saper 2002).

The anatomical organization of the CAN follows a hierarchical arrangement with four primary levels:

- Spinal level: At the spinal level, sympathetic and sacral parasympathetic neurons mediate segmental reflexes, such as blood flow regulation and bladder control. These spinal neurons function under the influence of higher centers (Benarroch 1993).
- Bulbopontine level (lower brainstem): The bulbopontine level contains several important nuclei. The nucleus of the solitary tract (NTS), ventrolateral medulla (VLM), and parabrachial nucleus (PBN) regulate fundamental autonomic functions. The VLM maintains cardiovascular reflexes, including baroreceptor responses to postural changes and blood volume alterations (Saper 2002). The PBN coordinates respiratory and cardiovascular control with feeding behavior patterns (Craig 2003).
- Pontomesencephalic level (upper brainstem): In the pontomesencephalic level, the periaqueductal gray (PAG) combines autonomic regulation with pain processing and stress responses. The PAG modifies autonomic functions during emotional states and coordinates adaptive responses through its connections with the hypothalamus and other brain regions (Critchley and Harrison 2013; Bandler et al. 2000).
- Forebrain level: At the forebrain level, the hypothalamus coordinates autonomic and endocrine responses for homeostatic maintenance. It processes both visceral and emotional inputs to regulate thirst, hunger, temperature, and stress responses. The anterior limbic circuit, including the insula, anterior cingulate cortex, and amygdala, processes

bodily sensations in relation to emotional states and goal-directed behaviors (Critchley 2005; Craig 2003).

The CAN processes convergent sensory information from various receptors, including nociceptors, thermoreceptors, and muscle receptors. This information undergoes processing at multiple levels, from the dorsal horn to forebrain structures, allowing the integration of interoceptive signals with emotional and cognitive information (Craig 2003; Critchley and Harrison 2013).

The neurotransmitter systems within the CAN include L-glutamate as the primary excitatory neurotransmitter, acting through AMPA and NMDA receptors. GABA provides inhibitory control via GABA-A and GABA-B receptors. Additional modulation occurs through cholinergic, monoaminergic, and peptidergic pathways, along with nitric oxide, purines, endocannabinoids, and neurosteroids (Benarroch 1993).

Although the functional anatomy of the CAN was initially characterized in experimental animal studies, advancements in functional neuroimaging have demonstrated that many of these areas are activated in humans during autonomic responses. Studies using functional MRI (fMRI) and positron emission tomography (PET) have revealed activation of regions such as the insular cortex, amygdala, and anterior cingulate during tasks that involve autonomic regulation, such as emotional processing or pain perception (Critchley 2005). The main central pathways regulating autonomic function are illustrated in Fig. 3.3.

Alterations in CAN function associate with various medical conditions, including autonomic failure, cardiovascular disease, and stress-related disorders (Critchley 2005; McAllen et al. 2011).

3.1.2 Peripheral Autonomic Nervous System

The peripheral autonomic nervous system regulates visceral organ functions through three major divisions: sympathetic, parasympathetic, and enteric nervous systems, along with visceral afferent pathways. The sympathetic and parasympathetic divisions operate via a sequential two-neuron arrangement, consisting of preganglionic neurons in the central nervous system and postganglionic neurons in autonomic ganglia that connect to the target organs (Porges 2007).

Acetylcholine functions as the primary neurotransmitter for all preganglionic neurons, producing rapid excitation of ganglion cells. The sympathetic preganglionic neurons are located within the intermediolateral cell column, spanning spinal cord segments T1–L2. These neurons extend their axons to postganglionic neurons situated in the paravertebral ganglia (sympathetic chain) and prevertebral ganglia while also connecting to chromaffin cells in the adrenal medulla (Jänig 2006). Most postganglionic sympathetic neurons release norepinephrine to influence blood vessels and visceral smooth muscles. Two exceptions exist: Sympathetic neurons innervating sweat glands use acetylcholine, and adrenal medullary cells release epinephrine into circulation. The sympathetic division maintains arterial pressure, modulates regional blood flow, regulates temperature, and coordinates responses to physical activity and stress (Low and Benarroch 2008; Valenza et al. 2019).

Parasympathetic preganglionic neurons originate from specific brainstem nuclei and sacral spinal cord segments (S2–S4). The vagus nerve (cranial nerve X) and pelvic nerves conduct the primary parasympathetic signals. Unlike sympathetic ganglia, parasympathetic ganglia position themselves adjacent to or within target organ walls, producing shorter postganglionic fibers. These neurons primarily release acetylcholine, though some utilize vasoactive intestinal polypeptide (VIP) or nitric oxide (NO) as neurotransmitters (Critchley 2005; Squire et al. 2008). Parasympathetic functions include the regulation of pupillary constriction, secretion from lacrimal and salivary glands, heart rate variability, gastrointestinal activity, bladder and rectal function, and erectile tissue responses (Jänig 2006).

The enteric nervous system, considered the most autonomous part of the autonomic nervous system, consists of the myenteric and submucosal plexuses, which control intestinal motility and

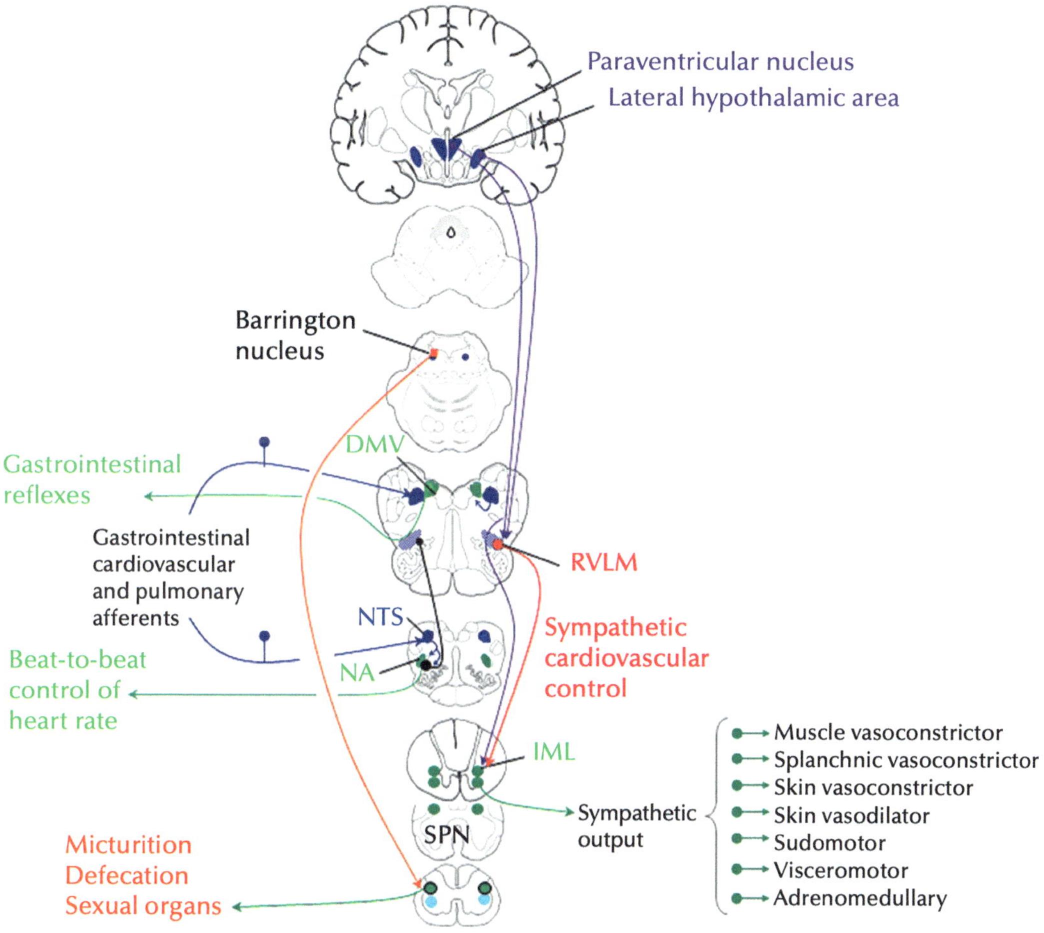

Fig. 3.3 Central pathways regulating autonomic function. Several pathways influence sympathetic preganglionic neurons in the intermediolateral cell column (IML) and preganglionic parasympathetic neurons located in the dorsal motor nucleus of the vagus (DMV), ventrolateral portion of the nucleus ambiguus (NA), and sacral parasympathetic nucleus (SPN). Sympathetic preganglionic neurons are functionally organized and controlled by parallel pathways originating from the rostral ventrolateral medulla (RVLM), rostral ventromedial medulla, paraventricular nucleus of the hypothalamus, and lateral hypothalamic area. Inputs from the Barrington nucleus, SPN, and Onuf nucleus (not shown) coordinate pelvic organ control. The DMV, NA, and RVLM act as effector regions for gastrointestinal, pulmonary, and cardiovascular reflexes mediated by afferents to the nucleus of the solitary tract (NTS). (Adapted with permission from Benarroch (2014))

secretion through local reflexes (Furness 2012). The enteric nervous system operates independently of the CNS but also interacts with sympathetic and parasympathetic fibers, allowing for coordination of digestive functions.

Visceral afferents are sensory fibers that originate from small neurons in the dorsal root ganglia, nodose ganglia, or jugular ganglia. These fibers convey visceral sensations, such as stretch or chemical stimuli, and mediate spinal, brainstem, and local reflexes. Visceral afferents play a critical role in sensorimotor feedback and autonomic regulation, as they help trigger responses that maintain homeostasis (Saper 2002; Craig 2003).

The combined actions of these systems ensure that the body can maintain internal balance and respond appropriately to external challenges. The primary pathways controlling the peripheral autonomic function are schematized in Figs. 3.4 and 3.5.

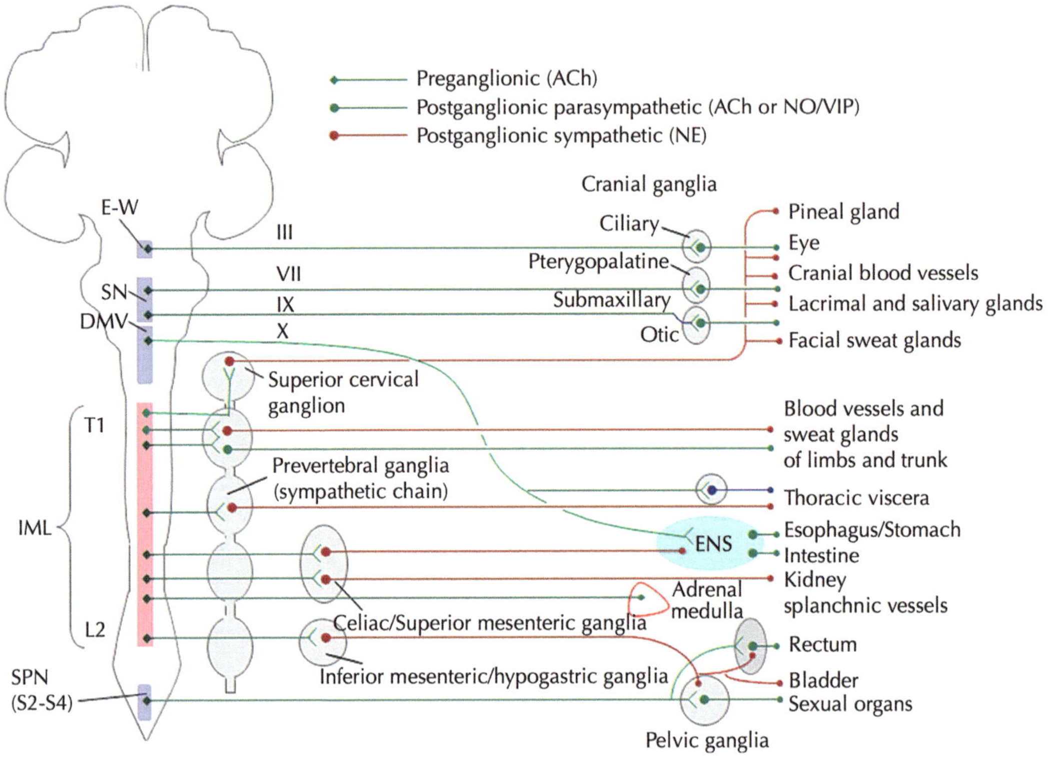

Fig. 3.4 General organization of sympathetic and parasympathetic pathways. Autonomic output includes preganglionic and postganglionic neurons. Preganglionic sympathetic neurons, primarily located in the intermediolateral cell column, activate neurons in paravertebral and prevertebral ganglia as well as adrenomedullary endocrine cells. Preganglionic parasympathetic neurons reside in the brainstem or sacral spinal cord and innervate ganglion cells near or within the walls of target organs. All preganglionic neurons are cholinergic. Abbreviations: ACh acetylcholine, ENS enteric nervous system, NE norepinephrine, NO nitric oxide, VIP vasoactive intestinal polypeptide. (Adapted with permission from Benarroch (2014))

3.1.2.1 Neurotransmitter Systems in Autonomic Neural Communication

The autonomic nervous system (ANS) employs specific neurotransmitters to mediate signal transmission between neurons and target organs. Acetylcholine functions as the primary neurotransmitter across all preganglionic neurons in both sympathetic and parasympathetic divisions, as well as in postganglionic parasympathetic neurons. The signaling occurs through two receptor types: nicotinic receptors, which mediate rapid synaptic transmission in ganglia, and muscarinic receptors, which produce prolonged effects in the target organs (Wehrwein and Joyner 2013).

Within the sympathetic division, postganglionic neurons primarily utilize norepinephrine for neurotransmission, except those innervating sweat glands, which release acetylcholine. Norepinephrine acts through adrenergic receptors, categorized into α and β subtypes. These receptors exhibit remarkable specificity, enabling varied physiological responses, including blood vessel constriction, cardiac rate modulation, and airway dilation (Critchley 2005; McCorry 2007; Porges 2007).

The autonomic nervous system also incorporates various neuropeptides that function as co-transmitters or neuromodulators. Substance P is concentrated in the intestinal tissue, where it regulates myenteric reflexes that control gastroin-

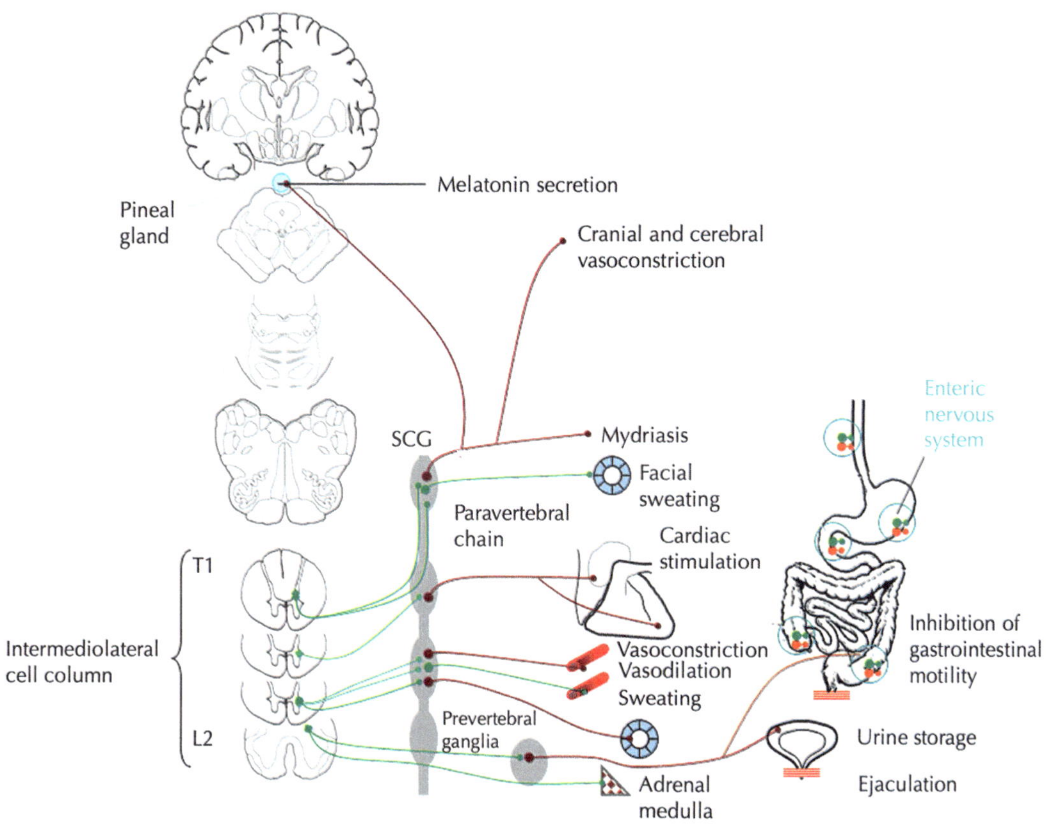

Fig. 3.5 Primary effectors and functions of the sympathetic system. Sympathetic effects are mediated by neurons in the paravertebral and prevertebral ganglia as well as by the adrenal medulla. Abbreviation: SCG superior cervical ganglion. (Adapted with permission from Benarroch (2014))

testinal movement patterns (Furness 2006). Vasoactive intestinal polypeptide (VIP), found predominantly in the sacral spinal cord and central nervous system, is released alongside acetylcholine at muscarinic junctions, contributing to blood vessel dilation and smooth muscle relaxation (Critchley 2005; Fahrenkrug 1989). Neuropeptide Y, which is released with norepinephrine, modifies blood flow patterns and vascular resistance, particularly during sympathetic vasoconstriction (Zukowska-Grojec 2015). The hypothalamus releases somatostatin, an inhibitory neuropeptide that regulates hormone and neurotransmitter release while influencing autonomic functions (Patel 1999; Benarroch 1993). Calcitonin gene–related peptide (CGRP), distributed throughout the central and peripheral nervous systems, participates in non-cholinergic and non-adrenergic vasodilation. CGRP maintains vascular tone and influences pain transmission, particularly in migraine disorders (Edvinsson 2019).

Despite the presence of these various neuropeptides in autonomic signaling pathways, clinical evaluation of autonomic function primarily focuses on acetylcholine- and norepinephrine-mediated responses, as these remain the predominant neurotransmitters in autonomic regulation (Low and Benarroch 2008) (Table 3.3).

3.1.2.2 Anatomy and Function of the Sympathetic Nervous System

The sympathetic nervous system displays a complex anatomical arrangement centered around bilateral sympathetic trunks. These ganglionated nerve cords extend from the cranial base to the coccyx along the vertebral column. The trunks

Table 3.3 Effects of autonomic nerve activity on effector organs and tissues

Tissue	Sympathetic receptor	Sympathetic stimulation	Parasympathetic stimulation
Eye			
Radial muscle of iris	α_1	Contraction (dilation of pupil; mydriasis)	–
Sphincter muscle of iris		–	Contraction (constriction of pupil; miosis)
Ciliary muscle	β_2	Relaxation for far vision	Contraction for near vision
Heart	β_1, β_2	↑ Heart rate	↓ Heart rate
		↑ Force of contraction	↓ Rate of conduction
		↑ Rate of conduction	
Arterioles			
Skin	α_1	Strong constriction	–
Abdominal viscera	α_1	Strong constriction	–
Kidney	α_1	Strong constriction	–
Skeletal muscle	α_1, β_2	Weak constriction	–
Spleen	α_1	Contraction	
Lungs			
Airways	β_2	Bronchodilation	Bronchoconstriction
Glands	α_1, β_2	↓ Secretion	↑ Secretion
Liver	α_1, β_2	Glycogenolysis	–
		Gluconeogenesis	–
Adipose tissue	β_3	Lipolysis	–
Sweat glands	Muscarinic;	Generalized sweating	–
	α_1	Localized sweating	–
Piloerector muscles	α_1	Contraction (erection of hair, goose bumps)	–
Adrenal medullae	Nicotinic	↑ Secretion of epinephrine, norepinephrine	–
Salivary glands	α_1, β_2	Small volume K^+ and water secretion	Large volume K^+ and water secretion; amylase secretion
Stomach			
Motility	α_1, β_2	Decreased	Increased
Sphincters	α_1	Contraction	Relaxation
Secretion			Stimulation
Intestine			
Motility	α_1, β_2	Decreased	Increased
Sphincters	α_1	Contraction	Relaxation
Secretion			Stimulation
Gallbladder	β_2	Relaxation	Contraction
Pancreas			
Exocrine	α	↓ Enzyme secretion	↑ Enzyme secretion
Endocrine (islets β cells)	α	↓ Insulin secretion	↑ Insulin secretion
Urinary bladder			
Detrusor muscle (bladder wall)	β_2	Relaxation	Contraction
Urethra sphincter		Contraction	Relaxation
Kidney	β_1	↑ Renin secretion	–

Adapted with permission from McCorry (2007)

connect to the spinal nerves through white and gray rami communicantes, with preganglionic axons entering through white rami and postganglionic axons exiting via gray rami to reach the target organs (Benarroch 1993; Standring 2020) (Fig. 3.6).

Anatomically, the sympathetic trunks occupy specific positions throughout their course. In the cervical region, they are positioned posterior to the carotid sheath and anterior to the cervical vertebrae's transverse processes. The cervical sympathetic ganglia are reduced to three due to fusion, with the superior cervical ganglion giving rise to the internal carotid nerve, which follows the internal carotid artery into the cranial cavity to supply the head (Porges 2007). The thoracic portion lies anterior to the rib heads, while the abdominal segment runs anterolateral to the lumbar vertebrae. The pelvic region shows the trunks positioned anterior to the sacrum and medial to the anterior sacral foramina, ultimately joining at the ganglion impar anterior to the coccyx (Standring 2020).

Preganglionic sympathetic neuron cell bodies are found in the lateral horn of the thoracic and upper lumbar spinal segments (T1–L2 or L3). These neurons have myelinated axons that exit the spinal cord via the ventral roots, joining the spinal nerves before branching off through the white rami communicantes into the sympathetic trunk (Critchley 2005; Robertson et al. 2012). Upon entering the sympathetic trunk, preganglionic fibers may:

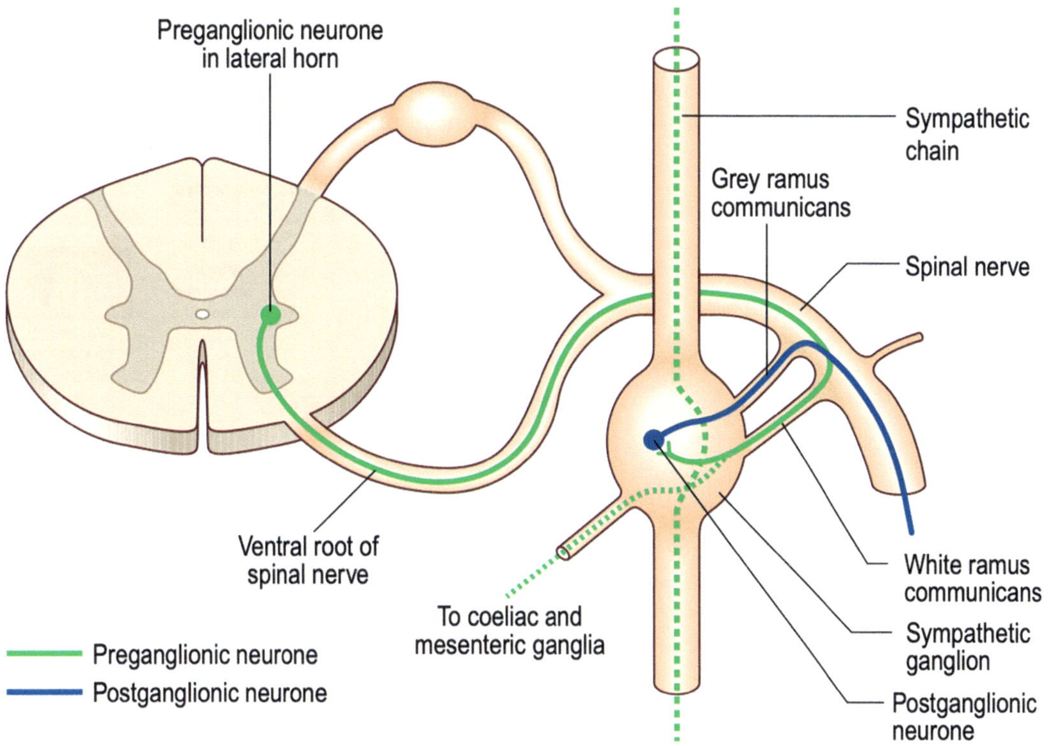

Fig. 3.6 Outflow of preganglionic sympathetic neurons from the lateral horn of the spinal cord gray matter. Preganglionic axons may synapse with a postganglionic neuron in a sympathetic ganglion at the same segmental level, traverse the sympathetic chain to synapse with a neuron in a more rostral or caudal ganglion, or extend to a ganglion within an autonomic plexus in the abdominal cavity, such as the celiac ganglion. (Adapted with permission from Standring (2020))

- Synapse in the nearest ganglion at their entry level
- Ascend or descend to synapse in other ganglia
- Bypass the sympathetic chain to synapse in autonomic plexuses, such as the celiac or mesenteric ganglia
- Directly innervate the adrenal medulla, stimulating the release of epinephrine and norepinephrine (Critchley 2005)

Postganglionic sympathetic neurons are located in the sympathetic trunk or peripheral autonomic ganglia. Their unmyelinated axons extend to the target organs via various routes. Some fibers return to the spinal nerves via gray rami communicantes to innervate blood vessels, sweat glands, and arrector pili muscles (Fig. 3.7). Others travel independently or along blood vessels to innervate organs such as the heart, lungs, and digestive tract (Mathias 2003; Porges 2007) (Fig. 3.8).

The sympathetic system employs norepinephrine as its primary postganglionic neurotransmitter, distinguishing it from the parasympathetic system's cholinergic transmission. Postganglionic fibers perform multiple functions, including vasoconstriction, sweat gland activation, and arrector pili muscle control. They may also produce vasodilation in the skeletal muscle vasculature (Critchley 2005; Mathias 2003).

The system's organization allows individual preganglionic neurons to influence multiple postganglionic neurons, enabling discrete control of different physiological processes. This arrangement supports independent regulation of functions like sweating and blood flow (Jänig 2006). Notably, the adrenal medulla is directly innervated by preganglionic neurons, bypassing the sympathetic chain to stimulate the release of stress hormones (Critchley 2005; Porges 2007).

The sympathetic response pathway begins in the hypothalamus, which connects to brainstem autonomic centers, including the periaqueductal gray matter, parabrachial region, and ventrolateral medulla. These areas project to the preganglionic sympathetic neurons in the spinal cord's intermediolateral cell column (Benarroch 2009).

The sympathetic system maintains a segmental organization pattern:

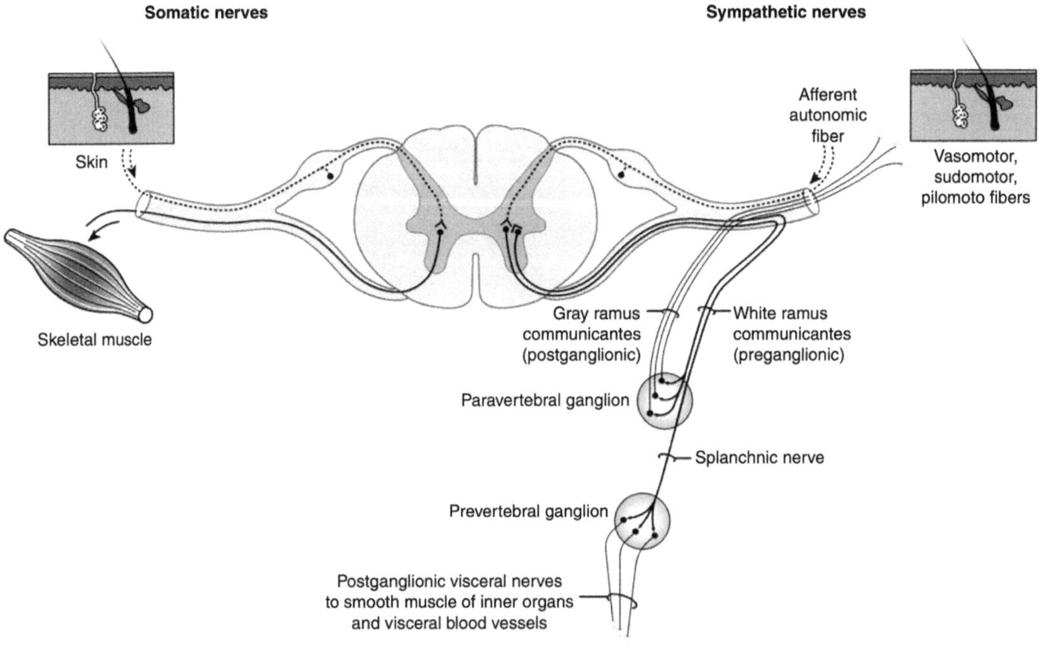

Fig. 3.7 Pathway of sympathetic outflow fibers originating from the spinal cord and roots. (Adapted with permission from Katirji et al. (2014))

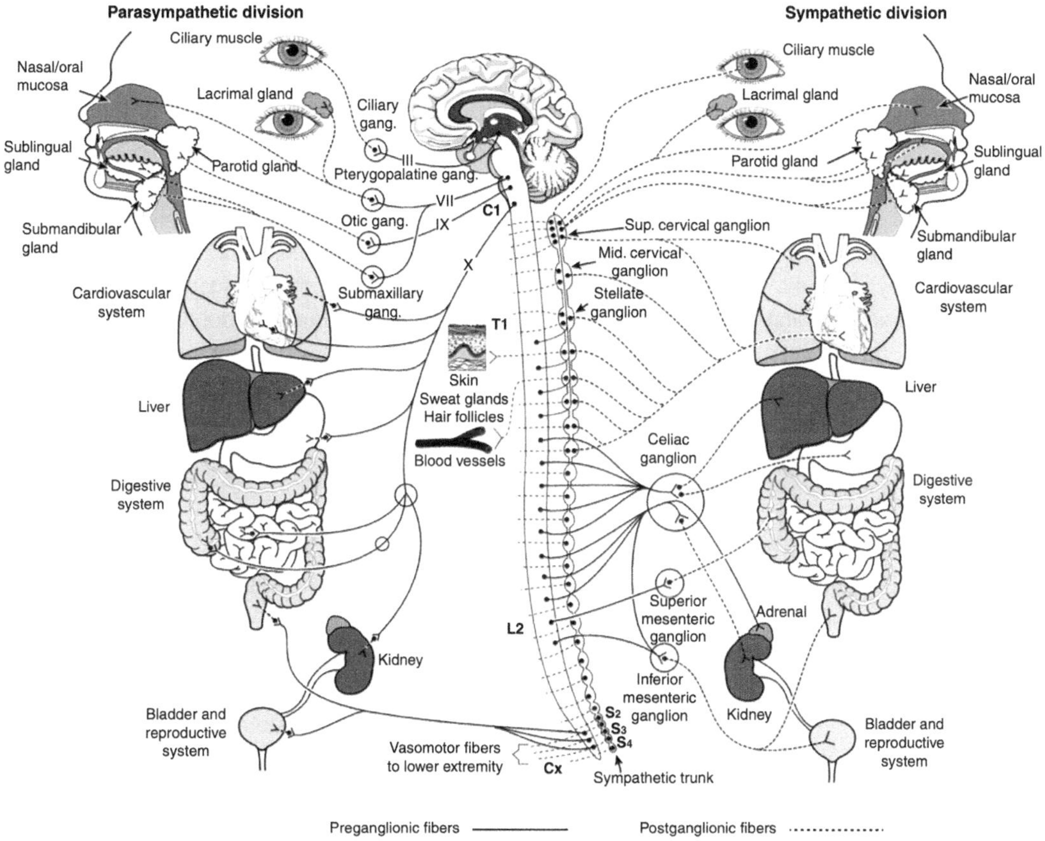

Fig. 3.8 Efferent pathways of the parasympathetic and sympathetic autonomic systems. (Adapted with permission from Katirji et al. (2014))

- T1 fibers innervate the head.
- T2 fibers supply the neck.
- T3–T6 fibers target the thorax.
- T7–T11 fibers innervate the abdomen.
- T12, L1, and L2 fibers supply the pelvis and lower limbs (Standring 2020).

This systematic arrangement enables precise control of physiological responses across body systems, from cardiovascular function to digestive and reproductive processes (Robertson et al. 2012).

Sympathetic Ganglia

The sympathetic nervous system contains three anatomically and functionally distinct ganglionic groups that form an integrated network for autonomic control. All these groups—paravertebral, prevertebral, and terminal ganglia—maintain specific roles in regulating the visceral function (Fig. 3.4). The functional organization of sympathetic ganglia reflects their anatomical distribution. Paravertebral ganglia primarily regulate the blood vessels, sweat glands, and piloerector muscles through segmental innervation. Prevertebral ganglia coordinate visceral functions, including digestion, absorption, and reproduction. Terminal ganglia fine-tune organ-specific responses based on local requirements (Benarroch 2014).

Each ganglion type processes neural information differently. Paravertebral ganglia often demonstrate divergent connections, where single preganglionic neurons influence multiple postganglionic cells. Prevertebral ganglia exhibit more complex integration, incorporating input from multiple sources to coordinate organ system responses. Terminal ganglia provide precise, localized control through direct organ innervation.

Paravertebral Ganglia

Paravertebral ganglia form bilateral chains adjacent to the vertebral column. These ganglia connect through longitudinal nerve strands to establish the sympathetic trunks, which extend from the superior cervical ganglion to the ganglion impar at the coccyx. The primary function of the paravertebral ganglia is to provide sympathetic innervation to all regions of the body except the abdominal, pelvic, and perineal organs. This system plays a critical role in regulating cardiovascular, sudomotor, and vasomotor functions across a broad range of tissues (Standring 2020). The cervical region contains three pairs of ganglia: superior, middle, and inferior cervical ganglia. The superior cervical ganglion, which represents the largest of the three, supplies sympathetic innervation to the head and neck through nerve fibers that accompany blood vessels (Jänig 2006).

The thoracic region typically includes 11 pairs of ganglia, while the lumbar and sacral regions contain four and four to five pairs, respectively. Each paravertebral ganglion connects to its corresponding spinal nerve through both white and gray rami communicantes. This arrangement allows preganglionic fibers to enter the sympathetic chain and postganglionic fibers to reach their peripheral targets (Robertson et al. 2012).

Postganglionic fibers originating from the paravertebral ganglia innervate the face, trunk, and limbs by following the paths of spinal nerves, blood vessels, or both. For example, axons from the superior cervical ganglion, which receives preganglionic input from the T1–T3 spinal segments (ciliospinal center of Budge), target multiple structures, including the eye (providing pupillary dilation), facial blood vessels, sweat glands, salivary glands, and the pineal, pituitary, and thyroid glands. These fibers reach their destinations by traveling along branches of the carotid artery (Standring 2020) (Table 3.4). Additionally, the superior cervical ganglion contributes to cerebral blood flow regulation by innervating cerebral blood vessels (Benarroch 1993).

Other paravertebral ganglia in the cervical, thoracic, and lumbar regions also supply sympathetic innervation to the trunk and extremities. These ganglia send postganglionic fibers that join peripheral spinal nerves, following a distribution pattern similar to the somatic nerves. The stellate ganglion, formed by the fusion of the inferior cervical and first thoracic ganglia, provides vasomotor and sudomotor innervation to the upper limb. The postganglionic fibers from this ganglion travel through the median and ulnar nerves, contributing to the regulation of blood flow and sweat production in the upper extremity (Critchley 2005).

In the lower limbs, postganglionic fibers from the lumbar paravertebral ganglia reach their targets primarily through the peroneal and tibial nerves, which innervate the vasculature, sweat glands, and arrector pili muscles in the lower extremity. These ganglia ensure appropriate regulation of skin blood flow and temperature control by modulating vasoconstriction and sudomotor activity (Benarroch 1993).

In addition to their role in the extremities, the paravertebral ganglia also provide innervation to the visceral organs. For instance, axons from the cervical and thoracic paravertebral ganglia innervate vital structures such as the heart, esophagus, and respiratory tract (Standring 2020). These postganglionic fibers help regulate the heart rate, bronchial diameter, and esophageal function, contributing to the sympathetic control of cardiovascular and respiratory systems (Cardinali 2018).

The paravertebral ganglia perform central functions in sympathetic regulation across multiple body systems. These ganglia establish an organized network for distributing sympathetic signals to diverse tissues throughout the body. Their anatomical placement along the vertebral column optimizes their ability to coordinate widespread physiological responses. The distribution pattern of sympathetic innervation from paravertebral ganglia follows a precise organizational scheme. Each ganglion connects to specific target tissues through postganglionic fibers,

Table 3.4 Peripheral innervation of sympathetic effectors

Segments	Output ganglia	Effectors	Effect (neurotransmitter)
T1–T3	Superior cervical	Eye	Pupil dilation (NE)
		Salivary glands	Contraction of tarsal muscle (NE)
		Face sweat glands	Increase in salivary protein (NE)
		Blood vessels	Facial sweating (ACh)
		Pineal gland	Vasoconstriction (NE)
			Melatonin secretion (NE)
T1–T8	Stellate and upper thoracic (upper limb, and thoracic viscera)	Sweat glands	Sweating (ACh)
		Skin blood vessels	Skin vasoconstriction (NE) or vasodilation (NO)
		Piloerector muscles	Piloerection
		Muscle blood vessels	Muscle vasoconstriction (NE) or vasodilation
		Periosteum, bone	Bone marrow activation (NE)
		Heart	Cardiac stimulation (NE)
		Respiratory tract	Bronchodilation (NE)
T5–T12	Celiac	Abdominal viscera	Inhibition of GI motility
	Mesenteric	Blood vessels	Vasoconstriction
	Adrenal medulla	Endocrine cells	Renin secretion
			Glucagon secretion
			Epinephrine secretion
T10–L2	Lumbar paravertebral	Sweat glands	Sweating (ACh)
		Skin blood vessels	Skin vasoconstriction (NE) or vasodilation (NO?)
		Piloerector muscles	Piloerection
		Muscle blood vessels	Muscle vasoconstriction (NE) or vasodilation
T2–L2	Inferior mesenteric	Pelvic viscera and blood vessels	Urine storage
	Hypogastric		Rectal inhibition
			Ejaculation

Abbreviations: *ACh* acetylcholine, *NE* norepinephrine, *NO* nitric oxide
Adapted with permission from Benarroch (2014)

allowing localized control of blood vessels, sweat glands, and other autonomic effectors. This arrangement enables the sympathetic nervous system to generate both broad and focused responses as physiological conditions require. The integration of signals through the paravertebral ganglia allows for coordinated responses across multiple organ systems. When environmental conditions change or internal homeostasis requires adjustment, these ganglia modify their activity patterns to generate appropriate physiological adaptations. This capability for integrated response helps maintain optimal body function under varying conditions (Porges 2007).

Prevertebral Ganglia

Prevertebral ganglia reside anterior to the vertebral column, positioned near major arterial branches from the abdominal aorta. The prevertebral ganglia play an important role in the autonomic innervation of abdominal and pelvic viscera, as well as their associated blood vessels (see Table 3.4). These ganglia include the celiac, superior mesenteric, and inferior mesenteric ganglia. The celiac ganglion receives input from the greater splanchnic nerves and coordinates upper abdominal organ function. The superior and inferior mesenteric ganglia obtain signals from lesser splanchnic nerves and regulate intestinal activity.

Additional small ganglia within this group support reproductive organ function (Mathias 2003). This system of ganglia is integral to understanding the potential complications and management strategies in spinal cord injury rehabilitation, particularly for injuries affecting the thoracic region.

Key prevertebral ganglia and their innervation patterns include the following:

- Celiac and superior mesenteric ganglia:
  - Receive inputs from spinal segments T5–T11 via the splanchnic nerves
  - Innervate abdominal viscera, mesenteric vessels, and renal vessels
  - Crucial for regulating gastrointestinal function and renal blood flow (Furness 2012)
- Inferior mesenteric ganglion:
  - Receives sympathetic output originating from spinal segments T10–L2
  - Gives rise to the hypogastric nerve
  - Along with neurons in the lumbar paravertebral chain, provides sympathetic innervation to pelvic organs (Standring 2020)
- Adrenal medulla:
  - Receives preganglionic inputs from the mid-thoracic region
  - Unique in that it receives direct preganglionic innervation, functioning as a modified sympathetic ganglion (Purves et al. 2018)

For specialists in spinal cord injury rehabilitation, understanding the organization of these prevertebral ganglia is necessary for several reasons:

- Predicting autonomic dysfunction: The level of spinal cord injury can help predict which prevertebral ganglia might be affected and, consequently, which organ systems may experience autonomic disturbances.
- Managing cardiovascular complications: Injuries above T6 can disrupt sympathetic control of the splanchnic vascular bed, potentially leading to orthostatic hypotension, autonomic dysreflexia, and other cardiovascular issues.
- Addressing gastrointestinal and genitourinary function: Injuries affecting T5–L2 can impact sympathetic regulation of abdominal and pelvic organs, necessitating specific management strategies for bowel and bladder dysfunction (Jänig 2006).
- Hormone regulation: Understanding the innervation of the adrenal medulla is important in managing stress responses and catecholamine release in patients with spinal cord injury (Benarroch 1997).

Terminal Ganglia

Terminal ganglia are located near or within the target organs, particularly in the heart, respiratory system, and pelvis. These ganglia process and modify neural signals before transmission to effector tissues. Their positioning allows rapid adjustments to local conditions while maintaining integration with broader autonomic control systems (Low and Benarroch 2008).

The anatomical location of the terminal ganglia provides significant physiological advantages. Their position near target tissues reduces signal transmission time, allowing rapid modifications of organ function in response to changing physiological demands. This arrangement supports precise temporal control of autonomic responses at the local level.

Terminal ganglia modify incoming neural signals through complex processing mechanisms before transmitting information to effector tissues. This processing includes signal integration, amplification, and modification based on local tissue conditions. The ganglia contain specialized neurons that respond to both central neural inputs and local molecular signals, enabling refined control of organ function. The physiological significance of terminal ganglia extends beyond local control. While these structures provide immediate regulation of organ function, they simultaneously maintain connections with broader autonomic control networks. This dual-level organization allows terminal ganglia to coordinate local responses with systemic autonomic regulation, ensuring appropriate organ function within the context of overall body homeostasis (McCorry 2007).

Preganglionic Sympathetic Units

The sympathetic preganglionic neurons are primarily located in the thoracolumbar region of the spinal cord, specifically within the T1–L2 segments. The majority of these neurons reside in the intermediolateral column (IML) of lamina VII, while others are found in regions such as the dorsolateral funiculus, the intercalate nucleus (positioned between the IML and the central canal), and the central autonomic nucleus in lamina X, dorsal to the central canal (Cabot 1996). Neurons within the IML are arranged in segmental clusters, with each cluster containing neurons that control specific targets. Each individual neuron within these clusters innervates a single, distinct target (Cabot 1996). These preganglionic neurons project to three main types of ganglia: paravertebral, prevertebral, and terminal ganglia, with the latter located near the target organ.

The sympathetic outflow is functionally diverse. Sympathetic preganglionic neurons are organized into distinct functional units, which include skin vasomotor, muscle vasomotor, secretomotor, and visceromotor units, among others. Each unit selectively innervates subpopulations of neurons in the sympathetic ganglia, thereby contributing to the specificity of sympathetic responses. These preganglionic units receive inputs from primary afferents originating in the dorsal root ganglia, as well as descending inputs from higher centers in the hypothalamus and brainstem (Benarroch 2014) (Fig. 3.9).

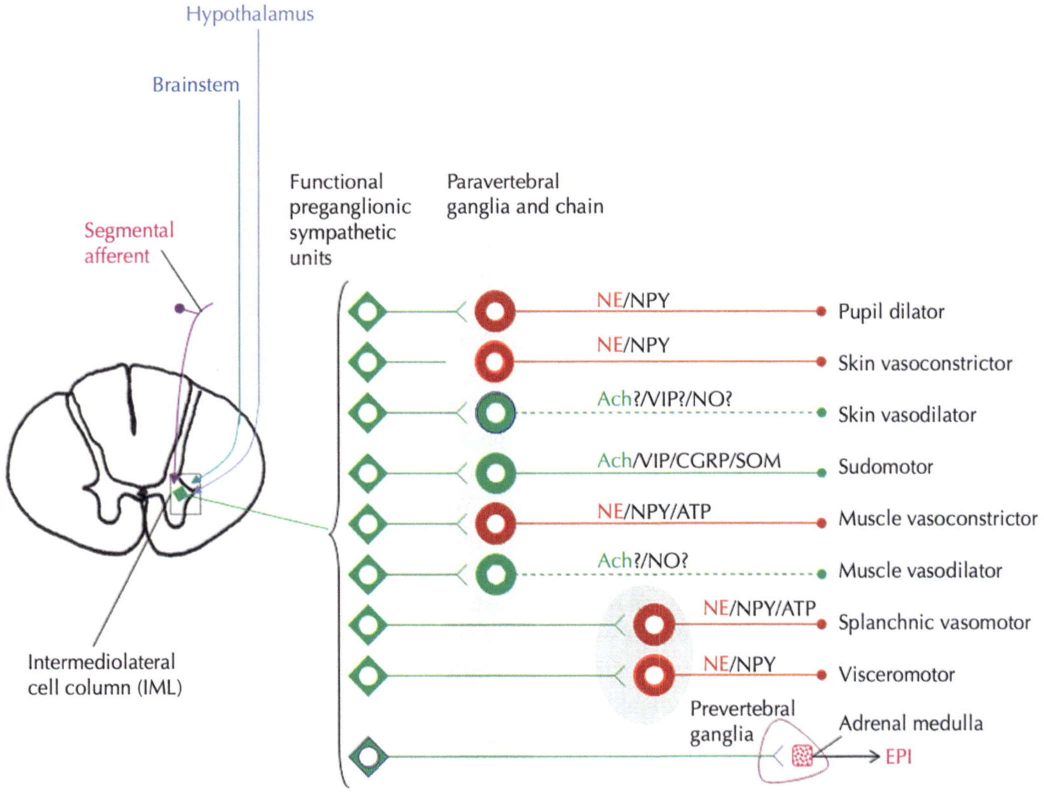

Fig. 3.9 Functional units of the sympathetic system. Preganglionic sympathetic neurons in the intermediolateral cell columns are organized into functional subunits that selectively regulate the activity of chemically coded, target-specific sympathetic ganglion neurons. Abbreviations: ACh acetylcholine, ATP adenosine triphosphate, ENS enteric nervous system, NE norepinephrine, NO nitric oxide, NPY neuropeptide Y, VIP vasoactive intestinal polypeptide. (Adapted with innervation from Benarroch (2014))

The main sources of premotor sympathetic inputs are located in various regions of the brainstem and hypothalamus, including the rostral ventrolateral medulla, the medullary raphe, the A5 noradrenergic group in the pons, the paraventricular nucleus of the hypothalamus, and the lateral hypothalamic area (Benarroch 2014). These premotor centers play a crucial role in coordinating the activity of preganglionic neurons. In response to specific stimuli, the preganglionic sympathetic units can be selectively activated by segmental afferents, which trigger somatosympathetic and viscerosympathetic reflexes. However, under normal physiological conditions, these units are recruited in a coordinated manner by premotor neurons in the brainstem and hypothalamus, enabling integrated autonomic responses across multiple organ systems (Saper 2002).

Preganglionic Input to the Sympathetic Ganglia

The sympathetic preganglionic output exhibits a segmental organization. Preganglionic axons, which are small myelinated fibers, exit the spinal cord via the ventral roots and travel through the white ramus communicans of the corresponding spinal nerves to reach the paravertebral sympathetic chain. At this point, some preganglionic axons synapse on paravertebral ganglion neurons at the same spinal level. However, a significant number of preganglionic fibers ascend or descend within the sympathetic chain to synapse on ganglia located at different levels (Standring 2020). Additionally, certain preganglionic axons form the splanchnic nerves, which innervate prevertebral ganglia and the adrenal medulla. Thus, while the sympathetic preganglionic neurons are organized in a segmental manner, there is an overlapping rostral-to-caudal arrangement that extends across multiple spinal segments, enhancing the distribution and coordination of autonomic output (Benarroch 2014).

Specific spinal segments provide inputs to different ganglia. For instance, the T1–T3 segments innervate the superior cervical ganglion, which controls cranial effectors, including the eye and facial blood vessels (Gibbins et al. 2003). The T1–T6 segments project to the stellate and upper thoracic ganglia, which are responsible for innervating the upper limb and thoracic structures. Meanwhile, the T5–T12 segments provide inputs to the adrenal medulla, celiac ganglia, and superior mesenteric ganglia, facilitating the regulation of abdominal viscera and blood vessels. The T10–L2 segments project to the inferior mesenteric ganglia and hypogastric ganglia, which innervate the pelvic viscera, as well as the lumbar prevertebral ganglia, which supply the lower limb (Gibbins et al. 2003).

The principal neurotransmitter released by the preganglionic sympathetic neurons is acetylcholine. Acetylcholine rapidly excites postganglionic neurons via nicotinic receptors of the ganglion type ($\alpha 3\beta 4$), which are important for fast synaptic transmission in sympathetic ganglia (Vernino et al. 2009). This cholinergic signaling mechanism is key to the initiation and regulation of sympathetic autonomic responses throughout the body.

Sympathetic Neurotransmitters

Norepinephrine is the principal neurotransmitter released by sympathetic neurons, necessary for sympathetic signaling. However, norepinephrine often does not work alone. Many sympathetic neurons also release adenosine triphosphate (ATP) and neuropeptide Y (NPY) alongside norepinephrine (Lundberg 1996). This combination of chemicals allows for the precise control of physiological systems, including blood pressure and metabolism.

While most sympathetic neurons use norepinephrine, there are some interesting exceptions. The neurons that make us sweat, for instance, use acetylcholine instead of norepinephrine (Weihe et al. 2005). Another special case is the adrenal medulla, which acts like a modified sympathetic ganglion. Instead of sending signals directly to the organs, it releases epinephrine (and some norepinephrine) into the bloodstream, allowing for body-wide sympathetic effects (Kvetnansky et al. 2009).

Sympathetic output can be categorized into three primary types:

- Sympathoneural adrenergic: This is the classic fight-or-flight response, causing blood vessels to constrict and heart rate to increase.
- Sympathoneural cholinergic: This regulates perspiration, essential for thermoregulation.
- Sympathoadrenal: This involves the adrenal medulla's release of epinephrine, preparing the whole body for action.

Understanding norepinephrine's role in blood vessel control helps in managing blood pressure issues common in spinal cord injury, like sudden high blood pressure (autonomic dysreflexia) or low blood pressure upon standing (orthostatic hypotension) (Krassioukov et al. 2012). The distinctive sweating control mechanism explains why individuals with spinal cord injuries may have difficulties in temperature management, an essential factor in the planning of therapy sessions and daily activities (Tate et al. 2004).

3.1.2.3 Anatomy and Function of the Parasympathetic Nervous System

The parasympathetic nervous system originates from specific regions of the CNS (Standring 2020). Preganglionic parasympathetic neurons are situated in select cranial nerve nuclei within the brainstem and the intermediate gray matter of the sacral spinal cord segments S2–S4 (Fig. 3.10). These neurons produce myelinated efferent fibers that exit the CNS via cranial nerves III (oculomotor), VII (facial), IX (glossopharyngeal), and X (vagus), as well as through the S2–S4 spinal nerves (Kiernan and Rajakumar 2014). These preganglionic neurons utilize acetylcholine as their neurotransmitter.

In contrast, the cell bodies of postganglionic parasympathetic neurons are primarily located outside the CNS. They form either discrete ganglia near their target organs or are dispersed within the walls of visceral structures (Furness 2006). The cranial component of the parasympathetic system includes four small peripheral ganglia: the ciliary, pterygopalatine, submandibular, and otic ganglia. It is important to note that these are efferent parasympathetic ganglia, distinct from the sensory ganglia associated with the trigeminal, facial, glossopharyngeal, and vagal nerves, which exclusively process afferent signals and house sensory neuron cell bodies (Standring 2020).

Although the cranial parasympathetic ganglia contain afferent fibers, postganglionic sympathetic fibers, and branchial efferent fibers in the case of the otic ganglion, these do not form synapses within these structures (Kiernan and Rajakumar 2014). Postganglionic parasympathetic fibers are typically unmyelinated and shorter than their sympathetic postganglionic fibers. This is due to the proximity of parasympathetic ganglia to their target organs, either within or near the innervated viscera. Similar to their preganglionic neurons, postganglionic parasympathetic neurons utilize acetylcholine as their principal neurotransmitter (Standring 2020).

The preganglionic segment of the parasympathetic nervous system consists of first-order neurons whose cell bodies are located within the CNS. These neurons originate from specific areas in the brainstem and spinal cord. The brainstem has numerous significant nuclei: the Edinger–Westphal nucleus, the salivary nucleus, the dorsal motor nucleus, and the nucleus ambiguus. Moving down to the spinal cord, preganglionic neurons are found in the ventral horns and intermediolateral columns of the sacral segments S2–S4 (Kiernan and Rajakumar 2014). From these origins, the preganglionic axons take different paths depending on their location. Cranial preganglionic axons travel along cranial nerves III, VII, IX, and X. These axons then form connections with short second-order neurons in parasympathetic ganglia. These ganglia are strategically positioned either in close proximity to or directly within the target organs. This arrangement allows for efficient communication between the nervous system and the organs it regulates (Furness 2006).

The sacral region of the parasympathetic nervous system has its own unique arrangement. Here, preganglionic axons extend through the pelvic nerve, ultimately reaching the inferior hypogastric plexus. Upon arrival, most of these axons form synapses with parasympathetic ganglia located within the target organs themselves.

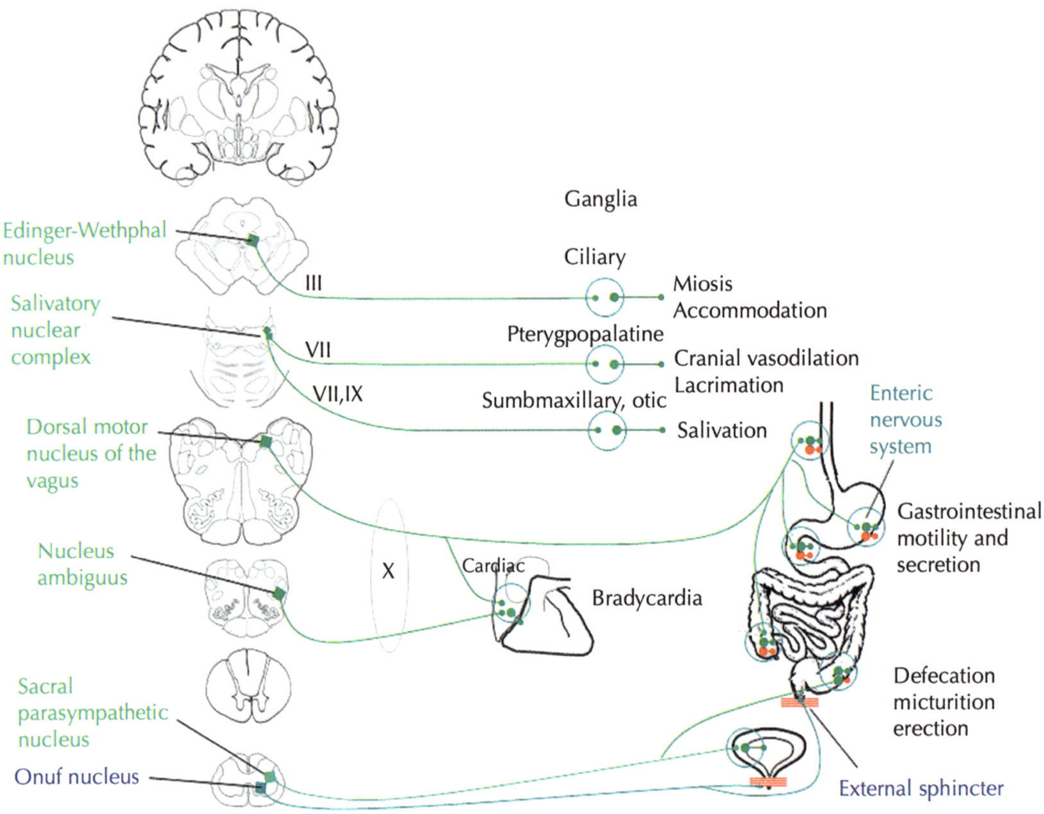

Fig. 3.10 Main effectors and functions of the parasympathetic system. Parasympathetic effects are mediated by neurons located near or within the walls of target effector structures. (Adapted with permission from Benarroch (2014))

However, some axons take a different route, connecting with neurons in the myenteric and submucosal plexuses of the enteric nervous system. Both pathways ultimately result in innervation by brief second-order neurons (Standring 2020).

Parasympathetic Ganglia and Their Associated Effector Organs

The parasympathetic nervous system's preganglionic fibers, which originate in the brainstem, form synapses with specific ganglia to innervate various effector organs (Standring 2020) (Fig. 3.8). This complex network of connections allows for precise control of various physiological functions:

- The ciliary ganglion receives input from the Edinger–Westphal nucleus in the midbrain via the oculomotor nerve (III). Its postganglionic fibers control the pupillary constrictor and ciliary muscle, regulating pupil size and lens accommodation (Kiernan and Rajakumar 2014).
- Parasympathetic fibers from the lacrimal nucleus in the pons travel with the facial nerve (VII) to the pterygopalatine ganglion. From here, postganglionic fibers extend to the lacrimal glands and various mucous glands in the palate, pharynx, and nasal cavities (Furness 2006).
- The submandibular ganglion is innervated by fibers from the salivary nucleus in the pons, also via the facial nerve. Its postganglionic axons supply the submandibular and sublingual glands, as well as the mucous glands in the oral cavity (Standring 2020).
- Fibers from the inferior salivary nucleus in the medulla travel with the glossopharyngeal nerve (IX) to the otic ganglion. Postganglionic fibers from this ganglion innervate the parotid

glands and the oral cavity mucous glands (Kiernan and Rajakumar 2014).
- The visceral ganglia receive input from the dorsal motor nucleus of X and nucleus ambiguus in the medulla via the vagus nerve (X). These ganglia provide extensive innervation to numerous organs, including the bronchi, lungs, heart, gastrointestinal tract (from esophagus to descending colon), liver, gallbladder, pancreas, kidneys, and intestines (Furness 2006).
- Parasympathetic fibers originating in the sacral spinal cord (S2–S4) travel through the pelvic nerve to the pelvic plexus, also known as the inferior hypogastric plexus. This plexus contains both pre- and postganglionic sympathetic and parasympathetic fibers. Postganglionic fibers from this plexus innervate the bladder, descending colon, rectum, and genitalia (Standring 2020).

The enteric nervous system, which some consider a third division of the autonomic nervous system, innervates the gastrointestinal tract, pancreas, and gallbladder. It consists of interconnected neurons forming two major plexuses: the myenteric (Auerbach) plexus and the submucous (Meissner) plexus. This system responds to changes in the gut's chemical environment and wall tension, functioning somewhat independently to maintain homeostasis. However, it remains largely under the influence of central nervous system reflexes mediated through both the parasympathetic and sympathetic nervous systems (Furness 2006).

Parasympathetic Output

The output of the parasympathetic nervous system is systematically organized, illustrating its function in preserving homeostasis and regulating specific organ functions (Standring 2020). The preganglionic neurons of this system are located in two principal areas of the central nervous system:

- The general visceral efferent column of the brainstem
- The sacral spinal cord segments S2–S4

Unlike their sympathetic ganglia, parasympathetic ganglia are positioned in close proximity to or within the walls of their target organs. This anatomical arrangement allows for more localized and precise control of organ function (Kiernan and Rajakumar 2014). Both preganglionic parasympathetic and sympathetic neurons utilize acetylcholine as their primary neurotransmitter. These neurons induce rapid excitation of parasympathetic ganglion cells through interaction with ganglion type $\alpha3/\beta4$ nicotinic receptors (Vernino et al. 2000).

A key distinction between the parasympathetic and sympathetic systems lies in their activation patterns. While the sympathetic system often activates in a broad, patterned fashion, the parasympathetic system mediates specific reflexes that control particular effectors. This targeted approach allows for fine-tuned regulation of various physiological processes (Furness 2006).

The parasympathetic output can be categorized into three main components:

- Output to cranial effectors: This includes control of structures in the head and neck region, such as the eyes, lacrimal glands, and salivary glands.
- Vagal output: This component regulates the heart, lungs, and gastrointestinal tract. The vagus nerve plays a key role in this aspect of parasympathetic function, influencing the heart rate, respiratory rate, and digestive processes.
- Sacral preganglionic output: This pathway controls the bladder, rectum, and sexual organs. It is responsible for regulating urination, defecation, and certain aspects of sexual function.

This organization allows the parasympathetic nervous system to exert precise control over specific organ systems, contributing to the maintenance of physiological balance and the body's "rest and digest" functions (Standring 2020).

Output to Cranial Effectors

The cranial parasympathetic system is a complex network of neurons and nerve fibers that play a key role in regulating various functions in the head and neck region. This system originates from neurons located in the general visceral efferent column of the midbrain, pons, and medulla. These neurons extend their preganglionic axons through four important cranial nerves: III (oculomotor), VII (facial), IX (glossopharyngeal), and X (vagus) (Standring 2020).

One of the key components of this system is the Edinger–Westphal nucleus, located in the midbrain. This nucleus gives rise to parasympathetic fibers that travel along the oculomotor nerve to reach the ciliary ganglion. This pathway is particularly important for visual function, as it controls pupil constriction and ciliary muscle contraction, both of which are essential for our ability to focus on objects at different distances (Wilhelm 2008). The facial and glossopharyngeal nerves share some interesting characteristics in their parasympathetic pathways. Both have their preganglionic neurons situated in the salivatory nuclear complex in the caudal pons. These pathways are interconnected through a network of intracranial structures, including the pterygopalatine, submandibular, and otic ganglia. Their primary functions are secretomotor and vasodilator, meaning they control glandular secretions and blood vessel dilation in their target areas (Kiernan and Rajakumar 2014). The pterygopalatine ganglion, which receives input from the facial nerve, has a wide-ranging influence. It innervates the lacrimal glands, controlling tear production, and also regulates blood flow in the eyes, nasal cavity, palate, and even the cerebral blood vessels. This ganglion plays a crucial role in maintaining the proper functioning and environment of these areas. Another important structure is the submandibular ganglion. It receives signals from the facial nerve via a roundabout path through the chorda tympani and lingual nerve. This ganglion is responsible for supplying the submandibular and major sublingual glands, which are key players in saliva production. The otic ganglion, on the other hand, receives its preganglionic axons from the glossopharyngeal nerve. These signals travel through the tympanic plexus and lesser petrosal nerve before reaching the ganglion. From here, postganglionic axons, primarily traveling via the auriculotemporal nerve, extend to innervate the parotid gland, another major salivary gland (Furness 2006).

It is also worth noting that there are numerous smaller ganglia, known as microganglia, associated with the ciliary muscle and salivary glands. These microganglia allow for even more refined local control of parasympathetic functions.

This complex organization of the cranial parasympathetic system allows for precise regulation of various functions in the head and neck. From controlling pupillary responses and tear production to managing salivation and regional blood flow, this system plays a crucial role in maintaining ocular, oral, and nasal homeostasis. Its complex network of pathways and ganglia ensures that these vital functions are fine-tuned to meet the body's needs in different situations (Standring 2020).

Vagal Output

The vagus nerve, also known as cranial nerve X, stands as a testament to the intricate design of the human nervous system. As the primary conduit for parasympathetic output to a vast array of visceral organs, it plays a crucial role in maintaining our body's delicate balance and regulating vital functions (Standring 2020).

The dorsal motor nucleus of the vagus nerve is essential for this system. This nucleus acts as the origin for the majority of the vagus nerve's preganglionic fibers, impacting practically all thoracic and abdominal organs. It is interesting to observe that only the distal colon and rectum are unaffected by its influence (Hopkins et al. 1996; Cheng et al. 1999). The structure of the dorsal motor nucleus of the vagus nerve exemplifies efficiency, with its viscerotropic and columnar arrangement facilitating precise regulation of various target organs.

When it comes to the heart, the vagus nerve's influence is particularly important. A cluster of neurons in the ventrolateral region of the nucleus ambiguus primarily governs cardiac innervation, whereas the dorsal motor nucleus of the vagus nerve additionally participates in this regulation. This dual input emphasizes the complexity of the vagal effect on cardiac function (McAllen and Spyer 1976).

The postganglionic neurons throughout the vagal pathways are strategically located in ganglia that create plexuses adjacent to their target tissues. These include the esophageal, pulmonary, and cardiac plexuses, as well as the myenteric (Auerbach) and submucosal (Meissner) plexuses of the enteric nervous system. This proximity allows for rapid, localized responses to physiological changes. These plexuses include a varied population of ganglion neurons. The majority use acetylcholine as their primary neurotransmitter, responsible for effects such as slowing the heart rate, contracting the smooth muscles, and stimulating secretory processes. Some neurons release vasoactive intestinal polypeptide (VIP) or nitric oxide (NO), eliciting smooth muscle relaxation and vasodilation (Furness 2006). This rich diversity of neurotransmitters and effects allows the vagus nerve to fine-tune its influence across various organ systems. The broad spectrum and complex architecture of the vagus nerve emphasize its key role in the parasympathetic regulation of essential organ systems (Kiernan and Rajakumar 2014).

Sacral Parasympathetic Output to Pelvic Organs

The sacral parasympathetic system is an important element in the neural system, regulating essential physical activities. Located in the S2–S4 regions of the spinal cord, the sacral parasympathetic nucleus functions as the control center for various vital actions (de Groat 2006; Fowler et al. 2008). These preganglionic neurons expand via the pelvic nerves, establishing complex connections with ganglion cells in the pelvic plexus and within the walls of the bladder, rectum, and reproductive organs. This strategic positioning facilitates accurate, localized control (Standring 2020).

The sacral parasympathetic output facilitates the contraction of the detrusor muscle, hence beginning urination and regulating bladder control. Simultaneously, it prompts the bladder neck to relax, facilitating a seamless flow. It is the stimulus for the rhythmic peristalsis in the distal colon and the final urge that facilitates rectal evacuation. By promoting relaxation of the smooth muscles in the erectile tissue, it plays a crucial role in the sexual response of both men and women (Kiernan and Rajakumar 2014).

Parasympathetic Neurotransmitters

The parasympathetic nervous system utilizes a variety of neurotransmitters to regulate its effects on various target organs. This mainly consists of acetylcholine, vasoactive intestinal polypeptide (VIP), and nitric oxide (NO), often used in various combinations to achieve specific physiological responses (Lundberg 1996; Burnstock 2004).

Acetylcholine serves as the principal neurotransmitter in the parasympathetic system. Its actions are mediated through various subtypes of muscarinic receptors found in target organs. The effects of acetylcholine are wide-ranging and include the following:

- Cardiac inhibition: Slowing the heart rate and reducing cardiac contractility
- Stimulation of gastric acid secretion: Promoting digestive processes
- Smooth muscle contraction: Affecting various organs, such as the bladder and gastrointestinal tract
- Exocrine gland secretion: Enhancing the production of saliva, tears, and other secretions

The diversity of muscarinic receptor subtypes allows for the fine-tuning of these responses in different target tissues (Standring 2020).

While acetylcholine often promotes contraction and secretion, VIP and NO play a different role. These neurotransmitters primarily elicit smooth muscle relaxation. This action is particularly important in certain physiological processes, such as:

- Vasodilation: Increasing blood flow to specific organs
- Bronchodilation: Relaxing the airways in the respiratory system
- Regulation of gastrointestinal motility: Coordinating the movement of digestive contents

The use of multiple neurotransmitters allows the parasympathetic system to exert refined control over its target organs. For instance, in the gastrointestinal tract, acetylcholine might promote contraction of certain muscles, while VIP or NO simultaneously relaxes others, allowing for coordinated peristaltic movements (Kiernan and Rajakumar 2014). The complexity of parasympathetic neurotransmission shows the complex nature of the nervous system. By utilizing different combinations of these chemical messengers, the parasympathetic system can fine-tune its effects, promoting the "rest and digest" state characteristic of parasympathetic dominance (Furness 2006).

3.2 Autonomic Regulation of Target Organs

The autonomic nervous system exerts precise control over various organ systems through intricate neural pathways and neurotransmitter mechanisms. This control system enables organs to respond appropriately to physiological demands while maintaining homeostasis. Understanding these regulatory mechanisms provides essential insights into normal physiological functions and pathological conditions. Each target organ system demonstrates unique patterns of autonomic innervation and response characteristics (Strominger et al. 2012) (Table 3.5).

3.2.1 Cardiovascular System

The autonomic nervous system regulates cardiovascular function through a complex interaction of sympathetic and parasympathetic influences. This balanced regulation maintains homeostasis and allows for rapid adaptations to physiological demands.

Sympathetic activity increases cardiac chronotropy and inotropy in both atrial and ventricular chambers. Concurrently, it induces widespread vasoconstriction, with notable exceptions in specific arteriovenous anastomoses of cutaneous and skeletal muscle beds. These latter structures receive cholinergic sympathetic vasodilatory input, allowing for localized blood flow regulation (Standring 2020).

The sinoatrial node, the heart's primary pacemaker, is under dual autonomic control. While both sympathetic and vagal efferents modulate its automaticity, tonic vagal activity predominates under basal conditions, providing beat-to-beat regulation of heart rate (Olshansky et al. 2008). Vagal input decelerates the heart rate by attenuating spontaneous sinoatrial node depolarization. Interestingly, vagal influence fluctuates with the respiratory cycle, diminishing during inspiration and intensifying during expiration. This phenomenon results in respiratory sinus arrhythmia, the magnitude of which inversely correlates with age in healthy individuals at rest.

Parasympathetic activation extends beyond chronotropic effects. It impedes atrioventricular conduction, prolongs ventricular refractoriness, and reduces contractility, primarily in the atria and, to a lesser extent, in the ventricles. A subset of cardiac ganglionic neurons receiving preganglionic vagal input utilizes vasoactive intestinal peptide (VIP) and nitric oxide (NO) as neurotransmitters. These chemical mediators promote relaxation and vasodilation, contributing to the nuanced modulation of cardiac function (Furness 2006).

Table 3.5 Effects of autonomic nervous activity on target organs and tissues

Structure	Sympathetic function	Parasympathetic function
Eye		
Radial muscle of iris	Dilates pupil (mydriasis)	
Sphincter muscle of iris		Contraction of pupil (miosis)
Ciliary muscle (accommodation)	Relaxation for far vision	Contraction for near vision
Glands of head		
Lacrimal gland		Stimulates secretion
Salivary glands	Scanty thick, viscous secretion	Profuse, watery secretion
Heart		
Rate	Increase	Decrease
Force of ventricular contraction	Increase	
Blood vessels	Generally constricts	Slight effect
Lungs		
Bronchial tubes	Dilates lumen	Constricts lumen
Bronchial glands		Stimulate secretion
Gastrointestinal tract		
Motility and tone	Inhibits	Stimulates
Sphincters	Stimulates	Inhibits (relaxes)
Secretion	May inhibit	Stimulates
Gallbladder and ducts	Inhibits	Stimulates
Liver	Glycogenolysis increase (blood sugar)	
Adrenal medulla	Secretion of epinephrine and norepinephrine	
Sex organs	Vasoconstriction, constriction of vas deferens, seminal vesicle, and prostatic musculature (ejaculation)	Vasodilation and erection
Skin		
Sweat glands	Stimulated	
Blood vessels	Constricted	Slight effect
Neurotransmitter at neuroeffector junction	Usually norepinephrine	Acetylcholine
Inactivation of transmitter	Slow and reuptake	Rapid
Reinforcement in body	Secretion of norepinephrine and epinephrine by adrenal medulla	

Modified with permission from Strominger et al. (2012)

The sympathetic influence on the heart exhibits notable lateralization. Right-sided sympathetic activation predominantly innervates the sinoatrial node, accelerating the heart rate and counteracting the vagal effects. In contrast, left-sided sympathetic stimulation primarily targets the atrioventricular node and ventricles. This lateralized innervation enhances atrioventricular conduction, augments His–Purkinje system excitability, and increases myocardial contractility and oxygen consumption (Goldstein et al. 2002).

Norepinephrine functions as the principal mediator of sympathetic cardiac actions. The interaction between the sympathetic and parasympathetic systems is characterized by significant antagonism, partially resulting from presynaptic interactions. For instance, sympathetic effects on the sinoatrial node become more pronounced when the basal vagal tone is elevated. This amplification occurs because norepinephrine-induced presynaptic inhibition of acetylcholine release contributes to heart rate acceleration. Conversely, vagal-induced cardiac responses, particularly in the ventricles, become more prominent during concurrent sympathetic activation. This phenomenon likely reflects vagally mediated presynaptic inhibition of sympathetic activity. Such interaction provides a significant protective mechanism by elevating the threshold for ventricular fibrillation (Coote 2013). The dynamic balance between sympathetic and parasympathetic influences allows for rapid cardiovascular adjustments. For example, at the onset of exercise, there is a swift decrease in vagal activity, permitting a rapid increase in the heart rate and cardiac output to meet the increased metabolic demands.

3.2.1.1 Autonomic Control of Cranial Vasculature

The blood vessels in the cranial region receive a complex network of innervation from multiple sources, each contributing to the fine-tuning of cerebral blood flow and cranial vascular tone. This intricate system involves sympathetic, parasympathetic, and sensory components.

Sympathetic innervation originates primarily from the superior cervical ganglion. These fibers induce vasoconstriction in both cranial and cerebrovascular beds, an effect mediated predominantly by norepinephrine. Neuropeptide Y (NPY) and adenosine triphosphate (ATP) also have a role in this vasoconstrictive effect, though to a lower degree (Golanov and Reis 1996). Interestingly, sympathetic inputs can also promote vasodilation in specific regions. For instance, they may dilate facial veins and actively expand the cutaneous vasculature of the face during states of heat stress or emotional arousal. However, the precise role of sympathetic vasoconstriction in intracranial vessels remains a subject of ongoing research and debate.

Parasympathetic control of cranial blood vessels stems from the pterygopalatine ganglion. This input elicits vasodilation in various cranial structures, including the eye, nasal cavity, palate, and intracranial circulation. Nitric oxide (NO) serves as the primary mediator of this vasodilatory effect, with acetylcholine and vasoactive intestinal peptide (VIP) playing supportive roles (Toda et al. 2009).

The sensory component of cranial vascular innervation involves branches of the trigeminal nerve, with cell bodies located in the trigeminal ganglion. These afferent fibers can induce antidromic vasodilation of cranial blood vessels through the release of substance P or calcitonin gene–related peptide. This process occurs during axon reflexes triggered by local inflammation, representing a key mechanism in neurogenic inflammation (Moskowitz 1984). Moreover, trigeminal afferents that reach the trigeminal nucleus caudalis can initiate trigemino-parasympathetic reflexes. These reflexes are mediated via the superior salivatory nucleus, which sends preganglionic axons to the sphenopalatine ganglion. The activation of this pathway results in a triad of responses: cranial vasodilation, lacrimation, and nasal congestion. This reflex arc forms the basis for several clinical phenomena, including certain types of headaches and autonomic symptoms associated with facial pain (Drummond 2006).

3.2.1.2 Role of Sympathetic Vasoconstriction in Blood Pressure Regulation

The sympathetic nervous system plays a crucial role in maintaining cardiovascular homeostasis, particularly through its vasoconstrictor effects. The distribution and function of sympathetic innervation exhibit significant regional variations, allowing for precise regulation of local blood flow (Gibbins et al. 2003).

Sympathetic noradrenergic fibers extensively innervate arterioles and veins in skeletal muscle and mesenteric beds. This innervation is fundamental for the short-term maintenance of arterial blood pressure, especially when transitioning to an upright posture. Upon standing, there is a reduction in the venous return, cardiac output, and blood pressure. This change unloads baroreceptors, triggering sympathetically mediated vasoconstriction. Dysfunction in this mechanism can lead to neurogenic orthostatic hypotension, a condition characterized by an inability to maintain blood pressure upon standing (Freeman et al. 2011).

Several methods exist to assess the integrity of this system. These include the following:

- Analyzing blood pressure responses during the Valsalva maneuver and head-up tilt tests
- Measuring plasma norepinephrine levels in supine and standing positions
- Recording muscle sympathetic nerve activity (used mostly in clinical research settings)

The last method provides direct insights into sympathetic outflow and has significantly advanced our understanding of sympathetic control in various physiological and pathological states (Wallin and Charkoudian 2007).

Interestingly, in humans, sympathectomy has a minimal effect on arterial pressure unless the splanchnic nerves are sectioned. This observation shows the particular importance of splanchnic vasoconstriction in blood pressure regulation. Correspondingly, individuals with spinal cord injuries typically do not develop significant orthostatic hypotension unless the lesion is above the level of splanchnic outflow (T6). It is notable that there is an age-dependent loss of splanchnic preganglionic neurons, estimated at approximately 5–8% per decade in adults (Low and Dyck 1978). This gradual decline may contribute to the increased prevalence of orthostatic intolerance in older populations.

The sympathetic noradrenergic system also extensively innervates various renal structures, including afferent and efferent arterioles, juxtaglomerular granular cells, and different segments of the nephron (proximal and distal tubules and loops of Henle). This renal sympathetic innervation serves three primary functions: renal vasoconstriction, enhanced tubular sodium reabsorption, and stimulation of renin secretion. These systems collectively facilitate the long-term regulation of arterial pressure, demonstrating the kidney's important involvement in blood pressure management (DiBona 2005). The renin–angiotensin–aldosterone system, activated by sympathetic stimulation, further amplifies these effects, leading to sustained increases in vascular tone and blood volume.

3.2.1.3 Autonomic Regulation of Cardiac Function

Cardiac function is tightly regulated by the interaction between sympathetic and parasympathetic (vagal) influences. This dual innervation allows for precise control of the heart rate, contractility, and conduction velocity, enabling rapid adaptations to varying physiological demands. Under resting conditions, there is a tonic vagal influence on the heart, which keeps the resting heart rate below the intrinsic rate of the sinoatrial node. During exercise or stress, sympathetic activation, coupled with vagal withdrawal, allows for rapid increases in the heart rate and contractility (Florea and Cohn 2014). The balance between sympathetic and parasympathetic influences is not simply additive but involves complex interactions. For example, vagal effects on the heart rate are more pronounced in the presence of background sympathetic tone, a phenomenon known as "accentuated antagonism" (Levy 1971).

Sympathetic Innervation

Sympathetic innervation of the heart originates from the thoracic spinal cord segments T2–T5. Preganglionic sympathetic fibers emerge from these segments and ascend in the paravertebral sympathetic chain. These fibers then synapse in either the cervical or upper thoracic (stellate) ganglia. Postganglionic noradrenergic fibers from these ganglia provide direct innervation to the heart (Jänig 2006).

Sympathetic stimulation results in the following:

- Increased heart rate (positive chronotropy)
- Enhanced contractility (positive inotropy)
- Accelerated conduction velocity (positive dromotropy)
- Increased relaxation speed (positive lusitropy)

These effects are primarily mediated by norepinephrine acting on β1-adrenergic receptors, although α-adrenergic effects also play a role, particularly in the coronary vasculature (Triposkiadis et al. 2009).

Parasympathetic (Vagal) Innervation

Vagal preganglionic fibers originate predominantly in the ventrolateral portion of the nucleus ambiguus, with a smaller contribution from the dorsal motor nucleus of the vagus. These fibers travel via the vagus nerve and reach the heart through the superior cervical, inferior cervical, and thoracic rami. Notably, these vagal fibers interconnect with cardiac sympathetic nerves to create the cardiac plexus, emphasizing the structural basis for complex sympathetic–parasympathetic interactions (Armour et al. 1997).

Vagal stimulation primarily results in the following:

- Decreased heart rate
- Reduced atrial contractility
- Slowed atrioventricular conduction

These effects are mediated by acetylcholine acting on muscarinic (M2) receptors.

Cardiac Ganglia

The majority of cardiac ganglion neurons are cholinergic, forming the final common pathway for vagal control of the heart. However, a subset of these neurons exhibits neurochemical diversity, synthesizing other neurotransmitters and neuromodulators, such as somatostatin, vasoactive intestinal peptide (VIP), or nitric oxide (NO), in various combinations (Armour 2008). This neurochemical complexity allows for the precise regulation of cardiac function beyond simple on/off control. Recent research has revealed that cardiac ganglia are not merely relay stations but can process and integrate information, possessing a degree of autonomy in cardiac control. This "little brain" in the heart can modulate cardiac function even in the absence of central nervous system input (Wake and Brack 2016).

3.2.2 Autonomic Regulation of Pupillary Function

The diameter of the pupil is dynamically controlled by the antagonistic actions of two smooth muscle groups: the radial dilator pupillae and the circular sphincter pupillae. These muscles are under the influence of the sympathetic and parasympathetic divisions of the autonomic nervous system, respectively.

The final pupil size at any given moment depends on the balance between these sympathetic and parasympathetic influences. This balance allows for the fine-tuning of pupil diameter in response to changing light conditions and visual tasks.

Several factors can influence this balance:

- Arousal state: Increased arousal or stress tends to shift the balance toward sympathetic dominance, resulting in pupil dilation.
- Circadian rhythms: There are diurnal variations in pupil size, with larger pupils typically observed in low light conditions and at night (Wilhelm et al. 1998).
- Age: The pupillary light reflex becomes less responsive with age, a phenomenon known as pupillary senescence (Bitsios et al. 1996).
- Pharmacological agents: Various drugs can affect the pupil size by modulating the autonomic function. For example, anticholinergics cause mydriasis, while cholinergic agonists cause miosis.

3.2.2.1 Sympathetic Control: Pupillary Dilation

Pupillary dilation (mydriasis) occurs through two primary mechanisms:

- Peripheral sympathetic activation: The radial smooth muscle of the pupil receives sympathetic innervation originating from the T1 and T2 segments of the spinal cord. These preganglionic fibers synapse in the superior cervical ganglion. Postganglionic noradrenergic fibers from the superior cervical ganglion innervate the dilator pupillae muscle. When activated, these fibers release norepinephrine, which acts on $\alpha 1$-adrenergic receptors to cause muscle contraction and subsequent pupillary dilation (McDougal and Gamlin 2015).
- Central inhibition: Dilation can also occur through the inhibition of the Edinger–Westphal nucleus, mediated by fibers originat-

ing from the posterior thalamus. This mechanism allows for pupillary dilation without direct sympathetic activation, demonstrating the complex central control of pupil size (Szabadi 2013).

3.2.2.2 Parasympathetic Control: Pupillary Constriction

Pupillary constriction (miosis) is primarily mediated by the parasympathetic nervous system and occurs in response to two main stimuli:

- Light reflex: Exposure to light triggers this reflex.
- Near vision (accommodation reflex): This occurs when focusing on close objects.

Both reflexes are mediated by the activation of parasympathetic preganglionic neurons in the Edinger–Westphal nuclei. These nuclei receive afferent input from the retina via the pretectal nuclei. Importantly, because these afferent pathways cross at the optic chiasm and again at the posterior commissure, illumination of one eye produces constriction of both pupils (consensual light reflex) (Loewenfeld 1999).

Parasympathetic postganglionic fibers originate in the ciliary ganglion and innervate the sphincter pupillae muscle. When activated, these fibers release acetylcholine, which acts on muscarinic (M3) receptors to cause muscle contraction and pupillary constriction.

3.2.3 Autonomic Regulation of Pelvic Organs

The pelvic organs are regulated by complex interactions between the lumbar sympathetic and sacral parasympathetic systems, with additional input from the sacral skeletomotor pathways to the external sphincters and pelvic floor. The brainstem pattern generators regulate the actions of various systems, integrating their functions for maximum physiological performance. Sympathetic innervation from T11–L2 spinal segments synapses in lumbar paravertebral or inferior mesenteric ganglia, influencing pelvic organs via hypogastric and pelvic nerves. This pathway plays a key role in storage functions and sexual responses. Parasympathetic input originates from the sacral preganglionic nucleus (S2–S3 segments), reaching the target organs through the pelvic nerve. This system is primarily involved in excretory and sexual functions. Pelvic ganglia mediate intricate interactions between preganglionic parasympathetic and postganglionic sympathetic efferents, allowing for fine-tuned control of pelvic organ function (Fowler et al. 2008).

The Onuf nucleus, located in the S2–S4 segments, contains somatic motoneurons innervating striated muscles of external urethral and anal sphincters and the pelvic floor. This nucleus is unique in its resistance to certain neurodegenerative processes and is central to the voluntary control of micturition and defecation. Gastrointestinal motility and secretion are controlled by multiple neural mechanisms: the enteric nervous system, extrinsic autonomic efferents that modulate enteric nervous system activity, visceral afferents that initiate reflexes at various levels, and somatic effects on specific striated muscles of the upper gastrointestinal tract and pelvic floor (Cersosimo and Benarroch 2008).

Sympathetic activation generally inhibits gastrointestinal motility by suppressing smooth muscle contraction and constricting sphincters. This system is predominantly active during periods of stress or digestion. Conversely, parasympathetic stimulation enhances motility and relaxes sphincters, promoting digestive processes and excretion (Furness 2012). This autonomic interplay is exemplified in bladder function, where sympathetic activity dominates during the storage phase, inhibiting detrusor contraction and facilitating urethral closure. During micturition, parasympathetic activation leads to detrusor contraction and internal sphincter relaxation, while somatic control via the Onuf nucleus allows for voluntary relaxation of the external urethral sphincter (Standring 2020).

3.2.3.1 Bladder

The neurological regulation of the bladder requires an intricate relationship between the sympathetic and parasympathetic systems. Sympathetic fibers, utilizing norepinephrine as a neurotransmitter, induce relaxation of the detrusor muscle while contracting the trigone, which includes the internal sphincter (bladder neck). Additionally, norepinephrine primarily inhibits synaptic transmission in parasympathetic ganglia. Conversely, parasympathetic fibers utilize three neurotransmitters: acetylcholine, ATP, and nitric oxide. Acetylcholine functions to contract the detrusor muscle and suppress norepinephrine release from sympathetic terminals. ATP contributes to bladder contraction, while NO facilitates relaxation of the urethral smooth muscle (Fowler et al. 2008).

Sacral preganglionic neurons provide parasympathetic output, activating cholinergic neurons situated in the microganglia around the bladder base and within its wall. During the bladder-filling phase (urine storage), lumbar sympathetic output promotes detrusor relaxation and bladder neck contraction. Concurrently, somatomotor output from the Onuf nucleus maintains tonic contraction of the external urethral sphincter. The process of micturition is initiated by sacral parasympathetic output via cholinergic pelvic ganglion neurons, which trigger detrusor contraction (de Groat et al. 2015).

Micturition requires a coordinated effort that includes detrusor muscle activation and external sphincter relaxation, regulated by the pontine micturition center. This intricate balance ensures efficient bladder emptying while maintaining continence during storage phases (Griffiths 2015).

3.2.3.2 Colon and Rectum

The anorectum and pelvic floor, similar to the lower urinary tract and external sphincter, receive innervation from parasympathetic, sympathetic, and somatic nervous systems. Sacral preganglionic parasympathetic output is conveyed via the pelvic nerve and plexus to neurons in the myenteric plexus of the descending colon and rectum, as well as to the ganglia distributed in the distal portion of the pelvic nerves (Bharucha 2006).

The sacral parasympathetic output comprises two distinct pathways: excitatory cholinergic and inhibitory nitroxidergic. The excitatory pathways regulate colonic propulsive activity during defecation, while the inhibitory pathways enable colonic volume adaptation to its contents and initiate colonic relaxation preceding the fecal bolus. The recto-anal inhibitory reflex, characterized by anal relaxation induced by rectal distention, is mediated by intrinsic nerves utilizing nitric oxide (NO) as a neurotransmitter. Tonic excitatory sympathetic input to the internal anal sphincter maintains closure of the anal canal at rest. The somatic innervation originating from the Onuf nucleus contributes significantly (25–50%) to the resting anal tone. The tonically active puborectalis muscle serves as an essential muscle for maintaining the resting anorectal angle, and its contraction is necessary for maintaining fecal continence (Rao 2004). This intricate neural control system ensures proper functioning of the colon and rectum, facilitating both defecation and continence as required. The balance between these various neural inputs allows for the complex coordination necessary for normal bowel function.

3.2.3.3 Sexual Organs

The neurological control of male sexual organs involves a complex relationship between sympathetic and parasympathetic systems, each fulfilling specific functions in different phases of sexual activity.

Smooth muscle tissues in the vas deferens, seminal vesicles, and prostate gland receive excitatory noradrenergic innervation from neurons located in the pelvic plexus. This sympathetic input is crucial for various aspects of male sexual function, including ejaculation and the maintenance of a non-erect state (Dean and Lue 2005). Parasympathetic input is relayed to neurons in local ganglia, which, upon activation, release a combination of neurotransmitters: acetylcholine, vasoactive intestinal peptide (VIP), and nitric oxide (NO). These neurotransmitters play significant roles in modulating blood flow and smooth muscle tone in the sexual organs (Andersson 2011).

Penile erection is primarily mediated by the vasodilation of penile arteries, a process triggered by NO. This neurotransmitter, released from both nerve endings and endothelial cells, initiates a cascade of events leading to smooth muscle relaxation and increased blood flow into the erectile tissues (Burnett 1997).

Sympathetic outflow serves multiple functions:

- It induces vasoconstriction and detumescence of the erectile tissue, contributing to the return to a flaccid state after erection.
- It stimulates the smooth muscle of the vas deferens, seminal vesicles, and prostate glands, which is necessary for ejaculation and the synthesis of seminal fluid (Giuliano and Clement 2005).

3.2.4 Salivary Glands

Salivary gland function is regulated by both the sympathetic and parasympathetic nervous systems, each contributing differently to salivary secretion. Sympathetic activation produces a relatively weak serous secretion, mainly from the submandibular gland. In contrast, parasympathetic stimulation elicits copious serous secretions from all salivary glands, including the parotid, submandibular, and sublingual glands (Proctor and Carpenter 2007). Parasympathetic control of lacrimal and salivary glands is mediated by neurons in the sphenopalatine (pterygopalatine), submandibular, and otic ganglia. These ganglia relay inputs from the salivatory nuclear complex, which includes superior and inferior portions, via the facial and glossopharyngeal nerves. When activated, these ganglion neurons trigger exocrine gland secretion and vasodilation through the release of acetylcholine, vasoactive intestinal peptide (VIP), and nitric oxide (NO) (Matsuo and Garrett 2019).

Sweat glands, unlike salivary glands, receive cholinergic inputs from sympathetic neurons of the superior cervical ganglion, which stimulates facial sweating. This arrangement, where sympathetic neurons use acetylcholine instead of norepinephrine as a neurotransmitter, is an exception to the general pattern of sympathetic neurotransmission (Sato et al. 1989).

A notable pathological condition can occur when lesions affect sympathetic axons. In such instances, postganglionic parasympathetic fibers from the otic and submaxillary ganglia may form connections with sweat glands. This atypical innervation can result in pathological sweating in response to parasympathetic reflexes triggered by the stimulation of structures in the mouth, a phenomenon known as gustatory sweating (Drummond 1994).

3.2.5 Gastrointestinal System

The gastrointestinal system is regulated by the intrinsic and extrinsic nervous systems. Sympathetic activation decreases gastrointestinal motility by inhibiting longitudinal and circular smooth muscle contraction and contracting sphincters. Conversely, parasympathetic activation increases motility by contracting these muscles and relaxing sphincters (Furness 2012).

The intrinsic enteric nervous system in the gut wall independently regulates gastrointestinal function, modulated by extrinsic sympathetic and parasympathetic nerves. The myenteric plexus (Auerbach plexus) lies between circular and longitudinal muscle layers, while the submucosal plexus (Meissner plexus) is under the mucosa (Fig. 3.11). These plexuses modulate motility via interstitial cells of Cajal, which generate slow waves for peristalsis. Interstitial cells respond to excitatory (acetylcholine and substance P) and inhibitory (VIP and NO) signals (Wood 2008). The enteric nervous system controls fasting motility, including the migrating motor complex, and basic sphincter tone. It also regulates mucosal circulation and neuroendocrine/immune cells. Food intake stimulates the enteric nervous system to modulate gastrointestinal activity (Furness 2012).

The extrinsic system connects the enteric nervous system to the CNS. Afferent signals from

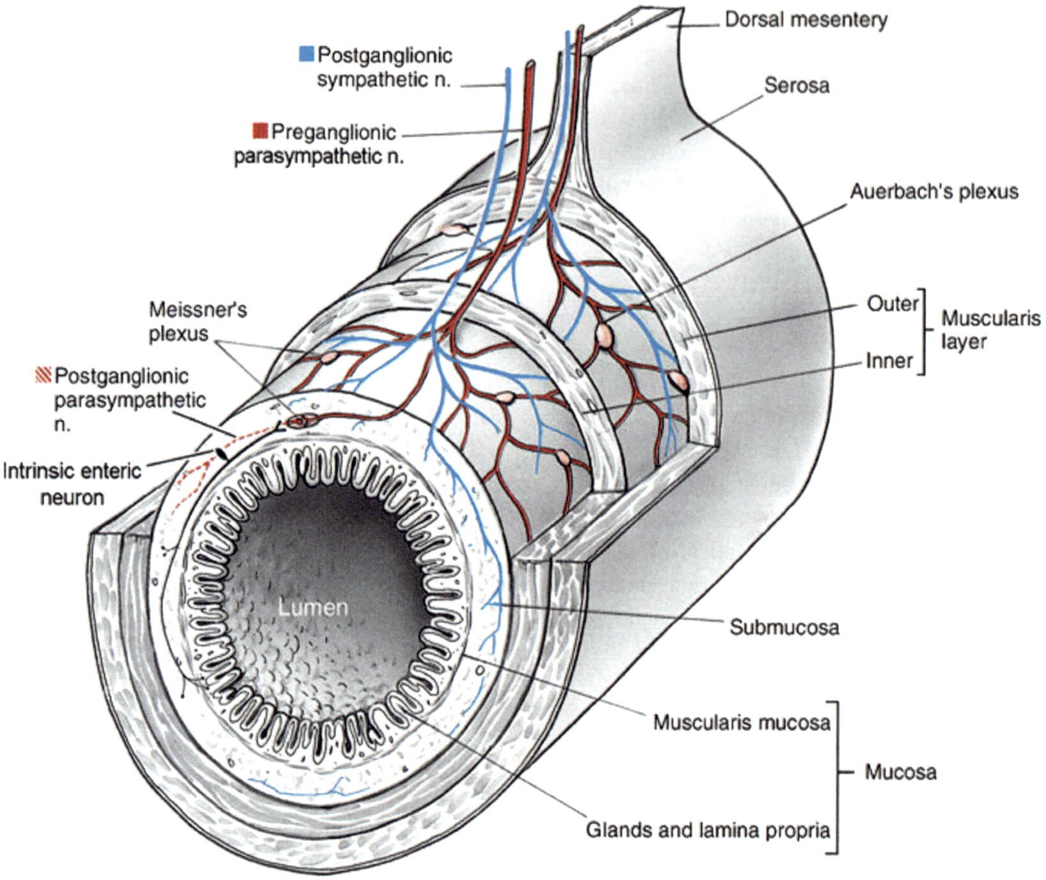

Fig. 3.11 Enteric division of the autonomic nervous system and intrinsic enteric nervous system, including myenteric plexus (Auerbach) and submucosal plexus (Meissner). (Adapted with permission from Strominger et al. (2012))

chemoreceptors and mechanoreceptors travel to the CNS, which initiates regulatory reflexes via autonomic pathways (Fig. 3.12). Cranial parasympathetic innervation occurs via the vagus nerve, with 80–90% sensory fibers. Afferents terminate in the solitary tract nucleus, projecting to regions like the hypothalamus. Efferents from the dorsal vagal nucleus innervate gastrointestinal ganglia (Powley 2000). The spinal autonomic system includes sympathetic splanchnic nerves (T1–L2) and parasympathetic pelvic nerves (S2–S4). Sympathetic innervation regulates vasomotor tone and blood flow redistribution. Parasympathetic innervation modulates motility, sphincter control, and secretion (Furness 2012).

Central autonomic network structures are involved in gastrointestinal autonomic innervation (Table 3.2). These structures modulate gastrointestinal autonomic function, influencing emotions, stress responses, and visceral sensations (Cersosimo and Benarroch 2008).

Gastrointestinal motility and secretion are controlled by (1) the enteric nervous system for local reflexes, (2) extrinsic autonomic efferents modulating enteric nervous system activity (Fig. 3.13), (3) visceral afferents initiating various reflexes, and (4) somatic effects on striated muscles of the upper gastrointestinal tract and pelvic floor (Cersosimo and Benarroch 2008).

3.2.5.1 Control of Gastrointestinal Motility

Gastrointestinal smooth muscle function is intricately controlled by the intrinsic (or enteric) nervous system through the release of peptides and transmitters. This intrinsic control is further mod-

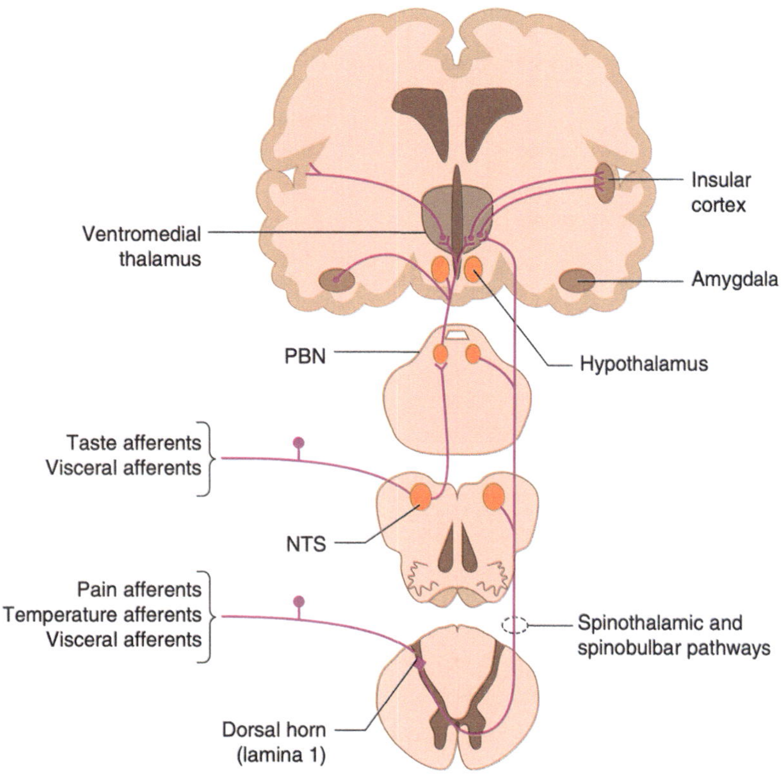

Fig. 3.12 Dorsal horn neurons in lamina I serve as the initial integration point, transmitting information to higher regulatory centers, including the nucleus of the solitary tract (NTS), medullary network, thalamus, and insular cortex. Abbreviations: NTS nucleus tractus solitarii, PBN parabrachial nucleus. (Adapted with permission from Struhal et al. (2017))

ulated by extrinsic autonomic nerves: the excitatory craniospinal parasympathetic input (vagus and S2, S3, and S4 nerves) and the predominantly inhibitory thoracolumbar sympathetic outflow, which is excitatory to sphincters (Fig. 3.14).

Gastrointestinal smooth muscle forms an electrical syncytium, allowing efficient transmission of contractile impulses across a sheet of sequentially linked cells in both transverse and longitudinal axes of the intestine. The interstitial cells of Cajal network serve as the pacemaker of this intestinal muscle syncytium, coordinating contraction circumferentially and longitudinally along the gut (Sanders et al. 2014).

The enteric nervous system in humans comprises a series of ganglionated plexuses, including the submucosal (Meissner), myenteric (Auerbach), deep muscular (Cajal), mucosal, and submucosal plexuses. These enteric nerves contain nearly 100 million neurons, comparable to the number of neurons in the spinal cord (Furness 2012).

At the diaphragm level, the vagus nerve is predominantly composed of afferent fibers. The current concept suggests that each vagal command fiber supplies an integrated circuit hardwired in the intestinal wall, resulting in specific motor or secretory responses. These circuits are required for many automated gastrointestinal responses, such as the peristaltic reflex, and continue functioning in a totally extrinsically denervated intestine (Powley 2000) (Fig. 3.15).

Pacemakers located on the greater curve of the gastric corpus and the duodenal bulb drive the maximum intrinsic contraction frequencies: three per minute in the stomach and 12 per minute in the small intestine. Similar to cardiac function, if the highest-frequency pacemaker malfunctions, the region with the next highest intrinsic contractile frequency takes over (Huizinga and Lammers 2009).

Disorders of extrinsic neural control of the gastrointestinal smooth muscle affect several clinically observed motility disorders, such as

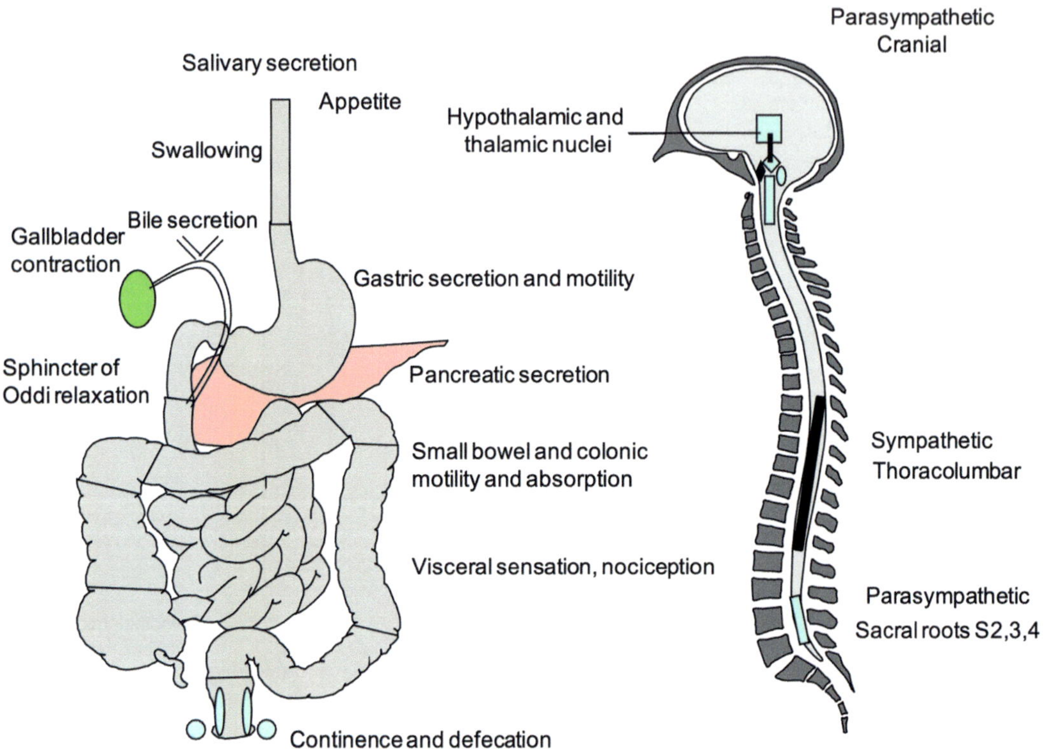

Fig. 3.13 Gastrointestinal physiology and functions regulated by extrinsic autonomic control. (Adapted with permission from Robertson et al. (2012))

diabetic gastroparesis and fecal incontinence resulting from obstetric trauma to the anal sphincters. Other diseases, such as achalasia or Hirschsprung's disease, result from disorders of enteric neural function (Knowles et al. 2013).

Enteric Nervous System

The prevertebral sympathetic ganglia innervating the gut function as convergence sites for preganglionic inputs via splanchnic nerves, afferents from cholinergic neurons of the enteric nervous system, and collaterals of primary dorsal root ganglion afferents. The sympathetic prevertebral reflex acts as a regulatory mechanism for motor activity in the gut. Enteric afferents, activated by gut distention, release acetylcholine, which stimulates prevertebral ganglion cells; these cells subsequently release norepinephrine at the myenteric plexus level, inhibiting gut motility (Furness et al. 2014).

The enteric nervous system consists of two ganglionated plexuses located within the walls of the gut: the myenteric plexus (of Auerbach) and the submucosal plexus (of Meissner) (Costa and Brookes 1994; Benarroch 2007; Furness 2006). This system regulates gastrointestinal motility and secretion largely independent of influences from the central nervous system. Neurons of the myenteric plexus control the activity of the smooth muscle of the gut, while those in the submucosal plexus modulate mucosal secretion and blood flow. The enteric nervous system manages gut motility and secretion through local reflexes triggered by local distention of the intestinal wall, distortion of the mucosa, and chemical contents in the lumen.

Most neurons in the enteric nervous system, including intrinsic primary afferent neurons, interneurons, and excitatory motoneurons that elicit smooth muscle contraction, utilize acetyl-

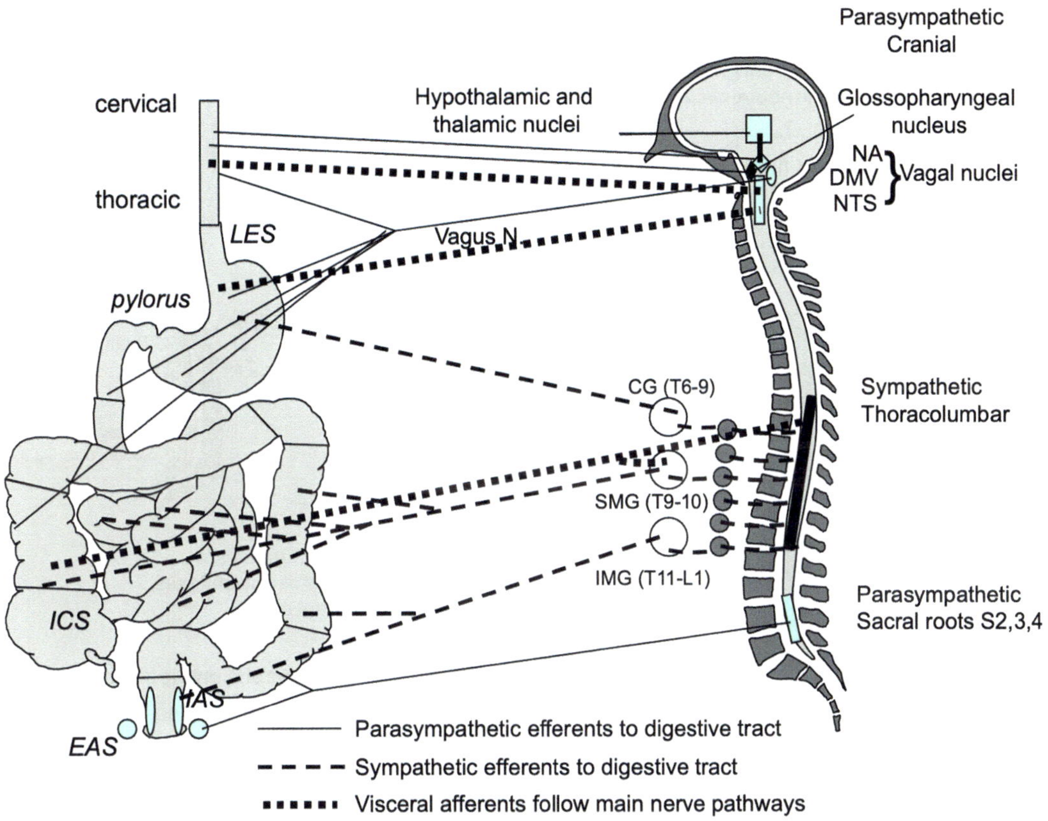

Fig. 3.14 Neural pathways of the sympathetic and parasympathetic nervous systems innervating the gastrointestinal tract. Abbreviations: ICS ileocolonic sphincter, IAS internal anal sphincter, EAS external anal sphincter, LES lower esophageal sphincter, CG celiac ganglion, SMG superior mesenteric ganglion, IMG inferior mesenteric ganglion, NA nucleus ambiguus, DMV dorsal motor nucleus of the vagus, NTS nucleus tractus solitarius, T thoracic. (Adapted with permission from Robertson et al. (2012))

choline as their primary neurotransmitter. The inhibitory motor neurons that promote smooth muscle relaxation utilize vasoactive intestinal peptide (VIP) or nitric oxide (NO). Secretomotor neurons use both acetylcholine and VIP (Anlauf et al. 2003).

Vagal Modulation of the Enteric Nervous System

Vagal preganglionic neurons originating from the dorsal motor nucleus of the vagus provide extensive cholinergic input to the enteric nervous system. This input affects various neuronal populations within the enteric nervous system, including the following:

- Excitatory cholinergic neurons that induce smooth muscle contraction (vagal excitatory pathway)
- VIP/NO neurons that promote smooth muscle relaxation (vagal inhibitory pathway)
- Submucosal acetylcholine/VIP neurons that stimulate intestinal secretion and vasodilation

The influence of vagal control on gut motility is most pronounced in the proximal gastrointestinal tract, particularly the esophagus and stomach. This vagal input plays a key role in several reflexes that regulate gastrointestinal function (Travagli et al. 2006):

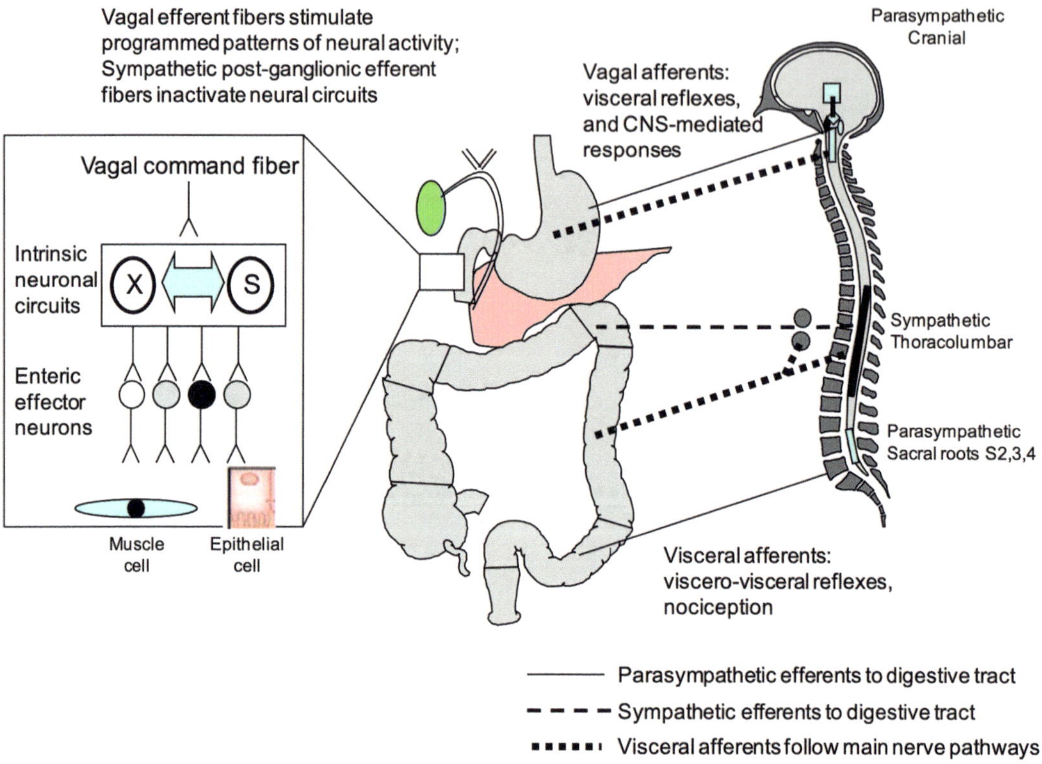

Fig. 3.15 Integration of extrinsic and intrinsic (enteric) neural control. Hardwired programs, such as peristalsis, are modulated by efferent inputs from the vagus and sympathetic nerves, which also carry afferent fibers mediating visceral sensation, nociception, and reflex responses. Abbreviation: CNS central nervous system. (Adapted with permission from Robertson et al. (2012))

- Esophageal motility
- Relaxation of the lower esophageal sphincter
- Gastric accommodation following meals
- Gastric emptying

As one progresses distally along the gastrointestinal tract, the significance of vagal control gradually diminishes. In the small bowel and colon, motility and secretion are predominantly regulated by local reflexes within the enteric nervous system, with less reliance on vagal input.

This differential distribution of vagal influence along the gastrointestinal tract reflects the functional specialization of different gut regions. The proximal gut, which is more involved in the initial stages of digestion and requires coordinated responses to food intake, relies more heavily on the central nervous system input via the vagus nerve. In contrast, the distal gut, which is more involved in absorption and waste elimination, operates more autonomously through local enteric reflexes (Powley et al. 2019).

Inhibitory Prevertebral Sympathetic Reflex
The prevertebral sympathetic ganglia, which innervate the gut, serve as a convergence point for several neural inputs:

- Preganglionic inputs via the splanchnic nerves
- Afferents from cholinergic neurons of the enteric nervous system
- Collaterals of primary dorsal root ganglion afferents

This convergence of inputs forms the basis of the sympathetic prevertebral reflex, a feedback mechanism that modulates motor activity in the gastrointestinal tract (Szurszewski and Linden 2012).

The reflex operates as follows:

- Gut distention activates enteric afferents.
- These activated afferents release acetylcholine.
- Acetylcholine excites prevertebral ganglion cells.
- The excited prevertebral ganglion cells release norepinephrine at the level of the myenteric plexus.
- Norepinephrine inhibits gut motility.

This reflex pathway provides a mechanism for fine-tuning gut motility in response to local conditions, particularly distention. It represents an important interface between the enteric nervous system and the sympathetic nervous system, allowing for rapid, localized adjustments to gut function without direct central nervous system involvement (Furness et al. 2014). The inhibitory nature of this reflex is particularly noteworthy. By reducing gut motility in response to distention, it may help prevent excessive pressure buildup within the gastrointestinal tract. This could be protective against potential tissue damage or discomfort associated with overdistention.

3.2.6 Thermoregulation and the Autonomic Control of Sweating

Sweating is a critical thermoregulatory mechanism mediated by the sympathetic nervous system. This process involves cholinergic activation of muscarinic M3 receptors in the eccrine sweat glands.

Sympathetic cholinergic fibers initiate sweat production. Acetylcholine stimulates sweat glands via muscarinic M3 receptors on epithelial cells. These fibers also contain vasoactive intestinal polypeptide, which enhances sweat production through its vasodilator effects. The cholinergic phenotype of these sympathetic fibers, unlike the typical adrenergic phenotype, results from a developmental switch. This switch is triggered by a signaling pathway involving cytokine receptor–like factor 1 (CRLF1) and cytokine-like factor 1 (CLF1), both expressed in sweat glands (Crisponi et al. 2007). In certain pathological states, such as chronic regional pain syndrome, the adrenergic influence on sweat glands may increase (Chemali et al. 2001).

The sympathetic output to the skin follows a segmental distribution:

- T1–T2 segments innervate the face.
- T2–T8 innervate the upper limbs.
- T4–T12 innervate the trunk.
- T10–L2 innervate the lower limbs.

The autonomic nervous system maintains a core body temperature around 37 °C, known as the thermoneutral zone. This requires a precise balance between heat production and loss. The temperature control network includes areas within the hypothalamus, periaqueductal gray, and medullary nucleus raphe pallidus. This network receives input from peripheral (skin) and central receptors, notably warm-sensitive neurons in the preoptic-anterior hypothalamus. Various factors influence warm-sensitive neuron activity, including circadian rhythm, hormones, plasma cytokines, glucose levels, and osmolarity.

Thermoregulation involves both autonomic and behavioral mechanisms. Autonomic thermoregulation includes heat production (shivering and non-shivering thermogenesis) and heat dissipation (sweating and vasomotor control). Behavioral thermoregulation involves environmental control and physical activity adjustments.

Sweating is humans' most effective heat dissipation method. Postganglionic sympathetic fibers innervate two to four million sweat glands segmentally. These glands are classified as apocrine or eccrine, with eccrine glands primarily responsible for thermoregulation. Sweat glands consist of a secretory coil in the lower dermis and a duct to the skin surface. Sweat is hypotonic relative to plasma due to sodium and chloride reabsorption in the duct.

Generalized hyperhidrosis often indicates an underlying condition, while focal primary hyperhidrosis is typically idiopathic, peaking in the second and third decades of life. Hypohidrosis frequently occurs as part of other autonomic disorders. Sweating patterns vary during sleep stages. It increases with deepening non-rapid eye movement (NREM) sleep but reduces during rapid eye movement (REM) sleep, except during emotionally charged dreams.

References

Andersson KE. Mechanisms of penile erection and basis for pharmacological treatment of erectile dysfunction. Pharmacol Rev. 2011;63(4):811–59.

Anlauf M, Schäfer MK, Eiden L, Weihe E. Chemical coding of the human gastrointestinal nervous system: cholinergic, VIPergic, and catecholaminergic phenotypes. J Comp Neurol. 2003;459(1):90–111.

Armour JA. Potential clinical relevance of the 'little brain' on the mammalian heart. Exp Physiol. 2008;93(2):165–76.

Armour JA, Murphy DA, Yuan BX, Macdonald S, Hopkins DA. Gross and microscopic anatomy of the human intrinsic cardiac nervous system. Anat Rec. 1997;247(2):289–98.

Bandler R, Keay KA, Floyd N, Price J. Central circuits mediating patterned autonomic activity during active vs. passive emotional coping. Brain Res Bull. 2000;53(1):95–104.

Barker RA, Barasi S. Neuroscience at a glance. Oxford: Blackwell Science; 1999.

Benarroch EE. The central autonomic network: functional organization, dysfunction, and perspective. Mayo Clin Proc. 1993;68(10):988–1001.

Benarroch EE. Central autonomic network: functional organization and clinical correlations. Armonk: Futura Pub. Co.; 1997.

Benarroch EE. Enteric nervous system: functional organization and neurologic implications. Neurology. 2007;69(20):1953–7.

Benarroch EE. Mechanisms of autonomic regulation: clinical implications. Mayo Clin Proc. 2009;84(10):885–98.

Benarroch E. Autonomic neurology. Oxford: Oxford University Press; 2014.

Bharucha AE. Pelvic floor: anatomy and function. Neurogastroenterol Motil. 2006;18(7):507–19.

Bitsios P, Prettyman R, Szabadi E. Changes in autonomic function with age: a study of pupillary kinetics in healthy young and old people. Age Ageing. 1996;25(6):432–8.

Brodal P. The central nervous system: structure and function. 4th ed. New York: Oxford University Press; 2010.

Burnett AL. Nitric oxide in the penis: physiology and pathology. J Urol. 1997;157(1):320–4.

Burnstock G. Cotransmission. Curr Opin Pharmacol. 2004;4(1):47–52.

Burnstock G. Purinergic signalling: therapeutic developments. Front Pharmacol. 2017;8:661.

Cabot JB. Some principles of the spinal organization of the sympathetic preganglionic outflow. Prog Brain Res. 1996;107:29–42.

Cardinali DP. Autonomic nervous system: basic and clinical aspects. Cham: Springer; 2018.

Cersosimo MG, Benarroch EE. Neural control of the gastrointestinal tract: implications for Parkinson disease. Mov Disord. 2008;23(8):1065–75.

Chemali KR, Gorodeski R, Chelimsky TC. Alpha-adrenergic supersensitivity of the sudomotor nerve in complex regional pain syndrome. Ann Neurol. 2001;49(4):453–9.

Cheng Z, Powley TL, Schwaber JS, Doyle FJ 3rd. Projections of the dorsal motor nucleus of the vagus to cardiac ganglia of rat atria: an anterograde tracing study. J Comp Neurol. 1999;410(2):320–41.

Coote JH. Myths and realities of the cardiac vagus. J Physiol. 2013;591(17):4073–85.

Costa M, Brookes SJ. The enteric nervous system. Am J Gastroenterol. 1994;89(8 Suppl):S129–37.

Costa M, Wiklendt L, Simpson P, Spencer NJ, Brookes SJ, Pinning PG. Neuromechanical factors involved in the formation and propulsion of fecal pellets in the Guinea-pig colon. Neurogastroenterol Motil. 2015;27(10):1466–77.

Craig AD. Interoception: the sense of the physiological condition of the body. Curr Opin Neurobiol. 2003;13(4):500–5.

Crisponi L, Crisponi G, Meloni A, Toliat MR, Nurnberg G, Usala G, et al. Crisponi syndrome is caused by mutations in the CRLF1 gene and is allelic to cold-induced sweating syndrome type 1. Am J Hum Genet. 2007;80(5):971–81.

Critchley HD. Neural mechanisms of autonomic, affective, and cognitive integration. J Comp Neurol. 2005;493(1):154–66.

Critchley HD, Harrison NA. Visceral influences on brain and behavior. Neuron. 2013;77(4):624–38.

de Groat WC. Integrative control of the lower urinary tract: preclinical perspective. Br J Pharmacol. 2006;147(Suppl 2):S25–40.

de Groat WC, Griffiths D, Yoshimura N. Neural control of the lower urinary tract. Compr Physiol. 2015;5(1):327–96.

Dean RC, Lue TF. Physiology of penile erection and pathophysiology of erectile dysfunction. Urol Clin North Am. 2005;32(4):379–95.

DiBona GF. Physiology in perspective: the wisdom of the body. Neural control of the kidney. Am J Physiol Regul Integr Comp Physiol. 2005;289(3):R633–41.

Drake RL, Vogl AW, Mitchell AWM. Gray's anatomy for students. 4th ed. Philadelphia: Elsevier; 2020.

Drummond PD. Sweating and vascular responses in the face: normal regulation and dysfunction in migraine, cluster headache and harlequin syndrome. Clin Auton Res. 1994;4(5):273–85.

Drummond PD. Mechanisms of autonomic disturbance in the face during and between attacks of cluster headache. Cephalalgia. 2006;26(6):633–41.
Edvinsson L. Role of CGRP in migraine. Handb Exp Pharmacol. 2019;255:121–30.
Fahrenkrug J. VIP and autonomic neurotransmission. Pharmacol Ther. 1989;41(3):515–34.
Florea VG, Cohn JN. The autonomic nervous system and heart failure. Circ Res. 2014;114(11):1815–26.
Fowler CJ, Griffiths D, de Groat WC. The neural control of micturition. Nat Rev Neurosci. 2008;9(6):453–66.
Freeman R, Wieling W, Axelrod FB, Benditt DG, Benarroch E, Biaggioni I, et al. Consensus statement on the definition of orthostatic hypotension, neurally mediated syncope and the postural tachycardia syndrome. Clin Auton Res. 2011;21(2):69–72.
Furness JB. The enteric nervous system. 1st ed. Oxford: Blackwell Publishing; 2006.
Furness JB. The enteric nervous system and neurogastroenterology. Nat Rev Gastroenterol Hepatol. 2012;9(5):286–94.
Furness JB, Callaghan BP, Rivera LR, Cho HJ. The enteric nervous system and gastrointestinal innervation: integrated local and central control. Adv Exp Med Biol. 2014;817:39–71.
Gibbins I. Functional organization of autonomic neural pathways. Organogenesis. 2013;9(3):122–8.
Gibbins IL, Jobling P, Morris JL. Functional organization of peripheral vasomotor pathways. Acta Physiol Scand. 2003;177(3):237–45.
Giuliano F, Clement P. Neuroanatomy and physiology of ejaculation. Annu Rev Sex Res. 2005;16:190–216.
Golanov EV, Reis DJ. Contribution of oxygen-sensitive neurons of the rostral ventrolateral medulla to hypoxic cerebral vasodilatation in the rat. J Physiol. 1996;495(Pt 1):201–16.
Goldstein DS, et al. Cardiac sympathetic dysautonomia in chronic orthostatic intolerance syndromes. Circulation. 2002;106(18):2358–65.
Goyal RK, Hirano I. The enteric nervous system. N Engl J Med. 1996;334(17):1106–15.
Griffiths D. Neural control of micturition in humans: a working model. Nat Rev Urol. 2015;12(12):695–705.
Hall JE. Guyton and Hall textbook of medical physiology. 13th ed. Philadelphia: Elsevier; 2015.
Hopkins DA, Bieger D, deVente J, Steinbusch WM. Vagal efferent projections: viscerotopy, neurochemistry and effects of vagotomy. Prog Brain Res. 1996;107:79–96.
Huizinga JD, Lammers WJ. Gut peristalsis is governed by a multitude of cooperating mechanisms. Am J Physiol Gastrointest Liver Physiol. 2009;296(1):G1–8.
Jänig W. The integrative action of the autonomic nervous system: neurobiology of homeostasis. Cambridge: Cambridge University Press; 2006.
Katirji B, Kaminski HJ, Ruff RL. Neuromuscular disorders in clinical practice. 2nd ed. New York: Springer; 2014.
Kiernan JA, Rajakumar N. Barr's the human nervous system: an anatomical viewpoint. 10th ed. Philadelphia: Wolters Kluwer; 2014.
Knowles CH, Lindberg G, Panza E, De Giorgio R. New perspectives in the diagnosis and management of enteric neuropathies. Nat Rev Gastroenterol Hepatol. 2013;10(4):206–18.
Krassioukov A, Biering-Sørensen F, Donovan W, Kennelly M, Kirshblum S, Krogh K, et al. International standards to document remaining autonomic function after spinal cord injury. J Spinal Cord Med. 2012;35(4):201–10.
Kvetnansky R, Sabban EL, Palkovits M. Catecholaminergic systems in stress: structural and molecular genetic approaches. Physiol Rev. 2009;89(2):535–606.
Levy MN. Sympathetic-parasympathetic interactions in the heart. Circ Res. 1971;29(5):437–45.
Loewenfeld IE. The pupil: anatomy, physiology, and clinical applications. Oxford: Butterworth-Heinemann; 1999.
Low PA, Benarroch EE. Clinical autonomic disorders. 3rd ed. Philadelphia: Wolters Kluwer; 2008.
Low PA, Dyck PJ. Splanchnic preganglionic neurons in man. III. Morphometry of myelinated fibers of rami communicantes. J Neuropathol Exp Neurol. 1978;37(6):734–40.
Lundberg JM. Pharmacology of cotransmission in the autonomic nervous system: integrative aspects on amines, neuropeptides, adenosine triphosphate, amino acids and nitric oxide. Pharmacol Rev. 1996;48(1):113–78.
Mathias CJ. Autonomic diseases: management. J Neurol Neurosurg Psychiatry. 2003;74(Suppl 3):iii42–7.
Matsuo R, Garrett JR. Neural and vascular control of salivary glands. In: Ekberg O, editor. Dysphagia. Springer; 2019. p. 255–72.
McAllen RM, Spyer KM. The location of cardiac vagal preganglionic motoneurones in the medulla of the cat. J Physiol. 1976;258(1):187–204.
McAllen RM, Salo LM, Paton JFR, Pickering AE. Processing of central and reflex vagal drives by rat cardiac ganglion neurones: an intracellular analysis. J Physiol. 2011;589(Pt 23):5801–18.
McCorry LK. Physiology of the autonomic nervous system. Am J Pharm Educ. 2007;71(4):78.
McDougal DH, Gamlin PD. Autonomic control of the eye. Compr Physiol. 2015;5(1):439–73.
Moskowitz MA. The neurobiology of vascular head pain. Ann Neurol. 1984;16(2):157–68.
Olshansky B, Sabbah HN, Hauptman PJ, Colucci WS. Parasympathetic nervous system and heart failure: pathophysiology and potential implications for therapy. Circulation. 2008;118(8):863–71.
Patel YC. Somatostatin and its receptor family. Front Neuroendocrinol. 1999;20(3):157–98.
Patestas MA, Gartner LP. A textbook of neuroanatomy. Oxford: Blackwell publishing; 2006.

Paxinos G. The human nervous system. 4th ed. San Diego: Academic Press; 1990.
Porges SW. The polyvagal perspective. Biol Psychol. 2007;74(2):116–43.
Powley TL. Vagal input to the enteric nervous system. Gut. 2000;47(Suppl 4):iv30–2.
Powley TL, Hudson CN, McAdams JL, Baronowsky EA, Phillips RJ. Vagal intramuscular arrays: the specialized mechanoreceptor arbors that innervate the smooth muscle layers of the stomach examined in the rat. J Comp Neurol. 2016;524(4):713–37.
Powley TL, Jaffey DM, McAdams J, Baronowsky EA, Black D, Chesney L, et al. Vagal innervation of the stomach reassessed: brain-gut connectome uses smart terminals. Ann N Y Acad Sci. 2019;1454(1):14–30.
Proctor GB, Carpenter GH. Regulation of salivary gland function by autonomic nerves. Auton Neurosci. 2007;133(1):3–18.
Purves D, Augustine GJ, Fitzpatrick D, Hall WC, LaMantia A-S, Mooney RD, et al. Neuroscience. 6th ed. Sunderland: Sinauer Associates; 2018.
Rao SS. Pathophysiology of adult fecal incontinence. Gastroenterology. 2004;126(1 Suppl 1):S14–22.
Robertson D, Biaggioni I, Burnstock G, Low PA, Paton JFR. Primer on the autonomic nervous system. 3rd ed. Amsterdam: Elsevier; 2012.
Sanders KM, Ward SM, Koh SD. Interstitial cells: regulators of smooth muscle function. Physiol Rev. 2014;94(3):859–907.
Saper CB. The central autonomic nervous system: conscious visceral perception and autonomic pattern generation. Annu Rev Neurosci. 2002;25:433–69.
Sato K, Kang WH, Saga K, Sato KT. Biology of sweat glands and their disorders. I. Normal sweat gland function. J Am Acad Dermatol. 1989;20(4):537–63.
Squire LR, Berg D, Bloom F, du Lac S, Ghosh A, Spitzer N. Fundamental neuroscience. 3rd ed. Amsterdam: Academic Press; 2008.
Standring S. Gray's anatomy: the anatomical basis of clinical practice. 42nd ed. London: Elsevier; 2020.
Strominger NL, Demarest RJ, Laemle LB. Noback's human nervous system: structure and function. 7th ed. New York: Humana Press; 2012.
Struhal W, Lahrmann H, Fanciulli A, Wenning GK. Bedside approach to autonomic disorders: a clinical tutor. Cham: Springer; 2017.
Szabadi E. Functional neuroanatomy of the central noradrenergic system. J Psychopharmacol. 2013;27(8):659–93.
Szurszewski JH, Linden DR. Physiology of prevertebral sympathetic ganglia. In: Johnson LR, editor. Physiology of the gastrointestinal tract. 5th ed. London: Academic Press; 2012. p. 583–627.
Tate DG, Forchheimer MB, Krause JS, Meade MA, Bombardier CH. Patterns of alcohol and substance use and abuse in persons with spinal cord injury: risk factors and correlates. Arch Phys Med Rehabil. 2004;85(11):1837–47.
Toda N, Ayajiki K, Okamura T. Cerebral blood flow regulation by nitric oxide: recent advances. Pharmacol Rev. 2009;61(1):62–97.
Travagli RA, Hermann GE, Browning KN, Rogers RC. Brainstem circuits regulating gastric function. Annu Rev Physiol. 2006;68:279–305.
Triposkiadis F, Karayannis G, Giamouzis G, Skoularigis J, Louridas G, Butler J. The sympathetic nervous system in heart failure physiology, pathophysiology, and clinical implications. J Am Coll Cardiol. 2009;54(19):1747–62.
Valenza G, Sclocco R, Duggento A, Passamonti L, Napadow V, Barbieri R, et al. The central autonomic network at rest: uncovering functional MRI correlates of time-varying autonomic outflow. Neuroimage. 2019;197:383–90.
Vernino S, Low PA, Fealey RD, Stewart JD, Farrugia G, Lennon VA. Autoantibodies to ganglionic acetylcholine receptors in autoimmune autonomic neuropathies. N Engl J Med. 2000;343(12):847–55.
Vernino S, Hopkins S, Wang Z. Autonomic ganglia, acetylcholine receptor antibodies, and autoimmune ganglionopathy. Auton Neurosci. 2009;146(1–2):3–7.
Wake E, Brack K. Characterization of the intrinsic cardiac nervous system. Auton Neurosci. 2016;199:3–16.
Wallin BG, Charkoudian N. Sympathetic neural control of integrated cardiovascular function: insights from measurement of human sympathetic nerve activity. Muscle Nerve. 2007;36(5):595–614.
Watson C, Kirkcaldie M, Paxinos G. The brain: an introduction to functional neuroanatomy. 1st ed. San Diego: Academic Press; 2010.
Wehrwein EA, Joyner MJ. Regulation of blood pressure by the arterial baroreflex and autonomic nervous system. Handb Clin Neurol. 2013;117:89–102.
Weihe E, Schütz B, Hartschuh W, Anlauf M, Schäfer MK, Eiden LE. Coexpression of cholinergic and noradrenergic phenotypes in human and nonhuman autonomic nervous system. J Comp Neurol. 2005;492(3):370–9.
Wilhelm H. The pupil. Curr Opin Neurol. 2008;21(1):36–42.
Wilhelm B, Wilhelm H, Lüdtke H, Streicher P, Adler M. Pupillographic assessment of sleepiness in sleep-deprived healthy subjects. Sleep. 1998;21(3):258–65.
Wood JD. Enteric nervous system: reflexes, pattern generators and motility. Curr Opin Gastroenterol. 2008;24(2):149–58.
Zukowska-Grojec Z. Neuropeptide Y: a novel sympathetic stress hormone and more. Ann N Y Acad Sci. 2015;771:219–33.

Recommended Additional Reading

Parent A. Carpenter's human neuroanatomy. 9th ed. Philadelphia: Lippincott Williams & Wilkins; 1996.
ten Donkelaar HJ. Clinical neuroanatomy: brain circuitry and its disorders. New York: Springer; 2011.

4 Anatomy and Function of the Skeletal Muscle, Skin, and Fascia

The human body's capacity for movement, self-protection, and structural integrity is fundamentally dependent on the complex interaction of skeletal muscles, skin, and fascia. This chapter examines these three fundamental tissue types, analyzing their structures, functions, and biomechanical characteristics.

Skeletal muscles form more than one-third of the body's total mass, serving as the primary drivers of movement and postural control. These muscles exhibit a diverse range of forms, from flat, sheet-like structures to compact, thick configurations. Their non-branching, striated fibers are enveloped by connective tissue that provides both support and flexibility. Understanding the skeletal muscle architecture is crucial, as the length and arrangement of muscle fibers directly influence their functional capabilities and force generation potential.

The skin, as the body's largest organ by surface area, provides an essential protective barrier while facilitating crucial functions like temperature regulation and sensory perception. Its complex structure, comprising multiple layers from the outermost epidermis to the deeper dermis, allows it to adapt to the body's constant movement while maintaining its protective properties. The skin's biomechanical characteristics become particularly important when considering pressure-related injuries and wound healing processes.

Fascia, often overlooked but critically important, creates an interconnected network throughout the body. This fibrous connective tissue exists in various forms, from the superficial layers just beneath the skin to the deep fascia that compartmentalizes and supports muscle groups. Fascia not only provides structural support but also plays essential roles in force transmission, protection of vital structures, and facilitation of fluid movement within the body.

Together, these three components form an integrated system that enables human movement, provides protection, and maintains bodily integrity. This chapter examines each component in detail, revealing how its individual properties and collective interactions contribute to overall body function and performance.

4.1 Structure and Functional Properties of the Skeletal Musculature

Skeletal muscles constitute approximately 40%–45% of the total human body mass, forming a fundamental component of the musculoskeletal system. These muscles exhibit a characteristic organization of non-branching, striated muscle fibers, which are enclosed by areolar connective tissue (Purslow 2020). The morphological features of skeletal muscles display considerable

H.-Y. Ko, *Practical Functional Anatomy for Spinal Cord Injury Rehabilitation*,
https://doi.org/10.1007/978-981-95-6457-6_4

variation throughout the body, including sheet-like arrangements, compact cylindrical formations, and elongated strap-like structures (Standring 2020).

The functional properties of skeletal muscles depend on their architectural design and fiber arrangement patterns. Research demonstrates that muscle fibers can decrease their resting length by 57%–63% during maximal contraction, with the exact percentage determined by the specific arrangement of muscle fibers and their pennation angles (Lieber and Fridén 2000; Lieber and Ward 2011). This contractile capability affects both the force generation and the mechanical efficiency of the muscle. Despite the variability in fiber orientation across different muscles, all skeletal muscle movements are fundamentally driven by contraction, which involves a reduction in the length of muscle fibers. This contractile process generates a tensile force that is transmitted across articular junctions, resulting in positional changes of the articulating bones and subsequently facilitating movement. The precise nature of this movement is determined by the muscle's origin and insertion and the type of joint it spans (Moore et al. 2018). The functional capacity of skeletal muscles is intricately linked to their physiological and biochemical properties. These properties include the muscle fiber type composition, mitochondrial density, and capillary network, which collectively influence the muscle's force production, endurance, and metabolic characteristics (Frontera and Ochala 2015).

At the cellular level, skeletal muscles contain multiple specialized components that determine their physiological capabilities. The sarcomere, as the basic functional unit, contains organized arrays of thick myosin and thin actin filaments. These filaments interact through cross-bridge formation during muscle contraction, following the sliding filament theory. The density and organization of these contractile proteins directly influence the muscle's force-producing capabilities (Herzog 2017).

The metabolic characteristics of skeletal muscles vary based on their fiber type composition. Type I fibers demonstrate high oxidative capacity and fatigue resistance, while Type IIa and IIx fibers exhibit progressively higher glycolytic capacity and force production. The distribution of these fiber types within a muscle reflects its primary functional role and adapts in response to different activity patterns (Egan and Zierath 2013).

Skeletal muscle tissues also participate in metabolic regulation through the secretion of myokines, which influence whole-body metabolism and tissue cross talk (Whitham and Febbraio 2016). Additionally, skeletal muscles demonstrate remarkable adaptability to various stimuli, including mechanical loading, neural activation patterns, and metabolic demands (Wilson et al. 2012).

4.1.1 Muscle Architecture, Fiber Arrangement, and Functions in the Skeletal Muscle

The arrangement of muscle fibers within skeletal muscles determines their mechanical and functional properties. Two main architectural patterns exist in muscle fiber organization: parallel and pennate arrangements. These patterns affect the biomechanical properties and functional capabilities of muscles according to their anatomical requirements (Lieber and Fridén 2000; Lieber et al. 2017).

Muscles with parallel fiber arrangement align their fibers along the muscle's line of pull. This pattern appears in two primary forms (Fig. 4.1). The first form presents a fusiform structure, as seen in the biceps brachii muscle. This configuration allows extensive shortening during contraction, enabling a wide range of motion, although it produces relatively lower force output compared to other arrangements (Standring 2020). The second form shows a flat, sheet-like arrangement, which appears in muscles such as the external oblique of the abdomen. This pattern provides extensive coverage and functions effectively in postural maintenance and body cavity stabilization (Moore et al. 2018).

Pennate muscles display fibers oriented at angles to the line of pull, allowing more fibers to

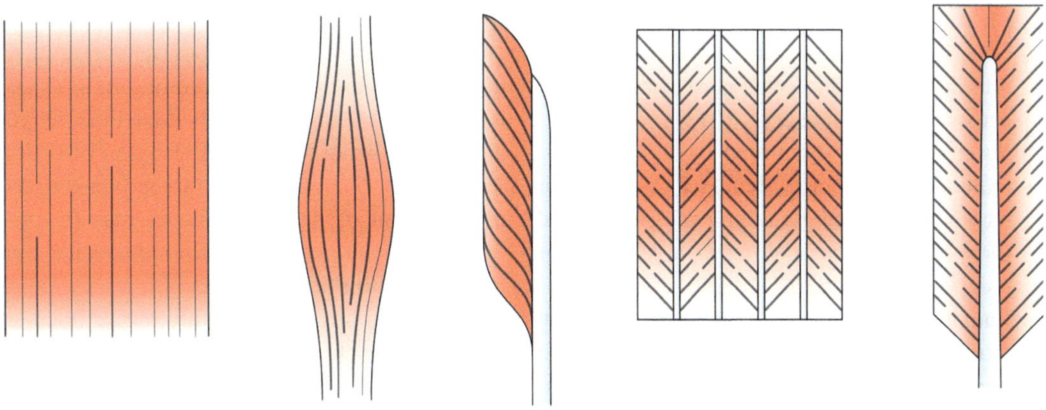

Fig. 4.1 Muscle shapes categorized into five distinct types. (Adapted from McGuinness (2018))

pack into a given volume. This results in enhanced force production, though with reduced contraction distance. Three main types of pennate arrangements exist (*see* Fig. 4.1). Unipennate muscles, such as the flexor pollicis longus and semitendinosus, attach their fibers to one side of the tendon. Bipennate muscles, including the rectus femoris and gastrocnemius, feature a central septum with fiber attachments on both sides, leading to a continuous central tendon. Multipennate muscles, exemplified by the deltoid, contain multiple intermediate septa with bipennate arrangements, maximizing the power output and force generation efficiency (Nishikawa et al. 2012; Wickiewicz et al. 1983).

The architectural variations in muscle fiber arrangements produce specific functional outcomes. Pennate arrangements generate greater force due to increased physiological cross-sectional area (CSA) (Gans and de Vree 1987). Parallel fiber arrangements permit larger movement ranges through extensive fiber shortening. Muscles with longer fibers, commonly found in parallel arrangements, achieve faster contraction velocities over greater distances. Pennate arrangements provide mechanical advantages through efficient force transmission to tendons.

Studies have shown that muscle architecture adapts to functional demands. These adaptations include changes in the pennation angle, fiber length, and physiological cross-sectional area in response to mechanical loading and activity patterns (Timmins et al. 2016).

4.1.1.1 Parallel Fiber Arrangements

The arrangement of skeletal muscle fibers plays a key role in their function, with parallel fiber patterns representing a foundational morphological structure. Some muscles, such as the intercostales interni, are relatively short, while others, like the pronator quadratus, rhomboids, and quadratus lumborum, are longer. At the extreme length are muscles such as the sartorius, gracilis, pectoralis minor, rectus abdominis, obliquus externus abdominis, and semitendinosus. Certain parallel-fibered muscles, including the rectus abdominis and semitendinosus, contain tendinous intersections, which serve as insertion points for muscle fibers; for example, the rectus abdominis typically has three such intersections. Muscle fascicles aligned along the long axis of the muscle exhibit five distinct forms, each adapted to meet specific functional demands (Hamill et al. 2015).

- Flat parallel muscles, such as the rectus abdominis and external oblique, are thin and broad, originating from sheet-like aponeuroses. This configuration allows for force distribution over a larger area, making them ideal for maintaining posture and stabilizing body cavities.
- Fusiform muscles like the biceps brachii, brachialis, and brachioradialis taper at both ends,

focusing force transmission to small bony sites. This shape is particularly effective for muscles that need to generate significant force across joints (Moore et al. 2018).
- Strap muscles, exemplified by the sartorius, maintain a uniform diameter along their length. This configuration enables targeted force transmission, making them suitable for precise movements.
- Radiate or convergent arrangement, seen in muscles like the pectoralis major and trapezius, combines aspects of flat and fusiform shapes. These muscles originate from broad aponeuroses and converge onto a tendon, allowing for multidirectional force generation—a valuable trait for muscles involved in complex movements of the shoulder girdle (Standring 2020).
- Circular muscles, such as the orbicularis oris surrounding the mouth, represent a unique concentric arrangement of strap-like fibers. Their primary function is to close these openings upon contraction, demonstrating how muscle architecture is finely adapted to specific physiological functions.

The biomechanical implications of parallel fiber arrangements are significant. In these muscles, fiber force aligns directly with the muscle's line of action, resulting in a greater range of shortening and higher movement velocity at the joint level. This is primarily because parallel muscles are often longer than other types, with muscle fibers extending farther than the tendon. For instance, in fusiform muscles like the biceps brachii, the fiber length can be equivalent to the muscle length (Wickiewicz et al. 1983).

4.1.1.2 Pennate Fiber Arrangements

In the pennate fiber arrangement, the muscle fibers are oriented diagonally, relative to a central tendon that extends along the length of the muscle. The overall shape of a pennate muscle resembles a feather, characterized by short fascicles angled in relation to the longitudinal axis of the muscle (Lieber and Fridén 2000). Due to the angular orientation of fibers in pennate muscles, the force generated by each fiber deviates from the direction of the line of pull of the muscle. The fibers in these muscles are shorter than the overall muscle length, resulting in a disparity between changes in individual fiber length and alterations in total muscle length (Wakeling et al. 2011). The primary factors influencing the development of pennate muscles may be predominantly geometrical: The total surface area of the skeleton is insufficient to accommodate the necessary origin sites for all muscle fibers in the body. Muscle fibers must secure their attachment to the limited skeletal surface area. This is accomplished by expanding the skeletal surface through muscle origin aponeuroses (Leijnse 1997). In pennate muscles, the individual muscle fibers apply force in parallel to the aponeurosis. However, the transmission of force along the aponeurosis occurs sequentially, with each fiber's force contributing to the cumulative force exerted by the preceding fibers. Consequently, sections of the aponeurosis (internal tendon) situated nearer to the external muscle tendon bear greater forces. This phenomenon accounts for the progressive thickening of the aponeurosis from the center of the muscle toward its periphery (Fig. 4.2).

Compared to parallel and fusiform muscles, pennate muscles offer both mechanical advantages and disadvantages. The main advantage lies in their ability to pack more muscle fibers into a given volume, allowing the total force generated by the muscle fibers to exceed that of parallel-fibered muscles of the same size. However, because the muscle fibers are oriented at an angle to the muscle's line of action, only a portion of their force contributes to the overall muscle force. Additionally, the muscle fibers in pennate muscles are shorter and contain fewer sarcomeres in series compared to parallel muscles of the same volume, resulting in reduced maximum displacement and contraction velocities.

Pennate muscles can be categorized into three subtypes based on the pattern of fiber attachment:

- Unipennate muscles have fibers running diagonally off one side of the tendon. Examples include the biceps femoris, extensor digitorum

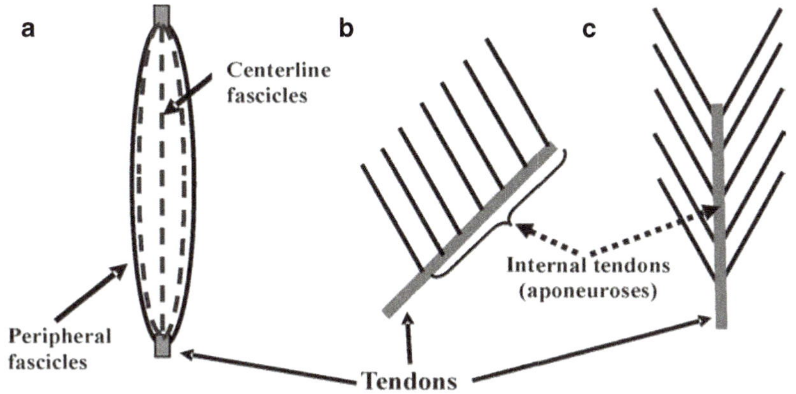

Fig. 4.2 Arrangements of muscle fibers. (**a**) Fusiform (spindle-shaped) muscle, (**b**) unipennate, and (**c**) bipennate configurations. In (**b**) and (**c**), external and internal tendons (aponeuroses) are illustrated. In fusiform muscles (**a**), fascicle length varies between the centerline and the periphery, with differences in curvature depending on the fascicle's proximity to the centerline. (Adapted from Zatsiorsky and Prilutsky (2012))

longus, flexor pollicis longus, semimembranosus, and tibialis posterior.
- Bipennate muscles feature fibers arranged on both sides of a central tendon. Examples include the rectus femoris, flexor hallucis longus, gastrocnemius, vastus medialis, vastus lateralis, and infraspinatus.
- Multipennate muscles contain fibers running diagonally from multiple central tendons, combining both unipennate and bipennate features. The deltoid and gluteus maximus are examples of multipennate muscles.

The structural features of pennate muscles influence their functional properties. Their shorter, diagonally oriented fibers typically produce slower movements with a reduced range of motion compared to fusiform muscles. However, this configuration allows for a larger physiological cross-sectional area, generally resulting in greater force production capabilities (Azizi et al. 2008).

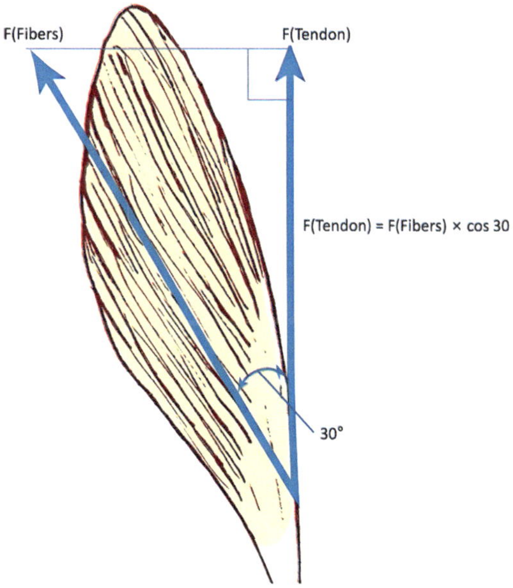

Fig. 4.3 The pennation angle, defined as the angle between the muscle fibers and the line of action of the muscle's pull. (Adapted with permission from Hamill et al. (2015))

Pennation Angle

The pennation angle refers to the angle formed between the fascicles and the line of action of the muscle (Lieber and Fridén 2000). This angle significantly influences the force transmission capabilities of pennate muscles (Fig. 4.3). As the magnitude of the pennation angle increases, the efficiency of force transmission to the tendon decreases. Notably, the pennation angle tends to increase during muscle contraction, which can lead to a reduction in the force-generating potential of the muscle.

The medial gastrocnemius, when acting on the ankle joint, exemplifies the impact of pennation angle on muscle function. When the knee is flexed to 90°, the pennation angle of the medial gastrocnemius can reach approximately 60°. In this configuration, only about half of the generated force is effectively transferred to the tendon, illustrating the mechanical disadvantage caused by large pennation angles (Narici et al. 1996). Conversely, muscles with lower pennation angles, such as the quadriceps, are less affected by this phenomenon. In these muscles, the pennation angle does not significantly impact force production and transmission (Fukunaga et al. 1997).

Muscle Volume and Cross-Sectional Area

The evaluation of muscle architecture requires several measurable parameters to assess its functional potential. These measurements can be obtained through cadaveric dissection studies, which provide direct access to muscle mass, length, and pennation angles. Modern imaging techniques, including ultrasonography and magnetic resonance imaging, now allow these parameters to be measured noninvasively in living subjects (Narici et al. 1989).

Muscle volume (MV) calculation employs a basic density-based formula, where MV (cm^3) = m/ρ, with m representing the muscle mass in grams and ρ representing the muscle density (1.056 g/cm^3). The cross-sectional area (CSA) can be derived from the muscle volume and fiber length using the formula CSA (cm^2) = MV/L, where L represents the fiber length in centimeters (Ward et al. 2009).

Research examining lower extremity muscles has documented specific volumetric measurements. The vastus lateralis shows the largest muscle volume in the thigh region at 1505 cm^3, while the soleus dominates the lower leg with 552 cm^3. Notable cross-sectional areas include the gluteus maximus at 145.7 cm^2 and the vastus medialis at 63 cm^2 (Ward et al. 2009).

Two types of cross-sectional measurements provide different architectural information. The anatomical cross section, measured perpendicular to the muscle's longitudinal axis, varies by measurement location. The physiological cross section represents the total fiber cross sections measured perpendicular to the fiber direction (Lieber and Fridén 2000).

The physiological cross-sectional area (PCSA) calculation incorporates multiple parameters: PCSA (mm^2) = $m \cos \theta/\rho L$. In this equation, m represents the muscle mass in grams, ρ denotes the muscle density, L indicates the muscle length in millimeters, and θ represents the pennation angle (Fig. 4.4).

Comparative analyses reveal significant variations in PCSA among muscles. The soleus demonstrates a PCSA of 230 cm^2, exceeding both the medial gastrocnemius (68 cm^2) and lateral gastrocnemius (28 cm^2) (Fukunaga et al. 1992). These measurements correlate directly with the force generation capacity. For example, the quadriceps femoris, with its large PCSA and short fibers, produces substantial force. In contrast, the hamstrings, characterized by smaller PCSA and longer fibers, excel in velocity-dependent actions.

4.1.1.3 Structures of an Individual Muscle

Skeletal muscles demonstrate a multilevel organizational structure that enables force generation and transmission. The muscle belly forms the main contractile area, with variations in its development across muscle groups (Lieber and Fridén 2000). The biceps brachii exhibits a pronounced belly, while others, like the wrist flexors and extensors, have less apparent ones (Winter and Challis 2010).

The epimysium, a connective tissue layer, covers the muscle and transfers tension from multiple generation points to the tendon and bone (Purslow 2010; Huijing 2009). Inside the epimysium, muscles contain numerous fibers—ranging from hundreds to tens of thousands—arranged in functional units called fascicles. Each fascicle, containing up to 200 fibers, has a perimysium covering that provides structural support and forms channels for blood vessels and nerves (Kjær 2004; Wang et al. 2012).

The epimysium and perimysium form a connective tissue network that determines the muscle's elastic characteristics. This structural arrangement allows muscle elongation under tension and return to resting length, making it

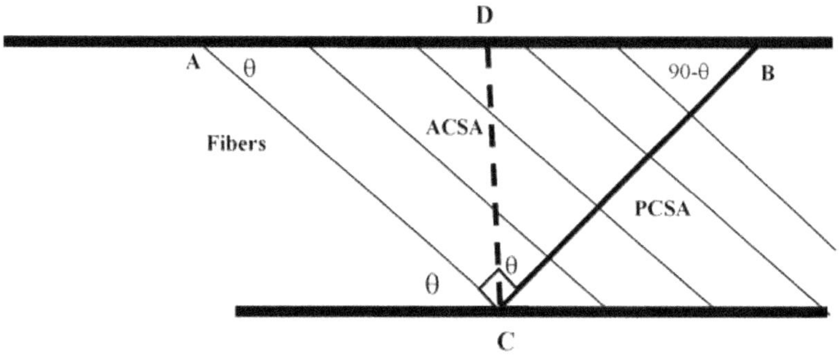

Fig. 4.4 Calculation of projected physiological cross-sectional area (PCSA). In a right triangle ABC, vertices A, B, and C are defined, where side AC represents a muscle fiber and side BC represents the physiological cross-sectional area (PCSA). Angle C is a right angle, and the pennation angle, θ, determines the remaining angles: θ and $90°–\theta$. From vertex C, an altitude is drawn to point D, forming triangle CDB. In this triangle, angle C is θ, and the projected PCSA is calculated as PCSA × cos θ. In the example, the projected PCSA equals the anatomical cross-sectional area (ACSA), though discrepancies may arise. (Adapted from Zatsiorsky and Prilutsky (2012))

relevant for flexibility training research and applications (Kjær 2004).

The cellular organization shows parallelly arranged fascicles containing cylindrical muscle fibers, which function as force-generating units. These fibers measure 10–100 μm in diameter and 15–30 cm in length, with the sarcolemma membrane and endomysium providing structural boundaries (Franchi et al. 2017; Gillies and Lieber 2011). Force transmission occurs through the basal lamina, which connects these structural elements (Fig. 4.5).

The endomysium contains capillaries and nerves that supply each muscle fiber. These neurovascular components enter near the muscle's midpoint and extend throughout the endomysial network. Research indicates that the endomysium provides electrical isolation for neural activity within the muscle (Frontera and Ochala 2015; Purslow 2020).

Microscopic examination reveals that muscle fibers contain myofibrils—rod-shaped structures extending the fiber length. These myofibrils, 1–2 μm in diameter, make up 80% of fiber volume. The remaining 20% contains organelles: mitochondria, sarcoplasm, sarcoplasmic reticulum, and T-tubules (Frontera and Ochala 2015; Rassier 2017) (Fig. 4.6). Myofibrils show alternating light and dark bands, reflecting the organization of sarcomeres—the basic contractile units. Z-membranes (Z-lines or Z-discs) mark sarcomere boundaries. Dark bands contain thick myosin filaments, while light bands hold thin actin filaments (Fig. 4.7). These sarcomeres form chains along each myofibril and generate muscle force through their structural arrangement (Herzog et al. 2015a, b; Huxley 1957).

Sarcomere function depends on precise protein arrangements and calcium regulation. Studies using electron microscopy and X-ray diffraction have revealed additional details about cross-bridge formation between actin and myosin filaments (Squire 2016). This molecular organization determines force production capabilities and influences muscle performance under different loading conditions (Rassier 2017).

4.1.2 Sensory Mechanisms and Reflexes in the Musculoskeletal System

The musculoskeletal system uses an extensive sensory network that provides continuous feedback about its physiological status and dynamic properties (Proske and Gandevia 2012). This network includes specialized receptors that monitor multiple parameters: muscle stretch, temperature variations, mechanical pressure, contractile tension, and nociceptive signals in the extremities. The sensory information collected by these receptors moves through primary afferent neu-

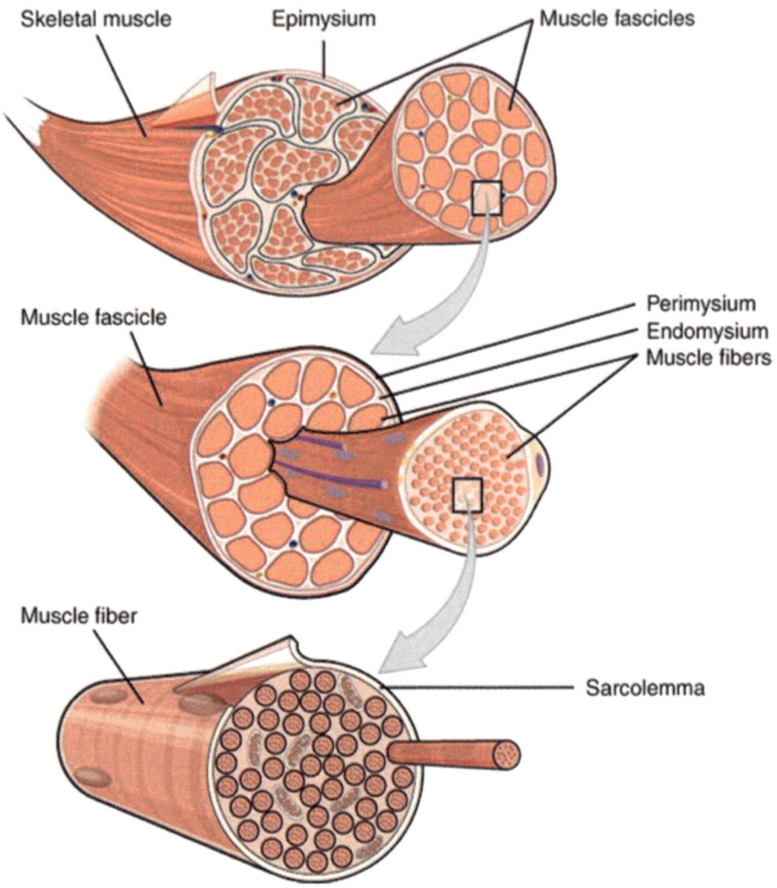

Fig. 4.5 Connective tissue layers in skeletal muscle. The epimysium encloses the entire muscle, the perimysium surrounds bundles of muscle fibers (fascicles), and the endomysium envelops each individual muscle fiber. (Adapted with permission from Xiao (2018))

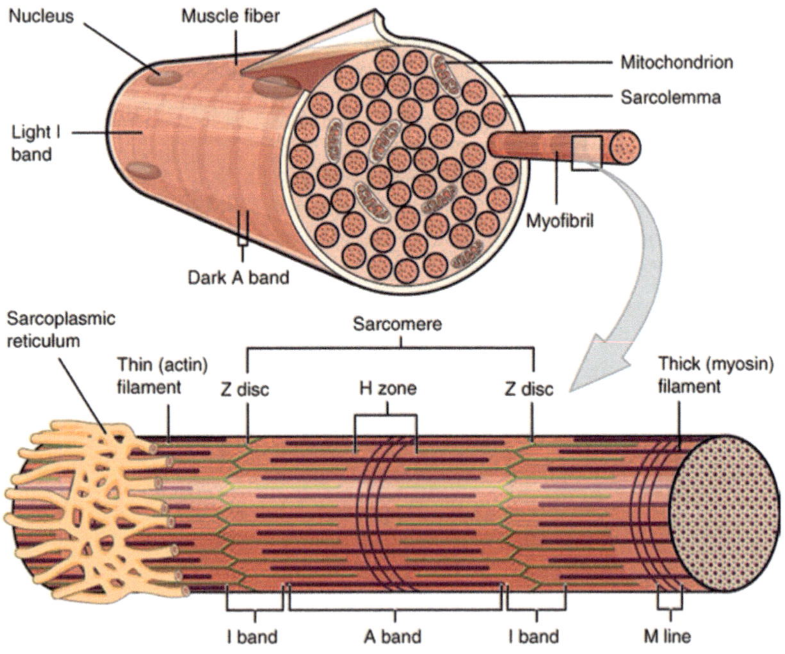

Fig. 4.6 Structure of a skeletal muscle fiber. The muscle fiber is enclosed by the sarcolemma (plasma membrane) and contains sarcoplasm, the cytoplasm of the muscle cell. The fiber is composed of numerous myofibrils arranged in a striated pattern, with each myofibril organized into repeating sarcomeres. (Adapted with permission from Xiao (2018))

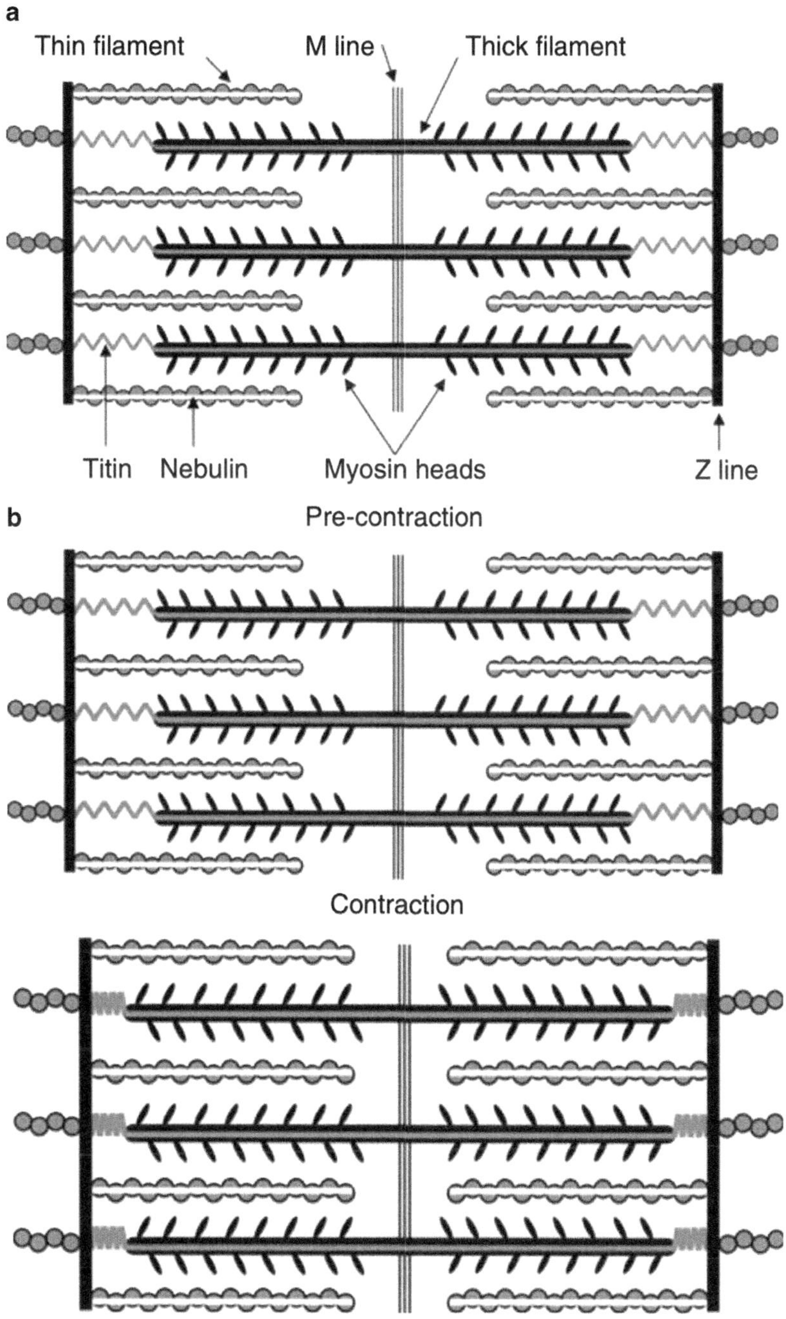

Fig. 4.7 (**a**) Schematic depiction of a sarcomere. Thin filament attach to the Z lines and facilitate the sliding of the thick filaments. Titin, a springlike protein, maintains the alignment of thick filaments within the sarcomere. (**b**) Sliding-filament model of contraction. The upper illustration shows a sarcomere in the resting state, while the lower illustration depicts sarcomere shortening during contraction, driven by myosin power stroke. The positions of the myosin heads shift, but the lengths of the thick and thin filaments remain constant. (Adapted with permission from Katirji et al. (2014))

rons to the spinal cord for initial processing before the central nervous system uses this information to generate or modify efferent signals to the muscles (Kandel et al. 2013).

Skeletal muscles contain multiple types of sensory receptors with specific physiological functions. Free nerve endings, which detect pain, show an irregular distribution pattern throughout the muscle tissue. These endings appear in high concentrations near the myotendinous junction, where mechanical stress accumulates during muscle contraction (Stecco et al. 2011). This anatomical arrangement allows complete monitoring of potential tissue damage and inflammatory responses throughout the muscle.

Golgi tendon organs (GTOs) function as mechanoreceptors within tendons. These structures contain axonal branches woven among organized collagen fiber bundles, with specialized connective tissue surrounding them. During muscular contraction, the deformation of collagen fibers activates these nerve terminals through mechanotransduction. This structural organization allows Golgi tendon organs to monitor muscle contraction intensity and force output with high precision (Jami 1992).

Muscle spindles appear as specialized proprioceptive structures within the muscle belly (Banks 2015). These receptors contain two types of modified muscle fibers, called intrafusal fibers, with specialized connective tissue around them. Unlike extrafusal fibers that produce contractile force, these intrafusal fibers divide into nuclear bag or nuclear chain fibers based on their nuclear morphology and distribution patterns. Group Ia neurons form annulospiral endings around the equatorial regions of both fiber types, while group II neurons make annulospiral terminations on nuclear chain fibers and flower-spray endings on nuclear bag fibers (Fig. 4.8). Table 4.1 summarizes the anatomical characteristics and primary functions of the intrafusal and extrafusal fibers, emphasizing their principal functional differences.

Muscle spindles respond through length-dependent mechanisms. Changes in muscle length modify the mechanical state of intrafusal fibers, which affects the activity of both Ia and II sensory neurons. This neuronal activity transmits information to the central nervous system about muscle length, its rate of change, and joint position. Different types of nuclear bag and chain fibers detect specific aspects of muscle length changes and their velocities (Prochazka and Ellaway 2012).

Muscle spindles receive efferent signals from γ-moto neurons, which form synapses with the polar regions of intrafusal fibers. When these neurons activate, they cause controlled contraction of the peripheral intrafusal fiber ends, which stretches the spindle's central sensory region and changes its mechanical sensitivity. This process allows the spindle to maintain appropriate sensitivity during various phases of muscle contraction or relaxation. Without γ-motoneuron activity, muscle spindles would only respond to muscle stretches at maximal extension (Kandel et al. 2013).

Sensory signals combine into coordinated motor responses through multiple reflex mechanisms at different nervous system levels. The patellar tendon reflex shows a monosynaptic reflex pattern where quadriceps muscle contraction matches the stimulus strength applied to the tendon. Polysynaptic reflexes include multiple sensory inputs with modification by descending and segmental factors. The flexor withdrawal reflex demonstrates this pattern, starting rapid withdrawal responses after noxious input while coordinating multiple muscle groups (Pierrot-Deseilligny and Burke 2012).

Propriospinal reflexes process information across multiple spinal segments and both sides of the spinal cord. The crossed extensor reflex responds to noxious stimuli by producing flexion in the affected limb with simultaneous extensor muscle activation in the opposite limb. This bilateral coordination maintains postural stability during protective responses (Jankowska 2008). The tonic neck reflex shows additional complexity, as head movements produce specific upper extremity responses: Head rotation results in ipsilateral arm extension with contralateral arm flexion, while sagittal head movements cause corresponding arm movements (Fukuda 1959).

Supraspinal reflexes involve extensive processing of spinal cord sensory information in brain centers before producing coordinated motor responses. The labyrinthine righting reflex responds to changes in body orientation by generating precise motor responses in the neck and limbs to maintain upright posture. This reflex combines information from multiple spinal levels and brain centers, using vestibular, visual, and proprioceptive information for postural control (Brodal 2010).

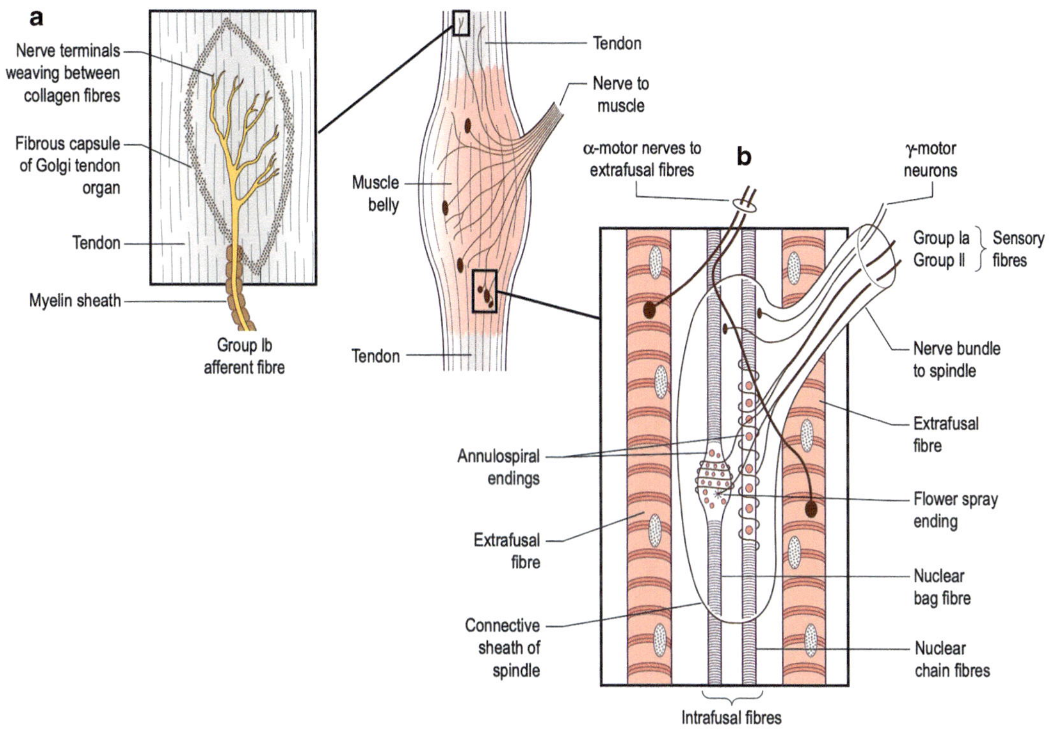

Fig. 4.8 Anatomy and positioning of sensory structures in muscle. (**a**) Golgi tendon organs, located at the junction between muscle and tendon. (**b**) Muscle spindles, embedded within the muscle fibers. (Adapted with permission from Palastanga and Soames (2012))

Table 4.1 Comparison of intrafusal and extrafusal fibers

Characteristic	Intrafusal fiber	Extrafusal fiber
Location	Within muscle spindles	Main body of the muscle
Primary function	Detect muscle length and changes (proprioception)	Muscle contraction and force generation
Innervation	Sensory (Ia, II) and motor (γ) neurons	Motor neurons (α)
Quantity	Few (dozens per muscle)	Many (comprise the bulk of the muscle)
Size	Small	Large
Types	Nuclear bag fibers, nuclear chain fibers	Type I (slow-twitch), type II (fast-twitch)
Contractile ability	Limited (for sensory function adjustment)	Strong (performs main muscle function)
Role in motor control	Provide muscle length information, regulate muscle reflexes	Execute actual movement

4.1.2.1 Muscle Spindle Organization and Function in Skeletal Muscles

Proprioceptors function as specialized sensory structures in the musculoskeletal system, detecting mechanical changes in muscle and joint position, muscle length, and tension. These structures convert mechanical stimuli into neural signals, which travel to the spinal cord to produce motor responses (Proske and Gandevia 2012). Among these, muscle spindles represent primary mechanoreceptors positioned parallel to muscle fibers within the muscle belly, maintaining connections to fascicles through connective tissue elements (Fig. 4.9).

The structural organization of muscle spindles comprises intrafusal fibers contained within a specialized capsule, setting them apart from surrounding extrafusal muscle fibers. Each spindle capsule holds up to 12 intrafusal fibers, classified

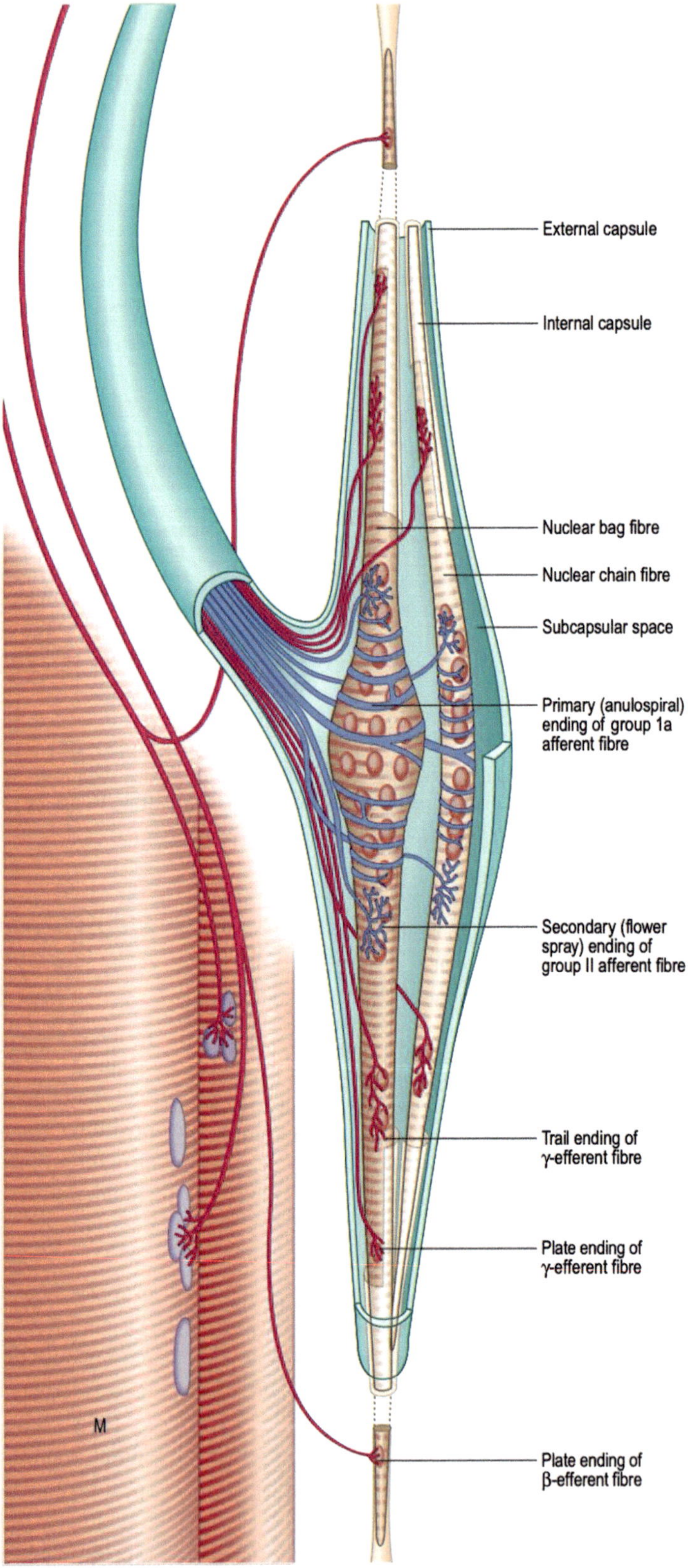

Fig. 4.9 Structure and function of the muscle spindle. The muscle spindle is arranged parallel to the surrounding muscle fibers. Each spindle capsule contains spindle fibers, classified as nuclear chain or nuclear bag fibers. Both types have contractile ends innervated by γ-motoneurons. Sensory signals generated in response to stretch are transmitted from the central region of both fiber types via type Ia sensory neurons and from the ends of nuclear chain fibers through type II sensory neurons. (Adapted with permission from Standring (2020))

into two main categories: nuclear bag and nuclear chain fibers (Bewick and Banks 2015; Manuel and Zytnicki 2011). Both types contain noncontractile central regions housing nuclei and sensory nerve endings, which send information to the central nervous system through the dorsal root of the spinal cord.

The distribution of muscle spindles varies across muscle groups. Higher concentrations appear in muscles requiring precise control, such as those in the eyes, hands, and upper back. In comparison, muscles like the latissimus dorsi and certain shoulder muscles contain fewer spindles. While most muscles contain spindles, specific type IIb fast-twitch glycolytic muscle fibers may not possess these proprioceptors (Banks 2015).

Nuclear bag fibers display a central nuclear cluster and a greater diameter, forming connections with both the spindle capsule and muscle connective tissue. In contrast, nuclear chain fibers show linearly arranged nuclei in their equatorial region and connect only with the spindle capsule (Bewick and Banks 2015).

γ-motoneurons innervate the contractile ends, initiating shortening upon neural input. These neurons, smaller than α-moto neurons, reside in the ventral horn of the spinal cord, with each γ-moto neuron innervating multiple muscle spindles (Manuel and Zytnicki 2011). Dynamic and static γ-efferents innervate nuclear bag and nuclear chain fibers, respectively, allowing precise adjustment of muscle spindle sensitivity to match movement requirements (Hamill et al. 2015) (Fig. 4.10).

Type Ia primary afferent neurons innervate the equatorial region of both nuclear bag and nuclear chain fibers, responding to changes in the spindle's central portion length. These sensory endings form connections with the dorsal horn through monosynaptic connections with α-moto neurons, resulting in contraction of the same muscle. The parallel arrangement of muscle spindles with extrafusal muscle fibers enables them to experience identical stretch forces (Proske and Gandevia 2012). Type II secondary afferent sensory neurons extend from the polar ends of nuclear chain fibers. These medium-sized neurons respond to muscle stretch at a higher threshold than type I sensory neurons (*see* Fig. 4.10). While most muscle spindles contain one or two type II sensory neurons, research indicates that 10%–20% of spindles and certain muscles lack them entirely (Banks 2015).

Primary afferents operate as velocity sensors, showing nonlinear sensitivity to muscle stretch rates. These neurons demonstrate high responsiveness to small, rapid length changes but reduced sensitivity to slower or larger alterations. Secondary afferents function primarily as muscle length sensors, maintaining some sensitivity to stretch rate changes (Manuel and Zytnicki 2011).

During muscle stretch, deformation of the intrafusal fibers' equatorial region initiates type I sensory neuron impulses to the spinal cord. These sensory action potentials interact with interneurons, generating excitatory local graded potentials that return to the stretched muscle. Sufficient stretch intensity activates muscle contraction through alpha moto neurons, forming a monosynaptic reflex arc (Bewick and Banks 2015). The stretch reflex (myotatic reflex) exemplifies this mechanism, where sensory neurons responding to muscle stretch increase motor input to the same muscle. This reflex pathway predominantly activates slow-twitch muscle fibers (Ellaway et al. 2015).

Type I sensory neurons transmit information to both cerebellar and cerebral sensory areas, providing comprehensive feedback about muscle length and velocity. Within the spinal cord, these signals connect with inhibitory interneurons, enabling reciprocal inhibition of antagonistic muscles. Additionally, connections form with excitatory interneurons linked to α-moto neurons of synergistic muscles, enhancing coordinated muscle activation (Proske and Gandevia 2012).

γ-moto neuron innervation prevents spindle discharge cessation during muscle shortening. The coactivation of α and γ systems maintains spindle tension, enabling continued position and length information transmission during muscle shortening. This mechanism functions primarily in slow movements and loaded conditions, while fast movements may rely more on antagonistic

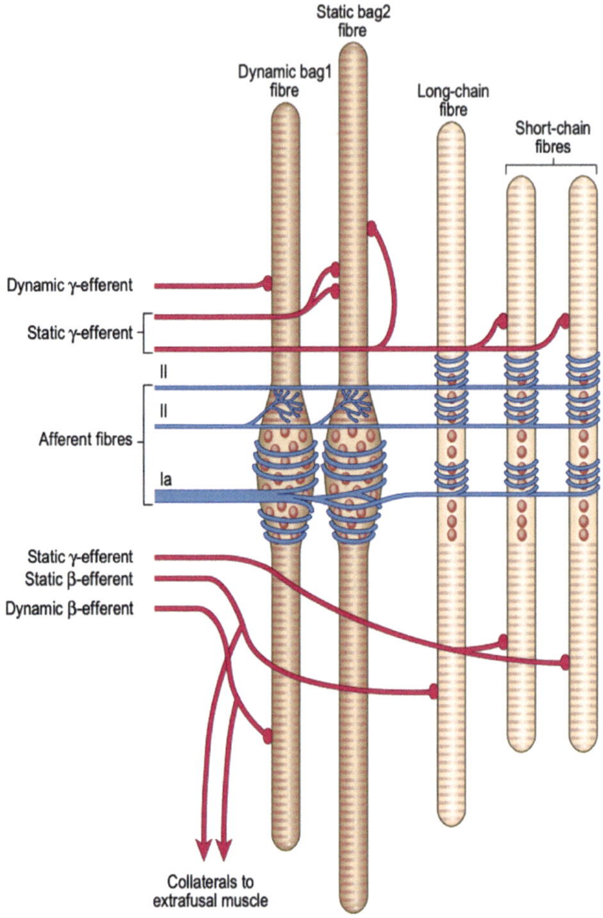

Fig. 4.10 Two afferent pathways bring information from the spindle into the spinal cord. Type Ia primary afferent pathway exiting from the equatorial regions of both the nuclear chain and bag fibers provides sensory information about muscle length and velocity of stretch. Type II secondary afferent pathway exiting from the ends of the nuclear chain fibers provides information about muscle length. Both fiber types receive motor innervation of the contractile ends via γ-efferent neurons. (Adapted with permission from Standring (2020))

muscle spindle stretch for positional feedback (Manuel and Zytnicki 2011).

In anticipation of lifting heavy objects, α- and γ-moto neurons establish specific system excitability levels. When lifting lighter-than-expected objects, the γ system reduces type I afferent output, rapidly adjusting α-motoneuron output and decreasing activated motor units. The lower activation threshold of γ-moto neurons enables them to initiate responses to postural changes by resetting spindle sensitivity and activating α output (Ellaway et al. 2015).

Modern molecular studies have identified specific ion channels and proteins in muscle spindle endings that mediate mechanotransduction. Advanced imaging techniques have revealed additional neural pathways connecting muscle spindles to motor planning areas and regions involved in proprioceptive processing. These findings expand our understanding of how muscle spindles integrate with the broader motor control system. The complex architecture of muscle spindles and their strategic placement within skeletal muscle tissue emphasize their importance in proprioception and motor control. This sensory apparatus contributes significantly to precise movement coordination and postural stability maintenance (Ellaway et al. 2015).

4.1.2.2 Golgi Tendon Organ Structure and Function in Skeletal Muscles

The Golgi tendon organ (GTO) serves as a specialized mechanoreceptor that plays a fundamental role in proprioception and motor control.

Located at the musculotendinous junction where muscle fibers transition into tendon tissue, GTOs exhibit a complex anatomical organization. Their structure consists of encapsulated collagen fascicles arranged in a spindle-shaped formation, with an outer capsule that extends internally to create distinct compartments (Jami 1992). The anatomical positioning of GTOs at this junction enables optimal force detection during muscular activities.

The microscopic architecture of GTOs reveals specialized type Ib afferent nerve endings that interweave between collagen fascicles. These sensory terminals establish synaptic connections with approximately 15–20 muscle fibers, creating a comprehensive force-monitoring network (Proske and Gandevia 2012). The collagen fascicles within GTOs demonstrate a specific alignment that optimizes their sensitivity to mechanical deformation during both active contraction and passive stretch conditions.

The mechanotransduction process in GTOs involves the compression of collagen fibers, which triggers action potentials in the type Ib afferent neurons. Research has demonstrated that these neural signals exhibit a linear relationship with both the magnitude and rate of load application, allowing precise force monitoring across various movement intensities (Proske and Gandevia 2012). The architectural arrangement of GTOs enables them to detect tension from multiple muscle fibers simultaneously, providing comprehensive feedback about force production within the muscle–tendon unit (Windhorst 2007).

GTOs demonstrate differential sensitivity patterns based on the nature of the mechanical stimulus. Their enhanced responsiveness to active tension compared to passive stretch stems from their unique structural organization. This specialized arrangement places GTOs in series with muscle fibers for active force detection while maintaining a parallel orientation to passive elements during stretch conditions. Consequently, muscle contractions typically activate GTOs at lower force thresholds compared to stretch stimuli (Dyck and Thomas 2004).

The neural circuitry involving GTOs generates complex inhibitory responses through spinal cord mechanisms. When activated, GTOs initiate an inverse stretch reflex through local graded potentials in the spinal cord (Proske and Gandevia 2012). This mechanism induces autogenic inhibition when the graded potential reaches sufficient magnitude, effectively modulating α-motoneuron output during high-velocity movements or elevated resistance conditions (Prochazka and Ellaway 2012). This inhibitory mechanism involves multiple interneuron populations within the spinal cord, creating complex modulatory circuits.

The functional capabilities of GTOs extend beyond simple force detection. These receptors maintain their sensitivity to minor tension fluctuations even during fatiguing conditions, enabling continuous force modulation throughout prolonged physical activities. Through the type Ib afferent pathway, GTOs can initiate rapid inhibitory responses to protect musculotendinous structures from excessive tension. Additionally, the functional output of GTOs demonstrates context-dependent plasticity. During locomotion, for instance, GTOs provide excitatory input to support muscles, facilitating extensor reflexes essential for maintaining posture and balance (Windhorst 2007).

The integration of GTO feedback occurs through a hierarchical processing system within the central nervous system. This includes primary integration at the spinal cord level, secondary processing in brainstem centers, and higher-order integration in cortical regions. This multilevel processing enables the combination of GTO input with other proprioceptive signals, particularly those from muscle spindles, creating a comprehensive representation of musculoskeletal dynamics (Riemann and Lephart 2002).

4.1.3 Structure and Function of the Neuromuscular Junction

The neuromuscular junction (NMJ) represents a specialized synapse that enables communication between motor neurons and skeletal muscle fibers, serving as the primary site for converting neural signals into mechanical force generation. This interface demonstrates structural and functional characteristics similar to central nervous system synapses, though with specific adaptations for rapid and reliable neurotransmission (Standring 2020).

The anatomical organization of NMJs begins with the terminal portions of both α-motoneurons and γ-motoneurons, which innervate extrafusal and intrafusal muscle fibers, respectively. As motor axons approach their target muscle cells, they undergo significant morphological changes. The myelin sheath terminates, and the axon develops a specialized terminal expansion that establishes contact with the muscle fiber membrane. This terminal region receives protection and electrical isolation through an encompassing Schwann cell sheath (Sanes and Lichtman 1999; Wilson and Deschenes 2005).

The region of muscle fiber membrane beneath the nerve terminal exhibits distinct structural modifications. The membrane forms a specialized postsynaptic apparatus characterized by elevated regions of accumulated sarcoplasm and cellular organelles. This structural arrangement creates the motor end plate, where the postjunctional membrane of the muscle fiber interfaces with the prejunctional membrane of the nerve terminal (Deschenes 2011; Slater 2017).

The presynaptic nerve terminal contains numerous synaptic vesicles filled with acetylcholine (ACh), the primary neurotransmitter at the NMJ. These vesicles cluster near specialized release sites called active zones, which contain voltage-gated calcium channels and other proteins essential for neurotransmitter release. When an action potential arrives at the nerve terminal, it triggers calcium influx, leading to the synchronized release of acetylcholine into the synaptic cleft (Nishimune et al. 2014; Slater 2015).

The synaptic interface demonstrates complex membrane specializations. The postjunctional membrane forms numerous invaginations, known as junctional folds, which interdigitate with corresponding projections of the prejunctional membrane. These formations create a synaptic cleft measuring 20–50 nanometers in width, containing specialized extracellular matrix molecules that facilitate synaptic transmission. The junctional folds significantly increase the surface area available for neurotransmitter reception and contain high concentrations of acetylcholine receptors (Burden et al. 2018; Hughes et al. 2006).

NMJ organization and function exhibit added complexity. The basement membrane within the synaptic cleft contains specific proteins that guide synapse formation and maintenance. Furthermore, perisynaptic Schwann cells actively participate in synaptic function by monitoring neurotransmission and releasing regulatory factors (Li et al. 2018; Wu et al. 2010).

4.1.4 Anatomy and Clinical Significance of Muscular Compartments

The compartmentalization of muscle groups represents a fundamental principle in musculoskeletal anatomy, where muscles are systematically organized within fascial boundaries. These compartments, formed by layers of dense connective tissue, create distinct anatomical spaces that house specific muscle groups, along with their associated neurovascular structures (Drake et al. 2020).

The anatomical arrangement of muscular compartments demonstrates remarkable consistency across different body regions. The thigh provides an exemplary model of this organization, with its three primary compartments, as shown in Fig. 4.11. The anterior compartment encompasses the quadriceps femoris muscle group, which consists of the rectus femoris, vastus lateralis, vastus medialis, and vastus intermedius. The posterior compartment contains the hamstring muscles, including the biceps femoris, semitendinosus, and semimembranosus. The medial compartment houses the adductor muscle group, comprising the adductor magnus, adductor longus, adductor brevis, and gracilis (Moore et al. 2018).

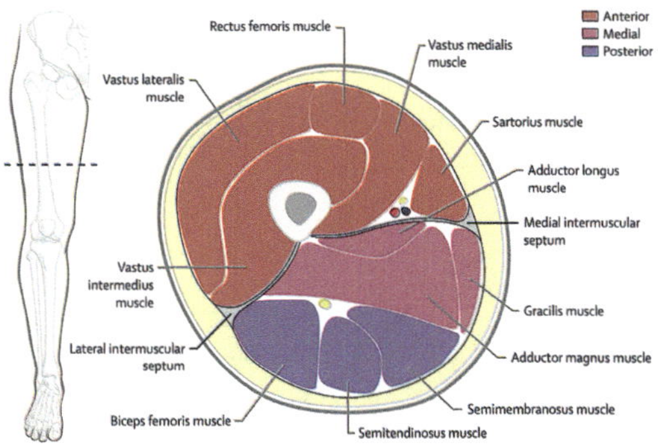

Fig. 4.11 Muscle compartments of the thigh. The thigh muscles are organized into compartments separated by fascial sheaths. Each compartment contains functionally similar muscle groups, categorized based on their actions, such as extensors, flexors, and adductors. (Adapted with permission from von Keudell et al. (2015))

Similar compartmental organizations exist throughout the body, particularly evident in the lower leg, as illustrated in Fig. 4.12. These compartments serve multiple physiological functions beyond simple anatomical organization. They facilitate coordinated muscle action by grouping functionally related muscles, provide mechanical advantage during contraction, and maintain optimal pressure relationships for muscle function (Standring 2020).

The fascial boundaries of these compartments demonstrate specific biomechanical properties. The fascia exhibits limited elasticity and maintains relatively constant volume within each compartment. Under normal physiological conditions, this arrangement supports proper muscle function and maintains appropriate pressure relationships. However, this same characteristic can become problematic when pathological conditions arise (Benjamin 2009).

Compartment syndrome represents a significant clinical manifestation of this anatomical arrangement. This condition occurs when increased pressure within a muscular compartment compromises local circulation and neural function. The anterior tibial compartment shows particular susceptibility to this condition due to its confined anatomical space and rigid fascial boundaries. Various factors can precipitate compartment syndrome, including trauma, excessive exercise, vascular injury, or postischemic swelling (Leversedge et al. 2011).

The pathophysiology of compartment syndrome involves a cascade of events. Initial pressure elevation leads to reduced capillary perfusion, resulting in tissue ischemia. This ischemia triggers an inflammatory response, causing further edema and additional pressure elevation. Without intervention, this cycle can lead to permanent tissue damage. Clinical manifestations include pain disproportionate to the injury, passive stretch pain, paresthesias, and, in advanced cases, paralysis (Frink et al. 2010; von Keudell et al. 2015).

Recent research has expanded the understanding of compartment pressure dynamics. Studies utilizing continuous pressure monitoring have identified pressure thresholds and temporal patterns critical for diagnosis and treatment. Additionally, investigation of fascial mechanical properties has revealed regional variations that influence susceptibility to compartment syndrome (Shadgan et al. 2015).

4.1.5 Essential Functions of the Skeletal Musculature

Skeletal muscle tissue serves as a fundamental component of human movement and postural control. Three primary functions directly related to human movement are contributing to the generation of skeletal movement, aiding in the stability of joints, and sustaining posture and body position (Neumann 2017).

4.1.5.1 Movement Generation and Control

The production of movement stands as a basic function of the skeletal muscular system. This process occurs through coordinated neural signals, muscular contraction, and force transmission. Skeletal muscles produce tensile forces that transfer through tendons to generate bone displacement and joint movement (Hamill et al. 2015).

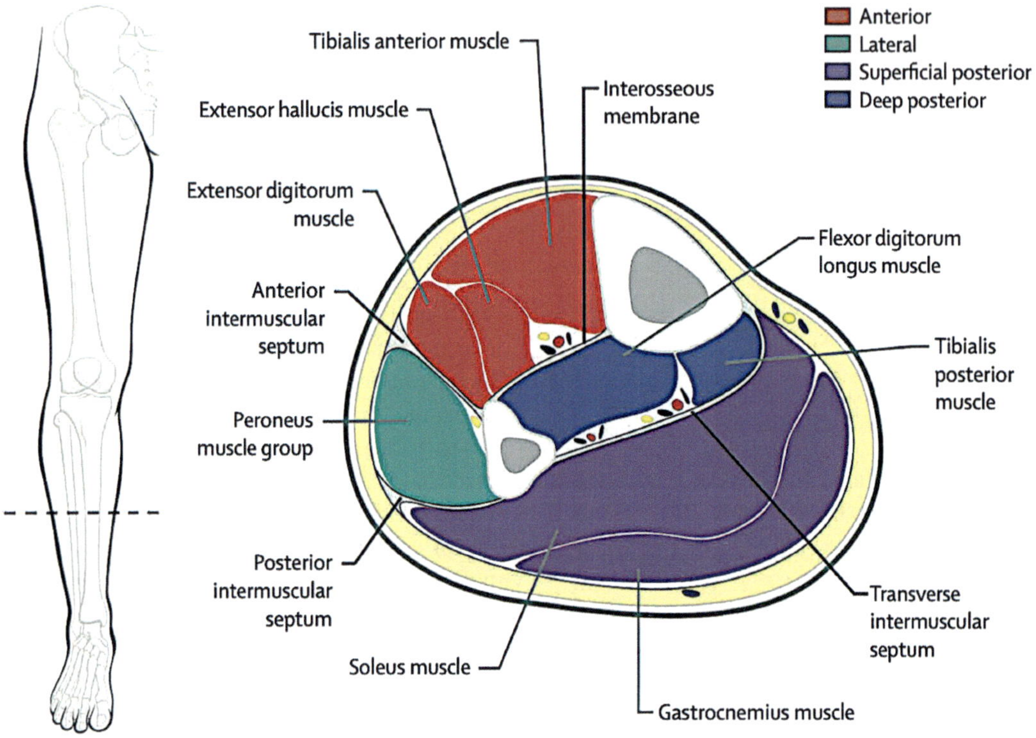

Fig. 4.12 Muscle compartments of the lower leg. The lower leg muscles are organized into compartments separated by fascial sheaths. Each compartment contains functionally similar muscle groups, classified based on their actions, such as extensors, flexors, evertors, and invertors. (Adapted with permission from von Keudell et al. (2015))

The mechanism of movement begins with neural activation of muscle fibers (Enoka and Duchateau 2016). Motor neuron stimulation of muscle tissue initiates a sequence of physiological events, leading to cross-bridge cycling between actin and myosin filaments. These molecular interactions generate contractile force within individual muscle fibers, resulting in tension throughout the muscle.

Force transmission occurs through specific anatomical structures (Herzog et al. 2015a, b). Tendons, made of dense regular connective tissue, connect muscles to bones. This arrangement allows muscles to transfer contractile forces from the muscle belly to skeletal attachment points. The mechanical configuration enables muscles to produce movements ranging from powerful athletic activities to precise motor tasks (Zajac et al. 2002).

Movement control requires coordination between multiple muscle groups (Winter 2009). Agonist muscles produce the primary force for intended movements, while antagonists modify and control these actions. This coordinated activity results in smooth, controlled movements and reduces the risk of injury or inefficient performance.

Accurate movement depends on continuous feedback from proprioceptive sensors within muscles and tendons (Proske and Gandevia 2012). This sensory information allows the nervous system to adjust muscle activation patterns as movement occurs. The combination of this feedback with motor commands forms a control system that adapts to changing movement requirements and environmental conditions.

4.1.5.2 Dynamic and Static Stabilization of Joints

The stability of joints comes from the combined effects of muscular forces and anatomical structures. Muscles generate tensile forces that transfer across joints through their tendons, providing dynamic stabilization during movement and static positions (Standring 2020).

Muscles perform a dual role in joint function by enabling movement while maintaining stability (Neumann 2017). The contractile forces of muscles compress joint surfaces together, which increases joint congruence and reduces excessive movement between articulating surfaces. This compression occurs through both active muscle contraction and passive muscle tension, maintaining joint stability across different positions and movement phases (Nordin and Frankel 2019).

The shoulder and knee joints demonstrate the importance of muscular stabilization. In the shoulder, the rotator cuff muscles maintain the humeral head's position within the glenoid fossa during arm movements. At the knee, the quadriceps and hamstring muscles work together to control joint position and movement while preventing excessive translation of the tibia relative to the femur (Neumann 2017).

This muscular stabilization adds to the support provided by ligaments and joint capsules. While ligaments offer passive restraint against excessive joint motion, muscles provide active control that adjusts to changing mechanical demands. The combination allows joints to maintain stability while permitting necessary movement for daily activities.

4.1.5.3 Postural Control and Balance

The maintenance of posture and body position requires continuous muscular activity that occurs through coordinated adjustments. This activity involves sustained contractions that maintain head position and balance body weight over the base of support (Kuo and Zajac 1993).

Postural control involves multiple muscular systems working in coordination. The axial muscles of the trunk and neck maintain spinal alignment, while the appendicular muscles of the lower extremities support body weight distribution (Winter et al. 1998). This muscular activity occurs at relatively low intensities but must persist throughout daily activities to maintain stability against gravitational forces.

Research has expanded our understanding of the neural control mechanisms underlying postural maintenance. The central nervous system processes information from visual, vestibular, and proprioceptive systems to generate appropriate muscular responses (Horak 2006). These responses occur through both anticipatory and reactive mechanisms, allowing adaptation to changing environmental conditions and task demands.

The antigravity muscles show particular importance in postural control. The soleus and gastrocnemius muscles in the lower leg maintain ankle position during standing, while the paraspinal muscles support spinal positioning. Additional muscular activity in the hip and thigh regions provides stability to the lower extremity joints (Kandel et al. 2013).

4.1.5.4 Supplementary Functions of Skeletal Muscles

- Beyond their primary roles in movement and posture, skeletal muscles serve several other essential functions in the human body:
- Protection and support: Muscles provide a protective layer for internal organs and tissues, shielding them from potential injury. They also offer structural support to the visceral components of the body.
- Pressure regulation: Muscular tension can modulate pressures within bodily cavities, playing a role in various physiological processes.

- Thermoregulation: Skeletal muscles contribute significantly to the maintenance of body temperature through heat production during contraction and metabolism.
- Sphincter control: Voluntary control over certain muscles enables the regulation of processes such as swallowing, defecation, and urination.

4.1.6 Properties and Characteristics of Skeletal Muscle Tissue

Skeletal muscle tissue demonstrates remarkable adaptability to mechanical demands through its defining physiological properties. The functional capacities of muscle tissue are derived from its specific structural and mechanical characteristics, which allow it to respond to varying loads and movement velocities. The four primary physiological properties that determine muscle performance include excitability, contractility, extensibility, and elasticity. These properties work together to enable precise movement control and force generation across different functional demands.

4.1.6.1 Excitability Properties

Skeletal muscle tissue demonstrates remarkable responsiveness to neural stimulation, ranking second only to nerve tissue in its sensitivity to stimuli. This property, known as excitability, allows skeletal muscles to respond rapidly and precisely to motor neuron signals through neurotransmitter release at the neuromuscular junction.

The excitation process begins when motor neurons release acetylcholine at the neuromuscular junction (Kandel et al. 2013). This chemical signal triggers a cascade of events in the muscle fiber membrane, leading to rapid changes in membrane potential. The speed and efficiency of this response system enable quick initiation of muscle contraction, when needed, for movement or postural adjustments.

The neural control of muscle excitability shows remarkable precision. The nervous system can selectively activate specific groups of muscle fibers and regulate the number of fibers recruited for any given task (Enoka 2008). This fine control allows for movements ranging from delicate manipulations requiring minimal force to powerful actions involving maximum muscle activation.

The high degree of excitability in skeletal muscle tissue supports its role in movement control. This property allows for rapid response to changing environmental demands and precise adjustment of muscle force output based on task requirements.

4.1.6.2 Contractile Properties

The fundamental property of contractility defines the ability of muscle tissue to generate force and reduce its length in response to stimulation. This property manifests through specific physiological mechanisms and operates within anatomical constraints that determine functional muscle performance (Lieber and Ward 2011).

Laboratory studies have demonstrated the extensive shortening capabilities of skeletal muscle tissue. Research shows that isolated muscles can reduce their length by 50%–70% from their resting state, with an average maximum shortening capacity of approximately 57% of the resting length (Gordon et al. 2020). These measurements provide important baseline data about the mechanical potential of muscle tissue under controlled conditions.

However, the functional shortening of muscles within the body differs significantly from these laboratory observations (Herzog 2018). Anatomical factors, including joint structure and the spatial relationships between body segments, limit the actual degree of muscle shortening during movement. The sartorius muscle illustrates this principle effectively—while capable of substantial shortening in isolated conditions, its functional range becomes limited by the hip joint architecture and the positional relationship between the trunk and thigh (Neumann 2017).

These anatomical constraints affect muscle function in several ways. The attachments of muscles to bones, their path across joints, and their relationships with surrounding structures all influence the practical expression of contractility (Zajac 1993). Understanding these limitations has proven valuable for analyzing movement patterns and developing effective rehabilitation strategies.

4.1.6.3 Extensibility

The extensibility of skeletal muscle represents a fundamental mechanical property that allows muscles to lengthen beyond their resting state. This passive property requires external forces or antagonist muscle action to achieve elongation (Kjaer 2004).

The structural basis for muscle extensibility resides primarily in the connective tissue framework that surrounds and interpenetrates muscle fibers (Purslow 2020). This framework includes the endomysium, perimysium, and epimysium, which contain collagen fibers arranged in specific patterns that permit controlled tissue elongation while maintaining structural integrity.

Practical applications of muscle extensibility appear in therapeutic and athletic contexts. During passive range of motion exercises, healthcare providers apply external forces to move limbs beyond their resting positions, demonstrating the natural capacity of muscles to elongate (Page 2012). This principle also manifests during activities such as stretching exercises, where muscles undergo controlled elongation to maintain or improve flexibility.

Biomechanical research has revealed specific characteristics of muscle extensibility. Studies show that the degree of possible elongation varies among different muscles and depends on factors such as fiber arrangement, connective tissue composition, and previous stretching history (Magnusson et al. 2019). The mechanical response to elongation follows a nonlinear pattern, with initial compliance followed by increasing resistance as the muscle approaches its extensibility limits.

4.1.6.4 Elasticity Properties

Skeletal muscle tissue possesses inherent elastic properties that enable it to recover its original length after stretching. This mechanical characteristic stems primarily from connective tissue components rather than the contractile elements of muscle fibers (Kubo et al. 1999).

The elastic behavior of muscle tissue performs multiple physiological functions. By returning stretched muscle to its resting length, elasticity maintains structural organization and baseline muscle tension. Additionally, this property enhances force production during movements that involve stretch–shortening cycles, improving mechanical efficiency during activities such as running and jumping.

A comparison with ligament tissue shows the distinctive elastic characteristics of muscles. Frontera and Ochala (2015) reveal that ligaments, mostly composed of collagen fibers, exhibit restricted elastic recovery following extension. When stretched beyond their resting length, ligaments may remain elongated, potentially resulting in joint instability. Conversely, muscle tissue exhibits uniform elastic recovery, reverting to its initial dimensions unless the exerted stresses are beyond tissue tolerances.

The elastic characteristics of muscles additionally aid in injury prevention. Proske and Morgan (2001) clarify the relationship of elasticity with other mechanical qualities in protecting muscle tissue during movement. This protective role is particularly important during dynamic activities that require rapid length alterations or significant stresses.

Research has revealed the molecular basis for muscle elasticity. The titin protein, along with other elastic elements in the extracellular matrix, provides the mechanical foundation for elastic recovery. These components store and release elastic energy during muscle lengthening and shortening, contributing to movement efficiency and tissue protection.

4.1.7 Force Generation in the Muscle

4.1.7.1 Muscle Contraction

The action potential from a motor neuron reaches a muscle fiber at a neuromuscular junction or motor end plate located near the center of the fiber. At this site, a synapse exists between the motor neuron and the muscle fiber membrane. When the action potential arrives at the synapse, a series of chemical reactions occurs, leading to the release of acetylcholine (ACh). ACh diffuses across the synapse, increasing the permeability of the muscle fiber membrane. ACh is rapidly broken down to prevent continuous muscle fiber stimulation. The speed at which the action potential travels along the membrane is known as the conduction velocity (Kandel et al. 2013).

The resting potential of the muscle fiber is −70–95 mV inside relative to outside. When the membrane potential reaches the threshold level (approximately −50 mV), a change in the potential of the muscle fiber membrane or sarcolemma occurs. The action potential is characterized by depolarization from the resting potential, causing the potential to become positive (approximately +40 mV). This is followed by a hyperpolarized state before returning to the resting potential through repolarization (Huxley 1957).

The depolarization wave of the action potential progresses along the nerve until it reaches the muscle fibers, where it spreads to the muscle membrane as calcium ions (Ca^{2+}) released into the area surrounding the myofibrils. These Ca^{2+} ions facilitate cross-bridge formation, resulting in an interaction between the actin and myosin filaments. When stimulation ceases, ions are actively removed from the area surrounding the myofibrils, releasing the cross-bridges. This process is known as excitation–contraction coupling. The calcium ions connect action potentials in a muscle fiber to contraction by binding to the filaments and initiating the interaction of actin and myosin to start contraction of the sarcomere (Sandow 1952).

Muscle force production is achieved through two primary mechanisms. First, muscle force can be increased by recruiting progressively larger motor units. Initially, during a muscle contraction, smaller motor units are activated. As muscle force increases, more and larger motor units are engaged. This is known as the size principle (Henneman et al. 1965). Second, a motor unit may be activated at various frequencies. A single action potential that activates a fiber will cause the force to increase and decrease, referred to as a twitch. If a second stimulus occurs before the initial twitch has subsided, another twitch builds upon the first. With subsequent high-frequency stimulations, the force continues to build. Eventually, the force builds to a level where no further increase in muscle force occurs. In a muscle contraction, both size recruitment and frequency of stimulation are simultaneously employed to increase the muscle force (Fuglevand et al. 1993).

Mechanisms of Muscle Contraction: The Sliding Filament Model

The process by which muscles generate tension has been a subject of extensive scientific inquiry. Among the various theories proposed, the sliding filament model, introduced by Huxley in 1957, provides the most widely accepted explanation for muscular contraction (Huxley 1957). This model provides a comprehensive framework for understanding the shortening of sarcomeres, the fundamental contractile units of muscle fibers (Frontera and Ochala 2015). Historically, muscle contraction was understood by establishing similarities to the process of blood coagulation, the elastic properties of rubber, a chain of circular elastic rings, and even sliding movements induced by opposing electrical charges within different filaments. However, these earlier theories have mostly been replaced by the sliding filament model.

According to Huxley's sliding filament theory, the contraction process is initiated by the release of calcium ions into the muscle in response to neurochemical stimulation (Huxley 2000). The sarcomere's contraction occurs as the myosin filament advances along the actin filament, forming cross-bridges between the myosin head and specific binding sites on the actin filament. In the fully contracted state, there is extensive overlap between the actin and myosin filaments along much of their lengths (Neumann 2017; Xiao 2018) (*see* Fig. 4.7).

The simultaneous sliding action of numerous sarcomeres arranged in series results in changes to both the length and force output of the muscle (Frontera and Ochala 2015). The magnitude of force generated by the muscle is directly proportional to the number of cross-bridges formed. As multiple sarcomeres, myofibrils, and fibers shorten concurrently, tension develops throughout the muscle and is transmitted to the bones at both attachment points, ultimately producing movement (Herzog 2017; Neumann 2017).

Muscle force is generated by the interaction between thick (myosin) and thin (actin) filaments within the sarcomeres. This process involves cyclic binding and releasing of myosin heads (cross-bridges) to the active sites on actin filaments. These cyclic interactions lead to both force production by the cross-bridges and relative sliding of the actin and myosin filaments (sliding). Importantly, while this sliding occurs, the actual lengths of the filaments themselves remain largely unchanged. During concentric muscle contraction (active shortening), actin filaments move closer to the A-bands, whereas in muscle elongation, they are drawn away from the myosin filaments (A-bands), still without altering their length. These relative actin–myosin displacements modify the extent of overlap between actin and myosin filaments (Fig. 4.13). If the load at the muscle ends permits shortening, actin filaments will slide along myosin filaments until the sarcomere reaches its active slack length (called the sarcomere active slack length)—the point at which no active force can be generated. It is noteworthy that the filaments exhibit significantly higher stiffness compared to the cross-bridges and exhibit minimal extensibility.

4.1.7.2 Transmission of Muscle Force to the Bone

Mechanism of Muscle Attachment: Direct, Tendinous, and Aponeurotic

Muscle attachment to bones or other tissues invariably occurs through connective tissue elements. Often, the muscle's perimysium and epimysium directly unite with the periosteum of the bone or the joint capsule. In cases where this connective tissue is not visibly distinct, the muscle exhibits a "fleshy attachment," leaving no evident mark on the bone, although the corresponding bone area may appear flattened or depressed. Frequently, the muscle's connective tissue elements coalesce to form a tendon, composed of collagen fiber bundles. Notably, there is no direct continuity between the muscle fibers and tendon fibers. Tendons display diverse forms, typically exhibiting high tensile strength, and may present as round cords, flattened bands, or thin sheets (the latter termed aponeurosis). Tendon attachments to the bone generally leave a smooth mark, except in instances of mixed fleshy and tendinous fiber attachments or long aponeurosis attachments, which roughen the bone surface.

The transfer of muscle force to skeletal structures occurs through three primary attachment mechanisms: direct bone insertion, tendinous connection, and aponeurotic attachment (Standring 2020). Muscle-to-bone force transmission occurs through the following distinct mechanisms:

- Direct bone insertion: In this method, muscle fibers attach directly to the periosteum of the bone. This attachment is characterized by a fusion between the epimysium and the bone surface (Benjamin 2009). An example of this type of attachment is observed in the trapezius muscle (Standring 2020).
- Tendinous connection: This mechanism involves the muscle attaching to the bone via a tendon, a dense connective tissue structure. The tendon integrates with the muscle fascia, forming a strong link between the muscle and the bone (Kjær 2004). This type of attachment is common in several muscle groups, including the hamstrings, biceps brachii, and flexor carpi radialis (Lieber and Fridén 2000).
- Aponeurotic attachment: An aponeurosis, often described as a flattened or sheet-like tendon, serves as the attachment medium in this mechanism (Purslow 2020). This fibrous sheath of tissue provides a broad surface area for muscle attachment. Notable examples of muscles utilizing aponeurotic attachments include the abdominal muscles and the trunk attachment of the latissimus dorsi (Benjamin 2009).

Each of these attachment types plays a critical role in the efficient transmission of force from the

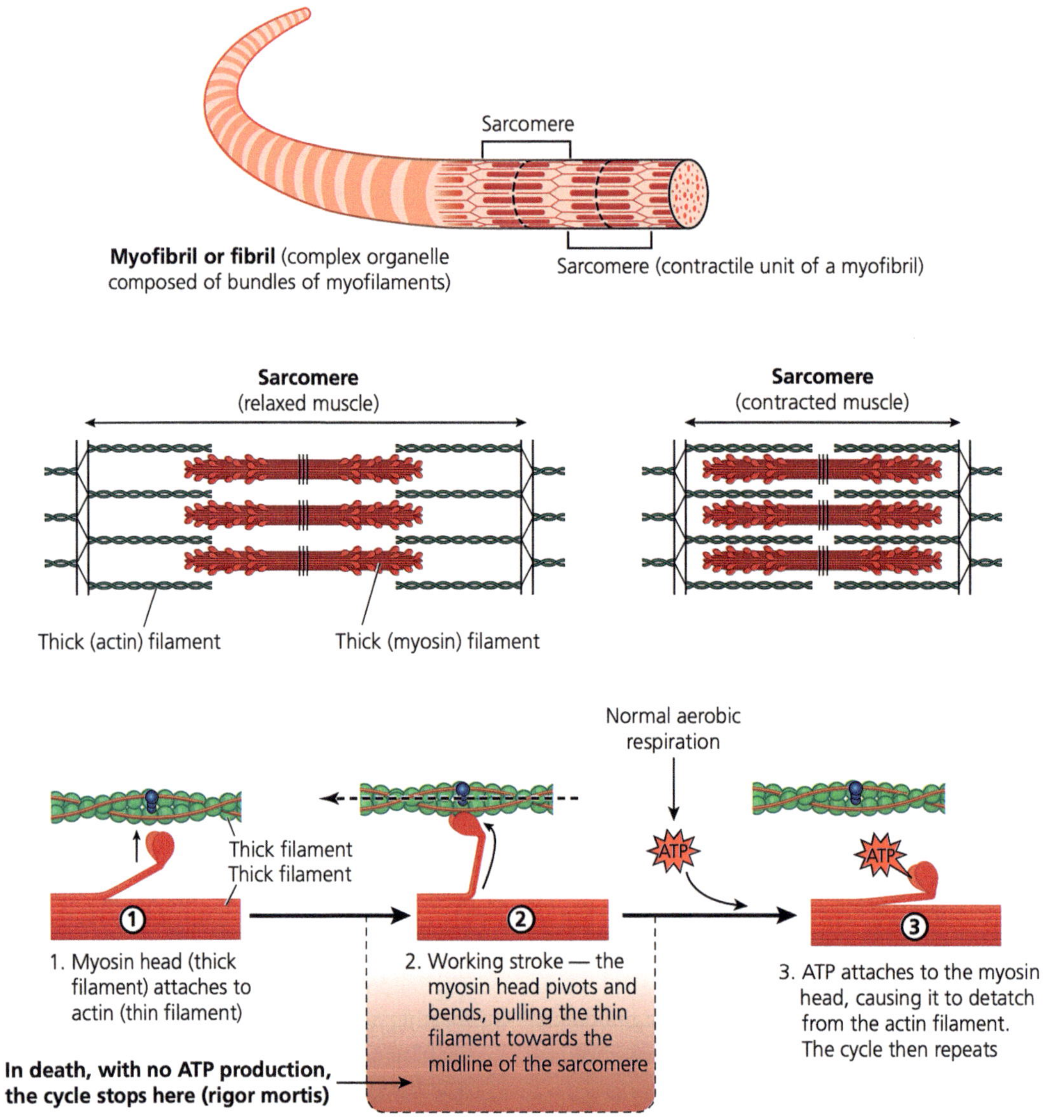

Fig. 4.13 Schematic representation of the mechanism of muscular contraction. (Adapted from McGuinness (2018))

muscle to the bone, contributing to the diverse range of movements and functions in the human musculoskeletal system (Kjær 2004). The specific type of attachment utilized by a muscle is often related to its functional requirements and anatomical location (Lieber and Fridén 2000).

Structural and Mechanical Properties of Tendons in Force Transmission

Tendons, the primary attachments for muscles, function as force transmitters from the muscle to the bone. The myotendinous junction serves as the interface where muscle fibers interweave with tendon collagen fibers. These structures exhibit considerable strength, transferring significant loads via perforating connections to bone surfaces. Tendons demonstrate stretch resistance while maintaining flexibility, allowing them to pass over cartilage, sesamoid bones, and bursae (Kannus 2000).

Tendons present in various forms—cord-like, strip-shaped, circular, oval, or flattened. Their structure consists predominantly of collagen fibers aligned parallel to the axis of force application. Although individual fibers are inelastic, tendons can exhibit elastic-like behavior through recoiling mechanisms and the elasticity of associated connective tissues. This characteristic allows tendons to withstand high tensile forces produced by muscles while displaying viscoelastic responses to loading. The Achilles tendon, for instance, has been shown to resist tensile loads comparable to or exceeding those of similarly dimensioned steel (Komi 1990).

The stress–strain relationship in tendons is characterized by viscoelasticity, manifesting as a nonlinear response with hysteresis. Tendons possess notable stiffness and strength relative to other anatomical structures, particularly when subjected to rapid loading. This stiff behavior is attributed to their high collagen content. Additionally, tendons exhibit marked resilience with minimal hysteresis, facilitating efficient energy storage and release (Wang 2006). These biomechanical properties play a key role in tendon function. The high stiffness and strength allow tendons to transmit force to the bone with minimal deformation. Their low hysteresis enables effective storage and release of elastic strain energy (Hamill et al. 2015; Neumann 2017). The differences in the strength and performance characteristics of tendons versus muscles and bones are presented in Fig. 4.14.

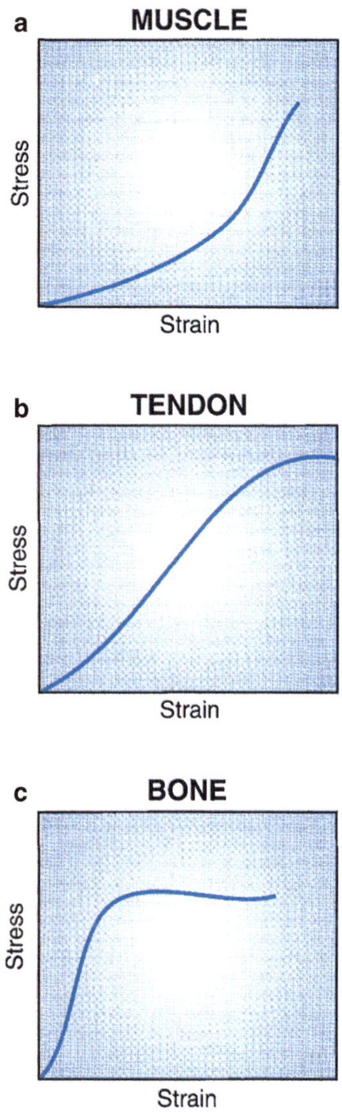

Fig. 4.14 The stress–strain curves for muscle, tendon, and bone tissue. Muscle is viscoelastic and thus deforms under low load and then responds stiffly (**a**). Tendon is capable of handling high loads. The end of the elastic limits of the tendon is also the ultimate strength level (no plastic phase) (**b**). Bone is a brittle material that responds stiffly and then undergoes minimal deformation before failure (**c**). (Adapted from Hamill et al. (2015))

The myotendinous junction features a multilayered interface where muscle fiber myofibrils merge with tendon collagen fibers (Tidball 1991). At the bone attachment site, tendons form a complex interface comprising fibrocartilage, mineralized fibrocartilage, and lamellar bone, connecting with the periosteum and subchondral bone (Benjamin et al. 2006).

In the musculoskeletal system, tendons and muscles operate in series to generate or absorb tension. This arrangement results in tendons experiencing the same tension as their associated muscles (Zajac 1989). The mechanical relationship between these structures is influenced by the magnitude of applied or generated force, the velocity of muscle action, and the degree of tendon slack.

The tendon structure is composed of parallel fibers that are not perfectly aligned, resulting in a wavy and wrinkled pattern. When muscle fibers generate tension while the tendon is slack, the tendon initially exhibits compliance as it straightens. Subsequently, it begins to recoil toward its original length. This recoiling action, which takes up the slack in the tendon, causes a delay in achieving the required tension level in the muscle fibers (Zajac 1989).

The recoiling property of the tendon reduces the muscle-shortening speed, consequently increasing the load a muscle can bear (Zajac 1989). In contrast, a stiff tendon with no recoil transmits tension directly to muscle fibers, leading to higher velocities but decreasing the muscle's load-bearing capacity. The stiff response in tendons facilitates rapid tension development in muscles, resulting in quick and precise movements.

Muscles and tendons are particularly vulnerable to injury during eccentric contractions. For instance, during the follow-through phase of throwing, the posterior rotator cuff stretches while contracting to decelerate the movement. Similarly, the quadriceps femoris muscle group lengthens and contracts during the support phase of running as knee flexion lowers the center of mass. The tendon absorbs the initial stretch of the relaxed muscle, and if the muscle contracts while being stretched, tension increases sharply in both structures (Zajac 1989).

The location of injury in tendons varies with the rate of tension development. At slow rates, the tendon–bone junction is more susceptible, while at faster rates, the tendon itself is more likely to fail. For the entire muscle–tendon unit, the muscle belly and the myotendinous junction are the common injury sites.

Many tendons traverse bony protuberances, which alter the muscle's pull angle and reduce tension. Examples include the quadriceps femoris muscles and patella and the hamstrings and gastrocnemius tendons as they pass over the femoral condyles. Some tendons are encased in synovial sheaths for protection and positioning. Tendon tension contributes to the formation of bony ridges and protuberances. Apophyses on bones develop due to tensile forces applied through tendons.

4.1.8 Neuromuscular Control and Muscular Function in Movement

The musculoskeletal system demonstrates precise control in motor task execution through selective activation of movement capabilities. During specific actions, such as raising an arm to comb the hair, the system utilizes approximately 20–30 degrees of freedom, though not all of them contribute directly to the intended task completion (Winter 2009). This selective muscular activation pattern represents a portion of the motor system's complete movement potential, enabling accurate control while reducing unnecessary motions.

Movement patterns become refined through coordinated activation of specific muscle groups, with each group performing defined functional roles (Shumway-Cook and Woollacott 2017). The coordination results in complex movements executed with minimal energy expenditure while maintaining effectiveness. During movement execution, different muscles function in specific roles: Primary movers produce the main action, while stabilizing muscles and antagonistic groups maintain movement precision and control (Neumann 2017).

The nervous system maintains continuous regulation of muscle activation patterns through integrated sensory feedback mechanisms (Kandel et al. 2013). This neuromuscular control system allows for moment-to-moment modifications of muscle activity based on specific task requirements and environmental conditions. The ongoing modulation of muscle activation patterns supports the development and enhancement of motor skills, enabling precise execution of complex tasks.

Research in motor control has shown that the central nervous system optimizes movement efficiency through learned patterns of muscle activation (Schmidt et al. 2019). These patterns develop through experience and training, leading to improved movement accuracy and reduced energy consumption. The neural circuits responsible for movement control continue to refine these patterns based on sensory feedback and task outcomes (Latash and Lestienne 2006a, b).

In clinical rehabilitation settings, understanding muscle activation patterns and motor control becomes valuable for developing therapeutic interventions (O'Sullivan et al. 2019). This scientific foundation guides the implementation of treatment strategies that focus on restoring functional movement patterns while considering the specific contributions of different muscle groups in motor task execution. Recent advances in neuroimaging and electromyography have enhanced our understanding of these motor control mechanisms, leading to more effective rehabilitation protocols (Enoka 2008).

4.1.8.1 Muscle Attachments: Origin and Insertion

The anatomy of muscle attachments exhibits specific characteristics related to their position and function. Muscles connect to bones at two primary sites, which were historically classified as "origin" and "insertion." However, modern anatomical understanding indicates that this classification requires modification, as muscles can generate movement at both attachment locations (Neumann 2017). The term "attachment" provides a more accurate description of these connection points.

Anatomically, the proximal attachment, traditionally known as the origin, appears broader and is positioned closer to the body's midline. The distal attachment, historically termed the insertion, frequently tapers into a tendon and is located further from the midline (Standring 2020). A functional perspective describes these attachments based on their relative mobility, comparing areas of lower mobility with those exhibiting higher mobility. Some muscles display multiple attachment points at either end, adding complexity to their mechanical function.

When muscle contraction occurs, forces are applied simultaneously to both skeletal connections, potentially resulting in movement at either or both attachment sites (Lieber and Ward 2011). The apparent stability of one bone during muscle action results from factors including the stabilizing influence of the surrounding musculature or mass differences between connected skeletal segments. Muscles that cross multiple joints demonstrate the capability to produce various movements across several segments (Winter 2009).

The psoas muscle exemplifies the adaptable nature of muscle function across different activities. During leg raises, it produces thigh flexion, while during curl-ups or sit-ups, it elevates the trunk. Similarly, the gluteus medius demonstrates functional variability, moving the pelvis during the stance phase and the leg during the swing phase of gait. The analysis of forces at all attachment sites, including those without visible movement, provides an understanding of stabilization requirements and actual forces transmitted to skeletal attachments.

4.1.8.2 Muscle Function and Attachment Angle

Muscles transfer tension to the bones through tendons and aponeuroses, with the attachment angle determining the distribution of force between rotational movement and joint loads. The angle of muscle insertion affects the proportion of force utilized for segment rotation versus joint compression or distraction (Zajac 1989).

In acute angle attachments or when tendons lie parallel to the bones, the muscular force primarily directs along the bone's longitudinal axis and into the joint. The biceps brachii demonstrates this principle during forearm extension, where its tendon attaches to the radius at a small angle. This mechanical arrangement requires increased muscle force for movement initiation, as a significant portion of force transmits into the elbow joint rather than producing rotation (Murray et al. 1995). The minimal resistance from forearm weight in extension allows movement despite reduced rotational force components.

Force analysis reveals two primary components: one parallel to the bone's length and another perpendicular to it, producing rotation. These components vary based on the attachment angle. While muscle tension may remain constant during motion, the rotational component changes with joint position (Lieber and Fridén

2000). Starting positions often exhibit reduced force output due to unfavorable insertion angles directing force along the bone. As joints move through midrange, increasing insertion angles typically direct more force into rotation (Veeger and van der Helm 2007).

Broad muscles with extensive attachment areas, such as the pectoralis major and trapezius, present additional mechanical complexity. Different sections may function independently due to varying attachment angles. The trapezius illustrates this concept, with its lower fibers attaching to the scapula at an angle opposite to the upper portion. During shoulder elevation and abduction, the lower trapezius may show minimal activity. This anatomical arrangement necessitates the analysis of multiple force vectors for accurate functional assessment (Ackland and Pandy 2009).

4.1.8.3 Muscular Contributions to Generating, Opposing, and Stabilizing Movements

Agonistic and Antagonistic Muscle Functions

Muscles that collaborate to produce a particular joint action are classified as agonists. In contrast, muscles that counteract or generate the opposite joint movement are termed antagonists (Lieber and Fridén 2000). For a movement to occur, antagonists must either relax or contract simultaneously with agonists to regulate or decelerate joint motion. Consequently, the most significant positional changes relative to other muscles are observed in the antagonists (Neumann 2017).

During the forward and upward swing of the thigh, the agonists facilitating the movement are the hip flexors, specifically the iliopsoas, rectus femoris, pectineus, sartorius, and gracilis muscles (Standring 2020). The hip extensors, including the hamstrings and gluteus maximus, function as antagonists, opposing hip flexion. The combined action of these antagonists, along with gravitational forces, serves to decelerate and ultimately terminate the hip flexion movement (Prilutsky 2000). Both agonistic and antagonistic muscle groups collaborate in the modulation and control of movement.

When a muscle serves as an antagonist, it becomes more prone to injury, particularly at its attachment sites or within the muscle fibers themselves (Garrett Jr. 1996). This increased vulnerability arises because the muscle contracts to decelerate the limb while simultaneously undergoing stretching.

Stabilizers and Neutralizers

Muscles also function as stabilizers, acting on one segment to facilitate specific movements in an adjacent joint. This stabilization is particularly significant in areas such as the shoulder girdle, where it enables smooth and efficient arm movements. Similarly, in the pelvic girdle and hip region, stabilization is necessary during gait. For instance, during walking or running, when one foot contacts the ground, the gluteus medius contracts to maintain pelvic stability, preventing a lateral drop (Neumann 2017).

Another significant muscular function is that of a synergist or neutralizer. In this role, a muscle contracts to counteract an unwanted joint action produced by another muscle (Kendall et al. 2005).

Force transfer between adjacent muscles can enhance the force in the target muscle (Neumann 2017). An illustrative example involves the gluteus maximus. When this muscle contracts at the hip joint to extend the thigh, it simultaneously tends to rotate the thigh externally. If external rotation is undesired, the gluteus minimus and tensor fasciae latae contract to generate an internal rotation force. This neutralizing action counterbalances the external rotation produced by the gluteus maximus, resulting in the intended extension movement (Kapandji 2007).

4.1.9 Functional Mechanics of Levers in Human Movement

The human body exemplifies biomechanical engineering, with its movements coordinated through an intricate system of levers. These levers, composed of bones, joints, and muscles, function together to produce the diverse pattern of motions observed in daily life. To comprehend the complexity of human movement, understanding the principles of these biological levers is necessary (Neumann 2017). In human biomechanics, three classes of levers facilitate movement, each with a unique role. These three classes of levers interact to facilitate movement, each defined by the arrangement of the fulcrum (pivot point), force arm (where force is applied), and load arm (where the load is located).

A first-class lever is used in balancing weight and/or changing the direction of pull. There is usually no gain in mechanical advantage. The first-class lever, less common in the human body, is demonstrated by standing on one leg. Here, the hip joint acts as the fulcrum, body weight to one side is the load, and the gluteus medius and minimus muscles on the same side provide the force (Levangie and Norkin 2005).

A second-class lever is exemplified when rising on one's tiptoes. The metatarsal heads at the ball of the foot act as the fulcrum, while the body weight transmitted through the tibia represents the load. The force is exerted by the contraction of the calf muscles. This lever arrangement provides a mechanical advantage, enabling the lifting of body weight with relatively less muscular force, however, at a reduced speed. The load arm is defined as the distance from the tibia to the metatarsal heads, while the force arm extends from the calf muscles' attachment on the calcaneus to the metatarsal heads (Kuo et al. 2005).

The third-class lever is the most common type in the human body. In this lever, the fulcrum is positioned at one end, the load at the other, and the force is applied between them. Although it operates at a mechanical disadvantage in terms of force production, it allows for greater speed and range of motion. This lever is illustrated by the action of the biceps brachii muscle across the elbow joint. When curling a dumbbell, the elbow joint is the fulcrum, the weight in the hand plus the forearm is the load, and the biceps muscle applies force between these two points. The load arm is the distance between the elbow and the forearm/hand's center of mass, while the force arm is the distance from the elbow joint to the biceps attachment on the forearm (Hamill et al. 2015; Oatis 2017) (*see* Fig. 2.17).

4.1.10 Mechanical Principles of Muscle Contraction Types

Muscle contractions exhibit specific mechanical characteristics based on length changes and tension development. During stimulation, muscles contract to reduce the distance between attachment points, generating force through different mechanical patterns (Enoka and Duchateau 2016). Isotonic contractions maintain relatively constant tension while allowing length changes. These contractions include two subtypes: concentric contractions, where muscles shorten while producing force, and eccentric contractions, where muscles lengthen while controlling movement against external forces. Eccentric contractions require less metabolic energy compared to concentric contractions while generating higher forces (Herzog et al. 2015a, b).

Isometric contractions occur when the muscle length remains unchanged despite active force production. These contractions develop increasing tension against immovable resistance or when opposing forces match muscle force output. Studies indicate that isometric force production shows unique neural control patterns compared to dynamic contractions (Duchateau and Baudry 2014).

Most functional movements incorporate combinations of these contraction types. The temporal sequencing of different contraction modes influences movement efficiency and force production capabilities. Advanced measurement techniques have shown that even during isometric conditions, some internal length changes occur within muscle fibers, affecting force generation mechanisms (Roberts and Azizi 2011).

4.1.10.1 Isometric Muscle Contractions

Muscle tension is developed against resistance to maintain position, lift a segment or an object, or lower or control a segment. When a muscle is active and develops tension without a visible or external change in the joint position, this is referred to as an isometric muscle contraction (Gordon et al. 1966). Figure 4.15 presents examples of isometric muscle actions. Consider bending over to 30° of trunk flexion and maintaining that position. The muscle action used to sustain this posture is termed isometric because no movement occurs. In this case, the back muscles contract isometrically to hold the trunk in a flexed position, counteracting the gravitational force that tends to further flex the trunk. Conversely, when the trunk is curled up to 30° and held in that position, an isometric muscle action of the trunk flexors occurs. This muscular effort resists the gravitational force that would otherwise extend the trunk (Neumann 2017). These examples illustrate how isometric contractions are fundamental in maintaining static postures against external forces, particularly gravity. This type of muscle action is prevalent in everyday activities and is a key component in postural control and stability (Hodges and Gandevia 2000).

4.1.10.2 Concentric Muscle Contractions

When a muscle actively generates tension while visibly shortening, this is classified as a concentric muscle action. In concentric joint actions, the net muscle forces producing movement align with the direction of joint angle change, indicating that the agonists are the controlling muscles (*see* Fig. 4.15). The limb movement resulting from a concentric muscle action is termed positive, as these joint actions often work against gravity or initiate the movement of a mass (Özkaya et al. 2017). Concentric muscle actions are responsible for numerous joint movements. For instance, flexion of the arm or forearm from a standing position is achieved through concentric action of the respective agonists or flexor muscles. Similarly, initiating horizontal adduction of the arm across the body involves concentric action of the horizontal adductors (Neumann 2017).

These muscle actions generate forces against external resistances. Common examples include lifting weights, pushing off the ground during locomotion, and propelling objects, such as in throwing activities (Zatsiorsky and Prilutsky 2012). The ability to perform concentric actions effectively is important for many athletic and daily living activities, as it allows for the generation of force and movement against resistance.

4.1.10.3 Eccentric Muscle Contractions

Eccentric muscle action occurs when an external torque exceeds the torque generated by the muscle, causing the muscle to lengthen. The external force producing this torque often originates from gravity or the action of antagonistic muscle groups (Enoka and Duchateau 2016). In eccentric joint action, the net muscular forces generating rotation oppose the change in joint angle, with antagonists functioning as the controlling muscles. The limb movement in eccentric muscle action is termed negative, as joint actions usually move downward with gravity or control rather than initiate mass movement. For example, during downhill walking, muscles act as shock absorbers, resisting downward movement while lengthening. Most downward movements, unless executed rapidly, are controlled by eccentric action of the antagonistic muscle groups. To illustrate, arm adduction from an abducted position involves eccentric contraction of the abductor muscles. Similarly, lowering into a squat position, which requires hip and knee flexion, is eccentrically controlled by the hip and knee extensors. Conversely, rising from a squat position involves concentric contraction of these same extensors to overcome gravity.

These examples indicate potential areas of muscular imbalance in the body. The extensors in the trunk and lower extremity are used for both lowering and raising segments. In contrast, the flexors in the upper extremity are used for both concentric raising and eccentric lowering of segments, resulting in more frequent use.

Eccentric contractions also decelerate movements. During rapid thigh flexion, as in kicking, the antagonists (extensors) eccentrically control and slow the joint action near the end of the range of motion. Athletes with impaired eccentric strength may face increased risk of injury during movements requiring rapid deceleration.

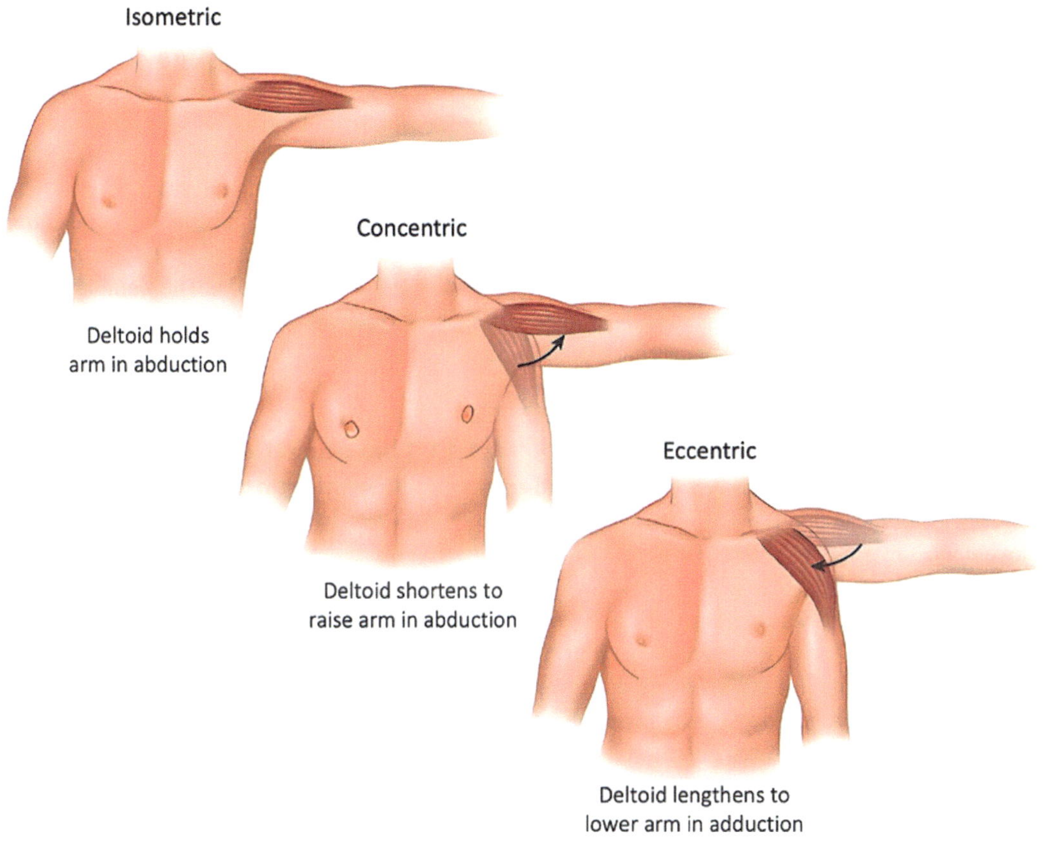

Fig. 4.15 Types of muscle actions. Isometric muscle action involves muscle tension without a change in joint position. Concentric muscle action occurs when the muscle shortens as it generates tension. Eccentric muscle action occurs when the muscle lengthens while resisting an external force. (Adapted with permission from Ko (2022))

When eccentric muscle contractions precede concentric ones, force output increases due to the contribution of elastic strain energy in the muscle. In throwing motions, for instance, the trunk, lower extremity, and shoulder internal rotators contract eccentrically during the windup, cocking, and late cocking phases. The elastic strain energy stored in these muscles subsequently enhances the concentric phase of the throw (Roach and Lieberman 2014).

4.1.10.4 Comparison of Isometric, Concentric, and Eccentric Muscle Contractions

Isometric, concentric, and eccentric muscle actions function in combination rather than isolation. Isometric actions stabilize body parts, while eccentric and concentric muscle actions occur sequentially to maximize energy storage and muscle performance. This natural sequence of muscle function, where an eccentric action precedes a concentric action, is known as the stretch–shortening cycle (Komi 2000).

These three muscle actions differ significantly in energy cost and force output. Eccentric muscle action can develop the same force output as the other two types with fewer activated muscle fibers . As a result, eccentric action is more efficient and can produce the same force output with less oxygen consumption than the others (LaStayo et al. 2003). Moreover, eccentric muscle action can generate greater force output using fewer motor units than isometric or concentric actions (Duchateau and Baudry 2014).

This occurs at the sarcomere level, where force increases beyond the maximum isometric force when the myofibril is stretched and stimulated (Herzog et al. 2015a, b).

Concentric muscle actions produce the lowest force output among the three types. Force relates to the number of cross-bridges formed in the myofibril. In isometric muscle action, the number of attached bridges remains constant. As the muscle shortens, the number of attached bridges decreases with increased velocity, reducing the force output generated by tension in the muscle fibers.

An additional factor contributing to noticeable force output differences between eccentric and concentric muscle actions appears during vertical movements. In this case, gravitational torques influence the force output in both concentric and eccentric actions (Enoka and Duchateau 2016). Gravitational force creates torque that contributes to the force output in an eccentric action as the muscles generate torque, controlling limb or body lowering. The total force output in a lowering action results from both muscular and gravitational torques (Winter 2009).

Gravity impedes upward limb movement, and before any movement occurs, the concentric muscle action must develop a force output exceeding the gravitational force acting on the limb or body. The total force output in a raising action is predominantly muscle force. This explains why concentric muscle action is more demanding than eccentric or isometric action (Enoka and Duchateau 2016).

This information is valuable when designing exercise programs for unconditioned individuals or rehabilitation programs (Kisner and Colby 2007). Even individuals with minimal strength may perform a controlled lowering of a body part or small weight but may struggle to hold or raise it. A program beginning with eccentric exercises, progressing to isometric, and then to concentric exercises may prove beneficial in strength progression or body part rehabilitation (Roig et al. 2009). For instance, a person unable to do a push-up should start at the extended position and lower into the push-up, receiving assistance on the up phase until developing sufficient strength for the concentric portion of the skill. When using eccentric exercises, factors to consider include controlling the speed of limb or weight lowering and managing the magnitude of the eccentrically imposed load, as muscle injury and soreness can occur more readily with eccentric muscle action under high-load and high-speed conditions (Proske and Morgan 2001). The comparison of the three types of muscle contractions is presented in Table 4.2.

4.1.11 Functional Considerations of Single- and Biarticular (Two-Joint) Muscles

The majority of muscles in the human body that act on a single joint exhibit two or three functions. However, some muscles demonstrate even greater versatility, possessing two to three times more functions than those that cross only one joint (Neumann 2017). A muscle that spans a single joint can exert influence on multiple aspects of joint movement. Specifically, it can generate torque around various axes of joint rotation. An illustrative example of this phenomenon is the anterior portion of the gluteus medius. Depending on the hip joint's position, this muscle can simultaneously contribute to flexion, abduction, and internal rotation of the femur around the hip's rotational axes (Standring 2020).

Muscles can exert forces that move one segment at a single point of attachment or affect two segments at both attachment sites. Importantly, a muscle has the capacity to accelerate and generate motion at multiple joints, regardless of whether it directly spans those joints. The soleus muscle exemplifies this phenomenon; while primarily functioning as an ankle plantar flexor, it can also promote knee extension despite not crossing the knee joint directly (Neumann 2017). This effect is observable in the standing position. As the soleus contracts to plantar-flex the ankle, the foot's contact with the ground necessitates a concurrent extension of the knee. Remarkably, the soleus accelerates the knee joint with twice

Table 4.2 Comparison of the three types of muscle contractions: isometric, concentric, and eccentric

Characteristic	Isometric muscle contraction	Concentric muscle contraction	Eccentric muscle contraction
Definition	Muscle generates force without changing length	Muscle shortens while generating force	Muscle lengthens while generating force
Primary function	Stabilization	Movement against resistance	Control of movement and deceleration
Force output	Intermediate	Lowest	Highest
Energy cost	Intermediate	Highest	Lowest
Oxygen consumption	Moderate	High	Low
Motor unit activation	Moderate	High	Low
Examples in activity	Holding a position (e.g., plank)	Lifting phase of a bicep curl	Lowering phase of a squat or downhill walking
Role in exercise programs	Used to maintain position or stabilize joints	Used to raise or lift a load	Used to control load during lowering or deceleration
Risk of injury	Low	Moderate	Higher risk, especially under high load or speed
Contribution of gravity	No direct contribution	Opposes movement	Assists movement
Cross-bridge attachment	Number remains constant	Decreases with velocity increase	Can exceed isometric maximum when stretched

the magnitude of its effect on the ankle, despite not physically traversing the knee.

Most human muscles cross a single joint, focusing their primary action on that specific joint. However, biarticular muscles present a distinct category, as they span two joints and allow for a variety of movements, often occurring in opposing sequences.

In the lower extremity, several biarticular muscles are present:

- Gastrocnemius (ankle plantar flexor and knee flexor or extensor in closed kinetic chain motion)
- Rectus femoris (knee extensor and hip flexor)
- Two-joint hamstrings (hip extensors and knee flexors): long head of biceps femoris, semitendinosus, and semimembranosus
- Gracilis (knee and hip flexors)
- Sartorius (hip and knee flexors)

The upper extremity has fewer biarticular muscles:

- Biceps brachii (elbow and shoulder flexors)
- Long head of triceps brachii (elbow and shoulder extensors)

Many distal muscles in the human limbs cross multiple joints, spanning not only proximal joints but also distal joints through long tendons and ligaments. Examples include the following:

- In the leg: plantaris and flexor hallucis longus
- In the arm: flexor carpi radialis and palmaris longus

An example of this complexity is the rectus femoris, which acts as both a hip flexor and a knee extensor. The function of a biarticular muscle is influenced by body positioning and external factors, such as ground reaction forces (Neumann 2017). For instance, during jumping, the rectus femoris mainly contributes to knee extension, but its effectiveness in hip flexion is limited due to the position of the hip, which aligns the muscle's force vector near the hip joint axis (Hamill et al. 2015).

The distance perpendicular to the muscle force vector from the joint center defines the moment arm. Muscle torque is calculated by multiplying the force by the moment arm. An increase in moment arm leads to greater joint torque, even with constant muscle force. For biarticular muscles, the primary action occurs at the joint with the larger moment arm or greater distance from the joint. The hamstring muscle group, for instance, primarily generates hip extension rather than knee flexion due to its larger moment arm at the hip. Similarly, the gastrocnemius produces ankle plantar flexion more effectively than knee flexion because of its greater moment arm at the ankle.

In sports like vertical jumping, the best performance is attained through a proximal-to-distal extension sequence, resulting in ankle plantar flexion. This sequence requires elevated joint moments and extension velocities at the ankle (Bobbert and van Ingen Schenau 1988). Biarticular muscles are important for this process. The gastrocnemius, which crosses both the knee and ankle joints, shows this. Knee extension while jumping enhances gastrocnemius length (Bobbert et al. 1986), enabling it to maintain a low contraction velocity as the ankle rapidly plantar-flexes. This reduced velocity allows the gastrocnemius to produce increased force throughout the jump.

The primary advantage of biarticular muscles in the lower limb is their ability to reduce the workload of single-joint muscles. They create a mechanical coupling between joints, facilitating rapid release of stored elastic energy. Biarticular muscles conserve energy by enabling positive work at one joint while simultaneously allowing negative work at the adjacent joint. During ankle plantar flexion (a concentric action involving positive work), the knee muscles can perform eccentric actions, storing elastic energy through negative work. Throughout the gait cycle, biarticular muscles work in coordinated pairs to facilitate movement: the sartorius and rectus femoris at heel strike; the hamstrings and gastrocnemius at mid-stance; the gastrocnemius and rectus femoris at toe-off; the rectus femoris, sartorius, and hamstrings during forward swing; and the hamstrings and gastrocnemius during foot descent (Zajac et al. 2002). At heel strike, the sartorius (hip and knee flexor) works in conjunction with the rectus femoris (hip flexor and knee extensor). The rectus femoris performs negative work, absorbing energy as the knee flexes, while the sartorius performs positive work as both a knee and hip flex with gravity (Zajac et al. 2002).

Biarticular muscles may experience functional limitations in certain joint positions. Passive insufficiency occurs when antagonistic muscle elongation is limited, preventing a full range of motion. An example is limited knee extension due to tight hamstrings. Conversely, active insufficiency happens when a muscle is excessively shortened, compromising its ability to generate maximum tension. This phenomenon is observed in the wrist, where finger flexors cannot produce maximum grip force when shortened by concurrent wrist flexion.

4.1.12 Determinants Affecting Force and Velocity Production in the Skeletal Muscle

Various physiological and biomechanical factors influence the ability of muscles to produce force and perform rapid contractions. These forces confront resistance from multiple components, including the muscle's own passive resistance, antagonistic muscle groups, surrounding soft tissues, gravitational forces, and the inertia of the load that is altered or controlled (Lieber and Ward 2011).

The key factors affecting force and velocity development in skeletal muscles are determined. These factors include the muscle cross-sectional area, muscle length, muscle fiber length, precontraction muscle preload, neural activation levels, muscle fiber types, and the aging of muscle tissue.

Each of these factors significantly influences the force–velocity characteristics of skeletal muscles, which has important implications for creating targeted rehabilitation protocols in patients with spinal cord injury (Gorgey and Dudley 2007). In the context of rehabilitation of spinal cord injuries, understanding how these factors interact becomes particularly relevant. For instance, changes in the muscle cross-sectional area due to atrophy can significantly impact force production capabilities. Similarly, alterations in neural activation patterns following spinal cord injury can affect both force generation and contraction velocity (Thomas et al. 2010). The length–tension relationship of muscles is another important consideration. In spinal cord injuries, changes in muscle length due to immobilization or altered postures can shift this relationship, affecting the muscle's ability to generate optimal force (Maganaris et al. 2006). Understanding the role of muscle fiber types is also valuable in the rehabilitation of spinal cord injuries. The shift in fiber type composition often observed after spinal cord injury can impact both force production and fatigue resistance, necessitating tailored exercise interventions.

4.1.12.1 Muscle Cross Section and Whole Length of Muscle Fibers

The architectural configuration of muscles significantly influences their capacity to generate force and alter length, which, in turn, affects movement velocity. The latter characteristic is particularly important, as the ability of a muscle to shorten is reflected in both its length changes and contraction speed, depending on the specific context. The size of a muscle is the primary factor determining its strength and force-generating potential. Typically, muscle tissue can produce a maximum contractile force ranging from 25 to 35 N/cm^2 of cross-sectional area, indicating that larger muscles are capable of producing greater force (Lieber and Fridén 2000).

In pennate muscles, the fibers are generally shorter and not aligned with the primary axis of force generation. This arrangement allows for an increased number of sarcomeres to be positioned in parallel, enhancing the muscle's force-producing capabilities. The increased physiological cross-sectional area in pennate muscles permits them to generate greater force than an equivalent mass of parallel fibers (Narici and Maganaris 2006). Muscles with parallel fibers and greater fiber lengths typically exhibit an extended functional range, resulting in increased range of motion and higher contraction velocities. This is due to the alignment of fibers parallel to the axis of force generation, which allows for a greater number of sarcomeres to be arranged in series. Consequently, this configuration leads to increased fiber lengths and an enhanced capacity for rapid shortening (Burkholder and Lieber 2001).

The ratio of the length of the muscle to the length of the tendon is a key factor in determining a muscle's potential for shortening. Muscles that attach to the bone via short tendons (e.g., the rectus abdominis) can undergo greater shortening distances compared to those with longer tendons (e.g., the gastrocnemius). Substantial shortening also occurs because skeletal muscles can reduce their length by approximately 30%–50% from their resting state. Additionally, muscles with less pennation can shorten over greater distances and achieve higher velocities (e.g., hamstrings and anterior tibialis). Conversely, muscles with greater pennation (e.g., the gastrocnemius) are capable of generating larger forces (Aagaard et al. 2001).

4.1.12.2 Length of Muscle Fibers

The force output of a muscle during contraction is significantly influenced by the length at which the muscle is maintained (Herzog et al. 1992). The length of a muscle can increase, decrease, or remain stable during contraction, depending on the external forces opposing its action. The anatomical structure of the region and the points of attachment to the bone establish limitations on the length of a muscle.

The maximum tension that a muscle fiber can produce is achieved when the muscle is activated at a length slightly exceeding its resting state, generally within a range of 80%–120% of the resting length. It is advantageous that the length of most muscles in the human body falls within this optimal force production range, enhancing efficient muscular function (Gordon et al. 1966).

At shorter lengths of the muscle, the increased overlap between actin and myosin filaments restricts optimal cross-bridge formation, leading to less force output. As the length of the muscle increases toward its resting length, filament overlap becomes more beneficial, allowing for maximum force production. Beyond the optimal length, the filament overlap decreases, leading to a reduction in the number of potential cross-bridges and, consequently, a decrease in force production (Rassier et al. 1999).

The passive component of muscle tension becomes increasingly significant as the muscle is stretched beyond its resting length. This passive tension arises primarily from the elastic properties of the connective tissues within the muscle, particularly the perimysium and epimysium. The combination of active and passive tension results in the characteristic length–tension curve observed in skeletal muscles (Lieber and Ward 2011).

4.1.12.3 Tension Development at Reduced Muscle Lengths

The ability of a muscle to generate tension decreases when it is activated at very short or excessively elongated lengths. At the level of the sarcomere, the optimal length correlates with the maximum overlap of myofilaments, hence enhancing the production of the highest number of cross-bridges (Lieber and Fridén 2000).

When a muscle has contracted to half of its resting length, its ability to generate additional contractile tension is significantly compromised. The decrease in tension at shorter lengths is due to the myofilaments exceeding their optimal overlapping state, resulting in incomplete activation of the cross-bridges due to a decreased number of potential formations (Herzog et al. 1992). As a consequence of this length–tension relationship, a muscle exhibits reduced strength and force-generating capacity at the extremes of a joint's range of motion or at the end point of a segment's movement (Gordon et al. 1966).

The reduced force production at shortened muscle lengths can be explained by the sliding filament theory. As the muscle shortens, the myosin filaments approach the Z-lines of the sarcomere, causing interference with cross-bridge formation. Furthermore, the actin filaments from opposite ends of the sarcomere begin to overlap, hence hindering the formation of effective cross-bridges (Rassier et al. 1999).

Understanding this aspect of muscle physiology can be helpful for rehabilitation professionals when designing exercise programs and assessing muscle function. For instance, strength testing performed at the end ranges of motion may not accurately represent the muscle's maximal force-generating capacity. Similarly, exercises that focus solely on shortened muscle lengths may not provide optimal stimulus for strength gains (Lieber and Ward 2011).

4.1.12.4 Tension at Elongated Muscle Lengths

When a muscle is elongated and subsequently activated, the initial tension in the muscle fibers is elevated due to the separation of cross-bridges after their initial attachment (Edman et al. 1978). This elevated tension continues until the length of the muscle slightly exceeds its resting state. Further elongation and contraction of the muscle result in a decline in tension generation, attributed to the slippage of cross-bridges, which leads to a reduction in the number of formed cross-bridges.

4.1.12.5 Neural Factors in Muscle Activation

The force output of a muscle is determined by the quantity of cross-bridges formed at the sarcomere level. Both the nature of motor unit stimulation and the types of motor units recruited influence the force output. As motor unit recruitment progresses from type I slow-twitch to type IIa and then type IIb fast-twitch fibers, force output increases from lower to higher levels. The recruitment of additional motor units or the engagement of fast-twitch fibers results in enhanced force production (Henneman et al. 1965).

4.1.12.6 Effects of Muscle Preloading on Contraction Force

When a concentric muscle action is preceded by a prestretch through eccentric muscle action, the subsequent concentric action can generate greater force. This phenomenon, known as the stretch–shortening cycle, involves an increase in muscle tension through the storage of potential elastic energy in the series elastic component of the muscle (Komi 1984). The stretching of a muscle results in a minor change in the length of the muscle and tendon (Kubo et al. 1999), accompanied by maximum accumulation of stored energy. Consequently, when a concentric muscle action follows, an enhanced recoil effect contributes to the force output through the muscle–tendon complex (Finni et al. 2000).

A concentric muscle action initiated immediately after a prestretch is further enhanced by the stored elastic energy in the connective tissue surrounding the muscle fibers. This stored energy results in a high force output in the initial phase of the concentric muscle action as these tissues return to their original length. The utilization of stored elastic energy is time-sensitive. If the shortening contraction of the muscle occurs within a reasonable time frame after the stretch (up to 0.9 s), the stored energy is recovered and utilized. However, if the stretch is maintained for an extended period before shortening occurs, the stored elastic energy dissipates by its conversion to heat (Komi 2000).

4.1.12.7 Effective Use of Muscle Prestretch

A brief, low-amplitude prestretch performed rapidly is the most efficacious technique for enhancing the performance of concentric muscle action. This enhancement is achieved through the reclamation of elastic energy and increased activation of the muscle (Bosco et al. 1981; Komi 2000). To maximize the energy return during the negative or eccentric phase, athletes should initiate the stretch rapidly but with limited range. It is important to promptly transition into concentric muscle action without stopping at the end of the stretch.

In jumping activities, for instance, a rapid counterjump from the anatomical position, characterized by a drop–stop–pop sequence and involving a descent of only 20–30 centimeters, proves significantly more effective than a jump initiated from a squat position or from a height that induces greater limb flexion (Bobbert et al. 1996).

The response to prestretch varies between slow-twitch and fast-twitch muscle fibers. Muscles predominantly composed of fast-twitch fibers benefit from a high-velocity prestretch over a limited distance due to their superior capacity for elastic energy storage (Komi 2000). The rapid cross-bridge formation in fast-twitch fibers enables them to accommodate a rapid stretch effectively. In contrast, the process of cross-bridge formation in slow-twitch fibers occurs at a slower rate (Bottinelli et al. 1996).

For slow-twitch fibers, a small-amplitude prestretch is less advantageous, as these fibers cannot store energy rapidly enough, and their cross-bridge formation is comparatively slower (Bottinelli et al. 1996; Komi 2000). Thus, slow-twitch fibers gain more advantage from a prestretch that is performed gradually and involves a wider range of motion. Athletes with a predominance of slow-twitch fibers should be recommended to utilize extended prestretches to maximize the advantages of the stretch. However, for most athletes, the preferred method continues to be a rapid prestretch within a limited range of motion. Understanding these fiber-specific responses to prestretch is necessary for customizing training and rehabilitation programs for particular athletes or patients. For instance, endurance athletes, characterized by a higher proportion of slow-twitch fibers, might benefit from different plyometric training protocols compared to sprinters or power athletes.

4.1.12.8 Muscle Fiber Types and Their Influence on Force and Velocity Generation

The force output of a muscle at varying movement speeds is closely associated with its fiber type composition (Lieber and Fridén 2000). During both lengthening (eccentric contraction) and shortening (concentric contraction), fast-twitch fibers demonstrate greater force production compared to slow-twitch fibers. Among all fiber types, type IIb fast-twitch fibers exhibit the highest maximal force generation capacity (Bottinelli et al. 1996) (Table 4.3).

In muscles containing a higher percentage of fast-twitch fibers, increased contraction velocities are observed at any given absolute force level. The increased contraction velocity in fast-twitch fibers is due to two primary mechanisms: rapid calcium ion (Ca^{2+}) release and increased adenosine triphosphatase (ATPase) activity (Schiaffino and Reggiani 2011). Conversely, slow-twitch fibers, which are recruited initially, predominate during low-load activities. These fibers exhibit a lower maximal shortening velocity compared to fast-twitch fibers, indicating their adaptation for prolonged, lower-intensity contractions (Pette and Staron 2000).

4.1.12.9 Age-Related Muscle Change and Sarcopenia

The age-related decline in muscle mass and quality, known as sarcopenia, leads to a reduction in muscle force production. This phenomenon affects bone density, overall function, glucose metabolism, and various other factors, contribut-

Table 4.3 Comparison between the three main types of skeletal muscle fibers

Characteristic	Red/slow (type I) slow-twitch fibers	Red/fast (type IIa) fast oxidative fibers	White/fast (type IIb) fast glycolytic fibers
Color	Red	Red	White
Contraction speed	Slow	Fast	Very fast
Oxidative capacity	High	High	Low
Resistance to fatigue	High	Medium (intermediate)	Low
Diameter (of muscle fiber)	Small	Medium (intermediate)	Large
Capillary density	High	Medium (intermediate)	Low
Mitochondrial density	High	High	Low
Glycogen reserves	Low	Intermediate	High
Myosin ATPase activity	Low	High	High
Main (metabolic) pathway for production of ATP	Aerobic cellular respiration—final stage: oxidative phosphorylation	Both aerobic and anaerobic metabolic pathways	*Only* anaerobic metabolism, esp. anaerobic glycolysis
Anaerobic enzyme content	Low	Medium	*High*
Force production (i.e., force produced by muscle)	Low	Medium-high	Very high
Example of typical use	Repeated low-level contractions, e.g., walking or low-intensity cycling for 30 mins	Used primarily for movements, such as walking (require more energy than postural control but less energy than sprinting). Activities involving speed, strength, and power	Used to produce rapid, forceful contractions to make quick, powerful movements. Short, fast, bursts of power such as heavy weight training, power lifting, and sprints
Examples of skeletal muscles with this type of fiber	Postural muscles of the neck and spine, leg muscles (type I and type IIa fibers)	Leg muscles (large quantities of both type I and type IIa fibers)	Arm muscles

Adapted with permission from Xiao (2018)

ing to functional limitations in older adults (Narici and Maffulli 2010).

Sarcopenia is characterized by both structural and biochemical alterations in aging muscle tissue. Anatomically, several changes occur, including the following:

- Diminished muscle mass and cross-sectional area
- Increased intramuscular fat and connective tissue
- Structural alterations at the sarcomere level
- Reduced number of motor units (Doherty 2003)
- Reduction in type II fiber dimensions
- Decreased number of both type I and II fibers

On a biochemical level, aging muscles exhibit the following:

- Decreased protein synthesis
- Modifications in muscle protein expression
- Alterations in enzyme activity

The decline in muscle force occurs at a rate of approximately 12%–15% per decade after the age of 50 years (Doherty 2003). This rate of strength loss increases with advancing age and is influenced by a complex interaction of anatomical, biochemical, nutritional, and environmental factors.

Research indicates that progressive resistance training is the most effective intervention to decelerate or potentially reverse the effects of aging on muscle tissue (Peterson et al. 2010). This finding has significant implications for rehabilitation specialists working with older adults, as it emphasizes the importance of tailored strength training programs in maintaining muscle function and overall health. Understanding the mechanisms of sarcopenia is particularly important in the context of rehabilitation. The overall effect of age-related muscular alterations and neurological impairment presents particular problems that require specialized strategies to maintain and improve muscle function in this population (Gorgey and Dudley 2007).

4.1.13 Mechanisms of Muscular Strength Enhancement

Muscular strength is defined as the maximum force a muscle or muscle group can generate at its skeletal attachment point (Kraemer and Ratamess 2004). Mechanically, this strength corresponds to the maximum isometric torque generated at a specific joint angle. While strength is often measured by determining the heaviest load that can be moved through a single repetition over a defined range of motion, it is important to note that this movement does not occur at a constant speed. Instead, joint movements typically involve varying velocities throughout the range of motion.

The assessment of strength is influenced by a multitude of factors. These include the type of muscle action—eccentric, concentric, or isometric—and the speed of limb movement and the complex interplay of length–tension relationships, force–angle characteristics, and force–time properties (Enoka 1996). These variables contribute to strength fluctuations across the range of motion, with measurements ultimately limited by the weakest joint position.

Strengthening training aims to achieve two primary objectives: increasing the muscle's cross-sectional area and enhancing the tension generated per unit of that area (Folland and Williams 2007). This principle is applicable to all age groups. The increased cross section, or hypertrophy, resulting from resistance training arises from several physiological adaptations. These include hypertrophy of individual muscle fibers, improved muscle capillarization, and an overall increase in the mean fiber area (MacDougall 1986; Jones et al. 1989). At the cellular level, the enlargement of muscle fibers is attributed to an increase in myofibril size and separation. While some researchers have postulated that muscle fibers might split as part of this process, this phenomenon remains unconfirmed in human studies (Jones et al. 1989).

The increase in tension per unit of cross-sectional area reflects the neural component of strength development (Sale 1988). In the early stages of strength training, a significant portion of strength gains can be attributed to nervous system adaptations. These include improvements in motor unit recruitment, firing rates, and synchronization (Häkkinen et al. 1985). As training progresses, hypertrophy becomes more prominent as the quality of muscle fibers improves.

4.1.13.1 Intensity in Enhancing Muscular Strength

The intensity of a training regimen is an important factor in the development of muscle strength. Research has consistently shown that strength improvements are closely associated with the tension generated within muscle fibers. For a muscle to react and adapt to training stimuli, it must be given stimuli exceeding a particular threshold, a principle referred to as overload (Kraemer et al. 2002). It is important to recognize that the fundamental factor for strength improvement is not solely the number of repetitions executed but rather the intensity of tension in the muscle.

The extent of overload is usually represented as a percentage of the maximal force that a muscle or muscle group can produce. Athletes and trainers usually aim to perform at the highest achievable percentage of maximal lifting capacity to enhance strength development. This method is based on the idea that training with low stress, especially when combined with large repetitions, may result in little strength gains, as the muscle may not be adequately stimulated for producing adaptive reactions. Research suggests that optimal strength gains are generally achieved when the muscle is worked near its maximum tension just before reaching a state of fatigue. This often involves performing sets of two to six repetitions (Campos et al. 2002). It is necessary to acknowledge that this method may require modification for individuals with spinal cord injuries or other neurological disorders, as their fatigue and recovery patterns may markedly differ from those of neurologically intact individuals (Gorgey and Dudley 2007).

The principle of increasing loading supports ongoing strength improvement. As muscles adapt to increased demands, a gradual increase in training load can result in beneficial adaptations in strength, power, and muscular endurance. This overload can be attained by multiple techniques, such as increasing the weight lifted, increasing the number of repetitions, modifying the speed of movements, decreasing rest intervals between exercises, or increasing the overall training volume.

For rehabilitation specialists, especially those treating spinal cord injury patients, a comprehensive understanding and careful application of the principles of intensity and progressive overload are necessary. It is extremely important to customize these methods to the specific requirements, abilities, and limitations of each patient. When used carefully, suitable intensity in strength training programs might potentially accelerate healing, enhance functional capacity, and significantly enhance the overall quality of life (Jacobs and Nash 2004).

4.1.13.2 Rest in Enhancing Muscular Strength

The efficacy and outcomes of a strength development program are closely associated with the rest intervals provided to the muscles. This rest is important at multiple levels: between sets, between training days, and before competitive events. The value of rest for skeletal muscles that have undergone resistive training stress is noteworthy, as these periods allow for recovery and rebuilding of muscle fibers (Schoenfeld et al. 2016a, b). During intense exercise, as skeletal muscles fatigue, their ability to develop tension decreases. This reduction in force production capability indicates that the muscle is no longer working under optimal overload conditions, which may limit the effectiveness of further training at that time (Willardson 2006).

The duration of rest intervals between sets can significantly influence the physiological responses to resistance exercise. Shorter rest intervals (30–60 s) tend to emphasize metabolic stress and may be more beneficial for muscular endurance and hypertrophy. However, longer rest periods (2–5 min) allow enhanced recovery of energy systems and may be more beneficial for maximal strength development (de Salles et al. 2009). Rest between training days is equally important. The frequency of training sessions for a particular muscle group should be balanced to allow adequate time for recovery and adaptation. While the optimal frequency can vary based on individual factors and training goals, most research suggests that training each muscle group two or three times per week provides a good balance between stimulus and recovery for most individuals (Schoenfeld et al. 2016a).

For individuals with spinal cord injuries, the concept of rest and recovery takes on additional importance. The altered neuromuscular function and potential changes in metabolic processes following spinal cord injury may require modified rest protocols. Rehabilitation specialists must carefully monitor fatigue levels and adjust rest periods accordingly to optimize strength gains while minimizing the risk of overtraining or secondary complications (Jacobs and Nash 2004).

4.1.13.3 Volume of Work in Enhancing Muscular Strength

The workload executed by a muscle is a significant factor in evaluating muscle recovery. The volume of work is determined by the total number of repetitions multiplied by the weight lifted (Kraemer and Ratamess 2004). This can be calculated on a weekly, monthly, or annual basis, including all main lifts and their frequencies. Two lifters could achieve the same weekly volume despite different routines: one may perform three sets of ten repetitions at 100 lb, resulting in a total volume of 3000 lb, while another may perform three sets of two repetitions at 500 lb, also totaling 3000 lb.

The optimal number of sets for strength development has been a topic of considerable debate. Some studies suggest that similar strength gains can be achieved with single sets as with multiple sets (Carpinelli and Otto 1998). However, a growing body of evidence supports significantly higher strength gains with three sets compared to single sets (Rhea et al. 2003). For heavy-resistance exercises with low repetitions, a rest period of 5–10 min between sets is necessary for energy system recovery (Kraemer and Ratamess 2004). For those engaging in heavy-resistance exercises with low repetitions, it is advisable to rest for 5–10 min between sets to allow for energy system recovery. Rest periods of less than 3 min may activate different energy systems, leading to lactic acid accumulation in the muscles.

At the outset of a weight-training regimen, the volume is typically high, characterized by more frequent sessions, a higher number of lifts per session, more sets per exercise, and greater repetitions per set. As training progresses, the volume is gradually reduced by decreasing the frequency of weekly sessions, reducing the number of sets, increasing the intensity of lifts, and lowering the number of repetitions. A common recommendation is to complete approximately 20,000 repetitions annually, which can be divided into monthly and weekly volumes depending on the required load adjustments. This allows for variations in volume across days and weeks to incorporate both higher- and lower-volume sessions.

Bodybuilders often employ short-rest, high-intensity training to enhance their muscle size, potentially compromising some strength gains that would otherwise be achieved with longer rest intervals. In cases where extended rest is impractical, incorporating high-repetition, low-resistance circuit training between high-resistance lifts may help mitigate lactic acid buildup. Bodybuilders typically work with loads ranging from six to twelve repetitions maximum (RM), which explains their lower strength levels compared to powerlifters or weightlifters who lift heavier loads.

Strength development for performance enhancement generally follows a structured plan, often involving periodization, as described in the literature for various sports and activities. This approach typically involves gradually increasing loads while decreasing the lifting volume over several months to avoid plateaus or strength losses due to slowed physical adaptations. As the competitive season approaches, lifting volume may be reduced by up to 60%, which can actually enhance muscle strength. Furthermore, if an athlete discontinues lifting immediately before a performance, strength levels can be maintained for at least 5 days and may even improve slightly following a short rest period.

4.1.13.4 Resistive Exercise Principles for Diverse Populations

The fundamental concepts of resistance exercise are often discussed in the context of athletes. These principles are equally useful in rehabilitation settings and for older adults, pediatric patients, and individuals with low fitness levels. Contemporary fitness guidelines increasingly utilize resistance training as an essential component of overall health maintenance. The American College of Sports Medicine recommends at least one set of resistance exercises twice weekly, including 8–12 repetition movements for adult populations (American College of Sports Medicine 2018). Individuals inexperienced in resistance training often exhibit remarkable progress compared to their trained counterparts, exhibiting positive responses to a wide range of protocols.

Resistance exercise has gained recognition as a beneficial modality for older adults. The age-associated decline in muscular strength is largely attributed to reduced physical activity levels. Consistent engagement in resistance training into advanced age may counteract bone tissue atrophy and potentially slow the progression of degenerative joint changes (Peterson et al. 2010). Research has also demonstrated the efficacy of eccentric training in enhancing strength among older adults (Roig et al. 2009). A comprehensive resistance training program for this demographic should prioritize specific muscle groups, including the cervical flexors, scapular stabilizers, core musculature, hip extensors, and quadriceps.

The primary variable in resistance training programs across different populations—from elite athletes to older adults and children—should be the magnitude of resistance applied. For instance, while a conditioned athlete might perform a dumbbell lateral raise with a 50-lb (22.7-kg) weight, an older adult may achieve a similar training effect by simply elevating the arm against gravity. It is imperative to implement high-resistance exercises with caution, particularly in pediatric and geriatric populations. In older adults, especially those with reduced bone density, excessive loading of the skeletal system through high-intensity lifting may lead to fractures (Beck et al. 2017).

In pediatric populations, the growth plates are particularly vulnerable to injury under high loads or improper technique. Consequently, high-intensity resistance programs are not advised for children. However, when proper safety measures are observed, children and adolescents can achieve significant strength gains through resistance training (Faigenbaum et al. 2009). Regular participation in a progressive resistance training program offers numerous potential benefits for young individuals, including enhanced bone strength, improved body composition, reduced injury risk, enhanced sports performance, and increased muscular endurance (Lloyd et al. 2014).

4.1.13.5 Kinetic Chain Exercise: Closed Versus Open Systems

While isokinetic assessment remains a common practice among rehabilitation specialists, there has been a shift in training methodologies. Many clinicians have transitioned from isokinetic training to closed-chain exercises, which incorporate both eccentric and concentric muscle actions utilizing the patient's body weight (Escamilla et al. 2012). A closed-chain exercise is defined as an isotonic exercise where the distal segment of the limb is fixed, such as when the foot or hand is in contact with a stable surface. An example of a closed-chain exercise targeting the quadriceps is the squat movement, performed with the feet remaining firmly on the ground.

There is a growing consensus among rehabilitation professionals that closed-chain exercises may offer superior benefits compared to open-chain exercises. For instance, a knee extension performed on an isokinetic dynamometer or a knee extension machine represents an open-chain exercise. The perceived advantages of closed-chain exercises include the utilization of body weight, preservation of muscle synergies, and enhanced transferability to functional activities of daily living (Stensdotter et al. 2003). Research has demonstrated that closed-chain kinetic exercises for the knee joint promote a more balanced activation pattern across the quadriceps muscle group compared to open-chain exercises (Escamilla et al. 2012). This finding has significant implications in the context of accelerated rehabilitation protocols following anterior cruciate ligament (ACL) reconstruction, where there is an increasing trend toward the implementation of closed-chain kinetic exercises in physical therapy regimens.

The question over open- and closed-chain exercises still remains unresolved. Notably, some studies have found no significant difference in the strains produced at the ACL during open- versus closed-chain exercises despite observations of increased anterior tibial translation during open-chain movements. This suggests that both exercise modalities may have their place in comprehensive rehabilitation programs, depending on the specific goals and stage of recovery. For instance, in patients with incomplete spinal cord injuries who have some lower extremity function, closed-chain exercises might be incorporated to promote weight-bearing and functional movement patterns. Conversely, open-chain exercises might be more appropriate in the early stages of rehabilitation or for patients with complete injuries, where isolated muscle strengthening is a primary goal.

4.1.13.6 Muscular Adaptations to Inactivity, Injury, and Immobilization

The effects of disuse or immobilization on muscle tissue can be profound. Atrophy, a primary indicator of limb immobilization, can manifest as a significant reduction in the muscle cross-sectional area, with studies reporting decreases of 20%–30% following 8 weeks of cast immobilization (Wall et al. 2013). This process of atrophy is precipitated by muscle remodeling, characterized by protein loss and alterations in muscle metabolism. The extent of atrophy appears to be muscle-specific, with lower extremity muscles experiencing more severe reductions in cross-sectional area compared to back or upper extremity muscles (Narici and de Boer 2011). The most significant changes occur during the initial weeks of disuse, emphasizing the importance of early intervention in rehabilitation and exercise protocols.

The process of muscle regrowth after inactivity or immobilization varies across age groups. Younger individuals generally exhibit more successful muscle regeneration compared to older adults, and the regrowth process differs between fast-twitch and slow-twitch muscle fibers (Suetta et al. 2013). It is significant that, even when the cross-sectional area of an atrophied muscle is successfully restored, the force output of the muscle often remains behind the structural recovery.

Muscular injury results in diminished force-producing capabilities. In response, compensatory mechanisms emerge, either through functional adaptations in other muscles or alterations in movement patterns to minimize reliance on the injured muscle. For instance, injury to a hip flexor can lead to a significant reduction in force production in the soleus due to its role in propelling the trunk forward during plantar flexion in gait (Neptune et al. 2001). Similarly, injury to the gluteus maximus can result in a redistribution of hip extension duties to the gluteus medius and hamstrings. It is important to recognize that the loss of function in one muscle can have far-reaching impacts across all joints in the kinetic chain, such as the lower extremity. Therefore, rehabilitation efforts should adopt a holistic approach, focusing on the entire musculoskeletal system rather than isolated muscle groups (Lepley and Palmieri-Smith 2015).

4.2 Structural and Biomechanical Characteristics of the Skin

The skin, the largest organ by surface area, performs multiple functions, including environmental protection, temperature regulation, and adaptation to body contours during movement. Its complex structure, comprising several layers, each with multiple components, results in intricate mechanical properties. Understanding the basic structure and biomechanical properties of the skin is particularly relevant in the context of pressure injury prevention and management. The skin forms a durable, flexible, and waterproof barrier that transitions smoothly into more delicate internal lining membranes at various body openings, such as the mouth, nose, eyelids, and urogenital and anal regions. The dermis, a key layer of human skin, primarily functions to nourish and mechanically support the epidermis, a thin cellular layer with limited mechanical strength on its own (Standring 2020).

In daily activities, the skin undergoes significant deformations and forces due to joint movements, minor trauma, and common actions like turning a door handle. This necessitates that the dermis allow for high elastic deformations with minimal force application while simultaneously maintaining high tensile strength (Silver et al. 2003). Beyond its protective role, the skin serves as a crucial sensory organ. It contains numerous nerve endings, making it the primary source of general sensations, including touch, pressure, temperature variations, and pain perception (McGlone and Reilly 2010).

The structure of human skin comprises several layers:

- Epidermis: A stratified, cellular outer layer.
- Dermis: An underlying layer of connective tissue.
- Dermal–epidermal junction: An undulating interface where epidermal ridges, known as rete ridges, project into the dermis. This junction provides mechanical support for the epidermis and acts as a partial barrier against the exchange of cells and large molecules.
- Subcutaneous layer: Below the dermis lies a fatty layer, the panniculus adiposus, commonly referred to as the "subcutaneous" layer (Kanitakis 2002).

The skin maintains the body's fluid balance through its waterproofing capability, largely facilitated by fatty secretions from sebaceous glands. These fatty secretions also contribute to vitamin D synthesis. Despite its efficient waterproofing, the skin possesses an absorptive capacity for certain drugs, vitamins, and hormones when applied appropriately. The skin also plays a role in excretion, particularly through sweating (Wong et al. 2016).

As a thermoregulatory organ, the skin helps maintain body temperature within a narrow range. Its blood supply and sweat glands facilitate heat loss through radiation, convection, and evaporation. In conjunction with the lungs, the skin accounts for over 90% of the body's total heat loss (Romanovsky 2014). The skin vasculature adjusts to environmental conditions by dilating to promote heat loss or constricting to conserve heat.

The skin's metabolic functions require a large surface area, approximately 1.8 m^2 in adults. Skin thickness varies with age and body region, ranging from 0.5 mm over the eyelids to thicker areas on the back of the neck, upper trunk, palms, and soles. Generally, the skin is thicker on posterior and extensor surfaces compared to anterior and flexor surfaces, with an average thickness of 1–2 mm (Sandby-Moller et al. 2003).

The skin demonstrates a balance between flexibility and firm adherence to underlying structures. In some regions, it is tightly bound, while in others, it is more loosely connected, allowing for ease of movement. Continuous friction prompts an increase in the thickness of the skin's superficial layers as a natural defense mechanism.

As people age, skin elasticity diminishes, leading to stretching in less firmly attached areas. The stretching pattern is influenced by the orientation of collagen fibers, which generally align perpendicularly to the stretch. In areas requiring high mobility, like joint flexion creases, the skin is securely anchored to underlying fibrous tissues to allow flexibility without interference from subcutaneous fat or mobile skin (Naylor et al. 2011).

Langer's lines, or cleavage lines, represent the direction of least tension in the skin, aligning with the natural orientation of dermal collagen fibers and, often, paralleling underlying muscle fibers. These lines are significant in surgical practice; incisions along these lines typically heal with minimal scarring, while cuts against them may lead to more prominent scarring due to internal stress (Borges and Alexander 1962).

Skin color varies based on melanin content and dermal blood flow. Temperature changes and blood oxygenation affect the skin color. Racial and individual skin color differences are determined by the distribution of melanin in the epidermal layers, which can increase with sun exposure (Rees 2003).

The skin's sensory capabilities are highly advanced, especially in the hands and digits. It also features unique patterns on the fingers, toes, palms, and soles, forming the basis for fingerprint identification. These patterns are due to the arrangement of dermal papillae, aiding grip and preventing slippage. Notably, these areas lack sebaceous glands and hair. Additionally, the skin is the origin of various appendages, including nails, hair, sebaceous glands, and sweat glands. The intricate pattern of skin creases forms distinct lines, with hair often emerging at the intersections of these creases.

4.2.1 Structure and Layers of the Skin: Epidermis and Dermis

The skin consists of two main layers: the superficial epidermis, which is of ectodermal origin, and the underlying dermis, derived from mesodermal tissue. The skin exemplifies remarkable biological engineering. Its complex structure allows it to perform numerous functions, from protection to sensation. However, in certain circumstances, such as chronic conditions, disabilities, or specific sports activities, the skin—particularly of the lower limbs—can face unusual mechanical stresses. These stresses may arise from extended periods of bed rest, wheelchair use, or the wearing of orthoses or prostheses, pushing the skin's adaptability to its limits (Gefen 2011).

To understand how the skin responds to these challenges, we need to consider its intricate structure. The skin is composed of three main layers: the epidermis, dermis, and hypodermis, or subcutis (Wong et al. 2016).

The epidermis is notably thin, measuring approximately 75–150 μm in thickness, and consists of four sub-layers: the stratum spinosum, stratum basale, stratum granulosum, and stratum corneum. Each of these is necessary for maintaining the integrity of the epidermal barrier (Kanitakis 2002). Beneath the epidermis lies the dermis, a dense fibrous tissue that gives the skin its strength and elasticity. Together with the epidermis, it forms the cutis, measuring about 1.5–2.5 mm in thickness. The dermis is moderately dense connective tissue containing three fibrous proteins: collagen, elastin, and small amounts of reticulin. These proteins are embedded in a supporting matrix, providing the skin with its remarkable ability to withstand mechanical stress (Silver et al. 2003). The hypodermis (subcutaneous fat, or subcutis) is primarily composed of adipose cells and varies significantly in thickness across different parts of the body. The hypodermis serves as both insulation and a cushion, accounting for about 10% of our body mass. This layer is particularly important in areas subjected to regular pressure or impact, such as the soles.

The integumentary system, with the skin as its primary component, demonstrates remarkable variability across different anatomical regions of the body. This heterogeneity in structure and function is a direct response to the diverse mechanical and environmental challenges faced by different body parts. The skin, originating from the mesodermal tissue, exhibits significant thickness variations throughout the body. Notably, the extensor surfaces of the limbs, including areas such as the buttocks, anterior thigh, and anterior lower leg, generally present thicker skin compared to other regions. An exception to this pattern is observed in the plantar surface of the foot, which exhibits exceptional thickness, a clear adaptation to its weight-bearing function (Holbrook and Odland 1974).

A comparative analysis of the skin on flexor and extensor aspects of the limbs reveals distinct characteristics. Flexor surfaces typically have a reduced hair follicle density and demonstrate enhanced sensitivity due to increased nerve innervation. In contrast, extensor surfaces not only have greater overall thickness but also possess a higher number of cellular layers. These structural differences contribute significantly to variations in skin resilience and susceptibility to mechanical damage across different body regions.

The epidermis, specifically its outer keratinized epithelium, shows variations in thickness and structural composition, influenced by local mechanical stresses and anatomical location. Beneath the keratinized surface lies the stratum corneum, a layer characterized by rapid keratin polymerization. Both the keratin layer and stratum corneum exhibit significantly increased thickness in plantar regions compared to other body areas, further illustrating the skin's adaptability to functional demands.

Functionally, the epidermis serves as a multifaceted protective barrier. It defends against mechanical trauma, provides protection from ultraviolet radiation, forms a relatively impermeable blockade against water and various substances, and constitutes the body's first line of defense against microbial invasion. These various functions demonstrate the critical role of the skin in maintaining overall health and homeostasis.

The skin's functional capabilities are further enhanced by its associated glands. Sebaceous glands produce lipid-rich secretions that soften the keratin layer, while eccrine glands contribute to skin hydration through their aqueous secretions. However, this hydration can be a double-edged sword. Excessive exposure to water can lead to maceration of the keratin layer, resulting in softening, swelling, pallor, and increased susceptibility to damage. This phenomenon, often referred to as "dishwasher's hands," is particularly relevant in contexts of prolonged water exposure (Sinnatamby 2011). The issues of perspiration and maceration extend beyond common scenarios and are of particular concern in specific populations. Athletes, individuals who regularly use orthoses or prostheses, and those who rely on wheelchairs for mobility face unique challenges. In these cases, extended contact between the skin and the equipment or seating surfaces can create microenvironments conducive to moisture accumulation, potentially leading to skin breakdown. This emphasizes the necessity for specialized care and prevention strategies in these populations (Sanders et al. 1995).

The overall thickness of the skin is predominantly influenced by two key elements: the stratum corneum and the dermal layer's dimensions. The dermis, frequently considered the essential layer of the skin, comprises dense, irregular connective tissue with abundant vasculature and cutaneous appendages. Its structural properties, including density, tensile strength, compliance, and thickness, exhibit significant regional variations. Anatomically, the dermis is subdivided into two distinct strata: the superficial papillary layer, characterized by a more loosely arranged structure, and the underlying reticular layer, which exhibits a more compact organization (Standring 2020).

The cutaneous vascular network is categorized, consisting of nutritive capillaries and thermoregulatory vessels, each fulfilling specific physiological roles. Nutritive capillaries form vertical loops that supply essential metabolites to the superior dermis and the avascular epidermis. In contrast, thermoregulatory vessels, which are larger in caliber and situated more deeply, course parallel to the cutaneous surface and contribute significantly to thermal homeostasis through heat dissipation. The hemodynamics and importance of nutritive capillaries, particularly in relation to their role in pressure ulcer pathogenesis, have been the focus of extensive scientific research, with emphasis on occlusion pressures and microcirculatory changes.

Hair structures originate from epidermal invaginations known as follicles. These obliquely oriented follicles penetrate deeply into the dermal layer, occasionally extending into the hypodermis. A notable characteristic of these follicular structures is their rich sensory innervation. The arrector pili muscles, smooth muscle fibers attached to the hair follicle sheaths and innervated by sympathetic nerve endings, are integral to the body's thermoregulatory response. Cold stimuli induce contraction of these muscles, resulting in cutis anserina, commonly referred to as "goosebumps." This physiological reaction manifests as a transient alteration in cutaneous topography, contributing to thermal regulation by modifying the skin's surface properties (Standring 2020).

Sebaceous glands are typically positioned at the interface of the hair follicle and the arrector pili muscle. These glands release their lipid-rich secretion upon muscle contraction. This physiological process is particularly pronounced during adolescence. In individuals with limb amputations, sebaceous glands present a unique concern: the potential development of epidermoid cysts on the residual limb, which may arise following infection of an obstructed sebaceous duct (Standring 2020).

The hypodermis, or subcutaneous adipose tissue, is a thick layer of connective tissue located beneath the dermis. It acts as a stabilizing layer between the dermis and the underlying deep fascia or, in some regions, directly the bone. This layer provides support to delicate structures like blood vessels and nerve endings while also cushioning the dermis from underlying tissues such as muscles and bones. This layer is notably prominent in the extremities, though its thickness and distribution exhibit considerable variability. Atrophy of this adipose layer, particularly in pressure-prone areas, reduces the protective cushioning around skeletal prominences. This reduction in natural padding may exacerbate the risk of pressure injury by intensifying localized compressive forces.

4.2.1.1 Epidermis: Composition and Function

The epidermis, a stratified squamous epithelium with a thickness ranging from 0.3 to 1.0 mm, is predominantly composed of keratinocytes. However, it also incorporates melanocytes and dendritic cells, including Langerhans cells. Keratinocytes perform multiple functions: They synthesize structural proteins such as epidermal keratin, natural moisturizing factors, and barrier lipids; proliferate to facilitate wound healing and replace exfoliated corneocytes (the outermost skin cells); transport water, glycerol, and urea via aquaporins; receive melanin from melanocytes; and contain antigen-presenting Langerhans cells, which serve as cutaneous protectors (Standring 2020).

Barrier lipids are produced in epidermal lamellar bodies, which also contain antimicrobial peptides, thus forming both permeability and antimicrobial barriers. Research has demonstrated that nerve fibers innervate all nucleated epidermal layers (Hilliges et al. 1995). The nucleated component of the epidermis comprises three distinct layers: the stratum basale (housing stem cells and transiently amplifying postmitotic cells), the stratum spinosum (also known as the spinous layer), and the stratum granulosum. The epidermal structure is characterized by multiple cellular strata, with deeper cells being metabolically active and proliferative. These cells gradually migrate superficially, undergoing cornification en route. Upon reaching the surface, they are shed through mechanical friction against clothing and other surfaces.

Despite being avascular, the epidermis contains sensory nerve endings and is intimately connected to the underlying dermis via rete ridges and corresponding dermal papillae. Lacking direct vascularization, nutrient transport within this layer occurs through diffusion via intercellular fluids, following passage across the selective barrier of the basement membrane that separates the epidermis from the vascularized dermis.

The epidermis is subdivided into several distinct layers, particularly pronounced in the thick skin of the palmar and plantar surfaces. These layers, arranged from deep to superficial, are as follows:

- *Stratum basale (basal cell layer)*: This monolayer of cells faces the dermoepidermal junction. The cells exhibit a cuboidal to columnar morphology, with a high nucleocytoplasmic ratio. They adhere to the basement membrane via hemidesmosomes and to adjacent cells via desmosomes. Together with the stratum spinosum, it plays a key role in producing new cells to replace those shed from the skin surface (Standring 2020). Keratinocytes originate in this layer, where they begin their life cycle. As new cells divide, they push neighboring cells upward to the skin surface. Melanin production also starts at this stage, providing a protective barrier against harmful ultraviolet radiation.

- *Stratum spinosum (prickle cell layer)*: Comprising multiple layers of irregularly shaped cells, the cells here are also cuboidal to columnar, with a high nucleocytoplasmic ratio. These become increasingly flattened as they approach the stratum granulosum. The cells maintain intercellular adhesion through desmosomes. The stratum basale and stratum spinosum collectively constitute the proliferative compartment, often termed the germinative zone, due to their role in cellular regeneration. Keratinization is initiated in this layer, where melanocytes, in response to sunlight, produce melanosomes and melanin. These are transferred to adjacent keratinocytes, offering protection against harmful ultraviolet radiation. As the melanin integrates into the keratinocytes, these cells continue their upward migration toward the stratum corneum, eventually undergoing desquamation.
- *Stratum granulosum (granular layer)*: In this layer, cells undergo progressive flattening and elongation, initiating the process of keratinization. These cells represent a transitional state, progressing toward cell death, making a significant step in the maturation process of the skin. As cells advance into the granular layer, several changes take place within the keratinocytes. Keratinization is finalized, and the hardened, granular structure of the keratinocytes helps prevent the absorption of various substances through the skin while also aiding in preserving the hydration of the deeper layers.
- *Stratum lucidum (clear layer)*: The stratum lucidum is composed of transparent cells that contain eleidin, a substance derived from keratohyalin. This layer is an important transitional phase in the formation of the outermost layer of the epidermis (stratum corneum). This layer represents a key transitional stage in the formation of the stratum corneum, where cells release lipids that form bilayers, acting as a "glue" to hold the corneocytes together and support the hydration of the epidermis. By the time the cells reach the clear layer, keratinization is fully completed.
- *Stratum corneum (horny layer)*: This superficial layer, from which corneocytes are shed, is the primary determinant of epidermal thickness. It plays a significant role in the skin's barrier function (Elias and Wakefield 2014). Upon reaching its final stage, the keratinocyte transforms into a corneocyte, serving as the body's first line of defense against injury and bacterial invasion and providing a waterproof barrier.

Within the deepest aspect of the epidermis reside epidermal melanocytes, the cellular source of cutaneous pigmentation. These dendritic cells synthesize and distribute melanin throughout the keratinocyte, contributing to the skin's overall chromatic properties and its variations across different anatomical regions and individuals (Standring 2020).

4.2.1.2 Dermis: Composition and Function

The dermis comprises about 90% of the skin's weight and forms its foundational layer. It is divided into two distinct zones:

- The superficial papillary dermis: This layer, just below the basement membrane zone, is characterized by a loose network of connective tissue with thin collagen fibrils.
- The reticular layer: Found deeper within the dermis, this layer consists of a dense connective tissue matrix with thick, regularly oriented collagen bundles (Standring 2020).

Collagen is the most abundant protein in the dermis, with types I, III, and V distributed throughout, and type IV present in the basement membrane. Notably, type III collagen is more prevalent in the papillary dermis, while type I collagen predominates in the reticular dermis.

The dermis, positioned deep into the epidermis, consists of a dense network of collagen and elastic fibers, contributing significantly to the skin's overall thickness. It is subdivided into two distinct regions: the superficial papillary layer, characterized by its fine texture, and the deeper, coarser reticular layer. While the papillary layer interconnects with the epidermis, establishing a clear demarcation, the reticular layer transitions seamlessly into the underlying hypodermis.

Dermal papillae, projections from the dermis, are necessary for cutaneous vascularity. These structures contain capillary networks that facilitate proximal blood flow to the epidermis, contributing to thermoregulation and physiological responses such as emotional flushing. Some papillae also contain tactile receptors, with their density correlating with the tactile sensitivity of different anatomical regions. For instance, these receptors are abundant in highly sensitive areas like the digits and labial region, while they are less concentrated in areas of reduced sensitivity, such as the dorsum.

The reticular layer of the dermis is characterized by a dense network of interwoven collagen and elastic connective tissue fibers, which provide the skin with its strength and elasticity. These fibers are predominantly oriented parallel to the cutaneous surface, with regional variations in alignment. This variation in fiber alignment gives rise to Langer's lines (cleavage lines), which exhibit anatomical site-specific patterns (Standring 2020).

The dermis is a highly complex tissue, including various structural and functional components. It contains an extensive network of vascular and lymphatic channels, neural elements, and sensory receptors. Furthermore, it contains a small amount of adipose tissue, pilosebaceous units, sudoriferous glands, and arrector pili muscles. The deep dermal surface exhibits invaginations from the hypodermis, allowing the entry of neurovascular structures into the skin. This complex organization shows the dermis's essential function in skin physiology and structural integrity (Palastanga and Soames 2012).

The dermis is distinguished by an abundant vascular supply, although no vessels traverse the dermoepidermal junction. Cutaneous motor innervation is autonomic, consisting of cholinergic fibers that supply eccrine sweat glands and adrenergic components innervating both eccrine and apocrine glands, smooth muscle fibers, arterioles, and arrector pili muscles. Sensory innervation of the skin is diverse, including free nerve endings, hair follicle–associated terminals, and specialized mechanoreceptors with expanded tips.

4.2.2 Biomechanics of the Skin in the Context of Pressure-Related Injuries

The biomechanical properties of the skin, including its rigidity, extensibility, and tensile strength, are significant in understanding pressure-related injuries, dermal senescence, and pathological mechanisms. These properties inform medical professionals about trauma processes, reconstructive surgical techniques, dermatological treatments, and scar tissue formation. Pressure-related injuries involve localized damage to the skin and/or underlying tissues, often occurring over osseous prominences or in association with medical devices. These injuries are influenced by both extrinsic and intrinsic factors. Extrinsic factors involve external conditions like excessive pressure, shear, friction, and tissue maceration, contributing to the injury's development. The damage arises from prolonged or intense pressure or a combination of pressure and shear forces.

The development of pressure-related injuries is inversely related to the magnitude and duration of applied force. High-intensity, short-duration pressure can be as detrimental as prolonged, lower-intensity pressure (Barnett and Ablarde 1995; Patterson and Fisher 1980). When tissue pressure surpasses capillary occlusion pressure, progressive ischemia occurs, leading to cellular necrosis. Pressure distribution is nonuniform, with the highest concentration occurring in tissues adjacent to bony structures. Muscle tissue exhibits greater susceptibility to pressure effects compared to cutaneous tissue.

Ischemia is a fundamental mechanism in the pathogenesis of pressure-related injuries. Sustained forces of sufficient magnitude can obstruct capillary blood flow, impairing cellular metabolism and resulting in tissue necrosis. Tissue injury occurs if the pressure applied exceeds the normal capillary perfusion pressure, which ranges from 14 mmHg on the venous side to 35 mmHg on the arterial side, with an overall capillary perfusion pressure of 20–30 mmHg. Notably, microscopic changes can be observed in the skin after just 1 h of continuous pressure at 60 mmHg (Patterson and Fisher 1980). Actual pressure necrosis may occur at sustained pressure levels as low as 30 mmHg, even in normal or denervated tissue.

Skin thickness, which varies according to the anatomical location, age, and sex, significantly influences its biomechanical properties. Comparative analysis reveals that forehead skin is thicker, more rigid, less taut, and less elastic than ventral forearm skin, reflecting region-specific structural variations. Age-related changes include dermal thinning due to reduced collagen and ground substance synthesis, with thickness decreasing by approximately 6% per decade. Male skin exhibits greater thickness than female skin in corresponding anatomical regions, attributed to higher collagen content.

4.2.3 Wound Healing Process of the Skin

The layers of normal skin function as a protective barrier against external threats. However, when this barrier is compromised through injury, the body initiates a process of wound healing. This complex process unfolds in four distinct yet overlapping phases: hemostasis, inflammation, proliferation, and maturation.

Phase 1: Hemostasis

Immediately following an injury, within the initial few minutes, the body prioritizes stopping the bleeding. Platelets (thrombocytes) rapidly accumulate at the injury site, forming a fibrin clot. This clot not only halts active bleeding but also establishes a temporary barrier against invading pathogens.

Phase 2: Inflammation

As the clot hardens and dries, it forms a protective scab over the wound. The scab provides a crucial layer of defense, allowing the skin beneath to start its repair process. During this phase, any premature removal of the scab can disrupt the newly regenerating tissue, leading to increased skin damage and a larger scar. The scab, although temporary, plays a vital role in protecting the underlying repair processes.

Phase 3: Proliferation

This phase begins approximately 1–24 days after the injury. It is characterized by the creation of a foundation for long-term wound repair. During this stage, fibroblasts deposit collagen, which strengthens the wound, while new cells migrate across the newly formed tissue to close the wound. The process of reepithelialization occurs as connective skin tissue forms to replace the damaged areas. The scab eventually loosens and falls off. One of the key aspects of this phase is wound contraction, where the wound size can decrease by as much as 40%–80%.

Phase 4: Maturation

The maturation phase occurs from 14 to 365 days post-injury. This final stage is when the tissues are remodeled and the dermis and epidermis connect and contract to fully close the wound. During maturation, scar tissue is formed as the body replaces the initial collagen with newly aligned collagen fibers. Unlike the random basket weave formation found in healthy skin, the collagen in scar tissue aligns in a single direction, giving scars a distinct appearance, texture, and reduced flexibility compared to normal skin.

4.3 Structural and Mechanical Properties of Fascia

According to medical terminology, fascia is defined as "a sheet or band of fibrous connective tissue that envelops, separates, or connects muscles, organs, and other soft structures of the body" (*American Heritage Stedman's Medical Dictionary* 2007). This definition restricts the term "fascia" to well-defined fibrous connective tissue layers. In the human body, from the cutaneous layer to the muscular plane, three primary fibrous connective layers are typically present: the superficial fascia, the deep fascia, and the epimysium, in addition to the various visceral fasciae. However, this layered arrangement is not always distinctly defined. In certain anatomical regions, one or more layers may be absent or tightly connected. For instance, in the palmar and plantar regions, the superficial and deep fasciae adhere to form a single connective layer, known as the palmar aponeurosis (Stecco et al. 2011) and plantar fascia/aponeurosis, respectively (Natali et al. 2010).

Historically, anatomists have differentiated between superficial and deep fascia, although many surgeons consider "fascia" to be synonymous with "deep fascia." The superficial fascia is conventionally described as a layer of areolar connective or adipose tissue located immediately beneath the skin. In contrast, the deep fascia is a more robust, dense connective tissue that is continuous with the superficial layer. The deep fascia commonly forms sheet-like structures and typically encases the underlying muscles and tendons like a stocking. An example of the deep fascia dividing compartments in the upper extremity is shown in Fig. 4.16.

Fasciae perform various functions, including acting as a soft tissue skeleton (ectoskeleton) that supports muscle attachments and providing protective coverings for underlying structures (Benjamin 2009). A notable example of fascia functioning as an ectoskeleton can be seen in muscles such as the gluteus maximus and tensor fascia lata, which attach predominantly to deep fascia (the iliotibial tract or band) rather than directly to the bone (Findley et al. 2012; Stecco et al. 2011) (Fig. 4.17).

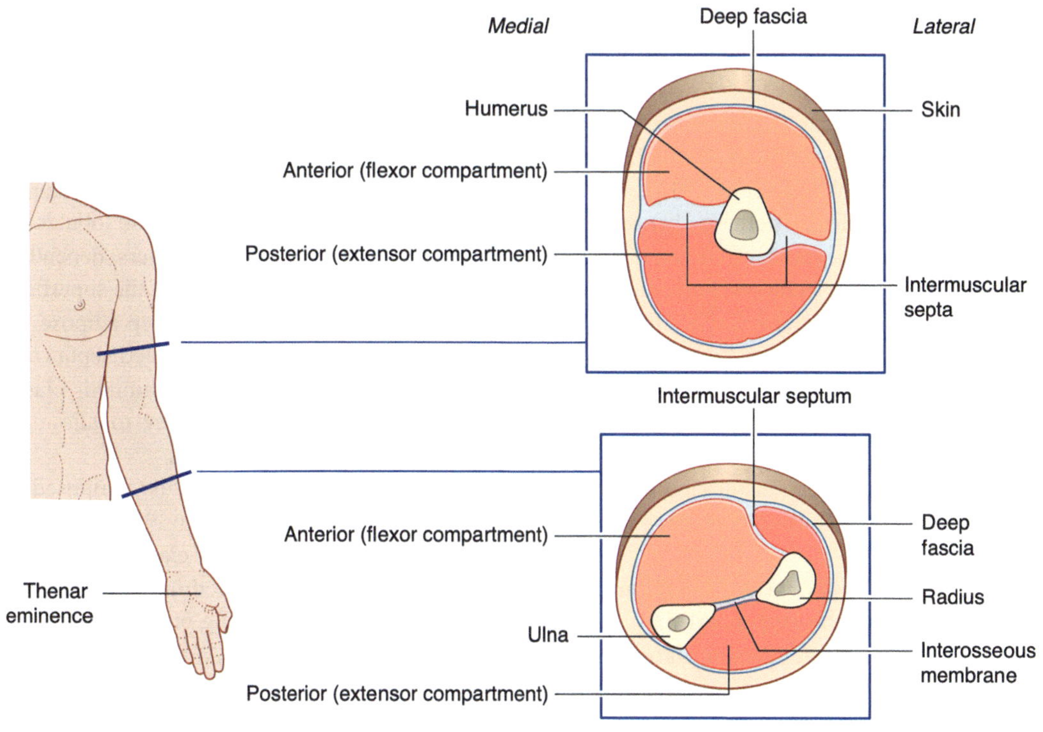

Fig. 4.16 Deep fascia dividing the compartments in the upper extremity. (Adapted with permission from Drake et al. (2020))

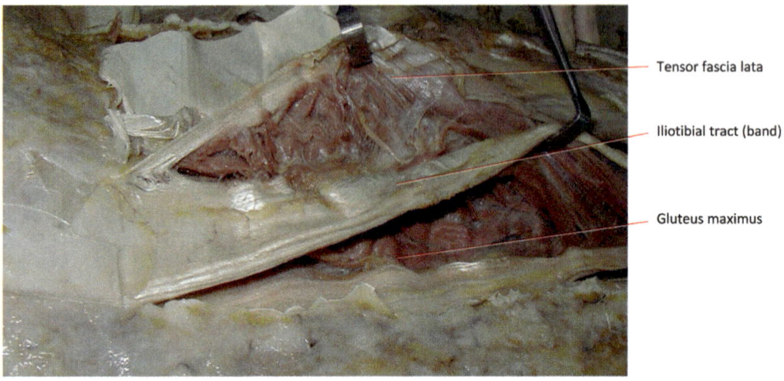

Fig. 4.17 Iliotibial tract with associated muscles. Tensor fascia lata attaches ventrally, while the gluteus maximus attaches dorsally to the iliotibial tract

4.3.1 Anatomical Forms of Fascia

The fascial system exhibits various anatomical forms, each with unique characteristics and functions. The superficial fascia is conventionally identified as a layer of loose connective or adipose tissue situated immediately beneath the dermis. This differs from the deep fascia, which is characterized by its tough and dense connective tissue composition, forming a continuum with the superficial layer (Standring 2020).

The deep fascia often appears in sheet-like configurations, encasing the underlying musculature and tendons similarly to a close-fitting stocking. This arrangement facilitates the structural integrity and functional dynamics of the musculoskeletal system (Willard et al. 2012).

The morphological differences between these fascial layers are significant for understanding their respective roles in biomechanics and pathophysiology. The superficial fascia, characterized by its looser structure, allows for enhanced flexibility and serves as a conduit for neurovascular structures. Conversely, the deep fascia, with its denser composition, provides increased mechanical support and force transmission capabilities (Stecco et al. 2011; Willard et al. 2012).

4.3.1.1 Superficial Fascia (Membranous Layer of the Hypodermis)

The superficial fascia is present throughout the body, with its arrangement and thickness varying by anatomical region, body surface, and sex. It is thicker in the lower extremities compared to the upper extremities and on the posterior aspect of the body rather than the anterior and is more pronounced in females than in males (Lancerotto et al. 2011). This fascial layer is a membranous connective tissue of variable thickness within the subcutaneous tissue, dividing it into superficial and deep adipose layers. Retinacula connect this membranous layer—referred to as the superficial fascia—to the skin and the deep fascia, creating a three-dimensional network between fat lobules (Abu-Hijleh et al. 2006).

The superficial adipose tissue consists of large fat lobules encased within fibrous septa, forming a honeycomb-like structure. It shows relatively consistent characteristics throughout the body. The septa (retinacula cutis superficialis) are well-defined, predominantly oriented perpendicular to the surface, and mechanically strong, anchoring the dermis to deeper planes. Adipose lobules are organized in single to multiple layers, dependent on fat content and the thickness of the superficial adipose tissue. In contrast, the deep adipose tissue features more obliquely oriented septa (retinacula cutis profundus), has limited elastic properties, and shows a tendency to allow the displacement of adipose lobules.

The superficial fascia itself is composed of interwoven collagen fibers, loosely packed and interspersed with abundant elastic fibers. Its thickness is more pronounced in the trunk compared to the limbs, becoming progressively thinner toward the extremities. At bony prominences and certain ligamentous folds, the superficial fascia adheres to the deep fascia (Standring 2020). Functionally, the superficial fascia contributes to maintaining skin integrity and supports subcutaneous structures, such as veins, by promoting their patency (Stecco et al. 2011).

The superficial fascia serves as a conduit for blood vessels and nerves, facilitating their passage to and from the skin. It also enhances movement between the skin and underlying structures. This function is particularly evident in areas overlying highly mobile joints and on the dorsum of the hand, where the skin exhibits considerable freedom of movement, allowing it to glide effortlessly over the extensor tendons during finger movements (Benjamin 2009). The mobility of the skin provided by the superficial fascia serves a protective function for both the skin and the deeper structures, shielding them from physical trauma. This mobility is facilitated by multiple sheets of collagen fibers, in conjunction with the presence of elastin (Kawamata et al. 2003). A consequence of the movement-promoting characteristics of the superficial fascia is that the blood vessels and nerves within it must follow a tortuous route. This arrangement allows them to adapt to changes in skin position relative to deeper structures (Standring 2020).

Fascial tissue also manifests in other forms throughout the body. Loose, areolar connective tissue surrounds skeletal muscle fibers, forming the endo- and epimysium, and creates thin films of tissue between adjacent muscles. In contrast, the deep fasciae in the limbs and back typically consist of dense connective tissue sheets characterized by large numbers of closely packed collagen fibers (Stecco et al. 2011).

4.3.1.2 Deep Fascia

The deep fascia is a fibrous membrane that forms a complex network, enveloping and separating muscles, creating sheaths for nerves and vessels, reinforcing ligaments around joints, and binding structures together into a cohesive mass. Although deep fascia surrounds all body muscles, its characteristics vary by region (Stecco et al. 2011). In the limbs, the deep fasciae appear as well-defined laminae of connective tissue, with an average thickness of 1 mm. Under this layer, muscles can glide freely due to their epimysium. A layer of loose connective tissue, rich in hyaluronic acid, lies between the epimysium and the deep fascia, aiding movement (McCombe et al. 2001). The deep fascia of the limbs is composed of two or three parallel layers of collagen fibers, forming a surrounding sheath. This fascia extends and connects with connective tissue layers that envelop muscle groups (such as the crural fascia in the leg), individual muscles (epimysium), muscle fascicles (perimysium), individual muscle fibers (endomysium), and bones (periosteum). These connective tissue layers are interlinked, creating a continuous network that integrates muscles, muscle compartments, and individual fibers. Additionally, tendons of numerous skeletal muscles anchor to the bone, both directly and via the fascia (Benjamin 2009; Stecco et al. 2007, 2011).

The deep fascia covering the extremities is a tough, membranous structure composed of dense, fibrous connective tissue that is mostly avascular. Its thickness varies across anatomical regions, but it consistently shows a resilient, irregular structure. The main functions of the deep fascia include offering cutaneous support and allowing smooth movement of the underlying musculature (Findley and Shalwala 2013).

In certain anatomical locations, the deep fascia acts as an attachment point for skeletal muscles. An example in the lower limb is the iliotibial tract, a significant part of the fascia lata, which provides insertion points for the tensor fasciae latae and gluteus maximus muscles. This arrangement allows these muscles to affect both the coxofemoral and tibiofemoral joints (Standring 2020).

The deep fascia contains many sensory fibers, making it highly responsive to stimuli. Interestingly, its innervation pattern matches that of the overlying skin rather than the underlying musculature. This unique feature influences pain perception and proprioception (Stecco et al. 2007).

The deep fasciae are strengthened around joints by retinacula, traditionally viewed as isolated structures functioning as pulleys to maintain tendon proximity to underlying bones during movement. Retinacula contribute to joint stability (Leardini and O'Connor 2002) and have roles in proprioception (Viladot et al. 1984) and peripheral motor coordination.

In the trunk, the structure of deep fasciae resembles that of limb fasciae in the thoracolumbar fascia, rectal sheath, and neck fasciae. However, the pectoral fascia and fasciae of the deltoid, trapezius, and gluteus maximus muscles exhibit different characteristics. This variation likely stems from the distinct embryological development of these muscles and fasciae. According to Sato and Hashimoto (1984), the superficial layer of trunk muscles develops within the superficial fascial layer, maintaining firm adherence through numerous intramuscular septa. Many muscle fibers also originate from the inner aspect of the surrounding fascia.

This structural organization of deep fasciae and their relationship with muscles and joints have implications for biomechanics and functional anatomy.

The reinforcement of fasciae around joints through retinacula suggests a role in maintaining joint integrity during movement. The connection between fasciae and muscle development in the trunk indicates a complex interplay between these structures, potentially influencing muscle function and force transmission.

4.3.2 Functional Properties of the Fascia

Fascial tissue provides mechanical stability between muscles, enabling force transmission and reducing friction during movement. This connective tissue transmits mechanical signals through its extensive network, contributing to proprioception and motor coordination (Stecco et al. 2011). The mechanical and sensory functions of fascia extend beyond local effects. Fascial continuity allows force transmission between distant body segments through myofascial chains. This anatomical arrangement modifies the traditional understanding of muscle function, as forces generated by muscle contraction can influence structures beyond immediate attachment sites (Wilke et al. 2016).

Fascia contains numerous mechanoreceptors, including Pacinian corpuscles and Ruffini endings, providing continuous feedback about tissue deformation and mechanical stress. These sensory inputs contribute to movement control and postural awareness. Studies indicate that fascial mechanoreceptors respond to various mechanical stimuli, including pressure, stretch, and vibration.

Studies have elucidated the function of fascia in fluid dynamics and tissue homeostasis. The fascial system functions in maintaining proper fluid balance and facilitating lymphatic drainage. Furthermore, fascia demonstrates adaptive properties, altering its structural organization in response to mechanical loading (Magnusson et al. 2010).

4.3.2.1 Protective Role

Fascia performs a protective function in specific body regions. Notably, dense sheets of connective tissue are present in the palms and soles. The palmar aponeurosis in the hand and the plantar aponeurosis in the foot shield the underlying vessels and nerves that course between the proximal and distal aspects of these extremities (Benjamin 2009).

These aponeuroses also connect the skin to the skeleton, regulating its movement during locomotion (Bojsen-Moller and Flagstad 1976). This arrangement is particularly important for the distribution of mechanical forces and the prevention of shear stress on neurovascular structures. In the palmar region, the aponeurosis forms a tough, fibrous sheet that overlies the flexor tendons, lumbricals, and neurovascular bundles. Its structure allows for the transmission of force from the flexor tendons to the skin, enhancing grip strength while protecting deeper structures (Standring 2020). Similarly, the plantar aponeurosis in the foot acts as a protective layer for the deeper soft tissues. It also contributes to the maintenance of the foot's arch and aids in the propulsive phase of gait. During toe extension, the plantar aponeurosis tightens through the windlass mechanism, elevating the arch and creating a rigid lever for push-off (Hicks 1954).

The protective role of fascia extends beyond these specific regions. Throughout the body, fascial layers provide a buffer against mechanical stress, reducing the risk of injury to underlying structures. This function is particularly evident in areas subjected to repeated impact or pressure, such as the gluteal region and the heel pad (Stecco et al. 2011).

4.3.2.2 Compartmentalizing Role

The firm nature of the deep fascia facilitates its role in segmenting and organizing muscle groups into defined regions referred to as "compartments." This fascial tissue unifies these compartments and mediates load transmission between them. As the compartmentalization function of deep fascia operates in conjunction with associated bones and intermuscular septa, these spaces are sometimes referred to as osteofascial compartments. Each limb segment—whether it be the arm, forearm, thigh, or lower leg—has characteristic compartments that distinguish functional groups of muscles based on their embryologic origin, vascular supply, and innervation. These compartments are generally named according to their anatomical location (anterior, posterior, medial, lateral, etc.) or based on the actions of their enclosed muscles.

Intermuscular septa, which are often attached to the periosteum, share a common developmental origin with the fascia, explaining why fascia seldom makes contact with the bone without also attaching to it. The tensile load on the deep fascia increases when tendons, particularly those in the posterior compartment of the lower leg, are stretched during eccentric muscle contraction. This stress is transmitted to the fascial attachment at the tibial crest.

The deep fascia in the lower leg is notably thick, corresponding with its structural role. Strong intermuscular septa extend inward from the deep fascia to attach to the periosteum of the tibia and fibula, dividing the limb into compartments for the dorsiflexor, plantar flexor, and peroneal (fibular) muscles. The anterior compartment, which contains the dorsiflexor muscles (including the tibialis anterior), is of clinical interest due to its association with compartment syndrome, a painful condition that can threaten limb viability. This syndrome occurs when pressure builds within the confined space of the compartment, impeding blood flow due to the restrictive nature of the deep fascia (Standring 2020).

4.3.2.3 Muscle Pumping Role

A significant function of deep fascia in the extremities is to provide a constraining sheath for underlying muscles. When these muscles contract against the resilient, thick, and resistant fascia, the thin-walled veins and lymphatics within the muscles undergo compression. The unidirectional valves in these vessels direct blood and lymph toward the heart. The prominence of deep fascia in the lower limb, compared to the upper limb, is partially due to the importance of muscle pumping for venous and lymphatic return, given the distance of the lower leg and foot below the heart.

The focus has shifted to deep vein thrombosis as a means of demonstrating the significance of the muscle pump. Comprehending the role of fascia in the muscle-pumping mechanism of the lower limb requires a thorough understanding of the venous system. A clear distinction is made between superficial and deep veins based on their position relative to the deep fascia. These two venous systems are connected by perforating veins, which traverse the deep fascia, linking the veins on both sides. Each of these three types of veins is equipped with valves that prevent backflow and divide the hydrostatic column of blood into segments (Meissner et al. 2007). Incompetent valves in the leg can lead to diminished muscle pump efficiency. Interestingly, perforating veins in the foot lack valves, permitting bidirectional flow. The calf muscle pump is the most significant in terms of volume and function, though it is initiated by the smaller muscle pumps in the foot. The influence of a similar pump in the thigh is considered minimal.

References

Aagaard P, Andersen JL, Dyhre-Poulsen P, Leffers AM, Wagner A, Magnusson SP, et al. A mechanism for increased contractile strength of human pennate muscle in response to strength training: changes in muscle architecture. J Physiol. 2001;534(2):613–23.

Abu-Hijleh MF, Roshier AL, Al-Shboul Q, Dharap AS, Harris PF. The membranous layer of superficial fascia: evidence for its widespread distribution in the body. Surg Radiol Anat. 2006;28(6):606–19.

Ackland DC, Pandy MG. Lines of action and stabilizing potential of the shoulder musculature. J Anat. 2009;215(2):184–97.

American College of Sports Medicine. ACSM's guidelines for exercise testing and prescription. 10th ed. Philadelphia: Wolters Kluwer; 2018.

American Heritage Stedman's medical dictionary. Boston: Houghton Mifflin Company; 2007.

Azizi E, Brainerd EL, Roberts TJ. Variable gearing in pennate muscles. Proc Natl Acad Sci USA. 2008;105(5):1745–50.

Banks RW. The innervation of the muscle spindle: a personal history. J Anat. 2015;227(2):115–35.

Barnett RI, Ablarde JA. Skin vascular reaction to short durations of normal seating. Arch Phys Med Rehabil. 1995;76(6):533–40.

Beck BR, Daly RM, Singh MA, Taaffe DR. Exercise and Sports Science Australia (ESSA) position statement on exercise prescription for the prevention and management of osteoporosis. J Sci Med Sport. 2017;20(5):438–45.

Benjamin M. The fascia of the limbs and back: a review. J Anat. 2009;214(1):1–18.

Benjamin M, Toumi H, Ralphs JR, Bydder G, Best TM, Milz S. Where tendons and ligaments meet bone: attachment sites ('entheses') in relation to exercise and/or mechanical load. J Anat. 2006;208(4):471–90.

Bewick GS, Banks RW. Mechanotransduction in the muscle spindle. Pflugers Arch. 2015;467(1):175–90.

Bobbert MF, van Ingen Schenau GJ. Coordination in vertical jumping. J Biomech. 1988;21(3):249–62.

Bobbert MF, Huijing PA, van Ingen Schenau GJ. An estimation of power output and work done by the human triceps surae muscle-tendon complex in jumping. J Biomech. 1986;19(11):899–906.

Bobbert MF, Gerritsen KG, Litjens MC, Van Soest AJ. Why is countermovement jump height greater than squat jump height? Med Sci Sports Exerc. 1996;28(11):1402–12.

Bojsen-Moller F, Flagstad KE. Plantar aponeurosis and internal architecture of the ball of the foot. J Anat. 1976;121(3):599–611.

Borges AF, Alexander JE. Relaxed skin tension lines, Z-plasties on scars, and fusiform excision of lesions. Br J Plast Surg. 1962;15:242–54.

Bosco C, Komi PV, Ito A. Prestretch potentiation of human skeletal muscle during ballistic movement. Acta Physiol Scand. 1981;111(2):135–40.

Bottinelli R, Canepari M, Pellegrino MA, Reggiani C. Force-velocity properties of human skeletal muscle fibres: myosin heavy chain isoform and temperature dependence. J Physiol. 1996;495(2):573–86.

Brodal P. The central nervous system: structure and function. 4th ed. Oxford: Oxford University Press; 2010.

Burden SJ, Huijbers MG, Remedio L. Fundamental molecules and mechanisms for forming and maintaining neuromuscular synapses. Int J Mol Sci. 2018;19(2):490.

Burkholder TJ, Lieber RL. Sarcomere length operating range of vertebrate muscles during movement. J Exp Biol. 2001;204(9):1529–36.

Campos GE, Luecke TJ, Wendeln HK, Toma K, Hagerman FC, Murray TF, et al. Muscular adaptations in response to three different resistance-training regimens: specificity of repetition maximum training zones. Eur J Appl Physiol. 2002;88(1–2):50–60.

Carpinelli RN, Otto RM. Strength training: single versus multiple sets. Sports Med. 1998;26(2):73–84.

de Salles BF, Simão R, Miranda F, Novaes JS, Lemos A, Willardson JM. Rest interval between sets in strength training. Sports Med. 2009;39(9):765–77.

Deschenes MR. Motor unit and neuromuscular junction remodeling with aging. Curr Aging Sci. 2011;4(3):209–20.

Doherty TJ. Invited review: aging and sarcopenia. J Appl Physiol. 2003;95(4):1717–27.

Drake RL, Vogl AW, Mitchell AWM. Gray's anatomy for students. 4th ed. Philadelphia: Elsevier; 2020.

Duchateau J, Baudry S. Insights into the neural control of eccentric contractions. J Appl Physiol. 2014;116(11):1418–25.

Dyck PJ, Thomas PK. Peripheral neuropathy. 4th ed. Philadelphia: Elsevier; 2004.

Edman KA, Elzinga G, Noble MI. Enhancement of mechanical performance by stretch during tetanic contractions of vertebrate skeletal muscle fibres. J Physiol. 1978;281:139–55.

Egan B, Zierath JR. Exercise metabolism and molecular regulation of skeletal muscle adaptation. Cell Metab. 2013;17(2):162–84.

Elias PM, Wakefield JS. Mechanisms of abnormal lamellar body secretion and the dysfunctional skin barrier in patients with atopic dermatitis. J Allergy Clin Immunol. 2014;134(4):781–91.

Ellaway PH, Taylor A, Durbaba R. Muscle spindle and fusimotor activity in locomotion. J Anat. 2015;227(2):157–66.

Enoka RM. Eccentric contractions require unique activation strategies by the nervous system. J Appl Physiol. 1996;81(6):2339–46.

Enoka RM. Neuromechanics of human movement. 4th ed. Champaign: Human Kinetics; 2008.

Enoka RM, Duchateau J. Translating fatigue to human performance. Med Sci Sports Exerc. 2016;48(11):2228–38.

Escamilla RF, Macleod TD, Wilk KE, Paulos L, Andrews JR. Anterior cruciate ligament strain and tensile forces for weight-bearing and non-weight-bearing exercises:

a guide to exercise selection. J Orthop Sports Phys Ther. 2012;42(3):208–20.

Faigenbaum AD, Kraemer WJ, Blimkie CJ, Jeffreys I, Micheli LJ, Nitka M, et al. Youth resistance training: updated position statement paper from the National Strength and Conditioning Association. J Strength Cond Res. 2009;23(5 Suppl):S60–79.

Findley TW, Shalwala M. Fascia Research Congress evidence from the 100 year perspective of Andrew Taylor Still. J Bodyw Mov Ther. 2013;17(3):356–64.

Findley T, Chaudhry H, Stecco A, Roman M. Fascia research–a narrative review. J Bodyw Mov Ther. 2012;16(1):67–75.

Finni T, Komi PV, Lepola V. In vivo human triceps surae and quadriceps femoris muscle function in a squat jump and counter movement jump. Eur J Appl Physiol. 2000;83(4–5):416–26.

Folland JP, Williams AG. The adaptations to strength training: morphological and neurological contributions to increased strength. Sports Med. 2007;37(2):145–68.

Franchi MV, Reeves ND, Narici MV. Skeletal muscle remodeling in response to eccentric vs. concentric loading: morphological, molecular, and metabolic adaptations. Front Physiol. 2017;8:447.

Frink M, Hildebrand F, Krettek C, Brand J, Hankemeier S. Compartment syndrome of the lower leg and foot. Clin Orthop Relat Res. 2010;468(4):940–50.

Frontera WR, Ochala J. Skeletal muscle: a brief review of structure and function. Calcif Tissue Int. 2015;96(3):183–95.

Fuglevand AJ, Winter DA, Patla AE. Models of recruitment and rate coding organization in motor-unit pools. J Neurophysiol. 1993;70(6):2470–88.

Fukuda T. The stepping test: two phases of the labyrinthine reflex. Acta Otolaryngol. 1959;50(2):95–108.

Fukunaga T, Roy RR, Shellock FG, Hodgson JA, Day MK, Lee PL, et al. Physiological cross-sectional area of human leg muscles based on magnetic resonance imaging. J Orthop Res. 1992;10(6):928–34.

Fukunaga T, Ichinose Y, Ito M, Kawakami Y, Fukashiro S. Determination of fascicle length and pennation in a contracting human muscle in vivo. J Appl Physiol. 1997;82(1):354–8.

Gans C, de Vree F. Functional bases of fiber length and angulation in muscle. J Morphol. 1987;192(1):63–85.

Garrett WE Jr. Muscle strain injuries. Am J Sports Med. 1996;24(6 Suppl):S2–8.

Gefen A. How do microclimate factors affect the risk for superficial pressure ulcers: a mathematical modeling study. J Tissue Viability. 2011;20(3):81–8.

Gillies AR, Lieber RL. Structure and function of the skeletal muscle extracellular matrix. Muscle Nerve. 2011;44(3):318–31.

Gordon AM, Huxley AF, Julian FJ. The variation in isometric tension with sarcomere length in vertebrate muscle fibres. J Physiol. 1966;184(1):170–92.

Gordon AM, Homsher E, Regnier M. Regulation of contraction in striated muscle. Physiol Rev. 2020;80(2):853–924.

Gorgey AS, Dudley GA. Skeletal muscle atrophy and increased intramuscular fat after incomplete spinal cord injury. Spinal Cord. 2007;45(4):304–9.

Häkkinen K, Komi PV, Alén M. Effect of explosive type strength training on isometric force- and relaxation-time, electromyographic and muscle fibre characteristics of leg extensor muscles. Acta Physiol Scand. 1985;125(4):587–600.

Hamill J, Knutzen KM, Derrick TR. Biomechanical basis of human movement. 4th ed. Philadelphia: Wolters Kluwer; 2015.

Henneman E, Somjen G, Carpenter DO. Functional significance of cell size in spinal motoneurons. J Neurophysiol. 1965;28(3):560–80.

Herzog W. Skeletal muscle mechanics: questions, problems and possible solutions. J Neuroeng Rehabil. 2017;14(1):98.

Herzog W. The multiple roles of titin in muscle contraction and force production. Biophys Rev. 2018;10(4):1187–99.

Herzog W, Kamal S, Clarke HD. Myofilament lengths of cat skeletal muscle: theoretical considerations and functional implications. J Biomech. 1992;25(8):945–8.

Herzog W, Leonard TR, Joumaa V, DuVall M. Mysteries of muscle contraction. J Appl Biomech. 2015a;31(1):1–13.

Herzog W, Powers K, Johnston K, Duvall M. A new paradigm for muscle contraction. Front Physiol. 2015b;6:174.

Hicks JH. The mechanics of the foot. II. The plantar aponeurosis and the arch. J Anat. 1954;88(1):25–30.

Hilliges M, Wang L, Johansson O. Ultrastructural evidence for nerve fibers within all vital layers of the human epidermis. J Invest Dermatol. 1995;104(1):134–7.

Hodges PW, Gandevia SC. Changes in intra-abdominal pressure during postural and respiratory activation of the human diaphragm. J Appl Physiol. 2000;89(3):967–76.

Holbrook KA, Odland GF. Regional differences in the thickness (cell layers) of the human stratum corneum: an ultrastructural analysis. J Invest Dermatol. 1974;62(4):415–22.

Horak FB. Postural orientation and equilibrium: what do we need to know about neural control of balance to prevent falls? Age Ageing. 2006;35(Suppl 2):ii7–11.

Hughes BW, Kusner LL, Kaminski HJ. Molecular architecture of the neuromuscular junction. Muscle Nerve. 2006;33(4):445–61.

Huijing PA. Epimuscular myofascial force transmission: a historical review and implications for new research. J Biomech. 2009;42(1):9–21.

Huxley AF. Muscle structure and theories of contraction. Prog Biophys Biophys Chem. 1957;7:255–318.

Huxley AF. Mechanics and models of the myosin motor. Philos Trans R Soc Lond Ser B Biol Sci. 2000;355(1396):433–40.

Jacobs PL, Nash MS. Exercise recommendations for individuals with spinal cord injury. Sports Med. 2004;34(11):727–51.

Jami L. Golgi tendon organs in mammalian skeletal muscle: functional properties and central actions. Physiol Rev. 1992;72(3):623–66.

Jankowska E. Spinal interneuronal networks in the cat: elementary components. Brain Res Rev. 2008;57(1):46–55.

Jones DA, Rutherford OM, Parker DF. Physiological changes in skeletal muscle as a result of strength training. Q J Exp Physiol. 1989;74(3):233–56.

Kandel ER, Schwartz JH, Jessell TM, Siegelbaum SA, Hudspeth AJ. Principles of neural science. 5th ed. New York: McGraw-Hill; 2013.

Kanitakis J. Anatomy, histology and immunohistochemistry of normal human skin. Eur J Dermatol. 2002;12(4):390–9.

Kannus P. Structure of the tendon connective tissue. Scand J Med Sci Sports. 2000;10(6):312–20.

Kapandji IA. The physiology of the joints: lower limb. 6th ed. Edinburgh: Churchill Livingstone; 2007.

Katirji B, Kaminski HJ, Ruff RL. Neuromuscular disorders in clinical practice. 2nd ed. New York: Springer; 2014.

Kawamata S, Ozawa J, Hashimoto M, Maeda T, Shimo S. Structure of the rat subcutaneous connective tissue in relation to its sliding mechanism. Arch Histol Cytol. 2003;66(3):273–9.

Kendall FP, McCreary EK, Provance PG, Rodgers MM, Romani WA. Muscles: testing and function with posture and pain. 5th ed. Baltimore: Lippincott Williams & Wilkins; 2005.

Kisner C, Colby LA. Therapeutic exercise: foundations and techniques. 5th ed. Philadelphia: F.A. Davis Company; 2007.

Kjaer M. Role of extracellular matrix in adaptation of tendon and skeletal muscle to mechanical loading. Physiol Rev. 2004;84(2):649–98.

Ko HY. Management and rehabilitation of spinal cord injuries. 2nd ed. Singapore: Springer; 2022.

Komi PV. Physiological and biomechanical correlates of muscle function: effects of muscle structure and stretch-shortening cycle on force and speed. Exerc Sport Sci Rev. 1984;12:81–121.

Komi PV. Relevance of in vivo force measurements to human biomechanics. J Biomech. 1990;23(Suppl 1):23–34.

Komi PV. Stretch-shortening cycle: a powerful model to study normal and fatigued muscle. J Biomech. 2000;33(10):1197–206.

Kraemer WJ, Ratamess NA. Fundamentals of resistance training: progression and exercise prescription. Med Sci Sports Exerc. 2004;36(4):674–88.

Kraemer WJ, Adams K, Cafarelli E, Dudley GA, Dooly C, Feigenbaum MS, et al. American College of Sports Medicine position stand. Progression models in resistance training for healthy adults. Med Sci Sports Exerc. 2002;34(2):364–80.

Kubo K, Kawakami Y, Fukunaga T. Influence of elastic properties of tendon structures on jump performance in humans. J Appl Physiol. 1999;87(6):2090–6.

Kuo AD, Zajac FE. Human standing posture: multi-joint movement strategies based on biomechanical constraints. Prog Brain Res. 1993;97:349–58.

Kuo AD, Donelan JM, Ruina A. Energetic consequences of walking like an inverted pendulum: step-to-step transitions. Exerc Sport Sci Rev. 2005;33(2):88–97.

Lancerotto L, Stecco C, Macchi V, Porzionato A, Stecco A, De Caro R. Layers of the abdominal wall: anatomical investigation of subcutaneous tissue and superficial fascia. Surg Radiol Anat. 2011;33(10):835–42.

LaStayo PC, Woolf JM, Lewek MD, Snyder-Mackler L, Reich T, Lindstedt SL. Eccentric muscle contractions: their contribution to injury, prevention, rehabilitation, and sport. J Orthop Sports Phys Ther. 2003;33(10):557–71.

Latash ML, Lestienne F. Motor control and learning. New York: Springer; 2006a.

Leardini A, O'Connor JJ. A model for lever-arm length calculation of the flexor and extensor muscles at the ankle. Gait Posture. 2002;15(3):220–9.

Leijnse JN. A generic morphological model of a muscle group—application to the muscles of the forearm. Acta Anat (Basel). 1997;160(2):100–11.

Lepley LK, Palmieri-Smith RM. Quadriceps strength, muscle activation failure, and patient-reported function at the time of return to activity in patients following anterior cruciate ligament reconstruction: a cross-sectional study. J Orthop Sports Phys Ther. 2015;45(12):1017–25.

Levangie PK, Norkin CC. Joint structure and function: a comprehensive analysis. 4th ed. Philadelphia: F.A. Davis Company; 2005.

Leversedge FJ, Moore TJ, Peterson BC, Seiler JG 3rd. Compartment syndrome of the upper extremity. J Hand Surg Am. 2011;36(3):544–59.

Li L, Xiong WC, Mei L. Neuromuscular junction formation, aging, and disorders. Annu Rev Physiol. 2018;80:159–88.

Lieber RL, Fridén J. Functional and clinical significance of skeletal muscle architecture. Muscle Nerve. 2000;23(11):1647–66.

Lieber RL, Ward SR. Skeletal muscle design to meet functional demands. Philos Trans R Soc Lond Ser B Biol Sci. 2011;366(1570):1466–76.

Lieber RL, Roberts TJ, Blemker SS, Lee SSM, Herzog W. Skeletal muscle mechanics, energetics and plasticity. J Neuroeng Rehabil. 2017;14(1):108.

Lloyd RS, Faigenbaum AD, Stone MH, Oliver JL, Jeffreys I, Moody JA, et al. Position statement on youth resistance training: the 2014 International Consensus. Br J Sports Med. 2014;48(7):498–505.

MacDougall JD. Morphological changes in human skeletal muscle following strength training and immobilization. In: Jones NL, McCartney N, McComas AJ, editors. Human muscle power. Champaign: Human Kinetics; 1986. p. 269–88.

Maganaris CN, Reeves ND, Rittweger J, Sargeant AJ, Jones DA, Gerrits K, et al. Adaptive response of human tendon to paralysis. Muscle Nerve. 2006;33(1):85–92.

Magnusson SP, Langberg H, Kjaer M. The pathogenesis of tendinopathy: balancing the response to loading. Nat Rev Rheumatol. 2010;6(5):262–8.

Magnusson SP, Vaagen P, Kjaer M. The impact of loading, unloading, ageing and injury on the human tendon. J Physiol. 2019;597(5):1283–98.

Manuel M, Zytnicki D. Alpha, beta and gamma motoneurons: functional diversity in the motor system's final pathway. J Integr Neurosci. 2011;10(3):243–76.

McCombe D, Brown T, Slavin J, Morrison WA. The histochemical structure of the deep fascia and its structural response to surgery. J Hand Surg Br. 2001;26(2):89–97.

McGlone F, Reilly D. The cutaneous sensory system. Neurosci Biobehav Rev. 2010;34(2):148–59.

McGuinness H. Anatomy and physiology. 5th ed. London: Dynamic Learning; 2018.

Meissner MH, Moneta G, Burnand K, Gloviczki P, Lohr JM, Lurie F, et al. The hemodynamics and diagnosis of venous disease. J Vasc Surg. 2007;46(Suppl S):4S–24S.

Moore KL, Dalley AF, Agur AMR. Clinically oriented anatomy. 8th ed. Philadelphia: Wolters Kluwer; 2018.

Murray WM, Delp SL, Buchanan TS. Variation of muscle moment arms with elbow and forearm position. J Biomech. 1995;28(5):513–25.

Narici MV, de Boer MD. Disuse of the musculo-skeletal system in space and on earth. Eur J Appl Physiol. 2011;111(3):403–20.

Narici MV, Maffulli N. Sarcopenia: characteristics, mechanisms and functional significance. Br Med Bull. 2010;95:139–59.

Narici MV, Maganaris CN. Adaptability of elderly human muscles and tendons to increased loading. J Anat. 2006;208(4):433–43.

Narici MV, Roi GS, Landoni L, Minetti AE, Cerretelli P. Changes in force, cross-sectional area and neural activation during strength training and detraining of the human quadriceps. Eur J Appl Physiol Occup Physiol. 1989;59(4):310–9.

Narici MV, Binzoni T, Hiltbrand E, Fasel J, Terrier F, Cerretelli P. In vivo human gastrocnemius architecture with changing joint angle at rest and during graded isometric contraction. J Physiol. 1996;496(1):287–97.

Natali AN, Pavan PG, Stecco C. A constitutive model for the mechanical characterization of the plantar fascia. Connect Tissue Res. 2010;51(5):337–46.

Naylor EC, Watson RE, Sherratt MJ. Molecular aspects of skin ageing. Maturitas. 2011;69(3):249–56.

Neptune RR, Kautz SA, Zajac FE. Contributions of the individual ankle plantar flexors to support, forward progression and swing initiation during walking. J Biomech. 2001;34(11):1387–98.

Neumann DA. Kinesiology of the musculoskeletal system: foundations for rehabilitation. 3rd ed. St. Louis: Mosby; 2017.

Nishikawa KC, Monroy JA, Uyeno TE, Yeo SH, Pai DK, Lindstedt SL. Is titin a 'winding filament'? A new twist on muscle contraction. Proc Biol Sci. 2012;272(1730):981–90.

Nishimune H, Stanford JA, Mori Y. Role of exercise in maintaining the integrity of the neuromuscular junction. Muscle Nerve. 2014;49(3):315–24.

Nordin M, Frankel VH. Basic biomechanics of the musculoskeletal system. 4th ed. Philadelphia: Lippincott Williams & Wilkins; 2019.

O'Sullivan SB, Schmitz TJ, Fulk GD. Physical rehabilitation. 5th ed. Philadelphia: F.A. Davis; 2019.

Oatis CA. Kinesiology: the mechanics and pathomechanics of human movement. 3rd ed. Philadelphia: Wolters Kluwer Health; 2017.

Özkaya N, Goldsheyder D, Nordin M. Fundamentals of biomechanics: equilibrium, motion, and deformation. 4th ed. New York: Springer; 2017.

Page P. Current concepts in muscle stretching for exercise and rehabilitation. Int J Sports Phys Ther. 2012;7(1):109–19.

Palastanga N, Soames R. Anatomy and human movement: structure and function. 6th ed. Philadelphia: Churchill Livingstone; 2012.

Patterson RP, Fisher SV. Pressure and temperature patterns under the ischial tuberosities. Bull Prosthet Res. 1980;10-34:5–11.

Peterson MD, Rhea MR, Sen A, Gordon PM. Resistance exercise for muscular strength in older adults: a meta-analysis. Ageing Res Rev. 2010;9(3):226–37.

Pette D, Staron RS. Myosin isoforms, muscle fiber types, and transitions. Microsc Res Tech. 2000;50(6):500–9.

Pierrot-Deseilligny E, Burke D. The circuitry of the human spinal cord: spinal and corticospinal mechanisms of movement. Cambridge: Cambridge University Press; 2012.

Prilutsky BI. Coordination of two- and one-joint muscles: functional consequences and implications for motor control. Mot Control. 2000;4(1):1–44.

Prochazka A, Ellaway P. Sensory systems in the control of movement. Compr Physiol. 2012;2(4):2615–27.

Proske U, Gandevia SC. The proprioceptive senses: their roles in signaling body shape, body position and movement, and muscle force. Physiol Rev. 2012;92(4):1651–97.

Proske U, Morgan DL. Muscle damage from eccentric exercise: mechanism, mechanical signs, adaptation and clinical applications. J Physiol. 2001;537(2):333–45.

Purslow PP. Muscle fascia and force transmission. J Bodyw Mov Ther. 2010;14(4):411–7.

Purslow PP. The structure and role of intramuscular connective tissue in muscle function. Front Physiol. 2020;11:495.

Rassier DE. Sarcomere mechanics in striated muscles: from molecules to sarcomeres to cells. Am J Physiol Cell Physiol. 2017;313(2):C134–45.

Rassier DE, MacIntosh BR, Herzog W. Length dependence of active force production in skeletal muscle. J Appl Physiol. 1999;86(5):1445–57.

Rees JL. Genetics of hair and skin color. Annu Rev Genet. 2003;37:67–90.

Rhea MR, Alvar BA, Burkett LN, Ball SD. A meta-analysis to determine the dose response for strength development. Med Sci Sports Exerc. 2003;35(3):456–64.

Riemann BL, Lephart SM. The sensorimotor system, part I: the physiologic basis of functional joint stability. J Athl Train. 2002;37(1):71–9.

Roach NT, Lieberman DE. Upper body contributions to power generation during rapid, overhand throwing in humans. J Exp Biol. 2014;217(12):2139–49.

Roberts TJ, Azizi E. Flexible mechanisms: the diverse roles of biological springs in vertebrate movement. J Exp Biol. 2011;214(3):353–61.

Roig M, O'Brien K, Kirk G, Murray R, McKinnon P, Shadgan B, et al. The effects of eccentric versus concentric resistance training on muscle strength and mass in healthy adults: a systematic review with meta-analysis. Br J Sports Med. 2009;43(8):556–68.

Romanovsky AA. Skin temperature: its role in thermoregulation. Acta Physiol (Oxf). 2014;210(3):498–507.

Sale DG. Neural adaptation to resistance training. Med Sci Sports Exerc. 1988;20(5 Suppl):S135–45.

Sandby-Moller J, Poulsen T, Wulf HC. Epidermal thickness at different body sites: relationship to age, gender, pigmentation, blood content, skin type and smoking habits. Acta Derm Venereol. 2003;83(6):410–3.

Sanders JE, Goldstein BS, Leotta DF. Skin response to mechanical stress: adaptation rather than breakdown–a review of the literature. J Rehabil Res Dev. 1995;32(3):214–26.

Sandow A. Excitation-contraction coupling in muscular response. Yale J Biol Med. 1952;25(3):176–201.

Sanes JR, Lichtman JW. Development of the vertebrate neuromuscular junction. Annu Rev Neurosci. 1999;22:389–442.

Sato T, Hashimoto M. Morphological analysis of the fascial lamination of the trunk. Bull Tokyo Med Dent Univ. 1984;31(1):21–32.

Schiaffino S, Reggiani C. Fiber types in mammalian skeletal muscles. Physiol Rev. 2011;91(4):1447–531.

Schmidt RA, Lee TD, Winstein CJ, Wulf G, Zelaznik HN. Motor control and learning: a behavioral emphasis. 6th ed. Champaign: Human Kinetics; 2019.

Schoenfeld BJ, Ogborn D, Krieger JW. Effects of resistance training frequency on measures of muscle hypertrophy: a systematic review and meta-analysis. Sports Med. 2016a;46(11):1689–97.

Schoenfeld BJ, Pope ZK, Benik FM, Hester GM, Sellers J, Nooner JL, et al. Longer interset rest periods enhance muscle strength and hypertrophy in resistance-trained men. J Strength Cond Res. 2016b;30(7):1805–12.

Shadgan B, Pereira G, Menon M, Jafari S, Darlene Reid W, O'Brien PJ. Risk factors for acute compartment syndrome of the leg associated with tibial diaphyseal fractures in adults. J Orthop Traumatol. 2015;16(3):185–92.

Shumway-Cook A, Woollacott MH. Motor control: translating research into clinical practice. 5th ed. Philadelphia: Wolters Kluwer; 2017.

Silver FH, Siperko LM, Seehra GP. Mechanobiology of force transduction in dermal tissue. Skin Res Technol. 2003;9(1):3–23.

Sinnatamby CS. Last's anatomy: regional and applied. 12th ed. Philadelphia: Churchill Livingstone; 2011.

Slater CR. The functional organization of motor nerve terminals. Prog Neurobiol. 2015;134:55–103.

Slater CR. The structure of human neuromuscular junctions: some unanswered molecular questions. Int J Mol Sci. 2017;18(10):2183.

Squire JM. Muscle contraction: sliding filament history, sarcomere dynamics and the two Huxleys. Glob Cardiol Sci Pract. 2016;2016(2):e201611.

Standring S. Gray's anatomy: the anatomical basis of clinical practice. 42nd ed. Edinburgh: Elsevier; 2020.

Stecco C, Gagey O, Belloni A, Pozzuoli A, Porzionato A, Macchi V, et al. Anatomy of the deep fascia of the upper limb. Second part: study of innervation. Morphologie. 2007;91(292):38–43.

Stecco C, Macchi V, Porzionato A, Duparc F, De Caro R. The fascia: the forgotten structure. Ital J Anat Embryol. 2011;116(3):127–38.

Stensdotter AK, Hodges PW, Mellor R, Sundelin G, Häger-Ross C. Quadriceps activation in closed and in open kinetic chain exercise. Med Sci Sports Exerc. 2003;35(12):2043–7.

Suetta C, Frandsen U, Jensen L, Jensen MM, Jespersen JG, Hvid LG, et al. Aging affects the transcriptional regulation of human skeletal muscle disuse atrophy. PLoS One. 2013;7(12):e51238.

Thomas CK, Häger-Ross CK, Klein CS. Effects of baclofen on motor units paralysed by chronic cervical spinal cord injury. Brain. 2010;133:117–25.

Tidball JG. Myotendinous junction injury in relation to junction structure and molecular composition. Exerc Sport Sci Rev. 1991;19:419–45.

Timmins RG, Shield AJ, Williams MD, Lorenzen C, Opar DA. Architectural adaptations of muscle to training and injury: a narrative review outlining the contributions by fascicle length, pennation angle and muscle thickness. Br J Sports Med. 2016;50(3):1467–72.

Veeger HE, van der Helm FC. Shoulder function: the perfect compromise between mobility and stability. J Biomech. 2007;40(10):2119–29.

Viladot A, Lorenzo JC, Salazar J, Rodriguez A. The subtalar joint: embryology and morphology. Foot Ankle. 1984;5(2):54–66.

von Keudell AG, Weaver MJ, Appleton PT, Bae DS, Dyer GSM, Heng M, et al. Diagnosis and treatment of acute extremity compartment syndrome. Lancet. 2015;386(10000):1299–310.

Wakeling JM, Blake OM, Wong I, Rana M, Lee SS. Movement mechanics as a determiner of muscle structure, recruitment and coordination. Philos Trans R Soc Lond Ser B Biol Sci. 2011;366(1570):1554–64.

Wall BT, Dirks ML, van Loon LJ. Skeletal muscle atrophy during short-term disuse: implications for age-related sarcopenia. Ageing Res Rev. 2013;12(4):898–906.

Wang JH. Mechanobiology of tendon. J Biomech. 2006;39(9):1563–82.

Wang JH, Guo Q, Li B. Tendon biomechanics and mechanobiology—a minireview of basic concepts and recent advancements. J Hand Ther. 2012;25(2):133–41.

Ward SR, Eng CM, Smallwood LH, Lieber RL. Are current measurements of lower extremity mus-

cle architecture accurate? Clin Orthop Relat Res. 2009;467(4):1074–82.
Whitham M, Febbraio MA. The ever-expanding myokinome: discovery challenges and therapeutic implications. Nat Rev Drug Discov. 2016;15(10):719–29.
Wickiewicz TL, Roy RR, Powell PL, Edgerton VR. Muscle architecture of the human lower limb. Clin Orthop Relat Res. 1983;179:275–83.
Wilke J, Krause F, Vogt L, Banzer W. What is evidence-based about myofascial chains? Arch Phys Med Rehabil. 2016;97(3):454–61.
Willard FH, Vleeming A, Schuenke MD, Danneels L, Schleip R. The thoracolumbar fascia: anatomy, function, and clinical considerations. J Anat. 2012;221(6):507–36.
Willardson JM. A brief review: factors affecting the length of the rest interval between resistance exercise sets. J Strength Cond Res. 2006;20(4):978–84.
Wilson MH, Deschenes MR. The neuromuscular junction: anatomical features and adaptations to various forms of increased or decreased neuromuscular activity. Int J Neurosci. 2005;115(6):803–28.
Wilson JM, Loenneke JP, Jo E, Wilson GJ, Zourdos MC, Kim JS. The effects of endurance, strength, and power training on muscle fiber type shifting. J Strength Cond Res. 2012;26(6):1724–9.
Windhorst U. Muscle proprioceptive feedback and spinal networks. Brain Res Bull. 2007;73(4–6):155–202.
Winter DA. Biomechanics and motor control of human movement. 4th ed. Hoboken: Wiley; 2009.
Winter SL, Challis JH. The force-length curves of the human rectus femoris and gastrocnemius muscles in vivo. J Biomech. 2010;43(16):3067–72.
Winter DA, Patla AE, Prince F, Ishac M, Gielo-Perczak K. Stiffness control of balance in quiet standing. J Neurophysiol. 1998;80(3):1211–21.
Wong R, Geyer S, Weninger W, Gunsalus CK, Wang K. The dynamic anatomy and patterning of skin. Exp Dermatol. 2016;25(2):92–8.
Wu H, Xiong WC, Mei L. To build a synapse: signaling pathways in neuromuscular junction assembly. Development. 2010;137(7):1017–33.
Xiao J. Muscle atrophy. Singapore: Springer; 2018.
Zajac FE. Muscle and tendon: properties, models, scaling, and application to biomechanics and motor control. Crit Rev Biomed Eng. 1989;17(4):359–411.
Zajac FE. Muscle coordination of movement: a perspective. J Biomech. 1993;26(Suppl 1):109–24.
Zajac FE, Neptune RR, Kautz SA. Biomechanics and muscle coordination of human walking: part I: introduction to concepts, power transfer, dynamics and simulations. Gait Posture. 2002;16(3):215–32.
Zatsiorsky VM, Prilutsky BI. Biomechanics of skeletal muscles. Champaign: Human Kinetics; 2012.

Recommended Additional Reading

Alberstone CD, Benzel EC, Najm IM, Steinmetz MP. Anatomic basis of neurologic diagnosis. New York: Thieme; 2009.
Buschmann J, Bürgisser GM. Biomechanics of tendons and ligament: tissue reconstruction and regeneration. Buxford: Woodhead Publishing; 2017.
Ellis H. Clinical anatomy: a revision and applied anatomy for clinical students and junior doctors. 11th ed. Oxford: Blackwell Publishing; 2006.
Gilroy AM. Anatomy in your pocket. New York: Thieme; 2018.
Latash ML, Lestienne F. Motor control: theories, experiments, and applications. New York: Springer; 2006b.

Vascular Anatomy of the Extremities and Lungs

5

The vascular system of the extremities and lungs consists of a complex network of vessels that maintains circulation and supports physiological functions through organized anatomical arrangements. This system includes both arterial and venous components, each showing specific structural and functional characteristics based on their anatomical locations (Standring 2020).

In the extremities, the vascular system follows a hierarchical branching pattern, with major arterial trunks dividing into progressively smaller vessels that supply specific anatomical regions. The upper extremity vasculature originates with the subclavian artery, which becomes the axillary artery at the lateral border of the first rib, continues as the brachial artery at the inferior border of the teres major muscle, and divides into the radial and ulnar arteries in the cubital fossa (Moore et al. 2018). This arterial network pairs with a detailed venous drainage system of superficial and deep components, including the cephalic and basilic veins, which return blood to the heart (Drake et al. 2020).

The lower extremity vascular system exhibits similar organizational principles, with modifications for its functional requirements. The femoral artery acts as the main arterial conduit, with its branches forming extensive anastomotic networks for tissue perfusion. The venous system contains multiple valves and uses the muscular pump mechanism to move blood against gravity (Meissner et al. 2007).

The pulmonary vasculature functions as a specialized system for gas exchange, operating under different physiological parameters than systemic circulation. This system works in a low-pressure environment, with mean pulmonary arterial pressures of 8–20 mmHg, lower than systemic arterial pressures (Glenny 2009). The pulmonary trunk divides into right and left pulmonary arteries, which branch into lobar and segmental vessels, forming capillary networks around the alveoli. This structure allows gas exchange while preventing pulmonary edema (West and Luks 2016).

Understanding the anatomy and relationships of these vascular systems improves clinical practice, as these structures frequently exhibit pathological conditions and provide access points for diagnosis and medical interventions, such as deep vein thrombosis and pulmonary embolism in patients with spinal cord injuries. This chapter discusses vascular structure and its clinical implications for the effective care of relatively immobile individuals, especially those with spinal cord injuries.

H.-Y. Ko, *Practical Functional Anatomy for Spinal Cord Injury Rehabilitation*,
https://doi.org/10.1007/978-981-95-6457-6_5

5.1 Vascular Structure of the Upper Extremity Vasculature

5.1.1 Shoulder Girdle Complex Vasculature

The shoulder girdle maintains an extensive blood supply through a network of anastomosing vessels that allows sufficient collateral circulation (Moore et al. 2018). This vascular framework includes multiple arterial branches from the subclavian and axillary arteries, with additional contribution from the descending aorta (Fig. 5.1). The main arterial supply to the shoulder joint flows through several vessels. The suprascapular artery emerges from the thyrocervical trunk of the subclavian artery and delivers blood to the superior aspect of the shoulder. The thoracoacromial artery contributes through its acromial branch, while the anterior and posterior circumflex humeral arteries branch directly from the axillary artery and supply the glenohumeral joint and surrounding tissues (Standring 2020). This vascular arrangement provides adequate blood flow to meet the metabolic requirements of shoulder musculature and maintains the joint capsule integrity (Drake et al. 2020; Moore et al. 2018).

The venous system follows the arterial pattern, with corresponding vessels sharing similar names. The venous drainage pathway begins with the suprascapular, thoracoacromial, and circumflex humeral veins. These vessels merge into the larger axillary and subclavian veins, connecting to the external jugular vein system (Gosling et al. 2017; Standring 2020). This organization allows efficient venous return from the shoulder region.

The lymphatic system of the shoulder follows specific drainage patterns related to the venous structures. Lymph fluid from the joint and surrounding tissues moves through lymph nodes in the axillary region (Netter 2019). These nodes form specific groups: the pectoral nodes drain the anterior shoulder region; the subscapular nodes process posterior drainage; and the central nodes receive fluid from both anterior and posterior groups (Clemente 2011). The lymphatic fluid progresses through the apical nodes before enter-

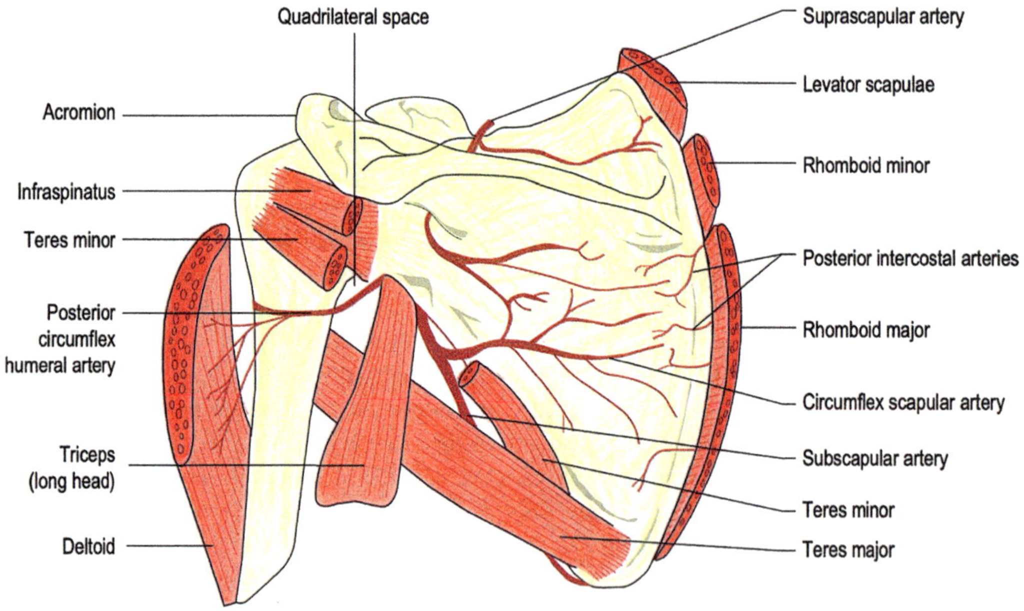

Fig. 5.1 Vessels supplying the shoulder region and those contributing to the scapular collateral anastomosis. (Adapted with permission from Palastanga and Soames (2012))

ing the subclavian lymph trunk, where it returns to the venous circulation (Agur and Dalley 2009; Netter 2019).

5.1.2 Upper Extremity Arterial Supply

The primary arterial conduit of the upper limb changes its nomenclature as it progresses through various anatomical segments, reflecting its path from the cervical region to the forearm. The vessel begins as the subclavian artery and changes to the axillary artery when it crosses the lateral border of the first rib, marking the beginning of the axillary region. As the axillary artery moves through the axilla, it provides blood supply to the shoulder area and upper limb structures. The vessel then becomes the brachial artery at the inferior border of the teres major muscle (Moore et al. 2018; Standring 2020).

The brachial artery descends through the arm along the medial aspect of the humerus, passing several anatomical structures, including the coracobrachialis muscle and the medial epicondyle. When the artery reaches the cubital fossa at the anterior elbow region, it divides into two branches: the radial and ulnar arteries. These vessels continue distally to supply blood to the forearm and hand structures (Drake et al. 2020; Moore et al. 2018) (Fig. 5.2). This arterial pathway maintains consistent blood flow, with each arterial segment corresponding to the anatomical organization and physiological needs of specific upper limb regions (Clemente 2011; Netter 2019).

5.1.2.1 Subclavian Artery

The right subclavian artery originates from the brachiocephalic trunk within the neck root, while the left subclavian artery begins from the aortic arch in the superior mediastinum and ascends to the neck. Both arteries transform into the axillary artery as they pass laterally over the first rib. The subclavian artery contains three segments defined by their relationship to the scalenus anterior muscle: medial, posterior, and lateral to the muscle (Standring 2020; Drake et al. 2020).

The subclavian artery forms an arch from the superior sternoclavicular joint margin to the clavicle's midpoint in the cervical region. This anatomical arrangement permits manual compression against the first rib for pulse assessment or bleeding control. The position makes the artery susceptible to compression from anatomical variations, including cervical ribs or muscle enlargement, potentially resulting in thoracic outlet syndrome with associated neurovascular symptoms (Jones et al. 2019; Moore et al. 2018; Netter 2019).

Multiple major branches emerge from the subclavian artery to supply various anatomical regions. The vertebral artery ascends through the cervical vertebrae's transverse foramina, supplying blood to the brain's posterior region and spinal cord via spinal arteries. The thyrocervical trunk branches into vessels, including the inferior thyroid and transverse cervical arteries, providing blood to the neck, shoulder, and thyroid gland. The internal thoracic artery courses down the anterior thoracic wall, supplying the intercostal spaces and offering alternative circulation pathways for cardiac perfusion when coronary arteries become occluded (Drake et al. 2020; Standring 2020).

Variations in subclavian artery anatomy, such as an anomalous right subclavian artery arising from the aorta after the left subclavian artery, carry implications for surgical procedures in the neck and thoracic areas (Drake et al. 2020; Netter 2019).

5.1.2.2 Axillary Artery

The axillary artery is a continuation of the subclavian artery, beginning at the lateral border of the first rib and extending to the lower edge of the teres major muscle. At this point, it transitions into the brachial artery near the lateral end of the posterior axillary fold. The axillary artery is conventionally divided into three segments based on its relationship with the pectoralis minor muscle: the first segment lies proximal to the muscle, the second is posterior to it, and the third is distal to it. The length of these segments can vary depending on the position of the upper limb, as shoulder abduction or adduction affects the muscle's

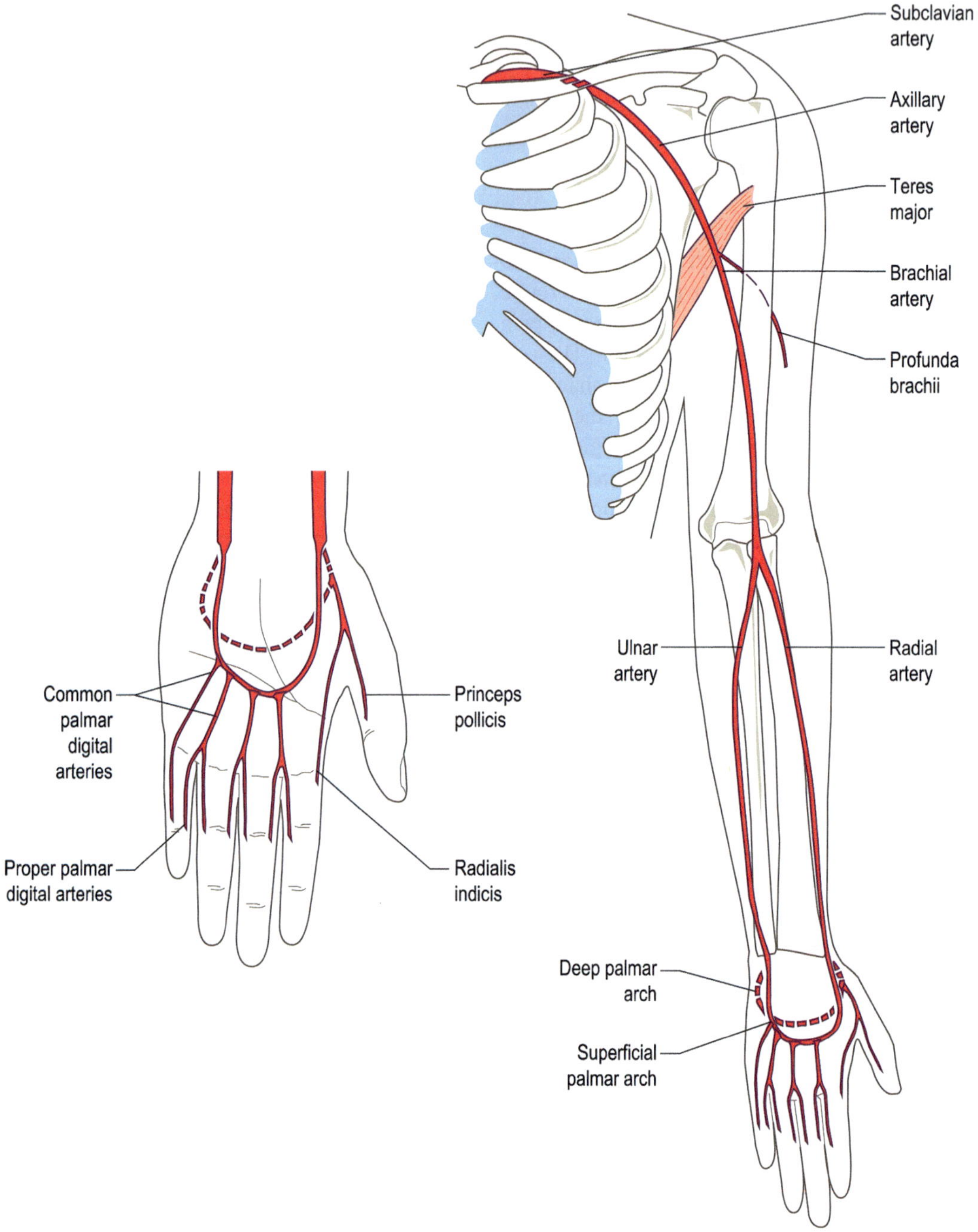

Fig. 5.2 Arteries of the upper extremity (ventral view) and arteries of the hand (palmar view). (Adapted with permission from Palastanga and Soames (2012))

placement relative to the artery (Moore et al. 2018; Standring 2020).

The brachial plexus cords are named according to their spatial relationship to the second segment of the axillary artery, with the lateral, medial, and posterior cords situated around the artery (Netter 2019). This anatomical relationship is important for clinical procedures, such as nerve blocks and vascular surgeries, where proximity to the artery and surrounding nerves must be considered to avoid complications.

The path of the axillary artery can be represented by a line that begins at the midpoint of the clavicle, travels below the coracoid process, and ends at the medial lip of the intertubercular groove of the humerus. This path forms a curved course with the concave side oriented inferomedially. Along this route, the axillary artery passes posterior to the pectoralis major and anterior to the subscapularis, latissimus dorsi, and teres major muscles (Sinnatamby 2011).

Clinically, the pulse of the axillary artery is palpable along the lateral wall of the axilla, where it lies posterior to the coracobrachialis muscle. This location serves as a useful pressure point for controlling bleeding in the upper limb. However, compression of the artery should be applied cautiously to avoid compressing adjacent structures, such as the median, ulnar, and radial nerves, which can cause numbness or tingling sensations in the arm (Drake et al. 2020).

The axillary artery has several branches that provide blood supply to the shoulder, chest wall, and lateral thoracic region. These branches include the superior thoracic artery, thoracoacromial artery, lateral thoracic artery, subscapular artery, and the anterior and posterior circumflex humeral arteries. These vessels participate in anastomoses with branches of the subclavian artery, including the transverse cervical and suprascapular arteries, as well as the posterior intercostal arteries originating from the descending thoracic aorta (Wineski 2019). In cases of aortic coarctation distal to the origin of the left subclavian artery, these anastomotic networks may become enlarged, establishing a collateral circulation that helps bypass the narrowed aortic segment, thus maintaining blood flow to the lower body (Drake et al. 2020).

5.1.2.3 Brachial Artery

The brachial artery, a continuation of the axillary artery, begins at the lower border of the teres major muscle and terminates in the cubital fossa near the neck of the radius, where it divides into the radial and ulnar arteries. Along its course, the brachial artery runs through the arm, traversing the long and medial heads of the triceps brachii muscle and passing anterior to the insertions of the coracobrachialis and brachialis muscles. Throughout its path, it is covered anteriorly by the medial border of the biceps brachii muscle. The median nerve crosses from lateral to medial at approximately the midpoint of the arm, closely associated with the artery (Drake et al. 2020; Moore et al. 2018).

Within the cubital fossa, the brachial artery is positioned deep to the bicipital aponeurosis, which separates it from the overlying median cubital vein. In this location, the artery is bordered medially by the median nerve and laterally by the biceps tendon. These anatomical relationships are clinically relevant during blood pressure measurement and cannulation of the brachial artery. The artery's course provides an accessible site for palpation of the brachial pulse, which can be felt by applying pressure against the humerus, especially along the medial aspect of the arm. The optimal site for auscultating Korotkoff sounds during blood pressure measurement is just medial to the bicipital aponeurosis, near the humeral medial epicondyle (Pickering et al. 2005).

Anatomical variations in the brachial artery are relatively common. The artery may divide into the radial and ulnar arteries at any point between the axilla and the cubital fossa, a condition known as high bifurcation. In such cases, the two arteries usually run parallel to each other, following the typical path of the brachial artery before reaching their usual destinations in the forearm (Rodríguez-Niedenführ et al. 2001; Standring 2020).

The profunda brachii, a significant branch of the brachial artery, travels alongside the radial nerve in the spiral groove of the humerus, situated between the lateral and medial heads of the triceps muscle. This artery provides blood supply to the posterior compartment of the arm, contributing to the vascular network around the elbow joint. Both the brachial artery and the profunda brachii branch into smaller vessels that supply the muscles of the arm and form an anastomotic network around the elbow, ensuring adequate blood flow during various arm movements (Standring 2020; Moore et al. 2018).

5.1.2.4 Radial Artery

The radial artery begins in the cubital fossa, opposite the neck of the radius, as a continuation of the brachial artery. It terminates by forming the deep palmar arch in the hand, contributing significantly to the blood supply of the hand and digits (Standring 2020). The radial artery's path is typically divided into three segments: the forearm, wrist, and palm (Moore et al. 2018).

In the forearm, the radial artery runs along the lateral aspect. When the forearm is midpronated and the brachioradialis muscle is contracted, the artery's course can be traced by a slightly convex line that starts at the biceps tendon, extending downward along the medial edge of the brachioradialis muscle, and ends just medial to the radial styloid process anteriorly. The artery is relatively superficial in this segment, which makes it an accessible site for cannulation and pulse palpation (Netter 2019). Throughout its course, it gives off several branches, including the radial recurrent artery, which contributes to the anastomotic network around the elbow, and muscular branches that supply the radial forearm muscles.

As the radial artery approaches the wrist, it passes lateral to the flexor carpi radialis tendon and enters the "anatomical snuffbox," where it lies superficially over the radiocarpal ligament and the bones of the wrist, including the scaphoid and trapezium. The artery is crossed from lateral to medial by the tendons of the abductor pollicis longus and the extensor pollicis brevis and longus muscles, which form the boundaries of the snuffbox (Wineski 2019). This location provides a convenient site for assessing the radial pulse and is often used in clinical practice for measuring heart rate.

In the palm, the radial artery passes between the two heads of the first dorsal interosseous muscle and then deep to the adductor pollicis muscle. It completes its course by forming the deep palmar arch, which contributes to the blood supply of the hand. Before completing the arch, the radial artery gives rise to the princeps pollicis artery, which supplies the thumb, and the radialis indicis artery, which provides blood to the lateral aspect of the index finger (Drake et al. 2020) (Fig. 5.2).

The radial pulse can be palpated at several locations along the artery's course, most commonly at the distal border of the radius, lateral to the flexor carpi radialis tendon, and within the anatomical snuffbox against the scaphoid bone. Its superficial location makes it accessible for procedures such as arterial blood sampling and catheter insertion. In surgical and interventional settings, the radial artery is a preferred access site for coronary angiography due to its easily compressible nature and lower risk of bleeding complications compared to femoral access.

The branches of the radial artery serve important roles throughout their course. In the forearm, the radial recurrent artery and muscular branches support the elbow anastomosis and adjacent musculature. At the wrist, the artery provides branches that supply the carpal bones and the dorsum of the hand, including the dorsal carpal branch, which participates in the formation of the dorsal carpal arch. Within the palm, the princeps pollicis and radialis indicis arteries provide the primary blood supply to the thumb and lateral side of the index finger, respectively (Moore et al. 2018; Standring 2020).

5.1.2.5 Ulnar Artery

The ulnar artery, a terminal branch of the brachial artery, begins in the cubital fossa and extends to the pisiform bone, where it divides into deep and superficial palmar branches to form the superficial palmar arch. Its path follows a medially convex line from the biceps tendon to the pisiform, continuing to the hamate hook in the hand (Moore et al. 2018; Netter 2019).

The artery passes over several muscles during its course. Proximally, it travels superficial to the brachialis muscle, then over the flexor digitorum profundus, and crosses the flexor retinaculum at the wrist. In its proximal path, multiple structures cross anterior to the artery from superior to inferior: pronator teres, median nerve, flexor carpi radialis, palmaris longus, and flexor digitorum superficialis. The flexor carpi ulnaris muscle provides protection as it partially overlies the artery near the wrist (Standring 2020; Wineski 2019).

The common interosseous artery branches from the ulnar artery inferior to the radial tuberosity. This vessel splits into anterior and posterior interosseous arteries along the interosseous membrane, supplying the deep flexor and extensor compartment muscles. The anterior interosseous artery runs on the membrane's anterior surface, while the posterior interosseous artery supplies the posterior compartment after penetrating the membrane. Both vessels contribute to the elbow's vascular network and collateral circulation (Drake et al. 2020; Moore et al. 2018).

The ulnar artery supplies blood to the hand primarily through the superficial palmar arch, which connects with the radial artery's superficial branch. The pulse can be detected on the wrist's medial aspect, lateral to the flexor carpi ulnaris tendon. Anatomical variations in the ulnar artery's origin, course, and branches occur frequently, requiring consideration during arterial line placement, graft harvesting, and hand reconstruction procedures. A superficial course variation may increase injury risk (Netter 2019; Wineski 2019).

The artery supplies blood to the elbow and wrist joints through proximal and distal branches. Near the elbow, it forms anastomoses with the radial recurrent and superior ulnar collateral arteries, maintaining blood flow during joint movement. The artery's branches at the wrist contribute to dorsal and palmar carpal arches, supporting the carpal bones' blood supply (Standring 2020).

5.1.2.6 Superficial Palmar Arch

The palmar aspect of the hand features two main arterial arches: the superficial and deep palmar arches. These vascular networks are vital for maintaining adequate blood supply to the hand and fingers, ensuring proper function and resilience against ischemic conditions (Moore et al. 2018; Standring 2020).

The superficial palmar arch is primarily formed by the ulnar artery, which serves as its main contributor. It is completed by a smaller contribution from the superficial palmar branch of the radial artery. However, variations in the formation of the arch are common; in some cases, the radial artery may play a larger role or may be absent altogether in completing the arch. The superficial palmar arch lies deep to the palmar aponeurosis and superficial to the long flexor tendons (Netter 2019; Drake et al. 2020).

From the convex aspect of the superficial palmar arch, four common palmar digital arteries arise. The most medial of these arteries runs along the ulnar side of the fifth digit, providing blood supply to the little finger. The remaining three common palmar digital arteries travel distally, each dividing into two proper palmar digital arteries, which supply the adjacent sides of the fourth, third, and second fingers. These arteries anastomose with branches from the deep palmar arch, creating a rich vascular network that ensures adequate perfusion even during hand movements that could potentially compromise blood flow (Standring 2020; Netter 2019).

5.1.2.7 Deep Palmar Arch

The deep palmar arch forms predominantly from the radial artery with additional contribution from the ulnar artery's deep branch (Wineski 2019). The arch lies beneath the long flexor tendons and their synovial sheaths, positioned against the metacarpal bone bases. It generates the palmar metacarpal arteries. The arch's distal convexity sits approximately 2 cm distal to the distal wrist crease (Tank and Grant 2017).

The palmar metacarpal arteries originate from this arch and supply blood to the deep palmar tissues. These vessels connect with the superficial palmar arch branches and contribute to the rich vascular network of the hand (Standring 2020). The deep palmar arch also provides small perforating branches that pass between the metacarpal bones to join the dorsal metacarpal arteries (Moore et al. 2018). This arterial structure maintains consistent anatomical relationships with surrounding structures, including the deep branch of the ulnar nerve and the deep palmar branches of the median nerve (Drake et al. 2020). Understanding these relationships holds particular importance for surgical procedures in the palm (Netter 2019).

5.1.3 Upper Extremity Venous Drainage

The upper extremity venous drainage system comprises two primary components: superficial and deep veins (Fig. 5.3). Superficial veins course within the superficial fascia beneath the skin, while deep veins parallel the major arteries within deeper fascial compartments. Both venous networks contain valves that direct blood flow heartward by preventing retrograde flow, thus optimizing venous return (Drake et al. 2020; Standring 2020).

These venous networks establish multiple connections through perforating veins, which pierce the deep fascia to join the superficial and deep systems (Moore et al. 2018). This arrangement allows blood to flow between the two systems and provides alternative drainage pathways when needed. The superficial veins often display more variability in their paths compared to deep veins, which follow more consistent courses alongside their arterial counterparts (Netter 2019).

The valvular system within these veins plays an important role in venous return mechanics. These one-way valves open in response to normal blood flow toward the heart and close when gravity or muscle relaxation would otherwise cause backward flow (Tank and Grant 2017). The muscular pumping action of the upper extremity assists this venous return mechanism during movement (Wineski 2019).

5.1.3.1 Superficial Venous System of the Upper Extremity

The superficial venous drainage of the upper limb consists of the cephalic, basilic, and median antebrachial veins with their accompanying tributaries. These vessels originate from the dorsal venous network of the hand and continue proximally through the upper extremity (Moore et al. 2018; Standring 2020) (Fig. 5.4).

Basilic Vein

The basilic vein originates on the ulnar side of the dorsal venous network of the hand and follows a subcutaneous path on the posterior forearm before transitioning to the anterior aspect. It travels between the biceps brachii and the pronator teres muscles. Near the shoulder, the vein penetrates the deep fascia and continues as the axillary vein (Moore et al. 2018). The basilic vein is the preferred site for venous cannulation in some clinical settings due to its accessibility and relatively large size.

Cephalic Vein

Arising from the lateral side of the dorsal venous network, the cephalic vein follows the radial border of the forearm and runs subcutaneously along the lateral aspect of the biceps brachii in the arm. At the elbow, the median cubital vein, which crosses anterior to the cubital fossa, connects the cephalic vein to the basilic vein. The cephalic vein then ascends through the deltopectoral groove and terminates in the axillary vein just below the clavicle. Occasionally, an accessory cephalic vein is present, enhancing venous drainage (Netter 2019).

Median Cubital Vein

The median cubital vein is a short, superficial vein located in the anterior aspect of the elbow, typically running superiorly and medially across the bicipital aponeurosis. This anatomical arrangement places the vein just beneath the skin and superficial to the brachial artery, with the bicipital aponeurosis acting as a protective layer that separates the vein from the underlying artery. This structural feature reduces the risk of accidental arterial injury during venipuncture (Standring 2020; Netter 2019).

The vein originates from the connection between the cephalic and basilic veins, with its terminal point at the junction with the basilic vein just above the medial epicondyle of the humerus.

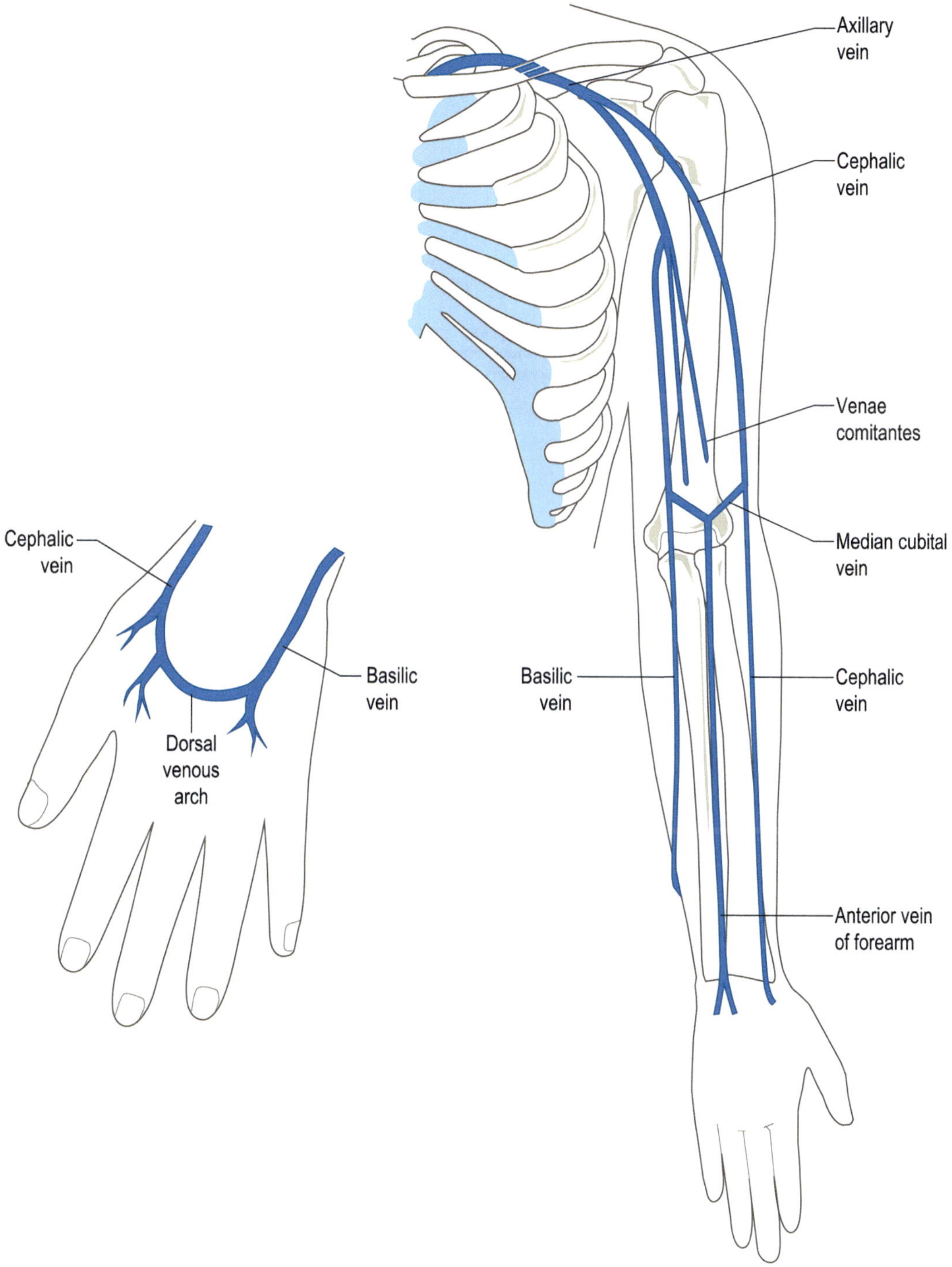

Fig. 5.3 Veins of the upper extremity (ventral view) and veins of the hand (dorsal view). (Adapted with permission from Palastanga and Soames (2012))

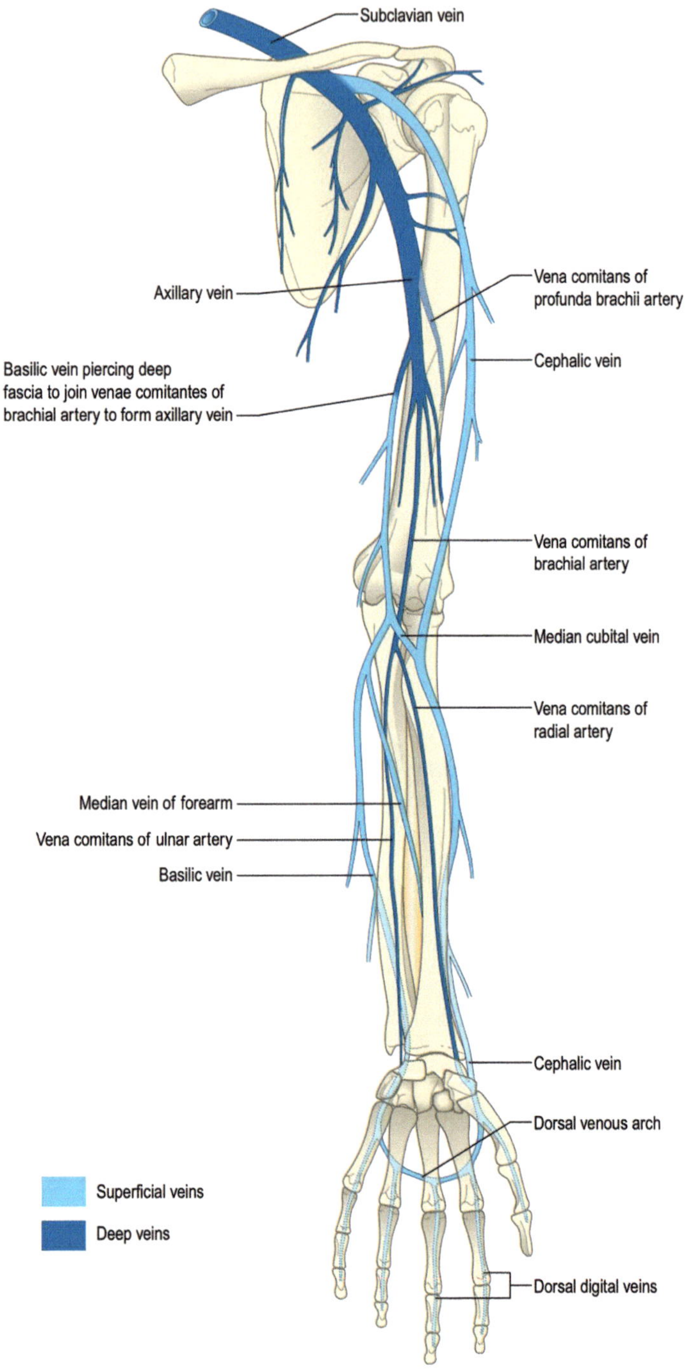

Fig. 5.4 Venous anatomy of the upper extremity. (Adapted with permission from Standring (2020))

The median cubital vein may show variations in its size, shape, and course, but it is generally a reliable landmark due to its consistent location (Drake et al. 2020).

Due to its relatively large diameter, superficial position, and predictable anatomical location, the median cubital vein is a preferred site for venipuncture, including blood draws and intravenous catheter placements. The ease of access and the minimal risk of complications, such as arterial puncture, make it a favored choice in clinical settings. Additionally, the median cubital vein's position above the bicipital aponeurosis provides some protection for deeper structures, making it less likely that a needle insertion will damage the brachial artery (Moore et al. 2018; Wineski 2019). Although the median cubital vein's anatomy is generally consistent, variations can occur. Some individuals may have a more tortuous course, multiple branches, or different patterns of communication with the cephalic and basilic veins. In certain cases, the median cubital vein may be absent or less prominent, requiring clinicians to use alternative veins for venipuncture (Netter 2019).

Dorsal Venous Plexus

The term "dorsal venous plexus" more accurately describes this structure due to its rarely evident arch-like form. Located on the back of the hand, its configuration and location exhibit considerable variability among individuals (Standring 2020).

5.1.3.2 Deep Venous System of the Upper Extremity

The deep venous system of the upper limb primarily features paired veins (venae comitantes) that accompany the arteries. These paired vessels run along the arterial pathways, using arterial pulsations to assist venous return toward the heart. The axillary vein presents an exception to this pattern, existing as a single vessel. It begins as a continuation of the basilic vein at the teres major muscle's inferior border and extends to the first rib's lateral edge, where it transitions into the subclavian vein (Snell 2012; Tank and Grant 2017).

The organization of the deep venous system mirrors the arterial pattern in its segmental arrangement. While these veins have smaller diameters compared to superficial vessels, they perform an important function in draining blood from the deep tissues of the upper extremity (Drake et al. 2020). The venae comitantes typically flank their corresponding arteries, forming a three-vessel bundle contained within the same fascial compartment (Moore et al. 2018).

These deep veins contain numerous valves that direct blood flow toward the heart and prevent retrograde flow (Standring 2020). The muscular contractions of the upper limb compress these vessels, aiding in venous return. Their position between muscles provides protection and mechanical advantage for blood propulsion (Netter 2019).

In the Hand and Wrist

The deep veins of the hand include venae comitantes accompanying the superficial and deep palmar arterial arches. The superficial palmar venous arch drains blood from the common palmar digital veins, while the deep palmar venous arch collects blood from the palmar metacarpal veins. Together, these venous arches help ensure adequate venous return from the hand, supporting circulation during various hand movements (Standring 2020).

In the Forearm

In the forearm, the deep venous system is formed by companion veins that accompany the radial, ulnar, and interosseous arteries. These veins merge at the level of the elbow to form the brachial veins, which typically exist as a pair and follow the brachial artery proximally. The deep veins of the forearm assist in draining blood from the deep musculature and interosseous spaces, returning it to the brachial veins for further drainage (Drake et al. 2020).

Brachial Veins

The brachial veins accompany the brachial artery and receive tributaries from the deep venous structures of the upper arm, including veins draining the muscles and connective tissues. These veins typically merge with the basilic vein at the level of the teres major muscle to form the axillary vein. In some cases, the brachial veins may merge directly with the axillary vein without receiving the basilic vein, indicating anatomical variations in venous drainage patterns (Snell 2012).

Axillary Vein

The axillary vein begins as the continuation of the basilic vein at the lower border of the teres major muscle and ascends medially to the axillary artery. It extends up to the lateral edge of the first rib, where it transitions into the subclavian vein. The axillary vein also receives major tributaries such as the cephalic vein, which drains blood from the lateral aspect of the upper limb, and other smaller veins draining the shoulder region (Detton 2017).

Subclavian Vein

The subclavian vein is formed as a continuation of the axillary vein beyond the lateral border of the first rib. It travels anterior and inferior to the subclavian artery, joining the internal jugular vein at the medial border of the scalenus anterior muscle to form the brachiocephalic vein. The subclavian vein also receives venous tributaries, including the external jugular vein and dorsal scapular vein. On the left side, the subclavian vein receives the thoracic duct at its junction with the internal jugular vein, while on the right side, the right lymphatic duct or trunk drains into the vein (Drake et al. 2020; Standring 2020).

5.2 Lower Extremity Vasculature

5.2.1 Palpating Arteries of the Lower Extremity

Lower extremity arterial pulse assessment involves examination of four main vessels: the femoral, popliteal, dorsalis pedis, and posterior tibial arteries. Each location requires specific anatomical knowledge and palpation techniques.

The femoral pulse is detected at the midinguinal point, halfway between the anterior superior iliac spine and pubic symphysis within the femoral triangle. The artery lies lateral to the femoral vein and one finger breadth medial to the femoral nerve, positioned above the femoral head. The vessel courses from the midinguinal point toward the patella's center through the femoral triangle, bordered by the inguinal ligament, sartorius muscle, and adductor longus muscle. The artery then passes through the adductor canal before emerging through the adductor hiatus to become the popliteal artery (Drake et al. 2020; Standring 2020) (Fig. 5.5).

Popliteal pulse assessment requires prone positioning with slight knee flexion. The pulse

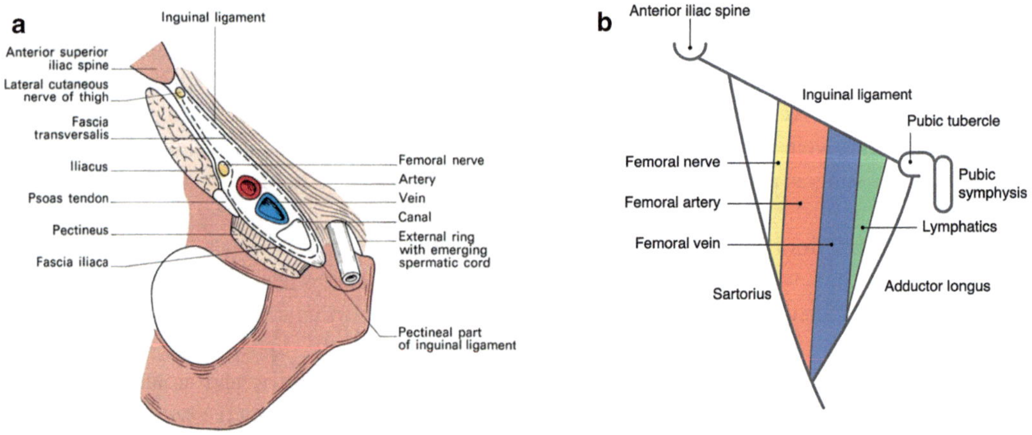

Fig. 5.5 Schematic illustration of the vascular and nerve structures of the femoral canal (**a**) and femoral triangle (**b**), including their surrounding anatomy as they pass beneath the inguinal ligament through right inguinal canal. (Adapted from Ellis (2006) and with permission from Drake et al. (2020))

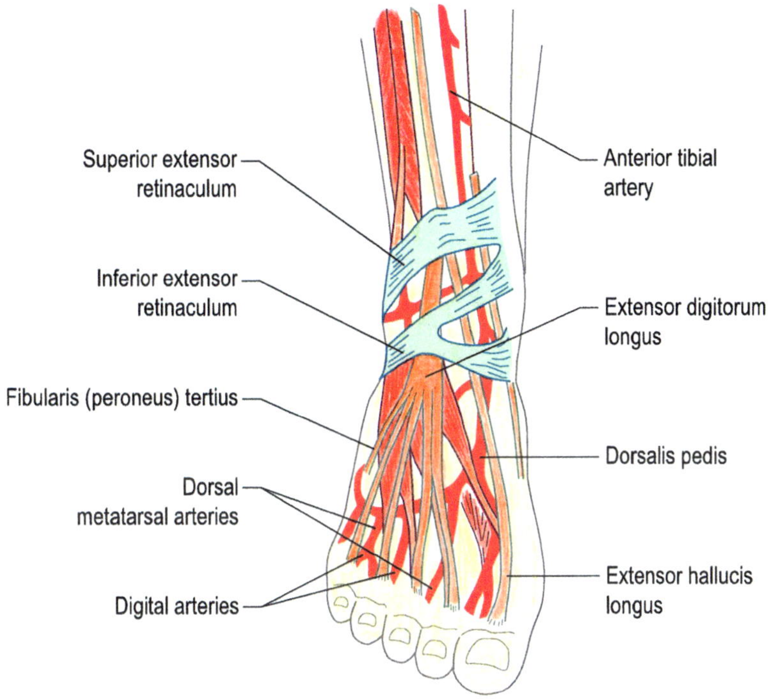

Fig. 5.6 Arteries supplying the dorsal aspect of the foot. (Adapted with permission from Palastanga and Soames (2012))

may be difficult to detect as the artery lies deep within the popliteal fossa, surrounded by the semimembranosus, semitendinosus, biceps femoris, and gastrocnemius muscles. Firm pressure against the posterior femur improves detection (Moore et al. 2018; Snell 2012).

The dorsalis pedis pulse is located on the foot's dorsum between the extensor hallucis longus and extensor digitorum longus tendons, along a line from the ankle's midpoint to the first intermetatarsal space (Fig. 5.6). This pulse may be absent in 10–15% of healthy individuals due to anatomical variations (Detton 2017; Netter 2019).

The posterior tibial pulse is found posterior to the medial malleolus in the tarsal tunnel, between the tibialis posterior and flexor digitorum longus tendons, accompanied by the tibial nerve and flexor hallucis longus tendon. Approximately 5% of individuals lack a palpable posterior tibial pulse due to anatomical variations, though adequate circulation occurs through collateral vessels, including the peroneal and dorsalis pedis arteries (Drake et al. 2020; Standring 2020).

5.2.2 Hip Joint Blood Supply

The hip joint is nourished by a rich network of blood vessels derived from multiple sources, forming a periarticular anastomosis that ensures adequate perfusion to the joint structures (Moore et al. 2018). The principal arteries involved in this vascular supply include the medial and lateral circumflex femoral arteries, the obturator artery, and the superior and inferior gluteal arteries (Standring 2020). Together, these arteries form a collateral network that provides continuous blood flow to the hip, which is important in maintaining joint health and promoting healing following trauma or surgery (Fig. 5.7).

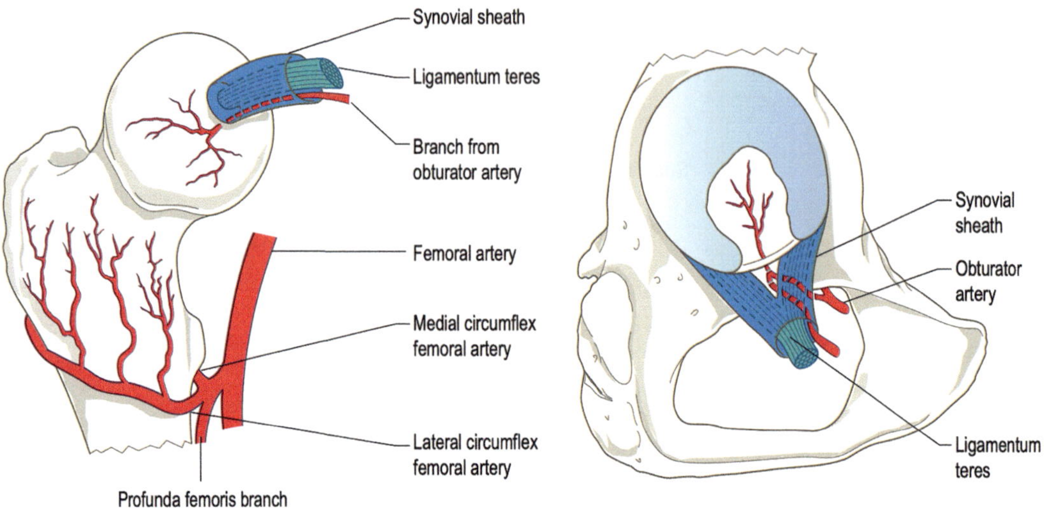

Fig. 5.7 Arterial supply of the hip joint. (Adapted with permission from Palastanga and Soames (2012))

The majority of the joint's blood supply is derived from the periarticular anastomosis. Branches from this network penetrate the joint capsule at its femoral attachment and course through the retinacula, situated between the capsule and synovial membrane, to supply the femoral neck and head (Drake et al. 2020). The integrity of this periarticular arterial network is particularly important for bone nutrition, especially for the proximal femoral epiphysis until ossification is complete, typically between ages 16 and 20 (Wineski 2019). Additionally, adequate blood supply is necessary for bone healing following fractures. Femoral neck fractures are often intracapsular, complicating the realignment of head and neck fragments. Moreover, rupture of the blood vessels supplying the head and neck can lead to ischemic degeneration of the femoral head, as the artery of the ligamentum teres often provides insufficient alternative blood supply. Consequently, early prosthetic replacement of the femoral component or total joint arthroplasty is frequently performed for subcapital fractures. Similarly, hip dislocations and slipped capital femoral epiphysis can potentially damage the retinacular blood vessels, further emphasizing the importance of understanding this vascular anatomy in trauma management.

5.2.3 Arterial System of the Lower Extremity

The principal vascular supply to the lower extremity is derived from the femoral artery, a continuation of the external iliac artery. The lower limb has been provided with a well-established collateral network that is capable of maintaining adequate tissue perfusion in a resting limb for extended periods during arterial compromise, despite the prominent role of the femoral artery (Moore et al. 2018; Standring 2020). This network includes connections through the profunda femoris artery and the genicular arteries, providing an alternate route for blood flow when the femoral artery is obstructed. Such collateral circulation is particularly beneficial during conditions of femoral artery occlusion, trauma, or when surgical procedures on major vessels are undertaken, offering a compensatory mechanism to prevent ischemia and tissue damage (Standring 2020).

The collateral pathways have an important role in maintaining limb viability during vascular interventions, such as bypass grafting or endovascular procedures. The compensatory hypertrophy of collateral vessels observed in patients with chronic arterial occlusive diseases serves to further emphasize the significance of these anastomoses.

5.2.3.1 Femoral Artery

The femoral artery is a direct continuation of the external iliac artery, which itself arises from the common iliac artery—a branch of the abdominal aorta (Netter 2019). Upon entering the thigh, the artery passes beneath the inguinal ligament and is enclosed within a funnel-shaped extension of the abdominal fascia called the femoral sheath (Drake et al. 2020) (Fig. 5.8). The femoral sheath contains the artery and vein in separate compartments; the femoral vein lies medially to the artery, while the femoral nerve is positioned laterally, outside of the sheath (Wineski 2019). Deep to the psoas major muscle, the femoral artery is easily palpable just below the inguinal ligament, midway between the anterior superior iliac spine and the pubic tubercle, where the femoral pulse can be detected.

Traversing through the femoral triangle, the artery moves medially over the femoral vein before entering the adductor (subsartorial) canal, which originates at the apex of the femoral triangle. The femoral artery then exits the anterior compartment of the thigh via the adductor hiatus, a tendinous opening within the adductor magnus muscle, and transitions into the popliteal artery as it reaches the popliteal fossa (Palastanga and Soames 2012).

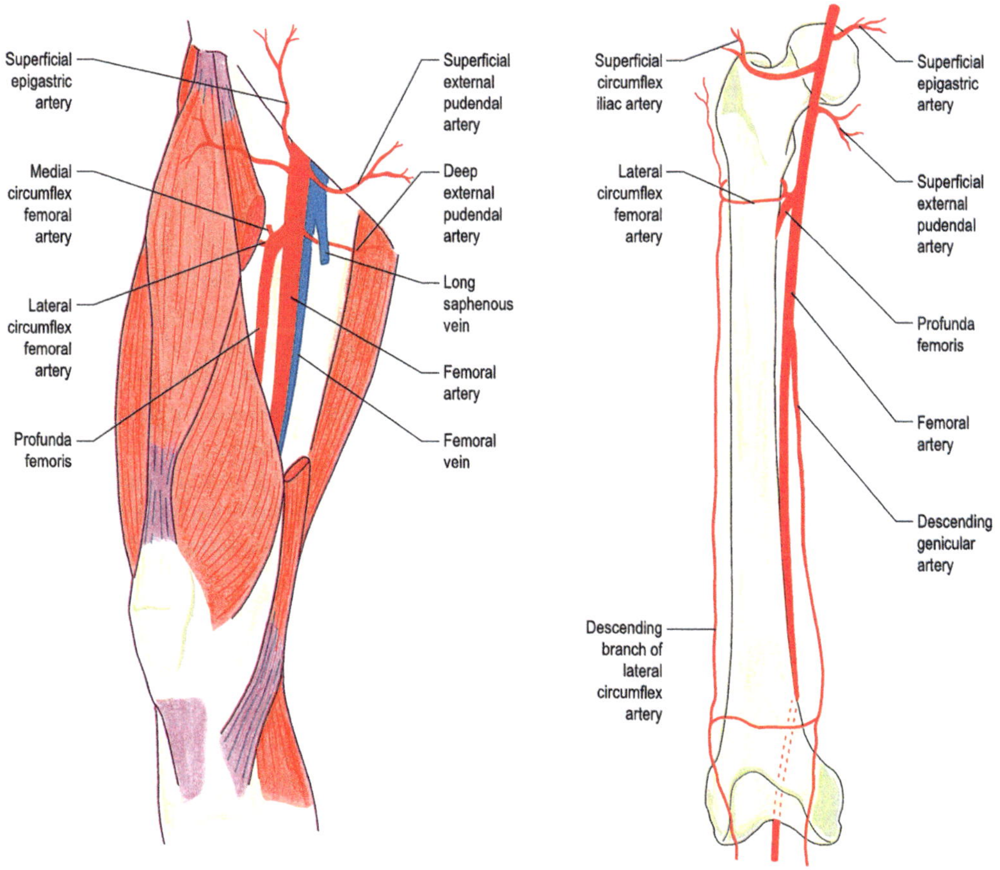

Fig. 5.8 Vessels within the femoral triangle and arterial supply of the thigh. (Adapted with permission from Palastanga and Soames (2012))

The femoral artery gives rise to several branches, each contributing to the vascular supply of different regions of the lower limb:

- *Superficial Circumflex Iliac Artery*: Provides blood flow to the lateral aspects of the iliac and gluteal regions.
- *Superficial and Deep External Pudendal Arteries*: Supply the external genitalia and perineum.
- *Descending Genicular Branch*: Participates in the formation of the anastomotic network around the knee, aiding in blood supply during knee flexion.
- *Muscular Branches*: Nourish surrounding muscles, including portions of the quadriceps.
- *Profunda Femoris (Deep Femoral Artery)*: Supplies most of the deep structures within the thigh, including the hamstrings and adductor muscles, and gives off perforating branches that contribute to the vascular network of the thigh (Standring 2020).

5.2.3.2 Profunda Femoris

The profunda femoris (deep femoral artery) is the largest branch of the femoral artery and has a similar diameter to the femoral artery itself. It originates from the lateral aspect of the femoral artery, approximately 5 cm inferior to the inguinal ligament (Moore et al. 2018; Standring 2020). The profunda femoris courses posteriorly to the femoral artery, exiting the femoral triangle between pectineus and adductor longus to reach the anterior surface of adductors brevis and magnus (Netter 2019).

In the distal thigh, the profunda femoris penetrates the adductor magnus as the fourth perforating artery, contributing to the anastomosis surrounding the knee joint (Drake et al. 2020). Shortly after its origin, it gives off several significant branches. These include the lateral circumflex femoral artery, which provides ascending, transverse, and descending branches that supply the gluteal region and hip joint, the quadriceps, and the knee joint, respectively. Additionally, the medial circumflex femoral artery anastomoses with the lateral circumflex femoral artery and supplies similar regions. The profunda femoris also emits numerous muscular branches and perforating arteries that pass above or through small openings in the adductor magnus, connecting with the descending branch of the lateral circumflex femoral artery (Wineski 2019).

5.2.3.3 Popliteal Artery

The popliteal artery is a continuation of the femoral artery after it emerges through the adductor hiatus (Palastanga and Soames 2012) (Fig. 5.9). Within the popliteal fossa, it descends vertically from the upper medial border, terminating by dividing into the anterior and posterior tibial arteries at the level of the tibial tuberosity (Ouriel 2001).

The popliteal artery is the deepest structure within the popliteal fossa. It provides cutaneous branches to the back of the leg, muscular branches to the adductor magnus and hamstring muscles, and articular branches that participate in the anastomosis around the knee joint. These articular branches are divided into three groups. The upper group includes the medial and lateral superior genicular arteries, which encircle the lower part of the femur just above the condyles. The middle genicular artery is smaller and pierces the posterior part of the joint capsule to supply the cruciate ligaments. The lower group consists of the medial and lateral inferior genicular arteries, which encircle the upper part of the tibial condyles just below the knee joint line. The superior and inferior arteries are interconnected by vertical arteries that run down either side of the patella.

5.2.3.4 Anterior Tibial Artery

The anterior tibial artery begins at the inferior border of the popliteus muscle and terminates anterior to the ankle joint, where it transitions into the dorsalis pedis artery. Its initial course involves passing anteriorly through an opening above the interosseous membrane between the tibia and fibula (Moore et al. 2018; Standring 2020).

As the artery passes over the membrane, it gives off recurrent arteries that ascend to partici-

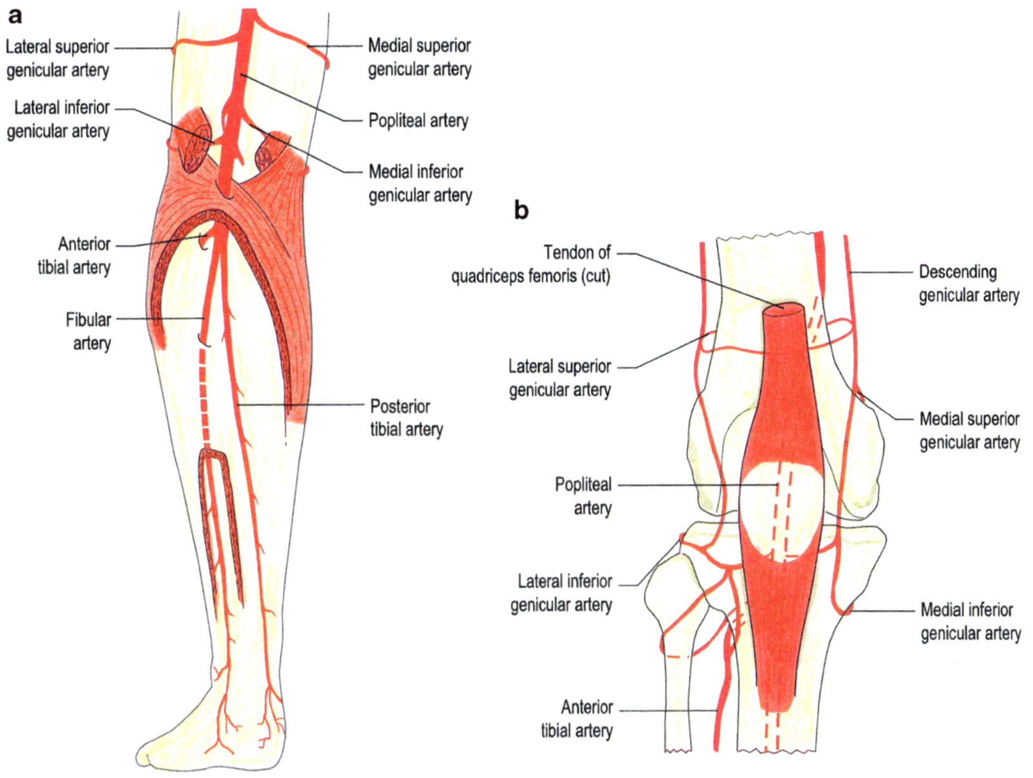

Fig. 5.9 (**a**) Arteries of the posterior aspect of the knee and leg, and (**b**) arteries surrounding the knee joint (anterior view). (Adapted with permission from Palastanga and Soames (2012))

pate in the anastomotic network surrounding the knee joint (Netter 2019). The anterior tibial artery then descends along the anterior surface of the interosseous membrane, becoming more superficial as it approaches the ankle (Drake et al. 2020). At the ankle joint, the artery passes deep to the extensor retinaculum, positioned between the tendons of extensor hallucis longus and extensor digitorum longus. This location allows for easy palpation of the anterior tibial pulse (Wineski 2019). The artery continues onto the dorsum of the foot as the dorsalis pedis artery, coursing toward the first interosseous space through which it passes to the plantar aspect of the foot. In this region, the dorsalis pedis pulse can be palpated.

Before its descent to the plantar surface, the dorsalis pedis artery gives rise to a significant branch, the arcuate artery. This vessel courses across the metatarsal bases, forming a connection laterally with the fibular (peroneal) artery. The arcuate artery gives origin to the dorsal metatarsal arteries, which bifurcate at the toe clefts into dorsal digital arteries, supplying adjacent toe sides. Typically, separate branches from the arcuate artery supply the medial aspect of the great toe and the lateral aspect of the little toe (Palastanga and Soames 2012).

5.2.3.5 Posterior Tibial Artery

The posterior tibial artery, the larger of the two terminal branches of the popliteal artery, originates at the lower border of the popliteus muscle. It travels downward along the posterior aspect of the leg, initially situated deep to the soleus and gastrocnemius muscles, while overlaying the flexor digitorum longus and flexor hallucis longus (Moore et al. 2018; Standring 2020). In the distal third of the leg, the artery becomes more accessible as it is covered only by deep fascia and skin, making it easier to palpate (Netter 2019). It runs posterior to the medial malleolus and deep to the flexor retinaculum before bifurcating into the medial and lateral plantar arteries immediately after crossing the ankle (Drake et al. 2020).

A significant branch arising from the posterior tibial artery is the fibular (peroneal) artery, which courses between the tibialis posterior and the flexor hallucis longus muscles. The fibular artery contributes to the vascularization of the lower leg, providing malleolar branches and a perforating branch that connects with the anastomotic network around the ankle, linking to branches from the anterior tibial and dorsalis pedis arteries. Clinically, the posterior tibial pulse, palpable in the retromalleolar region, serves as a critical indicator for assessing lower limb perfusion (Ouriel 2001; Wineski 2019).

The anatomical importance of the posterior tibial artery extends to conditions such as tarsal tunnel syndrome, where it may become compressed within the tarsal tunnel, leading to symptoms of pain and neuropathy. Understanding its precise anatomical course and relationships is crucial during surgical decompression to avoid iatrogenic injury.

5.2.3.6 Medial and Lateral Plantar Arteries

The plantar aspect of the foot receives its primary blood supply from the medial and lateral plantar arteries, which arise as terminal branches of the posterior tibial artery after it passes behind the medial malleolus. These vessels are responsible for the perfusion of the plantar tissues, including muscles, fascia, ligaments, and bones, and contribute significantly to the foot's complex vascular network (Standring 2020).

The medial plantar artery, the smaller of the two branches, runs anteriorly along the medial edge of the foot, closely accompanying the medial plantar nerve. It is located between the abductor hallucis and flexor digitorum brevis muscles, where it provides several branches that contribute to the arterial supply of the medial portion of the foot, including the plantar fascia, skin, and intrinsic foot muscles (Attinger et al. 2006). As it progresses distally, the artery divides into superficial and deep branches. The superficial branch supplies the skin and subcutaneous tissues, while the deep branch follows a deeper course alongside the muscles, eventually anastomosing with branches of the first plantar metatarsal artery. These connections play a role in maintaining adequate perfusion to the medial and plantar surfaces of the foot, especially in cases where collateral circulation is necessary due to arterial insufficiency (Attinger et al. 2006).

The lateral plantar artery, being larger, has a more complex trajectory. Initially, it courses anterolaterally between the flexor digitorum brevis and quadratus plantae muscles, paralleling the lateral plantar nerve. Upon reaching the base of the fifth metatarsal, the artery takes a sharp medial turn, traversing beneath the flexor digiti minimi muscle and forming the deep plantar arch (Standring 2020; Attinger et al. 2006). The plantar arch runs across the bases of the metatarsals, providing the primary vascular supply to the forefoot. This arch is a vital structure, as it anastomoses with the dorsalis pedis artery, creating a connection between the dorsal and plantar arterial networks, which ensures that if one vessel is occluded, collateral pathways can still maintain blood flow to the foot (Norgren et al. 2007).

The medial and lateral plantar arteries play essential roles in various clinical conditions. For instance, diabetic patients often suffer from peripheral arterial disease, leading to compromised blood flow in these vessels. Understanding the vascular anatomy of the foot, including the contributions of the plantar arteries, is essential for performing limb-salvage surgeries, such as revascularization procedures or flap reconstructions for nonhealing ulcers and ischemic wounds (Attinger et al. 2006; Norgren et al. 2007). The angiosome concept, which maps distinct vascular territories supplied by specific arteries, is particularly useful in planning surgical interventions and predicting wound healing outcomes. The medial and lateral plantar arteries correspond to specific angiosomes, and their anastomoses with the dorsal arteries provide multiple options for enhancing blood supply in reconstructive surgeries (Attinger et al. 2006).

5.2.4 Venous System of the Lower Extremity

The venous drainage of the lower extremities consists of three systems: the deep venous system parallels the tibia and femur; the superficial venous system lies within the superficial tissue layer between the deep muscular fascia and the skin; and the perforating or connecting veins connect these two systems. These perforating vessels received their name due to their characteristic penetration through anatomical barriers. Both the superficial and deep venous systems contain valves that facilitate unidirectional blood flow toward the heart, with deep veins containing more valves due to the higher pressures and volumes they must handle (Drake et al. 2020; Moore et al. 2018; Standring 2020). The venous system of the lower extremity displays significant anatomical variability between individuals and even between the left and right limbs. For rehabilitation specialists managing patients with spinal cord injuries, understanding the most common venous configurations is essential for effective treatment.

Superficial veins are generally larger in caliber and have fewer valves compared to deep veins. They lie within the superficial fascia and drain blood from the skin and subcutaneous tissues. Conversely, deep veins, known as venae comitantes, are typically paired with corresponding arteries and located within the muscular compartments (Meissner et al. 2007; Mozes and Gloviczki 2004) (Fig. 5.10). These veins have a higher density of valves to support blood return from deeper tissues, where muscle contractions aid in moving blood toward the heart (Standring 2020).

Extensive anastomotic connections exist between the superficial and deep venous systems, allowing for regulation of venous blood flow under different physiological conditions (Bergan and Bunke-Paquette 2014). In the foot, drainage mainly occurs through the dorsal venous network, whereas in the calf region, blood is directed from the superficial veins into the deep system via perforating veins (Meissner et al. 2007). These perforators, which traverse the deep fascia, contain valves that help maintain forward blood flow.

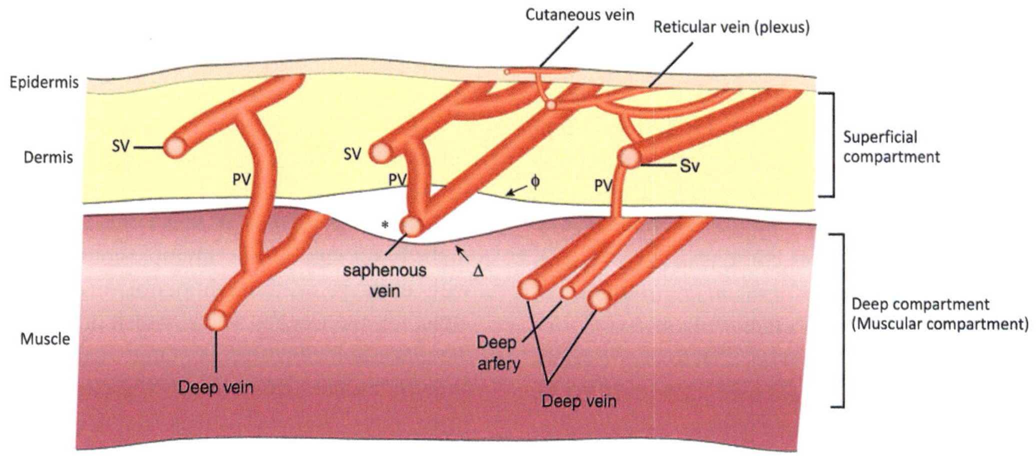

Fig. 5.10 Relationship between fascia and veins in the lower extremity. The fascia encases the muscles, separating the deep compartment from the superficial compartment. Superficial veins (SV) drain the subpapillary and reticular venous plexuses and are connected to deep veins through perforating veins (PV). Deep veins are located beneath the deep fascia (Δ), while perforating veins traverse the superficial fascia (Ø) and the saphenous space (*). The saphenous space, also referred to as the saphenous compartment, is a subcompartment of the superficial compartment. Small veins located just beneath the skin are termed reticular veins. (Adapted with permission from Sadick et al. (2013))

The calf muscle pump mechanism plays a significant role in venous return from the lower extremities. When the calf muscles contract, they compress veins within their fascial compartments, propelling blood into the deep veins. The strategic arrangement and orientation of valves in perforating veins ensure that blood moves in a single direction, thereby preventing backflow (Recek 2013).

In pathological conditions, such as chronic venous insufficiency, valvular incompetence in perforating veins may lead to retrograde blood flow into the superficial veins. This results in venous hypertension, causing the superficial veins to become distended and tortuous. Such changes often manifest as visible, bluish varicose veins, particularly along the posteromedial aspect of the calf and medial thigh (Eberhardt and Raffetto 2014).

5.2.4.1 Superficial Venous System of the Lower Extremity

The primary superficial veins of the lower extremity are the great (long or magna) saphenous vein and the small (lesser or parva) saphenous vein. These veins have numerous unnamed branches that connect the superficial and deep venous systems, facilitating blood drainage from the lower limb. The "great" and "small" saphenous veins are the most prominent components of the superficial venous system. International consensus has discouraged the use of terms such as "long," "greater," "short," and "lesser" to avoid confusion often caused by the use of abbreviations. Additionally, this consensus has standardized the nomenclature for the major tributaries of the saphenous veins (Sadick et al. 2013).

The superficial venous system is a key component of venous drainage, particularly visible on the dorsum of the foot, where it collects blood from the dorsal and plantar surfaces of the toes, as well as the lateral and medial borders of the foot (Standring 2020). This network is interconnected with the deep venous system via perforating veins that traverse the intermetatarsal spaces, allowing for efficient redistribution of blood between the superficial and deep compartments (Moore et al. 2018).

Great (Long, Greater, Magna, Internal) Saphenous Vein

The great saphenous vein is the longest vein in the human body, originating from the medial marginal vein of the foot and ending at the femoral vein near the saphenous opening below the inguinal ligament. Along its course, the great saphenous vein ascends anterior to the medial malleolus, follows the medial border of the lower leg, then curves behind the knee, and continues up the medial thigh. Anatomical variations such as duplication below the knee or connections with the small saphenous vein through the Giacomini vein are common (Netter 2019) (Fig. 5.11).

The vein passes obliquely along the posteromedial aspect of the calf, travels near the femoral and tibial condyles, and then runs superiorly, anteriorly, and laterally in the thigh. It penetrates the cribriform fascia at the saphenous opening in the femoral triangle to join the femoral vein. This point is located inferior to the midpoint of the inguinal ligament (Schuenke et al. 2020). Throughout its course, the great saphenous vein contains 8–20 bicuspid valves that help maintain unidirectional blood flow, working against gravity to facilitate venous return (Meissner et al. 2007). Venous connections along the pathway of the great saphenous vein in each region are as follows:

- At the ankle: The great saphenous vein arises from the medial marginal vein and travels upward along the medial aspect of the leg.
- In the lower leg: It forms multiple connections with deep veins through perforating veins, including the Cockett veins, which drain into the posterior tibial veins. Other tributaries include branches from the calf muscles.
- In the thigh: The great saphenous vein receives branches from the medial and posterior regions. It may also form an accessory saphenous vein, which runs parallel to the main vein.

The great saphenous vein anastomoses freely with the small saphenous vein and connects with deep intermuscular veins via perforating veins, especially near the knee and ankle joints (Bergan and Bunke-Paquette 2014). It also receives lateral tributaries before entering the saphenous opening:

- Superficial circumflex iliac vein (from the iliac region)
- Superficial external pudendal vein (from the external genitalia)
- Superficial epigastric vein (from the lower abdominal wall) (Alimi et al. 1994)

The structural properties and connections of the great saphenous vein make it a preferred vessel for grafts in coronary artery bypass surgery due to its length and accessibility.

Small (Short, Parva, External, Lesser) Saphenous Vein

The small saphenous vein originates from the lateral marginal vein of the foot and ascends along the posterior aspect of the leg. It begins by passing posterior to the lateral malleolus and runs upward along the lateral side of the tendocalcaneus to the calf's posterior surface. It eventually reaches the popliteal fossa, where it pierces the deep fascia and drains into the popliteal vein behind the knee joint (Schuenke et al. 2020) (Fig. 5.11).

As the small saphenous vein travels proximally, it is accompanied by the sural nerve, which runs parallel to the vein. The vein contains between 6 and 12 bicuspid valves that aid in

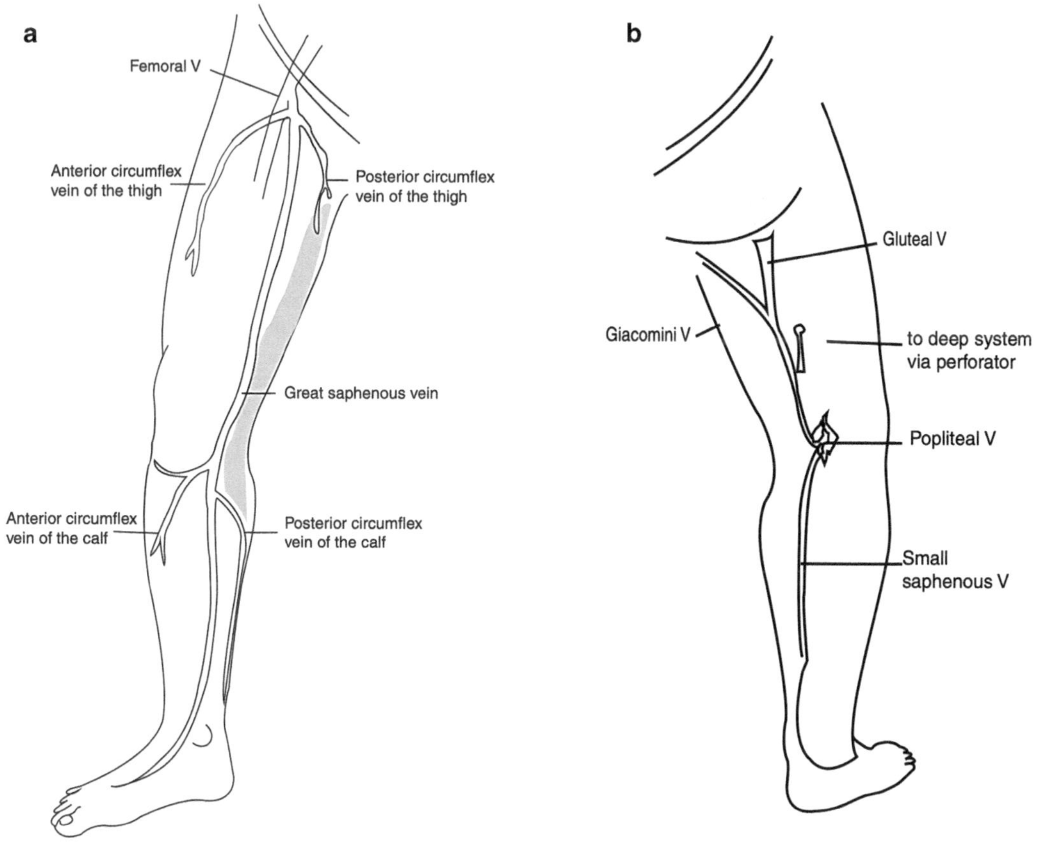

Fig. 5.11 The great saphenous vein and its primary tributaries. (**a**) Ventromedial view and (**b**) posterior view. In (**a**), vein segments shaded in grey are located with the saphenous compartment, while those shaded in blue are superficial to the superficial fascia in (**a**). (Adapted with permission from Sadick et al. (2013))

maintaining unidirectional blood flow. It receives cutaneous tributaries from the lateral regions of the ankle and leg, as well as small venous branches from the posterior calf. The vein's entry into the popliteal fossa occurs between the two heads of the gastrocnemius muscle, where it penetrates the roof of the fossa, formed by the deep fascia (Netter 2019).

The small saphenous vein frequently communicates with the great saphenous vein through various perforating veins along its course, forming an interconnected venous network that helps regulate blood flow and pressure. In some individuals, the small saphenous vein may continue as the Giacomini vein, extending beyond the popliteal region and connecting with the great saphenous vein or deep veins in the thigh (Bergan and Bunke-Paquette 2014).

The small saphenous vein is clinically significant in procedures such as sclerotherapy and vein harvesting for grafting. Its anatomical location and relationships with the sural nerve make it an important consideration in surgical interventions to avoid nerve injury (Meissner et al. 2007).

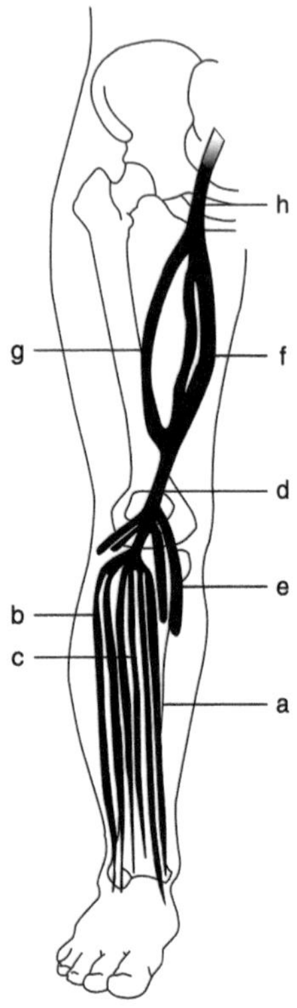

Fig. 5.12 Deep veins of the lower extremity, ventral view. (**a**) Paired posterior tibial veins. (**b**) Paired anterior tibial veins. (**c**) Paired peroneal veins. (**d**) Popliteal vein, which is frequently partially or completely duplicated. (**e**) Medial gastrocnemius veins. (**f**) Femoral veins (This vein is often duplicated; it is duplicated in this diagram). (**g**) Deep femoral vein, often duplicated (as shown in this diagram). (**h**) Common femoral vein. (Adapted with permission from Sadick et al. (2013))

5.2.4.2 Deep Venous System of the Lower Extremity

The deep veins of the lower extremity accompany their corresponding arteries and branches, running parallel throughout the leg. They are responsible for the majority of venous return from the lower limb, aided by muscle contractions and the presence of numerous valves that direct blood flow toward the heart, preventing backflow (Standring 2020; Moore et al. 2018).

The deep venous system comprises a network of vessels that play a critical role in venous return. This system is characterized by paired veins, known as venae comitantes, which accompany smaller arteries, sharing their nomenclature. These paired veins are found throughout the leg, except for larger vessels such as the popliteal, femoral, and profunda femoris veins, which are single rather than paired (Standring 2020) (Fig. 5.12). A key characteristic of the deep venous system is the presence of numerous bicuspid valves that allow unidirectional flow, which is particularly important in preventing gravitational effects (Meissner et al. 2007).

Deep veins are often involved in conditions such as deep vein thrombosis, where blood clots may form and potentially travel to the lungs, leading to a pulmonary embolism. The arrangement of these veins, including their connections to the superficial system via perforating veins, is needed for understanding the development of chronic venous insufficiency, which is characterized by poor venous return and venous hypertension (Meissner et al. 2007).

Understanding the anatomy and function of the deep venous system is essential for rehabilitation specialists, especially those working with patients who have spinal cord injuries. Reduced mobility and impaired muscle function in these patients can compromise venous return, increasing the risk of complications like deep vein thrombosis. Implementing preventive strategies, such as passive leg exercises and compression therapy, may help enhance venous circulation and reduce the likelihood of clot formation (Sezer et al. 2015; Nash et al. 1996).

In the Foot

The venous drainage of the foot starts with the plantar digital veins, which form from plexuses on the surface of the toes. These veins communicate with the dorsal digital veins, creating a network of interconnected venous channels. The plantar digital veins merge to form the four plantar metatarsal veins, which drain into the deep venous system of the foot (Netter 2019).

The plantar metatarsal veins connect with the veins on the dorsum of the foot through perforating veins, forming the deep plantar venous arch, which runs parallel to the plantar arterial arch. This arrangement is essential for efficient venous return and blood redistribution between the deep and superficial systems during standing and walking (Standring 2020; Caggiati et al. 2002).

The medial and lateral plantar veins emerge from the deep plantar venous arch, accompanying the corresponding arteries. These veins converge behind the medial malleolus to form the posterior tibial veins after communicating with the great and small saphenous veins (Netter 2019).

In the Leg

Posterior Tibial Veins

The posterior tibial veins run alongside the posterior tibial artery and drain the deep compartments of the posterior leg. They receive major tributaries, including the peroneal veins, which follow the peroneal artery, and perforating veins that connect with the superficial venous system. These veins join the anterior tibial veins to form the popliteal vein (Standring 2020; Moore et al. 2018).

Anterior Tibial Veins

The anterior tibial veins are paired veins accompanying the anterior tibial artery, responsible for draining the anterior compartment of the leg. They cross the interosseous membrane between the tibia and fibula and merge with the posterior tibial veins, forming the popliteal vein in the popliteal fossa (Netter 2019).

Popliteal Vein

The popliteal vein is formed at the inferior border of the popliteus muscle through the confluence of the anterior and posterior tibial veins. As it ascends within the popliteal fossa, it receives multiple tributaries, including the genicular veins, the small (lesser) saphenous vein, and other muscular branches such as the gastrocnemius veins. The vein travels from medial to lateral as it courses through the fossa, eventually crossing posterior to the popliteal artery (Drake et al. 2020; Moore et al. 2018).

After traversing the popliteal fossa, the vein enters the adductor canal and transitions to become the femoral vein upon passing through the adductor hiatus. The popliteal vein maintains close anatomical relationships with surrounding structures, which is significant for surgical procedures and the diagnosis of vascular pathologies such as deep vein thrombosis (Netter 2019).

The popliteal vein is a common site for deep vein thrombosis, where clot formation can obstruct venous return and potentially lead to pulmonary embolism. Its anatomical location and connection with the superficial and deep venous systems make it important in the management of venous disorders and surgical interventions (Caggiati et al. 2002).

In the Thigh

Femoral Vein (Superficial Femoral Vein)

The femoral vein is a continuation of the popliteal vein and accompanies the femoral artery as it ascends through the thigh. It becomes the external iliac vein at the level of the inguinal ligament, where it is sometimes referred to as the common femoral vein. The vein lies medial to the femoral artery within the femoral sheath and receives the great saphenous vein through the saphenous opening approximately 3 cm distal to the inguinal ligament (Standring 2020; Caggiati et al. 2002).

The femoral vein follows the path of the femoral artery, but its relationship with the artery changes as it ascends. In the adductor canal, it is located posteriorly relative to the artery but transitions from a lateral to a medial orientation as it progresses proximally. Upon entering the femoral triangle, it becomes positioned more medially (Schuenke et al. 2020).

The femoral vein receives multiple tributaries, including the following:

- Muscular veins from the thigh
- The vena profunda femoris (deep femoral vein), which drains deep structures and runs anterior to the arteria profunda femoris
- Lateral and medial circumflex femoral veins, which contribute to the venous return from the proximal thigh and hip regions (Bergan and Bunke-Paquette 2014)

The vena profunda femoris collects blood from muscular tributaries and perforating branches that establish anastomoses with the popliteal vein below and the inferior gluteal vein above, supporting collateral circulation.

5.2.4.3 Perforating Veins

Perforating veins connect the superficial and deep venous systems, with valves that typically direct blood from the superficial veins to the deep veins, preventing backflow. Valve failure in these veins may contribute to conditions such as chronic venous insufficiency and varicose veins.

- In the foot: Four consistent perforating veins are located:
  One 2.5 cm below the medial malleolus.
  One 3.5 cm below and anterior to the medial malleolus.
  Two positioned 3.0 cm below and anterior to the lateral malleolus.
- In the leg: Sixteen perforators include the following:
  Eight veins draining into the posterior tibial veins (Cockett and Boyd veins).
  Four draining into the peroneal veins.
  Four draining into the soleal and gastrocnemius veins.
- In the thigh: Two perforators, known as the Dodd perforators or Hunter's perforating veins, link the superficial veins to the deep femoral system.

5.3 Lung Vasculature

The pulmonary vascular system is necessary for the transportation of blood between the heart and lungs to enable efficient gas exchange. Oxygen-depleted blood from the right ventricle is directed to the lungs, where carbon dioxide is exchanged for oxygen within the alveolar-capillary networks. In addition to its respiratory function, the pulmonary vasculature also supplies nutrients and other essential substances to the lung tissues beyond the terminal bronchioles, ensuring the metabolic needs of the distal airways are met.

Similar to the systemic circulation, the pulmonary vascular network consists of a hierarchy of vessels, including arteries, arterioles, capillaries, venules, and veins. The pulmonary arteries carry deoxygenated blood from the right ventricle, bifurcating into progressively smaller branches until reaching the capillary beds surrounding the alveoli. Here, gas exchange occurs across the thin respiratory membrane. Following this, oxygenated blood is collected by the pulmonary venules and veins, which transport it to the left atrium for systemic circulation (West 2011; Weibel 1984).

The pulmonary vascular system differs from systemic circulation in terms of pressure and resistance. It performs in a low-pressure, low-resistance environment to accommodate the delicate structure of the alveoli and prevent fluid accumulation in the lungs. Pulmonary arterial pressures typically range from 8 to 20 mm Hg, significantly lower than systemic pressures, which allows for the efficient transfer of gases while avoiding vascular damage (Glenny 2009). Additionally, the pulmonary circulation's unique ability to accommodate varying blood volumes without significant changes in pressure reflects its high compliance, important for maintaining optimal gas exchange during various states of activity.

5.3.1 Arterial Vasculature of the Lung

The right ventricle of the heart propels deoxygenated blood into the pulmonary artery, which subsequently divides into the right and left branches just below the aorta. These branches follow the tracheobronchial tree in a posterolateral course, paralleling the branching pattern of the airways. The pulmonary arteries are defined by their elastic nature, allowing them to accommodate fluctuations in blood flow with minimal resistance. Initially, the walls contain limited smooth muscle, but the amount of smooth muscle progressively increases along the vessels as they extend to the peripheral regions of the lungs, with significant changes occurring at the level of the fifth or sixth arterial divisions (Standring 2020).

Pulmonary arterial branches measuring between 1.0 mm and 0.1 mm in diameter are mainly muscular, while those smaller than 0.1 mm consist largely of thin-walled endothelial tubes that lack muscular support. These smaller vessels form dense anastomotic networks that contribute to the alveolar capillary system, a primary structural element of the respiratory membrane. The capillary network enables efficient gas exchange by facilitating the transfer of oxygen and carbon dioxide between the alveoli and blood. This structural arrangement optimizes the interface for respiratory function and minimizes diffusion distance, allowing for rapid equilibration of gases (West 2011).

The pulmonary circulation is characterized by low resistance and low pressure, with mean pulmonary arterial pressures ranging between 8 and 20 mm Hg (average of 13 mm Hg). This low-pressure system is supported by the high compliance of the pulmonary vasculature, which allows the vessels to accommodate increases in blood volume with little impact on pressure. The resistance in the pulmonary circuit is approximately one-sixth of that in the systemic circulation, an important factor in preserving the integrity of the delicate alveolar structures and preventing fluid leakage into the alveolar spaces (Glenny 2009).

The ability of the pulmonary vascular bed to expand under increased blood flow, such as during exercise, is a key adaptive feature. The pulmonary circulation can handle a tripling of blood flow without significantly raising arterial pressure, primarily due to the recruitment and distention of additional capillaries. This adaptive reaction provides sufficient oxygen supply during elevated metabolic demand while preventing increased pressure that can impede gas exchange or result in pulmonary edema (Standring 2020).

Blood distribution within the lungs is as follows: approximately 30% of the total blood volume resides in the larger and smaller pulmonary arteries, while roughly 20% is present within the pulmonary capillary network. The remaining 50% is found in the pulmonary veins, which drain oxygenated blood back into the left atrium of the heart (Hlastala and Berger 2001). The extensive network of capillaries enveloping the alveoli maximizes surface area for gas exchange, while the thin endothelial walls of these vessels ensure efficient diffusion of respiratory gases.

Additionally, the distribution of blood flow within the lungs is influenced by both gravity and local factors such as alveolar oxygen tension. In an upright position, pulmonary blood flow is greater at the lung bases compared to the apices due to gravitational effects on perfusion. However, local regulation also occurs through mechanisms such as hypoxic pulmonary vasoconstriction, which directs blood flow away from poorly ventilated areas to optimize the ventilation–perfusion matching essential for effective gas exchange (West 2011).

5.3.1.1 Pulmonary Trunk

The pulmonary trunk serves as the conduit for deoxygenated blood from the right ventricle to the lungs for gas exchange. Measuring approximately 5 cm in length and 3 cm in diameter, it originates at the base of the right ventricle, positioned superiorly and to the left of the supraventricular crest. This vessel arises from the right ventricular outflow tract, or pulmonary conus, at the level of the pulmonary semilunar valves, where it channels blood into the pulmonary circulation. The trunk follows an ascending and posterior trajectory, lying anterior to the left side of the ascending aorta (Fig. 5.13). It is entirely enclosed within the pericardium, which provides structural support and minimizes friction during cardiac movements (Standring 2020).

The pulmonary trunk bifurcates into the left and right pulmonary arteries, which are similar in size and serve the corresponding lungs. The bifurcation angle of the main pulmonary artery can vary significantly, ranging from 100° to 180°, depending on individual anatomical differences. The diameter of the right pulmonary artery typically ranges from 17 to 30 mm, with an average measurement of 23.4 mm, while the main pulmonary artery itself has a diameter that spans from 20 to 30 mm, averaging around 26.4 mm (Ghaye et al. 2001b). Notably, the combined diameters of the left and right pulmonary arteries are larger than that of the pulmonary trunk, allowing for efficient distribution of blood to both lungs.

The bifurcation of the pulmonary trunk occurs just below the aortic arch, with the right pulmonary artery passing under the arch and traveling posteriorly toward the right lung, while the left pulmonary artery follows a more direct route to the left lung. This anatomical arrangement facilitates the even distribution of blood flow despite differences in the relative positions of the lungs

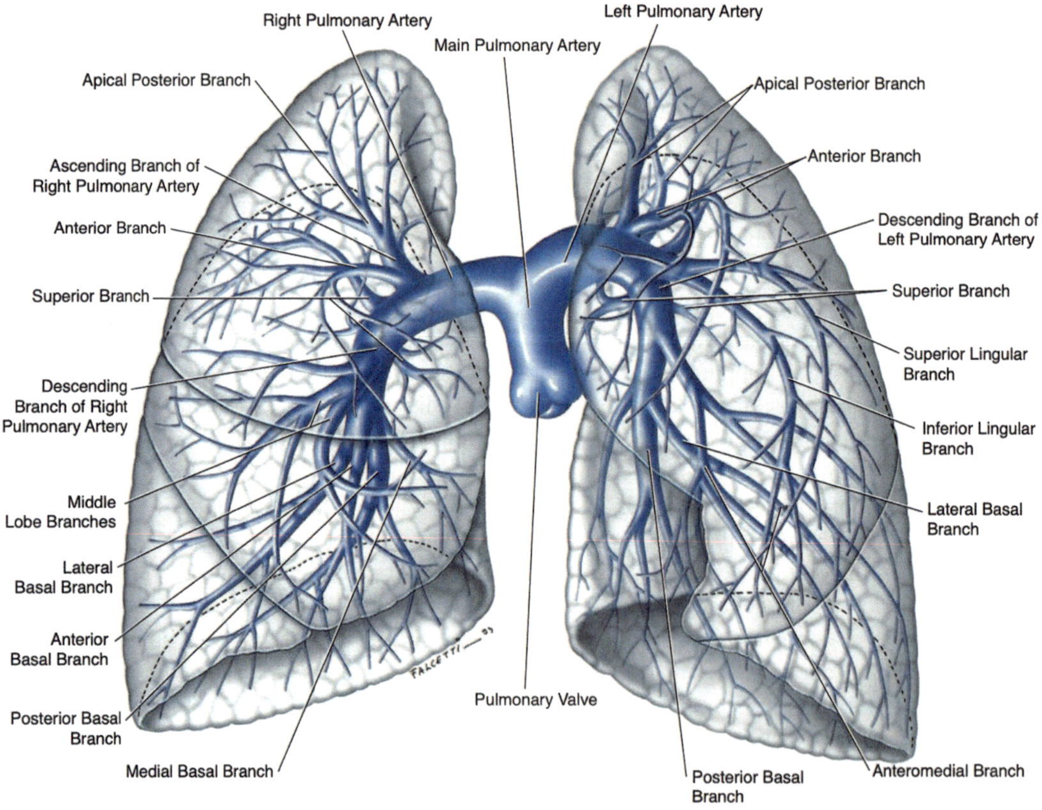

Fig. 5.13 Distribution of the pulmonary artery in both lungs. (Adapted from Uflacker and Guimaraes (2021))

and other thoracic structures. The right pulmonary artery is generally longer and courses laterally behind the superior vena cava and ascending aorta before reaching the right lung, while the left pulmonary artery maintains a shorter and more horizontal path.

Variations in the branching pattern and the diameters of the pulmonary arteries can influence hemodynamic properties, potentially affecting the efficiency of pulmonary circulation. In cases of pulmonary hypertension or congenital heart defects, structural alterations in the pulmonary trunk and its branches can lead to changes in blood flow dynamics, warranting careful evaluation through imaging techniques such as 3D volume-rendered CT (Ghaye et al. 2001a). These techniques provide a detailed view of the bifurcation and the relationships between the pulmonary arteries and surrounding structures, offering valuable insights into the anatomical and functional status of the pulmonary vasculature.

5.3.1.2 Right Pulmonary Artery

The right pulmonary artery, though slightly smaller in diameter than the main pulmonary artery, is a major conduit for blood supply to the right lung. It originates from the bifurcation of the pulmonary trunk just beneath the aortic arch and runs horizontally across the chest. Its path can be slightly downward in some individuals, extending toward the hilum of the right lung. On angiographic imaging, the right pulmonary artery appears to run behind the ascending aorta and the superior vena cava while lying anterior to the tracheal bifurcation and the esophagus. This anatomical arrangement places the artery in a unique position to maintain its relationship with both the cardiovascular and respiratory structures in the thoracic cavity (Standring 2020). At the level of the right lung hilum, the right pulmonary artery bifurcates into two main branches: the ascending branch, which supplies the right upper lobe, and the descending branch, which supplies the middle and lower lobes. This branching pattern closely follows the anatomical organization of the bronchopulmonary segments, allowing for the compartmentalized delivery of blood that matches the airway distribution.

Ascending Branch

The ascending branch of the right pulmonary artery travels upward to reach the right upper lobe, where it further divides into three segmental arteries: the apical, posterior, and anterior segmental arteries. Each of these segmental arteries serves a specific bronchopulmonary segment, ensuring an organized blood supply that supports efficient gas exchange.

- *Apical Segmental Artery*: The apical segmental artery supplies the apical segment of the right upper lobe. It divides into two major branches, the apical and posterior rami, which deliver blood to distinct subregions within the segment. This arterial division facilitates a more precise distribution of blood, enhancing perfusion to the ventilated areas (Ghaye et al. 2001a, b).
- *Posterior Segmental Artery*: Often arising in conjunction with the apical segmental artery, this artery branches into two primary divisions: the posterior ramus, which supplies the posterior part of the segment, and the lateral ramus, which extends toward the outer portions. The posterior segmental artery plays an important role in providing perfusion to the posterior bronchopulmonary segment of the right upper lobe, an area that can be variably shaped depending on individual anatomical differences (Standring 2020).
- *Anterior Segmental Artery*: The anterior segmental artery is the most inferior branch of the trifurcation from the ascending branch, supplying the anterior segment of the right upper lobe. It divides into two rami: the anterior ramus and the lateral ramus. These branches ensure that the anterior segment receives an adequate supply of oxygenated blood to meet metabolic demands during respiration (West 2011).

Descending Branch

The descending branch of the right pulmonary artery continues caudally from the hilum and provides blood to the right middle and lower lobes. It runs in close association with the bronchi, aligning itself with the airway branching pattern.

- *Middle Lobe Artery*: This artery arises from the descending branch, opposite the origin of the superior segmental artery of the right lower lobe. It travels anteriorly and inferiorly, dividing into two segmental arteries: the lateral and medial segmental arteries, which supply their respective bronchopulmonary segments in the right middle lobe (Glenny and Robertson 2011).
- *Superior Segmental Artery of the Right Lower Lobe*: This artery emerges from the descending branch and runs posteriorly, with an upward and lateral course, supplying the superior segment of the right lower lobe. Its pathway allows it to deliver blood to a region critical for ventilation, especially in the upright position, where gravitational effects influence perfusion distribution (Glenny and Robertson 2011; Standring 2020).
- *Medial Basal Segmental Artery*: The third major branch of the right pulmonary artery's descending branch arises distal to the origin of the superior segmental artery and directs blood to the medial basal segment of the right lower lobe.
- *Anterior Basal Segmental Artery*: Emerging from the anterolateral aspect of the descending branch, slightly beyond the origin of the medial basal segmental artery, it supplies the anterior basal segment (Ghaye et al. 2001a).
- *Posterior and Lateral Basal Segmental Arteries*: These vessels commonly arise as a bifurcation of the descending branch, with the posterior basal segmental artery reaching the most dependent portion of the lung and the lateral basal segmental artery supplying the lateral basal segment. Together, these arteries facilitate perfusion to the dependent regions of the right lower lobe, which are often more affected by conditions like aspiration pneumonia due to gravitational effects on fluid distribution.

The complex architecture of the right pulmonary artery's branches reflects the segmentation of the lung into distinct functional units. This anatomical arrangement allows for targeted blood flow, matching the distribution of air within each bronchopulmonary segment. Variations in the branching pattern, such as different points of origin for segmental arteries or atypical bifurcations, are clinically relevant for procedures such as bronchoscopy, pulmonary resection, and embolization therapy (Ghaye et al. 2001a; Weibel 1984).

5.3.1.3 Left Pulmonary Artery

The left pulmonary artery serves as a continuation of the main pulmonary artery, transporting deoxygenated blood from the right ventricle to the left lung. It is shorter than the one on the right side and follows a complex path, initially running upward, posteriorly, and to the left, before sharply turning left and descending toward the hilum of the left lung. Along its course, the left pulmonary artery passes anterior to the descending aorta and beneath the arch of the aorta. It is connected to the arch by the ligamentum arteriosum, a remnant of the ductus arteriosus, which provides a crucial landmark for both surgical and radiological procedures (Standring 2020). Upon reaching the hilum, the left pulmonary artery bifurcates into two primary branches: the ascending branch, which supplies the upper lobe, and the descending branch, which supplies the lower lobe. These branches match the bronchopulmonary segmental arrangement, enabling organized blood flow that aligns with the airway distribution.

Ascending Branch

The ascending branch of the left pulmonary artery arises approximately 2–4 cm from the origin of the main pulmonary artery and travels cranially toward the upper lobe. It divides into two segmental arteries: the posterior apical segmental artery and the anterior segmental artery, which supply specific regions of the left upper lobe.

- *Posterior Apical Segmental Artery*: This artery supplies the apical posterior bronchopulmonary segment of the left upper lobe and is the larger of the two segmental arteries. It bifurcates into two primary rami—the apical ramus, which runs medially, and the posterior ramus, which runs laterally. The bifurcation enables effective perfusion of the apical and posterior regions, which may differ in size and shape among individuals (Ghaye et al. 2001a).

- *Anterior Segmental Artery*: The anterior segmental artery, arising as the inferior branch of the ascending artery, courses anteriorly and bifurcates into the anterior and lateral rami. These branches supply the anterior bronchopulmonary segment of the left upper lobe, ensuring adequate blood flow to support the segment's respiratory function.

Descending Branch

The descending branch of the left pulmonary artery continues downward as a smooth extension of the main artery, supplying the left lung's lower regions. Along its path, it gives rise to several segmental branches, including the lingular artery, the superior segmental artery, and arteries supplying the basal bronchopulmonary segments of the left lower lobe.

- *Lingular Artery*: This artery emerges approximately 2 cm distal to the origin of the ascending branch and branches into two segmental arteries—the superior lingular segmental artery and the inferior lingular segmental artery. These arteries supply the corresponding segments of the lingula, a structure analogous to the right middle lobe in terms of anatomy and function. The lingular segments are often involved in conditions such as pneumonia and bronchiectasis, making the lingular artery a key focus during segmental interventions (Weibel 2019).
- *Segmental Arteries to the Left Lower Lobe*: The left lower lobe is usually divided into three bronchopulmonary segments, as opposed to the four found on the right. The segmental arteries include the following:
  - *Superior Segmental Artery*: This artery supplies the superior segment of the left lower lobe and arises posteriorly from the descending branch, just distal to the lingular artery. It follows a course similar to the superior segmental artery on the right side, delivering blood to the uppermost portion of the left lower lobe.
  - *Anteromedial Basal Segmental Artery*: Unlike the right side, where the anterior and medial basal arteries are separate, the left side commonly has a combined anteromedial basal segmental artery. This artery is larger than the one in the right lower lobe, providing blood to the anteromedial regions of the basal lung.
  - *Posterior and Lateral Basal Segmental Arteries*: These arteries typically arise together as a bifurcation from the descending branch. The posterior basal segmental artery courses inferiorly and posteriorly to supply the posterior bronchopulmonary segment. The lateral basal segmental artery follows an inferior and lateral path, delivering blood to the lateral basal segment. These arteries ensure perfusion to the dependent regions of the lung, which are especially significant in conditions that affect the lower lobes, such as aspiration pneumonia (Glenny and Robertson 2011).

5.3.1.4 Pulmonary Microcirculation

The pulmonary circulation serves as a principal filtering system for blood passing between venous and arterial circuits, where particles larger than 75 μm are typically retained in the pulmonary arterioles due to their narrow structure (Standring 2020). These arterioles decrease rapidly from a diameter of approximately 100–50 μm. The dimensions of pulmonary capillaries range from 8 to 9 μm in diameter, but their size can fluctuate based on factors such as gravitational forces and the individual's body position. Experimental evidence indicates that particles as large as 400 μm may traverse precapillary shunts between pulmonary arteries and veins, entering the venous side of the circulation, suggesting variability in microvascular filtering capabilities (West 2001).

The alveolar-capillary interface in adults comprises an extensive surface area of approximately 70–90 m^2, which allows for optimal gas exchange.

Oxygen and CO_2 are exchanged across the alveolar-capillary barrier primarily via diffusion, driven by concentration gradients. The capillary network is densely organized, with short cylindrical segments that create a hexagonal mesh-like structure within the alveolar walls. This network configuration is necessary for maximizing the surface area available for gas exchange, thereby improving respiratory efficiency (West et al. 1964).

Although the pulmonary artery accommodates a significant blood volume and flow rate, it does not contribute to direct lung tissue nourishment. Instead, this role is carried out by the bronchial circulation, a component of the systemic circulation. The bronchial and pulmonary capillary systems are interconnected, permitting blood drainage into the systemic veins through the azygos vein or into the left atrium via the pulmonary veins. These interconnections at the capillary level provide an important pathway that can alleviate elevated capillary hydrostatic pressures when imbalances in right or left atrial pressure arise (Standring 2020). This adaptive mechanism is particularly significant during pathological states such as pulmonary hypertension or congestive heart failure, where bronchial circulation can support collateral blood flow to maintain lung tissue viability when the pulmonary arterial supply is compromised.

Bronchial arteries, which emerge from the lung hilum, traverse two principal paths: along the bronchial tree and along the visceral pleura. These arteries maintain a substantial diameter until reaching the terminal bronchioles, where the pulmonary circulation assumes responsibility for nutrient delivery. However, smaller branches of the bronchial arteries may extend to the alveolar ducts and even penetrate the lung parenchyma surrounding the alveolar sacs, thereby contributing to local microvascular perfusion (West 2001). The bronchial arteries are essential for supplying blood to the walls of the bronchi, the smooth muscle layers, glandular tissues, and cartilaginous structures. They also provide the vasa vasorum to nourish the walls of the pulmonary arteries and the vasa nervorum for nerve supply.

The bronchial circulation is characterized by a dual venous drainage system. Blood from the bronchial arteries can drain into systemic veins, including the bronchial veins and tributaries of the azygos vein, or into the superior vena cava. Additionally, bronchial blood can enter the alveolar capillary network at precapillary, capillary, and postcapillary levels, subsequently draining into the pulmonary veins. This complex network of arteriolar branches and capillary beds is present throughout the bronchial tree, extending to the terminal bronchioles, with a gradual reduction in density as the airway caliber decreases (West et al. 1964). In the surrounding connective tissues of the pulmonary arteries, similar vascular structures are found, albeit with fewer vessels. Most of the blood from the juxtapleural bronchial arteries ultimately drains into the pulmonary capillaries. Within the mediastinal pleura, fine vascular networks establish connections between the bronchial venous plexus and the branches of the pulmonary vein.

Precapillary and postcapillary anastomoses exist between the bronchial and pulmonary arterial systems, with small venous branches linking bronchial venous plexuses to pulmonary veins. These vascular connections facilitate direct blood flow interactions between bronchial venous plexuses and adjacent alveolar capillaries through small venules, which are evident along both bronchi and bronchioles (Standring 2020). This anatomical arrangement supports adaptive responses to physiological stress and maintains the integrity of lung perfusion under various conditions.

5.3.2 Venous Vasculature of the Lung

The pulmonary veins originate from the capillary networks within the alveolar walls and pleura, forming an extensive vascular system crucial for oxygenated blood transport. The smaller pulmonary veins resemble arterioles histologically, displaying a thin wall structure predominantly composed of endothelial cells. As the veins

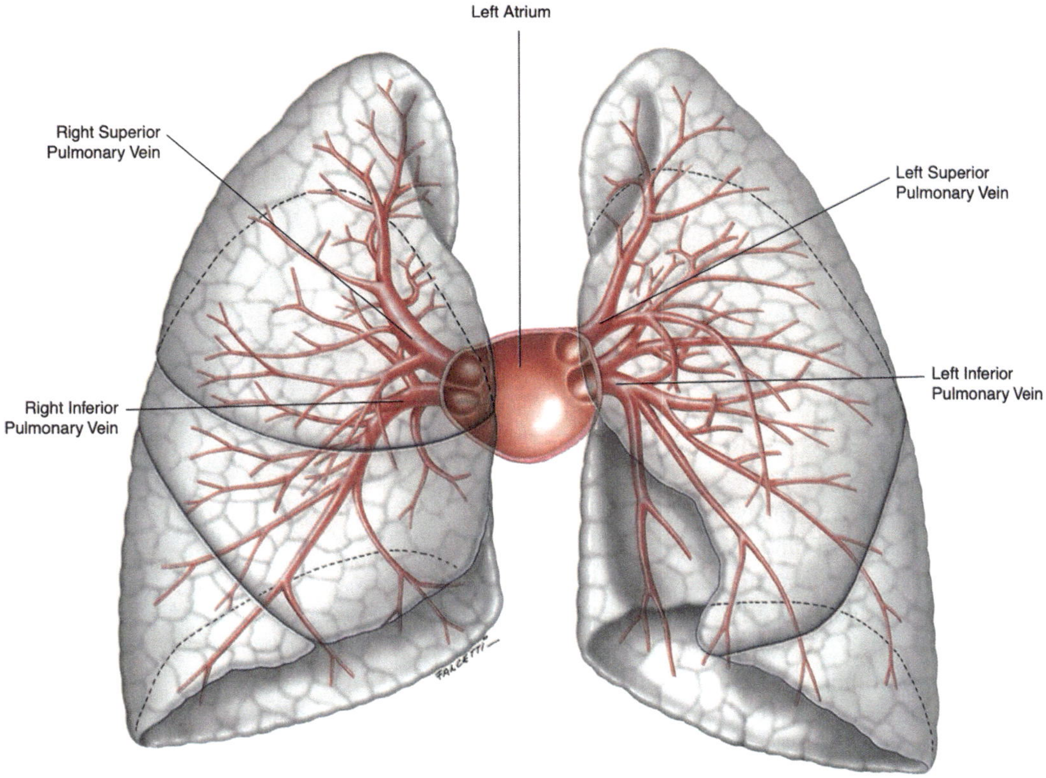

Fig. 5.14 Schematic illustration of the pulmonary veins and left atrium, showing the two superior pulmonary veins and the two inferior pulmonary veins. (Adapted from Uflacker and Guimaraes (2021))

increase in size to approximately 60–100 μm in diameter, distinct layers of smooth muscle cells and elastic fibers emerge, enhancing the structural integrity and aiding in maintaining venous tone. These features become increasingly prominent in larger pulmonary veins, which join the interlobular septa and continue toward the left atrium, carrying oxygen-rich blood (Gao and Raj 2005; Yuan et al. 2011) (Fig. 5.14).

The structural organization of the pulmonary veins differs from the bronchoarterial pathways, as these veins course along the interlobular septa rather than following the bronchial tree. This anatomical arrangement reduces the likelihood of interference with airflow, maintaining efficient respiratory function. Larger pulmonary veins also exhibit gradual thickening of the tunica media, reflecting an adaptation to the higher pressures encountered as blood approaches the heart (Huang et al. 1996).

Surrounding the bronchi, a dense microvascular network exists, composed of arterioles that terminate in bronchial capillaries and extensive bronchial venous plexuses. These plexuses feature irregular shapes and pathways, with some connecting directly to the smaller branches of the pulmonary veins, facilitating communication between the bronchial and pulmonary circulations. Such interconnections may serve as collateral pathways for blood flow during certain pathological conditions, such as pulmonary hypertension or embolism, where primary circulatory routes are compromised (Yuan et al. 2011; Huang and Yen 1998).

The dual role of the pulmonary venous system in both drainage and local blood flow regulation is supported by its complex anatomy and adaptive features. The presence of smooth muscle in the walls of the larger veins contributes to the modulation of venous return and the distribution of blood within the lung parenchyma. Additionally, the interconnected venous plexuses around the airways provide an alternative pathway for blood drainage, which can be significant in conditions that increase pulmonary vascular resistance (Gao and Raj 2005; Yuan et al. 2011).

The pulmonary venous system typically consists of four main veins: two superior and two inferior pulmonary veins. The superior veins drain the upper and middle lobes on the right and the upper lobe on the left, while the inferior veins drain the lower lobes. These veins carry oxygenated blood from the lungs to the left atrium of the heart, where it is subsequently pumped into systemic circulation. The pulmonary venous system accommodates approximately 50% of the lung's total blood volume at any given time, which is significant for its role in maintaining efficient gas exchange and regulating cardiac output. Imaging modalities such as CT are highly effective for visualizing the pulmonary veins and identifying any anatomical variations (Gao and Raj 2005).

Variations in pulmonary venous anatomy are relatively common. The most frequent anomaly on the right side involves three separate pulmonary veins, with an additional vein draining the right middle lobe. On the left, a common variant is a single trunk into which all the veins empty before reaching the left atrium. Complex arrangements, though rare, may occur and are usually regarded as normal anatomical variants, as they often do not affect the overall pathway of blood flow (Yuan et al. 2011).

The right pulmonary veins course through the hilum below the main pulmonary artery and posterior to the superior vena cava. They typically enter the left atrium as two separate veins. On the left side, the pulmonary veins cross in front of the descending aorta and may either enter the left atrium independently, like the right side, or merge into a common vein within the pericardial cavity before draining into the atrium (Yuan et al. 2011).

Understanding pulmonary venous anatomy has gained importance with the advent of interventional procedures for treating arrhythmias, such as atrial fibrillation. These arrhythmias often originate near the orifice of the pulmonary veins, where abnormal conduction pathways can develop. Treatment strategies involve the use of radiofrequency ablation to isolate these pathways by applying energy around the vein's orifice, thereby disrupting the aberrant electrical signals (Calkins et al. 2017).

References

Agur AMR, Dalley AF. Grant's atlas of anatomy. 12th ed. Philadelphia: Wolters Kluwer; 2009.

Alimi YS, Barthelemy P, Juhan C. Venous pump of the calf: a study of venous and muscular pressures. J Vasc Surg. 1994;20(5):728–35.

Attinger CE, Evans KK, Bulan E, Blume P, Cooper P. Angiosomes of the foot and ankle and clinical implications for limb salvage: reconstruction, incisions, and revascularization. Plast Reconstr Surg. 2006;117(7 Suppl):261S–93S.

Bergan JJ, Bunke-Paquette N. The vein book. 2nd ed. New York: Oxford University Press; 2014.

Caggiati A, Bergan JJ, Gloviczki P, Jantet G, Wendell-Smith CP, Partsch H. Nomenclature of the veins of the lower limbs: an international interdisciplinary consensus statement. J Vasc Surg. 2002;36(2):416–22.

Calkins H, Hindricks G, Cappato R, Kim YH, Saad EB, Aguinaga L, et al. 2017 HRS/EHRA/ECAS/APHRS/SOLAECE expert consensus statement on catheter and surgical ablation of atrial fibrillation. Heart Rhythm. 2017;14(10):e275–444.

Clemente CD. Clemente's anatomy dissector. 3rd ed. Philadelphia: Wolters Kluwer; 2011.

Detton AJ. Grant's dissector. 16th ed. Philadelphia: Wolters Kluwer; 2017.

Drake RL, Vogl AW, Mitchell AWM. Gray's anatomy for students. 4th ed. Philadelphia: Elsevier; 2020.

Eberhardt RT, Raffetto JD. Chronic venous insufficiency. Circulation. 2014;130(4):333–46.

Ellis H. Clinical anatomy: applied anatomy for students and junior doctors. 11th ed. Oxford: Blackwell Publishing; 2006.

Gao Y, Raj JU. Role of veins in regulation of pulmonary circulation. Am J Physiol Lung Cell Mol Physiol. 2005;288(2):L213–26.

Ghaye B, Szapiro D, Fanchamps JM, Dondelinger RF. Congenital bronchial abnormalities revisited. Radiographics. 2001a;21(1):105–19.

Ghaye B, Szapiro D, Mastora I, Delannoy V, Duhamel A, Remy J, et al. Peripheral pulmonary arteries: how far in the lung does multi-detector row spiral CT allow analysis? Radiology. 2001b;219(3):629–36.

Glenny RW. Determinants of regional ventilation and blood flow in the lung. Intensive Care Med. 2009;35(11):1833–42.

Glenny RW, Robertson HT. Determinants of pulmonary blood flow distribution. Compr Physiol. 2011;1(1):39–59.

Gosling JA, Harris PF, Humpherson JR, Whitmore I, Willan PLT. Human anatomy, color atlas and textbook. 6th ed. Philadelphia: Elsevier; 2017.

Hlastala MP, Berger AJ. Physiology of respiration. 2nd ed. Oxford: Oxford University Press; 2001.

Huang W, Yen RT. Zero-stress states of human pulmonary arteries and veins. J Appl Physiol. 1998;85(3):867–73.

Huang W, Yen RT, McLaurine M, Bledsoe G. Morphometry of the human pulmonary vasculature. J Appl Physiol. 1996;81(5):2123–33.

Jones MR, Prabhakar A, Viswanath O, Urits I, Green JB, Kendrick JB, et al. Thoracic outlet syndrome: a comprehensive review of pathophysiology, diagnosis, and treatment. Pain Ther. 2019;8(1):5–18.

Meissner MH, Moneta G, Burnand K, Gloviczki P, Lohr JM, Lurie F, et al. The hemodynamics and diagnosis of venous disease. J Vasc Surg. 2007;46(Suppl S):4S–24S.

Moore KL, Dalley AF, Agur AMR. Clinically oriented anatomy. 8th ed. Philadelphia: Wolters Kluwer; 2018.

Mozes G, Gloviczki P. New discoveries in anatomy and new terminology of leg veins: clinical implications. Vasc Endovasc Surg. 2004;38(4):367–74.

Nash MS, Montalvo BM, Applegate B. Lower extremity blood flow and responses to occlusion ischemia differ in exercise-trained and sedentary tetraplegic persons. Arch Phys Med Rehabil. 1996;77(12):1260–5.

Netter FH. Atlas of human anatomy. 7th ed. Philadelphia: Elsevier; 2019.

Norgren L, Hiatt WR, Dormandy JA, Nehler MR, Harris KA, Fowkes FG. Inter-society consensus for the management of peripheral arterial disease (TASC II). J Vasc Surg. 2007;45(Suppl S):S5–67.

Ouriel K. Peripheral arterial disease. Lancet. 2001;358(9289):1257–64.

Palastanga N, Soames R. Anatomy and human movement: structure and function. 6th ed. Edinburgh: Elsevier; 2012.

Pickering TG, Hall JE, Appel LJ, et al. Recommendations for blood pressure measurement in humans and experimental animals: part 1: blood pressure measurement in humans. Hypertension. 2005;45(1):142–61.

Recek C. Calf pump activity influencing venous hemodynamics in the lower extremity. Int J Angiol. 2013;22(1):23–30.

Rodríguez-Niedenführ M, Vázquez T, Nearn L, Ferreira B, Parkin I, Sañudo JR. Variations of the arterial pattern in the upper limb revisited: a morphological and statistical study, with a review of the literature. J Anat. 2001;199(5):547–66.

Sadick N, Khilnani N, Morrison N. Practical approach to the management and treatment of venous disorders. London: Springer; 2013.

Schuenke M, Schulte E, Schumacher U. Thieme atlas of anatomy: general anatomy and musculoskeletal system. 3rd ed. New York: Thieme; 2020.

Sezer N, Akkuş S, Uğurlu FG. Chronic complications of spinal cord injury. World J Orthop. 2015;6(1):24–33.

Sinnatamby CS. Last's anatomy: regional and applied. 12th ed. London: Churchill Livingstone; 2011.

Snell RS. Clinical anatomy by regions. 9th ed. Philadelphia: Lippincott Williams & Wilkins; 2012.

Standring S, editor. Gray's anatomy: the anatomical basis of clinical practice. 42nd ed. London: Elsevier; 2020.

Tank PW, Grant JC. Grant's dissector. 16th ed. Philadelphia: Wolters Kluwer; 2017.

Uflacker A, Guimaraes M. Uflacker's atlas of vascular anatomy: an angiographic approach. 3rd ed. Philadelphia: Wolters Kluwer; 2021.

Weibel ER. The pathway for oxygen: structure and function in the mammalian respiratory system. London: Harvard University Press; 1984.

Weibel ER. The pathway for oxygen: structure and function in the mammalian respiratory system. Cambridge: Harvard University Press; 2019.

West JB. Pulmonary physiology and pathophysiology: an integrated, case-based approach. 2nd ed. Baltimore: Lippincott Williams & Wilkins; 2001.

West JB. Respiratory physiology: the essentials. 9th ed. Philadelphia: Lippincott Williams & Wilkins; 2011.

West JB, Luks AM. West's respiratory physiology: the essentials. 10th ed. Philadelphia: Wolters Kluwer; 2016.

West JB, Dollery CT, Naimark A. Distribution of blood flow in isolated lung; relation to vascular and alveolar pressures. J Appl Physiol. 1964;19:713–24.

Wineski LE. Snell's clinical anatomy by regions. 10th ed. Philadelphia: Wolters Kluwer; 2019.

Yuan J, Garcia J, Hales C, Rich S, Archer SL, West JB. Textbook of pulmonary vascular disease. New York: Springer; 2011.

Recommended Additional Reading

Des Jardins T. Cardiopulmonary anatomy and physiology: essentials of respiratory care. 6th ed. Clifton Park: Delmar; 2013.

Anatomy and Function of the Spine

6

This chapter presents a comprehensive exploration of the spine's functional anatomy, biomechanics, and kinematics, providing an evidence-based foundation for the management of spinal cord injuries. While many texts focus on pathological conditions or surgical approaches, this chapter examines the underlying anatomical and biomechanical principles that determine spinal function and dysfunction. The complex anatomical organization of the vertebral column reflects its multiple functions: protecting the spinal cord and nerve roots, supporting body weight, and allowing movement across multiple segments. Understanding these structural relationships begins with embryological development, where early cellular migrations and tissue differentiation establish the basic vertebral architecture. This developmental perspective illuminates both normal anatomical variations and potential pathological conditions.

Surface anatomy provides the initial framework for clinical assessment, with specific anatomical landmarks guiding physical examination and procedural interventions. From this foundation, the text progresses through a detailed analysis of each spinal region—cervical, thoracic, and lumbar—examining their unique structural characteristics and biomechanical properties. This regional analysis includes comprehensive coverage of osseous elements, ligamentous structures, and neurovascular relationships.

Biomechanical principles receive particular attention, as they bridge the gap between static anatomy and dynamic function. The text examines load transmission patterns, movement coupling relationships, and the role of muscular systems in maintaining spinal stability. This biomechanical understanding proves valuable for analyzing both normal movement patterns and pathological conditions. While this chapter does not address spinal pain management directly, it provides the anatomical and biomechanical knowledge base required for informed clinical decision-making in spinal cord medicine. The material emphasizes practical application, allowing clinicians to translate theoretical concepts into effective rehabilitation strategies and improved patient outcomes. For medical professionals specializing in spinal cord injuries, this chapter offers an authoritative reference that integrates basic science with clinical relevance. The systematic presentation progresses from fundamental concepts to complex functional relationships, supporting both detailed study and quick reference during clinical practice.

H.-Y. Ko, *Practical Functional Anatomy for Spinal Cord Injury Rehabilitation*,
https://doi.org/10.1007/978-981-95-6457-6_6

6.1 Embryonic Development and Formation of the Vertebral Column and Spinal Cord

The formation of the vertebral column and spinal cord represents a complex developmental sequence that begins during the third week of human embryonic development. The process starts with neurulation, where the neural plate elevates and folds to form the neural tube and neural crest cells (Fig. 6.1). The neural tube closure initiates at the level of the fourth somite and proceeds bidirectionally in cranial and caudal directions (O'Rahilly and Müller 2010).

The paraxial mesoderm, positioned bilaterally along the developing neural tube, undergoes progressive segmentation to form epithelial somites (Fig. 6.2). These paired structures form in a rostral-to-caudal sequence. Each somite subsequently differentiates into two primary components: the dermomyotome dorsally and the sclerotome ventrally. The dermomyotome's lateral portions develop into the dermis and subcutaneous tissues of the back, while its primary mass differentiates into the deep back muscles, including the erector spinae muscle group.

The ventral sclerotome cells undergo an epithelial-to-mesenchymal transition and migrate ventromedially toward the notochord and neural tube (Fig. 6.3). These cells then form the vertebral bodies, vertebral arches, intervertebral discs, and associated ligamentous structures (Christ and Scaal 2008; Scaal 2016). Each sclerotome subdivides into cranial and caudal portions, with the cranial portion of one sclerotome joining the

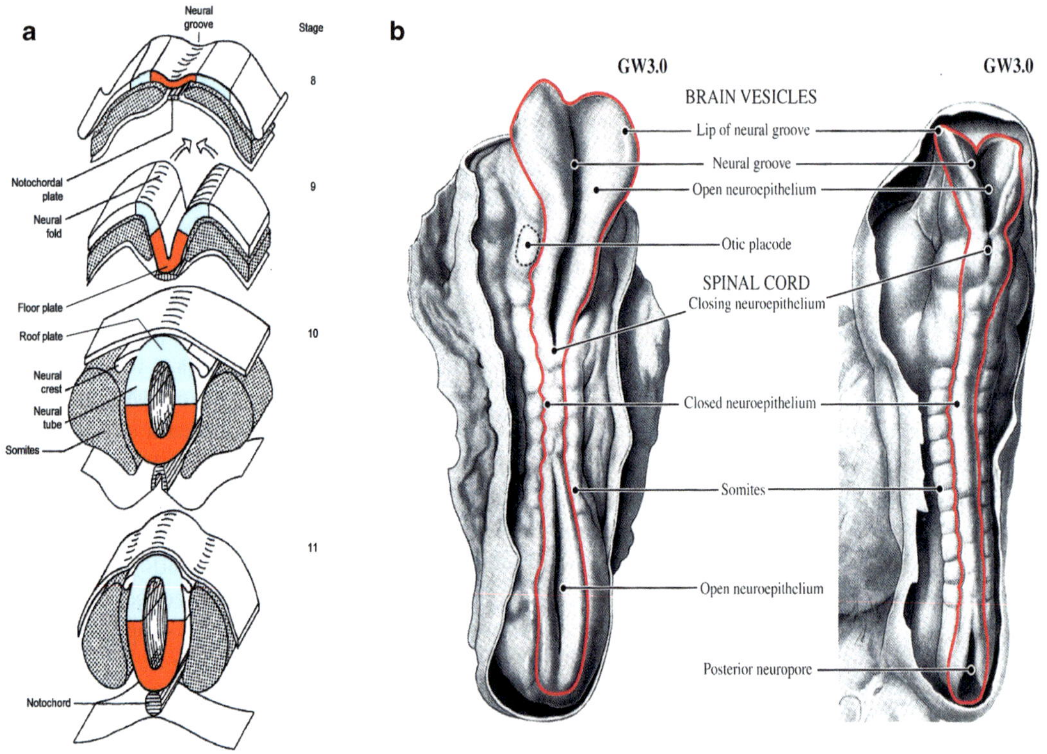

Fig. 6.1 GW 3 primary neurulation. The open neuroepithelium transitions into a closed neuroepithelium. By day 18, the neural plate can be distinguished from the surrounding ectoderm. Within 24 h of the first appearance of the neural plate, folds appear along the edges of the neural plate. Neural tube closure requires 4–6 days. Before complete closure, the tube's open ends form the anterior (cranial) and posterior (caudal) neuropore. (**a**) Cross-sectional diagram. (**b**) Dorsal view. (Adapted with permission from Altman and Bayer (2001))

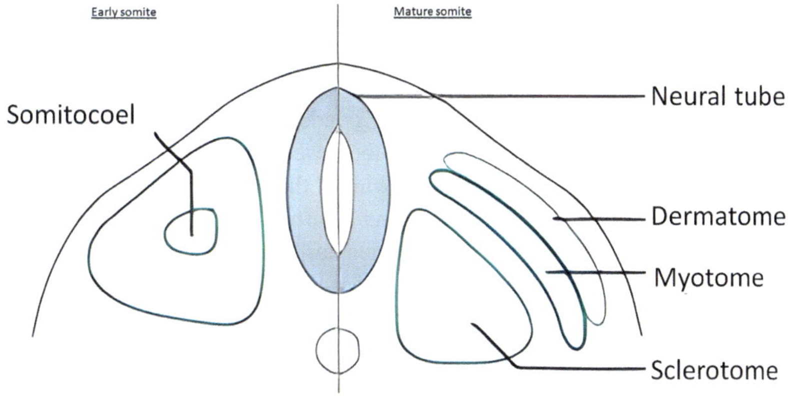

Fig. 6.2 Cell condensation around the somitocoel differentiates into dermamyotomes and sclerotomes. The dermamyotome further subdivides into the dermatome and myotome. (Adapted with permission from Webster and de Wreede (2016))

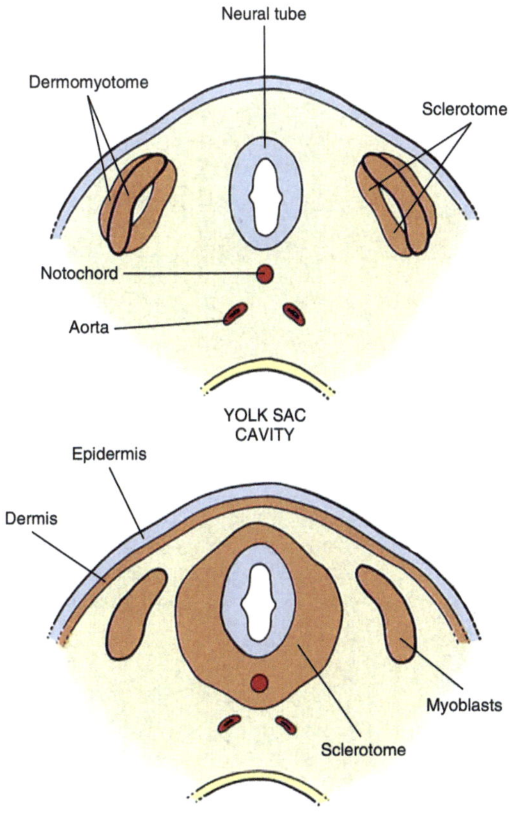

Fig. 6.3 Cross-sectional view illustrating two stages in somite differentiation: prior to (top) and following (bottom) the migration of dermis and sclerotome cells. (Adapted with permission from Cramer and Darby (2014))

caudal portion of the adjacent sclerotome to form a single vertebra—a process known as resegmentation.

Within the developing spinal cord, the neural tube differentiates into three distinct neuroepithelial layers. The innermost ependymal layer, surrounding the neural canal, contains actively dividing neuroepithelial cells. The intermediate mantle layer develops into the gray matter, containing neuronal cell bodies organized into the dorsal (sensory) and ventral (motor) horns (Butler and Bronner 2015). The peripheral marginal layer forms the white matter, comprising myelinated and unmyelinated axons arranged in specific ascending and descending tracts.

The neural crest cells, originating from the dorsal neural tube, undergo extensive migration and differentiation. These multipotent cells give rise to various structures, including the following (Le Douarin and Dupin 1993; Le Douarin and Kalcheim 1999):

- Dorsal root ganglia containing primary sensory neurons
- Sympathetic chain ganglia
- Schwann cells that myelinate peripheral nerves
- Melanocytes of the skin
- Components of the meninges

6.1.1 Vertebral Chondrification and Primary Ossification

The formation of vertebrae involves the condensation of sclerotomal cells, leading to the development of chondrification centers within the mesenchymal vertebrae (Cramer and Darby 2014). These centers, consisting of an anterior pair and at least one center in each posterior half, give rise to a cartilaginous model of the vertebra (Fig. 6.4). During this process, sclerotomal cells undergo controlled proliferation and differentiation to establish the cartilage template. Subsequently, each vertebra undergoes ossification through the emergence of three primary centers. The anterior center, situated in the future vertebral body, is termed the centrum. The remaining two centers are positioned laterally, surrounding the developing neural tube, and constitute the neural or posterior arch. Fusion of these lateral centers posteriorly results in the formation of the spinous process, with failure of this union leading to spina bifida (Cramer and Darby 2014).

On the anterior side, the left and right parts of the neural arch typically fuse with the centrum at a region called the neurocentral synchondrosis, situated in what will become the posterior part of the vertebral body (Cramer and Darby 2014).

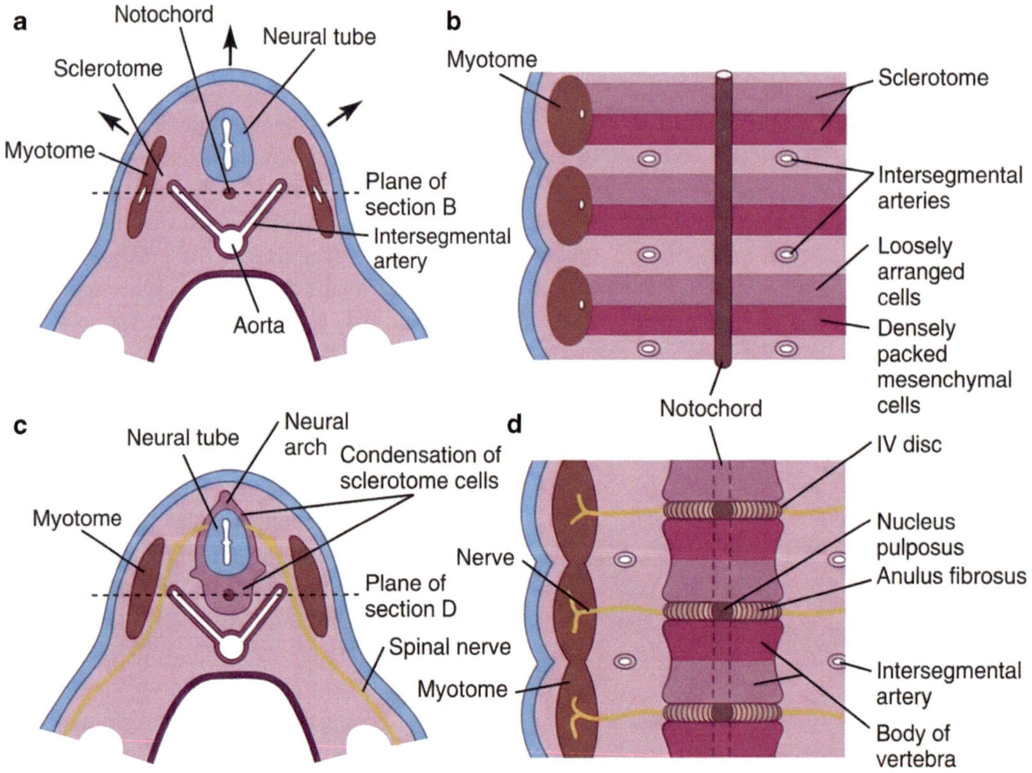

Fig. 6.4 Stages of vertebral development through somite differentiation. (**a**) Transverse section of a 4-week embryo, showing dorsal growth of the neural tube (arrows) and dorsolateral movement of the somite remnant, leaving a trail of migrating sclerotomal cells. (**b**) Diagrammatic frontal section of the same embryo, illustrating the condensation of sclerotomal cells around the notochord, consisting of a loosely packed cranial region and a densely packed caudal region. (**c**) Transverse section of a 5-week embryo, depicting the condensation of sclerotomal cells around the notochord and neural tube, forming a mesenchymal vertebra. (**d**) Diagrammatic frontal section showing the vertebral body forming from the cranial and caudal halves of two successive sclerotomal masses. Intersegmental arteries now cross the vertebral bodies, while spinal nerves lie between the vertebrae. The notochord is degenerating except in the region of the intervertebral disc, where it forms the nucleus pulposus. (Adapted with permission from Moore et al. (2013))

This fusion unites the neural arch's primary ossification centers with the centrum, resulting in the formation of the vertebral body from both the centrum and a small portion of the neural arch. This fusion results in a vertebral arch that is slightly smaller than its developmental precursor, the neural arch, while the vertebral body becomes somewhat larger than the original centrum (Cramer and Darby 2014).

The timing of the neurocentral synchondrosis fusion, where the neural arch meets the centrum, is still under research. Some studies, such as Maat et al. (1996), suggest that this fusion is completed by the age of 6, while others, such as Vital et al. (1989), propose that the neurocentral cartilage may persist until the age of 16. This cartilage plays a crucial role in the growth of the vertebral posterior arch. Premature fusion at this site is associated with the development of scoliosis (Vital et al. 1989).

Typically, the vertebral body forms from two chondrification centers, one on each side. The absence or failure of one center to develop leads to the formation of only half of the vertebral body, a condition known as a hemivertebra or cuneiform vertebra. This condition often causes a lateral spinal curvature, though it may be balanced by a similar anomaly on the opposite side at a different level (Cramer and Darby 2014). Vertebral bodies in development may present a wedge shape, being narrower at the anterior aspect than at the posterior aspect, which might mimic a compression fracture (Fesmire and Luten 1989). While such wedging across several vertebrae usually indicates a normal variant, the presence of a singular wedge-shaped vertebra with adjacent rectangular vertebrae should prompt evaluation for a possible compression fracture (Cramer and Darby 2014; Kyriakos et al. 2011).

6.1.2 Secondary Ossification Centers and Vertebral Column Maturation

The vertebral column undergoes significant development between the ages of 10 and 13, marked by the emergence of five secondary ossification centers (Cramer and Darby 2014; Standring 2020). These centers, described in anatomical literature as annular apophyses, ring apophyses, or annular epiphyses, are positioned on the cartilaginous end plates of a typical vertebral body. Secondary ossification centers also develop at the tips of the transverse processes and the single spinous process. The centers located on the transverse and spinous processes function as growth zones that enable rapid expansion of these structures during adolescent development (Bogduk 2005).

The annular apophyses, which comprise the two ossification centers associated with the peripheral rim of the superior and inferior surfaces of the vertebral bodies, do not contribute to the longitudinal growth of the vertebral bodies (Bogduk 2005). These centers integrate with the outer layers of the annulus fibrosus, which explains the bony attachment of these layers, while the more central layers maintain their attachment to the cartilage of the vertebral end plates. The process of fusion between all secondary ossification centers and their corresponding vertebral components occurs over an extended period between the ages of 14 and 25 (Bogduk 2005; Standring 2020). Following complete fusion, growth potential ceases. Medical professionals should note that during radiographic examination, these unfused centers may be misinterpreted as fracture lines (Cramer and Darby 2014).

6.1.3 Developmental Progression of Spinal Curvatures: Kyphoses and Lordoses

The human vertebral column exhibits four distinctive anteroposterior curves: two kyphotic and two lordotic curves. The primary curves, comprising the thoracic and pelvic curvatures, are established during early fetal development. The thoracic kyphosis extends from T2 to T12, characterized by thoracic vertebrae with greater posterior than anterior dimensions. The pelvic curve spans from the lumbosacral junction to the coccygeal apex, displaying an anteroinferior concavity due to the increased posterior dimensions of sacral segments (Standring 2020).

The secondary curves, cervical and lumbar lordoses, develop in response to postnatal functional demands. The cervical lordosis initiates during late intrauterine life but becomes morphologically distinct when infants develop head control in the prone position, typically at 3–4 months postnatally (Bagnall et al. 1977). This curvature becomes more defined as infants acquire sitting abilities and head stabilization for visual exploration, occurring around 9 months of age (Cil et al. 2005). The adult cervical curve's maintenance depends on the increased anterior thickness of the intervertebral discs.

The lumbar lordosis develops between 9 and 18 months, coinciding with the onset of bipedal locomotion. This curve forms as the erector spinae muscles align the lumbar spine for upright posture during walking, creating a posterior concavity. The lumbar lordosis extends from T12 to the lumbosacral junction, with more pronounced curvature in females. The curve demonstrates regional variation, with increased depth between L3 and the lumbosacral angle compared to the T12–L2 segment (Bernhardt and Bridwell 1989).

6.2 Surface Anatomical Features of the Back

Understanding surface anatomy represents a fundamental component of physical examination. Direct observation and palpation findings contribute significantly to patient assessment, as they allow clinicians to identify anatomical landmarks and their relationships to deeper structures. The correlation between surface landmarks and underlying anatomical structures provides approximate locations, and research indicates that practitioners can enhance their accuracy in identifying spinal structures through palpation with proper training and clinical experience (Byfield and Kinsinger 2002; Downey et al. 1999; Phillips et al. 2009). Multiple factors influence these anatomical relationships, including age, body habitus, gender, spinal alignment, and soft tissue composition. Additionally, respiratory mechanics modify structural positions through the movement of the thoracic cage and associated muscles, particularly affecting structures within the thoracic cavity. With the increasing integration of advanced imaging modalities in clinical settings, knowledge of normal anatomical relationships between internal organs, musculoskeletal structures, and the vertebral column has acquired greater clinical significance.

The median furrow represents a vertical depression extending along the back's midline from the external occipital protuberance superiorly to the gluteal cleft inferiorly, with its terminal point at the S3 spinous tubercle. This anatomical groove exhibits variable depths across different spinal regions due to the underlying muscle bulk and subcutaneous tissue distribution, showing minimal depth in the lower cervical area, where the nuchal ligament provides midline support and maximal depth in the lumbar region, where the well-developed paraspinal muscles create prominent lateral elevations. At its inferior aspect, the furrow expands to create an isosceles triangle, bounded superiorly by an imaginary line between the posterior superior iliac spines (PSISs) and inferiorly by the gluteal cleft. The PSISs manifest as paired surface depressions positioned 3–4 cm lateral to the midline, aligning with the S2 spinous tubercle, and mark the attachment points of the thoracolumbar fascia and various muscles. A horizontal cutaneous fold, the gluteal fold, extends from the midline laterally, marking the approximate inferior margin of the gluteus maximus muscle and corresponding to the level of the ischial tuberosity (Moore et al. 2013) (Fig. 6.5).

The trapezius presents as a large, triangular muscle with three distinct functional portions: superior (descending), middle (transverse), and inferior (ascending) fibers. It spans from the midline attachments at the external occipital protuberance, nuchal ligament, and cervicothoracic spinous processes (C1–T12) to its lateral insertion on the lateral third of the clavicle, acromion process, and scapular spine. Its superior muscle fibers contribute to the lateral neck–shoulder junction contour and function in scapular elevation. The latissimus dorsi muscle courses from its

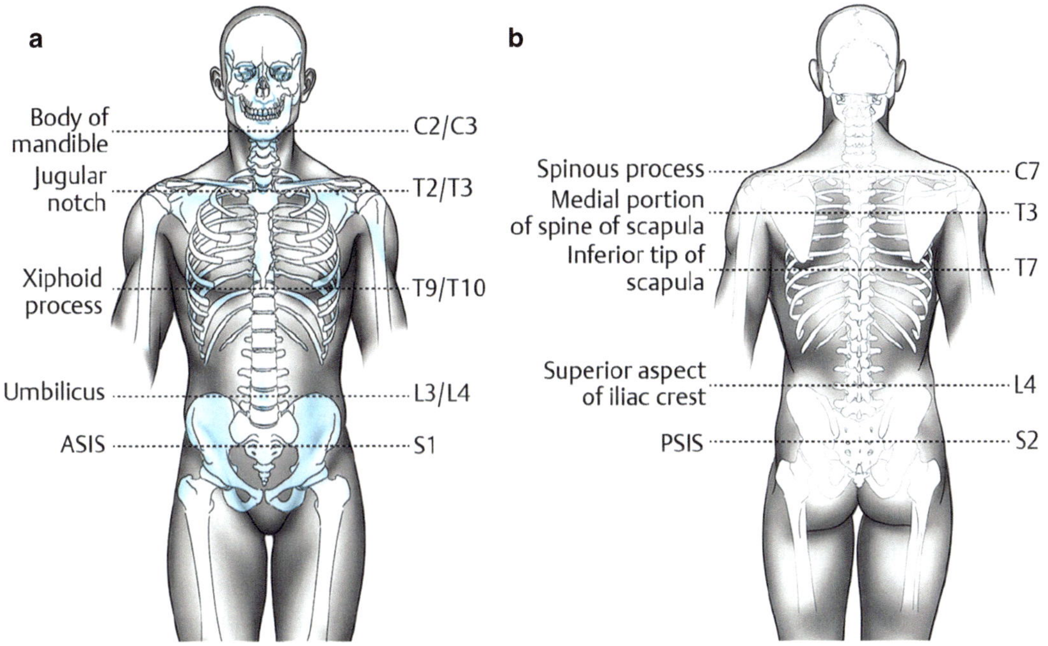

Fig. 6.5 Palpable surface landmarks for determining vertebral level. (**a**) Ventral view. (**b**) Dorsal view. (Adapted from Singh (2019))

broad origin at the thoracolumbar fascia, iliac crest, and lower thoracic spinous processes to the posterior axillary region, establishing the lateral border of the lower thoracic back. This muscle becomes particularly visible during resisted upper limb adduction and internal rotation. Between these muscles, the inferior scapular angle typically corresponds to the T8 spinous process level, serving as a reliable landmark for thoracic vertebral counting. Within the lumbar area, the paired erector spinae muscle groups (iliocostalis, longissimus, and spinalis) form prominent vertical elevations approximately 10 cm (one hand breadth) from the midline, increasing the median furrow depth (Moore et al. 2013; Standring 2020).

In the lower cervical spine, the C7 vertebra prominens exhibits a notably long and non-bifid spinous process, making it a primary landmark for vertebral level identification. During cervical flexion, both the C6 and T1 spinous processes may become apparent due to the stretching of overlying soft tissues. Studies demonstrate that 75% of individuals display C7 as the most prominent spinous process (Stonelake et al. 1988). Alternative patterns exist where C6 or T1 presents as the most prominent process in roughly 10% and 15% of cases, respectively. This anatomical variation holds clinical significance during cervical spine examination and procedural interventions. The C7 spinous process also marks the approximate level of several important anatomical structures, including the inferior border of the cricoid cartilage anteriorly and the junction between the cervical and thoracic portions of the spinal cord internally.

6.2.1 Anatomical Landmarks for Back Palpation

6.2.1.1 Palpatory Landmarks and Anatomical Correlations of the Cervical Spine

The external occipital protuberance (inion) occupies a central position within the occipital squama, marking a significant posterior cranial landmark. The superior nuchal line radiates laterally from this point, providing attachment for the upper fibers of the trapezius muscle and the occipital belly of the occipitofrontalis muscle. The transverse process of the atlas (C1) can be identified through palpation in a location slightly anterior and inferior to the mastoid process. The temporal bone's styloid process lies anterior to both the C1 transverse process and the great auricular nerve, with the latter ascending superficially within the fascial layers overlying the C1 transverse process (Stonelake et al. 1988).

Moving inferiorly from the inion, the axis (C2) spinous process presents as the first distinctly palpable posterior midline osseous structure. Continuing caudally, the next significant midline landmark encountered is the vertebra prominens, generally corresponding to the C7 spinous process. Research has documented that the vertebra prominens represents the most pronounced spinous process in 75% of the population, while the C6 or T1 spinous processes demonstrate greater prominence in 10% and 15% of cases, respectively (Stonelake et al. 1988). The C3 spinous process, characterized as the least prominent among cervical spinous processes, can be located through palpation at the level corresponding to the greater cornua of the hyoid bone (Moore et al. 2013; Standring 2020). The C6 spinous process holds particular clinical significance as it represents the most caudal cervical spinous process, maintaining unrestricted mobility during neck flexion and extension movements (White and Panjabi 1990) (Fig. 6.6).

The cervical zygapophysial (facet) joints can be palpated approximately 1.5 cm lateral to the posterior cervical midline. These bilateral joint sequences form the articular pillars. With the exception of the atlas (C1), individual cervical transverse process tips remain generally impalpable. However, the anterior aspects of these transverse processes can be located within the depression between the laryngeal structures and the sternocleidomastoid muscle (SCM). Optimal palpation of these structures may require gentle lateral displacement of the SCM muscle. The C6 vertebral level features notably prominent anterior tubercles on its transverse processes, recognized as the carotid tubercles due to their anatomical relationship with the common carotid artery.

From an anterior approach, the superior margin of the thyroid cartilage, which creates the laryngeal prominence (Adam's apple) in the anterior midline, corresponds to the horizontal plane of the C4 intervertebral disc. The C6 vertebral body demonstrates consistent alignment with the horizontal planes of both the cricoid cartilage and the first tracheal ring, providing reliable surface landmarks for vertebral level identification.

6.2.1.2 Palpatory Landmarks and Anatomical Correlations of the Thoracic Spine

When examining the posterior midline in a superior-to-inferior direction from the external occipital protuberance (EOP), the T1 spinous process represents the third prominent osseous landmark, following the spinous processes of C2 and C7. The spinous process of T3 demonstrates consistent alignment with the horizontal plane of the scapular spine root, providing a reliable reference point for vertebral level identification. At the apex of the thoracic kyphosis, the T4 spinous process typically manifests as the most prominent thoracic spinous process, making it a significant palpatory landmark in the upper thoracic region (Cooperstein et al. 2009).

The relationship between scapular position and vertebral levels varies with patient positioning. In standing or sitting positions with the upper extremities positioned adjacent to the trunk, the inferior angle of the scapula generally corresponds to the T8 spinous process level. This relationship changes when patients assume a prone position with upper extremities in flexion. In this position, scapular rotation occurs, resulting in more frequent alignment of the inferior scapular angle with the T9 spinous process (Cooperstein et al. 2015; Keogh and Ebbs 1984). Between the inferior scapular angle and the superior margin of the iliac crests, the T12 spinous process can be located at approximately the midpoint of this interval.

The thoracic vertebral architecture exhibits unique characteristics that affect surface landmark identification. The spinous processes demonstrate variable degrees of inferior angulation,

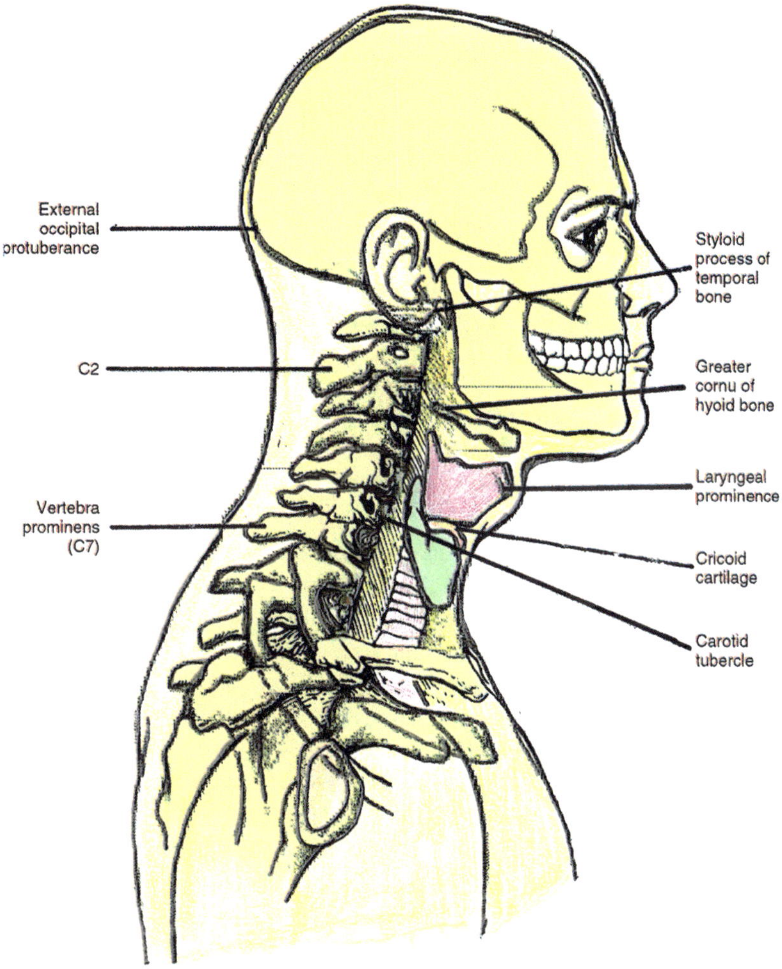

Fig. 6.6 Palpable landmarks of the lateral neck. (Adapted with permission from Cramer and Darby (2014))

creating a discrepancy between vertebral body position and the corresponding spinous process tip location (Cooperstein and Haneline 2007; Haneline et al. 2008). For T1–4 and T10–12 vertebrae, the transverse process tips are positioned one spinous interspace superior to the tip of the corresponding vertebral segment's spinous process. However, in T5–9 vertebrae, where spinous processes show more pronounced inferior angulation, the transverse process tips are located two spinous interspaces superior to their respective spinous process tips (Cooperstein et al. 2009). This anatomical arrangement holds particular significance for accurate vertebral level identification and clinical procedures targeting specific thoracic segments.

6.2.1.3 Anatomical Landmarks and Palpation of the Lumbosacral Spine

The lumbar vertebrae present distinctive anatomical features that influence their identification through surface palpation. In contrast to thoracic vertebrae, the posterior surfaces of lumbar spinous processes demonstrate a more horizontal orientation with flattened surfaces. Among the lumbar vertebrae, L4 and L5 possess notably shorter spinous processes, which affects their accessibility during palpation examinations. The L5 spinous process, in particular, presents significant challenges for accurate palpation due to its reduced prominence and deep position within the lordotic curve (Pysyk et al. 2010).

Clinical examination reveals that the L4 spinous process marks an important functional transition point, as it represents the most caudal spinous process that exhibits appreciable movement during sagittal plane trunk motions. This characteristic aids in distinguishing the L4 level from the more inferior L5 segment. Research utilizing ultrasound technology for vertebral level identification has established that the supracreistal plane, defined by the superior aspects of the iliac crests, intersects the vertebral column at the L3–4 intervertebral level in approximately 75% of healthy subjects. In the remaining population, this anatomical relationship demonstrates a bimodal distribution between the L2–3 and L4–5 levels (Pysyk et al. 2010).

The transverse processes of lumbar vertebrae maintain a relatively consistent lateral distance from the midline, positioned approximately 5 cm on either side. Despite their substantial size, these bony projections generally resist direct palpation due to the overlying soft tissue structures (Pysyk et al. 2010). Moving inferiorly to the sacral region, the S2 spinous tubercle, representing vestigial remnants of the second sacral spinous process, occupies a noteworthy position at the apex of the sacral kyphosis. This tubercle typically presents as the most prominent among the sacral spinous tubercles and establishes a reliable horizontal plane with the posterior superior iliac spines (PSISs). The PSISs can be readily palpated at points 3–4 cm lateral to the midline, creating important reference points for clinical examination.

The posterior aspect of the sacrum features four pairs of posterior sacral foramina, arranged in a systematic pattern. These foraminal pairs maintain consistent spatial relationships, positioned 2.5 cm lateral to the midline and separated by 2.5 cm intervals. However, these anatomical structures typically resist direct palpation due to their depth and overlying soft tissues. The sacral hiatus, representing the most inferior palpable midline depression on the posterior sacral surface, marks the caudal terminus of the vertebral canal and holds particular significance for caudal epidural procedures.

6.2.1.4 Anatomical Landmarks and Vertebral Correlations in the Anterior Neck and Trunk

The relationship between superficial anterior landmarks and their corresponding deeper structures provides fundamental information for comprehensive physical examination and clinical assessment (Table 6.1). These anatomical correlations assist medical practitioners in localizing specific vertebral levels and associated structures without relying on imaging studies.

In the anterior neck region, the laryngeal prominence serves as a significant midline landmark, demonstrating alignment with the C4 intervertebral disc. This anatomical feature displays sexual dimorphism, presenting with greater prominence in adult males due to the influence of androgens on laryngeal development (Moore et al. 2013). Proceeding inferiorly, the jugular notch, also termed the suprasternal notch, occupies the superior aspect of the sternal manubrium and corresponds to the T2 disc level or the T1–2 interspinous interval (Drake et al. 2020). The sternal angle, marking the junction between the manubrium and sternal body, maintains a consistent relationship with the T4 intervertebral disc level (Moore et al. 2013).

Moving down the anterior trunk, the xiphisternal junction, located near the sternum's inferior portion, demonstrates multiple anatomical correlations. This junction corresponds to the T9 vertebral body level, the inferior margin of the pectoralis major muscle, and the fifth costal cartilage (Drake et al. 2020). The costal margin's lowest component consists of the tenth costal cartilage, and the horizontal plane at this level, designated as the subcostal plane, intersects the L3 vertebral body (Standring 2020).

The transpyloric plane represents a clinically significant horizontal reference level situated equidistant between the superior margin of the pubic symphysis and the suprasternal notch. This plane intersects the L1 intervertebral disc and typically lies approximately one hand breadth (10 cm) inferior to the xiphisternal junction. At a more inferior level, the umbilicus generally corresponds to the L3 vertebral body position (Drake et al. 2020).

Table 6.1 Vertebral levels and clinically important structures

Vertebral level	Structure
C1	Transition of medulla oblongata into spinal cord
	Hard palate
	Anterior portion of soft palate
C2	Inferior border of free edge of soft palate
	Nasopharynx and oropharynx join
C2 disc	Superior cervical ganglion
C3	Epiglottis
	Oropharynx becomes laryngopharynx
C3 disc	Common carotid arteries split into internal and external carotid arteries
	Carotid sinus
C3 spinous	Greater cornua of hyoid bone
C4 disc	Laryngeal prominence
C5	Vocal folds
	Erb's point
	Superior margin of lobes of thyroid gland
C6	Middle cervical ganglion
	Cricoid cartilage
	First tracheal ring
	Transition of larynx to trachea
	Transition of laryngopharynx to esophagus
C7	Inferior cervical ganglion
T1	Stellate ganglion
	Inferior margin of thyroid gland
	Subclavian and internal jugular veins unite to form brachiocephalic veins
	Apices of lungs
T2	Brachiocephalic veins unite to form superior vena cava
T2 disc	Suprasternal notch
T4	Aortic arch
T4 disc	Sternal angle (of Louis)
T5	Pulmonary trunk divides into right and left pulmonary arteries
	Pulmonary artery and primary bronchus enter right lung
	Trachea divides into primary bronchi
	Posterosuperior ends of oblique fissures of lungs
T6	Base of heart
	Pulmonary artery and primary bronchus enter left lung
	Pulmonary veins exit right lung
	Superior vena cava enters right atrium
T7	Pulmonary veins exit left lung
	Inferior vena cava enters right atrium
	Horizontal fissure of right lung
T8	Caval hiatus (diaphragm)
	Superior extent of right hemidiaphragm
	Superior border of liver
T9	Xiphisternal junction
	Fifth costal cartilage
	Inferior border of pectoralis major muscle
	Inferomedial angles of posterior aspects of lungs
	Superior extent of left hemidiaphragm
	Superior pole of spleen
	Left extent of inferior border of liver

(continued)

Table 6.1 (continued)

Vertebral level	Structure
T10	Apex of heart
	Anteroinferior ends of oblique fissures of lungs
	Esophageal hiatus (diaphragm)
T11	Lowest extent of lungs (inferolateral angles of posterior aspects of lungs)
	Inferior extent of esophagus
	Cardiac orifice of stomach
	Left suprarenal gland
T12	Aortic hiatus (diaphragm)
	Costodiaphragmatic recesses
	Inferior pole of spleen
	Tail of pancreas
	Orifice of gallbladder
	Superior poles of kidneys (right slightly lower than left)
	Right suprarenal gland
L1	Pyloric orifice of stomach
	Superior horizontal (first) part of duodenum
	Left colic (splenic) flexure
L1 disc	Transpyloric plane
	Conus medullaris
	Hila of kidneys (right slightly below and left slightly above)
L2	Duodenal-jejunal junction
	Right colic (hepatic) flexure
	Head of pancreas
L3	Subcostal plane (lowest portion of costal margin made up from the tenth costal cartilage)
	Umbilicus (inconsistent)
	Inferior horizontal (third) part of duodenum
	Right extent of lower border of liver
	Inferior poles of kidneys (right slightly lower than left)
L4	Beginning of sigmoid colon
	Aorta divides into common iliac arteries
L5	Common iliac veins unite to form inferior vena cava
	Ileocecal junction
	Vermiform appendix arises from cecum
L5 disc	Anterior superior iliac spine
	Superolateral end of inguinal ligament
S3	Beginning of rectum
Lower sacrum	Superior extent of uterus
Coccyx	Pubic crest
	Ganglion impar
	Superior margin of pubic symphysis
	Inferomedial end of inguinal ligament
	Bladder (empty)

Adapted with permission from Cramer and Darby (2014)

The inguinal ligament establishes the proximal boundary of the thigh region, extending from its origin at the anterior superior iliac spine, which aligns with the L5 disc level, to its insertion on the pubic tubercle (Standring 2020). This ligamentous structure provides important surface landmarks for both musculoskeletal and vascular examinations of the lower extremity.

6.3 Structural Composition and Functional Significance of a Typical Vertebra

The fundamental vertebral structure comprises two primary components: the vertebral body and the vertebral (posterior/dorsal) arch. Both components display a distinctive anatomical organization consisting of an external cortical bone layer enclosing an internal trabecular or cancellous bone core (Skedros et al. 1994a, b). The internal trabecular architecture demonstrates specific orientation patterns that align with primary mechanical stress vectors. Within lumbar vertebral bodies, the trabecular organization has evolved to withstand axial compression forces during weight-bearing activities and locomotion. This architectural arrangement produces maximum resistance to compressive loads while maintaining structural integrity at the vertebral body-pedicle junction, where significant force transmission occurs during rotational and anteroposterior movements associated with normal ambulation (Table 6.2).

The vertebral body's cortical bone demonstrates regional thickness variations, with relatively thin layers along the discal surfaces and increased thickness throughout the vertebral arch and its processes. A specialized periosteal membrane covers this external cortical layer,

Table 6.2 Regional general characteristics of typical vertebrae

Structure	Cervical	Thoracic	Lumbar
Body	Rectangular	Heart-shaped	Kidney-shaped
	Uncinate processes	Taller posterior than anterior	Taller anterior than posterior
		Costal demifacets	
Pedicles	Small	Large, stout	Large, stout
	Superior and inferior vertebral notches	No superior vertebral notch	Superior (small) and inferior vertebral notches
Transverse processes	Anterior and posterior tubercles	Large	Large
	Foramen of transverse process	Transverse costal facet	
Articular processes (orientation [facing] of superior articular facet)	Posteriorly, superiorly, and medially 45° to vertical plane	Posteriorly, superiorly, and laterally 30° to vertical plane	Posteriorly and medially
			Biplanar
			Within vertical plane
Laminae	Short from superior-inferior	Tall from superior-inferior	Intermediate from superior-inferior
Spinous process	Bifid	Upper four: posteriorly directed	Spatulated in shape
		Middle four: long and inferiorly directed	
		Lower four: lumbarlike	
Intervertebral foramina	Oval-shaped: 35–50% filled with spinal nerve	Inverted pear-shaped: one twelfth filled with spinal nerve	Inverted pear-shaped: one third filled with spinal nerve
Vertebral canal	Trefoil-shaped: largest of vertebral column	Round in shape: smallest of vertebral column	Trefoil-shaped: intermediate in size between cervical and thoracic
Typical vertebrae	C3–6	T2–9	All

Adapted with permission from Cramer and Darby (2014)

containing nerve endings that transmit both nociceptive and proprioceptive signals (Edgar and Ghadially 1976; Higuchi and Sato 2002). The cortical bone structure incorporates numerous small vascular channels that permit the passage of nutritive vessels. Within the vertebral core, red marrow occupies the internal space, while the vertebral bodies contain specialized channels accommodating the basivertebral venous system.

Vertebral bone density demonstrates significant developmental patterns and individual variation. Research indicates that bone density increases substantially during pubertal development, reaching maximum levels in the third decade of life, concurrent with the fusion of secondary ossification center growth plates (Gilsanz 1998; Gilsanz et al. 1988). In pathological conditions such as osteoporosis, alterations in trabecular bone architecture occur, with disproportionate bone mass reduction in cancellous bone compared to the cortical shell (Feltrin et al. 2001; Mosekilde and Mosekilde 1986). These structural modifications reduce bone elasticity and increase skeletal fragility, leading to elevated vertebral fracture risk (Mosekilde and Mosekilde 1990; Feltrin et al. 2001).

6.3.1 Functional Anatomy of the Typical Vertebra

The vertebral structure demonstrates anatomical organization that supports both protective and biomechanical functions. Each vertebra incorporates multiple specialized components that work in concert to maintain spinal integrity and enable movement. The vertebral bodies provide primary axial support for the spine, while the posterior elements, comprising the pedicles and laminae, form a protective housing for the spinal cord (Fig. 6.7). The articular processes, through their specific facet orientations, modulate spinal motion patterns. The transverse and spinous

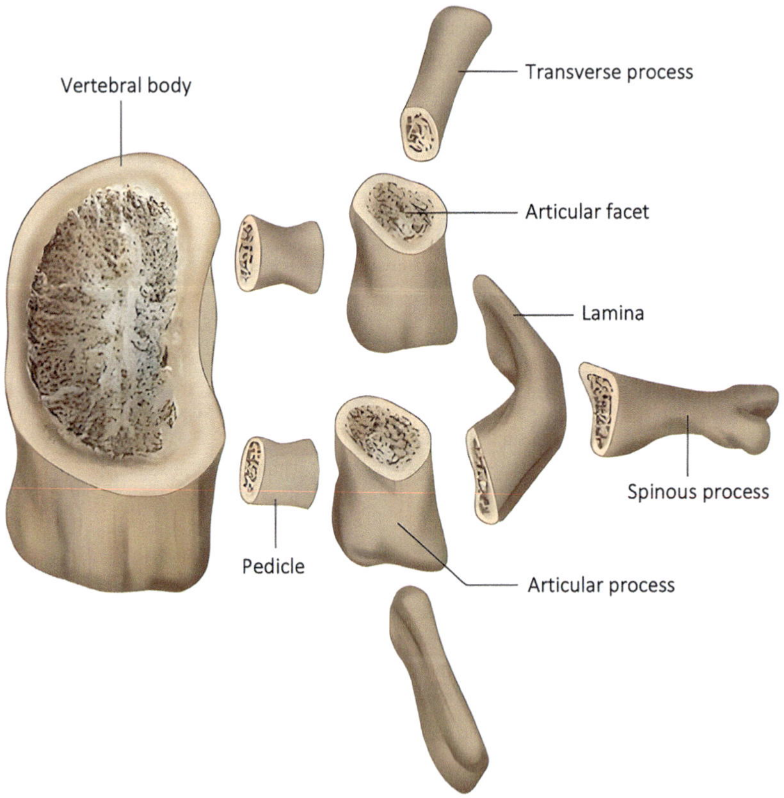

Fig. 6.7 Key functional components of a typical vertebra. The vertebral body provides support, while the pedicles and laminae protect the spinal cord in the cervical and thoracic regions, as well as the cauda equina below the L1 vertebra level. The spinous process, transverse processes, articular processes, and especially the articular facets function to control movement. (Adapted with permission from Ko (2022))

Fig. 6.8 A single spinal unit, also referred to as a motion unit or functional unit, consists of two adjacent vertebrae and the intervertebral disc between them. This unit forms the fundamental biomechanical segment of the spine, allowing controlled motion and providing stability while protecting the spinal cord and nerve roots. (Adapted with permission from Ko (2022))

processes function as attachment points for spinal musculature, providing mechanical advantage for vertebral movement. Biomechanically, the structure formed by two adjacent vertebrae and their intervertebral disc is referred to as a single spinal unit, motor unit, or functional unit of the spine (Fig. 6.8).

The posterior vertebral arch complex demonstrates significant weight-bearing capabilities beyond its protective role (Pal et al. 2001). Within the cervical region, the articular processes organize into bilateral support columns that participate in axial load distribution. The laminae of specific vertebral segments—C2, C7, T1, and T2—assume substantial weight-bearing responsibilities. This anatomical arrangement explains why surgical removal of laminae (laminectomy) at these levels significantly reduces cervical stability. In contrast, laminectomies performed between C3 and C6 typically present minimal risk to spinal stability, reflecting the reduced weight-bearing role of laminae in these segments (Pal et al. 2001).

The cervical spine demonstrates unique load transfer mechanisms through its pedicles, which facilitate bidirectional force transmission between the posterior arch and vertebral body (Pal et al. 2001). This pattern contrasts with thoracic vertebral mechanics, where load transfer occurs predominantly in a posterior-to-anterior direction, moving from the posterior arch toward the vertebral bodies. The biomechanical patterns in the upper lumbar region continue to undergo investigation. However, analysis of trabecular orientation patterns within L4 and L5 pedicles indicates that these segments exhibit primarily anterior-to-posterior load transfer, directing forces from the vertebral bodies toward the posterior arch structures (Pal and Routal 1987).

6.3.1.1 Anatomical Features and Biomechanical Adaptation of Vertebral Bodies

Vertebral bodies are interconnected through fibrocartilaginous intervertebral discs. Together, these structures form a flexible column that bears the weight of the trunk and head. These bodies adapt to additional loads exerted by the contraction of axial and proximal limb muscles (Issever et al. 2003). The shape and size of vertebral bodies vary across the different regions of the spine, with a progressive increase in transverse diameter from C2 to L3, likely to accommodate progressively greater loads. Variability in the width of the last two lumbar vertebrae is noted, with a gradual decrease from the first sacral segment to the coccyx's inferior tip (Standring 2020).

The internal architecture of vertebral bodies includes predominantly vertical trabeculae, reinforced by horizontal trabeculae. Prolonged and increased superior–inferior compressive loading stimulates an adaptive response, resulting in an increased number of both vertical and horizontal trabeculae within the vertebral bodies (Issever et al. 2003). In osteoporotic conditions, there is a marked reduction in the mass of horizontal trabeculae, disproportionately weakening the vertebral body relative to the reduction in bone mineral content—a 25% loss in bone mass can lead to a 50% decrease in load-bearing capacity (Mosekilde and Mosekilde 1986). Remodeling of vertebral bodies occurs following intervertebral disc degeneration, enhancing their ability to absorb compressive forces (Moore et al. 1992, 1996).

6.3.1.2 Blood Supply and Venous Drainage of the Vertebral Column

Vertebral bodies typically exhibit a concave shape posteriorly in the transverse plane, contributing to the formation of the vertebral foramina. Numerous small foramina, serving as passages for arteries and veins, are distributed across the anterior and lateral surfaces of the vertebral bodies. On the posterior aspect, posteriorly, small arterial foramina are present, along with one or two large, centrally located foramina that accommodate the exit of basivertebral veins (Standring 2020).

A series of large arteries penetrate the core of the vertebral bodies, encircling their entire circumference (Fig. 6.9). These principal nutrient arteries enter the vertebral bodies and establish an arterial plexus along the central horizontal plane. From this plexus, numerous smaller arterial branches extend vertically to supply the upper and lower edges of the vertebral bodies adjacent to the cartilaginous end plates of the intervertebral discs.

Arterial blood supply to the vertebral column's external aspect is derived from the deep cervical arteries, sourced from the costocervical trunks, and the ascending cervical arteries, originating from the inferior thyroid arteries for the cervical region (Cramer and Darby 2014). Additionally, the thoracic and lumbar regions are nourished by posterior intercostal arteries and lumbar segmental arteries, respectively. Conversely, the vertebral canal's internal aspect is supplied by segmental arteries branching from the vertebral arteries in the cervical area, the posterior intercostal arteries in the thoracic area, and the lumbar segmental arteries in the lumbar area to supply arterial blood (Griessenauer et al. 2015; Lazorthes et al. 1971; Ratcliffe 1980).

Upon entry into the intervertebral foramina (IVF), each segmental artery's spinal branch subdivides into three distinct branches. The posterior branch supplies the laminae and ligamenta flava; a prominent branch serves the spinous process base, extending posteriorly to its tip. This posterior branch also distributes smaller vessels to the extradural fat and the posterior spinal dura mater. Additionally, another major branch penetrates the lamina near the pedicle junction, splitting into ascending and descending branches, which traverse the superior and inferior articular processes to the articular facets' subchondral bone (Parke et al. 1981; Tveten 1976; Turnbull 1971). Another branch of the segmental artery's spinal ramus proceeds anteriorly, bifurcating into large offshoots supplying the vertebral body's posterolateral aspect before dividing into ascending and descending branches. The ascending branch crosses the intervertebral disc above, joining the level above's descending branch. The descending branch runs near the pedicle, supplying it with branches. Both ascending and descending branches nourish the vertebral bodies' posterior aspect and provide twigs to the posterior longitudinal ligament (Biglioli et al. 2004; Bosmia et al. 2015; Etz et al. 2011; Martirosyan et al. 2011). The spinal ramus's anterior branch bifurcates to target the vertebral body's posterolateral aspect before separating further into ascending and descending branches (Chakravorty 1971). These branches supply blood to the posterior vertebral bodies, with additional small vessels reaching the posterior longitudinal ligament (Crock and Yoshizawa 1977). Lastly, the neural branch, another division of each spinal ramus, extends to the spinal nerve. It splits into anterior and posterior radicular arteries, which nourish the anterior and posterior nerve roots and rootlets, respectively. The posterior radicular artery further provides branches to the dorsal root ganglion's distal and proximal poles (Griepp et al. 2012; Dommisse 1974).

The superior and inferior borders of the vertebral bodies are also sites for extensive venous drainage through numerous small veins. These veins feed into substantial horizontal tributaries positioned in close proximity to each superior and inferior vertebral border, collectively known as the horizontal subarticular collecting vein system (Crock and Yoshizawa 1977). Further, these tributaries merge into larger vertical venous channels that converge at the central horizontal plane of the vertebral body, forming a dense

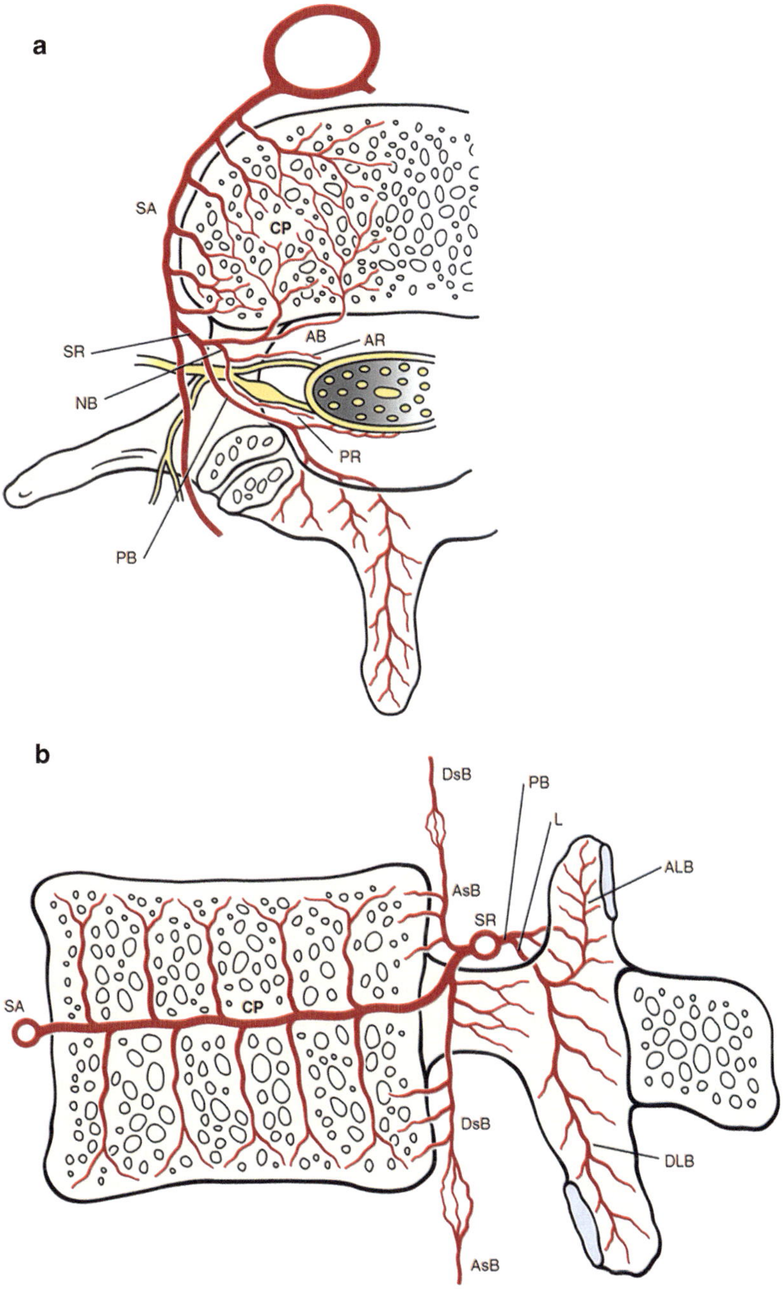

Fig. 6.9 Arterial supply to a typical vertebra and associated tissues. (**a**) Superior view of a horizontal section at the level of the vertebral body. (**b**) Midsagittal section. Abbreviations: SA, segmental artery; CP, central plexus; SR, segmental artery; AB, anterior branch; PB, posterior branch; NB, neural branch; AR, anterior radicular artery; PR, posterior radicular artery; AsB, ascending branch; DsB, descending branch; L, laminar artery; ALB, ascending laminar artery; DLB, descending laminar artery. (Adapted with permission from Cramer and Darby (2014))

venous network. This network typically drains through one and occasionally two basivertebral veins within a single vertebral body.

Additionally, the subarticular collecting vein system sends lateral tributaries that exit the vertebral body and merge into the external vertebral venous plexus (Groen et al. 1997). The vertebral venous plexus, situated externally to the vertebrae and vertebral canal, can be anatomically differentiated into two distinct components: the anterior external vertebral venous plexus, which encircles the vertebral bodies, and the posterior external vertebral venous plexus, which is related to the posterior arches of neighboring vertebrae (Groen et al. 1997). These plexuses establish connections with segmental veins at various

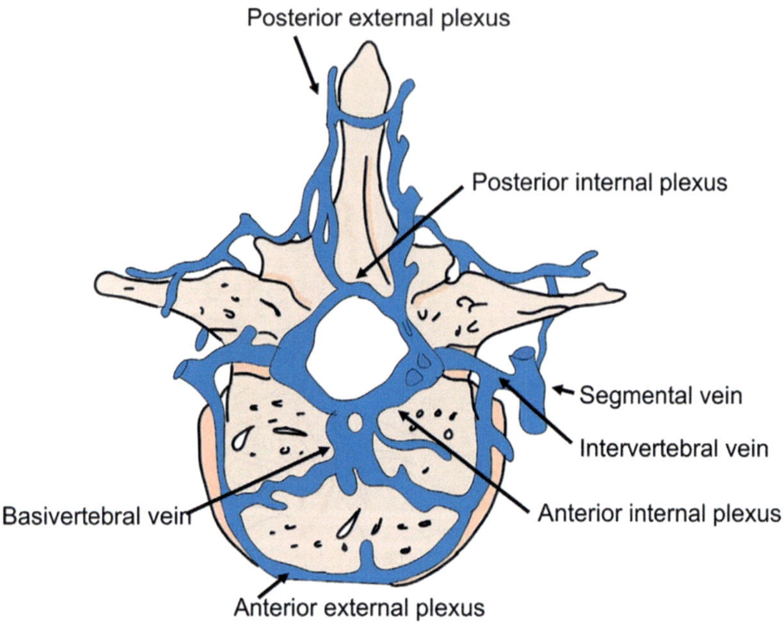

Fig. 6.10 The vertebral venous plexuses, illustrating the internal and external vertebral venous plexuses

spinal levels, including the deep cervical, intercostal, lumbar, and ascending lumbar veins, and they also maintain connections with the internal vertebral venous plexus within the vertebral canal (Demondion et al. 2000). Both the external and internal vertebral plexuses communicate through the intervertebral foramina, as well as directly through the vertebral bodies via the anterior plexus and through the ligamenta flava via the posterior plexus (Stringer et al. 2012).

The Batson venous plexus, also known as the valveless cerebrospinal system, establishes connections with the deep pelvic veins, thoracic veins, and the internal vertebral venous plexuses (Batson 1940; Tobinick and Vega 2006). It drains the lower regions of the urinary bladder, prostate, and breasts into the internal vertebral venous plexuses (intraspinal veins), which are located in the epidural space within the spinal canal (Groen et al. 1997; Nathoo et al. 2011). These plexuses consist of four interconnected vertical vessels: two positioned anteriorly adjacent to the posterior longitudinal ligament (anterior internal vertebral venous plexus) and two situated posteriorly on the anterior aspect of the vertebral arches and ligamentum flava (posterior internal vertebral venous plexus) (Groen et al. 1997) (Fig. 6.10). These plexuses form anastomoses with the posterior external venous plexus and are implicated in the potential spread of cancerous metastases to the vertebral column and brain due to their anatomical location and absence of valves (Batson 1940; Tobinick and Vega 2006; Nathoo et al. 2011).

6.3.2 Anatomical Structure and Development of Intervertebral Discs

The intervertebral discs (IVDs) are an important structure for clinical consideration, as they are a primary source of back pain and can lead to compression issues affecting posterior and anterior roots, spinal nerves, and the spinal cord itself. Moreover, pathological changes within these discs significantly influence spinal biomechanics (Humzah and Soames 1988).

Originating from the notochord and somatic mesenchyme (sclerotome), the IVDs consist of 12–20 layers forming the annulus fibrosus, while the central nucleus pulposus develops from the notochordal cells. Over time, notochordal cells in the vertebral bodies are replaced by osteoblasts and primarily by chondroblasts in the cartilage end plate, though residual notochordal cells can weaken the cartilage end plate, leading to potential herniation or intravertebral herniations known as Schmorl's nodes. These herniations often accelerate the degeneration of the IVD (Tanaka et al. 1993). Initially, during the fetal stage and shortly after birth, IVDs possess a rich vascular supply that gradually diminishes until the discs become nearly avascular by the second decade of life (Taylor 1975).

Located between adjacent vertebral bodies from C2 to L5–S1 and the sacrum and coccyx, IVDs form a symphysis joint and contribute significantly to the vertebral column's height, making up 20–33% of it (Standring 2020). The spine contains a total of 24 IVDs: 6 cervical, 12 thoracic, 5 lumbar (including the L5–S1 disc), and 1 between the sacrum and coccyx. Additional discs may be present between the first and second coccygeal segments or the fused sacral segments. IVDs are typically named using the two surrounding vertebrae (e.g., C4–5 disc) or the vertebra directly above (e.g., C6 disc). The functionality of these discs is vital for ensuring appropriate spinal flexibility and motion limitation between the segments, also aiding in load distribution across the spine. The strong connections between the IVD and the adjacent vertebral bodies form the fundamental components of the vertebral unit, also known as the motor or functional segment.

The shape of an IVD is determined by the contours of the adjacent vertebral bodies. IVDs vary in thickness depending on their spinal region, being thickest in the lumbar area and thinnest in the upper thoracic region. This variability helps form the lordoses of the cervical and lumbar regions, with these discs being thicker anteriorly. Conversely, thoracic discs display uniform thickness from anterior to posterior (Standring 2020).

The discs are anchored by the anterior and posterior longitudinal ligaments, with the posterior attachment being notably firm. While the anterior longitudinal ligament generally maintains a strong attachment to the vertebral body periosteum, its attachment to the IVD is relatively loose, though this varies regionally (Humzah and Soames 1988). In the thoracic area, the discs are additionally connected to the intraarticular ligaments, connecting them to the rib heads' crests from the second to the ninth ribs. IVDs attach to the anterior and posterior longitudinal ligaments.

6.3.2.1 Structure and Function of Annulus Fibrosus

In the lumbar and thoracic regions, the annulus fibrosus (AF) comprises numerous fibrocartilaginous layers, which are convex on the outside. These layers, known as lamellae, consist mainly of densely packed collagen fibers, complemented by a smaller proportion—about 10%—of elastic fibers (Bogduk 2005). Typically, the collagen fibers in each lamella align parallel to each other at an angle of roughly 65° relative to the vertical axis. The arrangement of fibers in adjacent lamellae results in a crisscross pattern, creating a 130° angle between them. Variability in fiber orientation among lamellae is common, differing both between individuals and between vertebrae (Humzah and Soames 1988). The outermost lamellae of the AF connect to the vertebral bodies via Sharpey's fibers, attaching to both the compact bone at the vertebral rim and the adjacent periosteum-covered vertebral body (Humzah and Soames 1988). The inner lamellae bond with the cartilaginous vertebral end plate.

In contrast to the lumbar discs, the cervical intervertebral discs exhibit distinct characteristics in the structural composition of the AF (Mercer and Bogduk 1999). In the cervical spine, the AF consists of a single, crescent-shaped structure of fibrocartilage, thicker anteriorly and tapering off laterally, and a thin, single lamella posteriorly. Similar to the lumbar region, the nucleus pulposus occupies the central portion of the cervical intervertebral discs, with cartilaginous end plates situated above and below the nucleus pulposus and AF. In contrast, cervical IVDs in children resemble those of the lumbar region across all age groups. The distinctive changes in the cervical IVDs begin as clefts in the AF (uncovertebral clefts) adjacent to the uncinate processes during adolescence. Over time, age-related shifts such as the formation of uncovertebral clefts adjacent to uncinate processes and the progressive narrowing of the posterior AF occur, predominantly due to cervical motion-related shearing forces (Taylor 1975). In terms of composition, the thoracic IVDs are largely similar to the lumbar structure.

The IVDs, under normal conditions, rarely face purely tensile stresses. Even during clinical spinal traction, there is some compressive load due to muscular tension. The AF is designed to handle tension in various directions dependent on the type of spinal movement or external condition: flexion primarily affects the posterior annulus, extension the anterior, lateral flexion the convex side, and axial rotation generates tension at about 45° to the disc plane. Additionally, during compression, the entire annulus undergoes tensile stress from the nucleus pulposus pushing outward (Humzah and Soames 1988). Despite the nucleus being removed experimentally, the AF retains a significant load-bearing role (Humzah and Soames 1988). Movements such as flexion cause the anterior AF to bulge outward, whereas extension results in an inward bulge toward the curve's concavity, and similar adjustments occur during lateral flexion. Notably, the anterior disc region is more robust than the posterolateral area, which is most susceptible to herniation risks.

The nerve supply to the AF varies by region: the most superficial lumbar lamellae and the outer cervical AF are innervated by both general somatic and visceral afferent nerves. The posterior AF receives its innervation from the recurrent meningeal nerve, while lateral and anterior regions are innervated by nerves from an anterior ramus and the sympathetic chain.

6.3.3 Structures of the Vertebral Arch

The vertebral arch, also known as the posterior arch, comprises several distinctive anatomical components, each with specific characteristics and functions. This includes the pedicles, laminae, superior and inferior articular processes, transverse processes, and spinous processes. The pedicles, which are thick, short processes, project posteriorly from the superior aspect of the vertebral body, connecting it to the arch (Standring 2020). The laminae are broad, flat plates that extend posteromedially from the pedicles, forming the roof of the vertebral foramen (Moore et al. 2013). The superior and inferior articular processes arise from the junctions of the pedicles and laminae, bearing articular facets that form synovial zygapophyseal joints between adjacent vertebrae, facilitating spinal motion and stability (Cramer and Darby 2014; Standring 2020). Transverse processes extend laterally from the arch, serving as attachment points for muscles and ligaments. The spinous process projects posteriorly from the midline junction of the laminae, varying in size and shape across spinal regions, and acts as a lever for muscle action (Bogduk 2012).

6.3.3.1 Anatomical Features and Composition of Vertebral Pedicles

The pedicles form the anterior segments of the vertebral arch and are positioned above the midpoint of a vertebral body. The smaller size of the pedicles compared to the vertebral bodies results in the formation of vertebral notches both above and below the pedicles, termed superior and inferior vertebral notches. The superior notch is notably shallower and less pronounced than the inferior notch.

Regionally, the pedicles exhibit varying compositions of cortical and cancellous bone, influenced by the degree of motion experienced by each segment of the spine (Pal et al. 2001). Areas subject to greater mobility, such as the middle cervical and upper lumbar regions, are characterized by a higher proportion of cortical bone, offering enhanced strength. Conversely, the thoracic region, which shows less movement, primarily consists of cancellous bone (Pal et al. 2001).

6.3.3.2 Morphological Characteristics of Vertebral Laminae

The vertebral laminae extend into the pedicles and comprise the substantial posterior part of the vertebral arch. These laminae arch posteromedially to join the spinous process, thereby completing the vertebral foramen. The vertebral canal is formed by the collective arrangement of all the vertebral foramina. Male laminae are generally slightly larger than those of females. Starting at C4, where they measure at their minimum (10.4 ± 1.1 mm), the laminae progressively increase in height, peaking at T11 (25.1 ± 2.5 mm). Beyond this point, there is a gradual decline in height from T12 through L4, with a sharper decrease at L5. Notably, the maximum breadth occurs at L5 (15.7 ± 2.0 mm), whereas the minimum is at T4 (5.8 ± 0.8 mm). While the cervical laminae are comparably wide, the thoracic ones, except at T11 and T12, are narrower, broadening progressively from T11 through L5. The thickness peaks at T2 (5.0 ± 0.2 mm) and is at its lowest at C5 (1.9 ± 0.6 mm), tapering from the upper to the lower thoracic segments. The cervical region at the lower end has the thinnest laminae, with the lumbar section displaying intermediate thickness (Xu et al. 1999).

6.3.3.3 Anatomical Characteristics and Orientation of Superior Articular Processes

The superior articular processes (zygapophyses) and their facets originate from the junction of the pedicles and laminae, similar to the transverse processes. Projecting superiorly, both the left and right superior articular processes possess articular surfaces that typically face posteriorly. The articular surface, or facet, of each superior articular process faces posteriorly; however, the exact orientation varies depending on the spinal region. In the cervical and lumbar regions, the facets are directed posteromedially, while in the thoracic region, they are oriented posterolaterally.

6.3.3.4 Anatomical Characteristics and Orientation of Inferior Articular Processes

Originating from the pediculolaminar junction, the left and right inferior articular processes, also termed zygapophyses, extend inferiorly. Each process's articular surface, or facet, is oriented anteriorly. The exact alignment of these facets transitions from anterolateral in the cervical region to anteromedial in the thoracic and lumbar regions. Adjacent zygapophyses constitute the zygapophysial joints, which permit limited motion. The degree of mobility at these joints varies significantly across different vertebral levels, and they contribute to forming the posterior boundary of the intervertebral foramina.

6.3.3.5 Anatomical and Functional Aspects of Zygapophysial Joints

Zygapophysial joints, formed by the articulation of superior and inferior articular facets between adjacent vertebrae, are small synovial, diarthrodial joints. The articular surface of each facet is covered by a 1- to 2-mm-thick layer of hyaline cartilage (Giles and Taylor 1987). These joints play a significant role in determining the direction and range of motion between vertebrae, as well as in load distribution, particularly during extension and rotation. Pain in zygapophysial joints may arise from direct injury, loss of motion, abnormal motion (Paris 1983), or degeneration of the hyaline cartilage (Lewinnek and Warfield 1986).

The zygapophysial joint capsule, a unique structure among synovial joints (Yamashita et al. 1996), surrounds the joint posterolaterally. This capsule includes an outer layer of dense fibroelastic connective tissue, a vascular middle layer of areolar and loose connective tissue, and an inner synovial membrane layer (Giles and Taylor 1987). The ligamentum flavum covers the anterior and medial parts of the joint.

Despite being thin and flexible, the capsules of the zygapophysial joints contribute to the stability of these joints during movement but are thought to be only minimally effective in limiting motion (Boszczyk et al. 2001; Onan et al. 1998). These joint capsules are attached at the margins of the adjacent superior and inferior articular facets and are generally characterized as thin and loose (Standring 2020). Superior and inferior external protrusions, known as recesses, bulge from the joint and contain adipose tissue, with the inferior recess being larger than the superior one. The capsules are longer and looser in the cervical region compared to the lumbar and thoracic regions.

The zygapophysial joint capsule is extensively innervated by sensory nerve fibers, as demonstrated by multiple studies (Cavanaugh et al. 1995; Vandenabeele et al. 1996). Moreover, Ahmed et al. (1993) discovered nociceptive fibers within the ligamentum flavum that are adjacent to these joints. The sensory nerve supply to each zygapophysial joint originates from the medial branch of the posterior primary division (posterior ramus) at the corresponding joint level, with additional innervations from the branches at the levels directly above and below. This extensive multilevel innervation may explain the broad and diffuse referral pain often associated with zygapophysial joint pathology.

The sensory receptors at these joints include three main types: free nerve endings, which are sensitive to pain (nociceptive), complex unencapsulated nerve endings, and encapsulated nerve endings (Yamashita et al. 1996; Beaman et al. 1993; Cavanaugh et al. 1995). Both the complex and encapsulated nerve endings are implicated in proprioception and in modulating protective muscle reflexes, whereas the free nerve endings are primarily involved in the pain response.

Innervation by mechanoreceptors is more concentrated in the cervical zygapophysial joint capsules compared to those in the thoracic and lumbar areas (McLain 1994; McLain and Pickar 1998). This higher density of mechanoreceptors in the cervical region might be linked to its greater range of motion, necessitating enhanced proprioceptive feedback to facilitate precise and smooth movements of the head, as well as to provide protection against injuries resulting from abrupt or improper motions. Furthermore, nociceptive innervation by free nerve endings is prevalent throughout all spinal regions (McLain and Pickar 1998).

6.4 Biomechanical Properties of Spinal Motion and Degenerative Changes

The human spine achieves significant mobility through combined movements of individual vertebral segments. These movements include flexion, extension, lateral flexion, rotation, and circumduction—a complex motion combining directional components (Pearcy and Tibrewal 1984; White and Panjabi 1990). Intervertebral discs both allow and limit movement at vertebral junctions. The cervical and lumbar regions display a greater range of motion due to their thicker intervertebral discs (Bogduk 2005). The morphology and alignment of articular facets determine the possible movement patterns between adjacent segments (Adams and Hutton 1983).

Intervertebral disc degeneration modifies spinal mobility patterns (Urban and Roberts 2003). During initial and middle stages (stages I–IV) of intervertebral disc degeneration, disc thinning leads to increased segmental motion and expanded joint mobility. The final phase (stage V) shows reduced segmental movement (Kirkaldy-Willis and Hayes 1978). These structural alterations affect the mechanical behavior of spinal segments through multiple mechanisms (Adams and Roughley 2006).

Changes in zygapophysial joints also affect spinal movement (Fujiwara et al. 2000). Initial degeneration phases (stages I–III) involve articular hyaline cartilage erosion, resulting in increased axial spine rotation. Progressive degeneration leads to additional bone formation beneath articular facets—subchondral sclerosis. This structural change reduces axial rotation in affected spinal segments below normal levels (Kirkaldy-Willis et al. 1978).

The spine's movement capabilities depend on multiple anatomical factors. Research has shown that age-related changes in disc composition alter mechanical properties and movement patterns (Antoniou et al. 1996; Urban et al. 2004). Additionally, facet joint orientation influences the direction and magnitude of possible movements between vertebrae (Adams and Hutton 1983; White and Panjabi 1990).

Understanding these biomechanical changes provides valuable information for treatment planning (Urban and Roberts 2003). The progression from increased mobility in early degeneration to reduced movement in advanced stages informs therapeutic approaches (Kirkaldy-Willis et al. 1978). Recent studies have demonstrated correlations between degenerative changes and specific movement limitations (Antoniou et al. 1996; Fujiwara et al. 2000).

6.4.1 Coupled Motions of the Spine and Regional Variations

In spinal biomechanics, the primary movements of flexion-extension, left–right axial rotation, and lateral bending occur in conjunction with secondary motions along other axes. These secondary movements, known as "coupled motions," represent consistent associations between movements about different axes (Haher et al. 1994). Research has documented variations in these motion patterns across different spinal regions.

The analysis of coupled motions reveals their occurrence across vertebral segments, with specific patterns emerging in different spinal regions (Ochia et al. 2006). When examining lateral flexion movements, researchers have observed more predictable patterns in the cervical and lumbar regions compared to other spinal areas (Harrison et al. 1998). These observations point to regional differences in biomechanical behavior.

The arrangement of superior and inferior articular facets influences the coupling between lateral flexion and axial rotation in the cervical and lumbar spine. These regions exhibit opposite rotational patterns and varying magnitudes of motion (Moroney et al. 1988). In the cervical spine, lateral flexion pairs with vertebral body rotation toward the concave side. During right lateral flexion of the cervical spine, the vertebral bodies rotate to the right while the spinous processes rotate to the left. This opposing movement occurs because spinous processes move in the direction opposite to vertebral bodies during rotation (White and Panjabi 1990). The lumbar spine demonstrates an inverse pattern, where lateral flexion pairs with vertebral body rotation toward the convex side of the lateral flexion arc. For example, left lateral flexion in the lumbar region produces right rotation of the vertebral bodies and left rotation of the spinous processes. The thoracic spine presents a graduated pattern of coupled motion that transitions between cervical and lumbar characteristics. The upper four thoracic vertebrae follow the cervical pattern, with vertebral body rotation toward the concavity during lateral flexion. The lower four thoracic vertebrae align with the lumbar pattern, rotating vertebral bodies toward the convexity. The middle four thoracic vertebrae show minimal coupled motion (White and Panjabi 1990).

6.4.2 Biomechanical Characteristics and Functions of Spinal Ligaments

The spinal ligaments demonstrate regional specialization from the cervical to the sacral areas. The ligamentum nuchae and anterior longitudinal ligament extend from the cervical spine, while the supraspinous ligament begins in the thoracic region. Tables 6.3a and 6.3b detail attachment points and motion-limiting functions of major spinal ligaments.

During normal spinal movement, ligaments support efficient motion patterns (Panjabi 1992). They maintain spinal cord protection by limiting excessive movement and absorbing impact forces (White and Panjabi 1990). The biomechanical effectiveness of these ligaments results from their structural properties and anatomical positioning relative to mobile vertebrae. Their fiber orientation optimizes tensile force resistance, although they buckle under compression (Bogduk 2005).

While individual ligaments show unique characteristics in stiffness, deformation limits, and failure thresholds, their overall biomechanical response patterns remain consistent, as demonstrated by similar load-deformation relationships. Ligament stiffness increases markedly as movement approaches physiological limits. During early spinal degeneration, increased mobility can place additional strain on ligaments. This condition increases susceptibility to sprains during various movements, including flexion, extension, axial rotation, and lateral flexion, making degenerative spines more vulnerable to ligamentous injury (Fujiwara et al. 2000; Iida et al. 2002).

Table 6.3a Ligaments of the upper cervical spine

Ligament	Attachment 1	Attachment 2	Motion Limited
Posterior atlanto-occipital membrane	Posterior arch of atlas	Posterior rim of foramen magnum	Flexion of occiput on atlas
Tectorial membrane	Posterior body C2	Anterior rim of foramen magnum (clivus)	Flexion and extension of atlas and occiput
Cruciform ligament			
Transverse ligament	Medial tubercle (colliculus atlantis) of lateral mass	Medial tubercle (colliculus atlantis) of contralateral lateral mass	Forms diarthrodial joint with odontoid process, allowing rotation
Superior longitudinal band	Transverse ligament	Anterior rim of foramen magnum (clivus)	Holds transverse ligament in place; may limit flexion and extension of occiput
Inferior longitudinal band	Transverse ligament	Body of C2	Holds transverse ligament in place; may limit flexion of occiput and atlas on axis
Alar ligaments	Posterior and lateral aspect of odontoid process	Medial surface of ipsilateral occipital condyle	Contralateral axial rotation
Apical ligament of odontoid process	Posterior and superior aspect of odontoid process	Anterior rim of foramen magnum (clivus)	Prevents some vertical translation and anterior shear of occiput
Anterior atlanto-occipital membrane	Superior aspect of anterior arch of atlas	Anterior margin of foramen magnum	Extension of occiput on C1

Table 6.3b Ligaments of the lower cervical, thoracic, and lumbar spine

Ligament	Attachment 1	Attachment 2	Motion Limited
Anterior longitudinal ligament	Occiput, anterior tubercle of atlas, and subsequent vertebral bodies	Sacrum	Extension of vertebral column
Posterior longitudinal ligament	Posterior body of C2 and subsequent intervertebral discs	Sacrum	Flexion of vertebral column; may help limit posterior intervertebral disc protrusion
Ligamenta flava (paired)	Anterior and inferior aspect of lamina of vertebra above	Posterior and superior aspect of lamina of vertebra below	May aid in extension of spine; slows last few degrees of spinal flexion
Supraspinous ligament	Posterior-most aspect of spinous process of vertebra above	Posterior-most aspect of spinous process of vertebra below	Flexion
Interspinous ligaments	Inferior aspect of spinous process of vertebra above	Superior aspect of spinous process of vertebra below	Flexion
Ligamentum nuchae	External occipital protuberance; spinous process of C7; dermis of posterior midline of neck	Cervical spinous processes (posterior tip and extending anteriorly to become continuous with cervical interspinous ligaments)	Flexion of cervical region
Intertransverse ligaments	Transverse process of vertebra above	Transverse process of vertebra below	Contralateral lateral flexion

Adapted with permission from Cramer and Darby (2014)

6.4.3 Anatomical Constraints on Spine Mobility

The vertebral column's mobility patterns are determined by specific osseous and ligamentous structures that work together to control movement while maintaining stability. These anatomical components create a system of mechanical limitations through bony stops and ligamentous brakes (Louis 1985; White and Panjabi 1990). The detailed anatomical structures that regulate spinal movement are presented in Table 6.4.

The biomechanics of spinal flexion involve complex interactions among multiple anatomical structures. During forward flexion, the anterior longitudinal ligament undergoes a controlled relaxation process, permitting controlled compression of the anterior portions of the intervertebral discs (Bogduk 2005). This movement creates increased intersegmental spaces between the vertebral laminae. Concurrent with these changes, the inferior articular processes of each vertebra demonstrate a superior gliding motion along the superior articular processes of the subjacent vertebra (Moore et al. 2013). The posterior longitudinal ligament, ligamentum flavum, and capsular ligaments experience tension during this motion, providing controlled resistance to excessive forward bending (Neumann 2017).

Table 6.4 Structures limiting spine motion

Motion	Structures limiting motion
Flexion	Posterior longitudinal ligament
	Ligamenta flava
	Interspinous ligament
	Supraspinous ligament
	Posterior fibers of intervertebral disc
	Articular capsules
	Tension of back extensor muscles
	Anterior surface of inferior articular facet against posterior surface of superior articular facet
Extension	Anterior longitudinal ligament
	Anterior aspect of intervertebral disc
	Approximation of spinous processes, articular processes, and laminae
Lateral flexion	Contralateral side of intervertebral disc and intertransverse ligament
	Approximation of articular processes
	Approximation of uncinate processes (cervical region)
	Approximation of costovertebral joints (thoracic region)
	Antagonist muscles
Rotation	Tightening of lamellar fibers of anulus fibrosus
	Orientation and architecture of articular processes

Adapted with permission from Cramer and Darby (2014)

Regional variations in spinal flexibility are notable across different vertebral segments. The lumbar and cervical regions exhibit enhanced flexion mobility compared to the thoracic segment (Standring 2020). This regional differentiation results from several anatomical factors: the thoracic spine contains relatively thinner intervertebral discs, experiences additional mechanical constraints from the thoracic cage architecture, and is influenced by the complex attachments of intercostal and paraspinal musculature (Bogduk 2005; Neumann 2017). The zygapophyseal joint orientation also contributes to these regional differences, with more sagittally oriented facets in the thoracic region limiting rotation and lateral flexion (Cramer and Darby 2014).

Extension movements demonstrate specific regional characteristics and limitations. The thoracic region shows particular restrictions during extension, attributed to its thin intervertebral discs, the structural configuration of thoracic vertebrae, and the mechanical influence of the costovertebral and costotransverse joints (Moore et al. 2013; White and Panjabi 1990). The posterior elements, including the spinous processes and laminae, can make contact during full extension, creating anatomical stops that prevent excessive backward bending (Bogduk 2005).

During lateral flexion, the ipsilateral aspects of the intervertebral discs undergo compression, while the contralateral sides experience tension. The cervical spine demonstrates the greatest capacity for lateral flexion, followed by the lumbar region, while the thoracic segment exhibits the most restricted lateral motion (Neumann 2017; Standring 2020). This variation relates to the orientation of the articular facets and the mechanical constraints imposed by the surrounding ligamentous and muscular structures (Mercer and Bogduk 1999).

6.5 Anatomy and Clinical Significance of the Cervical Spine

The cervical spine, a distinct section of the vertebral column, exhibits a high degree of structural complexity and functional significance. This relatively compact cylindrical region plays an important part in connecting the cranium and the thoracic cavity. It is known for its remarkable anatomical efficiency (Bogduk and Mercer 2000). The cervical spine has 76 distinct articulations, which is the highest degree of mobility among all spinal segments. This allows for a wide range of motion (Swartz et al. 2005). Nevertheless, this increased adaptability is contrasted with a high concentration of critical neuronal and circulatory structures, requiring careful examination in clinical assessments and therapeutic interventions (Swartz et al. 2005).

In order to accurately diagnose patients and develop effective treatment plans, rehabilitation professionals require a thorough knowledge of the complex structure of the cervical spine (Cramer and Darby 2014). By thoroughly examining the intricate details of the musculoskeletal, neurological, and vascular aspects of this region, medical professionals can gain an in-depth knowledge of how structure and function interact with each other. This information enables them to recognize small anomalies, forecast future consequences, and customize rehabilitative treatments to maximize outcomes for patients.

Therefore, a comprehensive examination of the structure of the cervical spine is necessary to establish an adequate basis for successful rehabilitation strategies. By understanding the detailed anatomy of this area, medical professionals are capable of managing the difficulties of cervical pathologies, leading to improved quality of treatment for patients with spinal cord injuries or related diseases.

6.5.1 Structural Overview of the First Cervical Vertebra (Atlas)

Named after the Greek Titan tasked with supporting the celestial sphere, the first cervical vertebra, or atlas, fulfills a similar role by supporting the globular head (Moore et al. 2013). This vertebra is distinctively structured with two arches, anterior and posterior, which are connected by lateral masses on either side. Each lateral mass extends outward, giving rise to a transverse process. Originating from three primary ossification centers—one in each lateral mass and another in the anterior arch—the atlas's ossification centers in the lateral masses appear first, typically around the seventh week postconception (Cramer and Darby 2014) (Fig. 6.11).

These ossification centers extend toward what will become the posterior arch, usually fusing by the age of four. In some instances, these centers do not fully unite, resulting in a persistent

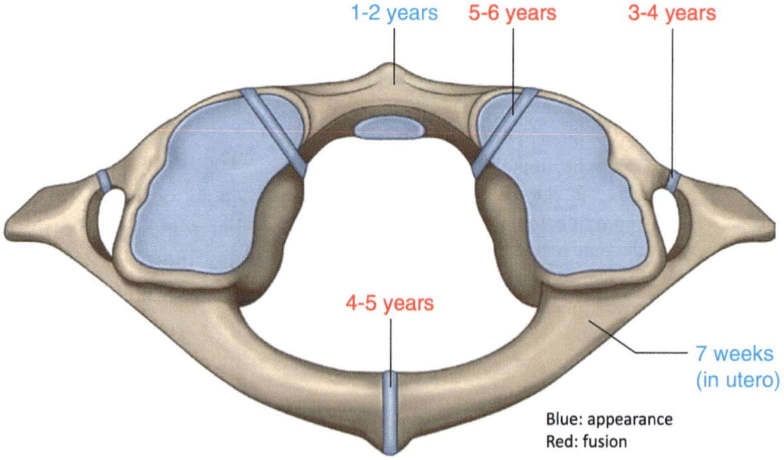

Fig. 6.11 Multiple ossification centers in atlas (C1) at birth. (Adapted with permission from Ko (2022))

cartilaginous bridge. On standard X-ray imaging, this bridge appears radiolucent, which can be mistaken for a fracture in cases of trauma.

6.5.1.1 Development and Structure of the Anterior Arch of the Atlas

In embryonic development, the anterior arch of the atlas is formed from the hypochordal arch, a connective tissue bridge linking the two lateral masses (Mekonen et al. 2015). While this arch is present throughout the spine during development, it uniquely persists into adulthood only in the anterior arch of the atlas. Typically, the primary ossification center for this arch manifests by the end of the first year after birth and consolidates with the lateral masses between the ages of 7 and 9 years (Piatt Jr and Grissom 2011). Smaller than its posterior arch, the anterior arch features a prominent elevation on its anterior surface, known as the anterior tubercle. This tubercle provides an attachment point for the anterior longitudinal ligament at the center and, more laterally, for the superior oblique fibers of the longus colli muscle (Schünke et al. 2010).

On its posterior surface, the anterior arch has a smooth, hyaline cartilage-covered articulating surface referred to as the facet for the dens, or odontoid process. This surface forms a diarthrodial joint with the anterior surface of the odontoid process, which is part of the axis vertebra. Due to the absence of a vertebral body in the atlas, the odontoid process substitutes as the analogous body region, enabling the atlas to pivot effectively around the odontoid process at this joint, facilitating significant rotational movement of the head (Joaquim et al. 2015).

6.5.1.2 Structure of the Posterior Arch of the Atlas

The atlas features a posterior arch that is notably larger than its anterior counterpart, comprising about two-thirds of the vertebral ring. There is a prominence called the posterior tubercle. The tubercle serves as a central point for attachment of the ligamentum nuchae and as the origin for the rectus capitis posterior minor muscle (Standring 2020).

Adjacent to the posterior arch, the first pair of ligamenta flava connects at its lower border. Notably thin, the superior lateral regions of this arch are recessed, forming grooves on both sides for the vertebral arteries. These grooves accommodate the vertebral arteries, veins, and the suboccipital nerves, which lie between the artery and the arch itself. The suboccipital nerve represents the posterior ramus of the C1 vertebra. Interestingly, in about 32–37% of individuals, these grooves are enclosed by a bone arch, leading to the development of what is sometimes termed an arcuate or arcual foramen (Taitz and Nathan 1986). The vertebral arteries reach their respective grooves by curving around the superior articular processes to rest atop the posterior arch. Furthermore, the posterior atlanto-occipital membrane, which may ossify at its lateral and inferior borders, is affixed to the groove sides, potentially contributing to arcuate foramen formation (Fig. 6.12) (Tubbs et al. 2006).

6.5.1.3 Structure of the Transverse Processes of the Atlas

The transverse processes of the atlas, both left and right, are notably larger compared to those of other cervical vertebrae and can be discerned between the mastoid process and the mandibular angle. These structures extend laterally from the lateral mass of the atlas, serving as a muscular attachment, thereby facilitating head rotation (Standring 2020). Not only are these processes substantial in size, making the atlas broader than most cervical vertebrae with the exception of C7, but they also display frequent asymmetry in their size and shape. Typically, the transverse processes of the atlas comprise a single lateral mass, as opposed to the anterior and posterior tubercles found in other cervical vertebrae. This configuration aligns closely with the posterior roots of other cervical vertebrae, essentially forming what could be considered a posterior root coupled with a minor segment of the intertubercular lamella (Standring 2020). Each transverse process also contains a foramen, the largest within the cervical spine, which accommodates the vertebral artery and veins as well as the sympathetic nerve plexus (Taitz et al. 1978).

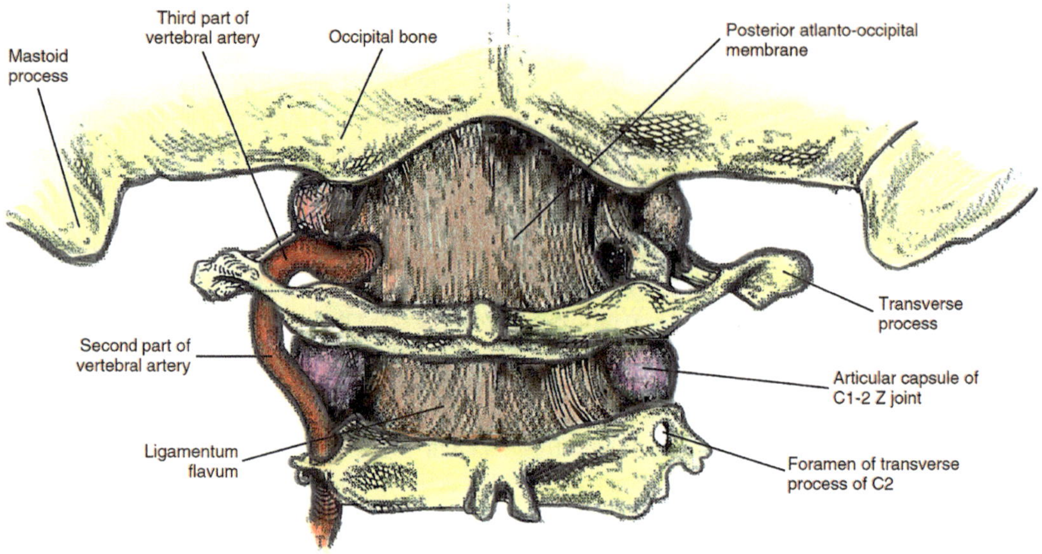

Fig. 6.12 Posterior ligaments of the upper cervical region. (Adapted with permission from Cramer and Darby (2014))

The atlas's transverse processes are sites for muscular attachment, engaging with muscles such as the rectus capitis lateralis, obliquus capitis superior and inferior, levator scapulae, splenius cervicis, and scalenus medius. These processes are also closely associated with the C1 spinal nerve. The posterior ramus of this nerve, known as the suboccipital nerve, primarily provides motor function to the suboccipital muscles. Conversely, the anterior ramus passes laterally around the lateral mass, staying close to the vertebral artery and beneath the rectus capitis lateralis muscle, before descending anterior to the atlas, where it merges with the ascending branch of the C2 anterior ramus.

6.5.1.4 Structure of the Vertebral Foramen of the Atlas

The vertebral foramen of the atlas typically exhibits a larger anteroposterior diameter compared to its transverse diameter. The spatial configuration within the C1 vertebral foramen is commonly segmented into thirds. These divisions consist of the odontoid process of the axis (C2) occupying one-third, the spinal cord situated in another third, and the remaining third comprising what is often referred to as "free space." However, this free space is not empty; it contains epidural fat, blood vessels, ligaments, meninges, and the subarachnoid space. This particular distribution within the vertebral foramen is frequently cited as Steel's rule of thirds (Foreman and Croft 2002) (Fig. 6.13). On rare occasions, the vertebral foramen of C1 may present with a reduction in size due to a hypoplastic posterior arch, a condition that can lead to spinal cord compression.

6.5.2 Structural Overview of the Second Cervical Vertebra (Axis)

The axis (C2) has distinctive morphological features compared with the other cervical vertebrae. It originates from ten ossification centers—five primary and five secondary. The primary centers include one in the body of the vertebra, one on each side of the neural arch, and two within the dens (refer to the odontoid process in the subsequent section). The secondary centers of ossification consist of one at the odontoid process, one at the lower part of the vertebral body, two at the transverse processes (one on each side), and one at the spinous process's posterior tip.

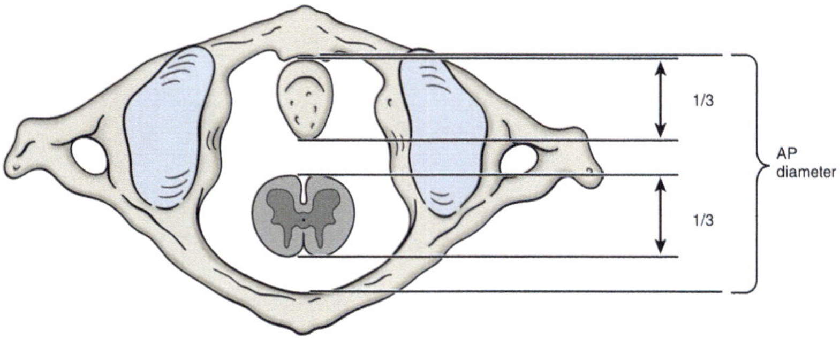

Fig. 6.13 Steel's rule of thirds. (Adapted with permission from Cramer and Darby (2014))

The most distinctive anatomical characteristics of the axis include the prominent dens or odontoid process, the superior articular processes, and the transverse processes (Standring 2020). The odontoid process arises from the superior aspect of the vertebral body. This structure plays an important role in the atlanto-axial joint, facilitating rotation of the head (Bogduk and Mercer 2000). The superior articular processes of the axis are uniquely adapted to articulate with the inferior articular processes of the atlas, forming a pivotal joint that allows for substantial rotational motion (Panjabi et al. 1988). Additionally, the transverse processes of the axis are notably pronounced compared to those of other cervical vertebrae. Another unique feature of the axis is its spacious vertebral foramen, designed to accommodate the descending spinal cord from the foramen magnum of the skull (White and Panjabi 1990). This enlarged space is important for protecting the neural structures and the spinal cord.

6.5.2.1 Structural Features of the C2 Vertebral Body

The C2 vertebral body exhibits distinct anatomical characteristics and serves as an attachment point for various ligaments and muscles. The body of C2 is characterized by a reduced amount of cancellous bone compared to its dens (Standring 2020). The inferior aspect of the body displays a notable indentation, while the posterior aspect features a less pronounced indentation (Panjabi et al. 2001). The anterior surface of the C2 vertebral body is concave, primarily due to the attachment site of the longus colli muscle. This feature is consistent throughout the cervical spine, where the anterior longitudinal ligament attaches to the lower margin of the vertebral bodies, closely associated with the attachment of the anterior fibers of the annulus fibrosus.

Like other cervical vertebrae, the lower or discal border of the C2 vertebra projects downward at its anterior side. The posterior surface of the vertebral body plays a crucial role as the attachment point for the posterior longitudinal ligament, extending upward to connect with the tectorial membrane, specifically anchoring to the posterior and inferior borders of the vertebral body. Additionally, the first intervertebral disc of the spine is located between the lower surface of the C2 vertebral body and the upper surface of the C3 vertebral body (Benzel 2017).

6.5.2.2 Developmental, Structural, and Vascular Features of the Dens (Odontoid Process)

The dens, also referred to as the odontoid process, arises from two primary ossification centers situated laterally and a secondary ossification center at the apex (Cramer and Darby 2014) (Fig. 6.14). The paired primary centers emerge during fetal development and typically achieve midline fusion by the eighth month of fetal development (Fesmire and Luten 1989). Following their union, they generally fuse to the inferior edge of the C2 vertebral body around 3–6 years. This fusion line between the odontoid process and C2's body remains detectable via X-ray until the age of 11, with one-third of people retaining this demarcation for life, often leading to misinterpretation as a fracture (Fesmire and Luten 1989). Additionally, a small disc at the juncture of the dens and C2 persists into late adulthood, frequently visible in sagittal MRI

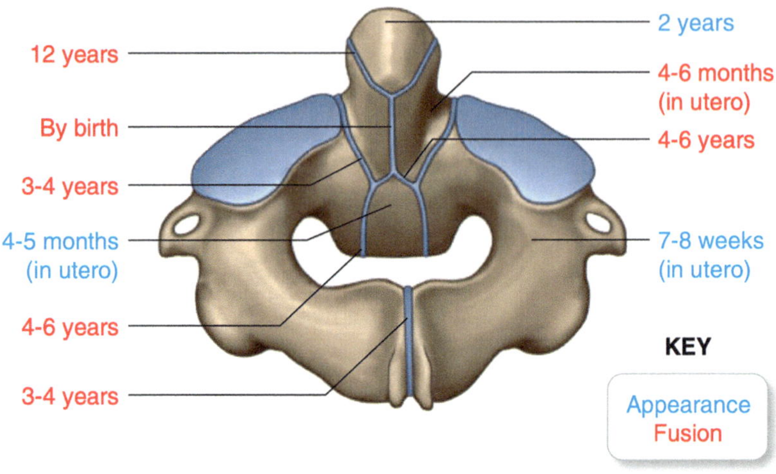

Fig. 6.14 Multiple ossification centers in axis (C2) at birth. (Adapted with permission from Ko (2022))

imaging. This joint is recognized as the subdental synchondrosis.

The apical secondary center of ossification typically emerges between the ages of 3 and 6, presenting as either V-shaped or cuneiform, thus creating a significant notch between the primary ossification centers (Cramer and Darby 2014). This center ordinarily merges with the rest of the odontoid process by age 12 (see Fig. 6.14). Prior to or in the absence of complete fusion, this fragment may appear on X-rays as a persistent ossiculum terminale, and distinguishing it from fractures is difficult (Gebauer et al. 2006).

The mature odontoid process resembles a peg and is topped with a curved surface. Observations indicate that, for 14–26% of individuals, this structure may lean slightly left or right (Pfirrmann et al. 2001; Koller and Robinson 2018). The anterior surface of the dens is covered with a hyaline-lined articular facet that joins with the facet on the posterior side of the atlas's anterior arch. At its base, the dens' posterior surface features a groove created by the transverse atlantal ligament, part of the cruciform ligament. This groove and the transverse ligament together form a synovial joint, contributing to the atlas and odontoid's complex of anterior and posterior joints, which are categorized as a trochoid (pivot) diarthrodial (synovial) joint. This arrangement permits the atlas to rotate around the odontoid process, allowing for approximately 45 degrees of movement in either direction. The lateral surfaces of the odontoid process, above the transverse ligament's groove, are flat and provide attachment points for the left and right alar ligaments, while the apical odontoid ligament connects to the odontoid's apex (Osmotherly et al. 2013).

Rich arterial networks supply the odontoid process. The vertebral arteries, on each side, bifurcate before traversing the transverse processes, sending an ascending artery to both the anterior and posterior surfaces of the odontoid process. These four arteries (left and right, anterior and posterior) converge at the apex to form an arterial arch over the odontoid. Additionally, prior to its entry into the carotid canal, the upper part of the extracranial internal carotid artery branches off to the odontoid's anterior surface and enhances the arterial supply from the ascending anterior artery of the axis (Schiff and Parke 1973). A remnant intervertebral disc (subdental synchondrosis) located between the odontoid process and the body of C2 inhibits significant vascular communication between these structures, contributing to the prolonged healing period of many type II odontoid fractures and the continued presence of os odontoideum (Althoff and Goldie 1977; Menezes and Traynelis 2008).

6.5.2.3 Structural Features of the C2 Superior Articular Processes

The superior articular processes of C2 can be visualized as flattened surfaces on the left and right pedicles, distinct from other cervical vertebrae, where the superior articular processes extend upward from the pediculolaminar junction (Panjabi et al. 1993). Unlike typical cervical vertebrae, where the superior articular processes extend upward from the pediculolaminar junction, those of C2 are level with the pedicles. This anatomical feature contributes to a distinctive configuration that, combined with a highly flexible articular capsule at this junction, permits significant axial rotation—up to 45° unilaterally—between C1 and C2 (Bogduk and Mercer 2000). The articular cartilage on the superior aspect of these processes is shaped convexly and features a transverse ridge from the medial to the lateral part of the process. This ridge allows the anterior and posterior aspects of the facet to slope inferiorly, facilitating more efficient rotational movement between C1 and C2 (Koebke and Brade 1982). Unlike other zygapophysial joints in the cervical spine, the articulation between the superior articular facet of C2 and the inferior articular facet of C1 is positioned more anteriorly, meaning that these facets do not contribute to the articular pillars found in the lower cervical vertebrae (Pal et al. 2001).

6.5.2.4 Structural Features of the C2 Transverse Processes

The transverse processes of the second cervical vertebra (C2) present unique anatomical characteristics that differentiate them from other cervical vertebrae. Unlike the transverse processes of cervical vertebrae C3–C7, but similar to C1, the C2 transverse processes are small and do not display well-defined anterior and posterior tubercles (Standring 2020; Taitz et al. 1978).

From a developmental perspective, research indicates that these processes correspond to the posterior components of typical cervical transverse processes. However, small representations of anterior tubercles appear at the junction where the transverse processes connect with the C2 vertebral body (Moore et al. 2013; Taitz et al. 1978). The C2 transverse processes extend in a superolateral orientation, and each process contains a foramen that forms an angular canal with two dural openings—one directed inferiorly and another laterally (Bogduk and Mercer 2000).

This anatomical configuration has significant implications for vertebral artery passage. The structure allows the vertebral artery to course laterally from the C2 transverse foramen toward the more laterally positioned foramen in the C1 transverse process (Cramer and Darby 2014). This arrangement accommodates the unique path of the vertebral artery as it traverses the upper cervical spine (Neumann 2017; Standring 2020).

Despite their relatively small dimensions, the C2 transverse processes serve as attachment sites for multiple cervical muscles. These include the scalenus medius, levator scapulae, splenius cervicis, and intertransversarii muscles (Moore et al. 2013). These muscular attachments contribute to the complex biomechanical function of the upper cervical spine (White and Panjabi 1990).

6.5.3 Anatomical Characteristics of the Typical Cervical Vertebrae (C3–C6)

The middle cervical vertebrae (C3–C6) demonstrate specific morphological features that set them apart from other vertebrae in the spinal column. These vertebrae, classified as typical cervical vertebrae, are smaller in size compared to other vertebral segments and display uniform structural patterns that reflect their functional requirements (Koller and Robinson 2018). The first and second cervical vertebrae (C1 and C2) are regarded as atypical, while the seventh cervical vertebra (C7) possesses unique features (Koller and Robinson 2018; Steinmetz and Benzel 2015).

6.5.3.1 Anatomical Features and Clinical Implications of the Vertebral Body (C3–C6)

From C2 to C7, both the transverse and anterior-to-posterior diameters of the cervical vertebrae increase, with the transverse diameter expanding more significantly than the anteroposterior diameter (Kwon et al. 2004). This dimensional increase supports the greater load demands on the lower cervical vertebrae. Notably, the transverse diameter exceeds the anteroposterior diameter except in the case of C7, where both dimensions are approximately equal at the superior and inferior vertebral margins. Additionally, the cortical thickness of the superior and inferior aspects of the vertebral bodies progressively increases from C3 to C7. The superior end plate of each vertebra from C3 to C7 is notably thicker posteriorly than anteriorly, a trait that is reversed in the inferior end plates. The anterior surface of these cervical vertebral bodies is convex laterally and concave vertically due to significant ridges at the superior and inferior borders created by the attachment of the anterior longitudinal ligaments (Panjabi et al. 2001). The attachments for the posterior longitudinal ligament are located on the posterior aspect of these bodies. Notably, the anterior aspects of the vertebral bodies are prone to developing osteophytes, which are asymptomatic in 20% to 30% of individuals but can cause complications such as dysphagia and dysphonia when they exert pressure on the esophagus or trachea (Kissel and Youmans 1992).

Projecting from the superior surface of each typical cervical vertebra are lateral lips known as uncinate processes, which become more pronounced in the lower cervical region (Dupuis et al. 1985). These structures form small synovial joints, termed the uncovertebral joints (of von Luschka), with indentations on the underside of the vertebra above, appearing around ages 9–10. These joints facilitate flexion and extension while limiting lateral flexion of the cervical spine. Notably, the uncinate processes are taller between C4 and C6, correlating with increased lengths of the posterior and anterior roots and spinal nerves extending from the dura mater to the intervertebral foramina, which could explain the higher prevalence of neural compression syndromes at these levels (Ebraheim et al. 1997a, b). Additionally, dense fibrous tissue connects to the uncinate processes, encasing the spinal nerve and vertebral artery in the intertransverse space, making the midcervical vertebral artery more susceptible to damage from osteophytes compared to its lower cervical counterpart (Ebraheim et al. 1997a, b).

6.5.3.2 Anatomical Features and Clinical Implications of the Transverse Processes

Each transverse process of a typical cervical vertebra features two roots—termed anterior and posterior. These roots extend laterally, terminating in tubercles (Standring 2020). Notably, at C5 or C6, it is not uncommon to find an anterior tubercle that is longer than those of adjacent vertebrae, often due to congenital fusion with the vertebral body above or below. The anterior and posterior tubercles are connected by an intertubercular lamella, sometimes referred to as a costotransverse lamella. At C1, the span between the lateral ends of the transverse processes is at its maximum, diminishing slightly and stabilizing from C2 through C6, and then markedly increasing at C7.

Each transverse process forms a neural groove between the anterior and posterior roots, providing a pathway for the emergence of the spinal nerve and its anterior primary ramus (Fig. 6.15). These grooves typically angle 50° anteriorly from the midline, except at C7, where the angle is more pronounced. The depth of these grooves progressively deepens from C3 to C7. The intertubercular lamellae at C3 and C4 descend obliquely from the anterior root toward the posterior, resulting in shorter anterior tubercles and deeper posterior grooves. The intertubercular lamellae at C6 are notably wider and shallower (Standring 2020).

A distinctive dural root sleeve encases each spinal nerve and its anterior ramus, anchored securely to the transverse process grooves by fibrous tissue. This attachment is particularly robust at the fourth through sixth cervical vertebrae (C4–C6) (see Fig. 6.15). This configuration limits the lateral movement of the dura, effectively preventing undue tension on the nerve roots and their connections to the spinal cord by anchoring the dura into the intervertebral foramen (Sunderland 1974a, b). The cervical dural root sleeves are richly innervated, primarily by fibers originating from the dorsal root ganglia rather than the recurrent meningeal nerves, a configuration that could lead to pain under significant traction or compression.

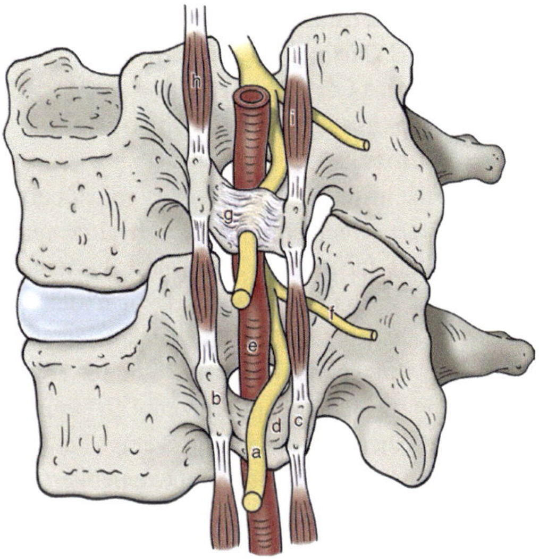

Fig. 6.15 Obliquely oriented cervical intervertebral foramina. The groove, or "gutter," for the spinal nerve is formed by the anterior and posterior tubercles and the intertubercular lamella of each transverse process in typical cervical vertebrae. (**a**) Anterior primary division (ventral ramus). (**b**) Anterior tubercle (of transverse process). (**c**) Posterior tubercle. (**d**) Intertubercular lamella. (**e**) Vertebral artery. (**f**) Posterior primary division (dorsal ramus). (**g**) Connective tissue surrounding the anterior primary division and anchoring it to the groove for the spinal nerve. (**h**) Anterior intertransversarius muscle. (**i**) Posterior intertransversarius muscle. (Adapted with permission from Cramer and Darby (2014))

Immediately after its formation by the confluence of posterior and anterior roots, each spinal nerve gives rise to a posterior ramus. This ramus, also known as the posterior primary division, initially progresses in a posterior and lateral direction along the zygapophysial joint. It provides this joint with sensory fibers (Bogduk 2012). Continuing posteriorly, the ramus distributes motor, nociceptive, and proprioceptive fibers to the deep muscles at the back of the neck and sensory fibers to the skin layers of the back (Drake et al. 2020).

In the cervical spine, the anterior transverse processes of vertebrae C4 to C6 end in distinct tubercles. The anterior tubercles act as attachment points for the scalenus anterior and the superior and inferior oblique fibers of the longus colli and longus capitis muscles (Moore et al. 2013). The posterior tubercles, which are generally positioned more laterally and slightly lower (with the exception of C6, where they are horizontally aligned), provide attachment sites for several muscles, including the splenius cervicis, longissimus cervicis, iliocostalis cervicis, levator scapulae, and the scalenus medius and scalenus posterior (Standring 2020).

Each cervical vertebra features a foramen within its transverse process, known as the transverse foramen. This foramen is bordered by four structures: the pedicle, anterior and posterior roots of the transverse process, and the intertubercular lamella (Benzel 2017). Asymmetry in size between the left and right foramina of a single vertebra is common, and occasionally a single transverse process may exhibit a double foramen. The vertebral artery typically enters the foramen of the transverse process at C6, ascending through the corresponding foramina from C5 to C1 (Tubbs et al. 2016). It then encircles the superior articular process of the atlas and forms a posteriomedial loop around the superior articular process before continuing cranially through the foramen magnum (Netter 2019). The anterior rami of C3 to C6 spinal nerves course directly posterior to the vertebral artery, often resting against the vessel as they emerge from the groove for the spinal nerve of the transverse process.

Accompanying the vertebral artery, the vertebral veins on each side also traverse the foramina of the transverse processes, originating in the atlanto-occipital region and descending through the foramina from C1 to C7 to drain into the subclavian vein (Winn 2017). These veins interconnect with the internal (epidural) and external vertebral venous plexuses (Cramer and Darby 2014). Accompanying the vertebral artery through the foramina of the transverse processes from C1 to C6 is a plexus of sympathetic nerves (Patestas and Gartner 2016).

6.5.3.3 Anatomical Features and Clinical Implications of the Cervical Vertebral Canal

The cervical vertebral canal demonstrates specific morphological characteristics that reflect its functional requirements. The vertebral foramen of typical cervical vertebrae presents a triangular (trefoil) configuration with adequate dimensions to contain the cervical enlargement of the spinal cord. The sequential arrangement of these foramina creates the vertebral canal, with structural contributions from both intervertebral discs and the ligamenta flava (Moore et al. 2013; Standring 2020).

The internal vertebral venous system within the canal shows complex anatomical organization. Within the epidural adipose tissue, the anterior internal vertebral venous plexus forms bilateral longitudinal channels. These channels maintain their position through delicate fibrous membranes located on each side of the canal. Retrocorporeal veins establish connections between these channels posterior to each vertebral body, coursing between the posterior longitudinal ligament and the vertebral bodies (Chaynes et al. 1998; Cramer and Darby 2014).

The upper cervical region presents specific vascular patterns with clinical implications. The anterior internal vertebral veins in this region, known as periodontal veins, form connections with retropharyngeal venous structures through pharyngovertebral veins. This anatomical arrangement creates potential pathways for infection transmission to the occipital condyles and atlas, which may lead to osteomyelitis in the upper cervical articulations (Bogduk 2005; Menezes and Traynelis 2008).

The cervical vertebral canal shows progressive dimensional changes from C3 to C6. While maintaining relatively constant transverse measurements, the canal exhibits a decrease in sagittal dimensions, resulting in an increasingly pronounced trefoil configuration in the caudal direction. At the C6 level, the spinal cord occupies approximately 75% of the canal's volume (Neumann 2017; White and Panjabi 1990).

Dynamic changes in the cervical subarachnoid space occur during spinal movements. From full extension to complete flexion, the cerebrospinal fluid volume between C2 and C7 undergoes significant variations. The fluid volume reaches its minimum during extension and maximum during flexion, with an approximate total volume change of 1.9 mL (Steinmetz and Benzel 2015).

6.5.3.4 Anatomical Features and Clinical Implications of the Cervical Articular Processes and Zygapophysial Joints

At birth, the articular processes of typical cervical vertebrae, which are lined with hyaline cartilage on their facets, display a more horizontal orientation with an average angle of about 34° relative to the horizontal plane. This orientation changes as they gradually shift to a more vertical position by the age of approximately 10 years, at which point the angle reaches an average between 45° and 53°. The facet joints in the upper cervical spine are positioned at roughly a 35° angle to the horizontal, while the zygapophysial joints in the lower cervical region align at about 65° (Oliver and Middleditch 1991). This alignment remains constant through later life stages. The seventh cervical vertebra exhibits a similar developmental pattern but retains a more vertical orientation throughout life compared to other cervical vertebrae. The horizontal orientation of articular processes in children, relative to adults, contributes to greater anterior-to-posterior translational movement among typical cervical vertebrae (Kasai et al. 1996). At birth, the articular processes of typical cervical vertebrae, which are lined with hyaline cartilage on their facets, display a more horizontal orientation with an average angle of about 34° relative to the horizontal plane. This orientation changes as they gradually shift to a more vertical position by the age of approximately 10 years, at which point the angle reaches an average between 45° and 53°. The facet joints in the upper cervical spine are positioned at roughly a 35° angle to the horizontal, while the zygapophysial joints in the lower cervical region align at about 65° (Oliver and Middleditch 1991). This alignment remains constant through later life stages. The seventh cervical vertebra exhibits a similar developmental pattern but retains a more vertical orientation throughout life compared to other cervical vertebrae. The horizontal orientation of articular processes in children, relative to adults, contributes to greater anterior-to-posterior translational movement among typical cervical vertebrae (Kasai et al. 1996).

The morphology of cervical zygapophysial joints evolves notably over the lifespan. In individuals younger than 20 years, the articular cartilage is uniformly smooth and maintains a thickness ranging from 1.0 to 1.3 mm, while the underlying subarticular bone demonstrates a consistent thickness. As aging progresses, the articular cartilage gradually becomes thinner. In adults, the cervical articular facets are typically characterized by a very thin cartilage layer, which lies over subarticular cortical bone that has become irregularly thickened.

The capsules of the zygapophysial joints are generally considered to have minimal impact on restricting normal movement in the cervical area (Onan et al. 1998). Nonetheless, the anterior parts of these capsules play a crucial role in limiting hyperextension. These capsules are notably thicker on the lateral sides than on the posterior, and they progressively increase in thickness from the posterior to the anterior along the lateral aspect, culminating in the anterolateral area being the thickest part of the capsule. Despite these variations, the articular capsules of the cervical zygapophysial joints are overall thin (Panjabi et al. 1989) and are longer and more flexible compared to those in the thoracic and lumbar regions.

When cervical vertebrae are interconnected, the articular processes on each side combine to form a structure known as the articular pillar, which protrudes laterally at the pediculolaminar junction of each vertebra (Standring 2020). This structure is prominently visible on lateral X-ray images. The cervical articular pillars are essential for supporting the head and neck's weight. As a result, the cervical region's weight-bearing capacity is distributed across three longitudinal columns: an anterior column through the vertebral bodies and two posterior columns via the right and left articular pillars (Louis 1985). A reduction in intervertebral disc height typically increases the load on the cervical articular facets (Hussain et al. 2010a, b, 2012). Articular pillar fractures are relatively common in the cervical spine and are often overlooked. These fractures are usually chip fractures of a superior articular facet and process. Affected patients may initially experience transient radicular pain, which often progresses to mild to intense neck pain. Persistent radiculopathy suggests that the displaced fractured facet and articular process impinge on the posterior root within the intervertebral foramen, causing ongoing symptoms (Czervionke et al. 1988).

Spondylolysis in a cervical vertebra involves a complete separation between the superior and inferior articular processes, a location analogous to the pars interarticularis in a lumbar vertebra. This condition may occur either unilaterally or bilaterally. While spondylolysis is a frequent finding in the lumbar spine, often regarded as a type of fracture, it is exceedingly rare in the cervical spine. The anomaly has been observed in every cervical vertebra except C1, with C6 being the most frequently affected, followed by C4 and C5 vertebrae in terms of occurrence (Paik 2010).

The cervical zygapophysial joints are innervated by the medial branches of two adjacent spinal segments, except for the C2–C3 joint, which is uniquely innervated solely by the third occipital nerve via a connection to the C2 posterior primary division (Bogduk 2005). This unique innervation pattern suggests the role of the zygapophysial joint capsules in transmitting information about joint position and tissue damage to the central nervous system (McLain 1994). Zygapophysial joints are implicated as a significant source of pain in many cases of both acute and chronic neck pain, with estimates suggesting that they contribute to pain in about 50% of individuals with chronic neck conditions (Aprill and Bogduk 1992; Bogduk 2005). The primary patterns of pain referral include neck pain and headaches originating from the C2–3 zygapophysial joints and neck and shoulder pain emanating from the C5–6 zygapophysial joints (Bogduk and Marsland 1988).

6.5.3.5 Anatomical Characteristic and Biomechanical Implications of Cervical Intervertebral Foramina

In the cervical spine, the intervertebral foramina are defined by the superior and inferior vertebral notches of adjacent cervical vertebrae. These foramina are oriented anteriorly in an oblique manner, angled approximately 45° from the midsagittal plane in the upper and middle cervical regions and about 55° in the lower cervical regions. Additionally, these openings are positioned inferiorly at a 10° angle relative to a horizontal plane that aligns with the superior vertebral end plates. Unique to the cervical spine, the uncinate processes form the anterior boundaries of these foramina and possess an almost oval configuration. The vertical dimension of the foramina progressively enlarges from the C3–4 to the C7–T1 levels. The size of the intervertebral foramina at C2–3 closely resembles those found in the middle to lower cervical areas (Ebraheim et al. 1996).

The dimensions of the intervertebral foramina are influenced by movement. For instance, flexion from a neutral position increases the vertical diameter of these foramina, whereas extension reduces their size (Yoo et al. 1992; Muhle et al. 2001). Additionally, applying therapeutic traction is known to enlarge the area and height of the intervertebral foramina—showing increases of 18.9% and 10.4%, respectively—when a traction weight of 15 kg is used (Liu et al. 2008). The internal pressure of the intervertebral foramina rises during extension, while shoulder abduction leads to a reduction in this pressure (Farmer and Wisneski 1994). Such a decrease in pressure, evident during upper extremity abduction, aligns with the positive Bakody's sign, which is the alleviation of distal upper extremity pain through arm abduction. Moreover, ipsilateral axial rotation at 20 and 40° can cause up to 15% and 23% narrowing of the foramina, respectively, whereas contralateral rotation at these angles results in foraminal expansion of up to 9% and 20%, respectively (Muhle et al. 2001).

The cross-sectional area of a typical cervical intervertebral foramen includes the spinal nerve along with its posterior and anterior roots and the dural sleeve, constituting 35–50% of its total area (Sunderland 1974a, b). In a neutral cervical spine position, these posterior and anterior roots are positioned in the lower section of the intervertebral foramen, at or below the level of the disc (Pech et al. 1985). Components such as epidural fat and blood vessels occupy the upper part of the foramen. Positioned posteriorly and slightly above the anterior root, the posterior root comes into contact with the superior articular process, while the dorsal root ganglion sits near a small notch on the anterior face of this process in the cervical area. The anterior root, meanwhile, aligns with the uncinate process. Separating the posterior and anterior roots, adipose tissue acts as a cushion within the intervertebral foramen (Pech et al. 1985).

6.5.3.6 Characteristic Features of the C6 Vertebra

The C6 vertebra, while generally typical, features distinct anterior tubercles on its transverse processes, known as carotid tubercles. These tubercles are particularly notable due to their proximity to the common carotid arteries on either side, allowing for potential compression of these arteries in the groove formed with the vertebral body of C6 (Standring 2020).

Moreover, the articular pillars of the C6 vertebra differentiate from those in other cervical vertebrae by three specific characteristics observable in both pediatric and adult vertebrae, with one feature unique to adults. Primarily, the superior articular processes and facets are positioned significantly more anterior compared to the inferior ones, which contrasts with the alignment seen in C3–C5, where they are nearly directly aligned. Secondly, the sulcus for the posterior primary division at C6 forms a pronounced incisure, visible from lateral and posterior aspects. Lastly, adults exhibit a notable "muscular process" at the lower end of the posterior edge of the C6 articular pillars, just below the incisure for the posterior primary division.

6.5.4 Characteristic Features of the C7 Vertebra

6.5.4.1 Structure of the C7 Transverse Processes

The seventh cervical vertebra (C7), also known as vertebra prominens, exhibits distinctive characteristics in its transverse processes. Notably, the anterior tubercle on each C7 transverse process is small and short, while the posterior tubercle's larger size results in an overall enlarged transverse process (Standring 2020). The anterior tubercle represents the costal element of C7 and arises uniquely from a separate primary ossification center, which typically fuses with the transverse process by the age of 5 or 6 years. Occasionally, this element may persist as a separate entity and develop into a cervical rib, a condition that can also occur at vertebrae C4 to C6, although less frequently. The area between the tubercles, known as the intertubercular lamella, often features a groove from the anterior ramus of the C7 nerve anteriorly and laterally to the transverse process foramen. Additionally, the posterior tubercle serves as an attachment point for the suprapleural membrane—also referred to as the cupola. This layer of connective tissue fortifies the apical pleura, protecting the lung's apex (Standring 2020).

In terms of vascular anatomy, the foramen within the C7 transverse processes, which is typically the smallest among the cervical spines, may occasionally present as a double foramen. These foramina are pathways for branches of the stellate ganglion, although commonly they only carry accessory arteries and veins, specifically from branches of the deep or ascending cervical arteries, along with their accompanying veins. The rest of the space within the C7 transverse process foramen is occupied by areolar connective tissue. Unlike at the C6 level and above, where the vertebral artery and accompanying venous and sympathetic structures typically pass, these structures traverse the C7 foramen approximately 5% of the time (Jovanovic 1990; Taitz et al. 1978).

6.5.4.2 Structure of the C7 Spinous Process

The seventh cervical vertebra, distinguished as the vertebral prominens, possesses a distinct spinous process that sets it apart from other cervical vertebrae. This spinous process exhibits the most pronounced projection in the cervical region, although in some cases the C6 spinous process may be more prominent (Kapandji 2008). C6 is notable as the most caudal cervical vertebra, demonstrating palpable motion during cervical flexion and extension maneuvers. Additionally, the first thoracic vertebra (T1) may occasionally present with a spinous process more prominent than that of C7 (Miele et al. 2012). The C7 spinous process is characterized by its direct posterior projection, unlike the usually bifid spinous processes found in other cervical vertebrae. The ligamentum nuchae's funicular part, which plays an important role in neck muscle attachment, connects at the singular posterior tip of the C7 spinous process (Miele et al. 2012). The prominent spinous process of C7, positioned at the cervicothoracic junction, functions as an important attachment site for numerous muscles, including the rhomboid minor, trapezius, and serratus posterior superior (Cramer and Darby 2014).

6.5.5 Biomechanically Important Ligaments of the Cervical Spine

6.5.5.1 Characteristics and Functional Significance of the Anterior Longitudinal Ligament

The anterior longitudinal ligament is a broad ligamentous structure that spans the anterior aspect of the vertebral bodies and intervertebral discs, extending from the occiput to the sacrum (Cramer and Darby 2014) (Fig. 6.16). At its superior attachment, the anterior longitudinal ligament forms a thickened, cord-like structure that fastens to the axis body and the atlas's anterior tubercle. Some of the atlantal fibers of the anterior longitudinal ligament spread laterally, attaching to the inferior aspect of the atlas's anterior arch. Superiorly, the anterior longitudinal ligament becomes continuous with the anterior atlanto-occipital membrane's medial portion (Cramer and Darby 2014). The width of the anterior longitudinal ligament at the C1–2 level is about 3.8 mm, broadening to 7.5 mm from C3 to T1 (Cramer and Darby 2014). In various spinal sections, the anterior longitudinal ligament exhibits differing attachment patterns. It typically adheres extensively to the vertebral bodies' superior and

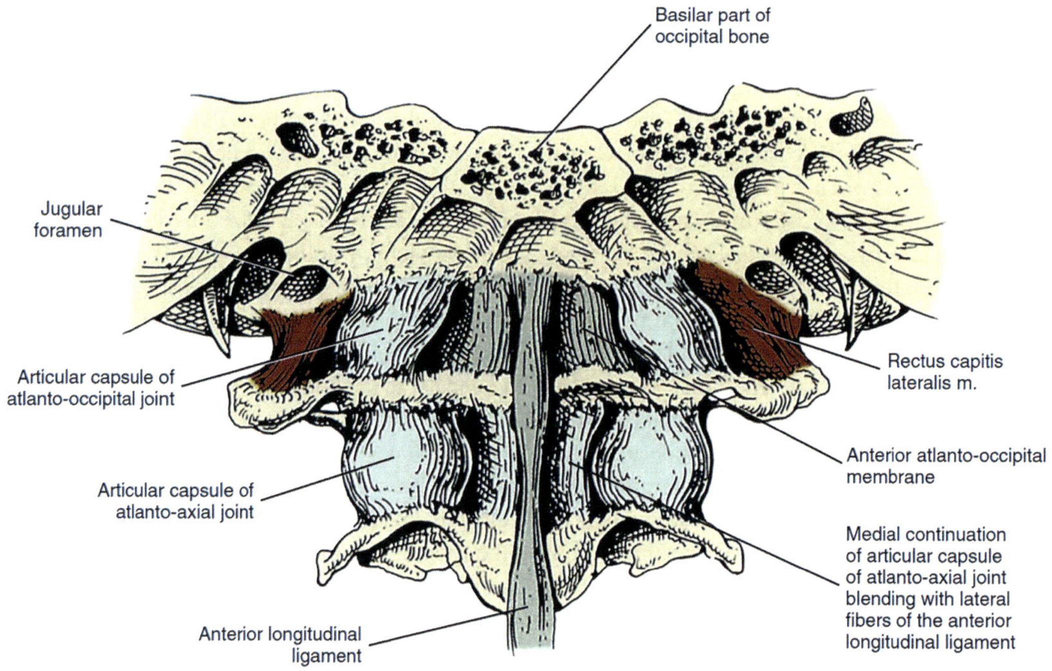

Fig. 6.16 Ventral view of the occiput, atlas, axis, and associated ligaments. The anterior longitudinal ligament narrows considerably between the atlas and the occipital bone, blending laterally with the anterior atlanto-occipital membrane. The articular capsules of the atlanto-occipital and lateral atlanto-axial joints are also clearly depicted. (Adapted with permission from Cramer and Darby (2014))

inferior bony end plates. Specifically, in the cervical area, it attaches firmly to the vertebral bodies' central region, less frequently in the center of the thoracic or lumbar vertebral bodies. The distinction between the ALL and the anterolateral fibers of the annulus fibrosus can be obscure laterally (Cramer and Darby 2014).

The anterior longitudinal ligament is composed of multiple layers; its superficial fibers span multiple vertebrae, whereas the deep fibers link adjacent vertebrae (Cramer and Darby 2014). The ligament generally exhibits greater thickness at the vertebral bodies than over the intervertebral discs, contributing to the anterior vertebral bodies' smoother contour. Primarily, the anterior longitudinal ligament restricts spinal extension and is often compromised during cervical extension injuries. In such injuries, the anterior annulus fibrosus tends to rupture before the anterior longitudinal ligament, given that the annulus fibrosus fibers are shorter and thus reach their tension limits earlier. Ossification of the anterior longitudinal ligament is another notable aspect, leading to decreased cervical flexibility. This ossification is frequently associated with the development of a fibrous mass behind the odontoid process, known as a retro-odontoid pseudotumor (Chikuda et al. 2009).

6.5.5.2 Characteristics and Functional Significance of the Posterior Longitudinal Ligament

The posterior longitudinal ligament extends as a downward continuation of the tectorial membrane, originating from the posterior side of the C2 vertebral body and extending to the sacrum and potentially to the coccyx (Behrsin and Briggs 1988). This ligament, along with the anterior longitudinal ligament, displays comparable tensile characteristics, each able to withstand similar loads on the spine; however, the anterior ligament restricts extension forces, whereas the posterior ligament counteracts flexion forces (Przybylski et al. 1996).

In its morphology, the posterior longitudinal ligament is broader and more uniformly shaped in the cervical and upper thoracic spine, exhibiting a thickness three to four times greater in the cervical area than in the thoracic and lumbar regions (Bland 1983). It comprises superficial fibers that stretch across multiple vertebrae, with deeper fibers running between adjacent vertebrae. Research by Panjabi et al. (1991a, b) and Bland (1983) has shown that the ligament is firmly anchored to the vertebral bodies and intervertebral discs in the cervical spine, suggesting its role in preventing posterior disc protrusion. Although firmly attached along the vertebral bodies in the cervical spine (Przybylski et al. 1998), it loosely connects in the central vertebral regions to facilitate the basivertebral veins' exit (Standring 2020). Notably, the ligament narrows over the vertebral bodies and broadens significantly over the intervertebral discs in the thoracic and lumbar regions.

Pathologically, the ligament may undergo ossification, predominantly in the cervical area and less frequently in the lumbar region. This ossification of the posterior longitudinal ligament (OPLL) is of clinical significance due to its potential to compress the spinal cord with minor trauma in the cervical area and cause radicular symptoms in the lumbar area (Chikuda et al. 2009). Primarily found in middle-aged and older Japanese men, the prevalence of OPLL, alongside ossifications of other spinal ligaments, is notably higher in this demographic, with a significant association also seen with diffuse idiopathic skeletal hypertrophy (DISH), a condition correlated with obesity and metabolic issues such as glucose intolerance (Matsunaga et al. 1999; Shingyouchi et al. 1996). Furthermore, high-salt, low-meat diets of Japanese and Taiwanese populations may contribute to OPLL (Wang and Thambuswamy 2011).

The posterior longitudinal ligament is richly innervated with nociceptive and vasomotor fibers, contributing to its sensitivity and involvement in regional blood flow, particularly following ligamentous injuries (Imai et al. 1995).

6.5.5.3 Characteristics and Functional Significance of the Ligamentum Flavum

The ligamenta flava consist of paired structures positioned bilaterally between the laminae of adjacent vertebrae, extending from C1–2 to L5–S1 (see Fig. 6.12). The posterior atlanto-occipital membrane serves as the ligamenta flava's equivalent at the occiput-C1 level. Originating as thin bands in the cervical region, these ligaments progressively thicken toward the lumbar area (Panjabi et al. 1993).

Positioned from the anterior-inferior edge of one vertebral lamina to the posterior superior edge of the lamina below, the ligamentum flavum increases in length from C2–3 to C7–T1, reflecting the increasing distance between laminae at these levels. Laterally, the ligament assists in stabilizing the anterior part of the zygapophyseal joint capsule. The ligamenta flava frequently merge with each other and with the interspinous ligament. This integration allows for the formation of small gaps between the left and right ligaments, facilitating the movement of veins connecting the internal and external vertebral venous plexuses. Notably, the ligamentum flavum between C1 and C2 is typically thinner and perforated by the C2 spinal nerve, as identified in a study by Panjabi et al. (1993) on six cervical spines, where this specific ligament was sometimes absent.

The ligamentum flavum, notable for its high elastin content, possesses significant elasticity essential for spinal flexion and extension (Yahia et al. 1990). This elasticity plays a critical role in managing the end range of spinal flexion and prevents the ligament from protruding into the spinal canal during extension movements (Kosaka et al. 2007). In its neutral position, the ligamentum flavum exhibits a "pretension" due to the abundance of elastin, resulting in the ligament being elongated by 15% compared with its length when untensioned (Nachemson and Evans 1968). Consequently, even in a neutral position, the ligament remains slightly stretched.

During fetal development and early childhood, the elastin percentage in the ligamentum flavum is at its peak (Postacchini et al. 1994).

As aging progresses, there is a notable decrease in this pretension, accompanied by a disorganization of elastic fibers (Kosaka et al. 2007). Aging also brings an increase in chondrocyte numbers and small calcified areas at the ligament's bone attachment sites. In cases of advanced age and degeneration, up to 80% of the elastic fibers are replaced with collagen and chondrocytes, significantly altering the ligament's composition and function (Postacchini et al. 1994; Yahia et al. 1990).

With advancing age or following trauma, the ligamentum flavum often thickens and may undergo calcification or fat infiltration (Ho et al. 1988). These alterations are likely to diminish the ligament's elasticity. Consequently, the ligament may buckle into the vertebral canal or the medial side of the intervertebral foramen, leading to a narrowing of these spaces. Consequently, the buckling of the affected ligamentum flavum can exacerbate the narrowing of these spaces, potentially causing compression of the neural structures within. Particularly in the thoracic and thoracolumbar spine, ossification of the ligamentum flavum frequently occurs and might compress the posterior portion of the spinal cord or the nerve roots exiting the spine (Hasue et al. 1983).

6.5.5.4 Characteristics and Functional Significance of the Cruciate Ligament

The cruciate ligament of the atlas, also known as the cruciform ligament, is aptly named due to its cross-shaped configuration. This ligament can be anatomically divided into three distinct components: a transverse ligament, a superior longitudinal band, and an inferior longitudinal band (Fig. 6.17).

Transverse Atlantal Ligament of the Cruciate Ligament

The transverse atlantal ligament, also referred to as the transverse component of the cruciate ligament, is regarded as the most important ligament in stabilizing the upper cervical spine (occiput-C1–C2 joint complex) (White and Panjabi 1990). This key structural component spans from the medial tubercle on one lateral mass of the atlas across to the corresponding tubercle on the opposite side, aligning in a horizontal plane. Notably, it forms an approximate 21° angle with the frontal plane, extending from its origin to where it curves behind the odontoid process (Panjabi et al. 1991a, b). This ligament courses about 5 mm below the point where the odontoid's apical and the alar ligaments attach (Ebraheim et al. 1997a, b). At the center, where it traverses behind the odontoid process, the ligament's width is maximized. A fine fibrocartilage layer anteriorly lines the transverse ligament of the atlas, facilitating the formation of a diarthrodial joint as it passes behind the odontoid (Ebraheim et al. 1997a, b).

The transverse atlantal ligament facilitates the pivoting motion of the atlas on the axis (Cramer and Darby 2014). Moreover, it maintains the atlas in its proper position, thereby preventing spinal cord compression during flexion of the head and neck. The transverse atlantal ligament fits into the groove on the posterior surface of the odontoid, ensuring the atlas remains in its correct position even when all other ligaments are severed (Standring 2020). Transection of the transverse atlantal ligament can result in an increase in the atlanto-odontal interspace from the normal 3 mm to as much as 8 mm (Menezes and Traynelis 2008). According to research by Panjabi et al. (1991a, b), the ligament is differentiated into superficial and deep layers, enhancing its structural integrity.

The area of the transverse atlantal ligament that passes posterior to the odontoid is particularly unique due to the encapsulating fibrous capsule around the fibrocartilage facet, forming a synovial joint with the odontoid's posterior articular facet. The biochemical composition of this fibrocartilage includes typical articular cartilage components like link protein, aggrecan, and type II collagen, which can become autoimmune targets in rheumatoid arthritis. This autoimmune reaction may lead to the erosion of the transverse ligament, contributing to the cervical instability commonly observed in many patients with this condition (Milz et al. 2001). Moreover, the reduced regenerative capacity of fibrocartilage compared to ordinary ligaments might explain

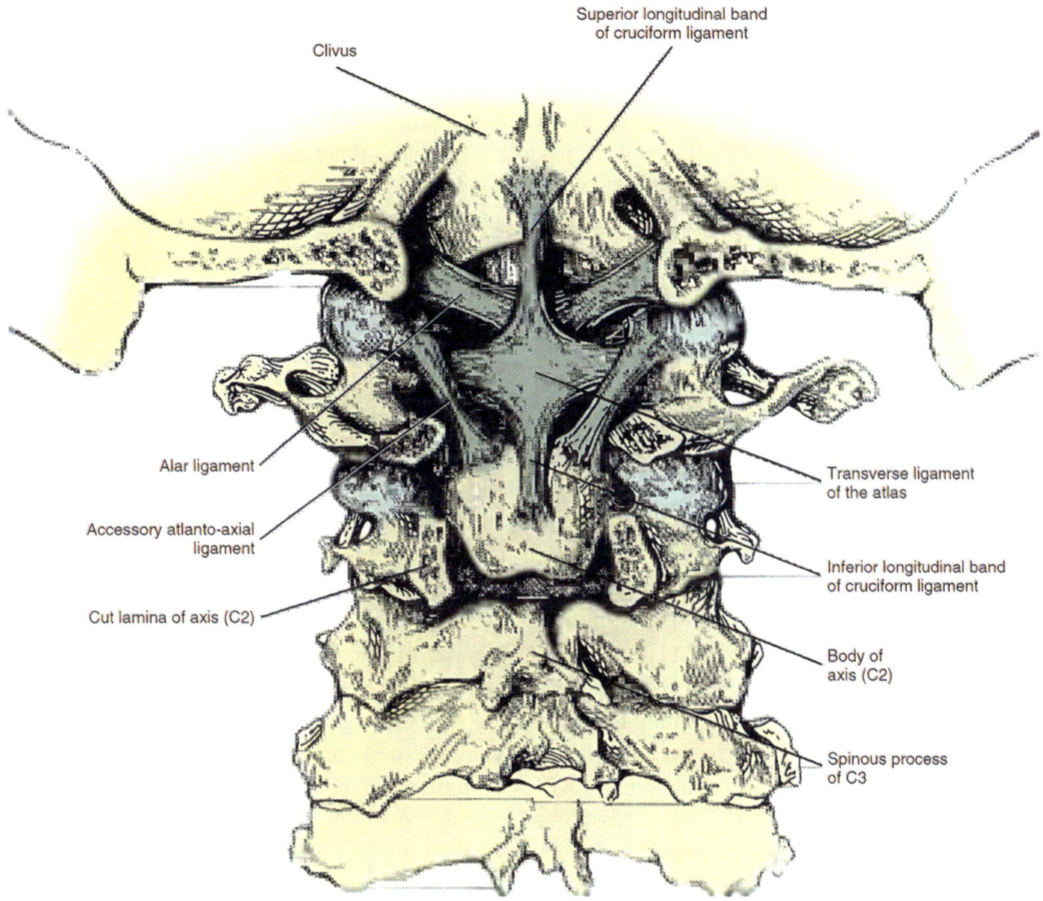

Fig. 6.17 Ventral view of the vertebral canal and foramen magnum, observed from a posterior perspective. The tectorial membrane has been removed, revealing several upper cervical ligaments can be seen. The centrally located cruciform ligament is visible, featuring narrow superior and inferior longitudinal bands and its stout transverse ligament. The alar and accessory atlanto-axial ligaments are also clearly displayed. (Adapted with permission from Cramer and Darby (2014))

the frequent occurrence of calcium deposits or ossification in the transverse atlantal ligaments of arthritic atlanto-axial joints.

Superior Longitudinal Band of the Cruciate Ligament

The superior longitudinal band, a component of the cruciate ligament, extends from the transverse atlantal ligament to the anterior margin of the foramen magnum, specifically attaching to the superior aspect of the basilar portion of the occipital bone (clivus) (Tubbs et al. 2011). The superior longitudinal band is situated between the apical ligament of the odontoid process anteriorly and the tectorial membrane posteriorly. While it contributes to limiting both the flexion and extension movements of the occiput, its principal role is to maintain the correct positioning of the transverse atlantal ligament (Dvorak and Panjabi 1987). Consequently, this ensures that the atlas is properly aligned with the odontoid process, which is important in maintaining optimal joint function and spinal stability (Saldinger et al. 1990).

The superior longitudinal band's unique anatomical position and attachments enable it to perform its stabilizing function within the complex ligamentous network of the craniocervical junction. The band's attachment to the clivus provides a stable anchor point, while its connection to the transverse ligament allows for the transmission of forces necessary to maintain the integrity of the atlanto-axial joint (Osmotherly et al. 2013). The relationship among the components of the superior longitudinal band, apical ligament, and tectorial membrane forms a balanced system that enables controlled movement and stability of the occiput and upper cervical spine (Dickman et al. 1995; Panjabi et al. 1991a, b).

Inferior Longitudinal Band of the Cruciate Ligament

The inferior longitudinal band, a component of the cruciform ligament, plays a significant role in maintaining stability at the craniocervical junction. This ligamentous structure connects the transverse ligament of the atlas to the anterior aspect of the C2 vertebral body (Tubbs et al. 2007). By anchoring the transverse atlantal ligament inferiorly, the inferior longitudinal band prevents excessive superior migration of the transverse ligament during cervical spine movements (Osmotherly et al. 2013). Furthermore, this band contributes to limiting excessive flexion of the occiput and atlas relative to the axis, working in conjunction with the superior band of the cruciform ligament and the transverse atlantal ligament itself.

6.5.5.5 Characteristics and Functional Significance of the Alar Ligaments

The alar ligaments, originating from the posterior and lateral aspects of the odontoid process and covering its entire posterior surface, extend laterally to insert onto the roughened areas of the medial surfaces of the occipital condyles on corresponding sides. The spatial arrangement of these ligaments varies, with certain fibers extending from inferior to superior, others maintaining a horizontal alignment, and, less frequently, taking a craniocaudal direction. This directional variation is influenced by the length of the odontoid process, with longer processes leading to more horizontal and subsequently craniocaudal orientations. The mean angle that results from the left and right alar ligaments ranges from 140° to 180°, and these ligaments typically exhibit asymmetry in their alignment with the coronal and horizontal planes (Pfirrmann et al. 2001). These pencil-width ligaments are characterized by their strength.

Functionally, the alar ligaments have an important role in limiting contralateral axial rotation of the cervical spine, although the exact function is still not completely understood (Dvorak and Panjabi 1987). Specifically, the left alar ligament restricts right rotation by the combined actions of fibers attached posterior to the axis of movement. These fibers operate together with the fibers of the right alar ligament, which are positioned anterior to this axis. This arrangement similarly restricts left rotation via right posterior and left anterior fibers (Standring 2020).

In addition to their role in motion limitation, the alar ligaments serve as secondary stabilizers of the upper cervical spine following damage to the tectorial membrane and cruciate ligaments. They are particularly susceptible to injury from movements combining axial rotation and flexion, as might occur in a motor vehicle accident when a driver looks back during a frontal collision. Damage to these ligaments, whether through tearing or stretching, leads to increased rotational and lateral displacement at the atlanto-occipital and atlanto-axial joints (Dvorak and Panjabi 1987).

6.5.5.6 Characteristics and Functional Significance of the Interspinous Ligaments

Interspinous ligaments are a group of ligaments connecting the spinous processes of adjacent vertebrae, extending from C2–3 to L4–5. In the cervical region, these ligaments are considered to be the anterior aspect of the ligamentum nuchae. Typically, in the cervical area, the interspinous ligaments are not well developed, often presenting as a thin, membranous, and translucent structure (Panjabi et al. 1993). In contrast, thoracic interspinous ligaments are characterized by their short height (superior to inferior) and broad width (anterior-to-posterior), while those in the lumbar region assume a more rectangular configuration

(Standring 2020). The thoracic interspinous ligaments are better developed, indicating adaptations to different mechanical demands across the spinal regions.

6.5.6 Vertebral Artery: Anatomical Characteristics and Variations

Originating as the initial branch of the subclavian artery, the vertebral artery enters the sixth cervical vertebra (C6) through its transverse foramen and ascends via the transverse foramina of the successive cervical vertebrae. The pathway of the nerve involves passing medially at C1 around the superior articular process of the atlas. It then goes beneath the posterior atlanto-occipital membrane before penetrating both the dura and arachnoid (see Fig. 6.12). The artery ascends via the foramen magnum and merges with another artery on the opposite side, forming the basilar artery.

The left and right vertebral arteries often exhibit asymmetry, with a typical diameter of 4.5 mm. The larger blood vessel is referred to as the "dominant artery," while the smaller one is called the "minor artery". A study reveals that there is a 59.2% prevalence of asymmetry in these arteries. Specifically, the left artery is dominant in 35.8% of cases, while the right artery is dominant in 23.4% of cases. A study conducted on 1013 Korean patients using CT angiography found that the asymmetry rate was reduced by 32.1%. The left vertebral artery was dominant in 22.3% of the patients, while the right vertebral artery was dominant in 9.8% (Hong et al. 2008). Research conducted by Peterson et al. (2010) indicates that females have a higher rate of asymmetry compared to males. Occasionally, the minor artery may be considerably smaller or completely missing, usually compensated for by a well-developed anastomosis.

6.5.6.1 The First Part of the Vertebral Artery: Anatomical Course and Relationships

The vertebral artery is segmented into four distinct parts, with the initial segment originating from the subclavian artery and extending to the transverse process of the C6 vertebra (Standring 2020). This segment (prevertebral segment) courses between the longus colli and the scalenus anterior muscles before arriving at the C6 transverse process. Half of the vertebral arteries exhibit some degree of tortuosity in their first part (Taitz and Arensburg 1989). Although there is ongoing discussion regarding whether vertebral artery tortuosity may reduce blood flow to the associated structures, no clinical consequences have been definitively related to mild or moderate tortuosity thus far. Notably, anomalies at the origin of the vertebral artery are uncommon, with the most frequent being an origin from the aortic arch, occurring in approximately 4% of cases. These anomalous arteries usually arise between the left common carotid and left subclavian arteries (Lemke et al. 1999).

Accompanying the first segment of the vertebral artery are multiple venous branches that converge to form the vertebral vein in the lower cervical area. Additionally, it is joined by a prominent branch and several smaller branches originating from the posteriorly situated inferior cervical ganglion or, when present (80% of cases), the cervicothoracic ganglion, also known as the stellate ganglion. These branches create a neural plexus surrounding the vertebral artery (Standring 2020).

6.5.6.2 The Second Part of the Vertebral Artery: Anatomical Course and Relationships

The vertebral artery's second segment advances superiorly via the transverse foramina of the C6 to C1 vertebrae. In the majority of instances (89.8%), it primarily enters the transverse foramen at C6, although there can be variations in entry locations, such as C5 at 6.3% and C7 at 3.0%. Usually, the entry is symmetrical on both sides in 85% of all cases and asymmetrical in 15%, with just a few cases when arteries enter more than two levels apart. Generally, the lower entry point is observed on the right side (Gluncic et al. 1999). Occasionally, when originating from the aortic arch, the artery tends to enter the transverse foramen at higher cervical levels, such as

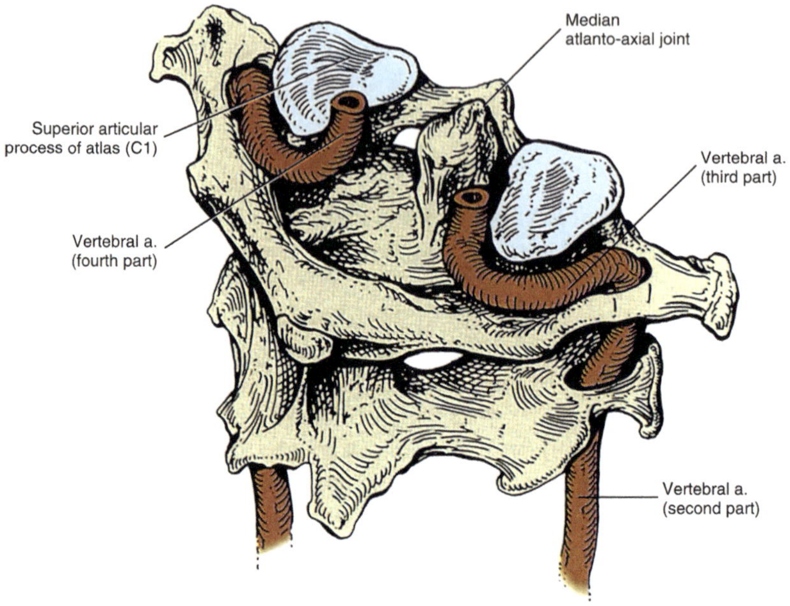

Fig. 6.18 The second, third, and fourth segments of the left and right vertebral arteries. Each vertebral artery courses laterally between C2 and C1, then passes posteriorly and medially at the level of C1. Finally, it ascends superiorly and medially to enter the foramen magnum. (Adapted with permission from Cramer and Darby (2014))

C5 or C4. As it ascends through these foramina, the artery is located roughly 1 mm posterior and lateral to the uncinate processes of the C2 to C6 vertebrae. The proximity of the vertebral artery to the uncinate processes decreases progressively from C6 upward to C3. Such anatomical closeness may make the vertebral artery vulnerable to compression by osteophytes at the uncinate processes within the mid-cervical area (Ebraheim et al. 1997a, b). Furthermore, the second segment of the vertebral artery courses anterior to the cervical spinal nerves and their anterior rami of C2–C6, which move from a medial to lateral direction in their respective cervical transverse process grooves.

The second segment of the vertebral artery often exhibits tortuosity between transverse processes, a condition that intensifies with advancing age. It remains unclear whether this feature is inherent, develops due to atherosclerosis, or results from a combination of factors. Severe intervertebral disc degeneration, occasionally associated with cervical spondylotic myelopathy, can lead to a reduction in the intervertebral space (Oga et al. 1996). This diminished distance between vertebrae may, in some cases, contribute to an increase in the tortuosity of the vertebral artery's second part (Oga et al. 1996). Occasionally, this condition may lead to the expansion of the transverse process foramen, although such cases are rare (Taitz and Arensburg 1989). In instances of severe cervical trauma, where an anterior dislocation of a vertebra may occur, there could be a total blockage of arterial flow through this section of the vertebral artery (Giacobetti et al. 1997).

Under normal conditions, the vertebral artery makes a significant lateral deviation—typically around 45° but reaching up to 90°—after it exits the foramen of the transverse process of the axis (Fig. 6.18). This deviation enables the artery to approach the laterally located transverse process of the atlas. Taitz and Arensburg (1989) observed considerable kinking or tortuosity in 11% (4 out of 36) of the vertebral arteries at this specific foramen.

The exact hemodynamics of blood flow through the left and right vertebral arteries within the atlanto-axial region remain incompletely understood. According to Taitz et al. (1978), it was believed that head extension coupled with rotation to one side typically compromised blood flow through the contralateral second part of the vertebral artery, primarily due to constriction between the atlas and axis. Haynes et al. (2002) found that atlanto-axial rotation generally does not result in the compression or narrowing of the vertebral artery. Li et al. (1999) observed a reduction in blood flow in both vertebral arteries during extreme extension and extension combined with rotation.

6.5.6.3 The Third Part of the Vertebral Artery: Anatomical Course and Relationships

The vertebral artery's third segment follows a characteristically tortuous path, commencing as it traverses the transverse process foramen of the atlas vertebra. At this point, it is situated posterior and medial to the rectus capitis lateralis muscle. The vessel then curves posteriorly and medially around the superior articular process of the atlas.

Upon reaching the posterior arch of the atlas, the vertebral artery lies in the groove specifically designed for it on the posterior arch. In this area, the posterior ramus of the first cervical nerve, also known as the suboccipital nerve, runs between the vertebral artery and the posterior arch of the atlas (Tubbs et al. 2011). Subsequently, the artery travels inferior to the posterior atlanto-occipital membrane. Occasionally, this membrane forms an ossified bridge over the artery, extending from the posterior arch of the atlas to the lateral mass (Muralimohan et al. 2009).

6.5.6.4 The Fourth Part of the Vertebral Artery: Anatomical Course and Relationships

The vertebral artery's fourth segment commences as the artery traverses beneath the bridge of the posterior atlanto-occipital membrane. From there, it progresses anteriorly, superiorly, and medially, ascending along the anterolateral aspect of the spinal dura mater (Standring 2020). Upon reaching the level of the foramen magnum, each vertebral artery penetrates the dura mater and arachnoid mater, continuing its ascent within the subarachnoid space. As the left and right vertebral arteries move medially along the anterior aspect of the medulla and inferior pons, they converge to form the basilar artery. This basilar artery ascends along the anterior aspect of the pons and midbrain, contributing to the cerebral arterial circle (circle of Willis) by bifurcating into the left and right posterior cerebral arteries (Waxman 2017).

Throughout its path, each vertebral artery generates several branches: muscular branches, osteoarticular branches, meningeal branches, and spinal rami. These spinal rami enter the intervertebral foramina, supplying the vertebral body, posterior arch, and soft tissue structures within the vertebral canal, including nerve roots and rootlets (Cramer and Darby 2014). Importantly, each vertebral artery forms a branch that, combined with its corresponding counterpart from the opposite side, forms a single anterior spinal artery that supplies the anterior aspect of the spinal cord along its entire length. Additionally, each vertebral artery produces a posterior spinal artery, which courses along the posterior aspect of the spinal cord. Moreover, each vertebral artery forms a posterior inferior cerebellar artery, supplying parts of the medulla and the inferior aspect of the cerebellum (Mtui et al. 2020).

Upon uniting at the level of the inferior pons, the right and left vertebral arteries form the basilar artery. This artery supplies the left and right posterior superior cerebellar arteries, which course posteriorly around the medulla to supply the medulla, inferior pons, and a significant portion of the cerebellum. The basilar artery also extends the left and right internal auditory arteries to the internal acoustic meatus of each side, supplying the middle and inner ear structures. As it ascends along the anterior pons, the basilar artery dispatches numerous small pontine branches to the pons, followed by the left and right superior cerebellar arteries. These arteries course posteriorly, supplying the superior pons and the inferior aspect of the cerebral peduncles of the midbrain before continuing to supply the superior aspect of the cerebellum. The basilar artery ends at the superior pons by dividing into the left and right posterior cerebral arteries.

These arteries join the circle of Willis and provide blood supply to the occipital lobes and the inferior portion of the temporal lobes of the cerebral cortex (Snell 2009).

6.5.6.5 Structural Differences Between the Intracranial and Extracranial Segments of the Vertebral Artery

Intracranial arteries have distinctly thinner walls compared to extracranial arteries of a similar size, primarily due to the differences in the media and adventitial layers, which are narrower and contain a lower quantity of elastic fibers (Walmsley 1983). Wilkinson (1972) observed that the vertebral artery's structure underwent significant changes just before it penetrates the dura mater. Before this penetration, the vertebral artery's structure was similar to that of most extracranial vessels. Beyond this point, its structure aligned more closely with intracranial vessels. Specifically, Wilkinson (1972) noted that the adventitia and media of the intracranial segment of the vertebral artery were thinner and contained fewer elastic fibrils. The external elastic lamina in this region was either entirely absent or consisted of only sparse, single elastic fibrils. In contrast, the intima and internal elastic lamina of the vertebral artery were similar in both its extracranial and intracranial segments. These observations represent the findings that reveal the variations in the structure between the intracranial and extracranial segments of the internal carotid artery.

The structural differences between intracranial and extracranial arteries have important implications for their function and susceptibility to pathological processes. The thinner walls and reduced elastic fiber content of intracranial arteries may render them more vulnerable to damage and less capable of withstanding hemodynamic stresses (Lee 1995). These structural features may also contribute to the higher incidence of certain cerebrovascular disorders, such as aneurysms and arterial dissections, in the intracranial arteries compared to their extracranial segments (Schievink 2001).

6.5.7 Anatomical Organization and Clinical Relevance of Cervical Nerve Roots

The cervical spine demonstrates distinct anatomical patterns in its nerve root organization. Within the cervical region, posterior and anterior rootlets emerge from the spinal cord and subsequently unite to form posterior and anterior roots. Studies have documented a progressive increase in nerve root length from the C4 to C8 levels (Yabuki and Kikuchi 1996).

The anatomical positioning of dorsal root ganglia within the intervertebral foramina (intraforaminal) shows considerable variation, with these structures occupying different positions along the proximal (intraspinal)–distal (extraforaminal) axis. This variable positioning holds clinical significance, as age-associated degenerative changes in the spine can result in compression of the dorsal root ganglia by surrounding anatomical structures.

After the convergence of posterior and anterior roots within the intervertebral foramen, they form the spinal nerve. This anatomical arrangement is characterized by the spinal nerve's brief course before its bifurcation into posterior and anterior rami. The concise length of the spinal nerve and its rapid division represent important anatomical features that influence the distribution of neural signals to specific regions of the cervical spine.

6.5.7.1 Characteristics of Cervical Rootlets and Dorsal Root Ganglia

Dorsal root ganglia (DRG) form ovoid swellings along dorsal spinal roots and contain cell bodies of primary afferent fibers. These structures demonstrate specific anatomical patterns in their distribution along the vertebral column. While most DRGs reside within intervertebral foramina, certain spinal levels show distinct anatomical variations (Standring 2020).

The cervical region presents unique anatomical features. The first cervical DRG, when present, appears small and lies on the atlas's posterior arch. The second cervical DRG occupies a position between the atlas's posterior arch and the C2 lamina, posterior and medial to the lateral atlanto-axial joint. This ganglion has an extradural location and fills 76% of the space between the atlas posterior arch and the C2 lamina. This anatomical arrangement increases its vulnerability to compression during prolonged cervical extension or from lateral atlanto-axial joint spondylosis (Lu and Ebraheim 1998). The C2 DRG contains sensory neurons that innervate the median and lateral atlanto-axial joints and supplies sensation to an extensive area of the neck and scalp (Bogduk 1982).

Lumbar DRGs show progressive size increases from L1 to L5. Measurements indicate width variations from 3.7 to 5.9 mm, length from 4.3 to 9.4 mm, and height from 4.3 to 8.3 mm (Hasegawa et al. 1995). L3 and L4 levels frequently display double ganglia formations. In the sacral region, upper DRGs maintain positions within intervertebral foramina, with S1 showing larger dimensions compared to other sacral ganglia (Ebraheim and Lu 1998).

DRG neurons exhibit pseudo-unipolar morphology, with single axons that divide into peripheral and central branches. The peripheral branch conducts sensory information from peripheral territories to the DRG neuron, while the central branch transmits signals to the spinal cord through dorsal roots. Quantitative analysis of the human L5 DRG reveals approximately 11,000 large neurons, 32,000 intermediate neurons, and 25,000 small neurons.

Two main histological categories define DRG neurons: "large, light" and "small, dark" cells. Large, light neurons contain abundant neurofilaments and generate myelinated Aα and Aβ fibers along with small myelinated mechanoreceptive Aδ fibers. Small dark neurons, characterized by lower neurofilament content, produce nociceptive Aδ and unmyelinated C fibers. Research indicates that large neurons process low-threshold mechanical and proprioceptive stimuli, while small neurons respond to high-threshold stimuli (Harper and Lawson 1985).

Small dark cells express multiple neuroactive substances, including calcitonin gene-related peptide, cholecystokinin, galanin, somatostatin, substance P, leu-enkephalin, nitric oxide, and vasoactive intestinal polypeptide (Ju et al. 1987). Human DRG neurons show higher substance P expression (44%) compared to rat neurons (20%). Additional neurochemical characteristics include nicotinic and muscarinic cholinergic receptors (Bernardini et al. 1999) and catecholaminergic properties in select neurons (Price and Mudge 1983).

The cervical region demonstrates significant variability in rootlet distribution, with anastomoses between adjacent segments occurring in 61% of cases, compared to 7% in the thoracic and 22% in the lumbar regions (Tanaka et al. 2000). These anatomical variations affect clinical presentations by altering traditional dermatomal patterns and influencing the manifestation of radicular symptoms.

6.5.7.2 Characteristics of Cervical Anterior Rami

In the cervical region, each anterior ramus emerges from its respective spinal nerve and exits the spine by passing posterior to the vertebral artery and between the anterior and posterior intertransversarii muscles (Bogduk 1985). These anterior rami provide innervation to the cervical spine's anterior musculature, which includes the longus capitis, longus colli, rectus capitis anterior, and rectus capitis lateralis muscles. The C1 anterior ramus innervates the atlanto-occipital joints, while the C2 anterior ramus supplies the median and lateral atlanto-axial joints (Bogduk 1985). In addition to muscular innervation, the cervical anterior rami contribute to the sensory supply of the vertebral bodies, the anterior longitudinal ligament, and the intervertebral disc's anterior aspect. These structures also receive sensory innervation from fibers originating in the sympathetic chain and autonomic fibers associated with the vertebral artery (Bogduk et al. 1988; Groen et al. 1990).

The cervical plexus (see Fig. 2.6) is composed of the anterior primary rami of the first four cervical nerves (C1–C4), forming a network of interconnecting loops between adjacent nerves. The C1 nerve emerges between rectus capitis anterior and rectus capitis lateralis, coursing downward anterior to the atlas transverse process to join the ascending branch of the C2 nerve. During this course, it dispatches a substantial branch to unite with the hypoglossal nerve. The C2, C3, and C4 cervical nerves each divide into upper and lower branches, with adjacent components forming connections. The lower branch of the C4 nerve may occasionally contribute to the brachial plexus (Standring 2020). The plexus's communicating loops are positioned in close proximity to the vertebral column, anterior to levator scapulae and scalenus medius, and posterior to the prevertebral muscles. Branches originating from these loops are situated posteromedial to the internal jugular vein, deep to the sternomastoid. For descriptive purposes, these branches can be categorized into superficial and deep groups. The superficial branches supply the skin of the inferior skull and neck regions, while the deep branches further subdivide into medial and lateral divisions, innervating muscles or forming communications with other nerves (Moore et al. 2013).

Four cutaneous nerves arise from the cervical plexus: the lesser occipital nerve, the greater auricular nerve, the transverse cervical cutaneous nerve, and the supraclavicular nerves. These nerves emerge in the posterior triangle above the midpoint of the sternomastoid's posterior border before penetrating the deep fascia to innervate the skin of the head and neck (Fig. 6.19). The deep branches, primarily muscular, originate alongside other plexus branches deep to the sternocleidomastoid. These can be further classified as lateral or medial branches based on their course through the posterior or anterior triangles of the neck, respectively (Drake et al. 2020).

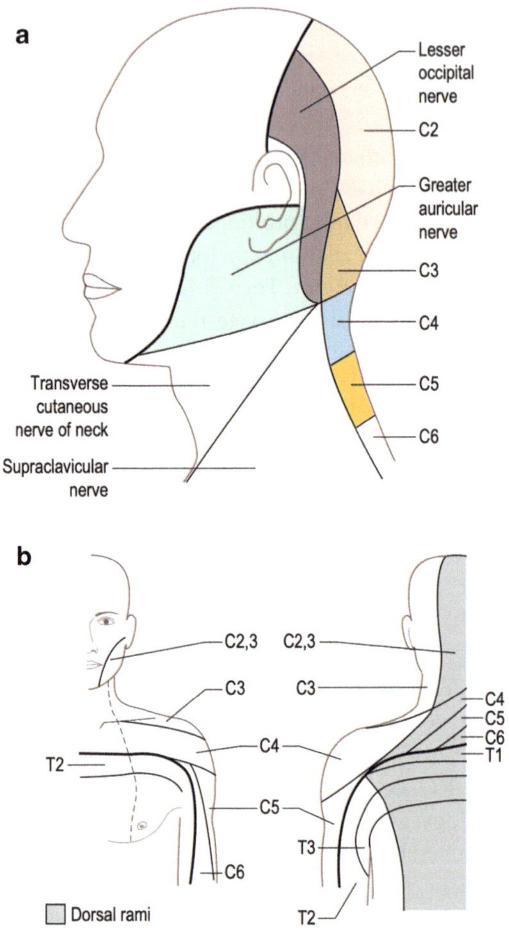

Fig. 6.19 (**a**) Cutaneous distribution of branches from the cervical plexus; (**b**) dermatome distribution. (Adapted from Palastanga and Soames (2012))

6.5.7.3 Characteristics of Cervical Posterior Rami

The posterior rami are often of smaller size in comparison to the anterior rami. After leaving the intervertebral foramen, the posterior ramus curves posteriorly, adjacent to the anterolateral aspect of the articular pillar (Fig. 6.20). Specifically, the posterior rami of C4 and C5 form an indentation on the lateral side of the articular pillars of the C4 and C5 vertebrae. After reaching the posterior and lateral sides of the superior articular process, each posterior ramus divides into medial and lateral branches.

The deep back muscles are among the primary structures innervated by the posterior rami. The medial branch of the posterior rami innervates the deeper, more segmentally oriented transversospinalis muscles. In contrast, the lateral branch innervates the longer, more superficial erector spinae muscles. Additionally, the medial branch provides innervation to the zygapophysial joints and the interspinous ligaments. The lateral branches of the upper cervical posterior rami (except for C1) extend posteriorly to innervate the erector spinae and splenius capitis and cervicis

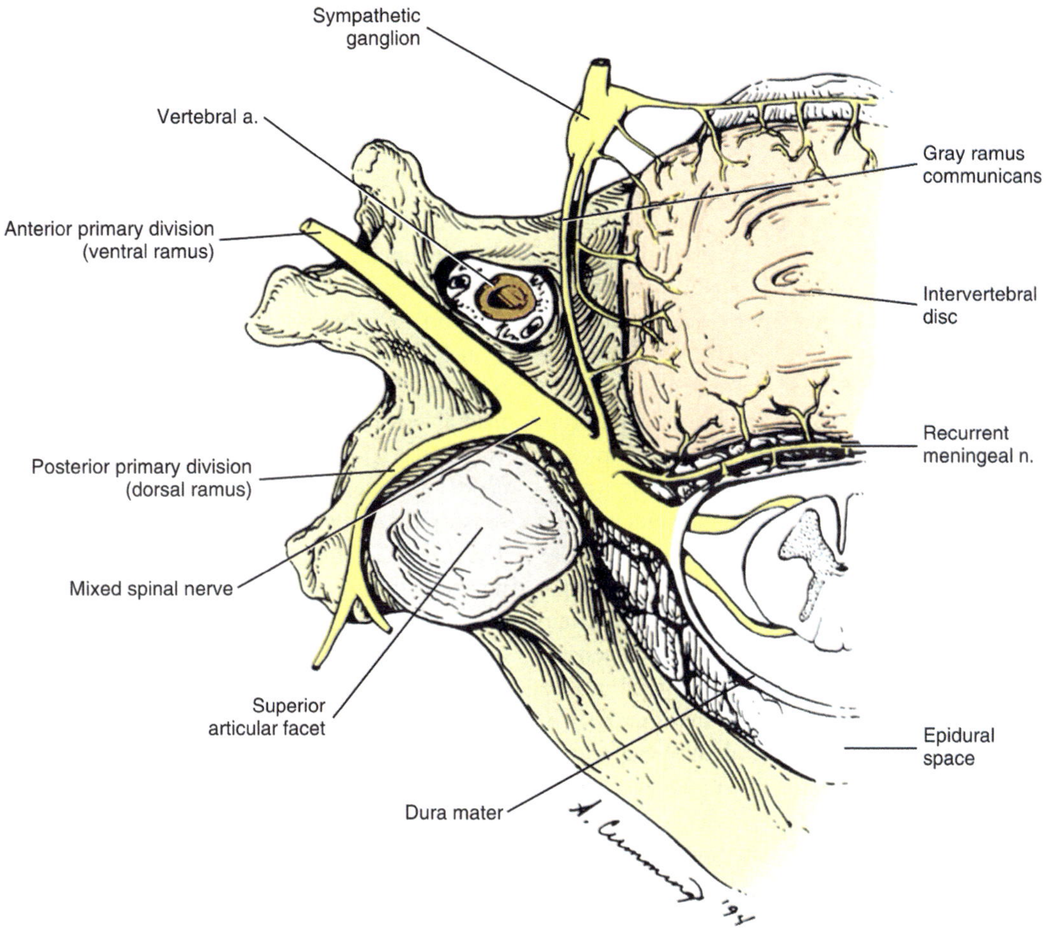

Fig. 6.20 Superior view of a typical cervical segment illustrating neural elements. The figure shows the dorsal and ventral roots, spinal nerve, and posterior and anterior primary divisions (dorsal and ventral rami). The posterior primary division divides into medial and lateral branches. The recurrent meningeal nerve is depicted entering the intervertebral foramen. Fibers from the middle cervical ganglion and the gray communicating ramus are also shown, supplying the anterior and lateral aspects of the intervertebral disc, vertebral body, and anterior longitudinal ligament. (Adapted with permission from Cramer and Darby (2014))

muscles before providing sensory innervation to the skin of the neck. The posterior rami of C6, C7, and C8 generally lack cutaneous branches.

The posterior ramus of the C1 spinal nerve is distinct. This nerve exits the vertebral canal by passing above the posterior arch of the atlas and quickly divides into anterior and posterior rami. The posterior ramus, known as the suboccipital nerve, travels between the posterior arch of the atlas and the vertebral artery. Unlike other posterior rami, it does not split into medial and lateral branches but instead curves superiorly for about 1 cm and terminates by providing motor innervation to the suboccipital muscles. Additionally, it sends a communicating branch to the posterior ramus of C2 (Standring 2020).

The C2 spinal nerve bifurcates into a posterior and anterior ramus posterior to the lateral atlanto-axial joint. The posterior ramus ascends around the inferior margin of the obliquus capitis inferior muscle, subsequently dividing into medial, lateral, superior, and inferior communicating branches, as well as a branch directed toward the obliquus capitis inferior muscle. The lateral branch of the posterior ramus of C2 provides motor innervation to the longissimus capitis, splenius capitis, and semispinalis capitis muscles (Bogduk 1982). The medial branch of the C2 posterior ramus, known as the greater occipital nerve, is notably large. It invariably penetrates the semispinalis capitis muscle and, in 20% of cases, the trapezius muscle as well. During its passage through the semispinalis muscle, the greater occipital nerve often accompanies the occipital artery (Janis et al. 2010). In cases where the greater occipital nerve does not pierce the trapezius muscle, it is believed to be shielded from compression during trapezius muscle contraction by passing through an aponeurotic sling associated with the trapezius and sternocleidomastoid muscle insertions onto the superior nuchal line (Bogduk 1982). After traversing the sling, the greater occipital nerve ascends and divides into several terminal branches, providing extensive sensory innervation from the occipital region medially to the area superior to the mastoid process and posterior to the ear laterally. Superiorly, these branches supply sensory innervation to the scalp, extending from the posterior occiput to the coronal suture of the skull (Bogduk 1982).

The C3 spinal nerve, which is the most rostral nerve to exit via an intervertebral foramen, branches into a posterior and anterior ramus within the lateral portion of the intervertebral foramen. The posterior ramus of C3 courses posteriorly between the transverse processes of C2 and C3, dividing into deep and superficial medial branches, a lateral branch, and a communicating branch with the posterior ramus of C2 (Bogduk 1982). The superficial medial branch, known as the third occipital nerve, traverses the lower segment of the C2–3 zygapophyseal joint from anterior to posterior, providing articular branches to this joint. The deep medial branch of the C3 posterior ramus innervates the uppermost multifidus muscles. The lateral branch of the C3 posterior ramus supplies the more superior layers of the neck muscles, including the longissimus capitis, splenius capitis, and semispinalis capitis. Additionally, the C3 posterior ramus innervates the C2–3 and C3–4 zygapophysial joints. The anterior rami of C1 and C2 innervate the atlanto-occipital joints and the median and lateral atlanto-axial joints, respectively (Bogduk 1982).

The spinal nerves from C4 to C8 exit through their corresponding intervertebral foramina. Their posterior rami quickly branch off, passing posteriorly medial to the posterior intertransversarii muscles, which they innervate. These rami then divide into medial and lateral branches. The medial branches of C4 and C5, and occasionally C6, further subdivide into superficial and deep branches. In contrast, the posterior rami of C6 (C7) and C8 do not divide, possessing only deep medial branches. The superficial branches provide innervation to the semispinalis cervicis and capitis muscles and send cutaneous fibers for sensory innervation to the posterior neck's skin. The deep medial branches innervate the multifidi muscles, specifically those muscle fibers attaching to the spinous process of the segmental level numbered one less than the nerve. The lateral branches of the posterior rami from C4 to C8 innervate muscles such as the semispinalis capitis, longissimus cervicis, splenius cervicis, and iliocostalis cervicis.

6.6 Anatomy and Clinical Significance of the Thoracic Spine

The thoracic spine demonstrates distinctive anatomical and functional properties that differentiate it from other spinal regions. Its restricted range of motion stems from its structural integration with the rib cage, where ribs connect anteriorly to the sternum (White and Panjabi 1990). This anatomical arrangement produces a more stable and rigid framework compared to cervical and lumbar regions.

The thoracic vertebrae exhibit a consistent pattern of dimensional increase from superior to inferior segments (Panjabi et al. 1991a, b). This progressive size augmentation reflects an adaptive response to the increasing axial loads that vertebrae must support along the spinal column. Biomechanical studies have demonstrated that the enhanced structural dimensions of lower thoracic vertebrae provide the necessary mechanical capacity to manage these elevated compressive forces (Stanley et al. 2004).

The thoracic region's relationship with the rib cage creates a biomechanical environment that prioritizes stability over mobility (Watkins 4th et al. 2005). This architectural design protects vital thoracic organs while maintaining appropriate spinal alignment. Analysis of thoracic spine mechanics has shown that the costovertebral joints significantly influence the region's movement patterns and load distribution (Willems et al. 1996). Understanding these anatomical characteristics proves valuable for clinicians evaluating thoracic spine disorders and planning therapeutic interventions.

6.6.1 Anatomical Characteristics of the Thoracic Vertebral Bodies

The thoracic vertebral bodies (T2–T8) exhibit greater size compared to those in the cervical region. The smallest thoracic vertebral body is found at the T3 level, with a gradual increase in size observed caudally (Masharawi et al. 2008). Thoracic vertebral bodies are characterized by a wedge-shaped configuration, with the anterior height being less than the posterior height. This wedging is most pronounced at T7 and progressively diminishes until L2 (Masharawi et al. 2008). The presence of mild lateral wedging may contribute to thoracic stability during normal spinal loading, with females demonstrating more pronounced left lateral wedging compared to males. This sex-related difference in lateral wedging could potentially explain the higher incidence of right-sided convex thoracic curves in adolescent idiopathic scoliosis among girls (Masharawi et al. 2008). Remnants of cervical uncinate processes can be observed on the posterior edge of the superior surface of upper thoracic vertebral bodies (Dupuis et al. 1985).

The T1 vertebra displays features typically associated with cervical vertebrae, such as uncinate processes and superior vertebral notches above the pedicles, along with a cervical-like vertebral body shape. When viewed from the lateral aspect, the T1 and T2 vertebral bodies exhibit a more rounded configuration, potentially facilitating the extensive flexion and extension movements occurring in the cervical region (Masharawi et al. 2008). The vertebral bodies of T9–T12 begin to acquire lumbar characteristics, with a greater increase in transverse dimension compared to anteroposterior dimension. Notably, the T12 vertebral body closely resembles a lumbar vertebra.

Typical thoracic vertebral bodies feature four small facets, known as costal demifacets, with two on each side for articulation with the heads of two adjacent ribs. The head of each rib articulates with the superior demifacet of the correspondingly numbered vertebra and the inferior demifacet of the vertebra above (e.g., the sixth rib head articulates with the superior demifacet of T6 and the inferior demifacet of T5). The T1 vertebral body usually has a full facet on the superior aspect for articulation with the first rib head, rather than a demifacet. The vertebral bodies of T10 and T11 each have a single facet on each side for articulation with the corresponding tenth and eleventh rib heads, respectively, with the facet on T11 being located on the pedicle. Additionally, T11 lacks an articular facet on the transverse process for articulation with the rib's articular tubercle. Similarly, T12 has a single facet on each side for articulation with the head of the corresponding twelfth rib.

6.6.2 Anatomical Characteristics of the Thoracic Articular Processes

The thoracic region demonstrates distinctive anatomical arrangements of articular processes that influence spinal mobility and stability. The superior articular processes and their facets exhibit a posterior, slightly superior, and lateral orientation, maintaining a more vertical alignment compared to their cervical counterparts. Correspondingly, the inferior articular processes and facets display an anterior, slightly inferior, and medial orientation, creating a reciprocal arrangement with the superior facets (White and Panjabi 1990).

This specific facet orientation significantly influences thoracic spine biomechanics. Research has demonstrated that this arrangement primarily restricts flexion and extension movements, while also imposing partial limitations on lateral flexion (Oda et al. 1996). The thoracic spine's rotational and lateral flexion capabilities are further constrained by the robust connections between thoracic vertebrae and the relatively immobile thoracic cage. These connections are established through costocorporeal and costotransverse articulations, which contribute substantially to thoracic spine stability.

The relationship between articular process orientation and thoracic cage attachments creates a biomechanical environment that emphasizes stability over mobility (Panjabi et al. 1991a, b). This architectural design supports the thoracic spine's role in protecting vital organs while maintaining appropriate spinal alignment. Understanding these anatomical relationships holds particular importance for clinicians evaluating thoracic spine disorders and planning therapeutic interventions.

6.6.3 Anatomical Characteristics of the Thoracic Spinous Processes

The thoracic spinous processes demonstrate distinctive anatomical patterns that vary by vertebral level. These bony projections perform significant functions as attachment sites for muscles and ligaments throughout the thoracic region (Standring 2020).

The upper thoracic vertebrae (T1–T4) feature spinous processes that extend in a predominantly posterior direction. Moving inferiorly, T5 through T8 demonstrate a more pronounced downward angulation, with the T8 vertebra exhibiting the most substantial process length in this group (Panjabi et al. 1991a, b). The lower thoracic vertebrae (T9–T12) show a gradual transition toward lumbar characteristics, marked by increased posterior projection and enhanced superior–inferior dimensions (White and Panjabi 1990).

Table 6.5 Attachments to thoracic spinous processes

Region	Structure(s) attached
Upper thoracic region	Ligaments: supraspinous, interspinous
	Muscles: trapezius, rhomboid major and minor, serratus posterior superior, deep back muscles (erector spinae and transversospinalis)
Lower thoracic region	Ligaments: supraspinous, interspinous
	Muscles: trapezius, latissimus dorsi, serratus posterior inferior, deep back muscles (erector spinae and transversospinalis)

Adapted with permission from Standring (2020)

The considerable length of the thoracic spine influences the distribution and characteristics of muscular and ligamentous attachments to these processes. This variation in attachments between upper and lower thoracic regions reflects the changing biomechanical demands along the thoracic spine (Jiang et al. 1994). The anatomical orientation and dimensions of these processes directly influence their capacity to provide effective attachment points for various spinal muscles and ligaments. Table 6.5 provides a detailed list of the attachments to the upper and lower thoracic spinous processes.

Understanding these anatomical characteristics holds particular relevance for clinical practice, as variations in spinous process morphology can affect surgical approaches and therapeutic interventions in the thoracic spine (Singer et al. 1988).

6.6.4 Anatomical Characteristics of the Thoracic Transverse Processes

The transverse processes of the thoracic vertebrae progressively shorten from T1 to T12. Consequently, the span between the tips of the left and right transverse processes is widest at T1 and narrows incrementally toward T12, where the transverse processes are minimal in size. This span widens again in the lumbar region (Standring 2020).

Each thoracic transverse process features a transverse costal facet, which articulates with the articular tubercle of the corresponding rib, contributing to the stability and movement of the thoracic spine (Moore et al. 2013).

6.6.5 Anatomical Characteristics of the Thoracic Intervertebral Foramina

The thoracic intervertebral foramina exhibit distinct anatomical features in contrast to those found in the cervical region. While cervical intervertebral foramina are oriented obliquely in an anterolateral direction, thoracic intervertebral foramina project directly laterally, resembling the orientation observed in the lumbar region (Sunderland 1974a, b). The first ten thoracic intervertebral foramina (T1–T10) are uniquely associated with their corresponding ribs, whereas the eleventh and twelfth ribs lack direct connections to intervertebral foramina.

Notably, the proportion of the intervertebral foramen occupied by the spinal nerve varies across different regions of the spine. In the thoracic region, merely one-twelfth of the foramen's cross-sectional area is filled by the spinal nerve. Conversely, the cervical region exhibits a significantly higher nerve-to-foramen ratio, with up to half of the intervertebral foramen's space occupied by the spinal nerve and its accompanying dural root sleeve (Sunderland 1974a, b). The lumbar region falls between these extremes, with approximately one-third of the intervertebral foramen's area taken up by the spinal nerve. These anatomical variations may contribute to the lower incidence of radiculopathy secondary to intervertebral disc protrusion in the thoracic region compared to the lumbar and cervical spine. The smaller nerve-to-foramen ratio in the thoracic spine could potentially reduce the likelihood of nerve compression by herniated disc material, as the nerve occupies a relatively smaller portion of the available space within the foramen.

6.6.6 Anatomical Characteristics of the Thoracic Anterior and Posterior Longitudinal Ligaments

The thoracic anterior longitudinal ligament shows specific anatomical characteristics that differ from its cervical and lumbar segments. In the thoracic region, this ligament presents increased anteroposterior thickness while maintaining reduced lateral dimensions (White and Panjabi 1990). The ligament attaches firmly to the superior and inferior vertebral end plates, though its connections to the intermediate thoracic vertebral bodies remain minimal.

The posterior longitudinal ligament in the thoracic spine maintains structural relationships with adjacent vertebral elements. Ossification of this ligament occurs less frequently in the thoracic region compared to the cervical spine. However, when present, these pathological changes may lead to neurological complications (Vera et al. 1997). This morphological alteration requires careful consideration during clinical evaluation of thoracic spine conditions.

The anatomical configuration of these ligamentous structures influences normal movement patterns and potential pathological processes in the thoracic spine. Recognition of these anatomical features supports accurate assessment and appropriate management of thoracic spine disorders (Tkaczuk 1968).

6.6.7 Anatomical Characteristics of the Thoracic Ligamentum Flavum

The thoracic ligamentum flavum comprises two distinct layers: a superficial (posterior) layer and a deep (anterior) layer. The superficial layer's fibers are obliquely oriented, while the deep layer's fibers are more organized and aligned in the sagittal plane. A potential gliding space separates these two layers (Viejo-Fuertes et al. 1998). The superficial layers of the left and right ligamentum flavum have a distinct border, as does the border between each superficial layer and the interspinous ligament located more medially. This region between the left and right superficial layers is termed the "zone of separation." In contrast, the deep layers of the left and right ligamentum flavum are continuous at the midline, with a prominent posterior midline indentation (cleavage separation) demarcating the division between the two sides.

The etiology of ligamentum flavum ossification remains unclear, and the condition itself is not well understood. Ossification occurs predominantly in the thoracic region. It is hypothesized that repetitive stress, such as chronic stretching of the ligament, may contribute to this ossification. This process is associated with the significant degeneration of elastic fibers, including their fragmentation and the disruption of elastic fiber bundles (Yayama et al. 2007). Although the T9–T10 and T10–T11 levels are most frequently affected, ossification can occur at all thoracic levels, with less frequent occurrences noted at the lower cervical or upper lumbar levels.

6.7 Characteristic Anatomy and Clinical Significance of the Lumbar Spine

The lumbar vertebrae demonstrate standard anatomical features, with the exception of the fifth lumbar vertebra. Studies of vertebral development have revealed that lumbar vertebrae grow at twice the rate of thoracic vertebrae between ages 2 and 16 years (Dimeglio and Canavese 2012). This difference in growth velocity relates to the directional curvature of each spinal region—thoracic kyphosis versus lumbar lordosis. Research indicates that thoracic vertebrae display accelerated growth in their posterior elements compared to vertebral bodies, while lumbar vertebrae show faster growth in their vertebral bodies relative to posterior elements (Taylor 1975).

The lumbar lordosis presents a more pronounced curvature than cervical lordosis in normal spinal alignment (Roussouly and Pinheiro-Franco 2011). This curve extends from the twelfth thoracic vertebra to the L5–S1 intervertebral disc space, with maximum curvature occurring between the third and fifth lumbar vertebrae. The development of lumbar lordosis results from two primary factors: increased anterior height of L4 and L5 vertebral bodies, and enhanced anterior dimensions of lumbar intervertebral discs (Bogduk 2005). Analysis has shown that intervertebral disc height variations contribute more significantly to lordotic curve formation than vertebral body height differences (Been and Kalichman 2014).

These developmental patterns and anatomical relationships influence spinal biomechanics and function. The specific growth characteristics and curvature formation mechanisms form a basis for understanding normal spinal development and potential pathological changes (Mac-Thiong et al. 2007).

6.7.1 Anatomical Characteristics of the Lumbar Vertebral Bodies

The lumbar vertebral bodies display specific dimensional characteristics and structural properties that influence spinal function. Volumetric analysis indicates sexual dimorphism, with males showing larger vertebral body volumes. A progressive volume increase occurs from L1 to L4, while L5 demonstrates smaller dimensions compared to L4 (Limthongkul et al. 2010). The bodies exhibit greater lateral width compared to anteroposterior dimensions. L4 and L5 vertebral bodies show increased anterior height, while L1 and L2 present greater posterior height. L3 maintains approximately equal anterior and posterior heights (Masharawi et al. 2008).

The vertebral bodies demonstrate specialized load-bearing capabilities through their internal architecture. The inner cancellous bone manages approximately 90% of compressive forces, with the outer cortical shell accommodating the remaining load (Silva et al. 1994). However, subsequent research suggests that this distribution might overestimate cancellous bone contribution (Cao et al. 2001). In conditions such as osteoporosis, the load distribution shifts, with the cortical shell assuming an increased proportion of compressive forces.

The vertebral end plates perform a significant mechanical function in load resistance. Research demonstrates that end plate removal reduces compression fracture resistance by 66% (Oxland et al. 2003). These structures show regional variations in strength, with maximum resistance occurring in posterolateral regions and minimum strength in central areas. The inferior end plates demonstrate greater strength compared to the superior ones, and sacral end plate strength parallels that of the inferior lumbar end plates (Grant et al. 2001).

6.7.2 Anatomical Characteristics of the Lumbar Transverse Processes

The lumbar transverse processes are quite long, with the transverse process of L3 being the longest. They possess neither articular facets (as do thoracic transverse processes) nor foramina of the transverse processes (as do cervical transverse processes). The lateral aspect of the anterior surface of the lumbar transverse processes is creased by a ridge that runs from superior to inferior. This ridge is created by the anterior layer of the thoracolumbar fascia. The middle layer of the thoracolumbar fascia attaches to the apex of the transverse processes. Table 6.6 lists structures that attach to the lumbar transverse processes.

Table 6.6 Attachments to lumbar transverse processes

Region	Structure(s) attached
Anterior surface	Psoas major and quadratus lumborum muscles
	Anterior layer of thoracolumbar fascia (separates psoas major and quadratus lumborum muscles)
	Medial and lateral arcuate ligaments (lumbocostal arches) (L1)
Apex	Middle layer of thoracolumbar fascia
	Iliolumbar ligament (L5, occasionally L4)
Superior border	Lateral intertransversarius muscle
Inferior border	Lateral intertransversarius muscle
Posterior surface	Deep back muscles (longissimus thoracis)

Adapted with permission from Standring (2020)

6.7.3 Anatomical Characteristics of the Lumbar Superior and Inferior Articular Processes

The lumbar superior articular facets exhibit consistent posterior and medial orientation across vertebral levels, while demonstrating level-specific angular variations (Fig. 6.21). Analysis reveals that L4 superior facets present a more sagittal orientation compared to L5 facets. The S1 superior facets, the L5–S1 zygapophysial joints, show a more coronal alignment relative to L4 and L5 facets (Van Schaik et al. 1985). This variable orientation affects the biomechanical properties of different vertebral segments.

Age-related changes in facet orientation occur most notably at the L4–L5 level, where the zygapophysial joints demonstrate progressive sagittal orientation. This alteration in joint alignment may explain the increased frequency of degenerative spondylolisthesis observed at this spinal level (Masharawi et al. 2004).

Anatomical studies indicate that left and right articular facets generally maintain symmetrical orientation throughout the lumbar spine. However, the L5–S1 articulation represents an exception to this pattern, as it frequently displays

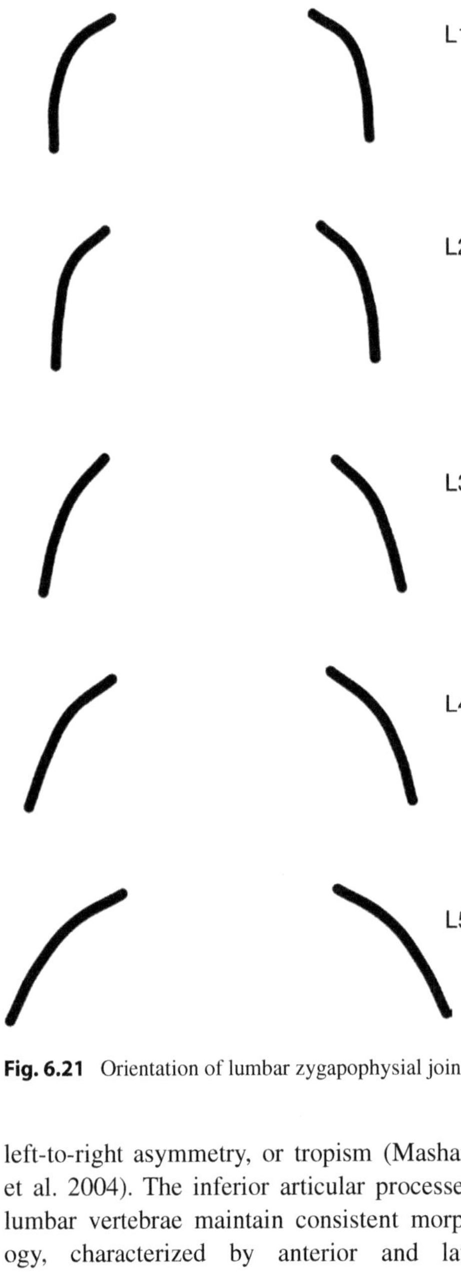

Fig. 6.21 Orientation of lumbar zygapophysial joints

left-to-right asymmetry, or tropism (Masharawi et al. 2004). The inferior articular processes of lumbar vertebrae maintain consistent morphology, characterized by anterior and lateral convexity.

These anatomical characteristics influence spinal mobility and stability. The varied orientation of articular facets at different levels affects the range and direction of movement possible at each segment, while also determining the distribution of mechanical forces through the spine.

6.7.4 Anatomical Characteristics of the Lumbar Zygapophysial Joints

The lumbar zygapophysial joints represent complex synovial articulations positioned in the vertical plane, with their structure determined by the configuration of superior and inferior articular facets (Standring 2020). During development, these joints demonstrate a notable transition from coronal orientation in childhood to acquiring a significant sagittal component with maturation (Masharawi et al. 2004).

The orientation and morphology of these joints directly influence spinal movement patterns. The sagittal component restricts rotational movement, while the coronal aspect limits flexion amplitude. The lumbar region permits approximately 60° of flexion before the bony architecture of articular processes prevents further movement. This range occurs despite the size of zygapophysial joints ultimately determining the flexion endpoint (Bogduk 2012).

The zygapophysial joint structure provides mechanical protection for intervertebral discs by significantly limiting rotation. During rotational movements, the articular facets undergo distraction on the ipsilateral side between adjacent lumbar articular facets (the superior facet of the vertebra below and the inferior facet of the vertebra above). Extension movements terminate when the inferior articular processes contact the pars interarticularis—the junction between the lamina and superior articular process of the vertebra below (Jaumard et al. 2011).

These biomechanical characteristics establish precise movement parameters while maintaining structural integrity. The joint architecture balances mobility requirements with protective mechanisms that prevent excessive movement and potential tissue damage (Kalichman and Hunter 2007).

6.7.5 Anatomical Characteristics of the Lumbar Intervertebral Foramina

Several features of the lumbar intervertebral foramina are unique. Many features of the region of the lumbar intervertebral foramina are different from those of the rest of the spine because of the unique characteristics of the lumbar and sacral spinal nerves. Because the spinal cord ends generally at the level of L1 to L2, the lumbar and sacral posterior and anterior roots must descend, sometimes for a considerable distance, within the subarachnoid space of the lumbar vertebral canal. This region of the subarachnoid space is known as the lumbar cistern.

The exiting nerves (posterior and anterior rootlets or roots) leave the lumbar cistern by entering a sleeve of dura mater. This usually occurs slightly inferior to the level of the intervertebral disc at the level above the intervertebral foramen that the roots eventually will occupy. On leaving the subarachnoid space of the lumbar cistern, the exiting posterior and anterior roots pass at an oblique inferior and lateral angle while retaining a rather substantial and very distinct covering of dura mater. This covering, known as the dural root sleeve, surrounds the neural elements and their accompanying radicular arteries and veins until they leave the confines of the intervertebral foramen. The largest lumbar dural root sleeve (proximal to the dorsal root ganglion) is L5 (5.6 mm diameter), and the smallest is L1 (3.5 mm diameter). On reaching the lateral edge of the intervertebral foramen, the dural root sleeve becomes continuous with the epineurium of the spinal nerve.

The anatomy of the L5 nerve root canal is unique, and because of this, the L5 roots and nerve are susceptible to compression in many different locations. This canal is longer and runs at a more oblique anterior angle than the rest of the lumbar nerve root canals. Also, the lateral recess of the L5 vertebra is the deepest laterally and often the narrowest from anterior to posterior of the entire spine. This narrow lateral recess may lead to compression of the L5 nerve in some instances. After leaving the lateral recess, the L5 roots and their dural root sleeve continue along the L5 nerve root canal by wrapping around the posterior and lateral aspect of the L5 vertebral body. The roots continue around the posterolateral aspect of the L5 intervertebral disc and unite to form the spinal nerve. The L5 nerve, which is the largest of the lumbar nerves, exits the lateral border of the L5 nerve root canal, which is the smallest intervertebral foramen of the lumbar spine (Olsewski et al. 1990). This makes the L5 nerve particularly susceptible to compression within its intervertebral foramen.

6.7.6 Anatomical Characteristics of the Iliolumbar Ligaments

The iliolumbar ligaments extend bilaterally from the L5 transverse processes, and infrequently from L4, to the sacrum and iliac crest on their respective sides. These ligaments display complex anatomical organization, with up to five distinct components (Bogduk 2005). Anatomical studies indicate that the most prominent structure consists of superior and inferior bands (Olsewski et al. 1990; Standring 2020).

The lumbosacral ligament demonstrates specific anatomical relationships, originating from the L5 transverse process and vertebral body and inserting into the sacral promontory. This ligamentous structure appears in a minimum of 3% of anatomical specimens (Olsewski et al. 1990).

The superior band of the iliolumbar ligament follows a more lateral course compared to the lumbosacral ligament, attaching to the iliac tuberosity anterior to the superior sacroiliac joint. This band maintains continuity with the quadratus lumborum muscle attachment at the L5 transverse process. Anatomical variations occur in its structure, which typically includes anterior and posterior components, though one component may be absent in some individuals. Historical anatomical studies have suggested that the iliolumbar ligament develops through metaplastic transformation of the quadratus lumborum muscle's epimysium. This developmental perspective adds to our understanding of the ligament's anatomical relationships and structural characteristics.

6.7.7 Anatomical Characteristics of the Lumbar and Sacral Dorsal Root Ganglia

The dorsal root ganglia (DRG) occupy positions proximal to the junction where posterior and anterior roots merge to form spinal nerves. A progressive increase in DRG diameter occurs from L1 to S1. These structures demonstrate positional variability, including potential left–right asymmetry, which holds clinical significance for diagnostic and therapeutic procedures.

Hasegawa et al. (1995) conducted morphometric analysis of L1–5 DRG locations, categorizing them into three positional variants: intraspinal (within vertebral canal), intraforaminal (within intervertebral foramen), and extraforaminal (lateral to intervertebral foramen). Their findings revealed specific distribution patterns. The L1–L3 DRG showed predominantly intraforaminal positioning (L1 92%, L2 98%, L3 100%). L4 and L5 demonstrated more varied distributions, with L4 showing 9.3% intraspinal, 86.1% intraforaminal, and 4.6% extraforaminal locations. L5 displayed 19.2% intraspinal, 72.8% intraforaminal, and 8.0% extraforaminal positions.

Sacral DRG exhibits distinct anatomical patterns. The S1 DRG maintains intraforaminal position (lateral to medial sacral pedicle border) in 55–60% of cases, with intracanalar location (medial to sacral pedicle medial aspect) in 40–45%. S2 DRG shows intraforaminal positioning in 15–50% of cases and intracanalar in 50–85%. S3 and S4 DRG consistently demonstrate intracanalar positions. No extraforaminal locations were observed in S1–S4 DRG. Among sacral DRGs, S1 presents the largest dimensions, with progressive size reduction occurring from S1 to S4.

6.8 Surface Anatomy of the Neck and Trunk

6.8.1 Posterior Aspect of the Neck

The superior boundary of the posterior surface of the neck is delineated by the superior nuchal line of the occipital bone, which extends medially to the external occipital protuberance and laterally to the mastoid process (Lumley 2008) (Fig. 6.22).

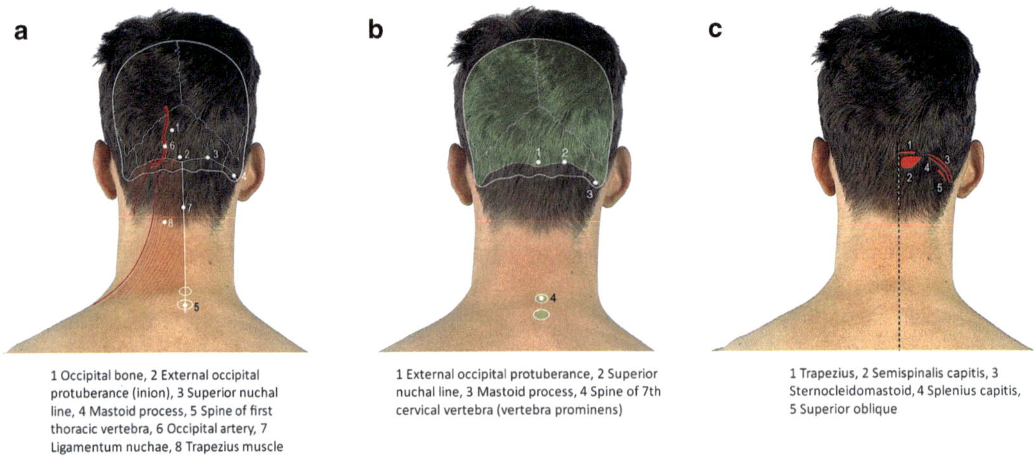

Fig. 6.22 Surface anatomy of of the posterior aspect of the neck. (**a**) Bones and soft tissues. (**b**) Palpable structures. (**c**) Muscle attachments. (Adapted with permission from Lumley (2008))

In the inferior neck region, structures are less distinguishable posteriorly due to the overlying musculature, including the trapezius, splenius, and the erector spinae group. The ligamentum nuchae runs along the midline between the muscles on both sides, attaching to the occipital bone and the cervical spinous processes. The spinous processes of the lower two or three cervical vertebrae and all thoracic vertebrae are palpable. However, the upper cervical spinous processes are not palpable and are covered by a median furrow. The C7 vertebra, known as the vertebra prominens, is readily identifiable due to its nonbifid spinous process, as is the first thoracic vertebra (T1) (Standring 2020).

The trapezius muscle, with its extensive attachment, forms the superficial layer (Fig. 6.23). Its superior attachment includes the medial third of the superior nuchal line, the ligamentum nuchae, and the spinous processes and interspinous ligaments of the cervical and thoracic vertebrae. Laterally, it attaches to the acromion and scapular spine, both of which can be palpated from the shoulder tip to the medial scapular border (Moore et al. 2013).

The sternocleidomastoid muscle attaches to the lateral third of the superior nuchal line and the mastoid process. Medial to this muscle, the occipital artery emerges and ascends over the cranium, where it can be palpated. The superior nuchal line corresponds to the location of the underlying transverse venous sinus. The posterior fossa, containing the hindbrain, is situated inferior to this level. Surgical access to the posterior fossa requires the detachment of the trapezius, erector spinae, and short occipital muscles from the occipital bone. Deep to these muscles, the vertebral artery crosses within the suboccipital triangle.

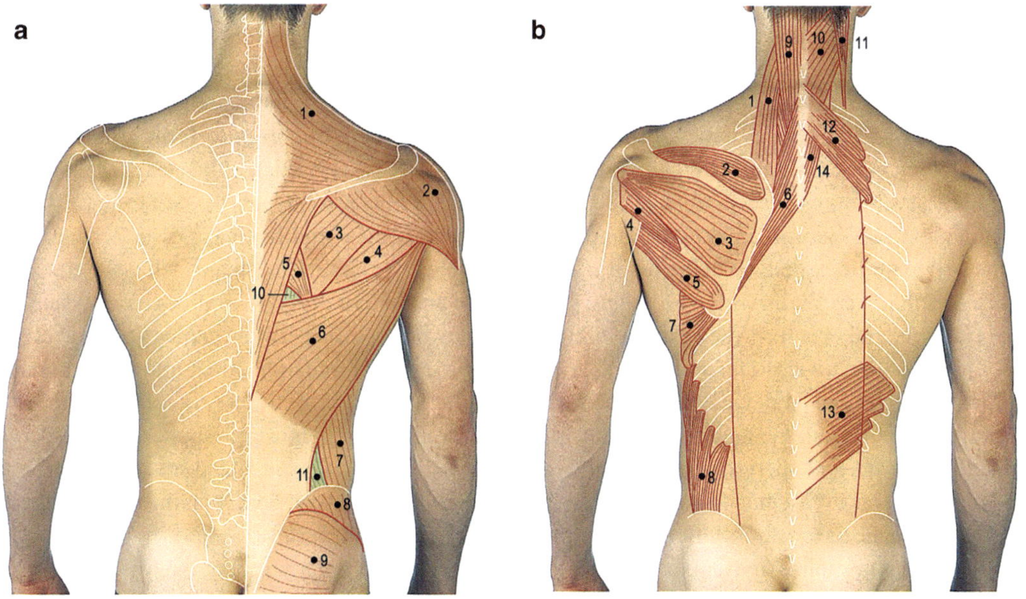

1 Trapezius, 2 Deltoid, 3 Infraspinatus, 4 Teres major, 5 Rhomboideus major, 6 Latissimus dorsi, 7 External oblique, 8 Gluteus medius, 9 Gluteus maximus, 10 Auscultatory triangle, 11 Lumbar triangle

1 Levator scapulae, 2 Supraspinatus, 3 Infraspinatus, 4 Teres minor, 5 Teres major, 6 Rhomboid muscles, 7 Serratus anterior, 8 External oblique, 9 and 14 Semispinalis capitis, 10 Splenius capitis, 11 Sternocleidomastoid, 12 Serratus posterior superior, 13 Serratus posterior inferior

Fig. 6.23 Surface anatomy of the posterior trunk. (**a**) Bones and superficial muscles. (**b**) Deep muscles of shoulder girdle. (Adapted with permission from Lumley (2008))

6.8.2 Posterior Aspect of the Trunk

The thoracic vertebrae serve as attachment points for the 12 rib pairs that constitute the thoracic cage, as well as the musculature of the shoulder girdle. While the terminations of the floating ribs and the margins of the costal region can be palpated, the majority of ribs are obscured by overlying muscular structures or the scapula, which covers ribs 2 through 7. These rib levels correspond to the fourth thoracic vertebral body and the ninth thoracic vertebral body, respectively (Standring 2020).

The caudally oriented thoracic spinous processes are all palpable. Their identification is facilitated by requesting the subject to flex forward, which accentuates their prominence. In contrast, the spinous processes of the lower thoracic and entire lumbar regions maintain a horizontal orientation and are easily discernible. A key anatomical landmark in this region is the spinous process of the fourth lumbar vertebra (L4), which aligns with the supracristal plane, an imaginary line connecting the superior aspects of the iliac crests (Moore et al. 2013).

The spinous processes of the sacrum, the sacral hiatus, and the coccyx can typically be palpated as they curve anteriorly. The anterior surface of these structures is accessible during rectal examination. The sacrum, an integral component of the pelvic girdle, articulates with the ilia bilaterally via the sacroiliac joints. Although these joints are synovial in nature, they exhibit minimal mobility and often undergo fusion.

The iliac crest can be palpated along its entire length, extending from the anterior superior iliac spine to the posterior superior iliac spine, the latter being marked by a cutaneous depression. In some individuals, the posterior inferior iliac spine and the sacroiliac joints may also be palpable. Through the gluteus maximus muscle, the posterior aspect of the acetabulum can sometimes be appreciated. The ischial tuberosities, located inferiorly, bear the body's weight in the seated position (Drake et al. 2020).

The female pelvis exhibits distinct anatomical characteristics compared to the male pelvis. It presents a relatively broader structure with increased capacity, while its osseous components are characterized by reduced thickness, enhanced smoothness, and diminished weight. Additional distinguishing features include a shallower acetabulum, a more vertically oriented ilium, and an obturator foramen that assumes a triangular shape rather than the oval configuration observed in males. Moreover, in females, the width of the acetabulum is less than the distance from its anterior rim to the pubic symphysis, whereas in males, this distance is greater. The inferior pubic angle also demonstrates greater width in females (Standring 2020).

The vertebral column's stability is maintained through a combination of factors, including the morphology of the vertebrae, the presence of physiological curvatures, intervertebral articulations, ligamentous structures, and powerful musculature. Intervertebral discs, positioned between the short cylindrical vertebral bodies, function as shock-absorbing elements that provide elasticity to the spinal column (Moore et al. 2013).

Typical vertebrae comprise an anteriorly situated body and a posterior osseous arch. The arch originates from short, stout pedicles on either side of the body, culminating posteriorly in the laminae, which converge to form a midline spinous process. The articular facets for the intervertebral synovial joints are located at the junction of each pedicle and lamina. These facets exhibit regional variations in shape, which dictate the type and extent of movement possible. While the range of motion between adjacent vertebrae is limited, the cumulative effect allows for considerable mobility of the vertebral column as a whole. Flexion is most pronounced in the cervical region, rotation in the thoracic region, and extension and lateral flexion in the lumbar region.

The first two cervical vertebrae, the atlas (C1) and axis (C2), are atypical of the typical vertebral structure. The atlas forms an oval ring of bone that articulates superiorly with the occipital condyles, facilitating flexion and extension of the head. Inferiorly, it articulates with the horizontal atlanto-axial facets of the axis, allowing for rotational movement of the head on the neck. The body of the atlas has fused with the axis, forming a distinct process known as the dens, which serves as a pivot for atlanto-axial rotation. The remaining cervical vertebrae primarily facilitate flexion and extension, with some degree of lateral flexion accompanied by slight rotation (Drake et al. 2020).

The thoracic spine's range of motion is constrained by the rib cage and the necessity for free respiratory motion. Some degree of rotation is possible, particularly in the lower thoracic vertebrae. In contrast, the lumbar spine's movement is primarily limited to flexion and extension. This can be visually demonstrated by marking the lumbar spinous processes and observing their longitudinal separation when the subject attempts to touch their toes (Neumann 2017).

Trunk flexion is primarily executed by the rectus abdominis muscle, with assistance from the prevertebral musculature. Lateral flexion is achieved through the action of the oblique abdominal wall muscles and the quadratus lumborum. Rotational movements are facilitated by the coordinated action of the internal and external oblique abdominal muscles. The body's center of gravity is typically located anterior to the second sacral segment. As the body moves, this center of gravity often shifts further anteriorly, necessitating a strong posterior muscle mass to counterbalance gravitational effects and restore upright posture. This function is primarily performed by the erector spinae muscle group (Moore et al. 2013).

The erector spinae, also known as the sacrospinalis, is a substantial muscle complex situated bilaterally along the vertebral column, extending from the sacrum to the skull. This term refers to a variety of small muscle groups that extend between neighboring vertebrae or across various levels of the spine. The muscle's bulk varies at different spinal levels. In the sacral region, it is relatively narrow, predominantly tendinous, and exhibits considerable strength. The lumbar region houses the most prominent portion of the erector spinae, where its cylindrical form is both visible and palpable (Standring 2020).

In the thoracic region, the erector spinae mass flattens and extends laterally to the angles of the ribs, to which it also attaches. This portion is covered by the scapula and the muscles of the shoulder girdle. In the cervical region, the cylindrical muscle masses are overlaid by the trapezius and splenius muscles. Although three parallel columns can be identified within each erector spinae muscle, it is most effectively conceptualized as a powerful extensor unit acting on both the vertebral column and the skull (Netter 2019).

6.9 The Role of Spinal and Trunk Musculature in Vertebral Column Function

The musculature of the spine and trunk plays an important part in the proper functioning of the vertebral column. These muscles are responsible for producing different movements of the spine and also play a role in maintaining posture (Cramer and Darby 2014). Moreover, they function as shock absorbers, distributing the loads that are exerted on the spine (Panjabi 1992).

Most muscles possess two attachment sites: the origin, which is typically more proximal and less mobile, and the insertion, which is usually more distal and more mobile. Occasionally, the origin and insertion sites can be reversed, with the origin being distal and the insertion proximal (Moore et al. 2013). The muscles of the back are categorized from superficial to deep, divided into six distinct layers (Bogduk 2012). Layer one represents the most superficial muscles, while layer six includes the deepest muscles.

6.9.1 First Layer Muscles: Trapezius and Latissimus Dorsi

The primary muscles constituting the first layer (superficial layer) of the back musculature include the trapezius and latissimus dorsi muscles. These muscles originate from the spine and occiput, inserting onto the shoulder girdle (scapula and clavicle) or the humerus (Moore et al. 2013).

The *trapezius muscle* (Fig. 6.24), the most superficial and superior muscle of the back, is a large and powerful muscle innervated primarily by the accessory nerve (CN XI) and by proprio-

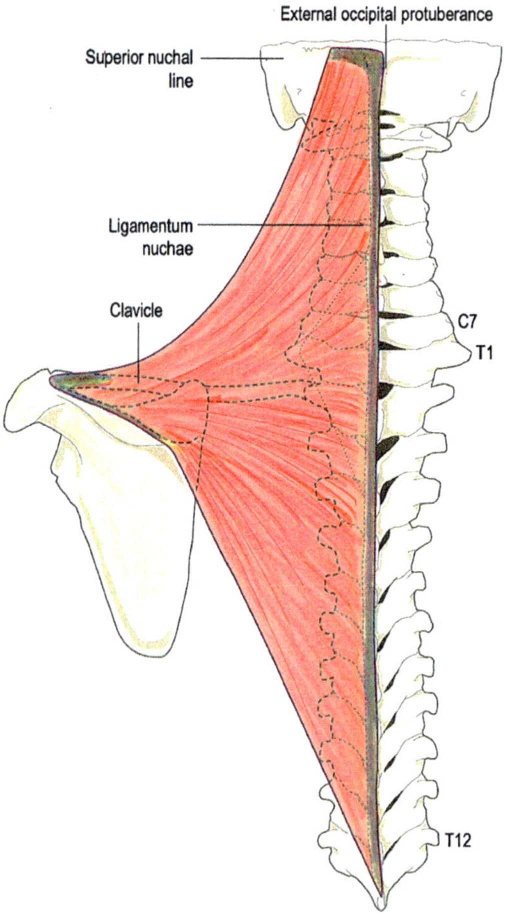

Fig. 6.24 Trapezius muscle. (Adapted with permission from Palastanga and Soames (2012))

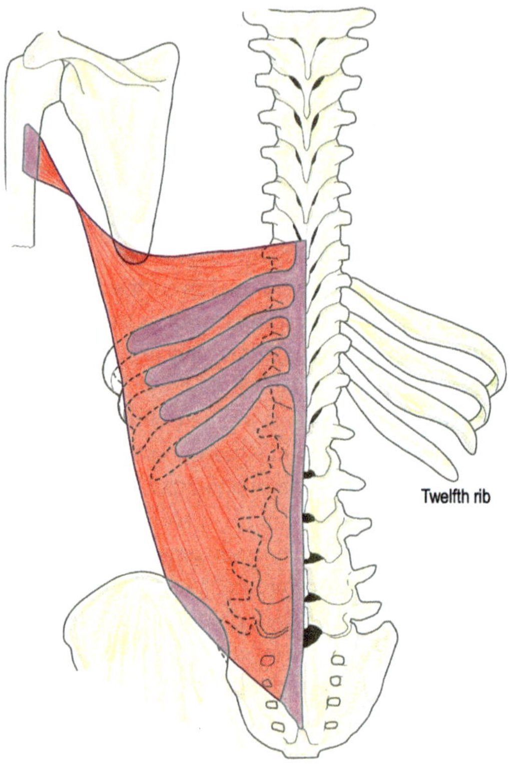

Fig. 6.25 Latissimus dorsi muscle. (Adapted with permission from Palastanga and Soames (2012))

ceptive fibers from the third and fourth cervical anterior rami (Standring 2020). Due to its extensive size, the trapezius has multiple attachment points, including the superior nuchal line, external occipital protuberance, ligamentum nuchae of the posterior neck, spinous processes of C7 to T12, and supraspinous ligament between C7 and T12. Its distal attachments include the spine of the scapula, acromion, and distal third of the clavicle (Standring 2020). The trapezius muscle's function varies based on the region that is contracting (upper, middle, or lower) (Neumann 2017). The middle fibers retract the scapula. The lower fibers depress the scapula while rotating it so the lateral angle moves superiorly. The actions of the upper trapezius fibers are influenced by whether the head, neck, or scapula is stabilized, as well as by whether the muscle is contracting unilaterally or bilaterally (Johnson et al. 1994).

The *latissimus dorsi muscle* (Fig. 6.25), situated inferiorly and laterally within the first layer of back muscles, is an expansive muscle with a broad proximal attachment site. The muscle originates from various structures, including the spinous processes and supraspinous ligament spanning from T6 to L5 vertebrae, despite the supraspinous ligament terminating between L2 and L4 (Bogduk 2005). Additionally, the latissimus dorsi muscle attaches to the thoracolumbar fascia, the posterior sacrum at the median sacral crest, the iliac crests, and the inferior four ribs. The muscle fibers ascend superiorly and laterally, inserting into the intertubercular groove of the humerus, positioned between the lateral attachment of the pectoralis major muscle and the medial attachment of the teres major muscle (Moore et al. 2013). Activation of the latissimus dorsi muscle facilitates adduction, medial rotation, and extension of the humerus. Innervation of this muscle is provided by the thoracodorsal nerve, a branch originating from the posterior cord of the brachial plexus (Standring 2020).

The *thoracolumbar fascia* (Fig. 6.26) is a fascial structure that spans from the thoracic region to the sacrum. In the thoracic region, it forms a thin covering over the erector spinae muscles. However, in the lumbar region, the thoracolumbar fascia is a robust structure composed of three distinct layers. The posterior layer of the thoracolumbar fascia attaches to the lumbar spinous processes, the interspinous ligaments between these processes, and the median sacral crest (Willard et al. 2012). This posterior layer is a strong sheet of dense connective tissue that connects two of the largest muscles in the body: the latissimus dorsi and the gluteus maximus. Consequently, the posterior layer plays a crucial role in transmitting forces between the spine, pelvis, and lower limbs. The superficial lamina of the posterior layer continues superiorly, merging with the rhomboid muscles, while the deep lamina of the posterior layer extends superiorly to become continuous with the splenius muscles (Barker and Briggs 2007).

6.9.2 Second Layer Muscles: Rhomboid Major, Rhomboid Minor, and Levator Scapulae

The second layer of back muscles, situated deep to the trapezius muscle, comprises three muscles: the rhomboid major, rhomboid minor, and levator scapulae muscles. All three muscles attach distally to the medial border of the scapula (Moore et al. 2013).

The *rhomboid major muscle* (Fig. 6.27) arises from tendinous fibers originating from the spinous processes of T2 through T5 and the supraspinous ligament at the corresponding levels. The muscle fibers course inferolaterally and attach via a tendinous band to the medial border of the scapula, between the root of the scapular spine and the inferior angle (Standring 2020).

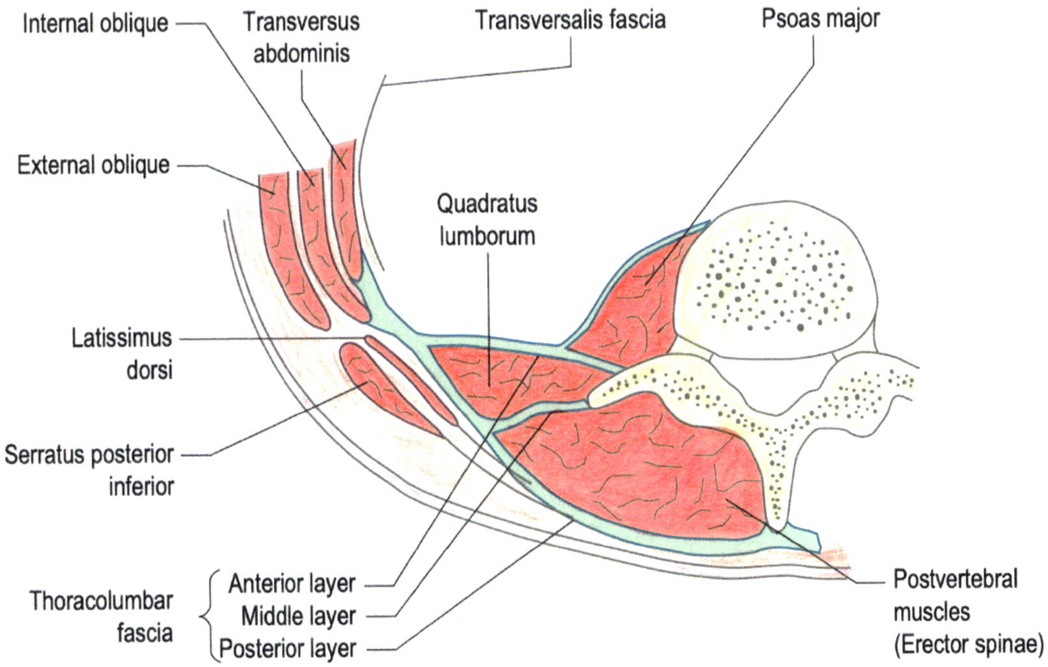

Fig. 6.26 Transverse section through the lumbar spine region, illustrating the formation and arrangement of the thoracolumbar fascia. (Adapted with permission from Palastanga and Soames (2012))

The *rhomboid minor muscle* is positioned immediately superior to the rhomboid major muscle, with its fibers also running in an inferolateral direction. The rhomboid minor muscle originates from the lower portion of the ligamentum nuchae and the spinous processes of C7 and T1 vertebrae, and it inserts on the medial border of the scapula at the level of the root of the scapular spine (Moore et al. 2013). Both rhomboid muscles receive innervation from the dorsal scapular nerve, which originates from the anterior primary division of the C5 spinal nerve.

The *levator scapulae muscle* (see Fig. 6.27) originates from tendinous slips arising from the transverse processes of the atlas and axis vertebrae and the posterior tubercles of the transverse processes of C3 and C4. The muscle fibers descend to attach to the medial border of the scapula, between the root of the scapular spine and the superior angle. Innervation is provided by branches from the anterior rami of the C3 and C4 spinal nerves, along with the dorsal scapular nerve (C5) (Standring 2020). When the cervical spine is stabilized, the levator scapulae muscle contributes to the elevation and rotation of the scapula, resulting in the inferior movement of the lateral angle (Paine and Voight 2013).

6.9.3 Third Layer Muscles: Serratus Posterior Superior and Serratus Posterior Inferior

The third layer of back muscles is composed of two thin, nearly quadrangular muscles: the serratus posterior superior and serratus posterior inferior (Moore et al. 2013).

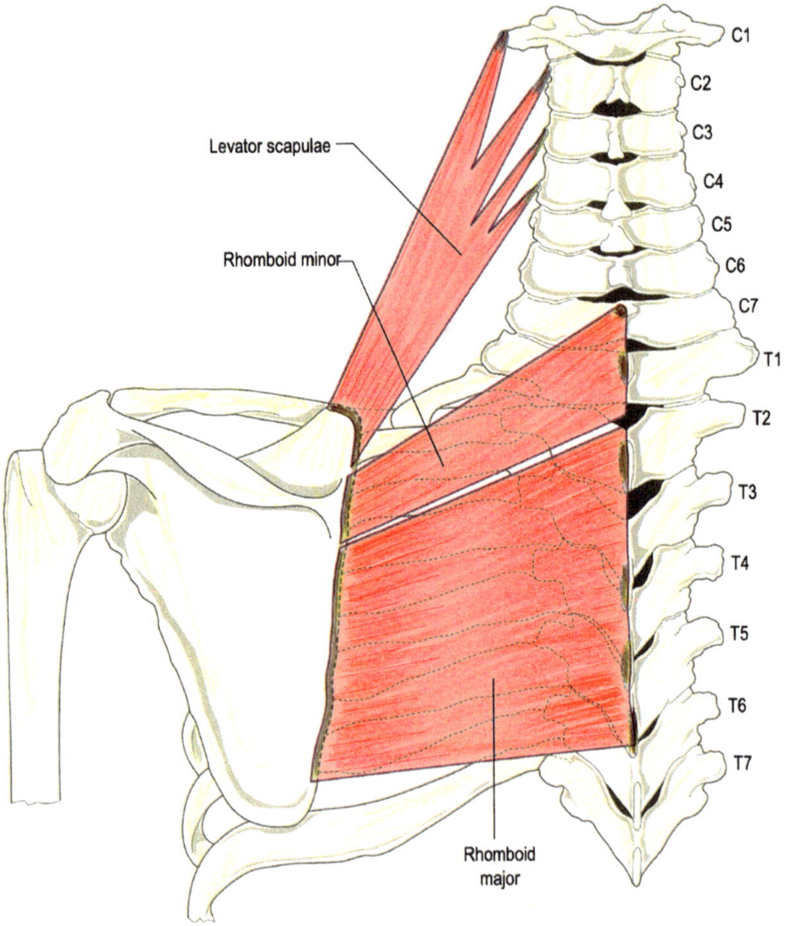

Fig. 6.27 Rhomboid major, rhomboid minor, and levator scapulae muscles. (Adapted with permission from Palastanga and Soames (2012))

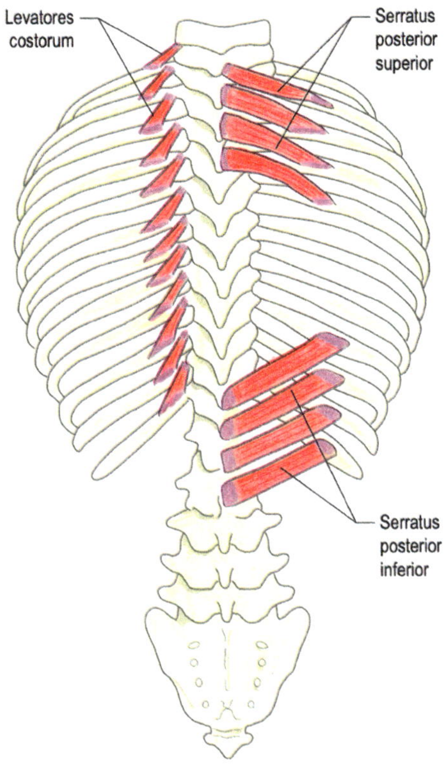

Fig. 6.28 Serratus posterior inferior, serratus posterior superior, and levator costorum muscles. (Adapted with permission from Palastanga and Soames (2012))

The *serratus posterior inferior muscle* (Fig. 6.28) originates from the spinous processes and the intervening supraspinous ligament, spanning from T11 to L2. The muscle fibers course in a superolateral direction to attach to the posterior and inferior surfaces of the lower four ribs (Standring 2020). Innervation of the serratus posterior inferior muscle is provided by the lower three intercostal nerves (T9 to T11) and the subcostal nerve, all of which are anterior primary divisions of their respective spinal nerves (Netter 2019). The primary function of the serratus posterior inferior muscle is to depress the ninth through twelfth ribs, potentially assisting in forced expiration (Neumann 2017).

The *serratus posterior superior muscle* (see Fig. 6.28) arises from the spinous processes of C7 through T3 and the supraspinous ligament connecting these processes. The muscle fibers run inferolaterally, attaching to the posterior and superior aspect of the second through fifth ribs (Moore et al. 2013). The serratus posterior superior muscle is involved in elevating the second through fourth ribs, potentially assisting with inspiration (Standring 2020).

6.9.4 Fourth Layer Muscles: Splenius Capitis and Splenius Cervicis

The fourth layer, the most superficial layer of the deep back muscles, consists of two muscles with fibers oriented in a superolateral direction: the splenius capitis and splenius cervicis (Moore et al. 2013).

The *splenius capitis muscle* originates from the inferior half of the ligamentum nuchae and the spinous processes spanning from C7 through T3 or T4 vertebrae. The muscle fibers course superolaterally to insert at the mastoid process of the temporal bone and the occiput, just below the lateral third of the superior nuchal line (Standring 2020).

The *splenius cervicis muscle* originates from the spinous processes of T3 through T6. Its fibers run parallel to those of the splenius capitis muscle and attach to the transverse processes of C1 and C2 and the posterior tubercles of the transverse processes of C3 and occasionally C4 (Kapandji 2008). These attachments are situated deep to the origins of the levator scapulae muscle. Both the splenius capitis and splenius cervicis muscles are innervated by lateral branches of the posterior primary divisions of the midcervical and lower cervical spinal nerves, respectively.

6.9.5 Fifth Layer Muscles: Erector Spinae

The fifth layer, comprising the largest group of back muscles, is known as the erector spinae or sacrospinalis muscle group (Fig. 6.29). These muscles form a series of longitudinal muscles that span the length of the spine, filling the groove lateral to the spinous processes. Their primary functions include extension and lateral flexion of

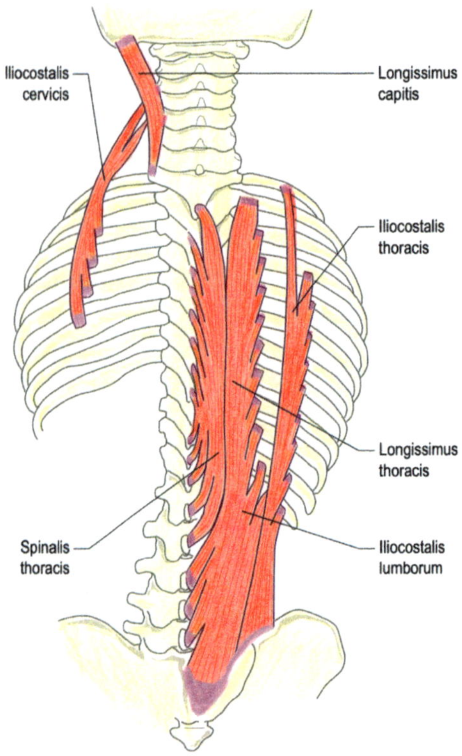

Fig. 6.29 Erector spinae muscles: iliocostalis, longissmus, and spinalis muscles. (Adapted with permission from Palastanga and Soames (2012))

the spine. The erector spinae muscles receive segmental innervation from the lateral branches of the posterior primary divisions (posterior rami) of the nearby spinal nerves. They are enveloped posteriorly in the thoracic and lumbar regions by the thoracolumbar fascia. This muscle group is subdivided into three groups from lateral to medial: the iliocostalis, longissimus, and spinalis muscles. Each group is further divided into three subdivisions based on their insertion points along the spine (e.g., lumborum, thoracis, cervicis, capitis) (Drake et al. 2020; Moore et al. 2013; Neumann 2017; Standring 2020).

The *iliocostalis muscles* originate inferiorly from the common origin of the erector spinae muscles, which includes the spinous processes and supraspinous ligament of T11 through L5 (supraspinous ligament terminates between L2 and L4), the median sacral crest, sacrotuberous ligament, the long posterior sacroiliac ligament, the lateral sacral crest, and the posteromedial iliac crest (Kapandji 2008). The *iliocostalis lumborum muscle*, the most inferior and lateral of the erector spinae muscles, courses superiorly to attach to the posterior and inferior surfaces of the angles of the lower six to nine ribs. It is innervated by lateral branches of the posterior primary divisions of the lumbar and lower thoracic spinal nerves (Standring 2020). The *iliocostalis thoracis* muscle arises from the superior aspect of the angles of the lower six ribs and ascends to attach to the angles of approximately the upper six ribs and the transverse process of the C7 vertebra. The *iliocostalis cervicis muscle* originates from the superior aspect of the angle of the third through sixth ribs and inserts superiorly onto the posterior tubercles of the transverse processes of C4 to C6 vertebrae (Moore et al. 2013).

The *longissimus muscles* lie medial to the iliocostalis group and comprise the thoracis, cervicis, and capitis divisions. The *longissimus thoracis muscle*, the largest of the erector spinae muscles, originates from the common origin of the erector spinae muscles and the transverse and accessory processes of the lumbar vertebrae. As the longest muscle of the back, the longissimus thoracis muscle inserts superiorly on the third through twelfth ribs, between their angles and tubercles, and attaches to the transverse processes of all 12 thoracic vertebrae (Moore et al. 2013; Standring 2020). The *longissimus cervicis muscle* originates from the transverse processes of the upper thoracic vertebrae (T1 to T5) and inserts on the posterior tubercle of the transverse processes and the articular processes of C3 through C6, as well as the posterior aspect of the transverse process and the articular process of C2. The *longissimus capitis muscle* arises from the upper thoracic transverse processes (T1 to T5) and articular processes of C4 through C7, terminating on the mastoid process of the temporal bone (Drake et al. 2020; Neumann 2017).

The *spinalis muscles*, consisting of thoracis, cervicis, and capitis divisions, originate from and insert onto spinous processes, with the exception of the spinalis capitis muscle (Kapandji 2008). The *spinalis thoracis muscle* arises from the spinous processes of the lower thoracic and upper lumbar vertebrae (T10 to L2) and inserts onto the spinous processes of the upper thoracic vertebrae (T2–T8) (Moore et al. 2013). The *spinalis cervicis muscle*, which is often absent or poorly developed, originates from the spinous processes of the lower cervical and upper thoracic vertebrae (C6 to T2) and inserts onto the spinous process of C2 (Standring 2020). The *spinalis capitis muscle* arises from the spinous processes of the lower cervical and upper thoracic vertebrae (C6 to T3) and the ligamentum nuchae. It inserts onto the occipital bone between the superior and inferior nuchal lines. This muscle is often indistinguishable from the semispinalis capitis (Drake et al. 2020).

6.9.6 Sixth Layer Muscles: Semispinalis, Multifidus, and Rotatores

The sixth layer of back muscles encompasses the deep back muscles, characterized by fibers that extend superiorly and medially. These muscles, often referred to as the transversospinalis group, typically originate from transverse processes and insert onto spinous processes. The muscles within this group are arranged in a manner where their lengths decrease from superficial to deep. The primary components of this group include the semispinalis, multifidus, and rotatores muscles (Cael 2010; Moore et al. 2013).

The *semispinalis muscles* are confined to the thoracic and cervical regions, subdivided into three distinct parts: semispinalis thoracis, semispinalis cervicis, and semispinalis capitis. The *semispinalis thoracis muscle* comprises slender muscular fasciculi interspersed between long tendons, attaching inferiorly to the transverse processes of the lower six thoracic vertebrae and superiorly to the spinous processes from C6 to T4 (Standring 2020). In contrast, the *semispinalis cervicis muscle* forms a thicker muscle mass, originating from the transverse processes of the upper five or six thoracic vertebrae and sometimes from the articular processes of the lower four cervical vertebrae. The *semispinalis cervicis* muscle primarily inserts onto the spinous process of the axis, with additional attachments to the spinous processes of C3 to C5 (Moore et al. 2013; Standring 2020). The *semispinalis capitis muscle*, the most robust and well-developed component of the semispinalis group, arises from the transverse processes of C7 to T6 and the articular processes of C4 to C6. It attaches superiorly to the medial region between the occiput's superior and inferior nuchal lines (Drake et al. 2020) (Fig. 6.30).

The *multifidus muscles* (see Fig. 6.30) are situated deep to the semispinalis muscles, occupying the space between the transverse and spinous processes of the vertebrae. This muscle group is composed of numerous muscular and tendinous fasciculi that originate from the posterior surface of the sacrum, posterior superior iliac spine, mammillary processes of the lumbar vertebrae, transverse processes of the thoracic vertebrae, and articular processes of the lower four cervical vertebrae. These fasciculi extend two to four, and

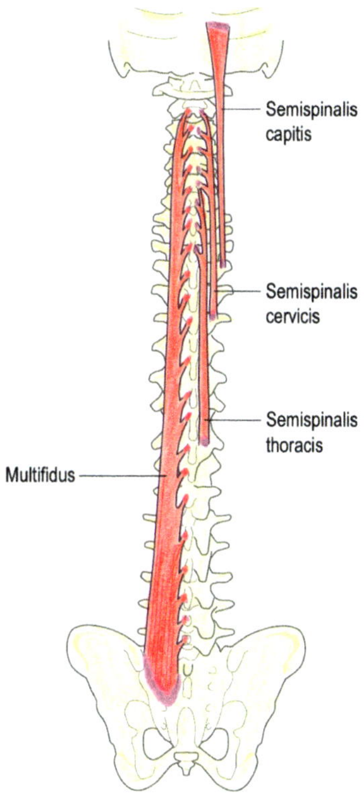

Fig. 6.30 Multifidus, semispinalis capitus, cervicis, and thoracis muscles. (Adapted with permission from Palastanga and Soames (2012))

occasionally five, vertebral segments before inserting onto a spinous process. The multifidus muscles attach to all vertebrae, with the exception of the atlas. The primary functions of the multifidus muscles are extension and contralateral rotation of the vertebral column (Cael 2010; Moore et al. 2013; Standring 2020).

Positioned deeper than the multifidus group, the *rotatores muscles* represent the deepest muscle fasciculi positioned in the groove between the spinous and transverse processes. This groove extends along the entire length of the vertebral column, from the sacrum to the axis (Standring 2020). The rotatores muscles are further categorized into rotatores brevis and rotatores longus. The *rotatores brevis* fascicles originate from the transverse processes and insert onto the root of the immediately superior vertebral spinous process. In contrast, the *rotatores longus* fascicles have similar origins but insert onto the root of the spinous process two vertebrae above. The rotatores longus muscles are positioned superficial to the rotatores brevis muscles (Drake et al. 2020; Moore et al. 2013). The rotatores muscles are most prominent in the thoracic region and are less developed in the cervical and lumbar regions, where they are often indistinguishable from the multifidus muscles. Unilateral contraction of the rotatores muscles contributes to the rotation of the spine, resulting in the movement of the vertebral bodies away from the contracted side (Cael 2010).

6.10 Muscles Involved in Movements of the Neck and Trunk

6.10.1 Muscles Involved in Movements of the Neck

The cervical region stands out as one of the most anatomically complex areas in the human body. Within this region, various compartments can be identified, each defined by specific muscle groups. Some of these compartments house critical structures, such as the carotid arteries. However, it's worth noting that these anatomical divisions often lack a clear functional basis (Standring 2020).

The neck is characterized by numerous small muscles, many of which cannot be palpated. Despite their size, these muscles play a crucial role in maintaining proper alignment and posture of the head and neck. While individual small muscles may exhibit specific actions, their primary function appears to be stabilizing the cranium on the vertebral column (Falla and Farina 2008).

In this discussion, the muscles have been categorized according to their primary actions, rather than their anatomical positions within the neck. The movements described include both the mobility of the cervical spine and the articulations at the craniovertebral junction. Consequently, the muscular groups are additionally classified into the following:

- Muscles acting exclusively on the cervical spine
- Muscles influencing both the head and neck simultaneously
- Muscles acting solely on the head in relation to the neck

6.10.1.1 Flexor Muscles of the Head and Neck: Sternocleidomastoid, Scalenus Anterior, Longus Colli, Longus Capitis, Rectus Capitis Anterior

Sternocleidomastoid (Innervated by the Spinal Part of the Accessory Nerve [CN XI]: C2, C3)

The sternocleidomastoid is a long, strap-like muscle that traverses obliquely along the lateral aspect of the neck, situated near the anterior midline (Standring 2020) (Fig. 6.31). It originates from two heads:

- The sternal head emerges as a narrow, cylindrical tendon from the superior portion of the anterior surface of the manubrium sterni.
- The clavicular head arises as a broad, muscular attachment from the superior surface of the medial third of the clavicle.

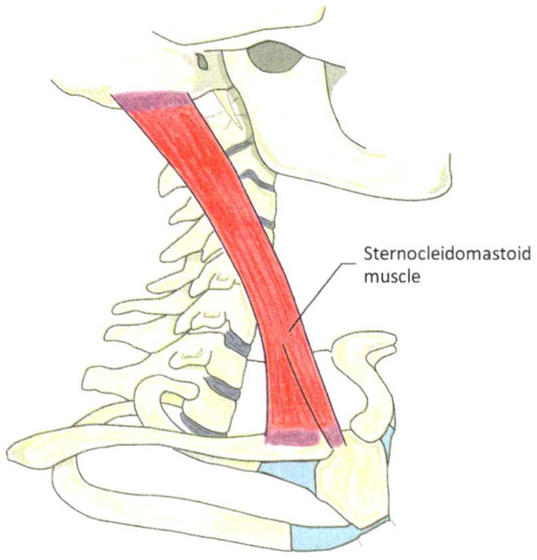

Fig. 6.31 Sternocleidomastoid muscle. (Adapted with permission from Palastanga and Soames (2012))

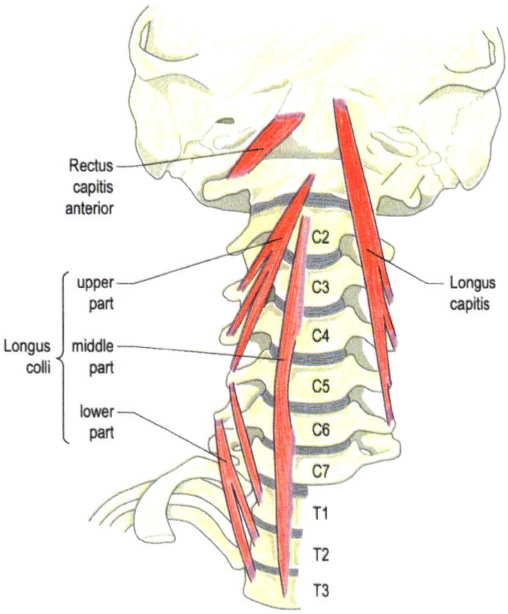

Fig. 6.32 Longus colli muscles. (Adapted with permission from Palastanga and Soames (2012))

Initially, a small gap separates these two heads, with the clavicular head positioned deep to the sternal head. As they ascend, they converge to form a relatively thick muscle belly. Superiorly, the sternocleidomastoid inserts via a short, strong, flattened tendon onto the lateral surface of the mastoid process of the temporal bone. Additionally, a thin aponeurosis attaches to the lateral third of the superior nuchal line of the occipital bone (Moore et al. 2013).

The spinal component of the accessory nerve enters the deep surface of the upper third of the sternocleidomastoid, exiting approximately midway along its lateral border. Sensory fibers may accompany the accessory nerve.

Unilateral contraction of the sternocleidomastoid results in ipsilateral head tilt and lateral flexion of the neck, accompanied by contralateral rotation of the head. The anterior fibers are thought to flex the head on the neck, while the posterior fibers (those attaching to the superior nuchal line) may extend the head at the atlanto-occipital joint (Netter 2019). When both muscles contract simultaneously, they produce flexion of the neck by drawing the head forward. If the head and neck are fixed, these muscles can potentially elevate the clavicle and manubrium sterni, consequently raising the ribs and thus functioning as accessory muscles of respiration during forced inspiration.

Longus Colli (Innervated by the Anterior Primary Rami of C3, C4, C5, C6)

The longus colli is a complex muscle composed of three distinct parts situated on the anterior and lateral aspects of the upper thoracic and cervical vertebral bodies (Standring 2020) (Fig. 6.32). Its anatomy can be described as follows:

- Inferior Portion: This segment originates from the anterior surface of the first through third thoracic vertebral bodies. It courses obliquely superiorly to insert on the anterior tubercles of the transverse processes of the fifth and sixth cervical vertebrae.
- Middle Portion: This segment runs vertically from the anterior aspect of the upper three thoracic and lower three cervical vertebral bodies. It attaches to the anterior surface of the second, third, and fourth cervical vertebral bodies.

- Superior Portion: This segment originates from the anterior tubercles of the transverse processes of the third, fourth, and fifth cervical vertebrae. It ascends to insert on the anterior tubercle of the atlas (C1).

The primary function of the longus colli, whether acting unilaterally or bilaterally, is cervical flexion (Neumann 2017). While it has been postulated that unilateral contraction of the inferior portion might contribute to ipsilateral lateral flexion and contralateral rotation, the central positioning of both its origin and insertion points renders this action questionable (Bogduk and Mercer 2000). The longus colli plays a significant role in maintaining cervical lordosis and providing anterior stability to the cervical spine. Its deep location and multi-segmental attachments make it an important contributor to fine motor control of cervical movements.

Longus Capitis (Innervated by the Anterior Primary Rami of C1, C2, C3, and Occasionally, C4)

The longus capitis, a slender and elongated muscle of the anterior neck, plays a significant role in cervical spine kinematics (Neumann 2017). This muscle originates from the anterior tubercles of the transverse processes of C3–C6 vertebrae, occasionally including C2 (Moore et al. 2013). It courses superiorly and medially, ultimately inserting on the basilar portion of the occipital bone, specifically lateral to the pharyngeal tubercle (Standring 2020).

The primary function of the longus capitis is to flex the head relative to the neck and the upper cervical spine. However, this movement typically occurs through the relaxation of the extensor muscles, with active flexion being necessary only when resistance is applied (Bogduk and Mercer 2000; Kapandji 2008).

Rectus Capitis Anterior (Innervated by the Anterior Primary Rami of C1 and C2)

The rectus capitis anterior, a short strap muscle, plays a significant role in the complex musculature of the neck (Standring 2020). This muscle is situated in a deep anatomical position, lying beneath the longus capitis. Its anatomical course originates from the anterior surface of the atlas's lateral mass and it extends upward and medially, ultimately inserting into the basilar part of the occipital bone. This insertion point is specifically located between the longus capitis and the occipital condyle (Standring 2020).

While the rectus capitis anterior contributes to the flexion of the head on the neck, its primary role appears to be more subtle yet crucial. Its main function is believed to be the stabilization of the atlanto-occipital joint during various head movements (Standring 2020).

6.10.1.2 Laterally Flexing Muscles of the Head and Neck: Scalenus Anterior, Scalenus Medius, Scalenus Posterior, Splenius Cervicis, Levator Scapulae, Sternocleidomastoid

Scalenus Anterior (Innervated by the Anterior Rami of C4, C5, and C6)

The scalenus anterior, the foremost of the three scalene muscles, occupies a significant position in the lateral neck region. As depicted in Fig. 6.33, this muscle is situated deep to the sternocleidomastoid and anterior to the scalenus medius (Moore et al. 2013). It originates from the anterior tubercles of the transverse processes of the third through sixth cervical vertebrae. From this point, the muscle fibers descend almost vertically, coursing deep to the prevertebral fascia. As they progress, these fibers converge to form a narrow tendon, which ultimately attaches to a prominent bony landmark known as the scalene tubercle, located on the inner border of the first rib.

When acting unilaterally, the scalenus anterior primarily laterally flexes the neck to the ipsilateral side, with a limited degree of contralateral rotation. Bilateral contraction of the scalenus anterior muscles results in neck flexion. Notably, when the upper attachments of these muscles are immobilized, they contribute to the mechanics of respiration. In the given situation, their function is to provide stability to the first rib throughout

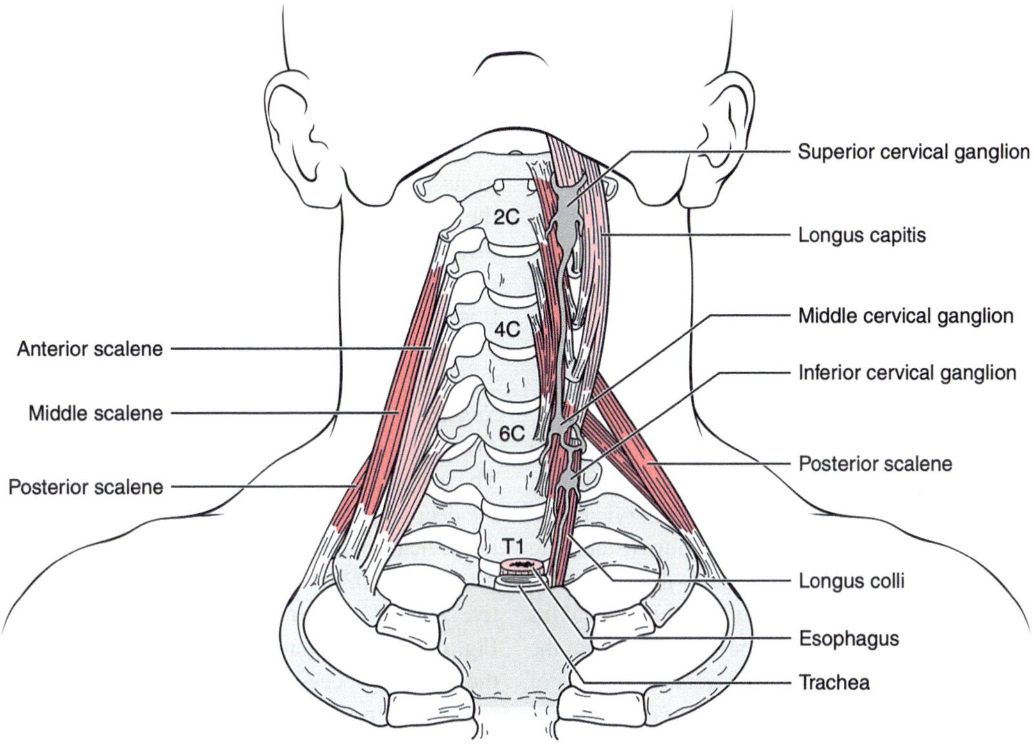

Fig. 6.33 Scalene muscles. (Adapted with permission from Jenkins (2009))

the process of breathing and may aid in its elevation, therefore facilitating the expansion of the upper thorax during inhalation (Neumann 2017). Application training involving reversed origin and insertion can be used for respiratory therapy in individuals with high tetraplegia.

Scalenus Medius (Innervated by the Anterior Primary Rami of C3–C8)

The scalenus medius, the middle and largest of the three scalene muscles, plays an important role in neck mechanics and respiratory function (Standring 2020). As illustrated in Fig. 6.33, this muscle occupies a significant portion of the lateral cervical region. The origins of the scalenus medius are extensive, arising from the transverse processes of the first and second cervical vertebrae, as well as from the posterior tubercles of the third to seventh cervical vertebrae (Moore et al. 2013). From these multiple points of origin, the muscle fibers course inferiorly and laterally, converging to insert onto a rough impression on the superior surface of the first rib. This insertion site is notably positioned posterior to the groove for the subclavian artery (Netter 2019).

The functional capabilities of the scalenus medius are significant and varied. When activated unilaterally, it serves as a powerful lateral flexor of the neck, drawing the cervical spine toward the side of contraction (Neumann 2017). This action is particularly important in activities requiring side-bending of the neck or in maintaining head position during unilateral load-bearing tasks. However, the role of the scalenus medius extends beyond mere movement production. When its superior attachments are fixed, this strong muscle becomes highly effective in stabilizing or elevating the first rib during respiratory efforts (Kendall et al. 2005). This respiratory function reveals the multifaceted nature of the scalenus medius and its significance in both postural control and breathing mechanics.

The anatomical position and substantial size of the scalenus medius make it a structure of particular interest in the context of thoracic outlet syndrome and cervical radiculopathy. Its close proximity to the brachial plexus and subclavian vessels means that hypertrophy or abnormal tension in this muscle can potentially contribute to neurovascular compression syndromes (Sanders et al. 2007). Therefore, a thorough understanding of the anatomy and function of the scalenus medius is important for physiatrists and rehabilitation specialists, particularly when addressing cervical spine disorders, respiratory dysfunction, or thoracic outlet syndrome.

Scalenus Posterior (Innervated by the Anterior Primary Rami of C6, C7, and C8)

The scalenus posterior, the smallest and most posteriorly situated of the three scalene muscles, plays a subtle yet significant role in cervical biomechanics and respiratory function (Standring 2020). As illustrated in Fig. 6.33, this muscle is located posterior to the other scalenes (Jenkins 2009). The scalenus posterior originates from the posterior tubercles of the transverse processes of the fourth, fifth, and sixth cervical vertebrae. From these points of origin, the muscle fibers course inferiorly and laterally, ultimately inserting onto the external surface of the second rib. This insertion site is notably positioned posterior to the attachment of the serratus anterior muscle (Moore et al. 2013).

Although less remarkable than those of its anterior and middle scalenes, the functional capacities of the scalenus posterior are also significant. Unilateral activation of this muscle results in lateral flexion of the neck, causing the cervical spine to be drawn toward the side of contraction (Neumann 2017). This action supplements the roles of the other neck muscles in generating and controlling lateral flexion motions of the head and neck. However, the role of the scalenus posterior extends beyond mere movement production. During respiratory efforts, particularly deep or forced breathing, this muscle may play a role in stabilizing the second rib (Kendall et al. 2005). This respiratory function, while secondary to its role in neck movement, underscores the multifaceted nature of the scalenus posterior and its integration into both the musculoskeletal and respiratory systems.

The anatomical position and attachments of the scalenus posterior make it a structure of interest in the broader context of cervical spine function and pathology. While less commonly implicated in thoracic outlet syndrome compared to its anterior and middle scalene counterparts, the scalenus posterior's proximity to the brachial plexus and its role in neck mechanics mean it cannot be overlooked in comprehensive evaluations of cervical and upper thoracic disorders (Sanders et al. 2007).

Splenius Cervicis (Innervated by the Posterior Primary Rami of C5, C6, and C7)

The splenius cervicis, a significant component of the deep muscles of the back, plays a crucial role in cervical spine movements and postural control (Standring 2020). This muscle exhibits a distinctive anatomical arrangement, as illustrated in Fig. 6.34. The splenius cervicis originates from the spinous processes of the third to sixth cervical vertebrae. From these points of origin, its muscle fibers course in a superior and lateral direction, ultimately inserting onto the posterior tubercles of the transverse processes of the upper three or four cervical vertebrae. Notably, this insertion occurs anterior to the attachment of the levator scapulae muscle (Moore et al. 2013).

The functional capabilities of the splenius cervicis are diverse and depend on whether it acts unilaterally or bilaterally. When activated

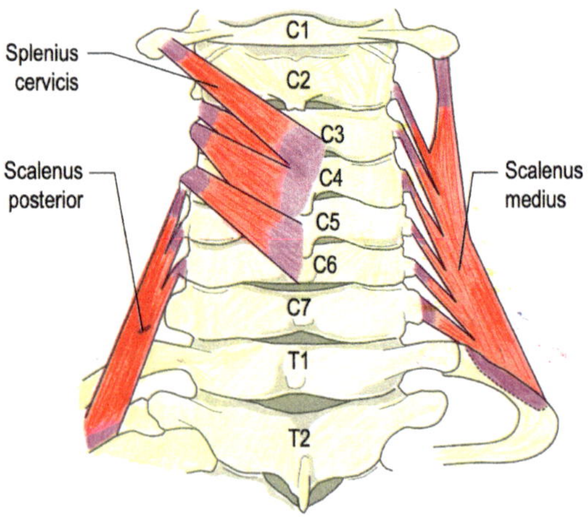

Fig. 6.34 Splenius cervicis muscle. (Adapted with permission from Palastanga and Soames (2012))

unilaterally, the splenius cervicis produces lateral flexion of the neck toward the same side, accompanied by a slight ipsilateral rotation (Neumann 2017). This combined action contributes to the complex three-dimensional movements of the cervical spine, allowing for precise control of head position. When both the right and left splenius cervicis muscles contract simultaneously, they work in synergy to extend the neck, pulling the cervical spine into a more upright position (Kendall et al. 2005).

The anatomical position and actions of the splenius cervicis make it a muscle of particular interest in the context of cervical spine biomechanics and pathology. Its role in both rotation and extension of the neck means that it can be implicated in a variety of cervical spine disorders, including cervicogenic headaches and chronic neck pain.

6.10.2 Muscles of the Movements of the Trunk and Thorax

The trunk, encompassing the lumbar and thoracic regions of the vertebral column, is capable of a diverse range of movements essential for daily activities and postural control. While the mobility between individual vertebrae is limited, the cumulative effect across multiple segments allows for significant flexion, extension, lateral bending, and rotation of the trunk (Neumann 2017; Standring 2020).

Flexion of the trunk involves the forward bending of the lumbar and thoracic spine, while extension refers to the opposite movement. These motions play a crucial role in many functional activities, particularly when coupled with hip flexion or extension (Magee 2014). However, it is important to note that when assessing trunk range of motion, care must be taken to isolate trunk movement from hip movement to ensure accurate measurement (Norkin and White 2016). Lateral flexion occurs when the shoulder level becomes inclined relative to the pelvis. This movement can be performed on both the left and right sides. During lateral flexion, adjacent vertebral bodies on the side of the bend approximate, while those on the opposite side separate (Moore et al. 2013). Rotation of the trunk involves a twisting motion to either the right or left while maintaining level shoulders. This movement results from the summation of individual vertebral rotations. It is worth noting that pelvic rotation around the hip joint can augment the overall rotational range of the shoulders. As with other trunk movements, accurate assessment of rotation requires stabilization of the pelvis (Kendall et al. 2005).

The thoracic cage, comprising the ribs articulating with the thoracic vertebrae and sternum, serves as an important protective structure for the underlying soft tissues. The muscles responsible for inspiration and expiration are intimately associated with this bony framework and are thus considered in this context (Tang and Bordoni 2024). The anterior abdominal wall muscles play a dual role. Not only do they contribute to trunk movements, but they also function to increase intra-abdominal pressure. This pressure increase aids in expiration and various straining activities such as micturition, parturition, coughing, and vomiting (Hodges et al. 2007).

6.10.2.1 Flexing Muscles of the Trunk: Rectus Abdominis, External Oblique, Internal Oblique, Psoas Major, Psoas Minor

Rectus Abdominis (Innervated by the Anterior Rami of the Lower Six or Seven Thoracic Nerves: T6 or T7 to T12)

The rectus abdominis, a prominent muscle of the anterior abdominal wall, plays a crucial role in trunk flexion and intra-abdominal pressure regulation (Neumann 2017). This paired muscle runs vertically on the front of the abdomen, enclosed within a fibrous structure known as the rectus sheath (Moore et al. 2013).

Originating from the anterior surface of the pubic symphysis and pubic crest via two tendons, the rectus abdominis ascends and widens, ultimately inserting onto the anterior surfaces of the xiphoid process and the costal cartilages of the fifth, sixth, and seventh ribs (Standring 2020). In lean individuals, the slightly convex lateral border of each muscle presents as a groove called the linea semilunaris, while the two rectus abdominis muscles are separated by a midline fibrous structure, the linea alba.

A unique feature of the rectus abdominis is the presence of transverse tendinous intersections, typically three in number, found in the anterior part of the muscle. These intersections are firmly attached to the rectus sheath and are positioned at specific anatomical landmarks: the lowest at the level of the umbilicus, the highest at the level of the xiphoid process, and the third approximately midway between these two (Netter 2019).

The rectus sheath, a complex fibrous envelope encasing each rectus abdominis muscle, is formed by the aponeuroses of the external oblique, internal oblique, and transversus abdominis muscles. The composition of this sheath varies at different levels of the abdomen (Kendall et al. 2005):

- Above the costal margin: The sheath is present only anteriorly, formed entirely by the aponeurosis of the external oblique.
- Between the costal margin and midway between the umbilicus and symphysis pubis: The anterior wall is formed by the aponeuroses of the external and the internal oblique, while the posterior wall consists of the aponeuroses of the internal oblique and transversus abdominis (Fig. 6.35a).
- Below the midpoint between the umbilicus and symphysis pubis: All three muscle aponeuroses pass anterior to the rectus abdominis, resulting in a deficient posterior wall (Fig. 6.35b). The inferior limit of the posterior layer is marked by a crescentic border called the arcuate line. Below this line, the rectus abdominis lies directly on the transversalis fascia, separated from extraperitoneal fat.

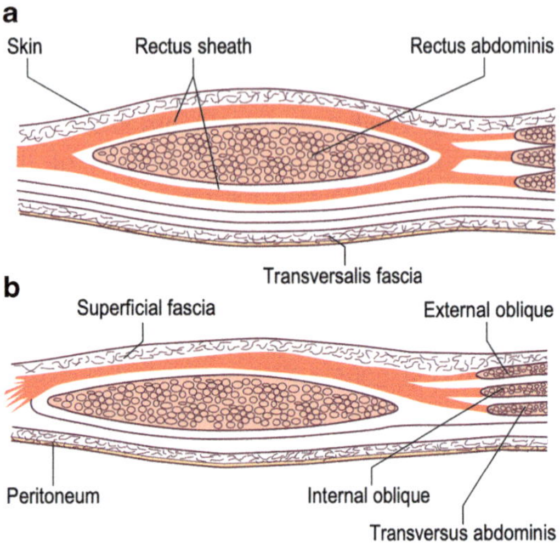

Fig. 6.35 Formation of the rectus sheath. The rectus abdominis is enclosed by aponeuroses of the external oblique, internal oblique, and transversus abdominis, which meet at the linea alba. (a) Above the arcuate line, the anterior wall is formed by external oblique and anterior internal oblique; the posterior wall by posterior internal oblique and transversus abdominis. (b) Below the arcuate line, all three aponeuroses pass anterior to rectus abdominis, leaving no posterior aponeurotic wall. (Adapted with permission from Palastanga and Soames (2012))

External Oblique (Innervated by the Anterior Rami of the Lower Six Thoracic Nerves: T7–T12)

The external oblique muscle is a key component of the anterolateral abdominal wall and is responsible for trunk rotation, lateral flexion, and regulation of intra-abdominal pressure (Neumann 2017). The fibers of the superficial layer of the anterior abdominal wall run in a distinct downward and medial direction, extending from the ribs toward the midline (Moore et al. 2013). The superior attachment of the external oblique is characterized by fleshy slips originating from the outer borders of the lower eight ribs and their costal cartilages. These origins form a complex interlocking pattern of muscle fiber through interweaving with the attachments of the serratus anterior superiorly and the latissimus dorsi inferiorly (Standring 2020). From these points of origin, the muscle fibers course inferiorly and medially, with those from the lower two ribs descending almost vertically to insert onto the outer lip of the anterior two-thirds of the iliac crest. This arrangement results in a free posterior border of the muscle extending between the 12th rib and the iliac crest. The remaining fibers of the external oblique give rise to a large aponeurosis, which is broader inferiorly than superiorly. This aponeurosis traverses the rectus abdominis muscle, contributing to the formation of the rectus sheath, and extends medially to fuse at the midline with the aponeurosis from the opposite side to form the linea alba. The linea alba, a fibrous raphe, runs from the tip of the xiphoid process of the sternum to the symphysis pubis, serving as a midline anchor for the abdominal muscles.

A significant anatomical feature associated with the external oblique is the inguinal ligament, formed by the lower free border of its aponeurosis. This ligament stretches between the pubic tubercle and the anterior superior iliac spine, creating a convex inferior border due to the pull of the fascia lata of the thigh, which attaches along its length (Tang and Bordoni 2024). The medial portion of the inguinal ligament expands posteriorly along the pecten pubis, forming the lacunar ligament. In the anatomical position, the lacunar ligament is nearly horizontal. A further extension of the lacunar ligament along the pecten is known as the pectineal ligament.

Superior to the pubic tubercle, a triangular cleft in the aponeurosis forms the superficial inguinal ring. This ring, with its base at the pubic crest and apex directed superolaterally, is bounded by medial and lateral crura. The superficial inguinal ring serves as an important landmark in inguinal hernia diagnosis and repair (Jenkins and O'Dwyer 2008). The external oblique, along with the internal oblique and transversus abdominis, forms the muscular corset of the abdomen. This arrangement not only facilitates movement but also plays a vital role in maintaining posture, supporting viscera, and assisting in respiratory and expulsive functions such as coughing, sneezing, and defecation (Hodges et al. 2007).

Internal Oblique (Innervated by the Anterior Primary Rami of T7–L1)

The internal oblique muscle, situated deep to the external oblique, forms the middle layer of the three sheets of abdominal wall musculature. This muscle is of significance in trunk rotation, lateral flexion, and the regulation of intra-abdominal pressure (Neumann 2017). The origins of the internal oblique are diverse and extensive. It arises from the lateral two-thirds of the inguinal ligament, the anterior two-thirds of the intermediate line of the iliac crest, and the thoracolumbar fascia (Moore et al. 2013). This broad origin allows the muscle fibers to fan out in a complex pattern, contributing to its versatile functional capabilities. The muscle fibers of the internal oblique demonstrate a variable orientation based on their location:

- The most posterior fibers ascend almost vertically to attach to the inferior borders of the lower four ribs.
- The more anterior and inferior fibers course superiorly and medially, transitioning into an aponeurosis along a line extending from the 10th costal cartilage to the body of the pubis (Standring 2020).

The aponeurosis of the internal oblique plays a complex role in the formation of the rectus sheath. It splits to envelop the rectus abdominis muscle before interlacing with the aponeurosis from the opposite side at the linea alba, contributing to the structural integrity of the anterior abdominal wall.

A notable feature of the internal oblique is the formation of the conjoint tendon. The portion of the muscle arising from the inguinal ligament courses medially and inferiorly, blending with the lower part of the transversus abdominis to form this tendon. The conjoint tendon then attaches to the pubic crest and pecten pubis, providing reinforcement to the inguinal region. An interesting anatomical relationship exists between the internal oblique and the male reproductive system. A few fibers from the inferomedial part of the muscle accompany the spermatic cord, forming the cremaster muscle. This muscle is responsible for the cremasteric reflex, which elevates the testis in response to various stimuli (Standring 2020).

The internal oblique, along with the external oblique and transversus abdominis, forms a dynamic muscular corset around the abdomen. This arrangement not only facilitates trunk movement but also plays a vital role in maintaining posture, supporting viscera, and assisting in respiratory and expulsive functions such as coughing, sneezing, and defecation (Hodges et al. 2007). The oblique muscles, including the internal oblique, play a significant role in the transfer of forces between the upper and lower body during functional activities. Their diagonal fiber orientation allows for the efficient transmission of rotational forces, making them key players in activities requiring trunk rotation, such as throwing or swinging motions.

The transversus abdominis, situated as the innermost of the four abdominal muscle layers, connects inferiorly to the inguinal ligament and iliac crest, posteriorly to the lumbar vertebrae (L1–L5), superiorly to the internal aspects of the lower seven ribs, and anteriorly to the linea alba. The contraction of its horizontally oriented fibers narrows the abdominal cavity. When the lumbar vertebrae are stabilized, it draws the abdominal wall inward. Conversely, with the anterior aponeurosis fixed, it contributes to the enhancement of lumbar lordosis.

6.10.2.2 Extending Muscles of the Trunk: Quadratus Lumborum, Multifidus, Semispinalis, Erector Spinae, Interspinales

Erector Spinae (Innervated by Adjacent Posterior Primary Rami)

The erector spinae, a powerful and complex group of muscles, plays a crucial role in spinal extension, lateral flexion, and postural control (Neumann 2017). This substantial muscular mass extends the entire length of the vertebral column, from the sacrum to the skull (see Fig. 6.29). In the lumbar region, the erector spinae presents as a broad, well-defined belly with a distinct lateral border. As it ascends, it differentiates into three parallel columns, each of which further subdivides into three parts based on its relative position (Moore et al. 2013). This intricate arrangement allows for fine control of spinal movements and contributes to the muscle group's versatility in function.

The inferior attachment of the erector spinae is characterized by a strong, thick, flat tendon that follows a U-shaped line encircling the origin of the multifidus muscle. This tendinous origin can be divided into medial and lateral components (Standring 2020). The medial limb of this tendon arises from the spinous processes of T11 to L5, extending to the supraspinous ligaments and the median sacral crest. The lateral limb attaches to the lateral sacral crest, sacrotuberous, sacrococcygeal, and posterior sacroiliac ligaments, as well as the posterior aspect of the iliac crest, medial to the internal oblique. Beneath the lateral limb, the erector spinae also has a fleshy attachment to the iliac tuberosity and the inner lip of the iliac crest. From these extensive origins, the muscle fibers ascend deep to the latissimus dorsi, splitting into the three distinct columns that contribute to its complex structure and function.

From this comprehensive origin, the muscle fibers of the erector spinae ascend deep to the latissimus dorsi, eventually dividing into three distinct columns. Each of these columns is further subdivided, contributing to the complex architecture of the posterior trunk musculature (Bogduk 2005). The erector spinae muscle group plays a vital role in maintaining upright posture, controlling spinal movements, and stabilizing the spine during various activities. Its extensive attachments and segmental innervation allow for both gross and fine motor control of the vertebral column. The erector spinae, along with the multifidus and other deep spinal muscles, forms part of the local muscle system of the spine. This system is responsible for providing segmental stability and control (Panjabi 1992).

Iliocostalis

The iliocostalis muscle, the most lateral column of the erector spinae group, plays a crucial role in spinal extension and lateral flexion (Neumann 2017). This muscle is characterized by its segmental arrangement, which allows for fine control of spinal movements and contributes to the overall stability of the vertebral column (Moore et al. 2013).

The iliocostalis can be divided into three parts based on its anatomical location and attachments:

- Iliocostalis lumborum: This most inferior portion of the muscle, inserts via six distinct slips into the inferior borders of the lower six ribs, near their angles (Standring 2020).

This arrangement allows the iliocostalis lumborum to exert a significant influence on lumbar spine extension and lateral flexion, as well as contributing to the control of rib cage movement during respiration.
- Iliocostalis thoracis: Arising medial to each insertion slip of the iliocostalis lumborum, the thoracic portion of the muscle ascends to insert near the angles of the upper six ribs and the transverse process of the seventh cervical vertebra (C7) (Netter 2019). This extensive attachment pattern enables the iliocostalis thoracis to play a role in thoracic spine extension and rotation, as well as contributing to the stabilization of the thoracic cage during various functional activities (Bogduk 2005).
- Iliocostalis cervicis: The most superior portion of the iliocostalis muscle originates medial to the insertion slips of the iliocostalis thoracis. It ascends to insert into the posterior tubercles of the transverse processes of the fourth to seventh cervical vertebrae (C4–C7) (Standring 2020). This arrangement allows the iliocostalis cervicis to contribute to cervical spine extension and lateral flexion, playing a role in head and neck posture and movement.

The iliocostalis, along with the other components of the erector spinae group, plays a vital role in maintaining an upright posture against gravity. Its activation is essential in activities that require spinal extension, such as lifting objects from the ground or maintaining an erect sitting posture. Moreover, its unilateral contraction contributes to lateral flexion of the spine, an action important in various functional and athletic activities (Neumann 2017).

Longissimus

The longissimus muscle, the intermediate column of the erector spinae group, is distinguished by its considerable length and thickness. The primary role of this muscle is to provide spinal extension, rotation, and stabilization between the vertebral column (Neumann 2017). The longissimus can be divided into three distinct parts based on its anatomical location and attachments:

- Longissimus thoracis: This portion of the muscle originates from the transverse and accessory processes of the lumbar vertebrae and the adjacent thoracolumbar fascia. It inserts via two sets of slips: one set attaches to the transverse processes of all 12 thoracic vertebrae, while the other set connects to the adjacent regions of the lower 10 ribs (Moore et al. 2013). This extensive attachment pattern allows the longissimus thoracis to exert significant influence over thoracic and lumbar spine extension and rotation, as well as contributing to rib cage stabilization during respiration.
- Longissimus cervicis: Arising from the transverse processes of the first six thoracic vertebrae (T1–T6), medial to the attachments of longissimus thoracis, this portion of the muscle ascends to insert into the posterior tubercles of the transverse processes of the second to sixth cervical vertebrae (C2–C6) (Standring 2020).
- Longissimus capitis: This most superior portion of the longissimus muscle has a complex origin, arising from two areas: (a) The transverse processes of the first five thoracic vertebrae (T1–T5), in common with longissimus cervicis; (b) The articular processes of the fourth to seventh cervical vertebrae (C4–C7) It inserts into the posterior aspect of the mastoid process of the temporal bone. This arrangement allows the longissimus capitis to contribute to head extension and rotation, playing a significant role in fine control of head positioning (Bogduk and Mercer 2000).

Spinalis

The spinalis muscle, the most medial and least developed column of the erector spinae group, plays a supportive role in spinal extension and stabilization (Neumann 2017). Despite its relatively minor size, the spinalis contributes to the overall function of the erector spinae complex.

Like its counterparts, the spinalis is divided into three parts:

- Spinalis thoracis: This is the most clearly defined portion of the spinalis muscle. It originates from the spinous processes of the eleventh thoracic to the second lumbar vertebrae (T11 to L2) and inserts into the spinous processes of the first to sixth thoracic vertebrae (T1 to T6) (Moore et al. 2013). This arrangement allows the spinalis thoracis to contribute to thoracic spine extension and stabilization.
- Spinalis cervicis and spinalis capitis: These portions are poorly developed and often blend with adjacent muscles, making their distinct actions difficult to isolate (Standring 2020). Their contributions to cervical spine and head movements are likely minor compared to the other erector spinae muscles.

When the three muscle columns of the erector spinae on both sides contract together, they extend the lumbar, thoracic, and cervical spines, as well as the head on the neck, making erector spinae the primary extensor of the trunk. Additionally, the erector spinae muscles are crucial for controlling trunk flexion. When the columns on one side contract together, they produce lateral flexion and rotation of the spine to the same side. During single-leg stance, the lower part of the erector spinae on the non-weight-bearing side contracts strongly to prevent pelvic drop. During walking, the erector spinae alternately contracts to stabilize the vertebral column on the pelvis. Due to the muscle's primary mass being in the lumbar region, particularly the longissimus thoracis, it plays a significant role in maintaining the secondary lumbar curvature during sitting and standing.

6.10.2.3 Rotating Muscles of the Trunk: Multifidus, Rotatores, Semispinalis, Internal Oblique, External Oblique

The rotational axis of the trunk is controlled by the coordinated action of several muscle groups. Left trunk rotation is achieved through the simultaneous contraction of the right external oblique and left internal oblique muscles. Conversely, right trunk rotation results from the coordinated activation of the left external oblique and right internal oblique muscles. These rotational movements may be accompanied by a degree of trunk flexion (Neumann 2017).

Multifidus (Innervated by the Posterior Primary Rami of Adjacent Spinal Nerves)

The multifidus muscle occupies a deep position relative to the semispinalis and erector spinae muscles, situated in the paravertebral groove between the spinous and transverse processes of the vertebrae throughout the spinal column. The multifidus originates laterally from several points, including the posterior aspect of the sacrum, the fascia overlying the erector spinae, the mammillary processes of the lumbar vertebrae, the transverse processes of the thoracic vertebrae, and the articular processes of the lower four or five cervical vertebrae (Standring 2020) (see Fig. 6.30).

From this extensive origin, the muscle fibers of the multifidus are arranged in three distinct layers as they course superiorly and medially to insert onto the spinous processes of all vertebrae from the fifth lumbar (L5) to the axis (C2). The deepest layer attaches to the immediately superior vertebra, the middle layer inserts onto the second or third vertebra above, and the superficial layer extends to the third or fourth vertebra above their origin (Bogduk 2005). The multifidus muscle plays a role in the rotational movement of the spinal segment on the opposite side. It attaches to the spinous processes from a place that is relatively laterally position.

Rotatores (Innervated by the Posterior Primary Rami of Adjacent Spinal Nerves)

The rotatores muscles (Fig. 6.36) are most prominent in the thoracic region, with less consistent presence in the lumbar and cervical segments. These muscles originate adjacent to the transverse process of one vertebra and insert on the lamina of the vertebra immediately superior (Standring 2020).

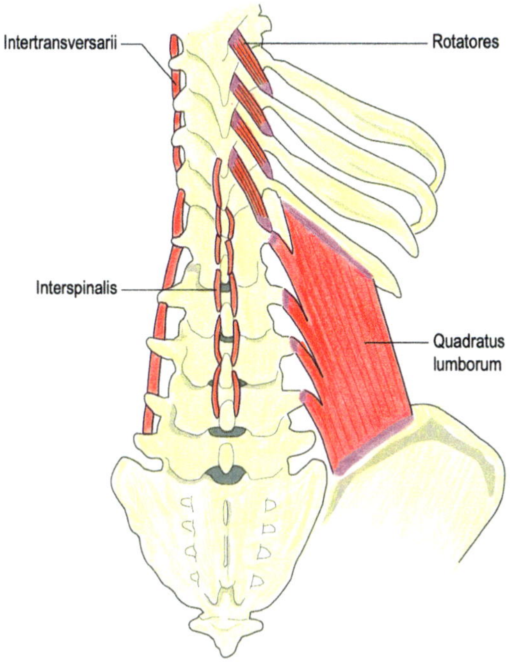

Fig. 6.36 Quadratus lumborum and intertransversalii muscles. (Adapted with permission from Palastanga and Soames (2012))

While multifidus can generate rotation, extension, and lateral flexion of the vertebral column at all levels, rotatores primarily produce rotation in the thoracic region. However, the functional significance of both muscle groups likely lies in their role as vertebral column stabilizers. They act as extensible ligaments, adjusting their length to stabilize adjacent vertebrae regardless of the overall position of the spine (Panjabi 1992).

Semispinalis (Innervated by the Posterior Primary Rami of Adjacent Spinal Nerves)

The semispinalis muscle group (see Fig. 6.30) spans from the lower thoracic region to the base of the skull and comprises three distinct parts:

- Semispinalis thoracis originates from the transverse processes of T6–T10 and inserts on the spinous processes of C6–T2.
- Semispinalis cervicis, a larger component, extends from the transverse processes of T1–T6 to the spinous processes of C2–C6.
- Semispinalis capitis, the largest part, originates from the transverse processes of T1–T6 and the articular processes of C4–C7, inserting on a medial impression between the superior and inferior nuchal lines at the skull base. The medial portion of the semispinalis capitis may occasionally be distinct and is referred to as spinalis capitis (Moore et al. 2013).

Bilateral activation of the semispinalis muscle group produces extension of the thoracic and cervical spine. Unilateral contraction results in rotation of the trunk and neck to the contralateral side (Neumann 2017).

6.10.2.4 Laterally Flexing Muscles of the Trunk: Quadratus Lumborum, External Oblique, Internal Oblique, Rectus Abdominis, Erector Spinae, Multifidus, Intertransversarii

Quadratus Lumborum (Innervated by the Anterior Primary Rami of T12–L4)

The quadratus lumborum is a large, flat, quadrilateral muscle situated in the posterior abdominal wall (see Fig. 6.36), extending between the pelvis and the 12th rib, positioned deep to the erector spinae. Its inferior attachment is to the iliolumbar ligament and the adjacent posterior portion of the iliac crest. The muscle fibers course superiorly and slightly medially to insert on the medial half of the 12th rib's inferior border. During its ascent, the medial border of the quadratus lumborum attaches to the lateral aspect of the anterior surface of all lumbar vertebrae's transverse processes. The anterior and middle layers of the thoracolumbar fascia encase this muscle (Standring 2020).

Unilateral contraction of the quadratus lumborum induces ipsilateral trunk lateral flexion. In a single-leg stance, it exhibits strong activation on the non-weight-bearing side to prevent pelvic drop. Additionally, it stabilizes the 12th rib during deep inspiration, providing a fixed origin for the diaphragm. Bilateral contraction assists in lumbar spine extension and contributes to lateral stability.

Intertransversarii (Innervated by the Anterior Primary Rami of Adjacent Spinal Nerves)
The intertransversarii are small muscular slips (see Fig. 6.36) bridging adjacent transverse processes in the cervical and lumbar regions. In the cervical spine, they are well developed, with each muscle comprising up to four slips extending between adjacent transverse processes from the atlas (C1) to T1. The lumbar intertransversarii exist as paired muscular slips: the lateral slip connects adjacent transverse processes, while the medial slip extends from the accessory process of one vertebra to the mammillary process of the vertebra above. The lateral slips may extend as far superiorly as the transverse process of T10. In the cervical region, these muscular slips are positioned both anterior and posterior to the emerging ventral ramus of the spinal nerve, whereas in the lumbar region, they lie posterior to the ventral ramus (Moore et al. 2013).

Unilateral activation of the intertransversarii produces ipsilateral lateral flexion in the lumbar and cervical regions. However, their primary function is to act as extensible ligaments, thereby stabilizing adjacent vertebral segments during trunk movements (Bogduk 2005).

References

Adams MA, Hutton WC. The mechanical function of the lumbar apophyseal joints. Spine (Phila Pa 1976). 1983;8(3):327–30.

Adams MA, Roughley PJ. What is intervertebral disc degeneration, and what causes it? Spine (Phila Pa 1976). 2006;31(18):2151–61.

Ahmed M, Bjurholm A, Kreicbergs A, Schultzberg M. Neuropeptide Y, tyrosine hydroxylase and vasoactive intestinal polypeptide-immunoreactive nerve fibers in the vertebral bodies, discs, dura mater, and spinal ligaments of the rat lumbar spine. Spine. 1993;18(2):268–73.

Althoff B, Goldie IF. The arterial supply of the odontoid process of the axis. Acta Orthop Scand. 1977;48(6):622–9.

Altman J, Bayer SA. Development of the human spinal cord: an interpretation based on experimental studies. 1st ed. New York: Oxford University Press; 2001.

Antoniou J, Steffen T, Nelson F, Winterbottom N, Hollander AP, Poole RA, et al. The human lumbar intervertebral disc: evidence for changes in the biosynthesis and denaturation of the extracellular matrix with growth, maturation, ageing, and degeneration. J Clin Invest. 1996;98(4):996–1003.

Aprill C, Bogduk N. The prevalence of cervical zygapophyseal joint pain: a first approximation. Spine (Phila Pa 1976). 1992;17(7):744–7.

Bagnall KM, Harris PF, Jones PR. A radiographic study of the human fetal spine. 1. The development of the secondary cervical curvature. J Anat. 1977;123(3):777–82.

Barker PJ, Briggs CA. Attachments of the posterior layer of lumbar fascia. Spine. 2007;32(17):1757–64.

Batson OV. The function of the vertebral veins and their role in the spread of metastases. Ann Surg. 1940;112(1):138–49.

Beaman DN, Graziano GP, Glover RA, Wojtys EM, Chang V. Substance P innervation of lumbar spine facet joints. Spine. 1993;18(8):1044–9.

Been E, Kalichman L. Lumbar lordosis. Spine J. 2014;14(1):87–97.

Behrsin JF, Briggs CA. Ligaments of the lumbar spine: a review. Surg Radiol Anat. 1988;10(3):211–9.

Benzel EC. Spine surgery: techniques, complication avoidance, and management. 4th ed. Philadelphia: Elsevier; 2017.

Bernardini N, Levey AI, Augusti-Tocco G. Rat dorsal root ganglia express m1–m4 muscarinic receptor proteins. J Peripher Nerv Syst. 1999;4(3–4):222–32.

Bernhardt M, Bridwell KH. Segmental analysis of the sagittal plane alignment of the normal thoracic and lumbar spines and thoracolumbar junction. Spine (Phila Pa 1976). 1989;14(7):717–21.

Biglioli P, Roberto M, Cannata A, Parolari A, Fumero A, Grillo F, et al. Upper and lower spinal cord blood supply: the continuity of the anterior spinal artery and the relevance of the lumbar arteries. J Thorac Cardiovasc Surg. 2004;127(4):1188–92.

Bland JH. The reversibility of osteoarthritis: a review. Am J Med. 1983;74(6A):16–26.

Bogduk N. The clinical anatomy of the cervical dorsal rami. Spine (Phila Pa 1976). 1982;7(4):319–30.

Bogduk N. The innervation of the vertebral column. Aust J Physiother. 1985;31(3):89–94.

Bogduk N. Clinical anatomy of the lumbar spine and sacrum. 4th ed. Edinburgh: Churchill Livingstone; 2005.

Bogduk N. Clinical and radiological anatomy of the lumbar spine. 5th ed. Toronto: Elsevier; 2012.

Bogduk N, Marsland A. The cervical zygapophysial joints as a source of neck pain. Spine (Phila Pa 1976). 1988;13(6):610–7.

Bogduk N, Mercer S. Biomechanics of the cervical spine. I: Normal kinematics. Clin Biomech (Bristol, Avon). 2000;15(9):633–48.

Bogduk N, Windsor M, Inglis A. The innervation of the cervical intervertebral discs. Spine (Phila Pa 1976). 1988;13(1):2–8.

Bosmia AN, Tubbs RS, Hogan E, Bohnstedt BN, Denardo AJ, Loukas M, et al. Blood supply to the human spinal cord: part I. Anatomy and hemodynamics. Clin Anat. 2015;28(1):52–64.

Boszczyk BM, Boszczyk AA, Putz R. Comparative and functional anatomy of the mammalian lumbar spine. Anat Rec. 2001;264(2):157–68.

Butler SJ, Bronner ME. From classical to current: analyzing peripheral nervous system and spinal cord lineage and fate. Dev Biol. 2015;398(2):135–46.

Byfield D, Kinsinger S. A manual therapist's guide to surface anatomy and palpation skills. 1st ed. Oxford: Butterworth-Heinemann; 2002.

Cael C. Functional anatomy: musculoskeletal anatomy, kinesiology, and palpation for manual therapists. Philadelphia: Lippincott Williams & Wilkins; 2010.

Cao KD, Grimm MJ, Yang KH. Load sharing within a human lumbar vertebral body using the finite element method. Spine. 2001;26(12):E253–60.

Cavanaugh JM, Kallakuri S, Özaktay AC. Innervation of the rabbit lumbar intervertebral disc and posterior longitudinal ligament. Spine. 1995;20(19):2080–5.

Chakravorty BG. Arterial supply of the cervical spinal cord (with special reference to the radicular arteries). Anat Physiol Curr Res. 1971;170(3):311–29.

Chaynes P, Verdié JC, Moscovici J, Zadeh J, Vaysse P, Becue J. Microsurgical anatomy of the internal vertebral venous plexuses. Surg Radiol Anat. 1998;20(1):47–51.

Chikuda H, Seichi A, Takeshita K, Shoda N, Ono T, Matsudaira K, Kawaguchi H, Nakamura K. Radiographic analysis of the cervical spine in patients with retro-odontoid pseudotumors. Spine (Phila Pa 1976). 2009;34(3):E110–4.

Christ B, Scaal M. Formation and differentiation of avian somite derivatives. Adv Exp Med Biol. 2008;638:1–41.

Cil A, Yazici M, Uzumcugil A, Kandemir U, Alanay A, Alanay Y, et al. The evolution of sagittal segmental alignment of the spine during childhood. Spine (Phila Pa 1976). 2005;30(1):93–100.

Cooperstein R, Haneline MT. Spinous process palpation using the scapular tip as a landmark vs a radiographic criterion standard. J Chiropr Med. 2007;6(3):87–93.

Cooperstein MA, Haneline MT, Young MD. The location of the inferior angle of the scapula in relation to the spinal level of prone patients. J Can Chiropr Assoc. 2009;53(2):121–8.

Cooperstein R, Haneline M, Young M. The location of the inferior angle of the scapula in relation to the spine in the upright position: a systematic review of the literature and meta-analysis. Chiropr Man Therap. 2015;23:7.

Cramer GD, Darby SA. Clinical anatomy of the spine, spinal cord, and ANS. 3rd ed. St. Louis: Elsevier; 2014.

Crock HV, Yoshizawa H. The blood supply of the vertebral column and spinal cord in man. Wien: Springer-Verlag; 1977.

Czervionke LF, Daniels DL, Ho PS, Yu SW, Pech P, Strandt J, et al. Cervical neural foramina: correlative anatomic and MR imaging study. Radiology. 1988;169(3):753–9.

Demondion X, Delfaut EM, Drizenko A, Boutry N, Francke JP, Cotten A. Radio-anatomic demonstration of the vertebral lumbar venous plexuses: an MRI experimental study. Surg Radiol Anat. 2000;22(3–4):151–6.

Dickman CA, Crawford NR, Brantley AG, Sonntag VK. Biomechanical effects of transoral odontoidectomy. Neurosurgery. 1995;36(6):1146–52.

Dimeglio A, Canavese F. The growing spine: how spinal deformities influence normal spine and thoracic cage growth. Eur Spine J. 2012;21(1):64–70.

Dommisse GF. The blood supply of the spinal cord: a critical vascular zone in spinal surgery. J Bone Joint Surg Br. 1974;56(2):225–35.

Downey BJ, Taylor NF, Niere KR. Manipulative physiotherapists can reliably palpate nominated lumbar spinal levels. Man Ther. 1999;4(3):151–6.

Drake RL, Vogl AW, Mitchell AW. Gray's anatomy for students. 4th ed. Philadelphia: Elsevier; 2020.

Dupuis PR, Yong-Hing K, Cassidy JD, Kirkaldy-Willis WH. Radiologic diagnosis of degenerative lumbar spinal instability. Spine (Phila Pa 1976). 1985;10(3):262–76.

Dvorak J, Panjabi MM. Functional anatomy of the alar ligaments. Spine (Phila Pa 1976). 1987;12(2):183–9.

Ebraheim NA, Lu J. Morphometric evaluation of the sacral dorsal root ganglia: an anatomic and clinical study. Surg Radiol Anat. 1998;20(2):105–8.

Ebraheim NA, An HS, Xu R, Ahmad M, Yeasting RA. The quantitative anatomy of the cervical nerve root groove and the intervertebral foramen. Spine. 1996;21(14):1619–23.

Ebraheim NA, Lu J, Biyani A, Brown JA, Yeasting RA. Anatomic considerations for uncovertebral involvement in cervical spondylosis. Clin Orthop Relat Res. 1997a;334:200–6.

Ebraheim NA, Xu R, Knight T, Yeasting RA. Morphometric evaluation of lower cervical pedicle and its projection. Spine (Phila Pa 1976). 1997b;22(1):1–6.

Edgar MA, Ghadially JA. Innervation of the lumbar spine. Clin Orthop Relat Res. 1976;115:35–41.

Etz CD, Kari FA, Mueller CS, Silovitz D, Brenner RM, Lin HM, et al. The collateral network concept: a reassessment of the anatomy of spinal cord perfusion. J Thorac Cardiovasc Surg. 2011;141(4):1020–8.

Falla D, Farina D. Neuromuscular adaptation in experimental and clinical neck pain. J Electromyogr Kinesiol. 2008;18(2):255–61.

Farmer JC, Wisneski RJ. Cervical spine nerve root compression. An analysis of neuroforaminal pressures with varying head and arm positions. Spine. 1994;19(16):1850–5.

Feltrin GP, Macchi V, Saccavini C, Tosi E, Dus C, Fassina A, et al. Fractal analysis of lumbar vertebral cancellous bone architecture. Clin Anat. 2001;14(6):414–7.

Fesmire FM, Luten RC. The pediatric cervical spine: developmental anatomy and clinical aspects. J Emerg Med. 1989;7(2):133–42.

Foreman SM, Croft AC. Whiplash injuries: the cervical acceleration/deceleration syndrome. 2nd ed. Baltimore: Williams & Wilkins; 2002.

Fujiwara A, Lim TH, An HS, Tanaka N, Jeon CH, Andersson GB, et al. The effect of disc degeneration and facet joint osteoarthritis on the segmental flexibility of the lumbar spine. Spine (Phila Pa 1976). 2000;25(23):3036–44.

Gebauer M, Lohse C, Barvencik F, Pogoda P, Rueger JM, Püschel K, et al. Subdental synchondrosis and anatomy of the axis in aging: a histomorphometric study on 30 autopsy cases. Eur Spine J. 2006;15(3):292–8.

Giacobetti FB, Vaccaro AR, Bos-Giacobetti MA, Deeley DM, Albert TJ, Farmer JC, et al. Vertebral artery occlusion associated with cervical spine trauma. A prospective analysis. Spine (Phila Pa 1976). 1997;22(2):188–92.

Giles LG, Taylor JR. Innervation of lumbar zygapophysial joint synovial folds. Acta Orthop Scand. 1987;58(1):43–6.

Gilsanz V. Bone density in children: a review of the available techniques and indications. Eur J Radiol. 1998;26(2):177–82.

Gilsanz V, Gibbens DT, Carlson M, Boechat MI, Cann CE, Schulz EE. Peak trabecular vertebral density: a comparison of adolescent and adult females. Calcif Tissue Int. 1988;43(4):260–2.

Gluncic V, Ivkic G, Marin D, Percac S. Anomalous origin of both vertebral arteries. Clin Anat. 1999;12(4):281–4.

Grant JP, Oxland TR, Dvorak MF. Mapping the structural properties of the lumbosacral vertebral endplates. Spine. 2001;26(8):889–96.

Griepp EB, Di Luozzo G, Schray D, Stefanovic A, Geisbüsch S, Griepp RB. The anatomy of the spinal cord collateral circulation. Ann Cardiothorac Surg. 2012;1(3):350–7.

Griessenauer CJ, Raborn J, Foreman P, Shoja MM, Loukas M, Tubbs RS. Venous drainage of the spine and spinal cord: a comprehensive review of its history, embryology, anatomy, physiology, and pathology. Clin Anat. 2015;28(1):75–87.

Groen GJ, Baljet B, Drukker J. Nerves and nerve plexuses of the human vertebral column. Am J Anat. 1990;188(3):282–96.

Groen RJ, Groenewegen HJ, van Alphen HA, Hoogland PV. Morphology of the human internal vertebral venous plexus: a cadaver study after intravenous Araldite CY 221 injection. Anat Rec. 1997;249(2):285–94.

Haher TR, O'Brien M, Dryer JW, Nucci R, Zipnick R, Leone DJ. The role of the lumbar facet joints in spinal stability. Identification of alternative paths of loading. Spine (Phila Pa 1976). 1994;19(23):2667–70.

Haneline MT, Cooperstein R, Young MD, Ross J. Determining spinal level using the inferior angle of the scapula as a reference landmark: a retrospective analysis of 50 radiographs. J Can Chiropr Assoc. 2008;52(1):24–9.

Harper AA, Lawson SN. Conduction velocity is related to morphological cell type in rat dorsal root ganglion neurons. J Physiol. 1985;359:31–46.

Harrison DE, Harrison DD, Troyanovich SJ. Three-dimensional spinal coupling mechanics: part II. Implications for chiropractic theories and practice. J Manip Physiol Ther. 1998;21(3):177–86.

Hasegawa T, An HS, Haughton VM, Nowicki BH. Lumbar foraminal stenosis: critical heights of the intervertebral discs and foramina. A cryomicrotome study in cadavera. J Bone Joint Surg Am. 1995;77(1):32–8.

Hasue M, Kikuchi S, Sakuyama Y, Ito T. Anatomic study of the interrelation between lumbosacral nerve roots and their surrounding tissues. Spine (Phila Pa 1976). 1983;8(1):50–8.

Haynes MJ, Cala LA, Melsom A, Mastaglia FL, Milne N, McGeachie JK. Vertebral arteries and cervical rotation: modeling and magnetic resonance angiography studies. J Manip Physiol Ther. 2002;25(6):370–83.

Higuchi K, Sato T. Anatomical study of lumbar spine innervation. Folia Morphol (Warsz). 2002;61(2):71–9.

Ho PS, Yu SW, Sether LA, Wagner M, Ho KC, Haughton VM. Ligamentum flavum: appearance on sagittal and coronal MR images. Radiology. 1988;168(2):469–72.

Hodges PW, Sapsford R, Pengel LH. Postural and respiratory functions of the pelvic floor muscles. Neurourol Urodyn. 2007;26(3):362–71.

Hong JT, Park DK, Lee MJ, Kim SW, An HS. Anatomical variations of the vertebral artery segment in the lower cervical spine: analysis by three-dimensional computed tomography angiography. Spine (Phila Pa 1976). 2008;33(22):2422–6.

Humzah MD, Soames RW. Human intervertebral disc: structure and function. Anat Rec. 1988;220(4):337–56.

Hussain M, Gay RE, An KN. Reduction in disk and fiber stresses by axial distraction is higher in cervical disk with fibers oriented toward loading direction: a finite element model analysis. J Manip Physiol Ther. 2010a;33(4):252–60.

Hussain M, Natarajan RN, An HS, Andersson GB. Motion changes in adjacent segments due to moderate and severe degeneration in C5-C6 disc: a poroelastic C3-T1 finite element model study. Spine (Phila Pa 1976). 2010b;35(9):939–47.

Hussain M, Natarajan RN, An HS, Andersson GB. Progressive disc degeneration at C5–C6 segment affects the mechanics between disc heights and posterior facets above and below the degenerated segment: a poroelastic C3-T1 finite element model study. Spine (Phila Pa 1976). 2012;35(16):E753–60.

Iida T, Abumi K, Kotani Y, Kaneda K. Effects of aging and spinal degeneration on mechanical properties of lumbar supraspinous and interspinous ligaments. Spine J. 2002;2(2):95–100.

Imai S, Hukuda S, Maeda T. Dually innervating nociceptive networks in the rat lumbar posterior longitudinal ligaments. Spine (Phila Pa 1976). 1995;20(19):2086–92.

Issever AS, Burghardt A, Patel V, Laib A, Lu Y, Ries M, et al. A micro-computed tomography study of the trabecular bone structure in the femoral head. J Musculoskelet Neuronal Interact. 2003;3(2):176–84.

Janis JE, Hatef DA, Reece EM, McCluskey PD, Schaub TA, Guyuron B. Neurovascular compression of the greater occipital nerve: implications for migraine headaches. Plast Reconstr Surg. 2010;126(6):1996–2001.

Jaumard NV, Welch WC, Winkelstein BA. Spinal facet joint biomechanics and mechanotransduction in normal, injury and degenerative conditions. J Biomech Eng. 2011;133(7):071010.

Jenkins DB. Hollinshead's functional anatomy of the limbs and back. 9th ed. St. Louis: Elsevier; 2009.

Jenkins JT, O'Dwyer PJ. Inguinal hernias. BMJ. 2008;336(7638):269–72.

Jiang H, Raso JV, Moreau MJ, Russell G, Hill DL, Bagnall KM. Quantitative morphology of the lateral ligaments of the spine. Assessment of their importance in maintaining lateral stability. Spine (Phila Pa 1976). 1994;19(23):2676–82.

Joaquim AF, Ghizoni E, Tedeschi H, Appenzeller S, Riew KD. Radiological evaluation of cervical spine involvement in rheumatoid arthritis. Neurosurg Focus. 2015;38(4):E4.

Johnson G, Bogduk N, Nowitzke A, House D. Anatomy and actions of the trapezius muscle. Clin Biomech (Bristol, Avon). 1994;9(1):44–50.

Jovanovic MS. A comparative study of the foramen transversarium of the sixth and seventh cervical vertebrae. Surg Radiol Anat. 1990;12(3):167–72.

Ju G, Hökfelt T, Brodin E, Fahrenkrug J, Fischer JA, Frey P, et al. Primary sensory neurons of the rat showing calcitonin gene-related peptide immunoreactivity and their relation to substance P-, somatostatin-, galanin-, vasoactive intestinal polypeptide- and cholecystokinin-immunoreactive ganglion cells. Cell Tissue Res. 1987;247(2):417–31.

Kalichman L, Hunter DJ. Lumbar facet joint osteoarthritis: a review. Semin Arthritis Rheum. 2007;37(2):69–80.

Kapandji IA. The physiology of the joints, Volume 3: the spinal column, pelvic girdle and head. 6th ed. Philadelphia: Churchill Livingstone; 2008.

Kasai T, Ikata T, Katoh S, Miyake R, Tsubo M. Growth of the cervical spine with special reference to its lordosis and mobility. Spine (Phila Pa 1976). 1996;21(18):2067–73.

Kendall FP, McCreary EK, Provance PG, Rodgers MM, Romani WA. Muscles: testing and function with posture and pain. 5th ed. Baltimore: Lippincott Williams & Wilkins; 2005.

Keogh B, Ebbs S. Normal surface anatomy. London: William Heinemann Medical Books; 1984.

Kirkaldy-Willis WH, Hayes WC. Pathology and pathogenesis of lumbar spondylosis and stenosis. Spine. 1978;3(4):319–28.

Kirkaldy-Willis WH, Wedge JH, Yong-Hing K, Reilly J. Pathology and pathogenesis of lumbar spondylosis and stenosis. Spine (Phila Pa 1976). 1978;3(4):319–28.

Kissel P, Youmans JR. Posttraumatic anterior cervical osteophyte and dysphagia: surgical report and literature review. J Spinal Disord. 1992;5(1):104–7.

Ko HY. Management and rehabilitation of spinal cord injuries. 2nd ed. Singapore: Springer; 2022.

Koebke J, Brade H. Morphological and functional studies on the lateral joints of the first and second cervical vertebrae in man. Anat Embryol (Berl). 1982;164(3):265–75.

Koller H, Robinson Y. Cervical spine surgery: standard and advanced techniques. Cham: Springer; 2018.

Kosaka H, Sairyo K, Biyani A, Leaman D, Yeasting R, Higashino K, et al. Pathomechanism of loss of elasticity and hypertrophy of lumbar ligamentum flavum in elderly patients with lumbar spinal canal stenosis. Spine (Phila Pa 1976). 2007;32(25):2805–11.

Kwon BK, Song F, Morrison WB, Grauer JN, Beiner JM, Vaccaro AR, et al. Morphologic evaluation of cervical spine anatomy with computed tomography: anterior cervical plate fixation considerations. J Spinal Disord Tech. 2004;17(2):102–7.

Kyriakos M, El-Khoury GY, McDonald DJ. Tumors and tumor-like lesions of bone. In: Pope TL, Bloem HL, Beltran J, Morrison WB, Wilson DJ, editors. Musculoskeletal imaging. 2nd ed. Philadelphia: Elsevier Saunders; 2011. p. 1017–140.

Lazorthes G, Gouaze A, Zadeh JO, Santini JJ, Lazorthes Y, Burdin P. Arterial vascularization of the spinal cord. J Neurosurg. 1971;35(3):253–62.

Le Douarin NM, Dupin E. Cell lineage analysis in neural crest ontogeny. J Neurosci. 1993;24(2):146–61.

Le Douarin NM, Kalcheim C. The neural crest. 2nd ed. Cambridge: Cambridge University Press; 1999.

Lee RM. Morphology of cerebral arteries. Pharmacol Ther. 1995;66:149–73.

Lemke AJ, Benndorf G, Liebig T, Felix R. Anomalous origin of the right vertebral artery: review of the literature and case report of right vertebral artery origin distal to the left subclavian artery. AJNR Am J Neuroradiol. 1999;20(7):1318–21.

Lewinnek GE, Warfield CA. Facet joint degeneration as a cause of low back pain. Clin Orthop Relat Res. 1986;213:216–22.

Li YK, Zhang YK, Lu CM, Zhong SZ. Changes and implications of blood flow velocity of the vertebral artery during rotation and extension of the head. J Manip Physiol Ther. 1999;22(2):91–5.

Limthongkul W, Karaikovic EE, Savage JW, Markovic A. Volumetric analysis of thoracic and lumbar vertebral bodies. Spine J. 2010;10(2):153–8.

Liu J, Ebraheim NA, Sanford CG Jr, et al. Quantitative changes in the cervical neural foramen resulting from axial traction: in vivo imaging study. Spine J. 2008;8(4):619–23.

Louis R. Spinal stability as defined by the three-column spine concept. Anat Clin. 1985;7(1):33–42.

Lu J, Ebraheim NA. Anatomic considerations of C2 nerve root ganglion. Spine (Phila Pa 1976). 1998;23(6):649–52.

Lumley JSP. Surface anatomy: the anatomical basis of clinical examination. 4th ed. Philadelphia: Churchill Livingstone; 2008.

Maat GJ, Matricali B, van Meerten EL. Postnatal development and structure of the neurocentral junction. Its relevance for spinal surgery. Spine (Phila Pa 1976). 1996;21(6):661–6.
Mac-Thiong JM, Labelle H, Roussouly P. Sagittal spino-pelvic balance in normal children and adolescents. Eur Spine J. 2007;16(2):227–34.
Magee DJ. Orthopedic physical assessment. 6th ed. St. Louis: Saunders Elsevier; 2014.
Martirosyan NL, Feuerstein JS, Theodore N, Cavalcanti DD, Spetzler RF, Preul MC. Blood supply and vascular reactivity of the spinal cord under normal and pathological conditions. J Neurosurg Spine. 2011;15(3):238–51.
Masharawi Y, Rothschild B, Dar G, Peleg S, Robinson D, Been E, et al. Facet orientation in the thoracolumbar spine: three-dimensional anatomic and biomechanical analysis. Spine. 2004;29(16):1755–63.
Masharawi Y, Salame K, Mirovsky Y, Peleg S, Dar G, Steinberg N, et al. Vertebral body shape variation in the thoracic and lumbar spine: characterization of its asymmetry and wedging. Clin Anat. 2008;21(1):46–54.
Matsunaga S, Yamaguchi M, Hayashi K, Sakou T. Genetic analysis of ossification of the posterior longitudinal ligament. Spine (Phila Pa 1976). 1999;24(10):937–8.
McLain RF. Mechanoreceptor endings in human cervical facet joints. Spine (Phila Pa 1976). 1994;19(5):495–501.
McLain RF, Pickar JG. Mechanoreceptor endings in human thoracic and lumbar facet joints. Spine (Phila Pa 1976). 1998;23(2):168–73.
Mekonen HK, Hikspoors JP, Mommen G, Köhler SE, Lamers WH. Development of the ventral body wall in the human embryo. J Anat. 2015;227(5):673–85.
Menezes AH, Traynelis VC. Anatomy and biomechanics of normal craniovertebral junction (a) and biomechanics of stabilization (b). Childs Nerv Syst. 2008;24(10):1091–100.
Mercer S, Bogduk N. The ligaments and annulus fibrosus of human adult cervical intervertebral discs. Spine (Phila Pa 1976). 1999;24(7):619–26.
Miele VJ, Panjabi MM, Benzel EC. Anatomy and biomechanics of the spinal column and cord. Handb Clin Neurol. 2012;109:31–43.
Milz S, Schlüter T, Putz R, Moriggl B, Ralphs JR, Benjamin M. Fibrocartilage in the transverse ligament of the human atlas. Spine (Phila Pa 1976). 2001;26(16):1765–71.
Moore RJ, Osti OL, Vernon-Roberts B, Fraser RD. Changes in endplate vascularity after an outer anulus tear in the sheep. Spine (Phila Pa 1976). 1992;17(8):874–8.
Moore RJ, Vernon-Roberts B, Osti OL, Fraser RD. Remodeling of vertebral bone after outer anular injury in sheep. Spine (Phila Pa 1976). 1996;21(18):2108–14.
Moore KL, Dalley AF, Agur AMR. Clinically oriented anatomy. 8th ed. Philadelphia: Wolters Kluwer; 2013.
Moroney SP, Schultz AB, Miller JA, Andersson GB. Load-displacement properties of lower cervical spine motion segments. J Biomech. 1988;21(9):769–79.
Mosekilde L, Mosekilde L. Normal vertebral body size and compressive strength: relations to age and to vertebral and iliac trabecular bone compressive strength. Bone. 1986;7(3):207–12.
Mosekilde L, Mosekilde L. Sex differences in age-related changes in vertebral body size, density and biomechanical competence in normal individuals. Bone. 1990;11(2):67–73.
Mtui E, Gruener G, Dockery P. Fitzgerald's clinical neuroanatomy and neuroscience. 8th ed. Edinburgh: Saunders/Elsevier; 2020.
Muhle C, Resnick D, Ahn JM, Südmeyer M, Heller M. In vivo changes in the neuroforaminal size at flexion-extension and axial rotation of the cervical spine in healthy persons examined using kinematic magnetic resonance imaging. Spine. 2001;26(13):E287–93.
Muralimohan S, Pande A, Vasudevan MC, Ramamurthi R. Suboccipital segment of the vertebral artery: a cadaveric study. Neurol India. 2009;57(4):447–52.
Nachemson AL, Evans JH. Some mechanical properties of the third human lumbar interlaminar ligament (ligamentum flavum). J Biomech. 1968;1(3):211–20.
Nathoo N, Caris EC, Wiener JA, Mendel E. History of the vertebral venous plexus and the significant contributions of Breschet and Batson. Neurosurgery. 2011;69(5):1007–14.
Netter FH. Atlas of human anatomy. 7th ed. Philadelphia: Elsevier; 2019.
Neumann DA. Kinesiology of the musculoskeletal system: foundations for rehabilitation. 3rd ed. St. Louis: Mosby; 2017.
Norkin CC, White DJ. Measurement of joint motion: a guide to goniometry. 5th ed. Philadelphia: F.A. Davis Company; 2016.
Ochia RS, Inoue N, Renner SM, Lorenz EP, Lim TH, Andersson GB, et al. Three-dimensional in vivo measurement of lumbar spine segmental motion. Spine (Phila Pa 1976). 2006;31(18):2073–8.
Oda I, Abumi K, Lü D, Shono Y, Kaneda K. Biomechanical role of the posterior elements, costovertebral joints, and rib cage in the stability of the thoracic spine. Spine (Phila Pa 1976). 1996;21(12):1423–9.
Oga M, Yuge I, Terada K, Shimizu A, Sugioka Y. Tortuosity of the vertebral artery in patients with cervical spondylotic myelopathy. Spine (Phila Pa 1976). 1996;21(9):1085–9.
Oliver J, Middleditch A. Functional anatomy of the spine. Oxford: Butterworth-Heinemann; 1991.
Olsewski JM, Simmons EH, Kallen FC, Mendel FC, Severin CM, Berens DL. Morphometry of the lumbar spine: anatomical perspectives related to transpedicular fixation. J Bone Joint Surg Am. 1990;72(4):541–9.
Onan OA, Heggeness MH, Hipp JA. A motion analysis of the cervical facet joint. Spine (Phila Pa 1976). 1998;23(4):430–9.

O'Rahilly R, Müller F. Developmental stages in human embryos: revised and new measurements. Cells Tissues Organs. 2010;192(2):73–84.
Osmotherly PG, Rivett DA, Mercer SR. Revisiting the clinical anatomy of the alar ligaments. Eur Spine J. 2013;22(1):60–4.
Oxland TR, Grant JP, Dvorak MF, Fisher CG. Effects of endplate removal on the structural properties of the lower lumbar vertebral bodies. Spine. 2003;28(8):771–7.
Paik NC. Bilateral cervical spondylolysis of C7. Spine J. 2010;10(11):e10–3.
Paine R, Voight ML. The role of the scapula. Int J Sports Phys Ther. 2013;8(5):617–29.
Pal GP, Routal RV. Transmission of weight through the lower thoracic and lumbar regions of the vertebral column in man. J Anat. 1987;152:93–105.
Pal GP, Routal RV, Saggu SK. The orientation of the articular facets of the zygapophyseal joints at the cervical and upper thoracic region. J Anat. 2001;198(4):431–41.
Palastanga N, Soames R. Anatomy and human movement: structure and function. 6th ed. Edinburgh: Elsevier; 2012.
Panjabi MM. The stabilizing system of the spine. Part I. Function, dysfunction, adaptation, and enhancement. J Spinal Disord. 1992;5(4):383–9.
Panjabi M, Dvorak J, Duranceau J, Yamamoto I, Gerber M, Rauschning W, et al. Three-dimensional movements of the upper cervical spine. Spine (Phila Pa 1976). 1988;13(7):726–30.
Panjabi MM, Duranceau JS, Oxland TR, Bowen CE. Multidirectional instabilities of traumatic cervical spine injuries in a porcine model. Spine (Phila Pa 1976). 1989;14(10):1111–5.
Panjabi M, Dvorak J, Crisco J 3rd, Oda T, Hilibrand A, Grob D. Flexion, extension, and lateral bending of the upper cervical spine in response to alar ligament transections. J Spinal Disord. 1991a;4(2):157–67.
Panjabi MM, Oxland TR, Parks EH. Quantitative anatomy of cervical spine ligaments. Part II. Middle and lower cervical spine. J Spinal Disord. 1991b;4(3):277–85.
Panjabi MM, Oxland T, Takata K, Goel V, Duranceau J, Krag M. Articular facets of the human spine. Quantitative three-dimensional anatomy. Spine (Phila Pa 1976). 1993;18(10):1298–310.
Panjabi MM, Crisco JJ, Vasavada A, Oda T, Cholewicki J, Nibu K, et al. Mechanical properties of the human cervical spine as shown by three-dimensional load-displacement curves. Spine (Phila Pa 1976). 2001;26(24):2692–700.
Paris SV. Anatomy as related to function and pain. Orthop Clin North Am. 1983;14(3):475–89.
Parke WW, Gammell K, Rothman RH. Arterial vascularization of the cauda equina. J Bone Joint Surg Am. 1981;63(1):53–62.
Patestas MA, Gartner LP. A textbook of neuroanatomy. 2nd ed. Hoboken: Wiley Blackwell; 2016.
Pearcy MJ, Tibrewal SB. Axial rotation and lateral bending in the normal lumbar spine measured by three-dimensional radiography. Spine. 1984;9(6):582–7.
Pech P, Daniels DL, Williams AL, Haughton VM. The cervical neural foramina: correlation of microtomy and CT anatomy. Radiology. 1985;155(1):143–6.
Peterson C, Phillips L, Linden A, Hsu W. Vertebral artery hypoplasia: prevalence and reliability of identifying and grading its severity on magnetic resonance imaging scans. J Manip Physiol Ther. 2010;33(3):207–11.
Pfirrmann CW, Binkert CA, Zanetti M, Boos N, Hodler J. MR morphology of alar ligaments and occipitoatlantoaxial joints: study in 50 asymptomatic subjects. Radiology. 2001;218(1):133–7.
Phillips DR, Barnard S, Mullee MA, Hurley MV. Simple anatomical information improves the accuracy of locating specific spinous processes during manual examination of the low back. Man Ther. 2009;14(3):346–50.
Piatt JH Jr, Grissom LE. Developmental anatomy of the atlas and axis in childhood by computed tomography. J Neurosurg Pediatr. 2011;8(3):235–43.
Postacchini F, Gumina S, Cinotti G, Perugia D, DeMartino C. Ligamenta flava in lumbar disc herniation and spinal stenosis. Light and electron microscopic morphology. Spine (Phila Pa 1976). 1994;19(8):917–22.
Price J, Mudge AW. A subpopulation of rat dorsal root ganglion neurones is catecholaminergic. Nature. 1983;301(5897):241–3.
Przybylski GJ, Carlin GJ, Patel PR, Woo SL. Human anterior and posterior cervical longitudinal ligaments possess similar tensile properties. J Orthop Res. 1996;14(6):1005–8.
Przybylski GJ, Patel PR, Carlin GJ, Woo SL. Quantitative anthropometry of the subatlantal cervical longitudinal ligaments. Spine (Phila Pa 1976). 1998;23(18):1944–50.
Pysyk CL, Persaud D, Bryson GL, Lui A. Ultrasound assessment of the vertebral level of the palpated intercristal (Tuffier's) line. Can J Anaesth. 2010;57(1):46–9.
Ratcliffe JF. The arterial anatomy of the adult human lumbar vertebral body: a microarteriographic study. J Anat. 1980;131(1):57–79.
Roussouly P, Pinheiro-Franco JL. Biomechanical analysis of the spino-pelvic organization and adaptation in pathology. Eur Spine J. 2011;20(5):609–18.
Saldinger P, Dvorak J, Rahn BA, Perren SM. Histology of the alar and transverse ligaments. Spine (Phila Pa 1976). 1990;15(4):257–61.
Sanders RJ, Hammond SL, Rao NM. Diagnosis of thoracic outlet syndrome. J Vasc Surg. 2007;46(3):601–4.
Scaal M. Early development of the vertebral column. Semin Cell Dev Biol. 2016;49:83–91.
Schievink WI. Spontaneous dissection of the carotid and vertebral arteries. N Engl J Med. 2001;344:898–906.
Schiff DC, Parke WW. The arterial supply of the odontoid process. J Bone Joint Surg Am. 1973;55(7):1450–6.
Schünke M, Schulte E, Schumacher U. Thieme atlas of anatomy: neck and internal organs. New York: Thieme; 2010.
Shingyouchi Y, Nagahama A, Niida M. Ligamentous ossification of the cervical spine in the late middle-aged Japanese men. Its relation to body mass index

and glucose metabolism. Spine (Phila Pa 1976). 1996;21(21):2474–8.
Silva MJ, Wang C, Keaveny TM, Hayes WC. Direct and computed tomography thickness measurements of the human lumbar vertebral shell and endplate. Bone. 1994;15(4):409–14.
Singer KP, Breidahl PD, Day RE. Variations in zygapophyseal joint orientation and level of transition at the thoracolumbar junction. Preliminary survey using computed tomography. Surg Radiol Anat. 1988;10(4):291–5.
Singh K. Spine essentials handbook. A bulleted review of anatomy, evaluation, imaging, tests, and procedures. New York: Thieme; 2019.
Skedros JG, Bloebaum RD, Mason MW, Bramble DM. Analysis of a tension/compression skeletal system: possible strain-specific differences in the hierarchical organization of bone. Anat Rec. 1994a;239(4):396–404.
Skedros JG, Mason MW, Bloebaum RD. Differences in osteonal micromorphology between tensile and compressive cortices of a bending skeletal system: indications of potential strain-specific differences in bone microstructure. Anat Rec. 1994b;239(4):405–13.
Snell RS. Clinical neuroanatomy. 7th ed. Philadelphia: Lippincott Williams & Wilkins; 2009.
Standring S. Gray's anatomy: the anatomical basis of clinical practice. 42nd ed. Edinburgh: Churchill Livingstone; 2020.
Stanley SK, Ghanayem AJ, Voronov LI, Havey RM, Paxinos O, Carandang G, et al. Flexion-extension response of the thoracolumbar spine under compressive follower preload. Spine (Phila Pa 1976). 2004;29(22):E510–4.
Steinmetz MP, Benzel EC. Benzel's spine surgery: techniques, complication avoidance, and management. 4th ed. Philadelphia: Elsevier; 2015.
Stonelake PS, Burwell RG, Webb JK. Variation in vertebral levels of the vertebra prominens and sacral dimples in subjects with scoliosis. J Anat. 1988;159:165–72.
Stringer MD, Restieaux M, Fisher AL, Crosado B. The vertebral venous plexuses: the internal veins are muscular and external veins have valves. Clin Anat. 2012;25(5):609–18.
Sunderland S. Mechanisms of cervical nerve root avulsion in injuries of the neck and shoulder. J Neurosurg. 1974a;41(6):705–14.
Sunderland S. Meningeal-neural relations in the intervertebral foramen. J Neurosurg. 1974b;40(6):756–63.
Swartz EE, Floyd RT, Cendoma M. Cervical spine functional anatomy and the biomechanics of injury due to compressive loading. J Athl Train. 2005;40(3):155–61.
Taitz C, Arensburg B. Erosion of the foramen transversarium of the axis. Anatomical observations. Acta Anat (Basel). 1989;134(1):12–7.
Taitz C, Nathan H. Some observations on the posterior and lateral bridge of the atlas. Acta Anat (Basel). 1986;127(3):212–7.
Taitz C, Nathan H, Arensburg B. Anatomical observations of the foramina transversaria. J Neurol Neurosurg Psychiatry. 1978;41(2):170–6.
Tanaka M, Nakahara S, Inoue H. A pathologic study of discs in the elderly. Separation between the cartilaginous endplate and the vertebral body. Spine (Phila Pa 1976). 1993;18(11):1456–62.
Tanaka N, Fujimoto Y, An HS, Ikuta Y, Yasuda M. The anatomic relation among the nerve roots, intervertebral foramina, and intervertebral discs of the cervical spine. Spine (Phila Pa 1976). 2000;25(3):286–91.
Tang A, Bordoni B. Anatomy, thorax, muscles. In: StatPearls [Internet]. Treasure Island (FL): StatPearls Publishing; 2024.
Taylor JR. Growth of human intervertebral discs and vertebral bodies. J Anat. 1975;120(1):49–68.
Tkaczuk H. Tensile properties of human lumbar longitudinal ligaments. Acta Orthop Scand. 1968;Suppl 115:1–69.
Tobinick E, Vega CP. The cerebrospinal venous system: anatomy, physiology, and clinical implications. MedGenMed. 2006;8(1):53.
Tubbs RS, Salter EG, Oakes WJ. The intracranial entrance of the atlantal segment of the vertebral artery in crania with occipitalization of the atlas. J Neurosurg Spine. 2006;4(4):319–22.
Tubbs RS, Kelly DR, Humphrey ER, Chua GD, Shoja MM, Salter EG, et al. The tectorial membrane: anatomical, biomechanical, and histological analysis. Clin Anat. 2007;20(4):382–6.
Tubbs RS, Hallock JD, Radcliff V, Naftel RP, Mortazavi M, Shoja MM, et al. Ligaments of the craniocervical junction. J Neurosurg Spine. 2011;14(6):697–709.
Tubbs RS, Shoja MM, Loukas M, editors. Bergman's comprehensive encyclopedia of human anatomic variation. Hoboken: Wiley Blackwell; 2016.
Turnbull IM. Microvasculature of the human spinal cord. J Neurosurg. 1971;35(2):141–7.
Tveten L. Spinal cord vascularity. Acta Radiol Diagn (Stockh). 1976;17(3A):289–98.
Urban JP, Roberts S. Degeneration of the intervertebral disc. Arthritis Res Ther. 2003;5(3):120–30.
Urban JP, Smith S, Fairbank JC. Nutrition of the intervertebral disc. Spine (Phila Pa 1976). 2004;29(23):2700–9.
Van Schaik JP, Verbiest H, Van Schaik FD. The orientation of laminae and facet joints in the lower lumbar spine. Spine. 1985;10(1):59–63.
Vandenabeele F, Creemers J, Lambrichts I. Ultrastructure of the human spinal arachnoid mater and dura mater. J Anat. 1996;189(2):417–30.
Vera CL, Cure JK, Naso WB, Gelven PL, Worsham F, Roof BF, et al. Paraplegia due to ossification of ligamenta flava in X-linked hypophosphatemia. A case report. Spine (Phila Pa 1976). 1997;22(6):710–5.
Viejo-Fuertes D, Liguoro D, Rivel J, Midy D, Guerin J. Morphologic and histologic study of the ligamentum flavum in the thoraco-lumbar region. Surg Radiol Anat. 1998;20(3):171–6.

Vital JM, Beguiristain JL, Algara C, Villas C, Lavignolle B, Grenier N, et al. The neurocentral vertebral cartilage: anatomy, physiology and physiopathology. Surg Radiol Anat. 1989;11(4):323–8.

Walmsley JG. Vascular smooth muscle orientation in curved branches and bifurcations of human cerebral arteries. J Microsc. 1983;131(3):377–89.

Wang MY, Thambuswamy M. Ossification of the posterior longitudinal ligament in non-Asians: demographic, clinical, and radiographic findings in 43 patients. Neurosurg Focus. 2011;30(3):E4.

Watkins R 4th, Watkins R 3rd, Williams L, Ahlbrand S, Garcia R, Karamanian A, et al. Stability provided by the sternum and rib cage in the thoracic spine. Spine (Phila Pa 1976). 2005;30(11):1283–6.

Waxman SG. Clinical neuroanatomy. 28th ed. New York: McGraw-Hill; 2017.

Webster S, de Wreede R. Embryology at a glance. 2nd ed. West Sussex: Wiley Blackwell; 2016.

White AA, Panjabi MM. Clinical biomechanics of the spine. 2nd ed. Philadelphia: Lippincott-Raven; 1990.

Wilkinson IM. The vertebral artery. Extracranial and intracranial structure. Arch Neurol. 1972;27:392–6.

Willard FH, Vleeming A, Schuenke MD, Danneels L, Schleip R. The thoracolumbar fascia: anatomy, function and clinical considerations. J Anat. 2012;221(6):507–36.

Willems JM, Jull GA, Ng JK. An in vivo study of the primary and coupled rotations of the thoracic spine. Clin Biomech (Bristol, Avon). 1996;11(6):311–6.

Winn HR, editor. Youmans and Winn neurological surgery. 7th ed. Philadelphia: Elsevier; 2017.

Xu R, Burgar A, Ebraheim NA, Yeasting RA. The quantitative anatomy of the laminae of the spine. Spine (Phila Pa 1976). 1999;24(2):107–13.

Yabuki S, Kikuchi S. Positions of dorsal root ganglia in the cervical spine. An anatomic and clinical study. Spine (Phila Pa 1976). 1996;21(13):1513–7.

Yahia LH, Garzon S, Strykowski H, Rivard CH. Ultrastructure of the human interspinous ligament and ligamentum flavum. A preliminary study. Spine (Phila Pa 1976). 1990;15(4):262–8.

Yamashita T, Minaki Y, Ozaktay AC, Cavanaugh JM, King AI. A morphological study of the fibrous capsule of the human lumbar facet joint. Spine (Phila Pa 1976). 1996;21(5):538–43.

Yayama T, Uchida K, Kobayashi S, Kokubo Y, Sato R, Nakajima H, Takamura T, Bangirana A, Itoh H, Baba H. Thoracic ossification of the human ligamentum flavum: histopathological and immunohistochemical findings around the ossified lesion. J Neurosurg Spine. 2007;7(2):184–93.

Yoo JU, Zou D, Edwards WT, Bayley J, Yuan HA. Effect of cervical spine motion on the neuroforaminal dimensions of human cervical spine. Spine. 1992;17(10):1131–6.

Recommended Additional Reading

Anderson DG, Vaccaro AR. Decision making in spinal care. 2nd ed. New York: Thieme; 2013.

Benzel EC. Biomechanics of spine stabilization. 3rd ed. New York: Thieme; 2015.

Calais-Germain B. Anatomy of movement. Revised ed. Seattle: Eastland Press; 1999.

McGuinness H. Anatomy & physiology. 5th ed. London: Dynamic Learning; 2021.

7 Functional Anatomy of the Upper Extremity: Nerve Supply and Musculature

Among individuals with spinal cord injury, the restoration of upper limb function represents the highest priority for patients with tetraplegia, as it directly impacts their independence and quality of life (Harvey 2008; Mulroy et al. 2011). While the functional capabilities of the upper limb post-injury are primarily determined by the neurological level of the lesion, significant functional improvements can be achieved through comprehensive rehabilitation strategies (Behrman et al. 2006). These strategies integrate detailed anatomical and biomechanical knowledge with goal-directed therapeutic interventions, complemented by appropriate assistive devices and adaptive equipment (Mulroy et al. 2004).

The optimization of upper limb function requires a thorough understanding of its complex functional anatomy (Moore et al. 2018; Neumann 2017). This understanding encompasses not only the structural components but also the intricate neuromuscular control mechanisms and biomechanical principles that govern upper limb movement (Hamill et al. 2015). The upper limb can be conceptualized as three interconnected functional units: the shoulder complex, the elbow complex, and the wrist–hand unit, each with distinct yet interdependent roles (Levangie and Norkin 2005).

The shoulder complex, comprising multiple joints and sophisticated muscle systems, serves as the foundation for upper limb function (Culham and Peat 1993; Terry and Chopp 2015). Its remarkable mobility allows the hand to access a wide workspace essential for activities of daily living, while simultaneously providing the stability necessary for both power-based tasks and precise manipulative activities (Ludewig and Reynolds 2018). This dual capacity for mobility and stability is crucial for activities ranging from basic self-care to complex occupational tasks (Paine and Voight 2013).

The elbow complex functions as a critical intermediate joint, working in concert with the shoulder to optimize hand positioning (Morrey et al. 1981). Through its flexion-extension and pronation-supination capabilities, the elbow enables precise spatial positioning of the hand while contributing to the overall stability and control of upper limb movements (Stroyan and Will 1993). This positioning is essential for activities requiring both gross motor control and fine manipulation (Safran and Baillargeon 2005).

The wrist-hand unit represents the terminal component of this kinetic chain, distinguished by its exceptional dexterity and complex sensorimotor control (Kaufmann et al. 2007; Moran and Berger 2003). This segment's unique anatomical architecture, combined with its rich sensory innervation and specialized motor control, enables complex grasping patterns, fine manipulation, and precise force modulation necessary for functional tasks (Long et al. 1970; Taylor and Schwarz 1955).

H.-Y. Ko, *Practical Functional Anatomy for Spinal Cord Injury Rehabilitation*,
https://doi.org/10.1007/978-981-95-6457-6_7

The functional integration of these three complexes is fundamental to upper limb performance (Kapandji 2007). Their coordinated action, regulated by precise neural control mechanisms, enables the smooth execution of varied motor tasks ranging from simple daily activities to complex skilled movements (Skirven et al. 2011). Understanding these components as a unified functional system is essential for developing effective rehabilitation strategies and optimizing functional outcomes in patients with tetraplegia (Mulroy et al. 2011).

7.1 Anatomical Organization and Distribution of Upper Extremity Neural Components

The innervation of the upper extremity is exclusively derived from the brachial plexus, a network of nerves originating in the cervical spinal cord. These peripheral nerves innervating the upper limb originate from cervical spinal nerve roots and various components of the brachial plexus, such as trunks and cords (Ferrante 2004) (Fig. 7.1).

The brachial plexus is formed by the ventral rami of the lower four cervical spinal nerves (C5–C8) and the first thoracic nerve (T1), with occasional contributions from C4 or T2, or both. These ventral rami, or anterior divisions of the spinal nerves, which form just external to the intervertebral foramina and are situated between the scalenus anterior and scalenus medius muscles, are collectively referred to as the roots of the plexus (Johnson et al. 2010; Standring 2020). Each spinal nerve root also receives autonomic fibers. C5 and C6 receive gray rami communicantes from the middle cervical ganglion, while C7, 8 and T1 receive them from the inferior or cervicothoracic ganglion.

The brachial plexus is structured such that the upper roots (C5 and C6) merge to form the upper

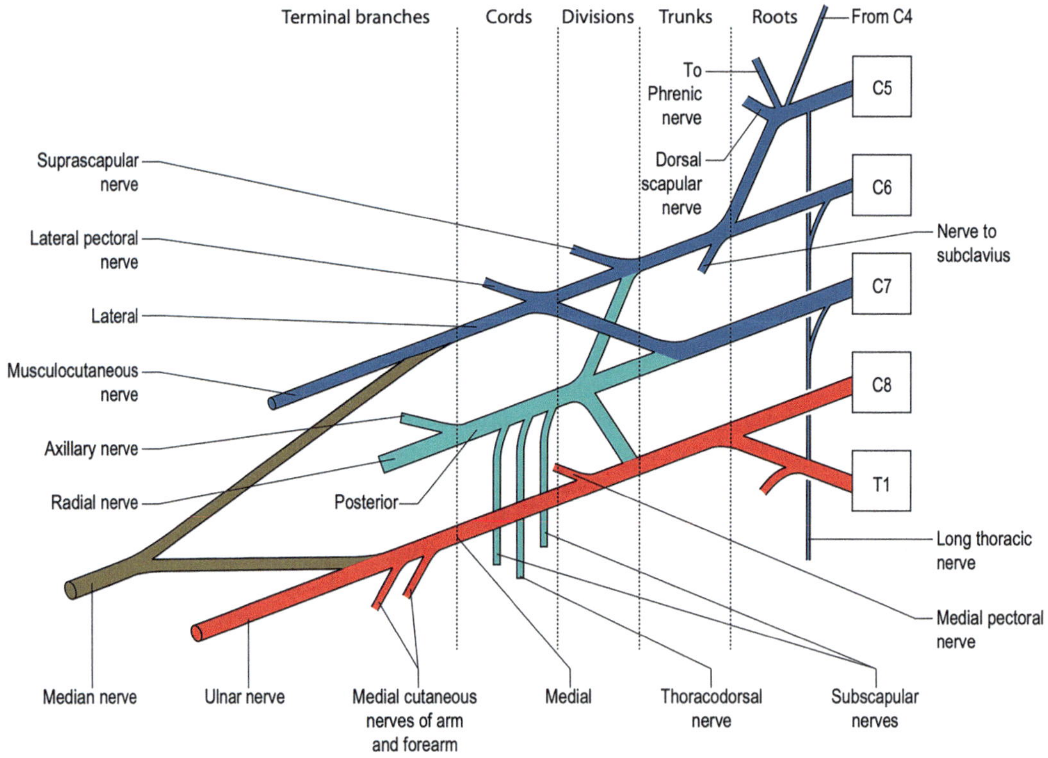

Fig. 7.1 Schematic diagram illustrating the brachial plexus. (Adapted with permission from Palastanga and Soames (2019))

trunk, the C7 root continues as the middle trunk, and the lower roots (C8 and T1) combine to create the lower trunk. These trunks are located between the scalene muscles and the upper border of the clavicle in the posterior triangle of the neck. The lower trunk often grooves the upper surface of the first rib behind the subclavian artery, with the T1 root in contact with the rib (Johnson et al. 2010; Standring 2020).

Each trunk splits into anterior and posterior divisions just above the clavicle. The anterior divisions of the upper and middle trunks unite to form the lateral cord, while the anterior division of the lower trunk continues as the medial cord. The three posterior divisions converge to form the posterior cord. These cords descend into the axilla, initially positioned posterolateral to the axillary artery and then reaching their respective positions around the second part of the axillary artery, posterior to the pectoralis minor muscle. The axillary artery is covered by the axillary sheath, an extension of the prevertebral fascia, which continues into the axilla (Ferrante 2004; Johnson et al. 2010; Standring 2020).

The upper limb musculature is categorized into four compartments. In the arm, the anterior compartment includes the coracobrachialis, biceps brachii, and brachialis, all innervated by the musculocutaneous nerve. The posterior compartment includes the triceps and anconeus, supplied by the radial nerve. In the forearm, the anterior compartment contains the wrist flexors, the long digital flexors, and the pronator teres and pronator quadratus, innervated by the median and ulnar nerves. The posterior compartment, supplied by the radial nerve, contains wrist and digit extensors, abductor pollicis longus, supinator, and brachioradialis. The hand's intrinsic muscles receive innervation from the median and ulnar nerves.

7.1.1 Branches from the Cervical Nerve Roots (Plexus Roots)

The cervical plexus generates two anatomically significant branches from its roots: the dorsal scapular nerve and the long thoracic nerve. These neural structures follow precise anatomical pathways and maintain specific innervation patterns that influence upper limb and thoracic function (Drake et al. 2020; Kim et al. 2004).

7.1.1.1 Dorsal Scapular Nerve

The dorsal scapular nerve emerges as a pure motor branch from the anterior ramus of the C5 root. After its origin, it pierces and passes through the substance of the scalenus medius muscle. The nerve then adopts a vertical trajectory along the vertebral border of the scapula, positioned deep to the levator scapulae muscle. During its course, it provides motor innervation to three muscles: the rhomboid major, rhomboid minor, and levator scapulae. These muscles function synergistically to maintain scapular stability and generate controlled movements of the shoulder girdle. Clinical studies have shown that dysfunction of this nerve can result in medial scapular winging and altered scapulothoracic mechanics (Johnson et al. 2010; Standring 2020).

7.1.1.2 Long Thoracic Nerve

The long thoracic nerve exhibits a more complex anatomical arrangement, receiving neural contributions from the anterior rami of C5, C6, and C7 roots. These root components unite to form the main trunk, which then courses inferior to the brachial plexus and deep to the subscapularis muscle. The nerve continues its descent along the lateral thoracic wall, positioned on the external surface of the serratus anterior muscle. Anatomical dissection studies have revealed a systematic pattern of innervation to the serratus anterior muscle, with distinct root levels corresponding to specific muscular segments (Bertelli and Ghizoni 2005).

The segmental distribution follows a precise pattern: the C5 root fibers innervate the upper segment (first and second digitations), C6 supplies the middle segment (third and fourth digitations), and C7 provides neural input to the lower segment (fifth through eighth digitations). This organized neural distribution enables fine control of various serratus anterior muscle functions, including scapular protraction, upward rotation, and stabilization against the thoracic wall during upper limb movements (Bertelli and Ghizoni 2005; Martin and Fish 2008; Shin et al. 2005).

7.1.2 Branches from the Trunks of the Brachial Plexus

Two significant branches emerge from the trunks of the brachial plexus: the nerve to the subclavius muscle and the suprascapular nerve, both primarily originating from the C5 and C6 nerve roots.

7.1.2.1 Nerve to the Subclavius Muscle

The nerve to the subclavius, a relatively minor branch from the upper trunk of the brachial plexus, courses anteriorly across the distal portion of the plexus and the subclavian artery and vein, primarily innervating the subclavius muscle.

7.1.2.2 Suprascapular Nerve

The suprascapular nerve, also arising from the upper trunk, traverses the posterior triangle of the neck to approach the suprascapular notch. It passes beneath the transverse scapular ligament, extends deep to the supraspinatus muscle, and wraps around the lateral margin of the scapular spine, ultimately innervating the infraspinatus fossa. This nerve innervates both the supraspinatus and infraspinatus muscles, playing an essential role in shoulder mobility and stability (Standring 2020).

7.1.3 Branches from the Cords of the Brachial Plexus

The cords of the brachial plexus give rise to several notable nerves, including the pectoral, subscapular, thoracodorsal, axillary, medial brachial cutaneous, and medial antebrachial cutaneous nerves.

7.1.3.1 Pectoral Nerves

The pectoral nerves consist of the lateral and medial pectoral nerves, which originate from the C5–C8 and T1 nerve roots. The lateral pectoral nerve, emerging from the lateral cord, innervates the pectoralis major muscle. The medial pectoral nerve arises from the medial cord and targets the pectoralis minor muscle.

7.1.3.2 Subscapular Nerve and Thoracodorsal Nerve

Originating from the C5 and C6 nerve roots and the posterior cord of the brachial plexus, the subscapular nerve innervates the subscapularis and teres major muscles. The thoracodorsal nerve, which arises from the C6–C8 nerve roots and the posterior cord, courses along the posterior wall of the axilla, deep to the latissimus dorsi muscle, which it innervates.

7.1.3.3 Axillary Nerve

The axillary nerve, another branch of the posterior cord, provides motor innervation to the deltoid and teres minor muscles. Additionally, it supplies sensory branches to the skin overlying the distal deltoid and articular filaments to the anterior capsule of the shoulder joint. This nerve courses over the subscapularis insertion and exits the axilla through the quadrilateral space, which is bounded by the surgical neck of the humerus, teres major and minor muscles, and the long head of the triceps.

7.1.3.4 Medial Brachial Cutaneous Nerve and Medial Antebrachial Cutaneous Nerve

The medial brachial cutaneous nerve, originating from the C8 and T1 nerve roots and the medial cord, innervates the skin of the medial arm, extending its coverage to the epicondyle and olecranon. Similarly, the medial antebrachial cutaneous nerve, also derived from the C8 and T1 nerve roots and medial cord, supplies sensory innervation to the skin over the biceps muscle, reaching the elbow and continuing along the ulnar aspect of the forearm to the wrist.

7.1.4 Terminal Branches of the Brachial Plexus

The primary terminal branches of the brachial plexus include the musculocutaneous (C5–C7), median (C6–C8, T1), radial (C5–C8, possibly T1), and ulnar (C8, T1) nerves.

7.1.4.1 Musculocutaneous Nerve and Lateral Antebrachial Cutaneous Nerve

The musculocutaneous nerve, originating from the lateral cord, bifurcates from this cord, with one branch aiding in the formation of the lateral fibers of the median nerve. Initially, the musculocutaneous nerve penetrates the coracobrachialis muscle and proceeds along the lateral side of the arm. This nerve is primarily responsible for motor functions, innervating the coracobrachialis, biceps brachii, and most of the brachialis muscle. Before reaching the elbow, it pierces the deep fascia near the biceps tendon. Continuing into the forearm, it transforms into the lateral antebrachial cutaneous nerve, providing sensory innervation to the posterior and radial regions of the forearm. The nerve's terminal branches establish connections with the superficial branch of the radial nerve and the palmar cutaneous branch of the median nerve near the thenar eminence (Ferrante 2004; Standring 2020).

7.1.4.2 Median Nerve

The median nerve originates from the medial and lateral cords of the brachial plexus. This nerve provides innervation to the radial aspect of the forearm's flexor compartment and the hand. The median nerve passes between the two heads of the pronator teres muscle in the forearm. Near the elbow, motor branches emerge to innervate the majority of superficial anterior forearm muscles, including the pronator teres, flexor carpi radialis, palmaris longus, and flexor digitorum superficialis, with the exception of the flexor carpi ulnaris.

The anterior interosseous nerve, a branch of the median nerve, supplies the deep anterior forearm muscles: pronator quadratus, flexor digitorum profundus, and flexor pollicis longus. This branch is vulnerable to compression as it courses along the anterior interosseous membrane's surface, situated between the flexor pollicis longus and flexor digitorum profundus (Moore et al. 2018).

As it passes through the carpal tunnel, the median nerve is prone to compression and trauma due to its superficial position in the carpal tunnel. Upon emerging from the carpal tunnel, it divides into muscular and digital branches, lying beneath the skin and palmar aponeurosis. The hand's motor branch leaves the nerve after the carpal tunnel, innervating the thenar eminence muscles, specifically the abductor pollicis brevis, opponens pollicis, and the superficial head of the flexor pollicis brevis. Notably, the deep head of the flexor pollicis brevis is not innervated by this branch.

Cutaneous innervation by the median nerve is provided through its palmar branch, supplying sensation to the palm's skin and thenar eminence and digital nerves. These digital nerves provide sensation to the thumb, index, middle, and the radial side of the ring fingers, including the posterior surface of their terminal phalanges. Additionally, these nerves provide motor branches to the first, second, and third lumbrical muscles (Ferrante 2004; Standring 2020).

7.1.4.3 Radial Nerve

The radial nerve, a terminal branch of the posterior cord, originates from the C5–C8 nerve roots and frequently incorporates fibers from the T1 root. It traverses the arm in a spiral pattern, lying in close proximity to the humerus between the medial and lateral heads of the triceps muscle. This anatomical relationship with the triceps exposes the nerve to potential compression, particularly significant as it supplies motor innervation to the triceps and anconeus muscles. Additionally, the radial nerve provides motor input to the brachialis, brachioradialis, and extensor carpi radialis muscles.

The nerve bifurcates distally into the posterior interosseous nerve, a predominantly motor branch, and the superficial radial nerve, which is purely sensory. The posterior interosseous nerve, responsible for innervating the extensor muscles of the forearm, including the extensor carpi ulnaris, is susceptible to entrapment as it penetrates the supinator muscle. The superficial radial nerve courses along the lateral forearm deep to the brachioradialis muscle. This branch provides sensory innervation to the ulnar side of the thumb, adjacent areas of the index and middle fingers, and through anastomosis with the posterior branch of the ulnar nerve and also innervates the adjacent side of the middle and ring fingers (Latef et al. 2018).

7.1.4.4 Ulnar Nerve

The ulnar nerve, which originates from the C8–T1 roots and the medial cord of the brachial plexus, descends along the medial aspect of the triceps muscle toward the elbow. At the elbow, it becomes superficial, positioned between the olecranon and the medial epicondyle. The nerve in this area is susceptible to compression and injury because it is only covered by the skin and fascia (Drake et al. 2020).

As it progresses, the ulnar nerve provides motor innervation to the flexor carpi ulnaris at the ulnar groove. Just distal to this point, it supplies the flexor digitorum profundus for the fourth and fifth digits (Spinner and Amadio 2003). The nerve's palmar branch traverses the ulnar side of the wrist and enters Guyon's canal. This canal is situated between the volar carpal ligament and the transverse carpal ligament, with the pisiform and hamate hook forming its boundaries. The anatomical positioning within this canal exposes the nerve to potential trauma (Mackinnon 2015).

Within Guyon's canal, the ulnar nerve divides into superficial and deep branches. The superficial branch innervates the palmaris brevis muscle and provides sensory innervation to the hypothenar eminence's skin. It further subdivides into digital branches supplying the little and ring fingers. The deep branch innervates the hypothenar muscles (flexor, abductor, and opponens digiti minimi), follows the deep palmar arch, and supplies the interossei, medial two lumbricals, adductor pollicis, and the flexor pollicis brevis's deep head (Eversmann Jr 1983; Woo et al. 2015).

The ulnar nerve's two cutaneous branches, the palmar cutaneous and dorsal palmar branches, do not pass through Guyon's canal. The palmar cutaneous branch arises in the mid-forearm, providing sensory innervation to the skin of the palm, while the dorsal cutaneous branch supplies sensation to the adjacent sides of the little and ring fingers, including their dorsal surfaces (Dy and Mackinnon 2016).

7.2 Surface Anatomy of the Upper Extremity

7.2.1 The Shoulder and Arm Regions

7.2.1.1 Anterior Aspect of the Shoulder and Arm

The shoulder girdle comprises the scapula and clavicle, which serve to attach the humerus to the axial skeleton (Moore et al. 2018) (Fig. 7.2). This structure is supported by powerful proximal musculature and transmits forces from the upper limb through the humerus, scapula, and clavicle to the axial skeleton (Neumann 2017).

The clavicle forms articulations medially with the sternum and first rib at the sternoclavicular joint, where the costoclavicular ligament functions as an effective shock absorber (Standring 2020). Laterally, it connects with the scapula at

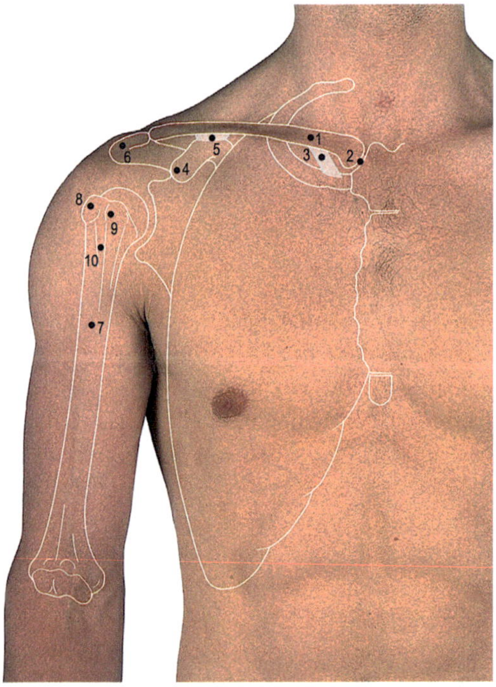

Fig. 7.2 Anterior aspect of the shoulder and arm: bony structure. (1) Clavicle, (2) sternoclavicular joint, (3) costoclavicular ligament, (4) coracoid process, (5) coracoclavicular ligaments, (6) acromion, (7) humerus, (8) greater tuberosity, (9) lesser tuberosity, (10) bicipital groove. (Adapted with permission from Lumley (2008))

the acromioclavicular joint, reinforced by sturdy coracoclavicular ligaments (Levangie and Norkin 2005). The anterior clavicular surface, acromion, and scapular spine are subcutaneous and readily palpable structures (Netter 2019).

The trapezius muscle originates from the lateral third of the clavicle, acromion, and scapular spine (Clarkson 2013) (Fig. 7.3). Its primary actions include scapular elevation, lateral rotation, and medial retraction, with innervation provided by the accessory nerve (cranial nerve XI). The deltoid muscle, which forms the rounded contour of the shoulder, attaches medially to the lateral quarter of the clavicle, acromion, and scapular spine, and inserts laterally onto the deltoid tuberosity of the humerus (Kendall et al. 2005). It functions as the primary abductor of the arm, with its anterior fibers contributing to flexion and posterior fibers to extension. The axillary nerve supplies this muscle (Magee 2014).

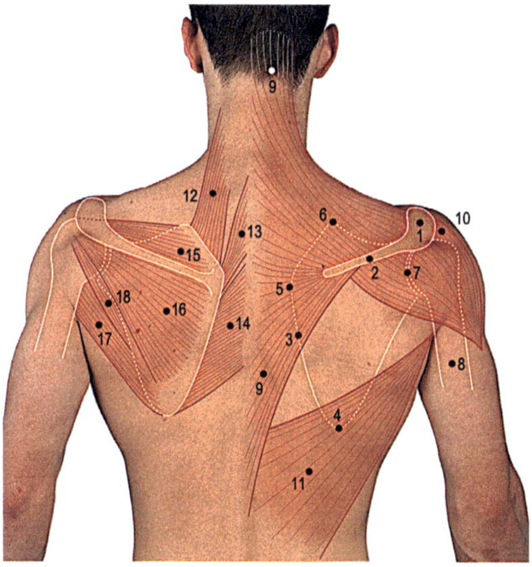

Fig. 7.3 Posterior aspect of the shoulder and arm: bones and muscles of the shoulder girdle. Superficial muscles are shown on the right, and deep muscles are shown on the left. (1) Acromion, (2) spine of scapula, (3) medial border of scapula, (4) inferior angle of scapula, (5) medial angle of scapula, (6) superior angle of scapula, (7) glenoid fossa, (8) humerus, (9) trapezius, (10) deltoid, (11) latissimus dorsi, (12) levator scapulae, (13) rhomboideus minor, (14) rhomboideus major, (15) supraspinatus, (16) infraspinatus, (17) teres major, (18) teres minor. (Adapted with permission from Lumley (2008))

The pectoralis major muscle has two heads: the clavicular head, which attaches to the medial two-thirds of the clavicle, and the sternocostal head, which attaches to the anterior surface of the sternum, the upper five to seven costal cartilages, and the superior part of the external oblique aponeurosis (Drake et al. 2020; Gilroy and MacPherson 2016). These two heads converge laterally into a narrow tendon that passes beneath the deltoid muscle to insert on the lateral lip of the bicipital groove on the humerus. The superior head overlaps the inferior, forming the bulk of the anterior axillary fold. The pectoralis major is a powerful adductor and medial rotator of the arm, with the clavicular head working with the deltoid's anterior fibers to flex the arm (Neumann 2017). Conversely, the sternocostal head acts as a strong extensor when the arm is in a flexed position, functioning in conjunction with the latissimus dorsi during activities such as pulling the body up on an overhead bar or during the follow-through of a tennis serve. This muscle's action can be clinically demonstrated by pressing the hands on the hips (Hamill et al. 2015; Hoppenfeld 1976; Moore et al. 2018). It is innervated by the medial and lateral pectoral nerves from the brachial plexus.

The cephalic vein courses through the deltopectoral groove, situated medial to the anterior deltoid fibers (Netter 2019). It penetrates the clavipectoral fascia, a layer interposed between the pectoralis minor and clavicle, before joining the axillary vein (Standring 2020).

The axillary artery and vein can be accessed through an incision that passes through the skin and superficial fascia, separating the two heads of the pectoralis major muscle and dividing the clavipectoral fascia (Hansen 2019). The axillary vein transitions into the subclavian vein at the outer border of the first rib, posterior to the middle of the clavicle, maintaining a close relationship with the bone's posterior surface (Drake et al. 2020). This anatomical arrangement allows for central venous access by inserting a needle inferior to the clavicle and directing it superomedially along the bone's posterior aspect.

The brachial artery runs along the arm within the groove between the biceps and brachialis muscles and can be palpated by applying lateral pressure against the humerus (Moore et al. 2018). The median nerve crosses anterior to the brachial artery in the mid-upper arm, shifting to its medial side within the cubital fossa (Netter 2019). Surgical approach to the shoulder joint can be accomplished through an anterior incision along the deltoid's anterior border, detaching its clavicular origin and the subscapularis tendon to expose the joint capsule and anterior articular surface (Miller and Thompson 2014).

7.2.1.2 Posterior Aspect of the Shoulder and Arm

The posterior shoulder region is characterized by prominent bony landmarks and muscular structures that are important to the function of the upper limb (Moore et al. 2018). The acromion and scapular spine are palpable subcutaneous structures, while the deltoid muscle creates the smooth contour of the shoulder, enveloping the proximal humerus. The posterior deltoid fibers contribute significantly to shoulder extension (Fig. 7.4).

The trapezius, a broad triangular muscle, originates medially from the medial aspect of the superior nuchal line on the occipital bone, the ligamentum nuchae, and the spinous processes and interspinous ligaments of the lower cervical and all thoracic vertebrae (Drake et al. 2020). Its lateral attachment spans the lateral third of the clavicle, the entire acromion, and the scapular spine. This broad medial origin facilitates a diverse range of scapular movements, including elevation, lateral rotation, and medial retraction.

The latissimus dorsi muscle exhibits a similarly broad medial attachment, originating from the lumbar spinous processes, thoracolumbar fascia, and the posterior half of the iliac crest (Gilroy and MacPherson 2016). Its fibers course superolaterally, inserting into the floor of the bicipital groove on the humerus. Additionally, the muscle attaches to the inferior angle of the scapula, facilitating medial rotation and stabilizing this aspect of the scapula against the thoracic wall during shoulder movements (Hamill et al. 2015). The latissimus dorsi functions as a powerful arm

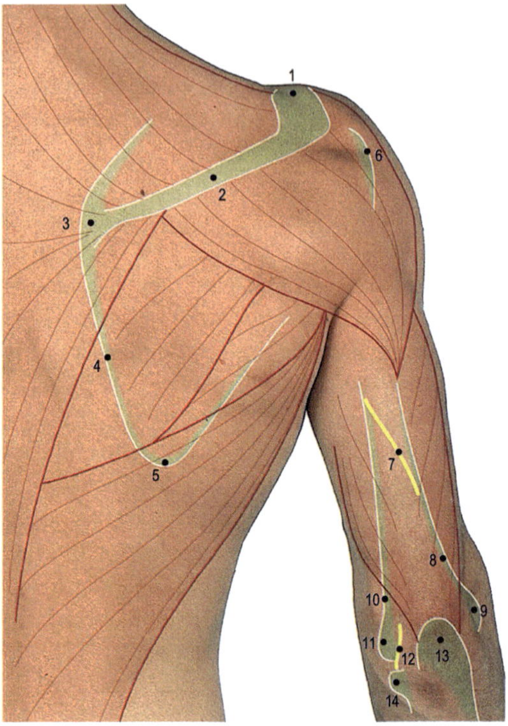

Fig. 7.4 Posterior aspect of the shoulder and arm: palpable structures. (1) Acromion, (2) spine of scapula, (3) medial angle, (4) medial border, (5) inferior angle of scapula, (6) greater tuberosity, (7) radial nerve deep to fibers of triceps, (8) lateral supracondylar ridge, (9) lateral epicondyle, (10) medial supracondylar ridge, (11) medial epicondyle, (12) ulnar nerve, (13) olecranon process, (14) head of ulna. (Adapted with permission from Lumley (2008))

adductor, with its narrow tendon wrapping around the teres major muscle in the posterior axillary fold (see Fig. 7.3). Innervation is supplied by the thoracodorsal nerve, a branch of the posterior cord of the brachial plexus, which descends along the medial axillary wall before entering the muscle within the axilla (Standring 2020).

The lower portion of the posterior axillary fold is primarily composed of the teres major muscle, which originates from the dorsal surface of the inferior angle of the scapula and inserts into the medial lip of the humeral bicipital groove (Moore et al. 2018). The lateral border of the scapula can be palpated through the teres major muscle bulk. Superiorly, the subscapularis muscle contributes to the formation of the posterior axillary wall. The term "rotator cuff" encompasses four muscles: supraspinatus, infraspinatus, teres minor, and subscapularis. These muscles are closely associated with the glenohumeral joint, playing a key role in maintaining its stability (Standring 2020).

The teres major and subscapularis muscles course anterior to the glenohumeral joint, with the latter attaching to the lesser tuberosity. Both muscles function as medial rotators of the shoulder joint. Conversely, the infraspinatus and teres minor muscles pass posterior to the joint, inserting on the greater tuberosity, and act as lateral rotators (Neumann 2017). The infraspinatus originates from the posterior scapular surface inferior to its spine and is palpable, while the teres minor attaches along the lateral border of the scapula.

The supraspinatus muscle, which can be palpated deep to the trapezius, originates from the posterior aspect of the scapula superior to its spine. It courses laterally beneath the acromion to insert on the humeral greater tuberosity. Beyond its role in joint stabilization, the supraspinatus is crucial for initiating abduction, as the deltoid lacks sufficient mechanical advantage to initiate this movement from the adducted position (Hamill et al. 2015). In cases of supraspinatus dysfunction, patients may need to employ compensatory mechanisms, such as swinging or flicking the arm laterally from the hip, to enable the deltoid to engage in abduction. While not primarily a rotator, the supraspinatus is integral to the rotator cuff complex (Magee 2014).

During mid-abduction of the arm, the supraspinatus tendon is compressed between the greater tuberosity of the humerus and the acromion, necessitating lateral rotation of the arm to achieve full abduction. This can be observed by holding the medial and lateral epicondyles of the humerus throughout the full range of abduction (Neumann 2017). Innervation of the shoulder musculature follows a specific pattern: the suprascapular nerve, originating from the upper trunk of the brachial plexus, supplies the supraspinatus and infraspinatus muscles. The subscapular nerves, derived from the posterior cord, innervate the teres major and subscapularis muscles. Notably, all intrinsic scapular muscles receive innervation from the fifth and sixth cervical spinal nerve roots (Moore et al. 2018). The rotator cuff muscles, including the supraspinatus, play a key role in stabilizing the glenohumeral joint anteriorly, posteriorly, and superiorly (Fig. 7.4).

The muscular contour of the posterior upper arm is primarily formed by the triceps brachii, with a minor contribution from the anconeus near the elbow. The triceps has three heads: the lateral and medial heads attach to the posterior humeral shaft above and below the radial groove, respectively, while the long head originates inferior to the glenoid fossa on the scapula. Distally, the triceps inserts onto the palpable olecranon process of the ulna, functioning as a powerful elbow extensor. The radial nerve provides innervation to this muscle (Standring 2020).

Clinical considerations in trauma to this region include:

- Clavicular fractures: Often resulting from falls on an outstretched hand, these typically occur at the junction of the middle and lateral thirds of the clavicle (Netter 2019).
- Shoulder dislocations: The glenohumeral joint is most vulnerable inferiorly. Dislocations in this direction commonly result from a blow to the upper lateral aspect of the abducted humerus, potentially compromising the axillary nerve. Assessment of sensation over the distal deltoid attachment (innervated by the upper lateral cutaneous branch of the axillary nerve) is preferable to testing deltoid function in acute post-traumatic scenarios (Magee 2014).
- Humeral fractures: These are typically caused by direct trauma. Proximal fractures through the surgical neck may endanger the axillary nerve. Shaft fractures are often spiral in nature, risking injury to the radial nerve as it courses through the posterior humeral groove. Radial nerve injury manifests as wrist drop, while triceps function remains intact due to its proximal innervation (Levangie and Norkin 2005).

7.2.2 The Elbow Region

The cubital fossa, a clinically significant area in the anterior elbow region, is demarcated superiorly by an imaginary line connecting the palpable medial and lateral epicondyles of the humerus. The lateral border of brachioradialis and the medial border of pronator teres define its boundaries (Moore et al. 2018) (Fig. 7.5).

Attachment of the brachialis to the anterior surface of the humerus is broad, with its tendon inserting onto the coronoid process of the ulna in the floor of the cubital fossa. This arrangement complicates palpation of the anterior structures of the elbow joint. The prominent tendon of the biceps traverses the center of the fossa, inserting onto the bicipital tuberosity of the radius on its medial aspect (Standring 2020). The firm and thin bicipital aponeurosis extends from the medial side of the biceps tendon across the roof of the fossa, attaching to the subcutaneous surface of the ulna. This aponeurosis hinders palpation of the underlying brachial artery, which has the median nerve on its medial side. For effective palpation of the artery, full extension of the elbow allows compression of the vessel against the joint. Bifurcation of the brachial artery into radial and ulnar arteries occurs at the apex of the fossa (Netter 2019).

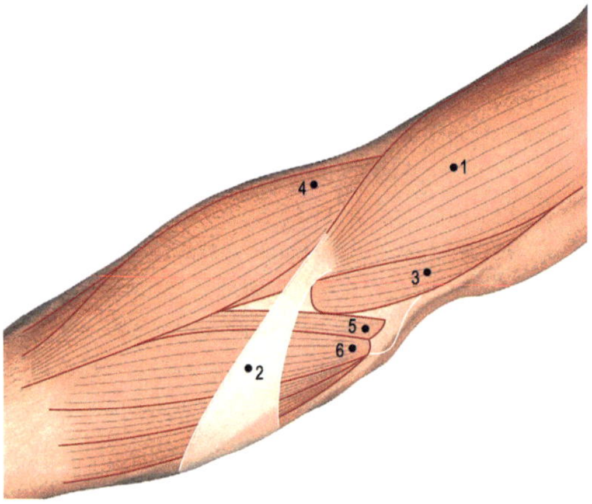

Fig. 7.5 Cubital fossa: soft tissue structures. (1) Biceps, (2) bicipital aponeurosis, (3) brachialis, (4) brachioradialis, (5) pronator teres, (6) common flexor origin. (Adapted with permission from Lumley (2008))

The brachial artery is a key site for arterial blood sampling and catheter insertion in various cardiac and arterial investigations. The median nerve exits the cubital fossa by passing between the humeral and ulnar heads of the pronator teres muscle and then courses deep to the superficial muscles of the forearm (Neumann 2017). In the roof of the cubital fossa, the cephalic, basilic, and median cubital veins are often utilized for venous sampling (Magee 2014).

7.2.3 The Forearm Region

7.2.3.1 Anterior Aspect of the Forearm

The anterior forearm region comprises several key muscles crucial for upper limb function. The attachment of the pronator teres is proximal to the common flexor origin on the anterior surface of the medial epicondyle. The common flexor origin provides attachment for flexor carpi radialis, palmaris longus, flexor digitorum superficialis, and flexor carpi ulnaris (Moore et al. 2018). Pronator teres has an additional head originating from the coronoid process of the ulna, while flexor carpi ulnaris has a second origin from the posterior subcutaneous border of the ulna. Distally, the pronator teres inserts into the lateral aspect of the mid-shaft of the radius, and it functions to pronate the forearm (Standring 2020).

Tendons of the other common flexor origin muscles extend to the hand. These muscles also contribute to weak elbow flexion. Innervation of the flexor carpi ulnaris is by the ulnar nerve, while the median nerve supplies the remaining muscles. Flexor pollicis longus, flexor digitorum profundus, and pronator quadratus are deep, impalpable muscles of the forearm. The anterior interosseous branch of the median nerve innervates all three, with flexor digitorum profundus receiving additional innervation from the ulnar nerve (Netter 2019).

The attachment of the brachioradialis is proximal to the upper two-thirds of the lateral supracondylar ridge of the humerus and distally to the lateral aspect of the distal radius. Its function is to pronate or supinate the forearm into the midprone position, where elbow flexion is most powerful.

The brachioradialis lies over the radial nerve and its posterior interosseous branch within the cubital fossa and is innervated by the radial nerve. Additionally, the radial nerve supplies the extensor carpi radialis longus muscle, which originates from the lower third of the lateral supracondylar ridge of the humerus. Although this muscle is deeply situated in the forearm, its tendon can be palpated at the wrist.

7.2.3.2 Posterior Aspect of the Forearm

The posterior aspect of the forearm presents distinct anatomical features relevant to rehabilitation medicine. Palpation of bony structures forming the elbow joint is more readily accomplished posteriorly. Medial and lateral epicondyles of the humerus, easily grasped between the finger and the thumb, serve as useful indicators of humeral position during shoulder rotation (Moore et al. 2018).

The posterior border of the ulna remains subcutaneous along its entire length. Triceps brachii forms the primary muscular bulk of the posterior arm, with its insertion on the palpable olecranon process. Through the musculature attached to the common extensor origin on the posterior aspect of the lateral epicondyle, the head of the radius, encased in the annular ligament, can be palpated (Standring 2020). Muscles originating from the common extensor origin include extensor carpi radialis brevis, extensor digitorum, extensor digiti minimi, and extensor carpi ulnaris. These muscles contribute to weak elbow extension and receive innervation from the posterior interosseous nerve (Netter 2019). The supinator muscle, also attached to the common extensor origin at a deeper level, has an additional head originating from the lateral aspect of the proximal ulna. It wraps around the proximal radius, inserting on its anterior surface. This muscle, along with other deep impalpable muscles on the posterior forearm, is supplied by the posterior interosseous nerve. Tendons of these muscles become palpable at the wrist (Neumann 2017).

The course of the ulnar nerve is posterior to the medial epicondyle, where it can be palpated and rolled over the bone before passing between the humeral and ulnar heads of flexor carpi ulnaris. Compression or percussion of the nerve at this point typically elicits paresthesia in the forearm (Magee 2014).

7.3 Anatomy and Function of the Shoulder Complex

The shoulder complex is an assembly of seven joints: the glenohumeral, suprahumeral, acromioclavicular, scapulocostal, sternoclavicular, costosternal, and costovertebral joints (Cailliet 1981). Among these, the suprahumeral and scapulocostal joints are considered functional rather than anatomical joints. Nevertheless, their roles are essential in the comprehensive evaluation of the shoulder area, facilitating the full range of movement through coordinated action of all involved joints.

The glenohumeral joint, classified as a ball-and-socket type, allows movement across numerous axes due to the spherical shape of the humeral head. The wide range of motion is accompanied by a reduction in stability. The glenoid fossa, being significantly smaller in surface area compared to the humeral head, contributes to this fundamental instability (Culham and Peat 1993). Enhancement of joint stability is achieved through the fibrocartilaginous labrum, which effectively increases the contact area between the humeral head and the glenoid fossa to 75% in the vertical axis and 56% in the transverse axis, thereby improving stability (Saha 1973).

The long head of the biceps traverses the humeral head within the glenohumeral joint's capsule and passes beneath the transverse humeral ligament into the bicipital groove. It merges with the short head, originating from the coracoid process. The long head's synovial sheath extends from the glenohumeral joint's synovium. Inflammation in the joint's synovium, such as that affecting the rotator cuff's adjacent synovium, can also affect the tendon. This interaction may contribute to the common occurrence of tendinitis in both the rotator cuff and the long head of the biceps (Neviaser 1980; Neviaser 1987).

In a resting position, the glenoid fossa is inclined downward at an angle of 5°. The scapular plane is positioned about 30° anterior to the coronal plane of the body (Perry 1983; Perry 1988). Additionally, there is a 30° retroversion of the humeral head relative to its shaft.

7.3.1 Scapulohumeral Rhythm Dynamics

Scapulohumeral rhythm is a term that refers to the coordinated movement between the glenohumeral joint and the scapula against the thorax, which is essential for lifting the arm. This synchronized motion is crucial for the humerus to move beneath the coracoacromial arch effectively. The initial phase of arm elevation demonstrates notable variability in scapular gliding between individuals, with its presence ranging from absent to minimal, reversed, or in rare cases, oscillatory (Doody et al. 1970; Inman et al. 1944). The scapula does not actively participate in the movement until reaching a 60° flexion and 30° abduction threshold. Research has identified the ratio of humeral to scapular motion as varying between 1.25:1 and 3:1 across different studies (Doody et al. 1970; Freedman and Munro 1966; Inman et al. 1944; Poppen and Walker 1976; Saha 1973), with a noted decrease in scapular movement as the arm elevates beyond 120°. Additionally, the first 30° of arm elevation includes a 5° rotation at the acromioclavicular joint. Clavicular elevation at the sternoclavicular joint constitutes the primary motion component associated with scapular movement until the arm reaches an elevation of 120°. The completion of full scapular rotation is dependent on both the acromioclavicular joint and clavicular rotation along its longitudinal axis beyond this point.

7.3.2 Mechanisms of Shoulder Joint Stability

The stability of the shoulder joint is primarily secured through the combined efforts of ligaments and muscular actions, crucial for maintaining joint integrity (Matsen et al. 1996). Ligaments are the key static stabilizers, with the shoulder joint being supported by the superior glenohumeral ligament, middle glenohumeral ligament, inferior glenohumeral ligament complex, and the coracohumeral ligament. The inferior glenohumeral ligament complex is further detailed into an anterior band, a posterior band, and an axillary pouch, all contributing to the reinforcement of the articular capsule's anterior wall (Warner et al. 1990).

Different phases of abduction see varied contributions from these ligaments for anterior stability: at 0° and 90° of abduction, the anterior restraint is provided by the subscapularis and the inferior glenohumeral ligament, respectively (O'Connell et al. 1990). When the arm is abducted to 45°, the stability is ensured through the coordinated action of the subscapularis, middle glenohumeral ligament, and the anterior band of the inferior glenohumeral ligament complex (Ovesen and Nielsen 1986). The superior glenohumeral ligament's function is to prevent inferior movement of the shoulder in an adducted state. The inferior glenohumeral ligament complex is the principal structure for posterior stability when the shoulder is abducted to 90°. The posterior rotator cuff muscles are also noted to provide static stabilization across all abduction positions (Ovesen and Nielsen 1986).

Dynamic stability, in contrast, is predominantly provided by the rotator cuff muscle tendons. They achieve this by pressing the humeral head into the glenoid cavity, effectively acting as a dynamic restraint against the humeral head's translation in any direction, be it anterior, posterior, or along the superior-inferior axis. The long head of the biceps also contributes to this stability mechanism, especially during the forearm's flexion and supination movements (Kumar et al. 1989).

7.3.3 Biomechanics and Muscular Coordination of the Shoulder Girdle Musculature

The shoulder girdle musculature demonstrates complex patterns of activation and coordination. Individual muscle actions, insertions, and nerve

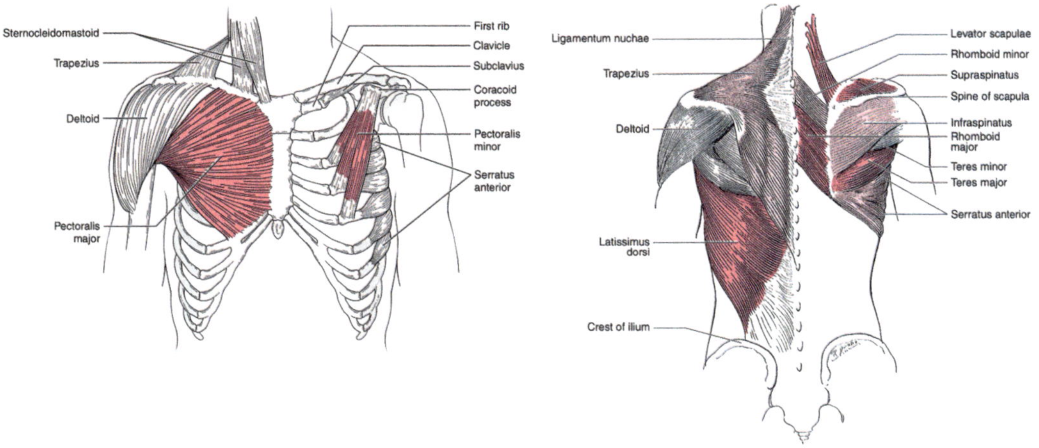

Fig. 7.6 Muscles acting on the shoulder joint and shoulder girdles: (**a**) Anterior aspect and (**b**) posterior aspect. (Adapted with permission from Jenkins (2009))

supply patterns for the shoulder joint and girdle are detailed in (Fig. 7.6). The muscles in this region perform dual functions of movement production and stabilization.

In shoulder abduction and flexion, the deltoid produces approximately 50% of the elevation force. The deltoid's contribution increases progressively with greater abduction angles, showing maximum activation between 90° and 180°. Research indicates that deltoid fatigue resistance peaks between 45° and 90° of abduction, establishing this range as optimal for therapeutic exercise protocols (Escamilla et al. 2019; Neumann 2017).

During arm elevation, the rotator cuff muscles (teres minor, subscapularis, infraspinatus, and supraspinatus) contribute significantly, as the deltoid cannot effectively elevate the arm without proper humeral head stabilization. The rotator cuff collectively generates about 50% of the force typically produced in elevation movements. The interaction between the deltoid and rotator cuff during abduction and flexion is illustrated in Fig. 7.7. In early elevation phases, the deltoid's vertical pull receives assistance from the supraspinatus, which simultaneously produces abduction and compresses the humeral head, counteracting superior translation forces. The rotator cuff muscles work as a unit to maintain proper glenohumeral positioning through compression forces. The teres minor, infraspinatus, and subscapularis provide stability through downward force application. The latissimus dorsi contracts eccentrically for additional humeral head stabilization, with increasing activity at higher elevation angles. Beyond 90° of flexion or

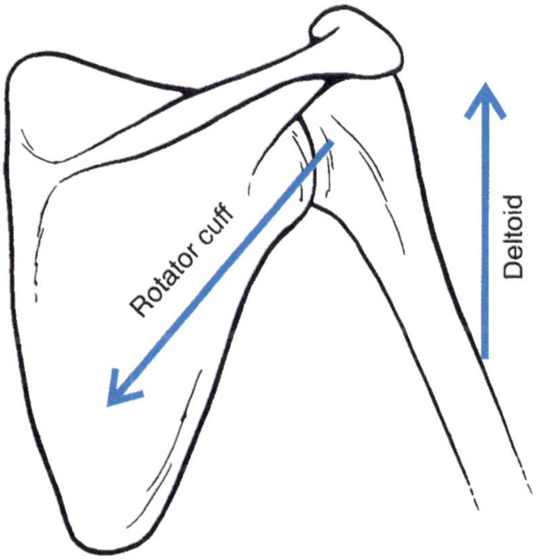

Fig. 7.7 Coordination of the deltoid muscle and rotator cuff for efficient arm flexion or abduction. During the initial stages of flexion and abduction up to 90°, the rotator cuff applies a stabilizing force to depress and stabilize the humeral head within the joint, while the deltoid muscle generates the force necessary to elevate the arm. (Adapted from Hamill et al. (2015))

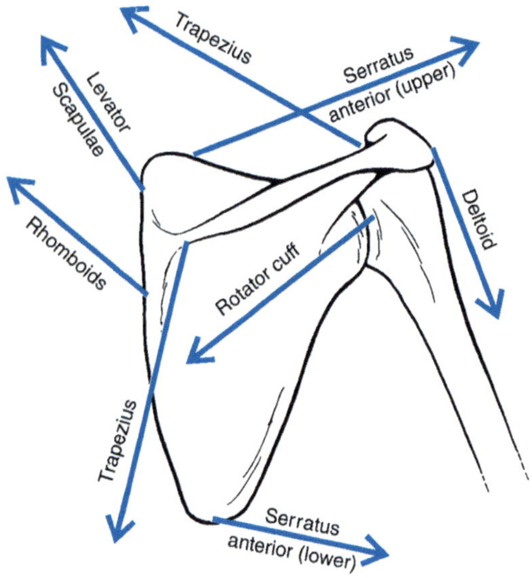

Fig. 7.8 Direction of pull of various shoulder girdle muscles, including the deltoid and rotator cuff, for the resting arm. The figure highlights the line of pull of the trapezius and serratus anterior, which coordinate to produce abduction, elevation, and upward rotation of the scapula required for arm flexion or abduction. Additionally, the levator scapulae and rhomboids are shown contributing to scapular elevation. (Adapted from Hamill et al. (2015))

abduction, rotator cuff force generation diminishes, increasing joint vulnerability. However, the supraspinatus maintains a substantial contribution above 90°. In this range, the deltoid's pull creates potential inferior subluxation forces. Elevation from 90° to 180° requires concurrent external rotation. When external rotation exceeds 20°, the biceps brachii contributes to abduction.

Scapular positioning during arm elevation requires coordinated movements, including protraction/abduction, elevation, and upward rotation with posterior clavicular rotation. As shown in Fig. 7.8, the serratus anterior and trapezius form a force couple generating lateral, superior, and rotational scapular motions. These actions initiate after the deltoid and teres minor begin arm elevation, continuing through 180°, with peak activity between 90° and 180°. The serratus anterior maintains scapulothoracic contact throughout these movements.

During controlled arm lowering (adduction/extension) with accompanying retraction, depression, and downward rotation of the shoulder girdle with forward clavicular rotation, the previously active muscles operate eccentrically. However, forceful lowering or movement against external resistance, such as during weight training, involves concentric activation patterns.

In concentric adduction or extension against resistance, as seen in swimming strokes, the primary movers include the latissimus dorsi, teres major, and the sternal portion of the pectoralis major. While the teres major activates primarily under resistance conditions, the latissimus dorsi maintains activity even in unloaded movements (Reinold et al. 2018).

The shoulder girdle's return movement involves retraction, depression, and downward rotation with forward clavicular rotation. The rhomboid muscle produces downward scapular rotation and functions with the teres major and latissimus dorsi to regulate arm and scapular motion during lowering. Additional scapular control comes from the pectoralis minor, which depresses and downwardly rotates the scapula, while the middle and lower trapezius portions work with the rhomboid to retract the scapula during resistance-based movements (Boettcher et al. 2020).

Rotational movements of the arm—internal and external rotation—become particularly important in athletic activities and arm elevation beyond 90°. External rotation, performed by the infraspinatus and teres minor, plays a key role in the preparatory phase of overhand throws. Both muscles show increased activation with progressive external rotation. The infraspinatus's dual role in rotation and stabilization makes it prone to early fatigue during elevated arm activities. Internal rotation involves the subscapularis, latissimus dorsi, teres major, and portions of the pectoralis major. The teres major contributes actively only against resistance. Despite the substantial force-generating capacity of these internal rotators, most upper extremity actions require minimal internal rotation force production.

Shoulder girdle adjustments during rotational movements vary with arm position. Elevated arm positions require the previously described abduction- and flexion-related girdle movements. Rotation in the neutral position needs minimal girdle assistance and allows a full 180° rotational range. This position-dependent variation occurs because elevation requires rotator muscles to maintain humeral head stability, limiting rotation in upper ranges. Internal rotation becomes particularly restricted in elevated positions due to greater tuberosity compression under the acromion.

Horizontal plane movements—flexion/adduction and extension/abduction—combine elements of elevated arm positions. The pectoralis major and anterior deltoid predominantly produce horizontal flexion, which is essential for power-based upper extremity activities. Horizontal extension utilizes the infraspinatus, teres minor, and posterior deltoid, commonly seen in backswing and preparatory phases of upper extremity skills.

7.3.4 The Shoulder Girdle and Its Musculature

The shoulder girdle comprises the scapula and clavicle, which articulate at the acromioclavicular joint. This structure forms a connection between the upper extremity and the axial skeleton through the glenohumeral and sternoclavicular joints, respectively. The scapula, a flat, triangular bone, serves as an attachment site for numerous muscles that either secure the shoulder girdle to the thorax or regulate upper limb positioning. The clavicle primarily functions as a support, holding the upper limb away from the trunk (Moore et al. 2018; Standring 2020).

Scapular positioning and movement are regulated by the activity of its attached musculature. Individual muscles, when functioning in various combinations, can contribute to multiple shoulder girdle movements. Therefore, each muscle's major action will be detailed individually. The fascial layers covering adjacent muscles enable gliding and sliding movements between the scapula and thorax (Neumann 2017).

From the anatomical position, the scapula lies obliquely over the second to seventh ribs on the posterior thoracic wall, with the coracoid process oriented anteriorly. The following movements are characterized (Dutton 2017; Hamill et al. 2015; Kapandji 2007; Kisner and Colby 2007; Levangie and Norkin 2005) (Fig. 7.9):

- Retraction: Scapular movement maintaining vertical alignment, with the medial border approaching the vertebral column. This action orients the glenoid fossa more laterally.
- Protraction: Forward scapular movement around the thoracic wall, as in rounding the shoulders. This movement may include some lateral rotation, resulting in a more anterior orientation of the glenoid fossa.
- Elevation: Upward movement of the shoulder girdle, as in shoulder shrugging.
- Depression: Downward movement of the shoulder girdle.
- Upward (lateral) rotation: A complex movement where the inferior angle of the scapula moves laterally around the thoracic wall. Simultaneously, the clavicle's strut-like action causes an upward scapular movement, causing the glenoid fossa to turn superiorly.
- Downward (medial) rotation: The return of the scapula to its resting position from lateral rotation. This movement involves the scapula rotating so that the inferior angle moves medially and downward, toward the spine, while the glenoid cavity moves downward.

Muscles involved in the movements of the shoulder girdle are listed in Table 7.1.

7.3.4.1 Retracting Muscles of the Shoulder Girdle: Rhomboid and Trapezius

Rhomboid Minor (Innervated by the Dorsal Scapular Nerve: C5)

The rhomboid minor is a small quadrilateral muscle situated in the upper thoracic region. Its fibers originate from the spinous processes of C7 and T1, as well as the supraspinous ligament situated between these vertebrae and the inferior portion of the ligamentum nuchae. These fibers course obliquely in an inferolateral direction to

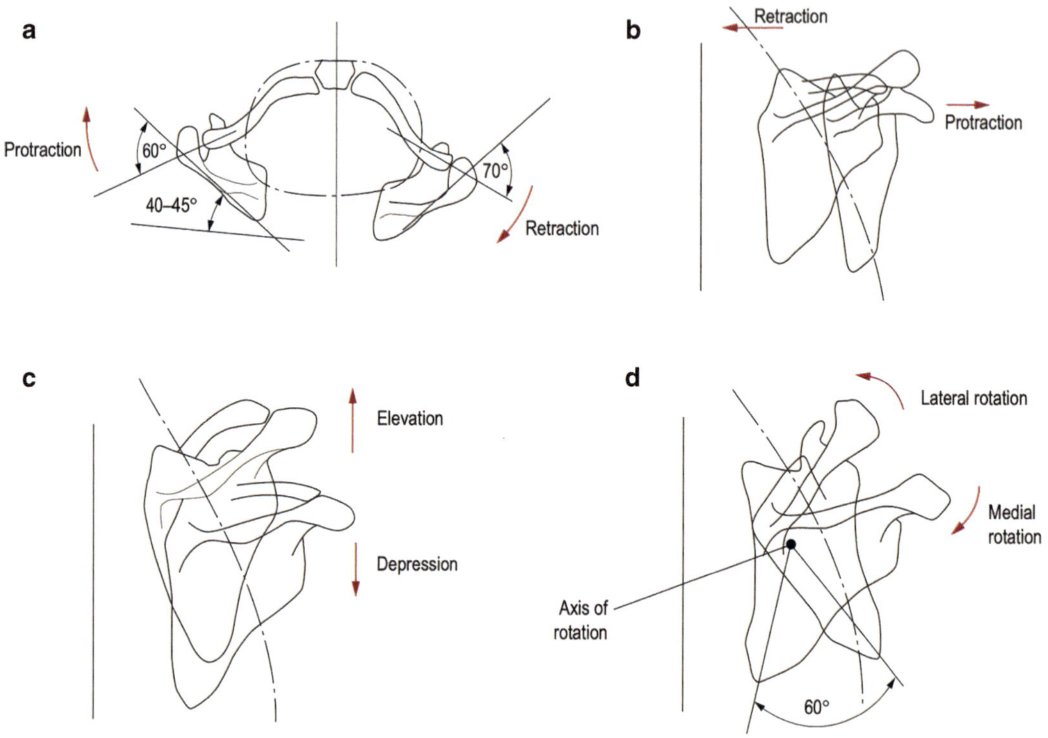

Fig. 7.9 Movement of the shoulder girdle: (**a**) protraction and retraction views from above, (**b**) protraction and retraction viewed from behind, (**c**) elevation and depression, (**d**) axial rotation. (Adapted with permission from Palastanga and Soames (2019))

Table 7.1 Movements of the shoulder girdle and the associated muscles

Movement	Muscles
Protraction	Serratus anterior Pectoralis minor
Retraction	Rhomboid major and minor Trapezius
Elevation	Trapezius (upper fibers) Levator scapulae
Depression	Pectoralis minor Trapezius (lower fibers)
Lateral rotation	Trapezius Serratus anterior
Medial rotation	Rhomboid major and minor Levator scapulae Pectoralis minor

Adapted with permission from Palastanga and Soames (2019)

insert onto the medial border of the scapula, specifically at the smooth triangular area located at the base of the scapular spine (Moore et al. 2018; Standring 2020) (Fig. 7.10).

This muscle's anatomical position and fiber orientation contribute to its primary function of scapular retraction, working in synergy with other posterior shoulder girdle muscles to stabilize and control scapular movement. The rhomboid minor's action is particularly significant in maintaining proper scapular alignment during upper extremity activities and postural control (Neumann 2017; Sahrmann 2002).

Rhomboid Major (Innervated by the Dorsal Scapular Nerve: C5)

The rhomboid major, larger than the rhomboid minor and often continuous with it, originates from the spinous processes of T2–T5 and the intervening supraspinous ligament spanning these vertebral levels. The muscle fibers progress in an oblique inferolateral trajectory, ultimately inserting along the medial border of the scapula, specifically in the region between the base of the

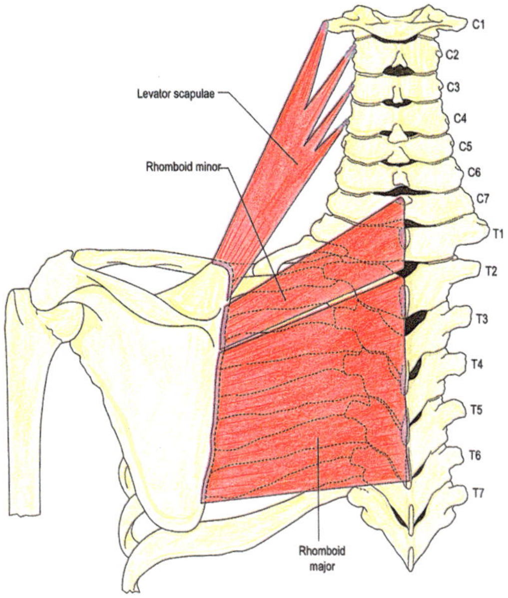

Fig. 7.10 Posterior thoracic view showing the attachment sites of the rhomboid major, rhomboid minor, and levator scapulae muscles. (Adapted with permission from Palastanga and Soames (2019))

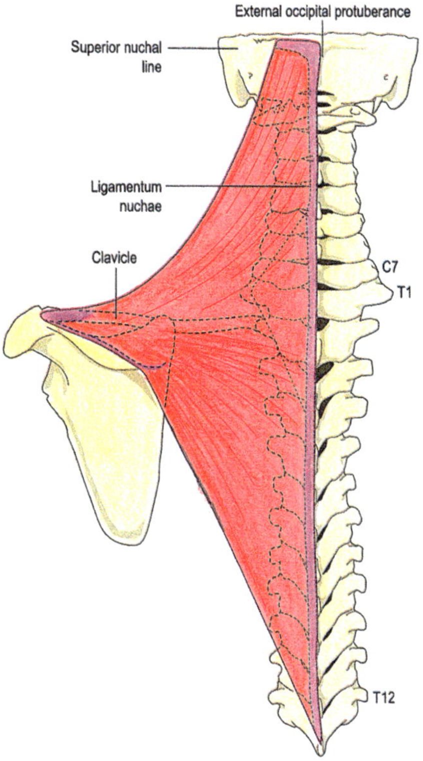

Fig. 7.11 Posterior view of the thorax illustrating the attachment site of the trapezius muscle. (Adapted with permission from Palastanga and Soames (2019))

scapular spine and the inferior angle (Moore et al. 2018; Neumann 2017; Standring 2020).

Both rhomboid muscles are situated superficial to the deep back musculature. The trapezius muscle overlays these structures, with the exception of the inferior border of the rhomboid major, which forms the floor of the "triangle of auscultation" (Clemente 2011). This anatomical arrangement has clinical significance in physical examination and auscultation of the posterior thorax.

The primary function of the rhomboids is scapular retraction. Additionally, these muscles contribute to medial rotation of the shoulder girdle. A key role of the rhomboids lies in their capacity as scapular stabilizers when other muscle groups are active (Levangie and Norkin 2005; Sahrmann 2002).

***Trapezius* (Innervated by the Accessory Nerve, XI)**

The trapezius is a large, flat, triangular muscle sheet that extends from the skull and vertebral column medially to the shoulder girdle laterally (Moore et al. 2018; Standring 2020) (Fig. 7.11). The muscle's medial attachment originates from multiple sites: the medial third of the superior nuchal line, the external occipital protuberance of the occipital bone, the ligamentum nuchae, and the spinous processes of C7 to T12, including the intervening supraspinous ligament. While most attachments are direct muscular insertions, there exists a triangular aponeurosis in the region between C6 and T3 (Clemente 2011; Neumann 2017).

The muscle fibers extend in different directions from this extensive medial attachment to the clavicle and scapula. The fiber orientation varies: upper fibers descend laterally, middle fibers run horizontally, and inferior fibers run upward and laterally. The upper fibers attach to the posterior border of the lateral third of the clavicle. The middle fibers insert on the medial border of the acromion and the superior border of the scapular spine crest, separated from the smooth area on the medial part of the spine by a small bursa. The lowermost fibers converge into a tendon that attaches to a tubercle on the inferior edge at the medial end of the scapular spine (Sahrmann 2002).

The upper free edge of the trapezius forms the posterior border of the neck's posterior triangle. The inferior free border delineates the medial boundary of the triangle of auscultation, an area of the thoracic wall minimally obstructed by the scapula and thinly covered by musculature. This triangle is further bounded inferiorly by the upper border of latissimus dorsi and laterally by the medial border of the scapula (Netter 2019).

The trapezius muscle plays a significant role in scapular stabilization during upper extremity movements. Its middle fibers, oriented horizontally, enable scapular retraction toward the spinal column. This action may be augmented by the simultaneous contraction of the upper and lower fibers, generating a resultant force directed medially (Johnson et al. 1994). Elevation of the shoulder girdle and maintenance of shoulder height against gravitational forces or during manual load-bearing are primarily accomplished by the upper trapezius fibers. Bilateral contraction of these fibers contributes to cervical extension, while unilateral activation results in ipsilateral cervical lateral flexion (Neumann 2017; Standring 2020).

The lower trapezius fibers are responsible for depressing the scapula's medial aspect, thereby lowering the shoulder, particularly when counteracting resistance, such as during chair transfers using upper limb support. Coordinated activation of the upper and lower fibers induces scapular lateral rotation around an axis approximating the base of the scapular spine (Hislop et al. 2007).

In cases of trapezius muscle paralysis, especially affecting its superior portion, this leads to anterior displacement of the scapula along the thoracic wall, with concomitant medial deviation of the inferior angle. This pathological state often manifests as a pronounced angulation in the typically smooth contour extending from the occiput to the acromion (Paine and Voight 2013).

7.3.4.2 Protracting Muscles of the Shoulder Girdle: Serratus Anterior, Pectoralis Minor

Serratus Anterior **(Innervated by the Long Thoracic Nerve: C5, C6 and C7)**

The serratus anterior is a large, flat muscular sheet situated on the lateral thoracic wall, positioned between the ribs and the scapula. Loose fascia exists between the deep surface of the muscle and the ribs or intercostal fascia, and also between its superficial surface and the subscapularis muscle, facilitating the free movement of the scapula. The deep fascia separates the muscle from the ribs and intercostal fascia, while the superficial fascia interfaces with the subscapularis muscle (Moore et al. 2018; Neumann 2017; Standring 2020). The serratus anterior forms the medial border of the axilla. Its inferior lateral aspect is partially obscured by breast tissue, while the upper digitations lie posterior to the clavicle. The latissimus dorsi crosses this lower border of the serratus anterior (Netter 2019).

The serratus anterior muscle originates from digitations that attach to the external surfaces of the upper eight or nine ribs, located just anterior to the midaxillary line, and the intercostal fascia in between. The initial digitation arises from the first and second ribs, with subsequent digitations originating from individual ribs. Notably, the lower four digitations intertwine with the costal attachments of the external oblique abdominis muscle.

The muscle fibers extend posteriorly from this broad origin to insert along the costal surface of the scapula's medial border, spanning from the superior to the inferior angles. However, the insertion pattern is not uniform. The uppermost digit follows a near-horizontal trajectory to the superior angle of the scapula. The lower four digitations converge to attach at the inferior angle, with the intermediate digitations distributed along the medial border (Kapandji 2007).

The serratus anterior is a main protractor of the shoulder girdle, playing a key role in forward-directed movements such as thrusting, pushing, and punching. These actions involve the anterior displacement of the scapula, which consequently propels the upper limb forward. The muscle's hypertrophy in boxers exemplifies its functional importance in these motions (Kim et al. 2014; Neumann 2017).

A critical function of the serratus anterior is scapular stabilization during upper limb movements. It exerts a strong contraction to maintain the scapula's medial border against the thoracic wall during arm flexion or when bearing weight anteriorly. Paralysis of this muscle results in "winging" of the scapula, where the scapula's medial border protrudes from the thoracic wall, significantly compromising upper limb function and mobility (Phadke et al. 2009; Martin and Fish 2008).

The lower digitations of the serratus anterior, in conjunction with the trapezius, facilitate lateral rotation of the scapula (Levangie and Norkin 2005). This action orients the glenoid fossa superiorly and anteriorly. Paralysis of the serratus anterior impairs this rotational capacity, limiting shoulder abduction to approximately 90° and substantially reducing upper extremity functional range (Martin and Fish 2008).

There is debate regarding the role of the serratus anterior as an accessory respiratory muscle during respiratory distress. The fiber orientation of most digitations, except possibly the superior two and potentially the most inferior, does not align with rib elevation. In fact, the muscle's action may be more conducive to rib depression (De Troyer and Estenne 1988; Ratnovsky et al. 2008).

Pectoralis Minor **(Innervated by the Medial Pectoral Nerve: C6, C7, and C8)**

The pectoralis minor is a thin, flat, triangular muscle located on the anterior thoracic wall, positioned deep to the pectoralis major (Moore et al. 2018; Neumann 2017; Standring 2020). It attaches inferiorly to the outer surfaces of the third, fourth, and fifth ribs near their costal cartilages and the intervening intercostal fascia. Occasionally, there may be additional attachments to the second or sixth rib, or more rarely, to both. The muscle fibers converge to form a short, flat tendon that passes superolaterally to attach to the upper surface and medial border of the coracoid process of the scapula (Borstad and Ludewig 2005).

The pectoralis minor muscle exerts a significant force on the coracoid process, facilitating anterior and inferior scapular displacement during pushing and striking motions. Additionally, when an individual leans on their hands, the pectoralis minor aids in transferring the weight of the trunk to the upper limb. Its attachment to the coracoid process allows the pectoralis minor to medially rotate the scapula against resistance (Levangie and Norkin 2005; Neumann 2017) (Fig. 7.12). This rotational component is integral to maintaining proper scapulohumeral rhythm and shoulder function (Kapandji 2007).

Furthermore, when the scapula and upper limb are fixed, the pectoralis minor can act as an accessory muscle of inspiration during respiratory distress, aiding in the elevation of the ribs to assist in breathing (De Troyer and Estenne 1988). During periods of respiratory distress, it may act as an accessory muscle of inspiration, augmenting the thoracic cavity's expansion (Ratnovsky et al. 2008). This dual functionality as both a shoulder girdle stabilizer and a potential respiratory assistant is particularly noteworthy in the context of spinal cord injury rehabilitation, where it is necessary to improve both upper extremity function and respiratory capacity.

7.3.4.3 Elevating Muscles of the Shoulder Girdle: Upper Fibers of the Trapezius, Levator Scapulae

Levator Scapulae **(Innervated by the Dorsal Scapular Nerve: C5)**

The levator scapulae muscle, innervated by the dorsal scapular nerve (C5), is anatomically positioned in the cervical posterior region. Its upper segment is concealed by the sternocleidomastoid muscle, while the lower portion lies deep to the trapezius (see Fig. 7.10). The intermediate section contributes to the posterior triangle's floor (Palastanga and Soames 2019; Standring 2020).

Superficial to the cervical extensor musculature, the levator scapulae originates via tendinous insertions on the transverse processes of the upper three or four cervical vertebrae. Specifically, it attaches to the posterior tubercles of the lower two vertebrae, posterior to the scalenus medius attachment. The muscle fibers descend laterally, terminating at the scapula's medial border between the superior angle and the base of the scapular spine (Leonello et al. 2007; Netter 2019).

Functionally, the levator scapulae coordinates with the trapezius to elevate and retract the shoulder girdle, as well as to counteract downward forces during manual load-bearing. Bilateral contraction of these muscles facilitates cervical extension, while unilateral activation induces ipsilateral cervical lateral flexion. Additionally, the levator scapulae contributes to scapular stabilization and participates in scapular medial rotation (Neumann 2017).

7.3.4.4 Medially Rotating Muscles of the Shoulder Girdle: Rhomboid Major, Rhomboid Minor, Pectoralis Minor, Levator Scapulae

The rhomboid major and minor originate from the vertebral column and insert on the medial border of the scapula. The pectoralis minor extends from the anterior chest wall to the coracoid process. The levator scapulae originates from the upper cervical vertebrae and inserts at the superior angle of the scapula (Moore et al. 2018).

The medial rotation of the shoulder girdle involves the movement of the inferior angle of the scapula toward the vertebral column. This movement is often facilitated by gravity and controlled by the eccentric activity of the lateral rotators, such as the trapezius and serratus anterior. However, during specific actions requiring resistance, such as when performing a chin-up, especially when moving from a hanging position to a full chin-up, the medial rotators work collectively against resistance to medially rotate the shoulder girdle (Neumann 2017).

The axis of rotation for medial rotation is located just below the spine of the scapula toward its base. This axis allows for coordinated movement where the pectoralis minor pulls down on the lateral side via its attachment to the coracoid process, while the rhomboids and levator scapulae pull up on the medial side, resulting in effective medial rotation of the shoulder girdle (Oatis 2017).

These muscular actions result in the movements depicted in Fig. 7.12a. The coordinated efforts of these muscles are particularly significant in activities requiring scapular stabilization and controlled movement against resistance (Levangie and Norkin 2005).

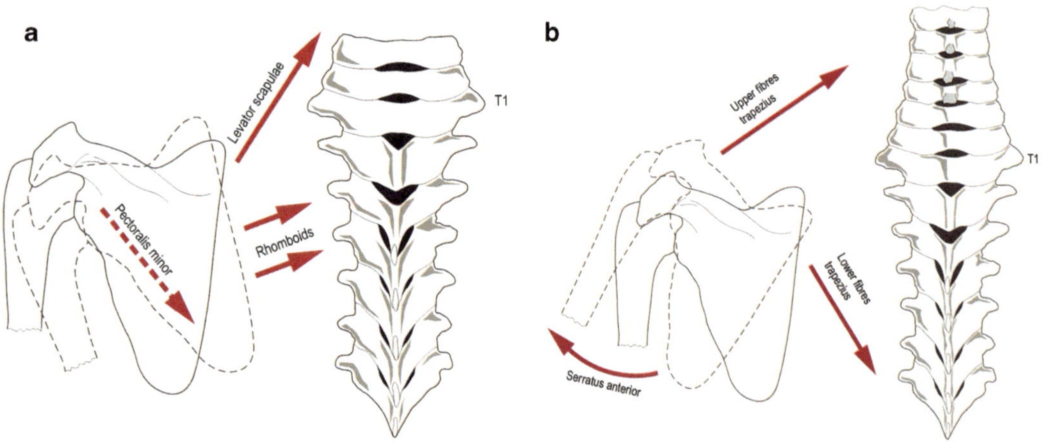

Fig. 7.12 Diagrammatic representation of scapular movements: (**a**) medial rotation and (**b**) lateral rotation, showing the direction of pull of the primary muscles involved. (Adapted with permission from Palastanga and Soames (2019))

7.3.4.5 Laterally Rotating Muscles of the Shoulder Girdle: Trapezius, Serratus Anterior

The trapezius and serratus anterior are positioned to enable lateral movement of the inferior angle of the scapula along the thoracic wall. This movement is restricted by the clavicle, which functions as a support structure by limiting motion at the acromion, resulting in the elevation of the acromion and lateral displacement of the inferior angle, allowing the glenoid fossa to orient more directly upward. This shoulder girdle movement significantly enhances the range of motion, particularly for abduction and flexion of the upper limb at the shoulder joint (Levangie and Norkin 2005; Neumann 2017).

The trapezius muscle contributes to this rotational movement through a coordinated action of its fibers. The upper fibers contract to elevate the lateral aspect of the clavicle and the acromion, while the lower fibers simultaneously exert a downward force on the medial end of the scapular spine. This differential action creates a rotational moment around the axis of the scapula, which is located just inferior to the base of the scapular spine (Moore et al. 2018) (Fig. 7.12b).

Among these two muscles, the serratus anterior plays a more dominant role in scapular lateral rotation. It exerts a strong lateral pull on the inferior angle of the scapula, where most of its fibers insert. This action effectively rotates the scapula around the thoracic wall (Oatis 2017).

7.3.5 The Shoulder Joint

The shoulder joint, or glenohumeral joint, exhibits a wide range of motion across multiple planes. The movement of the shoulder joint includes abduction and adduction in the coronal plane, flexion and extension in the sagittal plane, and medial and lateral rotation around the longitudinal axis of the arm (Neumann 2017).

7.3.5.1 Abducting Muscles of the Arm at the Shoulder Joint: Deltoid, Supraspinatus

Supraspinatus (Innervated by the Supraspinatus Nerve from the Upper Trunk of the Brachial Plexus: C5 and C6)

The supraspinatus muscle originates from the medial two-thirds of the supraspinous fossa and the deep surface of its overlying fascia (Fig. 7.13). Its course takes it laterally, deep to the trapezius, acromion, and coracoacromial ligament, ultimately traversing the superior aspect of the shoulder joint (Oatis 2017).

The tendon of the supraspinatus, which develops within the muscle belly, integrates with the deep surface of the shoulder joint capsule before inserting into the upper facet of the greater tubercle of the humerus.

The supraspinatus muscle initiates abduction at the shoulder joint and is particularly significant during the early phase of this movement. During the initial stages, the supraspinatus performs a dual function: it stabilizes the humeral head against the glenoid fossa, preventing upward shearing, and facilitates abduction (Levangie and Norkin 2005; Neumann 2017). After approximately 20° of abduction, the stronger deltoid muscle takes over, with the supraspinatus continuing to maintain the humeral head's position against the glenoid fossa. Following the first 20° of abduction, as the more powerful deltoid takes over the primary abduction role, the supraspinatus transitions to primarily maintaining the humeral head's position against the glenoid fossa (Moore et al. 2018; Oatis 2017).

The supraspinatus muscle is a component of the rotator cuff, a group of four muscles that form a musculotendinous structure encircling the humeral head (Standring 2020). This anatomical arrangement plays a significant role in maintaining glenohumeral joint stability during various shoulder movements (Moore et al. 2018). The primary function of the rotator cuff, including the supraspinatus, is to maintain the position of the

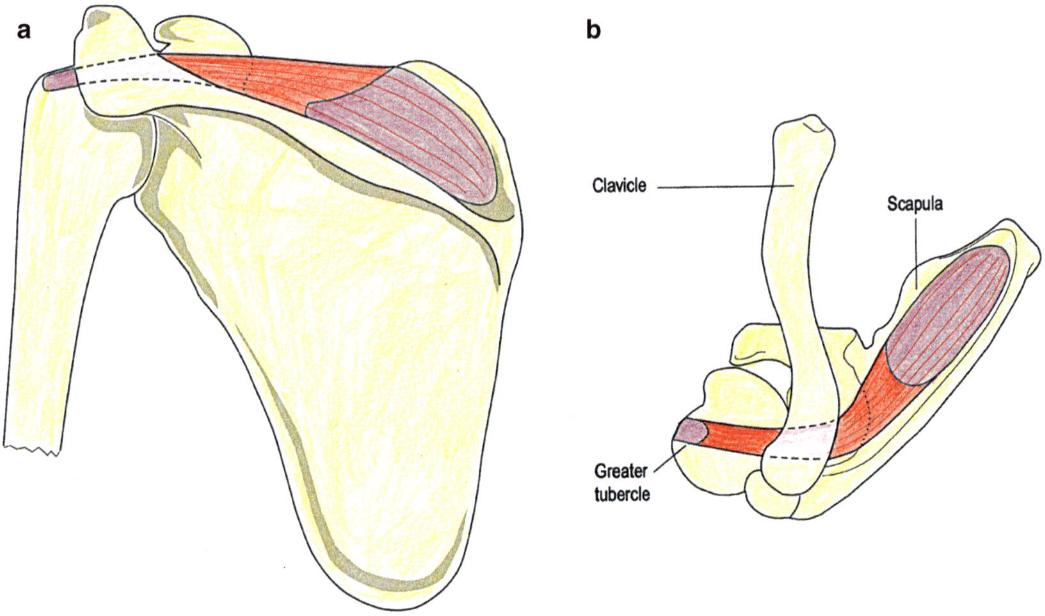

Fig. 7.13 The attachment of the supraspinatus: (**a**) posterior view and (**b**) superior view. (Adapted with permission from Palastanga and Soames (2019))

humeral head within the glenoid fossa throughout the range of shoulder motion (Neumann 2017). This stabilization is important for achieving appropriate shoulder biomechanics and preventing subluxation or dislocation (Kisner and Colby 2007).

Deltoid (Innervated by the Axillary Nerve from the Posterior Cord of the Brachial Plexus: C5 and C6)

The deltoid is a thick, triangular muscle that forms the rounded contour of the shoulder (Fig. 7.14). Anatomically and functionally, the deltoid is divided into three parts: anterior, middle, and posterior, with only the middle portion being multipennate (Moore et al. 2018). This muscle has an extensive origin from the shoulder girdle. Anteriorly, its fibers attach to the anterior border of the lateral third of the clavicle, while posteriorly, they attach to the lower lip of the crest of the scapular spine. Both the most anterior and posterior fibers run obliquely, in a continuous manner, to the deltoid tuberosity on the lateral surface of the humeral shaft (Neumann 2017).

The middle portion of the deltoid exhibits a more intricate architecture due to its multipen-

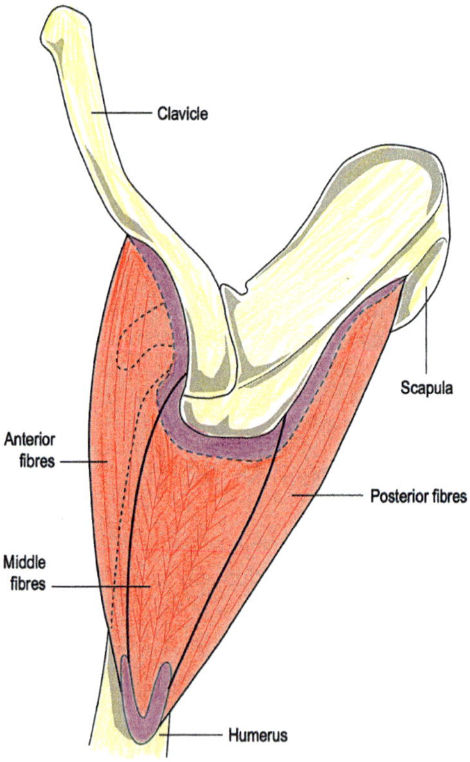

Fig. 7.14 Superior view of the deltoid, highlighting its attachment. (Adapted with permission from Palastanga and Soames (2019))

nate arrangement (see Fig. 7.14). This part comprises shorter oblique fibers that originate from four tendinous slips attached to the lateral margin of the acromion. These fibers then converge onto three intersecting tendinous slips, which ultimately insert into the humeral deltoid tuberosity. Despite working at a mechanical disadvantage, this intricate arrangement significantly enhances the strength of this particular muscle segment (Levangie and Norkin 2005).

The deltoid muscle serves as the primary abductor of the arm at the glenohumeral joint, with its middle multipennate fibers being the main effectors of this action. However, this deltoid-mediated abduction is effectively performed after the initial abduction by the supraspinatus muscle (Neumann 2017). During abduction, the deltoid's action tends to produce a superior translation of the humeral head. This potential instability is counteracted by the rotator cuff muscles: subscapularis anteriorly, teres minor and infraspinatus posteriorly, and supraspinatus superiorly (Fig. 7.15). These muscles work synergistically to maintain the humeral head centered in the glenoid fossa (Levangie and Norkin 2005).

The anterior portion of the deltoid functions as a potent flexor and medial rotator of the humerus. Conversely, the posterior portion acts as a strong extensor and lateral rotator. Additionally, the posterior deltoid aids in transferring the weight of heavy objects held in one's hand to the shoulder girdle (Oatis 2017). Interestingly, the posterior deltoid also plays a role during arm adduction. In this context, it counterbalances the medial rotation produced by the pectoralis major and latissimus dorsi, thereby maintaining optimal joint positioning (Standring 2020).

The deltoid muscle demonstrates a multifaceted functional profile in the biomechanics of the upper limb. Its middle fibers play a primary role in arm abduction through concentric contraction

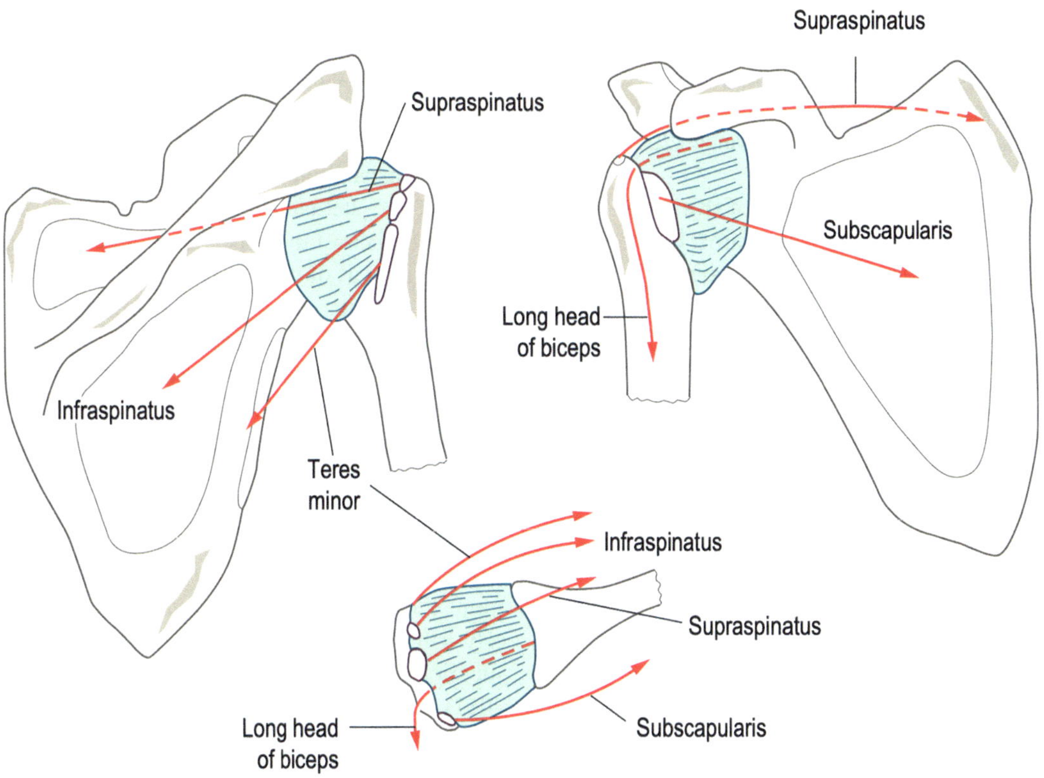

Fig. 7.15 The role of the rotator cuff muscles in stabilizing the shoulder joint. (Adapted with permission from Palastanga and Soames (2019))

(Standring 2020). However, the significant growth and multipennate architecture of this muscle may have developed to facilitate an extended period of maintaining a fixed position of the upper limb (Moore et al. 2018).

Many daily activities necessitate maintaining the arm in an abducted position for extended periods. During such tasks performed with the arms positioned anterior to the trunk, the middle fibers of the deltoid engage in isometric contraction. Subsequently, these fibers facilitate the controlled lowering of the arm back to its resting position through eccentric contraction (Neumann 2017). In the context of rehabilitation, exercises targeting the deltoid should consider its various contraction modes. Concentric strengthening can improve active arm elevation, while eccentric training may enhance control during lowering tasks. Isometric exercises can be beneficial for improving endurance in sustained arm positions, which is often required in activities of daily living (Kisner and Colby 2007). Furthermore, the deltoid's interaction with other shoulder girdle muscles, such as the rotator cuff group, is important to consider in maintaining shoulder stability and preventing impingement syndromes during rehabilitation protocols (Sahrmann 2002).

7.3.5.2 Flexing Muscles of the Arm at the Shoulder Joint: Pectoralis Major, Anterior Fibers of Deltoid, Long Head of Biceps, Coracobrachialis

***Pectoralis Major* (Innervated by the Medial and the Lateral Pectoral Nerves: Clavicular Part by C8 and T1, Sternocostal Part by C5, C6, and C7, Respectively)**

The pectoralis major is a thick, triangular muscle situated on the superior half of the anterior thoracic wall (Fig. 7.16), plays a significant role in shoulder joint flexion (Moore et al. 2018). Structurally, the pectoralis major comprises two parts: clavicular and sternocostal. While these sections may occasionally be separated by a groove, they typically form a continuous muscular unit. As the muscle fibers course toward their humoral insertion, they undergo a torsional arrangement, forming the characteristic rounded anterior axillary fold (Neumann 2017).

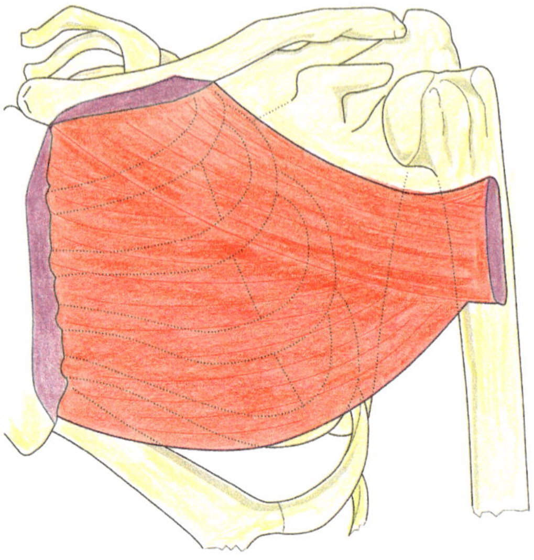

Fig. 7.16 Anterior view showing the attachment of the pectoralis major muscle. (Adapted with permission from Palastanga and Soames (2019))

The smaller clavicular portion originates from the medial half of the anterior surface of the clavicle, while the larger sternocostal portion has a more extensive origin, including the anterior surfaces of the sternal manubrium and body, the anterior aspects of the upper six costal cartilages, the anterior part of the sixth rib, and the aponeurosis of the external oblique abdominal muscle (Levangie and Norkin 2005).

The muscle converges laterally, inserting via a laminated tendon into the lateral lip of the intertubercular groove of the humerus. This tendon exhibits a U-shaped cross-sectional structure. The anterior lamina, formed by the clavicular part, attaches to the inferior portion of the humeral insertion site. The sternocostal part forms the posterior lamina, which passes superiorly behind the anterior lamina to attach to the superior portion of the humeral insertion (Oatis 2017). Topographically, the pectoralis major overlies the pectoralis minor, ribs, and serratus anterior. In females, the breast tissue covers this muscle, with fibrous septa of the breast attaching to the overlying deep fascia. The deltopectoral groove, containing the cephalic vein and branches of the thoracoacromial artery, separates the pectoralis major from the deltoid muscle (Standring 2020).

The pectoralis major muscle exhibits significant functional versatility in upper-body movements. As a primary climbing muscle, it demonstrates considerable strength when the upper limbs are fixed in an elevated position. The contractile force of the pectoralis major can effectively elevate the trunk. This action is complemented by the synergistic activity of the latissimus dorsi muscle (Moore et al. 2018; Neumann 2017). During dynamic movements such as pushing, punching, and throwing, the pectoralis major propels the humerus forward with substantial force. Concurrently, the serratus anterior and pectoralis minor muscles facilitate protraction of the shoulder girdle, enhancing the overall power and range of these movements (Standring 2020).

The functional capacity of the pectoralis major is particularly evident in specific exercises like the "push-up." During the ascent phase of this exercise, the muscle undergoes concentric contraction, effectively elevating the body. Conversely, in the descent phase, it contracts eccentrically, controlling the lowering of the body (Kendall et al. 2005).

7.3.5.3 Extending Muscles of the Arm at the Shoulder Joint: Latissimus Dorsi, Teres Major, Posterior Fibers of Deltoid, Pectoralis Major, Long Head of Triceps

Latissimus Dorsi (Innervated by the Thoracodorsal Nerve from the Posterior Cord of the Brachial Plexus: C6, C7, and C8)

The latissimus dorsi is a large triangular muscle spanning from the trunk to the humerus. This muscle forms key anatomical landmarks: its superior border forms the lower boundary of the triangle of auscultation, while its lateral border defines the medial boundary of the lumbar triangle (Moore et al. 2018).

The extensive origin of the latissimus dorsi includes the posterior layer of the thoracolumbar fascia, which attaches to the spinous processes of the inferior six thoracic vertebrae, all lumbar and sacral vertebrae, and associated supraspinous and interspinous ligaments. The portion arising from the six lower thoracic vertebrae is overlapped by the trapezius muscle (Neumann 2017). Additional attachments include the posterior aspect of the outer lip of the iliac crest, with direct muscular slips laterally. As the muscle fibers course superolaterally across the lower thorax, they attach to the external surfaces of the lower three or four ribs and via fascia to the inferior angle of the scapula (Levangie and Norkin 2005).

The convergence of these fibers results in a thin, flattened tendon that winds around the inferior border of the teres major. This tendon inserts into the intertubercular groove of the humerus, anterior to the teres major's tendon, separated by a bursa. Notably, the muscle undergoes a 180° twist, such that the anterior surface of the tendon is continuous with the posterior surface of the rest of the muscle. As a result, the fibers that start at the lowest level on the trunk have the highest attachment on the humerus (Oatis 2017).

The latissimus dorsi, a prominent muscle of the back, exhibits diverse functional capabilities depending on the fixed point of action. When the humerus is immobilized relative to the scapula, this muscle retracts the pectoral girdle. Additionally, it serves as a strong adductor and medial rotator of the humerus at the shoulder joint (Standring 2020).

From a functional perspective, the latissimus dorsi plays a significant role in climbing activities. When the arms are fixed above the head, it works with the pectoralis major to lift the trunk. This muscle is also actively involved in rowing motions and the downward stroke during swimming (Neumann 2017). Its attachment to the ribs contributes to its involvement in forceful expiration. During coughing or sneezing, the latissimus dorsi can be palpated as it compresses the thoracic and abdominal cavities. Its insertion on the inferior angle of the scapula allows it to assist in maintaining scapular stability against the thoracic wall during upper limb movements (Moore et al. 2018).

When the humerus is fixed, such as when using crutches, the latissimus dorsi can pull the trunk forward relative to the arms, simultaneously lifting the pelvis. This functionality is particularly relevant in patients with lower-body paralysis. As the latissimus dorsi attaches to the pelvis and retains its innervation, it can be utilized to generate movement of the pelvis and trunk (Kendall et al. 2005).

These patients can achieve a modified gait pattern by using lower limb orthoses and crutches to contract each latissimus dorsi alternately, resulting in a hip hitching motion.

Teres Major **(Innervated by the Lower Subscapular Nerve from the Posterior Cord of the Brachial Plexus: C6 and C7)**

The teres major muscle is located in the posterior part of the axilla and contributes to the formation of the inferior boundary of the upper triangular and quadrangular spaces (Moore et al. 2018) (Fig. 7.17). This substantial muscle also contributes to the formation of the axilla's posterior fold in conjunction with the latissimus dorsi (Standring 2020).

The teres major originates from an oval-shaped area on the dorsal surface of the scapula, proximal to the inferior angle, and from the fascia separating it from adjacent musculature. Its fibers, particularly those adjacent to the latissimus dorsi, course superolaterally, culminating in a broad, flat tendon (Neumann 2017). The insertion of the teres major is along the medial lip of the intertubercular groove of the humerus. A bursa separates its tendon from that of the latissimus dorsi, which largely overlaps the teres major (Clarkson 2013).

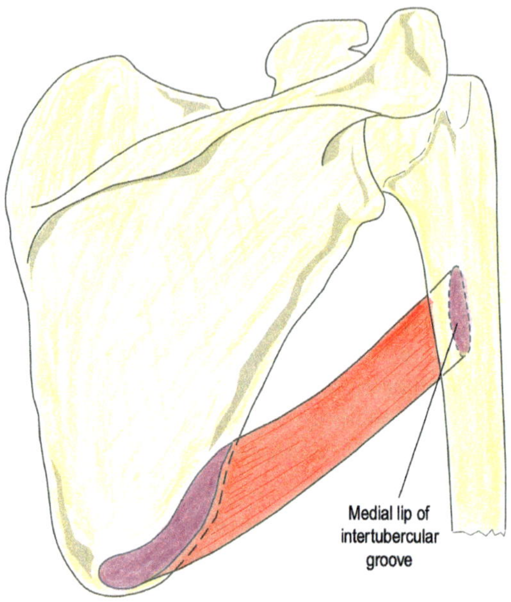

Fig. 7.17 Posterior view illustrating the attachment of the teres major muscle. (Adapted with permission from Palastanga and Soames (2019))

7.3.5.4 Adducting Muscles of the Arm at the Shoulder Joint: Coracobrachialis, Pectoralis Major, Latissimus Dorsi, Teres Major

Coracobrachialis **(Innervated by the Musculocutaneous Nerve from the Lateral Cord of the Brachial Plexus: C6 and C7)**

The coracobrachialis muscle is the sole true adductor muscle group in the arm (Moore et al. 2018). This muscle originates via a rounded tendon from the apex of the coracoid process of the scapula, sharing this origin with the short head of biceps brachii (Standring 2020). The insertion of the coracobrachialis is characterized by a flat tendon attaching to the medial aspect of the humeral shaft at its midpoint, positioned between the triceps and brachialis muscles. Some muscle fibers may extend into the medial intermuscular septum of the arm (Neumann 2017). Functionally, the coracobrachialis primarily acts as an adductor of the arm at the glenohumeral joint. Additionally, it contributes to weak flexion of the arm (Clarkson 2013).

7.3.5.5 Externally (Laterally) Rotating Muscles of the Arm at the Shoulder Joint: Supraspinatus, Infraspinatus, Teres Minor, Posterior Fibers of Deltoid

Infraspinatus (Innervated by the Suprascapular Nerve from the Upper Trunk of the Brachial Plexus: C5 and C6)

The infraspinatus is a key component of the rotator cuff complex (Moore et al. 2018). This muscle originates from the medial two-thirds of the infraspinous fossa of the scapula, including tendinous intersections attached to ridges within the fossa, and the thick fascia enveloping the muscle (Standring 2020) (Fig. 7.18).

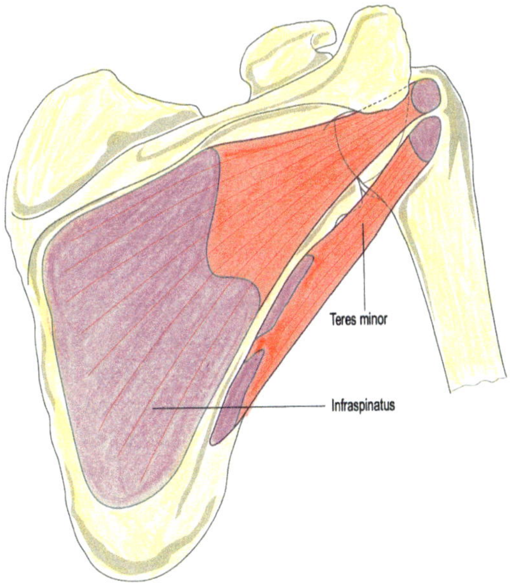

Fig. 7.18 Posterior view showing the attachment of the infraspinatus and teres minor muscles. (Adapted with permission from Palastanga and Soames (2019))

The muscle fibers of the infraspinatus converge to form a narrow tendon, which inserts onto the middle facet of the humeral greater tubercle and the posterior aspect of the shoulder joint capsule. A bursa, occasionally communicating with the glenohumeral joint, separates the muscle from the scapular neck. While the superior portion of the infraspinatus is deep to the trapezius, deltoid, and acromion process, its inferior part is superficial (Neumann 2017).

Infraspinatus, in conjunction with teres minor, plays a significant role in the biomechanics of full arm abduction. During the terminal phase of this movement, these muscles facilitate external rotation of the humerus, allowing the greater tubercle to clear the coracoacromial arch. This action enables complete contact between the humeral head and the glenoid fossa, permitting full abduction (Kisner and Colby 2007).

The infraspinatus, along with teres minor, supraspinatus, and subscapularis, forms the musculotendinous "rotator cuff." This structure functions as a dynamic stabilizer of the glenohumeral joint. The proximity of their tendons to the joint enhances their stabilizing effect. During humeral head movements on the glenoid fossa, the interaction between these muscles minimizes sliding and shearing forces. Additionally, when carrying weights in the hand, these muscles brace the humeral head against the glenoid fossa, maintaining joint integrity (Sahrmann 2002).

Teres Minor (Innervated by the Axillary Nerve from the Posterior Cord of the Brachial Plexus: C5 and C6)

The teres minor muscle, when observed from a posterior view, delineates the upper boundary of both the triangular and quadrangular spaces in the scapular region (Moore et al. 2018). It originates from the upper two-thirds of the lateral border of the scapula and the fascia between it and adjacent muscles, specifically teres major below and infraspinatus above. The muscle arises from two heads, separated by a groove accommodating the circumflex scapular artery (Standring 2020).

The muscle fibers of the teres minor course in a superolateral direction, converging to form a narrow tendon. This tendon inserts onto the inferior facet of the humeral greater tubercle and the bony surface immediately distal to it. Notably, the tendon integrates with and reinforces the inferior posterior aspect of the shoulder joint capsule, contributing to the overall stability of the glenohumeral joint (Neumann 2017).

In the context of shoulder biomechanics, teres minor, as a component of the rotator cuff, plays a significant role in maintaining the dynamic stability of the glenohumeral joint. It acts synergistically with other rotator cuff muscles to control the position of the humeral head within the glenoid fossa during various arm movements (Kisner and Colby 2007).

7.3.5.6 Internally (Medially) Rotating Muscles of the Arm at the Shoulder Joint: Subscapularis, Teres Major, Latissimus Dorsi, Pectoralis Major, Anterior Fibers of Deltoid

Subscapularis (Innervated by the Upper and Lower Subscapular Nerves from the Posterior Cord of the Brachial Plexus: C5, C6, and C7)

The subscapularis muscle forms the majority of the posterior wall of the axilla, positioned in close proximity to teres major and latissimus dorsi (Moore et al. 2018). Its anterior surface lies on the serratus anterior. From a frontal view, the subscapularis forms the upper boundary of the triangular and quadrangular spaces (Standring 2020).

As a component of the rotator cuff complex, the subscapularis performs a key function in preserving glenohumeral joint stability during various arm movements. It achieves this by maintaining the position of the humeral head within the glenoid fossa (Neumann 2017). Additionally, the subscapularis counteracts the superior translational forces on the humeral head when muscles such as the deltoid, biceps brachii, and the long head of the triceps are activated (Kisner and Colby 2007).

7.4 The Elbow Joint Complex

The elbow joint complex is composed of four articulations: the humeroradial, humeroulnar, proximal, and distal radioulnar joints. The first three of these are encapsulated within a single joint cavity. This configuration allows for a diverse range of movements essential for daily activities and complex motions (Moore et al. 2018).

The humeroradial joint, functioning as both a hinge and a pivot joint, enables flexion and extension in conjunction with the humeroulnar joint, as well as rotational movements together with the proximal radioulnar joint (Neumann 2017). Specifically, the humeroulnar joint, classified as a modified hinge joint, permits about 5° of internal and external rotation at the full range of flexion and extension, demonstrating the joint's slight rotational capability in addition to its primary motion (Stroyan and Will 1993). Stability in the elbow joint is significantly contributed by the congruent matching of the ulna's trochlea and the humerus's capitulum, which also restricts motion to primarily flexion and extension.

The cooperative function between the proximal and distal radioulnar joints, which operate independently from the other elbow joints, is pivotal for the pronation and supination of the forearm. This collaboration facilitates a range of motion encompassing 140° of flexion from a fully extended position and a rotation from 75° of pronation to 85° of supination. It is noted that the majority of daily living activities are achievable within a flexion range of 30–130° and with half the full capacity of pronation and supination (Morrey et al. 1981).

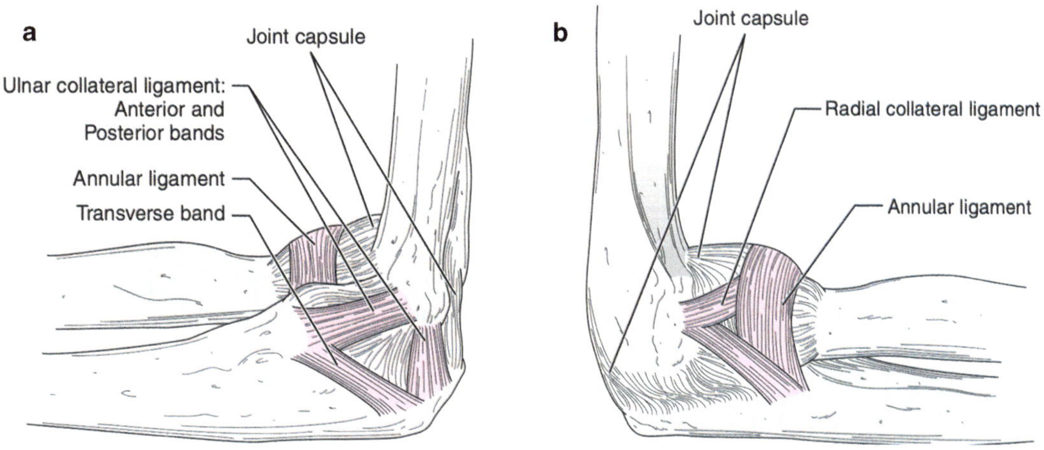

Fig. 7.19 Anatomy of the elbow ligament complexes: (**a**) medial ligament complex and (**b**) lateral ligament complex. (Adapted with permission from Jenkins (2009))

7.4.1 Elbow Joint Stability

The stability of the elbow joint is significantly enhanced by the medial and lateral collateral ligament complexes, which compensate for the articular capsule's relative insufficiency. The lateral side of the elbow features a complex comprising four important ligaments: the radial collateral ligament, the lateral ulnar collateral ligament, the accessory lateral collateral ligament, and the annular ligament (Fig. 7.19). On the other hand, the medial side is reinforced by the medial collateral ligament complex, which includes the anterior oblique ligament, the posterior oblique ligament, and the transverse ligament (Steinberg and Plancher 1995).

These collateral ligament complexes are fundamental in providing stability against lateral (varus) and medial (valgus) stresses. Specifically, the lateral complex guards against varus stress, while the medial complex is essential for valgus stability (Hotchkiss and Weiland 1987). Beyond these static stabilizers, dynamic stabilization is a critical component of elbow joint stability. The anconeus muscle acts as a dynamic stabilizer against varus stress. In contrast, the flexor-pronator muscle group serves as the primary dynamic defense against valgus stress (Safran 1995; Safran and Baillargeon 2005).

7.4.2 Biomechanics and Muscular Coordination of the Elbow Joint Musculature

The elbow represents a complex joint system where 24 muscles work in coordinated patterns to facilitate movement. While these muscles can potentially generate multiple actions across the elbow, wrist, and phalangeal joints, research has established that each muscle demonstrates a predominant function. Contemporary anatomical studies have identified four primary muscle groups that control elbow movement: the anterior flexor group, posterior extensor group, lateral extensor-supinator group, and medial flexor-pronator group (Neumann 2017).

Biomechanical analysis reveals that the mechanical efficiency of elbow flexors increases proportionally with the degree of flexion, primarily due to optimization of moment arm lengths (Basmajian and De Luca 2015). The brachialis

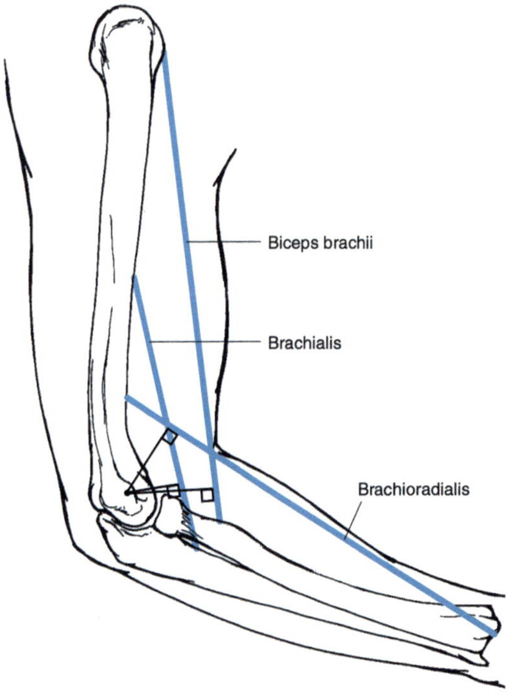

Fig. 7.20 The line of action of the three forearm muscles: The brachialis, despite its large size, has the smallest moment arm, resulting in the least mechanical advantage. The biceps brachii, with a large cross-sectional area, has a longer moment arm, while the brachioradialis, although smaller in cross section, possesses the longest moment arm, providing the greatest mechanical advantage in this position. (Adapted from Hamill et al. (2015))

muscle, despite having the most substantial cross-sectional area among flexors, demonstrates relatively limited mechanical efficiency. In comparison, the biceps brachii combines significant cross-sectional area with enhanced mechanical advantage. The brachioradialis, though smaller in cross section, exhibits optimal mechanical leverage (Fig. 7.20). Research indicates peak mechanical advantage occurs between 100° and 120° of flexion, with moment arms measuring 6 cm for the brachioradialis, 2.5–3.0 cm for the brachialis, and 3.5–4.0 cm for the biceps brachii (Palastanga and Soames 2019).

Each primary elbow flexor demonstrates specific functional characteristics based on joint position and mechanical advantage. The brachialis maintains consistent activity throughout forearm positions but experiences limitations due to suboptimal mechanical leverage, showing enhanced activity during pronation (Standring 2020). Biceps brachii function varies with both shoulder and radioulnar joint positions, with shoulder flexion reducing tension in the long head while extension increases it. Pronation significantly impacts biceps brachii tendon mechanics. The brachioradialis, characterized by minimal volume and extended fiber length, achieves remarkable efficiency through optimal mechanical positioning (Moore et al. 2018).

The triceps brachii, recognized as the primary elbow extensor, consists of three anatomically distinct portions. Shoulder position influences the long head function due to its proximal attachment. The medial head functions as the primary extensor component, maintaining consistent activity across various movement parameters. The lateral head, while possessing the greatest force-generating capacity, primarily activates during resistance-based activities.

The medial flexor-pronator muscle group, forming a common tendon on the medial epicondyle, includes the pronator teres and three wrist muscles (flexor carpi radialis, flexor carpi ulnaris, and palmaris longus). The pronator quadratus exhibits consistent activation patterns regardless of position or movement characteristics, while the pronator teres shows increased recruitment during rapid or high-resistance pronation, with peak activation at 60° of forearm flexion (Drake et al. 2020). The lateral epicondyle provides attachment for the extensor-supinator group. Electromyographic studies indicate that the supinator muscle predominantly controls slow, unresisted supination across all forearm positions, while the biceps brachii contributes during rapid or resisted movements in elbow flexion (Floyd 2012; Levangie and Norkin 2005).

7.4.3 Functional Muscle Groups in Elbow Movement

The muscles responsible for the movement of the elbow are categorized into four primary functional groups based on their actions: elbow flexors, elbow extensors, flexor-pronators, and extensor-supinators. This categorization aids in understanding the complex mechanics involved in elbow motion and its stabilization during various activities (Table 7.2).

Table 7.2 Elbow and forearm muscles and their respective functions

Muscle	Insertion	Innervating nerve	Flexion	Extension	Pronation	Supination
Anconeus	Lateral epicondyle of humerus to olecranon process on ulna	Radial nerve; C7, C8		Asst		
Biceps brachii	Supraglenoid tubercle and corocoid process to radial tuberosity	Musculocutaneous nerve; C5, C6	PM			PM
Brachialis	Anterior surface of lower humerus to coronoid process on ulna	Musculocutaneous nerve; C5, C6	PM			
Brachioradialis	Lateral supracondylar ridge of humerus to styloid process of radius	Radial nerve; C6, C7	PM			
Extensor carpi radialis brevis	Lateral epicondyle of humerus to base of third metacarpal	Radial nerve; C6, C7	Asst			
Extensor carpi radialis longus	Lateral supracondylar ridge of humerus to base of second metacarpal	Radial nerve; C6, C7	Asst			
Extensor carpi ulnaris	Lateral epicondyle of humerus to base of fifth metacarpal	Posterior interosseous nerve; C6–C8		Asst		
Flexor carpi radialis	Medial epicondyle of humerus to base of second, third metacarpal	Median nerve; C6, C7	Asst			
Flexor carpi ulnaris	Medial epicondyle to pisiform, hamate base, and base of fifth metecarpal	Ulnar nerve; C8, T1	Asst			
Palmaris longus	Medial epicondyle to palmar aponeurosis	Median nerve; C6, C7	Asst			
Pronator quadratus	Distal anterior surface of ulnar to distal anterior surface of radius	Anterior interosseous nerve; C8, T1			PM	
Pronator teres	Medial epicondyle of humerus, coronoid process of ulna to midlateral surface of radius	Median nerve; C6, C7	Asst		PM	
Supinator	Lateral epicondyle of humerus to upper lateral side of radius	Posterior interosseous nerve; C5, C6				PM

PM primary mover, *Ass* assistive mover

7.4.3.1 Flexing Muscles of the Elbow Joint: Biceps Brachii, Brachialis, Brachioradialis

The elbow flexor group, composed of the biceps brachii, brachialis, and brachioradialis, plays a pivotal role in the complex mechanics of elbow movement. These muscles, which extend across the elbow and, for some, the glenohumeral joint, each contribute distinctively to the action of elbow flexion.

Biceps Brachii (Innervated by the Musculocutaneous Nerve from the Lateral Cord of the Brachial Plexus: C5 and C6)

The biceps brachii is a prominent fusiform muscle occupying the anterior compartment of the arm (Moore et al. 2018) (Fig. 7.21). This muscle exhibits a unique anatomical arrangement, featuring two distinct origins and a complex distal attachment. It originates from two tendinous heads at its upper end and attaches to a tendinous and one aponeurotic insertion at its lower end. The short head arises from the coracoid process apex of the scapula, sharing its origin with the coracobrachialis muscle. The long head originates from the supraglenoid tubercle and adjacent glenoid labrum, with its tendon traversing the glenohumeral joint encased in a synovial sheath before entering the intertubercular groove deep to the transverse humeral ligament (Standring 2020).

The muscle's proximal portion is overlaid by the deltoid and pectoralis major muscles, while its main body is superficial, covered only by skin and subcutaneous adipose tissue. The two muscle bellies converge distal to the arm's midpoint, forming a single muscular unit (Neumann 2017). Distally, the biceps brachii forms a flattened tendon that undergoes a 90° rotation before inserting onto the posterior aspect of the radial tuberosity. A bursa interposes between this tendon and the remainder of the tuberosity. Additionally, a prominent aponeurotic extension, the bicipital aponeurosis (see Fig. 7.21), emerges from the tendon's lateral aspect. This structure courses inferomedially across the cubital fossa, anterior to the brachial artery and median nerve, before integrating with the deep fascia on the ulnar side of the forearm (Kisner and Colby 2007).

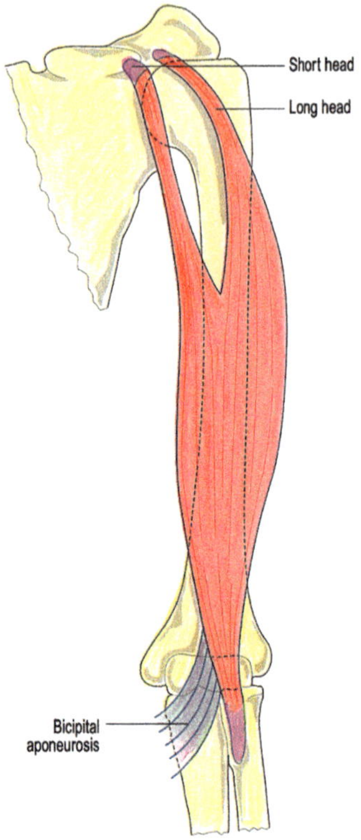

Fig. 7.21 Anterior view illustrating the attachment of the biceps brachii muscle. (Adapted with permission from Palastanga and Soames (2019))

The biceps brachii exhibits a multifaceted role in upper limb function, serving as both a primary elbow flexor and a potent forearm supinator, especially when met with resistance (Moore et al. 2018). The biceps brachii is a key flexor, particularly when the forearm is either supinated or in a neutral position, reaching its peak torque within an elbow flexion range of 80–100°. Its flexion efficacy, however, is reduced in conditions where the elbow is fully extended or when both the shoulder and elbow are flexed concurrently, due to the biceps shortening over both joints (Neumann 2017). In addition to its actions at the elbow, the biceps brachii also contributes to shoulder flexion. The long head's traversal of the superior aspect of the glenohumeral joint provides a significant stabilizing effect on this articulation (Standring 2020).

The biceps brachii's functional versatility is exemplified in activities requiring sequential supination and flexion, such as cork extraction. During this motion, the ulnar head may experience medial displacement due to the force transmission from biceps contraction to the ulna's posterior border via the bicipital aponeurosis (Kisner and Colby 2007).

In cases of deltoid paralysis, which is a potential complication in certain spinal cord injuries, the long head of the biceps can be therapeutically trained to assist in shoulder abduction. This adaptation is facilitated by laterally rotating the humerus at the glenohumeral joint, thereby optimizing the long head's line of action for this compensatory movement (Harvey 2008).

Brachialis (Innervated by the Musculocutaneous Nerve from the Lateral Cord of the Brachial Plexus: C5 and C6)

The brachialis muscle, situated deep to the biceps brachii in the distal half of the anterior compartment of the arm (Fig. 7.22), plays a key role in elbow flexion (Moore et al. 2018). This muscle originates from the distal two-thirds of the anterior surface of the humerus, extending laterally and medially to the respective intermuscular septa (Standring 2020). The brachialis is partially separated from the lateral intermuscular septum by the brachioradialis and extensor carpi radialis longus muscles, with potential fusion between these structures. The muscle fibers converge distally, forming a strong tendon that makes up the floor of the cubital fossa. This tendon attaches to the ulnar tuberosity and the lower part of the coronoid process, notably at the brachialis impression (Neumann 2017).

Some deeper fibers of the brachialis insert into the elbow joint capsule. This arrangement serves a protective function during elbow flexion, retracting the capsule to prevent its impingement between articulating bones (Kisner and Colby 2007).

Unlike the biceps brachii, the brachialis is the primary elbow flexor. While it functions to flex the elbow, it also plays a crucial role in controlling the extension produced by gravity. The flexors of the elbow regulate the movement in this scenario through an eccentric contraction, which helps to manage the speed and smoothness of extension (Sahrmann 2002).

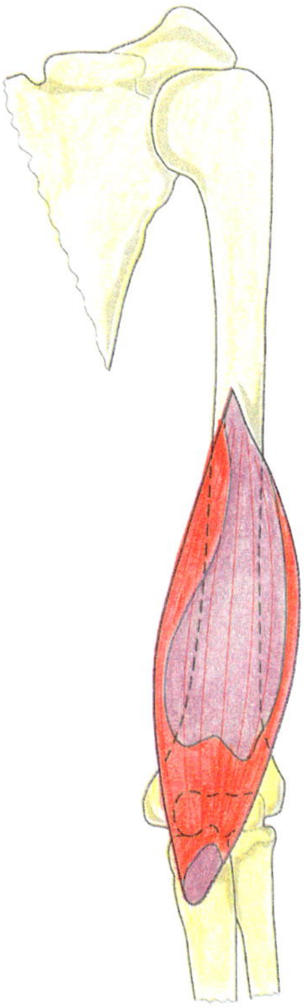

Fig. 7.22 Anterior view showing the attachment of the brachialis muscle. (Adapted with permission from Palastanga and Soames (2019))

Brachioradialis (Innervated by the Radial Nerve from the Posterior Cord of the Brachial Plexus: C5 and C6)

The brachioradialis is a superficial muscle on the lateral aspect of the forearm, extending distally to the wrist (Moore et al. 2018) (Fig. 7.23). This muscle delineates the lateral boundary of the cubital fossa and is partially overlapped proximally by the brachialis (Standring 2020).

The muscle originates from the superior two-thirds of the lateral supracondylar ridge of the humerus and the adjacent portion of the lateral intermuscular septum. Its fibers descend, forming a long, narrow, and flattened tendon along the middle third of the lateral aspect of the forearm. This tendon is intersected by those of the abductor pollicis longus and extensor pollicis brevis

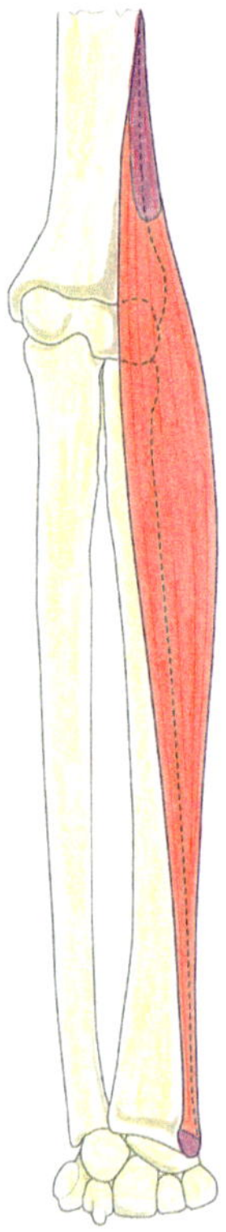

Fig. 7.23 Anterior view illustrating the attachment of the brachioradialis muscle. (Adapted with permission from Palastanga and Soames (2019))

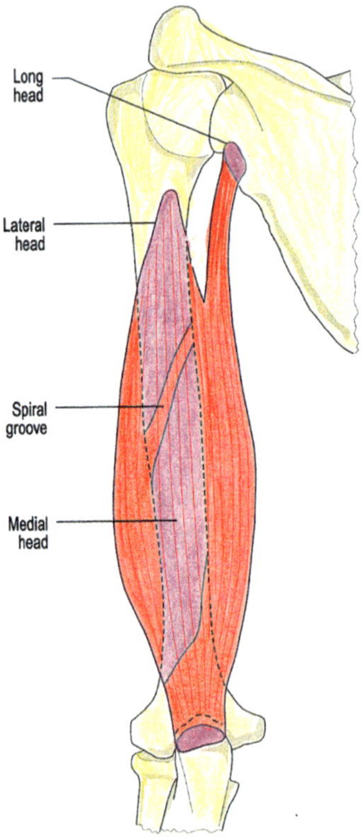

Fig. 7.24 Posterior view showing the attachment of triceps muscle. (Adapted with permission from Palastanga and Soames (2019))

before inserting onto the lateral surface of the radius, just proximal to the styloid process (Neumann 2017).

Functionally, the brachioradialis primarily acts as an elbow flexor, with its mechanical advantage optimized when the forearm is in a neutral position between pronation and supination. It also aids in repositioning the forearm to this neutral position from either extreme of rotation (Basmajian 1971; Kapandji 2007; Kisner and Colby 2007).

7.4.3.2 Extending Muscles of the Elbow Joint: Triceps Brachii, Anconeus

The primary extensors of the elbow are the triceps brachii and the anconeus. The long head of the triceps brachii, originating from the infraglenoid tubercle of the scapula, spans both the shoulder and elbow joints, rendering it less effective when the shoulder is hyperextended during elbow extension. The triceps brachii is most potent and generates the highest torque at 90° of elbow flexion. The anconeus muscle aids in initiating elbow extension and contributes to elbow stability during pronation and supination, although its efficacy diminishes when the elbow is fully extended (Pauly et al. 1967).

Triceps Brachii (Innervated by the Radial Nerve from the Posterior Cord of the Brachial Plexus: Lateral Head from C6 to C8, Long and Medial Heads from C7 and C8)

The triceps brachii, a prominent muscle of the posterior arm, is characterized by its three distinct heads: the long, lateral, and medial heads (Moore et al. 2018) (Fig. 7.24). It attaches via a tendon to the olecranon process of the ulna.

The long head of the triceps originates from the infraglenoid tubercle of the scapula and the adjacent glenoid labrum, integrating with the inferior aspect of the glenohumeral joint capsule. It courses distally as the most medial component, superficial to the medial head, before joining the common insertion tendon. During its descent, the long head traverses anterior to teres minor and posterior to teres major, delineating the medial boundary of the quadrangular space and triangular interval, while forming the lateral border of the triangular space (Standring 2020). The lateral head arises superior and lateral to the spiral groove on the posterior surface of the humerus, between the attachments of teres minor and deltoid. Its fibers descend to merge with the medial head, overlying the spiral groove. The medial head, the largest and deepest of the three, originates from the posterior humeral surface inferior and medial to the spiral groove, extending distally to the olecranon fossa (Kisner and Colby 2007; Neumann 2017). Notably, some fibers from the medial head insert into the posterior elbow joint capsule, functioning to retract it during joint extension, preventing capsular impingement.

The triceps brachii exhibits a complex functional profile in various upper limb activities. During elbow extension from a flexed position, gravity often provides the primary force, with the elbow flexors engaging in eccentric contraction to modulate the movement (Neumann 2017). The triceps brachii becomes active when the speed of extension is important (Moras et al. 2008).

In pushing and punching activities, the triceps contracts forcefully. During exercises like push-ups, it contracts concentrically to lift the body and eccentrically to lower it. This muscle functions similarly when using the arms to rise from or lower into a chair with armrests, or when using crutches or parallel bars to offload body weight from the legs during ambulation. In wheelchair use, the triceps brachii exerts considerable force to push the wheels and propel the chair forward (Mulroy et al. 2004).

Anconeus **(Innervated by the Radial Nerve: C7 and C8)**

The anconeus, a small triangular muscle located posterior to the elbow joint, exhibits a close anatomical relationship with the triceps brachii. Its origin is on the posterior aspect of the lateral epicondyle of the humerus and the adjacent elbow joint capsule. The muscle fibers course medially and distally, inserting on the lateral surface of the olecranon and the superior quarter of the posterior surface of the ulna, as well as the overlying fascia (Standring 2020).

Due to its extensive attachment on the ulna, the anconeus is believed to facilitate lateral movement (abduction of the ulna) and extension at the bone's distal end. These actions are particularly important during pronation, such as when using a tool like a screwdriver. During pronation, the ulna's movement relative to the radius adjusts the pronation-supination axis, enabling the forearm to rotate about a single axis rather than tracing an arc. This action enables the transfer of rotational motion from the forearm to the screwdriver, which is subsequently transmitted to the head of the screw.

7.4.3.3 Extensor-Supinators of the Elbow and Forearm

The extensor-supinators group consists of the extensor carpi radialis longus and brevis, supinator, extensor digitorum communis, extensor digiti minimi, and extensor carpi ulnaris. The supinator muscle, distinct from the biceps, efficiently performs supination regardless of the elbow's flexion angle, initiating and completing the movement. The biceps brachii requires resistance to activate its supination capability, while the other muscles in this group support various aspects of elbow extension and forearm supination.

Supinator **(Innervated by the Posterior Interosseous Branch of the Radial Nerve: C5 and C6)**

The supinator muscle is situated deep within the proximal forearm. This muscle is located on the superior portion of the radius and is concealed by the superficial muscles (Fig. 7.25). The supinator originates from multiple structures: the inferior aspect of the lateral epicondyle of the humerus, the radial collateral ligament, the annular ligament, and the supinator crest and fossa of the

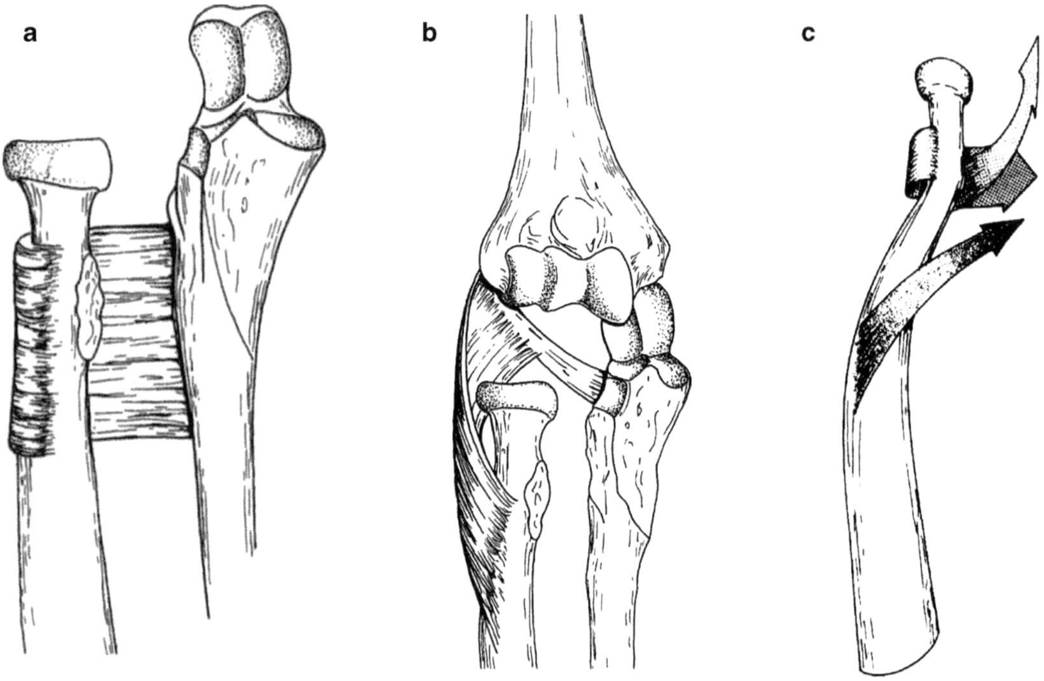

Fig. 7.25 The two layers of supinator muscle and their actions: (**a**) deep layer (ulnar layer), (**b**) superficial layer (humeral layer), and (**c**) combined supinating action of both layers with the biceps brachii. (Adapted from Calais-Germain (1999))

ulna (Moore et al. 2018; Standring 2020). For conceptual simplicity, the supinator can be considered as having two heads: humeral and ulnar. The posterior interosseous nerve traverses between these heads to access the extensor compartment of the forearm. From this extensive origin, muscle fibers descend laterally, enveloping the proximal third of the radius. They insert into the posterior, lateral, and anterior aspects of the radius, extending anteriorly to the margin between the neck and pronator teres attachment.

As implied by its nomenclature, the primary function of the supinator is forearm supination, characterized by an anterolateral rotation of the distal end of the radius around the ulna, resulting in parallel alignment of both bones (Neumann 2017). The supinator typically acts as the primary mover in this action unless a particularly powerful supinatory force is necessary, in which case the biceps brachii is also activated for additional power. However, it is important to note that the supination function of the biceps brachii is compromised when the elbow is fully extended; thus, forceful supination movements are optimally performed with the elbow flexed to approximately 120° (Morrey et al. 1981).

Extensor Carpi Radialis Longus (Innervated by the Radial Nerve from the Posterior Cord of the Brachial Plexus: C6 and C7)

The extensor carpi radialis longus, innervated by the radial nerve from the posterior cord of the brachial plexus (C6 and C7), is positioned on the lateral aspect of the posterior forearm compartment. This muscle is partially overlaid by the brachioradialis and originates from the anterior portion of the lower third of the lateral supracondylar ridge of the humerus and the adjacent intermuscular septum. In some cases, it may also attach to the lateral epicondyle via the common extensor tendon (Moore et al. 2018; Standring 2020).

The origin of the extensor carpi radialis longus is primarily from the anterior portion of the distal third of the lateral supracondylar ridge of the humerus and the adjacent intermuscular

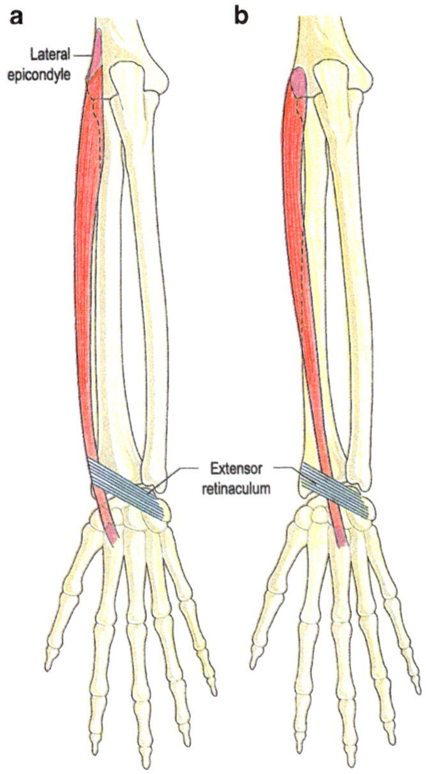

Fig. 7.26 Posterior view illustrating the attachment of (**a**) extensor carpi radialis longus and (**b**) extensor carpi radialis brevis. (Adapted with permission from Palastanga and Soames (2019))

septum. In some instances, an additional attachment point may be observed at the lateral epicondyle via the common extensor tendon (Netter 2019) (Fig. 7.26). As the muscle descends, it transitions into a flattened tendon approximately at the forearm's midpoint. This tendon continues distally, traversing the lateral surface of the radius. In the distal third of the forearm, the tendons of the abductor pollicis longus and extensor pollicis brevis cross over both the extensor carpi radialis longus tendon and that of the extensor carpi radialis brevis (Standring 2020). The extensor carpi radialis longus and extensor carpi radialis brevis tendons share a common synovial sheath as they pass deep to the extensor retinaculum. These tendons collectively form a groove on the posterior aspect of the radial styloid process. The extensor carpi radialis longus tendon ultimately inserts onto the posterior surface of the second metacarpal base (Drake et al. 2020).

Extensor Carpi Radialis Brevis **(Innervated by the Posterior Interosseous Branch of the Radial Nerve: C6 and C7)**

The extensor carpi radialis brevis is a significant muscle in the posterior compartment of the forearm, receiving innervation from the posterior interosseous branch of the radial nerve, specifically from spinal segments C6 and C7 (Moore et al. 2018). This muscle is positioned adjacent to and partially beneath the extensor carpi radialis longus, with which it may exhibit partial fusion (Standring 2020).

The extensor carpi radialis brevis originates from three primary sites: the lateral epicondyle of the humerus via the common extensor tendon, the lateral ligament of the elbow, and the surrounding fascia (Netter 2019). As the muscle descends, it forms a tendon at the midpoint of the forearm, which then courses alongside the extensor carpi radialis longus tendon. Both tendons pass deep to the abductor pollicis longus and extensor pollicis brevis before traversing beneath the extensor retinaculum within a shared synovial sheath. The extensor carpi radialis brevis tendon ultimately inserts onto the posterior aspect of the third metacarpal base (Drake et al. 2020).

From a functional perspective, the wrist extensors play a significant role in the biomechanics of gripping. The synergistic action of the extensor carpi radialis longus, extensor carpi radialis brevis, and extensor carpi ulnaris is fundamental to this process. By maintaining the wrist in an extended position, these muscles prevent wrist flexion that would otherwise occur due to the action of flexor digitorum superficialis and profundus. This mechanism allows these flexor muscles to exert their action primarily on the fingers. Without this extension, the flexor tendons cannot shorten sufficiently, resulting in ineffective movement at the interphalangeal joints, a condition known as active insufficiency (Neumann 2017). The significance of wrist extensor function in hand grip strength and dexterity is made apparent by this biomechanical theory (Neumann 2017; Skirven et al. 2011).

Extensor Carpi Ulnaris (Innervated by the Posterior Interosseous Branch of the Radial Nerve: C7 and C8)

The extensor carpi ulnaris is a significant muscle in the posterior compartment of the forearm, contributing to wrist extension and ulnar deviation (Moore et al. 2018). This muscle has a dual origin: it arises from the lateral epicondyle of the humerus via the common extensor tendon and the adjacent fascia, while also having an attachment to the posterior border of the ulna through a shared aponeurosis with flexor digitorum profundus and flexor carpi ulnaris (Standring 2020).

As the extensor carpi ulnaris descends, it forms a tendon near the wrist joint. This tendon traverses beneath the extensor retinaculum, occupying its own synovial sheath and compartment within a groove adjacent to the ulnar styloid process (Netter 2019). The tendon then attaches to a tubercle on the medial side of the base of the fifth metacarpal bone (Drake et al. 2020).

Functionally, the extensor carpi ulnaris operates together with the extensor carpi radialis longus and brevis to enable wrist extension. This synergistic action is particularly important in the biomechanics of gripping, as previously discussed in the context of the extensor carpi radialis muscles. The maintenance of wrist extension during gripping allows for optimal function of the digital flexors, preventing active insufficiency and enhancing grip strength. In addition to its role in wrist extension, the extensor carpi ulnaris collaborates with the flexor carpi ulnaris to produce adduction (ulnar deviation) at the wrist joint (Neumann 2017).

7.4.3.4 Flexor-Pronators of the Elbow and Forearm

This group includes the pronator teres, flexor carpi radialis, palmaris longus, flexor carpi ulnaris, and the flexor digitorum superficialis, all originating from the medial epicondyle's common flexor tendon. They enhance dynamic stability to the medial elbow and serve as modest flexors of the elbow. The pronator teres and pronator quadratus primarily affect forearm pronation, with the flexor carpi radialis playing a supportive role in this action.

Pronator Teres (Innervated by the Median Nerve from the Medial and Lateral Cords of the Brachial Plexus: C6 and C7)

The pronator teres muscle, which forms the medial boundary of the cubital fossa at the elbow, is the most lateral muscle of the superficial muscles within the flexor compartment of the forearm (Fig. 7.27). This muscle originates from two heads: the humeral and ulnar heads (Moore et al. 2018; Standring 2020).

The humeral head originates from multiple structures: the inferior portion of the medial supracondylar ridge, the adjacent intermuscular septum, the common flexor origin on the medial epicondyle of the humerus, and the overlying fascia (Netter 2019). The ulnar head arises from the pronator ridge of the ulna, which extends inferiorly from the medial aspect of the coronoid process. These two heads converge at the deep

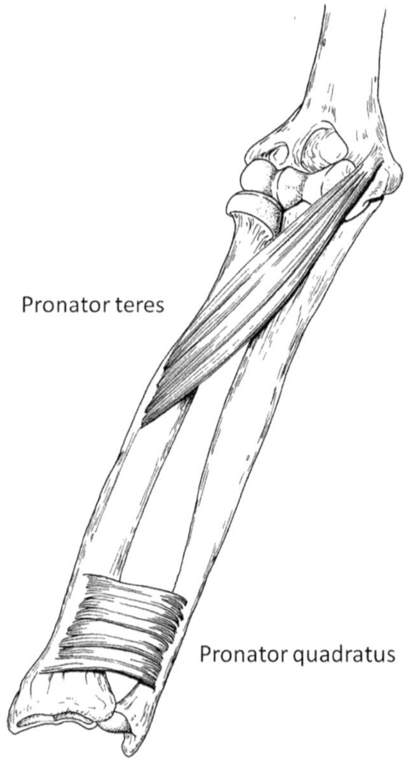

Fig. 7.27 Pronator teres, the primary pronator of the forearm, innervated by the median nerve (C6 and C7), and pronator quadratus, innervated by the anterior interosseous nerve (C8, T1). (Adapted from Calais-Germain (1999))

surface of the humeral head, with the median nerve passing between them. The muscle fibers of pronator teres course inferiorly and laterally, attaching via a flattened tendon to a roughened oval area on the middle portion of the lateral surface of the radius.

The primary function of the pronator teres is forearm pronation, which it accomplishes by generating an anteromedial movement of the distal end of the radius across the ulna, consequently rotating the hand (Neumann 2017). This action is particularly important in activities of daily living and many occupational tasks. Additionally, it serves as a weak flexor of the elbow (Morrey et al. 1981).

Flexor Digitorum Profundus (Innervated by the Anterior Interosseous Branch of the Median Nerve and the Ulnar Nerve: C7, C8, and T1)

The flexor digitorum profundus is a deep muscle situated on the medial aspect of the forearm, positioned beneath the flexor digitorum superficialis (Moore et al. 2018) (Fig. 7.28). The flexor digitorum profundus originates from multiple sites: the medial coronoid process of the ulna, the anterior and medial surfaces of the upper three-quarters of the ulna, and the medial third of the anterior surface of the interosseous membrane (Standring 2020). Additionally, it arises from an aponeurosis connecting the flexor carpi ulnaris to the posterior border of the ulna (Netter 2019). The portion of the muscle originating from the interosseous membrane forms a separate tendon midway down the forearm, which extends to the index finger. The remaining muscle fibers usually merge together to form tendons just proximal to the flexor retinaculum (Drake et al. 2020). These tendons traverse beneath the flexor retinaculum, positioned deep to the flexor digitorum superficialis tendons but sharing a common synovial sheath (Seradge et al. 1984).

Upon entering the palm, the four flexor digitorum profundus tendons diverge toward their respective digits. Initially, they course deep to the flexor digitorum superficialis tendons but subsequently become superficial as they pass through the bifurcation of the flexor digitorum superficialis tendon at the metacarpophalangeal joint level (Gosling et al. 2017). The flexor digitorum profundus tendons ultimately insert into the palmar aspect of the distal phalanx base after traversing a fibro-osseous tunnel (Detton 2017). Similar to the flexor digitorum superficialis tendons, the flexor digitorum profundus tendons are supported by vincula tendinum, which provide vascular supply and mechanical support.

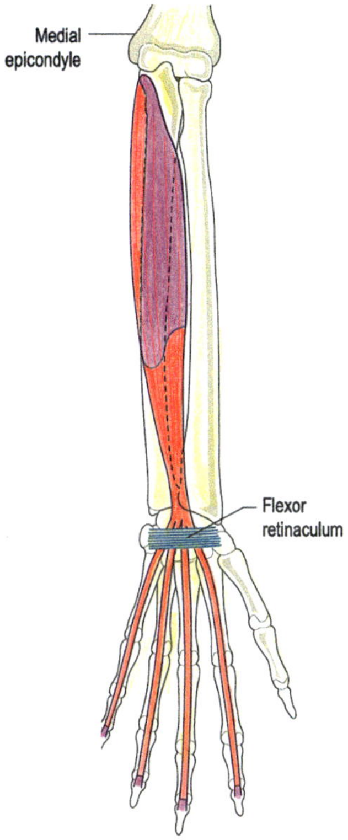

Fig. 7.28 Anterior view showing the attachment of flexor digitorum profundus. (Adapted with permission from Palastanga and Soames (2019))

Flexor Digitorum Superficialis (Innervated by the Median Nerve: C7, C8, and T1)

The flexor digitorum superficialis is situated superficial to the flexor digitorum profundus and flexor pollicis longus, while being deep to the pronator teres, palmaris longus, and flexor carpi radialis and ulnaris muscles (Moore et al. 2018; Standring 2020). The flexor digitorum superficialis has a bifid origin. The medial (humeroulnar) head arises from three distinct areas: the medial epicondyle of the humerus via the common flexor tendon, the anterior aspect of the ulnar collateral ligament, and the tubercle on the upper medial portion of the ulnar coronoid process (Netter 2019). The lateral (radial) head originates from the anterior border of the radius, specifically the upper two-thirds, extending downwards and laterally from the radial tuberosity (Drake et al. 2020).

As the muscle progresses distally, it narrows and divides into four separate tendons approximately midway down the forearm. These tendons pass deep to the flexor retinaculum, entering the hand in two pairs. The superficial pair is destined for the middle and ring fingers, while the deep pair serves the index and little fingers. Within the carpal tunnel, the flexor digitorum superficialis tendons lie superficial to those of the flexor digitorum profundus, sharing a common synovial sheath (Gosling et al. 2017).

Upon entering the palm, the tendons diverge toward their respective digits, maintaining their superficial position relative to the flexor digitorum profundus tendons. At the metacarpophalangeal joint level, each flexor digitorum superficialis tendon undergoes a longitudinal split. The resulting parts encircle the flexor digitorum profundus tendon, twisting to form a groove through which the flexor digitorum profundus tendon passes (Detton 2017). This unique arrangement creates a tunnel allowing the flexor digitorum profundus tendon to become superficial, effectively increasing the lever arm of the flexor digitorum superficialis at the proximal interphalangeal joint. This anatomical configuration facilitates a powerful digital grip. Prior to its final insertion on either side of the palmar surface of the middle phalanx base, the flexor digitorum superficialis tendon splits again. Additionally, the flexor digitorum superficialis tendons provide attachment points for the vincula tendinum, which supply blood vessels to the tendon.

7.5 The Wrist and Hand Complex

The hand exhibits anatomical characteristics on its palmar and dorsal surfaces. The palmar aspect is characterized by skin firmly adhered to the palmar aponeurosis via multiple strong fasciculi, resulting in limited mobility (Standring 2020). In contrast, the dorsal hand skin demonstrates greater flexibility (Moore et al. 2018).

A significant anatomical feature is the predominant lymphatic drainage through the areolar tissue on the dorsum of the hand. This anatomical arrangement often leads to the manifestation of lymphedema in this region (Lampe 1969). The surface of the hand is marked by transverse creases, which reflect the direct connection between the skin and underlying fascia, with no intervening adipose layer (Drake et al. 2020).

The architecture of the hand includes both proximal and distal palmar creases, aligned with the longitudinal fibers of the palmar fascia. These creases are often the initial sites affected by Dupuytren's disease, a condition characterized by proliferative fibrosis of the palmar and digital fascia, predominantly impacting the fourth and fifth fingers (Gosling et al. 2017; Strickland et al. 1990). The only transverse crease that aligns with an underlying joint is the middle digital crease. The distal digital crease is positioned just proximal to the distal interphalangeal joint (DIP), while the proximal digital crease overlies the midportion of the proximal phalanx (Netter 2019).

The hand consists of one longitudinal and two transverse arches (carpal and metacarpal) that are necessary for effective grasping of objects. This architectural design enhances the hand's adaptability in manipulating objects and increasing grip strength.

7.5.1 Skeletal Structure and Mechanics of the Wrist and Hand

The wrist joint is formed by the articulation between the radial head and the proximal carpal row. This ellipsoidal articulation permits a range of movements, including adduction, abduction, flexion, extension, and circumduction (Kaufmann et al. 2007). The midcarpal joint, located between the carpal rows, functions similarly to a hinge, allowing one degree of freedom in motion. The distal radioulnar joint, classified as a pivot-type articulation, is formed by the connection between the ulnar head and the ulnar notch of the radius. This joint is required for the supination and pronation movements of the wrist (Nakamura et al. 1996).

A significant anatomical structure in the wrist is the triangular fibrocartilage complex. This complex serves dual functions: it acts as a cushion for the ulnar aspect of the wrist and provides stability to the distal radioulnar joint, making it a key component in wrist biomechanics (Melone Jr and Nathan 1992). The intricate interaction between these articular components enables the complex movements of the hand while maintaining stability. This balance between mobility and stability is important to the hand's diverse functional capabilities, allowing for both precise manipulations and forceful grips (Palmer and Werner 1984).

The hand and wrist contain 27 bones, comprising 19 long bones, including five metacarpals and fourteen phalanges, and eight carpal bones. These elements form five polyarticulated rays, with each ray composed of metacarpals connected to phalanges. The skeletal architecture of the hand can be categorized into stationary and mobile components. The stationary elements include the distal carpal row and the second and third metacarpals, while the mobile elements encompass the phalanges, thumb, and ulnar two metacarpals (Littler 1960; Taylor and Schwarz 1955).

The interphalangeal joints, both proximal and distal, are classified as condylar articulations functioning as hinge joints. These joints permit flexion and extension movements (Kuczynski 1968). At the metacarpophalangeal joints, the collateral ligaments exhibit a unique property: they are lax during extension and hyperextension, allowing for considerable medial and lateral deviation. This characteristic necessitates assessment of ligamentous functional integrity when the joints are in a flexed position (Belliappa and Scheker 1993).

The capacity of the metacarpophalangeal joint for lateral mobility and rotation significantly enhances the grasping abilities of the hand. The carpometacarpal joints of the fingers are categorized as plain synovial joints, characterized by limited flexion and extension movements. In contrast, the thumb's metacarpotrapezial joint possesses a saddle configuration, providing two planes of motion. This arrangement accommodates flexion, extension, abduction, adduction, and minor rotational movements (Moran 1989). Interphalangeal joints function as hinge joints, restricting their motion to flexion and extension. This limitation in movement contributes to the stability and precision of digit control (Kapandji 2007). The diverse articular configurations within the hand, ranging from the complex saddle joint of the thumb to the simpler hinge joints of the interphalangeal articulations, collectively contribute to the hand's remarkable dexterity and functional versatility.

A notable anatomical feature is the orientation of the first metacarpal, which forms an approximate 45° angle with the second metacarpal in the sagittal plane (Kapandji 2007). This configuration is fundamental to the thumb's oppositional capability relative to the other digits. The ability of the thumb to perform flexion, extension, abduction, and adduction significantly enhances the functional capacity of the hand, with opposition representing a complex integration of these movements. The process of thumb opposition starts with extension, progresses through abduction, and ends with flexion and adduction (Cailliet 1994). The extensive range of motion exhibited by the thumb contributes substantially to the hand's overall functional capacity, accounting for approximately half of its capabilities (Kapandji 2007). This remarkable mobility and versatility of the thumb play a central role in the hand's ability to perform intricate tasks and manipulations.

7.5.2 Functional Dynamics of the Hand

The oblique orientation of the ulnar digits enables their movement toward the scaphoid tubercle during flexion at the metacarpophalangeal (MCP) and proximal interphalangeal (PIP) joints. The carpal bones are arranged in proximal and distal rows. The proximal row includes the pisiform (a sesamoid bone), scaphoid, lunate, and triquetrum. This row articulates with the radius, forming the radiocarpal joint, which is characterized by significant mobility (Tubiana et al. 1996).

The scaphoid, the largest bone in the proximal row, is important for its vascular anatomy. It lacks intraosseous arterial anastomoses between the arteries supplying the scaphoid tuberosity and those supplying the remainder of the bone. This anatomical feature makes the proximal scaphoid susceptible to avascular necrosis following a fracture. Similar risks of posttraumatic avascular necrosis exist for the lunate, capitate, and pisiform bones (Amadio and Taleisnik 1993). The distal row, which includes the trapezium, trapezoid, capitate, and hamate, is distinguished by its stability and rigidity, in contrast to the proximal row's flexibility. The capitate, the largest carpal bone, functions as the keystone of the distal row. The interaction between the proximal and distal rows facilitates wrist movements, supported by the dense arrangement of the carpal bones and the interconnecting interosseous ligaments (Garcia-Elias 1997; Moran and Berger 2003).

The Guyon canal, an anatomical passage formed by the pisiform and the hook of the hamate, with the flexor retinaculum as its floor and the expansion of the flexor carpi ulnaris as its roof, can potentially compress various branches of the ulnar nerve (Tubiana et al. 1996).

7.5.3 Ligament Structure in the Hand

The ligamentous framework of the wrist is classified into two primary categories: intrinsic and extrinsic ligaments. Intrinsic ligaments establish connections between the carpal bones, while extrinsic ligaments form attachments between the carpal bones and the radius, ulna, and metacarpals (Taleisnik 1976). The volar ligaments are notably thick and strong, functioning to prevent posterior instability. In contrast, the posterior ligaments display a comparatively thinner and weaker structure (Berger 2001). This anatomical arrangement contributes to the overall stability and functional integrity of the wrist joint.

At the metacarpophalangeal joints of the fingers, the four volar plates are interconnected by the transverse metacarpal ligament. This configuration plays a role in maintaining the collective stability of these joints (Bowers et al. 1980). For both the metacarpophalangeal and interphalangeal joints, lateral stability is provided by collateral ligaments. These ligaments resist abnormal lateral movements, thereby maintaining joint integrity during various hand movements. The fibrocartilaginous volar plates provide anterior support to these joints by preventing hyperextension and promoting the smooth articulation of the joint surfaces (Taleisnik 1976).

The intricate arrangement of these ligamentous structures enables the complex movements of the hand while maintaining joint stability. This balance between mobility and stability is necessary for the diverse functional capabilities of the hand, allowing for both precise manipulations and forceful grips (Garcia-Elias et al. 1991).

7.5.4 Wrist Movement Mechanics

The process of wrist flexion involves a synergistic action of multiple muscles. The flexor carpi radialis, flexor carpi ulnaris, and flexor digitorum superficialis work in coordination. Notably, the flexor digitorum profundus does not participate in this action (Tubiana et al. 1996).

Wrist extension, on the other hand, is achieved through the coordinated effort of the three primary extensors of the wrist—extensor carpi radialis longus, extensor carpi radialis brevis, and extensor carpi ulnaris—along with the extensor digitorum communis. The extensor carpi ulnaris acts as a ulnar deviator during both pronation and supination of the forearm. Long et al. (1970) conducted comprehensive studies elucidating the functional dynamics of both extrinsic and intrinsic hand muscles. Their research encompassed various finger positions, power exertion scenarios, and precision tasks, providing valuable insights into the complex interaction of these muscular structures during different hand activities. The efficacy of the extensor is most pronounced when the forearm is in a supinated position (Skirven et al. 2011).

7.5.5 Anatomy of Wrist Compartments

The extensor retinaculum of the wrist forms a protective layer over six distinct extensor compartments, as depicted in Fig. 7.29. Each of these compartments is characterized by a synovial sheath that extends beyond the wrist proximally and distally (Doyle and Botte 2003). The first posterior compartment is of particular clinical significance due to its susceptibility to tenosynovitis, commonly manifesting as de Quervain syndrome. However, it is important to note that tenosynovitis resulting from repetitive strain or injury can occur in any of the other compartments as well (Kiefhaber and Stern 1992).

On the volar aspect, the flexor retinaculum is situated inferior to the radial joint, forming the roof of the carpal tunnel, as illustrated in Fig. 7.29. Unlike other flexor tendons, the flexor carpi ulnaris and palmaris longus do not have synovial sheaths and do not pass through the carpal tunnel (Tubiana et al. 1996). The confined space of the carpal tunnel is susceptible to various pathological conditions. For instance, flexor tenosynovitis can lead to compression of the median nerve within this space, potentially resulting in carpal tunnel syndrome (Skirven et al. 2011).

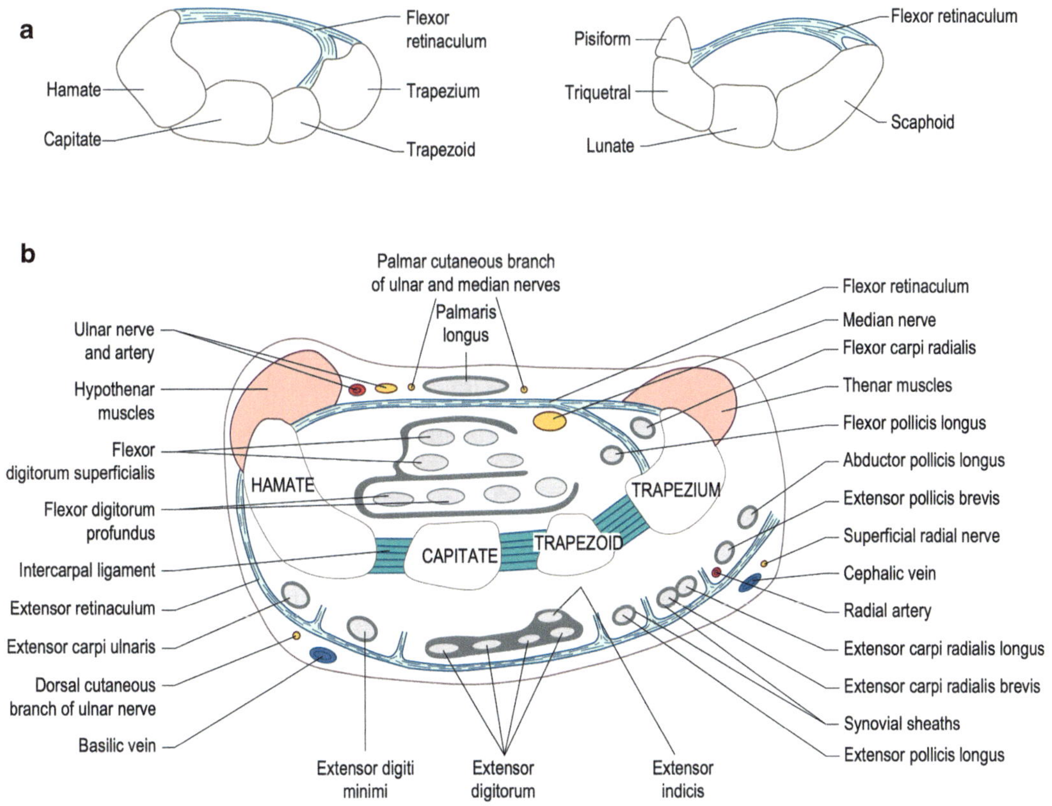

Fig. 7.29 (**a**) Arrangement of the carpal bones and the attachment sites of the flexor retinaculum, and (**b**) transverse section of the wrist region showing the spatial relationships of structures passing into the hand. (Adapted with permission from Palastanga and Soames (2019))

7.5.6 Biomechanics and Muscle Coordination of the Wrist and Hand Musculature

The wrist and hand operate through a sophisticated network of extrinsic and intrinsic muscles working in concert to enable precise movements (Standring 2020). The extrinsic muscles, which originate near the elbow joint, extend into the hand as tendons, with some reaching the distal fingertips. The extensor and flexor retinacula—transverse fibrous bands across the distal forearm and wrist—maintain these tendons in position during movement (Neumann 2017). Thirty-nine muscles control wrist and hand function, operating in agonist-antagonist pairs during all movements (Moore et al. 2018).

Within the hand, intrinsic muscles control movement at the metacarpophalangeal (MCP) and interphalangeal (IP) joints. The thenar eminence, a muscular region in the palm, contains four intrinsic thumb muscles. The hypothenar eminence, located on the palm's ulnar aspect, consists of three intrinsic muscles (Palastanga and Soames 2019) (Fig. 7.30).

The wrist flexors—flexor carpi ulnaris, flexor carpi radialis, and palmaris longus—demonstrate fusiform structures originating near the medial humeral epicondyle, transitioning to tendons approximately halfway along the forearm. The

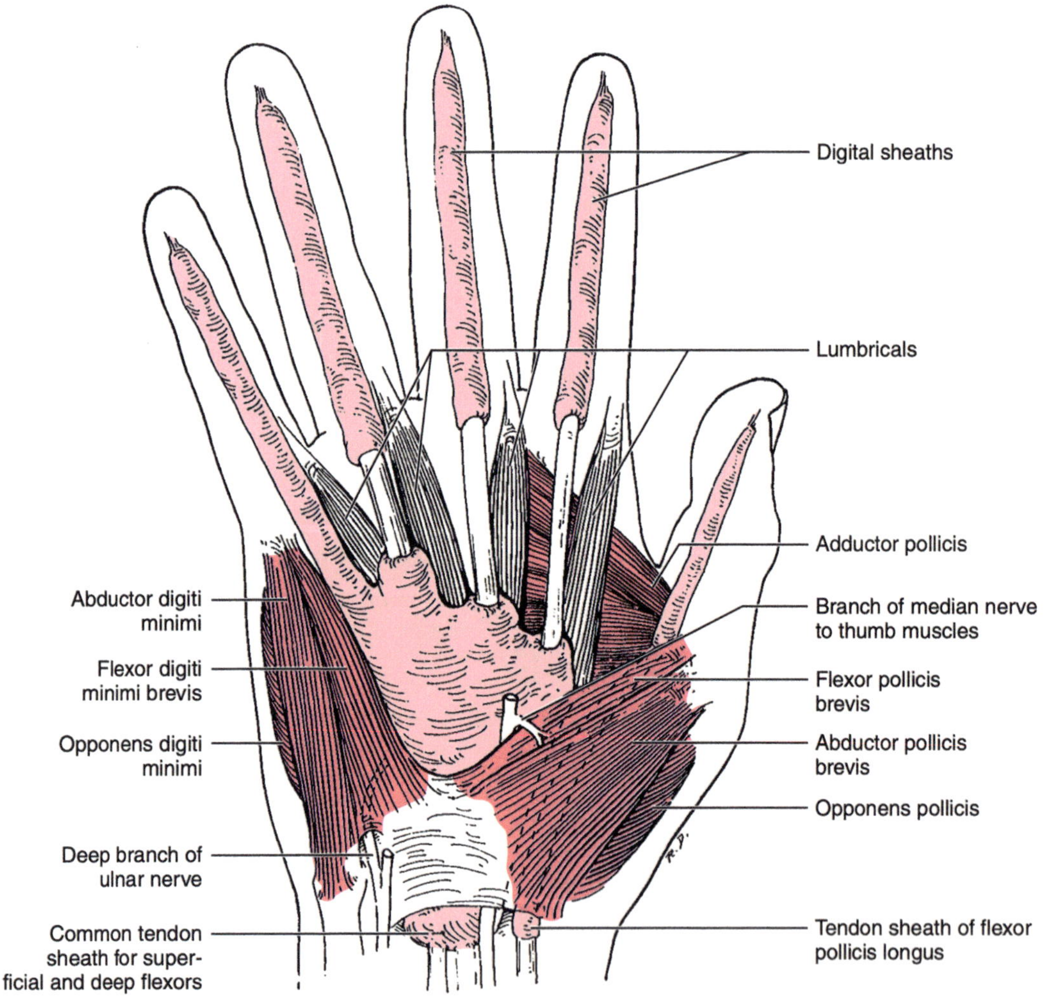

Fig. 7.30 Intrinsic muscles of the hand illustrated. (Adapted with permission from Jenkins (2009))

flexor carpi ulnaris and radialis perform the majority of wrist flexion. The palmaris longus may appear as a tendon or be absent in approximately 13% of individuals. The flexor carpi ulnaris, recognized as the strongest flexor, utilizes the pisiform bone as a sesamoid bone to optimize mechanical advantage (Skirven et al. 2011).

The wrist extensors—extensor carpi ulnaris, extensor carpi radialis longus, and extensor carpi radialis brevis—originate near the lateral epicondyle, forming tendons in the proximal third of the forearm. These muscles influence both wrist and elbow movements. The extensor carpi radialis longus and brevis contribute to elbow flexion, while the extensor carpi ulnaris produces elbow extension. Wrist extension provides support during gripping actions through finger flexion (Neumann 2017).

Coordinated muscle actions produce ulnar and radial deviation. The flexor and extensor carpi ulnaris generate ulnar deviation, while the flexor carpi radialis and both radial extensors perform radial deviation. Radial deviation, though having half the range of ulnar deviation, maintains wrist stability in racket sports through close-packed positioning (Moore et al. 2018).

The flexor digitorum profundus and flexor digitorum superficialis, originating near the medial epicondyle, function as the primary finger flexors. The flexor digitorum profundus has limited independent control over individual fingers, as the middle, ring, and little fingers flex simultaneously due to their shared tendinous origin. The index finger maintains independent flexion through its separate muscle and tendon arrangement (Standring 2020). At the proximal interphalangeal (PIP) joint, the flexor digitorum superficialis permits independent flexion of each finger, though this independence does not extend to the distal interphalangeal (DIP) joint. The flexor digiti minimi brevis adds to the little finger flexion capacity. The lumbricals and interossei, intrinsic muscles located in the palm and between metacarpals, produce finger flexion at the MCP joint while simultaneously extending the IP joints through their connection to the dorsal extensor hood. Studies indicate that complete flexion of all finger joints requires the long flexors to overcome the extension component of these intrinsic muscles, an action enhanced by slight wrist extension (Palastanga and Soames 2019).

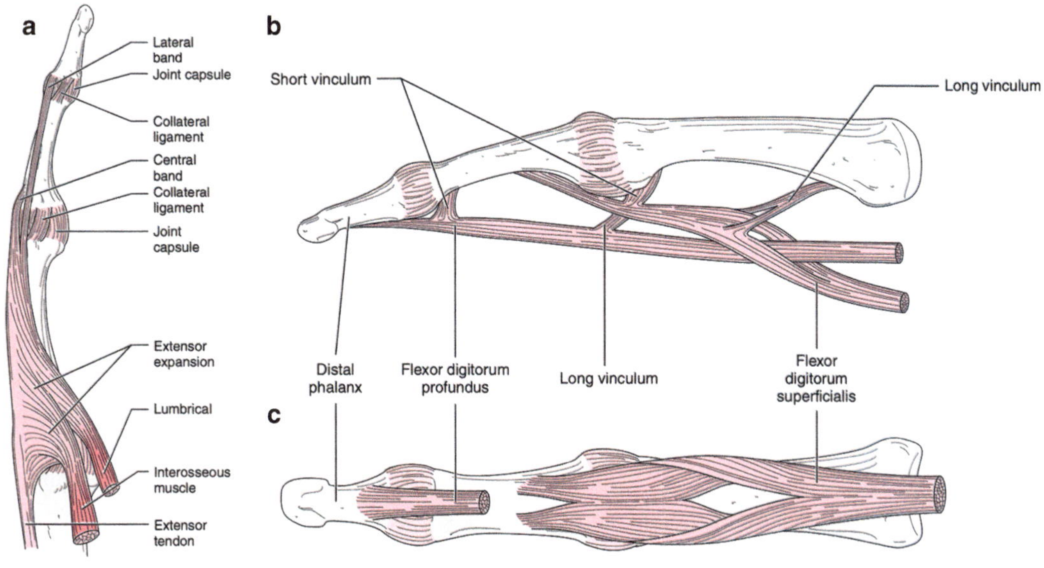

Fig. 7.31 Metacarpophalangeal and interphalangeal joints: (**a**) lateral view of the metacarpophalangeal joint showing the extensor tendon and extensor expansion, (**b**) lateral view of the finger flexor tendons, and (**c**) palmar views of the finger flexor tendons. (Adapted with permission from Jenkins (2009))

The extensor digitorum, which originates at the lateral epicondyle, directs finger extension through four tendon slips that divide at the MCP joint (Fig. 7.31). Each tendon develops a main slip inserting into the extensor hood and collateral slips connecting to adjacent fingers. The extensor hood, formed by the extensor digitorum tendon and fibrous tissue, spans the dorsal phalangeal surface to the distal phalanx. The lumbricals and interossei support PIP and DIP joint extension through their hood connection, with their action enhanced during extensor digitorum contraction (Neumann 2017).

The dorsal interossei, consisting of four intrinsic muscles between the metacarpals, control abduction of the second, third, and fourth digits. These muscles attach to the lateral aspects of the second and fourth digits and both sides of the third digit. The abductor digiti minimi brevis manages fifth digit abduction. Three palmar interossei muscles on the medial aspects of the second, fourth, and fifth digits perform adduction, while the third digit employs the dorsal interossei for adduction. Finger flexion restricts abduction due to collateral ligament tension and interossei length–tension relationships (Moore et al. 2018). Eight muscles control the thumb's extensive movement range, with opposition being fundamental for manipulative tasks. Opposition enables pinching, grasping, and gripping through thumb-to-finger contact. Multiple thenar muscles contribute to opposition; the opponens pollicis initiates this movement, with the opponens digiti minimi supporting little finger opposition movements (Skirven et al. 2011).

7.5.7 Musculature of the Hand

7.5.7.1 Hand Extrinsic Muscles

The extrinsic extensor musculature of the hand includes specialized mechanisms for the index and little fingers, facilitated by the extensor indicis proprius and extensor digiti minimi, respectively, in addition to the extensor digitorum communis. The tendons of these specialized muscles are positioned ulnar to the extensor digitorum communis tendon, enabling independent extension of these digits (Tubiana et al. 1996).

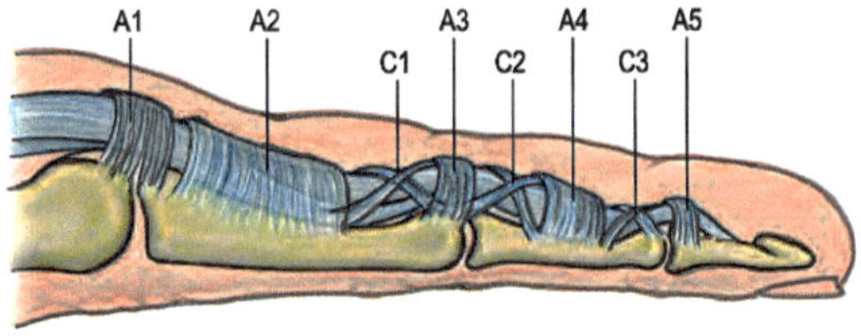

Fig. 7.32 The annular and cruciate pulleys of the fingers. (Adapted with permission from Standring (2020))

Each finger is actuated by two primary flexor tendons: the flexor digitorum superficialis and the flexor digitorum profundus. A complex support system, comprising five annular (A) pulleys and four cruciate (C) pulleys, maintains the integrity of these flexor tendons across the joints, as depicted in Fig. 7.32. Chronic stress on the flexor tendon can result in ligament sheath hypertrophy and nodule formation, typically manifesting as trigger finger. This condition most commonly affects the A1 pulley, particularly in the middle or ring fingers (Doyle 1988).

The flexor digitorum profundus of the index finger demonstrates independent function, unlike the interconnected movement of the other three fingers due to their shared muscular and tendinous structures. The flexor digitorum superficialis can be selectively activated by extending the adjacent fingers, thereby isolating the action from the flexor digitorum profundus. In unloaded finger flexion, the flexor digitorum profundus acts as the primary mover, while the flexor digitorum superficialis is recruited to provide additional force during loaded movements (Long et al. 1970).

7.5.7.2 Hand Intrinsic Muscles

The intrinsic muscles of the hand are defined by their origin and insertion sites, which are located distal to the wrist. This group includes the thenar and hypothenar muscles, interossei, and lumbricals, each contributing to the hand's detailed movements and functions (Tubiana et al. 1996).

The four dorsal interossei muscles enable finger abduction away from the central axis of the hand. Conversely, the three palmar interossei are responsible for adducting the index, ring, and small fingers toward the middle finger. The efficacy of these muscles in acting upon the interphalangeal joints is significantly influenced by the metacarpophalangeal (MCP) joint position. When the MCP joint is extended, the interosseous hood tightens, enabling extension of both proximal interphalangeal (PIP) and distal interphalangeal (DIP) joints. Conversely, a flexed MCP joint causes the hood to pass distal to the proximal phalanx, aiding in DIP and PIP joint flexion (Mentari et al. 1976).

The four lumbrical muscles, originating from the radial aspect of the flexor digitorum profundus tendons, primarily extend the DIP and PIP joints, regardless of MCP joint position. While they may contribute to MCP joint flexion, their influence is minimal compared to the profundus muscles (Brand and Hollister 1992). Lumbrical tendons run along the palmar aspect of the MCP joint axis, preventing excessive extension of this joint.

7.6 Phalangeal Dynamics

7.6.1 Phalangeal Flexion Dynamics

The process of finger joint flexion follows a specific sequence, initiating at the proximal interphalangeal (PIP) joint, progressing to the metacarpophalangeal (MCP) joint, and finally the distal interphalangeal (DIP) joint to achieve complete flexion (Tubiana 1981). This intricate mechanism involves the coordinated action of multiple muscle groups and ligamentous structures. At the onset of flexion, both the flexor digitorum profundus and the extensor digitorum communis are activated simultaneously. The extensor digitorum communis acts as a modulator, regulating DIP joint flexion to prevent hyperextension. The flexor digitorum profundus exerts its influence on the A4 pulley, initiating PIP joint flexion.

As the PIP joint flexes, tension in the extensor mechanism via the oblique retinacular ligament gradually diminishes, facilitating full flexion of the DIP joint. PIP joint flexion also stretches the lumbrical and interosseous tendons, which, through passive viscoelastic forces, initiate flexion of the MCP joint. Following this, the interossei muscles can actively contribute to MCP joint flexion (Belliappa and Scheker 1993). It is noteworthy that during this unopposed flexion process, both the flexor digitorum superficialis and the lumbricals remain inactive. This selective muscle activation pattern contributes to the smooth and coordinated flexion of the finger joints (Long et al. 1970).

7.6.2 Phalangeal Flexion Dynamics

The process of finger extension begins at the metacarpophalangeal (MCP) joint, driven by the extensor digitorum communis through its connection to the sagittal bands. Lumbrical muscle contraction is integral in preventing MCP joint hyperextension, thereby maintaining joint stability.

As extension progresses, the proximal interphalangeal (PIP) joint extends through the pull on the central slip of the extensor mechanism. Concurrently, the oblique band of the retinacular ligament tightens. Subsequently, the distal interphalangeal (DIP) joint extends, coordinated by the combined actions of the extensor digitorum communis and interossei muscles.

In cases of low ulnar nerve palsy, lumbrical paralysis results in an inability to stabilize the MCP joints. This instability allows extensor muscle action to hyperextend the MCP joints, while passive tension from the flexor digitorum profundus induces flexion in the DIP and PIP joints. This combination of factors leads to the development of claw hand deformity. Notably, the degree of clawing is diminished in the case of high ulnar palsy or tendon disruption, where the flexor digitorum profundus is also compromised. This reduced involvement of the flexor digitorum profundus results in a less prominent deformity (Tubiana et al. 1996).

References

Amadio PC, Taleisnik J. Fractures of the carpal bones. In: Green DP, editor. Operative hand surgery. 3rd ed. New York: Churchill Livingstone; 1993.

Basmajian JV. Muscles alive: their functions revealed by electromyography. 4th ed. Baltimore: Williams & Wilkins; 1971.

Basmajian JV, De Luca CJ. Muscles alive: their functions revealed by electromyography. 5th ed. Baltimore: Williams & Wilkins; 2015.

Behrman AL, Bowden MG, Nair PM. Neuroplasticity after spinal cord injury and training: an emerging paradigm shift in rehabilitation and walking recovery. Phys Ther. 2006;86(10):1406–25.

Belliappa PP, Scheker LR. Functional anatomy of the hand. Emerg Med Clin North Am. 1993;11(3):557–83.

Berger RA. The anatomy of the ligaments of the wrist and distal radioulnar joints. Clin Orthop Relat Res. 2001;383:32–40.

Bertelli JA, Ghizoni MF. Long thoracic nerve: anatomy and functional assessment. J Bone Joint Surg Am. 2005;87(5):993–8.

Boettcher CE, Ginn KA, Cathers I. Standard tests of shoulder strength for comparison with functional activity. J Rehabil Med. 2020;42(4):350–7.

Borstad JD, Ludewig PM. The effect of long versus short pectoralis minor resting length on scapular kinematics in healthy individuals. J Orthop Sports Phys Ther. 2005;35(4):227–38.

Bowers WH, Wolf JW Jr, Nehil JL, Bittinger S. The proximal interphalangeal joint volar plate. I. An anatomical and biomechanical study. J Hand Surg Am. 1980;5(1):79–88.

Brand PW, Hollister A. Clinical mechanics of the hand. 2nd ed. St. Louis: Mosby; 1992.

Cailliet R. Shoulder pain. 2nd ed. Philadelphia: FA Davis; 1981.

Cailliet R. Hand pain and impairment. 4th ed. Philadelphia: FA Davis; 1994.

Calais-Germain B. Anatomy of movement. Revised ed. Seattle: Eastland Press; 1999.

Clarkson HM. Musculoskeletal assessment: joint range of motion and manual muscle strength. 3rd ed. Philadelphia: Lippincott Williams & Wilkins; 2013.

Clemente CD. Anatomy: a regional atlas of the human body. 6th ed. Philadelphia: Lippincott Williams & Wilkins; 2011.

Culham EC, Peat M. Functional anatomy of the shoulder complex. J Orthop Sports Phys Ther. 1993;18(1):342–50.

De Troyer A, Estenne M. Functional anatomy of the respiratory muscles. Clin Chest Med. 1988;9(2):175–93.

Detton AJ. Grant's dissector. 16th ed. Philadelphia: Wolters Kluwer; 2017.

Doody SG, Freedman L, Waterland JC. Shoulder movements during abduction in the scapular plane. Arch Phys Med Rehabil. 1970;51(10):595–604.

Doyle JR. Anatomy of the finger flexor tendon sheath and pulley system. J Hand Surg Am. 1988;13(4):473–84.

Doyle JR, Botte MJ. Surgical anatomy of the hand and upper extremity. Philadelphia: Lippincott Williams & Wilkins; 2003.

Drake RL, Vogl AW, Mitchell AWM. Gray's anatomy for students. 4th ed. Philadelphia: Elsevier; 2020.

Dutton M. Dutton's orthopaedic examination, evaluation, and intervention. 4th ed. New York: McGraw-Hill Education; 2017.

Dy CJ, Mackinnon SE. Ulnar neuropathy: evaluation and management. Curr Rev Musculoskelet Med. 2016;9(2):178–84.

Escamilla RF, Yamashiro K, Paulos L, Andrews JR. Shoulder muscle activity and function in common shoulder rehabilitation exercises. Sports Med. 2019;39(8):663–85.

Eversmann WW Jr. Compression and entrapment neuropathies of the upper extremity. J Hand Surg Am. 1983;8(5 Pt 2):759–66.

Ferrante MA. Brachial plexopathies: classification, causes, and consequences. Muscle Nerve. 2004;30(5):547–68.

Floyd RT. Manual of structural kinesiology. 18th ed. New York: McGraw-Hill; 2012.

Freedman L, Munro R. Abduction of the arm in the scapular plane: scapular and glenohumeral movements. J Bone Joint Surg Am. 1966;48A:1503–10.

Garcia-Elias M. Kinetic analysis of carpal stability during grip. Hand Clin. 1997;13(1):151–8.

Garcia-Elias M, An KN, Berglund L, Linscheid RL, Cooney WP 3rd, Chao EY. Extensor mechanism of the fingers. I. A quantitative geometric study. J Hand Surg Am. 1991;16(6):1130–6.

Gilroy AM, MacPherson BR. Atlas of anatomy. 3rd ed. New York: Thieme; 2016.

Gosling JA, Harris PF, Humpherson JR, Whitmore I, Willan PLT. Human anatomy: color atlas and textbook. 6th ed. Philadelphia: Elsevier; 2017.

Hamill J, Knutzen KM, Derrick TR. Biomechanical basis of human movement. 4th ed. Philadelphia: Wolters Kluwer; 2015.

Hansen JT. Netter's clinical anatomy. 4th ed. Philadelphia: Elsevier; 2019.

Harvey L. Management of spinal cord injuries: a guide for physiotherapists. 1st ed. Edinburgh: Elsevier; 2008.

Hislop HJ, Avers D, Brown M. Daniels and Worthingham's muscle testing: techniques of manual examination. 8th ed. St. Louis: Elsevier; 2007.

Hoppenfeld S. Physical examination of the spine and extremities. New York: Pearson; 1976.

Hotchkiss RN, Weiland AJ. Valgus stability of the elbow. J Orthop Res. 1987;5(3):372–7.

Inman VT, Saunders JB, Abbott LC. Observations on the function of the shoulder joint. J Bone Joint Surg Am. 1944;26(1):1–30.

Jenkins DB. Hollinshead's functional anatomy of the limbs and back. 9th ed. St. Louis: Saunders; 2009.

Johnson G, Bogduk N, Nowitzke A, House D. Anatomy and actions of the trapezius muscle. Clin Biomech. 1994;9(1):44–50.

Johnson EO, Vekris M, Demesticha T, Soucacos PN. Neuroanatomy of the brachial plexus: normal and variant anatomy of its formation. Surg Radiol Anat. 2010;32(3):291–7.

Kapandji AI. The physiology of the joints, volume 1: upper limb. 6th ed. Philadelphia: Churchill Livingstone; 2007.

Kaufmann RA, Kozin SH, Mirarchi A, Holland B, Porter S. Biomechanical analysis of flexor digitorum profundus and superficialis in grip-strength generation. Am J Orthop (Belle Mead NJ). 2007;36(9):E128–32.

Kendall FP, McCreary EK, Provance PG, Rodgers MM, Romani WA. Muscles: testing and function with posture and pain. 5th ed. Baltimore: Lippincott Williams & Wilkins; 2005.

Kiefhaber TR, Stern PJ. Upper extremity tendinitis and overuse syndromes in the athlete. Clin Sports Med. 1992;11(1):39–55.

Kim DH, Murovic JA, Tiel RL, Kline DG. Management and outcomes in 353 surgically treated sciatic nerve lesions. J Neurosurg. 2004;101(1):8–17.

Kim SH, Kwon OY, Kim SJ, Park KN, Choung SD, Weon JH. Serratus anterior muscle activation during knee push-up plus exercise performed on static stable, static unstable, and oscillating unstable surfaces in healthy subjects. Phys Ther Sport. 2014;15(1):20–5.

Kisner C, Colby LA. Therapeutic exercise: foundations and techniques. 5th ed. Philadelphia: FA Davis; 2007.

Kuczynski K. The proximal interphalangeal joint. Anatomy and causes of stiffness in the fingers. J Bone Joint Surg Br. 1968;50(3):656–63.

Kumar VP, Satku K, Balasubramaniam P. The role of the long head of biceps brachii in the stabilization of the head of the humerus. Clin Orthop Relat Res. 1989;244:172–5.

Lampe EW. Surgical anatomy of the hand with special reference to infections and trauma. Clin Symp. 1969;21(3):66–109.

Latef TJ, Bilal M, Vetter M, Iwanaga J, Oskouian RJ, Tubbs RS. Injury of the radial nerve in the arm: a review. Cureus. 2018;10(2):e2199.

Leonello DT, Galley IJ, Bain GI, Carter CD. Brachial plexus anatomy. J Bone Joint Surg Am. 2007;89(5):1051–8.

Levangie PK, Norkin CC. Joint structure and function: a comprehensive analysis. 4th ed. Philadelphia: FA Davis; 2005.

Littler JW. The physiology and dynamic function of the hand. Surg Clin North Am. 1960;40:259–66.

Long C, Conrad PW, Hall EA, Furler SL. Intrinsic-extrinsic muscle control of the hand in power grip and precision handling: an electromyographic study. J Bone Joint Surg Am. 1970;52(5):853–67.

Ludewig PM, Reynolds JF. The association of scapular kinematics and glenohumeral joint pathologies. J Orthop Sports Phys Ther. 2018;39(2):90–104.

Lumley JSP. Surface anatomy: the anatomical basis of clinical examination. 4th ed. Philadelphia: Churchill Livingstone; 2008.

Mackinnon SE. Nerve surgery. New York: Thieme; 2015.

Magee DJ. Orthopedic physical assessment. 6th ed. St. Louis: Elsevier; 2014.

Martin RM, Fish DE. Scapular winging: anatomical review, diagnosis, and treatments. Curr Rev Musculoskelet Med. 2008;1(1):1–11.

Matsen FA III, Thomas SC, Rockwood CA Jr, Wirth MA. Glenohumeral instability. In: Rockwood Jr CA, Green DP, Bucholz RW, Heckman JD, editors. Rockwood and Green's fractures in adults. 4th ed. Philadelphia: Lippincott-Raven; 1996.

Melone CP Jr, Nathan R. Traumatic disruption of the triangular fibrocartilage complex. Clin Orthop Relat Res. 1992;275:65–73.

Mentari A, Johnson RJ, Yu J. Interosseous muscles in claw finger. Arch Phys Med Rehabil. 1976;57(10):458–62.

Miller MD, Thompson SR. DeLee & Drez's orthopaedic sports medicine. 4th ed. Philadelphia: Saunders; 2014.

Moore KL, Dalley AF, Agur AMR. Clinically oriented anatomy. 8th ed. Philadelphia: Wolters Kluwer; 2018.

Moran CA. Anatomy of the hand. Phys Ther. 1989;69(12):1007–13.

Moran SL, Berger RA. Biomechanics and hand trauma: what you need. Hand Clin. 2003;19(1):17–31.

Moras G, Buscà B, Peña J, Rodríguez S, Vallejo L, Tous-Fajardo J, et al. A comparative study between serve mode and speed and its effectiveness in a high-level volleyball tournament. J Sports Med Phys Fitness. 2008;48(1):31–6.

Morrey BF, Askew LJ, An KN, Chao EY. A biomechanical study of normal functional elbow motion. J Bone Joint Surg Am. 1981;63(6):872–7.

Mulroy SJ, Farrokhi S, Newsam CJ, Perry J. Effects of spinal cord injury level on the activity of shoulder muscles during wheelchair propulsion: an electromyographic study. Arch Phys Med Rehabil. 2004;85(6):925–34.

Mulroy SJ, Thompson L, Kemp B, Hatchett PP, Newsam CJ, Lupold DG, et al. Strengthening and optimal movements for painful shoulders (STOMPS) in chronic spinal cord injury: a randomized controlled trial. Phys Ther. 2011;91(3):305–24.

Nakamura T, Yabe Y, Horiuchi Y. Functional anatomy of the triangular fibrocartilage complex. J Hand Surg Br. 1996;21(5):581–6.

Netter FH. Atlas of human anatomy. 7th ed. Philadelphia: Elsevier; 2019.

Neumann DA. Kinesiology of the musculoskeletal system: foundations for rehabilitation. 3rd ed. St. Louis: Mosby; 2017.

Neviaser RJ. Lesions of the biceps and tendinitis of the shoulder. Orthop Clin North Am. 1980;11(2):343–8.

Neviaser TJ. The role of the biceps tendon in the impingement syndrome. Orthop Clin North Am. 1987;18(3):383–6.

O'Connell PW, Nuber GW, Mileski RA, Lautenschlager E. The contribution of the glenohumeral ligament to anterior stability of the shoulder joint. Am J Sports Med. 1990;18(6):579–84.

Oatis CA. Kinesiology: the mechanics and pathomechanics of human movement. 3rd ed. Philadelphia: Wolters Kluwer; 2017.

Ovesen J, Nielsen S. Anterior and posterior shoulder instability. A cadaver study. Acta Orthop Scand. 1986;57(4):324–7.

Paine R, Voight ML. The role of the scapula. Int J Sports Phys Ther. 2013;8(5):617–29.

Palastanga N, Soames R. Anatomy and human movement: structure and function. 7th ed. Edinburgh: Churchill Livingstone; 2019.

Palmer AK, Werner FW. Biomechanics of the distal radioulnar joint. Clin Orthop Relat Res. 1984;187:26–35.

Pauly JE, Rushing JL, Scheving LE. An electromyographic study of some muscles crossing the elbow joint. Anat Rec. 1967;159(1):47–53.
Perry J. Anatomy and biomechanics of the shoulder in throwing, swimming, gymnastics, and tennis. Clin Sports Med. 1983;2(2):247–70.
Perry J. Biomechanics of the shoulder. In: Rowe CR, editor. The shoulder. New York: Churchill Livingstone; 1988. p. 1–16.
Phadke V, Camargo P, Ludewig P. Scapular and rotator cuff muscle activity during arm elevation: a review of normal function and alterations with shoulder impingement. Rev Bras Fisioter. 2009;13(1):1–9.
Poppen NK, Walker PS. Normal and abnormal motion of the shoulder. J Bone Joint Surg Am. 1976;58A:195–201.
Ratnovsky A, Elad D, Halpem P. Mechanics of respiratory muscles. Respir Physiol Neurobiol. 2008;163(1-3):82–9.
Reinold MM, Gill TJ, Wilk KE, Andrews JR. Current concepts in the evaluation and treatment of the shoulder in overhead throwing athletes. Sports Health. 2018;2(1):39–50.
Safran MR. Elbow injuries in athletes. Clin Orthop Relat Res. 1995;310:257–77.
Safran MR, Baillargeon D. Soft-tissue stabilizers of the elbow. J Shoulder Elb Surg. 2005;14(1 Suppl S):179S–85S.
Saha AK. Mechanics of elevation of the glenohumeral joint. Acta Orthop Scand. 1973;44(6):668–78.
Sahrmann S. Diagnosis and treatment of movement impairment syndromes. 1st ed. St. Louis: Mosby; 2002.
Seradge H, Seradge E, Kleinert HE. Pathophysiologic role of the flexor tenosynovium in carpal tunnel syndrome. J Hand Surg Am. 1984;9(2):200–4.
Shin AY, Spinner RJ, Steinmann SP, Bishop AT. Adult traumatic brachial plexus injuries. J Am Acad Orthop Surg. 2005;13(6):382–96.
Skirven TM, Osterman AL, Fedorczyk JM, Amadio PC. Rehabilitation of the hand and upper extremity. 6th ed. Philadelphia: Elsevier; 2011.
Spinner RJ, Amadio PC. Compressive neuropathies of the upper extremity. Clin Plast Surg. 2003;30(2):155–73.
Standring S. Gray's anatomy: the anatomical basis of clinical practice. 42nd ed. London: Elsevier; 2020.
Steinberg BD, Plancher KD. Clinical anatomy of the wrist and elbow. Clin Sports Med. 1995;14(2):299–313.
Strickland JW, Idler RS, Creighton JC. Dupuytren's disease. Indiana Med. 1990;83(6):408–9.
Stroyan M, Will KE. The functional anatomy of the elbow complex. J Orthop Sports Phys Ther. 1993;17(6):279–88.
Taleisnik J. The ligaments of the wrist. J Hand Surg Am. 1976;1(2):110–8.
Taylor CL, Schwarz RJ. The anatomy and mechanics of the human hand. Artif Limbs. 1955;2(2):22–35.
Terry GC, Chopp TM. Functional anatomy of the shoulder. J Athl Train. 2015;35(3):248–55.
Tubiana R. The hand. Philadelphia: WB Saunders; 1981.
Tubiana R, Thomine JM, Mackin E. Examination of the hand and wrist. St. Louis: Mosby; 1996.
Warner JJP, Michel J, Arslanian LE, et al. Patterns of flexibility, laxity, and strength in normal shoulders and shoulders with instability and impingement. Am J Sports Med. 1990;18(4):366–75.
Woo A, Bakri K, Moran SL. Management of ulnar nerve injuries. J Hand Surg Am. 2015;40(1):173–81.

Recommended Additional Reading

Freivalds A. Biomechanics of the upper limb: mechanics, modeling, and musculoskeletal injuries. 2nd ed. Boca Raton: CRC Press; 2011.
Hirt B, Seyhan H, Wagner M, Zumhasch R. Hand and writ anatomy and biomechanics: a comprehensive guide. New York: Thieme; 2017.
Kahle W, Frotscher M. Nervous system and sensory organs. New York: Thieme; 2001.
Mesplié G. Hand and wrist rehabilitation: theoretical aspects and practical consequences. Cham: Springer; 2015.
Zatsiorsky VM, Prilutsky BI. Biomechanics of skeletal muscles. Champaign: Human Kinetics; 2012.

Functional Anatomy of the Lower Extremity: Musculature and Nerve Supply

8

The human lower extremity represents an example of evolutionary adaptation for bipedal locomotion and weight-bearing. Unlike the upper limb, which evolved for mobility and manipulation, the lower limb developed specialized features for support and movement (Moore et al. 2018). During embryological development, the lower limb undergoes a significant medial rotation of approximately 180°, establishing the characteristic alignment where extensor surfaces face anterolaterally and flexor surfaces orient posteromedially (Drake et al. 2020).

The neural control of the lower extremity emerges from an intricate network formed by the lumbar and lumbosacral plexuses, spanning from L1 to S4 spinal segments. This complex arrangement of nerve roots and their branches provides precisely coordinated motor control while establishing systematic patterns of sensory innervation through dermatomes (Standring 2020). The dermatomes' organization, though seemingly complex, follows a logical pattern when understood in the context of embryological development and functional requirements (Perry and Burnfield 2010).

The structural foundation of the lower extremity begins with the connection between the pelvic girdle and sacrum through the sacroiliac joint. This stable base contrasts with the relatively mobile attachment of the upper limb to the axial skeleton (Neumann 2017). The hip joint exemplifies this emphasis on stability, featuring a deep acetabular socket complemented by strong ligamentous support. Moving distally, the knee joint relies more heavily on muscular and ligamentous contributions for its stability, while the ankle and foot complex demonstrates remarkable adaptability through its multiple articulations (Kapandji 2007).

There is notable functional specialization in the muscle system that controls movement in the lower extremity. The hip musculature, anchored by the powerful gluteal complex, provides both stability and mobility crucial for bipedal stance and locomotion (Dicharry 2010). The knee's muscular control, dominated by the quadriceps anteriorly and hamstrings posteriorly, enables the precise regulation of weight-bearing and movement (Powers 2003). At the ankle and foot, the complex arrangement of muscles facilitates both powerful propulsion and fine control of foot positioning.

The foot's intrinsic musculature, including the often-overlooked lumbricals and interossei, plays a crucial role in maintaining the foot's arched structure and controlling precise toe movements. This architectural arrangement, combined with the complex system of tendons and ligaments, creates a dynamic platform capable of both shock absorption and powerful propulsion (Kelikian 2011).

H.-Y. Ko, *Practical Functional Anatomy for Spinal Cord Injury Rehabilitation*,
https://doi.org/10.1007/978-981-95-6457-6_8

8.1 Developmental Sequences of the Lower Extremity

The ontogenetic development of the lower extremities reveals fundamental aspects of their adult anatomical organization and functional capabilities (Moore et al. 2013). During early embryological stages, beginning at approximately week 4 of gestation, the lower extremity limb bud emerges from the lateral plate mesoderm and projects perpendicularly from the trunk wall at approximately 90° (Sadler 2012). The initial formation involves the proliferation of mesenchymal cells beneath the surface ectoderm, forming a paddle-shaped prominence. At this initial phase, the anatomical borders and surfaces correspond closely to those observed in the upper extremity anlage, reflecting their shared developmental origins from the somatic layer of lateral plate mesoderm (Schoenwolf et al. 2015).

As development progresses through weeks 7–8, the lower extremity undergoes a comprehensive medial rotation of approximately 180°, representing a significant morphological transformation (Fig. 8.1). This rotational process results in the repositioning of the extensor surfaces to assume an anterolateral orientation, while the flexor components become situated posteromedially (Moore et al. 2018). The ventral nerve roots supply the flexor musculature, while the dorsal divisions innervate the extensors, maintaining their relationships despite the rotation. The rotational modifications continue beyond birth, demonstrating the dynamic nature of lower extremity development (Standring 2020).

Human lower extremity development exhibits marked differences compared to precocial species. While precocial animals demonstrate immediate post-birth ambulatory capabilities, human neonates maintain their lower extremities in a characteristic flexed position for several months after birth. This flexed position includes hip flexion, knee flexion, and relative external rotation of the entire limb. This extended developmental timeline correlates with the specialized requirements for achieving stable bipedal locomotion in humans, allowing for the gradual development of neuromuscular control and skeletal maturation.

The developmental progression through early childhood involves continued extension and medial rotation of the lower extremity complex. These anatomical adjustments gradually modify the newborn's inverted foot position toward an everted orientation more appropriate for weight-bearing functions (Drake et al. 2020).

The torsional changes occur at multiple levels, including femoral anteversion and tibial torsion, which undergo progressive modifications during growth. The subhip rotation maintains the alignment of hip joint mobility and muscular orientation with the shoulder's cardinal planes, creating a biomechanical arrangement that differs from the organizational patterns observed at the knee, ankle, and foot levels (Netter 2019).

The ossification patterns of the lower extremity skeleton follow a precise temporal sequence, with primary ossification centers appearing prenatally and secondary centers developing throughout childhood and adolescence (Standring 2020). This prolonged skeletal maturation period allows for the fine-tuning of anatomical relationships and the development of appropriate joint surfaces for bipedal locomotion. The process involves complex interactions between mechanical forces, genetic factors, and hormonal influences that collectively shape the definitive adult morphology of the lower extremity (Moore et al. 2018).

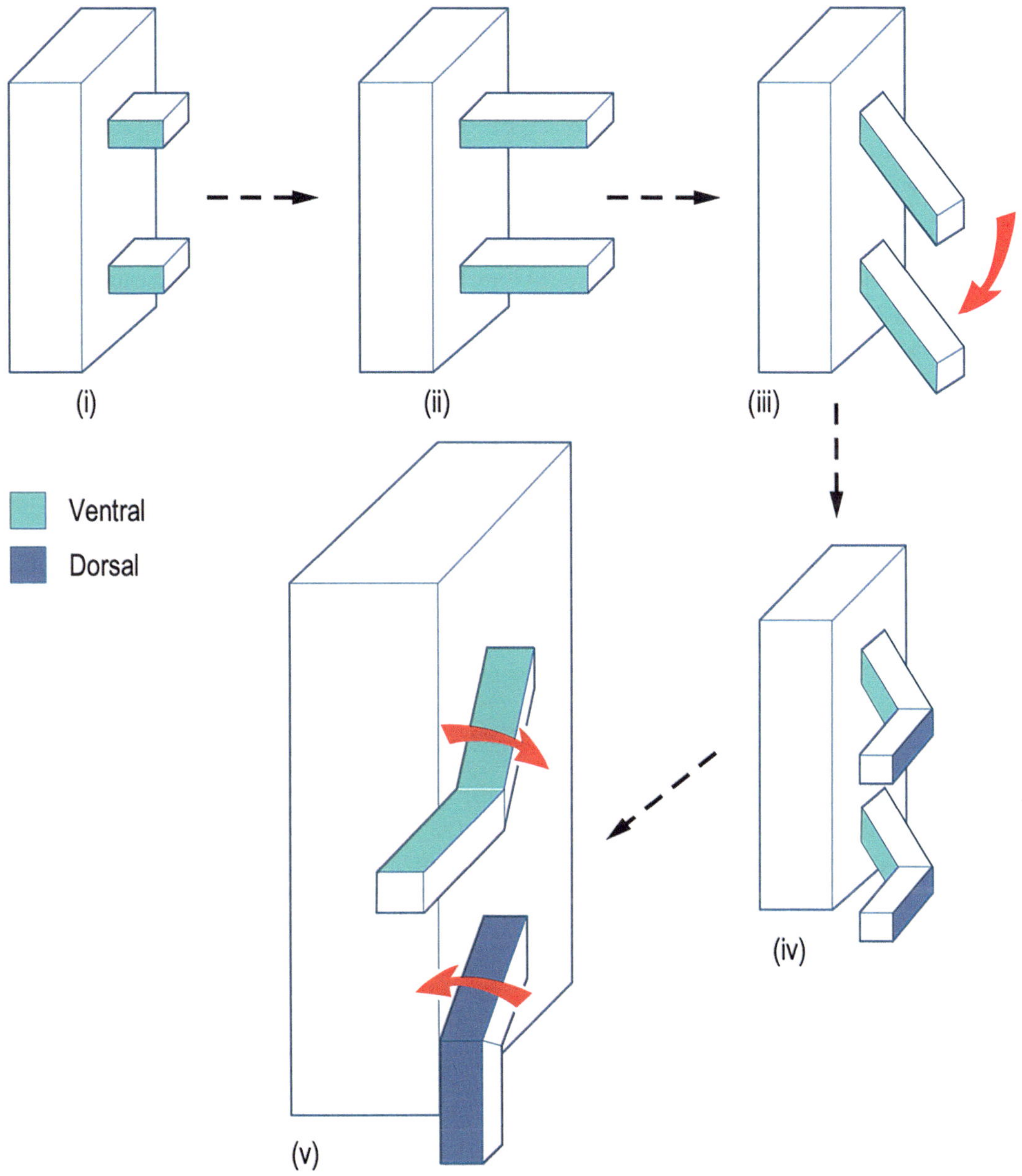

Fig. 8.1 Limb bud development and the directions of limb rotation. (Adapted with permission from Palastanga and Soames (2019))

8.2 Lower Limb Dermatomes

The cutaneous innervation pattern of the lower extremity demonstrates a complex but systematic organization that becomes more apparent through detailed dermatome analysis. A dermatome represents a specific area of skin innervated by a single spinal nerve through its dorsal and ventral rami, with contributions primarily from spinal segments L1 to S2 (Drake et al. 2020; Mancall and Brock 2011). While considerable overlap exists between adjacent dermatomes, certain regions show minimal overlap due to the distant primary innervation of neighboring nerves, particularly at the watershed zones between major nerve territories (Fig. 8.2).

The dermatomal distribution pattern becomes more evident when examining these innervation zones across the limb surface, independent of specific nerve trajectories. The L1 dermatome typically covers the inguinal region and upper anterior thigh, L2 supplies the anterior and upper lateral thigh, L3 innervates the anterior and lateral knee region, L4 extends down the anteromedial leg to the medial malleolus, and L5 follows the lateral leg to the dorsum of the foot (Standring 2020; Drake et al. 2020).

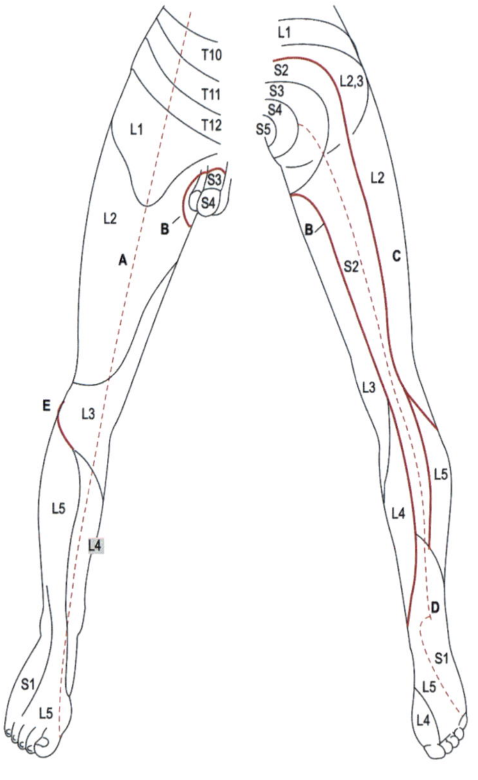

Fig. 8.2 Metamerism of the dermatome of the lower limb. (**a**) Preaxial border (line), (**b**) ventral axial line, (**c**) dorsal axial line, (**d**) postaxial line

Understanding the embryological development of the lower limb is fundamental, particularly the rotational changes during development. The lower limb bud appears around the fourth week of development and undergoes a 180° medial rotation. The original anterior surface shifts medially and then posteriorly, while the posterior surface rotates laterally and then anteriorly (Sadler 2012; Schoenwolf et al. 2015). This embryological process explains the adult innervation pattern where the tibial nerve, originating from the anterior divisions of the ventral rami (L4–S3), provides innervation to the posterior leg and plantar foot. Conversely, the common fibular nerve, derived from the posterior divisions of the ventral rami (L4–S2), innervates the anterolateral aspects of the leg (Moore et al. 2018; Netter 2019).

The dermatomes form continuous bands that wrap around the limb from posterior to anterior along its lateral aspect, revealing a more organized distribution than initially apparent. The sacral dermatomes (S1–S2) predominantly supply the posterior thigh, posterior calf, and lateral foot aspects, while also contributing to the plantar surface. When considering the cutaneous supply from the dorsal primary rami, most spinal roots provide innervation to a continuous strip of skin extending from the spinal region to the lower extremity (Standring 2020; Parent 1996).

These dermatomal strips vary in width throughout their course, with some becoming notably narrow in their proximal segments, appearing almost linear, before expanding to their main distribution areas more distally on the limb. This pattern is particularly evident in the L5 dermatome, which narrows significantly in the lateral thigh before broadening over the lateral leg and dorsum of the foot (Lee et al. 2008; Palastanga and Soames 2019).

The organization of cutaneous nerves arising from the lumbar and lumbosacral plexuses demonstrates specific innervation patterns to different areas of the limb. The femoral nerve (L2–L4) provides cutaneous innervation through its anterior cutaneous branches to the anterior thigh and, via the saphenous nerve, to the medial leg and foot. The lateral femoral cutaneous nerve (L2–L3) supplies the lateral thigh, while the obturator nerve (L2–L4) innervates the medial thigh. These nerves receive their fibers from the rami that form their parent nerves.

8.3 Anatomical and Functional Differences Between Upper and Lower Extremities

The upper and lower extremities exhibit significant structural differences that reflect their specialized functional adaptations. The lower limbs are primarily adapted for weight-bearing, support, and bipedal locomotion, while the upper limbs are structured for mobility and manipulation (Drake et al. 2020; Moore et al. 2018). These functional requirements have led to distinct anatomical characteristics in both appendicular segments.

The connection between the appendicular and axial skeleton demonstrates fundamental differences. The pelvic girdle forms a strong, rigid connection with the sacrum through the sacroiliac joints, which are amphiarthrodial joints reinforced by anterior and posterior sacroiliac ligaments. In contrast, the pectoral girdle maintains a more mobile attachment to the axial skeleton primarily through muscular connections, with only one synovial joint (the sternoclavicular joint) connecting it to the trunk (Palastanga and Soames 2019; Standring 2020).

The developmental fusion of the three pelvic components—ilium, ischium, and pubis—at the triradiate cartilage forms the acetabulum, creating a unified structure by late adolescence. The pubic bones connect anteriorly at the pubic symphysis, a fibrocartilaginous joint that provides additional stability. This arrangement significantly differs from the pectoral girdle's structure, where the scapula remains free-floating and supported by muscular attachments (Sadler 2012; Schoenwolf et al. 2015).

The hip joint demonstrates notable structural adaptations for stability. The acetabulum forms a deep socket that encompasses approximately 40% of the femoral head, creating inherent osseous stability. This articulation is enhanced by the acetabular labrum, which deepens the socket and increases the contact surface area. The joint capsule is reinforced by three major ligaments—iliofemoral, pubofemoral, and ischiofemoral—which provide additional stability while allowing controlled movement (Moore et al. 2018; Netter 2019).

In contrast, the glenohumeral joint of the shoulder prioritizes mobility over stability. The glenoid fossa is relatively shallow, covering only about 25–30% of the humeral head. While the glenoid labrum increases joint congruency, the primary stability comes from the rotator cuff muscles (supraspinatus, infraspinatus, teres minor, and subscapularis) rather than osseous or ligamentous structures. This arrangement allows for a greater range of motion but with reduced inherent stability compared to the hip joint (Drake et al. 2020).

8.4 Surface Anatomy and Clinical Landmarks of the Lower Extremity

The surface anatomy of the lower extremity provides important landmarks for clinical examination and medical procedures. These surface features correspond to underlying anatomical structures and help identify key anatomical relationships (Moore et al. 2018; Standring 2020).

8.4.1 Hip and Thigh Regions

8.4.1.1 Femoral Triangle Region

The anterior pelvic region presents several palpable bony landmarks, including the symphysis pubis, pubic body, pubic crest, pubic tubercle, and the anterior superior iliac spine (ASIS)

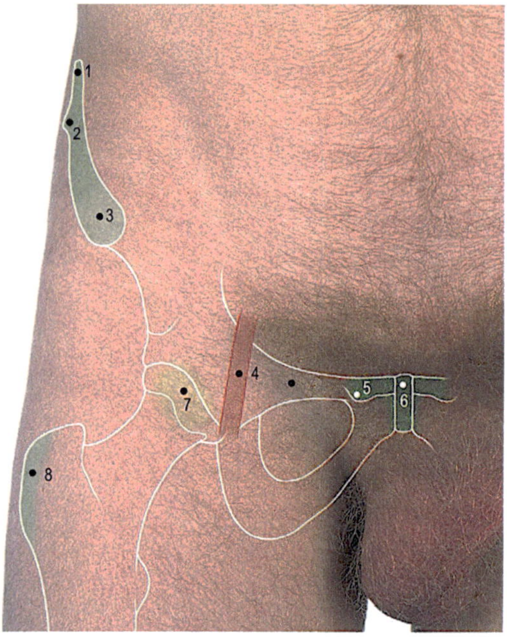

Fig. 8.3 Pelvis and anterior thigh: palpable structures. (1) Iliac crest, (2) tubercle of ilium, (3) anterior superior iliac spine, (4) femoral artery, (5) pubic tubercle, (6) symphysis pubis, (7) head of femur, (8) greater trochanter. (Adapted with permission from Lumley (2008))

(Drake et al. 2020) (Fig. 8.3). The inguinal ligament, a key anatomical structure, extends from the ASIS to the pubic tubercle, constituting approximately 75% of the base of the femoral triangle, with the remaining portion continuing across the pubic body (Netter 2019) (Fig. 8.4).

The lateral border of the femoral triangle is delineated by the sartorius muscle, which originates at the ASIS and courses distally and medially across the thigh, inserting on the subcutaneous surface of the proximal tibia. This muscle, referred to as the "tailor's muscle," is that which enables the cross-legged position. It is innervated by the femoral nerve (Moore et al. 2018). The adductor longus muscle forms the medial border of the triangle, originating from the pubic body and inserting along the linea aspera of the femur. This muscle receives its innervation from the obturator nerve (Standring 2020). Enveloping the thigh like a stocking, the fascia lata forms the roof of the femoral triangle. This fascial structure attaches superiorly to the inguinal ligament, iliac crest, inferior pubic ramus, and sacrum. The saphenous opening, a significant feature of this fascia, is located approximately 4 cm inferior and slightly lateral to the pubic tubercle. This opening overlies the femoral vein and allows passage of its major branch, the great saphenous vein (Drake et al. 2020).

The femoral triangle floor is composed of the iliopsoas and pectineus muscles, both of which are situated anterior to the hip joint. The femoral head can be palpated deep within this region. Iliopsoas muscle originates from ilium's medial aspect and lumbar vertebrae's lateral surfaces, inserts distally into the lesser trochanter of the femur. It functions as a potent hip flexor and medial rotator and receives innervation from the lumbar plexus via the second and third lumbar roots (Drake et al. 2020; Standring 2020). The pectineus muscle attaches proximally to the superior pubic ramus and distally just inferior to the iliopsoas tendon on the femur. The femoral nerve supplies this muscle (Moore et al. 2018).

The femoral artery traverses vertically through the triangle from the midinguinal point–equidistant between the anterior superior iliac spine and the symphysis pubis–and extends to the apex of the triangle. Its deep (profunda femoris) branch courses posteriorly between the adductor longus and pectineus muscles (Netter 2019). Medial to the femoral artery lies the femoral vein, both enclosed within a common sheath, and the femoral canal is positioned further medially. The femoral nerve lies lateral to the artery, external to the femoral sheath (Drake et al. 2020).

8.4.1.2 Lateral Aspect of the Hip

The greater trochanter, a prominent bony landmark, is palpable on the upper lateral thigh aspect. Its position is posterior and distal to the ASIS (Drake et al. 2020).

In the seated position, body weight distribution occurs primarily on the ischial tuberosities. This weight-bearing area is cushioned by the gluteus maximus muscle mass and buttock adipose tissue (Standring 2020) (Fig. 8.5).

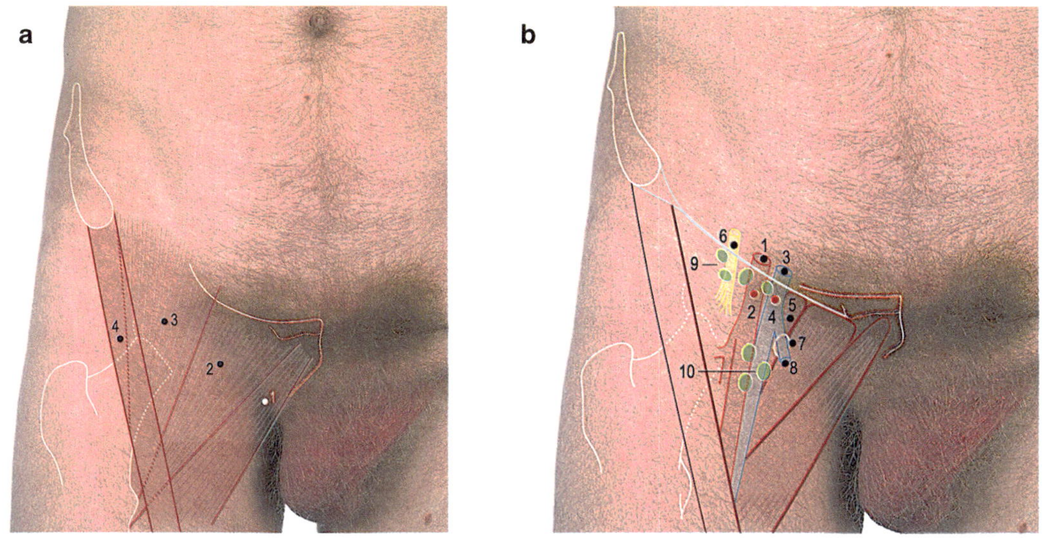

Fig. 8.4 (**a**) Femoral triangle: muscles. (1) Adductor longus, (2) pectineus, (3) iliopsoas, (4) sartorius. (**b**) Femoral triangle: vessels and nerves. (1) Femoral artery, (2) point of access to femoral artery, (3) femoral vein, (4) point of access to femoral vein, (5) femoral canal, (6) femoral nerve, (7) saphenous opening, (8) great saphenous vein, (9) horizontal and (10) vertical groups of inguinal nodes. (Adapted with permission from Lumley (2008))

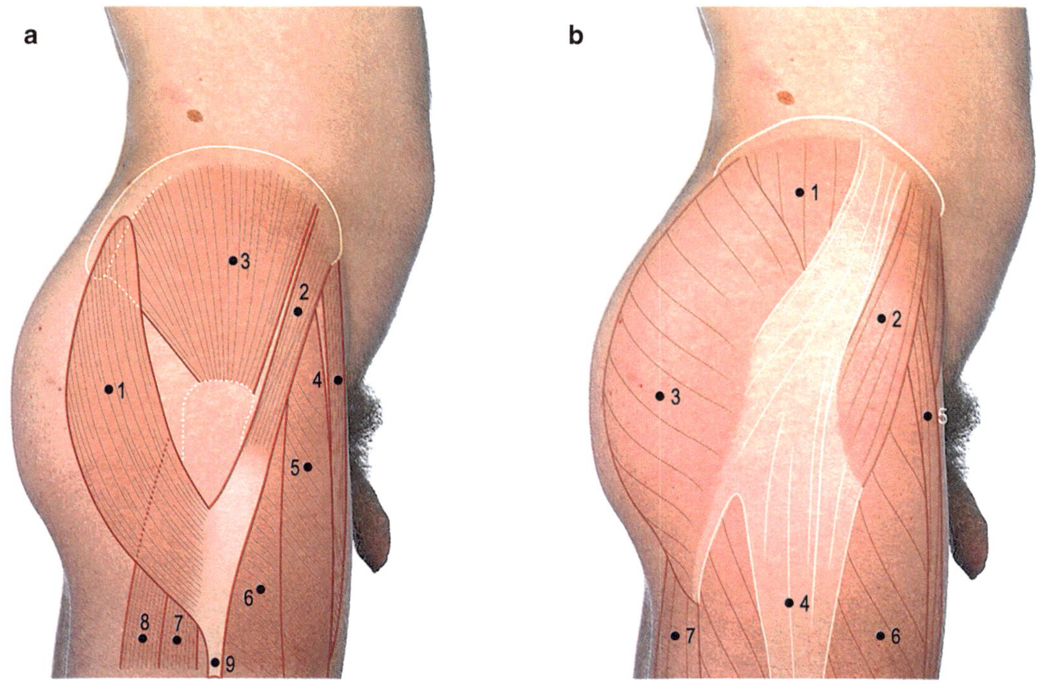

Fig. 8.5 (**a**) Lateral aspect of the hip joint: superficial muscles. (1) Gluteus maximus, (2) tensor fascia lata, (3) gluteus medius, (4) sartorius, (5) rectus femoris, (6) vastus lateralis, (7) long head of biceps femoris, (8) semitendinosus, (9) iliotibial tract. (**b**) Lateral aspect of the hip joint: superficial structures. (1) Gluteus medius, (2) tensor fascia lata, (3) gluteus maximus, (4) iliotibial tract, (5) sartorius, (6) vastus lateralis, (7) hamstring muscles. (Adapted with permission from Lumley (2008))

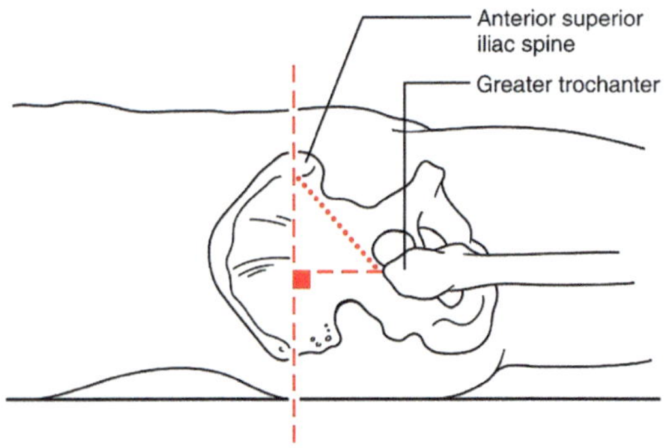

Fig. 8.6 Bryant's triangle. (Adapted with permission from Magee (2014))

Bryant's triangle, a clinical assessment tool, is utilized in diagnosing femoral neck fractures (Fig. 8.6). A reduction in the vertical distance of this triangle indicates a potential fracture, which can be confirmed by comparing the affected side with the contralateral normal side.

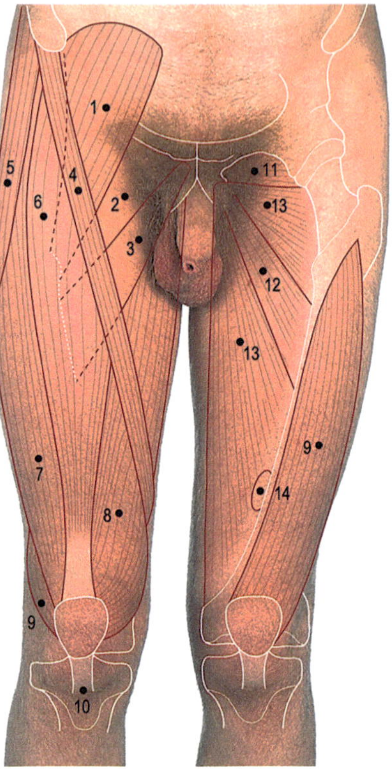

Fig. 8.7 Anterior and medial aspect of the thigh: superficial (right limb) and deep (left limb) muscles. (1) Iliopsoas, (2) pectineus, (3) adductor longus, (4) sartorius, (5) tensor fascia lata, (6) rectus femoris, (7) vastus lateralis, (8) vastus medialis, (9) vastus intermedius, (10) patellar tendon, (11) obturator externus, (12) adductor brevis, (13) adductor magnus, (14) adductor hiatus. (Adapted with permission from Lumley (2008))

8.4.1.3 Anterior Aspect of the Thigh

The anterior thigh is predominantly composed of the substantial quadriceps femoris muscle group (Fig. 8.7). This muscle complex comprises four distinct components with varying proximal attachments (Drake et al. 2020):

- Rectus femoris: Originates via two heads—a straight head from the anterior inferior iliac spine and an oblique head superior to the acetabulum.
- Vastus lateralis and medialis: Attach to their respective aspects of the intertrochanteric line, encircling the subtrochanteric femur before converging on the corresponding sides of the femoral linea aspera.
- Vastus intermedius: Originates from the anterior femoral shaft.

Distally, the quadriceps muscle attaches to the patella and continues beyond it as the strong patellar ligament, which is anchored to the tibial tuberosity. Both the patella and the tibial tuberosity are subcutaneous and easily palpable (Standring 2020). On either side of the patellar tendon, the knee joint capsule is reinforced by fibrous expansions of the quadriceps tendon, known as retinacula. These structures facilitate the muscle's attachment to the tibial condyles. The quadriceps femoris receives its innervation from the femoral nerve (Moore et al. 2018). While the proximal quadriceps components may not be readily visible in recumbent or relaxed standing positions, they become prominently defined when the entire lower limb is elevated against gravity with an extended knee.

The patellar tendon and the inferior horizontal fibers of the vastus medialis, which insert on the medial aspect of the patella, are particularly noticeable. These vastus medialis fibers are susceptible to rapid atrophy during periods of inactivity. The medial and lateral femoral condyles, along with the upper margin of the tibia, are palpable across most of the anterior, medial, and lateral aspects of the knee joint.

8.4.1.4 Medical Aspect of the Thigh

The adductor muscle group occupies the medial thigh region. The superficial layer, consisting of the pectineus and adductor longus muscles, has been previously described in the context of the femoral triangle. The adductor brevis extends from the inferior pubic ramus to the linea aspera, positioned between these muscles and the deeper, substantial adductor magnus (Drake et al. 2020) (Fig. 8.7). The adductor magnus, a powerful muscle, originates along the ischiopubic ramus and ischial tuberosity. It inserts laterally along the linea aspera and medial supracondylar ridge, with a robust tendon attaching to the adductor tubercle of the femur, which is palpable (Moore et al. 2018).

The femoral vessels traverse through an aperture in the adductor magnus's asperal attachment, known as the adductor hiatus, located approximately 10 cm proximal to the knee joint. At this point, they transition into the popliteal vessels. The gracilis muscle, a thin, flat structure, overlays the adductor muscles. It originates from the inferior pubic ramus and inserts on the proximal medial aspect of the tibia (Standring 2020).

The obturator nerve provides innervation to the gracilis and adductor muscles, which primarily function in hip joint adduction. The adductor magnus exhibits dual innervation; its portion arising from the ischial tuberosity receives supply from the sciatic nerve. This segment is functionally and developmentally associated with the hamstring muscle group, contributing to hip extension (Cramer and Darby 2014).

8.4.1.5 Posterior Aspect of the Hip and Thigh

Gluteal Region

The contour of the buttock is largely shaped by the gluteus maximus, a large quadrilateral muscle, enveloped by a variable, thick layer of fibrofatty superficial fascia (Standring 2020). This muscle has an extensive origin, including the lateral surface of the ilium posterior to the posterior gluteal line, the sacrum, coccyx, and sacrotuberous ligament. Its fibers course inferolaterally, inserting into the iliotibial tract and the gluteal tuberosity of the femur (Drake et al. 2020). Functionally, the gluteus maximus acts as a powerful extensor and lateral rotator of the hip joint. Additionally, through its connection to the iliotibial tract, it contributes to extension and stabilization of the knee joint (Moore et al. 2018). It is important to note that the inferior border of gluteus maximus does not align with the horizontal gluteal fold; instead, this fold represents the 'extensor line' or posterior skin fold associated with the hip joint (Agur and Dalley 2016).

The inferior gluteal nerve provides innervation to the gluteus maximus. This muscle overlays the hip joint, short articular muscles, and several nerves exiting the pelvis via the greater and lesser sciatic foramina, including the sciatic, gluteal, and pudendal nerves. To avoid potential injury to these deep structures, intramuscular injections in the buttock should be administered in the superolateral quadrant (Maruyama et al. 2001).

The tensor fasciae latae originates from the anterior quarter of the outer lip of the iliac crest and inserts distally into the iliotibial tract. It functions synergistically with the gluteus maximus in knee joint stabilization, extension, and internal rotator of the hip. The superior gluteal nerve innervates this muscle.

During a unilateral stance, the gluteus medius and minimus muscles elevate the non-weight-bearing side of the pelvis, maintaining the contralateral leg off the ground. These muscles originate from the anterolateral aspect of the ilium and insert on the greater trochanter. They receive innervation from the superior gluteal nerve and can be palpated contracting through the gluteus maximus during this action (Neumann 2017).

Thigh Region

The muscular mass of the posterior thigh is primarily composed of the hamstring group, which originates from the ischial tuberosity

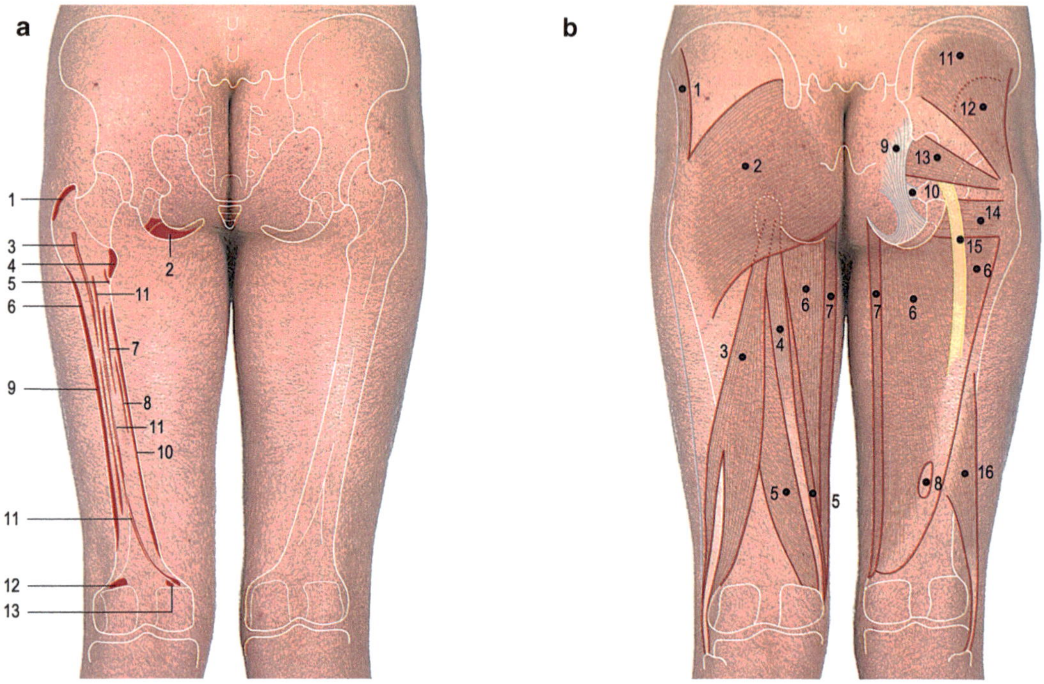

Fig. 8.8 (**a**) Gluteal region and posterior aspect of the thigh: muscle attachments. (1) Gluteus medius, (2) hamstring muscles, (3) gluteus maximus, (4) iliopsoas, (5) pectineus, (6) vastus lateralis, (7) adductor brevis, (8) adductor longus, (9) short head of biceps femoris, (10) vastus medialis, (11) adductor magnus, (12, 13) gastrocnemius. (**b**) gluteal region and posterior aspect of the thigh: superficial (left limb) and deep (right limb) muscles. (1) Tensor fascia lata, (2) gluteus maximus, (3) long head of biceps femoris, (4) semitendinosus, (5) semimembranosus, (6) adductor magnus, (7) gracilis, (8) adductor hiatus, (9) sacrotuberous ligament, (10) sacrospinous ligament, (11) gluteus medius, (12) gluteus minimus (deep to medius), (13) piriformis, (14) quadratus femoris, (15) sciatic nerve, (16) short head of biceps femoris. (Adapted with permission from Lumley (2008))

(Drake et al. 2020) (Fig. 8.8). The semimembranosus muscle inserts distally into a groove on the posteromedial aspect of the tibial condyle. Additionally, it has tendinous expansions to the lateral femoral condyle, forming the oblique popliteal ligament, and extends inferiorly to the soleal line of the tibia, creating the popliteal fascia overlying the popliteus muscle (Moore et al. 2018).

The semitendinosus muscle attaches distally to the upper subcutaneous medial surface of the tibia. In contrast, the biceps femoris inserts onto the fibular head, with its short head originating from the linea aspera (Standring 2020). Functionally, the hamstring muscles act as potent hip joint extensors and knee joint flexors. The sciatic nerve provides innervation to this muscle group (Neumann 2017).

The biceps femoris laterally, and the semimembranosus and semitendinosus medially, form an inverted V-shape that delineates the superior boundary of the popliteal fossa (Agur and Dalley 2016). This anatomical arrangement is significant for clinical assessments and surgical approaches in the posterior thigh region.

The sciatic nerve emerges from the pelvis through the greater sciatic foramen, maintaining close relationships with the ischium and the small posterior articular muscles of the hip joint. It courses deep to the gluteus maximus and subsequently beneath the hamstring muscles before bifurcating into its terminal branches: the tibial and common peroneal nerves (Netter 2019). Due to its intimate association with the hip joint, the sciatic nerve is susceptible to injury in cases of posterior hip dislocation.

8.4.2 Knee Region

8.4.2.1 Popliteal Fossa

The popliteal fossa is a diamond-shaped depression posterior to the knee joint. Its superior boundaries are delineated by the hamstring tendons, while the inferior margins are formed by the medial and lateral heads of the gastrocnemius muscle (Drake et al. 2020). The popliteal fascia, a thickened portion of the fascia lata, serves as the roof of this space (Moore et al. 2018) (Fig. 8.9).

The floor of the popliteal fossa is composed of three distinct structures, arranged in a superior-to-inferior sequence: the posterior surface of the femur, the knee joint, and the popliteus muscle

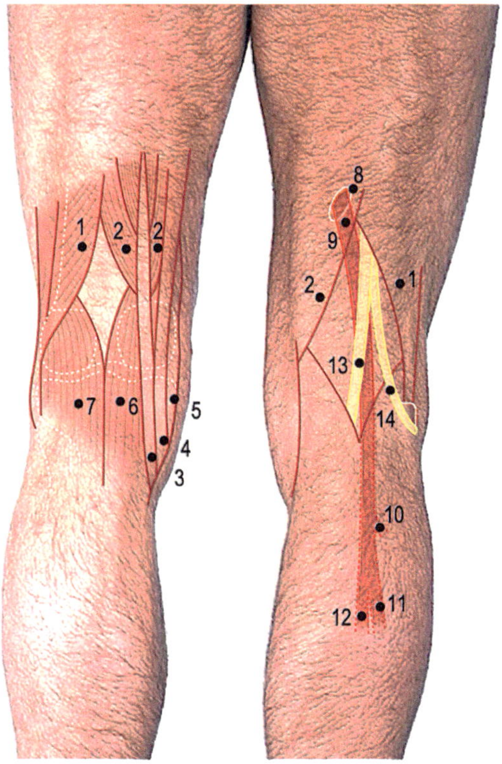

Fig. 8.9 Popliteal fossa: soft tissues. (1) Biceps femoris, (2) semimembranosus (on either side of semitendinosus tendon), (3) semitendinosus, (4) gracilis, (5) sartorius, (6) gastrocnemius, medial head, (7) gastrocnemius, lateral head, (8) adductor hiatus, (9) popliteal artery, (10) anterior tibial artery, (11) peroneal artery, (12) posterior tibial artery, (13) tibial nerve, (14) common peroneal nerve. (Adapted with permission from Lumley (2008))

overlying the proximal tibia (Standring 2020). This anatomical arrangement provides a complex interface between the thigh and leg regions. Within the popliteal fossa, several important neurovascular structures include the popliteal artery and vein, the tibial and common peroneal nerves, adipose tissue, and a small number of lymph nodes (Netter 2019). The popliteal artery, situated deep to its accompanying vein, maintains an intimate relationship with the knee joint capsule (Agur and Dalley 2016).

The popliteal artery can be palpated along the length of the fossa through deep compression. However, palpation is most readily accomplished with the knee in slight flexion, particularly over the proximal tibia (Neumann 2017). This positioning relaxes the overlying structures, facilitating easier access to the deep-lying artery.

8.4.2.2 Lateral Aspect of the Knee

The lateral aspect of the knee presents several palpable structures of clinical significance. When the knee is flexed to 90°, the biceps femoris tendon becomes distinctly visible and palpable, extending to its insertion on the fibular head (Drake et al. 2020) (Fig. 8.10). This positioning facilitates accurate assessment of tendon integrity and function.

The bony margin of the lateral tibial condyle is readily palpable. Adjacent to this, the fibular (lateral collateral) ligament can be felt as a taut, cord-like structure. It originates from the lateral femoral condyle and inserts on the fibular head, remaining separate from the joint capsule (Moore et al. 2018). This anatomical arrangement allows for specific ligament testing in clinical examinations.

Positioned medial to the tibial margin lies the lateral meniscus. While typically not palpable, it may become tender and occasionally discernible upon palpation when injured (Standring 2020). This clinical finding can aid in the diagnosis of meniscal pathologies.

The common peroneal nerve follows a superficial course in this region. Initially overlapped by the biceps femoris tendon, it becomes subcutaneous as it traverses the fibular neck. At this location, the nerve is palpable and vulnerable to

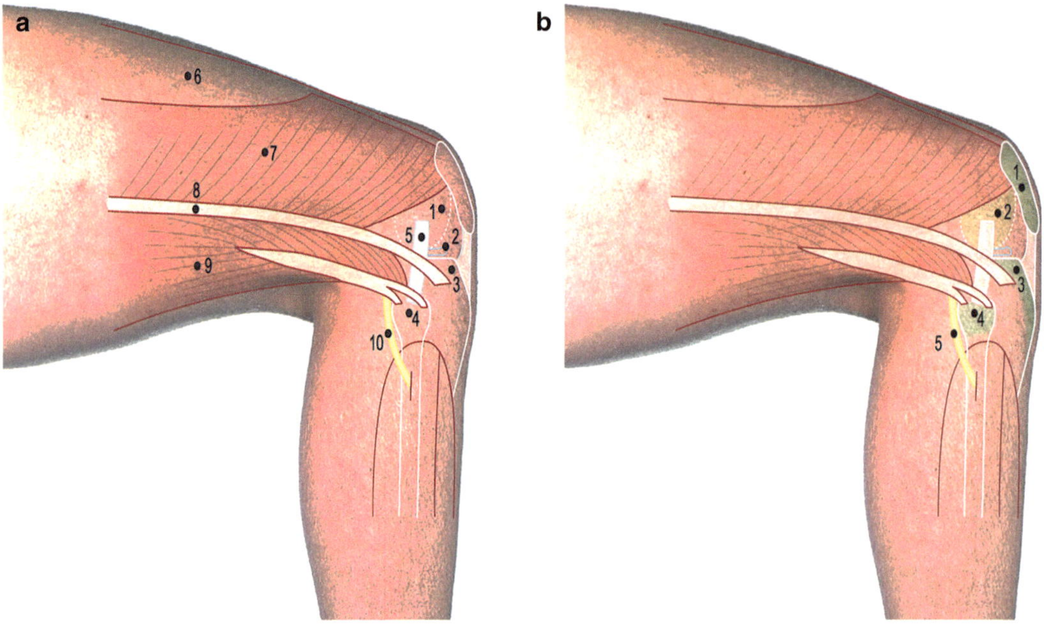

Fig. 8.10 (**a**) Lateral aspect of the flexed knee: bones and soft tissues. (1) Lateral femoral condyle; (2) lateral meniscus, (3) lateral tibial condyle, (4) head of fibula, (5) lateral collateral ligament, (6) rectus femoris, (7) vastus lateralis, (8) iliotibial tract, (9) biceps femoris, (10) common peroneal nerve. (**b**) Lateral aspect of flexed knee: palpable structures. (1) Patella, (2) lateral femoral condyle, (3) lateral tibial condyle, (4) head of fibula, (5) common peroneal nerve. (Adapted with permission from Lumley (2008))

injury (Netter 2019). Subsequently, it divides into deep and superficial branches, innervating the anterior and lateral leg compartments, respectively (Agur and Dalley 2016).

8.4.2.3 Medial Aspect of the Knee

The medial collateral ligament of the knee joint is a strong, broad, and flat structure that integrates with the joint capsule. It extends from the medial epicondyle of the femur to the upper portion of the tibial shaft (Fig. 8.11). The medial portion of the medial meniscus is situated between the femoral and tibial condyles. While typically impalpable, damage to this structure may result in tenderness along the joint line and potential protrusion at a tear site (Grood et al. 1981).

Attachment sites of the sartorius, gracilis, and semitendinosus tendons are located on the subcutaneous upper surface of the tibia, arranged in an anterior-to-posterior sequence. These structures become more readily identifiable when the knee is flexed to 90° (Basso et al. 2001).

8.4.3 Lower Leg Region

8.4.3.1 Anterior Aspect of the Lower Leg

The musculature of the lower leg is categorized into three distinct groups: anterior, lateral, and posterior (Moore et al. 2018). These compartments are delineated by anatomical structures, including the tibia, fibula, the interosseous membrane, and fascial components (anterior and posterior intermuscular septa) extending from the fascia lata to the fibula. The anterior intermuscular septum serves as a boundary between the anterior compartment, containing dorsiflexor and long digital extensor muscles, from the lateral compartment, which primarily contains the extensor muscles (Standring 2020) (Fig. 8.12).

The anterior border and medial surface of the tibia are largely subcutaneous, with the border extending from the tibial tuberosity and the surface extending from the medial condyle down to the medial malleolus (Netter 2019). Among the

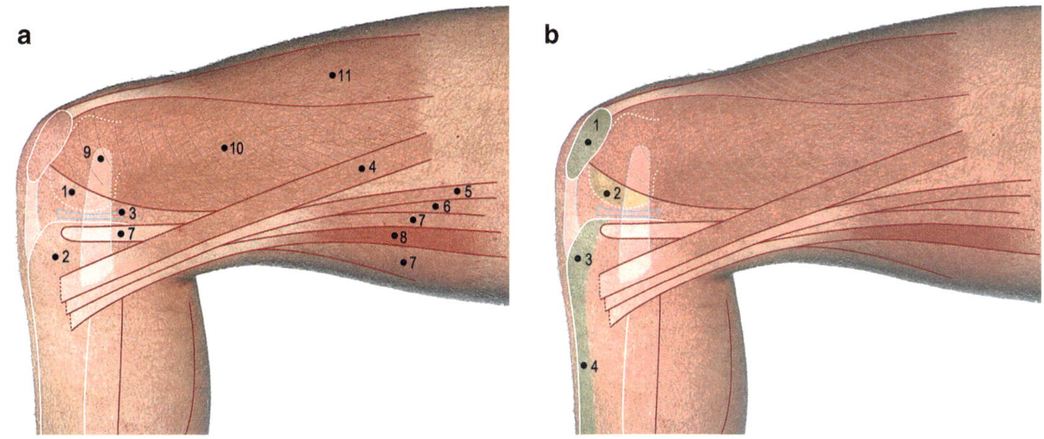

Fig. 8.11 (**a**) Medial aspect of the flexed knee: bones and muscles. (1) Medial femoral condyle, (2) medial tibial condyle, (3) medial meniscus, (4) sartorius, (5) gracilis, (6) adductor magnus, (7) semimembranosus, (8) semitendinosus, (9) medial collateral ligament, (10) vastus medialis, (11) rectus femoris. (**b**) Medial aspect of knee: palpable structures. (1) Patella, (2) medial femoral condyle, (3) medial tibial condyle, (4) subcutaneous anterior border of tibia. (Adapted with permission from Lumley (2008))

four muscles in the anterior compartment, only the tibialis anterior is directly attached to the tibia, anchoring to the upper two-thirds of the lateral surface of the tibia and the adjacent interosseous membrane. The distal tendon of the tibialis anterior inserts into the medial cuneiform and the base of the first metatarsal bones (Drake et al. 2020) (Fig. 8.13).

The extensor hallucis longus originates from the middle half of the fibula and interosseous membrane, while the extensor digitorum longus attaches to the upper two-thirds of the fibula and adjacent intermuscular septum (Moore et al. 2018). These tendons of the muscles extend to the distal phalanges of the toes, with extensor digitorum longus specifically targeting the lateral four toes and forming extensor expansions similar to those found in the hand. The peroneus tertius, a smaller muscle, connects the lower fibula to the fifth metatarsal body (Standring 2020).

Functionally, the anterior compartment muscles are responsible for ankle dorsiflexion. Additionally, extensor hallucis longus and extensor digitorum longus extend (dorsiflex) to the toes. Innervation for all four muscles in this compartment is provided by the deep peroneal nerve (Drake et al. 2020).

The anterior muscle group in the distal leg and ankle region is secured by two extensor retinacula (Standring 2020). The superior retinaculum, a thickening of the deep fascia, spans between the tibia and fibula approximately 3 cm proximal to the ankle joint. The inferior retinaculum, bifurcate in structure, originates from the superior aspect of the calcaneus and traverses the ankle. Its upper limb extends to the medial malleolus, while the lower limb connects with the plantar fascia (Moore et al. 2018). Beneath the inferior retinaculum, tibialis anterior and extensor hallucis longus possess individual synovial sheaths. The tibialis anterior's sheath extends proximally beneath the superior retinaculum. The extensor digitorum longus and peroneus tertius share a common synovial sheath (Drake et al. 2020).

The anterior tibial artery courses over the anterior aspect of the ankle joint, approximately midway between the medial and lateral malleoli. Medially, it is surrounded by the tendons of tibialis anterior and extensor hallucis longus, while laterally, it is bounded by the tendons of extensor digitorum and peroneus tertius. At this juncture, the artery transitions into the dorsalis pedis artery (Netter 2019). The dorsalis pedis artery pro-

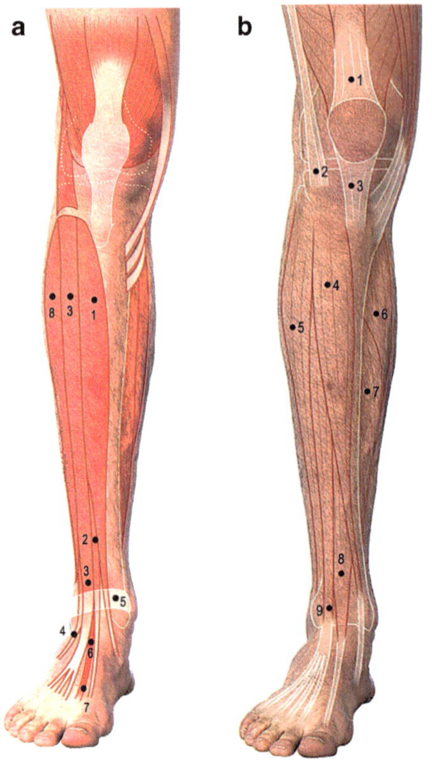

Fig. 8.12 (**a**) Anterior aspect of the lower leg: muscles. (1) Tibialis anterior, (2) extensor hallucis longus, (3) extensor digitorum longus, (4) peroneus tertius, (5) superior extensor retinaculum, (6) dorsalis pedis artery, (7) first dorsal metatarsal artery, (8) peroneus muscles. (**b**) Anterior aspect of right lower leg: superficial structures. (1) Quadriceps tendon, (2) iliotibial tract, (3) patellar tendon, (4) tibialis anterior, (5) peroneal muscles, (6) gastrocnemius, medial head, (7) soleus, (8) extensor hallucis longus, (9) extensor digitorum longus. (Adapted with permission from Lumley (2008))

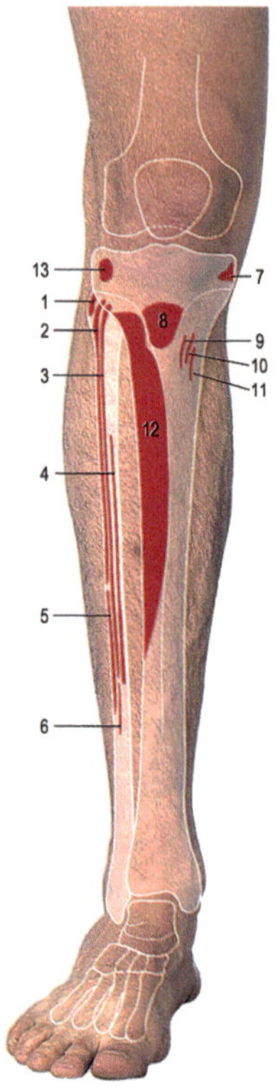

Fig. 8.13 Anterior aspect of the lower leg: muscle attachments: (1) Biceps, (2) peroneus longus, (3) extensor digitorum longus, (4) extensor hallucis longus, (5) peroneus brevis, (6) peroneus tertius, (7) semimembranosus, (8) ligamentum patellae, (9) sartorius, (10) gracilis, (11) semitendinosus, (12) tibialis anterior, (13) iliotibial tract. (Adapted with permission from Lumley (2008))

gresses distally, lateral to the extensor hallucis longus tendon. It then penetrates between the heads of the first dorsal interosseous muscle at the proximal end of the first intermetatarsal space, entering the plantar region. The distal continuation of the artery is termed the first dorsal metatarsal artery (Moore et al. 2018).

Clinically, the dorsalis pedis artery can be palpated by applying posterior pressure against the talus, navicular, or intermediate cuneiform bones in this area. Additionally, a perforating branch of the peroneal artery may occasionally be palpable as it descends over the anterior aspect of the lateral malleolus (Drake et al. 2020).

8.4.3.2 Medial Aspect of the Lower Leg

The medial tibial surface, exposed subcutaneously along its entire length, is susceptible to various injuries, from superficial lacerations to bone fractures. The tibial distal portion expands into the medial malleolus, which articulates laterally with the talus. Along the medial aspect of the foot, several bony landmarks are palpable: the posterior medial calcaneus, sustentaculum tali, navicular tubercle, and the medial aspects of the first metatarsal and its phalanges (Kelikian 2011).

The upper subcutaneous tibial surface serves as an attachment site for sartorius, gracilis, and semitendinosus muscles. The tendons of the deep posterior muscles of the calf run along grooves near the lower end of the tibia. While not all are readily palpable, tibialis posterior and flexor digitorum longus can be detected just above and medial to the posterior tibial artery. This artery's surface projection at the ankle joint lies between the medial malleolar prominence and the medial calcaneal eminence. It is palpable via compression on the talus prior to its deep course into the plantar region (Moore et al. 2018).

A deep fascial thickening, the flexor retinaculum, covers these deep tendons, extending from the medial malleolus to the calcaneal medial tubercle. Beneath this retinaculum, each tendon possesses a distinct sheath (Fig. 8.14).

The sustentaculum tali, a significant clinical landmark for tendon identification, can be palpated with firm pressure approximately 2 cm inferior to the medial malleolar apex. The tibialis posterior tendon courses inferiorly and anteriorly above the sustentaculum. The flexor digitorum longus crosses its medial surface, occasionally creating a groove. The flexor hallucis longus occupies a deep groove on the sustentaculum's inferior aspect (Kapandji 2007).

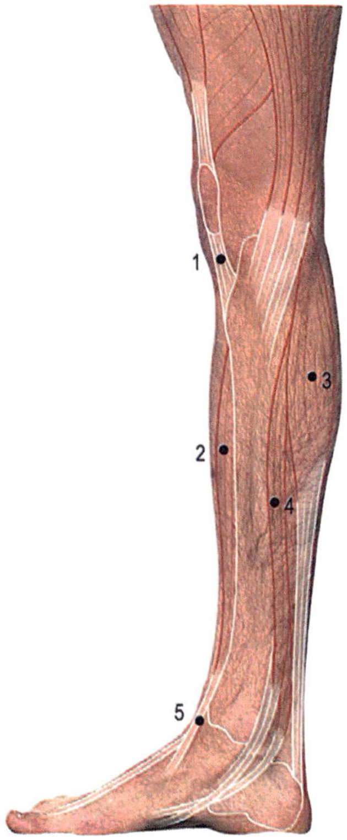

Fig. 8.14 Medial aspect of lower leg and foot: superficial structures. (1) Patellar tendon, (2) medial head of gastrocnemius, (3) soleus, (4) muscles of anterior compartment of leg, (5) tendon of tibialis anterior. (Adapted with permission from Lumley (2008))

8.4.3.3 Lateral Aspects of the Lower Leg

The fibular head is palpable inferior to the knee joint, while its shaft is surrounded by muscle tissue. The distal end of the fibula expands to form the lateral malleolus, which extends approximately 1 cm lower than the medial malleolus and contributes to the ankle joint mortise (Moore et al. 2018) (Fig. 8.15). Palpable bony landmarks on the lateral surface include the lateral calcaneal surface, peroneal tubercle, the cuboid, and the tuberosity of the fifth metatarsal base.

The lateral compartment contains two eversion muscles: peroneus longus and peroneus brevis. These muscles attach to the fibula's lateral surface, with peroneus longus occupying the superior two-thirds and peroneus brevis the inferior third (Netter 2019). The tendons of these muscles course posterior to the lateral malleolus, with peroneus brevis positioned anteriorly. They then traverse the lateral calcaneal surface, the brevis tendon superior to the peroneal tubercle, and the longus inferior. The peroneus longus tendon continues its path, grooving the inferior cuboid surface before attaching to the first metatarsal base and adjacent medial cuneiform. Conversely, the peroneus brevis tendon inserts into the tuberosity of the fifth metatarsal base (Standring 2020).

These muscles are involved in foot eversion and plantar flexion, receiving innervation from the superficial peroneal nerve. The superior and inferior peroneal retinacula, thickenings in the

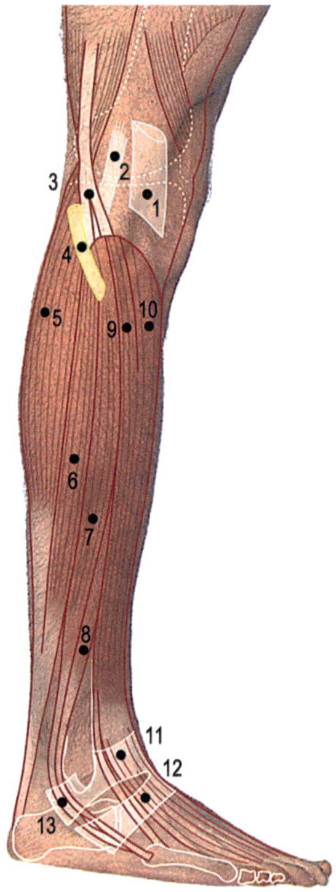

Fig. 8.15 Lateral aspect of the lower leg and foot: muscles. (1) Iliotibial tract, (2) lateral collateral ligament, (3) biceps femoris tendon, (4) common peroneal nerve, (5) gastrocnemius, (6) soleus, (7) peroneus longus, (8) peroneus brevis, (9) extensor digitorum longus, (10) tibialis anterior, (11, 13) superior extensor retinaculum, (12) inferior extensor retinaculum. (Adapted with permission from Lumley (2008))

deep fascia, secure these tendons to the lateral malleolus and calcaneus. A common synovial sheath envelops both tendons, extending to their respective distal attachments (Levangie and Norkin 2005).

Lower leg fractures typically involve both tibia and fibula. Due to the subcutaneous position of the tibia, these fractures are often compound, with the fracture site exposed to the external environment through a skin laceration. Rotational injuries of the ankle frequently occur, primarily affecting the lateral and medial ligaments. In more severe cases, fractures of the lateral and medial malleoli may occur. Major trauma can result in fractures of the posterior tibial rim (posterior malleolus) and ankle joint dislocation.

8.4.3.4 Posterior Aspect of the Lower Leg

The posterior aspect of the lower leg is dominated by the gastrocnemius and soleus muscles, which collectively form the calf's bulk. These muscles converge distally to create the tendo calcaneus (Achilles tendon), which attaches to the midpoint of the calcaneus's posterior surface (Moore et al. 2018) (Fig. 8.16). A bursa and adipose tissue pad separate this tendon from the superior posterior calcaneal surface, providing a cushioning effect.

The gastrocnemius, positioned superficial to the soleus, comprises medial and lateral heads that originate from their respective femoral condyles. In contrast, the soleus exhibits a continuous superior attachment spanning from the fibular head, across a tendinous arch overlying the posterior tibial vessels and tibial nerve, to the tibial soleal line (Netter 2019). The soleus protrudes on both sides of the gastrocnemius within the calf region. The plantaris, a slender vestigial muscle, runs alongside the lateral head of the gastrocnemius. It's worth noting that the deep calf muscles are not palpable above the ankle region (Drake et al. 2020).

These muscular structures function as potent plantar flexors of the foot, playing a significant role in both postural maintenance and ambulation. The tibial nerve provides innervation to this muscle group (Standring 2020). Intramuscularly and between these muscles, a venous plexus exists. Muscle contraction facilitates proximal blood flow, a mechanism known as the muscle pump (Levangie and Norkin 2005).

For rehabilitation specialists focusing on spinal cord injuries, understanding the anatomical arrangement and functional significance of these posterior leg muscles is fundamental. This knowledge aids in developing targeted interventions to address gait abnormalities, postural instability, and venous insufficiency often associated with spinal cord injuries. Additionally, the calf muscles' role in the muscle pump mechanism underscores their importance in preventing complications such as deep vein thrombosis, a common concern in patients with limited mobility due to spinal cord injuries.

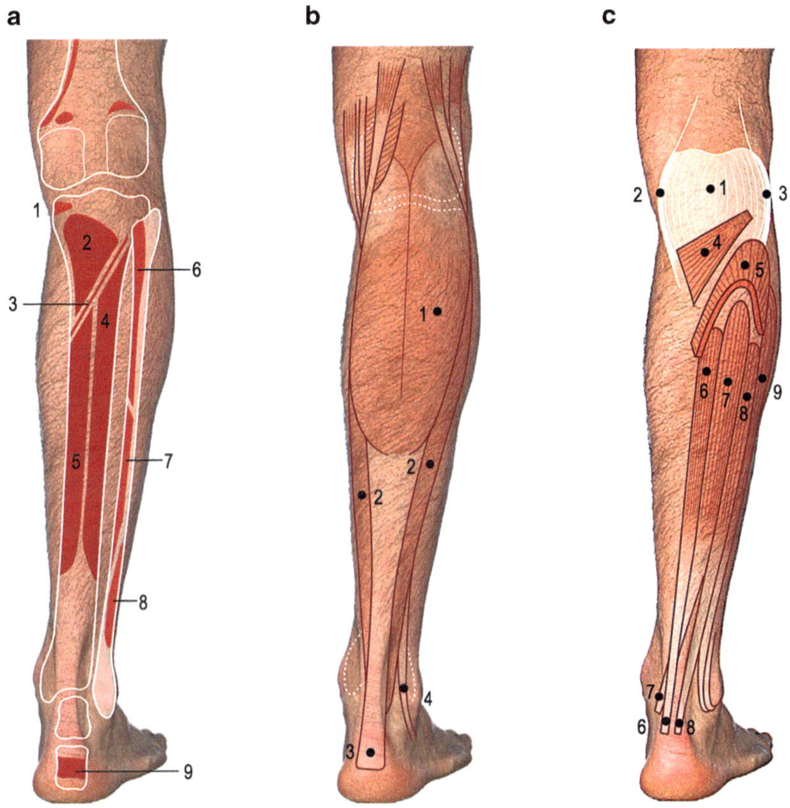

Fig. 8.16 (**a**) Posterior aspect of the lower leg and heel: muscle attachments. (1) Semimembranosus, (2) popliteus, (3, 6) soleus, (4) tibialis posterior, (5) flexor digitorum longus, (7) flexor hallucis longus, (8) peroneus brevis, (9) tendo calcaneus. (**b**) Posterior aspect of the lower leg and heel: superficial muscles. (1) Gastrocnemius, (2) soleus (3) tendo calcaneus, (4) peroneus longus. (**c**) Posterior aspect of the lower leg and heel: deep muscles. (1) Capsule of knee joint, (2) medial collateral ligament, (3) lateral collateral ligament, (4) popliteus, (5) soleus—partly excised, (6) flexor digitorum longus, (7) tibialis posterior, (8) flexor hallucis longus, (9) peroneal muscles. (Adapted with permission from Lumley (2008))

8.5 Identifying Bone and Joint Landmarks in the Lower Extremity

The lower extremity's surface anatomy features distinct landmarks that are critical for clinical examination and rehabilitation practices. The anterior superior iliac spine and the iliac crest are prominent structures that can be felt and sometimes seen. Located approximately a hand's width below the iliac crest, the greater tuberosity of the femur is palpable when the hip is abducted, facilitating a relaxed position for the hip abductors. The physical examination for patients with spinal cord injuries involves evaluating the Bryant's triangle (Fig. 8.17), by palpating the lateral hip area, especially relevant for those in wheelchairs. This assessment aims to identify possible fractures of the femoral neck or hip dislocations. The Bryant's triangle, or iliofemoral triangle, is delineated as a right-angled triangle extending from the anterior superior iliac spine to the apex to the greater trochanter (Magee 2014). The ischial tuberosity's function in weight-bearing when seated becomes palpable with hip flexion or when the patient is in a side-lying position, differing from when it is obscured by the gluteus maximus muscle in an upright position. Furthermore, the posterior superior iliac spines, situated just beneath the surface indentations at

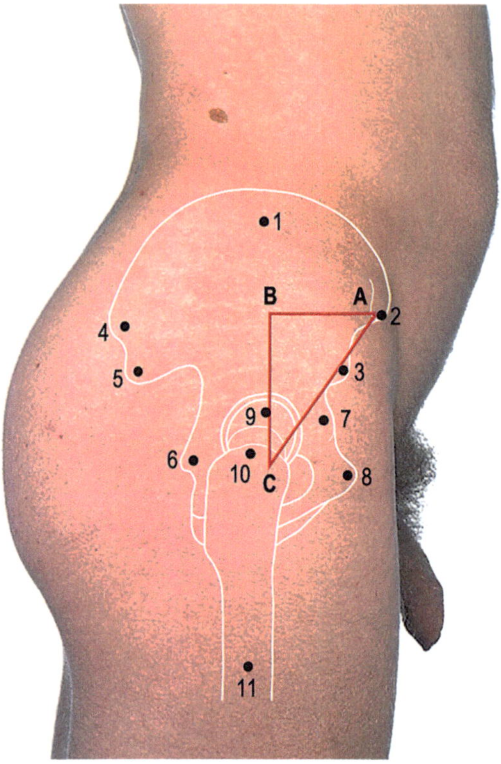

Fig. 8.17 Lateral aspect of the hip joint: bones. (1) Ilium, (2) anterior superior iliac spine, (3) anterior inferior iliac spine, (4) posterior superior iliac spine, (5) posterior inferior iliac spine ABC, Bryant's triangle, (6) ischial spine, (7) iliopubic eminence, (8) body of pubis, (9) head of femur, (10) greater trochanter, (11) shaft of femur. (Adapted with permission from Lumley (2008))

the upper region of the buttocks, are readily palpable landmarks (Ellis 1971).

Moving to the knee, the patella stands out as a significant landmark. Its lateral mobility is noticeable when the surrounding quadriceps muscle is in a state of relaxation. The femur and tibia's condyles, along with the fibula's head and the knee joint line, are palpable landmarks, especially when the individual is seated with the knee flexed, such as sitting on the edge of a table. The tibia's anterior and anteromedial edges are palpable throughout their length, just under the skin. In comparison, the fibula becomes superficial towards its end near the lateral malleolus, which is more protruded than the medial malleolus of the tibia.

At the ankle, the talus bone's head is palpable immediately before the malleoli. The navicular tuberosity, located about 1 inch ahead of the medial malleolus, serves as the primary attachment site for the tibialis posterior tendon. On the foot's lateral aspect, the base of the fifth metatarsal, where the peroneus brevis muscle attaches, can be located. The calcaneus, or heel bone, is also palpable easily. Below the lateral malleolus tip, the peroneal tubercle is felt about an inch down, and similarly, the sustentaculum tali is found 1 inch below the medial malleolus. These structures act as guides for the peroneus longus and flexor hallucis longus tendons, respectively.

8.6 Lower Extremity Joints

8.6.1 Pelvis and Hip Anatomy

The anatomical and functional features of the pelvis and hip regions differ significantly from those of the upper extremities. The pelvic girdle forms a rigid connection with the axial skeleton at the sacrum, without the need for muscular support to stabilize the sacroiliac joints. Muscular attachments, including those of the gluteal muscles and various short rotators, anchor to the upper portion of the femur, facilitating hip extension (via the gluteus maximus), abduction (through the gluteus medius and minimus), and lateral rotation (accomplished by the short rotators). Additionally, the proximal thigh is comprised of muscle groups responsible for flexion (the psoas, iliacus, and pectineus) and adduction. The rectus femoris and hamstrings, which are biarticular muscles, extend across the hip and primarily influence knee movement. The thigh's anterior neurovascular supply enters through the femoral triangle, while posterior access is via the greater sciatic notch, which accommodates the gluteal vessels and nerves.

The pelvis serves as a fundamental structural support for the upper body, establishing a connection between the spinal column and lower extremities. This structure is formed by a strong bony ring, comprising the pelvic bones (ilium,

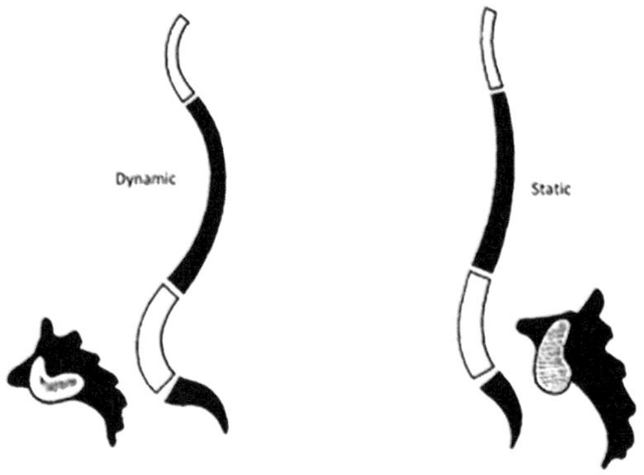

Fig. 8.18 Correlation between the degree of spinal curvatures and the morphology of the articular facet between the sacrum and ilium. When spine curvatures are more pronounced, the sacrum is positioned more horizontally, and the articular facet adopts an "L" shape. In contrast, with less spinal curvature, the sacrum and articular facet are oriented more vertically. These are considered normal variations. (Adapted from Shankar (1996))

pubis, and ischium), as well as the sacrum and coccyx, all of which are securely interconnected by strong ligaments, including the sacroiliac ligaments. The architecture of the pelvis includes three major joints: the hip, sacroiliac (SI) joint, and pubic symphysis. Notably, there is a significant variation in the size, cavity, and joint structures of the pelvis, both within and between genders. Particularly, the SI joints exhibit a considerable range of structural variability, which can be caused by variations in spinal curvature, as illustrated in Fig. 8.18.

Pelvic fractures are uncommon and usually result from severe trauma. Such fractures typically occur in more vulnerable points like the pubic rami, wing of the ilium, and the obturator foramen. Traumatic damage to the SI joints and pubic symphysis, while rare, primarily occurs under similar conditions of extreme force. In contrast, non-traumatic disorders such as infections (notably tuberculosis), tumors, or various forms of arthritis (osteoarthritis, psoriatic arthritis, Reiter's syndrome) are more frequent causes of complications within the SI joint. Conditions like sacroiliitis in ankylosing spondylitis are well documented.

Alternative diagnostic approaches have been developed to address the difficulties associated with physically evaluating the SI joints. Tests designed to apply pressure and induce movement in the joint, such as the pelvic rock test, Gaenslen's sign, and the Patrick/Faber test, are utilized to detect signs of disease within the SI joint (Kapandji 2007).

The hip joint is a ball-and-socket joint characterized by its spherical articulating surfaces. The femoral head, which is approximately two-thirds of a sphere, is supported by the neck of the femur, which connects to the shaft. The axis of the femoral neck is uniquely oriented, extending in a superior, medial, and anterior direction. In adults, the angle formed between this axis and the femoral shaft is approximately 125°. Variations in this angle can lead to different hip conditions. An increased angle may result in congenital hip dislocation (coxa valga), while a decreased angle may predispose individuals to femoral neck fractures or slippage of the femoral epiphysis (coxa vara). The femoral neck's axis also forms an angle with the frontal plane, known as the angle of anteversion, which typically ranges from 10° to 30° and points in a medial and anterior direction. During an upright posture, the coronal plane, which passes through the center of the femoral head and the axis of the femoral condyles, is positioned almost entirely anterior to the femoral shaft and the axis of rotation of the knee. The acetabulum, or socket, articulates with the femoral head and is oriented in a lateral, inferior, and anterior direction. The downward inclination of the acetabulum creates a prominent upper segment or roof, which bears the maximum compression forces against the femoral head. As a result, the articular cartilage of the acetabular roof and the superior portion of the femoral head is thickest. In cases of congenital hip dislocation, the upper lip of the acetabulum may not develop adequately, allowing the femoral head to dislocate posteriorly on the ilium (Kapandji 2007).

Hip fractures represent a significant health concern, leading to high rates of morbidity and mortality among the aged population. The majority of these fractures, exceeding 90%, occur in individuals older than 70 years. Depending on their location, fractures of the proximal femur can be categorized as occurring immediately below the head (subcapital), within the neck, between the trochanters (intertrochanteric), or beneath the trochanters (subtrochanteric). The specific site of the fracture is a critical factor in determining if the fracture will disrupt the blood flow to the femoral head.

The vascular supply to the hip joint is consistent with that of most synovial joints. In pediatric patients, the femoral shaft's nutrient artery ends at the metaphysis, meaning these arteries do not extend to the epiphysis, rendering the hip more susceptible to conditions like embolisms, infarctions, and osteomyelitis, which can lead to severe complications. However, in adults, the metaphysis becomes ossified, creating a network of rich anastomoses between the epiphysis and metaphysis, making these regions less prone to such pathologies.

At the hip specifically, the femoral head and the intracapsular portion of the neck receive their blood supply from two primary sources: a minor branch from the obturator artery that accompanies the ligament of the head along the neck, and retinacular arteries arising from the trochanteric anastomosis, ascending the neck. In younger adults, the connectivity between these vascular sources is strong, making hip fractures less likely to cause avascular necrosis of the femoral head. However, as individuals age, these anastomoses weaken, predominantly supplying the femoral head through smaller arteries from the trochanteric anastomosis that ascend the femoral neck. This shift results in the obturator artery's contribution diminishing. Consequently, in older adults, fractures at the femoral neck can sever the retinacular arteries, leading to a disruption in blood supply to the head. In contrast, fractures located externally to the joint capsule, including some intertrochanteric and all subtrochanteric fractures, do not typically affect these critical vessels.

The hip joint capsule is characterized by its looseness and exceptional strength, which is further reinforced by three distinct ligaments, each originating from a different component of the innominate bone: the iliofemoral, pubofemoral, and ischiofemoral ligaments. These ligaments exhibit a coiled or spiral structure, a result of the lower limb's rotation during the developmental process. The arrangement of the ligamentous fibers follows a clockwise pattern from the hip to the femur. Consequently, hip extension causes these ligaments to wind around the femoral neck, while flexion elicits the opposite effect. This phenomenon is particularly evident when an individual with complete paraplegia stands with their hips in extension, essentially "balancing" on the ligaments. During hip extension, all ligaments become taut as they wrap around the femoral neck, with the inferior band of the iliofemoral ligament experiencing the highest degree of tension (Kapandji 2008). It is important to note that the tautness is not limited to the iliofemoral ligament alone, as all ligaments contribute to the stability of the hip joint in this position.

The hip's primary flexion is accomplished through the action of two distinct muscles: the psoas major and the iliacus. These muscles are often collectively referred to as "iliopsoas," despite their considerable differences in structure, function, origin, innervation, and vascular supply. The iliacus, a broad and potent fan-shaped muscle, is situated below the durable iliacus fascia, originating from the iliac fossa. It receives its neural input from the femoral nerve, which travels toward the thigh's anterior compartment. Conversely, the psoas major—known in the culinary world as the filet mignon when referring to bovine anatomy—emerges from the lumbar vertebrae, extending from the abdominal cavity's posterior wall. Characterized by parallel fibers, the psoas major is designed for movement rather than strength. Its innervation is derived segmentally from the lumbar plexus, notably via the L2 and L3. Weakness of the psoas muscle, which can be caused by conditions such as a psoas abscess or hematoma, indicates a peripheral nerve lesion to a highly proximal location within the lumbar plexus or a radicular process.

The psoas major muscle has the ability to flex either the thigh or the spinal column during concentric contractions, depending on which section remains immobile. Understanding the neighboring abdominal structure is important. For example, when the appendix is positioned above the right psoas major, persons with appendicitis may experience difficulty in extending the right hip owing to the straining of the psoas, leading to pain. It is important to differentiate this symptom from the inclination to adopt a fetal posture to reduce peritoneal movement or stretching, which may be indicative of more serious conditions such as a perforated appendix.

Hip extension, which is largely facilitated by the gluteus maximus and hamstring muscles, performs an essential function in a range of everyday tasks and sporting endeavors. The gluteus maximus and hamstrings work to extend the hip joint while moving from a seated posture to standing, ascending stairs, or performing forceful jumps. Nevertheless, the relative effect of different muscle groups differs based on the level of intensity associated with the specific position. In activities characterized by low intensity, such as quiet walking or cycling on level terrain, the hamstrings undertake the predominant function of hip extension, whereas the gluteus maximus muscle exhibits comparatively lower levels of activity. In contrast, during more strenuous activities such as climbing stairs or hills, sprinting, or participating in competitive cycling, the gluteus maximus muscle, which is specifically developed to provide significant force, assumes the role of the primary hip extensor and external rotator. The functional distinction observed in the gluteus maximus and hamstrings demonstrates their distinct roles in various locomotor activities, showing their importance in both daily living locomotions and athletic endeavors.

The topic of the rotation of the hip remains a subject of controversy. The femoral shaft cannot rotate through its own axis, unlike the humerus. In contrast, the femoral neck undergoes backward and forward swinging actions during both internal and exterior rotation. It is indisputable that the force of lateral rotation is approximately three times greater than that of medial rotation. There is a considerable amount of consensus on the activities of the lateral rotators, which include the gluteus maximus, posterior fibers of the gluteus medius and minimus, and several short gluteal rotators. The medial rotators of the tensor fascia lata and gluteus medius and minimus muscles are the anterior fibers. The iliacus, psoas major, and pectineus muscles are subjects of ongoing controversy. The analysis of electromyographic data suggests that the pectineus muscle functions as a medial rotator, while the psoas and iliacus muscles exhibit less definitive evidence. The last analysis explores the potential role of the psoas and iliacus muscles as lateral rotators when the hip is flexed, and as medial rotators when the hip is in its anatomical position.

8.6.2 Thigh and Knee Anatomy

The thigh region is characterized by a substantial layer of subcutaneous fat, which houses the superficial veins, lymphatics, and lymph nodes, although these structures have limited clinical significance. Beneath the adipose tissue, the fascia lata, the deep fascia of the thigh, is notably thick and robust. Occasionally, the muscles beneath the fascia lata may protrude, causing a localized swelling that diminishes when the muscle contracts.

The lateral and medial intermuscular septa, attached to the deep surface of the fascia lata, divide it into extensor and flexor compartments. On the lateral aspect, the fascia lata thickens to form the iliotibial tract, which extends from the greater trochanter downward and serves as the insertion point for the tensor fascia lata and most of the gluteus maximus. The fibrous band of the iliotibial tract attaches to a small facet on the anterior surface of the tibial lateral condyle.

Overuse injuries, inflammation, sprains, and strains of the iliotibial tract are common among athletes, often leading to knee pain. Moreover, iliotibial tract contractures frequently occur in paralytic conditions, such as poliomyelitis and sports-related injuries, potentially causing secondary deformities, including leg length discrepancies and pelvic obliquities. The Ober test is employed to assess the presence of iliotibial tract contracture.

In the proximal thigh, the muscle groups are arranged according to their actions on the hip joint. Anteriorly, the flexor muscles are located, while the adductor muscles are situated medially. Posteriorly, the extensor muscles can be found, and the abductor muscles are positioned laterally. To better understand the spatial relationships between these muscle groups and the neurovascular structures, cross-sectional views are most informative. Due to the rotational changes that occur during development, the arrangement of the flexor and extensor muscles at the knee joint is opposite to that of the hip joint. In the mid and distal thigh, the extensor muscles are located anteriorly, while the flexor muscles are situated posteriorly.

The knee is structured as a diarthrodial joint that includes three compartments: the medial and lateral tibiofemoral articulations and the anterior patellofemoral articulation. It facilitates flexion, extension, and rotational movements along the longitudinal axis of the leg when it is flexed. By the 36th week of gestation, the knee's epiphyses are visible, and they generally ossify fully by the late teenage years, typically earlier in females. It is important to distinguish between the normal developmental variations of the patella, including different configurations and fragmentations, from actual fracture fragments during childhood.

The knee's stability is not significantly aided by its bony structures, especially in a flexed position; rather, it depends on robust connective tissues and muscular support. The tibial plateau features two condyles: the medial condyle is concave and extends longer in an anteroposterior direction than the lateral condyle. Positioned usually centrally, the tibial tubercle lies between these two condyles. Resting on the tibial plateau are the femoral condyles, with the medial condyle being noticeably larger and longer, featuring a wider circumference. These condyles are set at a 28° divergence angle, with a distinguishing notch found between the anterior one-third and posterior two-thirds of the lateral femoral condyle's circumference, which is visible on a true lateral radiograph.

The patellofemoral joint is composed of the femoral sulcus and the posterior surface of the patella. The sulcus exhibits a slight internal rotation relative to the femoral shaft, an anatomical detail significant for understanding knee mechanics (Tria and Klein 1992).

Flexion and extension movements of the knee occur in the sagittal plane around the transverse axis. Although these movements may appear simple, they involve a complex combination of rolling, shearing, and skidding motions of the femoral condyles on the tibial plateau. Additionally, there is a rotational component around the leg's long axis, which is crucial during both active and passive knee rotation. Active rotation is important in activities such as skiing and changing direction while running, while passive rotation plays a role in locking the knee in a fully extended position through the "screw-home" mechanism, which involves both ligamentous and bony factors.

As the knee extends fully, the anterior cruciate ligament becomes taut first. The quadriceps muscle then continues to extend the knee, causing the femur to medially rotate about the anterior cruciate ligament. The medial condyle, with its longer and more curved articular surface, glides around the anterior cruciate ligament, while the lateral condyle remains stationary. This action pulls the remaining knee ligaments taut, preventing further rotation and locking the knee. In this locked position, flexion cannot occur until the femur laterally rotates.

The popliteus muscle, a small muscle located at the floor of the popliteal fossa, originates from the tibia and inserts on the femur's lateral condyle and the posterior convexity of the lateral meniscus. This muscle plays a crucial role in unlocking the knee joint by laterally rotating the femur on the tibia, allowing flexion to occur. Additionally, the popliteus muscle retracts the lateral meniscus posteriorly, providing greater mobility and protection compared to the medial meniscus.

In many joints, including the knee, fibrocartilaginous discs called menisci divide the joint cavity into two separate compartments, allowing for different movements on each side and accommodating the joint's multiple degrees of freedom. The menisci play a crucial role in the knee's function by facilitating flexion and extension between the menisci and femoral condyles in the upper compartment, and rotation between the menisci and tibial condyles in the lower compartment.

In addition to their role in joint mechanics, the menisci contribute to joint lubrication and increase the contact area, enhancing joint stability. The importance of the menisci in maintaining joint stability is evident from the increased rotatory laxity observed after meniscectomy. Meniscal tears can result from either acute traumatic events or chronic degenerative processes. The medial meniscus is less mobile compared to the lateral meniscus due to its strong attachment to the medial collateral ligament and the absence of a muscle attachment similar to the popliteus muscle's attachment to the lateral meniscus. Meniscal injuries typically occur during weight-bearing with slight knee flexion, involving forced rotation and abduction of the tibia on the femur. In such cases, the medial meniscus experiences a sudden increase in both compressive and shear forces, leading to various types of tears. More forceful injuries can affect not only the medial meniscus but also the medial collateral ligament. In these cases, tenderness is present over one or both attachments of the ligament (to the femur and tibia) rather than the localized joint line tenderness associated with an isolated medial meniscus injury.

The stability of the patella is determined by the interaction of bony structures, ligaments, and muscular elements. During the final 30° of knee extension, there is a natural tendency for the patella to sublux laterally. This phenomenon is mostly attributed to the alignment of the lower limb and the produced force exerted by the quadriceps muscle. Three critical elements counteract this lateral pull: the prominent forward projection of the lateral femoral condyle, the medial patellar retinaculum, and the horizontally-oriented lower fibers of the vastus medialis, which attach directly to the medial side of the patella. Disruption in these anatomical relationships can lead to patellar tracking issues, which are typically manifested by symptoms such as pain, instability, swelling, catching, and a grinding sensation.

Particular anatomical variations, such as increased femoral neck anteversion, can heighten the risk of lateral subluxation of the patella. Additionally, a weakened vastus medialis can disrupt the balance of the quadriceps mechanism, further predisposing the patella to lateral subluxation. Anterior knee pain is often attributed to maltracking at the patellofemoral joint rather than complete subluxation or dislocation. Chronic damage to the hyaline cartilage within the patellofemoral joint, known as chondromalacia, is believed to result from improper patellar tracking. The pathogenesis of chondromalacia is surrounded by various theories concerning its causative and pathophysiologic mechanisms. Treatment approaches commonly recommend early intervention focusing on strengthening the vastus medialis, which is essential in preserving patellar stability and ensuring accurate tracking.

8.6.3 Lower Leg Anatomy

The leg is supported by two bones, the tibia and fibula, but only the tibia, which is the preaxial bone, contributes to the formation of the knee joint. This structural arrangement differs significantly from the upper extremity, where both the radius and ulna articulate with the elbow joint. Another notable distinction between the upper and lower limbs is the degree of movement between the two bones; while the forearm allows free supination and pronation, there is minimal movement between the tibia and fibula in the leg. The tibial shaft has a triangular shape, with its anterior border and anteromedial surfaces remaining subcutaneous along the entire length of the leg (Standring 2020).

In children, the proximal tibia is a common site for acute osteomyelitis, although involvement of the knee joint is uncommon (Peltola and Pääkkönen 2014). Atraumatic proximal leg pain in young children may also be caused by neoplasms such as Ewing's sarcoma, which can be challenging to differentiate from infection in the early stages. The tibia is the most frequently fractured long bone, especially in the exposed, subcutaneous, and slender distal third (Court-Brown and Caesar 2006).

Athletes often experience pain along the anterior border of the tibia, commonly referred to as "shin splints." This pain may be attributed to various conditions, including tibial stress fractures, overuse injuries of the anterior compartment or deep calf muscles (e.g., periostitis, tendinitis, tendinosis), or chronic compartment syndrome (Galbraith and Lavallee 2009). Pain characteristics can help differentiate between these conditions. Stress fracture pain is typically focal, occurs immediately upon impact, and is alleviated by avoiding weight-bearing. In contrast, pain from compartment syndrome develops gradually during activity and may take several hours to subside. Periostitis pain also develops over time during activity but may resolve within days (Wilder and Sethi 2004).

The leg is divided into three muscular compartments, each with specific actions at the ankle and foot: posterior, anterior, and lateral. To aid in memorizing the number of muscles in each compartment, the mnemonic "6-4-2" can be used, representing the calf, anterior, and peroneal compartments, respectively. The posterior compartment, or calf, contains six muscles: three superficial (gastrocnemius, soleus, and plantaris, collectively known as the triceps surae) and three deep (tibialis posterior, flexor hallucis longus, and flexor digitorum longus), separated by a robust deep transverse fascia (Drake et al. 2020). The anterior compartment consists of four muscles: tibialis anterior, extensor hallucis longus, extensor digitorum longus, and peroneus tertius (Moore et al. 2018). The lateral compartment contains the remaining two muscles, the peroneus brevis and longus. This classification excludes the popliteus muscle, which is better understood in relation to the knee joint (Standring 2020).

Overuse injuries of the leg are considered prevalent but not well understood. Approximately 25% of all athletes experience overuse injuries, many of which affect the leg tendons (Dugan and Bhat 2005). These injuries or syndromes have various names, such as repetitive motion disorder, repetitive stress syndrome, and activity-related musculoskeletal disorders, and have complex, multifactorial etiologies. Moreover, they can involve multiple tissues, including muscle, nerve, ligament, and tendon (Chéron et al. 2017). Although many terms suggest a well-defined pathophysiology, overuse injuries likely represent several distinct pathologic entities (e.g., tendinitis, tendinosis, and peritendinitis in tendons) associated with different causes, such as repetition, cumulative trauma, and sustained static loading (Khan et al. 1999). A practical definition of an overuse injury is one in which the summation of biomechanical loads exceeds the tolerance of the biological tissues (Barbe and Barr 2006).

8.6.4 Ankle and Foot Anatomy

The foot and ankle consist of a series of joints that enable the foot to adapt to ground irregularities and assume various positions in space. The ankle joint, located posteriorly, is a simple hinge joint primarily formed by the articulation between the talus and tibia, with its axis predominantly in the frontal plane. The articulating surfaces of the tibia's inferior aspect and the talus's superior aspect, along with the medial and lateral malleoli that enclose the talus, provide substantial bony stability to the ankle joint (Golanó et al. 2010). The lateral and medial malleoli differ in their characteristics, with the lateral malleolus being larger, more distally extended, and more posteriorly positioned than the medial malleolus (Standring 2020).

In addition to the bony structures, a robust capsule and ligaments further stabilize the ankle joint. The significance of these dense connective tissues in maintaining talocrural joint stability becomes evident in individuals who experience multiple ankle sprains and subsequently develop ankle instability (Hertel 2002). If the ligaments heal in an elongated position, abnormal shifting of the talus within the ankle mortise may occur, even during level ground walking (Hubbard and Hertel 2006). Medially, the strong deltoid ligament attaches the tibia to the calcaneus, navicular, and talus. High-force injuries may damage the medial ligament, most commonly avulsing the deltoid from the medial malleolus tip (Savage-Elliott et al. 2013). Laterally, the ankle is less robust and more susceptible to twisting inversion injuries, which are prevalent in sports such as basketball, rugby, and soccer. These injuries typically occur when the ankle and foot are in an inverted and plantarflexed position (Fong et al. 2007). The malleoli tightly grip the talar trochlea, allowing minimal lateral movement in the neutral and extended positions. However, the ankle is less stable in flexion when the narrower posterior trochlea lies within the wider anterior space between the malleoli. This mechanism strains the lateral ankle, injuring the ligaments from anterior to posterior (anterior talofibular, calcaneofibular, and posterior talofibular) (Golanó et al. 2010). Diagnostically, an increased anterior excursion of the ankle (anterior drawer) indicates a disruption of the anterior talofibular ligament (van Dijk et al. 1996).

Fracture-dislocations of the ankle joint are common, especially with twisting injuries involving forced external rotation and eversion of the ankle. The talus displaces the distal fibula laterally, leading to a fracture of the lateral malleolus (Pott's fracture). If the force is sufficiently strong, the deltoid ligament may be avulsed from the medial malleolus (Goost et al. 2014). Due to the poor blood supply of the distal fibula, this fracture may result in necrosis of the lateral malleolus (Bauer et al. 1985).

8.6.5 Foot Sole Anatomy

The foot sole and hand palm exhibit remarkable anatomical similarities, despite the foot's adaptations for locomotion and weight-bearing that result in noticeable morphological and proportional differences (Long et al. 2017). A comparative examination of these structures elucidates their similar features. In the foot, the skin and subcutaneous tissue exhibit exceptional thickness, toughness, and adherence to the underlying fascia. The substantial shock-absorbing pad overlying the calcaneus becomes evident during the dissection of the heel region. Deep to the heel pad lies the origin of the plantar aponeurosis, a specialized deep fascial layer analogous to the palmar aponeurosis of the hand. The plantar aponeurosis originates posteriorly from the calcaneal medial and lateral tubercles, which frequently develop osteophytes in conditions such as plantar fascitis, ankylosing spondylitis, Reiter's syndrome, and diffuse idiopathic skeletal hyperostosis (DISH, Forestier's disease) (Smith et al. 2005).

Plantar fasciitis, a prevalent condition among both athletic and sedentary individuals, is attributed to chronic strain on the plantar fascia. Clinical manifestations include pain and tenderness along the foot's plantar aspect, either at the calcaneal attachment of the plantar fascia or more distally from any osseous junction (Lee et al. 2015). The presence of a calcaneal spur may or may not accompany plantar fasciitis. As with many chronic overuse injuries involving dense connective tissues such as ligaments, tendons, and capsules, the precise pathophysiology of plantar fasciitis remains indefinable. Although often considered an inflammatory condition, plantar fasciitis likely encompasses a spectrum of traumatic, inflammatory, and degenerative disorders influenced by various intrinsic and extrinsic factors, including overuse, abnormal ankle and foot range of motion, muscular imbalances, genetic predisposition, footwear, and activity surfaces (Garcia et al. 2018).

The foot sole and hand palm exhibit remarkable muscular layer similarities (Table 8.1).

Table 8.1 The muscular layers of the foot

Layer	Muscles
Layer 1: short muscles	Abductor hallucis Flexor digitorum brevis Abductor digiti quinti
Layer 2: long tendons	Flexor hallucis longus Flexor digitorum longus Lumbricals are attached as in the hand Quadratus plantae
Layer 3: adductor layer	Abductor hallucis Two short flexors (flexor hallucis brevis and flexor digiti quinti)
Layer 4: interossei layer	Interossei muscles Two long tendons (peroneus longus and tibialis posterior)

Four distinct muscular layers can be identified: (1) short muscles, (2) long flexor muscle tendons originating in the calf, (3) adductor layer, and (4) interossei layer. Maintaining the balance between extrinsic and intrinsic muscles is crucial for preventing deformities, as evidenced by the intrinsic minus foot, which mirrors the intrinsic minus hand in form (Schieber and Santello 2004). When leg extrinsic muscle spasticity overpowers the foot's small, intrinsic muscles, clawed toes and an elevated medial longitudinal arch may develop (Smania et al. 2010).

The neurovascular plane in the foot sole, similar to that of the palm, is situated between the first and second layers, containing the tibial nerves and vessels within the neurovascular bundle. The tibial nerve bifurcates into medial and lateral plantar nerves within the tarsal tunnel, entering the foot sole deep to the abductor hallucis muscle (Netter 2019). The medial plantar nerve, analogous to the median nerve in the hand, innervates a larger cutaneous area but fewer muscles, including the abductor hallucis, flexor digitorum brevis, flexor hallucis brevis, and first lumbrical. Conversely, the lateral plantar nerve, equivalent to the ulnar nerve, supplies the majority of the foot's intrinsic muscles but a smaller cutaneous area (Singh 2017). Both plantar nerves terminate as digital nerves, which may be affected by Morton's neuroma, a painful neurofibroma most frequently occurring between the third and fourth metatarsal heads (Bencardino et al. 2000).

8.6.6 Foot Arches

The foot comprises three distinct arches—the medial longitudinal, lateral longitudinal, and transverse arches—which function as levers and bear weight during locomotor activities. These arches are architecturally analogous to engineered structures like stone bridges, relying on four key elements for their structural integrity: the morphology of the bones, especially the central keystone bone, ligaments that connect the inferior aspects of the bones, ligamentous and tendinous structures that span between the arch's endpoints, and supportive structures that suspend the arch's midpoint (Snell 1973).

Each foot arch exemplifies these architectural features. For instance, the medial longitudinal arch is supported by bony structures like the sustentaculum tali, which supports the talus, and the rounded talar head serving as the arch's keystone. The arch's inferior surface is stabilized by short plantar ligaments, preventing collapse under weight-bearing conditions. Static structures like the plantar aponeurosis and dynamic components such as the flexor hallucis longus and flexor digitorum longus tendons act as tie beams. The tibialis anterior and posterior muscles are prime examples of musculotendinous units that suspend the arch from above (Kapandji 2007).

The significance of these dynamic factors cannot be overemphasized, as arch collapse occurs in paralytic conditions. Arches may also fail due to overuse, extreme fatigue, prolonged standing, and chronic obesity. Pes planus, or flat foot, results from the collapse of the medial longitudinal arch, often accompanied by lateral and everted displacement of the forefoot (pronation). Conversely, pes cavus, or clawfoot, is characterized by an abnormally high medial longitudinal arch, typically resulting from muscular imbalances associated with flaccid or spastic paralysis (Sarrafian 1983).

8.6.7 Adaptive and Pathological Responses of Leg Tendons to Biomedical Loads

The leg tendons provide numerous examples of both adaptive and pathologic alterations. When tendons in the leg change direction, pass through narrow tunnels, traverse a pulley, or are compressed by adjacent structures, their anatomy and composition transform from regularly arranged dense connective tissue to fibrocartilage. This transformation has been observed in tendon regions subjected to compressive and frictional forces rather than pure longitudinal forces, such as the Achilles tendon passing over the calcaneus, the peroneus longus changing direction into the foot's sole, and the flexor hallucis longus hooking around the sustentaculum tali (Benjamin and Ralphs 1998). In essence, when a tendon experiences significant compression and friction, it adapts from a primarily fibrous tissue to a more fibrocartilaginous one (Vogel and Koob 1989).

Although initially adaptive, fibrocartilage has reduced vascularity and a diminished capacity to respond to repetitive microinjury compared to a typical tendon (Pufe et al. 2005). Over time, pathologic changes develop, leading to chronic overuse injuries. Numerous clinical examples of this problem exist in both the lower and upper extremities, such as Achilles tendinitis in runners and dancers and supraspinatus tears in pitchers (Maffulli et al. 2003; Seitz et al. 2011). Tendon injury within a relatively hypovascular fibrocartilaginous zone is likely one of the most significant concepts of chronic overuse injuries (Almekinders and Temple 1998).

The relationship between adaptive and pathologic tissue responses is highly dynamic, and many changes are potentially reversible (Sharma and Maffulli 2005). Consequently, treatment hypotheses can be developed to stimulate healing and repair by improving blood supply in the affected area and using active motion instead of immobilization. Further injury can be prevented by optimizing tissue and joint use, such as correcting muscle imbalances, reestablishing range of motion, and optimizing equipment, footwear, and environmental factors.

8.7 Nerve Supply to the Lower Extremity

Understanding the complex innervation patterns of the lower extremities is important for establishing effective rehabilitation strategies following spinal cord injuries. This exploration focuses on the lumbar and lumbosacral plexuses, demonstrating their important functions in motor control and sensory innervation. Notably, the discussion is mostly based on the segmental innervation, which includes dermatomes and myotomes, and is supported by key publications (Ellis 1971; Hollinshead 1969; Hollinshead and Rosse 1985; Last 1984; Snell 1973; Sunderland 1978; Warwick and Williams 1973).

Each nerve root associated with the lower extremities provides innervation to multiple muscles, establishing a network where muscles receive innervation from multiple nerve segments. This arrangement reveals the complexity of motor neuron organization within the spinal cord, which is divided into distinct movement centers. These centers demonstrate that each movement is controlled by motor neurons that extend spanning two adjacent levels. For instance, hip flexion is controlled by L2 and L3 motoneurons, demonstrating the precise coordination required for smooth movement. Furthermore, the close location of flexion and extension centers in the spinal cord, such as knee extension at L3, L4 and knee flexion at L5, S1, reveals the complex organization that enables limb movement. The hierarchical innervation pattern, wherein proximal joints receive innervation from more cranial motor centers compared to distal joints. For example, the hip is innervated starting at L2, followed by the knee at L3, and the ankle at L4.

The formation of the lumbar and lumbosacral plexuses represents an important development in vertebrate anatomy, allowing for complex neuromusculoskeletal behaviors. These plexuses, through the interchange of spinal nerve fibers, ensure that the main nerve branches receive input from multiple segmental spinal nerves. This structural and functional evolution not only enhances motor and sensory functions but also improves the limbs against injuries, demonstrating adaptation of vertebrate neuromuscular organization (Goldstein 1994).

The lower extremity receives its nerve supply from the ventral rami of spinal nerves originating from the L1 to the S4 spinal cord segments. These nerve roots exhibit varying courses within the vertebral canal before exiting their respective intervertebral foramina. The L1 nerve root originates approximately two vertebral levels superior to its exit point, while the S4 nerve root emerges from the spinal cord terminus, located between the L1 and L2 vertebrae (Drake et al. 2020; Standring 2020).

Upon exiting the intervertebral foramina, each nerve receives a gray ramus communicans, which connects it to the sympathetic trunk. Additionally, the L1 and L2 spinal nerve roots are linked to the sympathetic trunk by a white ramus communicans (Netter 2019).

The lumbar plexus is formed within the psoas major muscle by the ventral rami of the first three lumbar nerves and the superior portion of the L4 spinal nerve (Moore et al. 2018). The lower portion of the L4 spinal nerve and the ventral ramus of the L1 spinal nerve combine to form the lumbosacral trunk, which travels over the ala of the sacrum to join the ventral rami of the S1, S2, S3, and upper part of the S4 spinal nerves, creating the lumbosacral plexus. Finally, the S4 and S5 spinal nerves contribute to the formation of the sacral plexus (Wineski 2019).

The lumbar and lumbosacral plexuses have notable anatomical and functional uniformity among vertebrate species, ranging from fish to humans. It also demonstrates the important significance of these neural structures in limb function and movement.

Key characteristics of these plexuses include:

- Formation through convergence of multiple spinal nerves: The lumbar and lumbosacral plexuses are formed by the intricate intertwining, division, and recombination of spinal nerves originating from specific regions of spinal cord enlargements. This complex network formation is consistent across vertebrates, highlighting its evolutionary significance (Standring 2020).
- Functional segregation into anterior and posterior divisions: These plexuses consistently segregate into two primary divisions based on functionality and target muscle groups. The anterior divisions innervate flexor muscles and their corresponding surface areas, facilitating limb flexion. Conversely, the posterior divisions supply the extensor muscles and overlying body wall, enabling limb extension (Moore et al. 2018).
- Integration of sensory, motor, and sympathetic fibers: Each plexus serves as a comprehensive network, integrating sensory, motor, and sympathetic fibers. This integration enables extensive innervation of the extremities, supporting a wide range of sensory and motor functions. The complexity of these plexuses varies among species, reflecting evolutionary adaptations to different locomotor strategies.
- Compartmentalization and developmental origins: The lumbar and lumbosacral plexuses are divided into flexor and extensor compartments, demarcated by the preaxial and postaxial borders, identified by the great saphenous and small saphenous veins, respectively. The flexor compartment typically has a more extensive nerve supply, including a prominent caudal nerve primarily derived from the lumbosacral plexus (L1 to S4). Developmentally, the plexus begins as a single structure in the fetus and later divides due to pelvic girdle growth, resulting in the adult distinction between lumbar and lumbosacral plexuses (Drake et al. 2020).

8.7.1 The Lumbar Plexus

The lumbar plexus is formed by the ventral rami of the first three lumbar nerves and part of the L4 spinal nerve (see Fig. 2.8). Occasionally, the subcostal nerve (T12) may contribute to this plexus. The formation of the plexus occurs within the psoas major muscle, with the rami entering the muscle between its attachments to the vertebral body and transverse process (Drake et al. 2020; Standring 2020).

Typically, the L1 spinal nerve, sometimes joined by T12, divides into upper and lower branches. The upper branch further divides to form the iliohypogastric and ilioinguinal nerves. The lower part of the L1 spinal nerve combines with the upper portion of the L2 spinal nerve to form the genitofemoral nerve (Moore et al. 2018).

The lower part of the L2 spinal nerve, the L3 spinal nerve, and the upper portion of the L4 spinal nerve each divide into smaller anterior and larger posterior divisions. The anterior divisions of the nerves converge to form the obturator nerve, while branches from the third and fourth divisions may occasionally unite to form an accessory obturator nerve (Netter 2019). The posterior divisions from all three nerves join to form the femoral nerve, while branches from only the L2 and L3 spinal nerves form the lateral femoral cutaneous nerve. The psoas major (L1, 2, 3) and quadratus lumborum (L1–4) are innervated by separate direct muscle branches originating from the ventral rami (Wineski 2019).

The lumbar plexus emerges as a crucial neural network within the vertebrate anatomy, specifically located in the psoas major muscle on the posterior abdominal wall. It is intricately formed by five lumbar nerves, each exiting from intervertebral foramina and contributing to the plexus's composition. The branches of this plexus, which navigate anteriorly through the muscle, as well as laterally and medially, render it susceptible to various clinical conditions, including surgical interventions like abdominal hysterectomy, diseases such as psoas abscess, and pathological conditions including abdominal aortic aneurysm.

The lumbar plexus is particularly susceptible to adverse effects of anticoagulant therapy, such as retroperitoneal hemorrhage, which may compromise one or more of its nerves depending on the hematoma's placement. This plexus primarily provides nerve supply to the lower extremity and also extends motor innervation to the psoas (notably L2, L3) and quadratus lumborum muscles, as well as sensory innervation to the posterior parietal peritoneum. Among its five branches, three are notably significant for the applied anatomy of the lower extremity, emphasizing the importance of understanding this plexus in medical practice.

8.7.1.1 Branches from the Anterior Division of the Lumbar Plexus

Iliohypogastric Nerve

The iliohypogastric nerve arises from the L1 spinal nerve, with occasional contributions from T12. This nerve emerges from the superior lateral border of the psoas major muscle, traversing anterior to the quadratus lumborum and posterior to the kidney. As it approaches the midaxillary line, the nerve pierces the deep surface of the transversus abdominis muscle, giving rise to a lateral cutaneous branch. It then progresses anteriorly within the neurovascular plane situated between the transversus abdominis and internal oblique muscles, typically piercing the internal oblique approximately 2 cm medial to the anterior superior iliac spine (Moore et al. 2018; Netter 2019).

The iliohypogastric nerve then travels medially, deep to the external oblique aponeurosis, and becomes cutaneous at a point approximately 4 cm superior to the superficial inguinal ring. It innervates the muscles of the lateral abdominal wall and provides sensory innervation to a strip of skin extending from the upper lateral gluteal region to just above the pubis (Wineski 2019).

Ilioinguinal Nerve

The ilioinguinal nerve originates from the L1 spinal nerve, occasionally receiving fibers from T12. It emerges from the lateral border of the psoas major muscle, just below the iliohypogastric nerve, and travels obliquely downward within the abdomen, lying deep to the quadratus lumborum muscle. The nerve then pierces the internal oblique muscle, continues deep to the external oblique aponeurosis, and enters the inguinal canal. It ultimately reaches the skin through the superficial inguinal ring and external spermatic fascia. As the ilioinguinal nerve passes between the internal and external oblique muscles, it provides motor innervation to these abdominal wall structures. The sensory innervation from the ilioinguinal nerve covers the skin over the symphysis pubis and the superior medial aspect of the femoral triangle. Additionally, it supplies sensory fibers to specific areas of the external genitalia (Tagliafico et al. 2015).

Genitofemoral Nerve

The genitofemoral nerve is formed within the substance of the psoas major muscle from the L1 and L2 lumbar spinal nerves. It emerges anteriorly from the psoas muscle near its medial border, penetrating the fascia. The nerve's course then follows a caudal and anterior trajectory along the fascia, passing posterior to the ureter as it approaches the inguinal ligament.

At the level of the inguinal ligament, the genitofemoral nerve bifurcates into its terminal branches: the genital and femoral components. The genital branch enters the inguinal canal through the deep inguinal ring. Within the canal, it provides motor innervation to the cremaster muscle and sensory fibers to the cutaneous regions of the scrotum (or labia majora in females) and the adjacent thigh area (Rab et al. 2001). The femoral branch passes under the inguinal ligament, lateral to the femoral artery. It becomes superficial by traversing the saphenous opening and supplies sensory innervation to the skin over the upper part of the femoral triangle. Furthermore, the genitofemoral nerve's proximity to the psoas major muscle and its course through the inguinal region make it susceptible to entrapment or compression.

Obturator Nerve

The obturator nerve is formed by the anterior divisions of the L2, L3, and L4 spinal nerves, which unite within psoas major. It emerges from the medial border of the psoas major on the lateral sacral aspect. The nerve traverses the sacroiliac joint and obturator internus before entering the obturator canal, located below the superior ramus of the pubis and above the obturator membrane. After exiting the canal, the nerve lies superior to the obturator externus muscle and bifurcates into anterior and posterior branches (Standring 2020).

The anterior branch descends into the thigh, coursing anterior to the obturator externus and adductor brevis, while lying posterior to pectineus and adductor longus. Its terminal fibers are located between the adductor magnus and the medial intermuscular septum. This branch innervates the adductor longus, gracilis, and usually adductor brevis, with occasional innervation to pectineus. It also provides sensory innervation to the skin on the medial aspect of the thigh, with the distal third receiving input via the subsartorial plexus. Additionally, an articular branch reaches the hip joint through the acetabular notch (Moore et al. 2018).

The posterior branch pierces the obturator externus and descends between the adductor brevis anteriorly and adductor magnus posteriorly. It passes obliquely through adductor magnus and enters the popliteal fossa, terminating by piercing the oblique popliteal ligament to innervate the posterior part of the knee joint, including the cruciate ligaments. This branch also innervates the obturator externus and adductor magnus.

8.7.1.2 Branches from the Posterior Divisions of the Lumbar Plexus

Lateral Femoral Cutaneous Nerve

The lateral femoral cutaneous nerve originates from the posterior divisions of the L2 and L3 spinal nerves. It emerges from the lateral border of the psoas major muscle, then travels downward and forward onto the pelvic surface of the iliacus muscle (Grothaus et al. 2005). As it exits the pelvic cavity, the nerve passes medial to the anterior superior iliac spine, either passing below or through the inguinal ligament. It continues laterally, passing through or deep to the sartorius muscle and then penetrates the fascia lata to reach superficial tissues (Majkrzak et al. 2010).

Upon becoming superficial, the lateral femoral cutaneous nerve bifurcates into two primary branches. These branches extend distally along the lateral aspect of the thigh. The anterior branch provides sensory innervation to the anterolateral thigh surface, extending as far as the knee joint, while the posterior branch supplies sensory fibers to the lateral proximal two-thirds of the thigh below the greater trochanter (Aszmann et al. 1997). Compression or entrapment of this nerve can result in meralgia paresthetica, a condition characterized by sensory disturbances in the lateral thigh (Cheatham et al. 2013).

Femoral Nerve

The femoral nerve originates from the posterior divisions of the L2, L3, and L4 spinal nerves. It emerges from the lateral border of the psoas major muscle and runs in the groove between psoas and iliacus, situated deep to the iliac fascia. The nerve enters the thigh beneath the inguinal ligament, lateral to the femoral sheath, and into the femoral triangle (Standring 2020).

Here, it quickly divides into several branches that can be categorized into anterior and posterior divisions, based on whether they pass anterior or posterior to the lateral circumflex femoral artery. These divisions are anatomically defined by their relationship to the lateral circumflex femoral artery, with the anterior division passing anterior to the artery and the posterior division coursing posterior to it (Drake et al. 2020).

The anterior division of the femoral nerve provides motor innervation to sartorius and gives rise to the medial and lateral branches of the anterior cutaneous nerve of the thigh. These branches are responsible for sensory innervation of the anterior thigh region (Netter 2019). The posterior division innervates the quadriceps femoris muscle group. Additionally, it sends articular branches to both the hip and knee joints, contributing to proprioception and nociception in these articulations. It gives off the saphenous nerve, which supplies sensory innervation to the medial side of the leg and foot. The femoral nerve also provides motor innervation to the iliacus muscle.

Differentiating between femoral neuropathy, lumbar plexopathy, and L4 radiculopathy, particularly in cases of quadriceps weakness, presents diagnostic challenges. Femoral neuropathy is distinguished by its unique effects, affecting neither the lateral femoral cutaneous nerve of the thigh nor the obturator nerve, which are typically involved in lumbar plexopathies. Conversely, L4 radiculopathy affects the ankle extensors and the muscles innervated by the superior gluteal nerve, sparing the hip flexors and anterior thigh skin, which remain unaffected, unlike in femoral neuropathy.

8.7.2 The Lumbosacral Plexus

The lumbosacral plexus is situated on the posterior wall of the pelvis between piriformis and its fascia. It is composed of the ventral rami from the L4 to S4 spinal nerves (Fig. 8.19). The L5 spinal nerve and lower part of the L4 spinal nerve

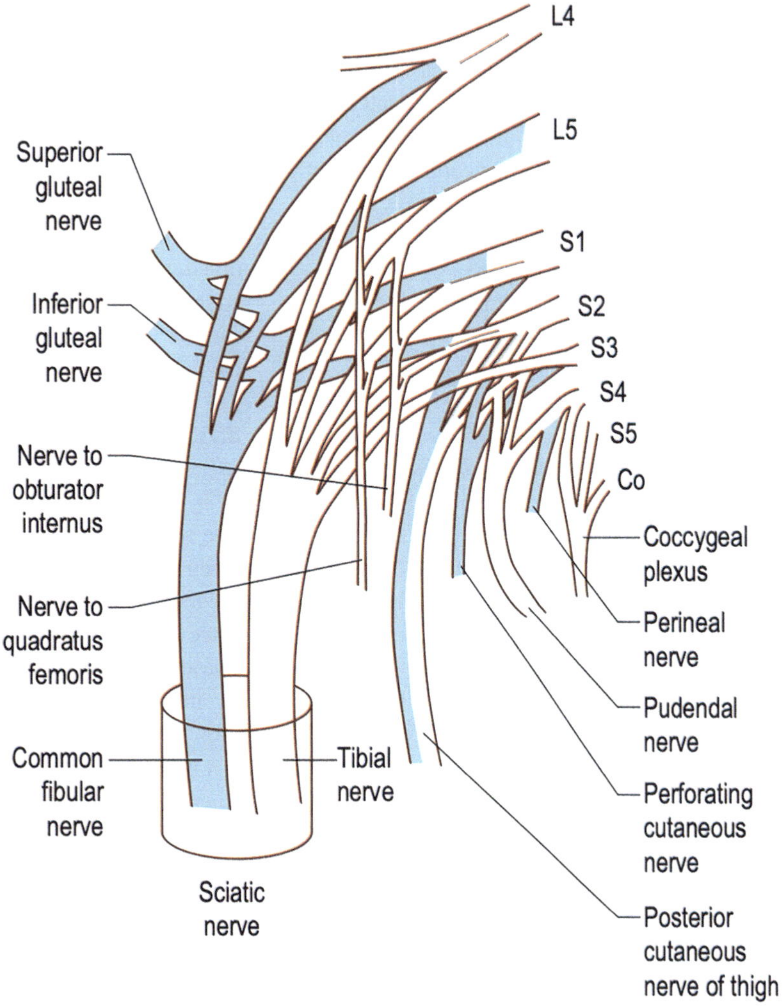

Fig. 8.19 Lumbosacral plexus: Illustration showing the structure and organization of the lumbosacral plexus, including its major branches and connections

combine to create the lumbosacral trunk, which passes over the sacral ala to join the laterally coursing ventral rami of the S1 to S4 spinal nerves (Netter 2019; Standring 2020).

Each ventral ramus receives a gray ramus communicans. Additionally, preganglionic parasympathetic fibers, known as pelvic splanchnic nerves, originate from the S2, S3, and S4 ventral rami. These fibers connect with the pelvic autonomic plexuses, innervating urogenital organs and the distal third of the gastrointestinal tract (Wineski 2019).

The lumbosacral plexus, like the brachial plexus, produces branches both from its roots and divisions, displaying a symmetry across its 12 nerves. Specifically, six nerves originate directly from the roots, while the remaining six emerge post-division. Notably, the division produces an equal split of nerves, with three originating from each of the anterior and posterior divisions. Each ventral ramus divides into anterior and posterior divisions, which converge at the greater sciatic foramen. The following nerves arise from the union of various anterior divisions (Drake et al. 2020):

- Nerve to quadratus femoris
- Nerve to obturator internus
- Pelvic splanchnic nerve
- Posterior femoral cutaneous nerve
- Pudendal nerve

Similarly, the posterior divisions give rise to:

- Branches to piriformis, coccygeus, and levator ani muscles
- Superior and inferior gluteal nerves
- Posterior femoral cutaneous nerve
- Perforating cutaneous nerve
- Perineal branch of S4

The sciatic nerve is the largest nerve formed by this plexus. It consists of the medially placed tibial nerve (anterior divisions L4 to S3) and the laterally placed common peroneal nerve (posterior division L4 to S2), bound together in a common sheath.

8.7.2.1 Nerve Branches from the Roots

The pudendal nerve and pelvic parasympathetic nerves play significant roles in pelvic and perineal innervation. However, when considering the lower limb, only two cutaneous nerves exert notable influence (Standring 2020). The perforating cutaneous nerve, while generally of limited clinical significance, innervates a small cutaneous area on the buttock's lower medial aspect. In contrast, the posterior cutaneous nerve of the thigh has a more extensive distribution. It innervates the buttocks' inferior portion via its inferior clunial branch, the posterior scrotum, and extends along the posterior thigh to the popliteal fossa (Moore et al. 2018). Interestingly, pain originating from prostatic or other cloacal-derived structures is thought to be transmitted through this nerve. It emphasizes the significance of distinguishing such pain from sciatica for accurate diagnosis and management in clinical practice (Drake et al. 2020).

Post-division nerve distribution reveals a complex anatomical arrangement. Six primary nerves emerge–three from the anterior division and three from the posterior division–each contributing to the formation of the sciatic nerve. The anterior division nerves primarily innervate the flexor and adductor muscle groups, while the posterior division nerves are responsible for the extensor muscles and their associated abductors (Netter 2019).

8.7.2.2 Branches from the Anterior Division of the Lumbosacral Plexus

The anterior divisions of the lumbosacral plexus give rise to several nerves, among which two are particularly noteworthy for their academic importance and utility in electromyographic studies. These include the nerves innervating the obturator internus and quadratus femoris, which also supply the gemelli muscles (Standring 2020).

Of significant clinical relevance is the tibial component, a major constituent of the sciatic nerve. This nerve plays a primary role in lower limb flexion, providing extensive innervation to the thigh's flexor musculature, with the exception of the short head of the biceps femoris. Its influence extends distally to innervate the flexor muscles of the leg and foot (Moore et al. 2018).

The hamstring muscle group, despite its biarticular nature allowing both knee flexion and hip extension, receives its innervation from the tibial nerve. This dual action of the hamstrings is a key consideration in gait analysis and rehabilitation strategies for spinal cord injury patients (Netter 2019).

Sciatic Nerve

The sciatic nerve is formed by the ventral rami of L4, L5, S1, S2, and S3. It exits the pelvis through the greater sciatic foramen, inferior to piriformis, entering the gluteal region. The nerve then descends along the posterior aspect of the thigh, deep to biceps femoris (Moore et al. 2018; Standring 2020). In its proximal course, the sciatic nerve overlies several structures, including superior gemellus, obturator internus, inferior gemellus, quadratus femoris, and adductor magnus. Typically, the nerve bifurcates into its terminal branches—the common peroneal and tibial nerves—approximately two-thirds down the thigh. However, this division can occur more proximally, or the two components may emerge separately from the pelvis. In such cases, the tibial nerve exits below piriformis, while the common peroneal nerve may exit above the piriformis or piece through the muscle (Netter 2019).

Surface anatomy of the sciatic nerve can be approximated by a line drawn from the midpoint between the ischial tuberosity and greater trochanter of the femur to a point two-thirds down the posterior thigh, where the medial and lateral hamstrings diverge.

The sciatic nerve provides articular branches to the knee joint and muscular innervation to the semitendinosus, semimembranosus, long head of the biceps femoris, and the hamstring portion of adductor magnus. The short head of biceps femoris receives innervation from the common peroneal component, while the tibial component supplies the other mentioned muscles.

Tibial Nerve

The tibial nerve, the medial terminal branch of the sciatic nerve, originates from the anterior divisions of the ventral rami of L4, L5, S1, S2, and S3. It continues the course of the sciatic nerve through the popliteal fossa, initially positioned lateral to the popliteal vessels before crossing superficially to the medial aspect to enter the leg under the tendinous arch of the soleus (Moore et al. 2013). The nerve descends obliquely inferomedially between the flexor digitorum longus and flexor hallucis longus muscles, passing behind the medial malleolus deep to the flexor retinaculum, situated between the tendons of these muscles. Upon entering the plantar aspect of the foot, the tibial nerve divides into its terminal branches, the medial and lateral plantar nerves (Moore et al. 2013; Standring 2020).

Within the popliteal fossa, the tibial nerve provides muscular branches to both heads of gastrocnemius, soleus, plantaris, popliteus, and tibialis posterior. Gastrocnemius receives innervation from its deep part, while soleus and plantaris are supplied from their superficial aspects. The nerve to popliteus has a unique course, descending over the muscle's superficial surface before curving around its inferior border to enter its deep surface (Standring 2020).

Articular branches from the tibial nerve supply the ankle, knee, and superior tibiofibular joints. The sural nerve, a cutaneous branch, descends between the two heads of gastrocnemius, piercing the deep fascia in the mid-calf region. It is joined by the peroneal communicating nerve and courses behind the lateral malleolus, extending along the lateral foot aspect. The sural nerve provides sensory innervation to the skin over the posterior and lateral lower third of the leg, the lateral foot border, and the fifth toe, excluding the distal phalanx (Moore et al. 2013; Netter 2019; Standring 2020).

In the lower leg, the tibial nerve supplies muscular branches to the deep part of soleus, flexor digitorum longus, flexor hallucis longus, and tibialis posterior muscles. Additional articular branches innervate the ankle joint, and a cutaneous branch supplies the heel and posterior foot region.

Pudendal Nerve

The pudendal nerve is the principal nerve of the perineal region, typically originating from the anterior division of the S2, S3, and S4 spinal nerves. This nerve exits the pelvic cavity via the lower medial segment of the greater sciatic foramen, accompanied by the pudendal vasculature and the nerve supplying the obturator internus muscle. Subsequently, it traverses the sacrospinous ligament, entering the lesser sciatic foramen to access the pudendal canal, which is situated on the deep aspect of the obturator fascia. The pudendal nerve provides motor innervation to the levator ani muscle and sensory innervation to the skin around the anus. Additionally, it supplies the genitalia and perianal cutaneous region (Shafik et al. 1995).

8.7.2.3 Branches from the Posterior Division of the Lumbosacral Plexus

The posterior division of the lumbosacral plexus gives rise to several key nerve structures, including the motor gluteal nerves, which are responsible for gluteal musculature innervation (Standring 2020). An anatomical mnemonic aids in recalling the innervation pattern: the superior positioning of the gluteus medius on the iliac crest, relative to the gluteus maximus, corresponds to its innervation by the superior gluteal nerve, which also supplies the gluteus minimus and tensor fasciae latae muscles (Moore et al. 2018).

The superior gluteal nerve primarily originates from the L4 segment, a notably higher level than the L5 segment from which the inferior gluteal nerve arises. This anatomical arrangement has implications for differential diagnosis in cases of nerve injury or spinal cord lesions.

Gluteal nerve injuries can occur through various mechanisms. Internally, damage may occur within the pelvic region, while external injuries can happen after the nerves traverse the greater sciatic notch into the gluteal region. Intrapelvic masses or traumatic events can affect the gluteal nerves concurrently with other lumbosacral plexus branches, such as the sciatic nerve. Externally, gluteal nerve injuries are predominantly the result of penetrating trauma, including iatrogenic causes during medical procedures or violent incidents like stabbings or gunshots (Drake et al. 2020).

A key clinical sign of superior gluteal nerve impairment is weakness in the gluteus medius and minimus muscles. This condition manifests as a positive Trendelenburg's sign during weight-bearing on the affected side (Fig. 8.20). This sign is characterized by a drop in the pelvis on the contralateral side during single-leg stance, indicating weakness in hip abduction.

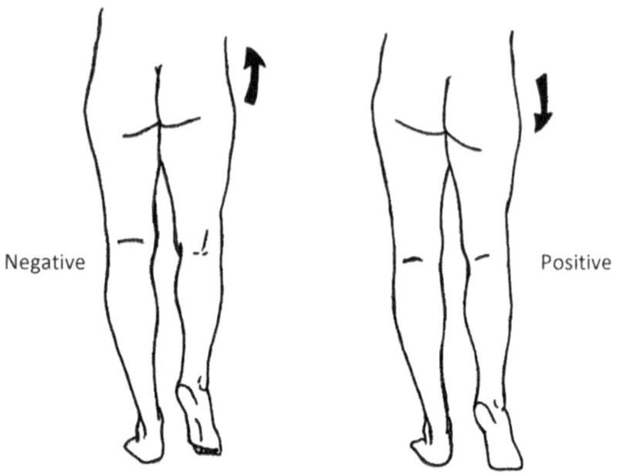

Fig. 8.20 Differential diagnosis of positive Trendelenburg's sign, which may result from hip adductor weakness (gluteus medius and minimus), hip dislocation, severe coxa vara, or pseudoarthrosis of the femoral head or neck. (Adapted from Shankar (1996))

The common peroneal nerve, which is enclosed with the tibial nerve within the sciatic sheath, branches in the thigh to innervate the short head of the biceps femoris. It separates from the tibial nerve in the upper popliteal fossa, coursing along the lateral border and then the medial border of the biceps femoris (Moore et al. 2018). Along its course, it gives rise to genicular branches for the knee and the lateral sural cutaneous nerve. The common peroneal nerve is particularly vulnerable to injury behind the fibular head due to direct pressure, prolonged squatting, sitting cross-legged, or orthopedic trauma. Within the peroneus longus muscle, it bifurcates into the deep and superficial peroneal nerves. These branches are susceptible to injury from fractures, compartment syndromes, tumors, or external pressure from tight clothing or footwear, each within its specific anatomical compartments—deep peroneal in the anterior compartment and superficial peroneal in the lateral compartment (Netter 2019).

The sciatic nerve, critical for lower limb innervation, is vulnerable to damage throughout its course. Potential causes include pelvic fractures, dislocations, hip surgeries, direct injuries, and improperly administered gluteal injections. Interestingly, sciatic nerve lesions often affect the common peroneal component more severely than the tibial component (Drake et al. 2020).

Within the thigh, the sciatic nerve is well-protected by surrounding soft tissue, making injuries there relatively rare. However, its vulnerability increases as it descends into the popliteal fossa. A key anatomical landmark is the sciatic nerve's bifurcation point into its terminal branches (tibial and common peroneal nerves), located approximately a hand's breadth above the knee joint at the apex of the popliteal fossa. Injuries at or above this division typically affect both branches, while those below predominantly impact the tibial nerve (Standring 2020).

Superior Gluteal Nerve

The superior gluteal nerve is formed by the convergence of the posterior divisions of the L4, L5, and S1 spinal nerves. This nerve exits the pelvic cavity in a posterolateral direction, traversing above the piriformis muscle alongside the superior gluteal blood vessels. The nerve then divides into superior and inferior branches, which proceed forward between the gluteus medius and gluteus minimus muscles. The superior branch specifically innervates the gluteus medius, while the inferior branch supplies both the gluteus medius and minimus, as well as the tensor fasciae latae muscle (Standring 2020).

Inferior Gluteal Nerve

The inferior gluteal nerve originates from the posterior divisions of the L5, S1, and S2 spinal nerves. It emerges from the pelvis inferior to the piriformis muscle, positioned superficial to the sciatic nerve. This nerve takes a direct course to the deep surface of the gluteus maximus muscle, which it exclusively innervates (Standring 2020).

Nerves to Piriformis, Coccygeus, and Levator Ani

The piriformis muscle receives its primary innervation from a nerve originating from the posterior division of the S2 spinal nerve, with occasional contributions from S1 (Windisch et al. 2007). This nerve follows a direct posterior trajectory to penetrate the anterior aspect of the piriformis muscle.

Descending branches from the S3 and S4 spinal nerves provide innervation to the coccygeus and levator ani muscles. Additionally, the perineal branch of S4 also innervates these muscles. This branch also innervates the external anal sphincter, along with the overlying cutaneous and fascial tissues (Barber et al. 2002).

Common Peroneal Nerve

The common peroneal nerve is the lateral terminal branch of the sciatic nerve. It is formed from the posterior divisions of the L4, L5, S1, and S2 spinal nerves. Its course begins along the superolateral aspect of the popliteal fossa, situated deep to the biceps femoris muscle and its tendon. The nerve then progresses to the posterior aspect of the fibular head. From this point, the common peroneal nerve advances anteriorly, encircling the fibular neck within the peroneus longus muscle. Here, it terminates by bifurcating into the superficial and deep peroneal nerves. Notably, the nerve is palpable posterior to the fibular head and as it winds around the fibular neck.

In terms of its distribution, the common peroneal nerve provides articular branches to both the knee joint and the superior tibiofibular joint. Additionally, it gives rise to the lateral cutaneous nerve of the calf, which innervates the posterolateral aspect of the proximal two-thirds of the leg. Typically, the peroneal communicating branch originates in conjunction with the lateral cutaneous nerve of the calf. This communicating branch subsequently unites with the sural nerve in the middle third of the lower leg.

Deep Peroneal Nerve

The deep peroneal nerve traverses inferomedially into the anterior compartment of the leg, passing deep to the extensor digitorum longus muscle. It then accompanies the anterior tibial vessels on the anterior aspect of the interosseous membrane. As it descends, the nerve is positioned on the interosseous membrane, deep to the extensor hallucis longus muscle and the superior extensor retinaculum.

At the level of the ankle joint, the deep peroneal nerve is situated deep to the inferior extensor retinaculum and the tendon of extensor hallucis longus, which crosses over it. Upon entering the dorsum of the foot, the deep peroneal nerve lies superficially between the tendons of extensor hallucis longus and extensor digitorum longus, where it bifurcates into medial and lateral branches.

The medial branch provides sensory innervation to the interdigital space between the first and second toes, extending to the distal interphalangeal joint. It also innervates the neighboring sides of these digits. The lateral branch, in contrast, innervates the extensor digitorum brevis muscle and numerous small joints of the foot, particularly those on the lateral aspect.

In the lower leg, the deep peroneal nerve provides muscular branches to extensor digitorum longus, tibialis anterior, extensor hallucis longus, and peroneus tertius. Additionally, it sends articular branches to the inferior tibiofibular and ankle joints. This detailed pathway and innervation pattern of the deep peroneal nerve are crucial for the motor and sensory functions of the anterior compartment of the leg and the dorsum of the foot, playing a significant role in foot dorsiflexion and toe extension.

Superficial Peroneal Nerve

The superficial peroneal nerve follows a near-vertical descent between the extensor digitorum longus and peroneus longus muscles, positioned anterior to the fibula. At approximately the midpoint of the lower leg, it emerges superficially by penetrating the deep fascia on the anterior surface. Subsequently, it divides into its terminal branches: the medial and intermediate dorsal cutaneous nerves. These branches traverse the anterolateral aspect of the ankle joint before entering the foot (Adkison et al. 1991). During its course, the superficial peroneal nerve provides motor innervation to the peroneus longus and brevis muscles. It also supplies cutaneous branches to the anterolateral aspect of the leg and the region surrounding the lateral malleolus.

The medial dorsal cutaneous branch innervates the medial aspect of the foot's dorsum, the great toe, and the adjacent sides of the second and third toes. In contrast, the intermediate dorsal branch supplies the dorsum of the foot and the adjacent sides of the third, fourth, and fifth toes. It is noteworthy that while the superficial peroneal nerve provides extensive cutaneous innervation to the dorsum of the foot, the skin overlying the distal phalanx receives innervation from the branches of the plantar nerves.

8.7.3 Autonomic Innervation of the Lower Extremity

The autonomic nervous system plays a significant role in lower extremity function through the integration of postganglionic sympathetic nerves with the roots of the lumbar and lumbosacral plexuses. These nerves emerge from the intervertebral foramina and accompany the lumbar and sacral plexuses, distributing sympathetic innervation throughout the lower limb (Standring 2020).

This intricate network regulates blood vessels, sweat glands, and erector pili muscles in the skin, with each dermatome receiving input via cutaneous nerves. The sympathetic ganglia, located along the lumbar and sacral portions of the vertebral column, serve as the origin for these postganglionic fibers (Moore et al. 2018).

The sympathetic innervation pathway to the lower limbs originates in the spinal cord, specifically in the thoracolumbar region (T11-L2). Preganglionic neurons, with cell bodies situated in the lateral horn of the spinal cord, project their axons through the anterior roots. These axons then travel down the sympathetic chain to synapse in the aforementioned ganglia (Netter 2019).

Beyond its physiological regulatory function, sympathetic innervation aids in diagnosing specific nerve damage by indicating lesion locations. Autonomic dysfunction in the lower limb can indicate that the lesion is distal to the point where postganglionic fibers merge with the spinal roots (Drake et al. 2020).

8.8 Musculature of the Lower Extremity

8.8.1 Muscles of the Hip Joint

8.8.1.1 Flexing Muscles of the Hip Joint: Psoas Major, Iliacus, Pectineus, Rectus Femoris, Sartorius

Psoas Major (Innervated by the Anterior Rami of L1, L2, and L3)

The psoas major is a powerful, large, and thick muscle predominantly located within the abdominal cavity (Fig. 8.21). The lumbar plexus is embedded within this muscle. The psoas major has several critical anatomical relationships: superiorly, it lies adjacent to the diaphragm and medial arcuate ligament, while inferiorly, it is closely associated with the kidney, psoas minor (if present), renal vessels, and ureter. On the right side, the muscle is overlapped by the inferior vena cava and the ileum, with the ascending and descending colon positioned laterally to the right and left psoas muscles, respectively. Medially, the lumbar vertebral column is in close proximity, while the transverse processes of the lumbar vertebrae are directly posterior to the muscle. The segmental lumbar nerves emerge from the intervertebral foramina behind the muscle (Neumann 2017).

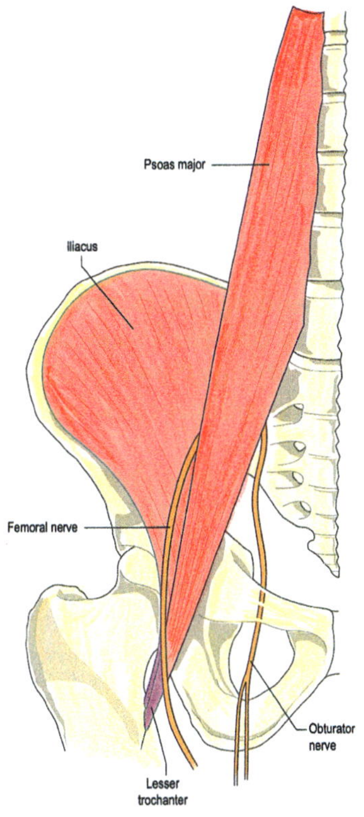

Fig. 8.21 Right psoas major and iliacus, anterior view. (Adapted with permission from Palastanga and Soames (2019))

The proximal attachment of the psoas major includes the adjacent margins of the vertebral bodies and the intervening intervertebral discs. Its uppermost attachment is at the inferior margin of the T12 vertebral body, while the lowermost attachment is at the superior margin of the L5 vertebral body. In addition, the psoas major attaches to the anterior medial aspect of each lumbar transverse process and originates from tendinous arches spanning the limited parts of the lumbar vertebral bodies (Moore et al. 2018).

The psoas major muscle fibers descend anteriorly towards the pelvic brim, with individual digitations from the vertebral column converging to form a thick muscle mass. This muscle gradually tapers as it traverses the pelvic brim beneath the inguinal ligament. At this juncture, the tendon alters its trajectory, assuming a more vertical orientation before coursing inferiorly, posteriorly, and laterally. A large bursa separates the muscle from the pubis and the hip joint capsule. Prior to crossing the pelvic brim, psoas is joined laterally by iliacus fibers, which continue to integrate with psoas even after it becomes tendinous, until the common tendon attaches to the tip and posterior aspect of the femoral lesser trochanter. Some iliacus fibers attach to the femur along a line extending inferiorly and anteriorly from the lesser trochanter (Neumann 2017).

Functionally, psoas major serves as a hip joint flexor. Due to its lumbar spine attachment, it can also flex the lumbar spine when the distal attachment is stabilized. The role of psoas in hip rotation has been a subject of debate. Initially, its attachment to the posterior and medial femur suggested a lateral rotation function of the thigh. However, when rotation is considered along an axis running through the femoral head and lateral condyle of the tibia, it should theoretically function as a medial rotator. Despite this, electromyographic studies have shown minimal activity during both medial and lateral rotation, leaving the question unresolved. If the psoas major of one side contracts independently with reversed origin and insertion, it induces lateral flexion of the lumbar spine to that side (Moore et al. 2018).

In terms of hip flexion, the actions of psoas major and iliacus are functionally combined. However, the psoas major acts independently on the lumbar spine when its distal end is fixed. For instance, during the movement of sitting up from supine position, both psoas muscles aid in lifting the trunk's weight, while the abdominal muscles work intensively to flex the trunk. Activating the abdominal muscles early is crucial to prevent excessive forward displacement of the lumbar spine before the trunk begins to rise. Elevating the head first helps prevent this undesirable and potentially harmful movement (Standring 2020).

Each lower limb constitutes approximately 15% of body weight. Therefore, lifting both legs simultaneously involves the hip flexors working against about 30% of the body's weight. During this initial phase, especially for the first 30° of leg elevation, the psoas major is primarily engaged. At this point, the lumbar spine is pulled forward due to the psoas muscle's action with reversed origin and insertion. This anterior displacement of the lumbar spine can cause significant strain and potential damage, especially in individuals with pre-existing degenerative changes in the region (Neumann 2017). Exercises such as raising both lower limbs simultaneously while in a supine position are frequently implicated in back issues and are, unfortunately, common in certain fitness regimens. It is essential to thoroughly understand the mechanics of this region before engaging in such exercises to prevent back problems rather than merely addressing them after they occur.

Iliacus (Innervated by the Femoral Nerve: L2 and L3)

The iliacus is a large, triangular muscle primarily located within the pelvic cavity (Fig. 8.21). Its primary superior attachment originates from the upper and posterior two-thirds of the iliac fossa, with additional fibers stemming from the sacral ala and anterior sacroiliac ligament. The muscle fibers descend anteriorly and medially, integrating with the lateral aspect of the psoas major. This muscular fusion persists over the pelvic brim, where the combined structure alters its course, proceeding inferiorly, posteriorly, and slightly laterally to insert into the lesser trochanter of the femur. At this point, it merges with the psoas major insertion at the tip of the lesser trochanter. Some fibers also attach to the hip joint capsule.

The iliacus exerts an influence on the hip similar to that of the psoas major. When the superior attachment serves as the fixed point, it anteriorly displaces the thigh, resulting in hip flexion. Conversely, if the inferior attachment is fixed, it pulls the pelvis forward, causing an anterior tilt. This action also constitutes hip flexion, but in this instance, the trunk is the moving segment (Neumann 2017).

In conjunction with the psoas major, the iliacus is active in all movements involving elevation of the lower limb anterior to the trunk, such as the forward swing of the leg during walking, running, and jumping. Additionally, it aids in anteriorly displacing the trunk from a supine to a sitting position.

8.8.1.2 Extending Muscles of the Hip Joint: Gluteus Maximus, Hamstrings (Semitendinosus, Semimembranosus, Biceps Femoris)

Gluteus Maximus (Innervated by the Inferior Gluteal Nerve: L5, S1, S2)

The gluteus maximus is the largest and most powerful of the gluteal muscles. It is situated on the posterior aspect of the hip joint. Evolutionary changes in human pelvic structure associated with bipedalism have transformed the primary function of this muscle from hip adduction in lower primates and early humans to hip extension in modern humans. This adaptation has been crucial in maintaining the erect posture, thereby liberating the upper limbs from weight-bearing roles and enabling their development into precise manipulative organs (Lieberman et al. 2006).

Morphologically, the gluteus maximus (Fig. 8.22) presents a quadrilateral shape with muscle fiber bundles aligned along its line of pull, resulting in a coarse surface appearance. The muscle is composed of two layers as it descends to its distal attachment. The proximal attachments of the gluteus maximus are extensive. It originates from the gluteal surface of the ilium, posterior to the posterior gluteal line, the posterior border of the ilium, and the adjacent portion of the iliac crest. Additional attachments include the lateral aspect of the coccyx, the posterior surface of the sacrum, and the superior part of the sacrotuberous ligament. The superior fibers attach to the aponeurosis of the sacrospinalis, while the deep anterior fibers originate from the fascia covering the gluteus medius.

The muscle fibers course inferiorly and anteriorly toward the proximal femur. Approximately

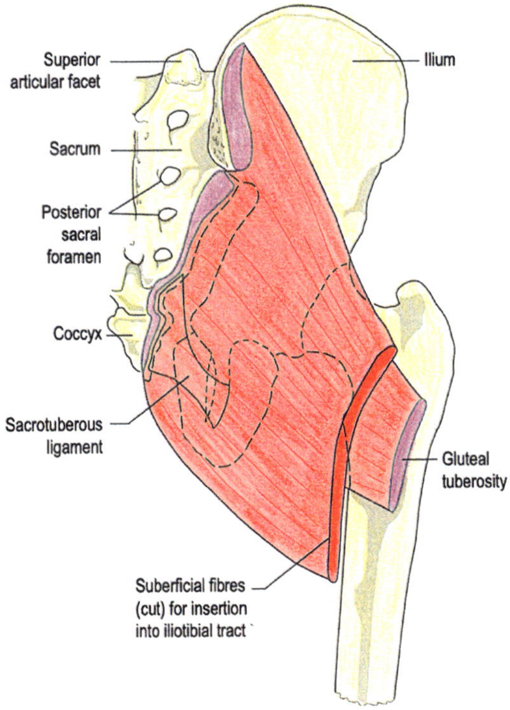

Fig. 8.22 Attachments of the right gluteus maximus (posterior view) with direction of fibers shown in diagrammatic form. (Adapted with permission from Palastanga and Soames (2019))

three-quarters of the superficial fibers form a lamina that narrows and inserts between the two layers of the fascia lata, contributing to the formation of the iliotibial tract. The remaining deep quarter of the fibers forms a broad aponeurosis that attaches to the gluteal tuberosity of the femur (Standring 2020).

When the gluteus maximus contracts with its superior attachment fixed, it draws the femoral shaft posteriorly, facilitating hip joint extension from a flexed position. Due to its lateral insertion on the thigh, it also induces lateral rotation during extension. The inferior fibers contribute to thigh adduction, while the superior fibers may assist in abduction (Neumann 2017).

The fibers inserting into the iliotibial tract (band) can indirectly extend the knee joint, as the tract's distal attachment is anterior to the knee's axis of movement on the lateral tibial condyle. Through this mechanism, the gluteus maximus provides significant lateral knee stability. When the femur is stabilized, contraction of the muscle causes posterior rotation of the ilium and pelvis around the hip joint, effectively lifting the trunk from a flexed posture.

The gluteus maximus is well adapted for forceful movements such as ascending stairs and running, due to its role as a strong hip extensor, especially when the hip is flexed. However, its activity is minimal during normal gait (Jonkers et al. 2003). The gluteus maximus works together with the hamstrings to raise the trunk from a flexed position. These muscle groups also control the forward flexion of the trunk, as this movement primarily occurs at the hip joint. During sitting, the gluteus maximus periodically relieves pressure on the ischial tuberosities through static or dynamic contractions, lifting these bony prominences from the supporting surface. This action is often alternated between sides to shift weight distribution. The gluteus maximus plays an important role in pelvic stabilization on the femoral heads, contributing to upright posture maintenance. Its ability to laterally rotate the femur when standing aids in elevating the medial longitudinal arch of the foot.

Paralysis of the gluteus maximus results in buttock flattening and impairs stair climbing and running abilities. While other muscles can compensate for hip extension, the movement becomes significantly weaker. In cases of quadriceps femoris weakness or paralysis, the gluteus maximus can be trained to produce functional knee extension, enabling limited weight-bearing during ambulation or standing.

8.8.1.3 Adducting Muscles of the Hip Joint: Adductor Magnus, Adductor Longus, Adductor Brevis, Gracilis, Pectineus

Adductor Magnus (Adductor Part, Innervated by the Posterior Division of the Obturator Nerve, L2 and L3; Hamstring Part, Tibial Division of the Sciatic Nerve, L4)

The adductor magnus is the largest and most posterior of the adductor muscles. It is positioned behind adductors brevis and longus (Fig. 8.23) and lies anterior to the semimembranosus and semitendinosus muscles (Fig. 8.24). This muscle comprises two distinct components: an adductor part and a hamstring part, forming a triangular muscular sheet with a thickened medial border (Neumann 2017). The proximal attachment originates from the femoral surface of the ischiopubic ramus, extending to the inferior surface of the ischial tuberosity laterally (Fig. 8.25). The anterior fibers attaching to the ischiopubic ramus form a twisted muscular sheet before its femoral insertion. Conversely, the posterior fibers, originating from the ischial tuberosity, descend vertically as a thickened cord (Moore et al. 2018).

Functionally, the adductor magnus primarily adducts the hip joint, with its posterior portion contributing to hip extension. While some researchers propose that this muscle, along with adductor longus, mediates hip internal rotation, though earlier views suggested external rotation. The rotational action likely depends on thigh

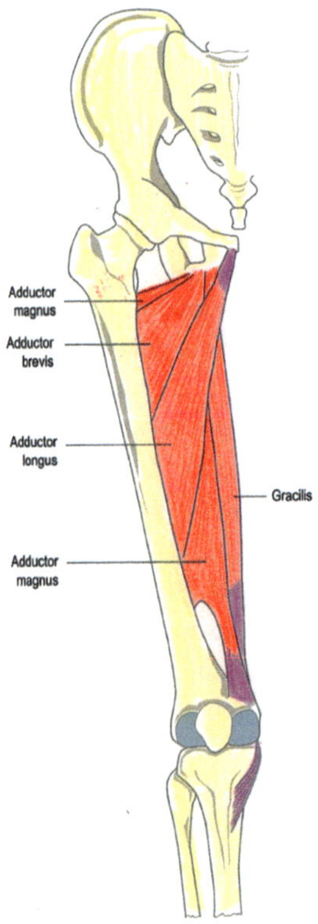

Fig. 8.23 Right adductor muscles, anterior view. (Adapted with permission from Palastanga and Soames (2019))

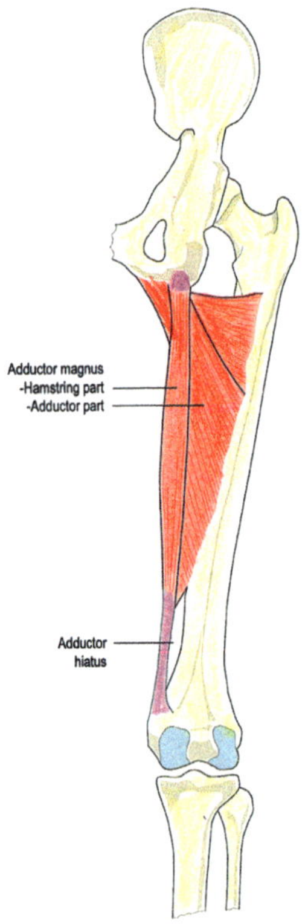

Fig. 8.24 Right adductor magnus, posterior view. (Adapted with permission from Palastanga and Soames (2019))

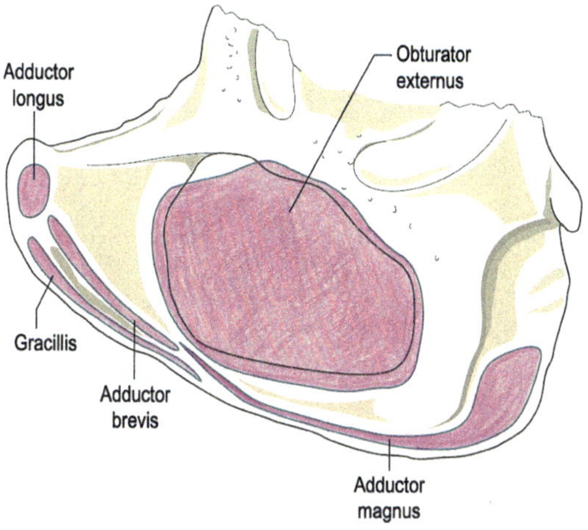

Fig. 8.25 Muscle attachments to the outer surface of the pubis, ischium, and obturator membrane. (Adapted with permission from Palastanga and Soames (2019))

positioning and the line of force relative to the mechanical axis of the femur. All adductor muscles play a significant role in maintaining lateral stability during the stance phase of gait (Palastanga and Soames 2019). Additionally, the medial collateral ligament of the knee joint appears to be a caudal continuation of the adductor magnus tendon. This anatomical relationship suggests that historically, the adductor magnus may have crossed the knee joint, potentially functioning as a knee flexor, similar to the gracilis muscle (Drake et al. 2020).

Adductor Longus (Innervated by the Anterior Division of the Obturator Nerve: L2, L3, L4)

Adductor longus, a triangular muscle characterized by its elongated and slender form (Fig. 8.26), occupies the medial thigh region, overlapping the middle portion of adductor magnus. Its proximal, narrower attachment originates from a small, roughened area situated immediately inferior to the medial end of the obturator crest on the anterior aspect of the pubic body (Fig. 8.25).

The muscle fibers of the adductor longus descend in a lateral direction, gradually expanding to insert along the middle two-quarters of the linea aspera. This insertion is positioned anterior to the adductor magnus inferiorly, the adductor brevis superiorly, and posterior to the vastus medialis (Moore et al. 2018).

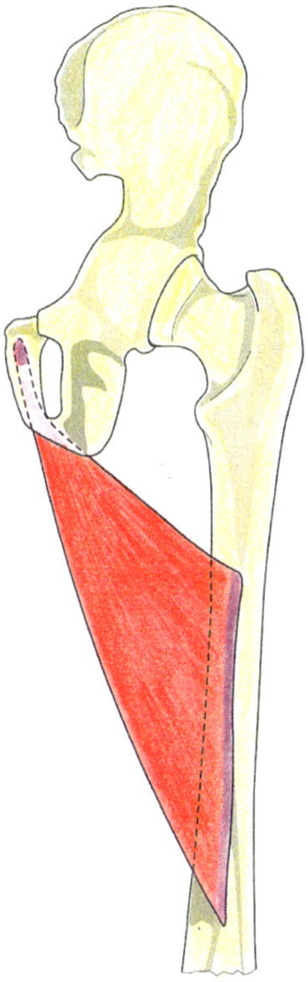

Fig. 8.26 Right adductor longus, posterior view. (Adapted with permission from Palastanga and Soames (2019))

Functionally, the adductor longus primarily serves as a thigh adductor. However, its role in thigh rotation remains a subject of debate among researchers. Additionally, this muscle assists in flexing an extended thigh and extending a flexed thigh, contributing to hip joint mobility in multiple planes (Neumann 2017).

Adductor Brevis (Innervated by the Anterior Division of the Obturator Nerve: L2, L3, L4)

The adductor brevis, a muscle with a triangular configuration, is situated within the medial aspect of the thigh. Its origin is localized to the anterior body's lateral segment and the pubic inferior ramus (Drake et al. 2020). The muscle's fibers traverse in a downward and lateral trajectory, with a slight posterior inclination, before inserting onto the superior half of the linea aspera, anterior to adductor magnus (Fig. 8.27). Topographically, the adductor brevis exhibits

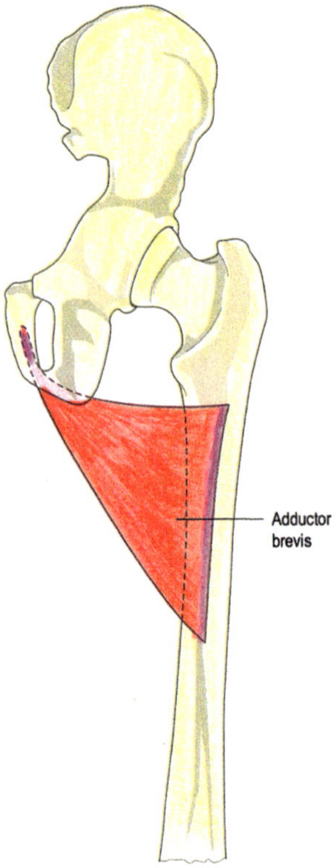

Fig. 8.27 Right adductor brevis, posterior view. (Adapted with permission from Palastanga and Soames (2019))

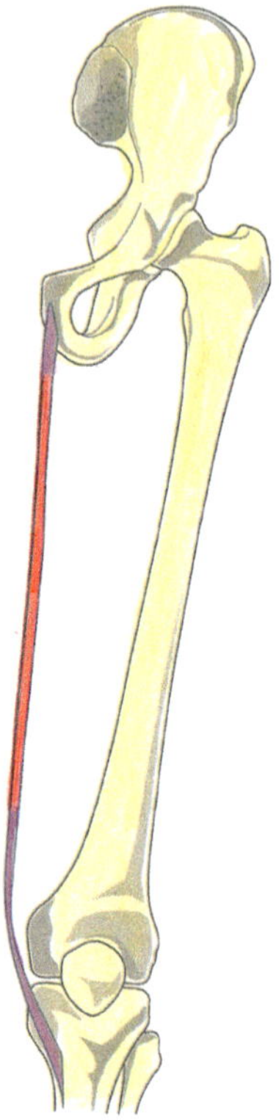

Fig. 8.28 Left gracilis, anterior view. (Adapted with permission from Palastanga and Soames (2019))

distinct relationships with adjacent muscles. Its superior portion is positioned deep to the pectineus muscle, while the inferior segment lies posterior to the adductor longus (Moore et al. 2018). This arrangement contributes to the complex layering of muscles in the medial thigh compartment, influencing biomechanical function and potential clinical considerations.

The anatomical orientation of adductor brevis, particularly its fiber direction and attachments, suggests a primary role in hip adduction. However, like other adductor muscles, it may contribute to hip flexion or extension depending on the initial position of the femur relative to the pelvis (Neumann 2017).

Gracilis (Innervated by the Anterior Division of the Obturator Nerve: L2, L3)

Gracilis, the most medially positioned muscle within the adductor group, serves dual functions as a thigh adductor and knee flexor. This slender, elongated muscle is situated along the medial aspect of the thigh (Fig. 8.28), representing the most superficial component of the adductor muscle complex. The proximal attachment of gracilis originates from the anterior surface of the pubic body and its inferior ramus, slightly extending onto the ischial ramus. As it descends between the semimembranosus posteriorly and sartorius anteriorly, gracilis develops a fusiform-shaped belly at its midpoint. The muscle becomes tendinous superior to the knee joint, crossing it before expanding to insert on a short vertical line on the proximal portion of the tibia's medial surface. This insertion is superior to that of the semitendinosus and posterior to, while merging with, the sartorius attachment. Bursae are present to separate the gracilis tendon from those of the sartorius and semitendinosus (Moore et al. 2018).

While it is evident that these muscles primarily contribute to thigh adduction, they appear to function most effectively when the hip joint is in the neutral, or anatomical position. The adductors also work synergistically during weight-bearing movements that involve simultaneous flexion and extension of the hip and knee. There remains some debate regarding their role in either medial or lateral rotation of the thigh (Neumann 2017). During ambulation, the adductors function to pull the supporting lower limb into adduction, thereby aligning the gravity line over the weight-bearing foot. They also play a key role in stabilizing the pelvis on the hip joint. As a group, the adductors are especially active when holding an object between the knees while seated, such as during horseback riding, where they are crucial for maintaining stability, particularly when the horse is in motion (Palastanga and Soames 2019).

In its capacity as a knee flexor, the gracilis augments hamstring function in basic flexion activities, such as initiating the swing phase of gait when knee flexion is required. It also aids in forceful flexion, as observed in the forward pull on a rowing boat's sliding seat. In horseback riding activities, gracilis is actively utilized in all its functions, assisting in gripping the horse while simultaneously helping maintain knee flexion control (Drake et al. 2020).

8.8.1.4 Abducting Muscles of the Hip Joint: Gluteus Maximus, Gluteus Medius, Gluteus Minimus, Tensor Fascia Lata

Gluteus Medius (Innervated by the Superior Gluteal Nerve: L4, L5, S1)

The gluteus medius is a significant muscle for hip and pelvic stability. This muscle exhibits a fan-shaped structure, located on the lateral and upper regions of the buttock, just below the iliac crest (Palastanga and Soames 2019) (Fig. 8.29). It broadens superiorly and narrows to its tendinous insertion inferiorly, occupying the space between

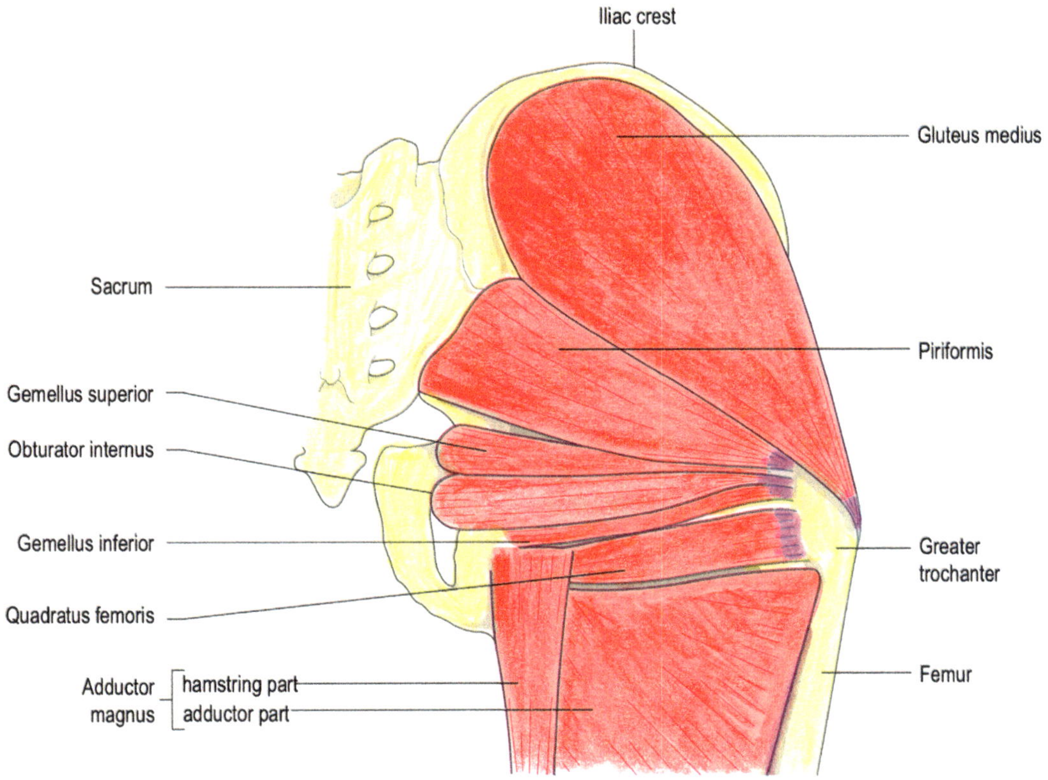

Fig. 8.29 Muscles responsible for lateral rotation of the hip joint: right side, posterior view. (Adapted with permission from Palastanga and Soames (2019))

the iliac crest and the greater trochanter of the femur. The gluteus maximus partially overlaps its posterior portion.

The muscle originates from an extensive area on the gluteal surface of the ilium, between the posterior and anterior gluteal lines. This attachment extends superiorly to the iliac crest and inferiorly almost to the greater sciatic notch (Moore et al. 2018). A strong fascial layer covers the muscle, providing a firm attachment for its fibers. This fascia is shared posteriorly with the gluteus maximus. The muscle fibers of the gluteus medius have distinct orientations: the posterior fibers descend obliquely forward, the middle fibers run vertically downward, and the anterior fibers course downward and backward. These fibers converge to form a flattened tendon, which inserts onto a roughened area along the superolateral aspect of the greater trochanter (Standring 2020).

When the pelvis is stabilized, the gluteus medius elevates the greater trochanter. However, as the hip joint acts as the fulcrum, this action causes lateral movement of the femoral shaft, resulting in hip abduction. With a fixed distal attachment, the muscle can lower the ipsilateral pelvis and raise the contralateral side. Additionally, the anterior fibers aid in femoral medial rotation when acting from a fixed pelvis, and when the femur is stable, these fibers rotate the contralateral side of the pelvis forward (Neumann 2017).

The gluteus medius plays a key role in walking, running, and single-limb weight-bearing. During gait, as the opposite limb lifts off the ground, the supporting side's gluteus medius contracts strongly to maintain or even elevate the contralateral pelvis, allowing smooth forward progression of the swing limb. If this muscle is weakened or paralyzed, the pelvis drops on the unsupported side, leading to the characteristic Trendelenburg gait (Perry and Burnfield 2010). Working with other muscles like the gluteus minimus and tensor fasciae latae, the gluteus medius also contributes to hip joint rotation during locomotion. With the femur as a relatively fixed point, it controls ipsilateral pelvic rotation. Impaired function of this muscle, due to paralysis or altered hip joint mechanics, can cause contralateral pelvic drop, known as the Trendelenburg sign (Mansfield and Neumann 2019).

8.8.1.5 Medially Rotating Muscles of the Hip Joint: Anterior Part of Gluteus Medius, Anterior Part of Gluteus Minimus, Tensor Fascia Lata, Psoas Major, Iliacus

Tensor Fascia Lata (Innervated by the Superior Gluteal Nerve: L4, L5)

The tensor fasciae lata is positioned anterolaterally to the hip joint and lies superficial to the gluteus minimus (Drake et al. 2020) (Fig. 8.30). Its proximal attachments include the anterior portion of the iliac crest's outer lip, including the iliac tubercle and anterior superior iliac spine. Its origin also includes the adjacent gluteal surface inferior to this area and the fascial layers between it and the gluteus minimus, as well as its superfi-

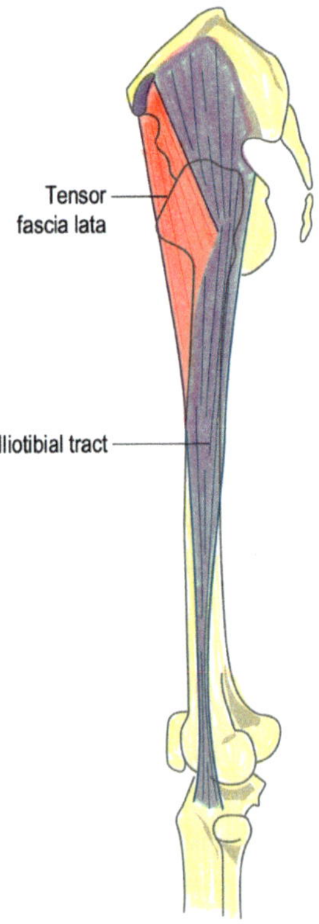

Fig. 8.30 Tensor fascia lata attaching anteriorly to the iliotibial tract. (Adapted with permission from Palastanga and Soames (2019))

cial covering. The inferior attachment is unique, inserting between the two layers of the iliotibial tract, distal to the level of the greater trochanter (Standring 2020).

Functionally, the tensor fasciae lata contributes to multiple hip joint actions. It assists in flexion, abduction, and medial rotation of the hip. Additionally, it counterbalances the posterior pull of the gluteus maximus on the iliotibial tract, providing a stabilizing effect (Neumann 2017). Working in conjunction with the superficial fibers of gluteus maximus, the tensor fasciae lata tightens the iliotibial tract. This action, coupled with the tract's attachment to the lateral tibial condyle, facilitates knee extension. The muscle's synergistic action with gluteus minimus enhances hip joint medial rotation, while its posterior fibers may contribute to thigh abduction (Palastanga and Soames 2019).

Functionally, the tensor fasciae lata and gluteus maximus create a linkage between the pelvis and tibia, stabilizing and controlling pelvic and femoral movements during weight-bearing activities. The muscle generates powerful medial rotation when the hip is extended, particularly during the 'toe-off' phase of walking, as it prepares the limb, pelvis, and trunk to absorb the force transmitted by the calf muscles (Perry and Burnfield 2010). In cases of quadriceps femoris paralysis, the tensor fasciae latae can be trained to produce knee extension sufficient for ambulation. However, it's important to note that this compensatory action is relatively weak and limited in range compared to normal quadriceps function.

8.8.1.6 Laterally Rotating Muscles of the Hip Joint: Gluteus Maximus, Piriformis, Obturator Internus, Gemellus Superior, Gemellus Inferior, Quadratus Femoris, Obturator Externus

Piriformis (Innervated by the Anterior Rami of the Sacral Plexus: L5, S1, S2)

The piriformis, a triangular muscle situated posterior to the hip joint and in alignment with the gluteus medius, originates in the pelvic cavity and terminates in the gluteal region (Fig. 8.29). Its proximal attachment includes the anterior aspect of the second through fourth sacral vertebrae, positioned lateral to the anterior sacral foramina. Additional attachments include the gluteal surface of the ilium and the pelvic surface of the sacrotuberous ligament as it exits the pelvis via the greater sciatic foramen. The muscle fibers course inferiorly, laterally, and anteriorly, converging into a tendon that inserts on the superior border and medial aspect of the greater trochanter of the femur. The fiber orientation follows a direct path from origin to insertion through the greater sciatic foramen (Drake et al. 2020).

In the anatomical position, despite its relatively superior location, the piriformis functions primarily as a lateral rotator of the thigh. However, when seated, the muscle plays a critical role in abduction, as its position is optimal for this movement. This muscle plays a critical role in maintaining femoral head stability within the acetabulum (Neumann 2017).

It is essential to recognize that muscle function often varies depending on body position. The piriformis exemplifies this principle, demonstrating different actions in standing versus sitting postures. This muscle is particularly significant in several functional activities:

- Abduction while seated, such as shifting from one chair to another without standing up
- Assisting in leg movement when exiting a vehicle in preparation for standing
- Stabilizing the pelvis during trunk rotation
- Maintaining pelvic balance control when standing on unstable surfaces, such as a moving bus

Given its diverse functional roles, strain of the piriformis muscle is not uncommon. However, due to its deep location and varied actions, such injuries may be overlooked in clinical assessments (Cassidy et al. 2012).

8.8.2 Muscles of the Knee Joint

8.8.2.1 Extending Muscles of the Knee Joint: Rectus Femoris, Vastus Medialis, Vastus Lateralis, Vastus Intermedius, Tensor Fascia Lata

The quadriceps femoris forms the prominent muscle mass on the anterior surface of the thigh. As its name suggests, it consists of four main components. One of these, the rectus femoris, originates above the hip joint, while the other three—vastus medialis, vastus lateralis, and vastus intermedius—originate from the shaft of the femur. All four muscles converge around the patella, forming a strong tendon known as the ligamentum patellae, which inserts into the tibial tuberosity (Moore et al. 2018).

While the rectus femoris is part of the quadriceps group, it also functions as a hip flexor by crossing anterior to the hip joint. This dual action distinguishes it from the other three components (Neumann 2017). The primary function of the quadriceps femoris is knee joint extension. However, each component appears to have a specific role in this action, often engaging at different phases of the movement. For instance, the vastus medialis demonstrates more pronounced activity during the terminal stage of extension. It is theorized to counteract the lateral displacement tendency of the patella, which results from the angular orientation of the femur (Powers 2003). Some clinicians advocate for targeted exercises to strengthen the oblique fibers of the vastus medialis. The rationale behind this approach is to influence patellar tracking, prevent the risk of dislocation, and potentially alleviate anterior knee pain in certain clinical conditions (Syme et al. 2009).

The rectus femoris exhibits particularly strong activation during straight leg raising exercises or in the combined movement pattern of hip flexion and knee extension. This muscle's unique positioning allows it to contribute significantly to both hip and knee joint movements.

The quadriceps femoris group is heavily recruited during stepping activities, such as ascending stairs and performing squats. The rectus femoris plays a crucial role specifically during the swing phase of gait, when the lower limb is advancing forward and the knee is extending. In the final stages of knee extension, the vastus medialis contributes to the joint's locking mechanism by facilitating medial rotation of the femur (Perry and Burnfield 2010).

Rectus Femoris (Innervated by the Femoral Nerve: L2, L3, L4)

The rectus femoris is a spindle-shaped bipennate muscle that is prominently visible on the anterior aspect of the thigh. This muscle has a unique dual-head origin: the straight head attaches to the anterior inferior iliac spine, while the reflected head originates from a roughened area superior to the acetabulum. These two heads form a continuous attachment point. Evolutionary biologists posit that the straight head is a relatively recent adaptation in human anatomy, associated with the development of upright posture (Lieberman et al. 2006).

From this unified origin, a single tendon descends, giving rise to the fleshy portion of the muscle. As it progresses distally, approximately two-thirds down the length of the thigh, the muscle narrows into a thick tendon that inserts onto the superior border of the patella. Some fibers extend around the patella, contributing to the formation of the patellar ligament.

The deep surface of the rectus femoris exhibits a smooth, tendinous structure. This anatomical feature facilitates uninhibited movement over a similarly smooth surface of the vastus intermedius, thereby allowing the rectus femoris to act independently on the hip joint (Neumann 2017). The unique anatomical configuration of the rectus femoris, particularly its dual-head origin and smooth deep surface, has significant functional implications. Its ability to act independently on the hip joint while simultaneously influencing knee extension makes it a key player in complex lower limb movements, especially those involved in bipedal locomotion and postural control (Perry and Burnfield 2010).

Vastus Medialis (Innervated by the Femoral Nerve: L2, L3, L4)

The vastus medialis, a component of the quadriceps femoris group, is positioned on the anteromedial aspect of the thigh, medial to the rectus femoris (Fig. 8.31). Its most prominent bulk is visible in the lower third of the thigh, just above the patella (Netter 2019). This muscle has an extensive linear origin that begins at the inferior medial end of the intertrochanteric line. The origin then courses inferiorly around the medial aspect of the proximal femoral shaft along the spiral line, continues along the medial lip of the linea aspera, and extends to the superior two-thirds of the medial supracondylar line. Additional points of origin include the medial intermuscular septum and the tendon of adductor magnus (Moore et al. 2018).

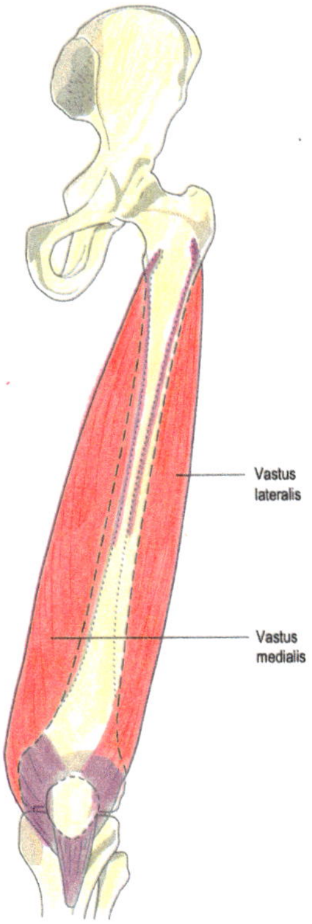

Fig. 8.31 Left vastus medialis and lateralis, anterior view. (Adapted with permission from Palastanga and Soames (2019))

The fiber orientation of the vastus medialis exhibits a distinctive pattern. The superior fibers primarily descend vertically, while the inferior fibers adopt a more horizontal, lateral trajectory. Some anatomists and biomechanists consider these two fiber groups to be anatomically and functionally distinct, with the oblique fibers specifically termed vastus medialis obliquus (Powers 2003).

The insertion of the vastus medialis is complex, attaching to the tendon of rectus femoris, the medial border of the patella, and the anterior aspect of the medial tibial condyle. Notably, the expansions that extend across the knee joint to attach to the tibia serve as a replacement for the joint capsule in this region, fusing with the deep fascia. This attachment also extends to the tibial tuberosity (Neumann 2017). The unique anatomical configuration of the vastus medialis, particularly the vastus medialis obliquus component, has significant implications for knee joint biomechanics and patellar tracking. The oblique fibers are thought to play a crucial role in maintaining proper patellar alignment during knee extension, potentially influencing patellofemoral joint function and stability (Syme et al. 2009).

Vastus Lateralis (Innervated by the Femoral Nerve: L2, L3, L4)

The origin of the vastus lateralis is characterized by an extensive linear attachment. It begins at the superior lateral portion of the intertrochanteric line and extends to the inferior border of the greater trochanter. Additional attachment points include the lateral aspect of the gluteal tuberosity and the superior half of the lateral lip of the linea aspera. The muscle also has connections to the fascia lata and the lateral intermuscular septum (Moore et al. 2018). From its origin, the muscle fibers of the vastus lateralis demonstrate a distinct orientation pattern. The superior fibers descend almost vertically, while the more inferior fibers adopt a downward and forward trajectory. This fiber arrangement contributes to the muscle's functional capabilities in knee extension and stabilization (Neumann 2017).

The bulk of the vastus lateralis is predominantly located in the upper half of the lateral thigh. As the muscle descends, it forms a broad tendon that narrows as it approaches the lateral aspect of the patella. This tendon inserts into the tendon of the rectus femoris and the base and lateral border of the patella. Some fibers extend to the anterior aspect of the lateral tibial condyle, blending with the iliotibial tract. This confluence contributes to forming an expansion that ultimately attaches to a line directed towards the tibial tuberosity. Notably, this portion of the attachment effectively replaces the knee joint capsule, reinforcing the lateral aspect of the knee (Standring 2020).

Vastus Intermedius (Innervated by the Femoral Nerve: L2, L3, L4)

The vastus intermedius, the deepest component of the quadriceps femoris complex, is positioned between the vastus lateralis and vastus medialis, lying deep to the rectus femoris. This muscle originates from the superior two-thirds of the anterior and lateral surfaces of the femur. The muscle fibers descend to form a broad tendon on its more superficial aspect. This tendinous structure has a complex insertion, attaching to the deep surface of the rectus femoris tendon, the tendons of the other vastus muscles, and the base of the patella (Moore et al. 2018).

The anatomical relationships of the vastus intermedius with its neighboring muscles are noteworthy. As the muscle descends farther, it becomes almost impossible to differentiate it from the vastus medialis, indicating the deep integration of these muscles in the quadriceps complex (Standring 2020).

The deep position and extensive origin of the vastus intermedius contribute significantly to its functional role. As a primary knee extensor, it works in synergy with the other quadriceps components to generate powerful extension forces at the knee joint. Its deep location may also provide additional stability to the femur during dynamic activities (Neumann 2017). The vastus intermedius, despite its deep location, plays a vital role in knee extension and stability. Its contribution to overall quadriceps function underscores the importance of considering all components of this muscle group in rehabilitation and strength training programs. Future research may further elucidate the specific contributions of the vastus intermedius to knee joint biomechanics and its potential role in patellofemoral disorders (Powers 2003).

8.8.2.2 Flexing Muscles of the Knee Joint: Hamstrings (Semitendinosus, Semimembranosus, and Biceps Femoris), Gastrocnemius Gracilis, Sartorius, Popliteus

It is important to note that, with the exception of the short head of biceps femoris, the three hamstring muscles span and act upon both the hip and knee joints, making their function highly complex. The specific contributions of these muscles to knee joint movement are essential to understanding their overall role (Neumann 2017).

While knee rotation by the hamstrings is typically described as occurring when the foot is off the ground, this perspective is somewhat oversimplified. While it may be easier to understand in theory, in practice, rotation typically occurs when the feet are fixed firmly on the ground and bearing weight. For instance, when moving laterally from one seat to another, the feet remain fixed while body weight is supported through them. In this scenario, the person remains seated but slightly lifts the buttocks off the seat, allowing the trunk to shift sideways. This movement is facilitated by the rotation of the femur on the tibial plateau, achieved through the coordinated action of the medial rotators on one knee and the lateral rotators on the other (Standring 2020).

The concurrent action of the hamstrings at both the hip and knee joints is particularly evident in athletic activities, such as when an athlete accelerates into a curve. In this situation, the hamstrings are responsible for raising the trunk to a more upright posture while simultaneously flexing the knee of the swing leg in preparation for the next stride. As the body navigates the curve, the hamstrings must also induce knee rotation to generate the turning force. The hamstrings are thought to function as a dynamic link between the posterior pelvis and the tibia, enabling precise adjustments in the alignment of these bones. This coordination is crucial during rapid posture changes in active movement, where additional forces arise from the acceleration of body segments. This dynamic interplay likely explains the high incidence of hamstring injuries among athletes (Lieberman et al. 2006; Opar et al. 2012).

Semimembranosus (Innervated by the Tibial Division of the Sciatic Nerve: L5, S1, S2)

The semimembranosus, along with the semitendinosus and biceps femoris, constitutes the hamstrings. Positioned on the posteromedial aspect of the thigh, semimembranosus lies deep to semitendinosus in its lower part. It originates from a strong membranous tendon attached to the upper lateral facet of the rough area on the ischial tuberosity (Fig. 8.32) and descends medially. As the muscle travels downward, it becomes fleshy on the medial side of the tendon, remaining deep to both semitendinosus and biceps femoris. From the lower portion of the muscle, a secondary aponeurotic tendon arises, narrowing as it inserts into a horizontal groove on the posteromedial surface of the medial tibial condyle. From this point, the fibers radiate in multiple directions, particularly upwards and laterally, forming the oblique popliteal ligament. Bursae are interposed between semimembranosus and the medial head of the gastrocnemius, as well as near its attachment to the tibia (Moore et al. 2018; Standring 2020).

Functionally, the semimembranosus exhibits versatility in its actions depending on whether it is working from its proximal or distal attachment. When acting from its distal attachment, it contributes to hip extension, particularly when the trunk is flexed forward. Conversely, when acting from its proximal attachment, it aids in knee flexion. In a position of knee semiflexion, it produces medial rotation of the knee. Interestingly, when the foot is fixed, the semimembranosus can function as a lateral rotator of the femur and pelvis on the tibia (Neumann 2017).

The complex functional characteristics of the semimembranosus show its importance in performing various biomechanical functions. Its involvement in both hip extension and knee flexion makes it a key element to the mechanics of walking and maintaining posture. The muscle's contribution to medial knee rotation is particularly relevant in activities requiring directional changes or stability in rotational movements (Powers 2003).

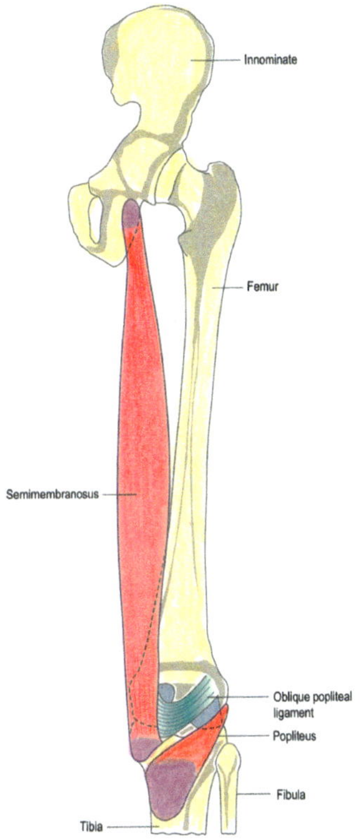

Fig. 8.32 Right semimembranosus and popliteus, posterior view. (Adapted with permission from Palastanga and Soames (2019))

Semitendinosus (Innervated by the Tibial Division of the Sciatic Nerve: L5, S1, and S2)

The semitendinosus originates from the lower medial facet of the lateral section of the ischial tuberosity. Notably, its proximal tendon is combined with that of the long head of biceps femoris, and these two muscles run in parallel for a short distance before diverging (Moore et al. 2018). The semitendinosus forms a fusiform muscle belly that quickly transitions into a long, slender tendon, which gives the muscle its descriptive name. This tendon courses inferiorly and medially, passing posterior to the medial femoral condyle. A small bursa separates the tendon from the medial collateral ligament, enhancing its smooth gliding action (Standring 2020). The distal attachment of semitendinosus is on a vertical line on the medial surface of the medial tibial condyle. This insertion is positioned just posterior to that of sartorius and posterior-inferior to the attachment of gracilis. The complex relationships between these tendons are facilitated by bursae: one separates semitendinosus from gracilis near their insertions, while another bursa intervenes between these two tendons and sartorius (Neumann 2017).

Functionally, the semitendinosus contributes to multiple actions at both the hip and knee joints. At the hip, it acts as an extensor and also assists in lateral rotation when the hip is flexed. At the knee, it serves as a flexor and, when the knee is flexed, contributes to medial rotation of the tibia on the femur. This dual-joint action makes semitendinosus a key player in complex lower limb movements, particularly in gait and sports-specific activities (Powers 2003).

The long tendon of semitendinosus provides a mechanical advantage, allowing for effective force transmission over a large range of motion. This characteristic is particularly beneficial in activities requiring rapid changes in limb position or in maintaining postural stability during dynamic movements (Opar et al. 2012). The complex relationship between the semitendinosus muscle and its adjacent muscles, enabled by the existence of multiple bursae, shows the significance of giving into consideration the combined role of the pes anserinus group (semitendinosus, gracilis, and sartorius) in knee biomechanics.

Biceps Femoris (Long Head, Innervated by the Tibial Division of the Sciatic Nerve; Short Head, by Common Peroneal Division: L5, S1, S2)

The biceps femoris, a key component of the hamstring muscle group, is unique in its dual-head structure and innervation pattern. The long head receives innervation from the tibial division of the sciatic nerve, while the short head is innervated by the common peroneal division, both deriving from spinal segments L5, S1, and S2 (Netter 2019).

Positioned on the posterolateral aspect of the thigh, the biceps femoris consists of two heads with different origins. The long head originates from the lower medial facet of the ischial tuberosity, sharing this attachment with semitendinosus, and extends onto the sacrotuberous ligament. Initially, these tendons descend together before separating into individual muscles. The long head forms a fusiform muscle that courses inferiorly and laterally across the posterior thigh, superficial to the sciatic nerve (Moore et al. 2018). The short head originates from the lower half of the lateral lip of the linea aspera, extending proximally to nearly reach the gluteus maximus attachment. Its origin continues onto the upper half of the lateral supracondylar line of the femur, with some fibers arising from the lateral intermuscular septum. In the lower third of the thigh, the short head merges with the narrowing tendon of the long head on its deep aspect (Standring 2020).

As the combined muscle approaches the knee, its tendon becomes palpable crossing the posterolateral aspect of the joint, directed toward the fibular head. The insertion of the biceps femoris is complex: the tendon splits around the fibular collateral ligament before attaching to the fibular head. Some fibers integrate with the ligament, while others attach to the lateral tibial condyle and the posterior aspect of the lateral intermuscular septum. A bursa facilitates smooth gliding between the tendon and the lateral collateral ligament (Neumann 2017).

Functionally, the biceps femoris contributes to multiple actions at both the hip and knee joints. At the hip, it acts as an extensor and lateral rotator. At the knee, it serves as a flexor and, when the knee is flexed, contributes to lateral rotation of the tibia on the femur (Powers 2003).

The dual innervation of biceps femoris presents unique considerations in neurological assessments and rehabilitation strategies, particularly in cases of sciatic nerve injuries or lumbar radiculopathies affecting L5-S2 segments.

The biceps femoris, in synergy with the other hamstring muscles, contributes to hip extension, particularly when elevating the trunk from a forward-bent position to an erect stance. During forward trunk flexion, all three hamstrings engage in eccentric contraction to regulate the movement. At the knee joint, the biceps femoris acts in concert with semimembranosus and semitendinosus to facilitate flexion. Additionally, when the knee is slightly flexed, the biceps femoris causes lateral rotation to the leg relative to the thigh or, if the foot is fixed in place, induces medial rotation of the thigh and pelvis on the leg (Koulouris and Connell 2003; Schache et al. 2012).

While knee flexion and joint stabilization are critical functions of the hamstrings, these actions alone do not necessitate such substantial muscle mass. Similarly, hip extension with the thigh as the moving segment could be accomplished by a smaller muscle group, especially considering the advantageous positioning of the gluteus maximus for this action. However, the considerable power required to elevate the trunk from a flexed position justifies the hamstrings' size, as they operate with a very short lever arm at the ischium and its ramus against the significant weight of the trunk acting on the contralateral side of the hip joint (Heiderscheit et al. 2010).

In static posture, the hamstrings contribute significantly to pelvic equilibrium, especially during upper trunk deviations from vertical alignment. Working in coordination with the abdominal muscles anterosuperiorly and the gluteus maximus posteroinferiorly, they modulate anteroposterior pelvic tilt, thereby influencing the degree of lumbar lordosis (Neumann 2010).

The unique biomechanical demands placed on the hamstrings may explain their susceptibility to injury during athletic activities. Sports-related hamstring injuries frequently occur in the initial 10–20 m of a sprint. While inadequate warm-up is often cited as a contributing factor, it is essential to consider the complex dual-joint action of these muscles during this phase of acceleration (Opar et al. 2012). At the commencement of a race, athletes adopt a forward-leaning posture to maximize forward propulsion, a position further accentuated by starting blocks. In this configuration, the hamstrings are maximally engaged, either to bring the trunk to an upright position or to maintain a posture that borders on forward collapse. Simultaneously, the lower limb is thrust forward with knee flexion to cover ground without premature foot contact. This combination of actions places immense strain on the hamstring muscles, potentially leading to tissue failure (Chumanov et al. 2011).

During gait, the hamstrings serve a crucial function in decelerating the forward swing of the tibia as the free leg extends, preventing hyperextension of the knee joint. This action is vital for maintaining smooth and controlled locomotion (Cappellini et al. 2006).

Gastrocnemius (Innervated by the Tibial Nerve: S1, S2)

The gastrocnemius muscle (Fig. 8.33) is primarily recognized for its role as a powerful plantarflexor (flexor) of the ankle joint. However, it also acts as a significant knee flexor in open-

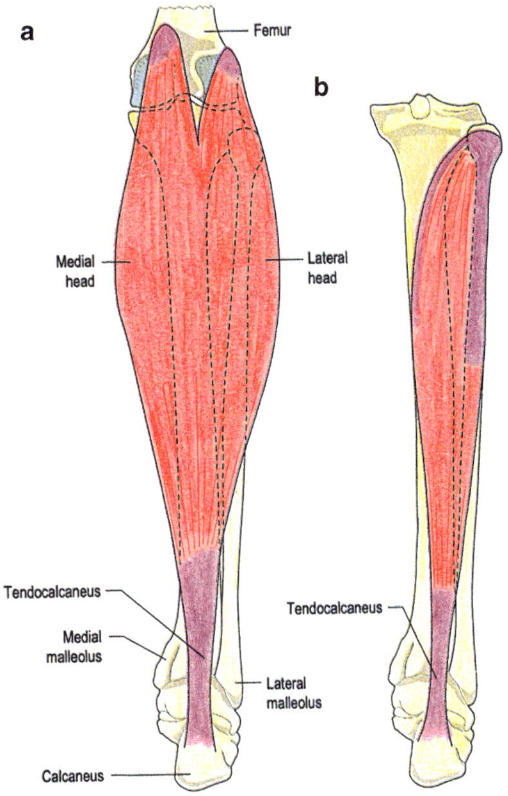

Fig. 8.33 (**a**) Right gastrocnemius and (**b**) right soleus. (Adapted with permission from Palastanga and Soames (2019))

kinetic chain movements and functions as a knee extensor in closed-kinetic chain movements.

The muscle's medial and lateral heads traverse the knee joint on their respective sides. When the foot is stabilized, the gastrocnemius becomes active, assisting in pulling the body forward over the feet. This motion is particularly evident during activities like sliding forward on a rowing seat or repositioning a fully reclined body. The muscle's role is crucial in movements such as turning or repositioning in bed or on a plinth, a key consideration given that individuals spend nearly a third of their lives in a supine position.

The gastrocnemius plays a role in various functional activities, particularly as a knee extensor during closed-chain movements when the foot is in contact with a surface (Harvey 2008; Ko 2022). Its dual action at both the ankle and knee joints enhances its versatility in locomotion and postural control (Neumann 2017).

Gracilis (Innervated by the Anterior Division of the Obturator Nerve: L2, L3)

The gracilis, a slender, elongated muscle, is positioned on the medial aspect of the thigh. It is the most superficial muscle of the adductor group. The muscle originates from the anterior surface of the pubic body and its inferior ramus, with a slight extension onto the ischial ramus. As it descends between the semimembranosus posteriorly and the sartorius anteriorly, the gracilis develops a fusiform-shaped belly approximately at its midpoint. The muscle becomes tendinous superior to the knee joint and crosses it before expanding to insert on a short vertical line on the proximal portion of the medial tibial shaft. This insertion is superior to that of the semitendinosus and posterior to, and blending with, the sartorius attachment. Bursae are present to reduce friction between the gracilis tendon and those of the sartorius and semitendinosus (Standring 2020).

Although the gracilis is categorized within the adductor muscle group, its contribution to hip adduction is secondary compared to its more prominent role at the knee. It primarily functions as a knee flexor and assists in medial rotation of the leg when the knee is semiflexed.

As a knee flexor, the gracilis supports the hamstring muscles during various functional activities, such as initiating knee flexion at the beginning of the swing phase during gait. The muscle is also used during activities requiring forceful knee flexion, such as pulling the body forward on a sliding seat in a rowing machine. In horse riding, the gracilis demonstrates its versatility by contributing to all its actions. When a rider grips the horse, the gracilis aids the adductor muscle group while simultaneously assisting in knee flexion control (Neumann 2017).

Sartorius (Innervated by the Femoral Nerve: L2, L3)

The sartorius, recognized as the longest muscle of the body, is a superficial strap-like structure in the anterior thigh compartment (Standring 2020). This muscle, named for its role in facilitating the cross-legged position historically adopted by tailors, exhibits a unique configuration with flattened tendons at both ends (Netter 2019).

Originating from the anterior superior iliac spine and its immediate inferior area, the sartorius courses medially and inferiorly across the thigh (Moore et al. 2018). Its distal attachment is on a vertical line along the medial tibial shaft, anterior to semitendinosus and gracilis insertions. The lower tendon partially merges with gracilis and extends fibers to the medial collateral ligament of the knee and the leg fascia. A bursa separates sartorius from gracilis distally, facilitating smooth movement. Topographically, the medial border of the sartorius defines the lateral border of the femoral triangle in its proximal third. The middle third constitutes the roof of the adductor canal (Schuenke et al. 2014).

Functionally, the sartorius acts on multiple joints, contributing to various lower limb movements. It aids in hip flexion, knee flexion, thigh abduction and lateral rotation, and medial rotation of the tibia on the femur (Neumann 2017). This action combination allows placement of the heel on the medial side of the opposite knee.

While assuming the cross-legged or tailor sitting posture is a common functional activity, the sartorius also assists in other movements that involve simultaneous knee and hip flexion with lateral rotation of the thigh. A notable example is during the breaststroke in swimming, where the muscle aids in drawing the lower limbs upward (Magee et al. 2016).

8.8.2.3 Medially Rotating Muscles of the Tibia at the Knee Joint: Semitendinosus, Semimembranosus, Gracilis, Sartorius, Popliteus

Popliteus (Innervated by the Tibial Division of the Sciatic Nerve)

The popliteus, a deep-seated muscle in the popliteal fossa, exhibits a triangular morphology and is located inferior and lateral to the knee joint (Palastanga and Soames 2019). Its origin is intracapsular, arising from a tendinous attachment on the anterior aspect of the lateral femoral condyle's groove, distal to the lateral epicondyle and the lateral collateral ligament insertion (Moore et al. 2018). The muscle's tendon follows a posteromedial course, traversing the joint line over the lateral meniscus's outer border, to which it attaches. This upper portion of the muscle is enveloped in a double layer of synovial membrane while it remains within the joint capsule. Upon exiting the capsule beneath the arcuate popliteal ligament, the muscle takes on a fleshy origin (Standring 2020). The popliteus then continues its course downward and medially, attaching via fleshy fibers to a triangular area on the posterior surface of the tibia above the soleal line, as well as to the fascia covering the muscle.

Functionally, when the foot is weight-bearing, the popliteus functions lateral rotation of the femur on the tibia. This action is critical in disengaging the knee from its close-kinetic position, enabling flexion (Neumann 2017). The muscle exerts a posterior force on the lateral femoral condyle's outer surface, causing it to rotate laterally around a vertical axis slightly medial to its center. This lateral rotation of the femur facilitates forward gliding of the medial condyle, thereby releasing the ligaments and muscles involved in maintaining the close-kinetic position (Kapandji 2007).

During forceful knee flexion, the popliteus activates to draw the tibia posteriorly relative to the femoral condyles. In non-weight-bearing conditions, it also assists the medial hamstrings in tibial medial rotation (Levangie and Norkin 2005). The popliteus plays a role in meniscal dynamics through its attachment to the lateral meniscus. During femoral lateral rotation, it pulls the meniscus posteriorly, preventing its entrapment between the articulating surfaces. Some researchers posit that this function may contribute to the lower incidence of lateral meniscus injuries compared to medial meniscus injuries (Jadhav et al. 2014; LaPrade et al. 2007).

8.8.2.4 Laterally Rotating Muscles of the Tibia at the Knee Joint: Biceps Femoris

As previously described, the biceps femoris is a key muscle responsible for the lateral rotation of the tibia at the knee joint. The muscle consists of two heads—the long head, which originates from the ischial tuberosity, and the short head, which originates from the lateral lip of the linea aspera of the femur. Both heads converge and insert onto the head of the fibula.

During knee flexion, particularly when the foot is off the ground, the biceps femoris contracts to produce lateral rotation of the tibia relative to the femur. This action plays a crucial role in activities requiring precise rotational control of the lower limb, such as pivoting and balancing movements.

Additionally, when the knee is extended, the biceps femoris assists in stabilizing the knee joint by counteracting excessive medial rotation, ensuring balanced movement during walking, running, and other dynamic activities.

8.8.3 Muscles of the Ankle Joint

8.8.3.1 Extending Muscles of the Ankle Joint: Tibialis Anterior, Extensor Digitorum Longus, Extensor Hallucis Longus, Peroneus Tertius

Tibialis Anterior (Innervated by the Deep Peroneal Nerve: L4, L5)

The tibialis anterior is a long, fusiform muscle located on the leg's anterior aspect, lateral to the tibial anterior border (Standring 2020) (Fig. 8.34).

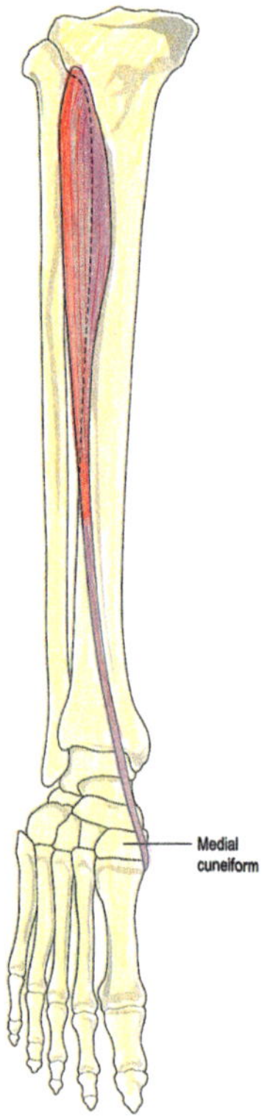

Fig. 8.34 Right tibialis anterior, anterior view. (Adapted with permission from Palastanga and Soames (2019))

This muscle is encased in a strong fascia and originates from the deep surface of this fascia, the upper two-thirds of the lateral surface of the tibia, and the adjacent interosseous membrane (Moore et al. 2018). In its distal third, the muscle transitions into a tendon that courses medially over the distal end of the tibia. This tendon traverses both superior and inferior extensor retinacula before inserting into the medial aspect of the medial cuneiform and the base of the first metatarsal bone. The insertion extends to the undersurface of both bones, blending with the peroneus longus tendon (Kahle et al. 1986).

Functionally, the tibialis anterior primarily dorsiflexes or extends the foot at the ankle joint. When acting in conjunction with the tibialis posterior, it inverts the foot, where the sole faces medially (Neumann 2017). In addition to its role in dorsiflexion, the tibialis anterior plays an important role in maintaining postural equilibrium. It collaborates with surrounding muscles to maintain body balance when the body weight shifts during upper-body activities (Magee 2014). During gait, the tibialis anterior serves multiple functions. It dorsiflexes the foot during the swing phase, preventing foot drop and toe drag. Additionally, it controls foot placement following initial heel contact.

Excessive tibialis anterior activity can lead to characteristic wear patterns on the posterolateral aspect of the heel due to frictional forces between the shoe and ground. Following heel contact, the rest of the foot gradually lowers in a controlled manner, adapting to surface irregularities (Whittle 1991).

The tibialis anterior, along with other dorsiflexors, is crucial in controlling forefoot lowering during walking or running, especially over uneven terrain. The anterior leg muscles are contained within a particularly taut fascia, limiting tissue expansion. This can lead to muscle compression during activity and strain on fascial attachments, particularly at bony interfaces, potentially resulting in "shin splints." Foot drop occurs when the tibialis anterior muscle becomes paralyzed, causing a lack of strength in the remaining dorsiflexors to lift the toes and prevent dragging of the foot on the ground. Patients may adjust for this condition by intentionally increasing the degree of knee flexion during walking, a phenomenon known as steppage gait (Gage et al. 2009).

8.8.3.2 Flexing Muscles of the Ankle Joint: Gastrocnemius, Soleus, Plantaris, Peroneus Longus, Peroneus Brevis, Tibialis Posterior, Flexor Digitorum Longus, Flexor Hallucis Longus

The calf muscles primarily function to plantarflex (flex) the foot at the ankle joint. Gastrocnemius and soleus play distinct roles in this action (Neumann 2017). Gastrocnemius acts as the primary propulsive force, mainly affecting the ankle while also contributing to knee flexion when strongly activated. Soleus, due to its anatomical position, functions more as a postural muscle. Its distal attachment serves as the fixed point, preventing anterior translation of the tibia under body weight, as the vertical projection of the center of gravity of the body typically falls anterior to the axis of rotation of the ankle joint.

The composition of these muscles reflects their functional specialization. Gastrocnemius consists predominantly of fast-twitch fibers, giving it a pale appearance, and thus is often referred to as "white" muscle. Conversely, soleus is composed mainly of slow-twitch fibers, resulting in a reddish appearance and thus termed "red" muscle (Scott et al. 2001).

Plantaris contributes minimally to ankle plantarflexion but gains clinical significance due to its susceptibility to injury. A condition known as "tennis leg" occurs when the plantaris tendon ruptures, often during activities like tennis. Patients typically report a sensation akin to being struck by a tennis ball on the posterior aspect of the calf. Complete tendon rupture may necessitate surgical intervention. The peroneus longus and brevis, while primarily evertors, also contribute to ankle plantarflexion. The tibialis posterior, primarily an invertor, assists in plantarflexion as well. The flexor digitorum longus and flexor hallucis longus, in addition to their actions on the toes, also aid in ankle plantarflexion (Moore et al. 2018).

Gastrocnemius (Innervated by the Tibial Nerve: S1, S2)

The gastrocnemius, a primary contributor to calf morphology, is situated on the posterior aspect of the leg, with its muscular bulk predominantly in the upper half (Standring 2020) (Fig. 8.33a). In conjunction with soleus and plantaris, it forms a composite structure known as the triceps surae. The two heads of gastrocnemius delineate the inferior boundaries of the popliteal fossa, a feature most apparent during knee flexion (Moore et al. 2018). The medial and lateral heads of gastrocnemius originate from distinct areas of the femur. The medial head arises from the popliteal surface of the femur, posterior to the medial supracondylar ridge and the adductor tubercle. The lateral head originates from the external surface of the lateral femoral condyle, superior and posterior to the lateral epicondyle. Both heads have additional attachments to the knee joint capsule and the oblique popliteal ligament (Kahle et al. 1986).

The muscle fibers from each head converge, without blending, to insert into the posterior surface of a broad, membranous tendon. This tendon fuses with the soleus tendon to form the proximal portion of the tendocalcaneus (Achilles tendon). The tendon narrows and becomes more cylindrical until approximately three fingers' breadth superior to the calcaneus, where it begins to broaden again before inserting into the middle portion of the posterior calcaneal surface (Doral et al. 2010).

The tendon fibers exhibit a helical arrangement from the myotendinous junction to their insertion, with the most medial fibers superiorly becoming posterior at the insertion site. This spiral configuration minimizes buckling when the tendon is lax and reduces deformation of individual strands under tension (Benjamin et al. 2008). A bursa is situated between the tendon and the superior aspect of the calcaneus, while a fat pad interposes between the tendon and the posterior ankle joint. Inferior to the insertion lies the fat pad of the heel (Theobald et al. 2005).

Gastrocnemius, in conjunction with soleus, serves as the primary plantarflexor of the ankle joint, providing the main propulsive force for locomotion (Neumann 2017). Its anatomical arrangement, crossing both the knee and ankle joints, allows it to function as a powerful knee flexor as well. However, this bi-articular nature introduces a biomechanical constraint: the muscle cannot exert maximal force at both joints simultaneously (Lichtwark and Wilson 2006). For instance, when the knee is flexed, gastrocnemius's ability to generate maximum force at the ankle joint is reduced, and similarly, its power at the knee is limited if the ankle is plantarflexed. This relationship is crucial in understanding the muscle's role in various functional activities and has implications for rehabilitation and performance optimization (Fukunaga et al. 2001).

In locomotor activities such as running, walking, and jumping, the gastrocnemius contributes significantly to the propulsive force. The triceps surae are recognized as one of the body's most powerful muscle groups because of the amount of strength it needs to move the body during these exercises. The gastrocnemius-soleus complex can generate forces several times body weight during explosive activities (Komi 1990).

The habitual use of high-heeled footwear can lead to adaptive shortening of gastrocnemius fibers, as this footwear chronically approximates the muscle's attachment points. This adaptation can result in limited ankle dorsiflexion, potentially causing discomfort or difficulty when walking in flat shoes or barefoot. Furthermore, the possibility of gastrocnemius shortening caused by footwear choices emphasizes the significance of educating patients and implementing preventive measures to ensure adequate function of the lower limbs (Cronin 2014; Csapo et al. 2010).

Soleus (Innervated by the Tibial Nerve: S1, S2)
Located deep to the gastrocnemius, the soleus is a broad, flat muscle, with a wider middle section that tapers as it descends (Fig. 8.33b). It originates from the soleal line on the posterior surface of the tibia, the posterior aspect of the upper third of the fibula, including the fibular head, and a fibrous arch connecting these bony attachments (Moore et al. 2018). The muscle fibers run downward, forming a belly that extends about halfway down the calf, leading to a membranous tendon on its deep surface. This tendon lies posteriorly and glides over a corresponding tendon on the deep surface of the gastrocnemius, allowing the muscles to move independently. The tendons of both muscles eventually merge to form the upper part of the tendocalcaneus (Achilles tendon), which travels behind the ankle joint and inserts into the middle portion of the posterior surface of the calcaneus (Doral et al. 2010).

The soleus muscle is essential for bipedal walking as it is one of the main muscles responsible for ankle plantarflexion. However, its function goes beyond simple locomotion. The soleus muscle plays an important role in maintaining posture by preventing the body from falling forward at the ankle while standing (Pandy and Andriacchi 2010). Due to its consistent and sustained activity, this muscle is highly suitable for actions that need endurance, which is why it is referred to as the "antigravity muscle."

Interestingly, the soleus also contributes significantly to cardiovascular health through its function as the "soleal pump." During standing and walking, intermittent contractions of this muscle facilitate venous return from the lower extremities. This action is made possible by the network of communicating vessels that traverse the muscle, connecting the deep and superficial venous systems (Meissner et al. 2007).

Understanding the soleus's anatomy and function is crucial for healthcare professionals involved in lower limb rehabilitation and sports medicine. Its role in postural control makes it a target for interventions aimed at improving balance and reducing fall risk, particularly in older adults (Onambele et al. 2006). The soleal pump mechanism is important for maintaining circulatory health and plays a significant role in managing venous insufficiency and preventing diseases such as deep vein thrombosis (Kahn et al. 2008).

8.8.3.3 Inverting Muscles of the Ankle and Foot: Tibialis Anterior, Tibialis Posterior

Tibialis Posterior (Innervated by the Tibial Nerve: L4, L5)
The tibialis posterior is the deepest muscle in the posterior compartment of the leg (Fig. 8.35). It originates from the upper half of the lateral aspect of the posterior surface of the tibia below the soleal line, the interosseous membrane, the posterior surface of the fibula between the medial crest and interosseous border, and the fascia covering its posterior aspect. The tendon of the tibialis posterior, which is enclosed in its own synovial sheath, passes behind the medial malleolus, where it grooves the bone, running medial to the tendons of the flexor hallucis longus and flexor digitorum longus. The tendon is superficial to the deltoid ligament and continues downward beneath the plantar calcaneonavicular ligament. It attaches primarily to the tubercle on the medial side of the navicular and the plantar surface of the medial cuneiform. Additional tendinous expansions extend to the plantar surfaces of all tarsal bones except the talus, including a strip that reaches the tip of the sustentaculum tali and the bases of the middle three metatarsals (Kelikian 2011).

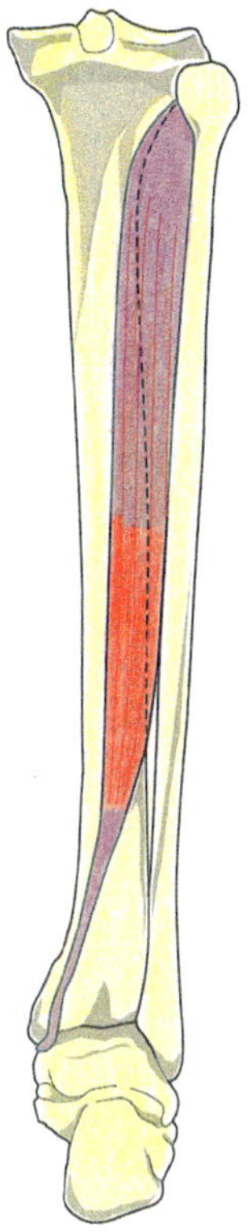

Fig. 8.35 Right tibialis posterior, posterior view. (Adapted with permission from Palastanga and Soames (2019))

It allows the tibialis posterior to exert a profound influence on foot mechanics. Primarily, it serves as the foot's principal invertor, working alongside the tibialis anterior to rotate the forefoot, orienting the sole medially (Maharaj et al. 2016). This action is crucial in activities ranging from the subtle adjustments of balance to the powerful pushoffs in sprinting.

While its role in plantarflexion is secondary to the calf muscles, the tibialis posterior shines in its ability to fine-tune foot position. It is a key player in controlling the midtarsal joint, where much of the foot's adaptive capability lies. By influencing the movement of the navicular and cuboid against the talus and calcaneus, it helps the foot conform to uneven terrain and absorb the shocks of locomotion (Blackwood et al. 2005). In addition, its contraction helps approximate the tibia and fibula, ensuring the ankle mortise maintains a firm grip on the talus during plantarflexion (Norkus and Floyd 2001). This action contributes to the ankle's stability, a critical factor in preventing injuries during dynamic activities.

The tibialis posterior has a significant but often overlooked role in supporting the arch structure of the foot. The multiple tendinous expansions in the foot provide a functional purpose by supporting and stabilizing the several arches of the foot. They play an important role in evenly distributing the body's weight and absorbing the impact of movement during walking or running (Kido et al. 2011). Dysfunction of this muscle, such as in tibialis posterior tendon dysfunction, can lead to significant alterations in foot biomechanics. The result can be an acquired flatfoot deformity, a condition that can cause chronic pain and significantly impact a person's mobility and quality of life (Kohls-Gatzoulis et al. 2004).

8.8.3.4 Everting Muscles of the Ankle and Foot: Peroneus Longus, Peroneus Brevis, Peroneus Tertius

Peroneus Longus (Innervated by the Superficial Peroneal Nerve: L5, S1)

The peroneus longus is a slender, fusiform muscle positioned on the lateral aspect of the leg (Fig. 8.36a). It is characterized by an elongated belly and an even longer tendon. This muscle is unique in that its tendon alters direction three times during its course to the insertion on the medial plantar surface of the foot (Moore et al. 2018).

The muscle originates from several sites: a limited area on the lateral condyle of the tibia (in conjunction with the extensor digitorum longus), the upper two-thirds of the lateral surface of the fibula, and the lateral aspect of the fibular head (Standring 2020). The lower half of its origin is situated posterior to the upper part of the peroneus brevis origin. Notably, a small region around the neck of the fibula remains unattached, allowing for the anterior passage of the common peroneal nerve. The muscle also attaches to the surrounding fascia and intermuscular septa both anteriorly and posteriorly (Netter 2019).

The tendon forms approximately a hand's breadth above the lateral malleolus and lies superficial to that of the peroneus brevis, sharing a common synovial sheath (Drake et al. 2020).

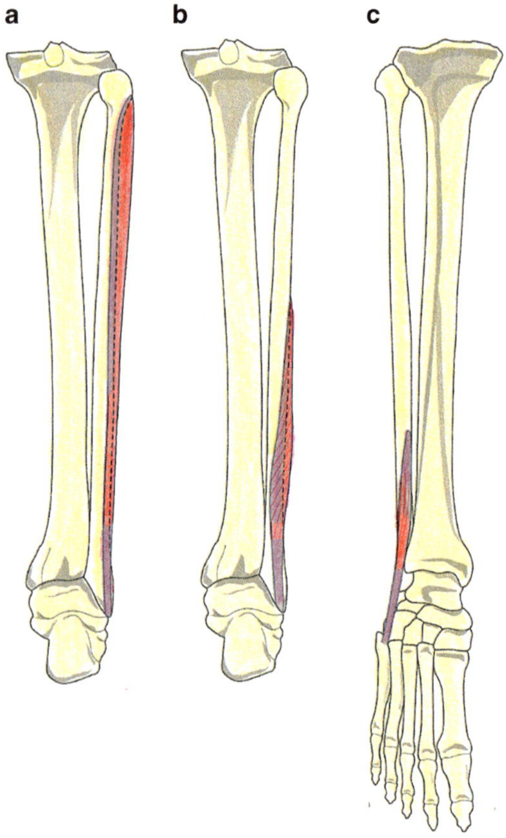

Fig. 8.36 (**a**) Right peroneus longus, (**b**) right peroneus brevis, and (**c**) right peroneus tertius. (Adapted with permission from Palastanga and Soames (2019))

It traverses a shallow groove posterior to the lateral malleolus, passing deep to the superior peroneal retinaculum. From this point, the tendon courses inferiorly and slightly anteriorly, passing inferior to the peroneal tubercle on the calcaneus, secured by the inferior band of the inferior peroneal retinaculum. At this juncture, the tendon is encased in a separate synovial sheath (Gilroy and MacPherson 2016). As the tendon reaches the inferolateral aspect of the cuboid bone, which it grooves, it changes direction to enter the groove on the inferior surface of the cuboid. This groove is transformed into a tunnel by fibers from the long plantar ligament and the tibialis posterior tendon (Agur and Dalley 2016). Within this tunnel, the tendon maintains its synovial sheath covering. The tunnel guides the tendon anteromedially across the foot to its insertion points on the plantar and lateral surfaces of the medial cuneiform and the base of the first metatarsal (Schuenke et al. 2014).

The peroneus longus functions as an evertor of the foot due to its origin on the lateral aspect of the leg and its course around the lateral foot (Neumann 2017). As it traverses from the posterior to the lateral malleolus to its insertion on the medial cuneiform and first metatarsal, it induces plantarflexion of the foot. This action results in a downward pull of the foot's medial aspect, contributing to pronation (Kapandji 2007).

Interestingly, the peroneus longus inserts into the same bones as the tibialis anterior, although the tibialis anterior approaches from the medial side. This arrangement is thought to create a supportive sling for the foot's arches, aiding in the regulation of arch height during dynamic activities (Standring 2020). The shared attachment of these muscles to the medial cuneiform and the base of the first metatarsal underscores the significance of medial foot control during various activities, particularly on uneven surfaces.

In the standing position, the peroneus longus, along with surrounding musculature, contributes to maintaining an upright posture. It stabilizes mediolateral balance by exerting pressure on the ground through the foot's medial aspect. This function becomes more apparent during single-leg stance, where the peroneus longus works intensively to maintain the leg's position over the foot and prevent contralateral body tilt. However, the primary functional role of the peroneus longus is observed during forceful foot actions, such as running, especially on irregular terrain. In these scenarios, its control over the foot's medial side and the first metatarsal (which bears the hallux) is crucial, working in conjunction with the tibialis anterior (Dicharry 2010).

Peroneus Brevis (Innervated by the Superficial Peroneal Nerve: L5, S1)

The peroneus brevis is located on the lateral aspect of the leg, sharing an osseofascial compartment with the peroneus longus (Fig. 8.36b). Its origin is from the lower two-thirds of the fibula's lateral surface, with the upper half situated anterior to the peroneus longus. The muscle also attaches to the adjacent intermuscular septa (Moore et al. 2018). The muscle's belly is short and fusiform, quickly transitioning into a tendon. This tendon accompanies that of the peroneus longus, passing posterior to the lateral malleolus within a shared synovial sheath. From there, it courses anteroinferiorly into a groove superior to the fibular tubercle on the calcaneus. The superior band of the inferior peroneal retinaculum converts this groove into a tunnel. The tendon then proceeds anteriorly to insert into the tubercle on the lateral aspect of the base of the fifth metatarsal bone (Drake et al. 2020). Proximal to the tubercle, the tendon is enveloped by a synovial sheath distinct from that of the peroneus longus. Typically, a slip from the tendon joins the long extensor tendon of the fifth toe. Additional separate slips may connect to the peroneus longus, calcaneus, or cuboid (Standring 2020).

The primary function of the peroneus brevis is foot eversion. Due to its anatomical course and attachments, its tendon's pull simultaneously produces ankle plantarflexion (Neumann 2017). This muscle is well-positioned to stabilize mediolateral sway during standing. In a single-leg stance, it helps prevent contralateral body tilt, effectively working with reversed origin and insertion. During gait or running, particularly on uneven terrain, it plays an important role in controlling foot position and should prevent excessive inversion (Dicharry 2010). However, this mechanism may not always function optimally. In some instances, the foot may over-invert, causing weight to be borne on the lateral foot aspect, further forcing the foot into inversion. This scenario can lead to severe damage or even rupture of the muscle tendon, and often affects the anterior talofibular ligament of the ankle joint as well (Hertel 2002).

Peroneus Tertius (Innervated by the Deep Peroneal Nerve: L5, S1)

Peroneus tertius is positioned on the lower lateral aspect of the leg (Fig. 8.36c). This muscle is thought to be a separate portion of the extensor digitorum longus. Its origin is from the anterior surface of the lower quarter of the fibula, in direct continuity with the extensor digitorum longus, with no intervening gap. It also arises from the intermuscular septum and adjacent fascia (Moore et al. 2018). The muscle fibers course inferolaterally, forming a tendon that passes deep to the superior extensor retinaculum and through the inferior extensor retinaculum. The tendon inserts into the medial and dorsal aspects of the fifth metatarsal's base (Drake et al. 2020).

Functionally, the peroneus tertius acts as a weak evertor and dorsiflexor of the foot at the ankle joint. The significance of this small muscle is challenging to determine, as its actions seem to be covered by other muscles with superior mechanical advantage. In fact, it is absent in some individuals (Standring 2020).

Notably, the muscle traverses the anterior talofibular ligament of the ankle joint, a structure frequently compromised in inversion injuries. This anatomical positioning suggests that the peroneus tertius may play a role in mitigating excessive inversion during athletic activities, potentially reducing injury incidence. However, during severe inversion events, the muscle itself is susceptible to tearing or complete rupture, resulting in significant pain and swelling (Neumann 2017). It has been postulated that with the evolution of bipedalism, the peroneus tertius may be assuming greater importance. This hypothesis is based on the observation that foot eversion is a distinctly human characteristic (Dicharry 2010).

Despite its seemingly redundant actions, the strategic location of the peroneus tertius over the anterior talofibular ligament suggests a potential protective function. This positioning could contribute to ankle stability during activities that risk excessive inversion, although this protective capacity may be compromised in cases of violent inversion (Hertel 2002).

8.8.4 Muscles of the Toes

8.8.4.1 Extending Muscles of the Toes: Extensor Hallucis Longus, Extensor Digitorum Longus, Extensor Digitorum Brevis, Lumricals

Extensor Hallucis Longus (Innervated by the Deep Peroneal Nerve: L5, S1)

The extensor hallucis longus is a unipennate muscle situated deep to and between the tibialis

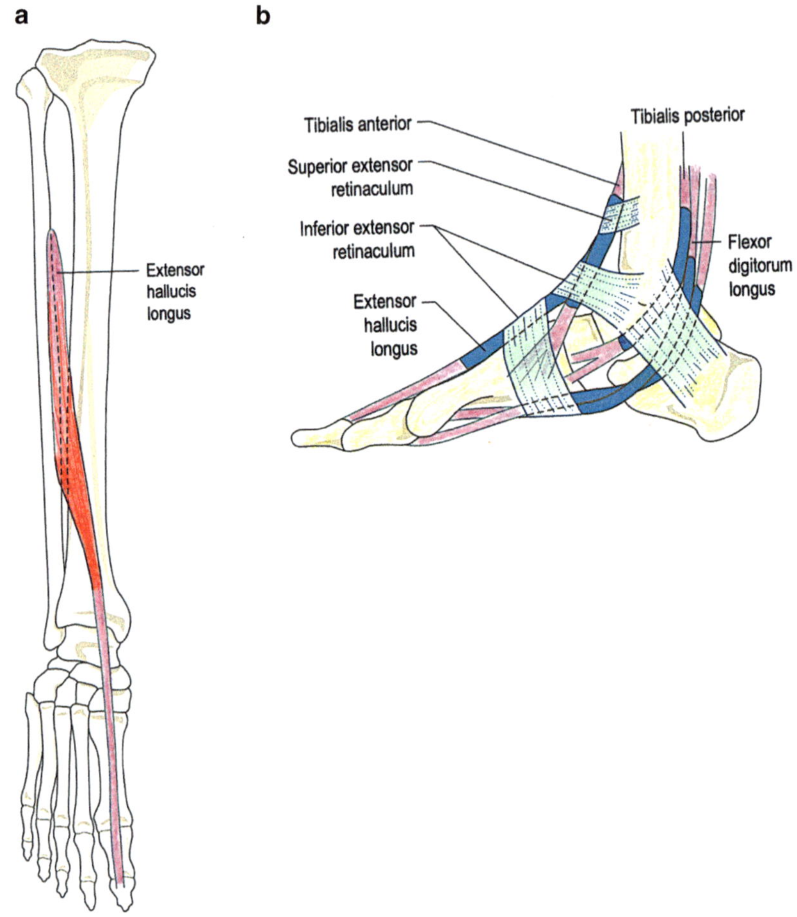

Fig. 8.37 (**a**) Extensor hallucis longus, anterior view, and (**b**) medial relations of the ankle joint. (Adapted with permission from Palastanga and Soames (2019))

anterior and extensor digitorum longus on the anterior aspect of the leg (Fig. 8.37). It originates from the middle half of the fibula's anterior surface and the adjacent interosseous membrane (Moore et al. 2018). The muscle fibers course inferomedially towards the tendon, which forms on its anterior surface. This tendon traverses beneath the superior extensor retinaculum and through the superior portion of the inferior extensor retinaculum in a distinct compartment, encased in its own synovial sheath. It then passes deep to the inferior band of the inferior extensor retinaculum as it approaches the base of the hallux (Drake et al. 2020). Typically, the tendon does not form a fully developed extensor hood. Instead, it attaches to the dorsal surface of the distal phalanx's base. Tendinous slips may extend to the dorsal aspect of the proximal phalanx's base and the first metatarsal (Standring 2020).

As its nomenclature suggests, the extensor hallucis longus extends all joints of the hallux, with primary action at the metatarsophalangeal joint. Additionally, it functions as a potent dorsiflexor of the foot at the ankle joint (Neumann 2017).

During running, the hallux is the final part of the foot to leave the ground, with the terminal propulsive force generated by the long flexors of the toes. Subsequently, the toe must return to an extended position while the foot simultaneously dorsiflexes and slightly inverts, preparing for heel strike in the next weight-bearing phase. The extension of the hallux and foot dorsiflexion also ensures ground clearance (Dicharry 2010). It's noteworthy that the hallux lacks associated lumbrical or interossei muscles. Consequently, extension of the interphalangeal joint relies solely on the extensor hallucis longus. Paralysis of this muscle results in joint flexion and toe buckling during the terminal phase of gait due to unopposed flexor muscle action (Hertel 2002).

Extensor Digitorum Longus (Innervated by the Deep Peroneal Nerve: L5, S1)

The extensor digitorum longus is a unipennate muscle situated on the anterior aspect of the leg, lateral to tibialis anterior and overlying extensor hallucis longus. It originates linearly from the upper two-thirds of the anterior surface of the fibula, the deep fascia, and the upper portion of the interosseous membrane, with some fibers extending laterally to the lateral condyle of the tibia, shared with the origin of the peroneus long (Moore et al. 2018).

The muscle fibers course downwards and medially, converging onto a tendon that becomes visible medially. This tendon passes over the anterior ankle joint, deep to the superior extensor retinaculum, and then through the inferior extensor retinaculum, accompanied by the peroneus tertius. At or immediately distal to the inferior extensor retinaculum, it divides into four tendons destined for the lateral four toes (Standring 2020). These four separate tendons share a common synovial sheath at the level of the inferior extensor retinaculum. On the dorsal surface of each proximal phalanx, the individual tendons form a triangular membranous structure known as the extensor hood or dorsal digital expansion. The lumbrical tendon joins this expansion medially, while the extensor digitorum brevis tendon attaches laterally for the second to fourth toes. The interossei of the foot do not contribute to the extensor hood (Kelikian 2011).

As it progresses over the proximal phalanx, the extensor hood divides into three components before reaching the dorsum of the proximal interphalangeal joint. The central portion attaches to the middle phalanx base, while the two lateral portions unite before inserting onto the distal phalanx base. Some studies have also described an attachment of the extensor hood to the dorsal aspect of the proximal phalanx (Riegger 1988).

Functionally, the extensor digitorum longus primarily extends the lateral four toes at the metatarsophalangeal joints and contributes to extension at the interphalangeal joints. However, it cannot perform the latter action independently, which is primarily executed by the lumbricals. In cases of lumbrical paralysis, extensor digitorum longus activation results in hyperextension of the metatarsophalangeal joint, accompanied by flexion of the interphalangeal joints. Due to its course across the anterior ankle joint, the extensor digitorum longus also aids in foot dorsiflexion (Neumann 2017). During gait and running, the extensor digitorum longus elevates the toes following their flexion prior to toe-off and maintains their clearance from the ground until the heel contact.

Extensor Digitorum Brevis (Innervated by the Deep Peroneal Nerve: L5, S1)

The extensor digitorum brevis is a slender muscle situated on the dorsal aspect of the foot, distal to the inferior portion of the extensor retinaculum. Its position is lateral to, and partially overlapped by, the tendons of peroneus tertius and extensor digitorum longus (Moore et al. 2018). The muscle originates from the anterior roughened area on the superior surface of the calcaneus, as well as from the deep fascia enveloping the muscle, which includes the stem of the inferior extensor retinaculum (Standring 2020). From its small belly, short tendons extend anteriorly and medially. The most medial of these tendons traverses the dorsalis pedis artery, inserting independently onto the dorsal aspect of the proximal phalanx base of the hallux. The other three tendons integrate with the lateral aspect of the extensor hood of the second, third, and fourth toes (Kelikian 2011). In some individuals, the most medial portion of the muscle may form a distinct belly, occasionally termed the extensor hallucis brevis.

Functionally, the medial segment of the muscle collaborates with the extensor hallucis longus to extend the great toe at the metatarsophalangeal joint. Concurrently, the remaining three tendons assist the extensor digitorum longus (Neumann 2017). Similar to the long extensor tendons, the extensor digitorum brevis contributes to the extension of the interphalangeal joints in conjunction with the lumbricals. However, it lacks the capacity to perform this action independently.

During ambulation and running, the extensor digitorum brevis functions in synergy with the extensor digitorum longus and extensor hallucis longus to extend the toes, ensuring they clear the ground. This coordinated action is essential for maintaining proper gait mechanics and preventing toe drag during the swing phase of the gait cycle.

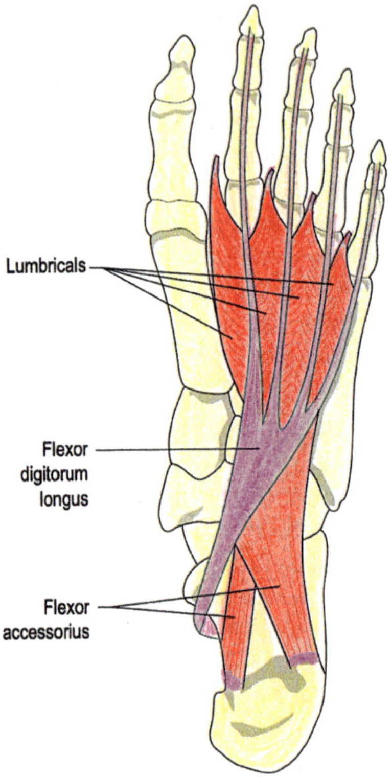

Fig. 8.38 Plantar aspect of the left foot showing the lumbricals, flexor digitorum longus, and flexor accessorius. (Adapted with permission from Palastanga and Soames (2019))

Lumbricals (Innervated the Medial Plantar Nerve for the Most Medial Muscle [S1, S2] and the Lateral Plantar Nerve for the Lateral Three Muscles [S2, S3])

The lumbricals are a set of four small muscles closely associated with the tendons of flexor digitorum longus, uniquely traversing from the flexor to the extensor compartment of the foot (Fig. 8.38). The most medial lumbrical originates from the medial aspect of the tendon of the second toe, adjacent to the flexor accessorius (quadratus plantae) attachment on the main longus tendon. The remaining three lumbricals have a distinctive origin pattern, each arising by two heads from the adjacent sides of two tendons: the second from the second and third toe tendons, the third from the third and fourth toe tendons, and the fourth from the fourth and fifth toe tendons (Moore et al. 2018). Each lumbrical muscle extends anteriorly, superficial to the deep transverse metatarsal ligament on the medial side of the toe. It then ascends obliquely to insert on the medial aspect of the extensor hood and the base of the proximal phalanx (Standring 2020).

Despite their small size, the lumbricals have been a subject of considerable debate regarding their functional role. They possess a relatively long muscle belly compared to their tendon length and serve as a unique bridge between the toe flexors and extensors. Their attachment to the proximal phalanx facilitates toe flexion at the metatarsophalangeal joint upon contraction. Concurrently, their insertion into the extensor hood enables extension of the interphalangeal joints, an action primarily attributed to the lumbricals rather than the long and short extensor tendons (Neumann 2017). A critical function of the lumbricals is preventing toe clawing during the propulsive phase of gait. In cases of lumbrical paralysis, the extensor muscles can pull the toes into hyperextension at the metatarsophalangeal joints, resulting in a clawed toe deformity even at rest.

Interestingly, the nerves innervating these muscles appear to contain a higher number of fibers than typically expected for such small muscles, with a significant proportion being sensory. This suggests that lumbricals may play a crucial role in providing proprioceptive feedback related to tension dynamics between the long flexor and extensor muscles. Such information is particularly valuable in locomotion, especially considering the lumbricals' attachment point is distant from the muscle bellies of the extensors and flexors (Riegger 1988). The complex relationship between the structure and function of lumbricals emphasizes their important function in maintaining optimal foot biomechanics.

8.8.4.2 Flexing Muscles of the Toes: Flexor Digitorum Longus, Flexor Digitorum Brevis, Flexor Hallucis Longus, Flexor Hallucis Brevis, Flexor Digiti Minimi Brevis, Lumbricals, Interossei

Flexor Hallucis Longus (Innervated by the Tibial Nerve: S1, S2)

The flexor hallucis longus is a powerful unipennate muscle located deep to the triceps surae, beneath the deep fascia of the calf. Its origin is from the inferior two-thirds of the posterior fibular surface and the adjacent fascia (Moore et al. 2018). The tendon descends, passing deep to the flexor retinaculum within its own synovial sheath, crossing the posterior aspect of the ankle joint lateral to flexor digitorum longus. During its course, it creates grooves in the distal tibia, the posterior talus (between medial and posterior tubercles), and the inferior surface of the sustentaculum tali, where a synovial-lined fibrous sheath forms a tunnel for its passage (Standring 2020).

In the plantar foot, the flexor hallucis longus tendon lies superficial to the plantar calcaneonavicular ligament, lateral to the flexor digitorum longus tendon. As it progresses anteriorly, the flexor hallucis longus tendon crosses deep to that of flexor digitorum longus, typically giving a slip to its medial two tendons. Subsequently, it traverses the fibrous digital sheath of the hallux, passing between the two sesamoid bones that surround the base of the proximal phalanx, and finally attaches to the plantar surface of the base of the distal phalanx (Kelikian 2011).

The flexor hallucis longus flexes all joints of the great toe, initially acting on the interphalangeal joint, followed by the metatarsophalangeal joint. As it crosses the ankle joint, it also contributes to plantarflexion of the foot. This muscle plays a significant role in locomotion, generating much of the final propulsive force from the foot during walking. At this stage of the gait cycle, the calf muscles have already exerted their maximum power, and the flexors of the lateral four toes are completing their peak contraction. Flexion of the hallux thus represents the terminal action before foot lift-off in preparation for the subsequent step. Additionally, the flexor hallucis longus contributes substantially to maintaining the medial longitudinal arch of the foot.

Flexor Digitorum Longus (Innervated by the Tibial Nerve: L5, S1, S2)

The flexor digitorum longus originates from the medial portion of the posterior tibial surface, inferior to the soleal line and the surrounding deep transverse fascia (Moore et al. 2018). The flexor digitorum longus tendon forms approximately three finger's breadth above the medial malleolus, positioned adjacent to the tibialis posterior tendon, which has crossed anterior to it and lies medially. The flexor digitorum longus tendon is situated medial to the tendon of extensor hallucis longus. It courses deep to the flexor retinaculum within its own synovial sheath along the medial aspect of the sustentaculum tali, occasionally creating a groove, before entering the plantar foot deep to the abductor hallucis (Standring 2020).

As it progresses anteriorly and laterally, the tendon crosses deep to the flexor hallucis longus tendon, typically receiving a slip that contributes to the medial two of its four branches. Approximately midway along the sole, on its lateral aspect, the tendon is joined by flexor accessorius (quadratus plantae). At this juncture, it divides into four individual tendons, one for each of the lateral four toes. The lumbrical muscles originate just distal to the attachment of flexor accessorius (Kelikian 2011).

Distal to the metatarsophalangeal joint, each tendon enters its respective fibrous sheath alongside the corresponding tendon of flexor digitorum brevis, which lies superficial to it. The brevis tendon then bifurcates, allowing the longus tendon to pass through and insert on the plantar surface of the distal phalanx base. Both tendons share a common synovial sheath.

Functionally, the flexor digitorum longus flexes the lateral four toes in a sequential manner: first flexing the distal interphalangeal joints, then the proximal interphalangeal joints, and finally the metatarsophalangeal joints. Its trajectory posterior to the medial malleolus enables it to contribute to ankle plantarflexion. However, when the ankle is plantarflexed, its toe flexion capacity is reduced (Neumann 2017). During the propulsive phase of locomotion, the flexor digitorum longus pulls the toes firmly towards the ground, maximizing grip and thrust during toe-off. In a standing posture, the toes tend to grip the ground, enhancing balance and stability.

Flexor Hallucis Brevis (Innervated by the Medial Plantar Nerve: S1, S2)

Flexor hallucis brevis is a short, deep muscle within the plantar aspect of the foot, positioned between abductor hallucis medially and flexor digitorum brevis laterally. Its origin is from the medial aspect of the plantar surface of the cuboid, posterior to the groove for peroneus longus, the adjacent surface of the lateral cuneiform, and the tendon of tibialis posterior (Moore et al. 2018).

The muscle fibers course anteriorly and medially towards the hallux, bifurcating into two bellies that lie on either side, deep to the tendon of flexor hallucis longus. Each belly's tendon inserts onto the corresponding side of the base of the proximal phalanx. The medial tendon unites with that of abductor hallucis, while the lateral tendon joins with adductor hallucis, forming common insertions. Small sesamoid bones develop within each tendon, articulating with shallow grooves on the head of the first metatarsal (Standring 2020).

The primary action of the flexor hallucis brevis is flexion of the metatarsophalangeal joint of the hallux. Additionally, it assists the flexor hallucis longus in generating the final propulsive force from the ground during locomotion. The collaborative insertion with abductor and adductor hallucis suggests a significant role in stabilizing the hallux during propulsion, likely to ensure maximal force generation (Neumann 2017). In cases of hallux valgus deformity, where the distal aspect of the hallux deviates laterally and the base medially, this propulsive thrust is compromised. Consequently, patients may experience difficulty in running or even walking, particularly at slower speeds. It is noteworthy that injuries to the sesamoid bones are not uncommon, especially in individuals who place considerable stress on the hallux. Such injuries can lead to inflammation at the articulation between the sesamoid bone and the metatarsal, resulting in significant pain and altered foot function (Boike et al. 2011).

Flexor Digitorum Brevis (Innervated by the Medial Plantar Nerve: S2, S3)

The flexor digitorum brevis is positioned in the plantar aspect of the foot, immediately deep to the central portion of the plantar aponeurosis. It is situated between the abductor hallucis medially and the abductor digiti minimi laterally (Moore et al. 2018). The muscle originates from the medial calcaneal tubercle, the deep surface of the central plantar aponeurosis, and the adjacent muscular septa. The flexor digitorum brevis fibers progress anteriorly through the mid-sole, diverging into four tendons destined for the lateral four toes. Just distal to the metatarsophalangeal joint, within their respective fibrous flexor sheaths, each tendon bifurcates to allow passage of the flexor digitorum longus tendon, which transitions from a deep to superficial position. The outer margins of each tendon's slips then reconverge after rotating nearly 180°, forming a shallow groove through which the flexor digitorum longus tendon glides. Upon crossing the proximal interphalangeal joint, the tendon splits again to insert bilaterally at the base of the middle phalanx (Standring 2020).

The primary action of flexor digitorum brevis is flexion of the proximal interphalangeal joints of the lateral four toes, followed by flexion at the metatarsophalangeal joints. This muscle plays a significant role, in conjunction with flexor digitorum longus, in generating propulsive force from the toes during locomotion (Neumann 2017).

8.8.4.3 Abducting Muscles of the Toes: Abductor Hallucis, Abductor Digiti Minimi, Dorsal Interossei

Abductor Hallucis (Innervated by the Medial Plantar Nerve: S1, S2)

Abductor hallucis is a powerful and functionally significant muscle located superficially on the medial aspect of the plantar foot, deep to the medial portion of the plantar aponeurosis (Moore et al. 2018). Its origins include the plantar aponeurosis, the plantar surface of the medial calcaneal tubercle, the flexor retinaculum, and the intermuscular septum separating it from flexor digitorum brevis. The muscle fibers progress anteriorly, forming a tendon that courses over the medial aspect of the metatarsophalangeal joint of the hallux. It inserts into the medial side of the proximal phalanx base, in conjunction with the tendon of flexor hallucis brevis (Standring 2020).

While hallux abduction itself is not typically a functional movement in daily activities, the muscle's strength and bulk suggest a more significant role in specific biomechanical functions. Its anatomical position along the medial foot, with attachments at both ends of the medial longitudinal arch, allows it to act as a dynamic support structure for the arch during propulsive activities (McKeon et al. 2015). The muscle's insertion on the medial aspect of the hallux contributes to maintaining the central alignment of this digit during flexion movements. When the abductor hallucis contracts forcefully, it not only moves the hallux medially but also positions the foot laterally, enhancing the alignment between the hallux and the medial foot.

References

Adkison DP, Bosse MJ, Gaccione DR, Gabriel KR. Anatomical variations in the course of the superficial peroneal nerve. J Bone Joint Surg Am. 1991;73(1):112–4.

Agur AMR, Dalley AF. Grant's atlas of anatomy. 14th ed. Philadelphia: Wolters Kluwer; 2016.

Almekinders LC, Temple JD. Etiology, diagnosis, and treatment of tendonitis: an analysis of the literature. Med Sci Sports Exerc. 1998;30(8):1183–90.

Aszmann OC, Dellon ES, Dellon AL. Anatomical course of the lateral femoral cutaneous nerve and its susceptibility to compression and injury. Plast Reconstr Surg. 1997;100(3):600–4.

Barbe MF, Barr AE. Inflammation and the pathophysiology of work-related musculoskeletal disorders. Brain Behav Immun. 2006;20(5):423–9.

Barber MD, Bremer RE, Thor KB, Dolber PC, Kuehl TJ, Coates KW. Innervation of the female levator ani muscles. Am J Obstet Gynecol. 2002;187(1):64–71.

Basso O, Johnson DP, Amis AA. The anatomy of the patellar tendon. Knee Surg Sports Traumatol Arthrosc. 2001;9(1):2–5.

Bauer M, Bergström B, Hemborg A, Sandegård J. Malleolar fractures: nonoperative versus operative treatment. A controlled study. Clin Orthop Relat Res. 1985;(199):17–27.

Bencardino J, Rosenberg ZS, Beltran J, Liu X, Marty-Delfaut E. Morton's neuroma: is it always symptomatic? AJR Am J Roentgenol. 2000;175(3):649–53.

Benjamin M, Ralphs JR. Fibrocartilage in tendons and ligaments—an adaptation to compressive load. J Anat. 1998;193(4):481–94.

Benjamin M, Kaiser E, Milz S. Structure-function relationships in tendons: a review. J Anat. 2008;212(3):211–28.

Blackwood CB, Yuen TJ, Sangeorzan BJ, Ledoux WR. The midtarsal joint locking mechanism. Foot Ankle Int. 2005;26(12):1074–80.

Boike A, Schnirring-Judge M, McMillin S. Sesamoid disorders of the first metatarsophalangeal joint. Clin Podiatr Med Surg. 2011;28(2):269–85.

Cappellini G, Ivanenko YP, Poppele RE, Lacquaniti F. Motor patterns in human walking and running. J Neurophysiol. 2006;95(6):3426–37.

Cassidy L, Walters A, Bubb K, Shoja MM, Tubbs RS, Loukas M. Piriformis syndrome: implications of anatomical variations, diagnostic techniques, and treatment options. Surg Radiol Anat. 2012;34(6):479–86.

Cheatham SW, Kolber MJ, Salamh PA. Meralgia paresthetica: a review of the literature. Int J Sports Phys Ther. 2013;8(6):883–93.

Chéron C, Le Scanff C, Leboeuf-Yde C. Association between sports type and overuse injuries of extremities in adults: a systematic review. Chiropr Man Therap. 2017;25:4.

Chumanov ES, Heiderscheit BC, Thelen DG. Hamstring musculotendon dynamics during stance and swing phases of high-speed running. Med Sci Sports Exerc. 2011;43(3):525–32.

Court-Brown CM, Caesar B. Epidemiology of adult fractures: a review. Injury. 2006;37(8):691–7.

Cramer GD, Darby SA. Clinical anatomy of the spine, spinal cord, and ANS. 3rd ed. St. Louis: Elsevier; 2014.

Cronin NJ. The effects of high heeled shoes on female gait: a review. J Electromyogr Kinesiol. 2014;24(2):258–63.

Csapo R, Maganaris CN, Seynnes OR, Narici MV. On muscle, tendon and high heels. J Exp Biol. 2010;213(15):2582–8.

Dicharry J. Kinematics and kinetics of gait: from lab to clinic. Clin Sports Med. 2010;29(3):347–64.

Doral MN, Alam M, Bozkurt M, Turhan E, Atay OA, Dönmez G, et al. Functional anatomy of the Achilles

tendon. Knee Surg Sports Traumatol Arthrosc. 2010;18(5):638–43.
Drake RL, Vogl AW, Mitchell AWM. Gray's anatomy for students. 4th ed. Philadelphia: Elsevier; 2020.
Dugan SA, Bhat KP. Biomechanics and analysis of running gait. Phys Med Rehabil Clin N Am. 2005;16(3):603–21.
Ellis H. Clinical anatomy: a revision and applied anatomy for clinical students. 5th ed. Oxford: Blackwell Scientific Publications; 1971.
Fong DT, Hong Y, Chan LK, Yung PS, Chan KM. A systematic review on ankle injury and ankle sprain in sports. Sports Med. 2007;37(1):73–94.
Fukunaga T, Kubo K, Kawakami Y, Fukashiro S, Kanehisa H, Maganaris CN. In vivo behaviour of human muscle tendon during walking. Proc Biol Sci. 2001;268(1464):229–33.
Gage JR, Schwartz MH, Koop SE, Novacheck TF. The identification and treatment of gait problems in cerebral palsy. 2nd ed. London: Mac Keith Press; 2009.
Galbraith RM, Lavallee ME. Medial tibial stress syndrome: conservative treatment options. Curr Rev Musculoskelet Med. 2009;2(3):127–33.
Garcia CA, Hoffman SL, Hastings MK, et al. Effect of metatarsal pads and orthotic devices on stress fractures and soft tissue injuries in sports: a meta-analysis. J Am Podiatr Med Assoc. 2018;108(2):116–27.
Gilroy AM, MacPherson BR. Atlas of anatomy. 3rd ed. New York: Thieme; 2016.
Golanó P, Vega J, de Leeuw PA, et al. Anatomy of the ankle ligaments: a pictorial essay. Knee Surg Sports Traumatol Arthrosc. 2010;18(5):557–69.
Goldstein B. Applied anatomy and electrodiagnosis of brachial plexopathies. Phys Med Rehabil Clin N Am. 1994;5:477–93.
Goost H, Wimmer MD, Barg A, Kabir K, Valderrabano V, Burger C. Fractures of the ankle joint: investigation and treatment options. Dtsch Arztebl Int. 2014;111(21):377–88.
Grood ES, Noyes FR, Butler DL, Suntay WJ. Ligamentous and capsular restraints preventing straight medial and lateral laxity in intact human cadaver knees. J Bone Joint Surg Am. 1981;63(8):1257–69.
Grothaus MC, Holt M, Mekhail AO, Ebraheim NA, Yeasting RA. Lateral femoral cutaneous nerve: an anatomic study. Clin Orthop Relat Res. 2005;437:164–8.
Harvey L. Management of spinal cord injuries: a guide for physiotherapists. Edinburgh: Elsevier; 2008.
Heiderscheit BC, Sherry MA, Silder A, Chumanov ES, Thelen DG. Hamstring strain injuries: recommendations for diagnosis, rehabilitation, and injury prevention. J Orthop Sports Phys Ther. 2010;40(2):67–81.
Hertel J. Functional anatomy, pathomechanics, and pathophysiology of lateral ankle instability. J Athl Train. 2002;37(4):364–75.
Hollinshead WH. Anatomy for surgeons. Vol. 3: the back and limbs. 2nd ed. New York: Harper & Row; 1969.
Hollinshead WH, Rosse C. Textbook of anatomy. 4th ed. New York: Harper & Row; 1985.
Hubbard TJ, Hertel J. Mechanical contributions to chronic lateral ankle instability. Sports Med. 2006;36(3):263–77.
Jadhav SP, More SR, Riascos RF, Lemos DF, Swischuk LE. Comprehensive review of the anatomy, function, and imaging of the popliteus and associated pathologic conditions. Radiographics. 2014;34(2):496–513.
Jonkers I, Stewart C, Spaepen A. The complementary role of the plantarflexors, hamstrings and gluteus maximus in the control of stance limb stability during gait. Gait Posture. 2003;17(3):264–72.
Kahle W, Leonhardt H, Platzer W. Color atlas of human anatomy, Vol. 1: Locomotor system. 3rd revised ed. New York: Thieme; 1986.
Kahn SR, Shrier I, Kearon C. Physical activity in patients with deep venous thrombosis: a systematic review. Thromb Res. 2008;122(6):763–73.
Kapandji AI. The physiology of the joints, Volume 2: the lower limb. 6th ed. Philadelphia: Churchill Livingstone; 2007.
Kapandji AI. The physiology of the joints, Volume 3: the spinal column, pelvic girdle and head. 6th ed. Philadelphia: Churchill Livingstone; 2008.
Kelikian AS, editor. Sarrafian's anatomy of the foot and ankle: descriptive, topographic, functional. 3rd ed. Philadelphia: Lippincott Williams & Wilkins; 2011.
Khan KM, Cook JL, Bonar F, Harcourt P, Åstrom M. Histopathology of common tendinopathies. Update and implications for clinical management. Sports Med. 1999;27(6):393–408.
Kido M, Ikoma K, Imai K, Maki M, Takatori R, Tokunaga D, et al. Load response of the tarsal bones in patients with flatfoot deformity: in vivo 3D study. Foot Ankle Int. 2011;32(11):1017–22.
Ko HY. Management and rehabilitation of spinal cord injuries. 2nd ed. Singapore: Springer; 2022.
Kohls-Gatzoulis J, Angel JC, Singh D, Haddad F, Livingstone J, Berry G. Tibialis posterior dysfunction: a common and treatable cause of adult acquired flatfoot. BMJ. 2004;329(7478):1328–33.
Komi PV. Relevance of in vivo force measurements to human biomechanics. J Biomech. 1990;23(Suppl 1):23–34.
Koulouris G, Connell D. Evaluation of the hamstring muscle complex following acute injury. Skeletal Radiol. 2003;32(10):582–9.
LaPrade RF, Morgan PM, Wentorf FA, Johansen S, Engebretsen L. The anatomy of the posterior aspect of the knee. An anatomic study. J Bone Joint Surg Am. 2007;89(4):758–64.
Last R. Anatomy: regional and applied. 7th ed. Edinburgh: Churchill Livingstone; 1984.
Lee MW, McPhee RW, Stringer MD. An evidence-based approach to human dermatomes. Clin Anat. 2008;21(5):363–73.
Lee WC, Wong WY, Kung E, Leung AK. Effectiveness of adjustable dorsiflexion night splint in combination with accommodative foot orthosis on plantar fasciitis: a randomized controlled trial. J Rehabil Res Dev. 2015;52(2):181–92.

Levangie PK, Norkin CC. Joint structure and function: a comprehensive analysis. 4th ed. Philadelphia: FA Davis; 2005.

Lichtwark GA, Wilson AM. Interactions between the human gastrocnemius muscle and the Achilles tendon during incline, level and decline locomotion. J Exp Biol. 2006;209(21):4379–88.

Lieberman DE, Raichlen DA, Pontzer H, Bramble DM, Cutright-Smith E. The human gluteus maximus and its role in running. J Exp Biol. 2006;209(11):2143–55.

Long C, Palastanga N, Soames R. Functional anatomy of the limbs and back. 10th ed. Philadelphia: Elsevier; 2017.

Lumley JSP. Surface anatomy: the anatomical basis of clinical examination. 4th ed. Philadelphia: Churchill Livingstone; 2008.

Maffulli N, Wong J, Almekinders LC. Types and epidemiology of tendinopathy. Clin Sports Med. 2003;22(4):675–92.

Magee DJ. Orthopedic physical assessment. 6th ed. St. Louis: Elsevier; 2014.

Magee DJ, Zachazewski JE, Quillen WS, Manske RC. Pathology and intervention in musculoskeletal rehabilitation. 2nd ed. Maryland Heights: Elsevier; 2016.

Maharaj JN, Cresswell AG, Lichtwark GA. The mechanical function of the tibialis posterior muscle and its tendon during locomotion. J Biomech. 2016;49(14):3238–43.

Majkrzak A, Johnston J, Kacey D, Zeller J. Variability of the lateral femoral cutaneous nerve: an anatomic basis for planning safe surgical approaches. Clin Anat. 2010;23(3):304–11.

Mancall EL, Brock DG. Gray's clinical neuroanatomy: the anatomic basis for clinical neuroscience. Philadelphia: Elsevier; 2011.

Mansfield PJ, Neumann DA. Essentials of kinesiology for the physical therapist assistant. 3rd ed. St. Louis: Elsevier; 2019.

Maruyama M, Feinberg JR, Capello WN, D'Antonio JA. The Frank Stinchfield Award: morphologic features of the acetabulum and femur: anteversion angle and implant positioning. Clin Orthop Relat Res. 2001;393:52–65.

McKeon PO, Hertel J, Bramble D, Davis I. The foot core system: a new paradigm for understanding intrinsic foot muscle function. Br J Sports Med. 2015;49(5):290.

Meissner MH, Moneta G, Burnand K, Gloviczki P, Lohr JM, Lurie F, et al. The hemodynamics and diagnosis of venous disease. J Vasc Surg. 2007;46(Suppl S):4S–24S.

Moore KL, Persaud TVN, Torchia MG. The developing human: clinically oriented embryology. 9th ed. Philadelphia: Elsevier; 2013.

Moore KL, Dalley AF, Agur AMR. Clinically oriented anatomy. 8th ed. Philadelphia: Wolters Kluwer; 2018.

Netter FH. Atlas of human anatomy. 7th ed. Philadelphia: Elsevier; 2019.

Neumann DA. Kinesiology of the hip: a focus on muscular actions. J Orthop Sports Phys Ther. 2010;40(2):82–94.

Neumann DA. Kinesiology of the musculoskeletal system: foundations for rehabilitation. 3rd ed. St. Louis: Mosby; 2017.

Norkus SA, Floyd RT. The anatomy and mechanisms of syndesmotic ankle sprains. J Athl Train. 2001;36(1):68–73.

Onambele GL, Narici MV, Maganaris CN. Calf muscle-tendon properties and postural balance in old age. J Appl Physiol. 2006;100(6):2048–56.

Opar DA, Williams MD, Shield AJ. Hamstring strain injuries: factors that lead to injury and re-injury. Sports Med. 2012;42(3):209–26.

Palastanga N, Soames R. Anatomy and human movement: structure and function. 7th ed. Edinburgh: Elsevier; 2019.

Pandy MG, Andriacchi TP. Muscle and joint function in human locomotion. Annu Rev Biomed Eng. 2010;12:401–33.

Parent A. Carpenter's human neuroanatomy. 9th ed. Baltimore: Williams & Wilkins; 1996.

Peltola H, Pääkkönen M. Acute osteomyelitis in children. N Engl J Med. 2014;370(4):352–60.

Perry J, Burnfield JM. Gait analysis: normal and pathological function. 2nd ed. Thorofare: SLACK Incorporated; 2010.

Powers CM. The influence of altered lower-extremity kinematics on patellofemoral joint dysfunction. J Orthop Sports Phys Ther. 2003;33(11):639–46.

Pufe T, Petersen WJ, Mentlein R, Tillmann BN. The role of vasculature and angiogenesis for the pathogenesis of degenerative tendons disease. Scand J Med Sci Sports. 2005;15(4):211–22.

Rab M, Ebmer J, Dellon AL. Anatomic variability of the ilioinguinal and genitofemoral nerve: implications for the treatment of groin pain. Plast Reconstr Surg. 2001;108(6):1618–23.

Riegger CL. Anatomy of the ankle and foot. Phys Ther. 1988;68(12):1802–14.

Sadler TW. Langman's medical embryology. 12th ed. Philadelphia: Wolters Kluwer; 2012.

Sarrafian SK. Anatomy of the foot and ankle: descriptive, topographic, functional. Philadelphia: J.B. Lippincott; 1983.

Savage-Elliott I, Murawski CD, Smyth NA, Golanó P, Kennedy JG. The deltoid ligament: an in-depth review of anatomy, function, and treatment strategies. Knee Surg Sports Traumatol Arthrosc. 2013;21(6):1316–27.

Schache AG, Dorn TW, Blanch PD, Brown NA, Pandy MG. Mechanics of the human hamstring muscles during sprinting. Med Sci Sports Exerc. 2012;44(4):647–58.

Schieber MH, Santello M. Hand function: peripheral and central constraints on performance. J Appl Physiol. 2004;96(6):2293–300.

Schoenwolf GC, Bleyl SB, Brauer PR, Francis-West PH. Larsen's human embryology. 5th ed. Philadelphia: Churchill Livingstone; 2015.

Schuenke M, Schulte E, Schumacher U. THIEME atlas of anatomy: general anatomy and musculoskeletal system. 2nd ed. New York: Thieme; 2014.

Scott W, Stevens J, Binder-Macleod SA. Human skeletal muscle fiber type classifications. Phys Ther. 2001;81(11):1810–6.

Seitz AL, McClure PW, Finucane S, Boardman ND 3rd, Michener LA. Mechanisms of rotator cuff tendinopathy: intrinsic, extrinsic, or both? Clin Biomech. 2011;26(1):1–12.

Shafik A, el-Sherif M, Youssef A, Olfat ES. Surgical anatomy of the pudendal nerve and its clinical implications. Clin Anat. 1995;8(2):110–5.

Sharma P, Maffulli N. Tendon injury and tendinopathy: healing and repair. J Bone Joint Surg Am. 2005;87(1):187–202.

Shankar K. Physiatric anatomic principles. Physical medicine and rehabilitation: state of the art reviews. Philadelphia: Hanley & Belfus, Inc. 1996;10(3).

Singh D. Netter's concise orthopaedic anatomy. 2nd ed. Philadelphia: Elsevier Saunders; 2017.

Smania N, Picelli A, Munari D, Geroin C, Ianes P, Waldner A, et al. Rehabilitation procedures in the management of spasticity. Eur J Phys Rehabil Med. 2010;46(3):423–38.

Smith DG, Michael JW, Bowker JH. Atlas of amputations and limb deficiencies: surgical, prosthetic, and rehabilitation principles. 3rd ed. Rosemont: American Academy of Orthopaedic Surgeons; 2005.

Snell RS. Clinical anatomy for medical students. Boston: Little, Brown; 1973.

Standring S. Gray's anatomy: the anatomical basis of clinical practice. 42nd ed. London: Elsevier; 2020.

Sunderland S. Nerves and nerve injuries. 2nd ed. London: Churchill Livingstone; 1978.

Syme G, Rowe P, Martin D, Daly G. Disability in patients with chronic patellofemoral pain syndrome: a randomised controlled trial of VMO selective training versus general quadriceps strengthening. Man Ther. 2009;14(3):252–63.

Tagliafico A, Bignotti B, Cadoni A, Perez MM, Martinoli C. Anatomical study of the iliohypogastric, ilioinguinal, and genitofemoral nerves using high-resolution ultrasound. Muscle Nerve. 2015;51(1):42–8.

Theobald P, Benjamin M, Nokes L, Pugh N. Review of the vascularisation of the human Achilles tendon. Injury. 2005;36(11):1267–72.

Tria AJ, Klein KS. An illustrated guide to the knee. New York: Churchill Livingstone; 1992.

van Dijk CN, Lim LS, Bossuyt PM, Marti RK. Physical examination is sufficient for the diagnosis of sprained ankles. J Bone Joint Surg Br. 1996;78(6):958–62.

Vogel KG, Koob TJ. Structural specialization in tendons under compression. Int Rev Cytol. 1989;115:267–93.

Warwick R, Williams PL. Gray's anatomy. 35th British ed. Philadelphia: WB Saunders; 1973.

Whittle MW. Gait analysis: an introduction. 1st ed. Oxford: Butterworth-Heinemann; 1991.

Wilder RP, Sethi S. Overuse injuries: tendinopathies, stress fractures, compartment syndrome, and shin splints. Clin Sports Med. 2004;23(1):55–81.

Windisch G, Braun EM, Anderhuber F. Piriformis muscle: clinical anatomy and consideration of the piriformis syndrome. Surg Radiol Anat. 2007;29(1):37–45.

Wineski LE. Snell's clinical anatomy by regions. 10th ed. Philadelphia: Wolters Kluwer; 2019.

Recommended Additional Reading

Calais-Germain B. Anatomy of movement. Revised ed. Seattle: Eastland Press; 1999.

Frost R. Applied kinesiology. A training manual and reference book of basic principles and practices. Berkeley: North Atlantic Books; 2002.

Hamill J, Knutzen KM, Derrick TR. Biomechanical basis of human movement. 4th ed. Philadelphia: Wolters Kluwer; 2015.

Jenkins DB. Hollinshead's functional anatomy of the limbs and back. 9th ed. St. Louis: Saunders; 2009.

Kahle W, Frotscher M. Nervous system and sensory organs. New York: Thieme; 2001.

Zatsiorsky VM, Prilutsky BI. Biomechanics of skeletal muscles. Champaign: Human Kinetics; 2012.

9 Functional Anatomy of the Lower Urinary Tract

The lower urinary system, comprising the urinary bladder and urethra, represents an essential component of the renal system. This anatomical network, which includes the kidneys and ureters, functions to filter metabolic waste from the bloodstream and maintain proper fluid balance in the body (Drake et al. 2020). Understanding the structural and functional aspects of this system is fundamental for comprehending various urological conditions and their management in spinal cord injuries.

The urinary bladder demonstrates a characteristic trilaminar wall structure consisting of the serosa (exterior), muscularis (middle), and mucosa (interior). Its anatomical position exhibits notable sexual dimorphism; in females, the bladder is situated inferior to the uterus and ileum and superior to the vagina and rectum, while in males, it is located superior to the prostate, anterior to the rectum, and inferior to the ileum and peritoneal cavity (Standring 2020).

Structurally, the bladder can be divided into two main regions: the body and the base. The body lies superior to the ureteral orifices, while the base consists of the trigone and bladder neck. These regions demonstrate characteristic neuromorphological and neuropharmacological properties that contribute to their specialized functions (el-Badawi and Schenk 1966). The bladder base exhibits a laminar architecture, featuring a superficial longitudinal layer beneath the trigone. A muscle layer located beneath this stratum integrates with the detrusor, predominantly exhibiting a circular arrangement (Dixon and Gosling 1987).

A significant sexual difference exists in the bladder neck structure. Males possess a complete ring of smooth muscle encircling the bladder neck, a feature absent in females (Gosling 1979). This structural difference has important functional implications, particularly in males, where the bladder neck facilitates antegrade ejaculation through its rich adrenergic innervation. The density of adrenergic innervation in the female bladder neck is comparatively lower.

The urethra, in conjunction with the pelvic floor musculature, constitutes the bladder outlet. Its composition includes both smooth and striated muscle components, with the latter forming the rhabdosphincter or external urethral sphincter. The blood supply to the bladder originates from the internal iliac arteries and veins, with additional contributions from the obturator and inferior gluteal arteries. This rich vascular network ensures adequate perfusion of the bladder tissue (Wein et al. 2015).

Understanding these anatomical relationships and structural variations is crucial for clinical practice, particularly in the context of urological disorders and their management. The complex interplay between anatomical structure and physiological function forms the basis for many therapeutic approaches in spinal cord injuries and related fields.

H.-Y. Ko, *Practical Functional Anatomy for Spinal Cord Injury Rehabilitation*,
https://doi.org/10.1007/978-981-95-6457-6_9

9.1 Anatomical and Functional Configuration of the Lower Urinary Tract

The lower urinary system, comprising the urinary bladder and urethra, is an integral component of the renal system. This anatomical network, which also includes the kidneys and ureters, functions to filter metabolic waste from the bloodstream (Drake et al. 2020). Understanding the structural and functional aspects of this system is fundamental for comprehending various urological conditions and their management.

The urinary bladder is a hollow, muscular organ with a trilaminar wall structure: the serosa (exterior), the muscularis (middle), and the mucosa (interior). Its anatomical position exhibits sexual dimorphism; in females, it is situated inferior to the uterus and ileum, and superior to the vagina and rectum. In males, the bladder is located superior to the prostate, anterior to the rectrum, and inferior to the ileum and peritoneal cavity (Standring 2020) (Fig. 9.1).

Structurally, the bladder is divided into two main regions: the body and the base. The body lies superior to the ureteral orifices, while the base consists of the trigone and bladder neck. These regions demonstrate distinct neuromorphological and neuropharmacological properties (el-Badawi and Schenk 1966). The bladder base exhibits a laminar architecture, featuring a superficial longitudinal layer beneath the trigone. A muscle layer located beneath this stratum integrates with the detrusor, predominantly exhibiting a circular arrangement (Dixon and Gosling 1987; Wein et al. 2015).

A notable sexual difference exists in the bladder neck structure. Males possess a complete ring of smooth muscle encircling the bladder neck, a feature absent in females (Gosling 1979). This structural difference has functional implications, particularly in males, where the bladder neck facilitates antegrade ejaculation through its rich noradrenergic innervation. Interestingly, the density of adrenergic innervation in the female bladder neck is comparatively lower (de Groat and Booth 1993).

The urethra, in conjunction with the pelvic floor musculature, constitutes the bladder outlet. Its composition includes both smooth and striated muscle components, with the latter forming the rhabdosphincter or external urethral sphincter (Fig. 9.2). The periurethral striated muscle is an integral part of the pelvic floor muscle complex, contributing to urinary continence mechanisms.

The blood supply to the bladder originates from the internal iliac arteries and veins, with additional contributions from the obturator and inferior gluteal arteries. This rich vascular network ensures adequate perfusion of the bladder tissue. The lymphatic drainage is managed by a series of lymph nodes, including the external iliac, internal iliac, and common iliac nodes (Kuru et al. 2013). This lymphatic system plays a

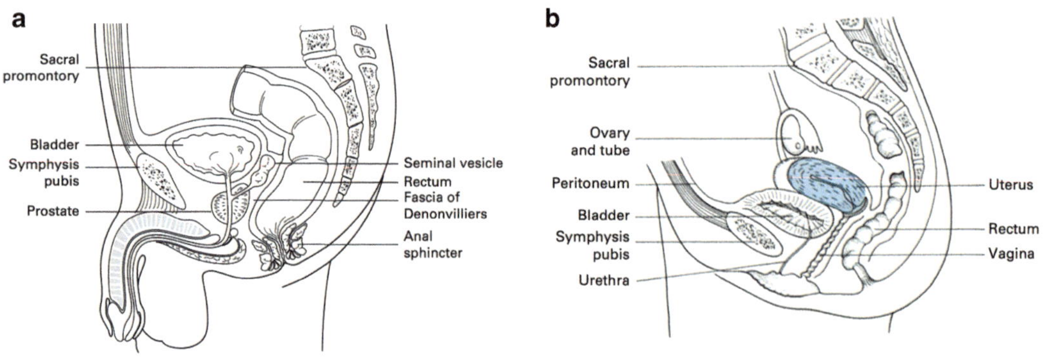

Fig. 9.1 Sagittal sections of the male (**a**) and female (**b**) pelvis. (**a**) In males, the bladder is positioned superior to the prostate, anterior to the rectum, and inferior to the ileum and the peritoneal cavity. (**b**) In females, the bladder is located inferior to the uterus and ileum, superior to the vagina, and anterior to the rectum. (Adapted from Wineski (2019))

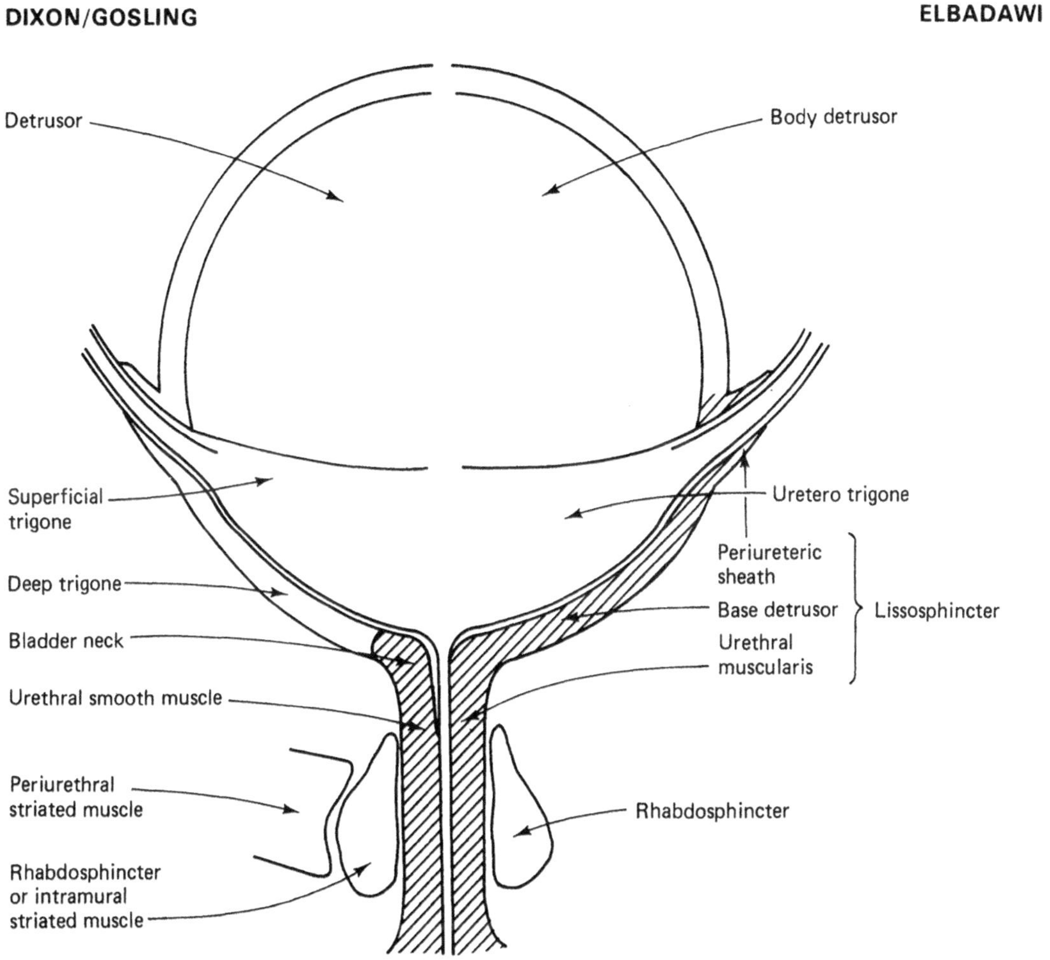

Fig. 9.2 Anatomy of the bladder and its outlet. This figure illustrates a comparison of terminology used by different authors, with definitions provided by Dixon and Gosling (left) and Elbadawi (right). (Adapted with permission from Torrens and Morrison (1987))

key role in immune function and fluid balance within the urinary system.

The ureters enter the bladder in a unique manner, approaching posteriorly and obliquely at the trigone level. The terminal 2 cm of each ureter is enveloped by the detrusor muscle wall in a fibromuscular tunnel, known as Waldeyer's sheath. The anatomical arrangement is functionally significant, as it contributes to the closure of the intramural ureter and ureteral orifices due to the natural pressure of the trigonal muscles (Tanagho and Pugh 1963). Under normal physiological conditions, ureteral peristalsis periodically propels a fluid bolus distally, overcoming the closing forces and facilitating urine expulsion. The dynamics at the ureteral-vesical junction are influenced by multiple factors, including detrusor stretch, intravesical pressure, and individual ureteral properties.

The relationship between bladder and ureteral pressures holds significant clinical relevance. Elevated detrusor pressure can precipitate various pathological conditions, including vesicoureteral reflux, hydronephrosis, and upper urinary tract injury. In cases of noncompliant or neurogenic bladders characterized by detrusor muscle hypertrophy, urine expulsion necessitates progressively increased ureteral pressures. This pathophysiological process can lead to a cascade of events, including ureteral wall dilation, impaired peristalsis, and hydroureteronephrosis (Abrams et al. 2002). Moreover, chronic obstruction at the bladder outlet can result in the formation of mucosal diverticula. These are herniations of the mucosa through the detrusor muscle layers, forming saccular structures outside the bladder. Such diverticula can become reservoirs for residual urine and potential sites for stone formation and tumor development (Wein et al. 2015).

The anatomical and functional configuration of the lower urinary system demonstrates the intricate balance necessary for normal urinary function. The complex relationships between structural elements, neural control, and pressure dynamics form the basis for understanding various urological conditions and their management.

9.1.1 Structure and Clinical Significance of the Urinary Bladder

The urinary bladder is a distensible, hollow, muscular, extraperitoneal organ functioning as a temporary urine reservoir. Located in the pelvis posterior to the pubic bone, it consists of two main portions: the dome and the base (Fig. 9.3). The dome, the upper part of the bladder, is characterized by its spherical shape, extensibility, and mobility. From its apex, the median umbilical ligament (urachus) ascends behind the anterior abdominal wall to the umbilicus, with the peritoneum behind it forming the median umbilical fold (Standring 2020).

In males, the peritoneum completely covers the superior surface of the dome, extending slightly to the base. It is in close proximity to the sigmoid colon and the terminal ileal coils. In females, the peritoneal reflection on the anterior uterine face creates the vesico-uterine pouch. The inferolateral part of the bladder lacks peritoneal covering in both sexes (Drake et al. 2020).

The base of the bladder, which forms its lower part, is fixed. The trigone, located at the posterior aspect of the base, is a triangular area defined by three orifices—two ureteral and one urethral (bladder neck) (Fig. 9.4). At the vesico-urethral junction or bladder neck, the specific arrangement of muscle fibers allows closure during the bladder-filling phase.

In adults, the bladder occupies a retropubic position and is palpable only during overdistention. However, at birth, it is relatively high and considered an abdominal organ, gradually descending to reach its adult position at puberty. In neonates and infants, the bladder, even when

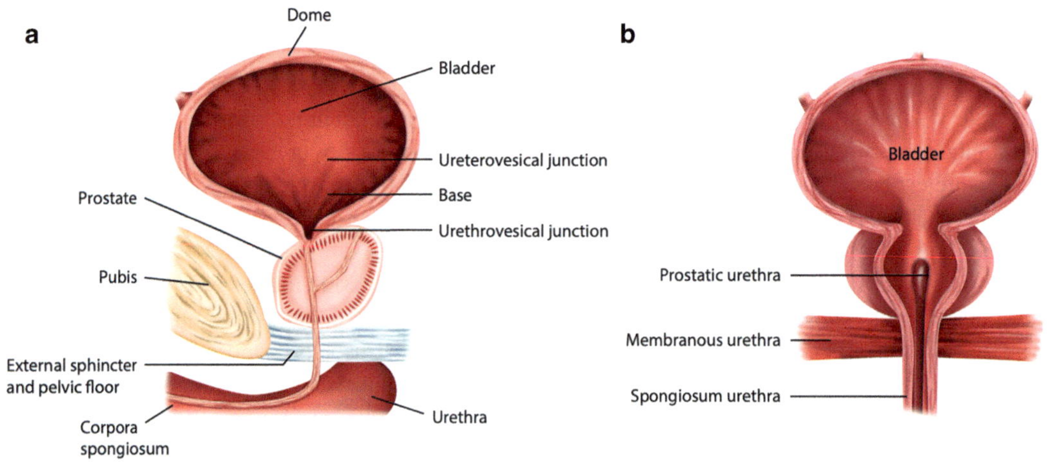

Fig. 9.3 Anatomy of the vesicosphincteric unit in males. (**a**) Sagittal view showing the spatial relationships of the bladder, urethra, and surrounding structures. (**b**) Frontal view highlighting the configuration of the vesicosphincteric unit. (Adapted from Corcos et al. (2016))

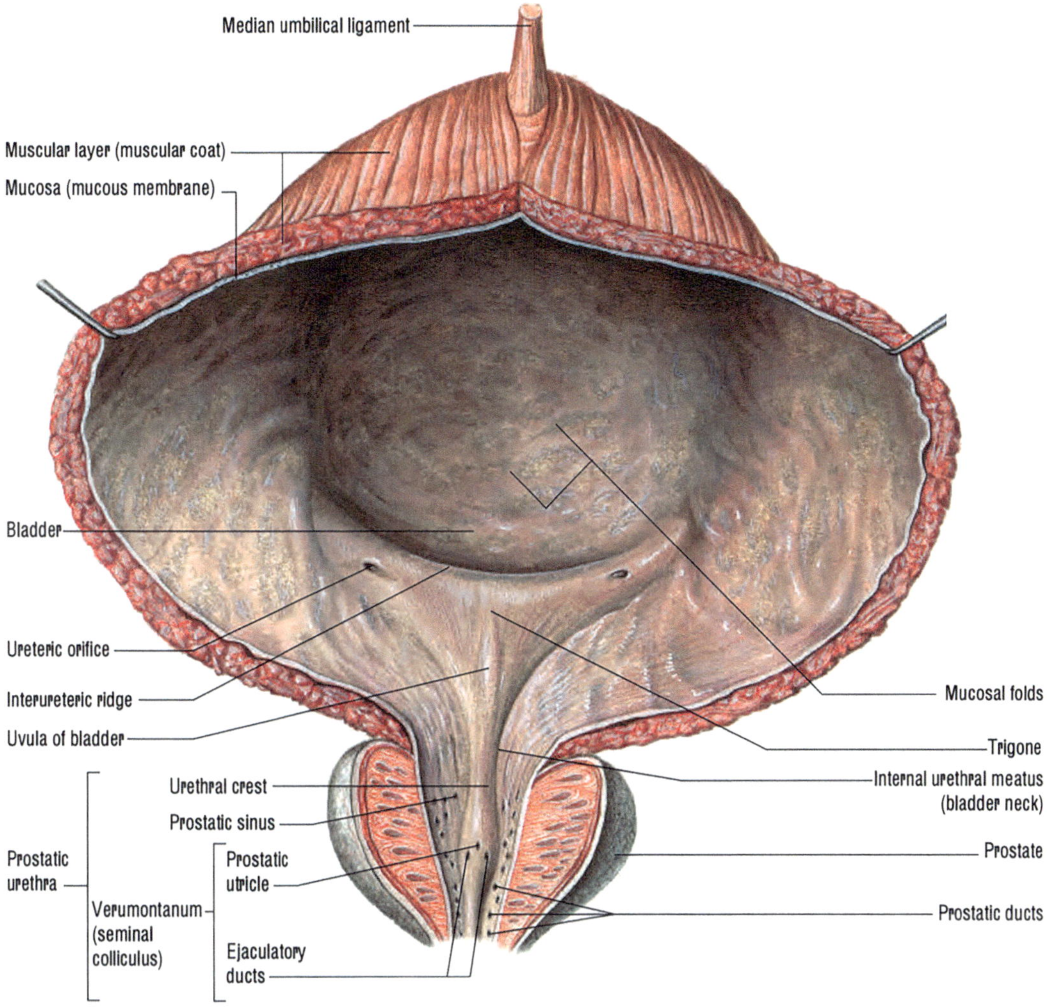

Fig. 9.4 Coronal section of the male urinary bladder. The appearance of mucosal folds varies based on the degree of bladder filling. (Adapted with permission from Paulsen and Waschke (2011))

empty, is an abdominal organ due to the minimal depth and volume of the pelvic cavity at birth. The bladder neck at this stage aligns with the upper edge of the pubic symphysis. The bladder has a tubular shape and contacts the posterior surface of the lower anterior abdominal wall. This anatomical feature is an important surgical consideration when performing lower abdominal laparotomy incisions in infants (Moore et al. 2018). The pelvic cavity acquires sufficient depth to accommodate the bladder only by age 7 or 8 years, and the bladder becomes truly intrapelvic in location around puberty, and then only when empty.

The empty bladder appears flattened from above downwards due to pressure from neighboring viscera but assumes an ovoid shape as it fills with urine. The empty, contracted adult bladder has a tetrahedral shape with four external surfaces: superior, posterior (also called the base), and two inferolateral (right and left). The borders between these surfaces are indistinct (Netter 2019). The anterior angle of the superior surface (where it meets the two inferolateral surfaces) forms the apex of the bladder. The median umbilical ligament, a fibrous, cord-like remnant of the embryological urachus, runs antero-superiorly from the apex to the umbilicus.

The bladder neck, the lowest part of the bladder, is where the posterior and inferolateral surfaces meet. It contains the internal urethral orifice, through which the bladder lumen connects with the urethra. In males, the bladder neck lies immediately above the upper surface (base) of the prostate. In females, it relates inferiorly to the pelvic floor and the pelvic fascia surrounding the upper urethra, and posteriorly to the anterior vaginal wall and fornix (Standring 2020).

The relationship of the bladder to the peritoneum has significant surgical and clinical implications. The superior surface (dome) of the bladder is completely covered by densely attached peritoneum. Anteriorly, the peritoneum leaves the superior surface and stretches upwards to meet the posterior surface of the anterior abdominal wall, blending with the fascia transversalis. This junction occurs 4–5 cm superior to the upper edge of the pubic symphysis (Standring 2020).

Three longitudinal folds in the peritoneum are observed as it extends from the bladder dome to the anterior abdominal wall. The median umbilical fold is the central fold, formed by the underlying median umbilical ligament, which is the remnant of the obliterated urachus. On either side of this are the medial umbilical folds, formed by the underlying medial umbilical ligaments, which are remnants of umbilical arteries). The peritoneum extends laterally from the superior surface of the bladder to the obturator fascia on the lateral pelvic wall.

In males, the peritoneum descends briefly from the superior bladder surface onto the posterior surface before turning posteriorly to reach the ventral rectal surface, forming the rectovesical pouch. In females, it extends directly backwards onto the uterine isthmus without dipping down the posterior bladder surface, forming the uterovesical pouch.

At the vesico-ureteral junction, the ureters traverse the bladder wall obliquely over a length of 1–2 cm. This anatomical arrangement creates a valve-like mechanism that prevents urine reflux towards the ureters when bladder pressure increases. The oblique path of the ureter through the bladder wall facilitates this process. As it passes through a hiatus in the detrusor (intramural ureter), compression and complete closure are achieved by detrusor contraction (Drake et al. 2020). The intravesical portion of the ureter lies immediately beneath the bladder urothelium, backed by a strong detrusor muscle plate. It is hypothesized that bladder filling results in passive occlusion of the ureter, functioning like a flap valve.

The urinary bladder functions as a reservoir, with its size, shape, position, and relationships varying based on its contents and the condition of adjacent organs. When empty, it lies entirely in the lesser pelvis, but as it distends, it expands anterosuperiorly into the abdominal cavity. An empty bladder has a tetrahedral shape with a base (fundus), neck, apex, a superior (dome) and two inferolateral surfaces. As the bladder fills with urine and distends, the demarcations between surfaces gradually disappear, and the bladder assumes a globoid or ovoid shape (Standring 2020).

9.1.1.1 Relations with the Adjacent Structures

The bladder's anatomical relationships vary between sexes and change with filling status, influencing clinical management strategies.

The base of the bladder, triangular in shape, is situated posteroinferiorly. In females, it closely relates to the anterior vaginal wall (Fig. 9.5). In males, it relates to the rectum, separated superiorly by the rectovesical pouch, and inferiorly by the seminal vesicles, vasa (ducti) deferentia, and Denonvilliers' fascia (Standring 2020) (Fig. 9.6).

The bladder neck, the most fixed part, is located 3–4 cm posterior to the lower pubic symphysis, just above the plane of the inferior aperture of the lesser pelvis. It maintains a constant position regardless of bladder and rectal variations. In males, it rests on and directly continues with the base of the prostate. In females, it relates to the pelvic fascia surrounding the upper urethra (Drake et al. 2020).

The apex of the bladder faces the upper pubic symphysis in both sexes. The median umbilical ligament (urachus) ascends from the apex to the umbilicus behind the anterior abdominal wall,

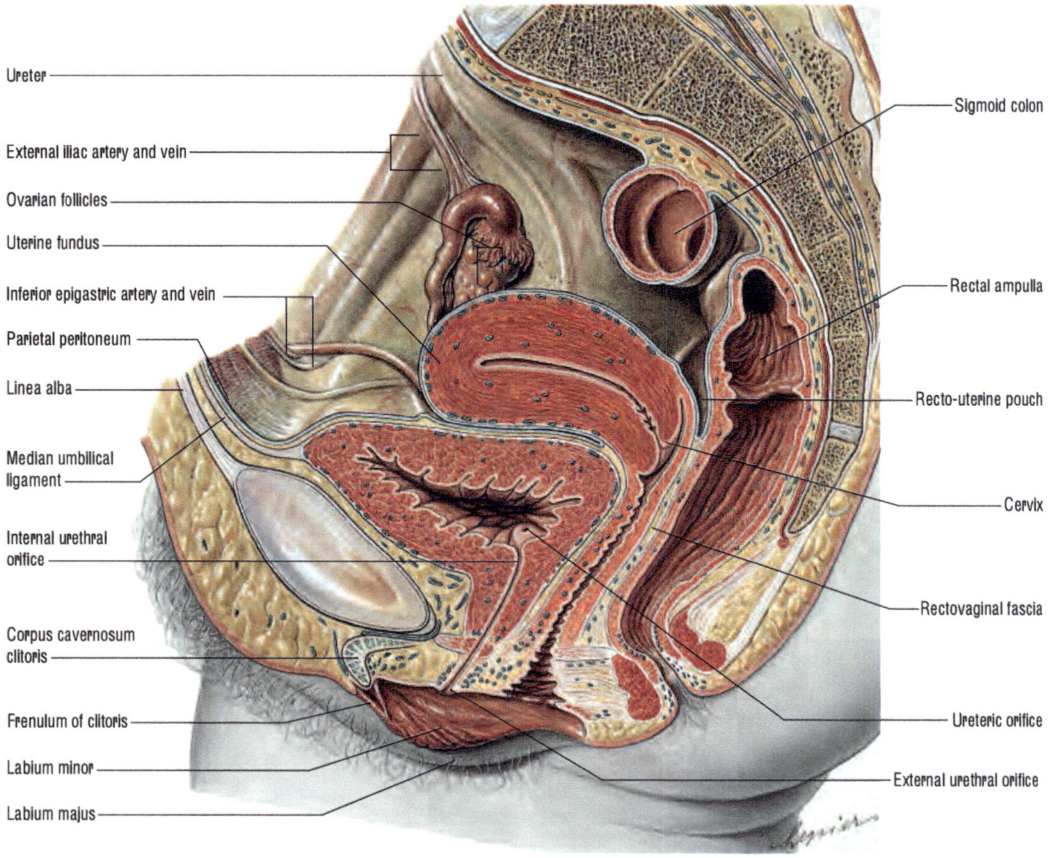

Fig. 9.5 Sagittal section of the female pelvis showing the anatomical relationship of the bladder to adjacent structures. (Adapted with permission from Paulsen and Waschke (2011))

covered by peritoneum forming the median umbilical fold (*see* Fig. 9.4). The anterior surface of the bladder is separated from the transversalis fascia by adipose tissue in the potential retropubic space (of Retzius) (Netter 2019).

In males, each inferolateral surface anteriorly relates to the pubis and puboprostatic ligaments. In females, the relations are similar, with pubovesical ligaments replacing puboprostatic ligaments. The inferolateral surfaces lack peritoneal covering (Standring 2020).

The triangular superior surface is bounded by lateral borders from the apex to the ureteric entrances, and a posterior border joining them. In males, peritoneum completely covers this surface, extending slightly onto the base and continuing posteriorly into the rectovesical pouch and anteriorly into the median umbilical fold. It contacts the sigmoid colon and terminal ileal coils (Standring 2020). The peritoneum in females predominantly covers the superior surface and reflects posteriorly onto the uterus at the level of the internal os, creating the vesicouterine pouch. The posterior region of the superior surface, lacking peritoneum, is distinguished from the supravaginal cervix by fibroareolar tissue. These relationships are important in managing bladder trauma. Extraperitoneal injuries often allow conservative management due to urine containment, while intraperitoneal injuries typically require surgical repair.

As the urinary bladder fills, it undergoes a remarkable transformation, shifting from a flattened structure to an ovoid shape. This metamorphosis brings about significant changes in its anatomical relationships, a process that holds

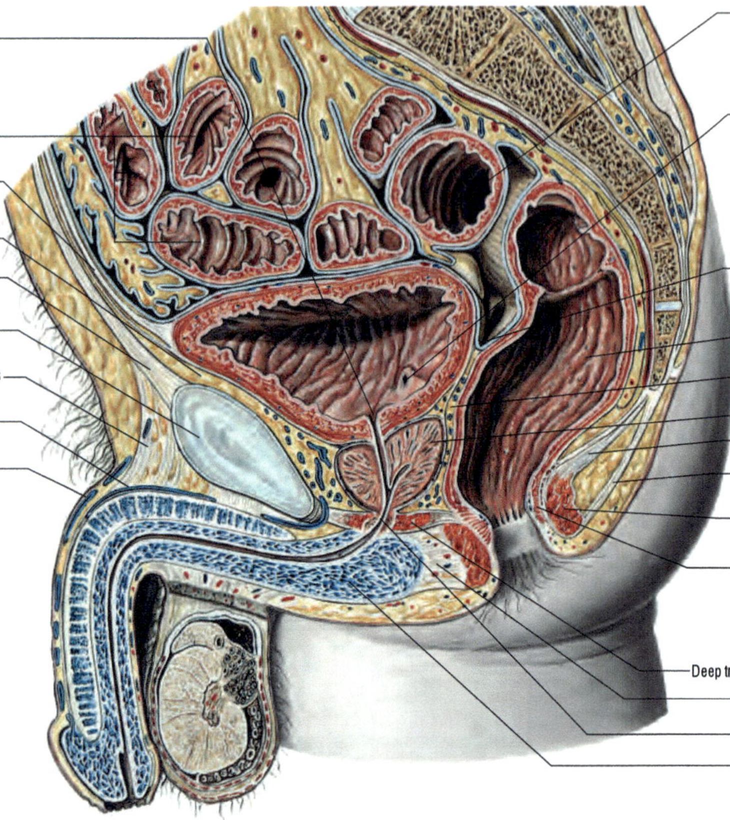

Fig. 9.6 Sagittal section of the male pelvis illustrating the anatomical relationships of the bladder with surrounding structures. Adapted with permission from Paulsen and Waschke (2011)

particular importance for rehabilitation specialists managing patients with neurogenic bladder following spinal cord injury. During the filling process, the bladder asserts itself in the anterior abdominal region, gently pushing the parietal peritoneum away from the suprapubic wall. The inferolateral surfaces of the bladder, once oriented to the sides, now rotate to face anteriorly. In this new position, they make direct contact with the abdominal wall, creating a peritoneum-free zone extending 5–7 cm above the pubic symphysis. This anatomical arrangement proves invaluable in clinical practice, allowing for the safe performance of suprapubic cystostomy.

The anterior and inferolateral anatomical relations of the bladder are similar in males and females. Inferolaterally, the bladder wall relates to the ipsilateral obturator internus and levator ani muscles, separated by loose connective tissue containing blood and lymphatic vessels. Anteriorly, the bladder neck and the anterior aspect of the bladder wall (where the two inferolateral surfaces meet) are related to the retropubic space, a fat-filled space which contains a plexus of veins. There are significant changes in the superior, posterior, and inferior relations. In females, the upper surface of the bladder often finds itself overlapped by an anteverted uterus, while its posterior surface develops intimate relationships with the anterior vaginal wall, the anterior vaginal fornix, and the front surface of the supravaginal cervix and uterine body. These structures are bound to the bladder's posterior surface by connective tissue. In males, the posterior surface of the bladder is associated with the seminal vesicles, as well as the terminal portions of the vas deferens and ureter.

These anatomical relationships are essential for managing bladder trauma, performing procedures like suprapubic cystostomy, and addressing complications in neurogenic bladder following spinal cord injury.

9.1.1.2 Interior of the Urinary Bladder

The internal surface of the urinary bladder is lined with transitional epithelium, also known as urothelium. This specialized tissue forms an uninterrupted barrier throughout the urinary system, extending from the renal pelvicalyceal complex through the ureters and bladder, and terminating in the proximal urethra (Drake et al. 2006). The mucosal layer of the bladder exhibits variable adhesion to the underlying detrusor muscle. While the majority of the mucosa maintains a loose connection, the trigonal region demonstrates a more secure attachment. In its non-distended state, the bladder mucosa presents numerous longitudinal folds, known as rugae. These folds disappear as the bladder expands with urine, a phenomenon readily observable during cystoscopic examination. The trigone, being the least extensible area, maintains a smooth surface regardless of bladder volume (Fowler et al. 2008).

The urinary bladder trigone represents an essential anatomical structure positioned on the inferior aspect of the posterior bladder wall. This region manifests as an inverted triangle, where its apex corresponds to the internal urethral orifice, and its base forms a line connecting the two ureteric orifices (hiatus) (Torrens and Morrison 1987) (Fig. 9.7). The external bladder wall in the trigonal region exhibits strong anatomical relationships with adjacent structures. In males, it is firmly attached to the superior surface of the prostate, while in females, it adheres to the anterior vaginal wall. This fixed positioning makes the trigone the least mobile portion of the bladder (Birder and Andersson 2013). In the human urinary system, muscle bundles extending craniolaterally from the base detrusor encompass the distal portion of each ureter. These extensions arrange themselves into superficial and deep periureteric sheaths. The superficial sheath encompasses three primary components: the Waldeyer sheath, which surrounds the juxtavesical (prehiatal) segment; muscle bundles of the ureteric hiatus, which enclose the intravesical (hiatal) portion; and transureteric bundles that traverse the submucosal (posthiatal) segments of the ureter (Andersson and Arner 2004; Torrens and Morrison 1987) (Fig. 9.8).

The muscular organization continues as the Waldeyer sheath invests the juxtavesical ureter and extends inferiorly to form the deep trigone, ultimately reaching the bladder neck. The ureteral musculature proceeds downward to estab-

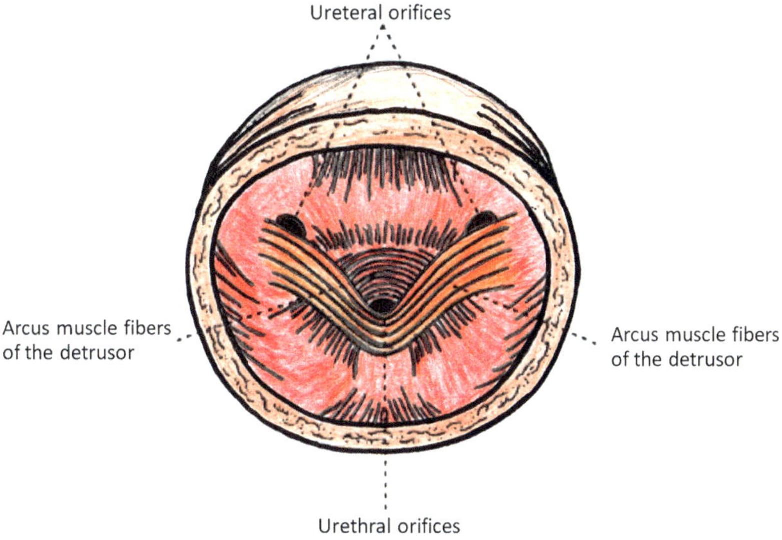

Fig. 9.7 Endovesical view of the trigon, showing its anatomical features and spatial orientation. (Adapted from Corcos et al. (2016))

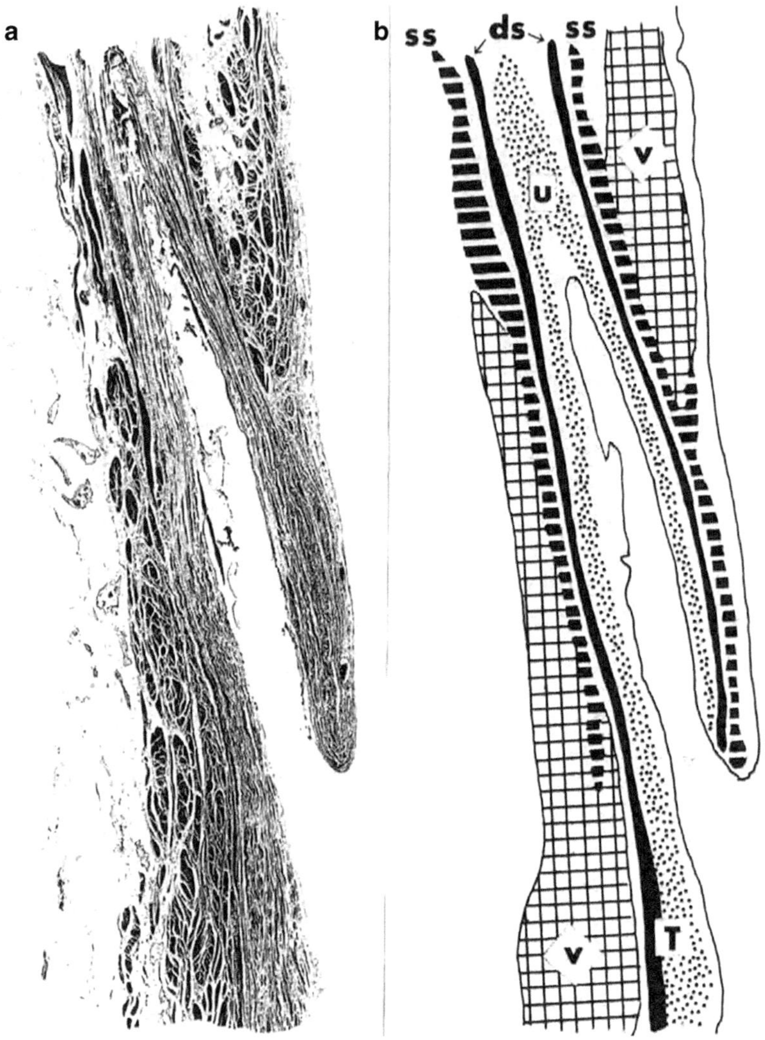

Fig. 9.8 Dual periureteric sheath at the ureterovesical junction, illustrated in a longitudinal section. (**a**) Photomicrograph of a histological section and (**b**) diagrammatic reconstruction based on muscle bundle tracings. Labels: u ureteric musculature, ds deep sheath, ss superficial sheath, v muscularis of bladder base, T trigonal region. (Adapted with permission from Torrens and Morrison (1987))

lish the superficial trigone, which extends to the verumontanum in males and terminates just proximal to the external meatus in females. Several muscle fibers connect Waldeyer's sheath to the detrusor muscle within the ureteral hiatus (Drake et al. 2010) (Fig. 9.9). The ureteric openings manifest as oblique slits with a characteristic separation of approximately 2.5–3 cm in the empty bladder state. This inter-orifice distance may increase to 5 cm during complete bladder distention. A submucosal elevation, known as the interureteric bar of Mercier, spans between these orifices (Andersson and Arner 2004; Moore et al. 2018).

Vesical Mucosa (Urothelium)

The vesical mucosa, also known as the urothelium, exhibits distinct anatomical characteristics throughout the urinary bladder (Standring 2020) (*see* Fig. 9.4). The majority of this mucosal layer maintains a loose connection to the underlying muscle tissue, allowing for significant adaptability in response to changes in bladder volume. This adaptability manifests as mucosal folds during bladder emptying, which subsequently flatten as the organ fills with urine (Apodaca 2004).

In contrast to the general mucosal pattern, the trigonal region, located superior and posterior to the internal urethral orifice, demonstrates a

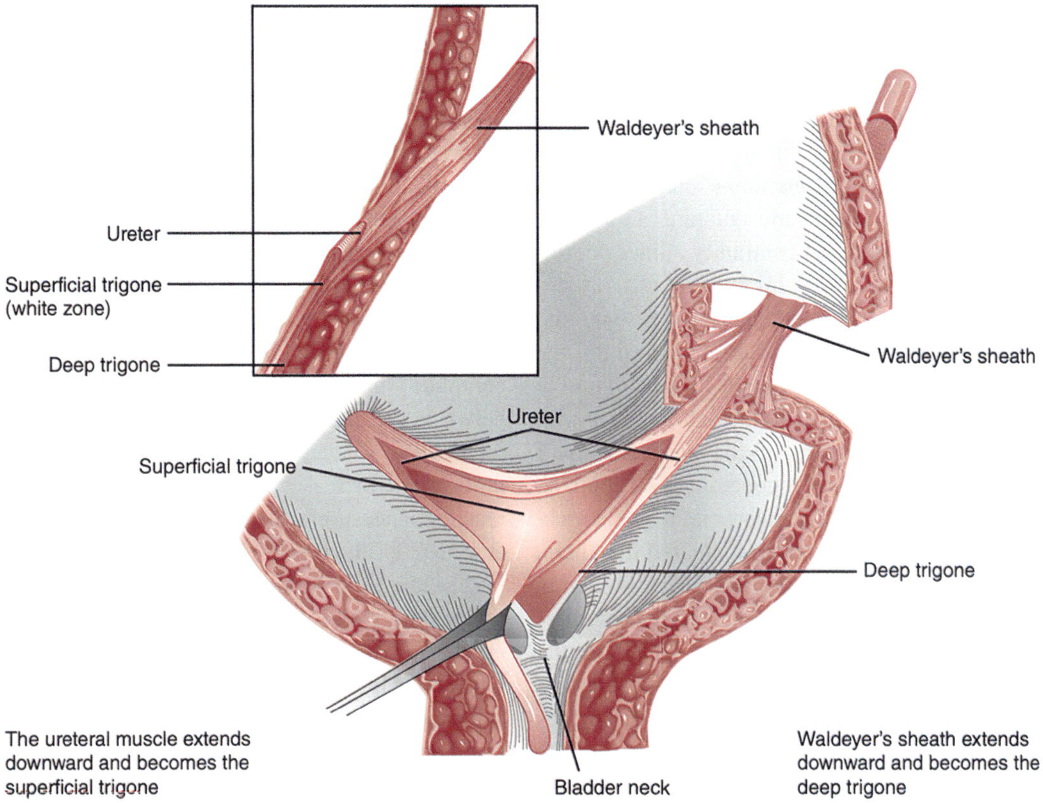

Fig. 9.9 Anatomy of the normal ureterotrigonal complex. The Waldeyer's muscular sheath surrounds the juxtavesical ureter and extends downward to form the deep trigone, reaching the bladder neck. The ureteral musculature forms the superficial trigone, which extends to the verumontanum in males and terminates just proximal to the external meatus in females. At the ureteral hiatus, Waldeyer's sheath connects to the detrusor muscle via a few muscular fibers. Inferior to the ureteral orifices, the sheath transitions into the deep trigone, while the descending ureteral musculature forms the superficial trigone. (Adapted from Doherty (2015))

unique anatomical arrangement. Here, the mucosa forms a tight adherence to the subjacent muscle layer, resulting in a consistently smooth surface regardless of bladder distention (Yoshimura and Chancellor 2003). The trigone itself is a triangular structure with well-defined boundaries. Its anteroinferior angle corresponds to the internal urethral orifice, while the posterolateral angles are formed by the ureteric orifices. The superior border of the trigone is demarcated by a slightly curved structure known as the interureteric bar. This bar represents the intravesical continuation of the ureters' internal longitudinal muscle and connects the two ureteric orifices (Elbadawi 1996).

Extending laterally beyond the ureteric openings are the ureteric folds. These structures are created by the terminal portions of the ureters as they traverse obliquely through the bladder wall. During cystoscopic examination, the interureteric crest appears as a pale band, serving as a valuable landmark for identifying the ureteric orifices (Tanagho and Lue 2008).

Vesical Trigone

The vesical trigone demonstrates a multilayered muscular organization that includes two anatomically and functionally significant components: the superficial trigonal muscle and the deep trigonal detrusor muscle (Drake et al. 2006). This anatomical arrangement supports multiple physiological functions related to urinary continence and micturition.

The deep trigonal detrusor muscle continues from the posteroinferior aspect of the detrusor muscle proper. Microscopic analysis reveals that the muscle cells in this layer share morphological similarities with those in other regions of the detrusor, suggesting functional integration with the broader bladder musculature (Elbadawi 1996). This anatomical continuity allows coordinated contraction during micturition.

The superficial trigonal muscle presents distinct morphological features compared to the deep layer. It contains smaller-diameter muscle bundles that maintain anatomical continuity with the intramural portions of the ureters at their proximal ends. Although generally characterized by its thin structure, the superficial trigonal muscle displays specific areas of increased thickness. These thickened regions appear prominently along the superior border, forming the interureteric ridge or bar, and at the lateral margins of the trigone (Gosling et al. 1999). The thickened interureteric ridge performs an important physiological role in preventing vesicoureteral reflux through its anatomical configuration and contractile properties (Wein et al. 2015).

Beyond the confines of the bladder, the superficial trigone muscle maintains anatomical continuity with the smooth muscle of the proximal urethra in both male and female anatomy. In male urogenital anatomy, this muscular continuation extends along the urethral crest to the level of the ejaculatory duct openings (de Groat and Yoshimura 2006; Yucel and Baskin 2004). The anatomical relationship between the superficial trigone and proximal urethral musculature indicates their combined role in maintaining bladder neck closure and regulating the micturition process (Fowler et al. 2008).

Ureteric Orifices

The ureteric orifices manifest as slit-like apertures positioned at the posterolateral angles of the vesical trigone. These anatomical structures exhibit dynamic spatial relationships that adapt to variations in bladder filling status (Fowler et al. 2008; McAninch and Lue 2013; Tanagho and Lue 2008). Their anatomical position and relationships play fundamental roles in maintaining urinary tract function and preventing retrograde flow.

In the empty bladder state, the ureteric orifices maintain an approximate distance of 3 cm from each other. Their position relative to the internal urethral orifice measures approximately 2.5–3 cm, establishing a triangular configuration that provides the anatomical basis for the term "trigone" (Standring 2020) (*see* Fig. 9.9). These spatial relationships demonstrate remarkable adaptability during bladder function. During complete bladder distention, the distances between the ureteric orifices, as well as their respective distances to the internal urethral opening, may increase substantially, potentially reaching measurements of up to 5 cm (Drake et al. 2020; Elbadawi 1996).

The morphological adaptations of the trigonal region during cycles of bladder filling and emptying produce significant physiological effects. The positional modifications of the ureteric orifices contribute to maintaining vesicoureteral junction competence, reducing the likelihood of vesicoureteral reflux during bladder filling phases (Birder and Andersson 2013). These dynamic changes in ureteric orifice positioning work in coordination with other anatomical mechanisms to maintain upper urinary tract protection.

The anatomical relationships of the ureteric orifices hold particular importance in clinical practice. During cystoscopic examination, understanding these measurements provides reference points for identifying and evaluating the ureteric orifices (Chapple et al. 2009). This knowledge becomes particularly relevant when assessing patients for vesicoureteral reflux or investigating upper urinary tract complications associated with neurogenic bladder conditions (Drake et al. 2020; Wein et al. 2015). The spatial relationships of the ureteric orifices also have practical implications for various diagnostic and therapeutic procedures in urological practice.

Internal Urethral Orifice

The internal urethral orifice occupies a precise anatomical position at the trigonal apex of the bladder. This positioning at the bladder's lowest point has direct implications for effective urinary drainage (Drake et al. 2020; Standring 2020). The orifice demonstrates a distinctive crescentic morphology in cross-sectional examination, which facilitates its role in regulating urine flow during micturition (Fowler and Griffiths 2010; Panicker et al. 2015; Yoshimura and de Groat 1997).

Male urogenital anatomy, particularly in advanced age, presents a notable anatomical relationship between the internal urethral orifice and the prostate gland. An elevation develops posterior to the orifice due to the median prostatic lobe (Tanagho and Lue 2008; Wein et al. 2015). This prostatic prominence exhibits considerable individual variation and undergoes age-related changes that may influence urinary function.

The anatomical architecture of the internal urethral orifice, including its crescentic configuration and relationships with adjacent structures such as the prostatic urethra in males, determines urodynamic parameters (Panicker et al. 2015). These morphological characteristics become particularly relevant in various clinical scenarios, including spinal cord injury and neurogenic bladder conditions (Birder et al. 2010; de Groat and Yoshimura 2015).

The structural and functional relationships of the internal urethral orifice provide fundamental considerations for therapeutic interventions. Understanding these relationships proves essential in managing various urological conditions, particularly those affecting micturition control (Yoshimura et al. 2014). The orifice's anatomical position and morphology also guide diagnostic approaches and surgical planning in contemporary urological practice.

Bladder Neck

The bladder neck, also referred to as the proximal or internal sphincter mechanism, is a specialized region of the lower urinary tract with unique anatomical and functional characteristics. This structure differs from the distal urinary sphincter mechanism and plays a significant role in maintaining urinary continence (Standring 2020). The smooth muscle of the bladder neck exhibits distinct histological, histochemical, and pharmacological properties that distinguish it from the detrusor muscle. These differences suggest that the bladder neck functions as a separate unit within the urinary system (Drake et al. 2006).

A notable feature of the bladder neck is its rich sympathetic noradrenergic innervation, which contrasts with the sparse parasympathetic cholinergic innervation. This innervation pattern differs markedly from that of the detrusor and other urethral smooth muscle, which are common to both sexes (Fowler et al. 2008).

The anatomical configuration of the bladder neck smooth muscle exhibits sexual variation: In males, the bladder neck is encircled by a collar of smooth muscle with specific adrenergic innervation. This collar extends distally to surround the preprostatic urethra. The muscle bundles forming this preprostatic sphincter are smaller than detrusor muscle bundles and are interspersed with a greater proportion of connective tissue rich in elastic fibers . In females, the smooth muscle of the bladder neck has a specific morphology (Fig. 9.10). Large-diameter fasciculi characteristic of the detrusor are replaced by small-diameter fasciculi that extend obliquely or longitudinally into the urethral wall. The female bladder neck typically rests above the pelvic floor, supported by pubovesical ligaments (Fig. 9.11), endopelvic fascia, and levator ani. This configuration aids in maintaining continence, particularly during increased intra-abdominal pressure. However, this arrangement may alter with age and after childbirth, potentially leading to stress incontinence (Klutke and Siegel 1995).

The ability of the bladder neck to maintain continence independently, even in cases of distal sphincteric incompetence, emphasizes its importance in urinary function. This capability has significant implications for managing urinary dysfunction in various clinical scenarios, including neurogenic bladder following spinal cord injury (de Groat and Yoshimura 2006). For rehabilitation specialists working with spinal cord injury patients, understanding the bladder neck's unique anatomy and innervation is essential. The predominance of sympathetic innervation in this region influences therapeutic approaches, particularly in managing detrusor-sphincter dyssynergia and other neurogenic bladder complications

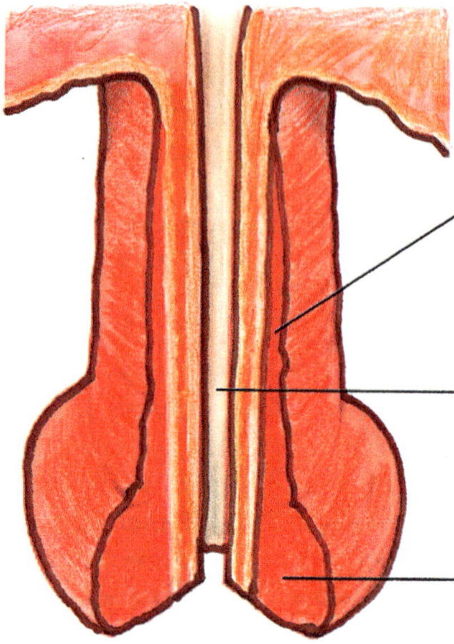

Fig. 9.10 Posterior view of the bladder neck in the female. (Adapted with permission from Standring (2020))

(Wein et al. 2015). Moreover, the sexual difference in bladder neck anatomy necessitates gender-specific considerations in the assessment and management of lower urinary tract dysfunction following spinal cord injury (Panicker et al. 2015).

9.1.1.3 Microstructure of the Bladder

The bladder is structurally composed of four layers: the urothelium (lining epithelium), lamina propria, muscularis propria, and serosa (Andersson and Arner 2004; Birder and Andersson 2013). This layered organization contributes to the organ's functional capabilities and protective mechanisms.

The innermost layer, the urothelium, consists of stratified epithelial cells, typically 3–6 cells thick. These cells are arranged in a specific pattern of basal, intermediate, and superficial zones. The superficial zone is particularly noteworthy as it forms an impermeable barrier, safeguarding the underlying structures from potentially harmful substances in the urine (Lewis 2000).

Adjacent to the urothelium lies the lamina propria, a supportive connective tissue layer. This layer is rich in capillaries, lymphatic vessels, and neuroreceptors, playing an important role in the bladder's sensory and vascular functions. A well-defined basement membrane separates the lamina propria from the urothelium, maintaining the structural integrity of these layers (Birder and Andersson 2013).

The muscularis propria, composed of the detrusor muscle, forms the third layer of the bladder wall. The detrusor is characterized by an interlacing network of multidirectional smooth muscle fibers, which gives the bladder its characteristic trabeculated appearance when viewed macroscopically. This muscle receives an abundant supply of parasympathetic, cholinergic fibers, enabling its contractile function for bladder emptying (Fry et al. 2010).

The trigonal region of the bladder exhibits a unique muscular arrangement. In this area, the musculature is organized into two histologically and biochemically different layers. The outer layer consists of parasympathetically innervated detrusor muscle fibers, while the inner layer, known as the superficial trigonal muscle, receives noradrenergic sympathetic innervation. This differential innervation pattern likely contributes to the specialized functions of the trigone in urine storage and voiding (Gosling et al. 1999).

The outermost layer, the serosa, provides a partial covering for the bladder, specifically related to the dome. Its external surface is lined

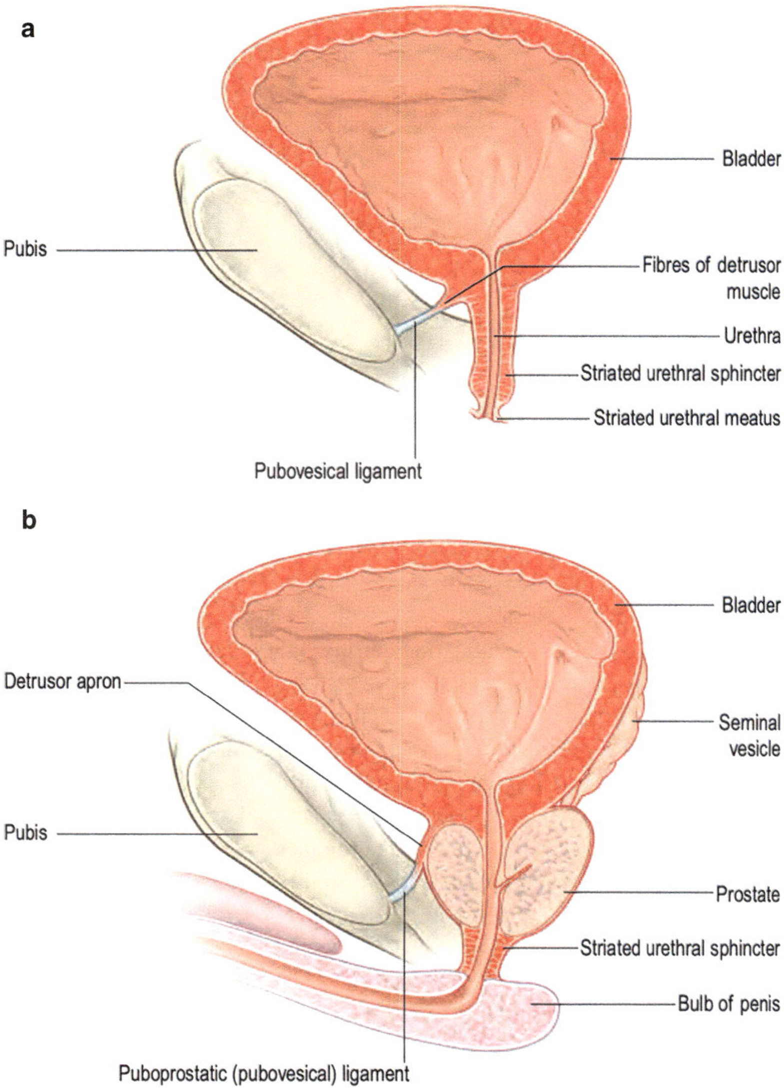

Fig. 9.11 Ligaments stabilizing the neck of the bladder and pelvic part of the urethra in relation to the pelvic bone: (**a**) In females, the ligaments support the bladder neck and urethra. (**b**) In males, the detrusor muscle extends over the anterosuperior aspect of the prostate and attaches to the pubic bone, condensing to form the pubovesical or puboprostatic ligaments. (Adapted with permission from Standring (2020))

by mesothelium, which facilitates the bladder's movement within the pelvic cavity. It is important to recognize that not all surfaces of the bladder are covered by serosa; the remaining areas are enveloped by a thin layer of non-serosal fibro-fatty connective tissue (Kunze et al. 2006).

Urothelium (Uroepithelium)

The urothelium, a specialized multilayered epithelium lining the urinary bladder, plays a pivotal role in the function and protection of the bladder. Its complex structure consists of basal, intermediate, and apical cell (umbrella cell) layers, each contributing to the unique properties of the urothelium (Jafari and Rohn 2022) (Fig. 9.12).

At the base, small cuboidal cells form the progenitor layer. Above this, the intermediate layer contains polygonal cells capable of stretching and flattening, allowing the bladder to accommodate varying urine volumes. The superficial "umbrella" cell layer, consisting of large, sometimes multinucleated cells, forms the uppermost boundary of the urothelium in direct contact with urine and microorganisms (Jafari and Rohn 2022; Lewis 2000) (see Fig. 9.12).

The thickness of the urothelium typically ranges from 4 to 7 cell layers, though it may

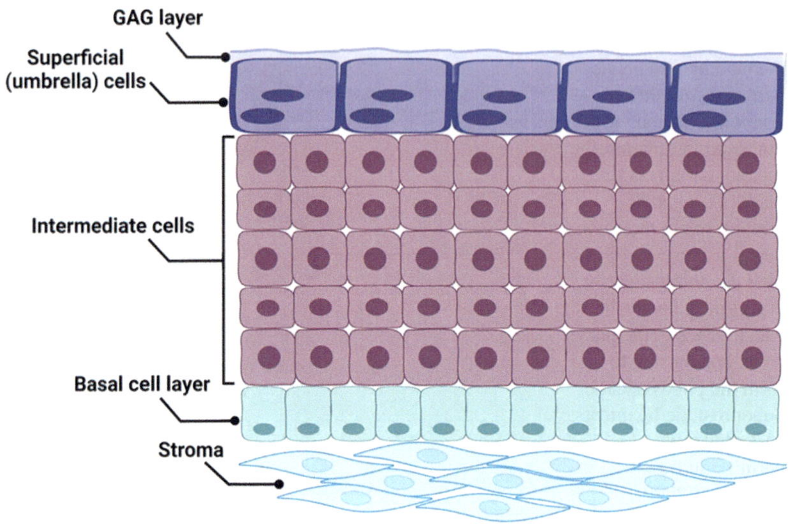

Fig. 9.12 Structures of the urothelium, which consists of three cell layers: basal cells, intermediate cells, and superficial (umbrella) cells. The umbrella cells are covered by an apical membrane plaque composed of uroplakin proteins on their luminal surface and a large pool of subapical vesicles. GAG glycosaminoglycan. (Adapted with permission from Jafari and Rohn (2022))

appear attenuated to 2–3 layers when the bladder is fully distended. This adaptability facilitates efficient expansion and contraction during the cycles of filling and voiding (Birder et al. 2012).

A unique feature of umbrella cells is the presence of 16 nm protein particles on their apical surface, arranged in hexagonal patterns to form two-dimensional crystals of asymmetric unit membranes. These asymmetric unit membranes are important in maintaining the permeability barrier of the bladder, preventing the reabsorption of urine into the bloodstream (Khandelwal et al. 2009).

The mucosa of the bladder, comprising the urothelium and underlying lamina propria, is folded when empty and loosely adherent to the submucosal tissue and detrusor muscle. However, it becomes more firmly attached over the trigone and around the neck of the bladder. This mucosal layer is richly vascularized and highly sensitive to various stimuli, including pain, distention, and temperature (Fowler et al. 2008).

The lamina propria, situated deep to the urothelium, forms a relatively thick layer of fibroelastic connective tissue. This layer allows for considerable distention of the bladder and contains numerous blood vessels. It also contains smooth muscle fibers forming a poorly defined muscularis mucosa, which may contribute to the fine motor control of the bladder during filling and emptying (Andersson and McCloskey 2014). Beneath the lamina propria lies the smooth muscle of the wall of the bladder, forming the detrusor muscle. This muscular layer is responsible for the contractile function of the bladder during micturition (Fry et al. 2010).

Lamina Propria

The lamina propria, a relatively thick layer of fibroelastic connective tissue, lies deep to the urothelium and is separated from it by a basement membrane. This layer plays a multifaceted role in bladder function, serving as more than just a supportive structure (Andersson and McCloskey 2014).

Structurally, the lamina propria is rich in capillaries, lymphatics, nerve endings, and elastic fibers. It facilitates considerable bladder distention and houses numerous blood vessels. A distinctive feature is the presence of the muscularis mucosae, a thin, poorly defined layer of smooth muscle fibers. The distribution of this layer within the bladder is variable, often lying close to the urothelium but sometimes found deeper within the lamina propria, where it may be well developed. It is essential to distinguish this layer from the muscularis propria (Andersson and McCloskey 2014).

The lamina propria serves several proposed functions:

- Acting as the capacitance layer of the bladder, influencing bladder compliance, and enabling adaptive changes to increasing volumes
- Supplying factors that affect the growth of the urothelium and/or detrusor
- Participating in signal transduction

Researchers have proposed that the lamina propria functions as the "functional center" for localized bladder control, coordinating urothelial and detrusor smooth muscle activities (Andersson and McCloskey 2014). This hypothesis is supported by the presence of a diffuse plexus of unmyelinated nerve fibers that connect with the urothelium, blood vessels, and detrusor smooth muscle. These nerve fibers express various receptors, including TRPV1 and P2X3, which are involved in sensory signaling (Yiangou et al. 2001).

In addition to nerve fibers, the lamina propria contains other important components, such as interstitial cells (myofibroblasts) and microvasculature. The myofibroblasts in the lamina propria are positioned to modulate physiologic interactions between the urothelium and detrusor smooth muscle. Recent research suggests these cells may play a role in spontaneous bladder activity and could be involved in various bladder disorders (Kanai et al. 2007).

The smooth muscle fibers in the lamina propria, forming the poorly defined muscularis mucosa, may contribute to the fine motor control of the bladder during filling and emptying cycles. This subtle muscular component adds another layer of complexity to bladder function, potentially allowing for localized adjustments in bladder wall tension (Andersson and McCloskey 2014).

Located underneath the lamina propria is the detrusor muscle, formed by the smooth muscle of the bladder wall. The detrusor, although separate from the lamina propria, interacts with it to execute the primary contractile function of the bladder during micturition (Fry et al. 2010).

The complex structure of the lamina propria, characterized by its varied cellular components and extensive innervation, shows its significance in bladder function. Its location between the urothelium and the detrusor muscle makes it ideally situated to facilitate communication between these layers and contribute to the integration of the bladder's sensory and motor functions.

Recent studies have also implicated the lamina propria in various bladder pathologies. For instance, changes in the properties of interstitial cells in this layer have been associated with overactive bladder syndrome, suggesting that the lamina propria may be a potential target for therapeutic interventions (Roosen et al. 2009).

Muscularis Propria (Detrusor Muscle)

The muscularis propria, commonly known as the detrusor muscle, forms the muscular wall of the bladder. Its structure and organization are fundamental to the dual functions of the bladder of urine storage and micturition. The detrusor muscle consists of three layers; however, they are not well delineated:

- An inner longitudinal layer
- An intermediate circular layer
- An outer longitudinal layer

Histological examination reveals that myofibrils are arranged into fascicles (bundles) in seemingly random directions. This architecture differs from the discrete circular and longitudinal smooth muscle layers found in the ureter or gastrointestinal tract. The individual cells within a bundle are interconnected to form a functional syncytium, allowing for coordinated contraction.

At the neck of the bladder, these layers become more organized, forming well-defined inner circular and outer longitudinal layers. This structural arrangement has significance for the bladder's functionality, particularly in preserving continence and enabling effective emptying.

Bladder smooth muscle cells differ from skeletal muscle fibers in that they do not possess cross-striations. Each detrusor smooth muscle cell is spindle-shaped and contains a single nucleus. When fully relaxed, these cells can measure several hundred micrometers in length with a maximum diameter of 5–6 μm (Smet et al. 1996). The cell membranes of detrusor muscle cells contain unique structures known as caveolae, which are flask-shaped invaginations often associated with components of the intracellular sarcoplasmic reticulum. These structures are involved in calcium signaling and other cellular processes essential for muscle contraction.

The primary motor innervation of the bladder smooth muscle comes from postganglionic parasympathetic nerve fibers. Intramural ganglia can also exist within the bladder wall, adding to the complexity of local control mechanisms. Nerve varicosities release various neurotransmitters, including acetylcholine and adenosine triphosphate (ATP). An important feature of detrusor muscle innervation is that not every smooth muscle cell receives direct synaptic contact. Instead, gap junctions between cells allow for the propagation of excitation throughout the smooth muscle syncytium. This arrangement ensures coordinated contraction of the entire detrusor muscle (Gillespie et al. 2006).

Detrusor smooth muscle cells express various postjunctional receptors, including muscarinic and purinergic receptors. When activated by their respective agonists, these receptors initiate the excitation-contraction cascade in the smooth muscle. Recent research suggests that the detrusor smooth muscle may also have afferent innervation, potentially mediating sensory signals related to muscle activity (Gillespie et al. 2006).

Conceptually, the detrusor muscle can be visualized as a sphere of smooth muscle bundles. Near the bladder neck, the three layers become more distinct, each playing a specific role in bladder function. In both men and women, the muscle fibers of the inner longitudinal layer extend down into the urethra in a funnel-shaped structure, contributing to continence and bladder emptying. However, there are notable differences between male and female anatomy in this region.

The middle circular layer in males forms a circular preprostatic sphincter at the bladder neck, essential for continence. The outer longitudinal layers are thickest posteriorly at the bladder base, providing strong trigonal support. Laterally, fibers from this sheet pass anteriorly and fuse to form a loop around the bladder neck, further contributing to the continence mechanism. The female bladder neck differs in that its sphincteric function is less pronounced, with some researchers questioning its existence as a distinct functional unit. This anatomical difference may contribute to the higher prevalence of stress urinary incontinence in women (Elbadawi 1996).

The intricate arrangement of muscle fibers and their innervation makes the detrusor susceptible to age-related changes and neurological disorders, which can significantly impact bladder function (Elbadawi et al. 1993).

Serosa

The serosa forms the outermost partial covering of the bladder. Its external surface is lined by mesothelium, a specialized epithelium common to body cavity organs. Beneath this mesothelial layer lies a variable amount of vascularized adipose tissue, which plays several important roles in bladder function (Khandelwal et al. 2009).

This adipose tissue is not confined to the serosa but often extends into the muscularis propria, intermingling with the detrusor muscle fibers. In some instances, it may even reach as deep as the lamina propria (Elbadawi 1996). This distribution pattern contributes to the bladder's ability to accommodate changes in volume and provides mechanical support.

The serosa typically covers the superior and posterolateral aspects of the bladder, while the anterior and inferior portions are covered by adventitia, a layer of connective tissue (Standring 2020). The mesothelial lining of the serosa produces a lubricating fluid that reduces friction between the bladder and adjacent organs during movement and volume changes. Additionally, the rich vascular network within the adipose tissue contributes to the blood supply of the bladder wall. Recent research suggests that the adipose tissue in the serosa may have endocrine and paracrine functions, potentially influencing bladder function through the release of various factors (Majesky 2007).

9.1.1.4 Bladder Distention and Anatomical Changes

As the bladder fills with urine, it undergoes significant anatomical changes. The initially flattened organ becomes progressively rounded and ascends above the pelvic brim into the abdominal cavity. This upward movement causes the bladder to strip the peritoneum away from the posterior surface of the anterior abdominal wall. Consequently, the inferolateral surfaces of the distended bladder directly contact the posterior surface of the abdominal wall, with no intervening peritoneum (Moore et al. 2018).

At this stage of distention, the bladder becomes palpable and can be identified through percussion. This anatomical arrangement allows for suprapubic access to the bladder without traversing the peritoneal cavity. A needle or cannula advanced perpendicular to the anterior abdominal wall in the midline, just above the pubic symphysis, can enter the distended bladder directly. This forms the anatomical basis for procedures such as suprapubic puncture, catheterization, and cystostomy (Netter 2019). The dome of the urinary bladder is its most mobile and least supported region. This anatomical characteristic has clinical implications, particularly in trauma situations. A powerful blunt injury to the lower abdomen of an individual with a tense, distended bladder may result in rupture of the bladder dome (Corriere Jr and Sandler 1986).

The location of bladder rupture significantly influences the path of urine leakage and, consequently, the management approach. A rupture of the bladder dome typically results in intraperitoneal urine leakage due to the close relationship between the peritoneum and this part of the bladder. In contrast, ruptures through other bladder walls are more likely to cause extraperitoneal urine leaks. The distinction between intra- and extraperitoneal bladder perforations is crucial, as their management strategies differ significantly (Gomez et al. 2004).

9.1.2 Structure and Function of the Urethra

The urethra serves as a muscular passage, transporting urine from the bladder to the external environment. It begins at the internal urethral orifice located at the bladder neck and terminates at the external urethral orifice. The position of this external aperture differs between males and females, indicating notable anatomical and functional differences (Drake et al. 2020).

In females, the external urethral orifice is situated within the vaginal vestibule. The female urethra, measuring approximately 3–4 cm, serves solely as a urinary passage. Conversely, the male urethra, with its external orifice at the glans penis tip, spans about 18–20 cm and performs a dual role, conveying both urine and seminal fluid (Moore et al. 2018; Standring 2020). This anatomical divergence leads to notable functional differences. The female urethra's exclusive urinary function contrasts with the male urethra's dual purpose, necessitating intricate neurological and muscular control to coordinate urination and ejaculation (Fowler et al. 2008). These structural and functional variations have important clinical implications. The shorter female urethra correlates with a higher incidence of urinary tract infections, while the longer, more complex male urethra is more prone to obstructive conditions like benign prostatic hyperplasia (Chapple et al. 2008).

9.1.2.1 Male Urethra

The male urethra, a muscular channel, extends from the neck of the bladder to the external meatus, serving as a conduit for both urine and seminal fluid. This complex structure, measuring 18–20 cm in an adult, comprises both striated and smooth muscle, and can be divided into four segments: the preprostatic, prostatic, membranous, and spongiose urethra (Drake et al. 2020) (Fig. 9.13). Alternatively, the male urethra is often described in three portions: the proximal or prostatic urethra; the membranous urethra (Both the prostatic and membranous portions are included in the posterior urethra); the anterior urethra (bulbar urethra, pendulous (or penile) urethra; fossa navicularis.

Histologically, the epithelial lining of the urethra changes along its course. The prostatic and membranous segments are lined with transitional epithelium, while the spongiose urethra transitions from pseudostratified columnar epithelium proximally to stratified squamous epithelium distally. The distal half of the spongiose urethra is

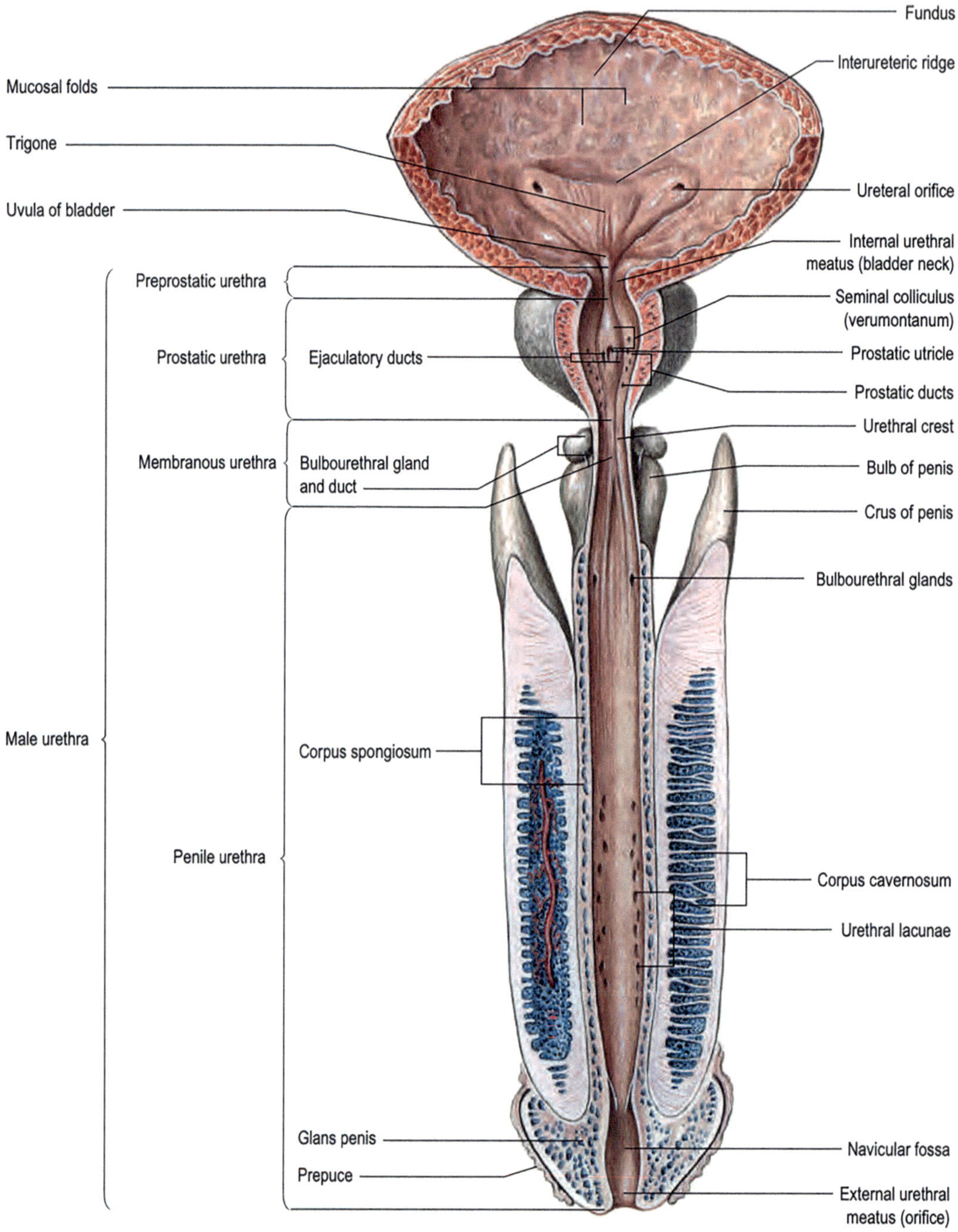

Fig. 9.13 Anatomy of the male urethra, illustrating its structural features and regional divisions. (Adapted with permission from Standring (2020))

dotted with openings of mucus-secreting periurethral glands of Littré (Hickey et al. 1982).

In a non-erect state, the urethra presents a double curve. The preprostatic, prostatic, and membranous segments run almost horizontally forward, while the bulbar portion of the urethra curves obliquely upwards and forwards. The penile portion then descends obliquely, creating this characteristic double curve (Moore et al. 2018).

Posterior Urethra

Preprostatic Urethra

The preprostatic urethra, a key segment of the male urinary tract, extends approximately 1 cm from the base of the bladder to the superior aspect of the prostate (Fig. 9.13). This short but important segment is important for urinary continence and ejaculatory function.

Anatomically, the preprostatic urethra is characterized by its stellate lumen, a feature readily observable during urethroscopy. It is surrounded by a distinct collar of smooth muscle fibers, which is structurally and functionally different from the detrusor muscle of the bladder (Standring 2020).

A key feature of the preprostatic urethra is its rich innervation by sympathetic noradrenergic fibers, in contrast to the predominantly parasympathetic innervation of the detrusor muscle. This unique innervation pattern is crucial for preventing retrograde ejaculation, as contraction of the preprostatic sphincter during ejaculation prevents seminal fluid from entering the bladder (Yucel and Baskin 2004).

The smooth muscle surrounding the preprostatic urethra, along with that of the bladder neck, forms what is known as the internal sphincter. This structure extends from the internal vesical meatus through the prostatic urethra to the level of the verumontanum, providing passive continence via sympathetic control (Drake et al. 2020). Interestingly, small periurethral glands located in this region may contribute to benign prostatic hyperplasia and symptoms of outflow obstruction in older men. These glands, while typically inconspicuous, can undergo hyperplasia and contribute to the overall enlargement of the prostate gland (Roehrborn 2008).

The preprostatic urethra transitions into the prostatic urethra, which continues through the prostate gland for another 3–4 cm. At this junction, the smooth muscle of the preprostatic urethra becomes continuous with the capsule of the prostate, forming a structurally and functionally integrated unit (Moore et al. 2018).

Understanding the anatomy and function of the preprostatic urethra is essential for comprehending various urological conditions, including benign prostatic hyperplasia, retrograde ejaculation, and certain types of urinary incontinence. Its unique structure and innervation make it a critical component in both urinary and reproductive functions.

Prostatic Urethra

The prostatic urethra, a key segment of the male urinary tract, traverses the entire height of the prostate gland. This segment, spanning 3.5–4 cm, is notable for being the widest and most dilatable portion of the male urethra (Moore et al. 2018).

Anatomically, the prostatic urethra tunnels through the prostate, positioned closer to the anterior than the posterior surface of the gland (Fig. 9.14). It connects superiorly with the preprostatic urethra and emerges slightly anterior to the prostatic apex inferiorly (Standring 2020).

A distinctive feature of the prostatic urethra is the urethral crest, a midline ridge projecting into the lumen along most of its posterior wall. This creates a crescentic appearance in transverse section. On either side of the crest lie shallow depressions called prostatic sinuses, where 15–20 prostatic ducts open (Drake et al. 2020) (Fig. 9.14). Midway along the urethral crest, an elevation known as the verumontanum or seminal colliculus is present. At this point, the urethra takes an anterior turn, typically at an angle of 30–35°, though this can vary with age and between individuals. The verumontanum contains the slit-like orifice of the prostatic utricle, a small cul-de-sac approximately 6 mm long that extends upwards and backwards into the prostate (Yucel and Baskin 2004).

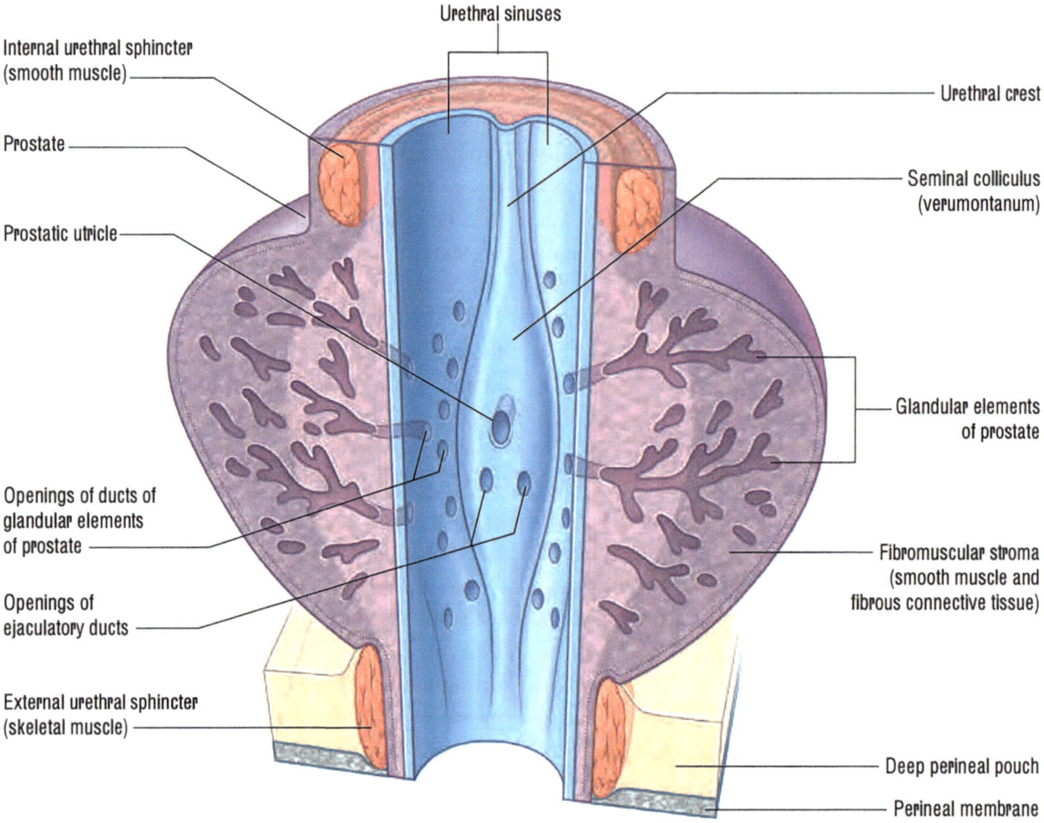

Fig. 9.14 The prostatic portion of the male urethra. The urethral crest features a raised area called the seminal colliculus, clinically referred to as the verumontanum. (Adapted with permission from Standring (2020))

The prostatic utricle, composed of fibrous tissue, muscular fibers, and mucous membrane, is pitted by openings of numerous small glands. Its embryological origin from the paramesonephric ducts or urogenital sinus has led to debates about its homology with female structures, being variously described as analogous to the vagina or uterus (Shapiro et al. 2004). On either side of the prostatic utricle, or sometimes within it, are the small openings of the ejaculatory ducts. These anatomical relationships are important in understanding both normal physiology and pathological conditions affecting this region (Moore et al. 2018).

The inferior portion of the prostatic urethra is anchored by the puboprostatic ligaments, rendering it immobile. This fixation is important in maintaining urethral stability and has implications for surgical procedures in this area (Walz et al. 2016).

Membranous Urethra

The membranous urethra, a critical yet often underappreciated segment of the male urinary tract, bridges the gap between the prostatic apex and the bulb of the penis. Despite its modest length of 1.5–2.5 cm, this portion plays a pivotal role in urinary continence and voiding control (Yucel and Baskin 2004). As it emerges from the prostatic apex, the membranous urethra descends with a gentle ventral concavity, threading its way through the perineal membrane. It lies approximately 2.5 cm posteroinferior to the pubic symphysis, a position that makes it vulnerable to injury in pelvic trauma but also accessible for certain surgical interventions (Standring 2020).

The wall of this urethral segment exhibits remarkable anatomical complexity. Beneath its epithelial lining lies a narrow layer of fibroelastic connective tissue, which in turn is surrounded by a muscular coat. This coat consists of two distinct layers: an inner layer of smooth muscle bundles that seamlessly continue from the prostatic urethra, and a prominent outer layer of circularly oriented striated muscle fibers. Together, these muscle layers form the external urethral sphincter, also known as the rhabdosphincter (Drake et al. 2020).

The external urethral sphincter, with its characteristic omega shape, encircles the urethra like a protective collar. Its unique composition of both smooth and striated muscle enables it to function as both a voluntary and involuntary unit, providing the fine control necessary for maintaining continence. Surrounding this sphincter, a layer of periurethral striated muscle fibers offers additional support, allowing for conscious control over urination, such as the ability to interrupt mid-stream (Moore et al. 2018).

Urinary continence at this level is not just the function of the sphincter. It is a cooperative effect involving radial folds of urethral mucosa, submucosal connective tissue, intrinsic smooth muscle, striated muscle fibers, and the pubourethral component of the levator ani muscle. This multifaceted approach for continence emphasizes the necessity of maintaining the integrity of this area throughout urological interventions (Walz et al. 2016).

The bulbourethral glands contribute to the complexity of this region (*see* Fig. 9.13). Situated within the sphincteric muscle, these glands discharge their secretions into the membranous urethra during sexual arousal, emphasizing the urethra's dual function in urine and reproductive systems (Shapiro et al. 2004).

Anterior Urethra

The anterior urethra, also known as the spongiose urethra, is the longest segment of the male urethral tract, extending approximately 15 cm in the flaccid penis. This portion is entirely encased within the corpus spongiosum, a spongiose erectile tissue that provides both protection and flexibility to the urethra (Standring 2020).

The spongiose urethra can be further divided into three distinct regions: the bulbar, penile, and fossa navicularis portions. Each of these segments has unique characteristics that contribute to the overall function of the urethra.

The anterior urethra starts with the bulbar segment, surrounded by the bulbospongiosus muscle and situated within the perineum. This section is the broadest portion of the urethra and represents the transition from the posterior to the anterior urethra. About 2.5 cm under the perineal membrane, the bulbourethral glands (Cowper's glands) discharge into this area, providing mucoid secretions that facilitate lubrication (Moore et al. 2018).

As the urethra continues its course, it curves downwards to form the penile urethra. This segment runs the length of the penile shaft and is characterized by its flexibility, allowing it to accommodate the changes in penile size and shape during erection. When empty, the penile urethra appears as a narrow, transverse slit, but it can expand to a diameter of about 6 mm during urination (Drake et al. 2020).

The final portion of the anterior urethra is the fossa navicularis, a dilated segment within the glans penis. This region terminates at the external urethral orifice, which is typically the narrowest part of the entire urethra. In adults, this orifice appears as a sagittal slit about 6 mm long, bordered on each side by small labia (Yucel and Baskin 2004).

Throughout its length, the anterior urethra is lined with epithelium that harbors numerous small mucous glands, particularly in the bulbar and distal penile segments. These glands, along with several small pit-like recesses called lacunae, contribute to the lubrication of the urethra. One notable lacuna, the lacuna magna, is larger than the rest and is located in the roof of the navicular fossa (Hickey et al. 1982).

The complex structure of the anterior urethra, with its varying diameters, glandular components, and flexible nature, allows it to perform its dual function of urine expulsion and semen transport effectively.

9.1.2.2 Female Urethra

This tubular passage begins at the bladder neck and terminates at the external meatus, coursing through the distal third of the anterior vaginal wall. In adult females, it measures approximately 4 cm in length and 6 mm in diameter (Drake et al. 2020). As it travels from the internal urethral orifice, it follows an anteroinferior path, nestling behind the pubic symphysis within the anterior vaginal wall (Yoshimura and Chancellor 2003) (Fig. 9.15).

Examining the urethral wall reveals a layered structure, each component contributing to its function. The outer muscular coat surrounds an inner mucosal lining. This mucosa comprises a stratified epithelium resting on a lamina propria rich in fibroelastic connective tissue and blood vessels (Standring 2020). Within the lamina propria, one finds a network of thin-walled veins and elastic fibers, oriented both circularly and longitudinally around the urethra.

The epithelial composition of the urethra varies from proximal to distal. The urothelium proximally lines the passage, resembling the bladder neck. This transitions to a non-keratinizing stratified squamous epithelium more distally. At the external meatus, the epithelium becomes keratinized, merging smoothly with the vestibular skin (Netter 2019).

The external urethral sphincter, also known as the rhabdosphincter, is a crucial component in urethral function. This structure, composed of slow-twitch muscle fibers, wraps around the ventral surface of the urethra in a horseshoe configuration. Innervated by the perineal branch of the

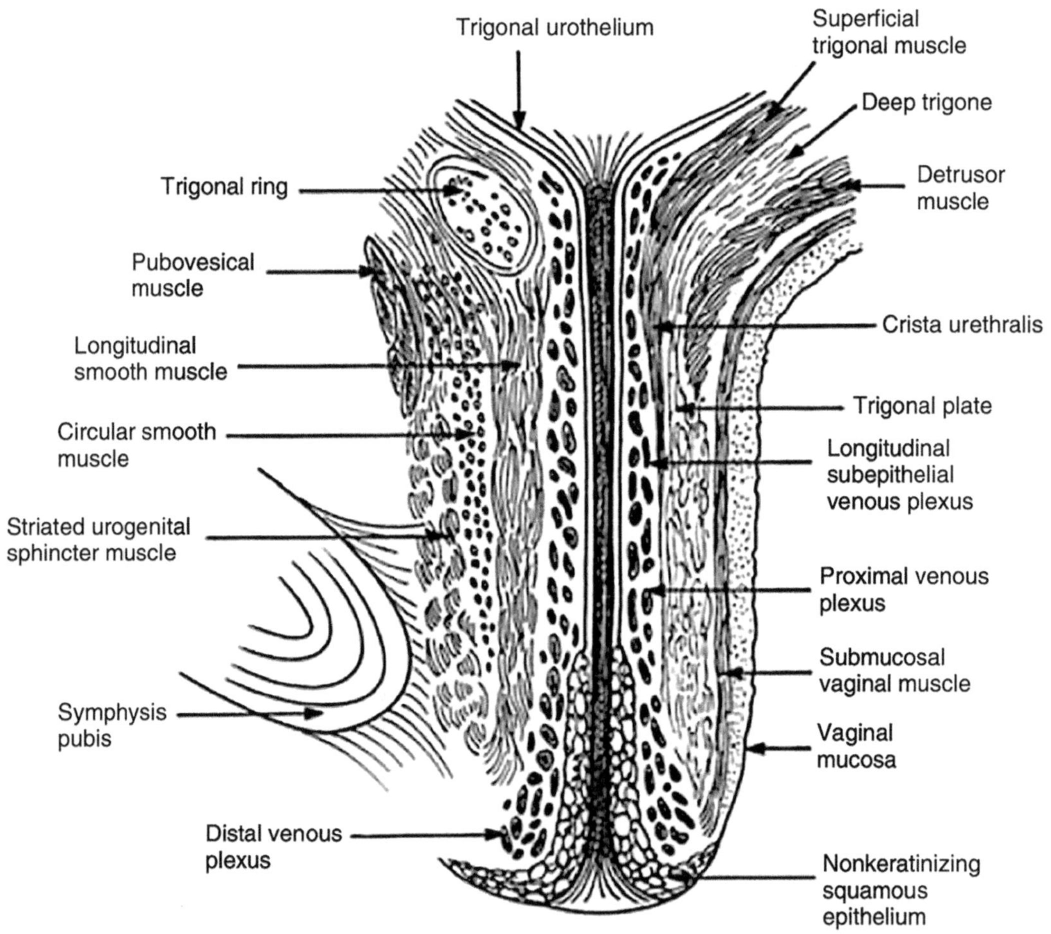

Fig. 9.15 Midsagittal section illustrating the anatomy of the female urethra. This figure emphasizes the contribution of multiple tissue structures to continence, including the striated sphincter, smooth sphincter, and subepithelial venous plexus. (Adapted with permission from Ashton-Miller and DeLancey (2007))

pudendal nerve, it allows for voluntary control of urination (Thor and de Groat 2010).

Urinary continence in females relies on three primary mechanisms. First, abdominal pressure transmits to the proximal urethra (Enhorning 1961). Second, the guarding reflex activates during stress events, causing the contraction of the external urethral sphincter (Constantinou and Govan 1982). Third, the "hammock hypothesis" suggests that abdominal pressure compresses the anterior urethral wall against its supported posterior wall (DeLancey 1994). Another aspect of urethral function has been proposed. Urethral attachments to the pubis and vaginal connections to pelvic muscles and fascia actively alter the position of the bladder neck and proximal urethra during voiding. This anatomical arrangement allows for urethral compression against the pubis during bladder filling and moments of strain (Ashton-Miller and DeLancey 2007).

The urethral structure also includes a network of glands. Small mucus-secreting glands populate the subepithelial tissue of the proximal urethra, their secretions flowing directly into the urethral lumen. Skene's glands are located in the distal urethra and discharge through paraurethral ducts adjacent to the external meatus (Pradidarcheep et al. 2011).

Innervation patterns of the urethral smooth muscle provide useful information about its function. These muscles receive extensive cholinergic parasympathetic innervation, with comparatively fewer noradrenergic inputs (Andersson and Wein 2004). The smooth muscle fibers predominantly align longitudinally, suggesting a role in shortening and widening the urethral lumen during micturition (Elbadawi 1996). The female urinary system lacks a distinct anatomical bladder neck sphincter. This absence raises questions about the mechanisms of continence at the bladder neck and proximal urethra. While one might expect smooth muscle activity to play a dominant role, evidence suggests otherwise. The competence of these regions likely relies more on the support provided by surrounding ligamentous structures than on smooth muscle action alone. The interplay between muscle orientation and innervation paints a picture of urethral function during micturition. The contraction of longitudinally oriented muscle fibers under parasympathetic stimulation results in the shortening and widening of the urethral lumen. This coordinated action facilitates urine passage through the urethra, emphasizing the role of urethral smooth muscle in the female micturition process.

9.1.2.3 Urethral Striated Muscle Fiber Composition

Striated muscles, including those in the urethral sphincter complex, are composed of distinct fiber types with varying functional and metabolic properties. These fibers are broadly categorized into slow-twitch and fast-twitch types, each playing a unique role in maintaining urinary continence (Gosling et al. 1981).

Slow-twitch fibers, also known as type I fibers, are characterized by their resistance to fatigue and ability to maintain tension over extended periods. These fibers predominantly rely on oxidative metabolism, containing numerous mitochondria and myoglobin, which gives them a red appearance (Padykula and Gauthier 1966). In the urethral sphincter, slow-twitch fibers are particularly well-suited for maintaining basal sphincter tone, a function crucial for long-term continence (Elbadawi 1982).

Fast-twitch fibers, or type II fibers, can be further subdivided into type IIa (fast oxidative) and type IIb (fast glycolytic) based on their metabolic profiles. These fibers are recruited rapidly and generate powerful contractions but fatigue quickly. They primarily utilize anaerobic metabolism and are rich in glycogen stores (Markwardt and Isenberg 1992). In the context of urinary function, fast-twitch fibers are essential for rapidly augmenting sphincter tone during sudden increases in intra-abdominal pressure, such as during coughing or sneezing (Tanagho et al. 1969).

The external urethral sphincter, also referred to as the rhabdosphincter, is a specialized skeletal muscle distinct from the periurethral muscles of the pelvic floor. It consists of two components with differing fiber type compositions: the periurethral striated muscle of the pelvic floor contains a mix of fast-twitch and slow-twitch fibers, while the striated muscle of the distal sphincter mechanism is predominantly composed of slow-twitch fibers (Elbadawi 1996). The distal component comprises more than 50% of static urethral resistance, emphasizing its importance in maintaining continence (Tanagho et al. 1969).

The fiber type distribution in the external urethral sphincter exhibits sexual dimorphism. In males, the rhabdosphincter comprises approximately 35% fast-twitch and 65% slow-twitch fibers. In contrast, females show a higher proportion of slow-twitch fibers, with a ratio of 87% slow-twitch to 13% fast-twitch fibers (Padykula and Gauthier 1966). This difference may reflect the different functional demands placed on the urethral sphincter in each sex.

At the molecular level, the contractile properties of these fiber types are governed by several factors. Fast-twitch fibers exhibit higher myosin ATPase activity, which catalyzes the actin-myosin interaction more rapidly, resulting in quicker contractions. They also express a fast isoform of the Ca^{2+}-ATPase in the sarcoplasmic reticulum, allowing for rapid calcium sequestration and, consequently, faster relaxation. Slow-twitch fibers, conversely, express a slow isoform of this enzyme (Markwardt and Isenberg 1992). Similar to smooth muscle, striated muscle fiber contraction is regulated by intracellular calcium through its interactions with the troponin-tropomyosin complex.

A research has revealed that a significant proportion of both fast-twitch and slow-twitch fibers in the human membranous urethral sphincter express nitric oxide synthase (NOS) in their sarcolemma. Specifically, the majority of fast-twitch fibers and approximately 25% of slow-twitch fibers show positive NOS staining (Ho et al. 2003). This finding suggests a potential role for nitric oxide in modulating sphincter function, although the precise mechanisms require further investigation.

The unique composition and properties of urethral striated muscle fibers have significant implications for urinary function and dysfunction. The predominance of slow-twitch fibers in the distal sphincter mechanism allows for sustained, low-level contraction necessary for maintaining urinary continence during daily activities. The fast-twitch fibers in the periurethral muscles provide quick contraction in response to abrupt increases in abdominal pressure, acting as an important backup mechanism to prevent stress urine incontinence (Gosling et al. 1981).

The external sphincter comprises smooth muscle components that receive noradrenergic innervation. Studies have shown that stimulation of the hypogastric nerve elicits myogenic potentials in the external urethral sphincter (Kakizaki et al. 1997). While the exact origin of this activity remains unclear, its persistence after α-adrenergic blockade suggests a contribution from striated muscle fibers.

9.2 Neuroanatomical Basis of the Lower Urinary Tract

The functioning of the lower urinary tract is systematically regulated by the autonomic nervous system and somatic motor nerves, which include the sympathetic and parasympathetic divisions as well as the pudendal nerve. This innervation system is essential for regulating urine storage and micturition (de Groat et al. 2015). Figure 9.16 illustrates the fundamental architecture of the autonomic nerve system in connection with the spinal cord.

Parasympathetic innervation, originating from the S2 to S4 spinal cord segments, initiates bladder contractions necessary for micturition. These fibers, known as pelvic splanchnic nerves or nervi erigentes, converge with sympathetic fibers from the T10 to L3 thoracolumbar spinal cord segments in the pelvis, forming the pelvic plexus (Fowler et al. 2008). This plexus extends branches to the bladder and urethra, providing both afferent and efferent innervation.

Within the detrusor muscle, parasympathetic preganglionic nerves synapse to become postganglionic nerves, predominantly innervating the muscle via cholinergic fibers. These fibers maintain a one-to-one relationship with individual muscle bundles, making complete detrusor denervation nearly impossible (Drake et al. 2010). This unique innervation pattern is particu-

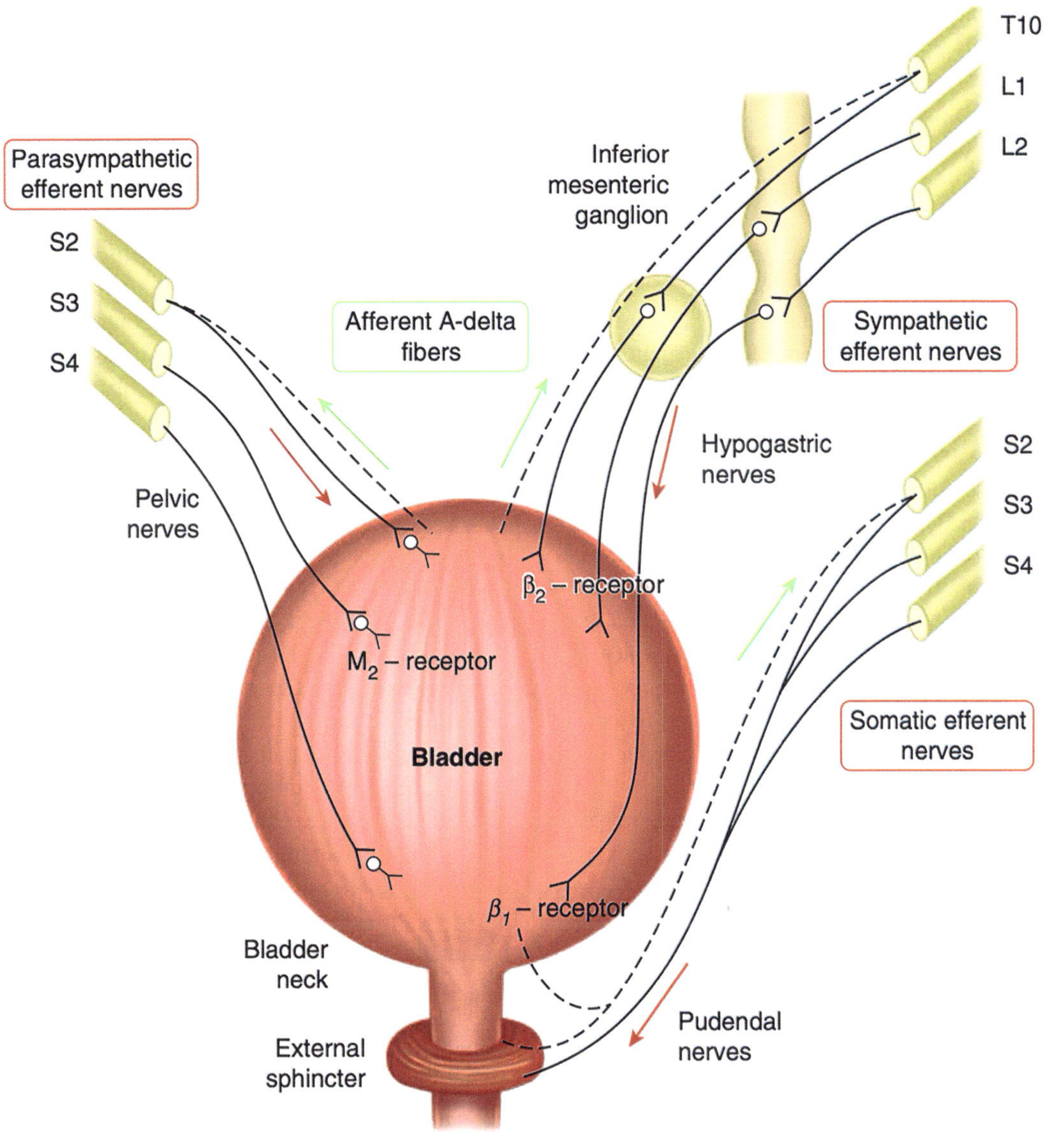

Fig. 9.16 Afferent and efferent neural pathways regulating lower urinary tract function, including the hypogastric nerve, pelvic nerve, and pudendal nerve. (Adapted with permission from Vignoli (2017))

larly relevant following suprasacral spinal cord injury, where the bladder may exhibit independent low-pressure contractions due to the synaptic architecture within the detrusor muscle.

Sympathetic nerve fibers descend through the lumbar splanchnic nerves to the hypogastric plexus, where they synapse to form postganglionic fibers. These hypogastric nerves merge with parasympathetic fibers in the pelvic plexus before reaching the bladder (Yoshimura and Chancellor 2003). Sympathetic innervation is distributed throughout the bladder, with a more pronounced influence on the bladder base and neck, particularly in modulating parasympathetic neuron activity at the preganglionic level.

The innervation of the bladder base and neck exhibits significant sexual dimorphism. In males, sympathetic innervation predominates, characterized by a lower density of cholinergic receptors and a high concentration of α-adrenergic fibers. These α-adrenergic fibers are crucial for maintaining tone in the bladder neck, proximal urethra, and prostate (Andersson and Wein 2004). This sympathetic dominance underlies the therapeutic use of α-antagonists to relax the bladder neck and prostatic smooth muscle in conditions such as benign prostatic hyperplasia and voiding dysfunction. Conversely, the female bladder neck and urethra have a limited adrenergic nerve supply and are primarily innervated by cholinergic fibers. This difference may be attributed to the embryological absence of a functional prostate and a corresponding structure to the membranous urethra in females (Thor and de Groat 2010). Understanding these neuroanatomical differences is important in developing targeted therapies for lower urinary tract dysfunction. For instance, the efficacy of α-blockers in male voiding dysfunction and their limited utility in females can be explained by these innervation patterns. Furthermore, the unique innervation of the detrusor muscle provides insights into the pathophysiology of neurogenic bladder conditions following spinal cord injury.

Somatic efferent motoneurons innervating the external striated urethral sphincter and pelvic floor muscles are located along the lateral edge of the ventral horn in the sacral spinal cord, an area often referred to as Onuf's nucleus. These sphincter motoneurons have transversely oriented dendritic bundles that extend laterally into the lateral funiculus, dorsally into the intermediate gray matter, and dorsomedially toward the central canal. Somatic nerve terminals release acetylcholine, which binds to skeletal muscle nicotinic receptors, triggering muscle contraction (Yoshimura and Chancellor 2003).

9.2.1 Suprapontine Control of Urinary Function

Micturition is a complex physiological process that integrates autonomic, voluntary, and emotional components under the control of the central nervous system (Fig. 9.17). This intricate regulation involves various suprapontine structures, each contributing to different aspects of urinary function.

The pontine micturition center, also known as Barrington's nucleus, serves as a critical integration point for signals from diverse brain regions, including the cerebellum, basal ganglia, thalamus, hypothalamus, and cerebral cortex (de Groat et al. 2015). The pontine micturition center's role as a central coordinator underscores the multifaceted nature of micturition control.

Voluntary control of micturition is primarily regulated by cortical centers in the frontal lobe, particularly in the paracentral lobule of the premotor area (Fowler and Griffiths 2010). This cortical involvement allows for conscious initiation and inhibition of voiding, a crucial aspect of social continence. The emotional aspects of micturition control involve a complex network within the limbic system, including the hypothalamus, hippocampus, cingulate gyrus, orbitofrontal cortex, amygdala, and several nonspecific thalamic nuclei (Griffiths and Tadic 2008). The exact role of the limbic system in relation to the pontine micturition center in humans is not yet completely understood; nonetheless, its connection shows the considerable effect of emotional states on urinary function and dysfunction.

At the brainstem level, particularly in the pons and medulla, the locus caeruleus complex and subcaeruleus nucleus complex play crucial roles in modulating vesicosphincteric function. Stimulation of these areas can induce either contraction or relaxation of the bladder and sphincter, with the anterior and dorsal parts of the pons being integral components of the Barrington center (Holstege 2005). The neurons within the subcaeruleus nucleus complex are predominantly noradrenergic, but a diverse array of neurotransmitters, including cholinergic, serotonergic, and enkephalinergic systems, are also involved in this regulation.

The subcaeruleus nucleus complex exerts its influence on micturition through both ascending and descending pathways. Ascending fibers contribute to the emotional and voluntary decision-making processes associated with micturition. Descending connections, originating from the

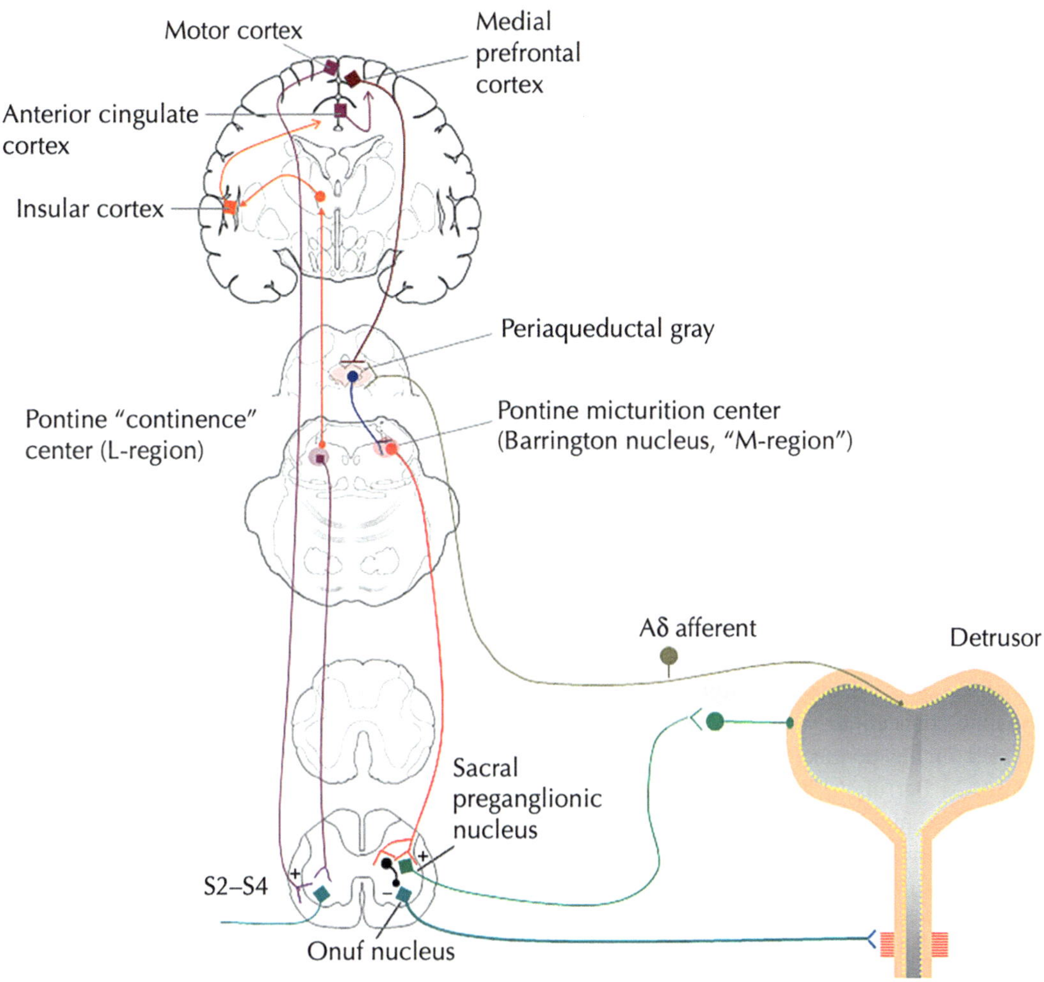

Fig. 9.17 Brainstem integration of bladder afferent input and the micturition reflex. The periaqueductal gray matter and the pontine micturition center coordinate the "on-off" switch mechanism regulating urine storage and micturition. (Adapted with permission from Benarroch (2014))

ventral part of the locus caeruleus complex, innervate extensive regions of the spinal cord. Two primary catecholaminergic pathways follow the intermediolateral column, ultimately innervating sympathetic thoracolumbar and parasympathetic sacral neurons (Yoshimura and Chancellor 2003).

Lesions located at or above the intracollicular level generally lead to detrusor overactivity and increased reflex responses (Fowler et al. 2008). This observation emphasizes the essential involvement of higher brain centers in modulating lower urinary tract function and provides insight into the pathophysiology of certain urological disorders.

Understanding these complex suprapontine control mechanisms is crucial for comprehending various urological disorders and developing targeted therapies. For instance, the involvement of frontal lobe areas in voluntary control of micturition explains why certain neurological conditions affecting these regions, such as stroke or dementia, can lead to urinary incontinence (Sakakibara et al. 2014).

The suprapontine control of urinary function represents a sophisticated interaction between autonomic, voluntary, and emotional processes. This complex regulation ensures the appropriate timing and social acceptability of micturition while maintaining lower urinary tract health.

9.2.1.1 Micturition Control of the Cerebral Cortex

The cerebral cortex is necessary for regulating lower urinary tract function, especially in managing bladder and detrusor muscle activity. This higher-level control is essential for the voluntary components of urination and continence. Principal cortical regions involved in this regulation are the superomedial frontal lobe, incorporating the medial prefrontal cortex and supplementary motor area, with the genu of the corpus callosum (Fowler et al. 2008) (Fig. 9.18).

Neural pathways originating from these cortical regions form a complex network that descends through subcortical structures, notably the basal ganglia. These pathways ultimately establish synaptic connections with detrusor motor nuclei located in the pontine micturition center, also known as Barrington's nucleus (de Groat et al. 2015). This hierarchical organization allows for both top-down control and bottom-up modulation of micturition reflexes.

The primary function of this cortical input is to provide an inhibitory effect on the micturition reflex, enabling conscious control over bladder function (Griffiths and Fowler 2013). This inhibition provide the basis for social continence, allowing individuals to suppress the urge to void when it is socially unacceptable. The voluntary control of bladder storage and voiding is influenced by both conscious decision-making processes and subconscious cerebral activities, demonstrating the intricate nature of this neural network (Michels et al. 2015).

Specifically, the medial portion of the sensorimotor cortex is responsible for regulating the external urethral sphincter (Holstege 2005). This area provides direct control over the voluntary muscles involved in maintaining continence and initiating micturition. The precise coordination between detrusor contraction and sphincter relaxation is necessary for appropriate urination.

Detrusor sphincter dyssynergia is generally seen in isolated cortical disorders; various cerebral lesions may lead to sphincter weakness or impaired control (Panicker et al. 2015). An important clinical consideration is the phenomenon of pseudodyssynergia, where voluntary sphincter contraction to maintain continence

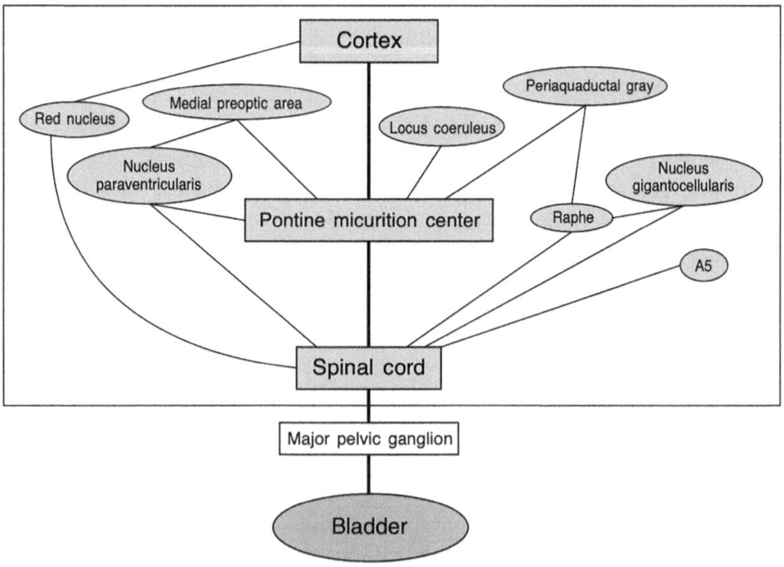

Fig. 9.18 Schematic representation of the central nervous system network involved in the control and modulation of bladder function. (Adapted with permission from Zermann et al. (1998))

against involuntary detrusor contractions can mimic true dyssynergia (Wein 2015). This indicates the importance of careful urodynamic assessment in patients with suspected neurogenic bladder dysfunction.

Cerebral pathologies, including strokes, neoplasms, and focal trauma, often present with a characteristic pattern of bladder dysfunction. Initially, these conditions may manifest as bladder areflexia, a state where the bladder fails to contract in response to filling. This is followed by the emergence of uninhibited detrusor contractions (Fowler et al. 2008). This progression is attributed to reduced suprapontine inhibition of the pontine micturition center, leading to disinhibition of the micturition reflex. However, as pontine-sacral pathways typically remain intact in such cases, the efficiency of bladder emptying is generally preserved once voiding is initiated (Griffiths et al. 2007).

Neuroimaging studies have further elucidated the cortical areas involved in micturition control. Functional magnetic resonance imaging (fMRI) has revealed activation in the anterior cingulate cortex, insula, and prefrontal cortex during bladder filling and voiding (Griffiths et al. 2007). These findings illustrate the complex interaction between various cortical regions in maintaining normal bladder function.

9.2.1.2 Basal Ganglia's Influence on Bladder Function

The neural control of micturition involves a complex interaction of various brain regions, with the basal ganglia playing a significant role in this process (Fowler et al. 2008). The basal ganglia, a group of subcortical nuclei including the caudate nucleus, subthalamic nucleus, putamen, globus pallidus, and substantia nigra, exert an inhibitory effect on bladder control, as evidenced primarily by animal studies.

Research has shown that electrical stimulation of specific areas within the basal ganglia, such as the putamen and globus pallidus, can suppress detrusor reflexes and induce urinary retention (Nishizawa et al. 1988). This finding shows the inhibitory nature of basal ganglia influence on bladder function. Conversely, damage to these structures can result in altered micturition control.

An example of this phenomenon is observed in feline models, where lesions in the substantia nigra have been demonstrated to cause detrusor overactivity. This overactivity is characterized by phasic, uninhibited contractions during the bladder filling phase (Yoshimura et al. 2003). Such observations provide valuable insights into the modulatory role of the basal ganglia in lower urinary tract function.

The implications of basal ganglia dysfunction on bladder control become particularly evident in neurological disorders affecting these structures. Conditions such as Parkinson's disease and multiple system atrophy (formerly known as Shy-Drager syndrome) are associated with alterations in micturition control (Sakakibara et al. 2012). These pathological states are hypothesized to disinhibit the micturition reflex from cortical regulation, resulting in detrusor overactivity and synergic sphincter activity.

9.2.1.3 Cerebellum Regulation of Urinary Tract Functions

The cerebellum, associated with motor control, balance, and coordination, also exerts a significant influence on lower urinary tract function (Dietrichs and Haines 2002). This structure acts as a neuromodulator, processing afferent signals from the bladder and pelvic floor while contributing to the maintenance of their basal tone.

The regulatory role of the cerebellum extends to the control of pelvic striated musculature, including the external urethral sphincter. It modulates both the intensity and frequency of contractions in these muscles, thereby fine-tuning their activity in relation to bladder function (Zermann et al. 1998).

Experimental studies using animal models have provided valuable insights into the specific contributions of cerebellar regions to lower urinary tract control:

- Fastigial nucleus: Electrical stimulation of this deep cerebellar nucleus has been shown to inhibit detrusor muscle activity (Bradley and Teague 1969). This finding suggests a poten-

tial role for the fastigial nucleus in preventing unwanted bladder contractions during the storage phase of micturition.
- Anterior vermis: Ablation experiments involving the anterior vermis have demonstrated a reduction in detrusor overactivity (Dietrichs and Haines 2002). This observation indicates that the anterior vermis may normally exert an excitatory influence on bladder function, and its removal can alleviate symptoms of overactive bladder.

These experimental findings demonstrate the complex and subtle function of the cerebellum in urinary tract regulation. The cerebellum's involvement in both inhibitory (via the fastigial nucleus) and potentially excitatory (via the anterior vermis) pathways indicates its importance in maintaining normal bladder function.

9.2.1.4 Neural Pathways and Spinal Cord Lesions

The complex mechanism of urine function is significantly dependent on the spinal cord, an essential pathway coordinating the interaction between the brain and the lower urinary tract. The spinal cord functions as both a conduit and a relay for signals between the pontine micturition center and the sacral micturition centers, enabling the exact regulation required for optimal bladder function.

Spinal cord lesions may dramatically alter urinary control, with the extent of disruption varying based on the lesion's location and severity. The body's remarkable capacity for neural plasticity and recovery also plays a role in determining the long-term impact of such injuries. One of the most significant complications arising from spinal cord damage between the pons and sacral regions is detrusor external sphincter dyssynergia. This condition manifests as a loss of coordination between the bladder's detrusor muscle and the external urethral sphincter. In detrusor external sphincter dyssynergia, these muscles contract simultaneously rather than in their normal, coordinated sequence (de Groat and Yoshimura 2006). Elevated bladder pressures may cause vesicoureteral reflux, potentially resulting in hydronephrosis. The risk of renal infections increases, and in severe cases, renal failure may occur.

The complexity of urinary function becomes even more apparent when considering the threefold innervation of the bladder detrusor and sphincter muscles. Parasympathetic, sympathetic, and somatic nervous systems each play essential functions in this complex interaction of regulation (Fowler et al. 2008).

Parasympathetic innervation, originating from cholinergic neurons in the sacral parasympathetic nucleus (S2–S4), is fundamental to the voiding process. These neurons activate cells in the pelvic and intramural ganglia via nicotinic receptors. The resulting release of acetylcholine, ATP, and nitric oxide has multiple effects. Acetylcholine, acting on M3 receptors, promotes detrusor muscle contraction, while simultaneously inhibiting norepinephrine release from sympathetic terminals through presynaptic M2 receptors. In contrast, sympathetic innervation, originating from preganglionic neurons in the T11–L2 spinal segments, primarily contributes to urine storage. Norepinephrine, the primary sympathetic neurotransmitter, induces detrusor muscle relaxation via β2- and β3-receptors, promotes bladder neck contraction through α1-receptors, and inhibits synaptic transmission in parasympathetic ganglia via presynaptic α2-receptors. The somatic nervous system's contribution comes from cholinergic motoneurons located in the Onuf nucleus at S2–S4 segments. These neurons, projecting through the pudendal nerve, activate the striated muscle of the external urethral sphincter via nicotinic acetylcholine receptors (Fig. 9.19).

In the storage phase of micturition, this complex system works together, with activation of sympathetic neurons, increased tonic activity of Onuf motoneurons, and inhibition of parasympathetic neurons. The voiding phase requires a precise reversal of this pattern: activation of parasympathetic output coupled with inhibition of both sympathetic and somatic outputs (Fowler et al. 2008).

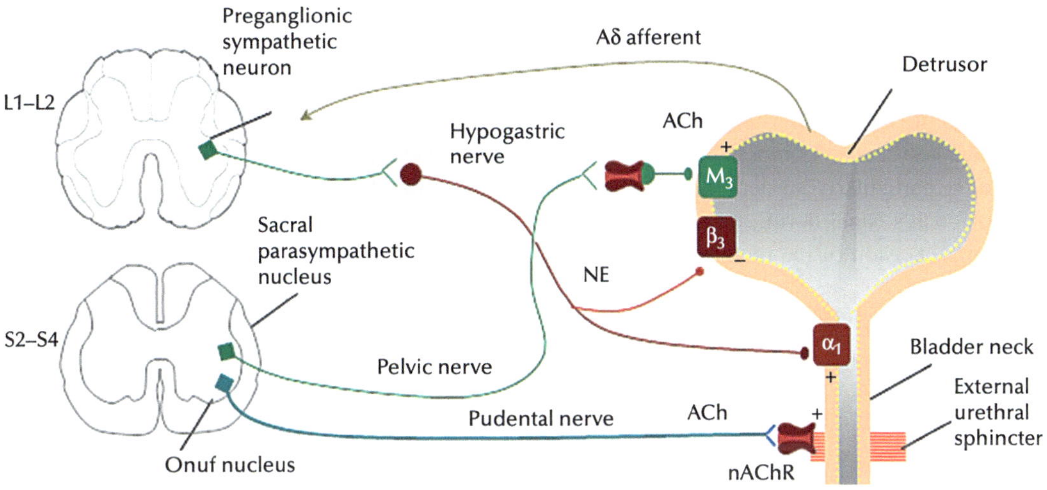

Fig. 9.19 Neural pathways involved in spinal output to the bladder. The detrusor and sphincter muscles are innervated by parasympathetic, sympathetic, and somatic pathways. Parasympathetic output originates from sacral parasympathetic nucleus neurons (S2–S4), traveling via the pelvic nerve, utilizing acetylcholine (ACh) to activate nicotinic receptors (nAChRs) in pelvic and intramural ganglia. ACh promotes detrusor contraction via M3 receptors and inhibits norepinephrine release through presynaptic M2 receptors. Sympathetic output arises from neurons in the L1–L2 spinal segments, projecting via the hypogastric nerve to ganglia and bladder tissues. Norepinephrine (NE) mediates detrusor relaxation through β2 and β3 receptors, bladder neck contraction via α1 receptors, and inhibition of parasympathetic ganglia through presynaptic α2 receptors. Somatic motor neurons in Onuf's nucleus (S2–S4) project via the pudendal nerve to activate the external urethral sphincter through nAChRs. During the storage phase, sympathetic activity and Onuf motor neuron activity increase, while parasympathetic output is inhibited. Voiding occurs with parasympathetic activation and suppression of sympathetic and somatic pathways. (Adapted with permission from Benarroch (2014))

9.2.2 Anatomical Configuration and Clinical Significance of the Pelvic Plexus

The pelvic plexus serves as a critical component in the neural regulation of pelvic viscera (Standring 2020). This neural network is formed by the confluence of pelvic parasympathetic nerves (originating from S2 to S4) and hypogastric sympathetic nerves (derived from T10-L2) in proximity to the rectum and, in females, the uterus (Baader and Herrmann 2003). A fascial sheath encapsulates each plexus, extending from the fundus of the bladder to the lateral pelvic wall, incorporating the ureter and branches of the hypogastric vessels (Raychaudhuri and Cahill 2008).

The anatomical positioning of the pelvic plexus varies between sexes. In females, it extends from the lateral aspect of the vagina to the cardinal ligament at the base of the broad ligament. In males, the plexus is situated between the anterior rectal wall and the posterior aspect of the prostate and seminal vesicles (Barber 2005). This complex neural structure comprises a heterogeneous collection of preganglionic parasympathetic and postganglionic sympathetic nerves (Thor and de Groat 2010).

Parasympathetic ganglia are located in close proximity to the bladder wall, offering a degree of protection against denervation injuries. However, it's important to note that decentralization remains a potential complication (Fowler et al. 2008). In contrast, sympathetic nerves within the pelvic plexus are predominantly postganglionic. Injury to these fibers results in degeneration and significantly impacts external sphincter function, often manifesting as a nonrelaxing deficit (Thor and de Groat 2010).

The clinical significance of the pelvic plexus becomes particularly evident in the context of surgical procedures. Operations such as radical hysterectomy and abdominoperineal resection pose a substantial risk to the integrity of this neural network (Moszkowicz et al. 2011). During rectal excision, the risk extends beyond direct trauma to include traction-induced damage. The pudendal nerve, a key component of pelvic innervation, is particularly susceptible as it emerges from Alcock's canal during perineal procedures involving excision of the rectum and anus (Rob et al. 2010).

The risk of pelvic plexus injury is somewhat decreased with radical hysterectomy; however, it continues to be a worry for both surgeons and patients. The risk increases if the surgical excision reaches below the cardinal ligaments or involves extensive removal of vaginal tissue (Rob et al. 2010).

9.2.3 Innervation of the Urinary Bladder

The urinary bladder receives its innervation from nerve fibers originating in the right and left inferior hypogastric plexuses, also referred to as pelvic plexuses (Standring 2020). These plexuses are mixed autonomic nerve networks, containing both sympathetic and parasympathetic components, each carrying efferent and afferent fibers (Baader and Herrmann 2003) (Fig. 9.20).

The sympathetic fibers reach the plexus via the right and left superior hypogastric nerves, which are derived from the superior hypogastric plexus. In contrast, the parasympathetic fibers arrive through the pelvic splanchnic nerves, emerging from the anterior sacral foramina (Raychaudhuri and Cahill 2008). The parasympathetic fibers in the inferior hypogastric plexus originate from the S2–S4 segments of the spinal cord, while the sympathetic fibers innervating the bladder originate from the T11–L2 segments of the spinal cord (Fowler et al. 2008).

Anatomically, the inferior hypogastric plexus is an elongated neural mesh, approximately 5–7 cm in length, positioned lateral to the rectum, internal genitalia, and bladder neck. This plexus provides all the parasympathetic and most of the sympathetic innervation to the pelvic and perineal viscera (Thor and de Groat 2010).

The clinical significance of these plexuses becomes evident in the context of radical pelvic surgery for cancers of the rectum, prostate, and cervix uteri. Disruption of these important neural networks can result in significant morbidity, manifesting as bladder dysfunction in both sexes and as erectile and/or ejaculatory dysfunction in males (Moszkowicz et al. 2011).

Functionally, the parasympathetic fibers provide the primary motor innervation to the detrusor muscle. The sympathetic fibers, on the other hand, supply motor innervation to part of the trigonal musculature (specifically the superficial trigonal muscle) and, in males, the pre-prostatic genital sphincter (Barber 2005).

Sensory innervation of the bladder is complex. The sensation of bladder distention is detected by stretch receptors in the lamina propria and conveyed via parasympathetic fibers to the posterior column of the spinal cord. However, pain sensations associated with bladder inflammation (due to calculi or infections) are transmitted by both sympathetic and parasympathetic fibers to the lateral spinothalamic tract (Rob et al. 2010).

Neural control of the bladder functions involves complex interactions between spinal and supraspinal centers. Urine storage depends on spinal reflexes mediated by lumbosacral sympathetic and motoneurons of the Onuf nucleus, located at the S2–S4 levels of the spinal cord. The pontine micturition center, also known as the pelvic organ stimulating center, coordinates the normal function of the bladder and other pelvic organs. It activates the sacral preganglionic parasympathetic output to the bladder detrusor while simultaneously inhibiting the Onuf nucleus (Benarroch 2010).

Higher brain centers also play a role in micturition control. The Onuf nucleus receives direct excitatory projections from the nucleus retroambiguus in the medulla. Behavioral and emotional influences affect the micturition reflex through integration of frontal lobe and bladder afferent inputs at the level of the periaqueductal gray, which projects to both the pontine micturition center and nucleus retroambiguus (Holstege 2005).

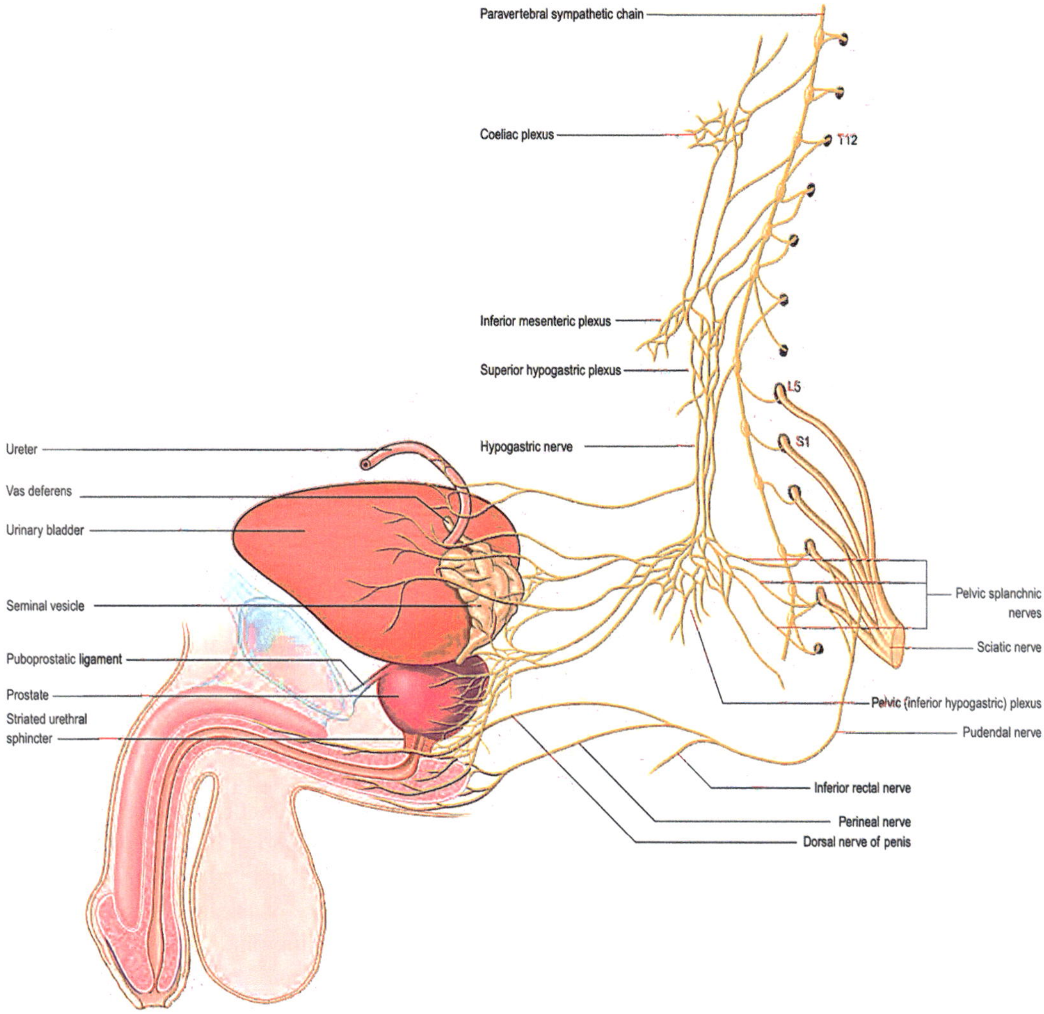

Fig. 9.20 Neural innervation of the lower urinary tract and male genitalia illustrating the contributions of parasympathetic, sympathetic, and somatic pathways. (Adapted with permission from Standring (2020))

9.2.3.1 Bladder Sensation

The urothelium, the superficial lining of the lower urinary tract, functions as mechanosensory transduction through bidirectional interactions with afferent and efferent nerves (Birder and de Groat 2007). During the process of bladder filling, urothelial cells release a variety of chemical signals, including acetylcholine, nitric oxide (NO), and adenosine triphosphate (ATP). These chemical mediators modulate the excitability of afferent fibers, thus influencing bladder sensation (Fowler et al. 2008).

Muscarinic and nicotinic receptors are thought to play a significant role in neural-urothelial interactions. Their activation triggers the release of ATP from urothelial cells, further amplifying the sensory signal (Birder and de Groat 2007). This complex relationship between the urothelium and neural elements reveals the importance of the bladder's epithelial lining in sensory function.

Afferent innervation of the bladder and urethra is primarily composed of two types of nerve fibers: small myelinated Aδ afferents and unmyelinated C afferents (de Groat and Yoshimura

2009). The Aδ afferents function as tension receptors, responsible for the physiological sensation of bladder fullness. These fibers are instrumental in triggering the spinobulbospinal micturition reflex, which is essential for normal bladder emptying (Thor and de Groat 2010).

In contrast, C afferents have a higher activation threshold and are mostly involved in mediating a spinal micturition reflex. This reflex becomes unmasked following spinal cord transection, emphasizing the adaptability of bladder innervation in pathological states (Holstege 2010). Interestingly, a subset of C fibers remains "silent" under normal conditions but becomes activated by chemical mediators released during bladder inflammation. This mechanism may contribute to altered sensation and urgency in various bladder pathologies (Raychaudhuri and Cahill 2008).

The afferent fibers from the bladder travel through the pelvic and hypogastric nerves, terminating in specific regions of the sacral and lumbar spinal cord. These areas include the superficial dorsal horn, dorsal commissure, and intermediate gray matter (de Groat and Yoshimura 2009). A distinct cell group in the sacral cord, known as Gert's nucleus, serves as a relay station, forwarding pelvic organ information to the periaqueductal gray (Holstege 2010). This pathway is required for the integration of bladder sensory information with higher brain centers. Many bladder afferents also project directly to the sacral parasympathetic nucleus. Here, they form synapses with both preganglionic neurons and local interneurons, facilitating segmental spinal reflexes (Barber 2005). This local circuitry allows for rapid responses to changes in bladder volume and pressure.

The central processing of bladder sensation involves multiple brain regions. Neurons receiving bladder and urethral afferent input send projections to the parabrachial nucleus. From here, sensory information is relayed via the thalamus to the insular cortex (Sugaya et al. 2005). This cortical representation of bladder sensation is necessary for conscious awareness of bladder fullness and the voluntary control of micturition.

9.2.3.2 Efferent Innervation of the Urinary Bladder

The efferent innervation of the urinary bladder comprises a complex network of parasympathetic and sympathetic fibers, each playing distinct roles in bladder function (Standring 2020).

Parasympathetic fibers originate from the second to the fourth sacral segments of the spinal cord. These fibers enter the pelvic plexuses on the posterolateral aspects of the rectum as the pelvic splanchnic nerves (Baader and Herrmann 2003). In contrast, sympathetic fibers derive from neuronal cell bodies in the lower three thoracic and upper two lumbar segments of the spinal cord. These sympathetic fibers form the celiac and mesenteric plexuses around the great vessels in the abdomen, from where the hypogastric plexuses descend into the pelvis as discrete nerve bundles within the extraperitoneal connective tissue posterior to the ureter on each side (Raychaudhuri and Cahill 2008).

The anterior part of the pelvic plexus, known as the vesical plexus, contains small groups of autonomic neurons. These multipolar intramural neurons, rich in acetylcholinesterase, occur in ganglia consisting of up to 20 neuronal cell bodies throughout the bladder wall (Fowler et al. 2008). The majority of preganglionic nerve terminals correspond morphologically to presumptive cholinergic fibers. Noradrenergic terminals also relay on cell bodies in the pelvic plexus, although it remains unclear whether similar nerves synapse on intramural bladder ganglia (Thor and de Groat 2010).

The detrusor muscle, including the trigonal portion, is profusely innervated, forming a dense plexus among the muscle cells. Most of these nerves contain acetylcholinesterase and are abundant throughout the muscle coat of the bladder (Barber 2005). Electron microscopy reveals that axonal varicosities adjacent to detrusor muscle cells possess features typical of cholinergic nerve terminals, containing clusters of small agranular vesicles, occasional large granulated vesicles, and small mitochondria (Moszkowicz et al. 2011). The human detrusor muscle also possesses a sparse but definite supply of sympathetic noradrenergic nerves, generally accompanying the vascular supply and rarely extending among the myocytes. Non-adrenergic, non-cholinergic nerves have been identified, and various neurotransmitters or neuromodulators, including somatostatin, have been detected in intramural ganglia (Rob et al. 2010).

9.2.4 Innervation of the Urethro-Vesical Unit

The superficial trigonal muscle exhibits a unique innervation pattern, with more noradrenergic (sympathetic) fibers than cholinergic (parasympathetic) nerves. This difference supports the view that the superficial trigonal muscle should be considered "ureteric" rather than "vesical" in origin (Holstege 2010). However, it is important to note that this muscle forms a minor part of the total muscle mass of the bladder neck and proximal urethra in both sexes, likely playing a limited role in the physiological control of these regions.

In males, the smooth muscle of the bladder neck is predominantly oriented obliquely or circularly. It receives sparse cholinergic (parasympathetic) innervation but possesses rich noradrenergic (sympathetic) innervation. This pattern extends to the smooth muscle of the prostate gland, seminal vesicles, and vasa deferentia (Benarroch 2010). Sympathetic stimulation causes contraction of smooth muscle in the wall of the genital tract, resulting in seminal emission and closure of the preprostatic sphincter, preventing reflux of ejaculate into the bladder. In females, the bladder neck's smooth muscle receives relatively few noradrenergic nerves but is richly supplied with presumptive cholinergic fibers. This sparse sympathetic supply likely relates to the absence of a functioning 'genital' portion of the female urethral wall (de Groat and Yoshimura 2009).

The lamina propria of the bladder fundus and inferolateral walls is predominantly lacking autonomic nerve fibers, except for some noradrenergic and occasional presumptive cholinergic perivascular nerves. However, the density of nonvascular nerves increases closer to the urethral orifice. At the bladder neck and trigone, a nerve plexus of cholinesterase-positive axons extends throughout the lamina propria, independent of blood vessels (Sugaya et al. 2005).

9.2.4.1 Hypogastric Nerve

The hypogastric nerve, an important element in bladder control, consists of both motor and sensory fibers. Its origin can be traced to preganglionic spinal neurons located in the thoracolumbar intermediolateralis cord, specifically at the level of T11 to L1 or L2 (Standring 2020).

The neural pathway of the hypogastric nerve follows a complex route. Preganglionic axons initially extend to the paravertebral sympathetic ganglionic chain, where they form synapses with ganglionic neurons. Subsequently, postganglionic axons traverse the superior hypogastric plexus to reach the vesical or inferior hypogastric plexus (Baader and Herrmann 2003).

The adrenergic innervation provided by these nerves exhibits regional specificity within the bladder. β-Adrenergic receptors predominate in the dome region, while α1-adrenergic receptors are more prevalent at the bladder base and neck, including the superficial trigone (Fowler et al. 2008). This differential distribution of receptor types results in distinct physiological effects: the dome experiences relaxation, while the bladder neck undergoes contraction in response to adrenergic stimulation (Thor and de Groat 2010).

While the hypogastric nerves are primarily adrenergic in nature, they also contain cholinergic and peptidergic components. However, the precise functions of these non-adrenergic elements remain to be fully elucidated (Raychaudhuri and Cahill 2008).

Notably, there is a marked sexual dimorphism in the innervation of the bladder neck. In males, this region receives rich sympathetic innervation. In contrast, the female bladder neck is predominantly supplied by cholinergic fibers, with significantly less adrenergic innervation (Barber 2005). This disparity in nerve supply may be attributed to the primary genital function ascribed to the bladder neck in males, a role that is less pronounced in females (Holstege 2010). Moreover, the sexual dimorphism in bladder neck innervation has important clinical implications. For instance, it may explain the different prevalence rates of certain urological disorders between males and females, and could inform sex-specific treatment approaches (de Groat and Yoshimura 2009).

9.2.4.2 Pelvic Nerves

The pelvic nerves constitute the primary parasympathetic component of bladder innervation. These neural structures originate from the S2 to S4 segments of the spinal cord and converge at the vesical plexus, from which branches extend to innervate the bladder (Standring 2020).

Predominantly, the fibers of the pelvic nerves are cholinergic in nature. However, it is worth noting that some noradrenergic fibers also contribute to the composition of the pelvic plexus, adding complexity to its neurochemical profile (Fowler et al. 2008).

The bladder, including the trigone region, receives extensive innervation from a dense neural network that interweaves among the detrusor muscle fibers. This extensive innervation reflects the high degree of neural control required for proper bladder function (Thor and de Groat 2010).

Cholinergic nerves comprise the majority of this neural plexus. These nerves typically follow the course of the vascular supply. Interestingly, they seldom extend among the nonstriated muscle components of the bladder neck and urethra, suggesting a specific pattern of innervation that may be relevant to bladder physiology (Raychaudhuri and Cahill 2008).

The predominance of cholinergic innervation in the bladder provides a rationale for the effectiveness of anticholinergic medications in treating certain bladder disorders. However, the presence of noradrenergic fibers suggests that adrenergic signaling may also play a role in bladder function, potentially offering additional therapeutic targets (de Groat and Yoshimura 2009).

9.2.4.3 Pudendal Nerves

The pudendal nerves play a significant role in both motor and sensory functions of the pelvic region. These nerves originate from the spinal motoneurons of Onuf's nucleus, located at the base of the anterior horn of S2–S4 (Standring 2020).

The axons of these motoneurons traverse the pudendal plexus, which is formed by the S2–S4 nerves. These axons then coalesce to form the pudendal nerves, which are responsible for innervating all the striated muscles of the pelvic floor, including the external urethral and anal sphincters (Fowler et al. 2008). Electrical stimulation of the pudendal nerve was found to inhibit involuntary detrusor contractions associated with bladder overactivity or hyperreflexia. This effect was attributed to the influence of stimulated pudendal afferents on both inhibitory and excitatory pathways in the spinal cord (Thor and de Groat 2010). Specifically, the pudendal afferents appear to modulate the inhibitory effects of the hypogastric or pelvic nerves on the bladder dome. Additionally, they influence the excitatory effects of the hypogastric nerve on the bladder neck, mediated by α-adrenergic receptors (Raychaudhuri and Cahill 2008).

9.2.5 Regulation of Urinary Storage and Micturition

Knowledge of bladder anatomy and physiology is fundamental for the diagnosis and management of urologic conditions arising from neurological injuries. Bladder dysfunction is typically classified into two main categories: issues with storage and problems with voiding (Standring 2020).

Lesions affecting the standard afferent or efferent pathways of the nervous system are potential sources of urodynamic disorders. These pathways form an intricate network that controls the filling and emptying of the bladder, and any disruption can lead to significant functional impairment (Fowler et al. 2008).

The process of urinary storage involves complex neural mechanisms. During this phase, the bladder must accommodate increasing volumes of urine while maintaining low intravesical pressure. This is primarily achieved through the activation of sympathetic pathways, which promote relaxation of the detrusor muscle and contraction of the bladder neck and external urethral sphincter (Thor and de Groat 2010). Conversely, micturition requires the coordinated activation of parasympathetic pathways, leading to detrusor contraction, along with the inhibition of sympathetic and somatic pathways, resulting in relaxation of the bladder outlet. This coordinated action allows for efficient bladder emptying (Raychaudhuri and Cahill 2008).

Neurological injuries can disrupt these finely tuned mechanisms in various ways. For instance, suprapontine lesions may result in detrusor overactivity due to the loss of inhibitory control from higher brain centers. Spinal cord injuries, depending on their level and completeness, can lead to a range of bladder dysfunctions, from detrusor-sphincter dyssynergia to acontractile bladder (Benarroch 2010).

Understanding these regulatory mechanisms is essential for developing targeted therapeutic approaches. For example, anticholinergic medications are often used to treat detrusor overactivity by reducing parasympathetic-mediated bladder contractions. Similarly, alpha-blockers may be employed to reduce bladder outlet resistance in cases of sympathetic overactivity (de Groat and Yoshimura 2009).

9.2.5.1 Bladder Compliance and Neural Control in Urine Storage and Voiding

The capacity of the bladder to hold substantial volumes of urine at low intravesical pressures characterizes its function during the storage phase. This ability arises from the bladder's compliance, allowing extensive passive filling without activating sensory nerves (Standring 2020).

Studies in cats have shown that most bladder sensory nerves are unmyelinated. Sacral sensory nerves, essential for proprioception of bladder fullness, are evenly distributed throughout the bladder's muscle layer. In contrast, lumbar sensory nerves, which detect external sensations like pain and temperature, are primarily located in the urothelium covering the posterior trigone and anterior bladder neck (Fowler et al. 2008).

Proprioception is further supported by muscle spindles in the periurethral striated muscle, which transmit information to the sacral spinal cord. The bladder's detrusor muscle is monitored by sensory nerves carried by Aδ fibers within the pelvic nerve, responding to passive distention. These nerves activate at intravesical pressures of 1–5 mmHg, corresponding to the initial awareness of bladder filling during cystometric analysis (Thor and de Groat 2010).

Animal studies have demonstrated that these sensory pathways enter the spinal cord and proceed directly to the lateral pontine micturition center without intermediate synapses. For conscious recognition of bladder filling, information must be transmitted from the detrusor and pelvic nerve to the sensorimotor cortex (Raychaudhuri and Cahill 2008).

C fibers, responsive to irritants and cold, follow the spinothalamic tracts to the sensorimotor cortex, potentially decreasing bladder storage capacity and promoting urination. The external urethral sphincter and pelvic floor muscles respond to bladder filling pressure. Sensory feedback from these muscles, via muscle spindles and Golgi tendon organs, is transmitted through the pudendal nerve to the sacral spinal cord and continues through the posterior columns to interact with γ- and α-motoneurons (Benarroch 2010).

During urine storage, sympathetic efferent activity predominates, maintaining continence. Pontine connections extend to the intermediolateral gray columns of the thoracolumbar cord, facilitating sympathetic outflow. This activates $\alpha 1$-adrenergic receptors in the bladder neck smooth muscle, causing contraction and closure of the bladder outlet. Sympathetic activity also suppresses parasympathetic ganglia, inhibiting detrusor muscle contraction (de Groat and Yoshimura 2009). The sympathetic nervous system targets the bladder body with predominantly β-receptors, generally associated with detrusor muscle relaxation. In humans, the exact role of these β-receptors remains unclear, with varied functions observed in animal studies (Birder and de Groat 2007).

Pontine linkage to the pudendal α-motor nucleus (segments S2–S4) via reticulospinal tracts results in external sphincter contraction, enhancing continence. As the bladder fills, external sphincter activity progressively increases, ensuring the bladder neck stays closed and urethral pressure exceeds bladder pressure (Reitz et al. 2003).

The coordination between brain's subcortical and cortical areas with the pons in regulating bladder storage is not fully understood. In felines, the cerebrum and midbrain are thought to contribute to voluntary inhibition of the pontine micturition center, allowing continued urine storage and filling. Inputs from the cerebellum, basal ganglia, thalamus, and hypothalamus may also play a role in suppressing the pontine micturition center, further supporting the storage phase (Holstege 2010).

9.2.5.2 Voiding Dynamics

Urinary voiding is a multifaceted process involving various sensory and motor pathways, including voluntary control by the brain. The pontine micturition center, located within the dorsalis tegmentolateralis nucleus of the pons, serves as a primary coordinator in managing the interaction between the bladder's detrusor muscle and the uretheric sphincters (Standring 2020).

As bladder pressure increases, sensory signals are conveyed to the brain, generating conscious awareness of the need to urinate. The decision to urinate activates specific brain regions, notably the anterior, inferior, and medial areas of the bilateral prefrontal cortex. These regions communicate through corticoreticular pathways to the pontine micturition center (Fowler et al. 2008). This process results in the inhibition of sympathetic and somatic pudendal nerve activities, leading to relaxation of both the internal and external sphincters. Simultaneously, the pontine micturition center stimulates parasympathetic nerves via reticulospinal connections to the sacral intermediolateral column, initiating a contraction of the detrusor muscle, which facilitates bladder emptying (Thor and de Groat 2010).

Research has shown that the prefrontal cortex plays a significant role in the voluntary control of micturition. Functional imaging studies have demonstrated increased activity in this region during both the decision to void and the act of voiding itself. This suggests that the prefrontal cortex is involved not only in initiating the voiding process but also in modulating ongoing micturition (Benarroch 2010).

The parasympathetic activation leading to detrusor contraction is mediated primarily through cholinergic pathways. Acetylcholine released from parasympathetic nerve endings acts on muscarinic receptors in the detrusor muscle, triggering contraction. This mechanism forms the basis for the use of anticholinergic medications in treating overactive bladder syndrome (de Groat and Yoshimura 2009).

Interestingly, recent studies have also identified a role for non-adrenergic, non-cholinergic (NANC) neurotransmission in the voiding process. Substances such as ATP and various neuropeptides have been implicated in modulating detrusor contraction and sensory signaling in the bladder. This expanding understanding of bladder neurophysiology opens new avenues for therapeutic interventions in voiding disorders (Birder and de Groat 2007).

The external urethral sphincter, controlled by the pudendal nerve, also plays a critical role in voiding dynamics. Its relaxation, coordinated with detrusor contraction, is necessary for efficient bladder emptying. Disorders affecting pudendal nerve function can lead to voiding difficulties and urinary retention (Reitz et al. 2003).

9.2.5.3 Controlled Interruption of Voiding

The brain's higher cortical areas possess the capability to interrupt the micturition reflex mediated by the pontine center. This interruption involves the suppression of signals that promote urination while simultaneously enhancing signals that support urine storage (Standring 2020).

The voluntary cessation of urination is a complex process that engages specific brain regions.

When an individual decides to stop urinating, there is activation of the deep medial frontal lobe's motor strip, which exerts control over the periurethral striated muscles, including the external urethral sphincter (Fowler et al. 2008). From this cortical origin, motor signals follow a specific neural pathway. They proceed via the pyramidal pathways, undergoing a crossing at the medullary pyramids. Subsequently, these signals travel down the corticospinal tracts, taking an anterolateral route to reach their destination in the somatic pudendal nucleus within the sacral spinal cord (Thor and de Groat 2010).

The culmination of this neural cascade results in the contraction of the external sphincter, effectively halting the process of urination, albeit temporarily. This mechanism demonstrates the remarkable level of voluntary control that can be exerted over what is fundamentally an autonomic process (Raychaudhuri and Cahill 2008).

Recent neuroimaging studies have provided further insights into the cortical areas involved in this process. Functional MRI studies have shown increased activity in the prefrontal cortex, anterior cingulate cortex, and insula during attempts to voluntarily control urination. These findings suggest a complex network of brain regions involved in the conscious control of micturition (de Groat and Yoshimura 2009).

The role of the external urethral sphincter in this process is particularly noteworthy. Unlike the internal sphincter, which is under involuntary control, the external urethral sphincter is composed of striated muscle and is subject to voluntary control. This dual sphincter system provides a robust mechanism for maintaining continence and allows for the fine control of urination (Birder and de Groat 2007).

9.2.6 Urethral Innervation

The innervation of the urethra involves a network of autonomic and somatic nerves, each playing specific roles in urethral function (Standring 2020). The prostatic plexus, which supplies the smooth muscle of the prostate and prostatic urethra, originates from the pelvic plexus on each side. This plexus is situated on the posterolateral aspect of the seminal vesicle and prostate. The cavernous nerves, both major and minor, penetrate the bulb of the corpus spongiosum to innervate the cavernous bodies of the penis (Fowler et al. 2008).

Sympathetic autonomic nerves supplying the internal urethral sphincter are derived from the pelvic plexus as it descends in the pelvis adjacent to the inferior prostatovesical pedicle and prostate. These nerves play a role in preventing retrograde ejaculation (Thor and de Groat 2010).

Parasympathetic preganglionic axons originate from neuronal cell bodies in the S2–S4 segments. These axons run in the pelvic splanchnic nerves and synapse in the vesical plexus in or near the bladder wall. Postganglionic fibers are then distributed to the smooth muscle of the urethral wall (Raychaudhuri and Cahill 2008).

The innervation of the external urethral sphincter has been a subject of debate. The prevailing view is that it receives innervation from neurons in Onuf's nucleus and from perineal branches of the pudendal nerve located on the perineal aspect of the pelvic floor. In both cases, the axons originate from neurons in the S2–S4 segments. Fibers from Onuf's nucleus (somatic) travel with the pelvic plexus on each side before branching off to run on the pelvic aspect of the pelvic floor and enter the membranous urethra (Benarroch 2010).

Somatic fibers innervating the striated muscle are also derived from the S2 to S4 segments. These fibers run in the pelvic splanchnic nerves but do not synapse in the vesical plexus (de Groat and Yoshimura 2009). Sensory fibers follow the path of the pelvic splanchnic nerves to the S2–S4 segments of the spinal cord. These afferent pathways are essential for conveying information about urethral tension and flow, contributing to the neural control of micturition (Birder and de Groat 2007).

In females, postganglionic sympathetic fibers arise from the plexus surrounding the vaginal arteries. This arrangement reflects the close anatomical and developmental relationship between the urogenital structures (Reitz et al. 2003).

9.3 Intra-Abdominal Pressure Raising Muscles (External Oblique, Internal Oblique, Rectus Abdominis, Transversus Abdominis, Cremaster)

The eight muscles of the abdominal wall (four on each side) collectively contribute to the function of raising intra-abdominal pressure. This action is accomplished by the 'sheet' muscles exerting tension on the rectus sheath via their aponeuroses, resulting in a flattening of the abdomen and compression of the abdominal viscera (Standring 2020).

When the diaphragm maintains its tone and resists upward displacement, the subsequent increase in intra-abdominal pressure plays a significant role in various physiological processes. These include:

- Micturition: In conjunction with appropriate sphincter relaxation, the increased pressure on the bladder facilitates urination.
- Defecation: The elevated pressure on the rectum assists in the elimination of feces.
- Emesis: Increased pressure on the stomach contributes to the process of vomiting.
- Parturition: In the final stages of childbirth, the compressive force generated by these muscles aids in the expulsion of the fetus from the uterus

When the diaphragm is in a relaxed state, the increase in intra-abdominal pressure causes the abdominal viscera to press against its lower surface, resulting in an upward displacement of the diaphragm. This action elevates the pressure within the thorax, such that when the glottis opens, air is forcefully expelled from the lungs, producing a vigorous cough or sneeze. The coughing action is further enhanced by the abdominal muscles exerting a downward pull on the lower ribs.

The coordinated action of the abdominal muscles and the diaphragm can also create a 'muscular corset' effect, which serves to stabilize the abdominal viscera. This function can be intensified during activities such as lifting, where it forms a type of pneumatic cushion anterior to the lumbar spine. This protective mechanism is often observed when individuals hold their breath, thereby anchoring the diaphragm, prior to and during the movement of heavy objects.

Understanding the function of these muscles has significant clinical implications. For instance, weakness or dysfunction of the abdominal muscles can lead to various issues, including:

- Urinary incontinence: Inadequate intra-abdominal pressure may compromise bladder control.
- Constipation: Reduced ability to generate pressure can impair defecation.
- Lower back pain: Insufficient abdominal support may increase stress on the lumbar spine.

Moreover, the role of these muscles in generating intra-abdominal pressure is relevant in the context of pelvic floor disorders. Excessive or chronic increases in intra-abdominal pressure, such as those associated with obesity or chronic coughing, can contribute to the development of pelvic organ prolapse and stress urinary incontinence.

9.3.1 Rectus Abdominis (Innervated by the Anterior Rami of the Lower Six or Seven Thoracic Nerves: T6 or T7 to T12)

The rectus abdominis, a paired muscle prominently situated in the anterior abdominal wall, serves dual primary functions: facilitating trunk flexion and regulating intra-abdominal pressure (Neumann 2017).

The rectus abdominis originates inferiorly via two tendons from the anterior surface of the pubic symphysis and pubic crest. As it ascends, the muscle broadens and ultimately inserts onto the anterior surfaces of the xiphoid process and the costal cartilages of ribs five through seven

(Standring 2020). This superior-inferior orientation contributes to its efficacy in trunk flexion movements.

A distinguishing feature of the rectus abdominis is the presence of transverse tendinous intersections, typically three in number, located in the anterior portion of the muscle. These intersections are firmly adherent to the rectus sheath and occupy specific anatomical positions (Netter 2019):

- The lowest intersection is situated at the umbilical level.
- The highest intersection aligns with the xiphoid process.
- The third intersection is approximately midway between the aforementioned two.

These tendinous intersections not only provide structural reinforcement but also contribute to the muscle's segmental function, potentially allowing for more refined control of trunk movements.

The rectus abdominis is encased within a complex fibrous envelope known as the rectus sheath. This sheath is formed by the aponeuroses of three lateral abdominal muscles: the external oblique, internal oblique, and transversus abdominis (Moore et al. 2018) (*see Fig. 6.35*). The composition of the rectus sheath varies at different levels of the abdomen, reflecting the functional demands placed on the abdominal wall at these locations (Kendall et al. 2005):

- Superior to the costal margin: The sheath is present only anteriorly, composed entirely of the external oblique aponeurosis.
- Between the costal margin and the midpoint between the umbilicus and symphysis pubis:
  - Anterior wall: Formed by aponeuroses of external and internal oblique muscles.
  - Posterior wall: Composed of aponeuroses from internal oblique and transversus abdominis muscles.
- Inferior to the midpoint between the umbilicus and symphysis pubis: All three muscle aponeuroses pass anterior to the rectus abdominis, resulting in an absent posterior wall. The inferior limit of the posterior layer is demarcated by a crescentic border known as the arcuate line. Below this line, the rectus abdominis rests directly on the transversalis fascia, separated from the extraperitoneal adipose tissue.

In individuals with low body fat, the slightly convex lateral border of each rectus abdominis muscle manifests as a depression known as the linea semilunaris. The two rectus abdominis muscles are separated by a midline fibrous structure called the linea alba. These surface landmarks are not only aesthetically notable but also serve as important reference points for clinical examination and surgical procedures (Drake et al. 2020).

The rectus abdominis plays a role in respiratory function, particularly during forced expiration. In high-level spinal cord injuries affecting inspiratory muscles, the ability to recruit abdominal muscles for expiration can be crucial for effective cough and secretion clearance (Terson de Paleville et al. 2011).

9.3.2 Superficial Lateral Abdominal Musculature: External Oblique (Innervated by the Anterior Rami of the Lower Six Thoracic Nerves: T7–T12)

The external oblique represents a principal muscular component within the anterolateral abdominal wall, functioning in axial rotation, side-bending of the trunk, and modulation of intraabdominal pressure (Neumann 2017). This muscle displays a distinctive fiber orientation in the superficial layer of the anterior abdominal wall, with fibers running obliquely from superior-lateral to inferior-medial aspects relative to the midline (Moore et al. 2018) (Fig. 9.21). The muscle originates through digitations from the external aspects of ribs 5–12 and their cartilaginous components, demonstrating an intricate interdigitating pattern with serratus anterior muscle superiorly and latissimus dorsi muscle inferiorly (Standring 2020).

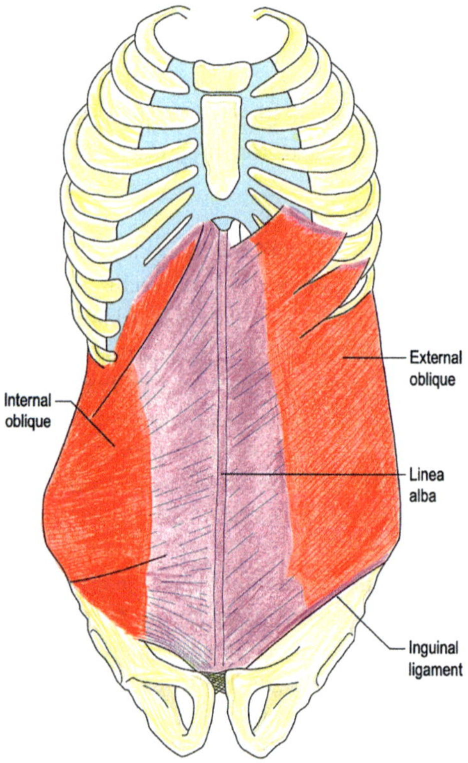

Fig. 9.21 Anterior view illustrating the right internal oblique and left external oblique muscles, emphasizing their orientation and anatomical relationships. (Adapted with permission from Palastanga and Soames (2012))

The muscle's architectural arrangement shows regional variations in fiber direction and insertion patterns. The fibers originating from the inferior two ribs follow a near-vertical trajectory, attaching to the lateral aspect of the anterior two-thirds of the iliac crest, producing an unattached posterior margin between the twelfth rib and iliac crest. The remaining muscular fibers transition into an expansive aponeurosis, characterized by greater width in its inferior portion (Netter 2019). This aponeurotic expansion crosses the rectus abdominis muscle, participates in rectus sheath formation, and merges at the midline with its contralateral aponeurosis to establish the linea alba. This median fibrous structure extends from the xiphoid process to the pubic symphysis, providing attachment for the bilateral abdominal musculature.

The inguinal ligament, a significant anatomical derivative of the external oblique, forms from the inferolateral margin of its aponeurosis. This ligamentous structure spans from the pubic tubercle to the anterior superior iliac spine, exhibiting a downward convexity due to fascial attachments from the thigh (Fan et al. 2018). The medial aspect of this ligament extends posteriorly along the pecten pubis as the lacunar ligament, maintaining a predominantly horizontal orientation in anatomical position. The pectineal ligament constitutes a further continuation of the lacunar ligament along the pecten pubis (Drake et al. 2020).

A triangular aperture in the aponeurosis superior to the pubic tubercle constitutes the superficial inguinal ring, bounded by medial and lateral pillars, with its base corresponding to the pubic crest. This anatomical opening holds clinical significance in hernia evaluation and surgical intervention (Jenkins and O'Dwyer 2008). The external oblique functions synergistically with the internal oblique and transversus abdominis muscles to establish the muscular framework of the abdomen (Willard et al. 2012). This arrangement facilitates not only movement but also contributes to postural control, visceral support, and respiratory mechanics, including forceful expiratory actions such as coughing, sneezing, and defecation (Hodges et al. 2005).

9.3.3 Middle Layer of the Anterolateral Abdominal Wall: Internal Oblique (Innervated by the Anterior Primary Rami of the T7–L1)

The internal oblique muscle forms the intermediate stratum of the three-layered anterolateral abdominal wall musculature, positioned deep to the external oblique (Moore et al. 2018). Its anatomical basis includes multiple attachment sites: the lateral two-thirds of the inguinal ligament, the anterior two-thirds of the intermediate line of the iliac crest, and the thoracolumbar fascia. This wide origin enables a radiating fiber arrangement that supports multiple biomechanical functions (Neumann 2017). The muscle demonstrates regional variations in fiber orientation and architectural organization:

- The most dorsal fibers ascend almost vertically to attach to the inferior borders of the lower four ribs (9th–12th).
- The more anterior and inferior fibers course superiorly and medially, transitioning into an aponeurosis along a line extending from the 10th costal cartilage to the body of the pubis (Standring 2020).

The aponeurotic component of the internal oblique exhibits complex anatomical relationships in forming the rectus sheath. It divides to include the rectus abdominis muscle before merging with its opposite aponeurosis at the linea alba, reinforcing anterior abdominal wall stability. Recent microscopic studies have revealed the presence of mechanoreceptors within these aponeurotic expansions, suggesting a proprioceptive role in addition to their mechanical functions (Stecco et al. 2019).

A unique anatomical feature involves to the combined tendon formation, where muscle fibers originating from the inguinal ligament merge with the inferior transversus abdominis fibers. This tendinous structure attaches to the pubic crest and pecten pubis, strengthening the inguinal region (Seeras et al. 2024). The internal oblique also displays sexual dimorphism in its relationship with reproductive structures. In males, inferomedial fibers associate with the spermatic cord to form the cremaster muscle, which mediates testicular elevation during the cremasteric reflex (Netter 2019).

The internal oblique functions as an integral component of the abdominal muscular system, collaborating with the external oblique and transversus abdominis. This three-layered arrangement supports postural maintenance, visceral positioning, and participation in respiratory and expulsive mechanisms, including coughing, sneezing, and defecation (Hodges et al. 2005). The diagonal fiber architecture enables efficient force transmission between upper and lower body segments during functional activities, particularly those involving rotational components (Willard et al. 2012).

9.3.4 Deep Layer of the Anterolateral Abdominal Wall: Transversus Abdominis (Innervated by the Anterior Primary Rami of the T7–T12 and the L1)

The transversus abdominis represents the innermost layer of the anterolateral abdominal wall musculature, characterized by its horizontal fiber arrangement (Moore et al. 2018). This muscle establishes multiple attachment points: inferiorly at the lateral third of the inguinal ligament and anterior two-thirds of the iliac crest's internal lip, posteriorly via the thoracolumbar fascia, and superiorly along the internal aspects of costal cartilages 7–12, where it demonstrates interdigitation with diaphragmatic attachments (Standring 2020).

The muscle fibers extend horizontally around the abdominal circumference, transitioning into an aponeurotic layer that merges with the posterior component of the internal oblique aponeurosis before reaching the linea alba. This arrangement contributes significantly to rectus sheath formation and anterior abdominal wall structural integrity (Drake et al. 2020). Microscopic analyses have revealed specialized mechanoreceptive elements within these aponeurotic expansions, suggesting proprioceptive functionality beyond mechanical support (Stecco et al. 2019). The inferior muscle fibers originating from the inguinal ligament demonstrate a unique descending path, combining with internal oblique fibers to establish the conjoint tendon. This tendinous structure inserts into the pubic crest and pecten pubis, posterior to the superficial inguinal ring (Seeras et al. 2024).

The inferior margin of the transversus abdominis, spanning between its inguinal ligament attachment and pecten pubis insertion, displays a characteristic downward concavity. The associated transversalis fascia establishes contact with both oblique muscles and the inguinal ligament inferior to this border. A circular aperture within the lateral aspect of this fascia constitutes the deep inguinal ring, holding particular clinical relevance in hernia development and surgical repair (Jenkins and O'Dwyer 2008).

A biomechanical research has identified the transversus abdominis as a primary contributor to core stabilization. Its activity occurs before to limb movement during functional activities, suggesting a predicted function in postural control (Hodges and Richardson 2018). The muscle's horizontal fiber orientation facilitates compression of abdominal contents, contributing to intraabdominal pressure modulation and spinal stabilization (Neumann 2017).

9.3.5 Cremaster

The cremaster muscle represents a fascinating example of sexual dimorphism in human anatomy, manifesting as a series of loosely organized muscular loops that encircle the spermatic cord and testes. This muscle extends from its origin at the inferior margin of the internal oblique muscle and adjacent portions of the inguinal ligament, continuing downward to its insertion at the pubic tubercle. While prominently developed in males, the muscle appears notably sparse in females, reflecting its specialized role in male reproductive physiology (Standring 2020).

The neural control system of the cremaster is innervated by the genital branch of the genitofemoral nerve (L1–L2). Despite this innervation pattern, the muscle operates independently of voluntary control. A characteristic feature of cremaster function is the cremasteric reflex, which produces testicular elevation in response to stimulation of the medial thigh region. This reflex demonstrates marked age-related variations, showing particular robustness in infants but experiencing significant attenuation by puberty (Moore et al. 2018).

References

Abrams P, Cardozo L, Fall M, Griffiths D, Rosier P, Ulmsten U, et al. The standardisation of terminology of lower urinary tract function: report from the Standardisation Sub-committee of the International Continence Society. Neurourol Urodyn. 2002;21(2):167–78.

Andersson KE, Arner A. Urinary bladder contraction and relaxation: physiology and pathophysiology. Physiol Rev. 2004;84(3):935–86.

Andersson KE, McCloskey KD. Lamina propria: the functional center of the bladder? Neurourol Urodyn. 2014;33(1):9–16.

Andersson KE, Wein AJ. Pharmacology of the lower urinary tract: basis for current and future treatments of urinary incontinence. Pharmacol Rev. 2004;56(4):581–631.

Apodaca G. The uroepithelium: not just a passive barrier. Traffic. 2004;5(3):117–28.

Ashton-Miller JA, DeLancey JO. Functional anatomy of the female pelvic floor. Ann N Y Acad Sci. 2007;1101:266–96.

Baader B, Herrmann M. Topography of the pelvic autonomic nervous system and its potential impact on surgical intervention in the pelvis. Clin Anat. 2003;16(2):119–30.

Barber MD. Contemporary views on female pelvic anatomy. Cleve Clin J Med. 2005;72(Suppl 4):S3–11.

Benarroch EE. Neural control of the bladder: recent advances and neurologic implications. Neurology. 2010;75(20):1839–46.

Benarroch E. Autonomic neurology. New York: Oxford; 2014.

Birder L, Andersson KE. Urothelial signaling. Physiol Rev. 2013;93(2):653–80.

Birder L, de Groat WC. Mechanisms of disease: involvement of the urothelium in bladder dysfunction. Nat Clin Pract Urol. 2007;4(1):46–54.

Birder L, de Groat W, Mills I, Morrison J, Thor K, Drake M. Neural control of the lower urinary tract: peripheral and spinal mechanisms. Neurourol Urodyn. 2010;26(1):128–39.

Birder LA, Ruggieri M, Takeda M, van Koeveringe G, Veltkamp S, Korstanje C, et al. How does the urothelium affect bladder function in health and disease? ICI-RS 2011. Neurourol Urodyn. 2012;31(3):293–9.

Bradley WE, Teague CT. Cerebellar influence on the micturition reflex. Exp Neurol. 1969;23(3):399–411.

Chapple CR, Wein AJ, Abrams P, Dmochowski RR, Giuliano F, Kaplan SA, et al. Lower urinary tract symptoms revisited: a broader clinical perspective. Eur Urol. 2008;54(3):563–9.

Chapple CR, MacDiarmid SA, Patel A. Urodynamics made easy. 3rd ed. Elsevier; 2009.

Constantinou CE, Govan DE. Spatial distribution and timing of transmitted and reflexly generated urethral pressures in healthy women. J Urol. 1982;127(5):964–9.

Corcos J, Ginsberg D, Karsenty G. Textbook of the neurogenic bladder. 3rd ed. Boca Raton: CRC Press; 2016.
Corriere JN Jr, Sandler CM. Management of the ruptured bladder: seven years of experience with 111 cases. J Trauma. 1986;26(9):830–3.
de Groat WC, Booth AM. Synaptic transmission in pelvic ganglia. In: Maggi CA, editor. Nervous control of the urogenital system. Chur: Harwood Academic Publishers; 1993. p. 291–347.
de Groat WC, Yoshimura N. Mechanisms underlying the recovery of lower urinary tract function following spinal cord injury. Prog Brain Res. 2006;152:59–84.
de Groat WC, Yoshimura N. Afferent nerve regulation of bladder function in health and disease. Handb Exp Pharmacol. 2009;194:91–138.
de Groat WC, Yoshimura N. Anatomy and physiology of the lower urinary tract. Handb Clin Neurol. 2015;130:61–108.
de Groat WC, Griffiths D, Yoshimura N. Neural control of the lower urinary tract. Compr Physiol. 2015;5(1):327–96.
DeLancey JO. Structural support of the urethra as it relates to stress urinary incontinence: the hammock hypothesis. Am J Obstet Gynecol. 1994;170(6):1713–20.
Dietrichs E, Haines DE. Possible pathways for cerebellar modulation of autonomic responses: micturition. Scand J Urol Nephrol Suppl. 2002;36(4):16–20.
Dixon JS, Gosling JA. Structure and innervation in the human. In: Torrens M, Morrison JFB, editors. The physiology of the lower urinary tract. London: Springer-Verlag; 1987. p. 3–22.
Doherty GM. Current diagnosis and treatment surgery. 14th ed. New York: McGraw Hill; 2015.
Drake MJ, Fry CH, Eyden B. Structural characterization of myofibroblasts in the bladder. BJU Int. 2006;97(1):29–32.
Drake MJ, Fowler CJ, Griffiths D, Mayer E, Paton JF, Birder L. Neural control of the lower urinary and gastrointestinal tracts: supraspinal CNS mechanisms. Neurourol Urodyn. 2010;29(1):119–27.
Drake RL, Vogl AW, Mitchell AWM. Gray's anatomy for students. 4th ed. Philadelphia: Elsevier; 2020.
Elbadawi A. Neuromorphologic basis of vesicourethral function: I. Histochemistry, ultrastructure, and function of intrinsic nerves of the bladder and urethra. Neurourol Urodyn. 1982;1(1):3–50.
Elbadawi A. Functional anatomy of the organs of micturition. Urol Clin North Am. 1996;23(2):177–210.
el-Badawi A, Schenk EA. Dual innervation of the mammalian urinary bladder. Am J Anat. 1966;119(3):405–27.
Elbadawi A, Yalla SV, Resnick NM. Structural basis of geriatric voiding dysfunction. II. Aging detrusor: normal versus impaired contractility. J Urol. 1993;150(5 Pt 2):1657–67.
Enhorning G. Simultaneous recording of intravesical and intraurethral pressure. A study on urethral closure in normal and stress incontinent women. Acta Chir Scand Suppl. 1961;Suppl 276:1–68.
Fan C, Fede C, Gaudreault N, Porzionato A, Macchi V, Caro DER, et al. Anatomical and functional relationships between external abdominal oblique muscle and posterior layer of thoracolumbar fascia. Clin Anat. 2018;31(7):1092–8.
Fowler CJ, Griffiths DJ. A decade of functional brain imaging applied to bladder control. Neurourol Urodyn. 2010;29(1):49–55.
Fowler CJ, Griffiths D, de Groat WC. The neural control of micturition. Nat Rev Neurosci. 2008;9(6):453–66.
Fry CH, Meng E, Young JS. The physiological function of lower urinary tract smooth muscle. Auton Neurosci. 2010;154(1–2):3–13.
Gillespie JI, Markerink-van Ittersum M, de Vente J. Sensory collaterals, intramural ganglia and motor nerves in the Guinea-pig bladder: evidence for intramural neural circuits. Cell Tissue Res. 2006;325(1):33–45.
Gomez RG, Ceballos L, Coburn M, Corriere JN Jr, Dixon CM, Lobel B, et al. Consensus statement on bladder injuries. BJU Int. 2004;94(1):27–32.
Gosling JA. The structure of the bladder and urethra in relation to function. Urol Clin North Am. 1979;6(1):31–8.
Gosling JA, Dixon JS, Critchley HO, Thompson SA. A comparative study of the human external sphincter and periurethral levator ani muscles. Br J Urol. 1981;53(1):35–41.
Gosling JA, Dixon JS, Jen PY. The distribution of noradrenergic nerves in the human lower urinary tract. A review. Eur Urol. 1999;36(Suppl 1):23–30.
Griffiths D, Fowler CJ. The micturition switch and its forebrain influences. Acta Physiol (Oxf). 2013;207(1):93–109.
Griffiths D, Tadic SD. Bladder control, urgency, and urge incontinence: evidence from functional brain imaging. Neurourol Urodyn. 2008;27(6):466–74.
Griffiths D, Tadic SD, Schaefer W, Resnick NM. Cerebral control of the bladder in normal and urge-incontinent women. NeuroImage. 2007;37(1):1–7.
Hickey DS, Phillips JI, Hukins DW. Arrangements of collagen fibrils and muscle fibres in the female urethra and their implications for the control of micturition. Br J Urol. 1982;54(5):556–61.
Ho KM, Borja MC, Persson K, Brading AF, Andersson KE. Expression of nitric oxide synthase immunoreactivity in the human female intramural striated urethral sphincter. J Urol. 2003;169(6):2407–11.
Hodges PW, Richardson CA. Relationship between limb movement speed and associated contraction of the trunk muscles. Ergonomics. 2018;40(11):1220–30.
Hodges PW, Eriksson AE, Shirley D, Gandevia SC. Intra-abdominal pressure increases stiffness of the lumbar spine. J Biomech. 2005;38(9):1873–80.
Holstege G. Micturition and the soul. J Comp Neurol. 2005;493(1):15–20.
Holstege G. The emotional motor system and micturition control. Neurourol Urodyn. 2010;29(1):42–8.
Jafari NV, Rohn JL. The urothelium: a multi-faceted barrier against a harsh environment. Mucosal Immunol. 2022;15(6):1127–42.
Jenkins JT, O'Dwyer PJ. Inguinal hernias. BMJ. 2008;336(7638):269–72.

Kakizaki H, Fraser MO, de Groat WC. Reflex pathways controlling urethral striated and smooth muscle function in the male rat. Am J Phys. 1997;272(5 Pt 2):R1647–56.

Kanai A, Roppolo J, Ikeda Y, Zabbarova I, Tai C, Birder L, et al. Origin of spontaneous activity in neonatal and adult rat bladders and its enhancement by stretch and muscarinic agonists. Am J Physiol Renal Physiol. 2007;292(3):F1065–72.

Kendall FP, McCreary EK, Provance PG, Rodgers MM, Romani WA. Muscles: testing and function with posture and pain. 5th ed. Baltimore: Lippincott Williams & Wilkins; 2005.

Khandelwal P, Abraham SN, Apodaca G. Cell biology and physiology of the uroepithelium. Am J Physiol Renal Physiol. 2009;297(6):F1477–501.

Klutke CG, Siegel CL. Functional female pelvic anatomy. Urol Clin North Am. 1995;22(3):487–98.

Kunze A, Neuhaus J, Stolzenburg JU. Quantitative immunohistochemical study of the innervation of the Guinea-pig lower urinary tract. BJU Int. 2006;98(2):424–9.

Kuru TH, Roethke MC, Rieker P, Roth W, Fenchel M, Hohenfellner M, et al. Histology core-specific evaluation of the European Society of Urogenital Radiology (ESUR) standardised scoring system of multiparametric magnetic resonance imaging (mpMRI) of the prostate. BJU Int. 2013;112(8):1080–7.

Lewis SA. Everything you wanted to know about the bladder epithelium but were afraid to ask. Am J Physiol Renal Physiol. 2000;278(6):F867–74.

Majesky MW. Developmental basis of vascular smooth muscle diversity. Arterioscler Thromb Vasc Biol. 2007;27(6):1248–58.

Markwardt F, Isenberg G. Gating of maxi K^+ channels studied by Ca^{2+} concentration jumps in excised inside-out multi-channel patches (myocytes from Guinea pig urinary bladder). J Gen Physiol. 1992;99(6):841–56.

McAninch JW, Lue TF. Smith & Tanagho's general urology. 18th ed. McGraw-Hill; 2013.

Michels L, Blok BF, Gregorini F, Kurz M, Schurch B, Kessler TM, et al. Supraspinal control of urine storage and micturition in men--an fMRI study. Cereb Cortex. 2015;25(10):3369–80.

Moore KL, Dalley AF, Agur AMR. Clinically oriented anatomy. 8th ed. Philadelphia: Wolters Kluwer; 2018.

Moszkowicz D, Alsaid B, Bessede T, Penna C, Nordlinger B, Benoît G, et al. Where does pelvic nerve injury occur during rectal surgery for cancer? Color Dis. 2011;13(12):1326–34.

Netter FH. Atlas of human anatomy. 7th ed. Philadelphia: Elsevier; 2019.

Neumann DA. Kinesiology of the musculoskeletal system: foundations for rehabilitation. 3rd ed. St. Louis: Mosby; 2017.

Nishizawa O, Sugaya K, Noto H, Harado T, Tsuchida S. Pontine micturition center in the dog. J Urol. 1988;140(4):872–4.

Padykula HA, Gauthier GF. Cytological studies of fiber types in skeletal muscle. A comparative study of the mammalian diaphragm. J Cell Biol. 1966;28(2):333–54.

Palastanga N, Soames R. Anatomy and human movement: structure and function. 6th ed. Philadelphia: Elsevier Health Sciences; 2012.

Panicker JN, Fowler CJ, Kessler TM. Lower urinary tract dysfunction in the neurological patient: clinical assessment and management. Lancet Neurol. 2015;14(7):720–32.

Paulsen F, Waschke J. Sobotta atlas of human anatomy: internal organ. 15th ed. Munich: Urban & Fischer; 2011.

Pradidarcheep W, Wallner C, Dabhoiwala NF, Lamers WH. Anatomy and histology of the lower urinary tract. Urinary Tract. 2011;202:117–48.

Raychaudhuri B, Cahill D. Pelvic fasciae in urology. Ann R Coll Surg Engl. 2008;90(8):633–7.

Reitz A, Schmid DM, Curt A, Knapp PA, Schurch B. Afferent fibers of the pudendal nerve modulate sympathetic neurons controlling the bladder neck. Neurourol Urodyn. 2003;22(6):597–601.

Rob L, Halaska M, Robova H. Nerve-sparing and individually tailored surgery for cervical cancer. Lancet Oncol. 2010;11(3):292–301.

Roehrborn CG. Pathology of benign prostatic hyperplasia. Int J Impot Res. 2008;20(Suppl 3):S11–8.

Roosen A, Datta SN, Chowdhury RA, Patel PM, Kalsi V, Elneil S, et al. Suburothelial myofibroblasts in the human overactive bladder and the effect of botulinum neurotoxin type A treatment. Eur Urol. 2009;55(6):1440–8.

Sakakibara R, Tateno F, Kishi M, Tsuyuzaki Y, Uchiyama T, Yamamoto T. Pathophysiology of bladder dysfunction in Parkinson's disease. Neurobiol Dis. 2012;46(3):565–71.

Sakakibara R, Panicker J, Fowler CJ, Tateno F, Kishi M, Tsuyusaki Y, et al. Is overactive bladder a brain disease? The pathophysiological role of cerebral white matter in the elderly. Int J Urol. 2014;21(1):33–8.

Seeras K, Qasawa RN, Ju R, Prakash S. Anatomy, abdomen and pelvis: anterolateral abdominal wall. In: StatPearls [Internet]. Treasure Island: StatPearls Publishing; 2024.

Shapiro E, Huang H, McFadden DE, Masch RJ, Ng E, Lepor H, et al. The prostatic utricle is not a Müllerian duct remnant: immunohistochemical evidence for a distinct urogenital sinus origin. J Urol. 2004;172(4 Pt 2):1753–6.

Smet PJ, Jonavicius J, Marshall VR, de Vente J. Distribution of nitric oxide synthase-immunoreactive nerves and identification of the cellular targets of nitric oxide in Guinea-pig and human urinary bladder by cGMP immunohistochemistry. Neuroscience. 1996;71(2):337–48.

Standring S. Gray's anatomy: the anatomical basis of clinical practice. 42nd ed. London: Elsevier; 2020.

Stecco C, Macchi V, Porzionato A, Duparc F, De Caro R. The fascia: the forgotten structure. Ital J Anat Embryol. 2019;116(3):127–38.

Sugaya K, Nishijima S, Miyazato M, Ogawa Y. Central nervous control of micturition and urine storage. J Smooth Muscle Res. 2005;41(3):117–32.

Tanagho EA, Lue TF. Smith's general urology. 17th ed. New York: McGraw-Hill Medical; 2008.

Tanagho EA, Pugh RC. The anatomy and function of the ureterovesical junction. Br J Urol. 1963;35:151–65.

Tanagho EA, Meyers FH, Smith DR. Urethral resistance: its components and implications. I. Smooth muscle component. Inv Urol. 1969;7(2):136–49.

Terson de Paleville D, McKay W, Aslan S, Folz R, Sayenko D, Ovechkin A. Respiratory motor control disrupted by spinal cord injury: mechanisms, evaluation, and restoration. Transl Stroke Res. 2011;2(4):463–73.

Thor KB, de Groat WC. Neural control of the female urethral and anal rhabdosphincters and pelvic floor muscles. Am J Physiol Regul Integr Comp Physiol. 2010;299(2):R416–38.

Torrens M, Morrison JFB. The physiology of the lower urinary tract. Berlin: Springer-Verlag; 1987.

Vignoli G. Urodynamics: a quick pocket guide. Switzerland: Springer; 2017.

Walz J, Epstein JI, Ganzer R, Graefen M, Guazzoni G, Kaouk J, et al. A critical analysis of the current knowledge of surgical anatomy of the prostate related to optimisation of cancer control and preservation of continence and erection in candidates for radical prostatectomy: an update. Eur Urol. 2016;70(2):301–11.

Wein AJ, Kavoussi LR, Partin AW, Peters CA. Campbell-Walsh urology. 11th ed. Philadelphia: Elsevier; 2015.

Willard FH, Vleeming A, Schuenke MD, Danneels L, Schleip R. The thoracolumbar fascia: anatomy, function and clinical considerations. J Anat. 2012;221(6):507–36.

Wineski LE. Snell's clinical anatomy by regions. 10th ed. Philadelphia: Wolters Kluwer; 2019.

Yiangou Y, Facer P, Ford A, Brady C, Wiseman O, Fowler CJ, et al. Capsaicin receptor VR1 and ATP-gated ion channel P2X3 in human urinary bladder. BJU Int. 2001;87(9):774–9.

Yoshimura N, Chancellor MB. Neurophysiology of lower urinary tract function and dysfunction. Rev Urol. 2003;5(Suppl 8):S3–10.

Yoshimura N, de Groat WC. Neural control of the lower urinary tract. Int J Urol. 1997;4(2):111–25.

Yoshimura N, Kuno S, Chancellor MB, de Groat WC, Seki S. Dopaminergic mechanisms underlying bladder hyperactivity in rats with a unilateral 6-hydroxydopamine (6-OHDA) lesion of the nigrostriatal pathway. Br J Pharmacol. 2003;139(8):1425–32.

Yoshimura N, Ogawa T, Miyazato M, Kitta T, Furuta A, Chancellor MB, et al. Neural mechanisms underlying lower urinary tract dysfunction. Korean J Urol. 2014;55(2):81–90.

Yucel S, Baskin LS. An anatomical description of the male and female urethral sphincter complex. J Urol. 2004;171(5):1890–7.

Zermann DH, Ishigooka M, Doggweiler R, Schmidt RA. Central autonomic innervation of the lower urinary tract—neuroanatomy study. World J Urol. 1998;16:417–22.

Recommended Additional Reading

Andersson KE, Michel MC. Urinary tract. Berlin: Springer; 2011.

Chapple CR, Steers WD. Practical urology: essential principles and practice. London: Springer; 2011.

Dmochowski R, Heesakkers J. Neuro-urology. Cham: Springer; 2018.

Liao L, Madersbacher H. Neurourology: theory and practice. Dordrecht: Springer; 2019.

Shankar K, editor. Physiatric anatomic principles. State of the art reviews. Philadelphia: Hanley & Belfus, Inc.; 1996.

Functional Anatomy of the Lower Gastrointestinal System

10

The lower gastrointestinal (GI) system, particularly the anorectum, demonstrates complex neural control mechanisms that are frequently compromised following spinal cord injury. Understanding the complex anatomical and physiological relationships between these systems is crucial for comprehending neurogenic bowel dysfunction, one of the most challenging consequences of spinal cord injury that significantly impacts quality of life (Krogh et al. 2017).

The walls of the lower GI tract maintain a characteristic four-layered architecture consisting of the mucosa, submucosa, muscularis propria, and serosa/adventitia. While this basic structure remains consistent throughout the tract, the anorectum exhibits specialized adaptations crucial for maintaining continence and coordinating defecation (Standring 2020). These adaptations include distinct muscle groups—the internal and external anal sphincters—which receive different types of innervation and are differentially affected by various levels of spinal cord injury (Furness 2012).

Neural control of anorectal function occurs through an intricate interaction between the autonomic and somatic nervous systems. The enteric nervous system maintains considerable autonomy in regulating basic gut functions, yet optimal performance requires intact communication with higher centers through spinal pathways (Costa et al. 2000). After spinal cord injury, disruption of these pathways leads to neurogenic bowel, characterized by impaired colorectal motility and defecation (Krogh and Christensen 2009). The severity and presentation of neurogenic bowel dysfunction vary depending on the level and completeness of spinal cord injury. Injuries above the conus medullaris typically result in upper motor neuron bowel syndrome, characterized by increased colonic wall tone and anal sphincter spasticity. In contrast, injuries affecting the conus medullaris or cauda equina lead to lower motor neuron bowel syndrome, marked by reduced colonic motility and sphincter tone.

The complex neural network controlling anorectal function involves both conscious and unconscious mechanisms. While the internal anal sphincter receives autonomic innervation and maintains continuous tone, the external anal sphincter depends on somatic innervation for voluntary control. Following spinal cord injury, this delicate balance is disrupted, necessitating targeted rehabilitation strategies based on the specific pattern of neurological deficit (Browning and Travagli 2014).

Understanding these anatomical and physiological relationships is essential for healthcare professionals managing spinal cord injury patients with neurogenic bowel. The following sections examine in detail the structural components, neuroanatomy, and physiological aspects

H.-Y. Ko, *Practical Functional Anatomy for Spinal Cord Injury Rehabilitation*,
https://doi.org/10.1007/978-981-95-6457-6_10

of the lower GI system. This knowledge forms the foundation for developing effective bowel management programs and improving outcomes for individuals with spinal cord injury-related bowel dysfunction.

10.1 Overall Structure and Function of the Gastrointestinal System

The gastrointestinal (GI) system comprises a continuous tract extending from the oropharynx to the large intestine, with associated digestive organs including the liver, gallbladder, and pancreas (Standring 2020). The walls of these hollow organs display a characteristic four-layered architecture: mucosa, submucosa, muscularis propria, and serosa/adventitia. While the mucosal layer exhibits organ-specific cellular variations, the remaining layers maintain relative uniformity throughout the tract (Ross and Pawlina 2016; Standring 2020).

Neural control of gastrointestinal function occurs through dual mechanisms (Furness 2012). The enteric nervous system functions independently, yet external neural input remains necessary for optimal physiological performance (Costa et al. 2000; Furness 2012). The innervation pattern incorporates postganglionic sympathetic fibers and preganglionic parasympathetic neurons, with parasympathetic cell bodies located within two major nerve networks: the myenteric plexus of Auerbach and the submucosal plexus of Meissner (Browning and Travagli 2014; Costa et al. 2000).

The primary functions of the GI system involve nutrient processing, digestion, and absorption (Barrett et al. 2019). Daily intake typically consists of 800 gm solid matter and 1200 mL fluid, which combines with approximately 7 L of secretions from multiple sources–salivary glands, gastric mucosa, hepatobiliary system, pancreas, and intestinal glands (Barrett et al. 2019; Hall 2016). The system demonstrates remarkable efficiency in fluid management, absorbing 8.85 L and producing approximately 150 gm of solid waste (Hall 2016). This precise regulation depends on integrated neuroendocrine mechanisms, where alterations can significantly modify gastrointestinal operations. The following sections examine the detailed structure, neuroanatomy, and physiological aspects of the GI system.

10.1.1 Structure and Functional Organization of Gastrointestinal Musculature

The GI tract exhibits a complex muscular architecture that integrates both smooth and skeletal muscle cells, each contributing to specific physiological functions (Barrett et al. 2019) (Fig. 10.1). While skeletal muscle predominates in masticatory muscles and the upper esophagus, a gradual transition occurs through the esophageal length. The upper third contains exclusively skeletal muscle, followed by a mixed composition in the middle third, and finally transitions to purely smooth muscle in the lower third, establishing the characteristic layered structure of the GI tract

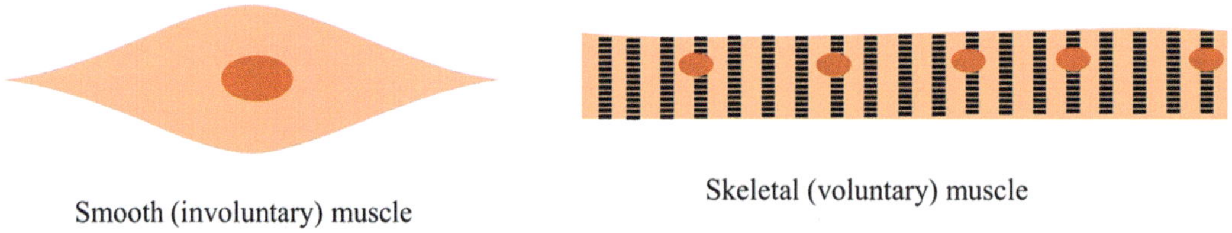

Fig. 10.1 Muscles of the gastrointestinal (GI) tract. The diagrams differentiate the two types of the GI tract muscles: smooth (non-striated), which possess intrinsic excitability, and skeletal (striated) muscles, which lack this characteristics. (Adapted with permission from Welcome (2018))

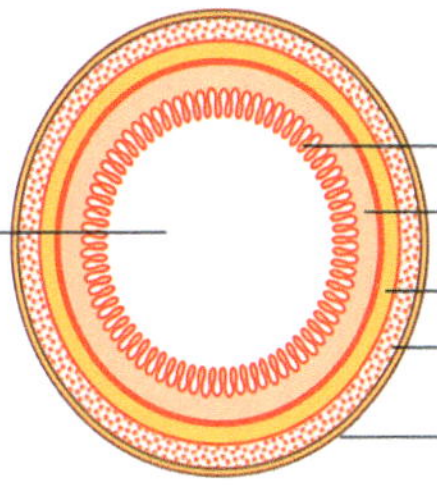

Fig. 10.2 Layers of the gastrointestinal (GI) tract, highlighting the location of smooth muscle. The GI tract generally has two muscle layers, except for the stomach, which contains three. The outer longitudinal muscle layer lies between the tunica serosa (outermost layer) and the myenteric (Auerbach's) plexus (not shown). The inner circular muscle layer is situated between the myenteric plexus and the submucosa. The Meissner's plexus (not shown) is positioned between the circular muscle and submucosa. These two layers, collectively termed the tunica muscularis externa, facilitate segmental contractions and peristaltic movements for mechanical breakdown and mixing of food with digestive enzymes. GI motor function is regulated by the autonomic nervous system, including sympathetic and parasympathetic inputs, and by intrinsic gut plexuses. The tunica mucosa, the innermost layer, consists of an epithelial layer, the lamina propria (containing blood vessels, mucous glands, lymph nodes, and sensory nerve endings), and the muscularis mucosae, a layer of smooth muscle that produces localized mucosal movements but does not contribute to food propulsion. (Adapted with permission from Welcome (2018))

walls (Hall 2016). This unique distribution reflects the evolutionary adaptation to different functional requirements along the digestive tract, from voluntary control in the upper regions to autonomous function in the lower segments. Smooth muscle cells are organized into two layers within the walls of the GI tract (Fig. 10.2).

The presence of smooth muscle extends well beyond the digestive system, forming integral components of various hollow organs and tubular structures throughout the body. These include blood vessels, respiratory airways, the urogenital tract, and erectile tissues (Ross and Pawlina 2016). Within the GI tract, smooth muscle cells aggregate into precise bundles, with individual cells measuring 20–500 μm in length and 2–5 μm in diameter. This organization enables the coordinated regulation of luminal dimensions and content flow essential for digestive processes (Huizinga and Lammers 2009). The contractile activity of these muscle bundles generates both mixing movements (segmentation) and propulsive movements (peristalsis), which are fundamental for proper digestion and absorption.

The smooth muscle cell membrane, or sarcolemma, reveals intricate organizational complexity (Fig. 10.3). Its structure incorporates a glycocalyx layer intimately associated with a protein-rich lipid bilayer (Ross and Pawlina 2016). The membrane surface features numerous caveolae, specialized microdomains containing specific proteins, including caveolin and the dystrophin complex, along with various signaling molecules such as G-protein coupled receptors, ion channels, and regulatory enzymes. These membrane components play important roles in signal transduction and calcium homeostasis (Gabella 1987). The extracellular matrix surrounding these cells contains an intricate mesh of types I and III collagen fibers, elastin, and glycoproteins, which collectively provide structural integrity and contribute to the tissue's viscoelastic properties. This matrix composition allows for the remarkable adaptability of smooth muscle to varying mechanical demands.

The neural control of GI smooth muscle demonstrates remarkable complexity, receiving dual innervation from both sympathetic and parasympathetic pathways (Furness 2012). Communication occurs through specialized varicosities that form neuromuscular junctions, with synaptic clefts measuring 50–200 nm. Neurotransmitter release at these junctions triggers complex intracellular signaling cascades that regulate muscle tone and contractility. A particularly fascinating aspect of this system involves the interstitial cells of Cajal, which establish communications with smooth muscle cells through gap junctions, creating an integrated network for coordinated function (Sanders et al. 2006). These cells act as pacemakers and intermediaries in neurotransmission, generating and coordinating the slow waves that drive GI motility patterns (Huizinga and Chen 2014) (Fig. 10.4).

This cellular communication network facilitates synchronized activity within "effector muscle bundles" through both mechanical and electrical coupling mechanisms. Intermediate

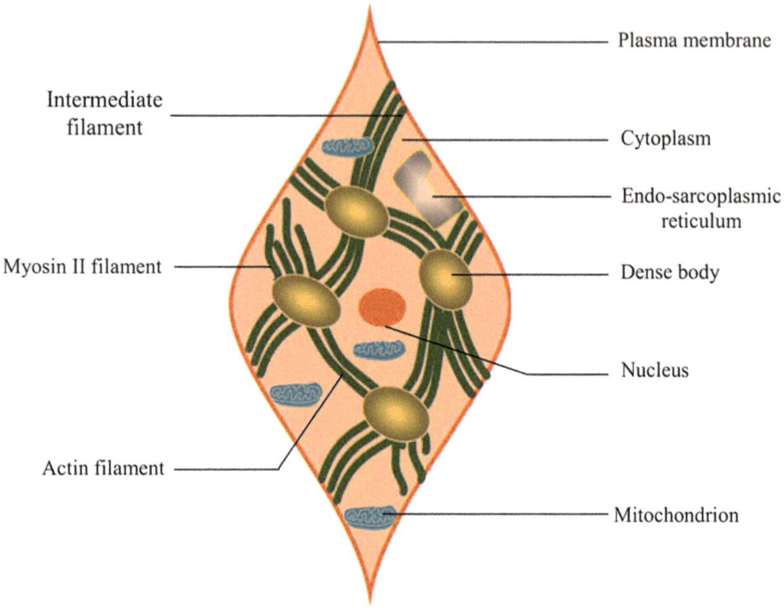

Fig. 10.3 Structure of smooth muscle cell. (Adapted with permission from Welcome (2018))

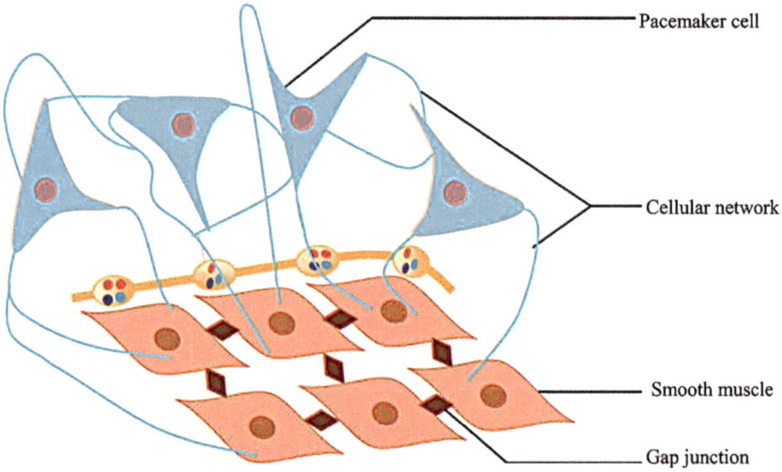

Fig. 10.4 Cellular network of pacemaker and gastrointestinal (GI) smooth muscle cells. Smooth muscle cells contain functional L-type calcium channels, which enhance conductivity upon activation by pacemaker cells. Notably, Cajal cells are also found in other regions of the body, including the brain, urinary tract, reproductive tract, and vascular system, where they generate pacemaker currents that drive information processing or contractile functions in these organs. (Adapted with permission from Welcome (2018))

junctions enable force transmission, while gap junctions permit the exchange of ions and small molecules between cells. This dual coupling system creates a functional syncytium essential for coordinated GI motility (Huizinga and Lammers 2009). The electrical coupling through gap junctions allows for the propagation of slow waves and action potentials, coordinating contractile activity across large regions of the GI tract. This synchronization is particularly important in maintaining proper motility patterns and preventing retrograde movement of luminal contents.

The cytoskeletal architecture of smooth muscle cells reveals further complexity, comprising myosin filaments, actin, and intermediate filaments (Small and Gimona 1998). The type II myosin found in smooth muscle contains subunits: two heavy chains, each associated with a pair of light chains. These myosin molecules display a distinctive side-polar arrangement of cross-bridges, contrasting with the bipolar arrangement found in skeletal muscle. This structural difference contributes to the unique contractile properties of smooth muscle, allowing for sustained contractions with lower energy expenditure, a phenomenon known as the "latch state" (Webb 2003).

The actin filaments, serving as the primary contractile proteins, interact with myosin through a calcium-dependent mechanism that differs significantly from that of skeletal muscle. The regulation of smooth muscle contraction involves phosphorylation of myosin light chains rather than the troponin-tropomyosin system found in skeletal muscle. This distinctive regulatory mechanism allows for fine-tuned control of contractile force and maintenance of tone (Sanders et al. 2006). Intermediate filaments, primarily composed of desmin and vimentin, provide structural anchoring points and contribute to mechanical stability. These filaments form a complex three-dimensional network that maintains cellular architecture and facilitates force transmission throughout the tissue (Huizinga and Chen 2014). The intermediate filament network also participates in mechanotransduction, converting mechanical stimuli into biochemical signals that influence cell behavior and function.

Recent investigations have uncovered additional complexity in smooth muscle regulation, demonstrating that non-muscle myosin isoforms can contribute to force generation (Webb 2003). This finding has important implications for understanding GI motility disorders and developing targeted therapies. Furthermore, studies have identified various mechanosensitive ion channels and receptors that allow smooth muscle cells to respond to mechanical stress and strain, contributing to the regulation of GI motility patterns.

10.1.2 Vascular Organization of the Gastrointestinal System

The GI vascular system exhibits a complicated organizational structure that can be analyzed from several viewpoints (Drake et al. 2020). Conventional classification separates arterial and venous circulations, whereas an organ-based approach provides regional blood flow patterns. The system functions within the extensive splanchnic circulation, originating from the Greek term "splanchnikos," meaning "inwards" (Standring 2020).

The splanchnic circulation has several parallel circulatory networks that supply specific organs: gastric, small intestine, large intestinal, hepatic, pancreatic, and splenic systems (Hall 2016). This circulatory framework receives blood through three major arterial branches of the abdominal aorta: the celiac, superior mesenteric, and inferior mesenteric arteries (Paulsen and Waschke 2011). These vessels create an intricate network that ensures adequate perfusion throughout the digestive system.

After tissue perfusion, blood collects in the portal venous system, which carries relatively deoxygenated blood enriched with absorbed nutrients to the liver. This unique arrangement, known as the hepatic portal system, allows hepatic processing of absorbed substances before they enter the systemic circulation (Rosenblum et al. 1997).

The splanchnic circulation has several unique characteristics that indicate its specialized functions:

- High blood flow capacity, receiving approximately 25–30% of cardiac output at rest
- Extensive collateral network development potential
- Remarkable adaptability to varying metabolic demands
- Specialized mucosal microcirculation supporting absorption

10.1.2.1 Arterial Distribution to the Gastrointestinal System

The arterial circulation of the GI tract originates from three principal branches of the abdominal aorta: the celiac artery, superior mesenteric artery, and inferior mesenteric artery (Standring 2020) (Fig. 10.5). These vessels create extensive interconnections through arterial trunk anastomoses, forming collateral networks that maintain blood flow and nutrient distribution to specific regions and organs (Standring 2020). This vascular organization provides alternative pathways for blood flow, which becomes particularly relevant in occlusive vascular diseases or during surgical procedures (Moore et al. 2018).

At the level of the twelfth thoracic vertebra (T12), the celiac trunk emerges from the thoracic aorta with a diameter of 6–8 mm (Netter 2019). This vessel delivers oxygenated blood to several abdominal organs, including the stomach, liver, pancreas, proximal duodenum, and abdominal esophagus. The celiac trunk divides into three primary branches: the splenic, hepatic, and left gastric arteries, although additional collateral vessels may arise from this trunk (Standring 2020). The distribution pattern of these vessels reflects their embryological development, which

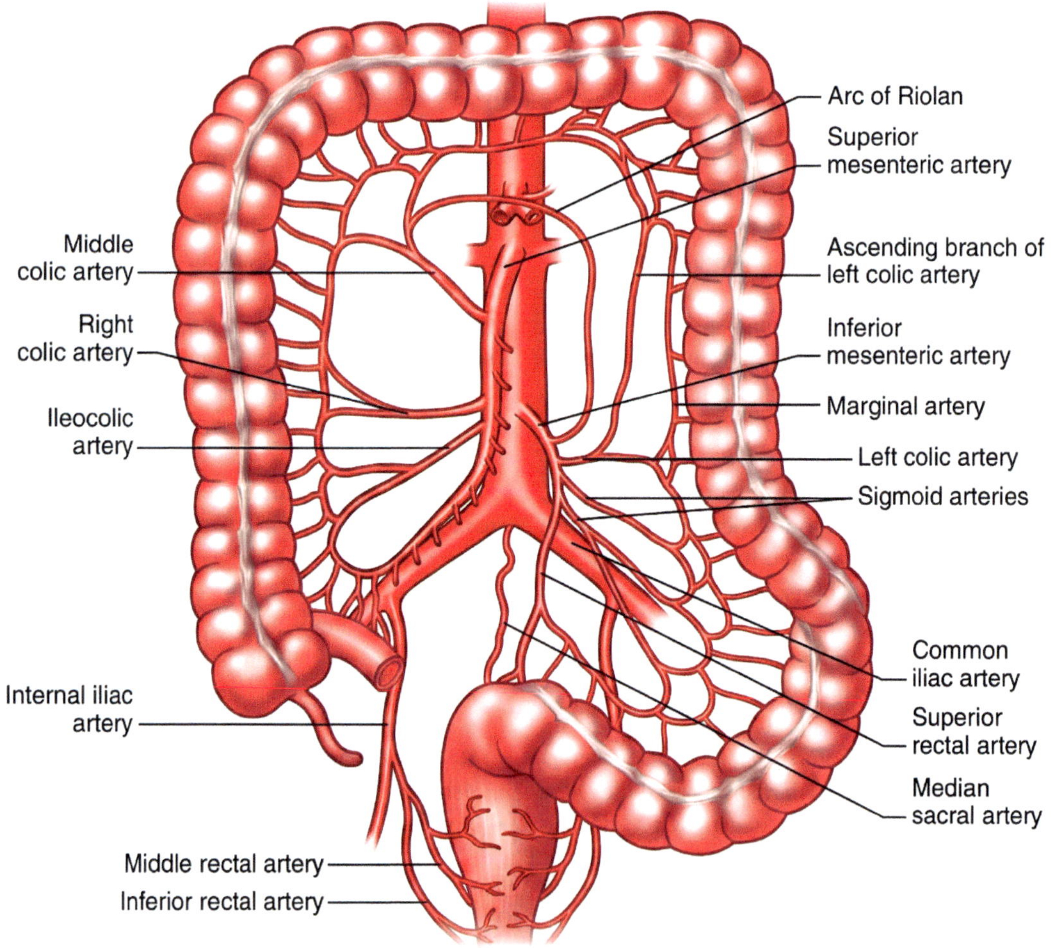

Fig. 10.5 Arterial anatomy of the colon and rectum. (Adapted with permission from Beck et al. (2019))

explains their segmental arrangement and frequent anatomical variations (Geboes et al. 2001).

The left gastric artery generates multiple esophageal and gastric branches. Complementary perfusion to the esophagus comes from small arteries originating from the thoracic aorta (Drake et al. 2020). These esophageal branches create longitudinal anastomoses, establishing a continuous vascular network along the esophageal length. This arrangement holds clinical significance in portal hypertension cases, where these vessels may dilate to form varices.

From the common hepatic artery arise three main divisions: the proper hepatic artery, right gastric artery, and gastroduodenal artery (Ellis and Mahadevan 2019). The splenic artery produces numerous branches supplying the pancreas, omentum, and spleen. Within the hepatic circulation, the hepatic artery provides 25% of the blood flow, delivering oxygen-rich blood, while the portal vein contributes 75%, transporting nutrient-rich blood from the intestines. Blood pressure in the hepatic artery maintains approximately 100 mmHg, similar to aortic pressure (Clemente 2011).

The superior mesenteric artery, measuring 7–8 mm in diameter, supplies blood to the right colon, portions of the transverse colon, and small intestine. In comparison, the inferior mesenteric artery, with a smaller diameter of 3–4 mm, perfuses the left colon (Wineski 2019). This difference in vessel caliber helps explain the increased vulnerability of the left colon to ischemic events.

Within the GI tract, the arterial circulation manifests in two distinct patterns. The intramural system comprises blood vessels within the walls of GI organs, arranged in specific layers corresponding to the muscular and mucosal components. The extramural system characterizes the vascular arrangement of the esophagus, where vessels penetrate from outside (Standring 2020).

Multiple arterial sources supply the rectoanal canal. The superior rectal artery branches from the inferior mesenteric artery, the middle rectal artery from the internal iliac artery, the inferior rectal artery from the internal pudendal artery, and additional branches arise from lower sacral arteries. This abundant vascular supply contributes to the remarkable healing capacity of this region (Barleben and Mills 2010) (Fig. 10.6).

Superior Mesenteric Artery and Its Branches

The superior mesenteric artery exhibits multiple branches along its course, with the ileocolic artery as its terminal component. The ileocolic artery originates from the right aspect of the superior mesenteric artery and progresses obliquely through the mesenteric tissues to reach the ileocecal junction. This vessel demonstrates a consistent presence and divides into superior and inferior branches (Michels 1963).

The right colic artery displays considerable anatomical variation in its origin, which may arise from one of three locations: the superior mesenteric trunk, the middle colic vessel, or the ileocolic artery. Some individuals show a complete absence of this vessel (Murono et al. 2015). The middle colic artery emerges from the superior mesenteric artery either posterior to the pancreas or at its inferior border. This vessel proceeds in a curved trajectory toward the hepatic flexure and splits into two components: a right branch forming an anastomosis with the ascending segment of the right colic artery, and a left branch creating an anastomotic connection with the ascending portion of the left colic artery (Kahn et al. 2001).

Superior mesenteric artery syndrome occurs when the distal (third) portion of the duodenum undergoes compression between the anterior surface of the abdominal aorta and the posterior aspect of the superior mesenteric artery (Rabie et al. 2015). This vascular compression syndrome affects the normal passage of intestinal contents, resulting in varied gastrointestinal manifestations (Merrett and Wilson 2009). The anatomical relationship of the superior mesenteric artery to the aorta creates an angle that ranges from 38 to 65 degrees in healthy individuals, while patients with this syndrome exhibit an angle less than 25 degrees (Welsch et al. 2007) (Fig. 10.7).

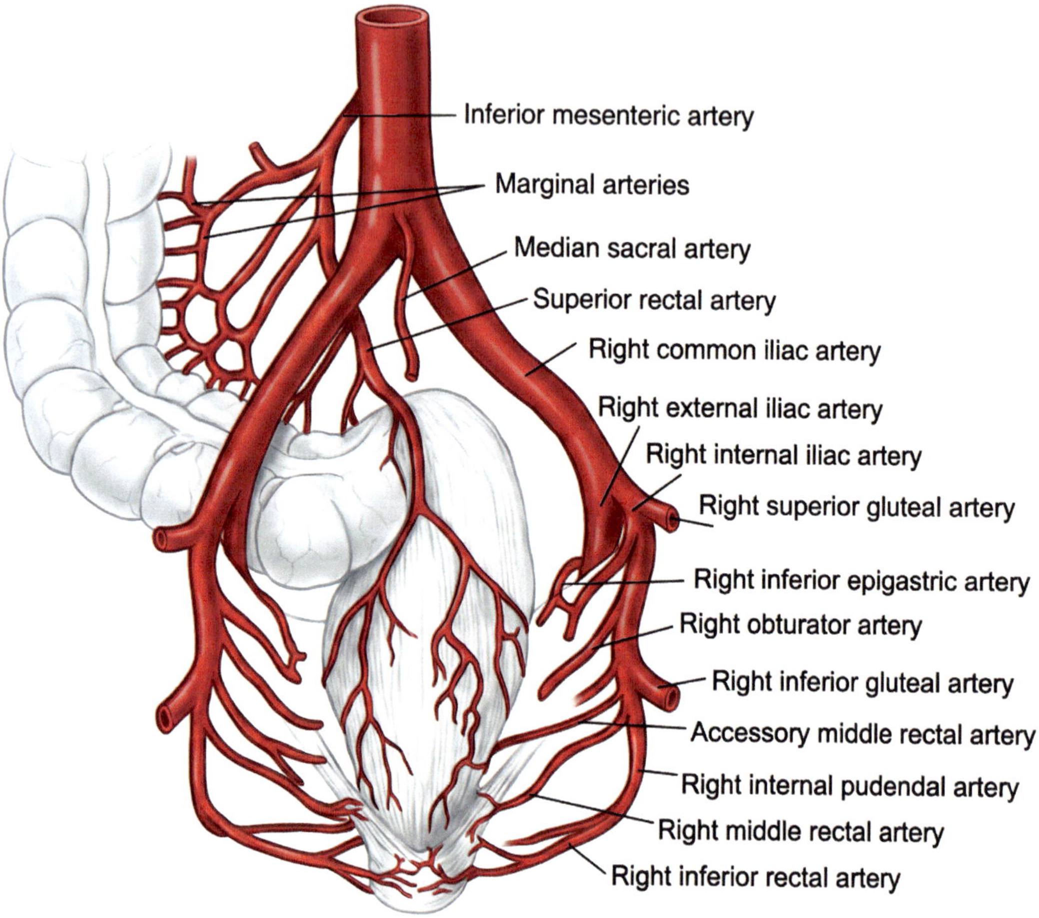

Fig. 10.6 Arterial blood supply of the anorectal region. (Adapted from Gordon and Nivatvongs (2007))

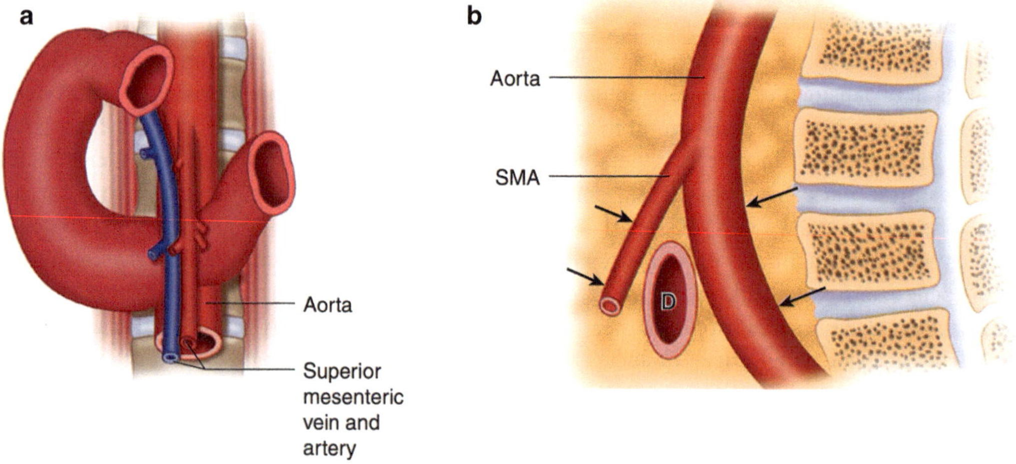

Fig. 10.7 (**a**) Diagram illustrating the anatomy of the third part of the duodenum and (**b**) its anatomical relationship with the aorta, vertebrae, and superior mesenteric artery. (Adapted with permission from Ko (2022))

Inferior Mesenteric Artery and Its Branches

The inferior mesenteric artery, the smallest of the three anterior branches of the abdominal aorta, originates from its anterior surface at the level of L3 vertebra, approximately 3–4 cm superior to the aortic bifurcation and 10 cm above the sacral promontory (Standring 2020). This position corresponds to approximately 3–4 cm inferior to the origin of the superior mesenteric artery (Moore et al. 2018).

The left colic artery, emerging as the initial branch 2.5–3 cm from the origin, demonstrates a consistent branching pattern. Upon bifurcation, its ascending component extends toward the splenic flexure, establishing anastomoses with the left branch of the middle colic artery. This anatomical arrangement forms part of the marginal artery of Drummond, a significant collateral pathway. The descending component creates anastomoses with the sigmoid arteries, maintaining the vascular supply to the left colon (Drake et al. 2020).

The sigmoid arteries exhibit notable anatomical variations in their origin and distribution. While they predominantly arise from the inferior mesenteric artery, the first sigmoid artery occasionally originates from the left colic artery. The number of sigmoid branches varies from one to six, with most individuals presenting two to four branches. These arteries form an interconnected network supplying the sigmoid colon, with multiple anastomoses creating a vascular arcade (Netter 2019).

As the inferior mesenteric artery descends, it crosses anterior to the left common iliac artery and vein, continuing into the base of the sigmoid mesocolon, where it becomes the superior rectal artery (*see* Fig. 10.6). This arterial continuation forms both rectosigmoid and upper rectal branches before dividing into left and right terminal branches. These terminal components follow a distinctive course, descending anteriorly around the inferior two-thirds of the rectum to the level of the levator ani muscle (Agur and Dalley 2009).

The middle rectal arteries demonstrate significant anatomical variability in their origin. Statistical analysis reveals that 67% originate from the internal pudendal arteries, while the remaining vessels arise from the inferior gluteal arteries (17%) and internal iliac arteries (17%). The inferior rectal arteries, branching from the internal iliac arteries via the pudendal artery within the Alcock canal (pudendal canal), traverse the ischioanal fossa. These vessels provide the primary blood supply to the anal canal and external anal sphincter muscles, forming anastomoses with superior rectal vessels (Standring 2020).

10.1.2.2 Gastrointestinal Venous Circulation

Figure 10.8 demonstrates the venous network of the GI tract. The portal vein functions as a principal component in digestive physiology, exhibiting specific anatomical characteristics: a valveless structure, low-pressure conditions (5–10 mm Hg), and minimal resistance properties. This vascular system transports nutrients from the intestines to the hepatic tissue through a defined pathway: originating in the villous core, proceeding through submucosal and serosal layers, continuing across the mesentery, and terminating in the portal vein via mesenteric venous tributaries (Vollmar and Menger 2009).

The formation of the portal vein results from the confluence of multiple venous channels, comprising the superior and inferior mesenteric veins and vessels draining the spleen, stomach, intestines, gallbladder, and pancreas. These vessels terminate in hepatic sinusoids, which are specialized capillary structures within the liver parenchyma. Venules represent approximately 40% of the mesenteric circulation, demonstrating their regulatory function in GI blood flow. Research indicates that the portal vein delivers 75–80% of hepatic blood flow, maintaining a total blood flow volume of 700–850 ml/min (Iwakiri et al. 2014). Portal venous flow fluctuates based on several variables: mesenteric arterial inflow, venous drainage velocity from abdominal organs, and intrahepatic vascular resistance patterns.

The portal system delivers nutrient-rich blood to the hepatic tissue, where reticuloendothelial and hepatic cells within the sinusoids process 50–80% of GI-absorbed substances. The hepatic

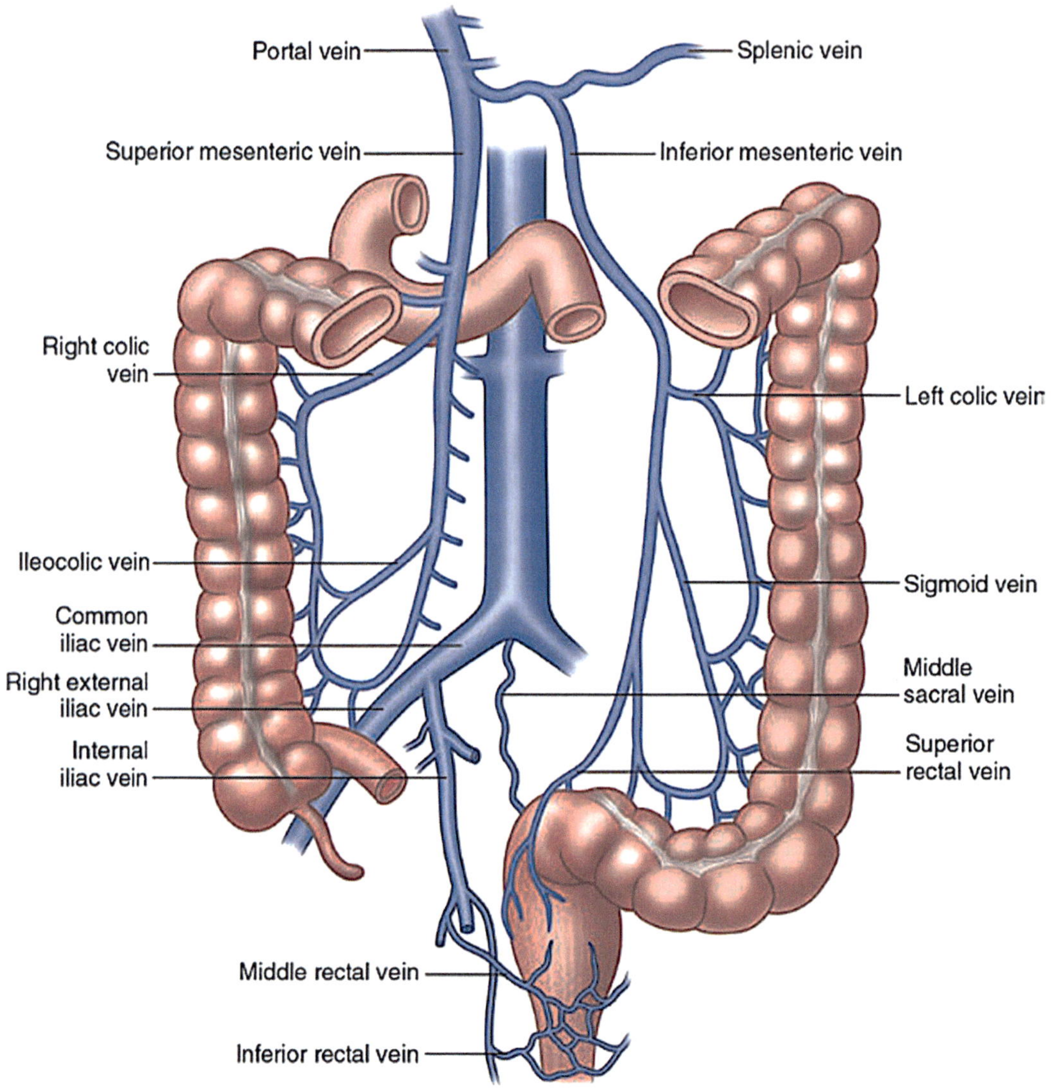

Fig. 10.8 Venous anatomy of the colon and rectum. (Adapted with permission from Beck et al. (2019))

vein subsequently conducts deoxygenated blood from the liver to the inferior vena cava, directing flow to the right atrium. Blood then moves through the pulmonary system for oxygenation before cardiac distribution throughout the body, including the GI tract (Garcia-Tsao et al. 2017).

The venous architecture of the intestines corresponds to its arterial organization, following identical nomenclature patterns.

Superior Mesenteric Vein

The superior mesenteric vein constitutes a major venous conduit receiving drainage from the right and transverse colon, maintaining an anatomical position anterior and right lateral to the superior mesenteric artery. This anatomical relationship holds significance during surgical procedures, particularly in pancreatic and colorectal interventions. The ileocolic veins consistently drain into the superior mesenteric vein, forming a predictable venous pattern (Sakaguchi et al. 2010).

Anatomical studies utilizing multidetector computed tomography demonstrate that the right colic vein exhibits two predominant drainage patterns: direct superior mesenteric vein junction in 56% of individuals, and gastrocolic trunk confluence in 44%. This anatomical variability requires careful preoperative evaluation, especially in laparoscopic right colectomy procedures (Sakaguchi et al. 2010).

The middle colic vein exhibits the highest degree of anatomical variation among mesenteric tributaries. Both the middle and right colic veins may form anatomical confluences with the right gastroepiploic vein and/or pancreaticoduodenal vein, creating the gastrocolic trunk of Henle. Morphometric analyses reveal that in 85% of cases, the middle colic vein establishes direct drainage into the superior mesenteric vein, while the remaining 15% utilize the gastrocolic trunk pathway.

Inferior Mesenteric Vein

The inferior mesenteric vein anatomically originates as a continuation of the superior rectal vein, establishing venous drainage from the left colon, rectum, and upper portion of the anal canal. The venous tributaries of the inferior mesenteric vein maintain consistent spatial relationships with their corresponding arterial counterparts, positioning slightly left lateral to the arterial vessels. This anatomical arrangement facilitates surgical identification and preservation during left-sided colonic resections (Silva et al. 2020).

At the anatomical level of the left colic artery, the inferior mesenteric vein deviates from its arterial companion, establishing an independent ascending course through the extraperitoneal space superior to the psoas muscle and lateral to the ligament of Treitz. The venous pathway continues in the retroperitoneal space, coursing posterior to the pancreatic body before its terminal confluence with the splenic vein. This anatomical course creates surgical considerations during procedures involving the pancreas, left colon, or splenic flexure mobilization (Murono et al. 2015).

10.2 Anorectal Structure and Function

10.2.1 Anatomy and Function of the Rectum

The rectum represents a specialized anatomical segment of the gastrointestinal tract, extending distally from the sacral promontory for 12–18 cm. This region exhibits distinctive morphological characteristics that differentiate it from the colonic segment, most notably the transformation of taeniae coli into a continuous outer longitudinal muscle layer and the absence of sacculations and appendices epiploicae. The rectal wall architecture incorporates two to three submucosal folds, known as Houston's valves, which create characteristic luminal curvatures that influence content progression.

The peritoneal relationships of the rectum demonstrate complex anatomical variations that hold significant clinical implications. The anterior surface maintains peritoneal coverage over its upper two-thirds, while lateral peritoneal investment extends only across the upper third. Recent morphometric analyses have established that the peritoneal reflection occurs at approximately 9 cm from the anal verge in females and 9.7 cm in males, though individual variation exists. The extraperitoneal portion receives support from the endopelvic fascia, specifically Denonvilliers fascia (Fig. 10.9), with additional stabilization provided by the lateral rectal stalks and the strong fascial attachment at the S4 level, termed Waldeyer's ring.

The biomechanical properties of the rectum demonstrate complex viscoelastic characteristics that permit its dual functions of storage and evacuation. These properties enable the maintenance of low intraluminal pressure during filling phases while preserving the capacity for effective evacuation. The rectal wall exhibits greater elasticity compared to the intra-abdominal colon, a feature particularly evident during surgical manipulation. This compliance mechanism becomes particularly relevant in pathological states, where inflammatory conditions can reduce wall elasticity and alter pressure-volume relationships.

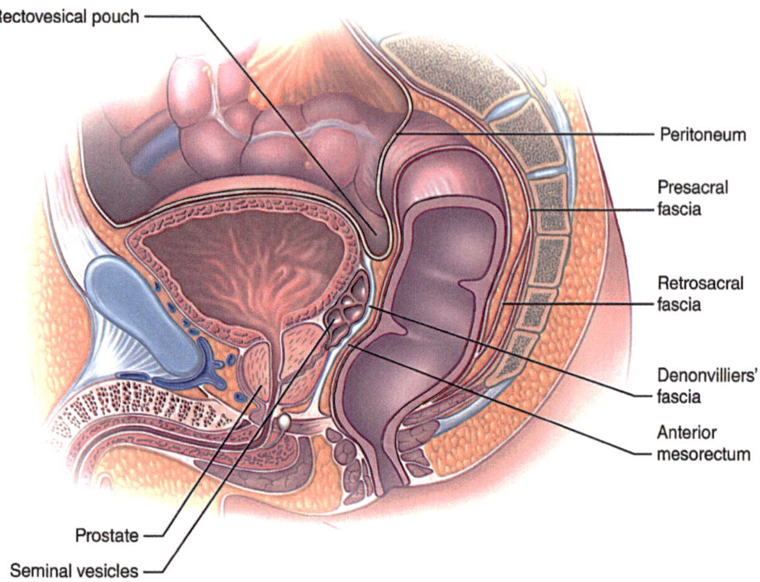

Fig. 10.9 Fascial relationships of the rectum, illustrating the key fascial layers and their anatomical connections. (Adapted with permission from Beck et al. (2019))

These alterations explain the superior functional outcomes observed with J-pouch reconstruction compared to straight anastomoses in reconstructive procedures.

The neurophysiological regulation of rectal function involves complex integration of various sensory pathways. Research indicates that the primary sensory mechanisms are predominantly located in extrarectal regions, as demonstrated by the preservation of defecatory sensation following complete rectal excision with subsequent reconstruction. The mechanoreceptors in the pelvic floor and puborectalis muscles contribute significantly to the sensory apparatus, while mucosal pathways provide complementary input. This multimodal sensory integration involves both autonomic and somatic neural networks, forming a critical control system for continence and evacuation (Moszkowicz et al. 2012).

The functional dynamics of the rectum encompass coordinated motor patterns that facilitate both storage and evacuation processes. Individual variation exists in the contribution of smooth muscle contraction to evacuation, with some individuals relying predominantly on voluntary muscular control through the Valsalva maneuver, while others demonstrate significant rectal wall contractility. The storage function depends on precise regulation of compliance and pressure thresholds, coordinated with pelvic floor activity through synchronized relaxation during defecation.

Recent advances in functional imaging and physiological monitoring have enhanced our understanding of rectal motor patterns. High-resolution manometry studies have revealed specific pressure zone modulations that correspond to different phases of rectal function. These findings have particular relevance in the context of defecatory disorders, where disruption of normal pressure patterns can lead to clinical symptoms. The clinical significance of rectal anatomical organization extends beyond basic structure-function relationships. Surgical segmentation of the rectum into lower (0–7 cm), middle (7–12 cm), and upper (12–15 cm) regions from the anal verge has evolved from therapeutic necessity, particularly in oncological management. This segmentation influences surgical approach selection, anastomotic technique determination, and functional outcome prediction.

Pathophysiological alterations in rectal compliance manifest in various clinical conditions. In inflammatory states such as ulcerative colitis and radiation proctitis, reduced wall compliance results in heightened intraluminal pressures at lower filling volumes, leading to symptoms of urgency and increased frequency. Understanding these mechanical alterations has influenced surgical reconstruction techniques, as evidenced by the development of compliance-enhancing procedures such as J-pouch construction (Kumar and Emmanuel 2017).

The neuromuscular integration controlling rectal function demonstrates remarkable complexity. Sensory dysfunction can manifest through various mechanisms, including elevated activation thresholds for the afferent neural cascade or excessive rectal compliance, both potentially lead to impaired distention detection. The preservation of sensory function following certain surgical procedures suggests redundancy in sensory pathways, a feature that may have evolved to maintain this critical physiological function (Moszkowicz et al. 2012).

Advanced physiological studies have revealed that rectal evacuation mechanisms vary significantly among individuals. While some rely primarily on increased intra-abdominal pressure through the Valsalva maneuver, others demonstrate substantial rectal smooth muscle contribution to evacuation. This variability in evacuation strategies suggests adaptive mechanisms that may compensate for individual anatomical or physiological variations.

10.2.2 Anatomy and Function of the Anal Canal

The anal canal extends approximately 2.5–5 cm from the levator ani muscle to the anal verge. This structure contains internal and external anal sphincter muscles, with the internal sphincter representing a continuation of the rectum's inner circular smooth muscle layer. Digital examination allows palpation of the puborectalis muscle, which forms the superior component of the external anal sphincter at the anorectal ring. The internal anal sphincter receives superior coverage from the levator ani muscle, followed distally by the superficial external sphincter muscle (an extension of the anococcygeal ligament), and the subcutaneous external striated anal sphincter muscle (Standring 2020) (Fig. 10.10).

The classification of the anal canal follows two primary definitions. The "anatomic" anal canal spans from the dentate line to the anal verge, based on embryological and histological characteristics. However, this definition holds limited practical value in surgical applications. Milligan and Morgan (1934) introduced the concept of the "surgical" anal canal, which extends from the anorectal ring to the anal verge (Milligan and Morgan 1934). The anorectal ring comprises the upper internal anal sphincter, conjoined longitudinal muscle, puborectalis, and external sphincter. This surgical definition accurately reflects the physiological mechanisms of anal continence (*see* Fig. 10.10).

Morphometric analysis reveals gender-specific variations in anal canal dimensions. Males demonstrate a longer surgical anal canal, measuring 4.4 cm posteriorly compared to 4.0 cm in females. Notably, anal canal length remains constant across different age groups. Magnetic resonance imaging studies have documented equivalent posterior anal canal lengths between genders; however, females exhibit significantly reduced anterior and posterior external anal sphincter lengths, excluding the puborectalis (DeLancey et al. 2007).

The rectum originates at the sacral promontory and terminates at its passage through the levator ani muscles. Unlike the colon, it features a complete outer longitudinal muscle layer and measures 12–15 cm in length, lacking sacculations and appendices epiploicae. The rectum presents three lateral curves: a superior and inferior curve convex to the right and a middle curve convex to the left. These infoldings manifest internally as Houston's valves, with the middle fold corresponding to the anterior peritoneal reflection (Standring 2020).

The anal canal's muscular architecture resembles a dual-tube configuration. The inner tube consists of autonomically innervated visceral smooth muscle, including the internal anal sphincter and longitudinal layer. The outer tube comprises somatically innervated skeletal muscles, including the puborectalis and external anal sphincter components. This outer muscular structure enables voluntary control of continence and

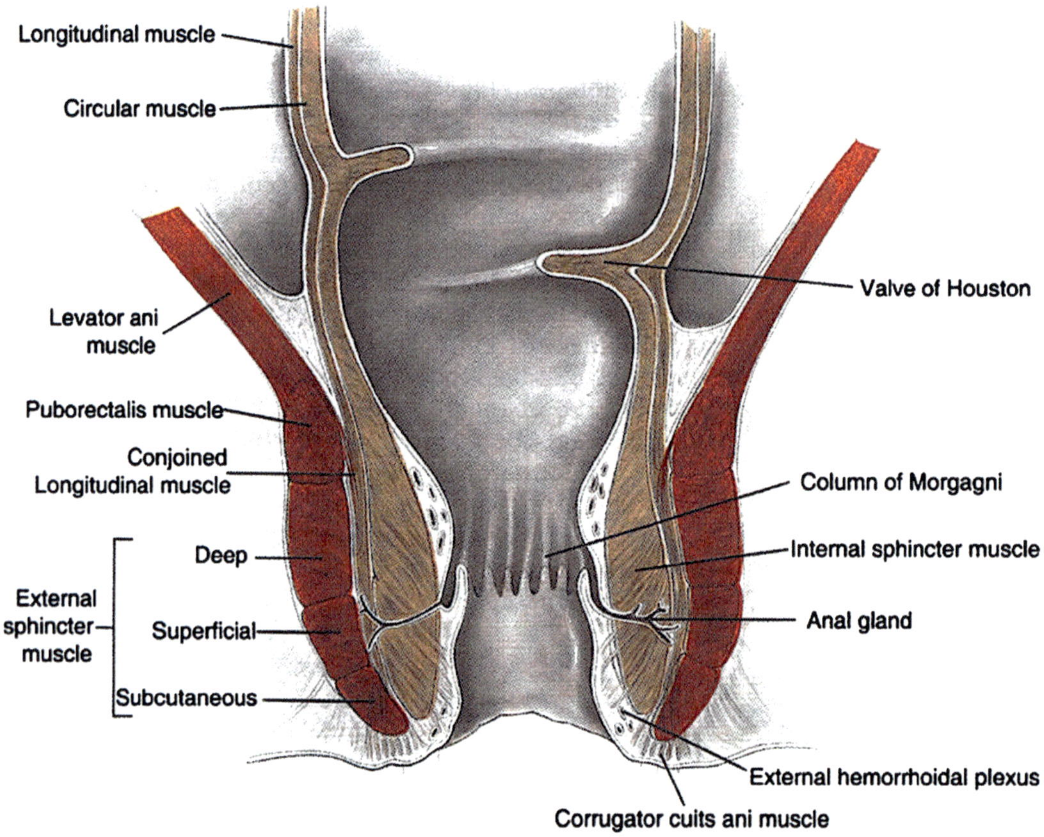

Fig. 10.10 Internal and external anal sphincters. (Adapted from Gordon and Nivatvongs (2007))

responds to Kegel exercise training (Jorge and Wexner 1997).

The anal canal's orientation shows a posterior direction due to the puborectalis component of levator ani exerting anterior traction on the rectum, creating the anorectal angle. It positions 2–3 cm anterior and slightly inferior to the coccygeal apex, adjacent to the prostatic apex in males. The anal verge marks the transition between squamous epithelium and perineal skin, with surrounding pigmentation approximately corresponding to the external anal sphincter's extent.

The structural composition includes an epithelial lining, vascular subepithelium, sphincter muscles, and fibromuscular support tissue with dense autonomic and somatic neural networks. This high-pressure zone ranges from 2 to 5 cm in adult subjects, with a shorter anterior wall compared to the posterior aspect. The canal forms an oval or triradiate configuration in the anteroposterior plane during rest (Bharucha and Rao 2014).

Anatomical relationships include anterior attachment to the perineal body, separating it from the membranous urethra in males and the lower vagina in females. The ischioanal fossae contain loose adipose tissue laterally and posteriorly, facilitating canal expansion while creating potential paths for infection spread. The anococcygeal ligament connects the posterior middle external anal sphincter to the coccyx (Shafik et al. 2005).

Digital examination enables palpation of the ischial spines laterally in the superior anal canal, where the pudendal nerves cross the sacrospinous ligament attachment. This anatomical relationship permits digital measurement of pudendal nerve motor terminal latency using modified electrodes (Stoker et al. 2002).

10.2.2.1 Muscles of the Anal Canal

The anal sphincter complex consists of the internal and external anal sphincters, which are anatomically separated by an intervening conjoint longitudinal muscle layer (Sinnatamby 2011). Recent anatomical investigations utilizing endoanal ultrasonography and magnetic resonance imaging have provided detailed documentation of this region's structural organization. The muscular components maintain a concentric arrangement, with each layer contributing specific functions to anorectal continence.

Advanced imaging techniques have revolutionized the understanding of anal sphincter anatomy, offering three-dimensional visualization of muscular relationships and supporting structures (Regadas et al. 2007). These imaging modalities permit precise assessment of sphincter integrity, which holds particular significance in rehabilitation medicine and surgical planning.

Internal Anal Sphincter

The internal anal sphincter constitutes a distal extension of the rectal circular smooth muscle, concluding in a distinctive rounded border positioned approximately 1 cm superior to the external anal sphincter's distal margin (Raizada and Mittal 2008). During digital rectal examination, practitioners can readily palpate the internal anal sphincter terminus, which manifests as the intersphincteric groove, providing a reliable anatomical landmark (Standring 2020).

Although the total volume is similar between genders, the anatomical distribution patterns exhibit significant differences. Female participants exhibit increased thickness in the medial and distal regions. Morphometric evaluations have recorded an average thickness of 2 mm and a length of 35 mm. Researchers recognize ongoing difficulties in delineating the proximal internal anal sphincter boundary due to its uninterrupted extension with the lower rectal wall structure.

The internal anal sphincter presents as a specialized white, thickened terminal component of the large intestine's inner circular muscle. Notable variations exist between populations, with increased thickness observed in male subjects and individuals with chronic constipation. The muscle's architectural arrangement reveals an intriguing pattern—rather than forming perfect circles, the fibers configure into a tight spiral formation that undergoes dimensional changes during relaxation, exhibiting both shortening and widening (Fritsch et al. 2002). Originating at the anorectal junction and terminating superior to the anal verge, the internal anal sphincter maintains specific relationships with surrounding structures. Its inferior border, palpable at the intersphincteric groove, corresponds precisely with the superior margin of the external anal sphincter's subcutaneous portion. The structural complexity increases as fibers from the conjoint longitudinal muscle traverse the internal anal sphincter medially, extending into the submucosa (DeLancey and Ashton-Miller 2004).

The physiological function of the internal anal sphincter exhibits remarkable complexity. While maintaining tonic contraction during rest periods, it exhibits precise reflex-mediated relaxation patterns, particularly evident during defecation. The upper portion of the internal anal sphincter responds to two primary stimuli: rectal distention, which triggers the recto-anal inhibitory reflex, and postprandial rectal contractions, initiating the sampling reflex. This coordinated relaxation mechanism permits distal rectal contents to enter the upper anal canal, enabling both conscious and subconscious assessment of physical properties. Concurrent with this process, the distal internal anal sphincter and external anal sphincter maintain continence through sustained contractile activity (Bharucha and Rao 2014).

The neurological control of these reflexes has significant consequences for clinical practice. The enteric nervous system predominantly regulates the recto-anal inhibitory reflex, though spinal pathways may offer modulatory effects. The absence of this reflex is diagnostically significant, providing as an indicator the signal of Hirschsprung's disease and providing valuable diagnostic information (Rao and Patcharatrakul 2016).

Conjoint Longitudinal Muscle

The conjoint longitudinal muscle represents an anatomical continuation of the rectum's outer longitudinal smooth muscle, positioned between the internal and external anal sphincters. In its superior portion, this muscle receives additional striated muscle fibers from the medial aspect of levator ani (Lunniss and Phillips 1992). Morphometric analysis indicates its thickness ranges from 0.5 to 2.0 mm. The muscle originates at the anorectal ring, extending caudally in conjunction with puborectalis muscle fibers.

Developmental studies reveal particularly pronounced muscle prominence during fetal life, where its thickness exceeds that of the internal anal sphincter. Age-related changes include progressive replacement of muscle tissue with connective tissue elements, resulting in marked thinning in elderly individuals, particularly evident in the distal region (Haas et al. 1984).

Recent anatomical investigations have challenged traditional interpretations of this structure. While classical descriptions present the conjoint longitudinal muscle as a combination of smooth longitudinal rectal fibers and skeletal levator ani components, contemporary immunohistochemical analyses of cadaveric specimens suggest an alternative arrangement. These studies indicate that the tissue between the internal and external anal sphincters consists predominantly of smooth muscle derived from longitudinal rectal fibers, with levator ani forming direct attachments to the longitudinal rectal muscle.

The muscle's fiber distribution follows three primary trajectories as it descends through the anal canal:

- Internal fibers penetrate the internal anal sphincter to reach the anal submucosa.
- Inferior fibers traverse the lower subcutaneous external anal sphincter to insert into perianal skin, with some fibers encircling the anal orifice.
- Lateral fibers extend through the superior external anal sphincter.

The most caudal portion includes fibers known as the corrugator cutis ani muscle, which penetrate the distal external anal sphincter to insert into perianal skin and ischiorectal fossa fat. Additional fibers pass obliquely through the internal anal sphincter, forming an intricate subepithelial network (Thomson 1975).

Autonomic innervation parallels that of the internal anal sphincter, with functional implications during defecation, where muscle contraction produces anal canal shortening and widening, accompanied by anal orifice eversion. Clinical significance extends to various pathological conditions. Thomson's hypothesis suggests that disruption of Treitz's muscles may contribute to anal cushion prolapse, vascular outflow obstruction, and hemorrhoidal complications.

The muscle's distal fibers create a honeycomb configuration within subcutaneous fat, establishing anatomical separation between superficial perianal space and the deeper ischioanal fossa. This architectural arrangement holds particular relevance for understanding infection spread patterns in anorectal pathology.

External Anal Sphincter

The external anal sphincter constitutes an elliptical tube of striated muscle that encircles the internal anal sphincter and conjoint longitudinal muscle. Its composition predominantly features type I slow-twitch muscle fibers, adapted for sustained contractile activity. The intersphincteric groove forms where this structure extends beyond the distal margin of the internal anal sphincter, with corrugator cutis ani muscle fibers from the conjoint longitudinal muscle traversing through to insert into perianal skin (Rociu et al. 2000).

The external anal sphincter muscle comprises three separate divisions arranged from proximal to distal: sphincter ani externus profundus, superficialis, and subcutaneus. Subsequent study, however, demonstrated that the external anal sphincter forms a continuous skeletal muscle sheet extending to the puborectalis and levator ani muscles. Contemporary literature acknowledges this continuity, however, anatomical descriptions often refer to proximal (deep), mid (superficial), and distal (subcutaneous) portions for practical purposes (Fritsch et al. 2004) (*see* Fig. 10.10).

The muscle's architectural relationships include posterior attachment to the coccyx via the anococcygeal ligament at its mid-portion, while the proximal region maintains continuity with the puborectalis muscle. Anteriorly, the proximal external anal sphincter integrates with the transverse perineal muscle to form part of the perineal body (Fig. 10.11). The uppermost fibers demonstrate complex relationships, blending with puborectalis' lower medial fibers and forming attachments to the anococcygeal raphe posteriorly and transverse perineal muscles anteriorly.

Modern imaging studies using magnetic resonance imaging and three-dimensional endoanal ultrasonography have revealed distinct morphological differences between males and females. Female subjects frequently display a natural anterior defect along the proximal sphincter length below the puborectalis sling, observed in 75% of nulliparous individuals. This anatomical feature correlates with anal manometry findings and holds clinical significance, particularly in interpreting endoanal ultrasound results and avoiding overdiagnosis of obstetric sphincter injuries. This anatomical variation provides scientific rationale for limiting anterior anal sphincterotomy procedures in female patients (Shafik et al. 2005).

Neural control of the external anal sphincter derives from bilateral innervation via the pudendal nerve's inferior rectal branch (S2 and S3) and the perineal branch of S4 (Fig. 10.12). The substantial bilateral overlap in pudendal innervation presents a physiological advantage, enabling partial reinnervation from the contralateral side following unilateral nerve injury. Like the levator ani and internal anal sphincter, this muscle maintains tonic contraction during rest through postural reflex mechanisms.

Female subjects typically demonstrate reduced sphincter length and thickness compared to males. The lower portion extends inferior to the internal anal sphincter, receiving traversing terminal fibers from the conjoint longitudinal muscle. Anterior attachments include connections to bulbospongiosus and bulbocavernosus muscles, contributing to the complex three-dimensional organization of the pelvic floor.

Vascular Supply and Innervation of the Anal Sphincter Complex

The vascular and neural anatomy of the anal sphincter complex exhibits intricate patterns of organization that directly influence sphincter function. Terminal branches of the superior and inferior rectal vessels provide blood supply to the internal anal sphincter. The autonomic nervous system maintains extrinsic control through specific spinal segment contributions: sympathetic innervation originates from the upper two lumbar segments, while parasympathetic fibers emerge from the S2–S4 segments. Both neural pathways reach the internal anal sphincter via the inferior hypogastric plexus (Drake et al. 2020).

Neurotransmitter-receptor interactions produce distinct physiological responses in the internal anal sphincter. Parasympathetic stimulation, mediated through cholinergic muscarinic receptors, results in internal anal sphincter relaxation concurrent with longitudinal anal muscle contraction. Conversely, sympathetic input via α-adrenergic receptor activation induces contraction in both the internal anal sphincter and longitudinal muscle components (Matsufuji and Yokoyama 2003).

The external anal sphincter receives its blood supply predominantly from the inferior rectal vessels, with supplementary contributions from the median sacral artery. Its neural control stems from bilateral innervation by the pudendal nerve's inferior rectal branch, which carries fibers from the S2–S4 spinal nerves (Bharucha and Rao 2014). This same nerve pathway transmits afferent signals from the anal canal lining and perianal skin. Additional motor innervation to the superior portion of the external anal sphincter may arrive via the nerve to levator ani, primarily containing ventral rami from the S3 and S4 segments.

10.2.2.2 Anal Canal Lining

The anal canal's muscularis mucosae demonstrates anatomical continuity with the upper anal canal. A distinctive feature involves fibers from the conjoint longitudinal muscle, which penetrate the internal anal sphincter before turning cranially to merge with the muscularis mucosae. This arrangement forms the submucosal muscle of the

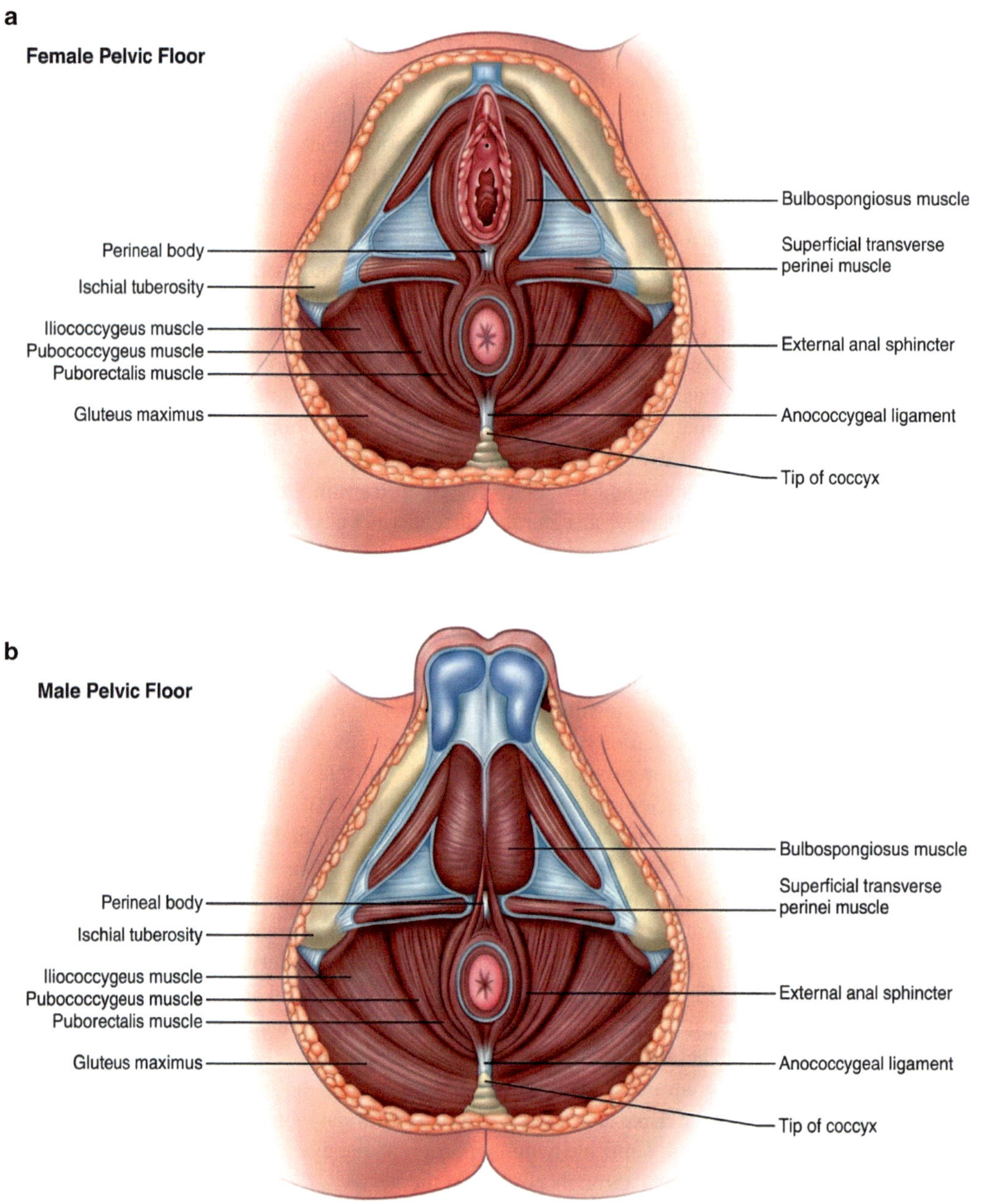

Fig. 10.11 Pelvic floor muscles of females (**a**) and males (**b**). (Adapted with permission from Beck et al. (2019))

anal canal, alternatively known as musculus submucosae ani, musculus canalis ani, or Treitz's muscle (Erlichman et al. 2019).

At the dentate line, longitudinal fibers establish particularly dense attachments to the submucosal muscle, forming Parks' mucosal suspensory ligament. This anatomical arrangement results in firm mucosal tethering and approximately demarcates the watershed between portal and systemic venous drainage of the hemorrhoidal plexuses. The interactions between longitudinal muscle fibers and anal canal muscularis mucosae create

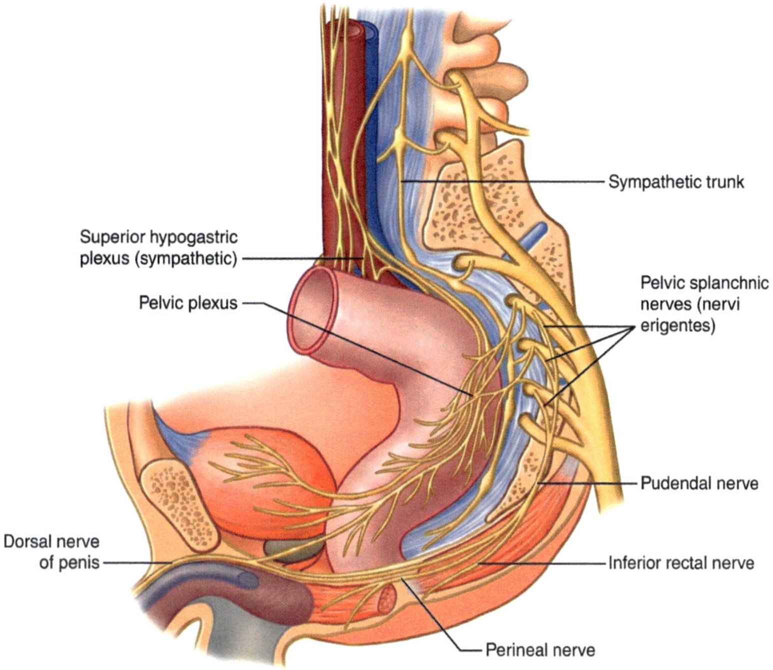

Fig. 10.12 Innervation of the rectum. (Adapted with permission from Beck et al. (2019))

three distinct submucosal spaces: a proximal submucous space, a marginal space extending from the dentate line to the intersphincteric groove, and a distal perianal space.

Histological analysis reveals three distinctive zones within the anal canal. The proximal zona columnaris, extending from the anorectal ring apex to the dentate line, displays simple columnar epithelium. The intermediate zona hemorrhagica, positioned between the dentate line and intersphincteric groove (Hilton's white line), features stratified squamous non-keratinized epithelium. Below the intersphincteric groove, the zona cutanea exhibits stratified squamous keratinized epithelium.

The anal crypts maintain connections with underlying anal glands through anal ducts (Fig. 10.13), structures implicated in the pathogenesis of most anorectal abscesses and fistulas. Anatomical studies document an average of six anal glands circumferentially distributed around the anal canal, ranging from three to twelve, with higher concentration in posterior quadrants. The anatomical arrangement exhibits variability—multiple glands may share a single crypt opening, while some crypts lack glandular connections. The anal gland ducts follow an inferolateral trajectory from the anal canal, entering the submucosa. Two-thirds of these ducts penetrate the internal anal sphincter, with half terminating in the intersphincteric plane (Seow-Choen and Ho 1994). Ductal obstruction precipitates anal fistula and abscess formation. Furthermore, the anatomical relationship between crypts and glands explains the typical location of "cryptoglandular" anal fistula internal openings at the dentate line (Sugrue et al. 2017).

The anal canal demonstrates distinct anatomical zonation with specialized epithelial characteristics. In its proximal region, the canal exhibits columnar epithelium with a characteristic pink appearance, morphologically similar to rectal mucosa (Walls 1958). This epithelium contains specialized secretory and absorptive cells organized into numerous tubular crypts, facilitating optimal mucosal function. The anal transition zone, alternatively termed the cloacogenic zone, initiates approximately 6–12 mm superior to the dentate line. This region manifests a distinctive purple hue, representing the interface between

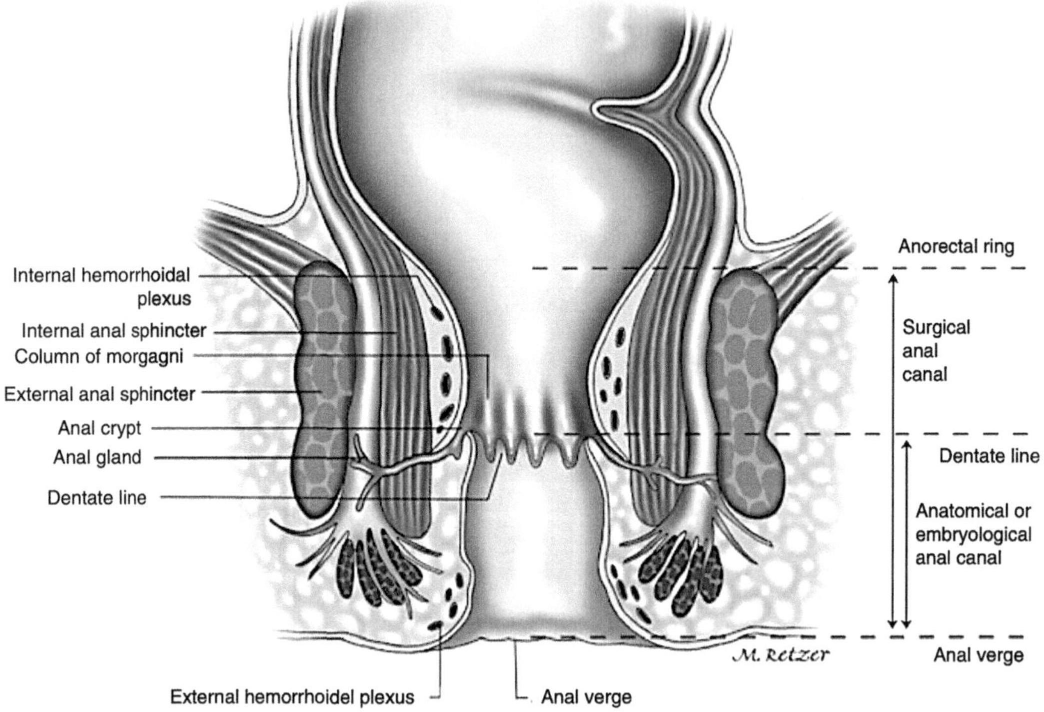

Fig. 10.13 Anatomy of the anal canal. (Adapted with permission from Beck et al. (2014))

columnar and squamous epithelia (Barleben and Mills 2010).

The columns of Morgagni constitute a significant anatomical feature, presenting as redundant mucosal elevations with anal crypts positioned at their base. These structures collectively form the undulating pattern characteristic of the dentate line. Advanced imaging techniques have demonstrated the precise architectural organization of these columns, including their rich neurovascular supply and lymphatic drainage patterns.

As the canal progresses distally, the epithelium undergoes transformation to cuboidal morphology, exhibiting increased pigmentation superior to the anal valves. The subepithelial connective tissues in this region demonstrate remarkable biomechanical properties, including significant compliance and distensibility. This tissue architecture accommodates an intricate network of submucosal vascular structures, which play fundamental roles in both physiological function and pathological processes (Enck et al. 2005; Walls 1958).

In the midanal canal, a notable feature includes 6–10 longitudinal mucosal elevations, termed anal columns (Standring 2020), with enhanced definition observed in pediatric populations versus adults (Gordon and Nivatvongs 2007). The superior region of the anal canal contains an internal hemorrhoidal venous plexus, characterized by dilated submucosal veins. These vascular structures exhibit unique arteriovenous anastomoses, which accounts for their elevated oxygen saturation and characteristic bright red appearance (Lunniss and Phillips 1992).

The vascular anatomy maintains a specific pattern, with each column typically housing a terminal branch of the superior rectal artery and vein, supplemented by variable contributions from middle and inferior rectal vessels. Notably, the submucosal vasculature displays particular prominence in three specific quadrants when viewed in the lithotomy position: left-lateral (3 o'clock), right-posterior (7 o'clock), and right-anterior (9 o'clock) positions (Milligan and Morgan 1934). These locations correspond to the formation of three distinct "anal cushions" where subepithelial tissue expansion occurs.

The anal cushions, despite their variable number and position, serve a fundamental physiological purpose. They facilitate anal canal closure and contribute significantly to maintaining continence against flatus and fluid (Jorge and Wexner 1997). Understanding the anatomy and function of these cushions becomes particularly relevant in clinical practice, as they represent a primary anatomical substrate in hemorrhoidal pathology (Rasmussen 1994).

The anatomical complexity extends to the inferior portion of the anal columns, where crescentic structures known as anal valves form, separated by small recesses termed anal sinuses (Standring 2020). Together, these structures constitute the dentate (pectinate) line, displaying a characteristic scalloped configuration with firm anchoring to the underlying submucosal connective tissue (Bharucha and Rao 2014). This region holds clinical significance due to the presence of anal glands, which demonstrate a branched pattern and contain stratified columnar epithelium. These glands drain into specialized depressions within the anal valves, known as anal crypts.

Clinical considerations should include the potential for pathological changes, particularly regarding the anal glands. These structures may develop cystic dilations capable of extending through the internal anal sphincter and occasionally reaching into the external sphincter muscle (Gordon and Nivatvongs 2007). Such anatomical variations and potential pathological developments bear particular relevance for rehabilitation specialists managing anorectal disorders.

The anal transition zone denotes the junction between columnar and squamous epithelium of the anal canal. This anatomical landmark exhibits variable length and position, commonly extending approximately 1 cm superior to the dentate line (Standring 2020) (Fig. 10.14). The submucosal region of the upper anal transition zone contains specialized nerve endings, notably thermoreceptors. These neural components constitute a refined "sampling" mechanism that aids continence by detecting the contents of the lower rectum during upper anal canal relaxation (Bharucha and Rao 2014).

The dentate line marks a significant embryological boundary, separating endoderm from ectoderm. The tissue superior to this line receives autonomic (sympathetic and parasympathetic) innervation, with vascular and lymphatic drainage connecting to the hypogastric vessels. Conversely, the inferior region receives somatic innervation and vascularization via the inferior hemorrhoidal system (Gordon and Nivatvongs 2007). This anatomical configuration explains two clinical observations: the metastatic spread of sub-dentate line malignancies to superficial inguinal lymph nodes, and the presence of pain in external hemorrhoids, which exclusively develop inferior to the dentate line (Rasmussen 1994).

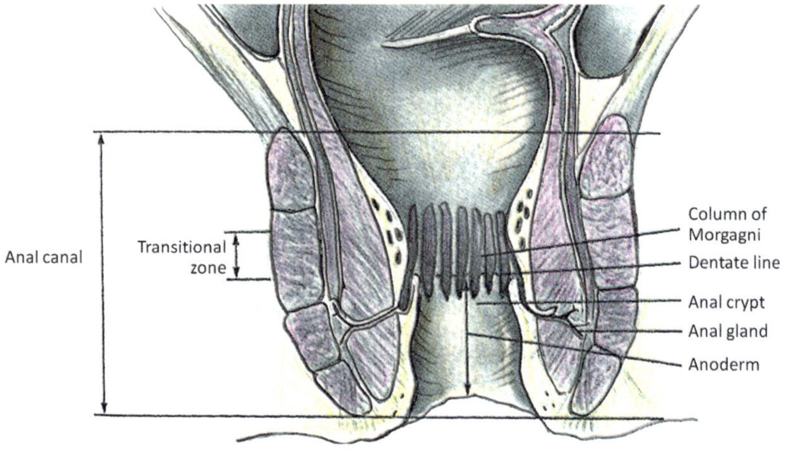

Fig. 10.14 Rectal and anal mucosa including the transitional zone. (Adapted from Gordon and Nivatvongs (2007))

The anoderm, a specialized epithelial region extending approximately 1.5 cm inferior to the dentate line, displays distinct histological features (*see* Fig. 10.14). This region presents as non-keratinized, stratified squamous epithelium lacking cutaneous appendages–hair follicles, sebaceous glands, and sweat glands (Lunniss and Phillips 1992). Despite this simplified structure, the anoderm contains numerous somatic sensory nerve endings that mediate touch, pain, and temperature sensations. This specialized epithelium continues to the intersphincteric groove, an anatomical depression that corresponds to the inferior margin of the internal sphincter (Milligan and Morgan 1934).

The most distal portion of the anal canal demonstrates further histological modification. Beyond the intersphincteric groove, the epithelium transitions to keratinized, stratified squamous type with hair follicles, displaying continuity with the perianal skin (Jorge and Wexner 1997). The subepithelial tissue inferior to the dentate line contains the external hemorrhoidal venous plexus, which appears as a circumferential ring at the anal margin during Valsalva maneuver (Standring 2020).

10.3 Anatomy of the Pelvic Floor

10.3.1 Perineal Body

The perineal body represents a crucial anatomical intersection where multiple muscles converge, including the external anal sphincter, superficial and deep transverse perinei muscles, and the bulbospongiosus muscle (Fig. 10.15). Investigations using high-resolution magnetic resonance and ultrasound imaging have revealed that the transverse perinei and bulbospongiosus muscles play a more significant role in maintaining anal continence than previously recognized (Stoker et al. 2002). Anatomical conceptions advocate for the collective designation of these muscles as the "external anal sphincter complex muscles," suggesting a purse-string configuration rather than the conventional donut-shaped model. This anatomical understanding has important clinical implications, particularly emphasizing the importance of perineal body repair during sphincter reconstruction procedures (DeLancey and Ashton-Miller 2004).

10.3.2 Anococcygeal Ligament

Detailed cadaveric analyses have enhanced our understanding of the anococcygeal ligament's structure, revealing its bilayered composition. The ventral layer, characterized by its thickness, extends from the presacral fascia to the conjoint longitudinal layer of the anal canal. In contrast, the dorsal layer presents as a thinner structure connecting the coccyx to the external anal sphincter. This anatomical arrangement carries significant surgical implications, particularly during intersphincteric proctectomy or low anterior resection procedures, which require division of the thick ventral layer. More extensive procedures, such as abdominoperineal resection, necessitate division of both layers (Tsukada et al. 2016).

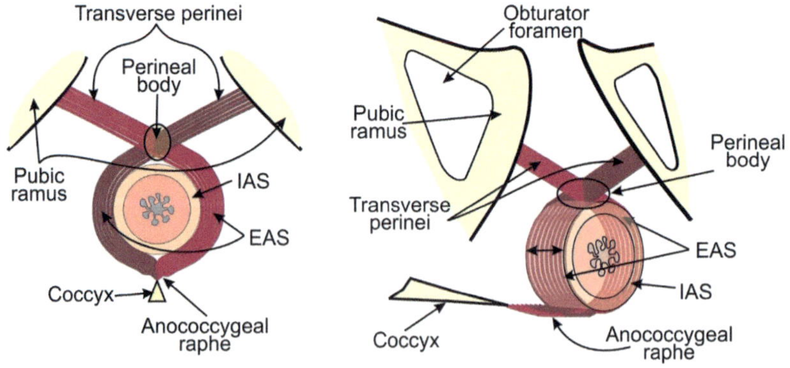

Fig. 10.15 Frontal and sagittal views of external anal sphincter (EAS) and its attachment to the bony pelvis, illustrating its anatomical orientation and connections. (Adapted from Mittal and Tuttle (2022))

The mechanical role of the anococcygeal ligament, particularly its dorsal component, appears less substantial than previously thought. The superficial portion's weak coccygeal insertion and undulating course suggest limited contribution to external anal sphincter stability. This anatomical characteristic may influence surgical planning and postoperative expectations (Rajasekaran et al. 2008).

10.3.3 Pelvic Floor Muscles: Focus on Levator Ani, Coccygeus

The pelvic floor muscles, particularly the levator ani complex and coccygeus muscle, serve as important components in maintaining pelvic organ support and continence. Their coordinated function relies on complex neural pathways that can be significantly impacted by spinal cord injury.

10.3.3.1 Levator Ani Complex (Innervated by the Anterior Rami of S3 and S4, and by a Branch from the Perineal Branch of the Pudendal Nerve [S4])

The levator ani muscles form an essential component of the pelvic floor, creating a dynamic supportive hammock across the lower pelvic cavity. These broad, thin muscles establish a natural separation between the pelvic cavity and perineum (Fig. 10.16). In males, the prostate separates the muscles at the midline, while in females, the urethra and vagina occupy this space (Drake et al. 2020).

The origin of each levator ani extends continuously from the inner pubic body anteriorly, following the obturator membrane, and reaching the pelvic surface of the ischial spine laterally. The muscle fibers course posteriorly, inferiorly, and medially toward the midline, with insertions into three key points: the perineal body, anal canal walls, and the anococcygeal raphe between the anal canal and coccyx (Standring 2020).

Anatomical studies have identified three distinct components of the levator ani: the pubococcygeus, puborectalis, and iliococcygeus. While some historical perspectives suggested the puborectalis belonged to the deep external anal sphincter, modern imaging and embryological evidence confirms its association with the levator ani complex. Advanced MRI studies in women have demonstrated distinct muscle fascicle orientations for each component, though histological examination shows no clear boundaries between these subdivisions (DeLancey et al. 2007).

The innervation pattern of the levator ani complex has been clarified through detailed cadaveric studies. The muscles receive dual innervation: branches of the pudendal nerve (perineal and inferior rectal branches) and direct sacral nerve contributions from S3 and/or S4 (known as the levator ani nerve). Specifically, the pudendal nerve primarily supplies the pubococcygeus and puborectalis, while direct sacral innervation predominates in the iliococcygeus (Barber et al. 2002).

The levator ani complex demonstrates constant activity, playing a vital role in pelvic visceral support, particularly crucial in females. Its contractile function helps regulate pelvic floor openings through both reflexive and voluntary mechanisms. During activities that increase intra-abdominal pressure, such as coughing, reflexive levator ani contraction assists in urethral and anal closure, preventing involuntary evacuation. This mechanism can be voluntarily engaged to resist inconvenient urges to micturate or defecate (Morgan et al. 2010).

In the female pelvis, the anatomical relationship between the levator ani and vagina proves particularly significant for uterine support. Childbirth or surgical procedures like episiotomy can cause excessive muscle stretching or trauma. Such injury may compromise both anal and urethral function, potentially leading to stress incontinence. This condition manifests as involuntary leakage during activities that increase intra-abdominal pressure.

Male patients generally experience stress incontinence less frequently, though it may develop following prostatectomy. Similar to female patients, pelvic floor exercise programs play a crucial role in restoring levator ani tone

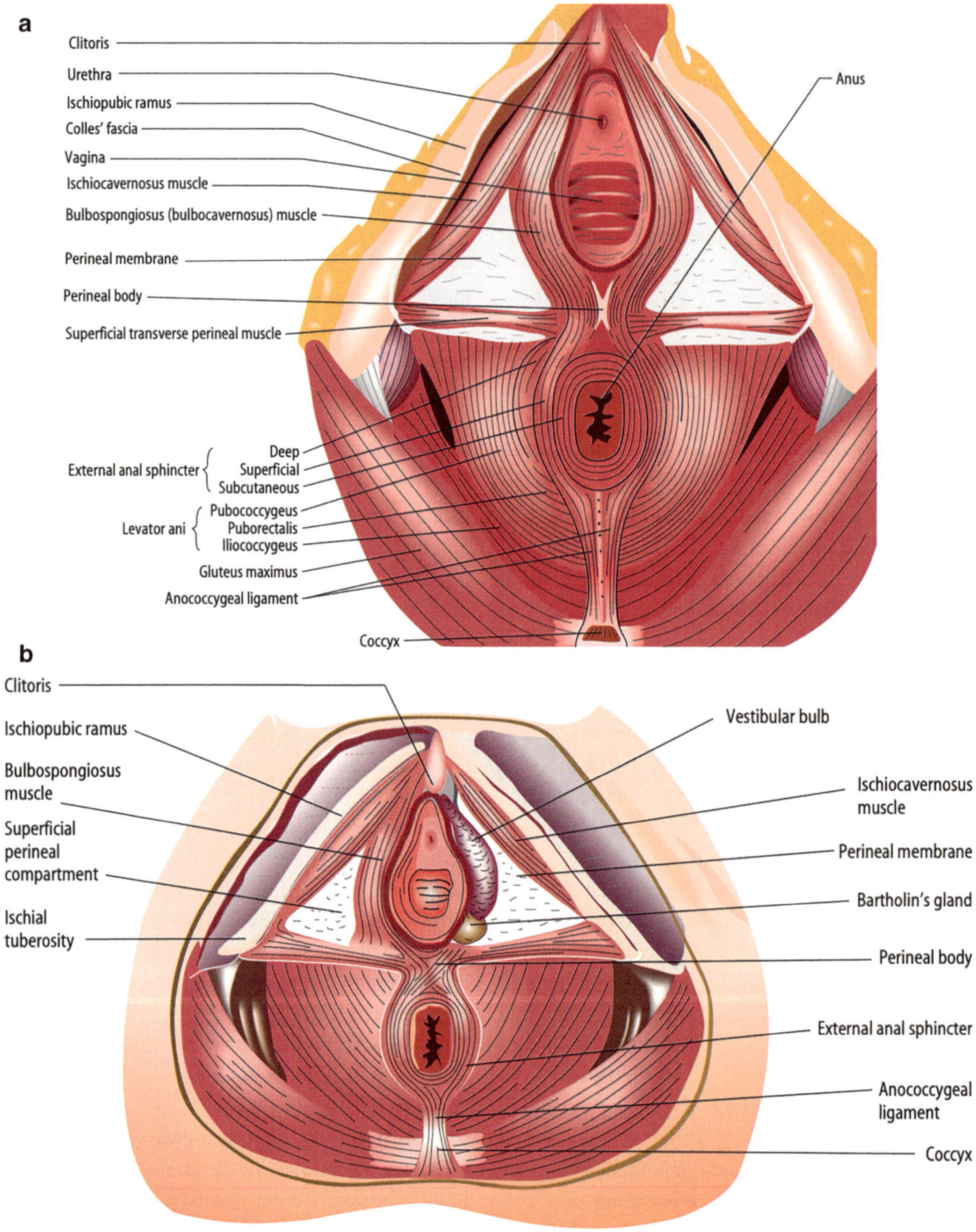

Fig. 10.16 The superficial compartment of the perineum, containing the superficial transverse perineal muscle, bulbospongiosus, and ischiocavernosus muscles, which form a triangular arrangement on either side of the perineum with the perineal membrane as the floor. (**a**) The left bulbospongiosus muscle is removed to reveal the vestibular bulb and Bartholin's gland. (**b**) Close-up view of the vestibular bulb and Bartholin's gland. (Adapted with permission from Sultan et al. (2007))

and function. Recent studies have demonstrated improved continence outcomes when pelvic floor rehabilitation begins preoperatively in men undergoing radical prostatectomy (Anderson et al. 2015).

Pubococcygeus

The pubococcygeus muscle exhibits gender-specific anatomical patterns in its anterior region, which originates proximal to the obturator canal. In male anatomy, the anterior muscle fibers extend posterior to the prostatic gland, whereas in females, these fibers encircle the vaginal wall, with both variations inserting into the perineal body. The middle portion of the muscle forms an anatomical sling surrounding the anorectal junction, maintaining its anterior positioning relative to the pubic symphysis. Select fibers merge with the longitudinal muscle of the rectum, extending through the external anal sphincter via fibroelastic attachments before reaching the perianal integument (Raizada and Mittal 2008).

Anatomically positioned medial to the puborectalis muscle, the pubococcygeus muscle originates from two primary sites: the superior posterior aspect of the pubic bone and the anterior half of the obturator fascia. These muscle fibers follow a posterior-inferior and medial trajectory, intersecting with their contralateral counterparts to establish the anococcygeal raphe (also termed the anococcygeal ligament). The muscle's insertion points include the inferior sacrum and coccygeal apex. Specific pubococcygeus muscle fibers contribute to forming the conjoined longitudinal muscle.

The pubococcygeus muscle creates an anatomical opening known as the "levator hiatus," which accommodates the inferior rectum, urethra, and gender-specific structures—the vaginal canal in females or the dorsal penile vein in males. A fascial condensation, designated as the "hiatal ligament," connects the levator hiatus to its contained organs. This ligament originates circumferentially from the hiatal margin, continuing from the fascia overlying the pelvic surface of the levator muscle. Clinical research has associated expansion of the levator hiatus with female pelvic organ prolapse (DeLancey et al. 2007). Studies have identified the pubococcygeus muscle as the primary component of the levator ani complex susceptible to injury during complicated vaginal deliveries (Dietz and Lanzarone 2005).

Iliococcegeus

The iliococcygeus originates from the obturator internus fascia and inserts into the anococcygeal body and coccygeal margins, partially overlapped by the posterior pubococcygeus. A research has revealed that levator ani injury appears in 55% of women with pelvic organ prolapse, compared to only 16% without prolapse, highlighting its crucial role in pelvic organ support (Dietz and Simpson 2008).

Puborectalis Muscle

The puborectalis muscle originates from the inferior region of the pubic symphysis and the superior fascia of the urogenital diaphragm. Its fibers extend laterally around the anorectal junction, uniting posterior to the rectum to create a muscular sling (Drake et al. 2020). The anatomical configuration known as the "anorectal ring" consists of the superior margins of both the internal anal sphincter and puborectalis muscle (Netter 2019).

The puborectalis muscle sling's contraction generates a horizontal force vector that results in two primary mechanical effects: closure of the pelvic diaphragm and reduction of the anorectal angle during squeeze maneuvers (Bharucha and Rao 2014). This biomechanical action represents the primary mechanism responsible for maintaining gross fecal continence (Raizada and Mittal 2008).

10.3.3.2 Coccygeus (Innervated by the Anterior Primary Rami of S4)

The coccygeus muscle presents as a flat, triangular configuration of muscular and fibrous tissue, positioned in an identical plane but posterior to the levator ani muscle. Its anatomical attachments extend from the ischial spine to the lateral borders of the coccyx and inferior portions of the sacrum (S4–S5) (Moore et al. 2018). On its gluteal aspect, the muscle's fibrous component forms the sacrospinous ligament (Standring 2020).

This muscle, receiving innervation from the anterior primary rami of S4, constitutes the posterior segment of the pelvic floor. The coccygeus functions synergistically with the levator ani to provide support for pelvic organs and sustain intra-abdominal pressure (Netter 2019). Additionally, it facilitates anterior coccygeal movement following posterior displacement during defecation or childbirth (Drake et al. 2020).

10.4 Neural Regulation of the Gastrointestinal System: From Central Integration to Peripheral Regulation

Neural control of gastrointestinal function involves a complex interaction between multiple nervous system components, where neural secretions–molecular messengers released by central and peripheral neurons–modulate diverse gut functions. These neural signals originate not only from extrinsic nerve supplies but also from the gut's own neurons and glial cells, creating a complex regulatory network (Grundy and Brookes 2012).

The gastrointestinal tract's nervous organization encompasses two major divisions: the extrinsic system, operating within the autonomic framework (including parasympathetic and sympathetic components) (Fig. 10.17), and the intrinsic system. Both systems function as integral elements of the peripheral nervous system, which establishes crucial connections between the central nervous system and body tissues through bundled axons and strategically positioned ganglia (Furness 2012).

Within this network, information flows bidirectionally through specialized neural pathways. Afferent (sensory) neurons transmit signals from gastrointestinal tissues to central processing centers, while efferent (motor) neurons convey regulatory commands from central regions to target organs. However, most neural pathways exhibit mixed characteristics, carrying both afferent and efferent signals simultaneously (Costa et al. 2000). These neural centers serve as sophisticated integration centers, capable of decoding incoming information and generating appropriate response signals via specific pathways.

The regulation of gastrointestinal functions, including food intake and body weight homeostasis, demonstrates remarkable complexity. The central nervous system integrates diverse information streams, incorporating signals from the peripheral nervous system alongside hormonal messages from multiple sources—the gut, liver, pancreas, and adipose tissue (Wood 2008). This comprehensive integration enables precise monitoring of luminal content characteristics and metabolite levels in the plasma, including glucose, lipids, and ketone bodies. Such processing proves essential for appropriate feeding behavior regulation and overall gastrointestinal homeostasis.

The enteric nervous system represents a neural network that extends throughout the gastrointestinal tract, from the esophagus to the anus, including associated organs such as the salivary glands, pancreas, and gallbladder. This intricate system consists of intrinsic neurons and glial cells that work in concert to regulate both gastrointestinal and extra-gut activities. While traditionally classified within the autonomic nervous system due to its involuntary nature, the enteric nervous system demonstrates unique characteristics that set it apart from other neural networks (Furness 2012).

One of the most intriguing aspects of this system lies in its ability to function independently while maintaining constant communication with the central nervous system. This communication flows bidirectionally, creating a dynamic relationship where the brain and spinal cord influence enteric neural activities, and conversely, the enteric nervous system shapes central nervous system functions. This autonomous yet integrated nature allows for precise control of digestive processes, even in the absence of external input from autonomic neurons (Costa et al. 2000).

The regulating activities of the enteric nervous system extend far beyond basic digestive functions. It analyzes and integrates inputs from several sources through complex neural reflexes, coordinating reactions that regulate motility,

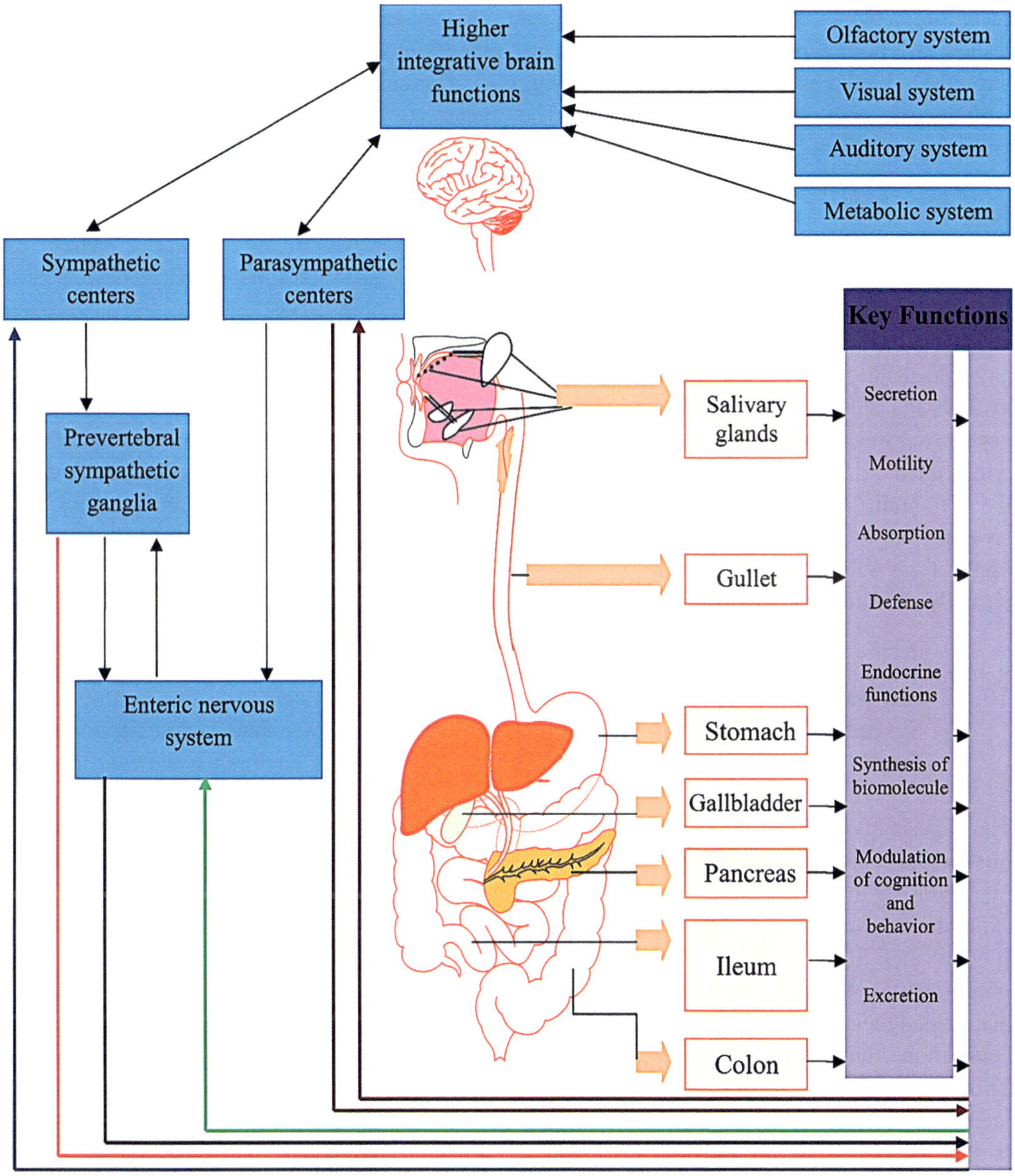

Fig. 10.17 Central schematic representation of the nervous system's influence on gastrointestinal (GI) function. Parasympathetic centers are located in the brainstem and sacral regions, with preganglionic efferent nerves, such as the vagus and pelvic nerves, projecting to the GI tract. The vagus nerves form synapses with intrinsic neurons in the esophagus, stomach, small intestine, colon, gallbladder, and pancreas, while the pelvic nerves connect with intrinsic neurons in the colon. These synaptic connections facilitate interaction between the extrinsic and intrinsic nervous systems, modulating secretion, motility, and absorption in the GI tract. Sympathetic centers are located in the thoracic and lumbar regions of the spinal cord (preganglionic ganglia). (Adapted with permission from Welcome (2018))

secretions, blood flow, and immunological responses. This comprehensive regulation occurs through a hierarchical organization involving both central and peripheral components. The brain and spinal cord serve as central regulatory centers, whereas peripheral control involves a complex network of ganglia, including prevertebral and paravertebral sympathetic ganglia, as well as enteric ganglia (Wood 2008).

Research has revealed further aspects of the enteric nervous system's influence on human physiology. In addition to its traditional role in digestion, this neural network participates in nutrient sensing, metabolic regulation, and endocrine function. Furthermore, its involvement in the gut–brain axis shows its significance in broader physiological processes and potential implications for various health conditions (Grundy and Brookes 2012).

10.4.1 Intrinsic (Enteric) Nervous System of the Gastrointestinal Tract

The genesis of our understanding of gastrointestinal neural control traces back to a groundbreaking discovery in 1902. Through their innovative experiments on canine intestinal segments, Bayliss and Starling observed a remarkable phenomenon: isolated sections of intestine maintained their motility patterns even after disconnection from central nervous system input. This observation, initially termed "the law of the intestine" and later known as the peristaltic reflex, laid the foundation for the field of neurogastroenterology (Bayliss and Starling 1902).

This autonomous neural network, known as the enteric nervous system or intrinsic nervous system of the gut, demonstrates remarkable complexity and independence. Within the gastrointestinal tract walls, from esophagus to colon, exists an intricate arrangement of enteric glial cells and neuronal cell bodies organized into ganglia. These structures form sophisticated plexuses through interconnected bundles of nerve processes, functioning without direct control from the brain or spinal cord, and operating independently of vagal or prevertebral ganglionic input (Furness 2012).

The human gut wall has around 100 million neuronal cells. These neurons communicate through neurotransmitter secretions, mimicking central nervous system functionality while maintaining operational autonomy (Costa et al. 2000). This extensive neural network orchestrates numerous gastrointestinal functions beyond basic motility control. The enteric nervous system modulates exocrine secretions from mucosal and submucosal tissues, regulates endocrine function, and manages absorption processes. Additionally, it maintains precise control over local microcirculation, coordinates immune responses, and modulates inflammatory processes within the gastrointestinal environment (Wood 2008).

The complexity of this "second brain" includes its sensory functions and its regulatory mechanisms. Through intricate neural pathways, the enteric nervous system processes gastrointestinal sensations and modulates tissue development, demonstrating its comprehensive role in maintaining digestive system homeostasis (Grundy and Brookes 2012). This multifaceted control system emphasizes the evolutionary significance of autonomous gastrointestinal regulation and its importance in maintaining overall physiological balance.

The parallels between the brain and the enteric nervous system extend beyond fundamental functional similarities, highlighting structural and organizational complexity within the gut's neural network. Embedded within the gut walls is a neural architecture that resembles the central nervous system, comprising ganglia, dense fiber networks, and a diverse array of neuronal populations. This complex arrangement supports the enteric nervous system's categorization as the "second brain" or "gut-brain," terms that indicate its significant autonomous and capacity for independent function (Furness 2012).

Similar to the central nervous system, the enteric nervous system demonstrates a high degree of self-sufficiency via pacemaker cells capable of generating spontaneous impulses independent of external stimuli. This autonomous function is supported by specialized neurons, including sensory neurons that detect changes in the local environment, interneurons that process sensory information, and motor neurons that regulate muscular contractions and secretory activity. Together, these neurons form an extensive network, with primary interganglionic tracts that branch into secondary and tertiary fibers. This network influences various effector systems—muscle layers, glandular secretions, blood vessels, and immune cells—enabling the enteric nervous system to independently modulate digestive processes (Costa et al. 2000).

The GI tract contains an intrinsic neural network—the enteric nervous system—which maintains autonomous control over digestive functions without external neural input (Furness 2012). This system incorporates two major neuronal networks: the myenteric plexus (of Auerbach) and the submucosal plexus (of Meissner), as shown in Fig. 10.18 (Costa et al. 2000). Located between longitudinal and circular muscle layers throughout the alimentary canal, the myenteric plexus directs propulsive activity. The submucosal plexus, positioned within the submucosa, modulates secretory functions and smooth muscle activity (Sanders et al. 2006).

Specialized pacemaker cells within these plexuses generate baseline electrical activity necessary for peristaltic movement (Huizinga and Lammers 2009). Multiple factors affect this electrical rhythm: mechanical stretch of the intestinal wall, endocrine mediators, and extrinsic innervation signals. This neural organization

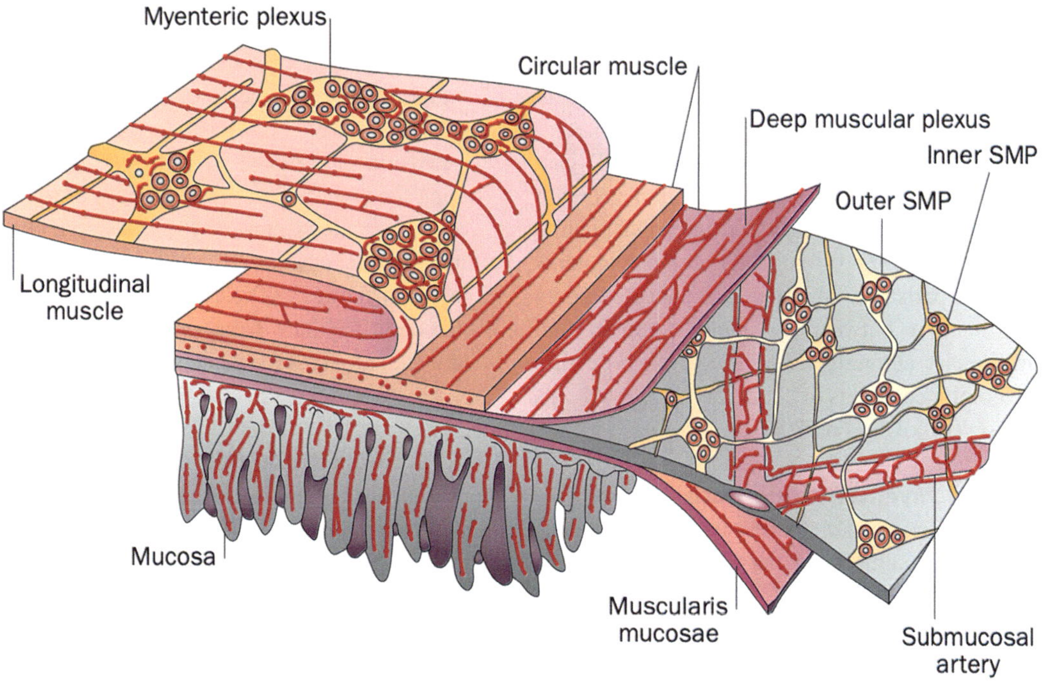

Fig. 10.18 Organization of the enteric nervous system (ENS). This diagram depicts the ENS in the small intestine of humans and medium- to large-sized mammals. The ENS includes two ganglionated plexuses: the myenteric plexus, located between the longitudinal and circular muscle layers, and the submucosal plexus (SMP), which has outer and inner components. Nerve fiber bundles connect the ganglia and form plexuses that innervate the longitudinal muscle, circular muscle, muscularis mucosae, intrinsic arteries, and mucosa. Axons of extrinsic origin also travel within these nerve fiber bundles. Innervation of gastroenteropancreatic (GEP) endocrine cells and gut-associated lymphoid tissue (GALT), not shown here, is also part of the ENS. (Adapted with permission from Lyte and Cryan (2014))

demonstrates the advanced capabilities of the enteric nervous system in maintaining digestive processes (Spencer et al. 2016).

The motor patterns of the GI tract show specific regional variations. Gastric contractility occurs at 3–5 cycles per minute, while duodenal activity presents at 10–12 cycles per minute. The pyloric sphincter acts as a regulatory point between these motor patterns (Sanders et al. 2006). Small intestinal motility progresses through three phases, characterized by slow waves every 5 s with superimposed contractions (Huizinga and Chen 2014). Both intrinsic and extrinsic neural pathways modify this motor pattern, along with various neuroactive substances. These include 5-hydroxytryptamine (serotonin), which affects intestinal motor function; endogenous opioids, which influence nociception and GI activity; and histamine, which modulates immune responses and acid production. Colonic motor activity exhibits irregular contractions regulated by the autonomic nervous system, which determines the frequency and amplitude of peristalsis. Several peptide mediators influence colonic motility: substance P, gastrin, enkephalin, and cholecystokinin increase motor activity, while vasoactive intestinal peptide (VIP) reduces it. Although the colon maintains inherent peristaltic abilities, defecation requires coordination between voluntary somatic and involuntary autonomic neural control, reflecting the combined effects of neural and hormonal regulation (Brookes et al. 2013).

10.4.1.1 Myenteric (Auerbach) Plexus

The myenteric plexus represents a complex neural network comprising nerve strands and ganglia, positioned between the longitudinal and circular muscle layers of the intestinal wall. This network maintains continuity both circumferentially and longitudinally throughout the gastrointestinal tract. While myenteric ganglia exhibit variations in morphology and orientation across species and intestinal regions, each major anatomical segment displays characteristic patterns (Wilson et al. 1981).

Ganglia, which function as nodal points within the plexus, connect through internodal strands or interganglionic connectives. In the guinea pig ileum, ganglion size ranges from 5 to over 200 neuronal cell bodies, with occasional solitary neurons located adjacent to nerve strands outside the primary network (Llewellyn Smith et al. 1985). The mechanical properties of these ganglia allow for deformation during intestinal movements, necessitating consideration of the intestinal wall's contractile state when analyzing their morphometric parameters.

The myenteric plexus comprises three anatomically and functionally integrated components (Fig. 10.19):

- Primary plexus: Consists of ganglia and internodal strands forming the main meshwork. Many nerve fibers bypass ganglia, typically coursing between the ganglion and longitudinal muscle before continuing within another internodal strand.
- Secondary plexus: Features finer nerve bundles that branch from primary internodal strands or ganglia. These components run parallel to circular muscle bundles, frequently intersecting internodal strands and extending along the inner aspect of the primary plexus.
- Tertiary plexus: Composed of fine nerve bundles that occupy spaces within the primary plexus meshwork (Richardson 1958). These elements originate from primary internodal strands, ganglia, and secondary strands.

10.4.1.2 Submucosal (Meissner) Plexus

The submucosal ganglionated plexus, first documented in the mid-nineteenth century by Meissner (1857) and Billroth (1857), represents a specialized neural network primarily found in the small and large intestines (Fig. 10.20). While the esophagus and stomach contain scattered submucosal ganglia, these structures do not develop into an organized plexus comparable to that observed in the intestines (Timmermans et al. 1997).

The anatomical characteristics of the submucosal plexus differ notably from those of the myenteric plexus, featuring more delicate interconnecting strands and more compact ganglia. This neural network maintains continuity both

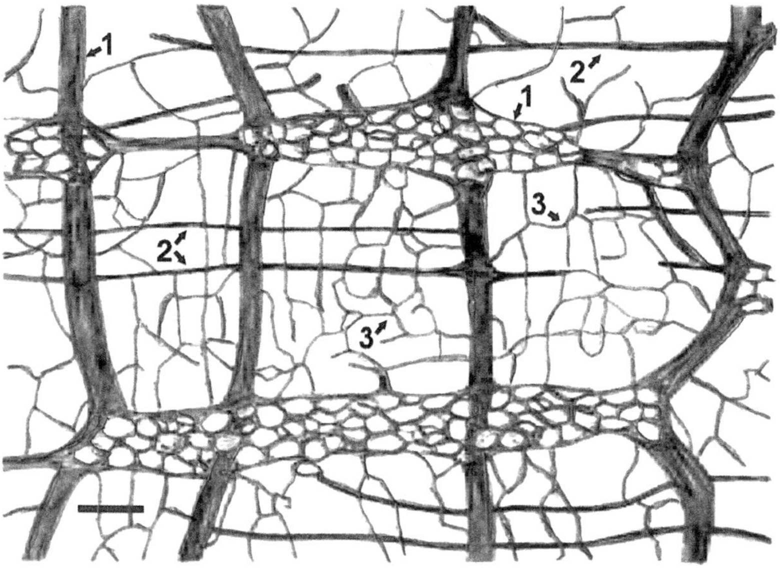

Fig. 10.19 Components of the myenteric plexus, illustrated using a wholemount from the guinea pig small intestine. (1) The primary component, common to all species, consists of ganglia and internodal strands (interganglionic connectives). (2) The secondary component includes nerve strands running parallel to the circular muscle. (3) The tertiary plexus, present only in areas where the longitudinal muscle is thin, contains sparse nerve fibers within the longitudinal layer. Neuron cell bodies are shown as white ovals within the ganglia. Calibration: 100 μm. (Adapted from Furness (2006))

around the circumference and along the length of the intestinal tract, forming an integrated system for local control of intestinal function.

The outer submucosal plexus, positioned closer to the circular muscle layer, contains neurons that extend their influence beyond the immediate submucosal region. These neurons provide innervation not only to the circular muscle but also reach the longitudinal muscle layer, as demonstrated through detailed anatomical studies (Furness et al. 2014; Sanders and Smith 1986). Additionally, this outer plexus maintains substantial projections to the mucosa, indicating its multifaceted role in coordinating intestinal function.

In contrast, the inner submucosal plexus, situated closer to the intestinal lumen, shows a more focused pattern of innervation. While it contains relatively few neurons that innervate the muscle layers, it demonstrates extensive connections with the mucosa. This arrangement suggests a specialized role in regulating mucosal function, possibly related to secretion and absorption.

The organization becomes notably simpler in small mammals, exemplified by studies in guinea pigs. These species typically display a single layer of submucosal ganglia, which predominantly contains secretomotor neurons rather than muscle-innervating motor neurons. Research has identified four primary types of neurons within these ganglia in the small intestine (Furness et al. 1984). Interestingly, this simpler arrangement bears a strong resemblance to the inner submucosal plexus observed in larger species, suggesting possible evolutionary relationships.

10.4.2 The Extrinsic Nervous System of the Gastrointestinal Tract

The GI tract receives extensive innervation from sources external to its walls through the extrinsic nervous system. This system mediates bidirectional communication between the central nervous system and digestive organs via the

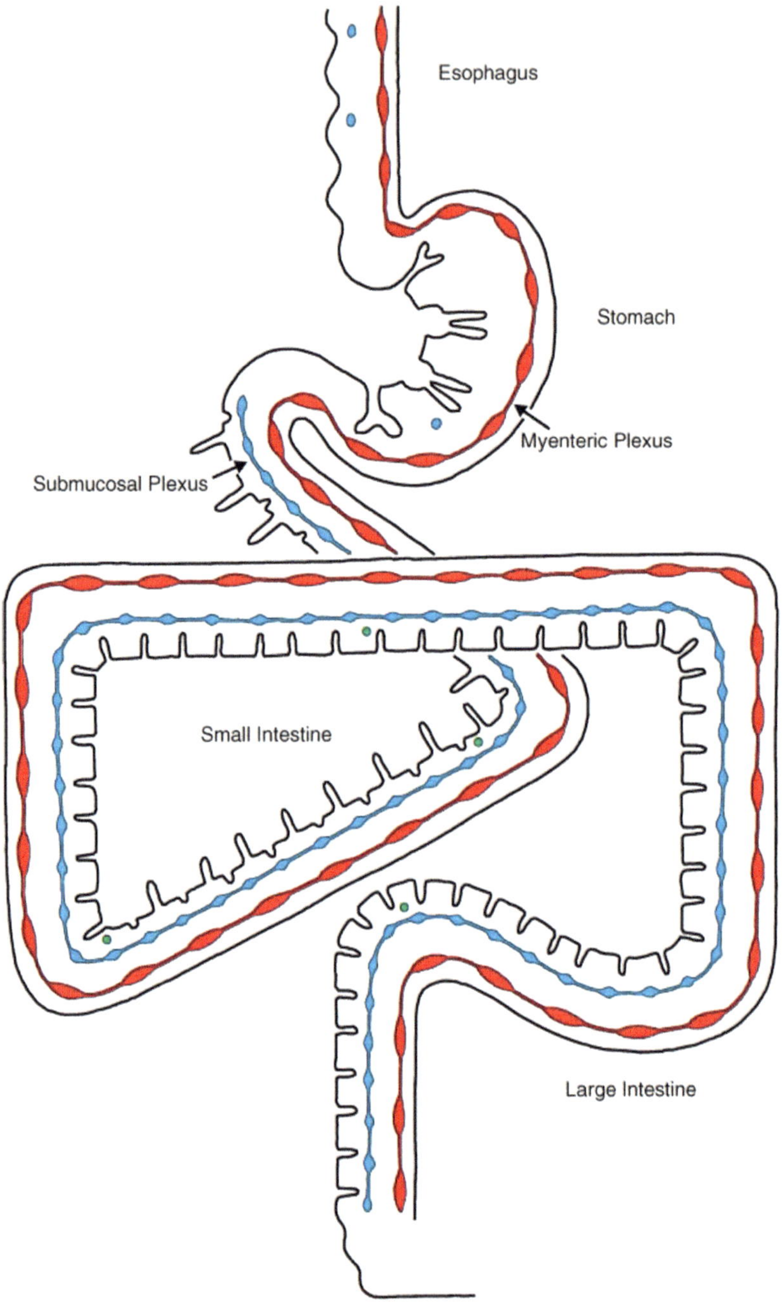

Fig. 10.20 Distribution of enteric ganglia in the tubular digestive tract. A schematic longitudinal section illustrates the myenteric ganglia, forming a continuous plexus from the upper esophagus to the internal anal sphincter, and the submucosal plexus, which is most prominent in the small and large intestines. Isolated ganglia are present in the gastric and esophageal submucosa as well as within the mucosa throughout the digestive tract. (Adapted from Furness (2006))

autonomic nervous system's parasympathetic and sympathetic divisions (Furness et al. 2014) (Fig. 10.21). The extrinsic innervation provides regulatory oversight of GI functions while allowing for considerable local autonomy through the enteric nervous system.

The parasympathetic division primarily promotes digestive functions through vagal and sacral pathways. The vagus nerve supplies the upper GI tract from the esophagus to the proximal two-thirds of the transverse colon, while sacral parasympathetic nerves innervate the distal colon

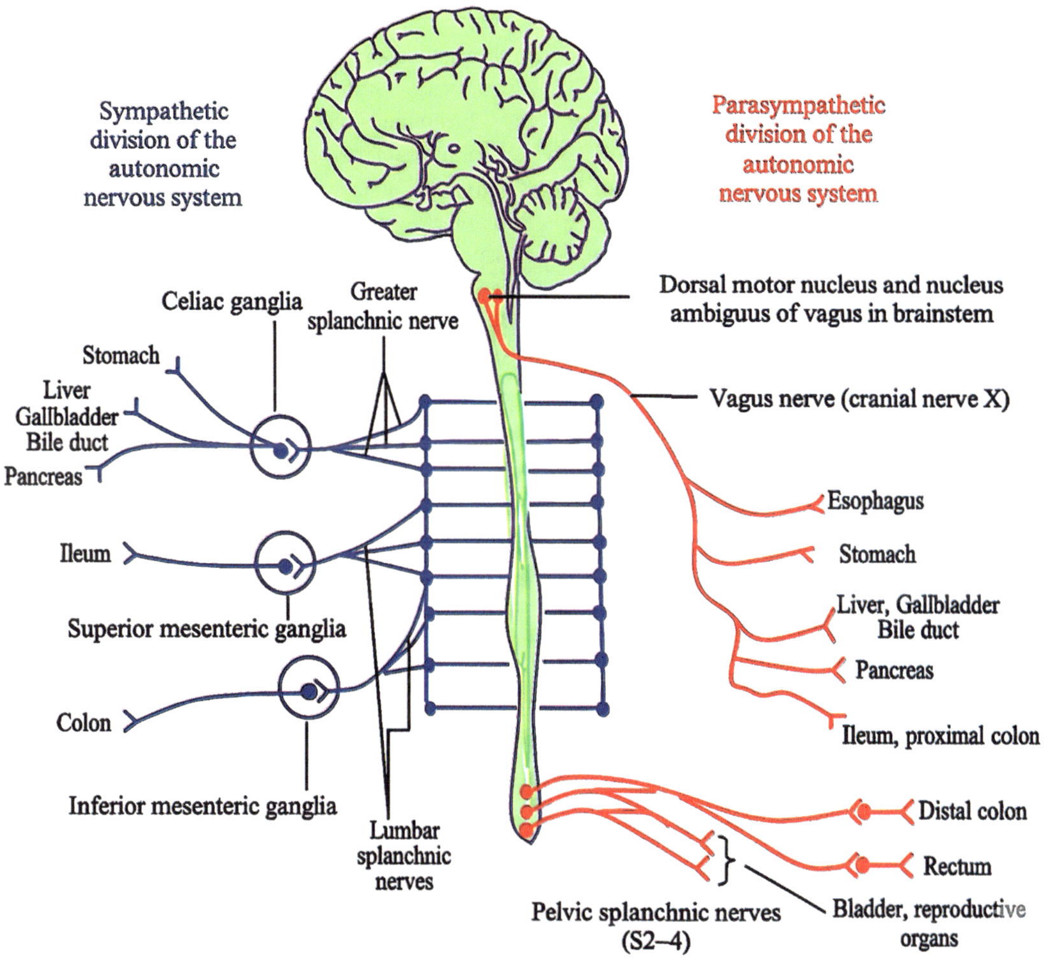

Fig. 10.21 Simplified schematic representation of the sympathetic and parasympathetic innervation of the enteric nervous system (ENS). Gut activity is regulated by both divisions of the autonomic nervous system. Parasympathetic preganglionic neurons originate from the dorsal motor nucleus of the vagus nerve in the brainstem and the intermediate gray zone in the sacral spinal cord. The vagus nerve innervates the esophagus, stomach, liver, pancreas, gallbladder, small intestine, and proximal colon, while the transverse colon, ileocecal sphincter, and ascending colon are innervated by the cruciform parasympathetic nucleus of the spinal cord. The distal rectum (colorectum) is innervated by pelvic plexuses, which also supply the bladder and reproductive organs. Sympathetic preganglionic neurons originate from the thoracolumbar spinal cord and modulate gut activity through the celiac, superior, and inferior mesenteric ganglia. (Adapted with permission from Welcome (2018))

and rectum (Browning and Travagli 2014). Parasympathetic preganglionic neurons synapse with postganglionic neurons in the enteric plexuses, forming an integrated network that modulates motility, secretion, and blood flow (Powley et al. 2019).

Sympathetic innervation originates from thoracolumbar segments of the spinal cord, with preganglionic fibers synapsing in prevertebral ganglia—including the celiac, superior mesenteric, and inferior mesenteric ganglia. Postganglionic sympathetic neurons generally reduce GI motility and secretion while controlling blood flow through vasoconstriction (Furness 2012). This system becomes particularly active during stress responses or when energy conservation is required.

The integration of these extrinsic inputs with the intrinsic enteric nervous system creates a complex control network. This arrangement allows for both precise local control of digestive functions and appropriate responses to broader physiological and environmental demands (Wood et al. 1999).

10.4.2.1 Central Processing of Gastrointestinal Visceral Signals

The human body maintains an intricate communication network between the GI system and the brain, facilitating a continuous flow of information that influences both digestive function and behavioral responses. This complex system relies on multiple neural pathways, with the vagus nerve serving as the primary communication conduit. As the most extensive sensory pathway in the human body, the vagus nerve transmits around 80% of afferent signals, with only 20% allocated for efferent communication (Browning and Travagli 2014).

The journey of visceral signals from the gut to the brain follows several routes. While some signals travel through the spinal cord via dorsal root ganglia, others take a more direct path through cranial nerves VII, IX, and X. Upon entering the central nervous system, these signals converge at an important processing area in the dorsomedial medulla—the nucleus tractus solitarius (NTS) (Fig. 10.22). This neural hub acts as a complex integration center, processing incoming information and coordinating appropriate physiological responses (Bonaz et al. 2018).

From the NTS, processed information flows to various brain regions, including the hypothalamus, amygdala, and thalamus, creating a complex network of communication that influences multiple aspects of bodily function. This distributed processing allows the brain to interpret and respond to gastrointestinal signals in ways that affect not only digestive functions but also emotional states and behavioral responses (Browning and Travagli 2014).

The vagus nerve's influence extends far beyond basic digestive control. Research has revealed its remarkable role in modulating mood, anxiety, and feeding behavior. Following food intake, vagal signaling triggers behavioral responses associated with satiety and sedation. Moreover, these neural pathways can influence emotional states through region-specific changes in neurotransmitter levels, particularly noradrenaline and GABA in the limbic system. Interestingly, these emotional effects occur independently of the hypothalamic-pituitary-adrenal axis, suggesting a direct pathway for visceral modulation of emotional states (Forsythe et al. 2014).

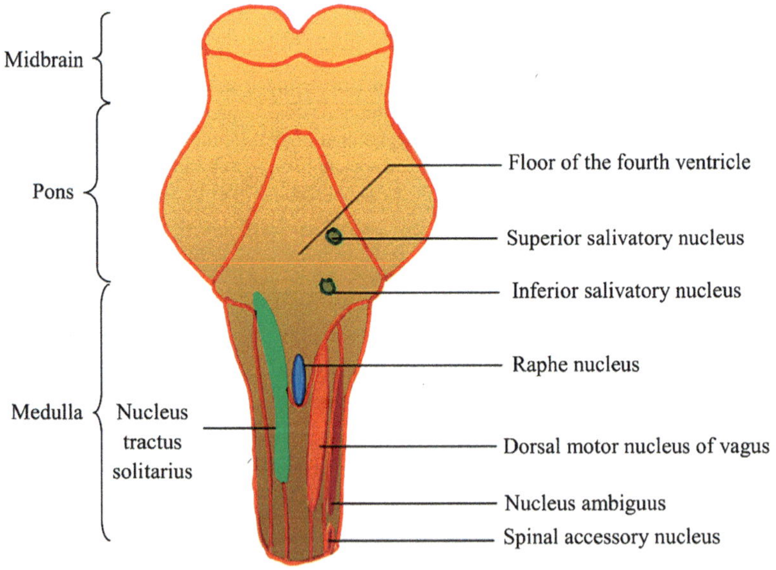

Fig. 10.22 Structure of the brainstem and the location of nuclei involved in gastrointestinal (GI) activity regulation. The raphe nuclei within the reticular formation play a significant role in modulating GI functions. These nuclei are a primary site of serotonin synthesis, a neurotransmitter crucial for various functions in the central and peripheral nervous systems, including mood regulation. (Adapted with permission from Welcome (2018))

The vagus nerve's control over feeding behavior operates through multiple mechanisms, including the regulation of gastric compliance, accommodation, and emptying. This neural pathway also influences pancreatic secretion and can trigger protective reflexes such as emesis when necessary. Beyond these digestive functions, vagal signaling plays a crucial role in modulating inflammation and pain perception, demonstrating the extensive reach of this neural network in maintaining bodily homeostasis (Yuan and Silberstein 2016).

Understanding these complex neural pathways has significant clinical implications. The vagus nerve's extensive influence on both physiological and psychological functions makes it a promising target for therapeutic interventions in various conditions, from inflammatory bowel disease to mood disorders. As research continues to unveil the intricacies of this gut–brain communication network, new therapeutic possibilities emerge for treating both gastrointestinal and neurological conditions (Bonaz et al. 2017).

Central Processing of Gastrointestinal Nociception

Gastrointestinal pain represents a progression that originates in the gut wall and culminates in conscious perception. When discomfort or pain arises from the gastrointestinal tract, specialized mechanoreceptors stationed next to sensory neurons first detect these signals. These primary sensory neurons, with their cell bodies located in dorsal root ganglia, act as the initial gatekeepers of pain input (Grundy and Brookes 2012).

The path that pain signals follow from the gut to the brain proves more complex than initially thought. While traditionally associated with the anterolateral pathway, researchers have discovered that pain information travels through both the anterolateral and dorsal column pathways. The dorsal column pathway may surpass the classical spinothalamic tract in several key aspects: it more effectively activates thalamic neurons, produces stronger behavioral responses to pain, and generates more increases in cerebral blood flow (Vermeulen et al. 2014).

The anterolateral system has various specialized neuronal pathways that transmit different elements of the pain sensation. The spinothalamic tract enables conscious pain perception, whereas its associated pathways—spinoreticular, spinomesencephalic, and spinohypothalamic—coordinate unconscious responses, including emotional reactions and behavioral responses to gastrointestinal discomfort (Brookes et al. 2013).

Inflammatory conditions can dramatically alter how pain signals travel and are interpreted, leading to unusual pain experiences known as visceral allodynia and hyperalgesia. In visceral allodynia, patients lose their normal ability to perceive pain appropriately, while hyperalgesia causes them to experience heightened sensitivity to painful stimuli. These disruptions commonly occur in conditions like inflammatory bowel disease, where multiple factors converge to alter pain processing–from changes in peripheral nerve sensitivity to modifications in how the spinal cord processes pain signals. Even our thoughts and emotions can influence how these signals are ultimately interpreted (Vanner et al. 2016).

The final part in this pain processing occurs in the somatosensory cortex, an intricate area of the brain that integrates information from various sensory receptors throughout the body. This cortical area, carefully organized into primary (Brodmann areas 1, 2 and 3) and secondary (Brodmann areas 5 and 40) regions, doesn't work in isolation but coordinates with other brain areas to create our complete experience of gastrointestinal sensations (Mayer and Tillisch 2011).

Descending Neural Control of Gastrointestinal Function

The brain's control over gastrointestinal function is a complex interaction of neural pathways from many brain regions, coordinating various aspects of digestive function. Through interconnected pathways, the brain monitors visceral activities through interconnected pathways, with the limbic system significantly influencing gastrointestinal functions via specialized tracts (Browning and Travagli 2014).

This descending control system originates from several key brain regions, including the hypothalamic periventricular gray, midbrain periaqueductal gray, nucleus raphe, and locus coeruleus. Higher brain centers, including the anterior cingulate and somatosensory cortices, also contribute to this control network. These pathways ultimately converge on the anterior column of the spinal cord, where they influence motor output to the gut (Bonaz et al. 2016).

The motor control system divides into two main categories: pyramidal and extrapyramidal tracts. The pyramidal system, comprising corticospinal and corticobulbar tracts, provides direct control through the medullary pyramids. In contrast, the extrapyramidal system takes a more indirect route, bypassing the pyramids to modulate anterior horn neurons. This dual system allows for both precise motor control through the pyramidal system and broader, coordinated movements via the extrapyramidal pathways (Brookes et al. 2013).

The periaqueductal gray emerges as a crucial control center for descending pain modulation. Located in the midbrain's tegmental core, it works in concert with the nucleus raphe to influence autonomic responses. A particularly significant inhibitory circuit involves the periaqueductal gray–rostral ventromedial medulla–dorsal horn pathway, which plays a vital role in pain processing (Vanner et al. 2016).

The locus coeruleus, positioned in the rostral pons, serves as the brain's primary source of norepinephrine and significantly influences gastrointestinal motor function. As demonstrated by Tsuruoka et al. (2010), this nucleus can modulate visceral motor responses through its widespread projections throughout the nervous system. Its influence extends through both excitatory and inhibitory pathways, mediated by different adrenergic receptor subtypes.

The vagus nerve provides the final common pathway for much of this descending control. Efferent fibers from the dorsal vagal nucleus and nucleus ambiguus carry processed information from higher brain centers to regulate various aspects of gastrointestinal function, including secretion, motility, and neuroendocrine activities. These signals ultimately influence multiple effector organs, including blood vessels, smooth muscles, enteric neurons, and gut-associated lymphoid tissues (Powley et al. 2019).

Neural Control of Autonomic Function in the Gastrointestinal System

Autonomic reflexes mostly originate in the spinal cord, although higher centers such as the medulla oblongata, pontine nuclei, hypothalamic regions, and some cerebral cortex areas offer conscious control of these responses (Furness et al. 2014). The gastrointestinal tract is innervated by both sympathetic and parasympathetic divisions, which cooperatively regulate digestive functions and peristaltic movement of food contents (Furness et al. 2014).

Sympathetic Modulation of Gastrointestinal Function

The sympathetic division of the autonomic nervous system modifies digestive activity through inhibitory effects on gastrointestinal motility and secretion. This inhibition occurs through direct neural mechanisms and indirect effects via reduced blood flow to the gastrointestinal tract (Standring 2020). The sympathetic preganglionic neurons originate in the intermediolateral cell column of the lower thoracic (T5–T12) and upper lumbar (L1–L2) spinal cord segments, constituting the thoracolumbar outflow (Drake et al. 2020). These neurons project their axons through the anterior roots into the peripheral nervous system. The preganglionic fibers leave the peripheral nerves via white (myelinated) rami communicantes to synapse in the paravertebral ganglia of the sympathetic chain (Fig. 10.23). These fibers follow three potential pathways: they may form synapses with postganglionic neurons at the same level, ascend or descend to different levels before synapsing, or transit through the chain to terminate in prevertebral ganglia proximal to visceral organs, such as the celiac and superior mesenteric ganglia, or in terminal ganglia, including the inferior mesenteric plexus (*see* Fig. 10.21).

The sympathetic paravertebral ganglia comprise 22 bilateral pairs adjacent to the vertebral column, forming two interconnected chains that

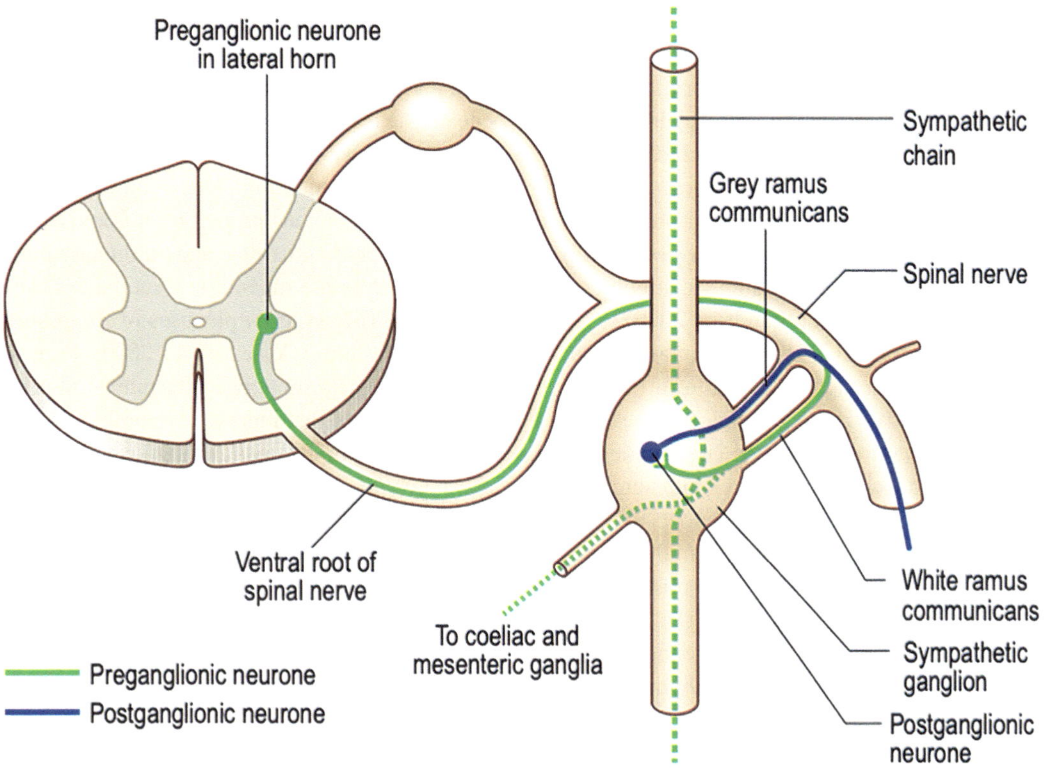

Fig. 10.23 Pathways of preganglionic sympathetic neurons originating in the lateral horn of the spinal cord. Their axons may synapse with postganglionic neurons in a sympathetic ganglion at the same spinal level, ascend or descend within the sympathetic chain to synapse in rostral or caudal ganglia, or project to prevertebral ganglia within autonomic plexuses of the abdominal cavity, such as the coeliac ganglion. (Adapted with permission from Standring (2020))

extend from the cranial base to the coccygeal region (Furness et al. 2014). These ganglia exhibit specific functional distributions: the superior cervical ganglion regulates cephalic glandular and smooth muscle activity; the middle cervical and stellate ganglia supply the esophagus; while the thoracic (10–12 pairs), lumbar (4 pairs), and sacral (4–5 pairs) ganglia provide innervation to the stomach, intestinal tract, and colonic segments. The postganglionic fibers exit the ganglia through gray (unmyelinated) rami communicantes to rejoin spinal nerves or form separate nerve branches (Moore et al. 2018).

The prevertebral sympathetic ganglia, situated near principal arterial vessels, coordinate sympathetic input to abdominal viscera. The celiac ganglion, positioned at the origin of the celiac artery, and the superior mesenteric ganglion, located near the superior mesenteric artery, regulate sympathetic output to the stomach, small intestine, hepatobiliary system, kidneys, and splenic tissue. These ganglia receive preganglionic fibers primarily from T5 to T12 spinal segments (Standring 2020). The terminal ganglionic structures, including the inferior mesenteric and hypogastric plexuses, are positioned adjacent to their target organs, with the inferior mesenteric plexus specifically mediating sympathetic regulation of colonic function through fibers originating from L1 to L2 segments.

The postganglionic sympathetic fibers primarily release norepinephrine, which interacts with α and β adrenergic receptors on target tissues. This neurotransmitter generally reduces gastrointestinal blood flow, decreases motility, and inhibits secretion, preparing the body for "fight or flight" responses by diverting resources away from digestive functions (Barrett et al. 2019).

Parasympathetic Innervation and Gastrointestinal Motility

The parasympathetic division originates from two main regions: the brainstem (yielding cranial nerves III, VII, IX, and X) and the sacral spinal cord (segments S2–S4). Its anatomical configuration exhibits characteristic long preganglionic and short postganglionic fibers, a structural pattern that differentiates it from sympathetic organization (Furness 2012; Standring 2020).

The craniosacral division maintains homeostatic balance through innervation of oral structures, digestive glands, and the complete gastrointestinal tract (Furness 2012). Integration centers reside within specific brainstem nuclei and sacral spinal cord segments (S2–S4). Preganglionic axons emerging from these centers form synaptic connections with postganglionic neurons that supply target tissues (Browning and Travagli 2014). Parasympathetic control of salivary glands occurs through postganglionic nerves originating from two primary ganglia. The otic ganglion supplies the parotid gland, while the submandibular ganglion provides innervation to the submandibular and sublingual glands (via cranial nerves VII or IX) (Furness 2012; Costa et al. 2000). Minor salivary glands receive postganglionic parasympathetic fibers from cranial nerve VII. The lingual nerve, a branch of the mandibular division of the trigeminal nerve (cranial nerve V), innervates various oral structures. This nerve transmits sensory information from the tongue to the brain and collects taste signals from the anterior two-thirds of the tongue, which then pass through the chorda tympani—a facial nerve branch (Browning and Travagli 2014). Parasympathetic ganglia, excluding those of the vagus nerve, reside in the head where synaptic connections between preganglionic and postganglionic neurons occur (Spencer and Hu 2020). Cranial nerve VII emerges from the pons and synapses at the submandibular ganglia, generating postganglionic fibers that supply the parotid gland. The submandibular ganglion lies lateral to the hyoglossus muscle, inferior to the lingual nerve (Grundy and Brookes 2012). Cranial nerve IX's preganglionic component originates in the medulla oblongata and projects to the otic ganglia, which emit postganglionic fibers to the submandibular and sublingual salivary glands. The otic ganglion sits medial to the mandibular nerve, beneath the foramen ovale.

The vagus nerve (CN X), with cell bodies situated in its dorsal motor nucleus, requires particular attention. Its axons project to multiple neural networks, including cardiac and pulmonary plexuses, as well as the myenteric (Auerbach's) and submucosal (Meissner's) plexuses of the gastrointestinal system (Costa et al. 2000; Furness 2012). These neural pathways terminate in postganglionic fibers that innervate smooth musculature and secretory glands throughout the upper gastrointestinal tract, extending to the splenic flexure of the colon (Standring 2020) (Fig. 10.24).

Within the gastrointestinal system, the vagus nerve stands as the primary source of parasympathetic innervation, establishing a two-way communication network between the brain and gut through its afferent and efferent components (Bonaz et al. 2017). Throughout the gastrointestinal tract, afferent nerve fibers form connections with various sensory receptors—mechanoreceptors, osmoreceptors, and chemoreceptors—which begin the process of signal transduction. These receptors reside on epithelial cells, smooth muscle cells, and other gastrointestinal tissues, creating an extensive sensory network (Browning and Travagli 2014). The elaborate network of specialized receptors continuously monitors and transmits information about the gastrointestinal environment to the dorsal vagal complex in the medulla oblongata through vagal afferent pathways. Simultaneously, additional sensory information travels to the central nervous system via lumbosacral spinal routes. This comprehensive sensory input includes monitoring food presence, volume, and chemical composition, while also detecting various sensations such as fullness, bloating, nausea, urgency, discomfort, and pain.

The sacral parasympathetic outflow (S2–S4), comprising the nervi erigentes, courses through pelvic nerves to reach ganglia embedded within the walls of the descending colon, rectum, and urinary bladder (Grundy and Brookes 2012). These neural elements form synapses with postganglionic neurons that regulate smooth muscle

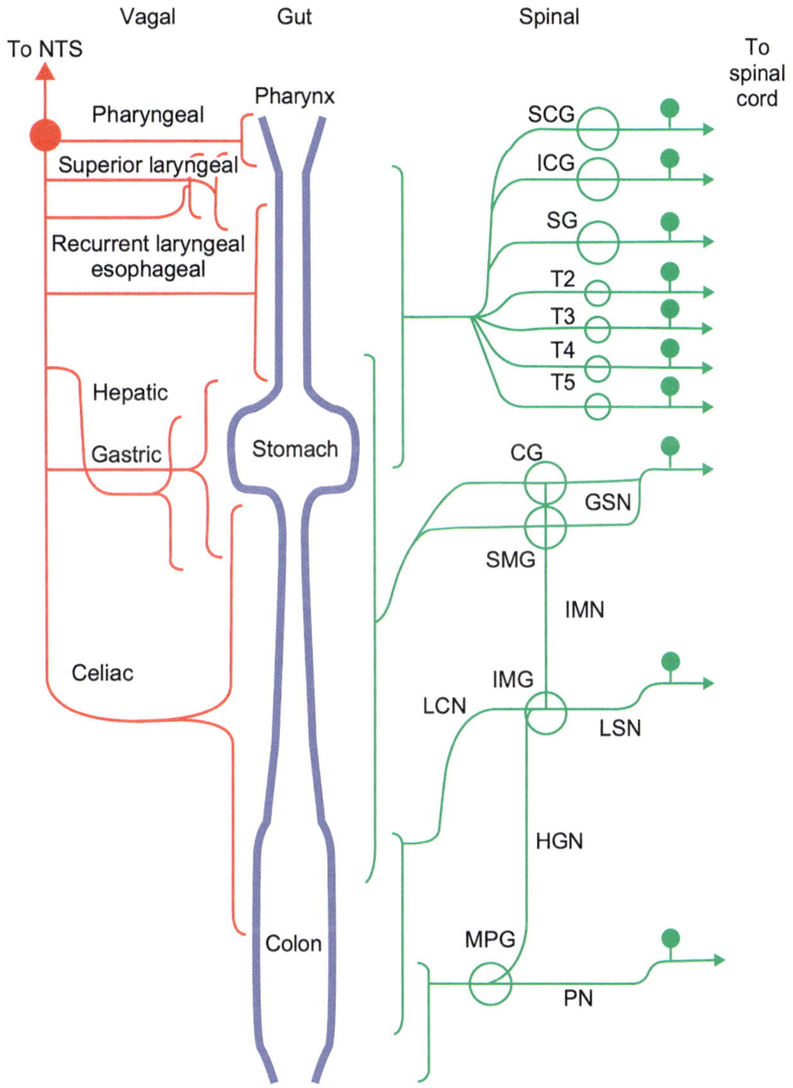

Fig. 10.24 Schematic representation of vagal and spinal innervation of the gastrointestinal (GI) tract. Abbreviations: CG celiac ganglia, GSN greater splanchnic nerve, HGN hypogastric nerve, ICG intermediate cervical ganglia, IMG inferior mesenteric ganglia, IMN intermesenteric nerve, LCN lumbar colonic nerves, LSN lumbar splanchnic nerve, MPG major pelvic ganglia, NTS nuclei tractus solitarii, PN pelvic nerve, SCG superior cervical ganglia, SG stellate ganglia, SMG superior mesenteric ganglia. (Adapted with permission from Johnson (2012))

function in these regions. Parasympathetic activation induces relaxation of the internal anal sphincter while promoting motility in the sigmoid colon and rectum (Costa et al. 2000; Spencer and Hu 2020). Pelvic afferent fibers penetrate multiple tissue layers—serosa, muscle, and mucosa—where they detect and transmit various stimuli, including nociception, mechanical distention from fecal matter, urgency sensations, and defecatory signals to regulatory centers (Grundy and Brookes 2012).

The descending colon and rectum receive innervation from pelvic splanchnic nerves originating from sacral spinal cord segments S2–S4 (Spencer and Hu 2020). These parasympathetic nerves conduct various mechanosensory and chemosensory signals from the colon to the spinal cord (Grundy and Brookes 2012). Pelvic nerve function maintains normal colorectal motility, and its dysfunction often results in fecal incontinence or constipation. These conditions commonly occur as sequelae of hysterectomy, colorectal surgery, or childbirth (Brierley et al. 2018; Spencer and Hu 2020).

When food enters the gastrointestinal tract, mechanoreceptors in smooth muscles detect these mechanical changes, triggering vagal afferents to relay signals to the central nervous system.

The brain processes this information and responds with appropriate motor commands (Powley et al. 2016). During normal peristaltic contractions, tension receptors become particularly active as distention levels increase (Spencer and Hu 2020).

10.4.3 Neural Innervation of the Specific Regions of the Gastrointestinal System

10.4.3.1 Neural Control and Motility of the Large Intestine

The large intestine begins at the cecum, continues through the colon, and concludes at the rectoanal region. The neural control of these anatomical segments involves an intricate network of regulatory systems: the enteric nervous system, autonomic pathways, and visceral sensory fibers (Brookes et al. 2013).

The physiology of the large intestine centers on three fundamental processes that maintain systemic homeostasis: the active absorption of sodium and water from the intestinal lumen, the coordinated movement of intestinal contents in an antegrade direction, and the regulated storage of fecal material (Gudsoorkar and Quigley 2014). The motor function of the colon shows distinct circadian patterns, with peak activity during daytime hours. This activity pattern receives modulation from neural mechanisms, including sympathetic activation upon awakening and the gastrocolic reflex following meals. Two primary contractile patterns characterize colonic motor function: prolonged tonic contractions that establish baseline muscle tone, and transient phasic contractions that direct content movement (Sanders and Smith 1986).

Colonic motility exhibits two forms of phasic contractions based on their propagation characteristics. The non-propagating type creates localized mixing through haustral compartmentalization, while propagating sequences produce bidirectional movement. In normal conditions, antegrade propagation predominates in distal colonic regions, regulating transit time. Conversely, proximal segments show increased frequency of retrograde sequences, enhancing absorption by prolonging contact between luminal contents and the mucosal surface.

High-amplitude propagated contractions represent distinct motor events occurring 4–6 times per day in healthy individuals. These contractions begin at various colonic locations, most commonly in proximal regions, and progress distally for variable distances. Manometric recordings demonstrate temporal correlation between these contractions and defecation or gas expulsion. The sigmoid colon shows characteristic rhythmic motor patterns controlled by intrinsic neural circuits, which regulate fecal delivery to the rectum (Spencer et al. 2016).

The enteric nervous system comprises two main plexuses: the submucosal (Meissner's) plexus and the myenteric (Auerbach's) plexus, located between longitudinal and circular muscle layers. These neural networks include excitatory and inhibitory motor neurons, plus intrinsic sensory neurons responsive to mechanical and chemical stimuli. Interneurons within these plexuses enable local reflex propagation, supporting coordinated motor function (Furness 2012).

Interstitial cells of Cajal serve as gastrointestinal pacemakers, producing rhythmic electrical signals that coordinate smooth muscle activity. These cells create networks that process mechanosensory and chemosensory information, enhancing intestinal motility regulation (Huizinga and Chen 2014). Recent research has revealed their role in maintaining the slow-wave electrical rhythm that underlies coordinated intestinal contractions.

The autonomic nervous system provides segmentally organized innervation to the large intestine. Sympathetic innervation originates from thoracic segments T5–T12 for midgut structures and lumbar segments L1–L2 for hindgut components. The postganglionic sympathetic neurons release norepinephrine, which influences intestinal motility through presynaptic inhibition of enteric neural circuits (Browning and Travagli 2014). The parasympathetic system supplies dual innervation: vagal fibers innervate midgut structures, while pelvic splanchnic nerves from segments S2–S4 supply hindgut regions. This parasympathetic innervation facilitates colonic propulsion and internal anal sphincter relaxation (Powley et al. 2019).

Advanced physiological research has identified a specialized high-pressure region in the distal sigmoid colon, known as the rectosigmoid sphincter. This area demonstrates unique contractile patterns in response to mechanical stimulation of both sigmoid and rectal segments. Electrophysiological studies show specific neural activation patterns in this region, suggesting its function in regulating colonic transit (Dinning et al. 2014).

The motor activity of the rectum manifests through coordinated myoelectric complexes initiated by two primary mechanisms: propagating contractions from proximal colonic regions and mechanical stimulation from sigmoid-derived contents. These complexes typically show retrograde propagation, an adaptation that maintains rectal emptiness between defecation events.

The sensory innervation follows multiple pathways throughout the large intestine. Midgut sensations travel via vagal routes, while hindgut sensitivity utilizes lumbar (L2–L3) and sacral (S1–S2) dorsal root ganglia. Blood vessels contain specialized nerve endings that detect colonic distention, providing ongoing information about wall tension and luminal pressure (Brookes et al. 2013). These afferent signals integrate with central nervous system pathways to regulate gastrointestinal function and generate conscious sensations (Costa et al. 2000).

10.4.3.2 Neural Control of the Rectum and Anus Region

The neural control of the rectum and anal region involves an organized network of sympathetic, parasympathetic, and somatic innervation that coordinates storage and evacuation functions. This organization enables precise regulation through multiple neural pathways and specialized receptors.

The sympathetic innervation originates from lumbar segments L1–L3, with nerve fibers traveling through sympathetic chains to join the preaortic plexus (Drake et al. 2020) (*see* Fig. 10.12). These fibers continue along the inferior mesenteric artery as the mesenteric plexus, providing innervation to the superior rectal region. The inferior rectal region receives innervation via presacral nerves from the hypogastric plexus. Two primary hypogastric nerves border the rectum laterally, transmitting sympathetic impulses from the hypogastric plexus to the pelvic plexus. This pelvic plexus occupies a position on the lateral pelvic wall adjacent to the inferior third of the rectum (Rajasekaran et al. 2008).

The pelvic plexus functions as an integrative center, processing both sympathetic and parasympathetic inputs to coordinate smooth muscle activity in the rectoanal region. The sympathetic system promotes storage function through several mechanisms: maintaining internal anal sphincter tone, inhibiting rectal smooth muscle activity, and modulating blood flow to the region. Conversely, parasympathetic activation facilitates evacuation by stimulating rectal contraction and internal sphincter relaxation. This reciprocal relationship allows for efficient switching between storage and evacuation modes (Spencer et al. 2011).

Parasympathetic innervation to the rectoanal complex derives from sacral segments S2–S4, with nerve fibers emerging through sacral foramina as the nervi erigentes. These nerves course laterally and anteriorly to merge with sympathetic hypogastric nerves, forming the pelvic plexus on the lateral pelvic wall. From this plexus, postganglionic fibers containing both parasympathetic and sympathetic components supply the rectum, reproductive organs, and anal canal. A specialized division of the pelvic plexus, the periprostatic plexus, innervates male reproductive and urinary structures including the prostate, seminal vesicles, corpora cavernosa, vas deferens, urethra, ejaculatory ducts, and bulbourethral glands (Kinugasa et al. 2011).

Recent neurophysiological studies have identified specialized mechanoreceptors within the rectal wall that respond to various stimuli, including distention, pressure changes, and chemical factors (Brookes et al. 2013). These receptors transmit information through dedicated afferent pathways to both spinal and supraspinal centers, contributing to conscious sensation and reflex control of anorectal function. The pudendal nerve provides additional sensory input from specialized receptors in the anal canal, enabling fine discrimination between solid, liquid, and gaseous contents. This sensory information integrates with autonomic reflexes to maintain continence and coordinate appropriate evacuation timing.

The internal anal sphincter receives dual autonomic innervation: sympathetic input from L5 and parasympathetic contributions from S2 to S4, following pathways similar to rectal innervation. The external anal sphincter demonstrates a different innervation pattern, receiving bilateral supply from the inferior rectal branches of the internal pudendal nerve (S2–S3) and perineal branches of S4 (*see* Fig. 10.12). The pudendal nerve mediates voluntary control of external sphincter contraction but does not participate in unconscious sphincter.

10.5 Anatomy and Physiology of Defecation

The process of defecation, representing the terminal phase of digestion, involves coordinated expulsion of undigested food residue and waste products through the anus. This physiological process requires precise synchronization between the rectum and anus, which form the most distal portions of the gastrointestinal tract (Barrett et al. 2019). The rectum demonstrates dual functionality as both a reservoir and propulsive organ. Its wall contains specialized mucosal folds, known as Houston's valves, which create mechanical barriers regulating fecal progression (Fig. 10.25). These valves enhance their barrier effect as fecal matter accumulates, and recent studies have revealed the presence of specialized mechanoreceptors within these structures that contribute to rectal filling sensation (Standring 2020).

The anal canal extends approximately 2–4 cm in length and displays a complex epithelial organization. The epithelium transitions from columnar in the upper region to stratified squamous in the lower portion through a specialized transitional zone (Drake et al. 2020). The muscular architecture of the anal wall incorporates both smooth and skeletal muscle components. The smooth muscle arranges in inner circular and outer longitudinal layers, with the circular layer expanding to form the internal anal sphincter. This sphincter maintains continence through its resting tone. The skeletal muscle component, comprising the levator ani complex, includes the pubococcygeus, iliococcygeus, and puborectalis muscles, which provide voluntary control of defecation (Raizada and Mittal 2008).

The defecation process involves complex physiological coordination between colonic motility, rectal compliance, and sphincter function (Spencer et al. 2011). High-amplitude propagating contractions in the colon initiate the defecatory urge by propelling intraluminal contents toward the rectum (Chen et al. 2017). The rectum exhibits adaptive relaxation to accommodate incoming material, while specialized sensory mechanisms in the upper anal canal perform periodic sampling to determine content nature (Knowles et al. 2010).

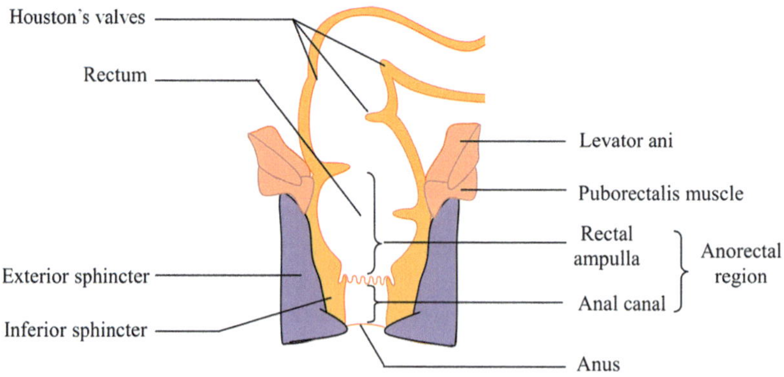

Fig. 10.25 Anatomy of the rectum, illustrating the position and features of Houston's valves. (Adapted with permission from Welcome (2018))

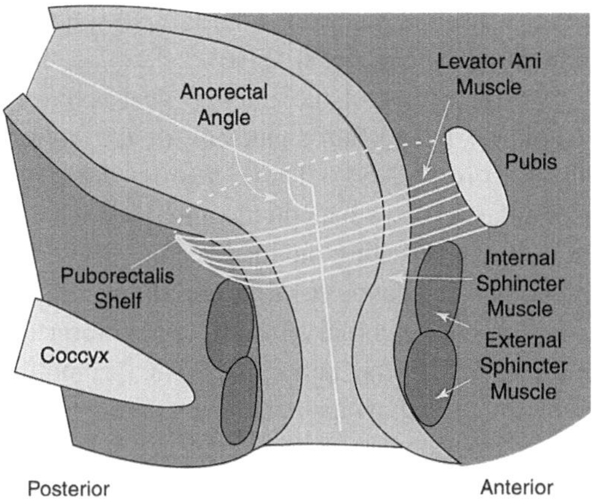

Fig. 10.26 Structure of the anorectum. The internal anal sphincter contributes 70% to 85% of resting sphincter pressure. The anorectal angle, approximately 90° at rest, becomes more obtuse during defecation. (Adapted with permission from Docimo and Brusciano (2023))

In situations where defecation is socially inappropriate, voluntary contraction of the puborectalis and external anal sphincter can return stool to the distal rectum. When defecation proceeds, multiple coordinated events occur, including progression of colonic propagating complexes, rectal contractions, increased intra-abdominal pressure through voluntary straining, and relaxation of both puborectalis and anal sphincters (Mayer and Tillisch 2011). The anorectal angle, which normally measures 90°–96° at rest, becomes more acute during squeeze maneuvers (70°–98°) and more obtuse during defecation (110°–155°) (Fig. 10.26). The anorectal angle exhibits no gender differences.

The completion of defecation initiates a series of restorative events, including external sphincter contraction, gradual recovery of internal sphincter tone, and re-expansion of the anal cushions (Dinning et al. 2013). This coordinated process ensures effective evacuation while maintaining continence between bowel movements. Various factors can influence the defecation process, including colonic transit patterns, stool characteristics, body position, and psychological state. Social behavior often modifies the timing of defecation, allowing it to occur at convenient times rather than immediately upon sensing the defecatory urge. This adaptation requires complex integration of both voluntary and involuntary neural mechanisms (Rao and Patcharatrakul 2016).

The neural control of defecation involves multiple pathways. The internal anal sphincter receives autonomic innervation, while the external sphincter responds to voluntary control through somatic nerves. Specialized mechanoreceptors within the rectal wall detect distention and pressure changes, sending signals to both spinal and supraspinal centers. This sensory information contributes to both conscious awareness and reflex control of anorectal function (Knowles et al. 2010).

Disorders of defecation may arise from various functional and mechanical abnormalities. Functional disorders include inadequate expulsive effort, paradoxical contraction of pelvic floor muscles (dyssynergia), and altered rectal sensation. These conditions often respond to biofeedback therapy and other conservative measures (Rao and Patcharatrakul 2016). Mechanical causes encompass structural abnormalities such as rectal intussusception, obstructing masses, and anal stenosis. Additional mechanical factors include rectocele formation and excessive perineal descent, which can alter the normal force vectors during defecation (Scott and Gladman 2008).

Advanced diagnostic technology have enhanced our understanding of defecation physiology. High-resolution anorectal manometry provides detailed pressure mapping of the anal canal and rectum, while magnetic resonance defecography allows dynamic visualization of the defecation process. These techniques have revealed complex patterns of muscular coordination and have identified previously unrecognized mechanisms of continence maintenance (Carrington et al. 2017).

10.5.1 Neural Control of Defecation

The neural regulation of defecation integrates both voluntary and autonomic components to coordinate multiple physiological functions. This control system focuses primarily on two critical regions of the gastrointestinal tract: the oropharyngeal–esophageal segment for food passage and the rectoanal region for fecal evacuation (Furness 2012; Standring 2020).

The intrinsic nervous system of the anorectum comprises enteric plexuses, with the myenteric plexus forming synaptic connections with extrinsic nerves. The extrinsic control involves complex interactions between parasympathetic pelvic nerves and sympathetic hypogastric innervation. The parasympathetic supply originates from sacral segments S2-S4, forming the inferior hypogastric plexus, which gives rise to superior, middle, and inferior rectal nerves. Parasympathetic stimulation inhibits internal anal sphincter activity. Conversely, sympathetic nerves, originating from lower thoracic ganglia and forming the superior hypogastric plexus, can both contract and relax the internal anal sphincter (Brookes et al. 2013).

Fecal continence depends on precise neural control of the external anal sphincter (Bharucha et al. 2005). The pudendal nerve, with roots from S2 to S4, provides voluntary control of this sphincter (Raizada and Mittal 2008). The levator ani complex receives innervation through specific pathways: pudendal nerve branches supply the puborectalis, while direct fibers from S3 and S4 innervate the pubococcygeus and iliococcygeus muscles. These muscles maintain the anorectal angle, crucial for continence (Liu et al. 2006) (Fig. 10.27).

The defecation sequence initiates when rectal mucosal mechanoreceptors activate, generating afferent signals to spinal and cerebral centers (Palit et al. 2012). These signals create conscious awareness and trigger efferent responses from pontine and sacral defecation centers. The pontine center coordinates voluntary and involuntary mechanisms (Brierley et al. 2018). Sensory information travels to the dorsal aspect of spinal segments S2–S4 via dorsal root ganglia.

In socially appropriate conditions, reduced cerebral inhibition combines with voluntary increases in intra-abdominal pressure through coordinated diaphragmatic and abdominal wall contractions (Brookes et al. 2013). Rectal distention activates autonomic pathways, producing rectal wall contraction and internal anal sphincter relaxation (Spencer et al. 2016).

Motor control of the external anal sphincter originates from Onuf's nucleus in the ventral sacral cord (S2–S4). These neurons form the pudendal nerves, which divide into inferior rectal branches innervating the anal region. The completion of defecation requires coordinated relaxation of both external sphincter and levator ani muscles, integrating parasympathetic (promoting defecation) and sympathetic (maintaining continence) components (Furness et al. 2014; Rao and Patcharatrakul 2016).

10.5.2 Mechanism of Defecation: Defecation Reflex

The defecation reflex represents a complex physiological process that begins with fecal accumulation in the rectal ampulla. This accumulation triggers stretch receptors in the rectal wall, initiating a cascade of neural signals that ultimately lead to rectal muscle activation. While this process operates through reflexive mechanisms, it remains subject to voluntary control through higher cortical centers (Rao and Patcharatrakul 2016).

The neural control of defecation operates through two synergistic routes. At the local level, the intrinsic pathway, regulated by the myenteric plexus, modulates the activity of smooth muscles in the anorectum and internal anal sphincter. Simultaneously, the extrinsic pathway involves parasympathetic sensory nerves that convey afferent signals to the dorsal spinal cord segments S2–S4. Upon processing this sensory information, these spinal segments generate inhibitory responses transmitted via parasympathetic motor fibers, resulting in internal anal sphincter relaxation (Brookes et al. 2013).

While both anal sphincters contribute to continence maintenance, the external anal sphincter emerges as the primary controller of defecation. This control occurs through precise cortical regulation of lower motor neurons that innervate the sphincter (Spencer et al. 2016). During normal defecation, the respiratory diaphragm and

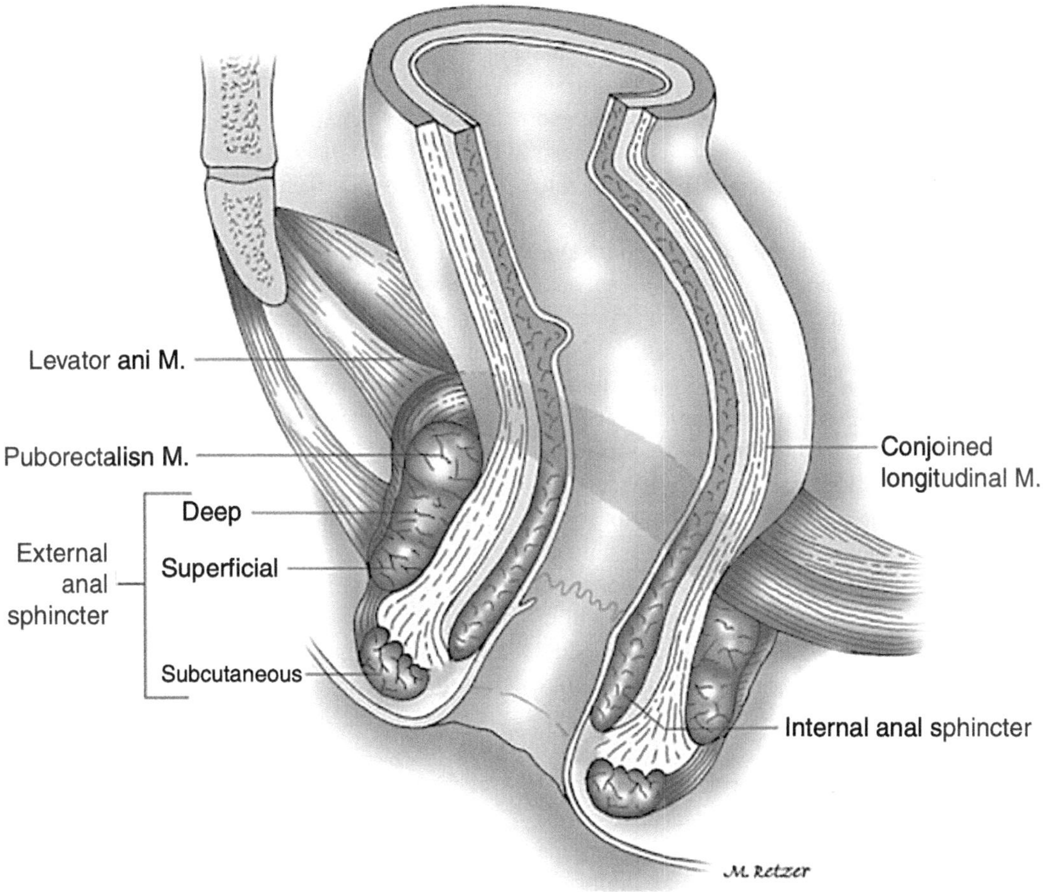

Fig. 10.27 Anatomy of the muscles of the anal canal. (Adapted with permission from Beck et al. (2014))

abdominal wall muscles contract in coordination, elevating both intra-abdominal and rectal pressure. Concurrent relaxation of pelvic floor and anal sphincter muscles facilitates rectal contraction and subsequent fecal expulsion (Palit et al. 2012).

The maintenance of continence relies on three interconnected anal reflexes, all mediated through pelvic and pudendal nerve pathways. The rectoanal inhibitory reflex functions to release sphincter muscle tone when appropriate, while the rectoanal contractile reflex prevents unexpected fecal release. The third component, the sensorimotor response, involves puborectalis muscle contraction in response to the sensation of defecatory urgency.

Contemporary medical science has developed numerous techniques for studying anorectal structure and function. These include magnetic resonance imaging, video X-ray defecography, and computed tomography for structural assessment, alongside functional studies using anorectal manometry, endoanal ultrasound, and electromyographic recordings. These diagnostic tools provide detailed information about both the static anatomy and dynamic function of the anorectum during defecation.

10.5.3 Internal Anal Sphincter in Defecation Control

The internal anal sphincter, as a specialized smooth muscle, maintains continuous maximal contraction through intrinsic myogenic properties and extrinsic autonomic neural control.

This sphincter provides the primary barrier against involuntary stool loss. Studies have shown that healthy adults typically demonstrate anal canal resting pressures between 50 and 70 mm Hg, with lower values observed in women and elderly individuals (Remes-Troche and Rao 2008).

The composition of anal canal resting tone reveals the predominant role of the internal anal sphincter, contributing 50–85% of total pressure. The external anal sphincter provides 25–30%, while anal cushion expansion accounts for the remaining 15%. Pressure measurements along the anal canal demonstrate a proximal-to-distal gradient, with peak pressures typically recorded 1–2 cm above the anal verge. This high-pressure zone, corresponding to concentrated smooth muscle fibers of the internal sphincter, exhibits gender-specific variations, measuring 2–3 cm in women compared to 2.5–3.5 cm in men. Notably, this gender difference persists even in nulliparous women, suggesting anatomical rather than obstetric factors (Liu et al. 2006).

When rectal distention occurs, the internal sphincter undergoes reflexive transient relaxation, allowing rectal contents to descend into the proximal anal canal. This rectoanal inhibitory reflex can be demonstrated experimentally through controlled balloon distention while monitoring anal pressure changes. The resulting exposure of the highly innervated anal canal to rectal contents enables discrimination between different material consistencies, particularly important for selective gas passage without stool leakage (Dinning et al. 2013).

The sampling reflex operates through the enteric nervous system, explaining its persistence after extrinsic denervation and its absence in Hirschsprung's disease. Following initial relaxation in response to rectal distention, the internal sphincter gradually recovers its tone as the rectum accommodates to the distending stimulus (Spencer et al. 2016). Clinical studies have identified significant internal sphincter dysfunction in approximately 25% of patients with idiopathic fecal incontinence. Particularly concerning is spontaneous sphincter relaxation without compensatory external sphincter activation, which may contribute significantly to incontinence development (Rao and Patcharatrakul 2016).

10.5.4 External Anal Sphincter and Defecation Control

Defecation represents a complex physiological process involving multiple integrated mechanisms under central nervous system regulation. The process begins when rectal filling from the sigmoid colon activates stretch receptors in the pelvic floor muscles, generating conscious awareness of defecatory urgency (Bharucha and Lacy 2020).

The initial rectal distention triggers the rectoanal inhibitory reflex, causing internal anal sphincter relaxation. This relaxation exposes the highly sensitive anal mucosa to rectal contents, enabling discrimination between flatus and stool through a specialized sampling mechanism that determines defecation urgency. During this sampling phase, the external anal sphincter reflexively contracts to maintain continence.

When social circumstances preclude immediate defecation, voluntary external anal sphincter contraction, combined with rectal compliance mechanisms, allows time for internal sphincter function to recover. This coordinated response prevents involuntary evacuation until appropriate conditions exist (Raizada and Mittal 2008). The act of defecation typically begins with assuming either a sitting or squatting position, which modifies the anorectal angle. This postural change, combined with increased intrarectal and intra-abdominal pressures, induces reflexive relaxation of the external anal sphincter, internal anal sphincter, and puborectalis muscle, often enabling defecation without active straining (Spencer et al. 2016).

The conjoint longitudinal muscle contracts during defecation, serving multiple functions. It retracts the vascular cushions alongside the anal wall while simultaneously shortening the anal canal. These actions, combined with pelvic floor descent and funneling, allow direct transmission of increased abdominal pressure through the relaxed pelvic floor, facilitating fecal expulsion.

The consistency of stool influences the evacuation pattern, determining whether emptying occurs through mass peristalsis of the left colon or intermittent stool passage. Following evacuation, a "closing reflex" involving transient external anal sphincter and puborectalis contraction restores internal anal sphincter tone and anal canal closure.

References

Agur AMR, Dalley AF. Grant's atlas of anatomy. 12th ed. Philadelphia: Wolters Kluwer; 2009.

Anderson CA, Omar MI, Campbell SE, Hunter KF, Cody JD, Glazener CM. Conservative management for postprostatectomy urinary incontinence. Cochrane Database Syst Rev. 2015;1(1):CD001843.

Barber MD, Bremer RE, Thor KB, Dolber PC, Kuehl TJ, Coates KW. Innervation of the female levator ani muscles. Am J Obstet Gynecol. 2002;187(1):64–71.

Barleben A, Mills S. Anorectal anatomy and physiology. Surg Clin North Am. 2010;90(1):1–15.

Barrett KE, Barman SM, Brooks HL, Yuan JX. Ganong's review of medical physiology. 26th ed. New York: McGraw-Hill; 2019.

Bayliss WM, Starling EH. The movements and innervation of the small intestine. J Physiol. 1902;26(3–4):125–38.

Beck DE, Steele SR, Wexner SD. Fundamentals of anorectal surgery. 3rd ed. Cham: Springer; 2019.

Beck DE, Wexner SD, Hull TL, Roberts PL, Saclarides TJ, Senagore AJ, et al. The ASCRS manual of colon and rectal surgery. 2nd ed. New York: Springer; 2014.

Bharucha AE, Fletcher JG, Seide B, Riederer SJ, Zinsmeister AR. Phenotypic variation in functional disorders of defecation. Gastroenterology. 2005;128(5):1199–210.

Bharucha AE, Lacy BE. Mechanisms, evaluation, and management of chronic constipation. Gastroenterology. 2020;158(5):1232–49.

Bharucha AE, Rao SSC. An update on anorectal disorders for gastroenterologists. Gastroenterology. 2014;146(1):37–45.

Billroth T. Einige Beobachtungen über das ausgedehnte vorkommen von nervenanastomosen im tractus intestinalis. Arch Anat Physiol Wiss Med. 1857:148–58.

Bonaz B, Bazin T, Pellissier S. The vagus nerve at the interface of the microbiota-gut-brain axis. Front Neurosci. 2018;12:49.

Bonaz B, Sinniger V, Pellissier S. Anti-inflammatory properties of the vagus nerve: potential therapeutic implications of vagus nerve stimulation. J Physiol. 2016;594(20):5781–90.

Bonaz B, Sinniger V, Pellissier S. Vagus nerve stimulation: a new promising therapeutic tool in inflammatory bowel disease. J Intern Med. 2017;282(1):46–63.

Brierley SM, Hibberd TJ, Spencer NJ. Spinal afferent innervation of the colon and rectum. Front Cell Neurosci. 2018;12:467.

Brookes SJ, Spencer NJ, Costa M, Zagorodnyuk VP. Extrinsic primary afferent signalling in the gut. Nat Rev Gastroenterol Hepatol. 2013;10(5):286–96.

Browning KN, Travagli RA. Central nervous system control of gastrointestinal motility and secretion and modulation of gastrointestinal functions. Compr Physiol. 2014;4(4):1339–68.

Carrington EV, Heinrich H, Knowles CH, Rao SS, Fox M, Scott SM. Methods of anorectal manometry vary widely in clinical practice: results from an international survey. Neurogastroenterol Motil. 2017;29(8):e13016.

Chen JH, Yu Y, Yang Z, Yu WZ, Chen WL, Yu H, et al. Intraluminal pressure patterns in the human colon assessed by high-resolution manometry. Sci Rep. 2017;7:41436.

Clemente CD. Anatomy: a regional atlas of the human body. 6th ed. Philadelphia: Wolters Kluwer; 2011.

Costa M, Brookes SJH, Hennig GW. Anatomy and physiology of the enteric nervous system. Gut. 2000;47(Suppl 4):iv15–9.

DeLancey JO, Ashton-Miller JA. Pathophysiology of adult urinary incontinence. Gastroenterology. 2004;126(Suppl 1):S23–32.

DeLancey JO, Morgan DM, Fenner DE, Kearney R, Guire K, Miller JM, et al. Comparison of levator ani muscle defects and function in women with and without pelvic organ prolapse. Obstet Gynecol. 2007;109(2 Pt 1):295–302.

Dietz HP, Lanzarone V. Levator trauma after vaginal delivery. Obstet Gynecol. 2005;106(4):707–12.

Dietz HP, Simpson JM. Levator trauma is associated with pelvic organ prolapse. BJOG. 2008;115(8):979–84.

Dinning PG, Wiklendt L, Gibbins I, Patton V, Bampton P, Lubowski DZ, et al. Low-resolution colonic manometry leads to a gross misinterpretation of the frequency and polarity of propagating sequences. Neurogastroenterol Motil. 2013;25(10):e640–9.

Dinning PG, Wiklendt L, Maslen L, Gibbins I, Patton V, Arkwright JW, et al. Quantification of in vivo colonic motor patterns in healthy humans before and after a meal revealed by high-resolution fiber-optic manometry. Neurogastroenterol Motil. 2014;26(10):1443–57.

Docimo L, Brusciano L. Anal incontinence. Clinical management and surgical techniques. Cham: Springer; 2023.

Drake RL, Vogl AW, Mitchell AWM. Gray's anatomy for students. 4th ed. Philadelphia: Elsevier; 2020.

Ellis H, Mahadevan V. Clinical anatomy: applied anatomy for students and junior doct ors. 14th ed. Oxford: Wiley-Blackwell; 2019.

Enck P, Hinninghofen H, Merletti R, Azpiroz F. The external anal sphincter and the role of surface electromyography. Neurogastroenterol Motil. 2005;17(Suppl 1):60–7.

Erlichman DB, Kanmaniraja D, Kobi M, Chernyak V. MRI anatomy and pathology of the anal canal. J Magn Reson Imaging. 2019;50(4):1018–32.

Forsythe P, Bienenstock J, Kunze WA. Vagal pathways for microbiome-brain-gut axis communication. Adv Exp Med Biol. 2014;817:115–33.
Fritsch H, Brenner E, Lienemann A, Ludwikowski B. Anal sphincter complex: reinterpreted morphology and its clinical relevance. Dis Colon Rectum. 2002;45(2):188–94.
Fritsch H, Lienemann A, Brenner E, Ludwikowski B. Clinical anatomy of the pelvic floor. Adv Anat Embryol Cell Biol. 2004;175:III-IX:1–64.
Furness JB. The enteric nervous system. Oxford: Blackwell Publishing; 2006.
Furness JB. The enteric nervous system and neurogastroenterology. Nat Rev Gastroenterol Hepatol. 2012;9(5):286–94.
Furness JB, Callaghan BP, Rivera LR, Cho HJ. The enteric nervous system and gastrointestinal innervation: integrated local and central control. Adv Exp Med Biol. 2014;817:39–71.
Furness JB, Costa M, Keast JR. Choline acetyltransferase- and peptide immunoreactivity of submucous neurons in the small intestine of the Guinea-pig. Cell Tissue Res. 1984;237(2):329–36.
Gabella G. Innervation of airway smooth muscle: fine structure. Annu Rev Physiol. 1987;49:583–94.
Garcia-Tsao G, Abraldes JG, Berzigotti A, Bosch J. Portal hypertensive bleeding in cirrhosis: risk stratification, diagnosis, and management. Hepatology. 2017;65(1):310–35.
Geboes K, Geboes KP, Maleux G. Vascular anatomy of the gastrointestinal tract. Best Pract Res Clin Gastroenterol. 2001;15(1):1–14.
Gordon PH, Nivatvongs S. Principles and practice of surgery for the colon, rectum, and anus. 3rd ed. New York: Informa Healthcare; 2007.
Grundy D, Brookes S. Neural control of gastrointestinal function. San Rafael: Morgan & Claypool Life Sciences; 2012.
Gudsoorkar VS, Quigley EM. Colorectal sensation and motility. Curr Opin Gastroenterol. 2014;30(1):75–83.
Haas PA, Fox TA Jr, Haas GP. The pathogenesis of hemorrhoids. Dis Colon Rectum. 1984;27(7):442–50.
Hall JE. Guyton and Hall textbook of medical physiology. 13th ed. Philadelphia: Elsevier; 2016.
Huizinga JD, Chen JH. The myogenic and neurogenic components of the rhythmic segmentation motor patterns of the intestine. Front Neurosci. 2014;8:78.
Huizinga JD, Lammers WJ. Gut peristalsis is governed by a multitude of cooperating mechanisms. Am J Physiol Gastrointest Liver Physiol. 2009;296(1):G1–8.
Iwakiri Y, Shah V, Rockey DC. Vascular pathobiology in chronic liver disease and cirrhosis—current status and future directions. J Hepatol. 2014;61(4):912–24.
Johnson LR. Physiology of the gastrointestinal tract. 5th ed. San Diego: Elsevier; 2012.
Jorge JMN, Wexner SD. Anatomy and physiology of the rectum and anus. Eur J Surg. 1997;163(10):723–31.
Kahn JK, Kolesov EI, Sankar A. Vascular anatomy of the gastrointestinal tract. Best Pract Res Clin Gastroenterol. 2001;15(1):1–14.
Kinugasa Y, Arakawa T, Abe S, Ohtsuka A, Suzuki D, Murakami G, et al. Anatomical reevaluation of the anococcygeal ligament and its surgical relevance. Dis Colon Rectum. 2011;54(2):232–7.
Knowles CH, De Giorgio R, Kapur RP, Bruder E, Farrugia G, Geboes K, et al. The London classification of gastrointestinal neuromuscular pathology. Gut. 2010;59(7):882–7.
Ko HY. Management and rehabilitation of spinal cord injuries. 2nd ed. Singapore: Springer; 2022.
Krogh K, Christensen P. Neurogenic colorectal and pelvic floor dysfunction. Best Pract Res Clin Gastroenterol. 2009;23(4):531–43.
Krogh K, Emmanuel A, Perrouin-Verbe B, Korsten MA, Mulcahey MJ, Biering-Sørensen F. International spinal cord injury bowel function basic data set (Version 2.0). Spinal Cord. 2017;57(7):692–8.
Kumar L, Emmanuel A. Internal anal sphincter: clinical perspective. Surgeon. 2017;15(4):211–26.
Liu J, Guaderrama N, Nager CW, Pretorius DH, Master S, Mittal RK. Functional correlates of anal canal anatomy: puborectalis muscle and anal canal pressure. Am J Gastroenterol. 2006;101(5):1092–7.
Llewellyn-Smith IJ, Costa M, Furness JB. Light and electron microscopic immunocytochemistry of the same nerves from whole mount preparations. J Histochem Cytochem. 1985;33(9):857–66.
Lunniss PJ, Phillips RK. Anatomy and function of the anal longitudinal muscle. Br J Surg. 1992;79(9):882–4.
Lyte M, Cryan JF. Microbial endocrinology: the microbiota-gut-brain axis in health and disease. New York: Springer; 2014.
Matsufuji H, Yokoyama J. Neural control of the internal anal sphincter motility. J Smooth Muscle Res. 2003;39(1–2):11–20.
Mayer EA, Tillisch K. The brain-gut axis in abdominal pain syndromes. Annu Rev Med. 2011;62:381–96.
Meissner G. Über die nerven der darmwand. Z Ration Med N F. 1857;8:364–406.
Merrett ND, Wilson RB. Superior mesenteric artery syndrome: diagnosis and treatment strategies. J Gastrointest Surg. 2009;13(2):287–92.
Michels NA. The variant blood supply to the small and large intestines in regional resections. J Int Coll Surg. 1963;39:127–70.
Milligan ETC, Morgan CN. Surgical anatomy of the anal canal: with special reference to anorectal fistulae. Lancet. 1934;224(5804):1150–6.
Mittal RK, Tuttle LJ. Anorectal anatomy and function. Gatroenterol Clin N Am. 2022;51:1–23.
Moore KL, Dalley AF, Agur AMR. Clinically oriented anatomy. 8th ed. Philadelphia: Wolters Kluwer; 2018.
Morgan DM, Cardoza P, Guire K, Fenner DE, DeLancey JO. Levator ani defect status and lower urinary tract symptoms in women with pelvic organ prolapse. Int Urogynecol J. 2010;21(1):47–52.
Moszkowicz D, Peschaud F, Bessede T, Benoit G, Alsaid B. Internal anal sphincter parasympathetic-nitrergic and sympathetic-adrenergic innervation: a

3-dimensional morphological and functional analysis. Dis Colon Rectum. 2012;55(4):473–81.
Murono K, Kawai K, Kazama S, Ishihara S, Yamaguchi H, Sunami E, et al. Anatomy of the inferior mesenteric artery evaluated using 3-dimensional CT angiography. Dis Colon Rectum. 2015;58(2):214–9.
Netter FH. Atlas of human anatomy. 7th ed. Philadelphia: Elsevier; 2019.
Palit S, Lunniss PJ, Scott SM. The physiology of human defecation. Dig Dis Sci. 2012;57(6):1445–64.
Paulsen F, Waschke J. Sobotta atlas of human anatomy. 15th ed. Munich: Elsevier; 2011.
Powley TL, Hudson CN, McAdams JL, Baronowsky EA, Phillips RJ. Vagal intramuscular arrays: the specialized mechanoreceptor arbors that innervate the smooth muscle layers of the stomach examined in the rat. J Comp Neurol. 2016;524(4):713–37.
Powley TL, Jaffey DM, McAdams J, Baronowsky EA, Black D, Chesney L, et al. Vagal innervation of the stomach reassessed: brain-gut connectome uses smart terminals. Ann N Y Acad Sci. 2019;1454(1):14–30.
Rabie ME, Ogunbiyi O, Al Qahtani AS, Taha SB, El Hadad A, El Hakeem I. Superior mesenteric artery syndrome: clinical and radiological considerations. Surg Res Pract. 2015;2015:628705.
Raizada V, Mittal RK. Pelvic floor anatomy and applied physiology. Gastroenterol Clin N Am. 2008;37(3):493–509.
Rajasekaran MR, Jiang Y, Bhargava V, Littlefield R, Lee A, Lieber RL, et al. Length-tension relationship of the external anal sphincter muscle: implications for the anal canal function. Am J Physiol Gastrointest Liver Physiol. 2008;295(2):G367–73.
Rao SS, Patcharatrakul T. Diagnosis and treatment of dyssynergic defecation. J Neurogastroenterol Motil. 2016;22(3):423–35.
Rasmussen OO. Anorectal function. Dis Colon Rectum. 1994;37(4):386–403.
Regadas FS, Murad-Regadas SM, Lima DM, Silva FR, Barreto RG, Souza MH, et al. Anal canal anatomy showed by three-dimensional anorectal ultrasonography. Surg Endosc. 2007;21(12):2207–11.
Remes-Troche JM, Rao SSC. Neurophysiological testing in anorectal disorders. Expert Rev Gastroenterol Hepatol. 2008;2(3):323–35.
Richardson KC. Electronmicroscopic observations on Auerbach's plexus in the rabbit, with special reference to the problem of smooth muscle innervation. Am J Anat. 1958;103(1):99–135.
Rociu E, Stoker J, Eijkemans MJ, Laméris JS. Normal anal sphincter anatomy and age- and sex-related variations at high-spatial-resolution endoanal MR imaging. Radiology. 2000;217:395–401.
Rosenblum JD, Boyle CM, Schwartz LB. The mesenteric circulation. Anatomy and physiology. Surg Clin North Am. 1997;77(2):289–306.
Ross MH, Pawlina W. Histology: a text and atlas with correlated cell and molecular biology. 7th ed. Philadelphia: Wolters Kluwer; 2016.
Sakaguchi T, Suzuki S, Morita Y, Oishi K, Suzuki A, Fukumoto K, et al. Analysis of anatomic variants of mesenteric veins by 3-dimensional portography using multidetector-row computed tomography. Am J Surg. 2010;200(1):15–22.
Sanders KM, Koh SD, Ward SM. Interstitial cells of Cajal as pacemakers in the gastrointestinal tract. Annu Rev Physiol. 2006;68:307–43.
Sanders KM, Smith TK. Motoneurones of the submucous plexus regulate electrical activity of the circular muscle of canine proximal colon. J Physiol. 1986;380:293–310.
Scott SM, Gladman MA. Manometric, sensorimotor, and neurophysiologic evaluation of anorectal function. Gastroenterol Clin N Am. 2008;37(3):511–38.
Seow-Choen F, Ho JM. Histoanatomy of anal glands. Dis Colon Rectum. 1994;37(12):1215–8.
Shafik A, Ahmed I, Shafik AA, El-Ghamrawy TA, El-Sibai O. Surgical anatomy of the perineal muscles and their role in perineal disorders. Anat Sci Int. 2005;80(3):167–71.
Silva NGOD, Barbosa ABM, Silva NA, Araújo DN, Assis TO. Anatomical variations of the superior mesenteric artery and its clinical and surgical implications in human. Arq Bras Cir Dig. 2020;33(2):e1508.
Sinnatamby CS. Last's anatomy: regional and applied. 12th ed. Edinburgh: Churchill Livingstone; 2011.
Small JV, Gimona M. The cytoskeleton of the vertebrate smooth muscle cell. Acta Physiol Scand. 1998;164(4):341–8.
Spencer NJ, Dinning PG, Brookes SJ, Costa M. Insights into the mechanisms underlying colonic motor patterns. J Physiol. 2016;594(15):4099–116.
Spencer NJ, Hu H. Enteric nervous system: sensory transduction, neural circuits and gastrointestinal motility. Nat Rev Gastroenterol Hepatol. 2020;17(6):338–51.
Spencer NJ, Nicholas SJ, Robinson L, Kyloh M, Flack N, Brookes SJ, et al. Mechanisms underlying distension-evoked peristalsis in Guinea pig distal colon: is there a role for enterochromaffin cells? Am J Physiol Gastrointest Liver Physiol. 2011;301(3):G519–27.
Standring S. Gray's anatomy: the anatomical basis of clinical practice. 42nd ed. Philadelphia: Elsevier; 2020.
Stoker J, Bartram CI, Halligan S. Imaging of the posterior pelvic floor. Eur Radiol. 2002;12(4):779–88.
Sugrue J, Nordenstam J, Abcarian H, Bartholomew A, Schwartz JL, Mellgren A, et al. Pathogenesis and persistence of cryptoglandular anal fistula: a systematic review. Tech Coloproctol. 2017;21(6):425–32.
Sultan AH, Thakar R, Fenner DE. Perineal and anal sphincter trauma: diagnosis and clinical management. London: Springer; 2007.
Thomson WH. The nature of hemorrhoids. Br J Surg. 1975;62(7):542–52.
Timmermans JP, Adriaensen D, Cornelissen W, Scheuermann DW. Structural organization and neuropeptide distribution in the mammalian enteric nervous system, with special attention to those components involved in mucosal reflexes. Comp Biochem Physiol A Mol Integr Physiol. 1997;118(2):331–40.

Tsukada Y, Ito M, Watanabe K, Yamaguchi K, Kojima M, Hayashi R, et al. Topographic anatomy of the anal sphincter complex and levator ani muscle as it relates to intersphincteric resection for very low rectal disease. Dis Colon Rectum. 2016;59(5):426–33.
Tsuruoka M, Wang D, Tamaki J, Inoue T. Descending influence from the nucleus locus coeruleus/subcoeruleus on visceral nociceptive transmission in rat spinal cord. Neuroscience. 2010;165(4):1019–24.
Vanner S, Greenwood-Van Meerveld B, Mawe G, Shea-Donohue T, Verdu EF, Wood J, et al. Fundamentals of neurogastroenterology: basic science. Gastroenterology. 2016;150:1280–91.
Vermeulen W, De Man JG, Pelckmans PA, De Winter BY. Neuroanatomy of lower gastrointestinal pain disorders. World J Gastroenterol. 2014;20(4):1005–20.
Vollmar B, Menger MD. The hepatic microcirculation: mechanistic contributions and therapeutic targets in liver injury and repair. Physiol Rev. 2009;89(4):1269–339.
Walls EW. Observations on the microscopic anatomy of the human anal canal. Br J Surg. 1958;45(193):504–12.
Webb RC. Smooth muscle contraction and relaxation. Adv Physiol Educ. 2003;27(1–4):201–6.
Welcome MO. Gastrointestinal physiology: development, principles and mechanisms of regulation. Cham: Springer; 2018.
Welsch T, Büchler MW, Kienle P. Recalling superior mesenteric artery syndrome. Dig Surg. 2007;24(3):149–56.
Wilson AJ, Furness JB, Costa M. The fine structure of the submucous plexus of the Guinea-pig ileum. II. Description and analysis of vesiculated nerve profiles. J Neurocytol. 1981;10(5):785–804.
Wineski LE. Snell's clinical anatomy by regions. 10th ed. Philadelphia: Wolters Kluwer; 2019.
Wood JD. Enteric nervous system: reflexes, pattern generators and motility. Curr Opin Gastroenterol. 2008;24(2):149–58.
Wood JD, Alpers DH, Andrews PL. Fundamentals of neurogastroenterology. Gut. 1999;45(Suppl 2):II6–16.
Yuan H, Silberstein SD. Vagus nerve and vagus nerve stimulation, a comprehensive review: part I. Headache. 2016;56(1):71–8.

Recommended Additional Reading

Fazio VW, Church JM, Delaney CP, Kiran RP. Current therapy in colon and rectal surgery. 3rd ed. Philadelphia: Elsevier; 2017.
Givel JC, Mortensen NJ, Roche B. Anorectal and colonic diseases: a practical guide to their management. 3rd ed. Berlin: Springer; 2010.
Reinus JF, Simon D. Gastrointestinal anatomy and physiology: the essentials. Hoboken: Wiley Blackwell; 2014.

Anatomical Organization of the Respiratory System

11

The respiratory system executes two fundamental gas exchange processes: external respiration between blood and atmospheric air, and internal respiration occurring between systemic capillaries and tissue cells (West and Luks 2016). This physiological mechanism requires precise coordination among multiple anatomical structures and neural pathways to maintain optimal respiratory function.

The respiratory tract's anatomical organization consists of upper and lower airways. The upper airway includes the nasal passages, their associated paranasal sinuses, pharynx, and larynx. The lower airway comprises the trachea, bronchial tree, and lung parenchyma (Moore et al. 2018). The mechanical aspects of breathing depend on the coordinated action of the pleurae, diaphragm, and thoracic wall musculature during inspiration and expiration (Drake et al. 2020).

Bronchopulmonary segmental anatomy maintains clinical significance across medical specialties. This anatomical organization, derived from embryological development patterns, establishes specific lobar and segmental divisions within each lung (Kumar and Clark 2012). The right lung contains three lobes divided into ten segments, while the left lung comprises two lobes with eight segments. Medical professionals, including physiatrists and spinal cord medicine specialists, apply this knowledge in diagnostic and therapeutic interventions (Standring 2020).

In spinal cord injury medicine, respiratory system anatomy holds exceptional relevance as respiratory complications remain a significant source of morbidity. The impact on respiratory function correlates with the neurological level and completeness of injury. Injuries above the C3 level may require immediate ventilatory support due to complete diaphragmatic paralysis. Lesions between C3 and C5 may preserve partial diaphragmatic function through phrenic nerve innervation, though respiratory compromise often occurs (Harvey 2016).

The thoracic and abdominal wall muscles receive innervation from thoracic segments and contribute to forced expiration and cough production. Impairment of these muscles in high thoracic injuries affects secretion clearance mechanisms and increases susceptibility to respiratory infections. The accessory respiratory muscles, including the scalenes and sternocleidomastoid muscles with cervical root innervation, may compensate when primary respiratory muscles are compromised.

Recent research has demonstrated that respiratory dysfunction in spinal cord injury extends beyond mechanical impairment. Changes in autonomic regulation, altered inflammatory

H.-Y. Ko, *Practical Functional Anatomy for Spinal Cord Injury Rehabilitation*,
https://doi.org/10.1007/978-981-95-6457-6_11

responses, and respiratory muscle plasticity contribute to the complex pathophysiology. Understanding these mechanisms has led to advanced therapeutic approaches, including respiratory muscle training protocols and novel ventilatory support strategies.

11.1 Osseous Components and Biomechanics of the Respiratory System

The thoracic skeletal framework functions as the structural foundation for respiratory mechanics, forming an integrated system of bones, joints, and articular surfaces. The thoracic cavity contains multiple osseous elements—the vertebral column, sternum, ribs, scapulae, and clavicles—which combine through precise anatomical connections to enable respiratory movements. These anatomical relationships allow specific biomechanical actions during breathing (Drake et al. 2020; Moore et al. 2018; West and Luks 2016).

In the posterior thoracic region, the thoracic vertebrae (T1–T12) establish specialized articulations with the ribs through two primary joint types: the costovertebral joints, which connect the rib heads to vertebral bodies, and the costotransverse joints, joining the rib tubercles to vertebral transverse processes. These double articulations provide stability while allowing controlled movement during respiration (Netter 2019; Standring 2020).

The anterior thoracic wall incorporates the sternum, composed of three anatomically distinct segments. The superior segment (manubrium) forms synovial joints with the first and second costal cartilages at the T3–T4 vertebral levels. The sternal body extends these articulations by connecting with costal cartilages two through seven at the T5–T8 vertebral levels. The manubriosternal junction forms the sternal angle (angle of Louis), positioned at the T4 vertebral level—an anatomical marker used for clinical reference points and respiratory assessment (Moore et al. 2018; Netter 2019; Standring 2020).

Rib anatomy demonstrates specific variations based on sternal relationships and biomechanical function. The first seven pairs (vertebrosternal or true ribs) establish direct sternal connections through costal cartilages. The eighth through tenth pairs (vertebrochondral or false ribs) form indirect connections through shared costal cartilage attachments. The eleventh and twelfth pairs (floating ribs) maintain posterior vertebral articulations while remaining unattached anteriorly, contributing to abdominal protection and diaphragmatic function (Drake et al. 2020; Kumar and Clark 2012).

The detailed anatomy of individual ribs supports specific respiratory functions. Each rib head forms multiple synovial articulations: with the vertebral body at its level, the intervertebral disc above, and the superior vertebra. The costotransverse joints provide additional stability through ligamentous attachments. These anatomical arrangements allow two primary respiratory movements: the 'bucket handle' motion of middle and lower ribs, increasing thoracic cavity transverse diameter, and the 'pump handle' motion of upper ribs, expanding anteroposterior diameter (West and Luks 2016; Ratnovsky et al. 2008) (Fig. 11.1).

Maximal inspiration produces quantifiable changes in thoracic dimensions through these coordinated movements. The anteroposterior diameter increases by 2–3 cm through anterior rib elevation, while lateral rib excursion expands the transverse diameter by 3–4 cm. These movements can be assessed during clinical examination when patients place their palms on the anterior sternum with adducted upper extremities, allowing observation of thoracic expansion patterns (Ratnovsky et al. 2008).

The scapula contributes additional anatomical complexity by overlying ribs 2–7 in the posterolateral thorax. Its position varies with muscular activity, while its anatomical features—including the subscapular fossa, acromion process, and coracoid process—provide attachment sites for respiratory and thoracic wall muscles. The scapular spine anatomically divides the dorsal surface into supraspinous and infraspinous fossae, areas significant for respiratory muscle attachment and function (Skandalakis 2004; Standring 2020).

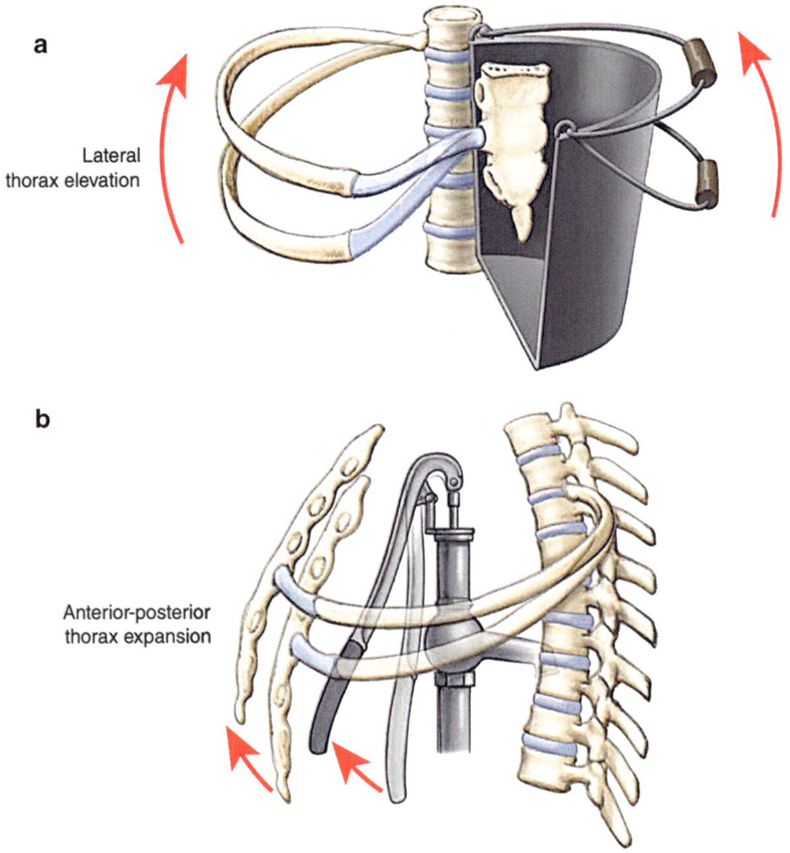

Fig. 11.1 Movement of the rib cage during breathing: (**a**) Bucket-handle motion, illustrating lateral expansion. (**b**) Pump-handle motion, illustrating anterior-posterior expansion. (Adapted with permission from Des Jardins (2013))

The clavicle completes the thoracic framework through its connections between the sternal manubrium medially and the scapular acromion laterally. Its morphology includes an anterior curve in the medial two-thirds and a posterior curve in the lateral third. The middle portion exhibits increased fracture susceptibility due to reduced muscular protection, affecting upper thoracic mobility (Moore et al. 2018).

Throughout respiratory cycles, this osseous framework demonstrates dynamic properties as anatomical landmarks shift relative to the vertebral column. The superior manubrial border maintains positional relationships with T2–T3 vertebrae, while the xiphisternal junction corresponds to the T9 vertebral level. These relationships undergo continuous modification during breathing, demonstrating thoracic adaptability (Standring 2020).

11.2 Anatomical and Functional Organization of the Airways

The respiratory system presents an organized network of airways extending from the external environment to the alveoli, where gas exchange takes place (Standring 2020). The conducting airways display anatomical and physiological differentiation into superior (upper) and inferior (lower) components, each with specific structural characteristics that support respiratory function (Drake et al. 2020; Moore et al. 2018).

The upper airways begin at the nasal cavity and extend through the pharynx to the larynx. The nasal passages contain specialized epithelium with a rich vascular supply, allowing for air warming and humidification. The pharyngeal region connects the nasal and oral cavities to the larynx, while maintaining separation of air and food passages through coordinated muscular actions (Netter 2019; Standring 2020).

Lower airways extend from the trachea through multiple bronchial generations to the terminal bronchioles. The trachea exhibits a C-shaped cartilaginous structure that maintains airway patency while allowing esophageal distention. Bronchi demonstrate progressive reduction in diameter and wall thickness as they divide, with a corresponding transition in supporting structures from cartilaginous rings to plates (Moore et al. 2018; West and Luks 2016).

The airway epithelium undergoes systematic changes that align with local functional requirements. The upper segments contain pseudostratified ciliated columnar epithelium with abundant goblet cells, transitioning to simple cuboidal epithelium in the terminal bronchioles (Proud 2008a) (Fig. 11.2). This epithelial organization supports specific functions including mucociliary clearance, particle filtration, and immune defense (Kumar and Clark 2012).

The conducting airways form an anatomical dead space of approximately 150 mL in adults. This volume represents the air that must move during each breath before fresh air reaches the respiratory zone, where gas exchange occurs (West and Luks 2016). The bronchial tree demonstrates specific branching patterns that optimize air distribution while minimizing resistance to flow. Each bronchial generation shows precise anatomical relationships with accompanying blood vessels and lymphatics, forming the bronchovascular bundle (Skandalakis 2004).

The airway smooth muscle arrangement changes throughout the bronchial tree. The larger airways contain discrete bundles of smooth muscle between cartilage plates, while smaller airways show a more continuous spiral arrangement. This anatomical organization allows for dynamic adjustment of airway caliber in response to various physiological and pathological conditions (Ratnovsky et al. 2008).

11.2.1 Anatomical and Functional Organization of the Upper Airways

The upper airways occupy the craniocervical region as the initial segment of the respiratory tract. This region includes the nasal cavity, oral cavity, pharynx, and larynx, forming an anatomical system that extends from the external environment to the thoracic cavity (Standring 2020). These structures perform multiple physiological functions that prepare inspired air for pulmonary gas exchange (West and Luks 2016).

The upper airway architecture displays specific epithelial and neurovascular characteristics. The respiratory mucosa primarily consists of pseudostratified ciliated columnar epithelium containing goblet cells, supported by submucosal glands. This cellular arrangement produces a protective mucus layer that works with ciliary movement to form a mechanical barrier against environmental particles and microorganisms (Whitsett and Alenghat 2015) (*see* Fig. 11.2). The submucosa contains dense vascular networks that regulate the temperature and humidity of inspired air before it reaches the lower airways (Drake et al. 2020).

Upper airway structures support both respiratory and non-respiratory functions. The olfactory neuroepithelium in the superior nasal conchae provides chemosensory capabilities, while the laryngeal structures work with the pharynx to control voice production and airway protection during swallowing (Moore et al. 2018). These multiple functions reflect an anatomical design that maximizes physiological efficiency.

The laryngeal region marks the transition between the upper airways and the tracheobronchial system, representing a change in mechanical properties. The trachea shows distinct anatomical features in its extrathoracic and intrathoracic portions, adapting to different mechanical forces and pressure changes (Kumar and Clark 2012).

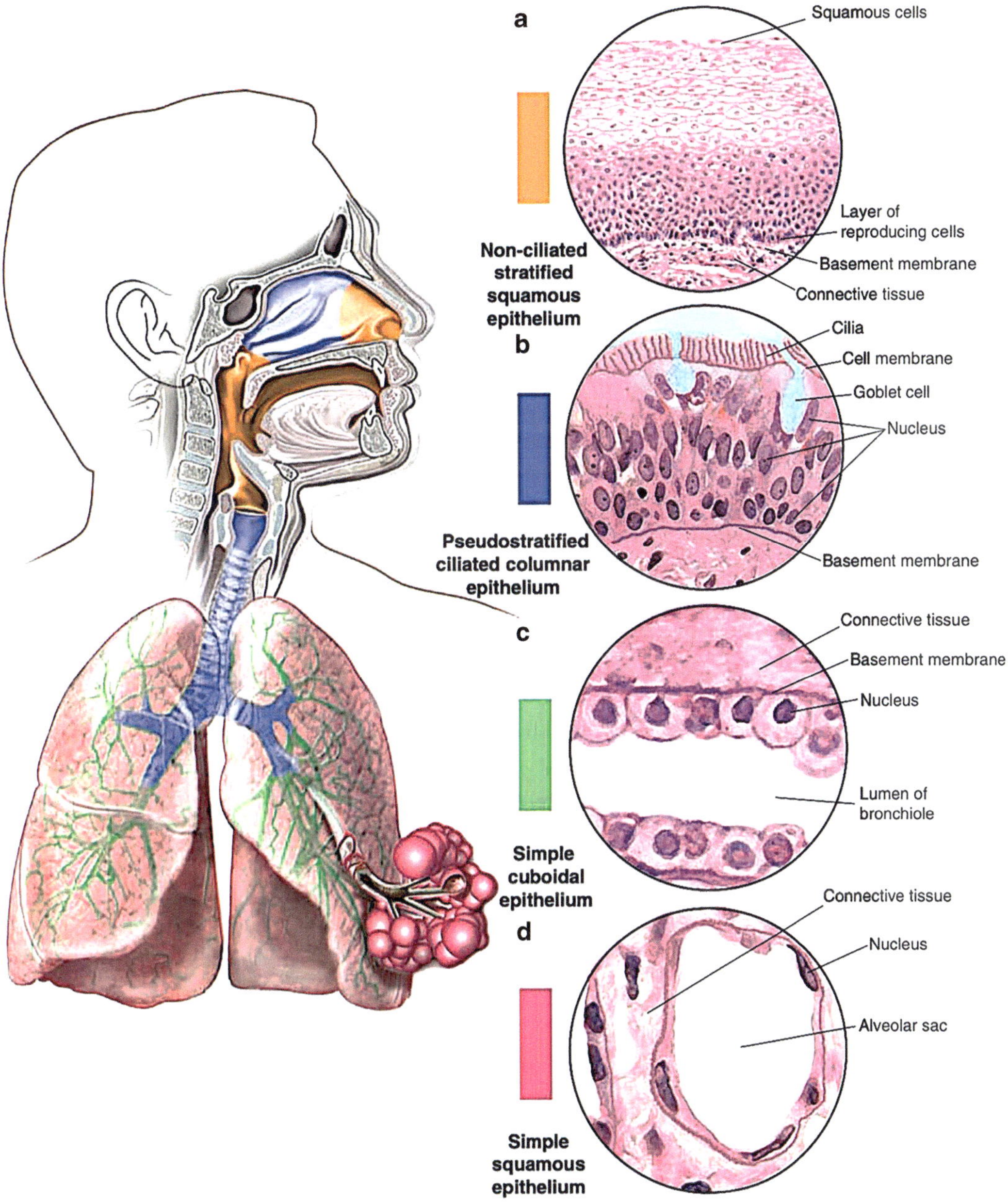

Fig. 11.2 Structural variations of the respiratory epithelium along the respiratory tract: (A) Stratified squamous epithelium comprises multiple cell layers and is located in the anterior nasal cavity, oral cavity, oropharynx, and laryngopharynx. (B) Pseudostratified ciliated columnar epithelium, with nuclei at varying levels, appears stratified and features cilia and goblet cells; it lines the posterior two-thirds of the nasal cavity and the tracheobronchial tree. (C) Simple cuboidal epithelium consists of a single layer of cube-shaped cells found in the bronchioles. (D) Simple squamous epithelium is a single layer of thin, flattened cells facilitating gas exchange and forming the alveolar walls and surrounding pulmonary capillaries. (Adapted with permission from Des Jardins (2013))

The extrathoracic segment contains specific cartilage arrangements and soft tissue components that accommodate these mechanical requirements.

The upper airway epithelium maintains specialized immunological functions through its cellular composition and molecular mechanisms. The respiratory mucosa contains immune cells and produces antimicrobial substances that provide additional protection against pathogens. This immune function works alongside mechanical barriers to maintain airway health.

Blood supply to the upper airways comes from multiple arterial branches that form extensive submucosal networks. This vascular arrangement supports both respiratory conditioning and local metabolic requirements. The lymphatic drainage parallels these vascular patterns, contributing to immune surveillance and fluid balance in the airway tissues (Netter 2019).

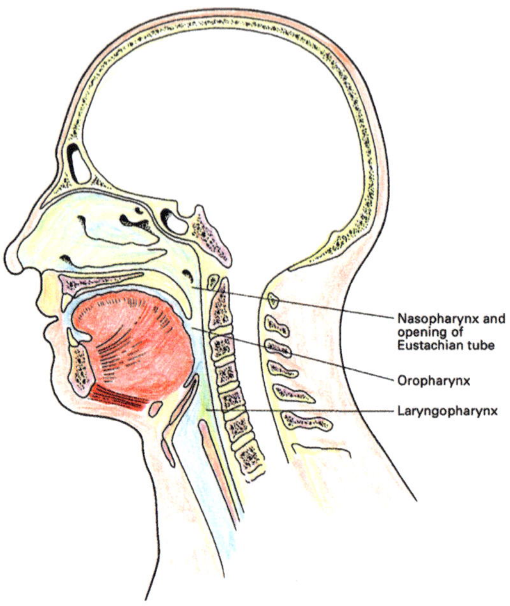

Fig. 11.3 Sagittal section of the head and neck illustrating the subdivisions of the pharynx. (Adapted from Ellis et al. (2004))

11.2.1.1 Pharynx

The pharynx represents a musculomembranous conduit that serves as a common pathway for both respiratory and digestive systems (Standring 2020). This anatomical structure extends from the basilar portion of the occipital bone superiorly to the esophageal origin at the sixth cervical vertebra (C6) inferiorly, maintaining intimate relationships with the prevertebral fascia and cervical vertebrae along its posterior aspect (Moore et al. 2018).

The pharyngeal complex demonstrates distinct regional specialization through three anatomically and functionally integrated segments (Drake et al. 2020): the nasopharynx, oropharynx, and laryngopharynx (Fig. 11.3). The nasopharynx, extending from the posterior nasal choanae to the soft palate, contains stratified cylindrical epithelium adapted for respiratory function. Its lateral walls house the pharyngeal ostia of the auditory (Eustachian) tubes, establishing communication with the middle ear cavities. This anatomical arrangement maintains pressure equilibrium across the tympanic membrane and provides mucociliary clearance for the middle ear cavity. The oropharynx spans from the soft palate to the vallecula, with its walls comprising the superior and middle pharyngeal constrictors. These muscular components contribute to deglutition and airway protection. The epithelial transition in this region reflects its exposure to both respiratory and digestive functions (Netter 2019). The laryngopharynx (hypopharynx) extends from the epiglottic tip to the esophageal inlet, featuring the middle and inferior pharyngeal constrictors in its wall composition. This region demonstrates complex neuromuscular coordination during both respiration and deglutition. The epithelial lining transitions to a stratified squamous type, protecting against mechanical and chemical stress.

Recent microanatomical studies have revealed specialized immunological zones within the pharyngeal mucosa, contributing to respiratory defense mechanisms. The pharyngeal wall architecture incorporates multiple muscle layers that coordinate to facilitate both respiratory and digestive functions while preventing their interference.

Pharyngeal Segments

Nasopharynx

The nasopharynx extends from the posterior nasal apertures (choanae) to the upper border of the soft palate (Standring 2020). This anatomical structure functions as the superior portion of the pharynx, positioned posterior to the nasal cavity (Drake et al. 2020). The region contains specialized mucosal lining composed of pseudostratified ciliated columnar epithelium, which maintains respiratory homeostasis through mucociliary clearance (Moore et al. 2018).

The posterior wall of the nasopharynx contains organized collections of lymphoid aggregates, known as pharyngeal tonsils or adenoids. These structures constitute a component of Waldeyer's ring—the ring of lymphoid tissue in the pharynx (Wineski 2019). During inflammatory conditions, these lymphoid structures may enlarge and result in complete obstruction of the nasopharyngeal airway, a condition particularly common in pediatric populations (Drake et al. 2020).

On the lateral walls of the nasopharynx lie the pharyngeal ostia of the eustachian tubes (tympanic tubes), establishing communication between the nasopharynx and the tympanic cavities (Standring 2020). These tubes regulate middle ear pressure through periodic opening during actions such as swallowing and yawning (Netter 2019). Various pathological conditions, including mucosal inflammation or excessive secretions, can impair the pressure-equalizing mechanism of these tubes, potentially leading to auditory deficits or middle ear effusion (Moore et al. 2018).

The nasopharyngeal mucosa possesses numerous mucous glands and maintains a rich vascular supply. These anatomical features support its physiological roles in air conditioning and immunological defense (Pawlina 2016).

Oropharynx

The oropharynx represents the middle pharyngeal segment, extending from the soft palate superiorly to the hyoid bone at the base of the tongue inferiorly (Drake et al. 2020). This anatomical region operates as a conduit for both deglutition and respiration, necessitating complex neuromuscular control (Standring 2020).

Two significant lymphoid structures reside within the oropharynx: the lingual tonsil at the posterior aspect of the tongue, and the palatine tonsils positioned between the palatopharyngeal and palatoglossal arches (Standring 2020). These lymphoid formations contribute to the immunological defense of the upper aerodigestive tract (Pawlina 2016).

The mucosal lining of the oropharynx differs from that of the nasopharynx, featuring nonciliated stratified squamous epithelium (*see* Fig. 11.2). This specialized epithelial structure responds to the mechanical forces experienced during swallowing and breathing (Wineski 2019). The morphological characteristics of this epithelium offer resistance against mechanical and chemical stimuli during food transit (Pawlina 2016).

The neural control of the oropharynx involves dual innervation: motor functions through the vagus nerve (Cranial Nerve X) and sensory input via the glossopharyngeal nerve (Cranial Nerve IX). This dual innervation pattern enables coordinated swallowing mechanisms and taste perception (Netter 2019). The complex neuromuscular organization allows for precise control of deglutition while maintaining airway protection (Moore et al. 2018).

Laryngopharynx

The laryngopharynx, also known as the hypopharynx, represents the inferior portion of the pharynx, extending from the base of the tongue to the entrance of the esophagus (Standring 2020). Its mucosal surface exhibits nonciliated stratified squamous epithelium, an adaptation that responds to regular mechanical stress during deglutition and respiration (Drake et al. 2020). Anterior to the laryngopharynx lies the epiglottis, which marks the superior boundary of the larynx. Mucosal extensions known as the aryepiglottic folds extend from the epiglottis around the laryngeal margins, creating a sphincteric mechanism that becomes active during swallowing (Moore et al. 2018). Medical professionals often examine these laryngopharyngeal structures through laryngoscopy while patients maintain a supine position (Wineski 2019).

The neural regulation of laryngopharyngeal function involves a complex interplay between sensory and motor components. The glossopharyngeal nerve (CN IX) provides sensory innervation, while the vagus nerve (CN X) supplies motor control to the musculature (Netter 2019). This neural configuration enables the pharyngeal reflex, commonly known as the gag reflex, which prevents aspiration of food and liquids. The same mechanism also prevents the base of the tongue from falling posteriorly and obstructing the laryngopharynx during sleep in the supine position (Wineski 2019).

Clinical management of airway complications often necessitates bypassing the upper airway through nasal or oral endotracheal intubation (Roberts and Hedges 2018). When implementing this intervention, medical practitioners must maintain appropriate gas conditioning through warming and humidification. Without proper conditioning, the tracheobronchial mucus layer becomes dehydrated, resulting in increased viscosity and decreased mobility. Viscous, stagnant secretions undergo stages of accumulation, resulting in partial obstruction with air-trapping, and potentially progressing to total obstruction with airway collapse (Gropper 2020). Medical practitioners must develop precise knowledge of laryngopharyngeal and laryngeal anatomical landmarks, including the vallecula, epiglottis, esophagus, vocal folds, and trachea. This knowledge becomes particularly significant during endotracheal intubation procedures and evaluation of dysphagia (Walls and Murphy 2012).

Pharyngeal Musculature

The pharyngeal muscles form a dynamic muscular tube that enables multiple physiological functions (Standring 2020). The classical comparison to nested flower pots effectively describes the arrangement of the superior, middle, and inferior constrictors, which create a complex contractile system along the pharyngeal length (Drake et al. 2020) (Fig. 11.4).

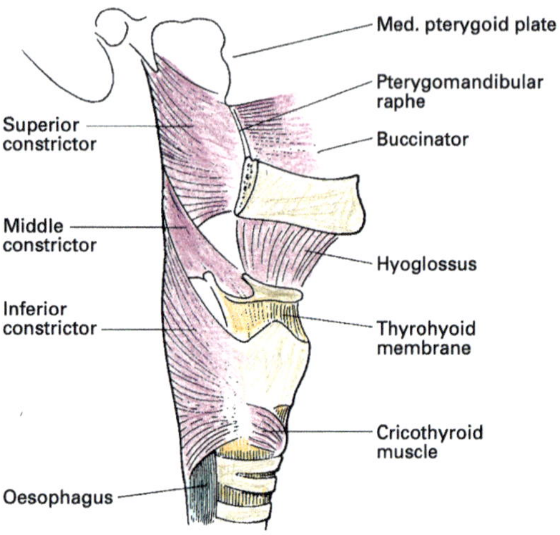

Fig. 11.4 The constrictor muscles of the pharynx. (Adapted from Ellis et al. (2004))

These constrictors demonstrate precise anatomical organization, originating from multiple bony and cartilaginous points before converging on a median raphe posteriorly. This raphe extends from the pharyngeal tubercle of the occipital bone superiorly to its integration with the esophageal wall inferiorly (Moore et al. 2018).

The superior constrictor begins its course from multiple points: the medial pterygoid plate's lower portion, pterygoid hamulus, pterygomandibular raphe, and the posterior mylohyoid line of the mandible. A notable anatomical gap exists between its upper margin and the skull base, creating a passage for the Eustachian tube into the nasopharynx (Netter 2019).

The middle constrictor displays a fan-like configuration, radiating from the hyoid bone's lesser horn, the superior aspect of its greater horn, and the stylohyoid ligament's inferior portion. This arrangement allows coordinated movement during swallowing and speech (Ellis and Mahadevan 2019).

The inferior constrictor, the most substantial of these muscles, demonstrates functional specialization through its two components. The lower cricoid-based portion (cricopharyngeus) acts as a sphincter with transverse fibers, while the superior thyroid-based segment facilitates propulsion through obliquely arranged fibers. This anatomical arrangement holds particular clinical significance, as dysfunction in the coordination between these components may result in pharyngeal pouch formation through Killian's dehiscence.

The neural control of these muscles flows through the pharyngeal nerve plexus, carrying accessory nerve fibers via the vagal pharyngeal branch. The inferior constrictor receives additional innervation from the superior laryngeal nerve's external branch and the recurrent laryngeal nerve, enabling fine coordination of swallowing and airway protection.

11.2.1.2 Larynx

The larynx represents a complex anatomical structure positioned between the base of the tongue and the superior aspect of the trachea. This remarkable organ has evolved from its primary function as a protective valve of the respiratory passages to incorporate the sophisticated mechanism of speech production (Standring 2020).

The larynx occupies a position opposite the fourth, fifth, and sixth cervical vertebrae (Fig. 11.5), separated from these vertebrae by the laryngopharynx. Its superficial coverage consists of investing deep fascia in the midline and thin strap muscles laterally, making it readily palpable. Women and children typically display a more superior positioning of the larynx (Moore et al. 2018).

The structural framework comprises articulating cartilages connected by ligaments, with nine distinct cartilaginous components: the hyoid bone, three unpaired cartilages (epiglottic, thyroid, and cricoid), and three paired cartilages (cuneiform, arytenoid, and corniculate) (Figs. 11.6 and 11.7). These cartilages demonstrate age-related variations, with hyaline composition in the thyroid, cricoid, and the majority of arytenoid cartilages, while elastic cartilage forms the epiglottic, corniculate, cuneiform, and vocal process of arytenoid in adults (Drake et al. 2020).

The larynx divides into three functional regions: the supraglottis (containing the epiglottis, aryepiglottic folds, false vocal cord folds, and

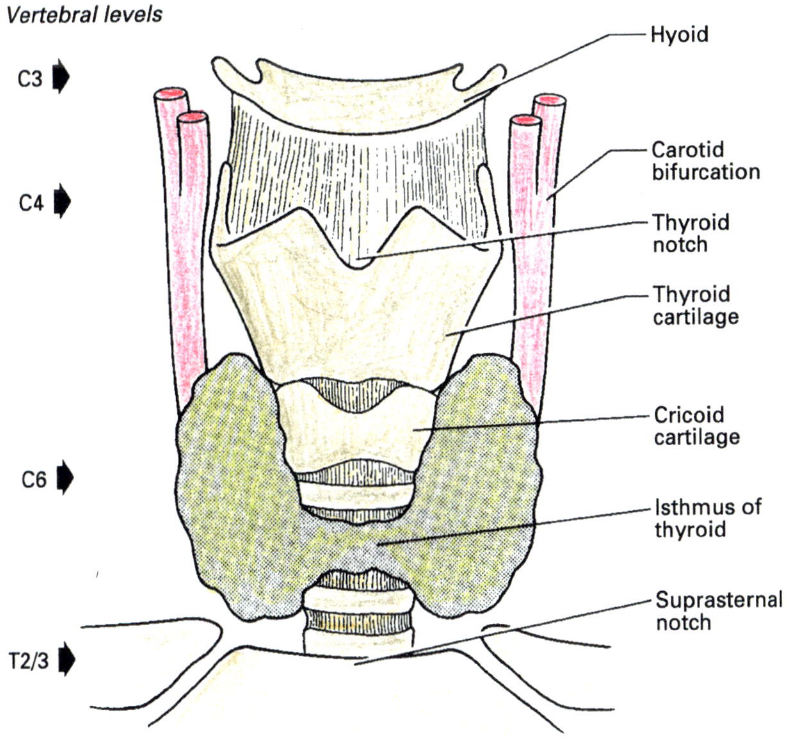

Fig. 11.5 Anterior view of the larynx illustrating key adjacent landmarks and their corresponding vertebral levels. (Adapted from Ellis et al. (2004))

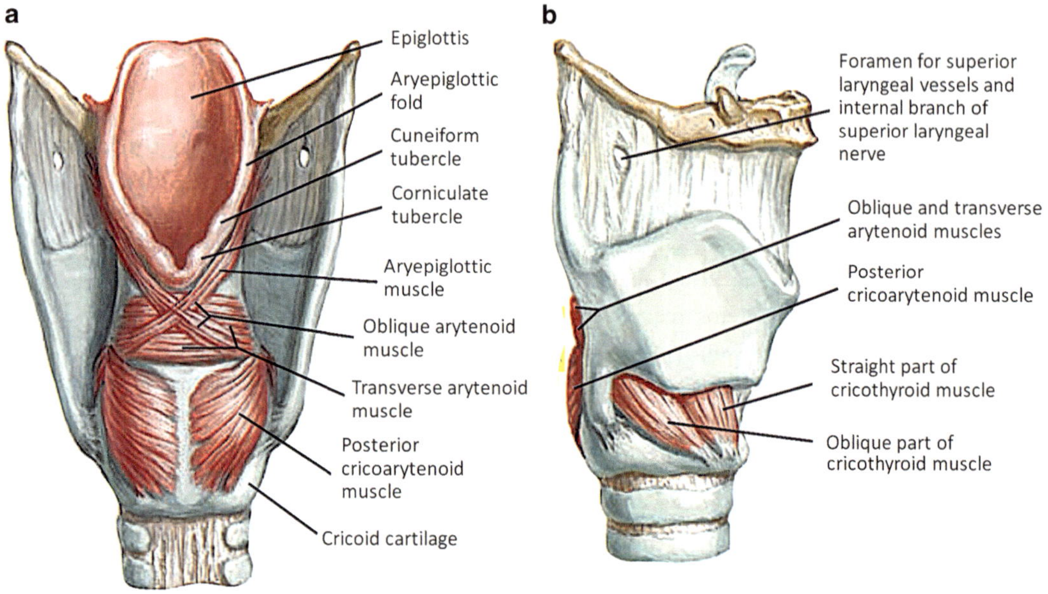

Fig. 11.6 Posterior (**a**) and lateral (**b**) views of the larynx. (Adapted with permission from McFarland (2015))

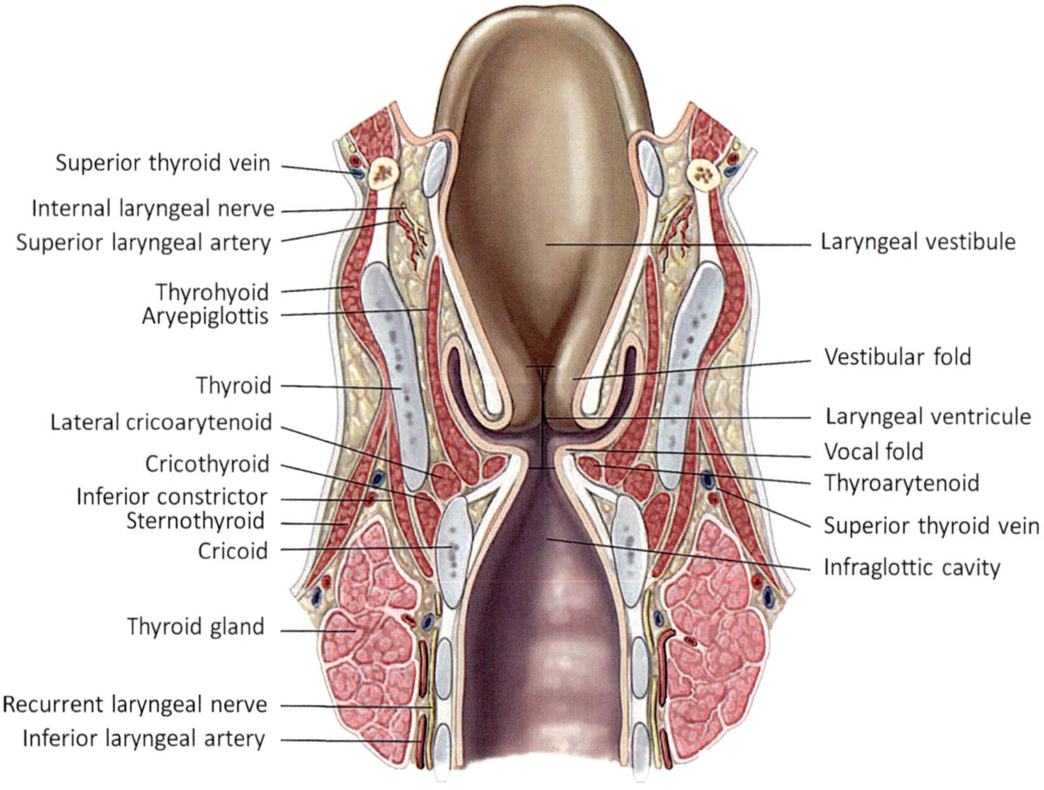

Fig. 11.7 Coronal section of the larynx, showing the vestibular and true vocal folds. (Adapted with permission from McFarland (2015))

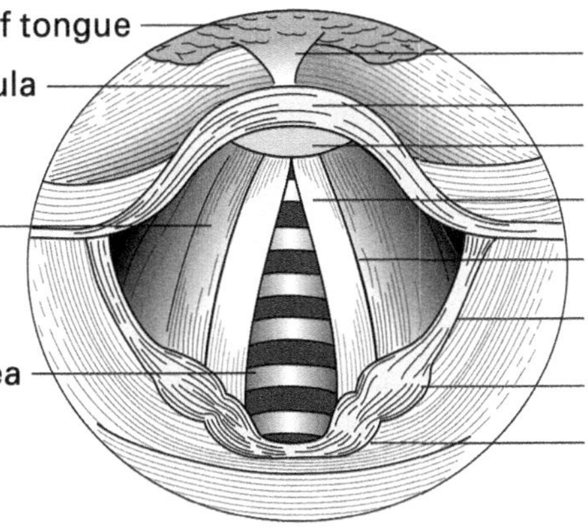

Fig. 11.8 Laryngoscopic view of the larynx. (Adapted from Ellis et al. (2004))

ventricle), the glottis (characterized by true vocal folds), and the subglottis (connecting to the trachea). The rima glottidis, positioned between the vocal folds, measures approximately 23 mm in males and 17–18 mm in females (Netter 2019).

During laryngoscopy, practitioners observe a sequence of structures: the base of the tongue, valleculae, anterior surface of the epiglottis, and the laryngeal aditus (Fig. 11.8). The vocal cords appear as pale, glistening ribbons extending from the thyroid cartilage angle to the arytenoids' vocal processes (Gropper 2020).

The laryngeal blood supply originates from superior and inferior thyroid arterial branches, with corresponding venous drainage into the superior thyroid and inferior thyroid veins. The vagus nerve provides both motor and sensory innervation through its superior and recurrent laryngeal branches. The laryngeal mucosa maintains continuity with the pharyngeal mucosa, featuring ciliated epithelium with intermittent goblet cells. Areas frequently contacting food display stratified squamous epithelium for enhanced protection (Roberts and Hedges 2018).

Laryngeal Cartilages

The laryngeal framework comprises nine cartilages that form an intricate structural assembly. This framework includes three unpaired cartilages (thyroid, cricoid, and epiglottis) and three paired cartilages (arytenoid, corniculate, and cuneiform), all maintained in position by a complex arrangement of ligaments, membranes, and muscles (Standring 2020).

Thyroid Cartilage

The thyroid cartilage, the largest of the laryngeal cartilages, presents as a shield-like structure with two laminae converging anteriorly. In males, this convergence creates the prominent laryngeal prominence or Adam's apple, while remaining less pronounced in females. Each lamina possesses superior and inferior cornua at its posterior border, with the inferior cornua featuring articular facets for cricoid cartilage articulation (Moore et al. 2018). The superior border displays a characteristic V-shaped thyroid notch, and the thyrohyoid membrane connects it to the hyoid bone above (Drake et al. 2020).

Cricoid Cartilage

The cricoid cartilage exhibits a distinctive signet ring configuration, with its broader posterior lamina connected anteriorly by a narrow arch. The lateral aspects of the lamina present two articular surfaces: one accommodating the thyroid cartilage's inferior horn and another supporting the arytenoid cartilage superiorly. This cartilage connects inferiorly to the first tracheal ring (Netter 2019).

Arytenoid Cartilages

These paired, pyramidal cartilages rest on the cricoid lamina's superolateral aspect. Each arytenoid features a lateral muscular process, receiving insertions from posterior and lateral cricoarytenoid muscles, and an anterior vocal process that anchors the vocal ligament. The apex articulates with the corniculate cartilage (Sinnatamby 2011).

Epiglottis

The leaf-shaped epiglottis attaches inferiorly to the thyroid cartilage via the thyro-epiglottic ligament. Its superior portion projects posterosuperiorly, overhanging the laryngeal inlet. The posterior surface features a tubercle, while the anterior surface forms median and lateral glosso-epiglottic folds. The valleculae, depressions flanking the median fold, hold clinical significance as common sites for foreign body impaction.

In neonates, the epiglottis displays distinctive characteristics, including deeper furrows and occasionally a V-shaped appearance during laryngoscopy. This elongated, deeply grooved configuration resembles aquatic mammalian anatomy, facilitating protection of the nasotracheal airway during suckling.

Accessory Cartilages

The corniculate cartilages present as small nodules at the arytenoid apices, while the cuneiform cartilages reside within the aryepiglottic folds, contributing to fold stability (Wineski 2019).

11.2.1.3 Interior Architecture of the Larynx

The internal aspect of the larynx displays an intricate arrangement of mucosal folds and specialized tissues that facilitate both protective and phonatory functions (Standring 2020). The mucosal lining forms two distinct pairs of folds that project into the laryngeal lumen: the superior false vocal folds and the inferior true vocal folds (Moore et al. 2018) (*see* Fig. 11.8).

Vocal Fold Architecture

The true vocal folds, fundamental to voice production, contain elastic tissue bands known as vocal ligaments along their medial borders. These folds establish anterior attachments to the posterior thyroid cartilage surface and posterior connections to the vocal processes of the arytenoid cartilages. The arytenoid cartilages' capacity for rotation around their vertical axes through the cricoarytenoid joints modulates vocal fold tension (Drake et al. 2020).

Glottic Region

The rima glottidis, or glottis, represents the interval between the true vocal folds. This anatomical space holds particular clinical significance as it marks the narrowest point of the adult larynx. However, in pediatric anatomy, the cricoid cartilage constitutes the narrowest segment (Netter 2019). The anatomical differences between adult and pediatric airways influence clinical manifestations of respiratory pathologies. Infants and young children commonly develop glottic and subglottic edema during viral or bacterial infections, manifesting as the croup syndrome (laryngotracheobronchitis and acute epiglottitis). This condition produces characteristic inspiratory stridor due to airway narrowing (Gropper 2020).

Mucosal Characteristics

The laryngeal mucosa demonstrates regional specialization in its epithelial composition. Superior to the vocal folds, the mucosa consists of nonciliated stratified squamous epithelium. In contrast, the subglottic region features pseudostratified ciliated columnar epithelium (Pawlina 2016). This differential epithelial distribution reflects the distinct functional requirements of each region (Wineski 2019).

11.2.1.4 Laryngeal Musculature

The laryngeal muscular system embodies a complex architectural design comprising extrinsic and intrinsic components, each contributing to precise movements essential for respiratory, phonatory, and deglutitive functions (Standring 2020). This complex neuromuscular organization enables the larynx to maintain airway patency, protect lower respiratory structures, and modulate vocal production through coordinated cartilaginous movements (Moore et al. 2018).

Extrinsic Muscles

The extrinsic musculature establishes a dynamic suspension system for laryngeal positioning. The infrahyoid group, including the sternothyoid, sternothyroid, thyrohyoid, and omohyoid muscles, facilitates vertical laryngeal displacement within the anterior neck. The sternothyroid muscle, originating from the posterior manubrial surface and inserting into the thyroid lamina's oblique line, receives innervation from the ansa hypoglossi. This neural connection enables precise laryngeal depression during complex functions such as voice modulation and swallowing (Drake et al. 2020).

The thyrohyoid muscle, receiving innervation from C1 fibers via the hypoglossal nerve, produces controlled laryngeal elevation. This action proves particularly significant during the pharyngeal phase of deglutition, where proper laryngeal elevation contributes to airway protection (Gropper 2020).

The suprahyoid muscle group—comprising the stylohyoid, mylohyoid, digastric, geniohyoid, and stylopharyngeus muscles—creates a complex force vector system. These muscles direct the hyoid bone in anterior, superior, and posterior directions, thereby influencing laryngeal position through the thyrohyoid membrane and associated ligamentous attachments (Netter 2019).

Intrinsic Muscles

The intrinsic muscles of the larynx exhibit specialized adaptations that enable precise control of airway patency and voice production (Drake et al. 2020; Moore et al. 2018) (Fig. 11.9). The posterior cricoarytenoid muscle serves as the primary abductor of the vocal folds, performing lateral rotation of the arytenoid cartilages. This action maintains airway patency during inspiration, preventing inspiratory stridor and ensuring adequate ventilation (Sinnatamby 2011; Standring 2020).

The lateral cricoarytenoid muscles function as primary adductors, facilitating medial rotation of the arytenoid cartilages to approximate the vocal folds. This mechanism operates in conjunction with the interarytenoid muscle, which stands out as the sole unpaired intrinsic muscle of the larynx (Pawlina 2016). The interarytenoid muscle comprises both transverse and oblique fiber components, with the oblique fibers extending into the aryepiglottic fold. This anatomical arrangement enhances the protective sphincteric function of the laryngeal inlet during swallowing and prevents aspiration (Norton 2017).

Voice production involves precise control of vocal fold tension through coordinated muscular actions. The thyroarytenoid muscle originates from the posterior aspect of the thyroid cartilage and contains specialized fibers that form the vocalis muscle (Wineski 2019). This structural organization enables fine-tuned adjustments of vocal fold tension, which is essential for modulating voice pitch. The cricothyroid muscle, positioned external to the laryngeal cartilaginous framework, acts as the primary tensor of the vocal folds through its effect on cricoid cartilage positioning (Moore et al. 2018; Netter 2019).

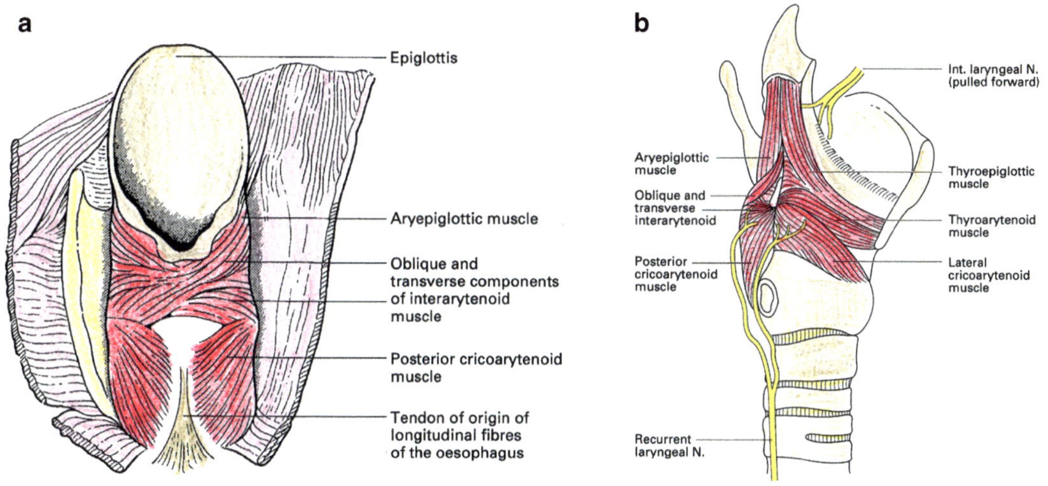

Fig. 11.9 Intrinsic muscles of the larynx: (**a**) posterior view and (**b**) lateral view. (Adapted from Ellis et al. (2004))

11.2.1.5 Innervation of the Laryngeal Structures

The vagus nerve (cranial nerve X) provides innervation to the larynx through its superior and recurrent laryngeal divisions, establishing both sensory and motor functions (Standring 2020).

The superior laryngeal nerve demonstrates specific anatomical relationships as it courses beneath the internal and external carotid arteries before bifurcating. Its external branch, approximately 2–3 mm in diameter, descends anteriorly along the lateral surface of the inferior pharyngeal constrictor muscle to innervate the cricothyroid muscle. This muscle functions as the primary tensor of the vocal folds, modulating voice pitch through vocal fold elongation (Sanders and Mu 1998). The larger internal branch (4–5 mm in diameter) penetrates the thyrohyoid membrane, accompanied by the superior laryngeal vessels. This branch provides sensory innervation to the laryngeal mucosa superior to the vocal folds, including the supraglottic and glottic regions. Anatomical studies confirm that this branch also transmits motor signals to the interarytenoid muscle, contributing to glottic closure.

The internal laryngeal nerve's anatomical position, situated under the piriform fossa mucosa, approximately 2–4 mm deep to the surface. This location permits effective topical administration of local anesthetic agents, particularly beneficial during diagnostic and therapeutic procedures such as laryngoscopy, bronchoscopy, and advanced airway management (Watkinson and Clarke 2018).

The recurrent laryngeal nerves exhibit asymmetric anatomical pathways. The right nerve diverges from the vagus at its intersection with the right subclavian artery (approximately at the level of T1–T2 vertebrae), forming a distinct loop under the artery before ascending through the tracheoesophageal groove. The left nerve follows a longer course, originating where the vagus crosses the aortic arch (typically at T4–T5 vertebral level), passing inferior to the arch, and ascending through the same groove. This anatomical variation accounts for the increased vulnerability of the left recurrent laryngeal nerve during thoracic surgical procedures (Wilson-Pauwels et al. 2002). Upon entering the cervical region, both nerves establish identical relationships, ascending in the tracheoesophageal groove approximately 0.5–1 cm lateral to the trachea. These nerves provide motor innervation to all intrinsic laryngeal muscles except the cricothyroid, while simultaneously conveying sensory information from the laryngeal mucosa inferior to the vocal folds.

The complex innervation pattern of the larynx necessitates careful consideration during surgical procedures in the anterior neck and superior mediastinum. Injury to these nerves can result in various voice disorders and airway protection deficits. Unilateral recurrent laryngeal nerve paralysis typically presents with hoarseness, while bilateral paralysis may cause severe respiratory distress requiring immediate medical intervention.

11.2.1.6 Respiratory Mechanics of the Larynx

The larynx, situated at the anterior aspect of the neck between C3 and C6 vertebral levels, executes two fundamental respiratory functions mediated by intrinsic laryngeal muscles (Standring 2020). During quiet respiration, the posterior cricoarytenoid muscles contract, causing the vocal folds to undergo lateral displacement (abduction), which expands the glottic aperture (Drake et al. 2020). This neuromuscular coordination occurs through vagal innervation, specifically the recurrent laryngeal nerve (Netter 2019). In the expiratory phase, these folds exhibit minimal medial movement (adduction) via thyroarytenoid muscle action while maintaining patent airways for continuous bidirectional airflow.

The second primary function involves glottic closure during expiration, known clinically as Valsalva's maneuver (Ludlow 2015). This physiologic mechanism manifests through synchronized contraction of the lateral cricoarytenoid and interarytenoid muscles, resulting in complete medial displacement of the laryngeal structures, incorporating both true and false vocal folds. The resultant hermetic seal of the laryngeal lumen facilitates increased intrathoracic pressure, reaching values of 40–100 mmHg during maximum effort. This pressure modulation becomes necessary during various physiologic activities: heavy lifting, forceful exertion, expectoration, pharyngeal clearance, emesis, micturition, defecation, and childbirth.

Research has demonstrated that patients with cervical spinal cord injuries above C5 may experience compromised laryngeal function due to interrupted descending neural pathways and altered respiratory muscle recruitment patterns (Berlowitz et al. 2016). The impairment predominantly affects the coordinated function of the posterior cricoarytenoid muscles, leading to reduced glottic aperture during inspiration. This pathophysiologic alteration may contribute to increased work of breathing and decreased respiratory efficiency (Brown et al. 2006).

Additionally, diminished expiratory muscle strength in cervical spinal cord injury patients can impair the effectiveness of the Valsalva maneuver, reducing the ability to generate adequate intrathoracic pressure for airway clearance (Schilero et al. 2014). This dysfunction may increase the risk of respiratory complications, particularly mucus retention and atelectasis, necessitating specialized respiratory rehabilitation protocols.

11.2.2 Anatomical and Functional Organization of the Lower Airways

The lower respiratory system initiates at the trachea, beginning at the inferior border of the cricoid cartilage, which anatomically corresponds to the sixth cervical vertebra (Standring 2020). This tubular structure descends through the superior mediastinum before reaching its terminus at the fifth thoracic vertebra, where it divides into right and left main bronchi (Drake et al. 2020). Throughout its course, the trachea maintains its patency through an ingenious design of 16–20 incomplete C-shaped cartilaginous rings interconnected by fibroelastic tissue, while the posterior aspect remains flexible due to the presence of the trachealis muscle (Netter 2019). The internal surface of the trachea presents a specialized ciliated pseudostratified columnar epithelium interspersed with goblet cells, creating an efficient mucociliary escalator system for airway clearance (Moore et al. 2018).

The main (primary) bronchi exhibit anatomical asymmetry. The right main bronchus adopts a more vertical trajectory, deviating approximately 25° from the sagittal plane, while maintaining a relatively shorter course. In contrast, its left main bronchus follows a more horizontal path, angling approximately 45° from the sagittal plane as it courses posterior to the aortic arch (Ellis and Mahadevan 2019). This anatomical arrangement explains the clinical observation of foreign bodies more frequently entering the right bronchial tree, creating a significant consideration in respiratory medicine.

As the airways progress deeper into the lungs, they undergo a series of precise divisions. The main bronchi first branch into secondary bronchi (lobar bronchi), supplying individual lobes—three on the right (upper, middle, lower) and two on the left (upper, lower). These further subdivide into tertiary bronchi (segmental bronchi), each responsible for a specific bronchopulmonary segment. The branching continues through multiple generations of increasingly smaller bronchioles, ultimately leading to respiratory bronchioles (terminal bronchioles), alveolar ducts, and the alveoli themselves, where the vital process of gas exchange occurs (West and Luks 2016).

The bronchial wall architecture demonstrates remarkable adaptability throughout this branching sequence. The prominent cartilaginous support gradually transitions from complete rings to scattered plates before disappearing entirely in the smaller airways. Simultaneously, smooth muscle becomes increasingly prominent, and the epithelial lining transforms from its initial pseudostratified columnar form to a simple cuboidal arrangement in the terminal bronchioles. This progressive modification in structure perfectly aligns with the changing functional demands at each level of the bronchial tree (Standring 2020) (*see* Fig. 11.2).

Through this intricate organization, the lower airways achieve their dual objectives with remarkable efficiency. The conducting airways, extending from the trachea through the terminal bronchioles, serve as complex conduits for air transport and conditioning. Meanwhile, the respiratory zone, encompassing the respiratory bronchioles through alveoli, provides the extensive surface area necessary for optimal gas exchange (West and Luks 2016).

11.2.2.1 The Tracheobronchial Tree

The tracheobronchial tree represents an intricate network of airways that progressively branch throughout the lungs following the passage of air through the larynx (Standring 2020). This system demonstrates a hierarchical branching pattern, with each subsequent generation becoming narrower, shorter, and more numerous (Drake et al. 2020). The airways differentiate into two fundamental categories: cartilaginous airways, which solely conduct air, and noncartilaginous airways, which facilitate both air conduction and gas exchange (Moore et al. 2018) (Table 11.1).

The branching pattern typically follows a dichotomous arrangement, with each division resulting in dimensional reduction of both length and diameter. The human tracheobronchial tree comprises 18–33 generations (average 23) and functionally divides into conducting and respiratory zones (West and Luks 2016). The conducting zone extends from the nasal cavity to the larger bronchioles, while the respiratory zone encompasses the smallest bronchioles to alveoli. The final 8–10 generations incorporate alveoli within their walls, marking the transition to gas exchange territories (Netter 2019).

Structural variations characterize different airway levels. Large and mid-sized bronchi contain substantial cartilaginous support, which progressively diminishes in smaller bronchi. Bronchioles, marking a significant transition point, lack cartilaginous components entirely. Terminal bronchioles represent the ultimate purely conducting airways before transitioning to respiratory bronchioles, where the epithelial lining becomes interrupted by alveolar presence (Levitzky 2007).

The pulmonary acinus, representing the fundamental respiratory unit, begins at the terminal bronchiole and includes successive generations of respiratory bronchioles, alveolar ducts, and alveolar sacs (Weibel 2017). Typically, three generations of respiratory bronchioles exist within each acinus. These structures, along with 2–9 generations of alveolar ducts and sacs, constitute the basic gas exchange unit. This architectural arrangement creates one of the human body's most extensive surface areas for facilitating respiratory function (Standring 2020) (Fig. 11.10).

The Cartilaginous Airways

As shown in Table 11.1, the conducting zone of the respiratory system encompasses the cartilaginous airways, which form a complex network beginning at the trachea and extending through the hierarchical bronchial divisions (Standring 2020). This anatomical continuum progresses through main bronchi, lobar bronchi, segmental

Table 11.1 Major structures and their corresponding generations within the tracheobronchial tree

Functional zone	Structures of the lungs	Generations	Characteristic
Conducting zone	Trachea	0	Cartilaginous airways
	Main stem bronchi	1	
	Lobar bronchi	2	
	Segmental bronchi	3	
	Subsegmental bronchi	4–9	
	Bronchioles	10–15	Noncartilaginous airways
	Terminal bronchioles	16–19	
Respiratory zone	Respiratory bronchioles	20–23	Sites of gas exchange
	Alveolar ducts	24–27	
	Alveolar sacs	28	

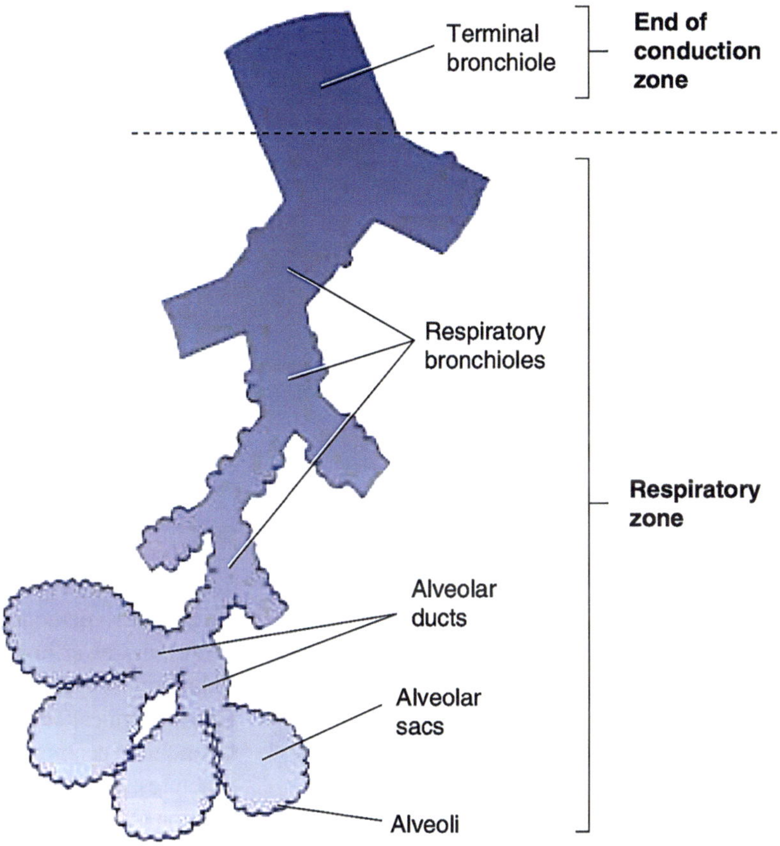

Fig. 11.10 Schematic illustration of structures distal to the terminal bronchioles, collectively termed the primary lobule. (Adapted with permission from Thiriet (2014))

bronchi, and ultimately subsegmental bronchi, each level maintaining structural characteristics while sharing fundamental cartilaginous support (Drake et al. 2020).

Trachea

The trachea is a well-structured passage that extends from the cricoid cartilage at the sixth cervical vertebra (C6) to its bifurcation at approximately the fifth thoracic vertebra (Standring 2020). In adults, this dynamic structure measures 11–15 cm in length, with approximately 5 cm positioned superior to the suprasternal notch—a distance that increases to nearly 8 cm during full neck extension. The tracheal diameter typically correlates with individual anthropometry, approximating the diameter of the subject's index finger (1.5–2.5 cm) (Drake et al. 2020).

The tracheal wall exhibits a specific regional structure, consisting of an anterolateral wall supported by 16–20 C-shaped cartilaginous rings and a posterior fibromuscular membrane (trachealis muscle). The cartilaginous rings, joined by fibroelastic tissue, terminate posteriorly at the junction of the trachea and esophagus (Moore et al. 2018). The carina, a unique keel-shaped cartilage, indicates the tracheal bifurcation and manifests as a noticeable sagittal ridge during bronchoscopic evaluation. The pathological flattening of this ridge frequently indicates hilar lymphadenopathy or structural distortion due to numerous illnesses (Netter 2019).

The microscopic structure of the tracheal wall exhibits significant specialization. The epithelial lining consists predominantly of ciliated cells interspersed with goblet cells in a 5:1 ratio, with cell heights ranging from 40 to 60 μm. These ciliated cells display hundreds of cilia on their apical surface, while goblet cells, found alone or in clusters, facilitate mucus secretion. Additional specialized cells, known as brush cells, feature characteristic apical microvilli measuring 1 μm

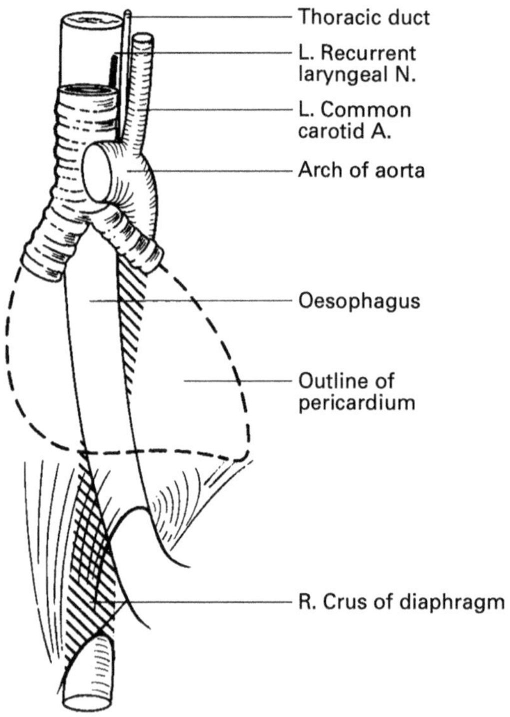

Fig. 11.11 The trachea and its surrounding anatomical structures. (Adapted from Ellis et al. (2004))

in length and 100 nm in diameter, potentially serving sensory or progenitor functions (Weibel 2017).

The trachea maintains intimate anatomical relationships with vital structures, including blood vessels, nerves, and the esophagus. Its anterior surface relates to the inferior thyroid veins, left brachiocephalic vein, and aortic arch, while laterally it associates with the inferior thyroid arteries and recurrent laryngeal nerves (Fig. 11.11). The right vagus nerve and brachiocephalic artery course along its right aspect, while the left recurrent nerve, aortic arch, and left common carotid and subclavian arteries parallel its left side (Standring 2020).

Clinical significance emerges in endotracheal intubation, where proper tube positioning requires placement approximately 2 cm superior to the carina, verified radiographically. Deep insertion beyond the carina typically results in right main stem bronchus intubation due to its more vertical orientation, potentially causing left lung atelectasis from inadequate ventilation (Levitzky 2007).

Main Bronchi

The tracheal bifurcation at the carina initiates the bronchial system, demonstrating anatomical asymmetry that significantly influences respiratory physiology and clinical interventions (Standring 2020). This bifurcation exhibits dynamic positioning: while typically aligned with the fourth thoracic vertebra (T4) in supine position, it may descend to the sixth thoracic vertebra (T6) during full inspiration in the erect position, reflecting the respiratory system's adaptability to positional changes (Drake et al. 2020).

The right main bronchus (Fig. 11.12) presents morphological characteristics: a shorter course (approximately 2.5 cm), wider luminal diameter (12–16 mm), and more vertical orientation (25° from vertical). In contrast, the left main bronchus (Fig. 11.13) demonstrates a longer trajectory (5 cm), narrower luminal diameter (10–14 mm), and more oblique course (40–60° from vertical). This anatomical asymmetry evolved as an adaptation to accommodate adjacent mediastinal structures, particularly the aortic arch's leftward position (Moore et al. 2018).

The neurovascular relationships of these structures reveal intricate three-dimensional arrangements essential for clinical consideration. The right main bronchus establishes significant relationships with the pulmonary artery, which transitions from an inferior to an anterior position relative to the bronchus. The azygos vein forms its characteristic superior arch, creating a distinctive anatomical landmark. The left main bronchus navigates a more complex course, passing beneath the aortic arch while maintaining posterior relationships with the esophagus, thoracic duct, and descending aorta. The left pulmonary artery demonstrates a superior-to-anterior positional transition relative to the bronchus (Netter 2019).

The histological organization of the main bronchi maintains continuity with tracheal architecture while demonstrating progressive specialization. The wall structure comprises incomplete cartilaginous rings united by fibroelastic tissue, with smooth muscle completing the posterior aspect. The mucosa presents pseudostratified ciliated columnar epithelium interspersed with

Fig. 11.12 Anterior view of the trachea and main bronchi. (Adapted from Ellis et al. (2004))

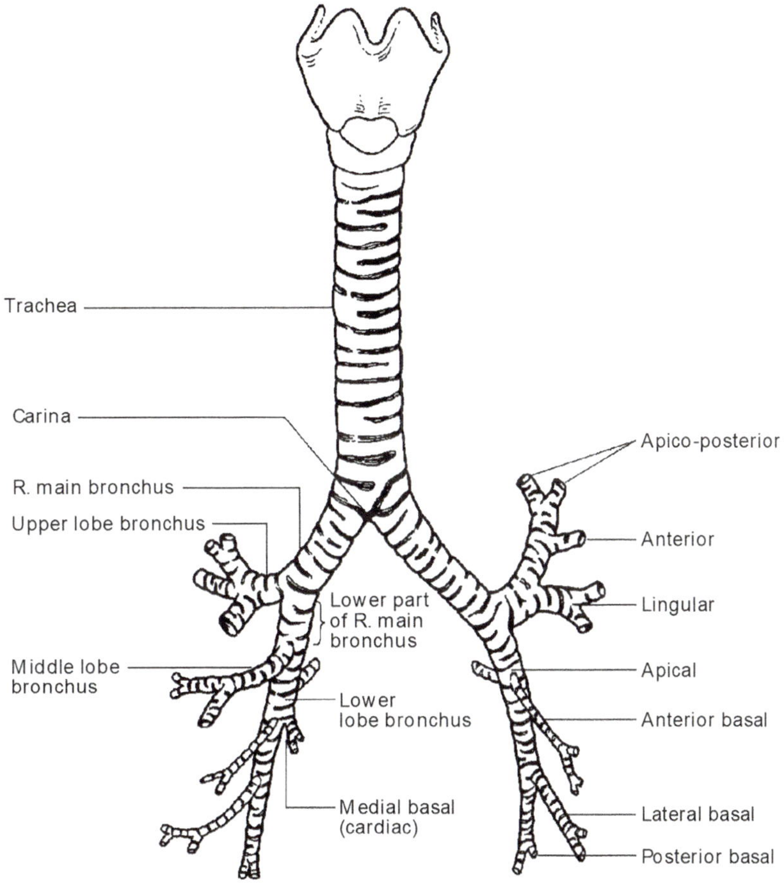

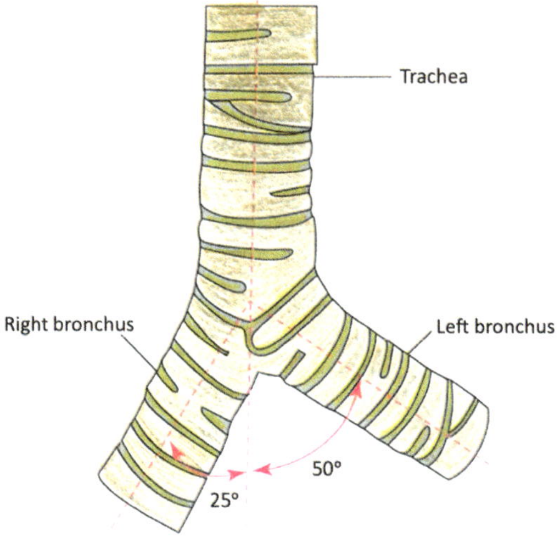

Fig. 11.13 Tracheobronchial angle illustrating the angle formed between the trachea and primary bronchi: 25° for the right bronchus and 50° for the left bronchus. (Adapted from Singh (2014a))

goblet cells, maintaining the respiratory tract's mucociliary clearance mechanism (Weibel 2017).

These anatomical considerations significantly influence clinical procedures, particularly during selective bronchial intubation and bronchoscopy. The right main bronchus's shorter length and early origin of its upper lobe bronchus (approximately 2.5 cm from the carina) necessitates specialized endotracheal tubes featuring lateral orifices to prevent upper lobe obstruction. Conversely, the left main bronchus's longer course (5 cm) between the carina and upper lobe bronchus origin readily accommodates standard endobronchial tubes without risking lobar obstruction (Ellis and Mahadevan 2019).

The branching patterns demonstrate systematic organization reflecting functional specialization: the right main bronchus divides into upper, middle, and lower lobar bronchi, opti-

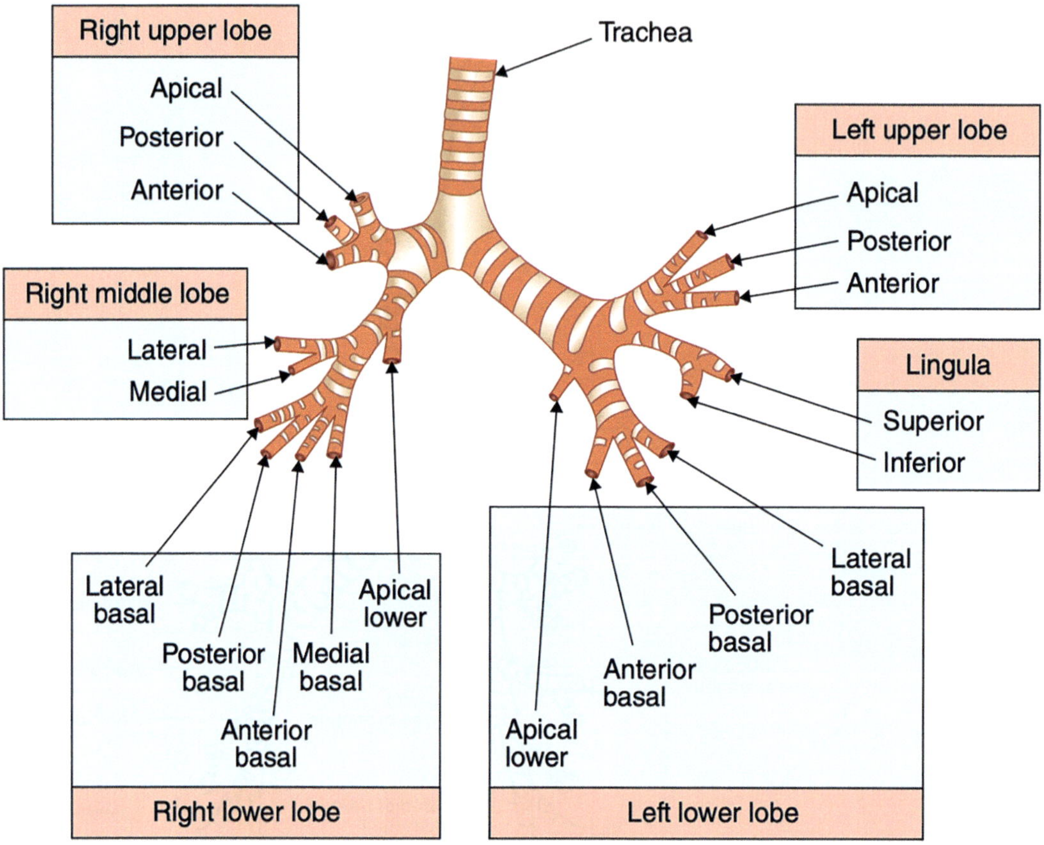

Fig. 11.14 Bronchopulmonary segments. (Adapted from Paramothayan (2019))

mizing ventilation distribution to the larger right lung. The left main bronchus supplies its corresponding lung through the upper and lower lobar bronchi, adapting to the reduced volume imposed by cardiac positioning. These lobar bronchi further subdivide into segmental bronchi, maintaining cartilaginous support characteristic of conducting airways while progressively reducing in diameter (Levitzky 2007) (Fig. 11.14).

From a developmental perspective, neonatal anatomy demonstrates different angular relationships, with both main bronchi forming approximately 55° angles with the trachea. This symmetrical configuration undergoes progressive modification during development, establishing the characteristic adult asymmetry essential for optimal respiratory function (Skandalakis 2004).

Lobar Bronchi

The lobar bronchi represent the second generation of the tracheobronchial tree, demonstrating asymmetric distribution patterns between right and left lungs that reflect fundamental differences in pulmonary architecture (Standring 2020). The right main bronchus divides into three lobar branches–superior (upper), middle, and inferior (lower)—corresponding to the trilobar organization of the right lung. In contrast, the left main bronchus bifurcates into superior and inferior lobar bronchi, aligning with the bilobar configuration of the left lung (Drake et al. 2020) (Table 11.2).

The structural composition of lobar bronchi demonstrates progressive modification from their parent bronchi. While maintaining cartilaginous support, these airways transition from the C-shaped rings characteristic of main bronchi to

Table 11.2 Lobes and segments of the human lungs

Right lung		Left lung	
Lobe	Segment	Lobe	Segment
Upper	1. Apical	*Upper*	1. Superior, apical
	2. Posterior		2. Superior, posterior
	3. Anterior		3. Superior, anterior
			4. Superior lingular
			5. Inferior lingular
Middle	4. Lateral		
	5. Medial		
Lower	6. Superior basal (apical)	*Lower*	6. Superior basal (apical)
	7. Medial basal		7. Medial basal
	8. Anterior basal		8. Anterior basal
	9. Lateral basal		9. Lateral basal
	10. Posterior basal		10. Posterior basal

a more complex arrangement of interconnected cartilaginous plates. This architectural adaptation provides both stability and flexibility necessary for accommodating respiratory movements and maintaining airway patency (Moore et al. 2018).

Segmental Bronchi

The third generation comprises segmental bronchi, which demonstrate precise anatomical organization reflecting bronchopulmonary segmentation. The right lung contains ten segmental bronchi distributed across its three lobes (Netter 2019) (Table 11.2):

- Superior (upper) lobe: apical, posterior, and anterior segments
- Middle lobe: lateral and medial segments
- Inferior lobe (lower): superior, medial basal, anterior basal, lateral basal, and posterior basal segments

The left lung presents eight segmental bronchi arranged within its two lobes (Ellis and Mahadevan 2019):

- Superior (upper) lobe: apicoposterior, anterior, and lingular (superior and inferior) segments
- Inferior (lower) lobe: superior, anteromedial basal, lateral basal, and posterior basal segments

This segmental organization holds particular clinical significance in diagnostic bronchoscopy, targeted medication delivery, and surgical resection. Each bronchopulmonary segment functions as a discrete respiratory unit with independent ventilation and perfusion, allowing selective therapeutic intervention or surgical removal without compromising adjacent pulmonary tissue (Levitzky 2007).

The histological architecture of segmental bronchi demonstrates further specialization from their parent airways. While maintaining cartilaginous support, these structures show increased smooth muscle content and more elaborate mucosal folding, enhancing their ability to regulate airflow distribution and participate in airway clearance mechanisms (Weibel 2017).

The Noncartilaginous Airways

The respiratory system exhibits an important progression from the conducting airways to the areas designated for gas exchange. The noncartilaginous airways arise between the tenth and fifteenth bronchial generations, indicating a significant transition in structure and function (Standring 2020). At that point, as airway widths diminish to less than 1 mm, the structure undergoes essential alterations that significantly affect respiratory physiology. The non-cartilaginous airways consist of the bronchioles and terminal bronchioles (Table 11.1).

Bronchioles

The bronchioles represent the transition from air conduction to the preparation for gas exchange. As these airways progress deeper into the pulmonary tissue, they shed their cartilaginous support, adopting instead a more dynamic structure where smooth muscle fibers arrange in distinctive spiral patterns. This architectural modification accompanies a transformation of the epithelial lining, which shifts from its previous pseudostratified columnar arrangement to a simpler cuboidal form. The surrounding tissue demonstrates equal complexity, as the lamina propria establishes direct connections with the pulmonary parenchyma, creating an integrated mechanical unit (Drake et al. 2020).

This evolutionary design, while optimizing the pathway toward gas exchange, introduces certain vulnerabilities into the respiratory system. Without the rigid support of cartilage, these airways become exquisitely sensitive to their mechanical environment. Their patency fluctuates with each breath, responding to the dynamic interaction between intra-alveolar and intrapleural pressures. This responsiveness extends to changes in lung volume and the contractile state of their muscular walls, creating a system that can rapidly adjust to varying respiratory demands (Moore et al. 2018).

The bronchioles typically generate three to four generations of increasingly refined passages before reaching the terminal bronchioles, each branch demonstrating progressive adaptation to its specialized role. This arrangement optimizes both airflow distribution and the preparation for gas exchange, yet its mechanical vulnerability often becomes central to respiratory pathology. Conditions affecting these airways can dramatically impact respiratory function, as seen in asthmatic bronchospasm, chronic bronchiolitis, and various forms of small airway disease (West and Luks 2016).

Terminal Bronchioles

At the threshold between conduction and gas exchange, terminal bronchioles emerge as complex structures marking a critical transition in respiratory physiology (Standring 2020). These airways, appearing between the sixteenth and nineteenth generations of the bronchial tree, demonstrate an average diameter of approximately 0.5 mm and represent the final purely conducting pathway before entering the gas exchange territory (Drake et al. 2020).

The terminal bronchioles exhibit unique architectural features reflecting their transitional position. Their walls progressively thin while maintaining specialized cellular elements, most notably the Clara cells, which project distinctive protoplasmic extensions into the airway lumen. These specialized cells likely serve multiple functions, including the production of surface-active materials and the metabolic processing of inhaled substances, though their complete physiological role remains under investigation (Moore et al. 2018).

A fascinating feature of terminal bronchioles lies in their development of Lambert's canals, microscopic channels establishing direct communications between the bronchiolar lumen and surrounding alveoli. These pathways may provide crucial alternative ventilation routes, particularly significant in pathological conditions such as chronic obstructive pulmonary disease, where standard airways may become compromised (Weibel 2017).

As terminal bronchioles transition into respiratory territories, they give rise to respiratory bronchioles, typically generating two to five generations before reaching the alveolar ducts. This branching pattern demonstrates remarkable variability, though Weibel's model typically describes three successive generations. Each respiratory bronchiole exhibits progressively increasing numbers of alveoli along its walls, creating a gradual transition into gas exchange regions (Levitzky 2007).

The organization culminates in the primary lobule or acinus, a fundamental functional unit encompassing all structures arising from a single terminal bronchiole. Each acinus, measuring approximately 3.5 mm in diameter, contains roughly 2000 alveoli and represents one of approximately 130,000 similar units distributed throughout the lungs. This arrangement optimizes gas exchange surface area, ultimately providing approximately 70 m^2 of interface between air and blood, comparable to a tennis court's dimensions (West and Luks 2016).

The epithelial transformation along this pathway reflects functional specialization, with progressive flattening from columnar to cuboidal cells and the gradual disappearance of ciliated cells and mucous glands. This architectural modification prepares for the intimate gas exchange relationships ahead in the alveolar regions, where pulmonary capillaries eventually cover 85–95 percent of alveolar surfaces (Netter 2019).

11.3 The Pleura

The pleura represents an example of developmental adaptation, arising through embryological invagination of the coelomic cavity around each lung (Standring 2020). This process creates a double-layered serous membrane that fundamentally influences respiratory mechanics. The inner visceral layer molds itself to every contour of the pulmonary surface, while the outer parietal layer provides a smooth lining to the thoracic cavity, together forming a continuous sheet that meets at the pulmonary hilum to create the pulmonary ligament (Drake et al. 2020).

The pleural margins trace a complex path across the thoracic wall, beginning approximately 4 cm above the medial clavicle. As they descend, they demonstrate intriguing asymmetry, particularly evident at the fourth costal cartilage, where the left pleura deviates laterally to accommodate the cardiac silhouette. This cardiac notch creates a unique anatomical arrangement on the left side, while the right pleural margin maintains a more vertical descent toward the costoxiphoid angle (Moore et al. 2018).

Following the thoracic contours, the pleural margins reach their lowest point at the midaxillary line, corresponding to the tenth rib. This inferior extent creates a physiologically significant region known as the costodiaphragmatic recess, where the pleural cavity extends beyond the lung margins during quiet breathing. This additional space becomes particularly meaningful during deep inspiration, allowing for pulmonary expansion. Surgeons must remain mindful of this anatomical relationship when approaching retroperitoneal structures, as inadvertent pleural entry may occur during procedures below the twelfth rib (Ellis and Mahadevan 2019).

A space exists between the visceral and parietal layers, maintained by a thin film of serous fluid. This fluid, far from being a simple lubricant, contains specific proteins and cellular components that maintain pleural homeostasis and support immune function. The microscopic thickness of this space reveals its physiological importance in facilitating smooth respiratory movements (Wang 1985).

11.4 Lungs

The lungs represent bilateral conical organs occupying the thoracic cavity, their morphology influenced by surrounding structures and the pleural space that contains them (Standring 2020). Much as a molded substance adopts the contours of its container, each lung's external configuration mirrors the pleural cavity's internal architecture. Sexual dimorphism manifests in pulmonary dimensions, with male lungs demonstrating greater mass and volume than female counterparts (Drake et al. 2020).

Each lung presents distinct anatomical surfaces and borders. The apex extends superiorly to the first rib level, while the broad, concave base conforms to the diaphragmatic dome. The anterior extent reaches the sixth rib at the xiphoid process, while posteriorly it descends to the eleventh rib (rib 11), positioned two intercostal spaces below the scapular angle. The mediastinal surface exhibits a characteristic concavity, accommodating cardiac and major vascular structures (Moore et al. 2018).

Asymmetry characterizes pulmonary anatomy, with the right lung measuring larger (average mass 620 g) compared to the left (570 g). This asymmetry extends to lobar organization. The right lung divides into three lobes through oblique and horizontal fissures. The oblique fissure separates upper and middle lobes from the lower lobe, while the horizontal fissure, extending to the fourth costal cartilage, differentiates the upper from the middle lobes (Netter 2019). Each lobe is further divided into bronchopulmonary segments.

The left lung demonstrates simpler organization with two lobes separated by an oblique fissure. This arrangement reflects the spatial requirements of the heart and its associated structures within the left hemithorax. Both lungs receive their neurovascular and bronchial components through the hilum, centrally positioned along the mediastinal border (Ellis and Mahadevan 2019).

The bronchopulmonary segmental organization creates functional units within each lobe, forming the basis for selective surgical resection and targeted therapeutic interventions. This segmental architecture maintains consistent patterns while allowing for individual anatomical variation (West and Luks 2016).

11.4.1 Lung Topography: Surface Projections and Anatomical Relationships

11.4.1.1 Anterior View

The pulmonary apex extends into the cervical structure, protruding about 2.5 cm superior to the medial third of the clavicle. This superior extension establishes significant anatomical relationships with cervical structures and affects clinical examination techniques (Drake et al. 2020).

The right lung's anterior border follows a defined path, descending posterior to the sternoclavicular joint before approaching the midline at the sternal angle. Its course continues behind the sternum to the sixth chondrosternal junction. The inferior border then traces a curved path laterally, intersecting the sixth rib at the midclavicular line and the eighth rib at the midaxillary line, before terminating at the tenth thoracic vertebral level posteriorly (Fig. 11.15). These relationships demonstrate respiratory variation, with inspiration moving these landmarks approximately two intercostal spaces inferiorly (Moore et al. 2018).

The anterior border of the left lung parallels that of the right lung until the fourth costal cartilage, where it diverges laterally to fit the cardiac silhouette, forming the cardiac notch. The inferior border of the left lung extends slightly lower than the right, indicative of asymmetric influence of subdiaphragmatic organs–chiefly the right-sided position of the liver (Netter 2019).

The fissural anatomy exhibits bilateral variation. The oblique fissure of the right lung, which divides the lower lobe from upper and middle lobes, terminates near the midclavicular line at the inferior border. The horizontal extends horizontally from the oblique fissure to the anterior border at the fourth costal cartilage level. The left lung generally presents only the oblique fissure, however, anatomical variants may include additional fissures, particularly between bronchopulmonary segments (Standring 2020).

The relationship between pulmonary margins and pleural reflections creates anatomically significant recesses. The costodiaphragmatic recess forms if the lung does not reach the inferior pleural limits, whereas the costomediastinal recess develops where the anterior pulmonary border does not meet the medial pleural boundary. These spaces vary with respiratory cycle and contain clinical significance in pleural fluid accumulation and thoracentesis (Moore et al. 2018).

Subdiaphragmatic relationships influence pulmonary position and movement. The right hemidiaphragm, elevated by the liver, affects right pulmonary excursion, while the left lung maintains relationships with the stomach and spleen. These relationships influence respiratory dynamics and pathological processes crossing the diaphragm.

11.4.1.2 Posterior View

The superior extent of the pulmonary apex corresponds to the vertebral terminus of the first rib, reaching the level of the first thoracic vertebra (T1) (Standring 2020). From this point, the lungs extend inferiorly to the diaphragmatic surface, where their bases conform to the superior diaphragmatic contour. The dome-like configuration of the diaphragm positions the maximum height of the right pulmonary base at the level of the eighth to ninth thoracic vertebrae (T8–T9), while the left pulmonary base sits slightly inferior to this level (Drake et al. 2020). Both pulmonary bases follow the diaphragmatic curvature to their respective inferior margins (Fig. 11.16).

Anatomically significant features include the oblique fissures of both lungs, which reach their maximum height posteriorly, approximately 2.5 cm from the midline at the level of the third to fourth thoracic vertebrae (T3–T4) (Netter 2019). The relationship between upper extremity posi-

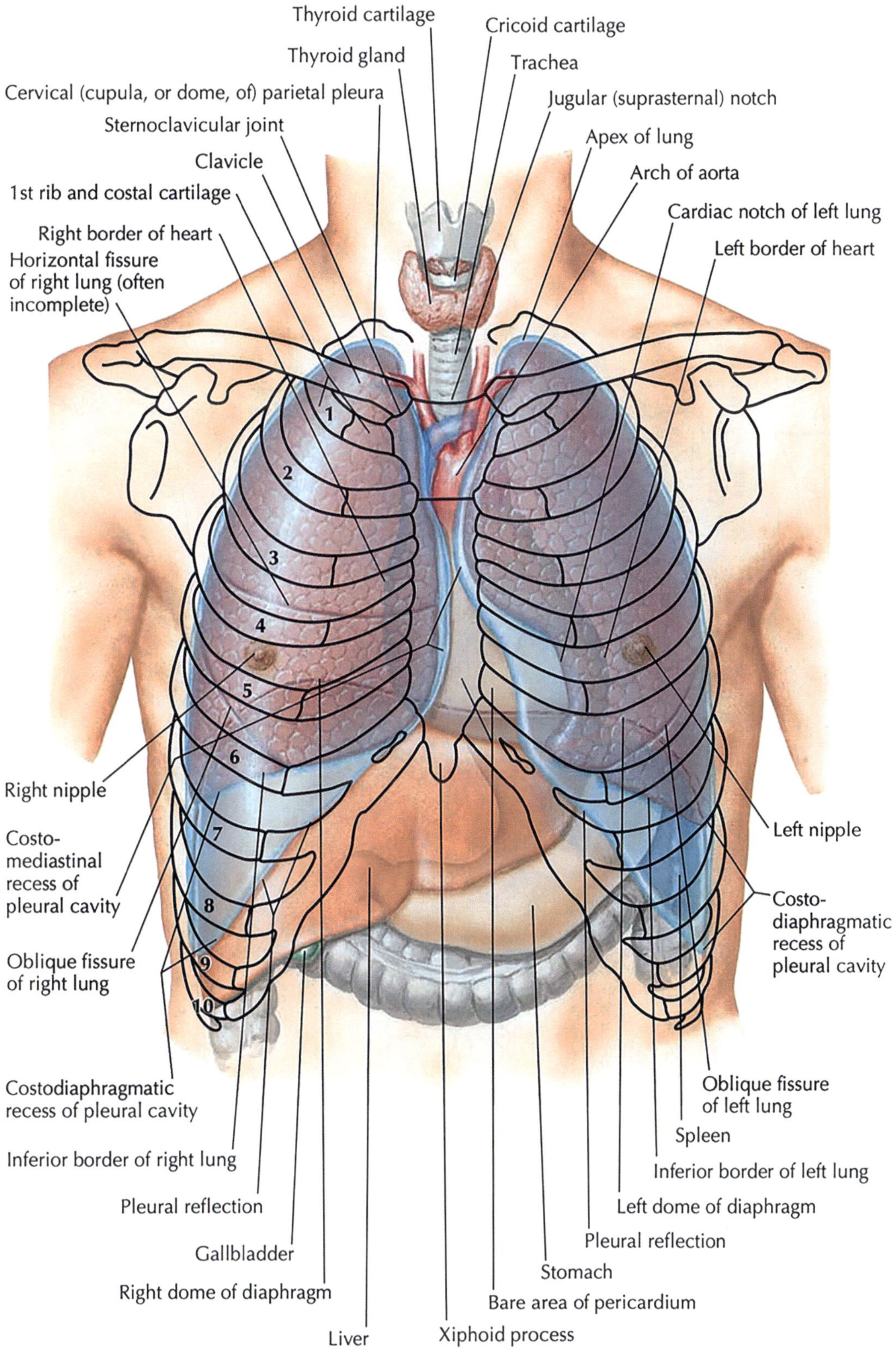

Fig. 11.15 Anterior view of lung topography. (Adapted with permission from Kaminsky (2011))

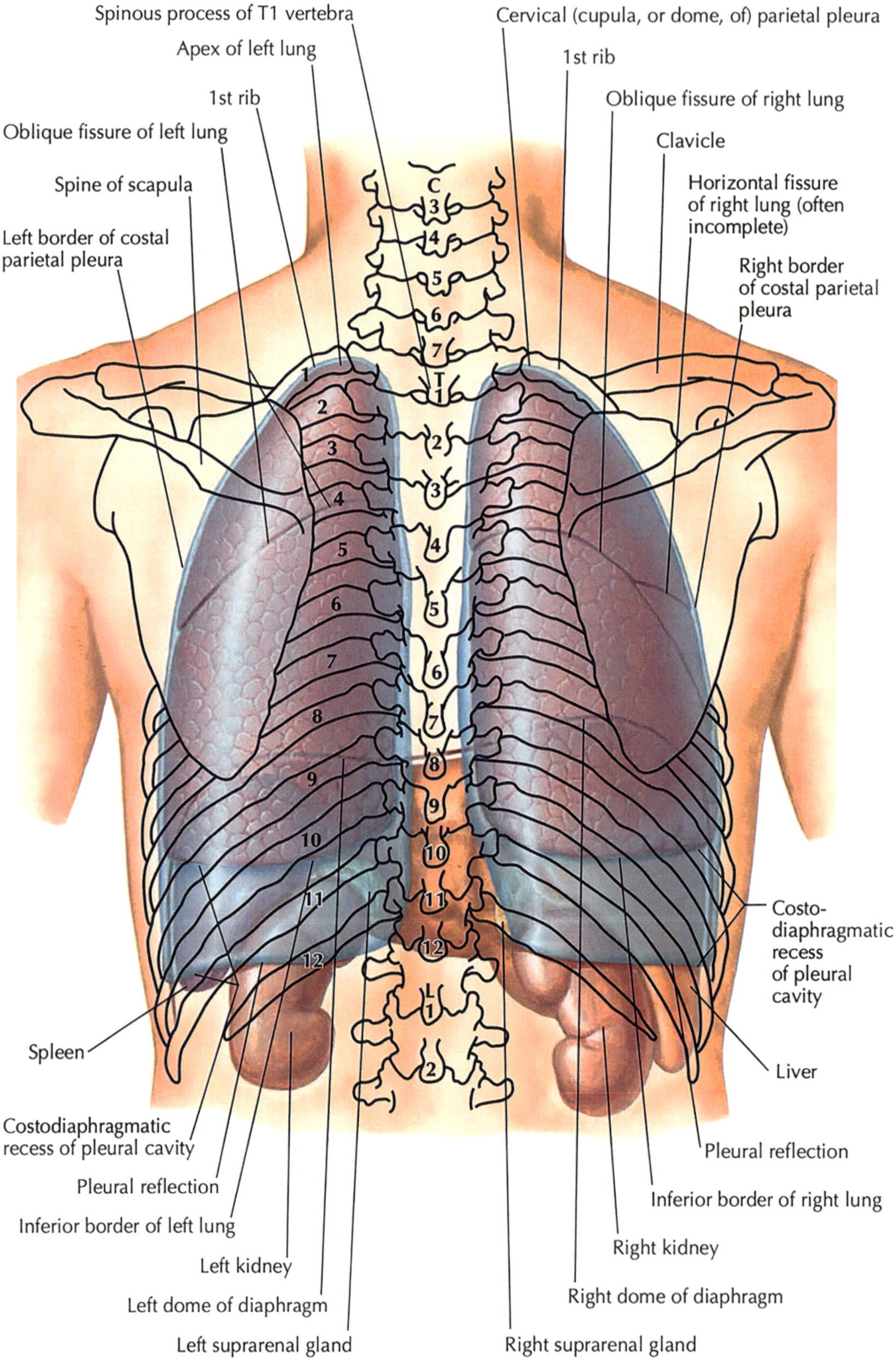

Fig. 11.16 Posterior view of lung topography. (Adapted with permission from Kaminsky (2011))

tioning and posterior thoracic examination merits clinical consideration. When the arm is elevated, the vertebral margin of the scapula approximates the oblique fissural position.

The pleural anatomy consists of parietal and visceral layers separated by a potential space containing minimal serous fluid under physiologic conditions (Standring 2020). The costodiaphragmatic recess, formed by the junction of costal and diaphragmatic components of the parietal pleura inferior to the pulmonary margin, facilitates inferior pulmonary expansion during inspiration. Pathologic conditions affecting this space may result in accumulation of air, excess serous fluid, blood, or purulent material, leading to pulmonary compression and respiratory compromise (Kumar et al. 2015).

Clinical significance extends to the anatomical relationships between pulmonary bases and adjacent organs. The left pulmonary base interfaces with gastrosplenic structures, where gastric distention can produce superior diaphragmatic displacement and potential respiratory restriction (Wineski 2019). On the right side, the diaphragm separates the pulmonary base from the hepatic surface. Hepatomegaly may cause diaphragmatic elevation with subsequent limitation of pulmonary expansion. In cases of hepatic abscess, trans-diaphragmatic extension may affect the adjacent pleural space and pulmonary tissue.

11.4.2 Anatomical Segmentation and Lobar Organization of the Lungs

The anatomical structure of the human respiratory system displays a complex organizational pattern through different structural levels. From a macroscopic viewpoint, each lung exhibits distinct anatomical divisions, characterized by asymmetric lobar distributions between the right and left sides. The right lung exhibits a trilobar configuration, while the left lung possesses a bilobar form. The lobar divisions arise from pleural folds that create anatomical fissures, comprising both horizontal and oblique fissures in the right lung, in contrast to a single oblique fissure in the left lung.

At a microscopic level, the pleural interface between adjacent lobes consists of specialized visceral pleural tissue. This anatomical arrangement facilitates respiratory mechanics by enabling smooth gliding movements between neighboring lobes during ventilation cycles. The biomechanical properties of these pleural surfaces resemble those observed in the parietal-visceral pleural interface of the entire lung within the thoracic cavity (Weinberger et al. 2019).

Each lung demonstrates a specific pattern of segmental organization. In the right lung, the superior lobe contains apical, anterior, and posterior segments. The middle lobe exhibits lateral and medial segments, while the inferior lobe comprises superior, anterior basal, lateral basal, posterior basal, and medial basal (cardiac) segments.

The left lung presents a modified segmental pattern that reflects its anatomical adaptations to accommodate cardiac structures. Its superior lobe includes two primary divisions: the upper division (containing apical, anterior, and posterior segments) and the lingular division (comprising superior and inferior segments). The left inferior lobe maintains five distinct segments: superior (also termed apical), anterior basal, medial basal, lateral basal, and posterior basal portions (Tortora and Nielsen 2017) (Table 11.2).

The spatial relationships and three-dimensional organization of these segments create a complex anatomical framework that supports optimal respiratory function. Each segment maintains its anatomical independence while contributing to the integrated function of the entire lung (Drake et al. 2020). This organizational pattern optimizes ventilation-perfusion matching and enables selective surgical intervention when necessary (Moore et al. 2018).

A detailed representation of the bronchopulmonary segments and their anatomical relationships is presented in Fig. 11.17 (Ellis and Mahadevan 2019; Singh 2014a). This illustration provides a comprehensive visualization of the

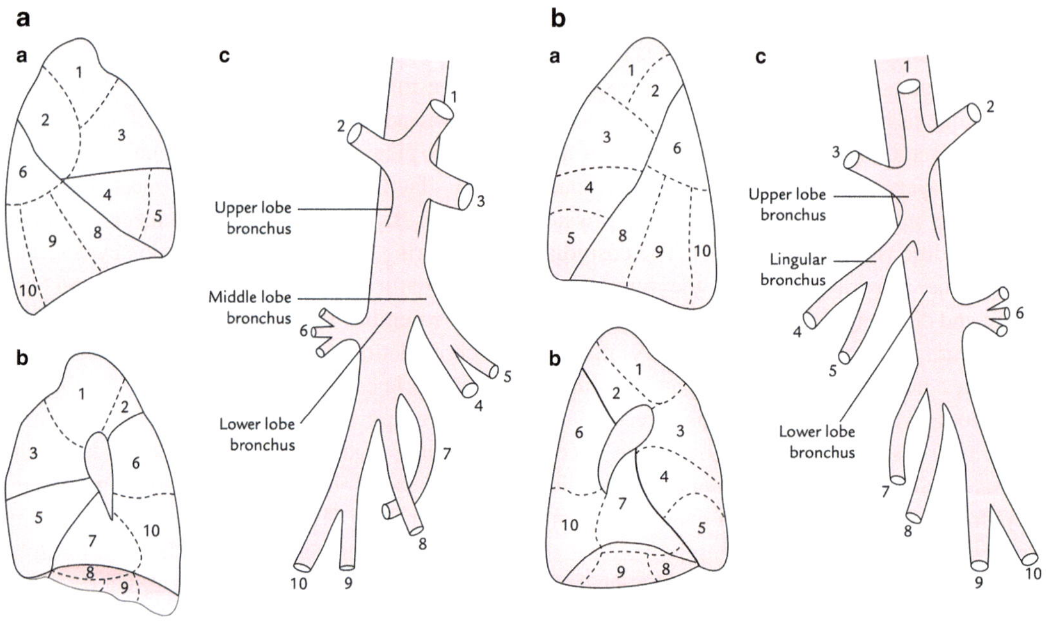

Fig. 11.17 Bronchopulmonary segments of the right lung (**a**) and left lung (**b**): A, lateral aspect; B, medial aspect; C, lobar and segmental bronchi. (Adapted from Singh (2014a))

segmental organization in both lungs, facilitating understanding of their three-dimensional relationships.

11.4.2.1 The Right Main Bronchus and Its Segmental Distribution

The right main bronchus exhibits a complex branching pattern with precise anatomical relationships (Moore et al. 2018). Following an initial course of approximately 2.5 cm, it produces the upper lobe bronchus at a right angle. This upper lobe bronchus extends for 1 cm before dividing into three segmental bronchi through trifurcation or closely spaced double bifurcation (Netter 2019).

The segmental bronchi of the upper lobe demonstrate specific directional patterns: the apical segment (1) ascends laterally, the posterior segment (2) courses posterolaterally with slight superior angulation, and the anterior segment (3) projects anterolaterally with minimal inferior deviation (Standring 2020).

Following the upper lobe bronchus, the main bronchus continues for 3 cm before producing the middle lobe bronchus, which projects anteroinferiorly. This middle lobe bronchus, measuring 1–1.5 cm, bifurcates into lateral (4) and medial (5) segments (Drake et al. 2020).

The bronchial segment between the upper and middle lobe branches, termed the "lower part of the right main bronchus," maintains anatomical significance due to its relationship with the right pulmonary artery. This relationship historically led to the designations "hyparterial bronchus" and "eparterial bronchus" for the respective segments (Ellis et al. 2015).

The apical segment bronchus of the lower lobe (6) originates near the middle lobe bronchus origin. This posterior-directed branch, approximately 1 cm in length, divides into superior, medial, and lateral branches. Its posterior orientation in supine position creates a potential site for foreign body or secretion accumulation.

The subsequent branching pattern includes the medial basal (cardiac) bronchus (7), followed by three additional basal segments: anterior basal (8), lateral basal (9), and posterior basal (10). The posterior basal segment represents the largest terminal branch, maintaining the directional course of the primary lower lobe bronchus.

The Right Upper Lobe Segments

The right upper lobe demonstrates segmental structure with defined anatomical boundaries and relationships (Standring 2020). The apical segment (S1) occupies the superior aspect of the right lung, extending superiorly into the cervical region to the level of the first rib's vertebral termination (Moore et al. 2018). This segment exhibits a characteristic lateral inclination between its posterior and anterior segments, with its boundary following an oblique course from the first rib anteriorly to approximately the second rib posteriorly (Drake et al. 2020).

The posterior segment (S2) of the right upper lobe establishes its anatomical territory from the inferior margin of the apical segment, extending downward to interface with the horizontal fissure laterally and the superior portion of the oblique fissure (Netter 2019). This anatomical arrangement creates specific surface relationships with the thoracic wall, providing important landmarks for clinical examination and surgical intervention (Standring 2020).

The anterior segment (S3) demonstrates a vertical orientation, originating from the inferior boundary of the apical segment and terminating at the horizontal fissure, corresponding approximately to the fourth rib level (Wineski 2019). This segmental organization facilitates precise localization during diagnostic imaging and therapeutic interventions (Ellis and Mahadevan 2019).

The Right Middle Lobe Segments

The middle lobe of the right lung exhibits a distinctive branching pattern characterized by the bifurcation of its bronchus into two primary segmental divisions (Standring 2020). This bifurcation establishes the foundation for two anatomically distinct segments: the lateral segment (S4) and the medial segment (S5) (Moore et al. 2018).

The anatomical separation between these segments occurs along a vertical plane that originates at the pulmonary hilum and extends laterally to reach the costal surface (Drake et al. 2020). This plane continues inferiorly until it reaches the anterior aspect of the oblique fissure's termination, establishing clear anatomical boundaries between the segments (Netter 2019).

Topographically, both segments maintain specific relationships with the anterior thoracic wall, corresponding to the fourth and fifth ribs and their associated costal cartilages (Standring 2020). This anatomical relationship provides important surface landmarks for clinical examination and diagnostic procedures (Wineski 2019). The precise delineation of these segments facilitates accurate localization during bronchoscopic procedures and surgical interventions (Ellis and Mahadevan 2019).

The Right Lower Lobe Segments

The right lower lobe demonstrates a complex segmental organization beginning with the superior segmental bronchus, which emerges from the lower lobe bronchus immediately inferior to the middle lobe bronchial origin (Moore et al. 2018). The superior segment (S6) occupies the upper portion of the lower lobe, extending from the superior aspect of the oblique fissure near the third rib's vertebral termination to the vertebral end of the fifth or sixth rib (Standring 2020).

Inferior to the superior (apical) segmental bronchus origin, the lower lobe bronchus divides into four basal segments (Drake et al. 2020):

- Medial basal segment (S7)
- Anterior basal segment (S8)
- Lateral basal segment (S9)
- Posterior basal segment (S10)

These basal segments collectively form the pulmonary base and maintain direct contact with the diaphragmatic surface (Netter 2019). A notable anatomical variation involves the medial basal segment, which occasionally presents with partial segmentation from adjacent basal segments due to an additional fissure. In such cases, this anatomically distinct portion has been historically referenced as the cardiac lobe of the lung (Wineski 2019).

11.4.2.2 The Left Main Bronchus and Its Segmental Distribution

The left main bronchus demonstrates anatomical characteristics, extending approximately 5 cm before producing the left upper lobe bronchus (Moore et al. 2018). Unlike the right bronchial tree, the left pulmonary artery loops this bronchus, eliminating the presence of an 'eparterial bronchus' configuration observed on the contralateral side (Standring 2020).

The upper lobe bronchus extends laterally for approximately 1 cm before bifurcating into superior and inferior (lingular) divisions (Drake et al. 2020). The superior division typically exhibits a unique branching pattern where the apical and posterior segments share a common origin, forming the apico-posterior bronchus (segments 1 and 2), while the anterior bronchus (segment 3) maintains independent origin (Netter 2019). However, anatomical variation may present with three independent bronchi, resembling the right-sided pattern (Standring 2020).

The lingular bronchus, representing the inferior division of the left upper lobe bronchus, projects anteroinferiorly and laterally for 1–2 cm before bifurcating into superior (segment 4) and inferior (segment 5) branches (Wineski 2019). This segmental arrangement contrasts with the medial-lateral orientation observed in the right middle lobe. The lingula's superior boundary frequently corresponds to an identifiable indentation along the upper lobe's anterior margin (Ellis and Mahadevan 2019).

The inferior directional course of the lingular bronchus creates anatomical conditions that may predispose this region to pathological processes, particularly in conjunction with left lower lobe involvement in infectious and bronchiectatic conditions (Moore et al. 2018). The left lower lobe bronchial distribution parallels its right lower lobe, with the notable exception of absent medial basal (cardiac) segmentation (Standring 2020).

The Left Upper Lobe Segments

The left upper lobe demonstrates unique anatomical characteristics that distinguish it from its right upper lobe, despite sharing some fundamental organizational principles (Standring 2020). The primary bronchus undergoes subdivision into superior and inferior divisions, with the latter forming the lingular component. Although these divisions roughly correspond to the right upper and middle lobes respectively, they manifest distinct anatomical features and lack the clear fissural separation observed on the right side (Moore et al. 2018).

When examining the superior division's segmental arrangement, anatomists have identified a more streamlined organization compared to the right lung. Instead of maintaining three distinct segments, the left superior division exhibits a merged apicoposterior segment, designated as S1 and S2, which effectively combines territories that remain separate in the right lung. Accompanying this merged segment, the anterior segment (S3) maintains its independent identity, paralleling its right-sided equivalent (Drake et al. 2020).

The lingular division, while functionally analogous to the right middle lobe, presents its own unique anatomical pattern. Rather than adopting the medial-lateral orientation characteristic of the right middle lobe, the lingula organizes into superior (S4) and inferior (S5) segments. This superior-inferior arrangement reflects the developmental and functional adaptations of the left lung (Netter 2019).

The absence of a separating fissure between these divisions creates a continuous anatomical territory, contrasting sharply with the more distinctly compartmentalized right lung. This anatomical continuity influences both the physiological function and pathological processes affecting these regions (Standring 2020).

The Left Lower Lobe Segments

The left lower lobe exhibits general similarities with its right lower lobe (Standring 2020). A significant anatomical variance is observed in the configuration of the basal segments, where the territories corresponding to the right anterior basal and medial basal regions merge into a single functional unit. This unified region, designated as the anteromedial basal segment (S8), receives its bronchial supply through a common stem, representing an efficient anatomical adaptation unique to the left lung (Moore et al. 2018).

Beyond this distinctive merger, the left lower lobe maintains recognizable segmental patterns that parallel the right side. The superior segment (S6) establishes the upper boundary of the lower lobe, while the lateral basal (S9) and posterior basal (S10) segments complete the basal portion of the lobe (Drake et al. 2020). This arrangement demonstrates how developmental factors influence the final anatomical organization, resulting in efficient space utilization within the left hemithorax.

11.5 Functional Anatomy of the Thoracic Musculature

The thoracic musculature comprises a complex neuromechanical system that modulates respiratory function through precisely coordinated actions of the intercostal muscles and diaphragm (Standring 2020). These muscles operate within a complex biomechanical system that optimizes ventilatory efficiency, with the diaphragm exhibiting phasic contractions synchronized with respiratory rhythm generation in the brainstem (De Troyer and Boriek 2011). This neural orchestration involves both pre-motor and motor neurons in the medulla oblongata, particularly the dorsal and ventral respiratory groups, which establish rhythmic breathing patterns (Butler et al.2014).

Biomechanical analyses have elucidated the three-dimensional dynamics of ribcage movement during respiration. The true ribs (1–7) demonstrate a "pump-handle" motion in the anterior-posterior plane, while the false ribs (9–12) exhibit a "bucket-handle" motion in the transverse plane. These movements are facilitated by specialized costochondral and costovertebral joint articulations, which permit specific degrees of freedom for rib rotation (Wilson et al. 2001).

The transpulmonary pressure gradient of 0.2 kPa generated during inspiration results from coordinated activation of the diaphragm and external intercostal muscles, with precise timing controlled by respiratory centers in the brainstem. This pressure differential induces alveolar expansion through complex mechanotransduction pathways involving surfactant-mediated surface tension modulation.

During forced respiration, accessory muscle recruitment follows a hierarchical pattern. Scalene and sternocleidomastoid muscles demonstrate early activation, followed by progressive recruitment of additional accessory muscles based on ventilatory demand. This sequential recruitment optimizes energy expenditure while maintaining adequate ventilation. The expiratory phase exhibits different mechanical characteristics based on breathing intensity. During quiet expiration, the passive elastic recoil of pulmonary tissue generates expiratory flow, while forced expiration involves active contraction of the internal intercostal and abdominal muscles.

11.5.1 Intrinsic Thoracic Musculature

The intrinsic muscles of the thorax constitute an integral component of the anterior thoracic wall, comprising three primary muscle groups: the external intercostal, internal intercostal, and transversus thoracis muscles (Moore et al. 2018). These muscles function cooperatively to facilitate respiratory mechanics and maintain thoracic wall stability.

The external intercostal muscles originate from each rib's inferior border and insert into the superior border of the subsequent rib. Their fibers demonstrate a characteristic oblique orientation, coursing inferiorly and medially (Netter 2019). These muscles extend from the tubercles of the ribs to the costal cartilages, where they transition into the anterior intercostal membranes. This anatomical arrangement optimizes their mechanical advantage during respiratory movements (Drake et al. 2020).

Internal intercostal muscles display a more complex architectural pattern, with origins at the inner lip and costal groove of the superior rib and its corresponding costal cartilage. Their insertion points are located along the superior border of the inferior rib. These muscles span from the sternum to the rib angles, continuing posteriorly as posterior intercostal membranes. The fiber orientation of internal intercostals runs inferiorly and laterally, contrasting with the external intercostals (Standring 2020).

Deep to the internal intercostals lie the innermost intercostal muscles, previously considered a component of the internal intercostals. These muscles attach to adjacent ribs' internal aspects, with fiber directionality matching that of the internal intercostals. All intercostal muscles receive innervation from their corresponding intercostal nerves, enabling precise control of respiratory movements (Ellis and Mahadevan 2019).

The transversus thoracis muscle presents as a thin sheet of muscular and tendinous fibers on the anterior thoracic wall's internal surface. It originates from the xiphoid process's posterior aspect, the lower third of the sternal body, and the adjacent costal cartilages. The muscle inserts via distinct slips onto the internal surfaces of the second or third through sixth costal cartilages (Moore et al. 2018) (Fig. 11.18).

An anatomical variant, the sternalis muscle, may occasionally appear superficial to the pectoralis major muscle's origin, running parallel to the sternum. This muscle exhibits variable attachment patterns, potentially connecting to costal cartilages, sternum, rectus sheath, and the sternocleidomastoid and pectoralis major muscles (Standring 2020).

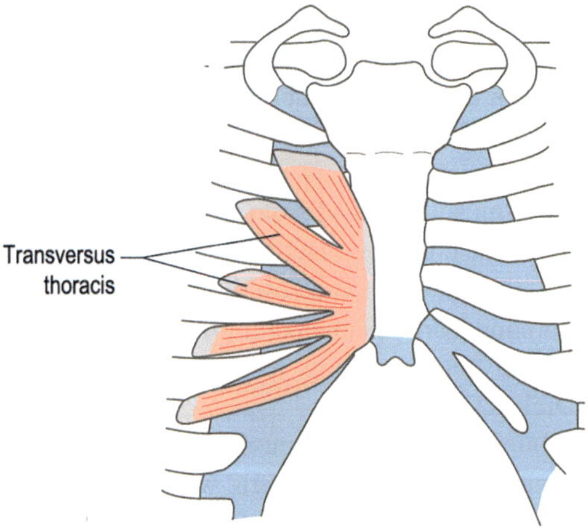

Fig. 11.18 Left transversus thoracis muscle located on the inner surface of the thoracic cage. (Adapted with permission from Palastanga and Soames (2019))

11.5.2 Anterior Thoracic Wall Musculature

The anterior thoracic region presents a complex layered architecture, with the outermost components consisting of cutaneous tissue and a specialized superficial fascial layer that contains mammary glands. The fundamental support structure is provided by the anterior skeletal elements of the thorax (Detton 2013). The muscular composition of this region demonstrates three different groups: muscles associated with upper limb function, muscles contributing to the anterolateral abdominal wall, and the thoracic musculature (Moore et al. 2018).

Among the upper limb-associated muscles, four primary muscular structures warrant detailed consideration. The most prominent is the pectoralis major, characterized by its distinctive fan-like morphology and tripartite origin pattern. Its superior attachment incorporates the anterior clavicular surface along its medial half. The intermediate portion establishes connections with the anterior aspects of both the manubrium and sternal body, extending to include the cartilaginous components of ribs one through six. A third, more variable component derives from the aponeurotic tissue of the external oblique muscle of the abdomen. These diverse origins unite to form a common insertion at the humeral greater tubercle (Standring 2020).

In the deeper anatomical plane, the pectoralis minor exhibits a triangular configuration. Its superior attachments arise from three consecutive ribs (third through fifth), specifically involving their superior margins and external surfaces near the costochondral junctions, while also incorporating the fascial investments of the adjacent intercostal muscles. The muscle's insertion is localized to the scapular coracoid process. Both pectoral muscles receive innervation from the medial and lateral anterior thoracic nerves, which emerge from the brachial plexus's corresponding cords (Drake et al. 2020).

The serratus anterior manifests as an expansive muscular formation that follows the thoracic contour. Its origin comprises multiple digitations from the initial eight or nine ribs, specifically their external surfaces and superior borders, along with the associated intercostal fascial layers. The muscle's insertion occurs along the anterior aspect of the scapula's vertebral border. Its neural supply is derived from the long thoracic nerve, which originates from cervical nerve roots five through seven and courses along the muscle's external aspect (Ellis and Mahadevan 2019).

The subclavius, despite its modest dimensions, occupies a significant anatomical position between the first rib and clavicle. This triangular muscle originates via a tendinous attachment at the first costochondral junction and finds its termination in a specialized groove near the lateral aspect of the clavicle's inferior surface. Its innervation is provided by a dedicated subclavian branch emerging from the brachial plexus (Netter 2019).

11.5.3 Dorsal Thoracic Wall Musculature

The posterior thoracic region exhibits a complex of tissue layers, beginning with cutaneous and superficial fascial components. The neural innervation of the cutaneous elements occurs through branches of the thoracic nerves' posterior primary divisions, with a distinctive distribution pattern: medial branches supply the upper six thoracic levels, while lateral branches innervate the lower six segments (Drake et al. 2020).

The superficial muscular layer comprises structures that establish connections between the upper extremity and vertebral column. This group includes five primary muscles: trapezius, latissimus dorsi, rhomboideus major and minor, and levator scapulae (Moore et al. 2018).

The trapezius demonstrates an extensive attachment pattern, originating from multiple points: the medial third of the superior nuchal line, the external occipital protuberance, the ligamentum nuchae's posterior border, and the spinous processes of C7 and all thoracic vertebrae with their associated supraspinous ligaments. Its insertion pattern shows regional variation: inferior fibers form an aponeurosis attaching to the scapular spine's medial triangle, middle fibers connect to the acromion's medial margin and scapular spine's superior border, and superior fibers terminate on the clavicle's lateral third posterior border. Neural input derives from the spinal accessory nerve (CN XI) and C3–C4 ventral rami. The muscle facilitates scapular medialization and rotation for shoulder elevation (Standring 2020).

The latissimus dorsi exhibits broad anatomical attachments, including a muscular component from the iliac crest's external lip and an extensive aponeurotic attachment to the lower six thoracic vertebrae, lumbar and sacral vertebrae, and associated ligaments. Its insertion occurs in the humeral intertubercular groove. Innervation comes from C6 to C8 via the thoracodorsal nerve. This muscle participates in shoulder extension, adduction, and medial rotation (Ellis and Mahadevan 2019).

The rhomboid muscles are often present as a continuous structure. Rhomboideus major originates from T2 to T5 spinous processes and associated ligaments, inserting into the scapular vertebral border through a tendinous arch. Rhomboideus minor attaches to T1 and C7 spinous processes and the ligamentum nuchae, inserting at the scapular spine's root. Both receive innervation from C5 to C6 via the dorsal scapular nerve and function in scapular retraction and elevation (Netter 2019).

The levator scapulae originates via four tendinous attachments to C1–C4 transverse processes, inserting along the scapula's vertebral border. Its primary innervation comes from C3–C4 ventral rami, facilitating scapular elevation and medial rotation (Detton 2013).

The intermediate layer of the posterior thoracic wall contains the serratus posterior muscle group (Fig. 11.19). The serratus posterior superior demonstrates aponeurotic origins from the ligamentum nuchae's inferior portion and the spinous processes with associated ligaments from C7 through T2–T3. This muscle establishes insertion points through distinct muscular digita-

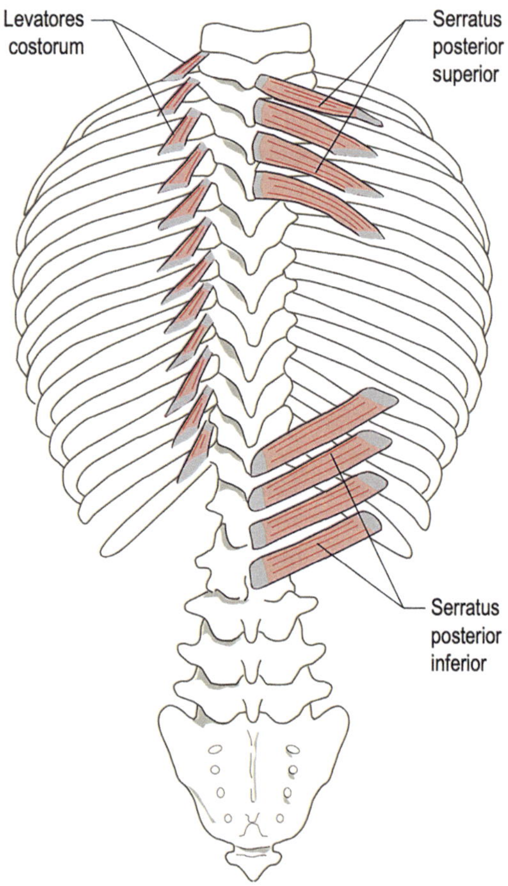

Fig. 11.19 Right serratus posterior superior and inferior muscles, and left levatores costarum. (Adapted with permission from Palastanga and Soames (2019))

tions into ribs 2–5, lateral to their angles. Its primary function involves thoracic cavity expansion through rib elevation (Standring 2020). The serratus posterior inferior originates via an aponeurotic sheet from the spinous processes and associated ligaments of T11–T12 and L1–L3. Its muscular digitations insert into the inferior borders of ribs 9–12, beyond their angles. This muscle facilitates downward and lateral displacement of the inferior ribs. Both serratus posterior muscles receive segmental innervation from the thoracic ventral rami corresponding to their anatomical positions (Moore et al. 2018).

Deep to the serratus posterior superior lie the thoracic components of the splenius muscle group. The splenius cervicis originates through tendinous attachments from T3 to T6 spinous processes and courses laterally to insert via distinct fasciculi into the transverse processes of C2–C3. The splenius capitis demonstrates origins from the inferior ligamentum nuchae and spinous processes of C7 through T3–T4, with insertions into the occipital bone inferior to the superior nuchal line's lateral third. These muscles facilitate posterior and lateral head and neck movement, with rotation toward the ipsilateral side. Their innervation derives from the middle and lower cervical posterior primary divisions (Drake et al. 2020).

The paravertebral groove, situated lateral to the thoracic spinous processes, contains the sacrospinalis muscle complex, enclosed by the thoracic portion of the lumbodorsal fascia. This muscular group represents the primary component of the deep back muscles, functioning in spinal extension and postural maintenance. The anatomical arrangement continues deep to include specialized vertebrocostal and intervertebral muscles, which contribute to segmental spinal control and stability (Ellis and Mahadevan 2019).

This complex muscular arrangement demonstrates the complicated biomechanical organization required for thoracic wall function, spinal stability, and upper extremity movement. The layered arrangement allows for coordinated action in maintaining posture, facilitating respiration, and enabling precise movement patterns of the trunk and upper limb (Detton 2013).

11.6 Respiratory Muscles

The respiratory system utilizes complex neuromuscular mechanisms for ventilation, with the diaphragm as the principal respiratory muscle (Polla et al. 2004). This dome-shaped musculofibrous structure, consisting of paired hemidiaphragms, demonstrates complex functional integration between thoracic and abdominal compartments (Standring 2020) (Fig. 11.20).

Each hemidiaphragm arises from three specific regions: the lumbar vertebrae, the costal margin, and xiphoid process. These muscle components converge at the central tendon, forming a unified structure that maintains anatomical continuity while allowing specialized regional func-

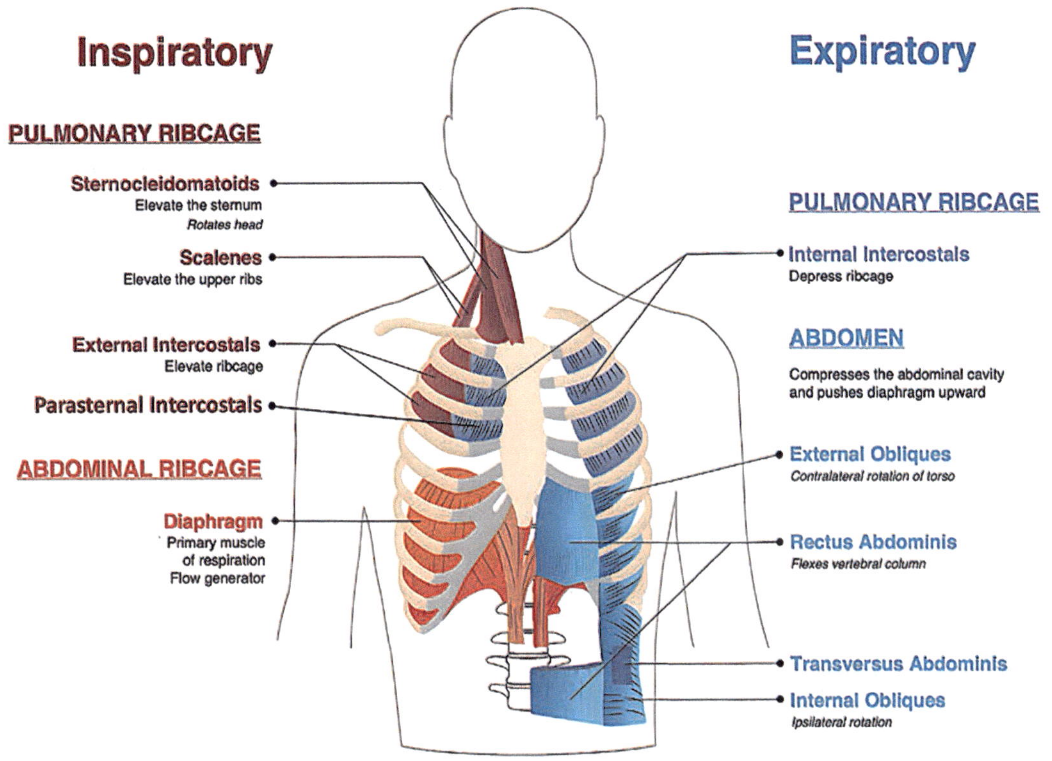

Fig. 11.20 Major respiratory muscles and their functions. Inspiratory muscles (red) and expiratory muscles (blue) are organized by compartments: pulmonary rib cages, abdominal rib cage, and abdomen. The paraspinal intercostal muscles, comprising the medial parasternal part of the internal intercostal muscles, act as inspiratory muscles. Primary respiratory actions and any secondary non-respiratory action (italicized) for each muscle or muscle group are detailed. (Adapted with permission from Ko (2023))

tion (Butler et al. 2014). The diaphragm has apertures for essential structures, such as the esophagus, aorta, various neural pathways, and the inferior vena cava (De Troyer and Estenne 1988).

The diaphragm receives primary motor innervation from phrenic nerve terminals, which originate from cervical segments C3–C5, with additional innervation provided by lower thoracic nerves (Wilson et al. 2001).

Biomechanical analyses have quantified diaphragmatic excursion during respiration. During inspiration, diaphragmatic contraction produces caudal displacement, simultaneously elongating the thoracic cavity while elevating and laterally expanding the inferior ribs. This mechanical action generates negative pressure gradients within the pleural and alveolar spaces, enabling ambient gas flow into the pulmonary system.

The external intercostal muscles contribute to thoracic expansion through two distinct mechanical actions: the "bucket-handle" motion, involving lateral ribcage elevation, and the "pump-handle" motion, producing anterior-posterior thoracic expansion (*see* Fig. 11.1). These movements optimize thoracic volume changes while minimizing muscular energy expenditure.

11.6.1 Inspiratory Muscles: Diaphragm, External Intercostals, Levatores Costorum, Serratus Posterior Superior

The respiratory system employs multiple muscle groups to execute inspiration efficiently. The diaphragm, which separates the thoracic and abdominal compartments, functions as the primary inspiratory muscle (West and Luks 2016). Its dome-shaped structure consists of a central tendinous region, with muscle fibers extending inferiorly to insert at specific anatomical locations: the xiphoid process anteriorly, the costal margin laterally, and the lumbar vertebrae posteriorly (Drake et al. 2020).

During inspiration, diaphragmatic contraction generates a caudal displacement of the central tendon, which increases the vertical dimension of the thoracic cavity. This action creates negative intrathoracic pressure, allowing atmospheric air to flow into the lungs (Feldman and Del Negro 2006). Although the diaphragm's attachment to the lower ribs could potentially reduce thoracic cavity expansion through costal elevation, accessory muscles prevent this mechanical disadvantage. The serratus posterior inferior and quadratus lumborum maintain rib position, optimizing diaphragmatic function (Standring 2020).

The thoracic cavity expansion occurs through the coordinated action of three muscle groups that elevate the ribs. The anterior muscle group includes the pectoralis minor, subclavius, sternocleidomastoid, and pectoralis major. Posteriorly, the serratus posterior superior and levatores costarum contribute to rib elevation, while the scalene muscles provide lateral force (Barrett et al. 2019). The external intercostal muscles, oriented obliquely between adjacent ribs, complete this mechanical sequence by approximating the ribs in an upward motion. This synchronized muscular activity results in anterolateral expansion of the rib cage, increasing both vertical and anteroposterior thoracic dimensions (Hall 2016).

The anterior thoracic wall presents a distinct muscular arrangement. In the region between the sternum and the chondrocostal junctions, the external intercostal muscles transition into a fibrous structure known as the anterior intercostal membrane. In this area, the internal intercostal muscles remain as the primary muscular component (De Troyer et al. 2005). The parasternal portion of the internal intercostal muscles exhibits unique anatomical and functional properties that differentiate it from the interosseous segment. These muscles, specifically termed "parasternal intercostals," occupy each intercostal space along the sternum (Standring 2020). Research has demonstrated that in humans, the parasternal intercostals function as inspiratory muscles across all intercostal levels (De Troyer and Boriek 2011).

11.6.1.1 Diaphragm (Innervated by the Left and Right Phrenic Nerves: C3, 4, 5)

The diaphragm represents a complex musculotendinous structure that forms a partition between thoracic and abdominal cavities. Its architecture centers around a trefoil-shaped central tendon, with muscle fibers radiating peripherally to attach at the thoracic margin (Moore et al. 2018).

The lumbar component of the diaphragm demonstrates bilateral asymmetry in its vertebral attachments. The right crus, being more extensive, originates from the vertebral bodies and intervertebral discs of L1–L3, while the left crus attaches to L1–L2 vertebrae and their intervening disc (Netter 2019). These crura form a crossing pattern anterior to the T12 vertebra, creating a passage for the descending aorta. A fibrous band, the median arcuate ligament, connects the crura at this intersection. The right crural fibers predominantly course leftward, forming a muscular sphincter around the esophageal hiatus before inserting into the central tendon (Drake et al. 2020) (Fig. 11.21).

Lateral to each crus, the diaphragm receives additional attachments via two significant ligamentous structures. The medial arcuate ligament, a fascial thickening overlying psoas major, extends from L2 vertebral body to L1 transverse process. The lateral arcuate ligament, derived from thoracolumbar fascia covering quadratus lumborum, spans between the L1 transverse process and the twelfth rib (Standring 2020).

The costal portion originates from the internal aspects of ribs 7–12 and their cartilages, where it interdigitates with the transversus abdominis muscle. The sternal component comprises two small muscular slips arising from the xiphoid process's posterior surface. All fibers proceed superiorly and medially toward the central tendon, creating an asymmetric dome configuration with the right side positioned slightly higher than the left.

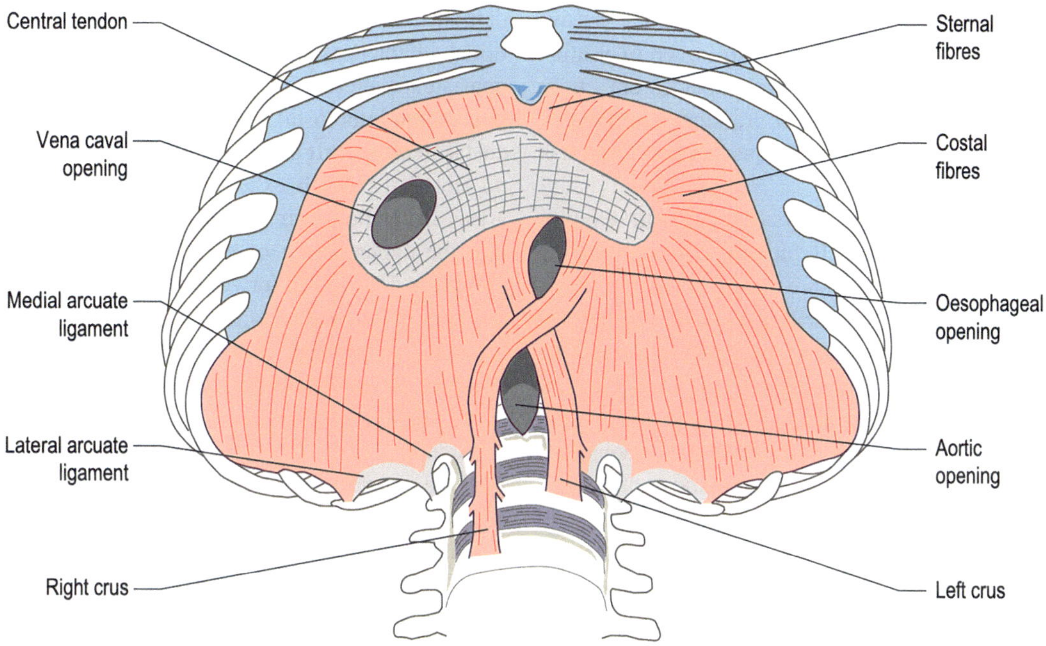

Fig. 11.21 Inferior view of the diaphragm. (Adapted with permission from Palastanga and Soames (2019))

The diaphragm's superior surface maintains intimate relationships with thoracic structures through parietal pleural coverage, while its inferior aspect relates to abdominal organs via peritoneal reflection. Several vital structures traverse the diaphragm through specific apertures: the caval opening at T8 level transmits the inferior vena cava and right phrenic nerve; the esophageal hiatus at T10 permits passage of the esophagus, vagal trunks, and left gastric vessels; and the aortic hiatus at T12 accommodates the aorta and thoracic duct (Ellis and Mahadevan 2019).

The diaphragm functions as the primary inspiratory muscle through a coordinated sequence of mechanical events. Its contraction initiates a series of anatomical changes that modify thoracic dimensions and facilitate pulmonary ventilation (West and Luks 2016). The respiratory cycle begins with contraction of the peripheral muscular portion against fixed ribs, causing flattening of the diaphragmatic domes and inferior displacement of the central tendon from its T8–T9 position (Hall 2016).

Further downward movement becomes limited by two factors: compression of abdominal organs and resistance from abdominal muscle tone, with additional restriction from pericardial tension. Following this initial phase, the central tendon becomes a stable point of reference. Subsequent muscular contraction produces specific rib cage movements: lateral expansion through upward and outward displacement of inferior ribs ("bucket-handle" movement) and anterior projection of the sternum via superior rib elevation ("pump-handle" movement). These modifications occur at multiple articular surfaces, including costovertebral, costotransverse, sternocostal, and interchondral joints (Barrett et al. 2019).

The diaphragm demonstrates variable excursion patterns correlated with respiratory intensity. During normal breathing, its displacement measures approximately 1.5 cm, increasing significantly to 7–13 cm during deep respiratory efforts. This vertical motion produces a piston-like effect, maximizing thoracic cavity expansion and contributing 60–75% of tidal volume (Hall 2016).

During inspiration, the domes exhibit greater displacement compared to the central tendon, while simultaneously elevating and everting the inferior costal margin. This dual action optimizes thoracic expansion. Expiration involves passive diaphragmatic relaxation, with forced expiration featuring active upward displacement from increased intra-abdominal pressure generated by anterior abdominal wall musculature (Feldman and Del Negro 2006).

The diaphragm's physiological role extends beyond respiration. It functions in maintaining intra-abdominal pressure during expulsive activities (defecation, micturition, emesis, and parturition), provides mechanical support to the lumbar spine during load-bearing activities, and contributes to gastroesophageal reflux prevention through right crural compression (Drake et al. 2020). Additionally, diaphragmatic motion creates pressure differentials between thoracic and abdominal compartments, facilitating venous return and lymphatic drainage from abdominal structures to the thorax (Standring 2020).

11.6.1.2 External Intercostal Muscles (Innervated by the Anterior Primary Rami of the Adjacent Intercostal [Thoracic] Nerves)

The intercostal muscles comprise three distinct layers occupying the spaces between adjacent ribs, with each layer receiving innervation from anterior primary rami of corresponding intercostal nerves (Moore et al. 2018) (Fig. 11.22). The external intercostal muscles lie in the superficial layer of the intercostal musculature. The internal intercostal muscle forms the intermediate layer, with a distribution pattern from the sternum anteriorly to the costal angles posteriorly. At this point, it continues as the internal intercostal membrane. The innermost intercostal muscle represents the deep layer, displaying incomplete coverage between the internal surfaces of adjacent ribs (Drake et al. 2020). This layer exhibits reduced development in superior intercostal spaces. While sharing directional orientation with the internal intercostals, these layers maintain separation by an intervening neurovascular plane containing intercostal nerves and vessels (Netter 2019).

The external intercostal muscles represent the superficial layer of thoracic wall musculature, demonstrating specific architectural and functional characteristics. These muscles exhibit a consistent pattern of fiber orientation in the first to eleventh intercostal spaces. The external intercostal muscles originate on ribs 1–11 and insert on ribs 2–12. Each external intercostal muscle arises from the lower border of a rib and insert into the upper border of the rib below, directed obliquely downward, forward, and medially. The external intercostal muscles are thicker than the internal intercostal muscles. The muscles involved in quiet and forced inhalation (Drake et al. 2020).

Anatomically, they extend from the costal tubercles posteriorly to the costochondral junctions anteriorly, where they transform into the external intercostal membrane, continuing to the sternum. In the inferior thoracic region, these muscles show anatomical continuity with the external oblique muscle. The fiber direction varies regionally, with medial orientation anteriorly and lateral orientation posteriorly (Standring 2020).

The primary mechanical function of external intercostals involves thoracic cage elevation during inspiration. This action increases thoracic dimensions in both lateral and anteroposterior planes, facilitating lung volume expansion. Additionally, these muscles prevent intercostal space retraction during forceful inspiration, functioning antagonistically to internal intercostal muscles (Barrett et al. 2019).

The clinical importance is evident in tetraplegia instances, when intercostal paralysis results in paradoxical movement of the intercostal spaces during respiration. This observation illustrates their typical stabilizing role in preserving thoracic wall integrity during respiration.

11.6.1.3 Levatores Costorum (Innervated by the Dorsal Rami of the Adjacent Thoracic Nerves)

The levatores costarum represent a series of small, well-defined triangular muscles extending from C7 to T11. Each side of the vertebral column contains twelve muscle units, displaying bilateral symmetry (Standring 2020) (*see* Fig. 11.19).

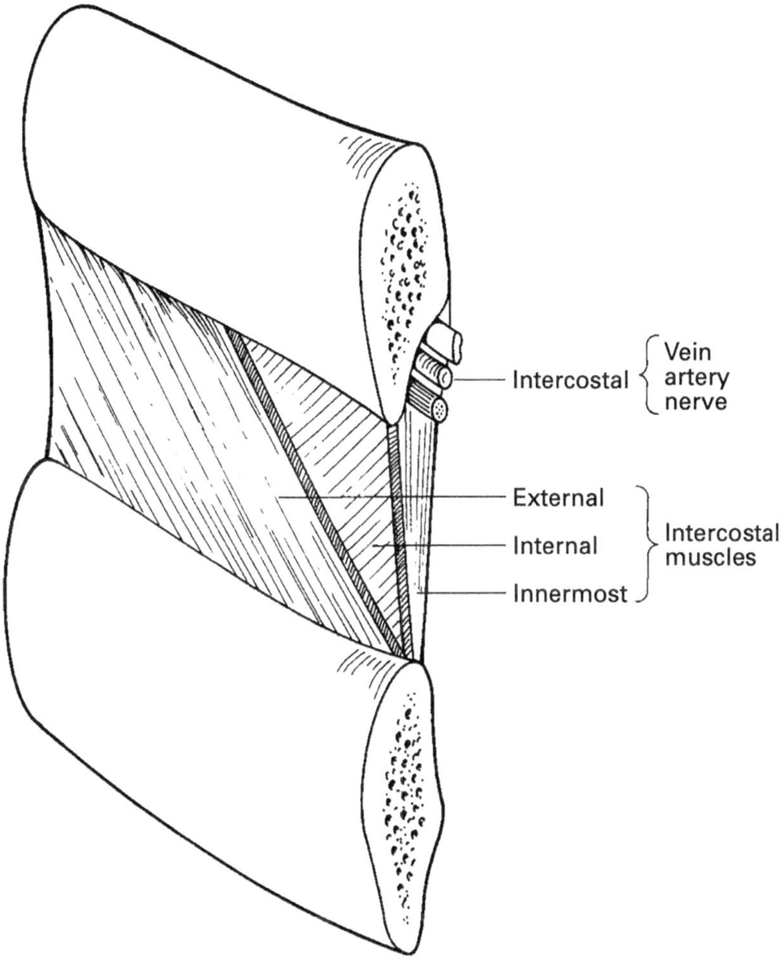

Fig. 11.22 Composition of an intercostal space. (Adapted from Ellis et al. (2004))

These muscles demonstrate a consistent anatomical pattern, originating from the transverse process tips of vertebrae and inserting into the superior border of the rib at the succeeding level, near its tubercle. Their fiber arrangement exhibits a characteristic fan-shaped configuration, expanding in width as they course inferolaterally (Drake et al. 2020; Moore et al. 2018).

The levatores costarum muscles perform multiple biomechanical functions. Their primary action contributes to the respiratory mechanism through rib elevation during inspiratory phases (Barrett et al. 2019). This mechanical action augments thoracic cavity expansion, complementing the actions of other respiratory muscles.

Their anatomical orientation and attachments enable additional functions beyond respiration. When activated unilaterally, these muscles facilitate lateral trunk flexion. Their fiber arrangement also permits generation of rotational forces in the thoracic spine, contributing to trunk rotatory movements (Neumann 2017).

11.6.1.4 Serratus Posterior Superior (Innervated by the Anterior Primary Rami of the Thoracic Nerves: T2–T6)

The serratus posterior superior presents as a thin, flattened muscular sheet positioned deep to the rhomboid muscles. Its anatomical origins include the inferior portion of the ligamentum nuchae and spinous processes spanning from the seventh cervical to the third thoracic vertebrae (Moore et al. 2018). The muscle fibers demonstrate an inferolateral trajectory, terminating at insertion points lateral to the costal angles of ribs two

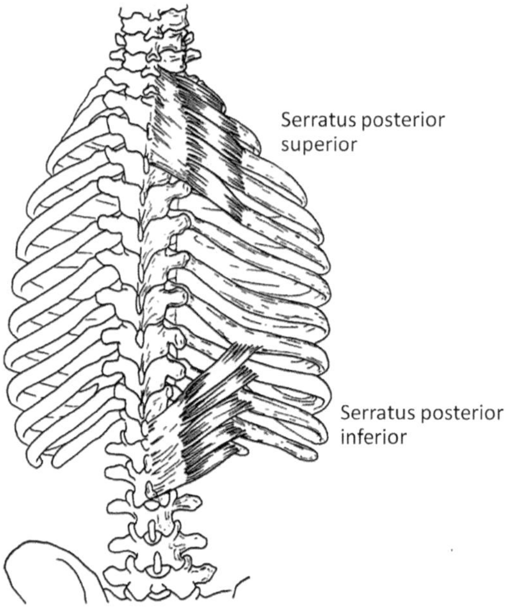

Fig. 11.23 Serratus posterior superior and inferior muscles. (Adapted from Calais-Germain (1999))

through five. This orientation creates a characteristic serrated appearance at its costal attachments (Drake et al. 2020) (Fig. 11.23).

The serratus posterior superior contributes to respiratory mechanics through its specific action on the ribs. During contraction, this muscle produces elevation of ribs two through five, participating in thoracic cage expansion during inspiration. This upward displacement of ribs augments the inspiratory phase by complementing the actions of primary respiratory muscles. Through this mechanical advantage, the muscle enhances thoracic volume expansion necessary for effective ventilation (West and Luks 2016).

11.6.2 Expiratory Muscles: External Oblique, Internal Oblique, Transversus Abdominis. Transversus Thoracis, Subcostals, Serratus Posterior Inferior, Latissimus Dorsi, Internal Intercostals

The process of expiration occurs through two primary mechanisms: passive recoil following inspiratory muscle relaxation or active muscular contraction (De Troyer et al. 2005). Passive expiration is adequate with silent respiration and appropriate speaking. However, active expiration becomes necessary during activities requiring prolonged phonation or increased ventilatory demands (Ratnovsky and Elad 2005).

The primary expiratory muscles comprise the anterolateral abdominal wall musculature. These muscles include the external oblique, internal oblique, transversus abdominis, and rectus abdominis (Hudson et al. 2011). When these muscles contract, they generate increased intra-abdominal pressure, resulting in superior displacement of abdominal contents and subsequent elevation of the diaphragm. This mechanical action, coupled with inspiratory muscle relaxation, reduces thoracic cavity volume and produces expiratory airflow (De Troyer and Boriek 2011).

The anatomical arrangement of expiratory muscles follows a specific pattern. The external oblique, internal oblique, and transversus abdominis originate from the pelvis and insert into the thoracic cage. The internal intercostal muscles, positioned deep to the external intercostals, contribute to active expiration by drawing adjacent ribs together and downward (Standring 2020). Additional expiratory muscles include the transversus thoracis, subcostals, serratus posterior inferior, and latissimus dorsi, which participate in forced expiration (Albert et al. 2008).

11.6.2.1 Abdominal Wall Musculature

The rectus abdominis muscle demonstrates specific anatomical characteristics, with its origin measuring 2.5 cm in width from the pubic crest and symphysis, and its insertion spanning 7.5 cm across the fifth, sixth, and seventh costal cartilages (Moore et al. 2018). The muscle exhibits developmental segmentation through three fibrous intersections: one at the umbilical level, another at the xiphoid level, and a third midway between these points. A fourth intersection may occur inferior to the umbilicus. These intersections, present only anteriorly, adhere to the anterior rectus sheath.

The rectus sheath's formation primarily results from the internal oblique aponeurosis splitting. The posterior aspect receives reinforcement from the transversus abdominis aponeurosis, while the anterior aspect is strengthened by the external oblique aponeurosis (Netter 2019).

The architectural arrangement varies along the length of the rectus sheath. Above the costal margin, the rectus abdominis directly contacts the costal cartilages, with only the external oblique aponeurosis forming the anterior sheath. In 5–8 cm below the costal margin, the transversus abdominis maintains its muscular form nearly to the midline. Below a level halfway between the umbilicus and pubis, demarcated by the arcuate line of Douglas, the aponeuroses of all three lateral muscles pass in front of the rectus. Here, then, posteriorly the rectus rests against the transversalis fascia, extraperitoneal fat, and peritoneum. Inferior to the arcuate line (semicircular line of Douglas), all three lateral muscle aponeuroses course anterior to the rectus, leaving only the transversalis fascia, extraperitoneal fat, and peritoneum as posterior relations (Drake et al. 2020) (Fig. 11.24).

The posterior rectus sheath exhibits four different zones from superior to inferior: cartilaginous, muscular, aponeurotic, and areolar. The anterior rectus sheath maintains consistent fascial resistance throughout its length, while the posterior wall's resistance varies, particularly above the pubis and in the epigastrium.

The linea alba, formed by the fusion of the rectus sheath aponeuroses, extends from the pubis to the xiphoid process. Its width increases superior to the umbilicus, while remaining relatively narrow in the inferior abdomen (Ratnovsky et al. 2008).

The lateral abdominal wall muscles demonstrate specific fiber orientations above the iliac crest: external oblique fibers course inferiorly and medially, internal oblique fibers run superiorly and medially, and transversus abdominis fibers maintain a horizontal orientation. Below the iliac crest, these muscles become aponeurotic, with fibers oriented inferiorly and medially,

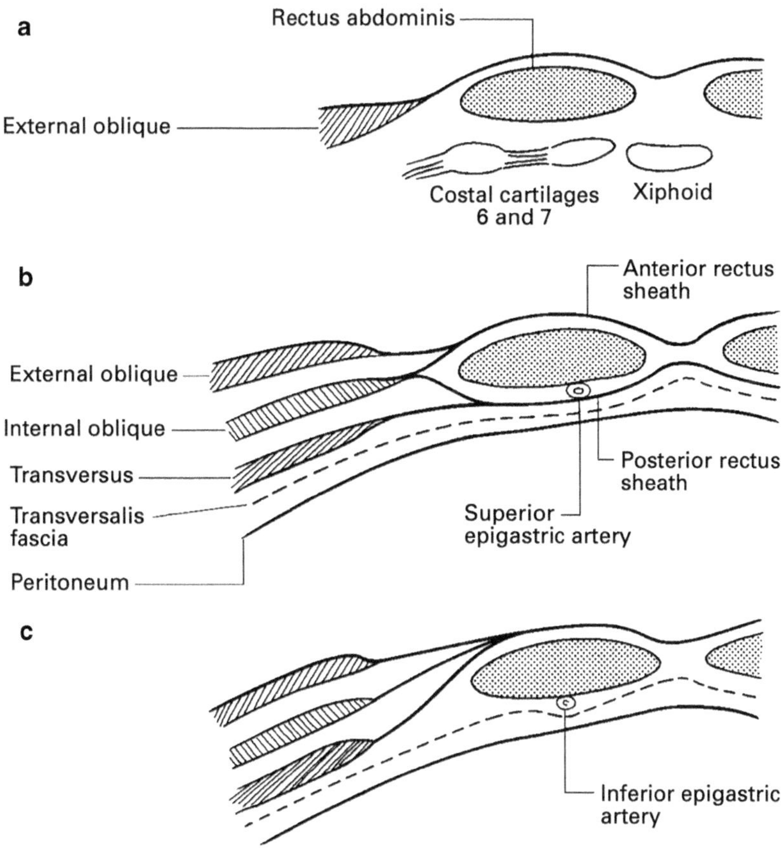

Fig. 11.24 Composition of the rectus sheath illustrated in transverse sections: (**a**) above the costal margin, (**b**) above the arcuate line, and (**c**) below the arcuate line. (Adapted from Ellis et al. (2004))

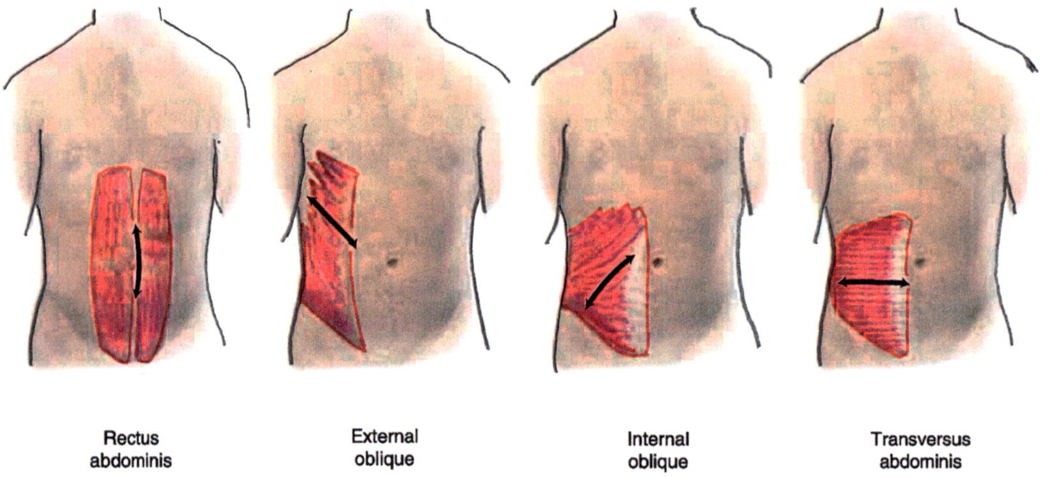

Fig. 11.25 Accessory muscles involved in expiration. (Adapted with permission from Des Jardins (2013))

contributing to the inguinal canal formation (Standring 2020). Functionally, these abdominal muscles participate in forced expiration and coughing by elevating intra-abdominal pressure and drawing the inferior ribs inferiorly and medially. Electromyographic analysis confirms their non-participation in inspiration (De Troyer et al. 2005) (Fig. 11.25).

Rectus Abdominis Muscle

The rectus abdominis represents paired muscular structures extending longitudinally through the anterior abdominal wall. Each muscle forms a vertical band approximately 10 centimeters in width, with the linea alba providing midline separation between the paired muscles (Moore et al. 2018). These muscles originate from the pubic symphysis and iliac crest, with superior attachments at the xiphoid process and the fifth through seventh costal cartilages (Drake et al. 2020; Singh 2014b) (Fig. 11.26).

The rectus abdominis contributes significantly to respiratory function through its role in active expiration. During contraction, these muscles generate increased intra-abdominal pressure, resulting in superior displacement of the diaphragm into the thoracic cavity. This mechanical action reduces thoracic volume and promotes expiratory airflow (Ratnovsky et al. 2008). The coordinated activity between the rectus abdominis and other expiratory muscles produces effective ventilation during increased respiratory demands (De Troyer and Boriek 2011).

External Oblique Muscle

The external oblique muscle presents as a broad, flattened structure positioned in the anterolateral abdominal wall. This muscle represents the most superficial component of the anterolateral abdominal musculature (Standring 2020). Its anatomical organization includes eight distinctive digitations originating from the external surfaces of the inferior eight ribs. The muscle fibers course inferiorly and medially, terminating in an extensive aponeurosis that inserts into the iliac crest and contributes to the formation of the linea alba (Netter 2019) (Fig. 11.27).

From a respiratory perspective, the external oblique exhibits significant expiratory function. During contraction, it increases intra-abdominal pressure, which promotes superior displacement of the diaphragm into the thoracic cavity. This mechanical action reduces thoracic volume and augments expiratory airflow (De Troyer et al. 2005). The muscle's oblique fiber arrangement also enables rotational torque production, which may enhance expiratory force during activities requiring increased ventilatory effort (Ratnovsky and Elad 2005).

Fig. 11.26 Origin and insertion sites of the rectus abdominis. (Adapted from Singh (2014b))

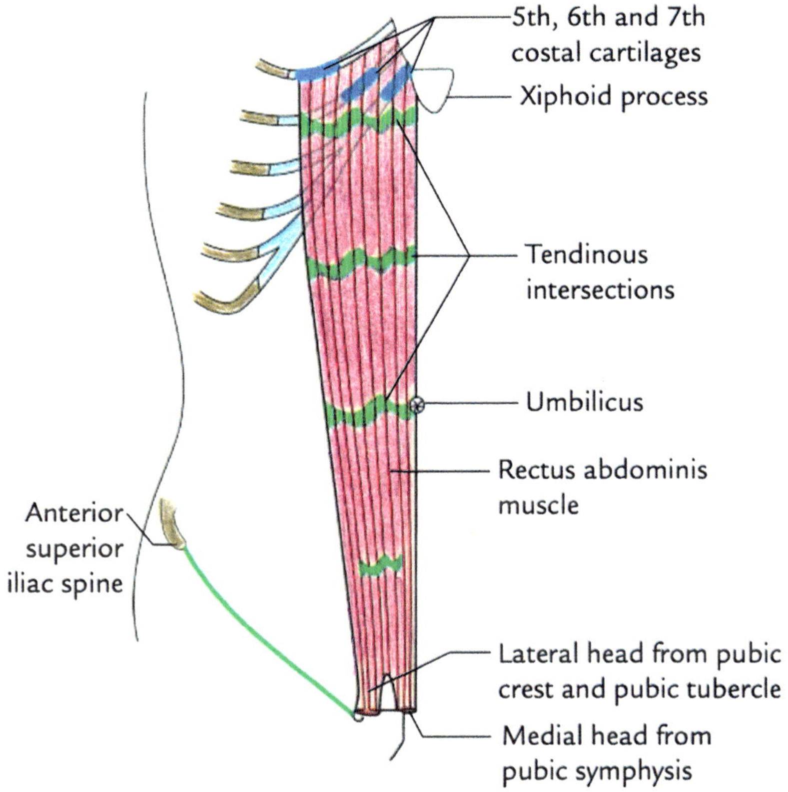

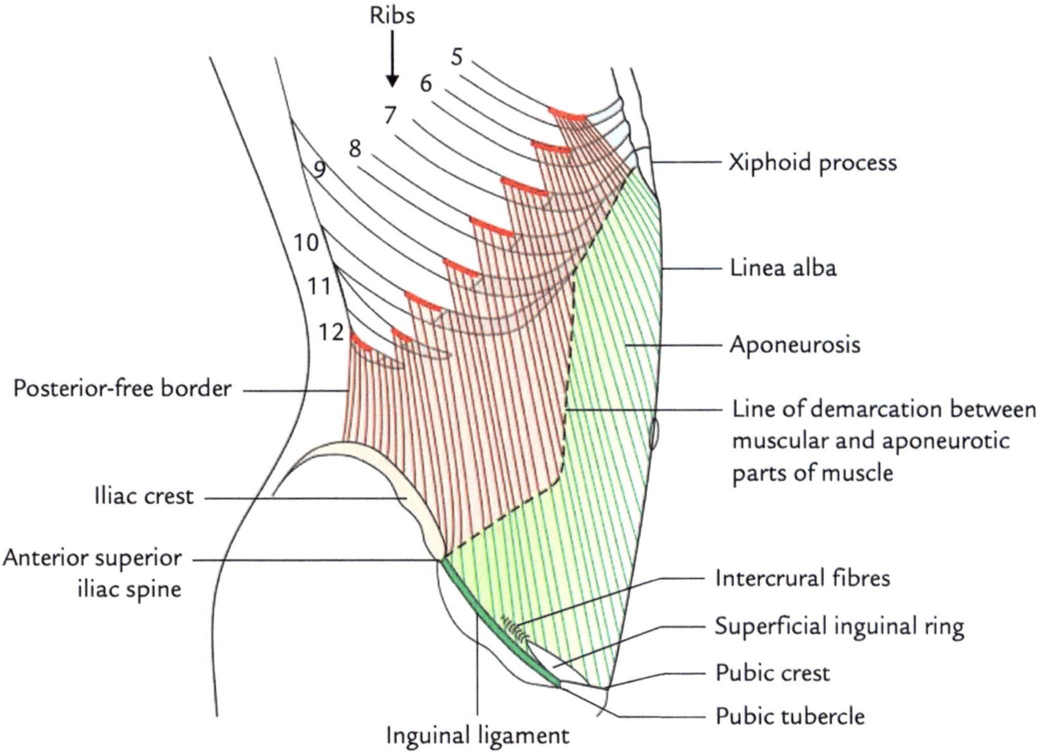

Fig. 11.27 Anatomy of the external oblique muscle. (Adapted from Singh (2014b))

Internal Oblique Muscle

The internal oblique muscle, situated deep to the external oblique, occupies the lateral and anterior regions of the abdominal wall. This muscle demonstrates reduced thickness and surface area compared to the external oblique muscle (Drake et al. 2020). The internal oblique originates from three areas: the inguinal ligament, the iliac crest, and the thoracolumbar fascia. Its fibers course superiorly and medially to insert into the inferior four costal cartilages and continue as an aponeurosis that contributes to the linea alba formation (Moore et al. 2018) (Fig. 11.28).

In respiratory physiology, the internal oblique functions as an active expiratory muscle. Through its contraction, it elevates intra-abdominal pressure, resulting in superior displacement of abdominal contents and subsequent diaphragmatic elevation. This mechanical action reduces thoracic cavity volume, promoting expiratory airflow (De Troyer and Boriek 2011). The muscle's fiber orientation allows it to work synergistically with other abdominal wall muscles during forceful expiration and various respiratory maneuvers (Ratnovsky et al. 2008).

Transversus Abdominis Muscles

The transversus abdominis represents the deepest layer of the anterolateral abdominal wall, positioned immediately deep to the internal oblique muscle. Its anatomical attachments include the inguinal ligament, iliac crest, thoracolumbar fascia, and inferior six costal cartilages. The muscle fibers maintain a horizontal orientation and merge into an aponeurosis that integrates into the linea alba (Standring 2020) (Fig. 11.29).

The respiratory function of the abdominal wall muscles involves coordinated action of all four paired muscle groups (rectus abdominis, external oblique, internal oblique, and transversus abdominis). Their synchronized contraction generates increased intra-abdominal pressure, resulting in superior displacement of the diaphragm into the thoracic cavity (De Troyer et al. 2005). This mechanical action produces two significant effects: reduction of thoracic cavity volume and elevation of intrapleural pressure. These changes enhance expiratory airflow during active expiration (Ratnovsky and Elad 2005).

The transversus abdominis demonstrates particular significance in respiratory mechanics

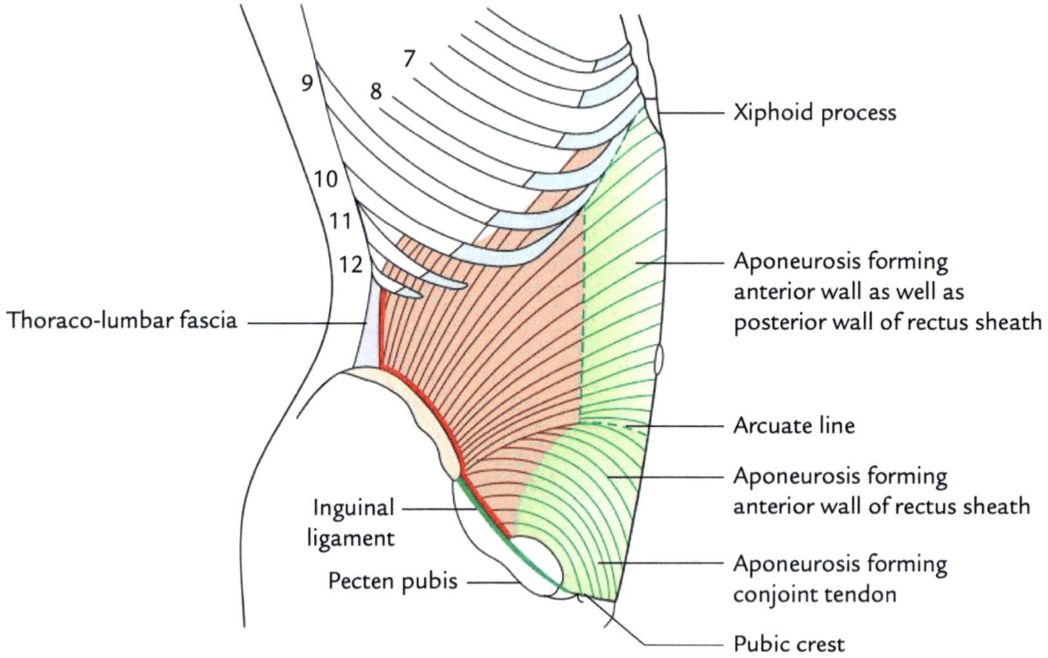

Fig. 11.28 Anatomy of the internal oblique muscle. (Adapted from Singh (2014b))

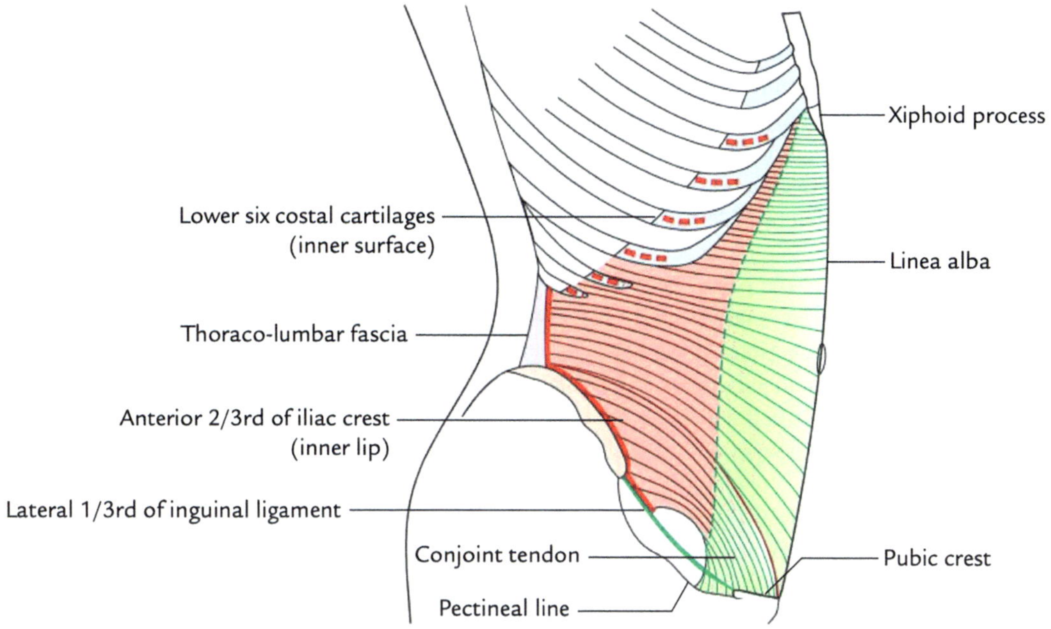

Fig. 11.29 Origin and insertion sites of the transversus abdominis muscle. (Adapted from Singh (2014b))

through its horizontal fiber arrangement, which provides efficient circumferential compression of abdominal contents. This anatomical configuration enables effective modulation of intra-abdominal pressure during respiratory maneuvers (Hudson et al. 2011).

11.6.2.2 Internal Intercostal Muscles

The internal intercostal muscles compose the intermediate layer of the intercostal muscle group, positioned deep to the external intercostals. These muscles extend from the sternum anteriorly to the costal angles posteriorly, where they transition into the posterior intercostal membrane (Standring 2020). Their fibers demonstrate an oblique orientation, coursing inferiorly and laterally in the anterior region, while adopting an inferomedial direction posteriorly (Moore et al. 2018).

The anatomical arrangement includes attachments from the upper border of each rib to the lower border of the adjacent upper rib. During active expiration, these muscles contract to produce inferomedial displacement of the ribs, reducing both the lateral and anteroposterior thoracic dimensions (Fig. 11.30). This action counteracts the external intercostal muscles and minimizes intercostal tissue displacement during forceful expiration (De Troyer and Boriek 2011).

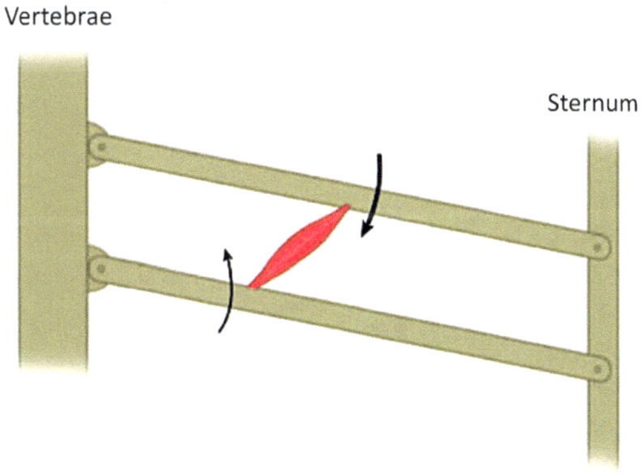

Fig. 11.30 Effect of internal intercostal muscle contraction: contraction generates a greater torque on the upper rib compared to the lower rib. (Adapted with permission from Ko (2022))

In the area between the sternum and chondrocostal junctions, the external intercostals transition into the anterior intercostal membrane, leaving the internal intercostals as the only muscular component. This anterior portion of the internal intercostal muscles, termed the "parasternal intercostals," exhibits unique anatomical and functional characteristics compared to its internal interosseous muscle (Fig. 11.31). Human parasternal intercostals function as inspiratory muscles across all intercostal levels (De Troyer et al. 2005).

The inner surface of the parasternal intercostals is covered by the triangularis sterni (transversus thoracis), which functions as an expiratory muscle. Although anatomically related, the triangularis sterni is not classified within the intercostal muscle group (Ratnovsky et al. 2008; De Troyer et al. 2005).

The internal intercostal muscles primarily activate during forceful expiration, such as during exercise or coughing, rather than during quiet breathing. This selective activation pattern reflects their role in augmenting expiratory effort when increased ventilatory demands exist (Hudson et al. 2011).

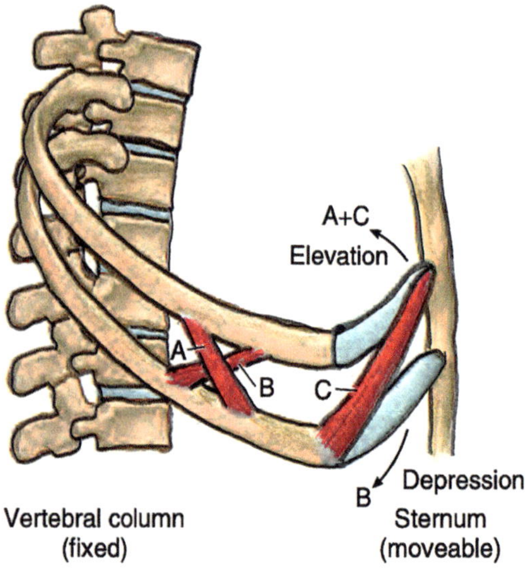

Fig. 11.31 A simplified model illustrating the action of the intercostal muscles. Contraction of the muscle fibers parallel to the slope of the ribs at a given point (fiber A and C) elevates the ribs and sternum, while contraction of muscle fibers oriented approximately perpendicular to the slope of the fibs (fibers B) results in rib depression. (Adapted from Moore et al. (2018))

11.6.2.3 Transversus Thoracis (Triangularis Sterni)

The transversus thoracis, also known as the triangularis sterni, occupies the internal surface of the anterior thoracic wall (Standring 2020). This muscle demonstrates specific anatomical attachments, originating from three regions: the posterior aspect of the xiphoid process, the inferior half of the sternal body, and the fourth through seventh costal cartilages (Drake et al. 2020) (*see* Fig. 11.18).

The muscle's fiber orientation exhibits a distinctive pattern. The inferior fibers maintain a horizontal trajectory, while the superior fibers course superolaterally to insert into the internal surfaces of the second through sixth costal cartilages (Moore et al. 2018). This anatomical configuration enables the transversus thoracis to generate inferomedial forces on the sternocostal junctions during contraction.

In respiratory mechanics, the transversus thoracis functions as an expiratory muscle by producing inferior displacement of the sternocostal cartilages. This action contributes to thoracic volume reduction during active expiration (De Troyer and Boriek 2011). The muscle's anatomical position and mechanical advantage make it particularly effective in modulating anterior chest wall compliance during expiratory maneuvers (Ratnovsky et al. 2008).

11.6.2.4 Subcostals

The subcostal muscles represent specialized muscular bands within the thoracic wall, characterized by their irregular arrangement and variable extent across intercostal spaces. These muscles demonstrate optimal development in the inferior thoracic region (Standring 2020). Their anatomical position includes attachments to the internal surface of the ribs near their angles, with fibers spanning one to two intercostal spaces (Moore et al. 2018) (Fig. 11.32).

The fiber orientation of the subcostal muscles parallels that of the innermost intercostal

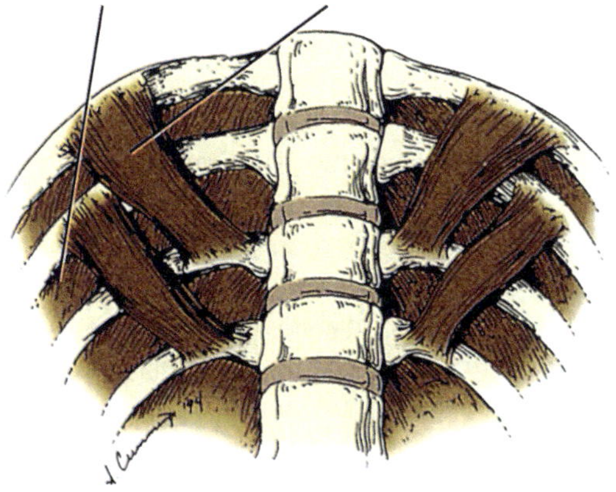

Fig. 11.32 Internal view of the posterior thoracic wall depicting several subcostal muscles. (Adapted with permission from Cramer and Darby (2014))

muscles, with which they may share continuity. This architectural arrangement suggests functional synergy between these muscle groups (Drake et al. 2020).

From a respiratory point of view, the subcostal muscles contribute to active expiration through their action of rib depression. Upon contraction, they contribute to thoracic volume reduction by drawing adjacent ribs closer together (De Troyer et al. 2005). Their anatomical position provides mechanical advantage for modulating local thoracic wall compliance during expiratory efforts (Ratnovsky and Elad 2005).

11.6.2.5 Serratus Posterior Inferior (Innervated by the Anterior Primary Rami of the Thoracic Nerves: T9–T11)

The serratus posterior inferior occupies a deep position relative to the latissimus dorsi muscle. Its anatomical attachments include proximal origins from the spinous processes of the eleventh thoracic through second lumbar vertebrae (T11–L2) and their associated supraspinous ligaments, with connections via the thoracolumbar fascia (Standring 2020) (*see* Fig. 11.23).

The muscle demonstrates a horizontal fiber orientation, extending laterally to insert into the external surfaces of the inferior four ribs at their angles (Moore et al. 2018). This architectural arrangement enables the muscle to generate posteroinferior forces on its costal attachments during contraction.

The serratus posterior inferior contributes to active expiration through its ability to produce posteroinferior rib displacement. The muscle receives innervation from the anterior primary rami of the ninth through eleventh thoracic nerves (T9–T11), reflecting its segmental organization (Drake et al. 2020). This innervation pattern supports coordinated activation with other expiratory muscles during increased ventilatory demands (De Troyer and Boriek 2011).

11.7 Biomechanics of Thoracic Cage Movement

The thoracic cage undergoes complex three-dimensional changes during respiration, involving modifications in vertical, transverse, and anteroposterior dimensions. These changes result from coordinated movements of the diaphragm, ribs, and sternum (Standring 2020).

The ribs and their associated costal cartilages function as lever systems, with movement patterns varying based on their length and sternal articulation patterns. The ribs articulating with the sternum demonstrate progressive lengthening in the superior-to-inferior direction, with anterior aspects positioned inferior to their posterior aspects. Although the ribs 8–9 show reduced length, their costal margin attachments maintain the progressive increase in thoracic transverse diameter through the rib 9 level (De Troyer et al. 2005).

Rib movement occurs around an axis extending through the costovertebral and costotransverse joints along the rib neck. This axis demonstrates posterolateral orientation, corresponding to vertebral transverse process inclination. The obliquity of these processes increases progressively in the inferior direction, resulting in distinct movement patterns between superior and inferior ribs (Moore et al. 2018).

The upper ribs (ribs 2–5) exhibit predominantly "pump-handle" movement, characterized by anterior elevation that raises the sternal body. The first rib demonstrates minimal anterior movement due to its shorter length and firm manubrial attachment. The elevation of longer superior ribs produces manubriosternal joint flexion, increasing anteroposterior thoracic diameter (Drake et al. 2020).

In contrast, ribs 8–10 demonstrate "bucket-handle" movement, characterized by superolateral displacement of their anterior aspects. This movement pattern increases the infrasternal angle and thoracic transverse diameter. The flat costotransverse joint surfaces in these regions permit combined rotational and translational movements (Ratnovsky et al. 2008).

The ribs 6 and 7 exhibit combined pump-handle and bucket-handle movements. The ribs 11 and 12, lacking anterior attachments, minimally affect thoracic transverse diameter but provide attachment sites for inferior diaphragmatic fibers and, with quadratus lumborum assistance, establish stable diaphragmatic anchoring points (De Troyer and Boriek 2011).

Thoracic vertebral column movements induce passive rib displacement patterns. Flexion approximates adjacent ribs, extension increases interrib spacing, lateral flexion produces rib approximation on the concave aspect and separation on the convex aspect, and rotation generates relative horizontal displacement between adjacent ribs.

11.8 Respiratory Mechanics in Speech Production

The respiratory system demonstrates dual functionality in maintaining life-sustaining processes and enabling speech production. Speech production usually utilizes only 13% of vital capacity, suggesting that baseline respiratory function generally supports normal speech requirements (Moore et al. 2018).

Respiratory patterns present in four principal forms: abdominal (diaphragmatic), thoracic, clavicular, and oppositional breathing. Diaphragmatic breathing, characterized by anterior abdominal wall displacement during diaphragm contraction, predominates in 65% of individuals. Gender-specific differences exist, with females demonstrating increased prevalence of thoracic breathing patterns. During quiet respiration and speech, diaphragmatic breathing often functions independently, while increased respiratory demands recruit additional thoracic muscle involvement (De Troyer et al. 2005).

The phrenic nerve provides primary motor innervation to the diaphragm. Clinical evidence suggests limited voluntary diaphragmatic control, with paralysis minimally affecting overall respiratory capacity. Speech-related respiratory control primarily involves thoracic musculature, allowing patients with diaphragmatic paralysis to maintain adequate respiratory and speech functions through thoracic breathing patterns. However, such patients may experience limitations during prolonged or intense physical activities requiring combined breathing patterns (Standring 2020).

Multiple respiratory muscle involvement presents more significant clinical challenges than isolated diaphragmatic paralysis. Unilateral respiratory muscle damage often permits effective compensation, with deficits becoming apparent only during strenuous activities. Bilateral involvement may necessitate clavicular breathing, primarily mediated by the sternocleidomastoid muscles, which receive innervation from the second and third cervical nerves (Ratnovsky and Elad 2005). Table 11.3 provides an overview of the primary respiratory muscles and their innervations.

In cervical trauma cases, respiratory function may depend solely on sternocleidomastoid activity. However, these muscles provide insufficient force for adequate thoracic expansion. Clavicular breathing patterns may also develop in conditions of extreme tension or generalized weakness, characterized by minimal thoracic and abdominal movement, subtle clavicular elevation, and visible sternocleidomastoid strain during inspiration (Drake et al. 2020). The inefficiency of clavicular breathing significantly impacts speech production, resulting in inadequate respiratory support, compromised voice quality, and deterioration during physical exertion. Clinical manifestations

Table 11.3 Summary of motor innervation of respiratory muscles

Muscle	Innervation
Muscles of inspiration	
Diaphragm	Phrenic nerve or plexus, C4, some input from C3 and C5
Pectoralis major	C5–8, T1
Pectoralis minor	C8, T1
Subclavius	C5, C6
Serratus posterior superior	T1–T4
Levator costarum	Intercostal (thoracic) nerves
Scalenes	C6, C7, C8
External intercostals	Intercostal (thoracic) nerves
Sternocleidomastoid	C3, C3
Muscles of expiration	
External obliques	Intercostal (T8–T12) nerves
Internal obliques	Intercostal (T8–T12) nerves
Rectus abdominus	Intercostal (T7–T12) nerves
Transverse abdominals	Intercostal *T7–T12) nerves
Serratus posterior inferior	T9–T12
Quadratus lumborum	T12, L1

Adapted and modified from Shankar (1996)

include excessive throat tension, irregular speech flow, and progressive voice weakness (Hudson et al. 2011).

11.9 Autonomic Innervation of the Tracheobronchial Tree and Lungs

The tracheobronchial tree and lungs receive autonomic innervation by three neural pathways that regulate pulmonary function. These pathways include autonomic afferent, parasympathetic efferent, and sympathetic efferent components, which coordinate to maintain optimal respiratory function (Standring 2020) (Fig. 11.33).

The autonomic nervous system presents a dual structure, with opposing sympathetic and parasympathetic pathways. This balanced neural control modulates respiratory and circulatory functions to achieve optimal ventilation-perfusion matching based on metabolic requirements. The system responds to input from peripheral sensory cells, which detect changes in various physiological parameters (Drake et al. 2020).

The autonomic control system integrates respiratory and cardiovascular functions through complex neural networks. These networks regulate bronchial tone, pulmonary blood flow, and secretory activity of the tracheobronchial tree. Sympathetic stimulation generally produces bronchodilation and vasoconstriction, while parasympathetic activation results in bronchoconstriction and increased secretory activity (Mazzone and Undem 2016).

Afferent pathways transmit sensory information from various receptors in the respiratory system, including mechanoreceptors, chemoreceptors, and nociceptors. This sensory input enables rapid adjustments in respiratory parameters to maintain homeostasis.

11.9.1 Autonomic Afferent Pathways in Respiratory Control

The respiratory system has various afferent pathways that transmit sensory information to the central nervous system. Pulmonary stretch receptors in the alveoli and irritant receptors in the airways convey signals through the pulmonary plexus, which surrounds the tracheal bifurcation and pulmonary hila, ultimately transmitting them to the central nervous system via the vagus nerve (Mazzone and Undem 2016).

The tracheal irritant receptors and laryngeal cough receptors also utilize vagal pathways for central signal transmission. These receptors respond to mechanical and chemical stimuli, initiating protective reflexes and modulating respiratory patterns. Peripheral chemoreceptors and baroreceptors demonstrate specific neural pathways. The carotid bodies and carotid sinus transmit afferent signals through the glossopharyngeal nerve, while the aortic bodies and arch utilize vagal pathways. These structures monitor blood gas composition and arterial pressure, contributing to respiratory and cardiovascular control (Standring 2020).

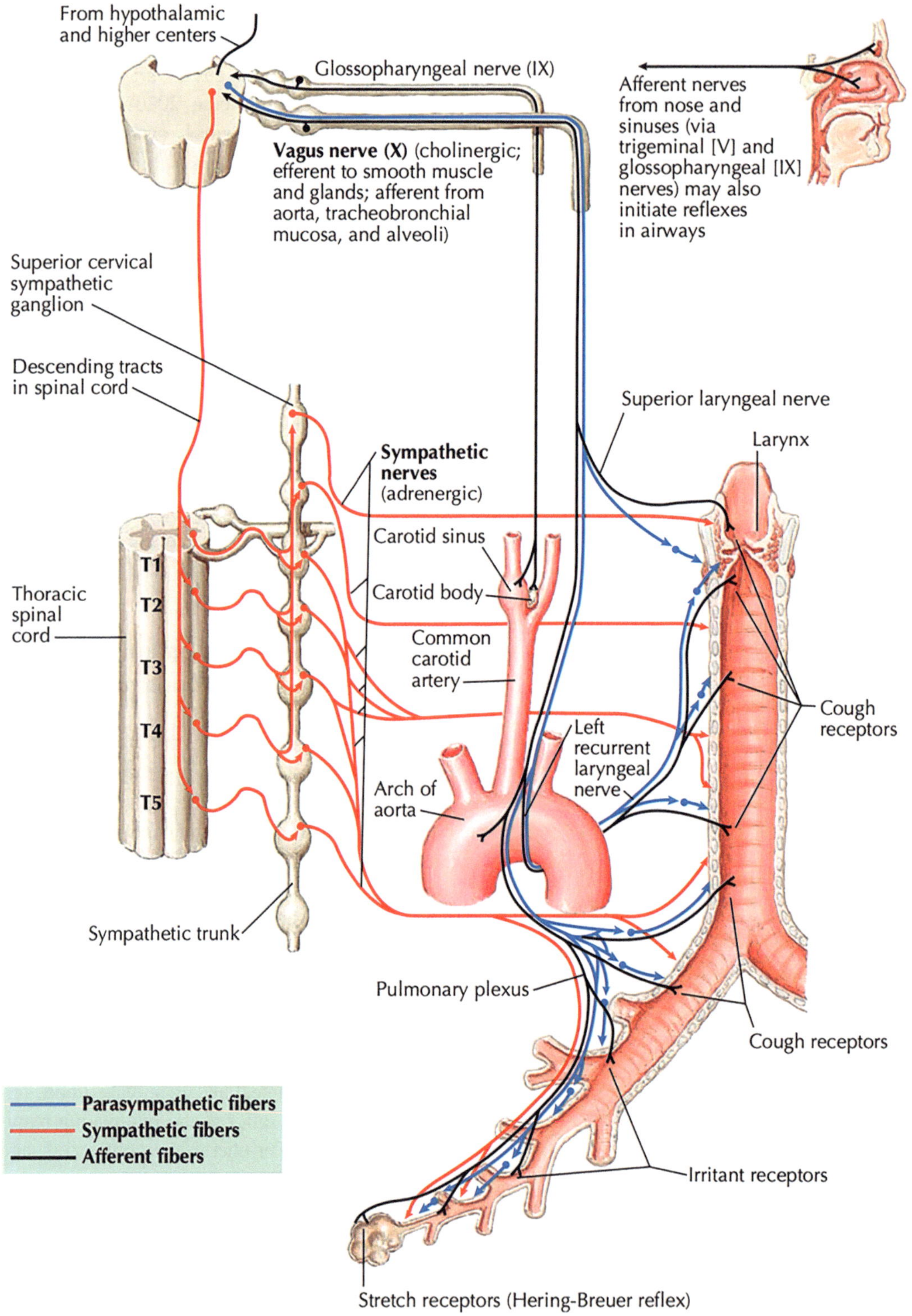

Fig. 11.33 Innervation pattern of tracheobronchial tree. (Adapted with permission from Jones et al. (2013))

Nasal and paranasal receptors transmit information through both trigeminal and glossopharyngeal pathways. Additionally, respiratory control centers receive modulatory input from the hypothalamus, higher cortical centers, and the reticular activating system, enabling integration of respiratory control with other physiological functions (Drake et al. 2020).

11.9.2 Parasympathetic Efferent Pathways

The parasympathetic nervous system regulates the tracheobronchial tree through a structured neural network. The vagus nerve is a principal pathway for all preganglionic parasympathetic efferent fibers. These fibers originate from cellular bodies located in the dorsal vagal nuclei, which maintain close association with the respiratory centers in the medulla oblongata. This anatomical arrangement ensures the smooth integration of respiratory control mechanisms (Standring 2020).

The neural signals follow a two-neuron pathway characteristic of parasympathetic innervation. As these preganglionic fibers approach their targets, they form synaptic connections with short postganglionic neurons. These secondary neurons are strategically positioned either in close proximity to or directly within the tracheobronchial wall, allowing for localized control of respiratory functions (Drake et al. 2020).

This parasympathetic network regulates various essential physiological responses in the respiratory system via cholinergic neurotransmission. The release of acetylcholine induces smooth muscle contraction in the bronchial walls, activates secretory functions in respiratory glands, and enables vasodilation of bronchial blood vessels. These coordinated actions have significance for maintaining normal airway function (Mazzone and Undem 2016).

11.9.3 Sympathetic Efferent Pathways

Sympathetic regulation begins in the thoracic spinal cord, where preganglionic fibers originate from segments T1(T2) to T5(T6). These fibers traverse the white rami communicantes to reach the sympathetic trunk, where they follow specific pathways based on their target tissues (Standring 2020). For the upper respiratory tract, including the larynx and upper trachea, preganglionic fibers ascend within the sympathetic trunk to synapse in cervical sympathetic ganglia. The remaining fibers form synaptic connections in upper thoracic ganglia, from which postganglionic fibers extend to innervate the inferior trachea and bronchial tree via the pulmonary plexus. These postganglionic terminals release norepinephrine as their primary neurotransmitter (Drake et al. 2020).

The physiological effects of sympathetic activity occur through interactions with specific adrenergic receptors. These receptors exhibit specific distribution patterns and functional characteristics throughout the respiratory system. α-Receptors are predominantly located in smooth muscle and exocrine glandular tissues, while β-receptors exhibit two subtypes with specific tissue distributions. β1-Receptors concentrate in cardiac tissue, whereas β2-receptors appear throughout various smooth muscle tissues, including the bronchial tree. The respiratory system's response to sympathetic stimulation includes bronchial smooth muscle relaxation, reduced glandular secretion, and vascular constriction. In pulmonary tissues, which exclusively express β2-receptors, sympathetic activation produces bronchodilation and may decrease mucus production. However, pharmacological stimulation of α-adrenergic receptors can induce bronchoconstriction, demonstrating the complex interaction between different receptor subtypes (Mazzone and Undem 2016).

References

Albert RK, Spiro SG, Jett JR. Clinical respiratory medicine. 3rd ed. Philadelphia: Elsevier; 2008.

Barrett KE, Barman SM, Brooks HL, Yuan JX. Ganong's review of medical physiology. 26th ed. New York: McGraw Hill; 2019.

Berlowitz DJ, Wadsworth B, Ross J. Respiratory problems and management in people with spinal cord injury. Breathe. 2016;12(4):328–40.

Brown R, DiMarco AF, Hoit JD, Garshick E. Respiratory dysfunction and management in spinal cord injury. Respir Care. 2006;51(8):853–68.

Butler JE, Hudson AL, Gandevia SC. The neural control of human inspiratory muscles. Prog Brain Res. 2014;209:295–308.

Calais-Germain B. Anatomy of movement. Revised ed. Seattle: Eastland Press; 1999.

Cramer GD, Darby SA. Clinical anatomy of the spine, spinal cord, and ANS. 3rd ed. St. Louis: Elsevier; 2014.

De Troyer A, Boriek AM. Mechanics of the respiratory muscles. Compr Physiol. 2011;1(3):1273–300.

De Troyer A, Estenne M. Functional anatomy of the respiratory muscles. Clin Chest Med. 1988;9(2):175–93.

De Troyer A, Kirkwood PA, Wilson TA. Respiratory action of the intercostal muscles. Physiol Rev. 2005;85(2):717–56.

Des JT. Cardiopulmonary anatomy & physiology: essentials of respiratory care. 6th ed. Clifton Park: Cengage Learning; 2013.

Detton AJ. Grant's dissector. 16th ed. Philadelphia: Wolters Kluwer; 2013.

Drake RL, Vogl AW, Mitchell AWM. Gray's anatomy for students. 4th ed. Philadelphia: Elsevier; 2020.

Ellis H, Mahadevan V. Clinical anatomy: applied anatomy for students and junior doctors. 14th ed. Oxford: Wiley-Blackwell; 2019.

Ellis H, Feldman S, Harrop-Griffiths W. Anatomy for anaesthetists. 8th ed. Oxford: Blackwell Publishing; 2004.

Ellis H, Logan BM, Dixon AK, Bowden DJ. Human sectional anatomy: atlas of body sections, CT and MRI images. 4th ed. Boca Raton: CRC Press; 2015.

Feldman JL, Del Negro CA. Looking for inspiration: new perspectives on respiratory rhythm. Nat Rev Neurosci. 2006;7(3):232–42.

Gropper MA. Miller's anesthesia. 9th ed. Philadelphia: Elsevier; 2020.

Hall JE. Guyton and Hall textbook of medical physiology. 13th ed. Philadelphia: Elsevier; 2016.

Harvey LA. Physiotherapy rehabilitation for people with spinal cord injuries. J Physiother. 2016;62(1):4–11.

Hudson AL, Butler JE, Gandevia SC, De Troyer A. Role of the diaphragm in trunk rotation in humans. J Neurophysiol. 2011;106(4):1622–8.

Jones HR, Burns TM, Aminoff MJ, Pomeroy SL. The Netter collection of medical illustration. Volume 7, Nervous system, Part II Spinal cord and peripheral motor and sensory systems. 2nd ed. Philadelphia: Elsevier; 2013.

Kaminsky DA. The Netter collection of medical illustrations: respiratory system. 2nd ed. Philadelphia: Elsevier; 2011.

Ko HY. Management and rehabilitation of spinal cord injuries. 2nd ed. Singapore: Springer; 2022.

Ko HY. A practical guide to care of spinal cord injuries: clinical questions and answers. Singapore: Springer; 2023.

Kumar P, Clark M. Clinical medicine. 8th ed. London: Elsevier; 2012.

Kumar V, Abbas AK, Aster JC. Robbins and Cotran pathologic basis of disease. 9th ed. Philadelphia: Elsevier; 2015.

Levitzky MG. Pulmonary physiology. 7th ed. New York: McGraw Hill; 2007.

Ludlow CL. Central nervous system control of voice and swallowing. J Clin Neurophysiol. 2015;32(4):294–303.

Mazzone SB, Undem BJ. Vagal afferent innervation of the airways in health and disease. Physiol Rev. 2016;96(3):975–1024.

McFarland DH. Netter's atlas of anatomy for speech, swallowing, and hearing. 2nd ed. St. Louis: Elesevier; 2015.

Moore KL, Dalley AF, Agur AMR. Clinically oriented anatomy. 8th ed. Philadelphia: Wolters Kluwer; 2018.

Netter FH. Atlas of human anatomy. 7th ed. Philadelphia: Elsevier; 2019.

Neumann DA. Kinesiology of the musculoskeletal system: foundations for rehabilitation. 3rd ed. St. Louis: Elsevier; 2017.

Norton NS. Netter's head and neck anatomy for dentistry. 3rd ed. Philadelphia: Elsevier; 2017.

Palastanga N, Soames R. Anatomy and human movement: structure and function. 7th ed. Edinburgh: Churchill Livingstone; 2019.

Paramothayan S. Essential respiratory medicine. Oxford: Wiley Blackwell; 2019.

Pawlina W. Histology: a text and atlas with correlated cell and molecular biology. 9th ed. Philadelphia: Wolters Kluwer; 2016.

Polla B, D'Antona G, Bottinelli R, Reggiani C. Respiratory muscle fibres: specialisation and plasticity. Thorax. 2004;59(9):808–17.

Proud D. The pulmonary epithelium in health and disease. West Sussex, UK: Wiley; 2008a.

Ratnovsky A, Elad D. Anatomical model of the human trunk for analysis of respiratory muscles mechanics. Respir Physiol Neurobiol. 2005;148(3):245–62.

Ratnovsky A, Elad D, Halpern P. Mechanics of respiratory muscles. Respir Physiol Neurobiol. 2008;163(1–3):82–9.

Roberts JR, Hedges JR. Roberts and Hedges' clinical procedures in emergency medicine. 7th ed. Philadelphia: Elsevier; 2018.

Sanders I, Mu L. Anatomy of the human internal superior laryngeal nerve. Anat Rec. 1998;252(4):646–56.

Schilero GJ, Radulovic M, Wecht JM, Spungen AM, Bauman WA, Lesser M. A center's experience: pulmonary function in spinal cord injury. Lung. 2014;192(3):339–46.

Shankar K, editor. Physiatric anatomic principles. State of the art reviews. Philadelphia: Hanley & Bells, Inc.; 1996.

Singh V. Textbook of anatomy: upper limb and thorax, vol. I. 2nd ed. New Delhi: Elsevier; 2014a.

Singh V. Textbook of anatomy: abdomen and lower limb, vol. II. 2nd ed. New Delhi: Elsevier; 2014b.

Sinnatamby CS. Last's anatomy: regional and applied. 12th ed. Edinburgh: Churchill Livingstone; 2011.

Skandalakis JE. Skandalakis' surgical anatomy: the embryologic and anatomic basis of modern surgery. Athens: Paschalidis Medical Publications; 2004.

Standring S. Gray's anatomy: the anatomical basis of clinical practice. 42nd ed. Philadelphia: Elsevier; 2020.

Thiriet M. Anatomy and physiology of the circulatory and ventilatory systems. New York: Springer; 2014.

Tortora GJ, Nielsen MT. Principles of human anatomy. 4th ed. Hoboken: Wiley; 2017.

Walls RM, Murphy MF. Manual of emergency airway management. 4th ed. Philadelphia: Wolters Kluwer; 2012.

Wang NS. Anatomy and physiology of the pleural space. Clin Chest Med. 1985;6(1):3–16.

Watkinson J, Clarke R. Scott-Brown's otorhinolaryngology head & neck surgery. 8th ed. Boca Raton: CRC Press; 2018.

Weibel ER. Lung morphometry: the link between structure and function. Cell Tissue Res. 2017;367(3):413–26.

Weinberger SE, Cockrill BA, Mandel J. Principles of pulmonary medicine. 7th ed. Philadelphia: Elsevier; 2019.

West JB, Luks AM. West's respiratory physiology: the essentials. 10th ed. Philadelphia: Wolters Kluwer; 2016.

Whitsett JA, Alenghat T. Respiratory epithelial cells orchestrate pulmonary innate immunity. Nat Immunol. 2015;16(1):27–35.

Wilson TA, Legrand A, Gevenois PA, De Troyer A. Respiratory effects of the external and internal intercostal muscles in humans. J Physiol. 2001;530(Pt 2):319–30.

Wilson-Pauwels L, Akesson EJ, Stewart PA, Spacey SD. Cranial nerves: function and dysfunction. 2nd ed. New York: People's Medical Publishing House; 2002.

Wineski LE. Snell's clinical anatomy by systems. 10th ed. Philadelphia: Wolters Kluwer; 2019.

Recommended Additional Reading

Aliverti A, Pedotti A. Mechanics of breathing: new insights from new technologies. 2nd ed. New York: Springer; 2014.

Martini FH, Ober WC, Nath JL, Bartholomew EF, Petti K. Visual anatomy & physiology. 3rd ed. New York: Pearson; 2018.

Proud D. The pulmonary epithelium in health and disease. West Sussex: Wiley; 2008b.

Functional Anatomy and Neuroanatomical Basis of Swallowing

12

The complex process of swallowing requires precise coordination of multiple anatomical structures and neurological pathways. Understanding these mechanisms becomes particularly relevant when examining dysphagia in various medical conditions. Dysphagia frequently occurs following neurological trauma, with increased prevalence in patients who have sustained moderate to severe traumatic brain injuries or spinal cord injuries (Wolf and Meiners 2003). The condition may also develop after cervical spine surgical interventions, presenting through various pathophysiological mechanisms, including mechanical obstruction or neurological dysfunction.

Several clinical factors correlate with increased dysphagia severity. Research has demonstrated higher rates of swallowing difficulties in patients requiring prolonged mechanical ventilation exceeding 14 days, those with decreased cognitive function as measured by Rancho Los Amigos scores below 7, and individuals with significant intracranial pathology requiring urgent surgical intervention (Mackay et al. 1999). Additionally, radiological findings of midline shift or brainstem involvement often indicate increased risk of dysphagia development (Bradley et al. 1983).

In the context of spinal cord injuries, specific patient characteristics and injury patterns influence dysphagia risk. Higher cervical spine trauma, particularly affecting C1–C4 levels, shows a strong association with swallowing dysfunction (Kirshblum et al. 1999). Additional risk factors include advanced age, presence of tracheostomy, and surgical interventions involving the cervical spine, particularly those utilizing anterior approaches or requiring occipitocervical fusion (Tian and Yu 2013).

Diagnostic evaluation of dysphagia commonly employs modified barium swallow studies, which provide real-time visualization of deglutition phases and help identify specific mechanical or neurological deficits. While therapeutic approaches exist for various causes of dysphagia, management of patients with concurrent cognitive and behavioral impairments requires specialized consideration and modified treatment strategies (Smith et al. 2000).

12.1 Normal Swallowing: Voluntary and Involuntary Stages

Normal swallowing (deglutition) represents one of the most complex neuromuscular processes in human physiology, requiring precise coordination among multiple anatomical structures. This process integrates the synchronized activation of muscles throughout the oral cavity, pharynx, larynx, esophagus, and stomach (Leopold and Kagel 1997). While most individuals perform these intricate movements effortlessly, even healthy persons may experience occasional, transient

H.-Y. Ko, *Practical Functional Anatomy for Spinal Cord Injury Rehabilitation*,
https://doi.org/10.1007/978-981-95-6457-6_12

swallowing difficulties that typically resolve through protective mechanisms such as coughing or throat clearing (Matsuo and Palmer 2008).

The upper digestive tract's remarkable adaptability allows it to support multiple physiological functions beyond swallowing, including respiration, phonation, and speech articulation. This anatomical region contains specialized valvular structures that modify their configuration based on specific functional demands. During swallowing, these valves undergo precise temporal adjustments to maintain airway protection while enabling efficient bolus transport (Logemann 1988). A study has improved our understanding of swallowing's temporal relationships. Martin-Harris et al. (2005a, b) demonstrated significant overlap between various phases of swallowing through detailed analysis of respiratory timing patterns. Their research identified four unique breathing-swallowing patterns that vary with age, challenging the traditional view of completely separate swallowing phases. They found that laryngeal closure and hyoid excursion form an initial functional unit that precedes pharyngoesophageal segment opening, while the hyoid typically returns to its rest position after breathing resumes.

The biomechanical aspects of normal swallowing involve coordinated pressure generation throughout the upper aerodigestive tract. This coordination optimizes bolus propagation while minimizing residue accumulation in oral and pharyngeal regions. The process requires precise temporal sequencing of muscle activation patterns, regulated by both cortical and brainstem mechanisms (Jean 2001). When neurological or muscular conditions affect the swallowing apparatus, the remarkable neuroplasticity of these structures often enables the development of compensatory mechanisms, allowing individuals to maintain safe swallowing with modified food and liquid consistencies (Robbins et al. 2008).

While traditionally described as separate oral preparatory, oral, pharyngeal, and esophageal phases, current evidence supports a more integrated view of swallowing physiology. These phases demonstrate significant temporal overlap, particularly at their transitions. Understanding these overlapping relationships becomes particularly relevant when considering the interaction between swallowing and respiratory functions. The temporal coordination between breathing and swallowing affects multiple aspects of bolus management, including transit time, retention patterns, and aspiration risk (Martin-Harris et al. 2005a, b).

The successful integration of voluntary and reflexive components requires intact neural pathways throughout multiple levels of the nervous system. This integration allows for both conscious control during the oral phase and automatic sequencing during the pharyngeal phase, creating a seamless transition between voluntary and involuntary components (Jean 2001).

12.1.1 Voluntary Stage of Swallowing: Oral Preparation and Transport

The oral phase of swallowing comprises a complex sequence of neuromuscular events that transform food from its initial state to a bolus ready for pharyngeal transfer. This process incorporates two primary stages: the oral preparatory stage and oral transport phase, each requiring precise anatomical and neurological coordination (Hiiemae and Palmer 2003).

The oral preparatory stage begins when food enters the oral cavity. During this phase, the oral structures cooperate to process food by mastication, simultaneously mixing it with saliva to create a cohesive bolus. This process demands intricate coordination among multiple anatomical components: the mandible executes rhythmic movements, the tongue continuously repositions food between dental surfaces, and the facial muscles—particularly the buccinators—prevent food accumulation in the lateral sulci. The successful formation of a proper bolus depends on both the mechanical action of these structures and appropriate salivary production (Matsuo and Palmer 2009).

As mastication nears completion, the process transitions to the oral transport phase. This stage features synchronized movement patterns that propel the prepared bolus posteriorly toward the pharynx. The anterior portion of the tongue progressively elevates against the hard palate, creating a wave-like motion that drives the bolus backward. Concurrent with this movement, the soft palate elevates to create a seal with the nasopharynx, preventing premature bolus entry into the pharyngeal space and protecting the nasal airways (Dodds et al. 1990).

The neural control of these stages demonstrates remarkable complexity. While both phases can be initiated voluntarily, they may also begin reflexively when specific oral and pharyngeal sensory regions receive appropriate stimulation. This dual control mechanism enables both conscious control over eating behavior and automatic responses to various food properties, including texture, temperature, and taste. The process requires intact sensorimotor integration, with continuous feedback from mechanoreceptors, proprioceptors, and taste receptors modifying motor output in real-time (Steele and Miller 2010).

For normal swallowing to occur, the anatomical structures of the upper aerodigestive system must maintain their structural integrity, and their sequential function must demonstrate appropriate timing. This coordination depends on both the motor and sensory components of the nervous system working in harmony. The real-time adjustments made during both stages rely on constant sensory feedback, allowing for modifications based on bolus characteristics and helping maintain optimal preparation and transport patterns (Palmer et al. 1993).

12.1.1.1 Oral Preparatory Stage

The oral preparatory stage of swallowing represents a complex integration of mechanical, sensory, and motor functions that transform food into a cohesive bolus suitable for swallowing. This process requires precise coordination among multiple anatomical structures and neural pathways (Matsuo and Palmer 2008).

The process begins with lip closure, maintained by the orbicularis oris muscle, which creates an anterior seal to prevent food loss. Simultaneously, the buccinator muscles maintain tension in the lateral walls of the oral cavity, preventing food accumulation in the vestibule. The soft palate initially adopts a forward position, closing the oral cavity posteriorly while maintaining nasal airway patency (Dodds et al. 1990).

Mastication involves both mechanical reduction and sensory processing. The teeth perform specialized functions: incisors provide cutting action, while molars execute grinding movements. The jaw muscles can generate remarkable force—up to 55 pounds on incisors and 200 pounds on molars. The primary muscles of mastication include the temporalis, masseter, and medial pterygoid for elevation, working in coordination with jaw depressors for rhythmic chewing movements (Miller 1982).

The tongue plays a central role in bolus formation through coordinated actions of its intrinsic and extrinsic muscles. The intrinsic muscles modify tongue shape, while extrinsic muscles alter its position within the oral cavity. These actions are interdependent—shape changes affect position and vice versa. For example, tongue flattening by the verticalis muscle results in forward displacement, while hyoglossus activation lowers both the body and sides of the tongue (Sanders et al. 1993) (Fig. 12.1).

The neural control of mastication involves multiple cranial nerves and central nervous system regions. The trigeminal nerve (CN V) provides both sensory and motor innervation to the masticatory muscles through its second and third divisions. Taste sensation is mediated by the facial nerve (CN VII) for the anterior two-thirds of the tongue and the glossopharyngeal nerve (CN IX) for the posterior third (Jean 2001).

The rhythmical chewing pattern emerges from brainstem reticular formation activity, modified by input from the hypothalamus, amygdala, and cortical regions associated with taste and smell. This pattern involves a reflexive sequence: food presence initially inhibits masticatory muscles, allowing mandibular depression. This movement triggers a stretch reflex, causing rebound contraction and jaw elevation. The resulting food compression against the oral mucosa reinitiated the cycle (Lund and Kolta 2006).

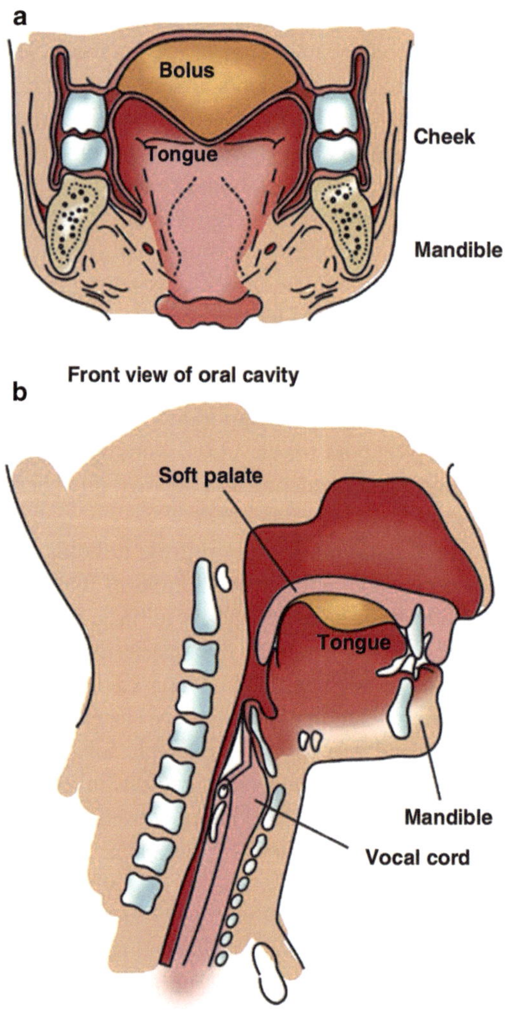

Fig. 12.1 Anterior (**a**) and lateral (**b**) views of the tongue holding a bolus just prior initiating the oral stage of swallowing. (Adapted with permission from Thankappan et al. (2018))

During bolus preparation, multiple sensory inputs, including taste, temperature, viscosity, and size, are continuously processed. This information modifies the activity of lips, tongue, buccal muscles, and dental manipulation to achieve optimal bolus consistency. The terminal phase of preparation involves the tongue gathering the processed food into a cohesive mass positioned against the hard palate, ready for transport (Palmer et al. 1993).

12.1.2 Involuntary Stages of Swallowing: Pharyngeal and Esophageal Phases

12.1.2.1 Pharyngeal Phase

The pharyngeal phase is an important phase in the swallowing process, representing the shift from voluntary control to an involuntary, meticulously organized sequence of protective actions. This phase begins when the bolus reaches the anterior tonsillar pillars, triggering a rapid sequence that often terminates within 1–2 s. The pharyngeal phase begins when the bolus is propelled posteriorly by the tongue into the pharynx, as depicted in Fig. 12.2.

Unlike earlier stages of swallowing that demonstrate flexibility in execution, the pharyngeal phase maintains strict requirements for completion. While one might bypass preparatory and oral phases through modifications in food consistency or head positioning, the pharyngeal phase remains steadfast in its execution and is essential for safe swallowing. The moment this phase begins, several protective mechanisms engage simultaneously. Diaphragmatic activity undergoes immediate inhibition, making concurrent breathing and swallowing physiologically impossible under normal circumstances. The soft palate rises to create a seal with the nasopharynx, preventing food from entering the nasal cavity. At the same time, the vocal cords move toward closure, establishing a crucial barrier for airway protection (Matsuo and Palmer 2008).

The complex interaction of the tongue, velopharynx, and larynx during the pharyngeal phase produces a precisely coordinated sequence. As shown in Fig. 12.3, the tongue elevates as velopharyngeal closure begins, triggering the forward motion of laryngeal structures to increase upper esophageal sphincter opening. The larynx also elevates, leading to relaxation of the cricopharyngeus musculature. When these actions maintain appropriate temporal integration, the bolus moves through the pharyngeal segment without penetration or aspiration into the airway (Kahrilas et al. 1992).

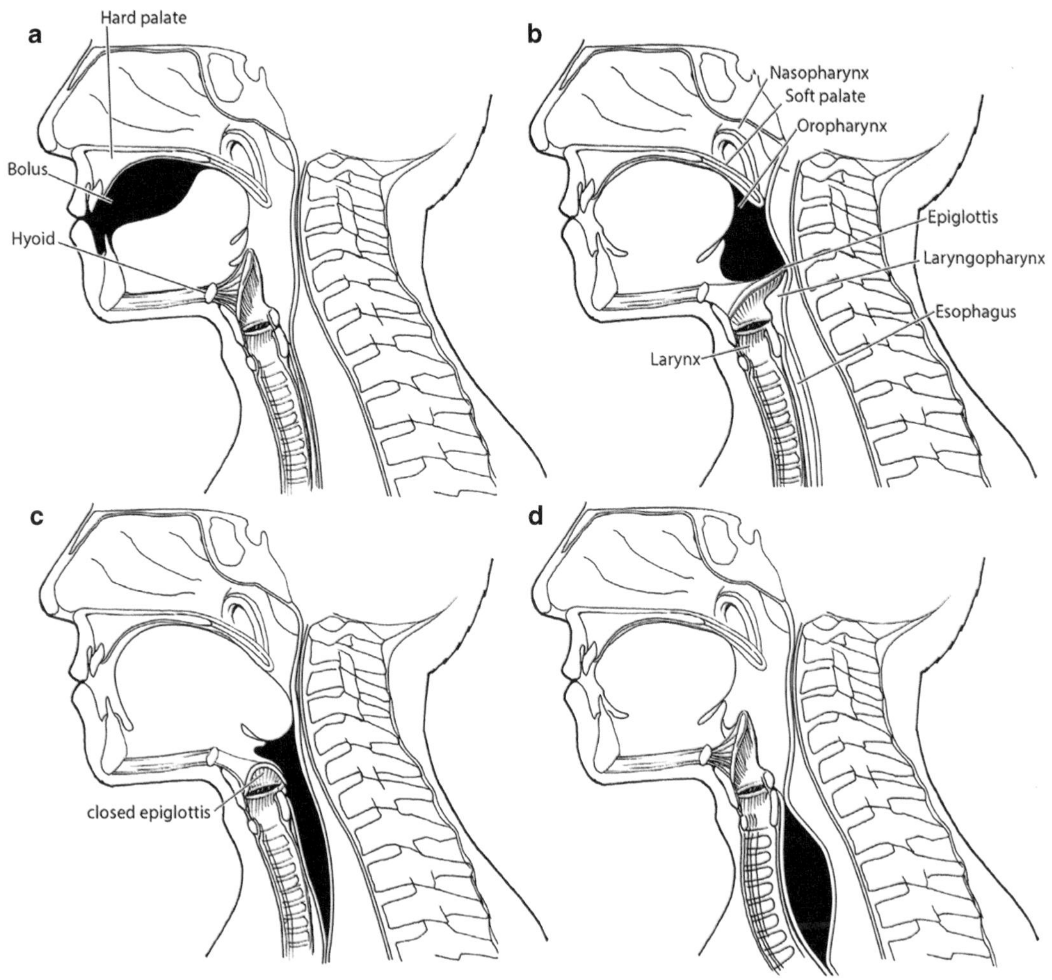

Fig. 12.2 Bolus propagation during swallowing: (**a**) Wave-like elevation of the tongue propels the bolus into the oropharynx. (**b**) Soft palate elevation seals the nasal cavity as the bolus passes into the pharynx. (**c**) The bolus is propelled through the pharynx past the closed epiglottis and enters the esophagus through the upper esophageal sphincter. (**d**) The bolus is transported through the esophagus to the stomach via peristaltic waves. The soft palate and tongue relax, and the epiglottis reopens to restore respiration. (Adapted with permission from Hennessy and Goldenberg (2016))

The effectiveness of pharyngeal swallowing varies with bolus characteristics. A well-formed, cohesive bolus might pass smoothly through the pharyngeal space, while a poorly organized one may require sustained engagement of pharyngeal mechanisms as it extends from the oral cavity into the oropharynx. Normal function of the pharyngeal phase depends on bolus consistency, size, and whether swallowing occurs as a single or continuous event (Dodds et al. 1990).

The neural control of these complex movements is predominantly located within brainstem circuits, specifically within the swallowing pattern generator network. This network coordinates muscle activation timing while processing continuous feedback from peripheral sensory receptors. These receptors monitor bolus characteristics and position, allowing real-time adjustments to maintain safe and efficient swallowing (Steele and Miller 2010).

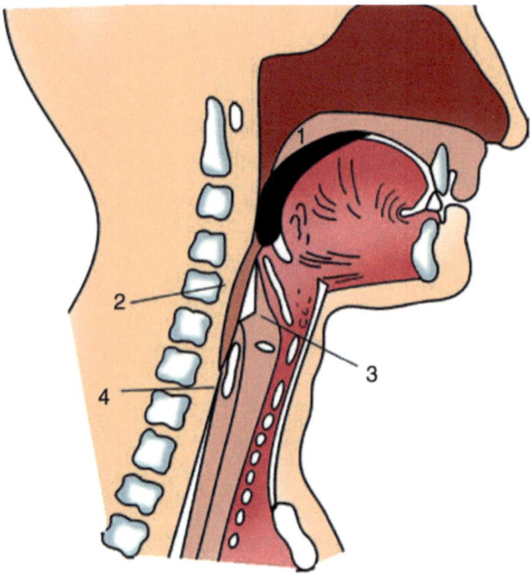

Fig. 12.3 Neuromuscular components of the swallowing reflex: (1) closure of the soft palate, (2) peristaltic motion, (3) elevation and closure of the larynx, and (4) relaxation of the cricopharyngeus muscle. (Adapted with permission from Thankappan et al. (2018))

Airway Protection Mechanisms During the Pharyngeal Phase of Swallowing

Protection of the airway during swallowing depends fundamentally on two different but interrelated components: elevation and closure. The elevation mechanism relies on systematic contraction of the strap musculature within the neck, which generates an upward and forward displacement of the larynx. This movement positions the laryngeal structure beneath the tongue base as it retracts during the final moments of the oral phase of swallowing (Fig. 12.4). Through this biomechanical action, the larynx moves superiorly and anteriorly, effectively displacing it from the path of the descending food bolus that travels over the tongue base (Logemann 1988).

Closure of the laryngeal inlet represents the second major protective component, accomplished through the coordinated sphincteric action of multiple laryngeal structures. These structures include the epiglottis, which fold posteriorly, the aryepiglottic folds, which constrict medially, and both the false and true vocal folds, which adduct to seal the airway. The epiglottis and aryepiglottic folds may contribute less to direct airway protection than previously thought. Instead, these structures appear to serve primarily as mechanical guides, functioning to direct the food bolus around the airway rather than acting as primary protective barriers (Matsuo and Palmer 2008).

The precise timing and coordination between these protective mechanisms create a multilayered defense system during swallowing. The superior and anterior movement of the larynx not only physically distances the airway from the bolus path but also contributes to upper esophageal sphincter opening through mechanical traction. This synchronization of protective mechanisms becomes particularly significant in populations with compromised swallowing function, where the integrity of these protective mechanisms may determine the risk of aspiration (Steele et al. 2011).

Modern imaging studies have further illuminated the dynamic nature of these protective mechanisms. Videofluoroscopic analyses demonstrate that laryngeal elevation and closure occur in a precisely timed sequence, with elevation slightly preceding complete closure. This temporal relationship ensures that the airway begins moving away from the bolus path even before the final protective barriers are fully engaged (Kim and McCullough 2008).

12.1.2.2 Esophageal Phase

The esophageal phase occurs within a muscular tube extending 20–25 cm in length, with an average diameter of 2 cm, from its origin just caudal to the cricopharyngeus muscle to its termination at the gastric cardia (Goyal and Chaudhury 2008). During swallowing, the esophagus actively shortens by approximately 10% through longitudinal muscle contractions. The esophagus functions as more than a simple passive conduit, demonstrating active roles in acid control and mucosal protection. In its resting state, the upper two-thirds remain collapsed while the lower third maintains a rounded configuration (Cook and Kahrilas 1999).

The hallmark of this phase involves peristalsis—the sequential contraction and relaxation of esophageal muscles that propel the bolus

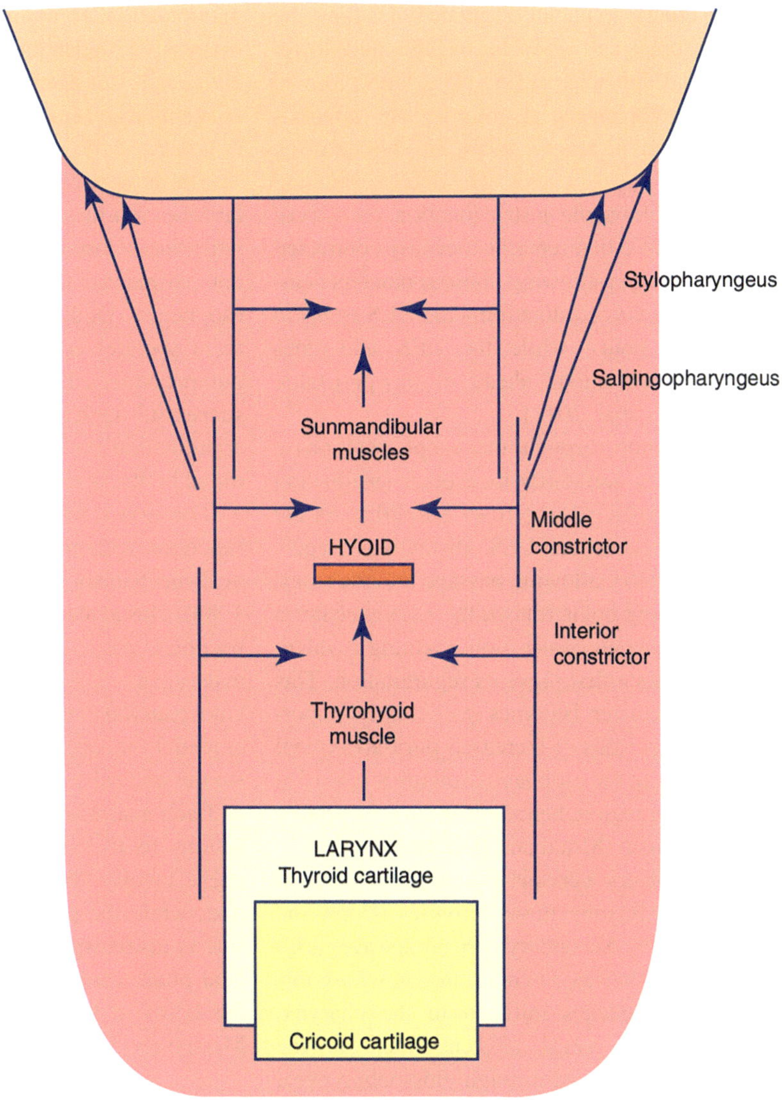

Fig. 12.4 Pharyngeal elevators (anterior view) and the directions of contractions during the swallow. (Adapted with permission from Thankappan et al. (2018))

toward the stomach. This process manifests in two characteristic forms: primary and secondary peristalsis. Primary peristalsis represents a continuation of the pharyngeal wave, propagating through the entire length of the esophagus in approximately 8–10 s. When primary peristalsis fails to clear all material, secondary peristaltic waves emerge in response to esophageal distension. These secondary waves, initiated without a swallow, continue until complete bolus clearance occurs (Mittal and Balaban 1997).

The anatomical composition of the esophagus varies along its length, influencing its neural control mechanisms. The upper third consists of striated muscle, controlled by glossopharyngeal and vagal nerve impulses. In contrast, the lower two-thirds comprise smooth muscle, primarily regulated through vagal innervation and the myenteric nervous system. This dual innervation ensures continued function even after significant neural disruption—for instance, even with brainstem paralysis, the myenteric plexus can generate effective peristaltic waves to propel tube-fed material into the stomach (Goyal and Chaudhury 2008).

Two crucial sphincters regulate bolus flow: the upper esophageal sphincter (UES) and lower esophageal sphincter (LES). The UES, formed by the cricopharyngeal muscle and approximately 3 cm in length, maintains high resting tone under vagal control. The LES, extending about 3 cm above the gastric junction, consists of circular muscle that remains tonically contracted at rest. During swallowing, these sphincters demonstrate precise coordination—the UES opens to accept the bolus, while the LES undergoes "receptive relaxation" ahead of the peristaltic wave (Miller et al. 2011).

Transit time through the esophageal phase shows greater variability than other swallowing stages, typically ranging from 8 to 20 s. As the bolus approaches the LES, distention triggers reflex relaxation, allowing passage into the stomach. Following bolus transit, the LES executes a strong, prolonged contraction, creating a valve-like mechanism that prevents regurgitation. This anti-reflux barrier becomes particularly important during periods of increased intra-abdominal pressure, since the esophagus collapses inward at the sphincter region, hence enhancing the protective mechanism (Mittal and Balaban 1997).

The myenteric nervous system plays a crucial role in coordinating these activities (Cook and Kahrilas 1999). Secondary peristaltic waves arise through both intrinsic neural circuits within this system and reflexes initiated in the pharynx, transmitted via vagal afferent fibers to the medulla, and returned through glossopharyngeal and vagal efferent pathways. This complex neural integration ensures efficient bolus transport while maintaining protective mechanisms against retrograde flow.

12.2 Anatomy of the Swallowing Mechanism

The swallowing mechanism comprises four principal anatomical divisions: oral, pharyngeal, laryngeal, and esophageal components (Matsuo and Palmer 2008). These structures function as an integrated system, with each component performing specific actions in the process of bolus manipulation, formation, and transport. A comprehensive understanding of each anatomical division is fundamental, as dysfunction in any area may result in deglutition disorders or aspiration risk.

The anatomical framework consists of muscular tubes reinforced by cartilaginous and osseous structures, forming three primary functional compartments: the oral cavity, pharynx, and esophagus (Dodds et al. 1990). Each compartment operates as an independent hydrodynamic pump, connected by specialized valves. However, successful deglutition requires precise temporal coordination among these compartments, orchestrated by the brainstem's swallowing center.

The larynx, while often considered separately, functions as an integral component of the pharyngeal mechanism during swallowing (Logemann 1988). This anatomical and functional integration becomes particularly relevant in radiological assessment of swallowing disorders. Understanding these intricate relationships is essential for accurate interpretation of diagnostic studies and subsequent clinical decision-making.

Each anatomical division contributes distinct actions to the swallowing sequence. The oral phase involves bolus containment and preparation, while the pharyngeal components manage safe bolus transport past the airway. The esophageal phase completes the process by delivering the bolus to the stomach (Dodds et al. 1990; Matsuo and Palmer 2008).

12.2.1 Functional Anatomy of the Oral Cavity

The oral cavity performs the preparatory and initiation phases of swallowing, involving food manipulation and bolus formation (Dodds et al. 1990). In these periods, the buccinators and orbicularis oris muscles establish labial competence, while synchronized soft palate lowering forms a posterior seal, preventing premature bolus entry into the oropharynx and maintaining airway protection (Matsuo and Palmer 2008).

The tongue's complex neuromuscular system, comprising both intrinsic and extrinsic muscles, controls bolus manipulation. Unilateral hypoglossal nerve (CN XII) injury can result in asymmetrical bolus control, potentially causing food to accumulate in the affected buccal space (Pauloski and Logemann 2000). The tongue base receives innervation from the vagus nerve (CN X), primarily through the glossopharyngeal muscle, while the oral portion remains under voluntary control through hypoglossal innervation.

Research has identified significant anatomical variations affecting swallowing dynamics. Studies demonstrate larger laryngeal and hypopharyngeal volumes in males compared to females, with progressive age-related volume reduction in both genders (Robbins et al. 1992). These anatomical differences, including decreased piriform sinus capacity with aging, require consideration during clinical assessment and therapeutic planning.

The tongue's sensory system includes distributed taste receptors detecting five primary taste modalities: sweet, sour, salty, bitter, and umami (zest) (Ertekin and Aydogdu 2003). Optimal tongue function depends on adequate moisture, primarily supplied by three pairs of major salivary glands—parotid, submandibular, and sublingual—complemented by approximately 300 minor salivary glands (Ship et al. 2002).

12.2.2 Functional Anatomy of the Pharynx and Larynx

The pharynx and larynx represent an evolutionarily refined system that manages both swallowing and respiration (Sasaki and Isaacson 1988). The primary function of these structures involves precise directional control of air and food, with particular emphasis on protecting delicate pulmonary tissues from contamination during swallowing (Ertekin and Aydogdu 2003).

The laryngeal complex has developed from its most primitive form as a fundamental sphincteric mechanism, primarily for airway protection, as demonstrated through comparative anatomical studies (Sasaki et al. 1977). This evolutionary development has resulted in a complex valvular system that efficiently manages both respiratory and alimentary functions (Logemann et al. 1992).

The effective functioning of this complex anatomical system depends on two fundamental components: structural integrity of the mobile elements and precise neurological timing of their movements (Dodds et al. 1990). This "time-sharing" mechanism requires coordinated action of multiple valves and chambers, controlled by intact neural pathways to maintain safe separation of respiratory and digestive functions.

The specialized protective mechanisms have evolved specifically to prevent aspiration of food and liquid into the airways, as pulmonary surfactant function remains particularly susceptible to contamination (Palmer et al. 2000). This protective function represents a critical evolutionary adaptation in maintaining respiratory system integrity.

12.2.2.1 Anatomical Organization and Clinical Significance of the Pharyngeal Complex

The pharyngeal structure extends posteriorly from the oral and nasal cavities to the esophagus, positioned behind the larynx and trachea, providing direct passage to the stomach (Matsuo and Palmer 2008). A series of constrictor muscles sequentially drives food inferiorly, with particular significance attributed to the cricopharyngeus muscle at the base of the pharyngeal constrictors. This muscle maintains contraction during respiration but exhibits controlled relaxation during swallowing to accommodate laryngeal elevation and food passage. Clinical implications arise when cricopharyngeal dysfunction occurs due to scarring, fibrosis, or nerve injury, potentially resulting in aspiration despite preserved oral and laryngeal function (Cook and Kahrilas 1999).

The pharyngeal structure demonstrates anatomical segmentation into the nasopharynx, oropharynx, and hypopharynx, as illustrated in Fig. 12.5. This complex extends from the skull base to precisely defined boundaries: the inferior border of the cricoid cartilage anteriorly and the inferior border of vertebra C6 posteriorly. The pharyngeal wall architecture comprises internal longitudinal muscles, including the

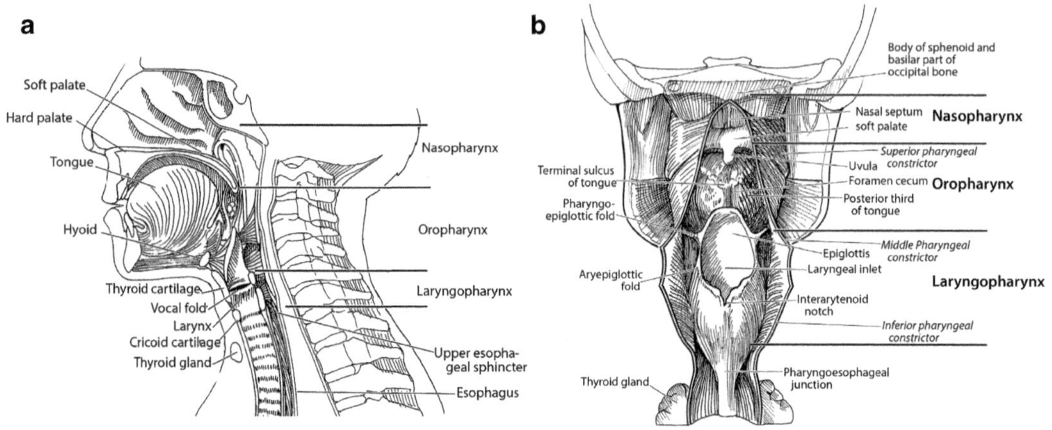

Fig. 12.5 Anatomy of the head and neck showing the oral cavity, pharynx, and larynx, with delineation of three major regions: nasopharynx, oropharynx, and hypopharynx. (**a**) Sagittal view and (**b**) posterior view. (Adapted with permission from Hennessy and Goldenberg (2016))

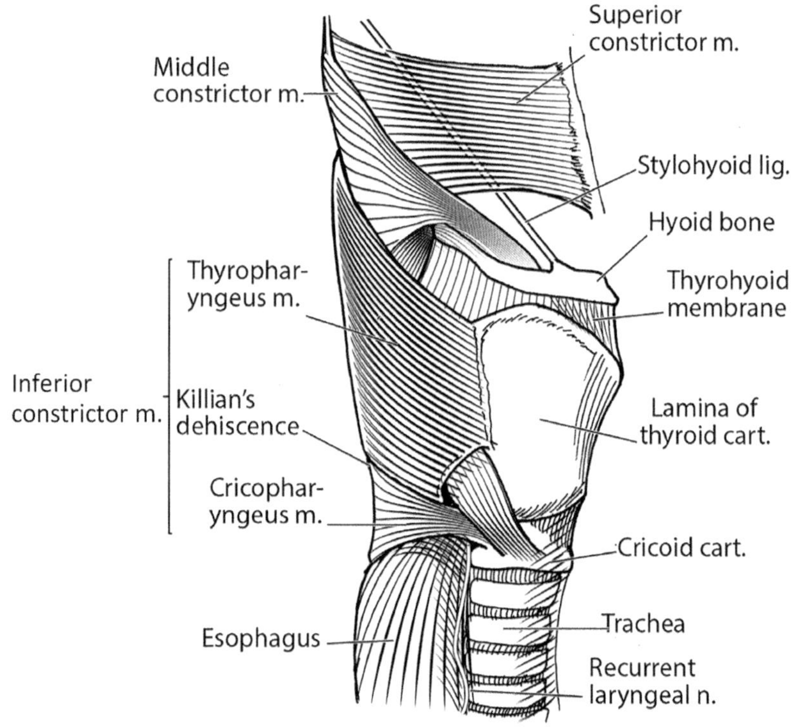

Fig. 12.6 The constrictor muscles of the pharynx, clearly showing the division of the inferior pharyngeal constrictor into the thyropharyngeus and cricopharyngeus components. (Adapted with permission from Hennessy and Goldenberg (2016))

palatopharyngeus, stylopharyngeus, and salpingopharyngeus, complemented by external circular muscles consisting of superior, middle, and inferior constrictors (Logemann et al. 1992) (Fig. 12.6). Of particular anatomical interest is the inferior pharyngeal constrictor, which represents the thickest component of the pharyngeal constrictors. As demonstrated in Fig. 12.7, this structure divides into the thyropharyngeus and cricopharyngeus components. Between these elements lies the Killian dehiscence, a structurally vulnerable region that, while anatomically

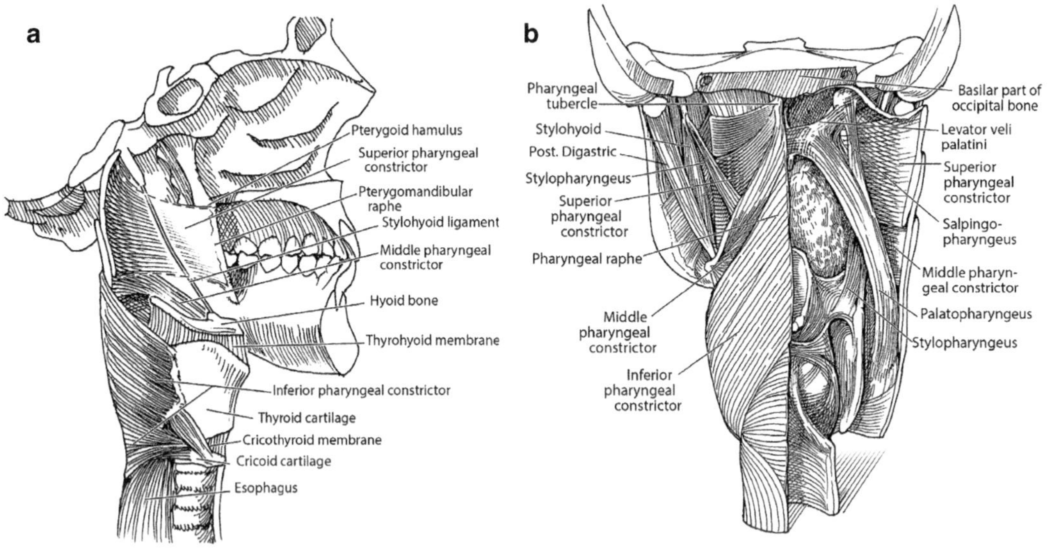

Fig. 12.7 The muscles of the pharynx. (**a**) Lateral view of the pharyngeal constrictor muscles. (**b**) Posterior view showing the muscles of the pharynx, with the pharyngeal raphe to reveal the internal longitudinal pharyngeal muscles. (Adapted with permission from Hennessy and Goldenberg (2016))

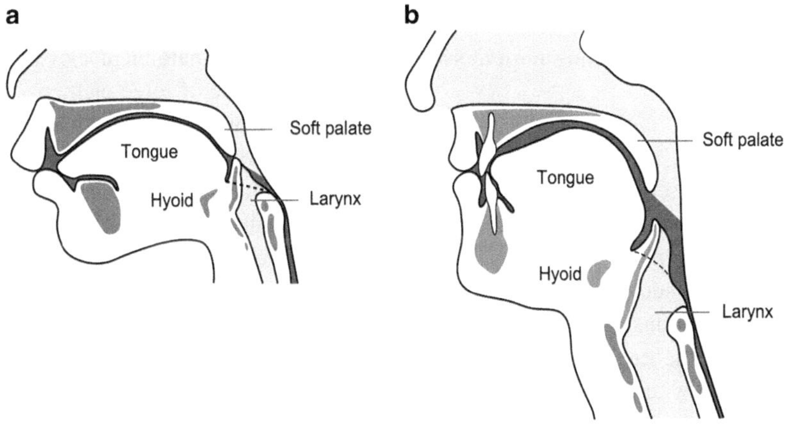

Fig. 12.8 Sagittal section of the head and neck in an infant (**a**) and an adult (**b**), illustrating the food passageway (dark gray) and airway (light gray): (**a**) In infants, the oral cavity is smaller, with a flatter tongue and palate. The epiglottis is nearly in contact with the soft palate, allowing the airway and food passageway to remain separate except during swallowing. (**b**) In adults, the larynx is positioned lower in the neck, resulting in the food passageway and airway crossing within the pharynx. (Adapted with permission from Matsuo and Palmer (2008), PMC)

normal, may predispose to diverticulum formation during instances of uncoordinated pharyngeal peristalsis (Cook et al. 1989a, b).

Human developmental processes significantly influence pharyngeal anatomy, as depicted in Fig. 12.8. The progressive elongation of the neck accompanies laryngeal descent, creating a longer vertical pharyngeal passage. While this anatomical adaptation enhances speech capabilities, it simultaneously introduces increased vulnerability to aspiration by establishing a shared pathway for both respiratory and alimentary functions (Sasaki et al. 1977). Table 12.1 provides detailed information regarding pharyngeal musculature, their attachments, and innervation patterns, underlining the complex integration of these

Table 12.1 Muscles of the pharynx

Muscle of pharynx		Cranial nerve supply	Main action
External circular layer	Superior pharyngeal constrictor	Vagus (pharyngeal branch of vague and pharyngeal plexus)	Constricts walls of pharynx during deglutition
	Middle pharyngeal constrictor	Vagus (pharyngeal branch of vagus, branches of external and recurrent laryngeal nerves and pharyngeal plexus)	Constricts walls of pharynx during deglutition
	Inferior pharyngeal constrictor (thyropharyngeus and cricopharyngeus)	Vagus (pharyngeal branch of vagus, branches of external and recurrent laryngeal nerves and pharyngeal plexus)	Constricts walls of pharynx during deglutition
Internal longitudinal layer	Salpingopharyngeus	Vagus (pharyngeal branch of vagus and pharyngeal plexus)	Elevate (shorten and widen) the pharynx and larynx during swallowing and speaking
	Stylopharyngeus	Glossopharyngeal nerve	Elevate (shorten and widen) the pharynx and larynx during swallowing and speaking
	Palatopharyngeus	Vagus (pharyngeal branch of vagus and pharyngeal plexus)	Elevate the larynx and pharynx while speaking and swallowing, depresses the soft palate and moves palatopharyngeal arch to the midline

Adapted with permission from Thankappan (2018)

anatomical elements in supporting normal swallowing function.

Nasopharynx

The nasopharynx represents the superior segment of the pharyngeal complex, exhibiting a distinctive anatomical arrangement that reflects its specialized functions (Matsuo and Palmer 2008). Its superior boundary demonstrates a characteristic downward slope in the craniocaudal direction, incorporating three key anatomical components: the basisphenoid, basiocciput, and the anterior aspects of the first two cervical vertebrae. The inferior margin of this region finds its demarcation at the level of Passavant's muscle and the hard palate, where it transitions into the oropharynx (Dua et al. 1997).

The lateral boundaries of the nasopharynx comprise the superior constrictor muscle and pharyngobasilar fascia, while maintaining anterior continuity with the nasal cavities. Within this architectural framework, several significant anatomical structures emerge: the pharyngeal tonsils (adenoids) occupy a midline position along the roof, while the Eustachian tube openings punctuate the posterolateral walls. The torus tubarius presents as an anatomical elevation formed by the convergence of three elements: the cartilaginous Eustachian tube, the levator veli palatini muscle, and their overlying mucosal coverage (Kahrilas et al. 1992).

A particular anatomical feature is the fossa of Rosenmüller, positioned slightly posterior and superior to the torus tubarius. The mucosal architecture of the nasopharynx demonstrates specific folding patterns that envelope the salpingopharyngeus and levator veli palatini muscles, contributing to the region's functional capabilities (Palmer et al. 2000).

Oropharynx

The oropharynx is an important part of the pharyngeal complex, extending posteriorly from the oral cavity and covering the inferior aspect of the soft palate to the upper margin of the epiglottis (Matsuo and Palmer 2008). This region maintains its anatomical relationship with the C2 and C3 vertebrae, which constitute its posterior boundary. The structural framework includes numerous key components: the posterior third of the tongue, bilateral palatine tonsils, soft palate, oropharyngeal mucosal lining, and the constrictor muscles, which work together to allow various activities (Logemann et al. 1998).

The lateral oropharyngeal wall displays two unique anatomical arches: anteriorly, the palatoglossal arch, supported by the palatoglossus muscle, and posteriorly, the palatopharyngeal arch, reinforced by the palatopharyngeus muscle. Between these parallel arches lies the tonsillar bed, a specialized region containing the palatine tonsils, which represent concentrated collections of lymphoid tissue (Kahrilas et al. 1993). The structural foundation of the tonsillar bed comprises the superior pharyngeal constrictor muscle and a delicate layer of pharyngobasilar fascia, which extends superiorly to merge with the cranial base periosteum, thereby establishing the pharyngeal wall's superior limits (Cook and Kahrilas 1999).

The epiglottic tip serves as an anatomical landmark, defining the oropharynx's inferior border and marking the transition to subsequent regions of the pharyngeal complex (Palmer et al. 2000). This precise anatomical organization facilitates the coordinated functions of swallowing while maintaining structural integrity during various physiological processes.

Hypopharynx

The hypopharynx represents a critical anatomical segment extending from the hyoid bone and valleculae superiorly to the cricopharyngeus muscle inferiorly, establishing continuity with the cervical esophagus at the level of the C6 vertebra (Matsuo and Palmer 2008). The posterior hypopharyngeal wall maintains direct continuation from the oropharynx, with the retropharyngeal space positioned immediately posterior to this structure. This architectural arrangement plays an essential role in the mechanics of swallowing (Logemann et al. 1998).

A distinctive anatomical feature of the hypopharynx includes the bilateral pyriform sinuses, which demonstrate important structural relationships with surrounding tissues. These sinuses communicate anteriorly with the posterior paraglottic space of the larynx, with their apices aligning with the true vocal cords. The lateral boundary of each pyriform sinus comprises the thyrohyoid membrane superiorly and thyroid cartilage inferiorly, while the aryepiglottic fold's lateral surface forms its medial wall (Cook and Kahrilas 1999). These sinuses function as protective channels during swallowing, directing the bolus toward the esophagus while serving as temporary reservoirs when lingual and pharyngeal peristalsis proves insufficient, thereby preventing airway compromise (Palmer et al. 2000).

The postcricoid region, forming the anterior wall of the lower hypopharynx, spans from the cricoarytenoid joints to the cricoid cartilage's inferior border. The hypopharyngeal wall demonstrates a complex four-layered structure: mucosa, pharyngobasilar fascia, muscular layer (comprising outer circular and inner longitudinal components), and buccopharyngeal fascia (Kahrilas et al. 1993). This multilayered organization provides both structural integrity and functional flexibility necessary for safe and efficient swallowing.

12.2.2.2 Cartilages of the Pharynx and Larynx

The anatomical foundation of swallowing mechanisms relies heavily on precise skeletal and cartilaginous relationships, particularly involving the mandible and hyoid bone (Matsuo and Palmer 2008). The medial surface of the mandible presents several attachment points for swallowing-related musculature. A prominent mylohyoid line, extending approximately 1 cm along the internal mandibular surface, provides insertion for the mylohyoid muscle. The posterior mandibular surface features distinct mental spines in the anterior midline region, serving as attachment points for both the geniohyoid and genioglossus muscles, which subsequently establish connections with the hyoid bone and intrinsic tongue structure, respectively (Dodds et al. 1990).

The hyoid bone exhibits a complicated anatomical structure with a central body with four protruding horns—two superior (lesser cornua) and two inferior (greater cornua). This structure gains additional stability through the stylohyoid ligament, a connective tissue band extending from each lesser cornu to the styloid process at the skull base. This ligamentous connection creates a dynamic suspension system that supports various phases of swallowing while maintaining anatomical relationships during swallowing movements (Palmer et al. 2000).

Thyroid Cartilage

The thyroid cartilage, representing the largest component of the laryngeal cartilaginous structure, exhibits a distinctive anatomical configuration formed by two quadrilateral laminae that converge anteriorly (Matsuo and Palmer 2008). These laminae form a characteristic anterior prominence with gender-specific angular variations: approximately 90° in adult males and 120° in females (Fig. 12.9). The superior aspect features a V-shaped notch where the laminae converge, while the posterior borders extend to form superior and inferior cornua.

The cartilage's external surface presents an oblique line extending from the superior thyroid tubercle to the inferior thyroid tubercle, providing attachment points for three crucial muscles: sternothyroid, thyrohyoid, and inferior constrictor. The superior border's concave configuration supports the thyrohyoid membrane attachment, with the lateral thyroid ligament forming a thickened region at the superior cornu. The superior cornua connect to the hyoid bone's greater cornua via the thyrohyoid ligament, while the inferior cornua establish synovial articulations with the cricoid cartilage, forming the cricothyroid joints.

The thyrohyoid membrane complex includes median and lateral components, featuring a lateral aperture that accommodates the passage of laryngeal neurovascular structures. Within the posterior-lateral aspect of the thyrohyoid ligament lies the triticeal cartilage, a small but notable anatomical component (Palmer et al. 2000). The thyroid cartilage, composed of hyaline cartilage, typically begins ossification around age 25, representing a significant developmental milestone in laryngeal structural maturation.

Cricoid Cartilage

The cricoid cartilage presents a signet ring configuration with significant anatomical and functional importance in the laryngeal framework (Fig. 12.10). This unique structure comprises two primary components: an anterior arcus, characterized by its thin profile, and a more substantial posterior lamina (Matsuo and Palmer 2008).

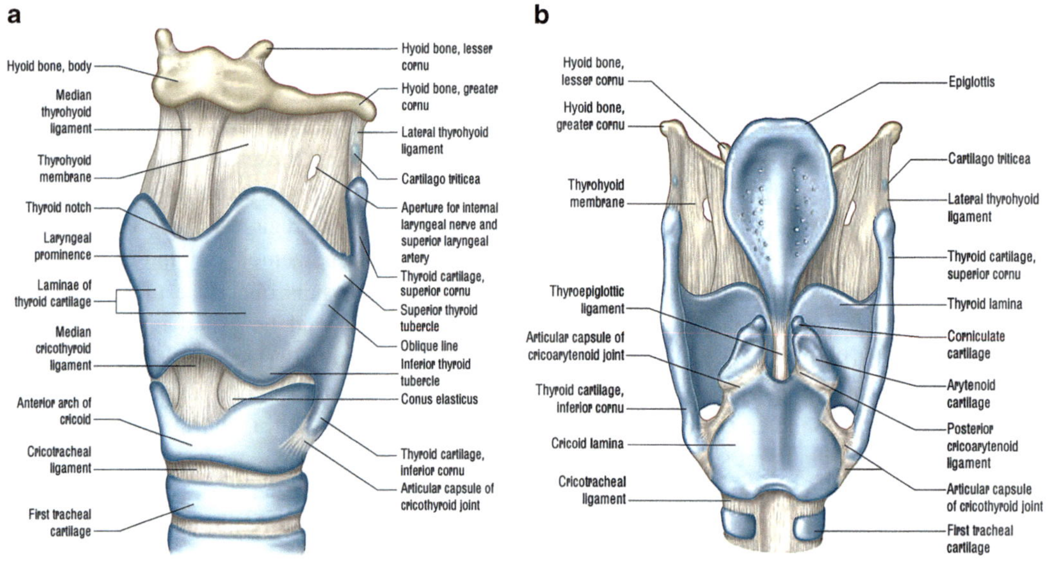

Fig. 12.9 Laryngeal cartilages and ligaments. (**a**) Anterolateral view of the larynx. (**b**) Posterior view of the larynx. (Adapted with permission from Standring (2020))

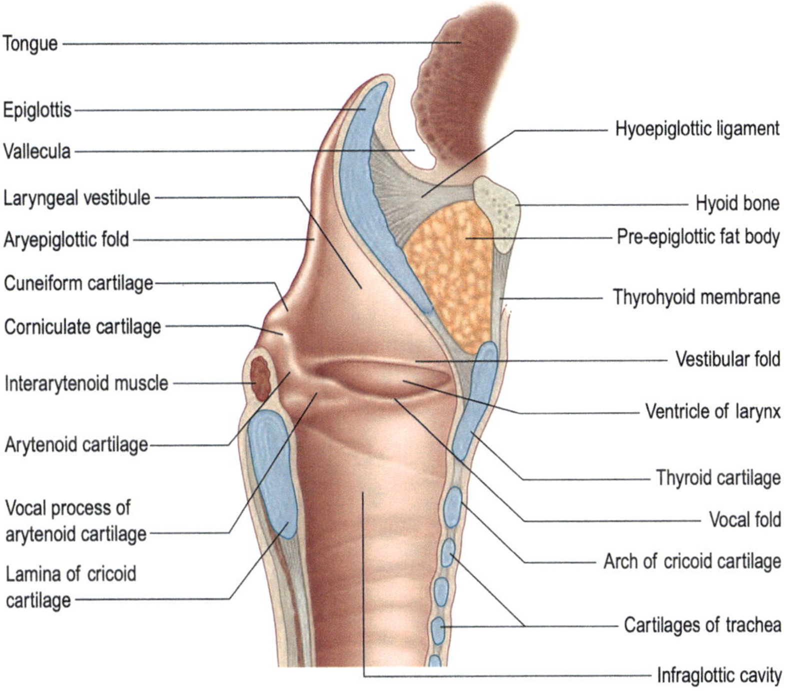

Fig. 12.10 Midsagittal section of the larynx illustrating fibroelastic membrane attachments. (Adapted with permission from Standring (2020))

Unlike its thyroid counterpart, the cricoid forms a complete circumferential ring, establishing a crucial foundation for laryngeal support.

The posterior lamina, measuring approximately 2–3 cm in vertical height, features a median vertical ridge bordered by shallow depressions that provide origin points for the posterior cricoarytenoid muscles. Its superior border presents two specialized oval convex facets that serve as articular surfaces for the arytenoid cartilages. At the junction where the lamina transitions into the anterior arch, bilateral articular facets accommodate the inferior cornua of the thyroid cartilage, forming the cricothyroid joints.

The cartilage maintains important structural relationships with adjacent components: superiorly connecting to the thyroid cartilage through the cricothyroid ligament, and inferiorly forming continuity with the trachea (Cook and Kahrilas 1999). The cricoid, made of hyaline cartilage, has varying degrees of ossification with advancing age, potentially influencing its biomechanical properties and function in swallowing and respiration.

Epiglottis

The epiglottis represents a characteristic anatomical structure composed of elastic cartilage that functions as an integral component of the upper respiratory system (Standring 2020) (Figs. 12.9, 12.10, and 12.11). This leaf-shaped cartilaginous structure demonstrates unique morphological features, with its superior portion exhibiting a broad, rounded configuration that transitions into a narrowed inferior segment known as the petiole (Drake et al. 2020).

The cartilaginous framework maintains specific anatomical relationships through various ligamentous attachments. The petiole connects to the internal aspect of the thyroid cartilage via the thyroepiglottic ligament, positioning itself in the midline immediately inferior to the thyroid notch (Standring 2020). The anterior surface of the epiglottis extends toward the hyoid bone through a fan-shaped ligamentous structure, which constitutes an anatomical continuation of the median glossoepiglottic ligament (Moore et al. 2018).

The posterior surface of the epiglottis displays a predominantly concave configuration, although it exhibits subtle convexities at both its superior

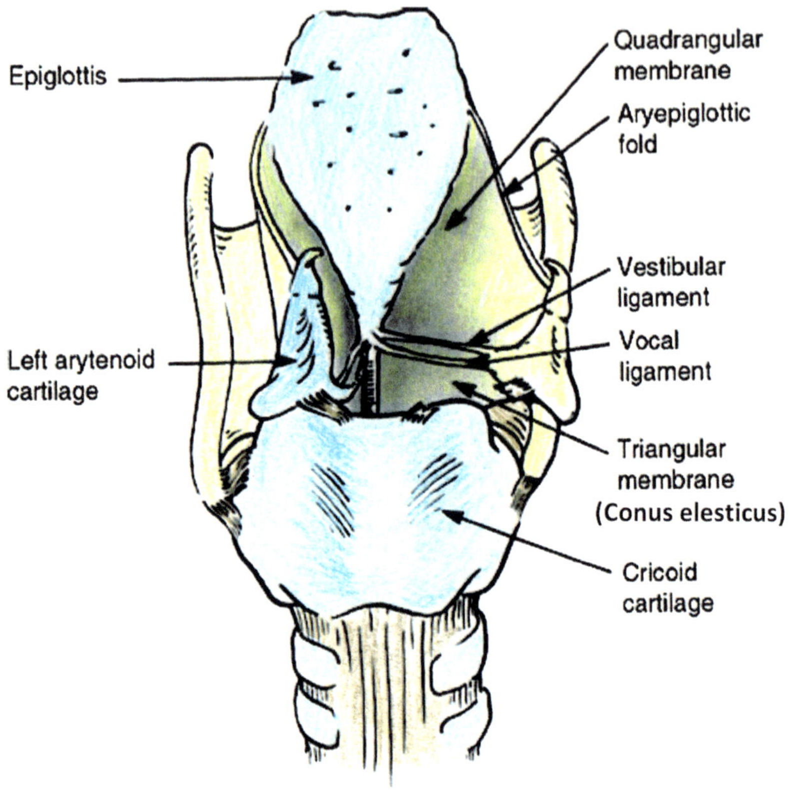

Fig. 12.11 Posterior view of the larynx showing fibroelastic membrane attachment. The arytenoid cartilage on the right is moved laterally to demonstrate the membrane's attachments to the ligament. (Adapted from Blitzer et al. (2009))

and inferior margins. A notable microscopic feature of this elastic cartilage includes the presence of multiple small depressions that accommodate mucous glands, contributing to the physiological functions of this structure (Netter 2019).

The anatomical orientation of the epiglottis positions its anterior surface in proximity to the base of the tongue, establishing a significant anatomical relationship that influences its physiological role in swallowing and airway protection (Lalwani 2020).

Arytenoid Cartilages

The arytenoid cartilages are paired pyramidal-shaped structures located at the posterior and superior aspects of the cricoid cartilage (Standring 2020) (Figs. 12.10 and 12.11). Each arytenoid cartilage terminates superiorly with a corniculate cartilage, forming an anatomical unit that contributes to laryngeal function (Drake et al. 2020).

The structural framework creates a continuous cartilaginous and ligamentous wall extending from the hyoid bone inferiorly to the anterior tracheal surface. This arrangement provides the foundation for muscular attachments and mucosal relationships (Moore et al. 2018).

Each arytenoid cartilage demonstrates specific anatomical features, including a base and three processes. The base presents a concave configuration that articulates with the cricoid cartilage. The vocal process extends anteriorly, providing attachment for the vocal ligament, while the muscular process projects posterolaterally, offering attachment points for the posterior and lateral cricoarytenoid muscles. The apex, oriented superiorly, posteriorly, and medially, forms an articulation with the corniculate cartilage (Netter 2019).

The arytenoid cartilages present three distinct surfaces: posterior, medial, and anterolateral. The posterior surface exhibits a smooth, triangular, and concave morphology. The medial surface maintains a narrow, smooth, and flat configuration, while the anterolateral surface shows a rough and convex appearance. A notable ridge extends from the vocal process to the apex on the lateral surface, creating two depressions. The superior depression receives the vestibular ligament, while the inferior depression provides attachment for the vocal ligament and partially accommodates the lateral cricoarytenoid and vocalis muscles, components of the thyroarytenoid muscle (Sasaki and Isaacson 1988).

The composition of each arytenoid cartilage primarily consists of hyaline cartilage, with elastic cartilage concentrated at the apex, reflecting the specialized mechanical requirements of these structures (Standring 2020).

Corniculate Cartilages

The corniculate cartilages are small, paired anatomical structures composed of elastic cartilage positioned at the apices of the arytenoid cartilages (Standring 2020). These cartilages project posteromedially and demonstrate a characteristic horn-like configuration (Sataloff et al. 2007). While they typically form synovial articulations with the arytenoid cartilages, anatomical variations exist where fusion between these cartilages occurs.

Located within the aryepiglottic folds, the cuneiform cartilages present as paired, elongated elastic cartilaginous elements that contribute to the structural framework of the laryngeal inlet (Netter 2019). Additionally, the lateral thyrohyoid membrane frequently contains the triticeal cartilage, a small elastic cartilaginous structure that exhibits considerable anatomical variability in its presence and dimensions (Figs. 12.9b and 12.10).

These anatomical components participate in various physiological functions, including voice production and airway protection during swallowing. Their elastic composition provides the necessary flexibility for optimal laryngeal function while maintaining structural integrity.

Cuneiform Cartilages

The cuneiform cartilages represent paired anatomical structures located within the laryngeal complex. These small, elongated cartilaginous elements demonstrate a characteristic club-shaped morphology and occupy positions anterior to the corniculate cartilages within the aryepiglottic folds (Standring 2020). The cartilages manifest as noticeable whitish prominences on the mucosal surface immediately anterior to the arytenoid cartilages (Singh 2014).

From a developmental perspective, these cartilages originate from the fourth pharyngeal arch and consist of elastic cartilage, which provides flexibility while maintaining structural integrity (Moore et al. 2018). Their average dimensions measure approximately 4–5 mm in length, though significant individual variation exists (Drake et al. 2020).

The cuneiform cartilages contribute to the dynamic function of the laryngeal inlet during phonation and swallowing. Their position within the aryepiglottic folds allows them to participate in laryngeal sphincter mechanisms. These cartilages serve to maintain the structural framework of the supraglottic larynx.

12.2.2.3 Muscles of the Pharynx

The pharyngeal region exhibits a specialized neuromuscular organization where striated muscle fibers compose the fundamental structure of the oral cavity, larynx, and pharynx (Standring 2020). Within this complex architecture, two muscular formations establish the tonsillar boundary: the palatopharyngeal muscle forms the medial arch, while the glossopalatine muscle creates the lateral arch. This anatomical configuration enables precise control of swallowing, vocalization, and respiratory functions (Matsuo and Palmer 2008).

The pharyngeal wall demonstrates a systematic layered arrangement, where fibrous fascia creates connections between the internal mucosal layer and external muscular components. Moving superiorly toward the skull base, the wall structure transforms as the muscular component diminishes. This superior region, approximately 2 cm in width, contains only mucosa and fascia, forming the pharyngeal fibrous layer (Table 12.1). The inferior region contains constrictor muscles, while elevator muscles occupy an internal position—an anatomical feature not found elsewhere in the gastrointestinal system.

The pharyngeal constrictors of the external circular layer are present in three anatomically and functionally distinct segments. The superior pharyngeal constrictor originates from four precise locations: the pterygoid process of the sphenoid bone, pterygomandibular raphe, mandibular mylohyoid line, and transverse lingual musculature. These components merge posteriorly, establishing the upper pharyngeal wall and meeting at the dorsal pharyngeal raphe (Dodds et al. 1990) (Fig. 12.7).

From the hyoid processes and stylohyoid ligament emerges the middle pharyngeal constrictor, forming a bridge between the skull base and the hyoid bone's minor processes. The muscle fibers form a posterosuperior plate, meeting their counterparts at the posterior midline's pharyngeal raphe (Miller 1982). The inferior pharyngeal constrictor demonstrates a tripartite origin from the cricoid cartilage, thyroid cartilage, and lateral thyrohyoid ligament. This muscle follows a superoposterior course around the pharynx, creating a superior convex arch at its inferior margin.

Among the internal longitudinal layer muscles (elevators), the stylopharyngeal muscle originates from the styloid process region, following an anteroinferior path through the interval between the superior and middle constrictors. Its fibers partially merge with contralateral components before inserting into the epiglottic margins and posterior thyroid cartilage border. This arrangement enables vertical pharyngeal movements during deglutition (Kahrilas et al. 1993). The palatopharyngeal muscle functions as the primary elevator, demonstrating multiple attachment points at the posterior border of the hard palate, palatine aponeurosis, and pterygoid process. The muscle descends to insert into the posterior thyroid cartilage and integrates within the constrictor musculature, enabling coordinated elevation during swallowing (Matsuo and Palmer 2008).

12.2.3 The Pharyngoesophageal Segment

The pharyngoesophageal region exhibits a complex muscular organization where the pharyngeal constrictors extend from near the skull base downward to merge with the esophagus. At the inferior extent of these constrictors lies the cricopharyngeal muscle, which displays three portions: oblique, transverse, and longitudinal (Goyal and Chaudhury 2008) (Fig. 12.12).

The oblique portion of the cricopharyngeal muscle extends superiorly and posteriorly from the lateral cricoid cartilage, positioning itself adjacent to the inferior constrictor. These oblique fibers typically converge at the pharyngeal raphe, functioning as a diminutive extension of the pharyngeal constrictors. The transverse, or semicircular portion, which comprises the majority of the muscle mass, courses posteriorly from the posterior and lateral aspects of the cricoid cartilage. Unlike the superior portions, the posterior midline junction of these muscles lacks a fibrous raphe (Kahrilas et al. 1988).

Two longitudinal muscles, termed esophageal elevators, originate from the inferior cricoid cartilage. These muscles course along the lateral aspects of the esophagus, integrating with the longitudinal esophageal musculature that originates from the cricoid cartilage's median lamina. A specific overlapping pattern characterizes this region: the inferior constrictor overlays the cricopharyngeal muscle, which in turn overlaps the esophageal circular muscle (Ekberg and Lindström 1987).

The anatomical structure presents two significant areas of structural vulnerability. Between the oblique and transverse components of the cricopharyngeal muscles lies a small triangular space known as Killian's opening or Laimer's triangle. This anatomical weakness provides a potential pathway for the development of Zenker diverticulum. A second vulnerable area exists laterally, inferior to the transverse portion and superior to the longitudinal cricoid muscle insertion. This region represents the site where Killian-Jamieson diverticula may develop.

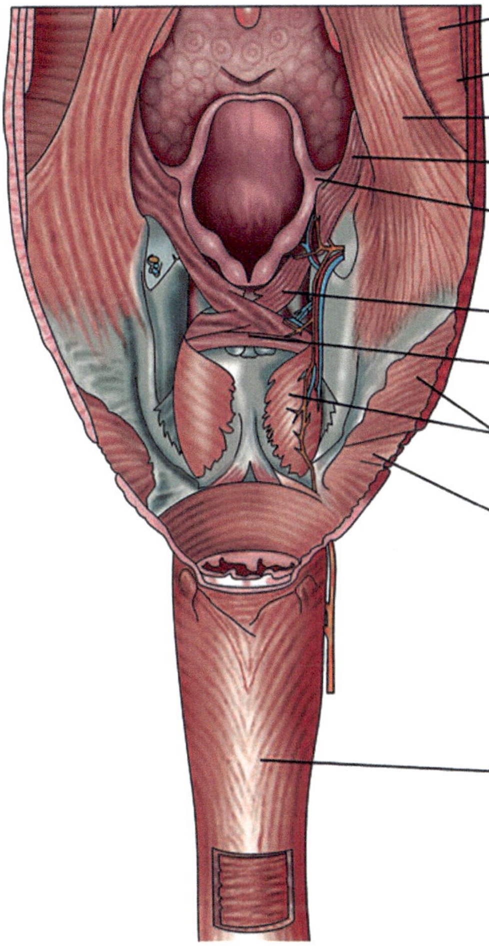

Fig. 12.12 Pharyngo-esophageal junction: posterior view with pharynx opened and mucosa removed. (Adapted with permission from Thankappan (2018))

12.2.4 Functional Anatomy of the Larynx

The larynx exhibits remarkable anatomical complexity, performing multiple physiological functions that extend well beyond its commonly recognized roles in respiration and phonation (Matsuo and Palmer 2008). As part of an advanced neuromuscular system, the larynx coordinates extensively with oral and pharyngeal mechanisms during swallowing. This coordination manifests in precise pressure modulation during various physiological activities, including Valsalva maneuvers associated with lifting and defecation (Leopold and Kagel 1997).

The biomechanical aspects of laryngeal function become apparent during the swallowing sequence, when the structure undergoes superior and anterior displacement. Medical professionals can evaluate this movement through palpation of the thyroid prominence during dry swallows. Various medical conditions may affect this displacement pattern, including post-radiation changes, surgical interventions, and neurological disorders. When laryngeal elevation becomes compromised, patients face increased aspiration risks, particularly with liquid consistencies. This risk becomes particularly significant when protective reflexes such as cough are simultaneously diminished (Steele and Miller 2010).

The neuromuscular events controlling laryngeal closure during swallowing follow a precise sequence. The epiglottis moves through two distinct phases, beginning with its transition from rest to a transverse orientation. This initial movement results from hyoid bone elevation, achieved through coordinated muscle activation including the stylohyoid, digastric, mylohyoid, and geniohyoid muscles. The thyrohyoid muscle simultaneously reduces the distance between the hyoid bone and thyroid cartilage (Miller 1982). In the subsequent phase, the thyroepiglottic muscle pulls the epiglottis downward over the arytenoid region, transforming its shape from a downward to an upward convex configuration. The aryepiglottic muscle then constricts the laryngeal inlet through a sphincter-like mechanism.

Vestibular closure occurs in different but integrated stages. The initial stage involves supraglottic space closure through superior thyroarytenoid muscle contraction. The next stage involves the compression of the subepiglottic space from an inferior approach, as the posterior epiglottic surface and its associated fat cushion compress against the arytenoid region. This sequential closure pattern generates a peristaltic-like mechanism that effectively eliminates bolus material from the vestibule (Steele and Miller 2010).

The median thyrohyoid fold represents a significant anatomical component in the swallowing process. This structure, comprising the epiglottic cartilage, preepiglottic fat cushion, and associated ligamentous connections (thyroepiglottic, median thyrohyoid, and hyoepiglottic), contributes to the posterior displacement of the superior-anterior vestibular wall. This anatomical arrangement maintains efficient vestibular clearance following the completion of deglutition.

Comparative anatomical research reveals fascinating evolutionary adaptations of the larynx across species. The most basic laryngeal function involves airway protection, as demonstrated by the progression from primitive air exchange structures in lungfish to more complex arrangements in terrestrial animals. The reptilian larynx developed active airway dilatation capabilities, while advanced vocalization features emerged in higher reptiles and mammals. The human larynx demonstrates specialized characteristics for complex phonation, including extended membranous vocal folds and modified resonating cavity dimensions (Cook et al. 1989a, b).

The clinical significance of laryngeal functional anatomy extends into multiple medical domains. Dysphagia specialists regularly assess laryngeal elevation and closure patterns when evaluating swallowing disorders. Post-surgical rehabilitation programs often focus on restoring normal laryngeal movement patterns. Following cerebrovascular events, healthcare providers monitor laryngeal function as an indicator of neurological recovery. Understanding these anatomical relationships helps medical professionals develop effective strategies for preventing aspiration pneumonia and restoring voice function.

12.2.4.1 Glottic Anatomy and Physiology

The glottis functions as a dynamic valve that modulates airflow for multiple physiological purposes, including airway protection, respiration, vocalization, and swallowing (Ludlow 2005). This valvular mechanism operates through a mobile, flexible mucosal-lined passage that rapidly changes configuration through synchronized actions of cartilages, ligaments, and muscles.

During inspiration, the glottis undergoes two primary modifications: lengthening and widening. The lengthening process occurs when the cricothyroid muscle contracts, creating distance between the posterior lamina of the cricoid cartilage and the thyroid cartilage. The paired vocal cords maintain specific anatomical relationships during this process—their anterior aspects remain fixed to the posterior midline of the thyroid cartilage, while their posterior portions attach to the vocal processes of the arytenoid cartilages (Sasaki and Weaver 1997).

The widening of the airway requires precise biomechanical adjustments in the posterior glottic structures. These modifications involve both rotational movement and lateral displacement of the arytenoid cartilages, which articulate with the superior edge of the cricoid lamina. The posterior cricoarytenoid (PCA) muscles, under recurrent laryngeal nerve (RLN) innervation, control these movements. The PCA demonstrates functional compartmentalization: its medial portion counteracts the thyroarytenoid and vocalis muscles' tethering effects by maintaining the arytenoid in an upright position. Meanwhile, the lateral portion generates both rotational and displacement movements of the arytenoid cartilage relative to the cricoid's superior surface through its attachment to the cartilage's muscular process (Sanders et al. 1993).

The neuromuscular control of these movements requires precise timing and coordination. Electromyographic studies have demonstrated that the activation patterns of laryngeal muscles vary with different respiratory and phonatory tasks. During quiet breathing, the PCA shows rhythmic activity that increases during inspiration and decreases during expiration. This pattern changes dramatically during phonation, when the thyroarytenoid and lateral cricoarytenoid muscles become the primary active muscles (Hillel 2001).

Understanding these anatomical relationships and physiological mechanisms has significant clinical implications. Disorders affecting the RLN can impair glottic function, leading to symptoms ranging from voice changes to breathing difficulties. Additionally, the complex interaction between these structures becomes particularly relevant during endotracheal intubation, where proper positioning must account for the natural biomechanics of the laryngeal structures (Noordzij and Ossoff 2006).

Arytenoid Motion and Vocal Function

Vocal fold mobility depends fundamentally on the cricoarytenoid joints, which function as true synovial joints and therefore remain susceptible to the same pathological conditions that affect other synovial joints throughout the body (Von Leden and Moore 1961). Rather than executing simple rotational movements, the arytenoids perform complex sliding motions across multiple dimensions on the cricoid lamina. During phonation, vocal fold approximation occurs through two primary mechanisms: medial rotation of the arytenoids and their medial displacement, achieved through interarytenoid muscle activation (Sataloff et al. 2007).

When rotation occurs without adequate medial displacement, excessive air escapes during phonation—a phenomenon readily observable during laryngoscopic examination while the subject whispers. This particular "glottic posture" frequently appears in hyperfunctional vocal disorders, representing an involuntary but inappropriate positioning of the arytenoids during phonation (Nagai et al. 2005).

The regulation of vocal pitch emerges from the complex interaction of multiple factors: vocal cord shape, tension, and precise arytenoid positioning (Sataloff et al. 2007). The positioning component relies heavily on two key mechanisms: cricothyroid muscle action, which creates distraction of the cricoid's posterior lamina, and posterior displacement of the arytenoids. The medial portion of the posterior cricoarytenoid muscle serves dual functions—it not only assists in pitch modulation by increasing tension on the membranous cord but also contributes to glottic aperture lengthening by maintaining proper arytenoid positioning during inspiration while the cricothyroid muscle contracts (Von Leden and Moore 1961). The significance of this function becomes particularly evident in cases of vocal fold paralysis, where anterior displacement of the arytenoid cartilage can be readily observed.

Glottic Closure

Vocal fold closure represents the primary intrinsic laryngeal function essential for airway protection, accomplished through rapid contraction of the thyroarytenoid and lateral cricoarytenoid muscles. The recurrent laryngeal nerve mediates the activity of all intrinsic laryngeal muscles except the cricothyroid (Ludlow 2005). The glottic closure reflex demonstrates remarkable speed, with a latency of less than 25 ms, making it one of the body's fastest reflexive responses.

During the swallowing sequence, glottic closure proceeds in a specific order: true vocal fold approximation occurs first, followed by false vocal fold approximation, and finally, the arytenoids approximate to the epiglottic petiole, establishing a third protective layer. This closure sequence initiates after respiratory cessation, typically during the expiratory phase and before laryngeal elevation (Logemann 1988).

Various pathological conditions can compromise glottic closure. Structural abnormalities may result from partial laryngeal surgery or age-related vocal fold atrophy (presbylarynx), characterized by vocal fold bowing. Neurological conditions frequently manifest as vocal fold paralysis, leading to glottic incompetence. Understanding these mechanisms proves particularly relevant in clinical practice, as therapeutic interventions targeting swallowing function invariably aim to enhance glottic closure efficiency (Steele et al. 2011).

Laryngeal Elevation and Depression

The vertical movement of the larynx represents a fundamental requirement for normal swallowing. This function often deteriorates in individuals with neurological conditions or following radiotherapy for head and neck cancer (Logemann et al. 2000). During swallowing, the superior and anterior displacement of the laryngeal complex serves to remove it from the bolus path while simultaneously aiding in the opening of the upper esophageal sphincter by distraction of the cricopharyngeus muscle.

The process can be conceptualized as the larynx "sliding up" along the bolus while coordinating with the downward peristaltic movement. This complex motion relies on an intricate suspension system involving multiple muscle groups (Palmer et al. 2000). The laryngeal complex maintains firm suspension from both the skull base and mandible through a complex network of muscular attachments. From the skull base, the posterior belly of the digastric muscle and the stylohyoid ligament and muscle insert into the hyoid bone. Anteriorly, the hyoid bone connects to the mandible through several muscles: the geniohyoid, mylohyoid, intrinsic tongue muscles, and the anterior belly of the digastric. The pharyngeal suspension involves both the active stylopharyngeus and the pharyngeal constrictors, while the thyrohyoid muscle connects the thyroid cartilage to the hyoid bone and assists in elevation (Cook et al. 1989a, b).

The depression mechanism of the larynx involves equally complex muscular arrangements. The larynx experiences inferior distraction through connections to both the sternum and scapula via the sternohyoid and omohyoid muscles. Additional depressor functions come from the thyrohyoid and sternothyroid muscles. These strap muscles demonstrate remarkable conservation across mammalian species and receive dual innervation from both the ansa hypoglossi and ansa cervicalis nerves (Pearson et al. 2012). This dual innervation and muscular redundancy ensures rapid elevation during swallowing, with approximation of the thyroid cartilage to the hyoid, followed by swift return to the lower "resting" position necessary for respiratory function.

Supraglottic Anatomy

The supraglottis includes all laryngeal tissues above the vocal cords, forming a vestibular tunnel leading to the glottis, a characteristic significantly interesting to individuals first studying laryngeal anatomy (Logemann et al. 1995) (Fig. 12.13). The functional significance of this mobile and flexible portion of the airway becomes especially apparent when studying swallowing patterns in patients who have undergone partial laryngeal surgery.

Popular misconception often attributes primary glottic protection to the epiglottis, viewing it as a simple "flapper valve" that seals the glottic aperture during swallowing. While the suprahyoid portion of the epiglottis and the aryepiglottic folds constitute the majority of this tract, research has demonstrated that its posterior movement during swallowing does not create an effective seal. Instead, the epiglottis functions more like a ship's prow, directing the bolus stream laterally into the pyriform sinuses during swallowing.

The epiglottis demonstrates unique structural properties, composed of elastic cartilage that provides significantly greater flexibility compared to the rigid hyaline cartilage of the thyroid and cricoid structures (Sasaki and Isaacson 1988). Its movement pattern, more accurately described as folding rather than rotation, results from several concurrent actions: laryngeal elevation (which reduces the hyoid-thyroid cartilage distance),

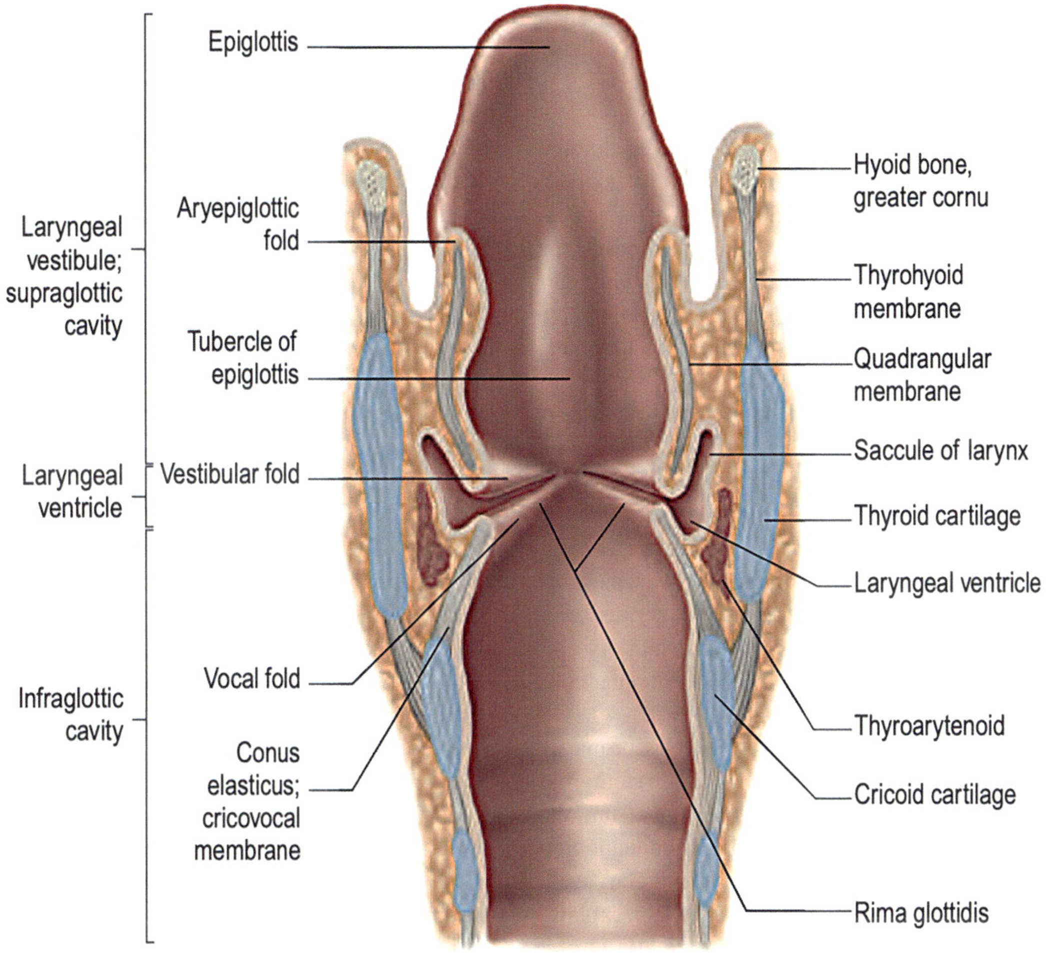

Fig. 12.13 Coronal section of the larynx and the cranial end of the trachea, posterior view, showing: supraglottic space (laryngeal vestibule) extends from the laryngeal inlet to the vestibular folds; transglottic space (laryngeal ventricle, intermediate laryngeal cavity) extends from the vestibular folds, across the laryngeal ventricle, to the vocal folds; infraglottic cavity (subglottic space) extends from the vocal folds to the inferior border of the cricoid cartilage. (Adapted with permission from Standring (2020))

hyoepiglottic ligament relaxation, increased pre-epiglottic fat space pressure, and the mechanical effect of the bolus weight itself. When epiglottic movement becomes compromised, it typically indicates broader swallowing dysfunction, such as post-radiation fibrosis or muscle weakness, rather than serving as the primary pathological focus.

The laryngeal vestibule, defined as the region superior to the vocal cords, comprises multiple structural components: the epiglottis, arytenoid and corniculate cartilages, interarytenoideus muscle, and the connecting aryepiglottic folds. Fluoroscopic examination during modified barium swallow studies reveals the functional significance of this "vertical pipe" configuration (Palmer et al. 1993). Thickened liquid can be observed flowing around the aryepiglottic folds into the pyriform sinuses while avoiding vestibular penetration. This protective mechanism becomes notably absent following partial laryngectomy, resulting in consistent vestibular penetration of ingested material. Successful rehabilitation after such procedures requires patients to develop compensatory strategies, though some degree of tracheal aspiration often persists during swallowing.

Subglottic Anatomy

The subglottic region, also termed the infraglottic space, comprises the portion of the larynx extending from the inferior surface of the true vocal cords to the lower border of the cricoid cartilage (Fig. 12.13). This anatomical region holds particular clinical significance due to its unique structure and position within the laryngeal framework.

The subglottis represents the narrowest segment of the laryngeal airway in children, though in adults the glottis typically forms the narrowest point. The mucosa in this region adheres firmly to the underlying cricoid cartilage, creating a relatively rigid tube that maintains airway patency. The subglottic mucosa contains numerous seromucous glands and is lined with respiratory epithelium—pseudostratified ciliated columnar cells interspersed with goblet cells.

This region's blood supply derives primarily from the subglottic branch of the cricothyroid artery, with additional contributions from the inferior laryngeal artery. The rich lymphatic network in this area drains into the prelaryngeal (Delphian) nodes and paratracheal chain, explaining patterns of tumor spread in laryngeal malignancies.

Laryngeal Cavities

The central tubular laryngeal cavity, lined throughout with mucosa, extends from the pharynx to the trachea. The laryngeal inlet opens into the pharynx, positioned below and behind the tongue, while its inferior aspect continues into the trachea. Within this complex structure, the cavity divides naturally into three major regions: the vestibule, the middle portion, and the infraglottic space (Fig. 12.14).

The cavity of the larynx communicates with the pharynx through the superior laryngeal aperture, which is framed by the epiglottis anteriorly, the aryepiglottic folds laterally, and the interarytenoid notch posteriorly (Sasaki and Isaacson 1988). Between these boundaries lies the vestibule, extending from the aperture to the vestibular folds, commonly known as false vocal cords. The vestibular folds themselves appear as thick mucosal pads covering the vestibular ligament, attaching to the thyroid cartilage just inferior to where the epiglottis connects.

On either side of the laryngeal cavity, between the vestibular and vocal folds, the mucosa creates lateral bulges known as the laryngeal ventricles. From these ventricles, the laryngeal saccules extend anterosuperiorly, positioning themselves between the vestibular fold and thyroid cartilage. These saccular walls contain essential mucous glands that maintain vocal fold lubrication (Palmer et al. 2000). The space between the false and true vocal cords houses these ventricles, with their associated saccules normally extending superiorly under the false vocal cord, sometimes reaching as far as the superior border of the thyroid cartilage.

The vocal folds and the space between them form the glottis, with the rima glottidis representing the gap between the vocal cords and vocal processes of the arytenoids. This space subdivides into a longer intermembranous portion between the cords and an intercartilaginous space between the processes. As the narrowest part of the laryngeal cavity, its shape and width continuously adapt during respiration and phonation.

The preepiglottic space presents as a wedge-shaped region anterior to the epiglottis, bounded by the thyrohyoid ligament and hyoid bone anteriorly. The hyoepiglottic ligament connects the hyoid bone to the epiglottis superiorly, while the space continues laterally as the paraglottic space. The thyroid cartilage creates its lateral boundary, with the conus elasticus and quadrangular membrane forming the medial border.

The mucosal lining shows regional specialization throughout the laryngeal cavity. While most areas feature pseudostratified columnar epithelium, the epiglottis and true vocal cords possess stratified squamous epithelium. Beneath this epithelial layer and superficial to the fibroelastic membrane lies Reinke's space. The posterior laryngeal surface, being part of the laryngopharynx, also displays stratified squamous epithelium. Throughout the mucosa, glands appear predominantly mucous in nature, with scattered

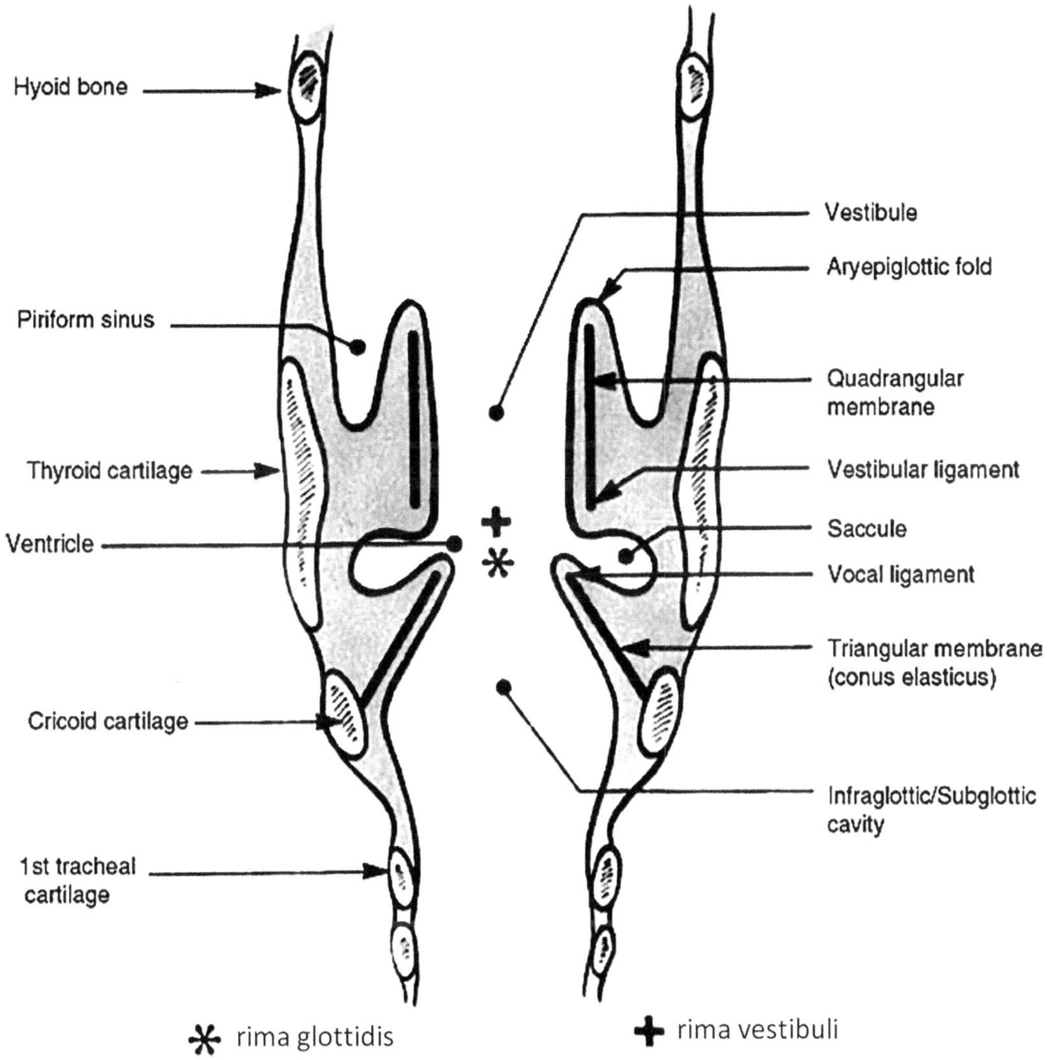

Fig. 12.14 Coronal section of the larynx illustrating laryngeal cavities: vestibule, middle portion, and infraglottic cavity. (Adapted from Blitzer et al. (2009))

serous cells, while the saccule contains concentrated glandular tissue specifically for vocal fold lubrication (Friedrich and Lichtenegger 1997).

12.2.4.2 Structural Composition and Function of the Laryngeal Cartilages

The larynx, positioned anterior to cervical vertebrae 4–6, comprises a complex cartilaginous framework that mediates respiratory and phonatory functions. The primary cartilaginous components include the thyroid, cricoid, paired arytenoids, and epiglottis, interconnected through ligamentous attachments and supported by intrinsic and extrinsic musculature (Drake et al. 2020) (Fig. 12.9).

The thyroid cartilage forms the anterior and lateral laryngeal boundaries through two laminae that converge anteriorly, creating the thyroid notch and laryngeal prominence. Posterior projections, the superior and inferior cornua, extend from each lamina, with the inferior cornua articulating with the cricoid cartilage. This articulation allows limited rocking and gliding movements between these structures (Standring 2020).

The cricoid cartilage exhibits a signet ring configuration, with an expanded posterior lamina. The arytenoid cartilages, pyramidal in shape, sit atop the posterior cricoid plate. These cartilages feature two processes: the muscular process for rotational movements and the vocal process for vocal fold attachment. The posterior arytenoid muscular attachments facilitate vocal fold mobility through precise adjustments.

The epiglottis connects anteriorly to the thyroid cartilage via ligamentous attachment, curving posteriorly behind the hyoid bone's anterior aspect. The aryepiglottic folds contain the cuneiform and corniculate cartilages, which contribute to fold stability and overall laryngeal function through muscular connections (Murry and Carrau 2018).

A mucous membrane covers the laryngeal framework, forming distinctive folds including the aryepiglottic folds and ventricular folds (false vocal folds). The latter, containing minimal muscle tissue, play a secondary role in phonation compared to the true vocal folds (Matsuo and Palmer 2008).

12.2.4.3 Laryngeal Muscles

The laryngeal musculature consists of two separate groups: extrinsic and intrinsic muscles. Extrinsic muscles control overall laryngeal position and individual cartilage movements, while intrinsic muscles modify vocal cord positioning and tension, thereby altering glottic configuration. These muscles work in precise coordination to support multiple laryngeal functions, from phonation to airway protection (Sasaki and Isaacson 1988).

Four primary extrinsic muscles—the digastric, mylohyoid, stylohyoid, and geniohyoid (collectively known as suprahyoid muscles)—accomplish laryngeal elevation by raising the hyoid bone (Fig. 12.15). The thyrohyoid muscle serves a specific function in approximating the hyoid bone and thyroid cartilage. Detailed anatomical descriptions of these laryngeal extrinsic and intrinsic muscles, including their attachments and innervation patterns, are presented in Tables 12.2 and 12.3.

The thyroarytenoid muscle demonstrates complex anatomical organization. It comprises two integrated components rather than separate entities: the vocalis and thyroepiglottic parts (Pearson et al. 2012). The vocalis portion lies deep and inferior, running parallel to the vocal ligament with its posterior attachment. The thyroepiglottic component, positioned superiorly, extends into the aryepiglottic fold and reaches the epiglottic margin. This muscular arrangement enables forward displacement of the arytenoid cartilages, resulting in vocal cord relaxation and shortening, while simultaneous inward rotation adducts the vocal folds and narrows the glottic inlet.

Glottic closure follows a precise inferior-to-superior sequence critical for airway protection during swallowing: true vocal folds close first, followed by false folds, and finally the aryepiglottic folds (Kochis-Jennings et al. 2012). The sphincteric closure of the aryepiglottic folds involves coordinated action of the lateral cricoarytenoid muscles, transverse and oblique arytenoid muscles, and aryepiglottic muscles. These structures form a triangular laryngeal opening when relaxed but contract to approximate the arytenoid cartilages to the epiglottis during function.

Extrinsic Muscles of the Larynx

The extrinsic laryngeal muscles (Table 12.2) establish a dynamic suspension that enables precise laryngeal control through two main muscle groups: y an essential role in anchoring the larynx and adjusting its posture throughout different physiological processes. These muscles play an important function in raising the larynx while swallowing and altering its position during vocalization, which directly affects changes in pitch. The extrinsic laryngeal muscles form a suspension system through two primary groups: the suprahyoid and infrahyoid (strap) muscles. This arrangement enables precise control of laryngeal positioning and movement during various physiological functions (Drake et al. 2020).

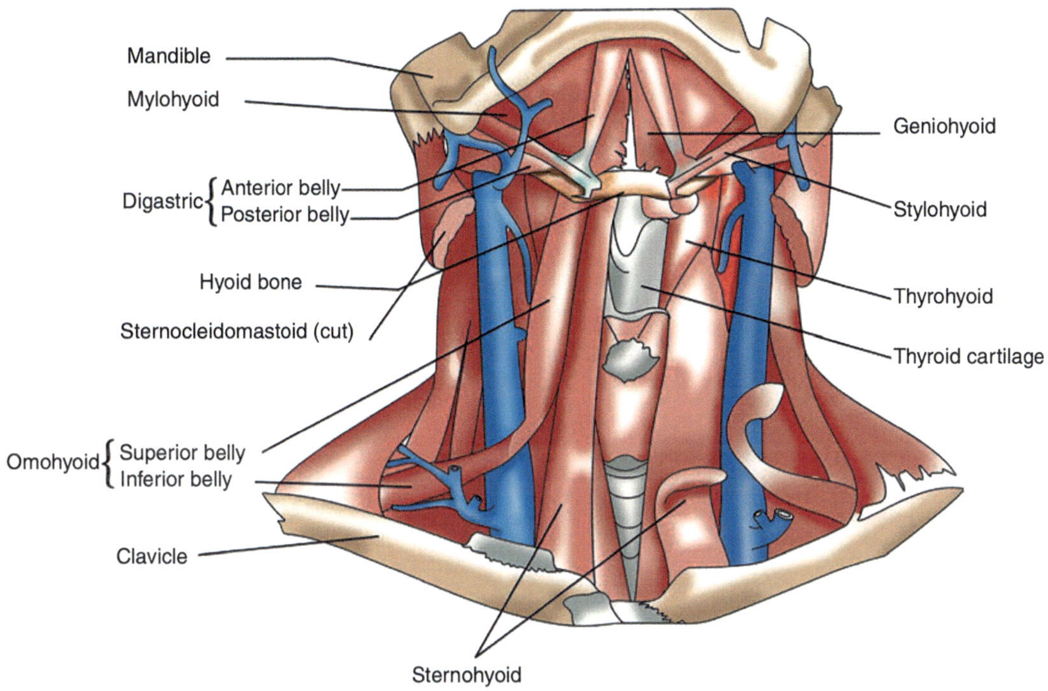

Fig. 12.15 Infrahyoid and suprahyoid muscles. (Adapted with permission from Thankappan et al. (2018))

The suprahyoid complex comprises several muscles—stylohyoid, mylohyoid, digastric, geniohyoid, and stylopharyngeus—creating a multidirectional force system. These muscles guide hyoid bone movement through anterior, superior, and posterior planes, affecting laryngeal position through thyrohyoid membrane connections. The palatopharyngeus and pharyngeal constrictors provide additional muscular support to this system (Netter 2019).

The thyrohyoid muscle merits particular attention due to its role in swallowing biomechanics. Receiving innervation from C1 fibers via the hypoglossal nerve, it facilitates controlled laryngeal elevation during the pharyngeal phase of swallowing, contributing significantly to airway protection.

The infrahyoid group includes the sternothyroid, sternohyoid, thyrohyoid, and omohyoid muscles. The sternothyroid muscle demonstrates unique anatomical characteristics, extending from the posterior manubrial surface to the thyroid lamina's oblique line. Its innervation through the ansa hypoglossi enables precise control of laryngeal depression during complex physiological tasks (Standring 2020).

Epiglottic movement relies on the coordinated action of the thyroepiglottic and aryepiglottic muscles. The aryepiglottic muscle courses within its corresponding folds from the arytenoid cartilage to the epiglottis's lateral border, facilitating epiglottic depression over the laryngeal inlet during deglutition (Sanders et al. 1993).

Intrinsic Muscles of the Larynx

The intrinsic laryngeal muscles, operating within a framework of nine cartilages, control vocal fold movements through precise manipulation of the arytenoid cartilages (Table 12.3). This cartilaginous framework consists of three unpaired cartilages (thyroid, cricoid, and epiglottis) and three paired cartilages (arytenoids, corniculates, and cuneiforms). The arytenoid cartilages maintain particular significance due to their role in muscular attachments and vocal fold mobility.

The posterior cricoarytenoid muscle functions as the primary vocal fold abductor. Its fibers course superolaterally from the posterior cricoid

Table 12.2 Extrinsic muscles of the larynx

Extrinsic muscles of the larynx		course	Cranial nerve supply	Main action
Suprahyoid muscles	Anterior belly of digastric	Down/back	Trigeminal nerve (via mandibular nerve)	Elevates hyoid bone, depresses mandible, fixes mandible and hyoid bone
	Posterior belly of digastric	Down/forward	Facial nerve	Elevates and retracts hyoid bone
	Mylohyoid	Anterior and mid fiber—lateral Posterior fibers—medial/down	Trigeminal nerve (via mandibular nerve)	Elevates hyoid bone, stiffens floor of mouth
	Stylohyoid	Down/forward	Facial nerve	Elevates hyoid bone
	Hyoglossus		Hypoglossal nerve	Retracts and depresses the tongue
	Geniohyoid	Back/slightly down	Ansa cervicalis (C1) and hypoglossal nerve	Pulls hyoid forward, fixes hyoid and mandible, depresses and retracts the mandible
Infrahyoid muscles	Sternohyoid	Up/medial	Ansa cervicalis from cervical plexus (C1–C3)	Depresses the hyoid bone and larynx during speech and swallowing
	Sternothyroid	Up	Ansa cervicalis from cervical plexus (C1–C3)	
	Omohyoid	Inferior belly—forward/up to course behind sternocleidomastoid, changes to obtuse angle and becomes tendinous, forms superior belly coursing vertical/up	Ansa cervicalis from cervical plexus (C1–C3)	
	Thyrohyoid	Up	Ansa cervicalis from cervical plexus (C1–C3) and hypoglossal nerves	

Modified with permission from Thankappan et al. (2018)

cartilage to the arytenoid cartilage's lateral and inferior aspects, enabling arytenoid rotation necessary for airway maintenance during inspiration. The lateral cricoarytenoid muscle complements this action, extending from the cricoid cartilage's lateral aspect in a superoposterior direction (Sataloff et al. 2007) (Fig. 12.19).

The thyroarytenoid muscle forms a substantial muscular plate along the laryngeal inlet. It originates from the thyroid cartilage's internal lamina and contains both lateral and vocal components. Its superior portion, the ventricularis muscle, contributes to rima glottidis closure and supraglottic space compression (Zhang 2016).

The cricothyroid muscle exhibits a dual-component structure. The anterior (vertical) portion extends from the cricoid ring to the thyroid lamina, while the posterior (oblique) segment spans from the lateral cricoid ring to the posterior thyroid lamina. This anatomical arrangement allows for complex vocal fold adjustments, including length, tension, and position modifications (Blitzer et al. 2009).

The hyoid bone, though not an intrinsic laryngeal component, provides essential support for laryngeal mobility. This U-shaped structure connects to the temporal bone through the stylohyoid ligament and to the larynx via the thyrohyoid

Table 12.3 Intrinsic muscles of the larynx

Intrinsic muscles of the larynx	Course	Cranial nerve supply	Main action
Anterior or vertical cricothyroid	Up/back/lateral	External branch of superior laryngeal nerve (branch from vagus)	Produces tension and elongation of the vocal ligament resulting in higher-pitch phonation
Posterior cricoarytenoid	Superior fibers—horizontal Mid fibers—oblique Inferior fibers—vertical	Recurrent laryngeal nerve	Abducts the vocal cord by rotating the arytenoid cartilages laterally
Lateral cricoarytenoid	Oblique up/back	Recurrent laryngeal nerve	Adducts the vocal cord by rotating the arytenoids medially, closes rima glottis
Transverse and oblique arytenoids	Transverse arytenoid—lateral Oblique arytenoid—oblique up/lateral	Recurrent laryngeal nerve (*nerves from both sides of the larynx innervate the muscle*)	Adducts the vocal folds via approximation of the arytenoid cartilages (closes the posterior rima glottis)
Thyroarytenoid	Back/lateral	Recurrent laryngeal nerve	Relaxes vocal ligament
Vocalis	Back/lateral	Recurrent laryngeal nerve	Relaxes posterior vocal ligament when maintaining or increasing the tension of anterior part

Modified with permission from Thankappan et al. (2018)

membrane, establishing a foundation for coordinated tongue and laryngeal movements (Matsuo and Palmer 2008).

Functional coordination among these muscles enables precise control of voice production and airway protection. The posterior cricoarytenoid muscles manage vocal fold abduction, while the lateral cricoarytenoid and arytenoid muscles control adduction. The thyroarytenoid and vocalis muscles, working in concert with the cricothyroid muscle, regulate vocal pitch through alterations in vocal fold length and tension (Logemann 1983) (Fig. 12.16).

12.2.5 Anatomy of the Esophagus

The esophagus functions as a flexible muscular passage within the digestive system, connecting the pharynx to the stomach. This anatomically complex structure measures approximately 23–25 cm, beginning at the inferior border of the cricoid cartilage (C6 vertebral level) and terminating at the cardiac opening of the stomach (T11 vertebral level) (Drake et al. 2020) (Fig. 12.17). The esophagus traverses the posterior mediastinum within the thoracic cavity, closely associated with maintaining intimate relationships with major structures such as the tracheobronchial tree, heart, and major vessels (Moore et al. 2018) (Fig. 12.18).

The esophageal wall exhibits a four-layered structure that aligned with its particular function. The exterior fibrous layer provides structural stability and allows contact with adjacent tissues. Below this is the muscular layer, consisting of an outer longitudinal layer and an inner circular layer that remains continuous with the inferior constrictor of the pharynx. The circular muscle layer is less thick cranially and progressively thickens distally. Between the two muscle layers are a number of neurons arranged in a plexus structure known as Auerbach's plexus. This contains both sympathetic and parasympathetic nerves. The vagus nerve is in close proximity to the esophagus, especially in the inferior region.

The submucosal layer has an extensive network of blood vessels, neuronal components, and mucous-secreting glands, enabling esophageal function. The innermost mucosal layer comprises

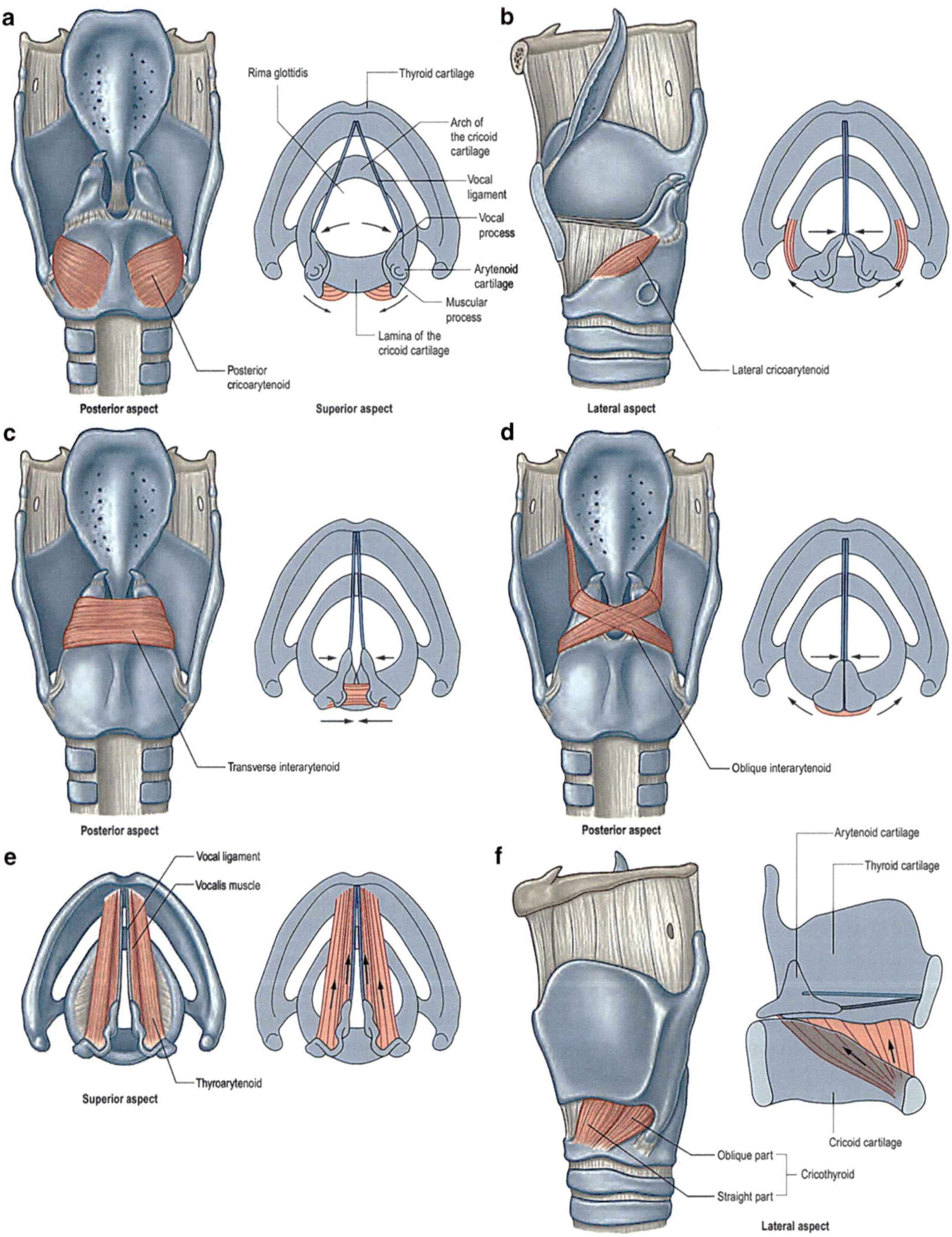

Fig. 12.16 The intrinsic muscles of the larynx shown in posterior (**a**, **c**, **d**) or lateral views (**b**, **f**), with their actions, illustrating in a superior view (E). Arrows indicate the direction of cartilage movement in each case. (Adapted with permission from Standring (2020))

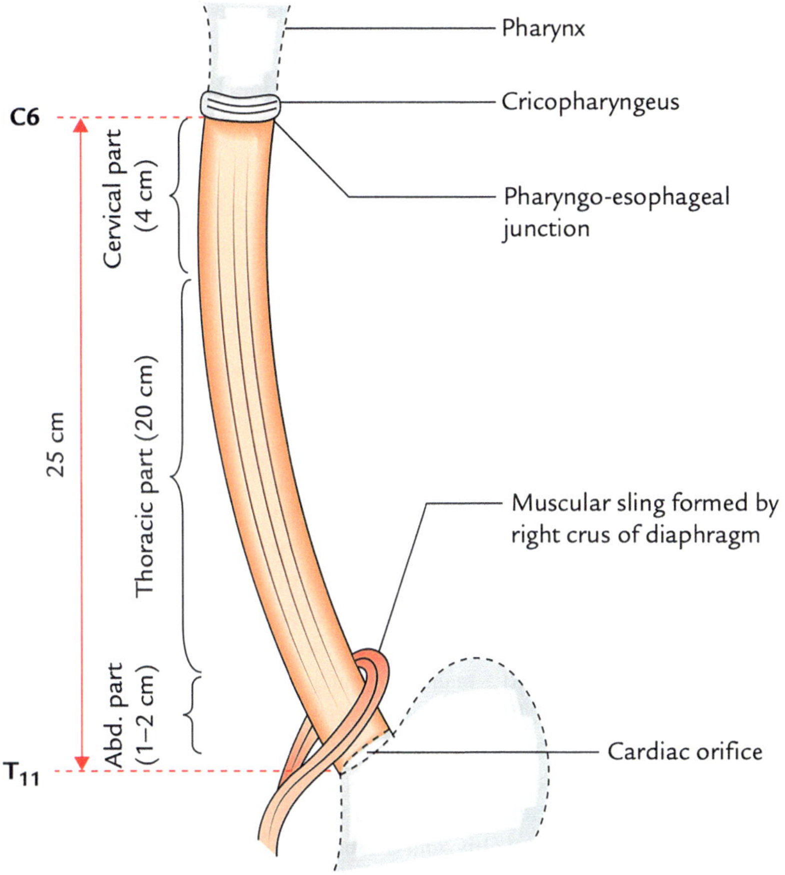

Fig. 12.17 The esophagus, divided into three parts: cervical, thoracic, and abdominal, extending from the C6 vertebral level and the T11 vertebral level. (Adapted from Singh (2014))

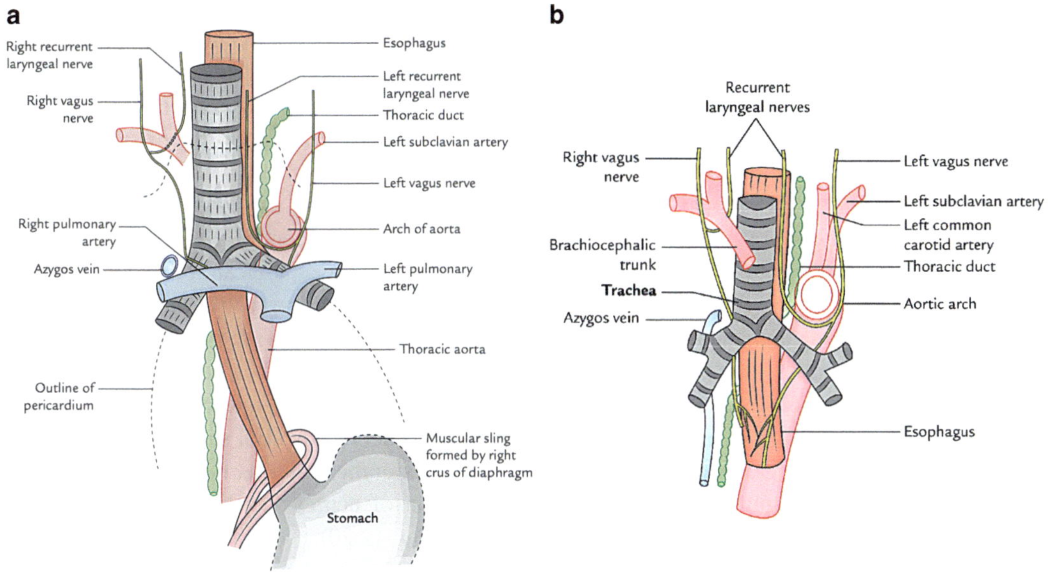

Fig. 12.18 (**a**) Anterior and lateral, and (**b**) posterior and lateral relations of the trachea, main bronchi, and major vessels. (Adapted from Singh (2014))

specialized stratified squamous epithelium with surface papillae, adapted to withstand mechanical stress during food transit (Standring 2020).

The muscularis mucosae, a layer of longitudinal non-striped muscular fibers, is situated between the submucosal and mucosal layers. This structure becomes progressively more prominent in the distal esophagus, influencing local motility patterns. The absence of a surrounding fascial envelope distinguishes the esophagus from other digestive organs, making it particularly susceptible to pathological processes affecting the mediastinum (Sinnatamby 2011).

Three anatomically significant constrictions characterize the esophageal course. The first occurs at the cricopharyngeal sphincter (15 cm from the incisors), representing a key point of physiological control. The second constriction appears where the aortic arch and left main bronchus cross the esophagus (23 cm from the incisors), while the third manifests at the diaphragmatic passage (40 cm from the incisors). These constrictions form natural points of physiological narrowing that influence bolus transit and may become clinically significant in various pathological conditions (Ellis and Mahadevan 2019) (Fig. 12.19).

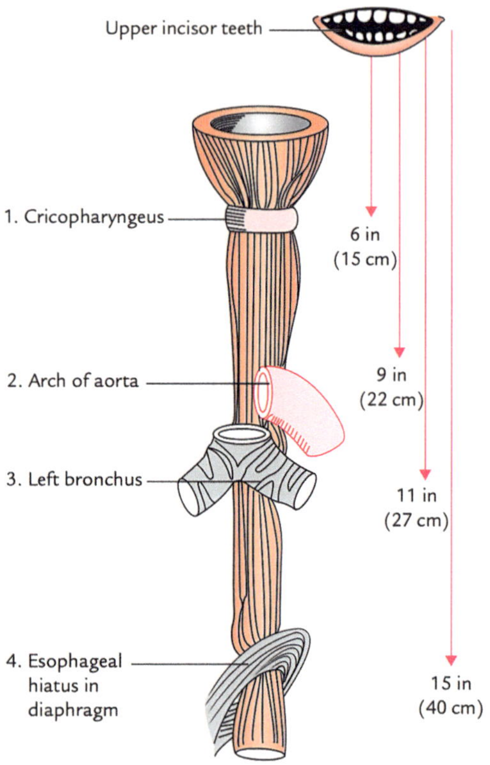

Fig. 12.19 Anatomical sites of esophageal constrictions, with their distances from the upper incisor teeth indicated on the right. (Adapted from Singh (2014))

The muscular structure of the esophagus demonstrates a transition from striated to smooth muscle, reflecting its position between voluntary and involuntary portions of the digestive tract. The upper third contains striated muscle fibers under voluntary control, while smooth muscle predominates in the lower two-thirds. This structural variation results in differential transit times: less than 1 s in the upper portion versus approximately 3 s in the lower segments. This specialized arrangement allows for both voluntary initiation of swallowing and autonomous continuation of peristaltic movements (Pawlina 2016).

Neural control of esophageal function involves an intricate network centered on the myoneural plexus. This complex integrates parasympathetic ganglia receiving vagal input with contributions from somatic segmental nerves. The sensory innervation pattern explains the characteristic referral of esophageal pain to specific dermatomes, typically manifesting as substernal discomfort. This neural organization allows for both local reflex control and central nervous system integration of esophageal function (Netter 2019).

Peristaltic activity in the esophagus manifests in three distinct patterns. Primary peristalsis represents normal physiological propulsive waves, initiated by voluntary swallowing and propagated through autonomic control. Secondary peristalsis includes non-physiological retrograde movements, often serving protective functions. Tertiary peristalsis appears as segmental contractions without a propulsive effect, becoming more prevalent with aging. Systemic conditions affecting peripheral nerve function, such as diabetes mellitus, can significantly alter these patterns through disruption of neural coordination (Young et al. 2014) (Fig. 12.20).

The coordinated muscular activity relies on the precise timing of circular muscle contraction and longitudinal muscle shortening. This syn-

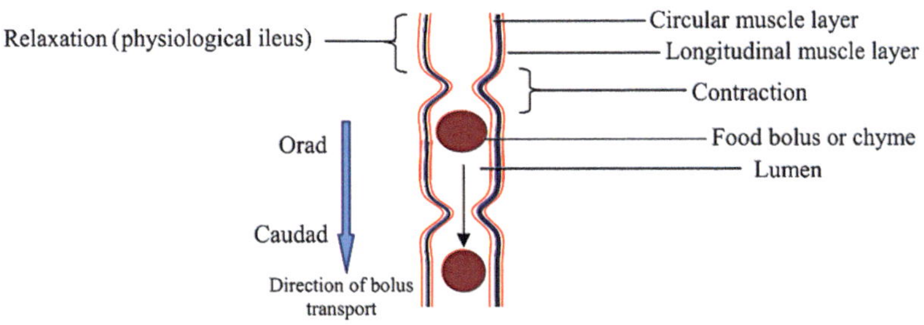

Fig. 12.20 Esophageal peristaltic contraction. The smooth muscle proximal to the bolus contracts to propel it downward, while the distal segment relaxes to facilitate its movement. This coordination is regulated by the excitatory and inhibitory activity of motor neurons innervating these muscle layers. Sensory neurons relay information to the medulla, where they synapse with both excitatory and inhibitory motor neurons. The contraction of smooth muscle above the bolus is mediated by excitatory motor neurons, while relaxation below the bolus is mediated by inhibitory motor neurons. (Adapted with permission from Welcome (2018))

chronization creates an effective propulsive force that moves the bolus at a relatively constant velocity, independent of gravitational effects. The combination of mechanical and neural elements allows the esophagus to function effectively in various body positions, maintaining efficient transport of ingested material from the pharynx to the stomach.

12.2.5.1 Upper Esophageal Sphincter

The upper esophageal sphincter (UES) extends 4–6 cm in length, comprising three muscular components: the cricopharyngeus muscle, the lower part of the thyropharyngeus muscle, and the upper esophageal muscle fibers (Lang et al. 2011) (Fig. 12.12). The cricopharyngeus muscle, forming the predominant component of the UES, presents as a C-shaped band that wraps around the anterior and lateral aspects of the pharyngoesophageal junction (Ertekin and Aydogdu 2003). This anatomical arrangement forms a high-pressure zone that maintains closure of the sphincter at rest, preventing air entry into the esophagus during respiration (Sivarao and Goyal 2000).

The coordinated contraction and relaxation of the sphincter facilitate bolus passage from the pharynx to the esophagus through a complex neuromuscular sequence. During the swallowing process, the UES undergoes three different phases: relaxation, opening, and closure. The sphincter relaxation precedes the arrival of the bolus by approximately 100–200 ms, demonstrating the precise temporal coordination of the swallowing mechanism (Mu and Sanders 2010). This relaxation phase coincides with anterior and superior movement of the larynx, which assists in sphincter opening and provides airway protection (Cook et al. 1989a, b).

The pharyngeal plexus provides primary innervation to the UES region, while the cricopharyngeus muscle receives dual innervation through both the recurrent laryngeal and superior laryngeal nerves. This dual innervation pattern enables precise laryngopharyngeal coordination during swallowing and represents a natural mechanism for airway protection (Mittal et al. 2013). The superior laryngeal nerve provides sensory information from the hypopharyngeal mucosa and motor innervation to the cricothyroid muscle, while the recurrent laryngeal nerve supplies motor innervation to the intrinsic laryngeal muscles and cricopharyngeus (German et al. 2009).

Electromyographic studies indicate that the UES maintains tonic contraction during rest, with pressure measurements generally between 30 and 110 mmHg. This baseline pressure decreases during sleep and increases during physical exertion or emotional stress, reflecting the dynamic nature of sphincter control. The complex integration of sensory and motor signals allows the UES to respond appropriately to various physiological conditions, including swallowing, belching, and vomiting (Kahrilas et al. 2015).

12.2.5.2 Lower Esophageal Sphincter

The lower esophageal sphincter (LES) is situated in a unique anatomical position between two pressure zones. The esophageal body lies in the thoracic cavity, which maintains subatmospheric pressure for respiratory function, while the distal esophagus and stomach lie within the higher-pressure peritoneal cavity. This pressure differential necessitates a functional sphincter mechanism that permits antegrade food passage while preventing retrograde flow (Boeckxstaens and Zaninotto 2012).

The LES maintains its normal resting tone through two primary mechanisms: the sphincteric function of circular muscular fibers and its anatomical position within the diaphragmatic crural folds (Mittal and Balaban 1997). This dual mechanism generates a constant resting pressure that physiologically decreases to subatmospheric levels immediately before bolus arrival. Dysfunction of this mechanism manifests in various pathological conditions. For instance, failed sphincter relaxation results in achalasia, preventing normal bolus transit into the stomach (Kahrilas et al. 2017).

The anatomical integrity of the LES depends significantly on ligamentous attachments maintaining its position within the diaphragmatic crura. Weakening of these attachments can lead to herniation of the distal esophagus and stomach into the thoracic cavity, primarily due to the trandiaphragmatic pressure gradient. This condition, known as hiatus hernia, disrupts normal LES positioning and function, potentially allowing gastric content regurgitation into the esophagus (Richter and Rubenstein 2018).

The esophageal mucosa demonstrates specific characteristics, featuring nonkeratinizing squamous epithelium extending from the upper sphincter to the gastric cardia (Orlando et al. 1981). During endoscopic examination, this presents as a clear demarcation between the pale esophageal squamous mucosa and the richly vascularized pink gastric mucosa. This junction represents a critical anatomical and physiological transition zone susceptible to pathological changes (Spechler and Souza 2014).

Exposure to gastric acid can trigger significant mucosal alterations at the gastroesophageal junction. These changes may manifest as ulceration, inflammation, dysplastic transformation, or metaplastic conversion of normal squamous mucosa to intestinal columnar epithelium, a condition known as Barrett's esophagitis (Sharma et al. 2016). This premalignant condition has caused considerable study interest, particularly due to its increasing prevalence in conjunction with esophageal cancer in Western countries (Wong and Fitzgerald 2005).

12.2.5.3 Esophageal Peristalsis: Mechanisms and Control of Bolus Transport

The coordinated movement of food and liquids through the gastrointestinal tract relies on peristalsis, a complex process of rhythmic muscular contractions. This unidirectional propulsion mechanism consists of two phases: an initial accommodative wave followed by a propulsive contractile wave. These movements result from synchronized contractions of longitudinal and circular muscles that enable efficient bolus transport (Goyal and Chaudhury 2008) (Fig. 12.20).

Esophageal peristalsis manifests in two distinct forms: primary and secondary. Primary peristalsis initiates with swallowing, triggered by sensory input as the bolus transitions from the pharynx into the cervical esophagus. This process maintains a consistent rate of travel independent of bolus movement and typically spans approximately 9 s. The contractile force generated during primary peristalsis exceeds that of secondary peristalsis in both magnitude and duration.

Secondary peristalsis develops in response to esophageal wall distention, where stretch receptors activate local reflexes to advance the bolus distally. This mechanism plays a significant role in acid clearance, explaining why patients with gastroesophageal reflux disease often display impaired secondary peristalsis. Manometric studies can assess various aspects of these contractions, including preload tension, contractile tension, and postload tension, providing insights into the strength, amplitude, and duration of peristaltic waves (Welcome 2018).

The neural control of esophageal peristalsis involves multiple systems. Brainstem lower motor neurons regulate skeletal muscle function in the proximal esophagus, while the central and enteric nervous systems coordinate smooth muscle activity in the distal portion. Additionally, the intrinsic properties of esophageal smooth muscle contribute to this complex regulatory mechanism (Mittal and Balaban 1997).

12.3 Neural Control of Swallowing

The neural control of swallowing requires precise coordination between respiratory and digestive functions, as the pharynx functions as a shared pathway for both processes (Jean 2001). This coordination system integrates multiple neural mechanisms to prevent aspiration through a complex network of sensory inputs, central processing, and motor outputs.

The initial sensory component involves multiple anatomical regions in the oropharynx that trigger the swallowing reflex. These regions include the soft palate, uvula, posterior tongue surface, pharyngeal surface of the epiglottis, pharyngoesophageal junction, and faucial pillars. Within these structures, specialized mechanoreceptors and chemoreceptors respond to various stimuli, including pressure, touch, and chemical properties of food boluses (Matsuo and Palmer 2008). The faucial pillars contain particularly specialized fluid and pressure receptors that demonstrate rapid adaptation to stimuli, contributing to the precise timing of swallow initiation (Lang et al. 2001).

The afferent pathway begins as primary sensory signals from the oropharyngeal mucosa travel through three major cranial nerves. The trigeminal nerve (CN V) transmits oral sensory information and controls masticatory muscles, while the glossopharyngeal nerve (CN IX) conveys taste information from the posterior third of the tongue and pharyngeal sensation. The vagus nerve (CN X) provides essential sensory information from the pharyngeal and laryngeal mucosa (Dodds 1989). These afferent signals converge in the solitary tract, where they synapse with second-order neurons in the nucleus tractus solitarius, forming the primary integration center for swallowing (Steele and Miller 2010).

Central processing of swallowing involves a distributed neural network. The nucleus tractus solitarius serves as the primary afferent portal, containing interneurons that execute complex levels of swallowing control. Recent research has identified a structurally defined central pattern generator within the brainstem, which coordinates the sequential activation of multiple muscle groups during swallowing (Ertekin and Aydogdu 2003). This central pattern generator receives modulatory inputs from higher cortical centers, particularly the region anterior to the precentral sulcus cortex, where stimulation induces sequential mastication and swallowing.

The efferent control system involves multiple cranial nerve nuclei. The ambiguus nucleus provides motor control to the pharyngeal and laryngeal muscles through the vagus nerve, while the hypoglossal nucleus controls the intrinsic and extrinsic muscles of the tongue through the hypoglossal nerve (CN XII) (Costa 2018). The facial nerve (CN VII) contributes to swallowing through control of the salivary glands and specific facial muscles, though its role is less prominent than that of the vagus and hypoglossal nerves.

These neural components regulate three phases of swallowing. The oral phase remains under voluntary control, involving conscious manipulation of the food bolus through intricate tongue movements coordinated by the hypoglossal nerve. The pharyngeal phase, initiated by stimulation of specific sensory regions, functions as an involuntary reflex that cannot be interrupted once initiated. This phase includes multiple protective mechanisms such as laryngeal closure and respiratory interruption (Kahrilas et al. 1991). The esophageal phase, regulated primarily by the enteric nervous system, demonstrates autonomic control with self-regulatory functions, including the ability to generate secondary peristaltic waves in response to retained or regurgitated material (Park 1999).

12.3.1 Central Nervous System Control of Swallowing

The brainstem maintains fundamental control of swallowing, sharing anatomical proximity with centers regulating respiration, blood pressure, and temperature regulation. However, successful feeding requires higher brain center involvement for food recognition, procurement, and oral placement. This hierarchical organization explains why patients with preserved basic swallowing function may still experience feeding difficulties due to cognitive impairments from various neurological conditions (Leopold and Daniels 2010).

The swallowing center resides in the brainstem, positioned adjacent to vagal sensory and motor nuclei. This location supports centrally driven motor patterns that respond to proprioceptive input. Multiple protective reflexes accompany the swallowing process, including glottic closure, which occurs with a latency period under 25 ms. This brief latency period indicates brainstem-level processing rather than higher center involvement (Sasaki and Isaacson 1988).

Research by Sasaki demonstrated that various physiological alterations affect laryngeal reflexes, including hypercapnia, hypoxia, and sedation (Sasaki et al. 2006). Central nervous system disruption, whether from focal injury or global impairment through anesthetic agents, can significantly affect swallowing function. This relationship manifests clearly in post-stroke dysphagia, where swallowing mechanisms become temporarily compromised.

The integration of swallowing control occurs at multiple levels within the central nervous system. While the brainstem coordinates the basic swallowing pattern, cortical and subcortical regions contribute significantly to voluntary aspects of feeding and swallowing initiation (Hamdy et al. 1999). This multi-level control system allows for both automatic protective responses and voluntary modification of swallowing patterns based on food consistency and volume (Ertekin and Aydogdu 2003).

12.3.2 Cranial Nerve Function in Swallowing

The process of swallowing relies on precise neural control through multiple cranial nerves that regulate sensory input, proprioception, and motor function. These neural pathways form an integral component of the swallowing mechanism, coordinating bolus movement through various anatomical chambers and valves (Miller 2008).

Each cranial nerve contributes specific functions to the swallowing process. The trigeminal nerve (CN V) provides sensory information from the oral cavity and controls the muscles of mastication. The facial nerve (CN VII) manages taste sensation from the anterior two-thirds of the tongue and regulates salivary gland function. The glossopharyngeal nerve (CN IX) transmits sensory information from the posterior third of the tongue and pharyngeal walls while providing taste sensation from the posterior tongue (Matsuo and Palmer 2009) (Table 12.4).

Table 12.4 Innervation of major muscles involved in swallowing

Cranial nerves	Muscles
Trigeminal nerve (V)	Masticatory muscles Mylohyoid Tensor veli palatini Anterior belly of digastrics
Facial nerve (VII)	Facial muscle Stylohyoid Posterior belly of digastrics
Glossopharyngeal nerve (IX)	Stylopharyngeus
Vagus nerve (X)	Levator veli palatine Palatopharyngeous Salpingopharyngeous Intrinsic laryngeal muscles Cricopharyngeus Pharyngeal constrictors
Hypoglossal nerve (XII)	Intrinsic tongue muscles Hyoglossus Geniohyoid Genioglossus Styloglossus Thyrohyoid

Adapted from Matsuo and Palmer (2008)

The vagus nerve (CN X) maintains a central role in swallowing physiology through its extensive innervation of the pharynx and larynx. Its branches, particularly the superior and recurrent laryngeal nerves, coordinate laryngeal function during swallowing. The spinal accessory nerve (CN XI) contributes to pharyngeal elevation through its innervation of the trapezius and sternocleidomastoid muscles. The hypoglossal nerve (CN XII) controls intrinsic and extrinsic tongue muscles, enabling precise bolus manipulation and propulsion (Dodds et al. 1990).

Dysfunction in any of these neural pathways can result in specific swallowing impairments. Clinical evaluation of swallowing disorders often begins with assessment of cranial nerve function, as patterns of dysfunction can indicate the location and extent of neurological impairment.

12.3.2.1 Vagus Nerve

The vagus nerve (CN X) serves as the principal neuronal conduit controlling swallowing physiology, offering extensive motor and sensory innervation to the palate, pharynx, larynx, esophagus, and stomach while simultaneously regulating respiratory and cardiovascular functions. The vagus nerve exits the skull through the jugular foramen, positioned adjacent to the internal jugular vein and cranial nerves IX and XI. At this location, the nerve expands to form the nodose ganglion, containing chemoreceptor cells that monitor blood carbon dioxide levels. These specialized cells organize into paraganglioma or glomus bodies, which may occasionally develop into functionally significant tumors despite their benign histology.

The vagal pathway continues alongside the carotid arteries, transitioning from a posteromedial to anterolateral position. This course makes it particularly vulnerable during carotid surgical procedures. Two major branches emerge along its cervical course: the superior laryngeal nerve, providing sensation to the supraglottis and anterior hypopharynx, and the recurrent laryngeal nerve, innervating the intrinsic laryngeal muscles (Sanders and Mu 1998).

Central components of the vagus include motor innervation originating from the nucleus ambiguus and sensory input processing through the nucleus solitarius. Approximately 80% of vagal fibers carry sensory information, while the remaining fibers conduct visceral preganglionic parasympathetic signals. This prevalence of sensory fibers indicates the nerve's essential function in monitoring and regulating swallowing physiology.

Multiple branches form the pharyngeal plexus by combining with glossopharyngeal nerve fibers, resulting in a complex network that innervates the soft palate and pharyngeal constrictors. This neural network traverses the esophagus, coordinating the complex muscular activities required for normal swallowing. Notably, anterior cervical spine surgical approaches may compromise this plexus, explaining the frequent occurrence of post-operative dysphagia.

Historical research by Kirchner demonstrated the vagus nerve's essential role through canine studies, where bilateral vagotomy at the skull base resulted in complete feeding inability (Kirchner 1958). Modern clinical observations of high vagal lesions consistently reveal significant swallowing impairment due to compromised palatal, pharyngeal, and laryngeal functions (Mistry and Hamdy 2008).

High vagal lesions, occurring above the superior laryngeal nerve branch point, produce particularly devastating effects, resulting in combined paralysis and anesthesia of ipsilateral pharyngeal and laryngeal structures. This clinical presentation emphasizes the importance of preserving vagal integrity during surgical procedures in the neck (Matsuo and Palmer 2008).

Beyond swallowing, the vagus maintains critical regulatory functions through baroreceptor innervation, influencing cardiac rate and blood pressure. Its extensive distribution continues below the recurrent laryngeal branch points, providing essential parasympathetic innervation to thoracic and abdominal organs, including the regulation of esophageal and gastric function.

Superior Laryngeal Nerve

The superior laryngeal nerve, a branch of the vagus nerve, emerges high in the neck and courses medially to the carotid artery and its branches, roughly parallel to the superior laryngeal branches of the superior thyroid artery (Standring 2020). This nerve divides into external and internal branches with specific functions in laryngeal innervation (Sanders and Mu 1998).

The external branch innervates the cricothyroid muscle. This muscle's contraction produces posterior movement of the cricoid ring relative to the thyroid cartilage, anterior approximation of these structures, and separation of the posterior cricoid lamina from the thyroid lamina. These actions expand the airway and increase vocal cord tension by lengthening the glottis's anterior-posterior dimension. The cricothyroid muscle participates in both respiratory function and voice pitch modulation.

The anatomical relationship between the external branch and the superior thyroid pole creates surgical complexities. The nerve may travel between superior thyroid artery branches or below the superior pole, making it vulnerable during thyroidectomy. While injury may produce minimal symptoms in many individuals, it significantly impacts professional singers by limiting vocal range (Cernea et al. 2009).

The internal branch provides sensory innervation to the supraglottic larynx, cricoarytenoid joints, posterior laryngeal region, and pyriform sinus mucosa (Sanders and Mu 1998). Damage to this branch causes supraglottic and pyriform sinus anesthesia, resulting in persistent dysphagia and aspiration from impaired sensory feedback. Research has demonstrated this branch's importance in normal swallowing physiology (Ludlow 2015).

Recurrent Laryngeal Nerve

The recurrent laryngeal nerve's unique course originates during mammalian embryogenesis. The pharyngeal structures' migration leads to nerve looping around fourth branchial arch derivatives—the aortic arch on the left and subclavian artery on the right (Standring 2020) (Fig. 12.21). This anatomical arrangement characterizes all mammals, though abnormal embryonic ciliary function may reverse this pattern in Kartagener syndrome, resulting in situs inversus with respiratory manifestations (Knowles et al. 2013).

The left recurrent laryngeal nerve follows an anteroposterior course around the aortic arch, passing beneath the ligamentum arteriosum, while the right recurrent laryngeal nerve typically encircles the right subclavian artery. This asymmetry creates differential nerve lengths requiring varied conduction velocities for synchronized vocal cord function. A striking example occurs in giraffes, where the left recurrent laryngeal nerve's length exceeds the right by approximately 20-fold.

Both nerves typically ascend in the ipsilateral tracheoesophageal groove, with the right nerve following a more lateral approach due to its shorter course. A significant anatomical variant occurs when the right fourth arch fails to develop into the right subclavian artery. This results in a non-recurrent right laryngeal nerve taking a lateral approach to the larynx, potentially complicating thyroid surgery. In these cases, the aberrant right subclavian artery may cause dysphagia lusoria through esophageal compression.

The recurrent laryngeal nerve provides motor innervation to all intrinsic laryngeal muscles except the cricothyroid. The posterior cricoarytenoid muscle, crucial for respiration as the sole glottic abductor, receives motor supply through this nerve (Sanders and Mu 1998). The nerve enters the larynx posterior to the cricothyroid joint, dividing into anterior and posterior branches. The anterior branch innervates the lateral cricoarytenoid and thyroarytenoid muscles, while the posterior branch supplies the posterior cricoarytenoid muscle.

Clinical implications of recurrent laryngeal nerve injury vary based on whether the damage is unilateral or bilateral. Bilateral injury prevents vocal cord abduction, creating an inadequate glottic airway that may require urgent intervention. During right-sided cervical spine surgery, laryngeal retraction may increase tension on the right recurrent laryngeal nerve, potentially explaining higher rates of right-sided nerve injury.

Beyond motor function, the recurrent laryngeal nerve provides sensory innervation to the subglottic larynx and trachea. Loss of this sensory function can impair swallowing mechanisms

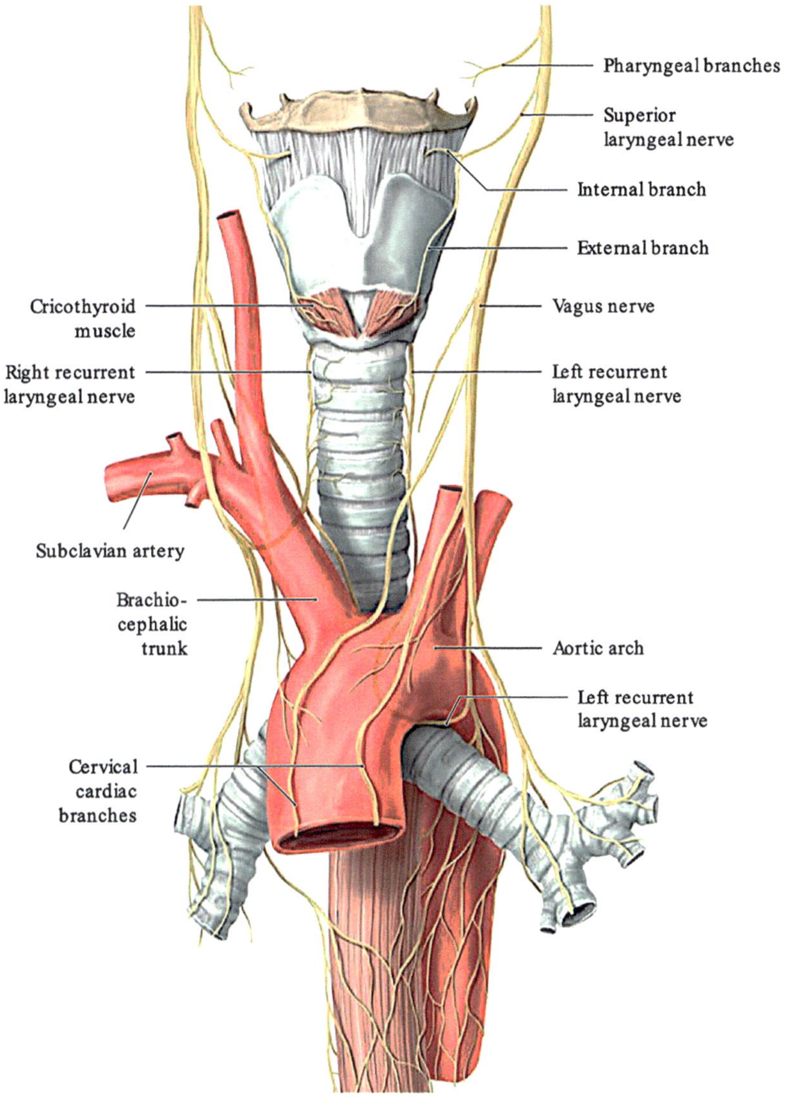

Fig. 12.21 The vague nerve in the neck, giving rise to four sets of branches: pharyngeal branches, superior laryngeal nerve, recurrent laryngeal nerve, and cervical cardiac branches. (Adapted from Schuenke et al. (2016))

and increase aspiration risk through reduced proprioception (Ludlow 2015). This dual motor-sensory function shows the nerve's clinical significance in maintaining airway protection and vocal function.

12.3.2.2 Glossopharyngeal Nerve

The glossopharyngeal nerve (CN IX) demonstrates intricate anatomical relationships and diverse functions, necessitating a detailed understanding for clinical practice. After exiting the jugular foramen with the vagus nerve, CN IX courses adjacent to the internal carotid artery and lateral to the pharyngeal constrictors before joining the pharyngeal plexus (Standring 2020).

This cranial nerve serves both sensory and motor functions. Its sensory component provides innervation to the soft palate and tonsil, while also conveying taste sensation from the posterior tongue (Sanders and Mu 1998). Additionally, CN IX carries parasympathetic fibers that synapse in the otic ganglion to regulate parotid gland secretion.

The motor functions of CN IX manifest through two primary pathways: participation in the pharyngeal plexus for middle pharyngeal constrictor innervation, and direct supply to the stylopharyngeus muscle for pharyngeal shortening. This dual motor role significantly influences swallowing biomechanics and pharyngeal function.

12.3.2.3 Trigeminal Nerve

The trigeminal nerve (CN V) functions as a primary neural component for facial sensation and motor control, with specific contributions to swallowing and oral functions (Standring 2020). Its three major divisions exhibit complex anatomical distributions, with the second and third divisions particularly involved in oral cavity functions. These divisions traverse the gasserian ganglion, located in Meckel's cave on the temporal bone's posterior aspect, where sensory neurons form synaptic connections. The sensory components extend through an elongated nuclear structure in the brainstem, continuing caudally into the spinal cord as the spinal tract of V (Kiernan and Rajakumar 2014).

The nerve carries sensory information from mucosal and proprioceptive receptors throughout the oral cavity, including the tongue, palatal structures, dentition, and buccal mucosa. Its motor component innervates the masticatory muscles and the tensor veli palatini. While the hypoglossal nerve controls tongue movements, the trigeminal nerve innervates the anterior digastric muscle, with the posterior portion receiving innervation from the facial nerve (CN VII) (Mtui et al. 2016).

The maxillary division (V2) exits through the foramen rotundum and distributes within the pterygopalatine fossa. It provides sensory innervation to both hard and soft palates, upper dentition, gingiva, and related structures. The nerve's branches access the nasal mucosa via the sphenopalatine foramen, accompanied by parasympathetic fibers. Palatine branches traverse the palatine canal, exiting via the greater and accessory palatine foramina (Standring 2020).

The mandibular division (V3) includes the lingual nerve, which supplies sensory innervation to the anterior two-thirds of the tongue, and the inferior alveolar nerve, which innervates lower dentition, gingiva, and lip. The lingual nerve's path makes it susceptible to injury during specific surgical procedures, potentially affecting bolus formation and swallowing function (Norton 2017).

Age-related changes in oral sensation can impact swallowing function, as proper sensory input remains necessary for effective bolus formation and swallow initiation (Robbins et al. 1992).

12.3.2.4 Hypoglossal Nerve

The hypoglossal nerve (cranial nerve XII) emerges from the caudal brainstem and follows a complex anatomical path through the skull base and neck. After exiting the hypoglossal canal superior to the occipital condyle, it maintains close proximity to cranial nerves IX, X, and XI at the jugular canal, making it susceptible to pathological processes in this region (Standring 2020).

The nerve's course includes an inferior and anterior trajectory lateral to the carotid arterial system and deep to the jugular vein. During its more horizontal path, it produces the ansa hypoglossal, a significant branch that provides motor innervation to the infrahyoid (strap) muscles. The nerve travels within a venous network that receives drainage from the lingual, anterior facial, and posterior facial veins. This anatomical arrangement creates potential surgical complications, as bleeding from these vessels during procedures may result in inadvertent nerve damage during hemostasis attempts (Norton 2017).

After passing beneath the posterior belly of the digastric muscle, the nerve ascends through the submandibular triangle to reach the tongue musculature. It branches to supply the mylohyoid muscle before distributing to both the extrinsic and intrinsic tongue muscles. This anatomical arrangement becomes particularly relevant in contemporary surgical management of obstructive sleep apnea. Therapeutic approaches involve selective nerve branch identification for stimulator placement, aiming to advance the tongue base anteriorly (Strollo et al. 2014).

The nerve's position deep to the submandibular gland and duct increases its vulnerability during surgical procedures addressing inflammatory conditions or malignancies of the submandibular gland. This anatomical relationship requires careful surgical planning and meticulous dissection techniques to avoid iatrogenic injury.

12.3.3 Autonomic Nervous Innervation of the Larynx

12.3.3.1 Sympathetic Innervation of the Larynx

The sympathetic nervous system exerts precise control over laryngeal functions through an intricate network of neural pathways. These pathways form a complex yet organized system that maintains homeostatic control and supports various laryngeal functions. A comprehensive understanding of this system provides valuable insights for clinical practice and research in laryngeal disorders (Drake et al. 2020).

The sympathetic innervation functions via two separate neuronal groups that cooperatively regulate laryngeal activity. The superior cervical ganglion contains the cell bodies of postganglionic neurons, serving as a principal integration and relay site. These neurons get input from preganglionic neurons originating in the gray matter of the upper thoracic spinal cord, specifically within segments T1 to T4. This configuration establishes an intricate relay system that facilitates exact regulation of laryngeal functions (Fitzgerald et al. 2012).

Sympathetic fibers originate in the thoracic spinal cord and begin on a highly coordinated anatomical pathway. They initially arise from the ventral root and combine with the spinal nerve. Thereafter, these fibers exit from the spinal nerve through the white ramus communicans, entering the sympathetic chain. The fibers ascend through the cervical sympathetic trunk before reaching their synaptic terminals in the superior cervical ganglion.

This neural organization supports multiple aspects of laryngeal function. The sympathetic system regulates blood flow through vasoconstriction of laryngeal vessels and controls secretory activity in laryngeal glands. Additionally, it modulates smooth muscle tone in laryngeal structures and contributes to protective reflexes. These functions collectively maintain optimal laryngeal performance and protect airways during various physiological states (Moore et al. 2018).

12.3.3.2 Parasympathetic Innervation of the Larynx

The parasympathetic nervous system exerts precise control over laryngeal secretory functions through a complex neural network organization. The structural arrangement positions postganglionic neuronal cell bodies within intralaryngeal ganglia, enabling these cholinergic neurons to provide direct regulation of vascular tone and glandular secretion (Kandel et al. 2013; Sanders et al. 1993).

Preganglionic neurons originate in the dorsal motor nucleus of the vagus nerve and project their axons to the larynx via vagal pathways. These pathways include both the superior and inferior (recurrent) laryngeal nerve branches, which serve as essential conduits for parasympathetic innervation. This anatomical arrangement allows for fine-tuned control of secretory functions at multiple levels within the laryngeal mucosa (Ludlow 2015; Mu and Sanders 2010).

The neurotransmitter acetylcholine (ACh) plays a fundamental role in signal transmission across several neural systems within the larynx. ACh mediates neurotransmission at multiple sites, including motor neuron synapses, parasympathetic pre- and postganglionic junctions, and specific sympathetic pathways. This neurotransmitter system ensures coordinated control of secretory responses essential for maintaining optimal laryngeal function and airway protection (Hillel 2001; Noordzij and Ossoff 2006).

12.4 Mechanisms for Airway Protection and Safe Swallowing

The coordination between respiratory function and swallowing requires complex neuroanatomical control mechanisms to maintain airway protection, particularly during the oropharyngeal phase of swallowing. These essential physiological processes share common anatomical pathways through the oropharyngeal structures, with integrated control mechanisms originating in the dorsal and ventral regions of the medullary centers within the brainstem. The neurological integration manifests through interneuronal networks connecting the nucleus tractus solitarius, nucleus ambiguus, and retrofacial nucleus, resulting in the temporary suspension of respiratory activity during swallowing. Respiratory pathologies frequently lead to compromised swallowing safety due to disruption of these neural pathways (Jean 2001; Leopold and Kagel 1997; Matsuo and Palmer 2009).

In individuals with intact neurological function, swallow apnea initiates prior to the oral phase of swallowing. While masticatory movements modify normal respiratory patterns through proprioceptive feedback mechanisms, complete respiratory cessation initiates when the bolus reaches the vallecular region, following a brief expiratory phase. This respiratory pause is mediated by inhibitory signals from the swallowing pattern generator to the ventral respiratory group neurons. After the bolus transitions through the pharyngoesophageal segment, the hyolaryngeal complex returns to its anatomical resting position as breathing resumes with expiration. This biomechanical sequence generates subglottic pressure, resulting in an audible expiratory burst detectable via auscultation at the level of the cricoid cartilage, serving as a protective mechanism against aspiration (Ertekin and Aydogdu 2003; Hind et al. 2001; Logemann 1983).

The temporal characteristics of swallow apnea demonstrate physiological variations ranging from 0.75 to 1.25 s, influenced by factors such as age, bolus volume, and viscosity. Larger boluses generally require extended apneic periods due to increased sensorimotor processing demands. Research involving post-laryngectomy patients has revealed that respiratory cessation during swallowing may operate independently of vocal fold movements through separate neural pathways, as these individuals demonstrate comparable apneic durations to those with intact laryngeal structures (Martin-Harris et al. 2005a, b).

The protective mechanism of the larynx involves three distinct sphincteric levels working in synchronized coordination. The superior level comprises the aryepiglottic folds at the laryngeal inlet, controlled by the aryepiglottic and oblique interarytenoid muscles innervated by the recurrent laryngeal nerve. The intermediate level consists of the ventricular folds (false vocal cords) and vestibular walls, regulated by thyroepiglottic muscles. The inferior level involves the true vocal cords, managed by the lateral cricoarytenoid and transverse interarytenoid muscles. The suprahyoid and infrahyoid muscle groups, including the thyrohyoid, stylohyoid, stylopharyngeus, digastric, and mylohyoid muscles, work synergistically to elevate and advance the hyolaryngeal complex toward the retracted tongue base (Mu and Sanders 2010; Pearson et al. 2012; Standring 2020).

The epiglottis follows a specific biomechanical sequence during deglutition, coordinated through sensory feedback from the internal branch of the superior laryngeal nerve. As the bolus head approaches, it initially tilts posteriorly against the posterior pharyngeal wall due to tongue base retraction. The subsequent hyolaryngeal elevation and anterior movement, facilitated by suprahyoid muscle contraction and regulated by the glossopharyngeal nerve (CN IX) and vagus nerve (CN X), position the epiglottis vertically. This movement directs the bolus around the laryngeal vestibule through the piriform sinuses. Following bolus passage, the epiglottis covers the laryngeal inlet through both passive mechanical displacement and active neuromuscular mechanisms, preventing aspiration of residual material during airway reopening (Matsuo and Palmer 2009; Miller 2008).

The passage of the bolus through the pharyngoesophageal junction occurs following cricopharyngeal muscle relaxation, mediated by vagal inhibitory signals through the myenteric plexus. While fluid boluses may transit passively under tongue-generated momentum, semi-solid and solid materials require coordinated peristaltic action through sequential contraction of the superior, middle, and inferior pharyngeal constrictor muscles. The esophageal transit process typically spans approximately 15 s, involving primary peristaltic waves initiated by the swallowing center and secondary peristaltic waves triggered by local stretch receptors. This process culminates in coordinated relaxation of the lower esophageal sphincter through vagovagal reflexes. The neuromuscular mechanism maintains its efficiency regardless of gravitational orientation, demonstrating the autonomic control of esophageal motility through the enteric nervous system (Goyal and Chaudhury 2008; Kahrilas et al. 2017).

References

Blitzer A, Brin MF, Ramig LO. Neurologic disorders of the larynx. 2nd ed. New York: Thieme; 2009.

Boeckxstaens GE, Zaninotto G. Achalasia and esophagogastric junction outflow obstruction: focus on the subtypes. Neurogastroenterol Motil. 2012;24(Suppl 1):27–31.

Bradley RM, Stedman HM, Mistretta CM. Superior laryngeal nerve response patterns to chemical stimulation of sheep epiglottis. Brain Res. 1983;276(1):81–3.

Cernea CR, Hojaij FC, De Carlucci D, Gotoda R Jr, Plopper C, Vanderlei F, et al. Recurrent laryngeal nerve: a plexus rather than a nerve? Arch Otolaryngol Head Neck Surg. 2009;135(11):1098–102.

Cook IJ, Kahrilas PJ. AGA technical review on management of oropharyngeal dysphagia. Gastroenterology. 1999;116(2):455–78.

Cook IJ, Dodds WJ, Dantas RO, Massey B, Kern MK, Lang IM, et al. Opening mechanisms of the human upper esophageal sphincter. Am J Phys. 1989a;257(5 Pt 1):G748–59.

Cook IJ, Dodds WJ, Dantas RO, Kern MK, Massey BT, Shaker R, et al. Timing of videofluoroscopic, manometric events, and bolus transit during the oral and pharyngeal phases of swallowing. Dysphagia. 1989b;4(1):8–15.

Costa MMB. Neural control of swallowing. Arq Gastroenterol. 2018;55(Suppl 1):61–75.

Dodds WJ. Physiology of swallowing. Dysphagia. 1989;3(4):171–8.

Dodds WJ, Stewart ET, Logemann JA. Physiology and radiology of the normal oral and pharyngeal phases of swallowing. AJR Am J Roentgenol. 1990;154(5):953–63.

Drake RL, Vogl AW, Mitchell AWM. Gray's anatomy for students. 4th ed. Philadelphia: Elsevier; 2020.

Dua KS, Ren J, Bardan E, Xie P, Shaker R. Coordination of deglutitive glottal function and pharyngeal bolus transit during normal eating. Gastroenterology. 1997;112(1):73–83.

Ekberg O, Lindström C. The upper esophageal sphincter area. Acta Radiol. 1987;28(2):173–6.

Ellis H, Mahadevan V. Clinical anatomy: applied anatomy for students and junior doctors. 14th ed. London: Wiley-Blackwell; 2019.

Ertekin C, Aydogdu I. Neurophysiology of swallowing. Clin Neurophysiol. 2003;114(12):2226–44.

Fitzgerald MJT, Gruener G, Mtui E. Clinical neuroanatomy and neuroscience. 6th ed. Philadelphia: Saunders; 2012.

Friedrich G, Lichtenberger G. Surgical anatomy of the larynx. J Voice. 1997;11(3):345–55.

German RZ, Crompton AW, Thexton AJ. Integration of the reflex pharyngeal swallow into rhythmic oral activity in a neurologically intact pig model. J Neurophysiol. 2009;102(2):1017–25.

Goyal RK, Chaudhury A. Physiology of normal esophageal motility. J Clin Gastroenterol. 2008;42(5):610–9.

Hamdy S, Rothwell JC, Brooks DJ, Bailey D, Aziz Q, Thompson DG. Identification of the cerebral loci processing human swallowing with H2(15)O PET activation. J Neurophysiol. 1999;81(4):1917–26.

Hennessy M, Goldenberg D. Surgical anatomy and physiology of swallowing. Oper Tech Otolaryngol. 2016;27:60–6.

Hiiemae KM, Palmer JB. Tongue movements in feeding and speech. Crit Rev Oral Biol Med. 2003;14(6):413–29.

Hillel AD. The study of laryngeal muscle activity in normal human subjects and in patients with laryngeal dystonia using multiple fine-wire electromyography. Laryngoscope. 2001;111(4 Pt 2 Suppl 97):1–47.

Hind JA, Nicosia MA, Roecker EB, Carnes ML, Robbins J. Comparison of effortful and noneffortful swallows in healthy middle-aged and older adults. Arch Phys Med Rehabil. 2001;82(12):1661–5.

Jean A. Brain stem control of swallowing: neuronal network and cellular mechanisms. Physiol Rev. 2001;81(2):929–69.

Kahrilas PJ, Dodds WJ, Dent J, Logemann JA, Shaker R. Upper esophageal sphincter function during deglutition. Gastroenterology. 1988;95(1):52–62.

Kahrilas PJ, Logemann JA, Krugler C, Flanagan E. Volitional augmentation of upper esophageal sphincter opening during swallowing. Am J Phys. 1991;260(3 Pt 1):G450–6.

Kahrilas PJ, Logemann JA, Lin S, Ergun GA. Pharyngeal clearance during swallowing: a combined manometric and videofluoroscopic study. Gastroenterology. 1992;103(1):128–36.

Kahrilas PJ, Lin S, Logemann JA, Ergun GA, Facchini F. Deglutitive tongue action: volume accommodation and bolus propulsion. Gastroenterology. 1993;104(1):152–62.

Kahrilas PJ, Bredenoord AJ, Fox M, Gyawali CP, Roman S, Smout AJ, et al. The Chicago Classification of esophageal motility disorders, v3.0. Neurogastroenterol Motil. 2015;27(2):160–74.

Kahrilas PJ, Bredenoord AJ, Fox M, Gyawali CP, Roman S, Smout AJPM, et al. Expert consensus document: advances in the management of oesophageal motility disorders in the era of high-resolution manometry: a focus on achalasia syndromes. Nat Rev Gastroenterol Hepatol. 2017;14(11):677–88.
Kandel ER, Schwartz JH, Jessell TM, Siegelbaum SA, Hudspeth AJ. Principles of neural science. 5th ed. New York: McGraw-Hill; 2013.
Kiernan JA, Rajakumar N. Barr's the human nervous system: an anatomical viewpoint. 10th ed. Philadelphia: Wolters Kluwer; 2014.
Kim Y, McCullough GH. Maximum hyoid displacement in normal swallowing. Dysphagia. 2008;23(3):274–9.
Kirchner JA. The motor activity of the cricopharyngeus muscle. Laryngoscope. 1958;68(7):1119–59.
Kirshblum S, Johnston MV, Brown J, O'Connor KC, Jarosz P. Predictors of dysphagia after spinal cord injury. Arch Phys Med Rehabil. 1999;80(9):1101–5.
Knowles MR, Daniels LA, Davis SD, Zariwala MA, Leigh MW. Primary ciliary dyskinesia: recent advances in diagnostics, genetics, and characterization of clinical disease. Am J Respir Crit Care Med. 2013;188(8):913–22.
Kochis-Jennings KA, Finnegan EM, Hoffman HT, Jaiswal S. Laryngeal muscle activity and vocal fold adduction during chest, chestmix, headmix, and head registers in females. J Voice. 2012;26(2):182–93.
Lalwani AK. Current diagnosis & treatment in otolaryngology. 4th ed. New York: McGraw Hill; 2020.
Lang IM, Medda BK, Shaker R. Mechanisms of reflexes induced by esophageal distension. Am J Physiol Gastrointest Liver Physiol. 2001;281(5):G1246–63.
Lang IM, Medda BK, Shaker R. Differential activation of medullary vagal nuclei caused by stimulation of different esophageal mechanoreceptors. Brain Res. 2011;1368:119–33.
Leopold NA, Daniels SK. Supranuclear control of swallowing. Dysphagia. 2010;25(3):250–7.
Leopold NA, Kagel MC. Pharyngo-esophageal dysphagia in Parkinson's disease. Dysphagia. 1997;12(1):11–8.
Logemann JA. Evaluation and treatment of swallowing disorders. San Diego: College-Hill Press; 1983.
Logemann JA. Swallowing physiology and pathophysiology. Otolaryngol Clin N Am. 1988;21(4):613–23.
Logemann JA, Kahrilas PJ, Cheng J, Pauloski BR, Gibbons PJ, Rademaker AW, et al. Closure mechanisms of laryngeal vestibule during swallow. Am J Phys. 1992;262(2 Pt 1):G338–44.
Logemann JA, Pauloski BR, Colangelo L, Lazarus C, Fujiu M, Kahrilas PJ. Effects of a sour bolus on oropharyngeal swallowing measures in patients with neurogenic dysphagia. J Speech Hear Res. 1995;38(3):556–63.
Logemann JA, Rademaker AW, Pauloski BR, Ohmae Y, Kahrilas PJ. Normal swallowing physiology as viewed by videofluoroscopy and videoendoscopy. Folia Phoniatr Logop. 1998;50(6):311–9.
Logemann JA, Pauloski BR, Rademaker AW, Colangelo LA, Kahrilas PJ, Smith CH. Temporal and biomechanical characteristics of oropharyngeal swallow in younger and older men. J Speech Lang Hear Res. 2000;43(5):1264–74.
Ludlow CL. Central nervous system control of the laryngeal muscles in humans. Respir Physiol Neurobiol. 2005;147(2–3):205–22.
Ludlow CL. Central nervous system control of voice and swallowing. J Clin Neurophysiol. 2015;32(4):294–303.
Lund JP, Kolta A. Generation of the central masticatory pattern and its modification by sensory feedback. Dysphagia. 2006;21(3):167–74.
Mackay LE, Morgan AS, Bernstein BA. Swallowing disorders in severe brain injury: risk factors affecting return to oral intake. Arch Phys Med Rehabil. 1999;80(4):365–71.
Martin-Harris B, Michel Y, Castell DO. Physiologic model of oropharyngeal swallowing revisited. Otolaryngol Head Neck Surg. 2005a;133(2):234–40.
Martin-Harris B, Brodsky MB, Michel Y, Ford CL, Walters B, Heffner J. Breathing and swallowing dynamics across the adult lifespan. Arch Otolaryngol Head Neck Surg. 2005b;131(9):762–70.
Matsuo K, Palmer JB. Anatomy and physiology of feeding and swallowing: normal and abnormal. Phys Med Rehabil Clin N Am. 2008;19(4):691–707.
Matsuo K, Palmer JB. Coordination of mastication, swallowing and breathing. Jpn Dent Sci Rev. 2009;45(1):31–40.
Miller AJ. Deglutition. Physiol Rev. 1982;62(1):129–84.
Miller AJ. The neurobiology of swallowing and dysphagia. Dev Disabil Res Rev. 2008;14(2):77–86.
Miller LS, Vegesna AK, Brasseur JG, Braverman AS, Ruggieri MR. The esophagogastric junction. Ann N Y Acad Sci. 2011;1232:323–30.
Mistry S, Hamdy S. Neural control of feeding and swallowing. Phys Med Rehabil Clin N Am. 2008;19(4):709–28.
Mittal RK, Balaban DH. The esophagogastric junction. N Engl J Med. 1997;336(13):924–32.
Mittal RK, Hong SJ, Bhargava V. Longitudinal muscle dysfunction in achalasia esophagus and its relevance. J Neurogastroenterol Motil. 2013;19(2):126–36.
Moore KL, Dalley AF, Agur AMR. Clinically oriented anatomy. 8th ed. Philadelphia: Wolters Kluwer; 2018.
Mtui E, Gruener G, Dockery P. Fitzgerald's clinical neuroanatomy and neuroscience. 7th ed. Philadelphia: Elsevier; 2016.
Mu L, Sanders I. Anatomy of the human internal superior laryngeal nerve. Anat Rec. 1998;252(4):646–56.
Mu L, Sanders I. Human tongue neuroanatomy: nerve supply and motor endplates. Clin Anat. 2010;23(7):777–91.
Murry T, Carrau RL, Chan K. Clinical management of swallowing disorders. 4th ed. San Diego: Plural Publishing; 2018.
Nagai H, Ota F, Connor NP. Effect of deficits in laryngeal sensation on laryngeal muscle biochemistry. Ann Otol Rhinol Laryngol. 2005;114(5):352–60.
Netter FH. Atlas of human anatomy. 7th ed. Philadelphia: Elsevier; 2019.

Noordzij JP, Ossoff RH. Anatomy and physiology of the larynx. Otolaryngol Clin N Am. 2006;39(1):1–10.
Norton NS. Netter's head and neck anatomy for dentistry. 3rd ed. Philadelphia: Elsevier; 2017.
Orlando RC, Powell DW, Carney CN. Pathophysiology of acute acid injury in rabbit esophageal epithelium. J Clin Invest. 1981;68(1):286–93.
Palmer JB, Kuhlemeier KV, Tippett DC, Lynch C. A protocol for the videofluorographic swallowing study. Dysphagia. 1993;8(3):209–14.
Palmer JB, Drennan JC, Baba M. Evaluation and treatment of swallowing impairments. Am Fam Physician. 2000;61(8):2453–62.
Park H, Conklin JL. Neuromuscular control of esophageal peristalsis. Curr Gastroenterol Rep. 1999;1(3):186–97.
Pauloski BR, Logemann JA. Impact of tongue base and posterior pharyngeal wall biomechanics on pharyngeal clearance in irradiated postsurgical oral and oropharyngeal cancer patients. Head Neck. 2000;22(2):120–31.
Pawlina W. Histology: a text and atlas. 7th ed. Philadelphia: Wolters Kluwer; 2016.
Pearson WG Jr, Langmore SE, Yu LB, Zumwalt AC. Structural analysis of muscles elevating the hyolaryngeal complex. Dysphagia. 2012;27(4):445–51.
Richter JE, Rubenstein JH. Presentation and epidemiology of gastroesophageal reflux disease. Gastroenterology. 2018;154(2):267–76.
Robbins J, Hamilton JW, Lof GL, Kempster GB. Oropharyngeal swallowing in normal adults of different ages. Gastroenterology. 1992;103(3):823–9.
Robbins J, Butler SG, Daniels SK, Gross RD, Langmore S, Lazarus CL, et al. Swallowing and dysphagia rehabilitation: translating principles of neural plasticity into clinically oriented evidence. J Speech Lang Hear Res. 2008;51(1):S276–300.
Sanders I, Wu BL, Mu L, Li Y, Biller HF. The innervation of the human larynx. Arch Otolaryngol Head Neck Surg. 1993;119(9):934–9.
Sasaki CT, Isaacson G. Functional anatomy of the larynx. Otolaryngol Clin N Am. 1988;21(4):595–612.
Sasaki CT, Weaver EM. Physiology of the larynx. Am J Med. 1997;103(5A):9S–18S.
Sasaki CT, Levine PA, Laitman JT, Crelin ES Jr. Postnatal descent of the epiglottis in man: a preliminary report. Arch Otolaryngol. 1977;103(3):169–71.
Sasaki CT, Yu Z, Xu J, Hundal J, Rosenblatt W. Effects of altered consciousness on the protective glottic closure reflex. Ann Otol Rhinol Laryngol. 2006;115(10):759–63.
Sataloff RT, Heman-Ackah YD, Hawkshaw MJ. Clinical anatomy and physiology of the voice. Otolaryngol Clin N Am. 2007;40(5):909–29.
Schuenke M, Schulte E, Schumacher U. THIEME atlas of anatomy: head, neck, and neuroanatomy. New York: Thieme; 2016.
Sharma P, Bergman JJ, Goda K, Kato M, Messmann H, Alsop BR, et al. Development and validation of a classification system to identify high-grade dysplasia and esophageal adenocarcinoma in Barrett's esophagus using narrow-band imaging. Gastroenterology. 2016;150(3):591–8.
Ship JA, Pillemer SR, Baum BJ. Xerostomia and the geriatric patient. J Am Geriatr Soc. 2002;50(3):535–43.
Singh V. Textbook of anatomy: upper limb and thorax. 2nd ed. New Delhi: Elsevier; 2014.
Sinnatamby CS. Last's anatomy: regional and applied. 12th ed. London: Churchill Livingstone; 2011.
Sivarao DV, Goyal RK. Functional anatomy and physiology of the upper esophageal sphincter. Am J Med. 2000;108(Suppl 4a):27S–37S.
Smith RV, Kotz T, Beitler JJ, Wadler S. Long-term swallowing problems after organ preservation therapy with concomitant radiation therapy and intravenous hydroxyurea. Arch Otolaryngol Head Neck Surg. 2000;126(3):384–9.
Spechler SJ, Souza RF. Barrett's esophagus. N Engl J Med. 2014;371(9):836–45.
Standring S. Gray's anatomy: the anatomical basis of clinical practice. 42nd ed. London: Elsevier; 2020.
Steele CM, Miller AJ. Sensory input pathways and mechanisms in swallowing: a review. Dysphagia. 2010;25(4):323–33.
Steele CM, Bailey GL, Chau T, Molfenter SM, Oshalla M, Waito AA, et al. The relationship between hyoid and laryngeal displacement and swallowing impairment. Clin Otolaryngol. 2011;36(1):30–6.
Strollo PJ Jr, Soose RJ, Maurer JT, de Vries N, Cornelius J, Froymovich O, et al. Upper-airway stimulation for obstructive sleep apnea. N Engl J Med. 2014;370(2):139–49.
Thankappan K, Iyer S, Menon JR. Dysphagia management in head and neck cancers: a manual and atlas. Singapore: Springer; 2018.
Tian W, Yu J. The role of C2-C7 and O-C2 angle in the development of dysphagia after cervical spine surgery. Dysphagia. 2013;28(2):131–8.
Von Leden H, Moore P. The mechanics of the cricoarytenoid joint. Arch Otolaryngol. 1961;73:541–50.
Welcome MO. Gastrointestinal physiology: development, principles and mechanisms of regulation. Cham: Springer; 2018.
Wolf C, Meiners TH. Dysphagia in patients with acute cervical spinal cord injury. Spinal Cord. 2003;41(6):347–53.
Wong DJ, Fitzgerald RC. Epidemiologic risk factors for Barrett's esophagus and associated adenocarcinoma. Clin Gastroenterol Hepatol. 2005;3(1):1–10.
Young B, Woodford P, O'Dowd G. Wheater's functional histology: a text and colour atlas. 6th ed. Philadelphia: Elsevier; 2014.
Zhang Z. Mechanics of human voice production and control. J Acoust Soc Am. 2016;140(4):2614–35.

Recommended Additional Reading

Ekberg O. Dysphagia: diagnosis and treatment. 2nd ed. Cham: Springer; 2019.

Fried MP, Tan M. Clinical laryngology: the essentials. New York: Thieme; 2015.

Groher ME, Crary MA. Dysphagia: clinical management in adults and children. 2nd ed. St. Louis: Elsevier; 2016.

Hisa Y. Neuroanatomy and neurophysiology of the larynx. Tokyo: Springer; 2016.

Jones B. Normal and abnormal swallowing: imaging in diagnosis and therapy. 2nd ed. New York: Springer; 2003.

Mankekar G. Swallowing—physiology, disorders, diagnosis and therapy. New Delhi: Springer; 2015.

Randolph GW. The recurrent and superior laryngeal nerves. Cham: Springer; 2016.

Saitoh E, Pongpipatpaiboon K, Inamato Y, Kagaya H. Dysphagia evaluation and treatment: from the perspective of rehabilitation medicine. Singapore: Springer; 2018.

Neuroanatomical Basis of Spasticity in Upper Motor Neuron Lesion

13

In rehabilitation medicine and spinal cord injury management, spasticity presents a substantial clinical issue that impacts patients' daily activities and participation in rehabilitation programs. The definition, understanding, and related anatomy and neuroanatomy of spasticity have evolved over time, reflecting progress in neuroscience and clinical research.

Lance first defined spasticity in 1980 as a velocity-dependent increase in tonic stretch reflex with hyperreactive tendon jerks, resulting from stretch reflex hyperexcitability (Lance 1980a, b). This initial physiological definition has since broadened. Pandyan and colleagues later characterized spasticity as "a sensorimotor control disorder resulting from upper motor neuron (UMN) lesions, appearing as intermittent or sustained involuntary muscle activation" (Pandyan et al. 2005). This expanded definition recognizes additional upper motor neuron syndrome manifestations, including flexor and extensor spasms, hyperreflexia, hypertonia, clonus, abnormal antagonistic muscle co-contraction, and pathological reflexes like the Babinski sign (Ko 2022a).

From a neuroanatomical perspective, spasticity involves the upper motor neuron system's neural pathways above the anterior horn cell, incorporating the spinal cord, brainstem, and cerebral structures. Clinical features include elevated muscle tone and reactivity, reduced precision in voluntary muscle control, and abnormal motor outputs, which vary according to temporal factors and movement patterns (Adams and Hicks 2005).

At the physiological level, spasticity represents an exaggeration of the myotatic reflex associated with spinal hyperexcitability, resulting from decreased inhibitory control from supraspinal descending pathways (Sheean 2002) (Fig. 13.1). Spasticity represents one component of widespread motor system dysfunction, as illustrated in Fig. 13.2.

Spinal plasticity following injury occurs through two main mechanisms:

- Axonal growth (collateral sprouting) from intact nerve fibers toward vacant synaptic sites
- Neuronal receptor hypersensitivity compensating for presynaptic afferent loss (Barnes and Johnson 2008a)

Clinically, spasticity emerges after spinal shock resolution, displaying velocity-dependent resistance during passive joint movement. Faster stretches lead to stronger antagonistic contractions due to enhanced Ia afferent activation. A characteristic "clasp-knife" phenomenon occurs when joint resistance initially increases, then abruptly decreases, attributed to an enhanced withdrawal reflex and specific segmental interneuron firing patterns (Cleland and Rymer 1990).

The dynamic characteristics of spasticity result in temporal variations and susceptibility to

H.-Y. Ko, *Practical Functional Anatomy for Spinal Cord Injury Rehabilitation*,
https://doi.org/10.1007/978-981-95-6457-6_13

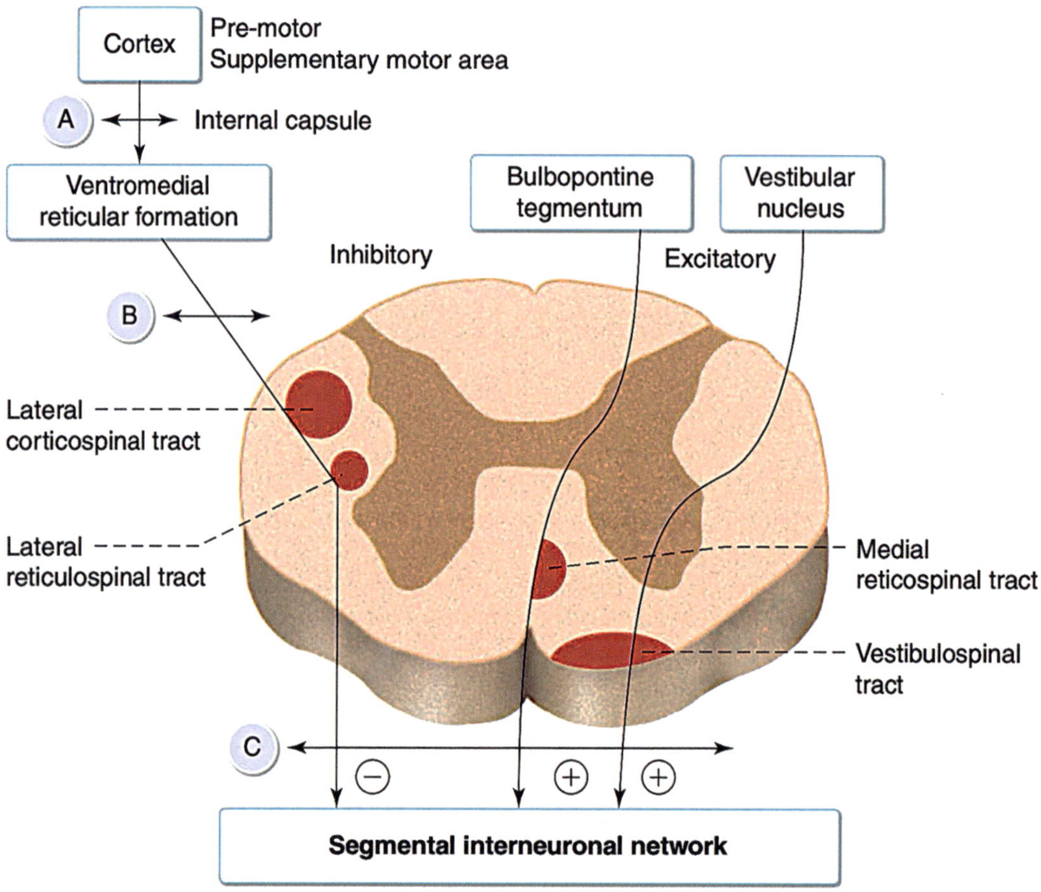

Fig. 13.1 Major descending systems exerting supraspinal control. The anatomical relations and the differences with respect to cortical control between the inhibitory and excitatory descending systems indicate that anatomical location of the upper motor neuron lesion plays a significant role in the determination of the resulting clinical pattern. A: Lesion affecting the corticospinal tract and the corticoreticular tract, which are facilitating structures of the main inhibitory system, namely, the lateral reticulospinal tract. B: Incomplete spinal cord lesion affecting the corticospinal fibers and the lateral reticulospinal tract. C: Complete spinal cord lesion affecting the corticospinal fibers, together with the lateral reticulospinal fibers, and the excitatory pathways, namely, the medial reticulospinal tract and the vestibulospinal tract. Plus symbol indicates excitatory pathway; minus symbol indicates inhibitory pathway. (Adapted with permission from Sindou et al. (2014))

factors beyond the initial spinal cord or brain injury. Its prevalence and severity relate to injury level, severity, pattern, and post-injury duration. Notable is the non-linear relationship between spasticity and spinal cord or brain injury severity (Adams et al. 2007). For instance, central cord syndrome at upper cervical levels may produce more significant upper extremity spasticity relative to lower extremities, while lower cervical lesions may exhibit an inverse pattern.

Upper motor neuron syndrome presents with positive and negative features that differentially impact functional ability. Positive features encompass enhanced tendon reflexes, reflex spread, Babinski sign, and clonus. While valuable for diagnosis, most positive features contribute minimally to disability (Burke et al. 2013). Clonus emerges as a notable exception among positive features. This rhythmic muscle contraction can significantly impair function, occurring during activities of daily living, such as traversing curbs, or occurring spontaneously at rest. The impact of clonus often necessitates specific therapeutic interventions beyond general spasticity

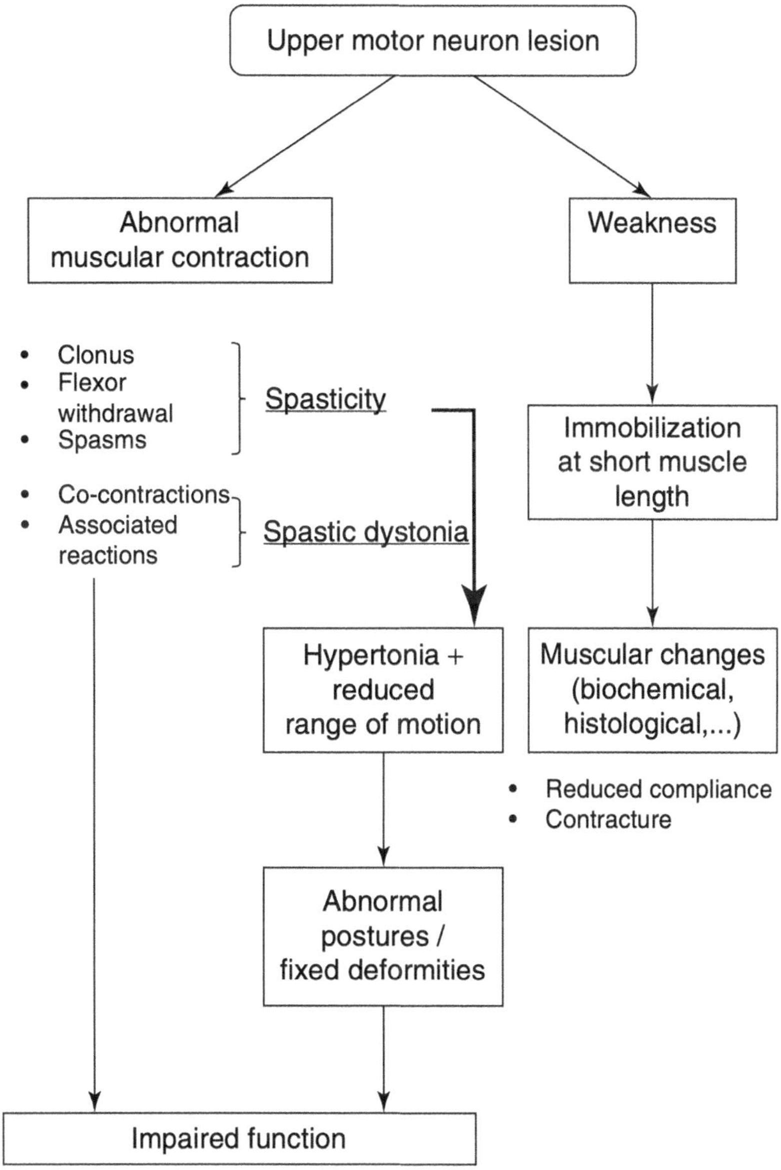

Fig. 13.2 Schematic representation of pathophysiology leading to impaired function in the upper motor neuron lesion. (Adapted with permission from Sindou et al. (2014))

management (Dietz and Sinkjaer 2007; Gracies 2005). Negative features manifest through reduced motor activation, including muscle weakness, impaired dexterity, and increased fatiguability. These symptoms typically create more substantial functional limitations than positive features and demonstrate greater resistance to rehabilitation interventions (Nielsen et al. 2007).

The upper motor neuron lesion presents with several clinical characteristics, with spasticity as a key element (Lance 1980a). This chapter focuses on the positive aspects of UMN syndrome, including spasticity and its related anatomy, as outlined in Table 13.1 (Sheean 2002). The positive manifestations of UMN syndrome primarily result from enhanced spinal reflexes, which operate under supraspinal regulation and receive additional modulation from segmental inputs (Nielsen et al. 2007). These spinal reflexes can be evaluated and quantified through electrophysiological methods, providing

Table 13.1 Clinical features of movement dysfunction in the upper motor neuron syndrome

Type	Neurological tonicity	Clinical signs
Positive symptoms	Spasticity	Increased muscle tone Exaggerated tendon reflexes (phasic stretch reflex) Stretch reflex spread to extensors Repetitive stretch reflex discharges (clonus) Babinski sign
	Released flexor reflexes	Massive synergy patterns
	Extensor spasms	Massive synergy patterns
	Dyssynergic patterns	Co-contraction during movement
Negative symptoms	Loss of dexterity	
	Weakness	Inadequate force generation Slow movement Fatigue
	Loss of selective control of muscles and limb segments	Loss of dexterity
Rheologic changes in spastic muscle	Stiffness Contracture Fibrosis Atrophy	

objective measurements of neural circuit function (Trompetto et al. 2014).

Understanding spasticity necessitates the analysis of multiple key components. The descending motor pathways, involving upper motor neurons, form the anatomical basis that, when impaired, results in upper motor neuron syndrome (Gracies 2005). Secondly, the spinal reflexes produce the evident clinical signs and symptoms. This includes the evaluation of non-reflexive components and biomechanical characteristics that affect clinical manifestation (Burke et al. 2013). Ultimately, those basic spinal mechanisms that account for increased reflex responses necessitate thorough examination to fully understand the pathophysiological context (Dietz and Sinkjaer 2007).

13.1 Anatomical Components and Pathophysiological Mechanisms in Spasticity

The spinal motor neuron receives continuous modulatory input from three primary neuronal sources: peripheral afferent fibers providing sensory feedback, local circuit interneurons mediating complex reflex patterns, and descending projections from supraspinal structures enabling voluntary motor control. Spasticity manifests as pathological enhancement of the monosynaptic stretch reflex, as illustrated in Fig. 13.3. Various central nervous system lesions can precipitate spasticity, particularly when they disrupt supraspinal descending control systems. The reticulospinal tracts appear most critically involved in this process, as their disruption significantly alters motor neuron excitability patterns (Young and Wiegner 1987).

The clinical manifestations of spasticity demonstrate considerable heterogeneity across multiple domains. The distribution patterns vary significantly among different muscle groups within individual limbs, often following characteristic anatomical patterns related to the site and extent of the causative lesion. Patients exhibit diverse responses to physiological stimuli, which may be exteroceptive (cutaneous), proprioceptive (muscle stretch), or interoceptive (visceral) in nature. The temporal progression also varies considerably regarding symptom onset following motor pathway lesions, ranging from immediate development to delayed manifestation over weeks to months. Furthermore, the condition shows marked diversity in its underlying etiology and the anatomical locations of precipitating lesions within the central nervous system (Mukherjee and Chakravarty 2010).

To investigate this clinical variability, researchers have developed multiple experimental models. Sherrington's classical decerebration model involves brainstem transection at the intercollicular level, resulting in immediate manifestation of decerebrate rigidity. Additional experimental approaches include complete spinal cord transection, which eliminates all supraspinal control, and selective ablation of motor cortical

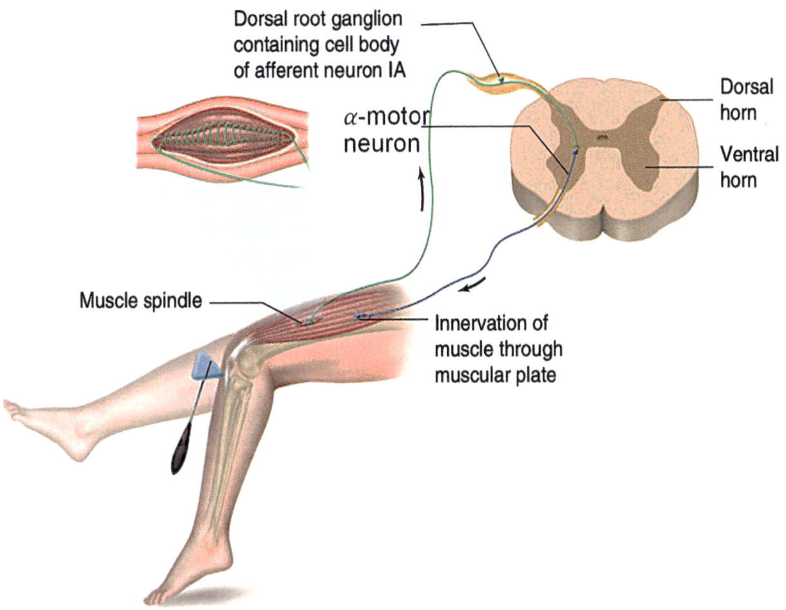

Fig. 13.3 Myotatic reflex. The myotatic reflex is a muscle's contraction in response to its stretching (for instance, when the patellar tendon is tapped with a tendon reflex hammer). This reflex provides automatic regulation of the muscle length. When a muscle is lengthened, its spindle, which is its sensory receptor, is stretched, increasing its nerve activity (IA fiber). The afferent volley in IA excites the α-motoneuron, causing the muscle to contract, thus resisting the stretching. (Adapted with permission from Sindou et al. (2014))

areas 4 and 6, which disrupts specific descending pathways. Spinal ischemia models have provided particular insight by producing selective interneuronal deficits that lead to motor neuronal hyperexcitability. These diverse experimental paradigms have significantly advanced our understanding of the pathophysiological mechanisms underlying spasticity (Dimitrijevic 1987; Dimitrijevic and Nathan 1967; Matthews 2004).

13.1.1 Components of Spasticity in the Spinal Cord

13.1.1.1 Alpha Motor Neurons in the Spinal Cord

Alpha motor neurons demonstrate a precisely ordered anatomical arrangement within lamina IX of the spinal cord ventral horn, exhibiting both segmental and longitudinal organization patterns. These lower motor neurons form discrete functional aggregations called motor nuclei, with neurons innervating the same muscle grouping into distinct anatomical assemblies termed motor neuron pools. These pools organize into longitudinal columns that typically extend across 2–4 spinal cord segments, with each column maintaining specific muscle innervation patterns.

The topographical arrangement of motor neuron pools follows consistent anatomical principles within the ventral horn gray matter. Neurons innervating proximal axial and girdle muscles occupy medial positions, while those supplying progressively more distal limb muscles are situated gradually laterally. This mediolateral organization reflects both developmental patterning and functional specialization of motor control (Romanes 1964). The dorsoventral distribution demonstrates additional organizational features, with flexor muscle motor neurons generally positioned dorsal to extensor motor neurons within each segment.

The somatotopic distribution of α-motoneurons demonstrates apparent precision, as illustrated in Fig. 13.4. This organization creates a three-dimensional map of muscle representation within the spinal cord, where specific locations correspond to particular muscle groups

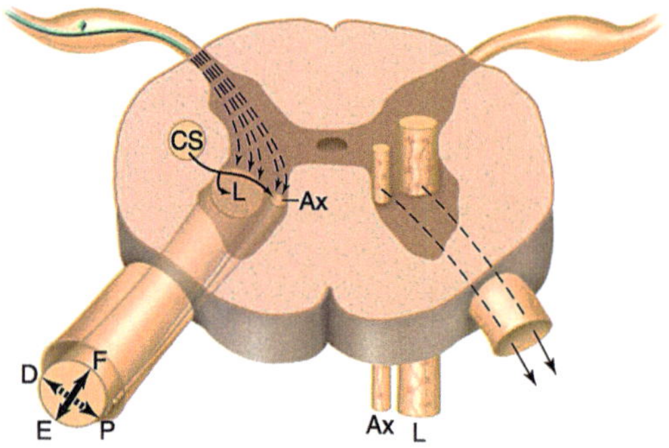

Fig. 13.4 Motor neurons in the ventral horn are organized somatotopically according to their destinations and functional roles into two longitudinal columns: a medial group (medial column), which innervates the axial musculature (neck and trunk), and a lateral group (lateral column), which innervates the extremities musculature (upper limb and lower limb). In this group, the motor neurons for extensor muscles are located more ventrally, whereas the motor neurons for flexor muscles are located more dorsally. Ax Axial musculature, L limbs, D for distal muscles of limbs, P for proximal muscles of limbs, F for flexor muscles, E for extensor muscles, CS corticospinal tract. (Adapted with permission from Sindou et al. (2014))

and movement patterns. The systematic relationship between motor neuron position and target muscle innervation allows for coordinated activation of muscle synergies during complex movements (Watson et al. 2009).

Alpha motor neurons function as essential integration sites for motor control, featuring extensive dendritic trees that receive approximately 50,000 synaptic inputs. Their cell bodies are among the largest in the central nervous system, measuring 30–70 μm in diameter. These neurons receive direct synaptic connections from descending motor pathways, including corticospinal, rubrospinal, and reticulospinal tracts, as well as extensive inputs from local circuit interneurons and proprioceptive afferents. The integration of these multiple input sources enables refined control of muscle activation patterns (Burke 2007). The intricate anatomical framework supports precise patterns of neuromuscular control through both direct and indirect pathways. Motor neurons receive monosynaptic excitation from Ia proprioceptive afferents and polysynaptic inputs from various interneuronal circuits, including Ia inhibitory interneurons, Renshaw cells, and reciprocal inhibitory pathways. This complex integration enables both voluntary movement control and automatic postural adjustments.

In spastic conditions, this precisely organized system becomes disrupted through multiple mechanisms. Changes in motor neuron excitability result from alterations in both descending control systems and local circuit function. These modifications affect voltage-gated ion channels, synaptic transmission properties, and intrinsic membrane characteristics of motor neurons. Additionally, changes in interneuronal circuits lead to reduced reciprocal inhibition and altered presynaptic inhibition of Ia afferents. These combined effects contribute to the characteristic features of spastic muscle activity, including velocity-dependent resistance to stretch and enhanced stretch reflexes (Heckman et al. 2009).

13.1.1.2 Primary Afferents and Neurotransmitters in Spinal Cord Circuitry

The spinal cord exhibits a highly organized laminar structure for primary afferent distribution, with direct input to all Rexed laminae except lamina VIII, which serves exclusively as an intercommissural exchange area. This organization demonstrates remarkable specificity in both anatomical arrangement and functional connectivity (Ralston 1979).

The spatial organization of primary afferents follows a precise pattern based on fiber diameter and function. Large-diameter fibers project to laminae III, IV, V, VI, and IX, while small-diameter fibers target both superficial laminae I and II and deeper laminae V and X. Among these, proprioceptive Ia fibers show a distinctive distribution pattern. While some directly synapse with lamina IX motor neurons to mediate monosynaptic reflexes, the majority terminate in lamina VI at the dorsal horn base. This region functions as a specialized "proprioceptive layer," where primary afferents interface with interneurons that initiate ascending pathways (Ralston 1979).

Human anatomical and physiological studies have revealed complex organization within the dorsal root entry zone. Fine fibers (Aδ and C), which primarily mediate nociception, cluster in the ventrolateral region before traversing the medial portion of Lissauer's tract to reach the dorsal horn's superficial layers. In contrast, large-diameter fibers (Aβ) occupy the dorsomedial region of the dorsal root entry zone and project to the dorsal column. These Aβ fibers carry epicritic tactile information and conscious proprioception to brainstem lemniscal pathways (Willis and Coggeshall 2004).

The neurotransmitter systems in these circuits show distinct patterns that correlate with fiber type and laminar distribution. Most primary afferents utilize excitatory amino acids, predominantly glutamate and aspartate. However, small-diameter afferents projecting to superficial laminae I and II employ inhibitory neurotransmitters, specifically GABA and glycine. The inhibitory circuitry shows further specialization: lamina I neurons receive exclusively glycinergic synaptic inhibition, while lamina II neurons display a mixed pattern. Approximately half of lamina II neurons receive pure GABAergic inhibition, while the remainder receive combined GABAergic and glycinergic inhibitory input (Todd 2010). This organization of neurotransmitter systems provides the basis for complex sensory processing and motor control (Todd and Spike 1993).

13.1.1.3 Neuroanatomical Organization of Spinal Interneurons

The spinal cord contains diverse populations of propriospinal interneurons that form complex neural circuits coordinating motor and sensory functions (Jankowska 2008). These interneurons are distributed throughout the gray matter and can be classified based on their axonal projections into short, intermediate, and long propriospinal neurons (Flynn et al. 2011).

The intermediate zone, particularly Rexed laminae VII and VIII, contains interneurons that regulate intersegmental spinal reflexes. Short propriospinal interneurons extend their axons through dorsolateral and ventrolateral tracts across 4–8 adjacent metameres, primarily forming synaptic connections with flexor motor neurons in the lateral portion of lamina IX. In comparison, intermediate propriospinal neurons project their axons through the ventrolateral tract to more distant metameres, predominantly targeting extensor motor neuron pools (Alstermark and Isa 2012).

Spinal interneurons integrate multiple inputs, including those from peripheral afferents and descending control systems, resulting in outputs that modify motor neuron activity through inhibitory or excitatory mechanisms (Jankowska 2001, 2008). The distribution of these interneurons follows a specific organizational pattern within the spinal gray matter, reflecting their specialized functions in motor control (Alvarez and Fyffe 2007).

Interneurons mediating Ib afferent-induced autogenic inhibition, which receive input from Golgi tendon organs, are concentrated in the lateral portions of laminae V and VI (Pierrot-Deseilligny and Burke 2005). These neurons contribute to force regulation during muscle contraction by providing negative feedback to homonymous and synergist motor neurons. Meanwhile, interneurons responsible for reciprocal Ia inhibition are primarily located in lamina VII, where they coordinate the activation of antagonist muscle pairs (Hultborn et al. 2004).

Renshaw cells represent a specialized class of inhibitory interneurons positioned in the ventromedial region of lamina VII. These cells form a recurrent inhibitory circuit with α-motoneurons, receiving excitatory input from motor axon collaterals and providing feedback inhibition to the same motor neuron pool and its synergists (Alvarez and Fyffe 2007). This arrangement creates a negative feedback loop that regulates motor output and contributes to the stability of motor patterns (Windhorst 2007).

The neurochemical properties of spinal interneurons vary according to their functional roles. Inhibitory interneurons primarily utilize glycine as their neurotransmitter, while others express GABA and enkephalins (Todd 2010). Renshaw cells, located in the ventral portion of lamina VII, primarily use glycine and occasionally GABA as neurotransmitters. These cells receive cholinergic input from motor axon collaterals and provide feedback inhibition to α- and γ-motoneurons, forming a regulatory circuit that is modulated by both peripheral and supraspinal inputs (Alvarez and Fyffe 2007).

GABAergic interneurons operate through multiple mechanisms, including presynaptic inhibition via axo-axonic synapses on primary afferents. These neurons utilize two main receptor types: GABA-A receptors, which are ionotropic and regulate chloride conductance, and GABA-B receptors, which are metabotropic and modulate potassium channels (Zeilhofer et al. 2012). Both receptor types produce membrane hyperpolarization, reducing motor neuron excitability. The activity of these inhibitory interneurons is regulated by primary afferents and descending monoaminergic pathways through α2-noradrenergic and 5-HT1 serotonergic receptors.

Enkephalinergic interneurons are concentrated in the superficial dorsal horn laminae and modulate nociceptive transmission through mu (μ), delta (δ), and kappa (κ) opioid receptors (Todd 2010).

13.1.1.4 Neuroanatomical Organization of Major Ascending Spinal Pathways

The spinocerebellar system comprises secondary neurons whose cell bodies are located in multiple regions, including Clarke's nucleus (nucleus dorsalis), Betcherew's nucleus within lamina VI, and additional neurons in laminae IV and IX (Matsushita and Hosoya 1989). This system forms two primary pathways that convey proprioceptive information to the cerebellum through the lateral spinal column.

The ventral spinocerebellar tract, also known as Gowers' tract, carries proprioceptive signals from the contralateral limbs. In contrast, the dorsal spinocerebellar tract, or Flechsig's tract, transmits ipsilateral proprioceptive information from the trunk musculature (Bosco and Poppele 2001). These pathways are fundamental for the cerebellum's role in motor coordination and postural control (Spanne and Jörntell 2013).

The ventral column contains the spinoreticular tract, which originates from neurons in the deeper layers of the spinal cord, specifically laminae VII, VIII, and X. This pathway maintains bilateral projections and extends to the reticular formation. While traditionally associated with nociceptive processing, the spinoreticular tract also significantly contributes to motor control through its connections with the reticular formation, which influences muscle tone and coordinated movement patterns.

13.1.2 Supraspinal Control of Motor Function: Upper Motor Neuron Systems

Contemporary understanding of the upper motor neuron system reveals a complex network of descending pathways that extends far beyond the traditional clinical focus on the corticospinal tract. This complex system modulates lower motor neuron activity through multiple pathways that coordinate voluntary movement, muscle tone, posture, reflexes, and autonomic functions. Among these descending systems, the corticospinal tract uniquely originates from the cerebral cortex, while other pathways emerge from the caudal brainstem segments (Lemon and Griffiths 2005).

The historical foundation of motor control research stems from extensive animal studies conducted during the late nineteenth and early twentieth centuries. Sherrington's work with the decerebrate cat model demonstrated immediate enhancement of extensor tone following precise lesioning between the superior and inferior colliculi. Nonetheless, the limitations of this paradigm in explaining human spasticity revealed the complexity of motor organization among species. Further studies involving primates, especially macaques and chimpanzees, yielded more therapeutically significant insights; yet, species-specific differences persisted as interpretative challenges (Pierrot-Deseilligny and Burke 2005).

Motor control originates from a precise cortical organization where corticospinal fibers emerge from both precentral and postcentral regions. The precentral component, comprising

60% of fibers, primarily originates from the primary motor cortex (Brodmann area 4, 40%) and premotor cortex (area 6, 20%), with each region serving distinct functions. Area 4 specializes in precise distal movements, while area 6 coordinates complex motor sequences and postural adjustments. The postcentral contribution, making up the remaining 40%, arises from the primary somatosensory cortex and posterior parietal cortex, primarily serving sensorimotor integration functions (Dum and Strick 2005).

Clinical observations from lesion studies have illuminated the intricate relationship between anatomical damage and motor dysfunction. Interestingly, isolated primary motor cortex lesions rarely cause spasticity but instead lead to reduced muscle tone and diminished reflexes. The development of spasticity typically requires involvement of both primary motor and premotor cortices, affecting pyramidal and parapyramidal pathways simultaneously. This understanding explains why subcortical lesions, particularly in the internal capsule where these pathways converge, more frequently result in spasticity compared to isolated cortical lesions (Nathan and Smith 1955).

Three primary descending motor pathways within the spinal cord interact to maintain motor control. The dorsal reticulospinal tract provides inhibitory control over spinal reflexes and muscle tone, while the medial reticulospinal tract facilitates motor activity and flexor reflexes. The vestibulospinal tract contributes primarily to extensor tone regulation and postural mechanisms. This intricate arrangement allows for precise control of movement and posture under varying conditions.

The development of spasticity reflects disruption of a finely balanced brainstem control system. Under normal conditions, the medullary reticular formation maintains precise control over spinal reflexes through the dorsal reticulospinal tract, while receiving modulatory input from cortical areas. Disruption of this inhibitory system, whether through cortical, subcortical, or spinal cord lesions, leads to unopposed excitatory influences from other pathways, ultimately resulting in spasticity (Burke et al. 2013).

Recent advances in our understanding of these systems have indicated the complexity of neurotransmitter interactions and molecular mechanisms that govern motor control. These insights continue to inform therapeutic approaches for managing motor disorders, ranging from pharmacological interventions to targeted physical therapies.

13.1.2.1 Corticospinal System

The corticospinal system represents a remarkably complex neural pathway that evolved relatively recently in mammalian evolution. In humans, this motor control system originates from extensive regions of the frontoparietal cortex, where layer V pyramidal neurons, particularly in Brodmann areas 4, 3a, 3b, 1, 2, 5, SII, and the supplementary motor area, give rise to corticospinal fibers (Lemon 2008).

The human corticospinal tract demonstrates impressive numerical complexity, containing over one million axons, of which approximately 700,000 are myelinated. The majority of these axons (80%) have diameters less than 2 μm, while the remainder range from 5 to 22 μm. A small but significant proportion (3.5%) exceeds 20 μm in diameter. The largest fibers typically originate from the giant pyramidal cells of Betz, measuring 10–20 μm in thickness and primarily located in the primary motor cortex (Porter and Lemon 1995).

The developmental trajectory of the corticospinal system parallels motor skill acquisition in human infants. Myelination begins in the internal capsule and peduncles between 10 and 14 days after birth and continues progressively through the second year of life, coinciding with the development of fine motor skills. This extended period of maturation reflects the system's role in complex, voluntary movement control (Eyre 2007).

As corticospinal fibers descend through the brain, they follow a precise anatomical course. After originating from layer V neurons, these fibers traverse the posterior limb of the internal capsule and continue through the cerebral peduncle in the midbrain before reaching the basilar pons. Diffusion tensor tractography studies have revealed that hand-related fibers occupy the anteromedial portion of the basilar pons, while leg-related fibers positioned posterolaterally, maintaining a typical somatotopic organization (Jang 2014).

A fascinating aspect of corticospinal organization involves the extensive collateralization of individual neurons. Nearly all corticospinal neurons possess multiple branches terminating across several spinal cord levels. This architectural feature explains why focal cortical lesions can affect multiple spinal segments and why cortical neurons labeled from a single spinal level appear distributed across broad cortical regions (Rosenzweig et al. 2010).

The majority of corticospinal fibers (approximately 90%) cross at the pyramidal decussation in the caudal medulla to form the lateral corticospinal tract. A smaller proportion (8–15%) remains uncrossed in adult humans and primates, though this percentage can reach 30% in some individuals. These uncrossed fibers form the ventral corticospinal tract, eventually crossing at the spinal level through the ventral white commissure. Notably, both crossed and uncrossed fibers originate from similar cortical regions and show substantial overlap in their terminal fields (Kuypers 1963).

Within the spinal cord, corticospinal projections demonstrate specific laminar termination patterns. Fibers originating from the motor cortex primarily target interneurons in the intermediate zone and motor neurons in the ventral horn, while those from parietal regions predominantly project to the dorsal horn. This organized distribution pattern underlies the tract's role in both motor control and sensory processing (Nathan and Smith 1955).

The evolutionary significance of the corticospinal system becomes apparent through its presence exclusively in mammals, with considerable variation across species. In rodents, the major corticospinal bundle occupies the dorsal column, while primates exhibit a more developed system predominantly situated in the lateral column. This evolutionary progression reflects the increasing demands for precise motor control in higher mammals (Lemon and Griffiths 2005).

The contribution of different cortical areas to the corticospinal tract shows species-specific variations that illuminate human motor specialization. In humans, approximately 80% of corticospinal fibers originate from Brodmann's areas 4 and 6, representing a higher proportion than observed in other primates. This increased contribution from motor areas, coupled with relatively sparse input from somatosensory regions, suggests evolutionary optimization for voluntary movement control (Porter and Lemon 1995).

The giant pyramidal cells of Betz, distinctive features of the primary motor cortex, show a non-uniform distribution that correlates with functional specialization. These cells appear most numerous in the leg area superiorly and decrease in number toward the face area inferiorly. Despite this organized distribution, the traditional concept of strict somatotopic organization requires revision. Clinical observations from electrical stimulation studies in conscious patients reveal that cortical activation typically produces coordinated muscle group movements rather than isolated muscle responses (Rothwell 2011).

Recent investigations have revealed intricate patterns in corticospinal terminal distribution within the spinal gray matter. Fibers entering through the intermediate zone (lamina VII) demonstrate complex branching patterns, reaching laminae IV, V, VII, and portions of lamina IX. The density of direct projections to lamina IX, containing α-motoneurons, peaks in the cervical and lumbar enlargements, particularly in regions innervating distal musculature. This arrangement underlies the system's crucial role in fine motor control of the extremities (Jankowska 2001).

The developmental timeline of the corticospinal system holds particular clinical significance. At birth, the tract remains unmyelinated and functionally immature. The two-year period required for complete myelination coincides with critical phases of motor development, including the acquisition of standing, walking, and fine motor skills. This extended maturation period may provide opportunities for therapeutic intervention following early nervous system injury (Eyre et al. 2001).

The extensive distribution of corticospinal collaterals carries important clinical implications. Single cortical neurons influence multiple spinal segments through branching axons, creating a distributed control system rather than a simple point-to-point network. This organization explains the widespread effects of focal cortical lesions and the potential for functional reorganization following injury. Understanding this architecture has profound implications for rehabilitation strategies after central nervous system damage (Rosenzweig et al. 2010).

13.1.2.2 Descending Brainstem Motor Control System

The brainstem gives rise to all descending spinal projections except the corticospinal tract. These pathways organize into two distinct functional groups, each with specific anatomical and physiological characteristics.

Group A pathways, including the vestibulospinal, medial reticulospinal, tectospinal, and interstitiospinal tracts, descend through the ventral and ventrolateral funiculi. These pathways exhibit extensive collateralization and form bilateral terminations in the ventromedial portions of laminae 5–7. Their main functions include postural control, integration of limb-trunk movements, coordination of synergistic muscle actions, and orientation of head and body movements (Peterson and Abzug 1975).

Group B pathways comprise the rubrospinal and lateral reticulospinal tracts, which course through the dorsolateral funiculus. In contrast to Group A, these pathways show limited collateralization and primarily target the dorsal and lateral regions of laminae 5–7, along with motor neurons innervating distal limb muscles. These projections enhance Group A's function by facilitating flexion-oriented movements of the limbs, particularly at the elbow and hand (Drew et al. 2004).

The lateral pontine and medullary reticular formation, specifically the gigantocellular nucleus, contribute significantly to these descending systems. Their projections, traveling in the dorsolateral column, establish contralateral connections with lamina I and the lateral portions of laminae V, VI, VII, and IX. This arrangement creates a complex control system for coordinated movement and postural adjustments (Matsuyama et al. 2004).

Understanding the organization of these brainstem pathways is clinically significant, as their impairment leads to various movement disorders. The complementary roles of Groups A and B provide both gross postural control and fine motor adjustments, establishing a hierarchical motor control system that functions in conjunction with the corticospinal system.

The Rubrospinal System in Motor Control

The evolutionary process of motor control systems demonstrates various adaptations among species, particularly demonstrated by the development of the rubrospinal tract. In cats and other quadrupedal mammals, this system functions as a key motor control system, promoting flexor activity while suppressing extensor motor neurons via complex interneuronal networks. Nevertheless, the human rubrospinal system presents a different evolutionary story, characterized by diminished neuronal populations and augmented rubro-olivary connections, indicative of our species' unique motor control requirements (Baker 2011; Onodera and Hicks 2009).

The red nucleus, located in the midbrain tegmentum, contains the neurons that originate this primordial motor pathway. These fibers, starting from big cells in the caudal region, traverse the midline and descend into the lateral funiculus, intermingling with corticospinal tract fibers. Although previous research focused on fixed somatotopic organization, current research demonstrates a more intricate arrangement, wherein individual neurons frequently impact several spinal segments via complex collateral networks (Kuypers 1981).

The red nucleus participates in a complex neuronal interaction, receiving precisely structured inputs from cortical and cerebellar origins. The ipsilateral corticorubral fibers from the motor cortex preserve somatotopic precision, whilst the crossing projections from the emboliform and globose cerebellar nuclei offer important modulatory effects. A combination of inputs establishes a special motor control circuit, wherein corticorubral and rubrospinal fibers constitute a nonpyramidal pathway that enhances the primary motor system (Kennedy 1990).

Electrophysiological investigations have unveiled the complex functional architecture of this system. Red nucleus stimulation triggers a coordinated pattern of neural activity, generating excitation in contralateral flexor α-motoneurons while simultaneously inhibiting their extensor counterparts. This complex regulation occurs indirectly via interneuronal circuits instead of direct motor neuronal connections, indicating the complex character of this motor control system (Lemon 2008).

The clinical significance of the rubrospinal system extends beyond its primary motor functions. The rubrospinal system has potential compensatory roles following corticospinal injury, offering promising implications for rehabilitation strategies. Moreover, understanding the evolutionary modifications of this system provides crucial insights into human motor control development and potential therapeutic approaches for movement disorders.

The Vestibulospinal System in Motor Control

The vestibular nuclei form a complex network in the lateral floor of the fourth ventricle, spanning the pons and medulla. These nuclei integrate inputs from the vestibular apparatus and cerebellar regions to maintain equilibrium and posture. The system's organization allows continuous adjustment of body position in three-dimensional space (Highstein and Holstein 2006).

The vestibulospinal tract originates predominantly from giant cells in the lateral vestibular nucleus, descending through the ventral portion of the lateral funiculus. While present throughout all spinal segments, its influence shows particular strength in cervical and lumbosacral regions. The tract's fibers terminate in the medial ventral horn, specifically within Rexed's laminae VII, VIII, and portions of IX, forming direct connections with extensor motor neuron somata and dendrites (Grillner et al. 1970). Individual axons frequently provide collaterals to multiple spinal segments, creating a more distributed influence than previously recognized. This arrangement enables coordinated control of posture and movement across different body segments (Pompeiano 1972).

The utricular macula's vestibular ganglion cells project specifically to the ventral lateral vestibular nucleus, suggesting a specialized role in responding to head tilt, gravity, and linear acceleration. This pathway works in conjunction with cerebellar inputs: inhibitory GABAergic projections from anterior lobe vermis Purkinje cells and excitatory glutamatergic inputs from the fastigial nucleus.

The system's importance becomes particularly evident in pathological conditions. Experimental lesions of the cerebellar anterior lobe result in increased extensor tone, demonstrating the critical role of cerebellar modulation in maintaining appropriate postural control. These findings have significant implications for understanding balance disorders and developing therapeutic strategies.

The Reticular Formation

The reticular formation forms the ancestral core of the brainstem, containing cell groups of varying sizes within an expansive fiber network. As the phylogenetically oldest part of the brainstem, it developed before the emergence of more specialized pathways. The reticular formation extends throughout the brainstem tegmentum, from the midbrain-diencephalon junction to the medullocervical boundary. Its neurons display extensive dendritic trees and branched axonal networks, creating its characteristic microscopic appearance. This structure occupies the spaces between cranial nerve nuclei and major nerve tracts, maintaining functional continuity between the spinal cord and higher brain centers (Sakai et al. 2009).

Spinal projections originate from the medial zones of both pontine and medullary regions. The medullary component sends mostly uncrossed fibers from cells dorsal to the inferior olivary complex, descending through the anterior lateral funiculus to reach interneurons in laminae VII and IX. Similarly, pontine reticulospinal fibers descend uncrossed in the ventral funiculus, terminating on interneurons in laminae VII and VIII. Unlike other descending pathways, reticulospinal projections lack somatotopic organization (Holstege 1982).

The system receives input from diverse sources, including sensory receptors, cerebellar pathways, and cortical regions. These inputs undergo processing that transforms specific sensory information into generalized control signals. The medullary region primarily provides inhibitory control over motor activity and muscle tone through γ-motoneuron regulation, while pontine and midbrain regions facilitate motor responses through polysynaptic pathways.

Clinical significance appears in conditions like cerebral palsy, where patients experience heightened startle responses and involuntary movements. These symptoms reflect disruption of the reticular formation's regulatory functions, affecting both descending motor control and ascending arousal systems (Drew et al. 1986).

Afferent Pathways in the Reticular Formation

The reticular formation processes neural signals from diverse brainstem pathways. Most ascending and descending pathways contribute collateral branches to the reticular formation, enabling integration of visceral and somatic information, whether sensory or motor in nature (Garcia-Rill et al. 2015).

The primary neural circuit organization involves initial synapsing in the lateral parvicellular nuclei, followed by signal transmission to the medial gigantocellular nuclei for broader reticular formation output distribution. The medial lemniscus pathway shows a unique pattern—while its discriminative sensory fibers from the dorsal columns bypass the reticular formation, the adjacent lateral spinothalamic and trigeminothalamic tracts provide substantial collateral input. The spinoreticular tracts also form direct projections to the reticular formation (Siegel and Sapru 2019).

Several pathways maintain precise topographic organization by bypassing the reticular formation, including the dorsal column-medial lemniscus pathway, retinogeniculocalcarine tract, and tonotopic auditory fibers. However, secondary projections from visual and auditory systems enable rapid responses to intense stimuli, as shown by startle reactions to sudden acoustic or visual input.

The reticular formation's extensive connectivity patterns allow it to modify arousal states and influence sensorimotor processing. This arrangement enables the reticular formation to respond to various sensory inputs while maintaining specific pathways for precise discriminative information (Haines and Mihailoff 2018).

Efferent Pathways in the Reticular Formation

The reticular formation generates output through two major pathways: the reticulobulbar and reticulospinal tracts, each with specific neurotransmitter systems. The reticulobulbar tracts extend to the cranial nerve nuclei, while the reticulospinal tracts descend through the ventrolateral spinal cord columns from the medial reticular formation nuclei (Purves et al. 2018).

The pontine reticulospinal tract originates in the pontine reticular nuclei and courses through the medial longitudinal fasciculus. These fibers terminate in spinal cord laminae VII and VIII across all segments, sharing terminal fields with vestibular nuclear projections. This arrangement allows control of axial musculature, particularly neck muscles, for postural control and muscle tone regulation (Carpenter and Reddi 2012).

From the nucleus reticularis gigantocellularis, the medullary reticulospinal tract emerges and terminates throughout the spinal cord in laminae VII and IX. These terminations share areas with corticospinal and rubrospinal tract endings. Most reticulospinal axons form synapses with interneurons in laminae VII to IX, while a smaller portion directly connects with ventral motor neurons.

This organization enables the reticular formation to modulate motor function through both direct and indirect pathways, affecting posture, movement, and muscle tone across multiple spinal segments (Watson et al. 2010).

13.1.3 Dual Control Systems of Spinal Reflexes in the Brainstem

The brainstem contains two balanced systems that control spinal reflexes: an excitatory and an inhibitory system (Brown 1994). These systems operate independently, with different anatomical locations and varying degrees of cortical control. The two systems maintain equilibrium in reflex control, though their organization and function reflect a simplified but accurate model of brainstem-spinal interactions (Rothwell 1994). Their anatomical separation and differential response to cortical input suggest specialized roles in motor control (Whitlock 1990).

13.1.3.1 Brainstem Excitatory System in Spinal Reflex Control

In the upper brainstem, a diffuse region facilitates spinal stretch reflexes. This system originates in the sub- and hypothalamus, with efferent pathways extending through the midbrain's central gray and tegmentum, pontine tegmentum, and medullary reticular formation. Experimental stimulation of this area enhances patellar reflexes in monkeys and increases reflex activity and extensor tone in spastic cats. Disruption of the bulbopontine tegmentum reduces spasticity (Schreiner et al. 1949).

While this system receives input from the somatosensory cortex and supplementary motor area (Whitlock 1990), motor cortex and internal capsule stimulation do not alter its facilitatory effects. This limited cortical influence differentiates it from the inhibitory system. The system projects through medial reticulospinal tracts in the ventromedial cord (Brown 1994).

The lateral vestibular nucleus, located near the pontomedullary junction, provides additional excitatory input. Its stimulation causes disynaptic excitation of extensor motor neurons (Rothwell 1994), with signals traveling via the lateral vestibulospinal tract. Though significant in decerebrate rigidity, its role in spasticity appears minimal. Single lesions here have a limited effect on feline spasticity but augment the antispastic impact of bulbopontine tegmentum lesions (Schreiner et al. 1949). Vestibulospinal tract lesions only temporarily reduce spasticity (Bucy 1938).

Both regions, despite their excitatory nature, inhibit flexor reflex afferents (Whitlock 1990). The lateral vestibulospinal tract specifically inhibits flexor motor neurons (Rothwell 1994).

13.1.3.2 Brainstem Inhibitory System in Spinal Reflex Control

The premotor cortex sends corticoreticular fibers (parapyramidal fibers) that activate an inhibitory region in the medulla's ventromedial reticular formation, located dorsal to the pyramids (Brown 1994). Electrical stimulation studies demonstrate this area's ability to inhibit patellar reflexes in intact cats. In decerebrate cats, stimulation reduces rigidity, while in cats with chronic cerebral lesions, it decreases spastic muscle tone.

Research on experimental poliomyelitis in monkeys revealed significant damage to this region during early spastic stages. The area's inhibitory effects extend to the tonic vibration reflex and flexor reflex afferents in intact cats (Whitlock 1990). Premotor cortex or internal capsule stimulation enhances these inhibitory effects, confirming cortical control. Additional input may come from the cerebellum (Lindsley et al. 1949). The dorsal reticulospinal tract, situated in the dorsolateral funiculus, carries the descending signals from this inhibitory region (Engberg et al. 1968).

13.2 Spinal Segmental Reflexes in Upper Motor Neuron Syndrome

Upper motor neuron syndrome is characterized by an increase in spinal segmental reflex activity, leading to exaggerated muscle activation patterns that contribute to motor impairments (Nielsen et al. 2007). These reflexes can be broadly classified into two categories based on their physiological mechanisms and pathways of activation (Trompetto et al. 2014).

The first category encompasses proprioceptive reflexes, including tonic and phasic stretch reflexes, as well as the positive supporting reaction. These reflexes are primarily mediated by muscle spindles and Golgi tendon organs, which regulate muscle length and tension under pathological conditions. Enhanced stretch reflexes, often observed in spasticity, contribute significantly to the resistance against passive joint movement (Trompetto et al. 2014). The second category includes nociceptive and cutaneous reflexes, such as flexor and extensor reflexes. These reflexes are activated in response to painful or tactile stimuli and are mediated through complex interactions between sensory input and motor output. The Babinski sign, a hallmark of corticospinal tract dysfunction, is an example of these reflexes, demonstrating an abnormal extension of the great toe following plantar stimulation (Li and Francisco 2015). The persistence of this reflex in adults indicates disruption in upper motor neuron pathways.

A notable phenomenon observed in spasticity is the clasp-knife response. This reflex displays features of both proprioceptive and nociceptive reflexes, with an initial resistance to passive joint movement that abruptly gives way. This response reflects the intricate interaction between different spinal reflex pathways, often involving altered afferent input and motor neuron excitability (Mukherjee and Chakravarty 2010). The study of such reflex patterns provides valuable insights into the underlying pathophysiology of upper motor neuron syndrome and informs targeted therapeutic strategies (Barnes and Johnson 2008a).

Clinicians evaluating upper motor neuron syndrome rely on recognizing these reflex abnormalities to estimate the severity of neurological impairment. Additionally, understanding the mechanistic basis of these reflexes aids in facilitating the development of interventions aimed at modulating hyperactive reflex circuits and improving functional outcomes in affected individuals (Nielsen et al. 2007; Trompetto et al. 2014).

13.2.1 Proprioceptive Reflexes

The proprioceptive system provides essential information regarding body position and movement, mediated by muscle spindles functioning as sensory receptors within the musculature (Proske and Gandevia 2012). Stretching of these muscle spindles activates their afferent fibers, which directly synapse with motor neurons in the spinal cord to regulate muscle contraction (Pierrot-Deseilligny and Burke 2005).

The stretch reflex pathway is foundational to deep tendon reflexes, commonly referred to as phasic stretch reflexes, characterized by their rapid response to brief stretches (Needle et al. 2014a, b). In contrast, tonic stretch reflexes manifest during prolonged muscle stretches, such as those encountered during assessments of muscle tone. Despite their temporal differences, both reflex types share the same underlying mechanisms (Mukherjee and Chakravarty 2010). The positive support reaction, another proprioceptive reflex, arises from proprioceptors in the foot and contributes to postural control and weight-bearing activities (Dietz and Duysens 2000; Nielsen et al. 2007).

13.2.2 Phasic Stretch Reflexes

Hyperexcitability of phasic stretch reflexes manifests clinically as exaggerated deep tendon reflexes, reflex irradiation, and clonus (Burke 1988). Although traditionally considered a monosynaptic response triggered by tendon percussion. Tendon percussion generates vibration waves throughout the limb, stimulating multiple muscle spindles and mechanoreceptors (Lance and De Gail 1965) (Fig. 13.5).

The tendon reflex response is driven by asynchronous firing of spindles and involves both monosynaptic and polysynaptic pathways. For instance, the rise in soleus motor neuron excitability following Achilles tendon percussion, occurring within 10 ms, suggests oligosynaptic and polysynaptic contributions (Burke et al. 1983).

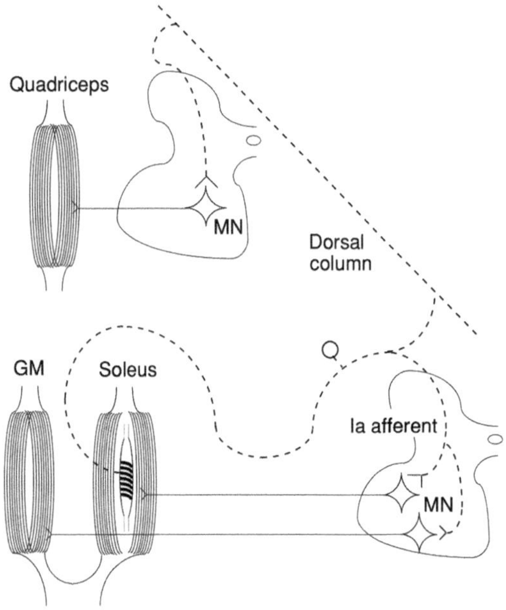

Fig. 13.5 Connections of monosynaptic Ia excitation. Ia afferents (dashed lines) originating from muscle spindle primary endings of the soleus have monosynaptic projections to homonymous soleus motor neurons (MNs), and to heteronymous MNs supplying its close synergist, the gastrocnemius medialis (GM), located at the same S1 segmental level and also, after running in the dorsal column, to MNs of the quadriceps located more rostrally in L4. (Adapted with permission from Pierrot-Deseilligny and Burke (2005))

Reflex irradiation, another phenomenon, occurs when tendon percussion elicits contractions in adjacent limb muscles. This phenomenon is attributed to mechanical stimulation of spindles in nearby muscles rather than newly formed synaptic connections (Burke 1988). In individuals with upper motor neuron syndrome, the hyperexcitable state of these stretch reflexes exacerbates such responses.

Clonus, characterized by rhythmic muscle contractions following rapid stretch, is most commonly observed at the ankle. Initially thought to stem from a central oscillator due to its consistent frequency (Dimitrijevic 1987; Walsh 1976), subsequent research demonstrated that clonus frequency varies with mechanical load, indicating that increased stretch reflex gain rather than central mechanisms predominates (Rack et al. 1984).

The mechanism underlying clonus involves rapid dorsiflexion of the ankle, which stretches the gastrocnemius-soleus muscle, triggering a reflex contraction. As the muscle relaxes under sustained dorsiflexion, the cycle repeats. The examiner's technique and the application of force significantly influence the elicitation and maintenance of clonus, with stronger loads often suppressing its activity (Burke 1988).

13.2.3 Tonic Stretch Reflexes

The clinical assessment of muscle tone involves examining resistance during passive joint movement. In upper motor neuron syndrome, this manifests as spasticity, a velocity-dependent form of hypertonia (Herman 1970). While slow movements may not reveal increased tone, faster movements elicit progressively greater resistance, correlating with electromyographic activity in the stretched muscle.

Thilmann et al. (1991) demonstrated that in normal subjects, stretch reflexes in relaxed biceps occur only with rapid movements exceeding 200° per second, with latencies of 61–107 ms. These responses appear to be phasic rather than tonic reflexes. Notably, at typical clinical testing velocities, normal relaxed muscles show no stretch reflex activity, indicating that baseline muscle tone derives from viscoelastic properties rather than reflex activity. In contrast, patients with spasticity exhibit stretch reflexes at much lower velocities (as low as 35° per second), with activity persisting throughout the stretching movement (Burke et al. 1970). This reflex activity shows a linear relationship with stretch velocity, confirming the velocity-dependent nature of spasticity (Powers et al. 1989).

The tonic stretch reflex exhibits both velocity and length dependency. In lower limbs, the reflex sensitivity decreases at longer lengths in ankle plantarflexors and quadriceps (Meinders et al. 1996). However, upper limb studies show opposite effects, with finger flexors displaying increased reflexes in shortened positions (Li et al. 2006).

Repeated stretching temporarily reduces spasticity through both biomechanical changes and decreased reflex activity (Nuyens et al. 2002). The mechanism may involve thixotropic changes in spindle sensitivity or habituation of central reflex pathways (Schmit et al. 1999).

13.2.4 Static Tonic Stretch Reflexes

While traditional views considered spastic stretch reflexes primarily dynamic, occurring only during joint movement, research has revealed more complexity (Thilmann et al. 1991). Some studies show EMG activity persisting during sustained muscle stretch, particularly in elbow flexors (Powers et al. 1989). The apparent absence of static reflexes in previous studies may reflect methodological factors (Powers et al. 1989). Length-dependent inhibition, as seen in the clasp-knife phenomenon, might explain its absence in the quadriceps. At the ankle, static stretch reflexes remain consistently unobserved (Berardelli et al. 1982; Herman 1970).

The mechanisms underlying static tonic stretch reflexes likely involve length-sensitive receptors. While primary muscle spindles (Ia afferents) respond mainly to velocity, secondary muscle spindles (group II afferents) maintain increased firing during sustained stretch. Pharmacological studies with baclofen and tizanidine in spinal cats suggest group II afferent involvement in spasticity (Skoog 1996). Research shows that L-DOPA, which selectively inhibits group II afferent transmission, reduces spasticity and related symptoms in spinal cord injury patients (Eriksson et al. 1996). However, Burke (1988) suggests that persistent EMG activity after movement might reflect other stimuli, such as cutaneous input, rather than pure stretch reflexes.

13.3 The Lower Motor Neuron

The anterior horn cells and their peripheral axons that connect to striated muscle form a structural and physiological unit known as the final common motor pathway, or the lower motor neuron (LMN). This system includes not only spinal motor neurons but also motor cranial nerve nuclei that direct head and neck muscle activity. At the molecular level, these neurons express choline acetyltransferase and utilize vesicular acetylcholine transporters for neurotransmitter packaging (Standring 2020).

Within the spinal cord, anterior horn cells organize into specific cell columns with precise somatotopic arrangements. A medial cell column extends along the spinal cord's length and contains cell groups that control long and short axial muscles. Adjacent to this lies the lateral cell column, which varies in size and function according to spinal level. In thoracic regions, this column remains compact and controls intercostal and anterolateral trunk muscles. The column expands significantly in cervical (C5–T1) and lumbosacral (L2–S2) areas to create multiple subgroups for extremity muscle control. Cells positioned anteriorly and peripherally direct extensor and abductor muscles, while those situated dorsally and centrally manage flexor and adductor muscle groups. This organization reflects evolutionary adaptations for precise limb control (Kandel et al. 2013).

The spinal gray matter displays a specific cytoarchitectural pattern known as Rexed's lamination, with motor neurons residing in lamina IX. These neurons appear as large multipolar cells (50–70 μm in diameter) with abundant Nissl granules and central nuclei. Their extensive dendritic arbors extend beyond lamina IX's boundaries, receiving synaptic inputs from various sources, including propriospinal neurons, descending pathways, and primary afferents. The motor neuron axons, measuring 12–20 μm in diameter and conducting at velocities of 70–120 m/s, exit through ventral roots and combine with dorsal root fibers beyond the dorsal root ganglion to form mixed spinal nerves. These nerves split into dorsal and ventral primary rami, which merge to create the brachial and lumbosacral plexuses in the limb regions (FitzGerald et al. 2012).

The anterior horn contains diverse neuronal populations. Internuncial neurons, smaller than motor neurons, form local circuits within the spinal cord and express various neurotransmitters, including glycine and GABA. The root cells that send axons outside the spinal cord exist in two primary varieties. Alpha motor neurons, comprising about 70% of ventral root fibers, innervate striated muscle fibers (extrafusal) and have soma diameters of 30–70 μm. Gamma motor neurons (soma diameter 15–30 μm) make up the remaining 30% and control muscle spindles (intrafusal). Additionally, the ventral roots in thoracic (T1–L2) and upper lumbar segments contain poorly myelinated preganglionic sympathetic fibers (diameter 1–3 μm), while sacral ventral roots (S2–S4) carry preganglionic parasympathetic fibers (Boron and Boulpaep 2017).

Alpha motor neurons utilize acetylcholine as their neurotransmitter, releasing it at specialized motor end-plates that contain high concentrations of nicotinic acetylcholine receptors (approximately 10,000 receptors/μm^2). Electrical stimulation triggers the synchronized release of approximately 50–200 acetylcholine quanta per action potential. After release, acetylcholinesterase, anchored in the synaptic cleft by collagen-Q, hydrolyzes acetylcholine within 100 μs. While α-motoneurons directly control muscle tension through this mechanism, γ-motoneurons modify muscle spindle sensitivity through similar but smaller end-plates on intrafusal fibers (Kandel et al. 2013).

The myotatic reflex exemplifies basic motor control principles. This monosynaptic reflex involves primary spindle afferents (group Ia fibers, conducting at 70–120 m/s) forming direct synapses with α-motoneurons. When muscle stretch activates these afferents, they release glutamate onto motor neurons, triggering an excitatory postsynaptic potential with a synaptic delay of approximately 0.5 ms. This reflex typically spans two or three spinal segments since most muscles receive input from multiple adjacent segments, creating a pattern similar to dermatomes in sensory innervation (Standring 2020).

Multiple sensory pathways modulate lower motor neuron function through complex neural circuits. Group Ia fibers provide monosynaptic excitation with a synaptic delay of 0.5–1.0 ms, while group Ib fibers from Golgi tendon organs create disynaptic inhibition through interneurons expressing glycine. Nociceptive pathways, using substance P and glutamate as neurotransmitters, terminate in Rexed's lamina I and II, triggering protective reflexes through polysynaptic circuits. These reflexes coordinate multiple muscle groups through reciprocal innervation, including crossed extensor responses that maintain postural stability during limb movement (FitzGerald et al. 2012).

13.4 Muscle Afferent Fibers

Historical research by Ruffini, Golgi, and Sherrington demonstrated that skeletal muscles contain numerous sensory receptors, with afferent axons exceeding the number of efferent axons in most muscles (Hunt 1990). These researchers analyzed conduction velocities to determine axonal diameters, identifying three main groups of myelinated fibers (Table 13.2).

Group I fibers, the largest diameter myelinated afferents in the peripheral nervous system, conduct impulses at the highest velocities. Group II and III fibers exhibit smaller diameters with corresponding slower conduction velocities. Group IV comprises unmyelinated axons with the slowest conduction speeds (Pfaff and Volkow 2016). Among muscle receptors, muscle spindles and Golgi tendon organs have undergone extensive research analysis. Muscle spindles receive dual innervation from Group I and II axons. In comparison, Golgi tendon organs receive innervation solely from Group I axons. The Group I afferents show further subdivision: Group Ia axons, which supply muscle spindles, possess marginally larger diameters than Group Ib axons innervating tendon organs (Banks et al. 2009).

13.4.1 Structure and Function of Muscle Spindles

Muscle spindles represent specialized mechanoreceptive sensory organs embedded within skeletal muscle tissue. Each spindle contains between

Table 13.2 Muscle afferents and efferents

Afferent axon type	Receptor	Sensitive to
Group Ia	Primary spindle ending	Muscle length and rate of change of length
Group Ib	Golgi tendon organ	Muscle tension
Group II	Secondary spindle ending	Muscle length (little rate sensitivity)
Group II	Nonspindle endings	Deep pressure
Groups III, IV	Free nerve endings	Pain, chemical stimuli, and temperature (important for physiologic response to exercise)
Efferent axon type	**Innervation**	**Function**
α-motoneurons (skeletomotor)	Extrafusal muscle fibers	Control lesion of muscle
β-fibers (skeletofusimotor)	Intrafusal muscle fibers (collaterals from α-motoneurons)	Control sensitivity of spindles (no independent control of spindle sensitivity)
γ-motoneurons (fusimotor)	Intrafusal muscle fibers	Control sensitivity of spindles (independently of α-motoneuron activation)

Adapted from Downey et al. (1994)

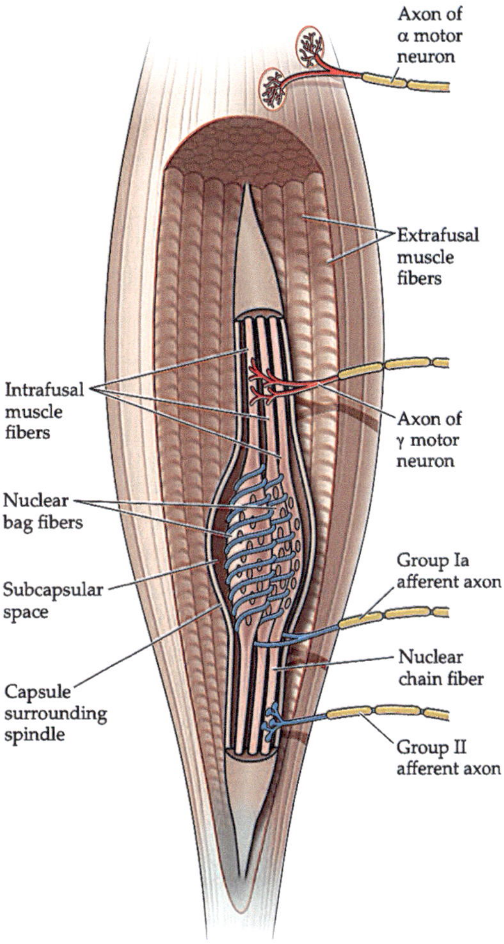

Fig. 13.6 Diagram of a muscle spindle, the sensory receptor that initiates the stretch reflex. (Adapted from with permission Purves et al. (2018))

2 and 25 intrafusal muscle fibers that anchor to the perimysium surrounding conventional extrafusal muscle fibers (Fig. 13.6). These receptors demonstrate precise parallel alignment with extrafusal fibers and reside within a distinct connective tissue capsule. This strategic orientation enables spindles to detect alterations in muscle length, thereby contributing to proprioception and muscle tone regulation (Banks 2006). Individual spindles demonstrate specific dimensional parameters, measuring up to 1 cm in length with diameters ranging from 100 to 200 μm.

The intrafusal fiber comprises two primary categories distinguished by their nuclear arrangement: bag fibers and chain fibers. Bag fibers demonstrate nuclear aggregation within their central (equatorial) region, while chain fibers exhibit linear nuclear distribution (Hunt 1990). The bag fiber category further subdivides into bag1 and bag2 variants, which differ in their density of elastic elements (Proske and Gandevia 2012). Though intrafusal fibers share the fundamental sarcomeric architecture of skeletal muscle, their contractile proteins concentrate in polar regions with reduced density in the equatorial zone. A typical human muscle spindle contains between 3 and 4 bag fibers and between 4 and 11 chain fibers (Banks et al. 1997).

Muscle spindles maintain distinct functional characteristics from Golgi tendon organs through their unique pattern of dual afferent and efferent innervation. While Golgi tendon organs respond specifically to tension, spindles detect changes in muscle length. Their discharge frequency decreases during extrafusal fiber contraction, reflecting their primary role in monitoring both static length and dynamic length changes. This response pattern occurs through mechanical deformation of intrafusal fibers.

The sensory innervation demonstrates precise organization, with single Group Ia afferent fibers branching to supply both bag and chain fibers, forming primary (annulospiral) endings within the equatorial region (Banks 1994) (Fig. 13.7). These Group Ia fibers share similar morphological and physiological properties with Group Ib afferents regarding myelination, diameter, and conduction velocity. They exhibit remarkable mechanosensitivity, activating with minimal stretch forces of approximately 2 g. Additionally, thinner Group II afferents with slower conduction velocities form secondary (flower spray) endings, predominantly on chain fibers. Individual spindles typically receive between 1 and 5 Group II axons. Human spindles generally receive innervation from single sensory axons, contrasting with the multiple innervation patterns observed in other mammals (Kucera et al. 1991).

The physiological response patterns of muscle spindles incorporate both tonic and phasic components. Through integrated Group Ia and II afferent systems, chain fibers generate tonic responses to sustained stretch independent of stretch velocity. The primary endings of Group Ia afferents demonstrate particular sensitivity to velocity changes in bag fibers, generating rapid discharge bursts during dynamic stretch. This dual response mechanism enables spindles to participate in stretch reflexes while simultaneously providing unconscious proprioceptive information regarding joint position and movement.

The intrafusal fiber length regulation occurs through γ-motoneurons, which innervate the polar regions. These neurons, characterized by smaller soma size and slower conduction velocities compared to α-motoneurons, differentiate into gamma-dynamic (γ1) and gamma-static (γ2) subtypes. This functional differentiation enables separate control of dynamic and static length change responses. In human muscle spindles, γ-motoneurons demonstrate preferential innervation patterns to bag1 fibers (Kucera and Walro 1992).

Upon γ-motoneuron activation, the discharge rates of both Group Ia and II fibers increase through elongation of the mechanoreceptor regions within bag and chain fibers. This sophisticated system modulates afferent discharge through two mechanisms: direct muscle stretch and intrafusal fiber contraction via γ-motoneuron activity. The spindle's response characteristics vary with both the rate and magnitude of length changes, maintaining optimal sensitivity throughout the muscle's physiological range of motion. The muscle spindle system incorporates additional sensory components, including Group III and IV afferent fibers originating from free nerve endings, which respond to intense or noxious stimuli. These fibers contribute to the flexor reflex pathway and may modulate the enhanced reflex responses observed in spastic conditions.

The anatomical distribution of muscle spindles throughout skeletal musculature demonstrates precise functional correlation. Muscles responsible for fine motor control contain higher spindle densities, particularly evident in cervical musculature. In contrast, large postural muscles such as the gluteus maximus exhibit lower spindle density per unit volume. Quantitative analyses reveal significant variations in spindle counts, ranging from 6 in the stylohyoideus muscle to approximately 1300 in the quadriceps femoris, correlating directly with each muscle group's control requirements (Proske 2005).

The structural organization of muscle spindles exhibits consistent dimensional parameters, ranging from 2 to 10 mm in length, with strategic positioning near intramuscular neurovascular structures. Their architecture incorporates multiple parallel intrafusal fibers within a specialized capsule, demonstrating central thickening, creating their characteristic fusiform morphology (Downey et al. 1994) (see Figs. 13.5 and 13.6).

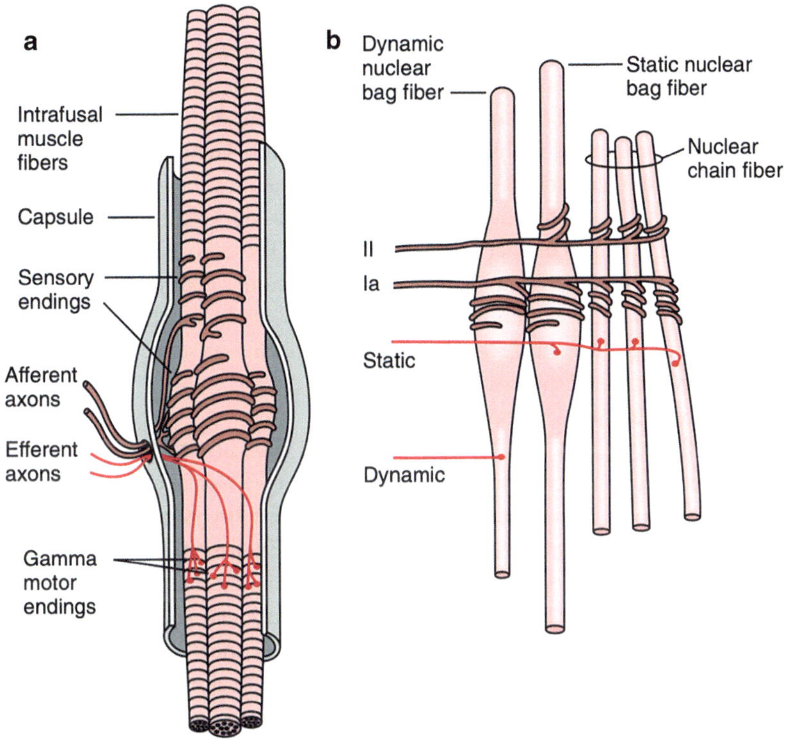

Fig. 13.7 (**a**) The main components of the muscle spindle are intrafusal muscle fibers, afferent sensory fiber endings, and efferent motor fiber endings. The intrafusal fibers are specialized muscle fibers; their central regions are not contractile. The sensory fiber endings spiral around the central regions of the intrafusal fibers and are responsive to stretch of these fibers. Gamma motor neurons innervate the contractile polar regions of the intrafusal fibers. Contraction of the intrafusal fibers pulls on the central regions from both ends and changes the sensitivity of the sensory fiber endings to stretch. (**b**) The muscle spindle contains three types of intrafusal fibers: dynamic nuclear bag, static nuclear bag, and nuclear chain fibers. A single Ia sensory fiber innervates all three types of fibers, forming a primary sensory ending. A group II sensory fiber innervates nuclear chain fibers and static bag fibers, forming a secondary sensory ending. Two types of motor neurons innervate different intrafusal fibers. Dynamic gamma motor neurons innervate only dynamic bag fibers; static gamma motor neurons innervate various combinations of chain and static bag fibers. (Adapted with permission from Vanderah and Gould (2016))

The capsule contains a fluid-filled space providing mechanical support and reducing friction during fiber movement (Matthews 1964).

Motor control of these specialized receptors occurs through γ-motoneurons innervating the contractile polar regions of intrafusal fibers. This arrangement enables central nervous system modulation of spindle sensitivity through selective activation of specific fiber types. The activation of γ-motoneurons induces polar region contraction, resulting in central region elongation and subsequent modification of sensory ending discharge rates. This complex control mechanism enables precise proprioceptive feedback during diverse movement patterns (Prochazka 1986; Proske and Gandevia 2018) (Fig. 13.7).

13.4.2 Structure and Function of Golgi Tendon Organs in Musculotendinous Junctions

Golgi tendon organs function as specialized mechanoreceptors measuring approximately 1 mm in length and 0.1 mm in diameter. These receptors are predominantly located at musculotendinous junctions, where collagen fibers form connections between tendon and muscle fibers, rather than within the tendon itself (Jami 1992).

The structural configuration of Golgi tendon organs comprises collagen fiber bundles arranged in series with 15–20 extrafusal muscle fibers. Within the capsule, these collagen fibers exhibit a braided arrangement. Each tendon organ is innervated by a single Group Ib axon, which loses its myelin sheath upon entering the capsule and subsequently branches into several fine terminals that intertwine with collagen fibers (Schoultz and Swett 1972) (Fig. 13.8).

The braided organization of collagen fibers enhances their capacity to influence nerve endings. Extrafusal muscle fiber contraction causes collagen fiber straightening, leading to nerve ending compression and stretching, thereby increasing their firing frequency. This anatomical configuration enables the detection of minimal changes in muscle tension. Furthermore, the connection of muscle fibers from multiple motor units to each tendon organ allows the afferent fiber to monitor motor unit recruitment patterns (Houk and Henneman 1967). Golgi tendon organs function beyond their traditional role as protective mechanisms against excessive muscle tension. These receptors participate actively in normal motor control by providing continuous feedback about muscle force production to the central nervous system (Proske and Gregory 1980).

13.4.3 Response Properties of Primary and Secondary Endings of Muscle Spindle Endings

The primary and secondary muscle spindle endings exhibit specific firing characteristics during muscle length alterations. These changes occur in

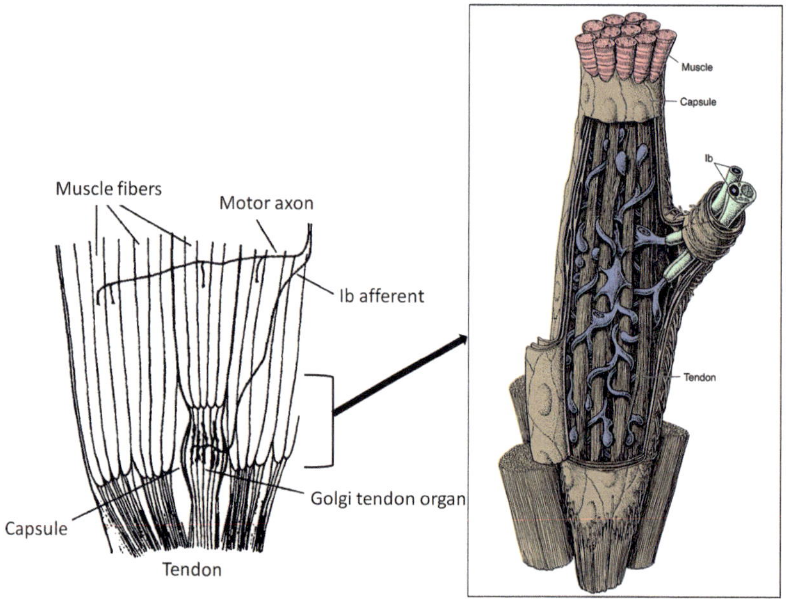

Fig. 13.8 Illustration of a Golgi tendon organ and its relationship to muscle fibers, with magnified illustration (right) of the area. A capsule surrounds those collagen fibers that connect a set of muscle fibers to the common tendon. A group of Ib afferent axon enters the capsule and splits into many fine, unmyelinated branches that intertwine among the collagen fibers. When the tendon is stretched, especially by a muscle contraction, these sensory endings are compressed and increaser their firing rate. Note that each tendon organ is connected in series to a small number of muscle fibers (10–20), but because these are typically from different motor units, the firing rate of the afferent fiber is sensitive to the effects of recruitment. (Adapted from Downey et al. (1994), and with permission from Vanderah and Gould (2016))

dynamic phases during length modification and static phases after achieving steady-state length (Matthews 1964).

Both primary and secondary endings maintain baseline discharge patterns reflecting muscle length during static conditions. Muscle stretching leads to increased firing rates in both endings to higher steady-state levels, while shortening results in decreased rates corresponding to reduced length (Hunt 1990).

Primary endings show characteristic responses during dynamic phases, generating high-frequency bursts during muscle stretch that exceed steady-state rates. These firing rates correspond to length change velocity, with higher rates during faster stretches. During rapid muscle shortening, primary endings typically pause before establishing new steady-state patterns. This velocity sensitivity makes primary endings responsive to vibration and muscle taps, while secondary endings show minimal response to these stimuli (Proske and Stuart 1985).

Primary endings exhibit maximum sensitivity to length changes below 0.1 mm, with diminished dynamic sensitivity for larger changes. They can modify their responsiveness after reaching new lengths, allowing detection of small changes from various positions. This adaptation creates complex relationships between muscle length, change rate, and firing patterns. The firing properties indicate primary endings mainly detect small movements and unexpected length-change rates rather than providing direct velocity measurements. This complexity prevents simple mathematical correlations between firing rates and muscle parameters (Banks 1994).

13.5 Dynamic and Static Gamma Motor Neurons in Muscle Spindle Control

The understanding of muscle spindle motor innervation emerged through research conducted from 1945 to 1955. Leksell's study revealed that the stimulation of small γ-motor axons increased spindle afferent discharge without significant changes in muscle tension. This finding, validated by single-fiber recordings, established the distinct fusimotor system for spindles, functioning concurrently with the skeletomotor (α-motoneuron) system (Hultborn 2006; Hunt 1951).

Gamma motor neurons modify spindle sensitivity through intrafusal fiber contraction. Two types of γ-motoneurons exist, each exerting distinct effects on spindle responses. Static γ-neurons enhance steady-state discharge from primary afferents and secondary endings during constant muscle length. Dynamic γ-neurons enhance high-frequency bursts during length changes while exerting minimal influence on static responsiveness.

The selective activation of these neuron types allows the central nervous system to adjust incoming sensory information. Static γ-neuron activation emphasizes muscle length information, while dynamic γ-neuron activation enhances detection of rapid length changes. This dual control system enables modification of proprioceptive feedback based on movement requirements (Matthews 1962). Gamma motor neurons receive extensive supraspinal input, allowing precise control of spindle sensitivity during various motor tasks. This control system plays a key role in movement coordination and postural stability (Burke 1988).

13.6 Reflex Pathways of Muscle Spindles and Tendon Organs

Muscle stretch reflexes represent fundamental neural mechanisms that generate automatic muscle contractions in response to passive muscle lengthening. The pioneering work of Sherrington provided essential insights into the spinal pathways underlying these reflexes through systematic studies involving dorsal and ventral root sectioning (Granit 1975; Burke et al. 1970). His research using decerebrate cat preparations demonstrated that stretch reflexes comprise two primary components: a rapid phasic response that occurs immediately following changes in muscle length, and a sustained tonic response that persists during maintained stretch.

The phasic component manifests clinically as the tendon jerk reflex, while tonic reflexes become apparent during active muscle contraction. Through detailed anatomical and physiological investigations, researchers identified that Ia afferent fibers establish direct monosynaptic connections with α-motoneurons in the spinal cord (Pierrot-Deseilligny and Burke 2005). This specialized anatomical arrangement enables rapid signal transmission and precise coordination of reflex responses across multiple muscle groups.

Sherrington's studies revealed the presence of coordinated activation patterns between different muscle groups, specifically the concurrent activation of synergist muscles accompanied by relaxation of antagonist muscles—a phenomenon termed reciprocal innervation (Burke et al. 1970; Enoka 2008). This organization allows for smooth, controlled movements and maintenance of posture. The stretch reflex system demonstrates remarkable plasticity, showing substantial adaptation through training and in response to injury (Matthews 1972; Rothwell 1994).

Current understanding indicates that stretch reflexes function as an integrated component of motor control, working in conjunction with descending motor signals from supraspinal centers. This integration enables precise movement control and postural stability. Multiple studies have shown complex interactions between spinal reflex circuits and higher-level control mechanisms in the modulation of these responses (Pierrot-Deseilligny and Burke 2005).

13.6.1 Spinal Pathways of Group Ia Afferents

Group Ia afferents enter the spinal cord through dorsal roots and branch extensively. Their collaterals form direct excitatory connections with α-motoneurons innervating the same muscle (homonymous) and similar (synergist) muscles. A single Ia afferent typically connects with most motor neurons supplying its muscle of origin, creating strong autogenic excitation.

The synaptic connections between Ia afferents and synergist motor neurons show variable strength based on mechanical action similarity. These connections create coordinated activation patterns across functionally related muscle groups. Additionally, Ia collaterals connect with inhibitory interneurons that project to antagonist muscle motor neurons, establishing disynaptic inhibition pathways (Pierrot-Deseilligny 1996).

Ia afferents also form connections with propriospinal and other interneurons, participating in complex reflex circuits. Some collaterals extend through the dorsal columns to brainstem nuclei, eventually reaching cortical regions. This ascending pathway contributes to conscious proprioception and movement control (Jankowska 1992; Hultborn 2006).

Mechanisms of the Monosynaptic Stretch Reflex

The monosynaptic stretch reflex functions through specific spinal pathways. When flexor muscles lengthen passively, Ia fiber discharge increases, leading to monosynaptic excitation of homonymous and synergist muscles. Simultaneously, antagonist motor neurons receive inhibition through interneurons, promoting coordinated muscle response (Baehr and Frotscher 2012; Pierrot-Deseilligny and Burke 2005) (Fig. 13.9).

While this reflex strongly maintains joint position in decerebrate animals, intact subjects show predominantly phasic responses. The primary endings' sensitivity to dynamic length changes requires rapid muscle stretch to elicit observable reflexes. Tendon taps generate vibration waves through muscle tissue, stimulating multiple primary endings simultaneously (Burke et al. 1983). The extensive convergence and divergence of Ia afferents generate temporal and spatial summation in motor neurons, resulting in characteristic muscle contractions. Despite being referred to as "deep tendon reflexes," these responses actually originate from muscle receptors rather than tendon organs. Tendon taps induce muscle vibrations that stimulate numerous primary endings (Hultborn 2006; Matthews 1964).

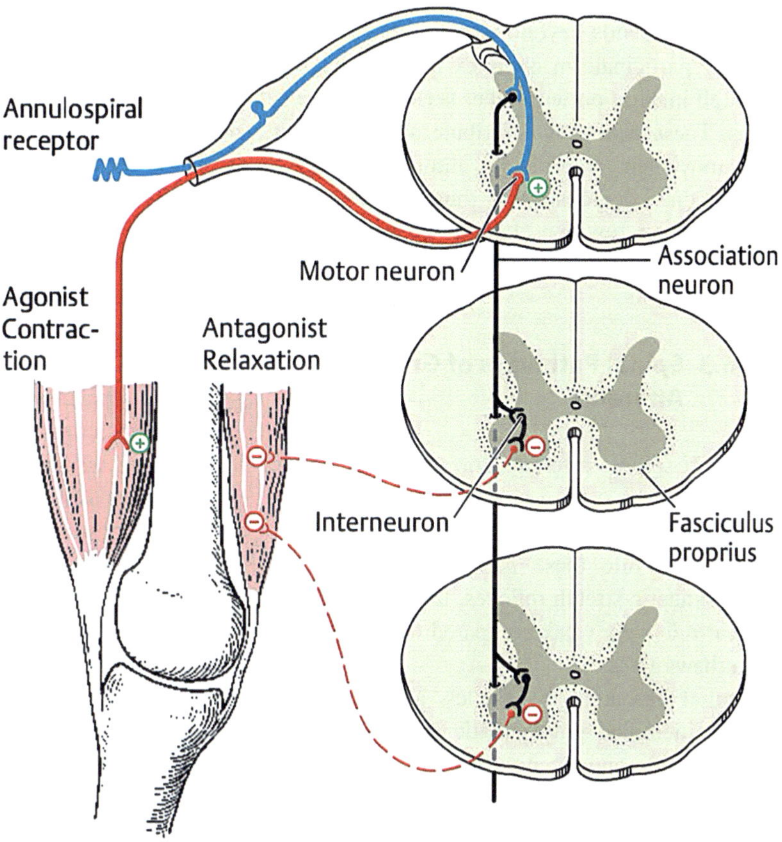

Fig. 13.9 Monosynaptic reflex with polysynaptic inhibition of antagonist muscles. (Adapted with permission from Baehr and Frotscher (2012))

13.6.2 Spinal Pathways of Group Ib Afferents from Golgi Tendon Organs

The neural circuitry of Group Ib afferents originating from Golgi tendon organs exhibits complex patterns of spinal connectivity that contribute significantly to motor control. These afferents establish polysynaptic connections within the spinal cord, primarily forming disynaptic or trisynaptic inhibitory pathways to homonymous motor neurons, a process known as autogenic inhibition. Simultaneously, they facilitate the activation of motor neurons innervating antagonist muscles. This organization creates a response pattern that functions oppositely to muscle spindle reflexes, leading to its classification as an inverse myotatic reflex (Jami 1992; Pierrot-Deseilligny and Burke 2005).

The integration of Ib afferent signals occurs through specialized interneurons that receive convergent inputs from multiple sensory receptors and descending motor pathways. These interneuronal networks extend beyond local circuits, influencing motor neurons of muscles throughout the limb. Group Ib pathways work collaboratively with group II afferents to regulate coordinated limb movements (Dietz 2002; Jankowska 2001; Pierrot-Deseilligny 2002).

Ib afferent pathways are important in maintaining postural stability and controlling locomotion. The activity patterns of these neural circuits show task-dependent modulation, adapting their responses based on specific movement phases and behavioral requirements (Eccles et al. 1957). This adaptive capability allows for precise force regulation during different motor tasks.

The functional significance of Ib afferent pathways extends beyond simple reflex actions, as they participate in complex motor behaviors through integration with other sensorimotor systems. These pathways contribute to both voluntary movement control and automatic postural adjustments, demonstrating their importance in overall motor function (Jankowska and McCrea 1983; Lundberg et al. 1987).

13.6.3 Spinal Pathways of Group II Afferents

Group II spindle afferents form complex spinal circuits that have been challenging to study due to overlap with other afferent fibers of similar diameter. While these afferents contribute to monosynaptic stretch reflexes, their connections appear relatively weak compared to Ia pathways (Matthews 1972).

Initial research by Eccles and Lundberg (1959) identified polysynaptic pathways where Group II afferents excited flexor motor neurons while inhibiting extensor motor neurons. However, subsequent studies revealed these effects partially originate from free nerve endings, forming part of the flexion reflex afferent system (Hultborn 2006; Jankowska 1992).

Lundberg's research suggests supraspinal control can modulate these pathways. These circuits participate in coordinated limb movements, particularly during locomotion, where they influence contraction timing and magnitude. Recent studies indicate these pathways show state-dependent modulation based on motor task requirements (Edgley 2001; Hultborn 2006; Pierrot-Deseilligny and Burke 2005).

13.7 Pathophysiological Mechanisms of Exaggerated Stretch Reflex Activity in Upper Motor Neuron Lesion

Spinal motor neurons serve a fundamental role in activating muscle fibers for posture and movement control. Following central motor pathway lesions, the spinal cord undergoes complex adaptations to maintain functional output despite reduced descending control. These changes occur as compensatory mechanisms when normal descending excitation diminishes .

The adaptive processes involve multiple levels of neuronal reorganization. At the cellular level, motor neurons exhibit increased membrane excitability through ion channel up-regulation and lowered action potential thresholds. Structurally, surviving descending fibers, sensory afferents, and local interneurons demonstrate sprouting, while motor neuron dendrites extend and form new synaptic connections.

Synaptic transmission undergoes significant modification through altered presynaptic inhibition mechanisms, enhancing the efficiency of remaining inputs. The interneuronal network shows widespread changes, including increased excitatory and decreased inhibitory influences on motor neurons. These adaptations represent coordinated responses aimed at preserving muscle activation patterns (Nielsen et al. 2007).

Research indicates these changes follow specific temporal sequences depending on lesion type and location. While these adaptations help maintain spinal output, they may not fully restore normal motor control. Understanding these mechanisms provides insights for developing targeted rehabilitation strategies and managing spasticity (Dietz and Sinkjaer 2007). This plasticity involves complex molecular signaling pathways and activity-dependent modifications. These findings suggest potential therapeutic interventions to optimize beneficial adaptations while minimizing maladaptive changes (Rossignol and Frigon 2011).

13.7.1 Motor Neuronal Changes

Neuronal membrane excitability undergoes dynamic regulation through receptors and ion channels controlled by intracellular signaling systems. After major excitatory input loss, significant molecular changes occur within cells (Heckman et al. 2008).

Voltage-dependent Ca^{2+} and Na^{+} persistent inward currents (PICs) regulate motor neuronal excitability. These currents generate plateau potentials when facilitated by serotonergic and noradrenergic input. PICs enhance excitatory synaptic inputs, enabling motor neurons to achieve firing frequencies necessary for movement control (Powers and Binder 2001).

Following spinal cord injury, PICs initially decrease during spinal shock but may return in chronic stages without neurotransmitter intervention. This recovery involves upregulation of monoaminergic receptors, particularly 5-HT(2B) and 5-HT(2C) subtypes. Similar adaptations likely occur in interneurons, though direct evidence remains limited (Murray et al. 2010).

Human studies suggest PICs contribute to spasticity and spasms after spinal cord injury. Motor units show altered recruitment patterns during spasms, requiring less synaptic drive for sustained activation. These changes appear linked to serotonergic receptor upregulation, primarily in spinal cord injury rather than supraspinal lesions, explaining the rarity of spasms in stroke patients (Gorassini et al. 2009). The specific role of PICs in different types of central nervous system lesions requires further investigation using novel experimental approaches (D'Amico et al. 2017).

13.7.2 Neuroplastic Changes and Synaptic Reorganization

After central nervous system injury, structural adaptations occur through dendritic outgrowth, collateral sprouting from surviving neurons, and the formation of new synapses. These changes affect descending pathways to spinal motor neurons, particularly following spinal cord injury (Raineteau and Schwab 2001). These adaptations contribute to motor control alterations, including spasticity and reduced movement specificity. Patients may experience difficulty performing precise movements, show involuntary associated movements, and develop spastic postures during activity. While muscle afferents likely participate in synaptic reorganization, direct evidence remains limited (Bareyre et al. 2004).

The reorganization process involves multiple mechanisms, including axonal sprouting, dendritic remodeling, and synapse formation. These changes follow specific temporal patterns and may be influenced by rehabilitation interventions (Fouad and Tetzlaff 2012; Weidner et al. 2001).

13.7.3 Presynaptic Regulation

Central nervous system injury triggers presynaptic modulation of synaptic transmission. While research emphasis remains on sensory afferent changes in spasticity, surviving descending fibers and interneurons undergo upregulation to maintain motor neuron activation (Rudomin 2009). Studies indicate a pathway-specific distribution of presynaptic inhibition. Presynaptic inhibition mediated by sensory afferents selectively preserves descending motor axons, whereas post-activation depression affects only sensory pathways (Nielsen et al. 2007).

Descending pathways exhibit distinct regulatory mechanisms contributing to motor control adaptation. These changes include altered neurotransmitter release probability and synaptic strength modification, though detailed mechanisms require further investigation (Alvarez and Fyffe 2007). Understanding pathway-specific presynaptic modulation advances knowledge of spasticity mechanisms. Recent research focuses on identifying molecular targets for selective regulation of specific pathways (Brownstone and Bui 2010).

13.7.4 Presynaptic Inhibition

Presynaptic inhibition of sensory afferents involves GABAergic interneurons receiving inputs from diverse sensory afferents and descending pathways. These interneurons suppress transmitter release through GABA-mediated primary afferent depolarization, reducing subsequent action potential amplitude and neurotransmitter release (Rudomin and Schmidt 1999).

Supraspinal centers regulate presynaptic inhibition during voluntary movement and locomotion. This modulation shapes sensory input to spinal motor neurons according to task requirements, facilitating specific muscle activation while preventing unwanted antagonist activation (Nielsen et al. 2007).

Following central nervous system injury, presynaptic inhibition shows varying changes. Spinal cord injury and multiple sclerosis patients exhibit reduced presynaptic inhibition correlating with spasticity development. However, stroke patients show decreased inhibition only in upper limb muscles, suggesting this mechanism may represent an epiphenomenon rather than direct spasticity causation (Kagamihara and Masakado 2005).

GABAergic medications effectively reduce spasticity, though this may not specifically indicate presynaptic inhibition's role. These drugs generally suppress synaptic transmission and motor neuron excitability through multiple mechanisms.

13.7.5 Recurrent Inhibition

Recurrent inhibition operates through Renshaw cells located in the spinal ventral horn. These cells receive motor axon collaterals and project to motor neurons and Ia inhibitory interneurons (see Fig. 2.4). Pierrot-Deseilligny developed techniques using H-reflex testing to study this system in humans (Katz and Pierrot-Deseilligny 1982).

Studies indicate recurrent inhibition remains largely normal in spasticity at rest. Only progressive paraparesis and amyotrophic lateral sclerosis show reduced inhibition, though this likely does not contribute significantly to spasticity development (Mazzocchio and Rossi 1989). Research suggests recurrent inhibition may demonstrate task-dependent modulation during movement, though its contribution to spasticity pathophysiology appears minimal. Understanding these mechanisms provides insights into motor control regulation in both healthy and pathological conditions (Hultborn and Nielsen 2007).

13.8 Pathophysiological Mechanisms of Clonus in Spasticity

Clonus manifests as rhythmic oscillations (3–8 Hz) at distal joints, particularly the ankle, following rapid muscle stretch. While commonly observed in spasticity with exaggerated stretch reflexes, many patients with severe spasticity and pronounced muscle stiffness do not exhibit clonus, suggesting different underlying mechanisms (Mukherjee and Chakravarty 2010).

Two principal mechanisms explain clonus generation. The traditional theory proposes alternating stretch reflexes in antagonist muscles, supported by observations that clonus frequency correlates with limb segment length. Studies demonstrate clonus cessation during sensory afferent blockade and its reset by peripheral nerve stimulation. Research reveals central pattern generator involvement. EMG recordings show synchronization between antagonistic muscles during clonus, independent of mechanical stretch. This mechanism explains observations of clonus triggered by cutaneous stimulation and its persistence despite mechanical loading modifications (Burke et al. 2013; Dietz and Sinkjaer 2007).

13.9 Pathophysiological Mechanisms of Spasm in Spasticity

Spasms represent involuntary muscle contractions lasting seconds to minutes, typically triggered by sensory stimuli. These contractions differ from spasticity through longer duration, greater strength, and preferential activation by cutaneous rather than muscle afferent inputs. Spasms can reach intensities causing significant pain (Sheean 2002).

The distribution of spasms shows characteristic patterns, occurring predominantly in spinal cord injury while rarely presenting in stroke patients, even those with severe stretch reflex exaggeration. This pattern suggests fundamentally different pathophysiological mechanisms from typical spastic responses (Ivanhoe and Reistetter 2004).

Research indicates spasms arise from increased transmission in flexor reflex afferent circuits. Studies reveal that upregulation of 5-HT receptors following loss of serotonergic innervation leads to enhanced persistent inward currents in motor neurons and flexor reflex afferent circuit interneurons. This mechanism explains the prevalence of spasms in spinal cord injury compared to stroke or cerebral palsy (Adams and Hicks 2005; Nielsen et al. 2007).

13.10 Anatomical Basis of Spastic Postures

The development of abnormal postures in spasticity and upper motor neuron syndrome results from various interacting mechanisms at the musculoskeletal level (Lance 1980a, b). These postural deviations present through various biomechanical patterns, resulting from alterations in muscle tone, alterations in viscoelastic characteristics, and changes in motor control mechanisms (Sheean 2002). The eventual postural configuration at any joint is determined by the mechanical interactions between antagonistic muscle groups, with relative strength, changes in muscle tone, and altered biomechanical characteristics being significant factors (Pandyan et al. 2005).

Research has shown that the force vectors generated by these muscular elements combine with modified tissue characteristics to produce specific postural patterns. While these patterns follow predictable biomechanical principles, individual variations exist based on the severity and distribution of neurological involvement. The mechanical output at each joint results from the combined action of primary movers and their antagonists, along with contributions from secondary muscle groups. This relationship provides the foundation for understanding how spasticity affects joint positioning (Ada et al. 2006).

When analyzing spastic postures, clinicians must consider the biomechanical aspects beyond simple force generation. The analysis needs to account for the muscles' moment arms and the influence of gravitational forces. This comprehensive approach requires thorough knowledge of joint architecture and associated muscle groups. The anatomical organization of muscles, particularly their attachment points and lines of action, enables clinicians to effectively analyze movement patterns and understand postural deviations (Burke et al. 2013).

In clinical practice, the examination of patients with spastic postures requires consideration of how altered neural input affects muscle function across different joint positions. Clinicians must evaluate the primary muscles contributing to observed postures while assessing how different muscle groups interact under varying conditions. The impact of gravity and positioning on postural maintenance becomes particularly relevant when planning therapeutic interventions. This understanding helps determine which muscles primarily contribute to specific postural patterns and how they might respond to various treatment approaches (Wissel et al. 2009).

The biomechanical analysis extends beyond individual muscle actions to include the complex interaction between muscle length–tension relationships and joint position. These relationships become particularly important when considering the long-term effects of spasticity on joint mobility and function. The sustained abnormal postures can lead to adaptive changes in muscle properties and joint structures, further complicating the clinical picture.

This section will describe spastic postures, focusing on the prevalent abnormal postures observed in patients with spinal cord injuries, particularly high-level tetraplegics, moving from distal to proximal joints in the upper and lower extremities.

13.10.1 Spastic Posture Patterns in the Upper Extremity

13.10.1.1 Spastic Posture Patterns of the Hand

Upper extremity spasticity manifests through characteristic patterns of hand involvement, reflecting complex interactions between neural control mechanisms and musculoskeletal anatomy (Trompetto et al. 2014). Understanding these patterns requires analysis of both neurophysiological changes and biomechanical consequences at the joint and muscle level (Li et al. 2019).

The clenched fist pattern represents a primary manifestation of hand spasticity, characterized by sustained finger flexion. This pattern emerges from abnormal activation of both extrinsic and intrinsic finger flexors, with variable wrist positioning depending on the relative involvement of wrist flexors and extensors (Sommerfeld et al. 2012). The sustained muscle activation leads to changes in muscle fiber composition and mechanical properties over time (Gracies 2005). These alterations can result in reduced sarcomere length and increased passive stiffness, resulting in the maintenance of the flexed posture. Clinical implications extend beyond functional limitations, as the sustained finger flexion may compromise palm hygiene, creating conditions conducive to skin maceration and ulceration in the palmar surface.

Flexion and Extension

A more complex pattern involves simultaneous flexion at the metacarpophalangeal (MCP) joints with extension at the interphalangeal (IP) joints. This pattern demonstrates the intricate interaction between neural control and musculoskeletal mechanics (Kamper and Rymer 2014). The posture results from coordinated activity between extrinsic finger flexors and extensors in the forearm, modulated by the unique action of lumbrical muscles (Fig. 13.10). These intrinsic hand muscles originate from the flexor digitorum profundus (FDP) tendons and insert into the extensor digitorum tendons, providing a mechanical advantage that simultaneously flexes the MCP joints while extending the IP joints (Santello et al. 2019).

Thumb Curling

Thumb involvement in spastic patterns presents particular characteristics due to its unique anatomical and functional properties. The thumb's metacarpophalangeal joint demonstrates greater mobility compared to other digits, allowing complex movements in multiple planes (van

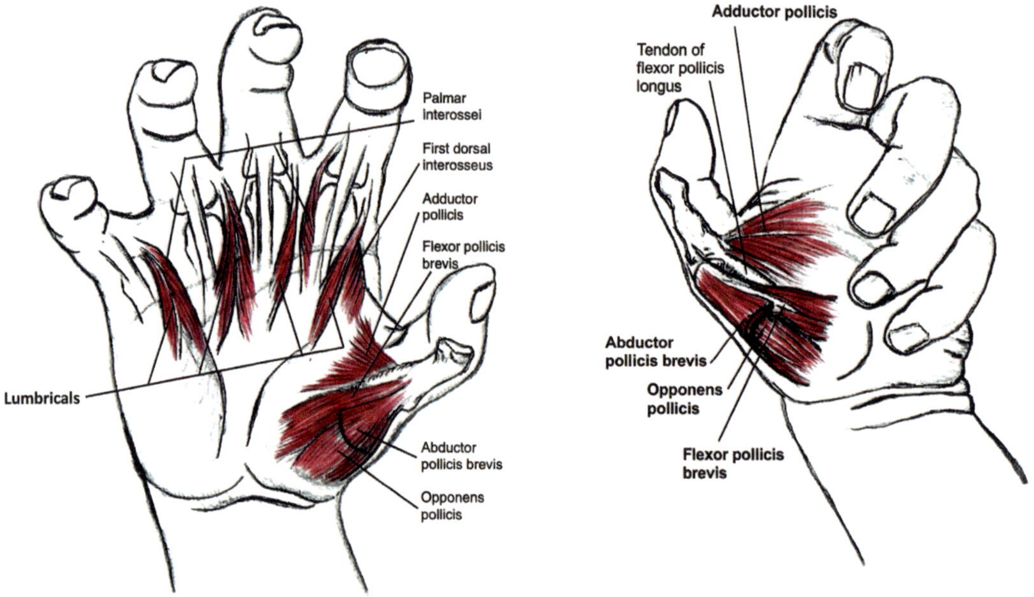

Fig. 13.10 Adductor pollicis brevis, flexor pollicis brevis, opponens pollicis, adductor pollicis, palmar interossei, dorsal interossei, and lumbricals. (Adapted with permission from Truong et al. (2009))

Oudenaarde et al. 1997). This mobility results from the unique articular surface geometry and ligamentous structure of the carpometacarpal and metacarpophalangeal joints.

In spastic conditions, the thumb typically assumes one of two primary positions: curled over the flexed fingers or adducted and flexed into the palm. These positions result from the combined action of multiple muscle groups with different mechanical advantages (Bhadra et al. 2001). The thenar muscles—flexor pollicis brevis, abductor pollicis brevis, and opponens pollicis—originate from the carpal bones and flexor retinaculum, providing the primary forces for thumb positioning. The first two muscles insert on the proximal phalanx, while the opponens attaches to the first metacarpal, creating a force couple that produces anterior and medial movement.

The adductor pollicis contributes additional forces through its unique anatomical position between the first two metacarpals. This muscle not only adducts the thumb but also influences MCP joint flexion through its line of pull. The flexor pollicis longus, extending from the anterior radius to the distal phalanx, completes the mechanical system by controlling IP joint position (Fig. 13.11). These combined actions form the characteristic thumb postures observed in spasticity, with implications for both function and therapeutic intervention.

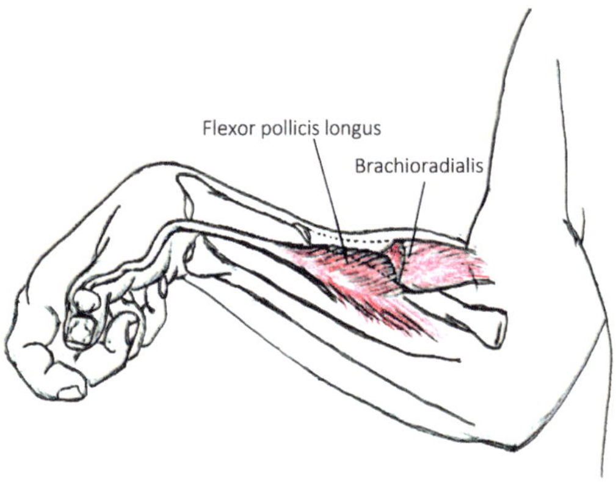

Fig. 13.11 Flexor pollicis longus and brachioradialis. (Adapted with permission from Truong et al. (2009))

13.10.1.2 Spastic Posture Patterns of the Wrist

Flexion

Wrist flexion represents the predominant pattern in spasticity, emerging from coordinated activation of multiple muscle groups (Sheean 2002). Research indicates that this pattern occurs in approximately 60–70% of patients with upper motor neuron lesions, affecting functional recovery and rehabilitation outcomes (Lundström et al. 2008).

The primary flexor system comprises three main muscles, each contributing unique mechanical advantages to wrist positioning. The flexor carpi radialis (FCR) originates from the common flexor tendon at the medial epicondyle of the humerus and follows a longitudinal course through the medial anterior forearm to insert into the second and third metacarpals. This anatomical arrangement generates both flexion and slight abduction moments at the wrist, with a mechanical advantage that varies throughout the range of motion (Lieber and Fridén 2019). Biomechanical studies demonstrate that the FCR generates its maximum force at approximately 15° of wrist extension (de Bruin et al. 2011).

The palmaris longus, while variable in its presence and morphology, contributes additional flexion force when present. This muscle shares its origin with the FCR and courses parallel to its medial border, ultimately merging with the palmar aponeurosis. Anatomical studies indicate that this muscle may be absent in approximately 15% of individuals or present in various structural forms. When present, it contributes approximately 2–8% of total wrist flexion force (Moore et al. 2018).

The flexor carpi ulnaris (FCU) adds complexity to wrist control through its unique dual-headed structure. Its humeral head arises from the common flexor tendon, while the ulnar head originates from the olecranon's medial surface. The FCU's insertion into the pisiform carpal bone creates a strong flexion moment arm, generating approximately 30–35% of total wrist flexion torque (de Bruin et al. 2011). Electromyographic studies demonstrate that

FCU activation patterns become significantly altered in spastic conditions, showing increased co-contraction with wrist extensors (Trompetto et al. 2014).

Extension

Extension patterns occur less frequently in spasticity, appearing in approximately 20–30% of cases, due to the biomechanical disadvantage of the extensor muscle group (Li et al. 2018). The wrist extension system involves three primary muscles working in a coordinated manner to produce controlled movement, with each muscle contributing specific directional components to the overall motion pattern.

The extensor carpi radialis longus demonstrates unique biomechanical properties due to its course across both the elbow and wrist joints. This muscle originates from the lateral supracondylar ridge and lateral epicondyle of the humerus, inserting into the second metacarpal base. Biomechanical studies indicate that it generates maximum tension at approximately 20° of wrist flexion and contributes about 40% of total wrist extension force (Gracies 2015).

The extensor carpi radialis brevis parallels this course, originating from the common extensor tendon and inserting into the third metacarpal base. These muscles work synergistically to produce wrist extension with an abduction component, generating approximately 35% of total extension torque. Their combined action creates a moment arm that varies with forearm position, reaching maximum efficiency in neutral forearm rotation (Li et al. 2018).

The extensor carpi ulnaris complements this system through its origin at the common extensor tendon and insertion along the fifth metacarpal's dorsal surface. This arrangement provides both extension and adduction moments at the wrist, contributing approximately 25% of the total extension force. Its activation patterns become particularly disrupted in spastic conditions, leading to altered wrist kinematics (Kollitz et al. 2015).

A clinically significant observation involves the mechanical coupling between wrist position and finger control. Spastic activation of wrist extensors often produces secondary MCP joint flexion due to passive tension development in the long finger flexors (tenodesis effect of the fingers). This mechanical interaction creates a force–length relationship that affects approximately 70–80% of patients with wrist extensor spasticity (Urban et al. 2010).

13.10.1.3 Spastic Posture Patterns of the Elbow

Joint Anatomy

The elbow's functional architecture enables four primary movement patterns: flexion, extension, pronation, and supination. The humeroulnar joint, functioning as a modified hinge mechanism, facilitates flexion–extension movements through the articulation between the humeral trochlea and the ulnar trochlear notch. The humeroradial joint, operating as a biaxial articulation, enables both rotational movement along the radial longitudinal axis and participates in flexion-extension motions (Stroyan and Wilk 1993). These joint structures work synergistically to produce coordinated upper extremity movements, with their alignment and configuration significantly influencing spastic posture development.

Flexion and Supination

Flexion combined with supination represents the predominant spastic pattern at the elbow due to the mechanical advantage and relative strength of these muscle groups (Li et al. 2019). The principal muscles involved demonstrate specific force-generating capabilities in different positions, creating a complex biomechanical system.

The biceps brachii, consisting of two heads that originate from different parts of the scapula (the short head from the coracoid process and the long head from the supraglenoid tubercle) (Fig. 13.12), exerts a maximum force during supination. The insertion onto the radial tuberosity and the formation of the bicipital aponeurosis, which extends into the medial forearm fascia (Fig. 13.13), generate effective flexion and powerful supination forces.

The brachialis, originating over the distal half of the anterior humerus and inserting onto the coronoid process and ulnar tuberosity, functions as a pure elbow flexor. Research indicates it generates approximately 40% of total elbow flexion torque, with maximum efficiency during pronation (Kholinne et al. 2018). Unlike the biceps, its anatomical arrangement prevents involvement in forearm rotation.

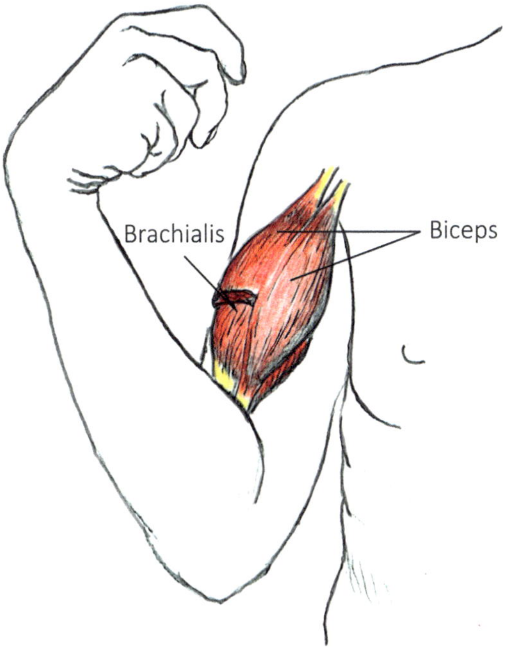

Fig. 13.12 Biceps and brachialis. (Adapted with permission from Truong et al. (2009))

The brachioradialis extends from the distal humerus to the radial styloid process (Fig. 13.14), demonstrating peak mechanical advantage in neutral forearm position. Biomechanical studies show it contributes approximately 15% of total flexion force (O'Sullivan et al. 2005).

Extension and Pronation

Extension patterns occur less frequently in spasticity, appearing in approximately 20–25% of cases. The triceps brachii, consisting of three heads (medial and lateral from the posterior humerus, long head from the infraglenoid tubercle), provides primary extension force through its insertion on the olecranon process. The anconeus,

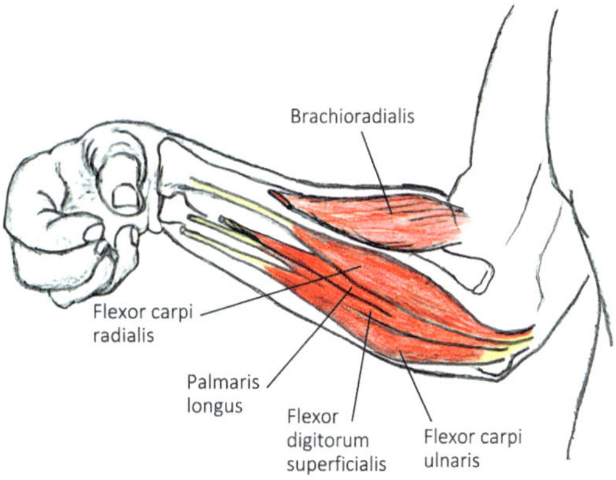

Fig. 13.14 Flexor digitorum superficialis, flexor carpi ulnaris, brachioradialis, flexor carpi radialis, and palmaris longus. (Adapted with permission from Truong et al. (2009))

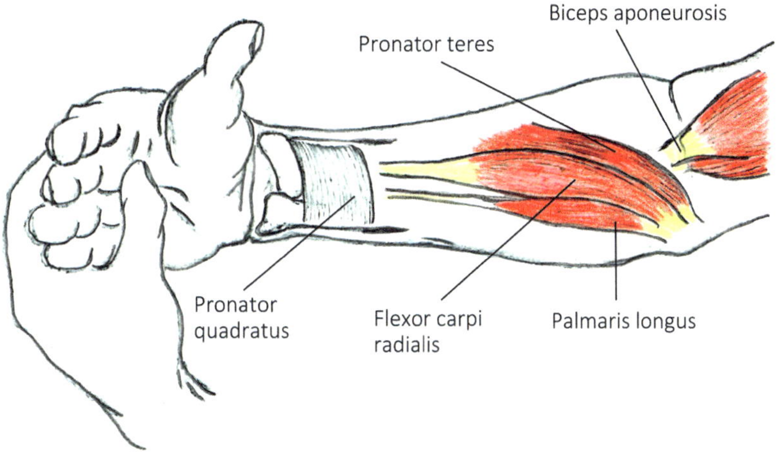

Fig. 13.13 Flexor carpi radialis, palmaris longus, pronator teres, and pronator quadratus. (Adapted with permission from Truong et al. (2009))

originating on the lateral humeral epicondyle and inserting on the proximal ulna, contributes to extension control and joint stability during rotational movements.

The pronator teres demonstrates a dual-headed structure: the humeral head from the medial epicondyle contributes to elbow flexion, while the ulnar head from the coronoid process generates pronation force. The pronator quadratus, positioned at the distal forearm between the ulna and radius, provides focused pronation force. Recent studies indicate these muscles work synergistically, with their relative contributions varying throughout the range of motion (Werner et al. 2014).

13.10.1.4 Spastic Postures of the Shoulder

Shoulder Adduction and Internal Rotation

The predominant spastic shoulder pattern involves adduction and internal rotation, reflecting the biomechanical advantage of these muscle groups (McPherson et al. 2018). This pattern occurs in approximately 50–70% of patients with upper motor neuron lesions and significantly impacts functional activities. The principal muscles contributing to this pattern include the pectoralis major, latissimus dorsi, teres major, and subscapularis, each providing specific force vectors that combine to form the characteristic posture. Studies demonstrate that this posture significantly affects activities of daily living, particularly in three key areas: dressing activities, mobility/transfer tasks, and axillary hygiene maintenance (Phadke et al. 2013).

Shoulder Subluxation

The initial flaccid phase following acute central nervous system injury creates a critical period of vulnerability for glenohumeral joint stability. During this period, which typically lasts 2–4 weeks, weakness in the rotator cuff and scapular stabilizing muscles may lead to inferior and/or anterior glenohumeral subluxation (Lo et al. 2003). This risk increases significantly during patient handling and transfers, as mechanical load on the paralyzed limb may exceed the impaired passive limitations of the joint.

Chronic subluxation may persist despite the development of spasticity, creating a complex clinical scenario. Research indicates that approximately 30–40% of patients develop persistent subluxation, which correlates with increased pain and reduced functional outcomes. When implementing interventions such as botulinum toxin for managing spasticity, clinicians must carefully consider the balance between reducing muscle overactivity and maintaining sufficient muscular support for joint stability. Particular attention should focus on preserving function in the rotator cuff muscles, deltoid, and pectoralis major and minor, as these muscles contribute significantly to dynamic joint stability.

Shoulder Hyperextension

While less common, occurring in approximately 15–20% of cases, shoulder hyperextension represents a specific pattern of spastic involvement (Datta et al. 1999). This pattern emerges when reduced activity in the subscapularis and pectoralis muscles allows unopposed action of the posterior muscle group, including the latissimus dorsi, teres major, posterior deltoid, and long head of triceps. Biomechanical studies demonstrate that this muscle imbalance creates a posterior force vector on the humerus.

The influence on shoulder flexion movement occurs through both mechanical and neurophysiological causes. EMG studies revealed altered activation patterns in the anterior shoulder muscles during attempted flexion, characterized by reduced acceleration and prolonged movement duration (Phadke et al. 2013).

13.10.2 Spastic Posture Patterns in the Lower Extremities

13.10.2.1 Spastic Posture Patterns of the Ankle and Foot

The ankle complex represents an intricate biomechanical system comprising two primary articulations: the talocrural joint (tibia-fibula-talus) and the subtalar joint (talus-calcaneus). The talocrural joint functions primarily in the sagittal plane, while the subtalar joint facilitates triplanar motion through its oblique axis orientation (Brockett and Chapman 2016). This anatomical arrangement allows four cardinal movements: plantar flexion (flexion), dorsiflexion (extension), inversion, and eversion, with the axis of rotation varying throughout the movement range (Kelikian 2011).

Plantar Flexion

The triceps surae muscle group generates approximately 80% of plantar flexion force through a complex force-coupling mechanism (Kay and Blazevich 2009). The gastrocnemius, comprising lateral and medial heads originating from the respective femoral condyles, exhibits type II fiber predominance and functions optimally at longer muscle lengths. The soleus, positioned deep to the gastrocnemius, contains predominantly type I fibers and maintains postural control through sustained low-level activation. These muscles converge into the Achilles tendon, which transmits force to the posterior calcaneus through a specialized enthesis structure. Biomechanical analyses reveal that the gastrocnemius components contribute 15–20% of knee flexion torque due to their bi-articular arrangement and moment arm variations throughout the range of motion (Barber et al. 2017).

Inversion

Multiple muscles contribute to foot inversion through complex force-coupling mechanisms. The triceps surae provides approximately 10% of inversion force during plantar flexion through its slight medial insertion vector (Hodges et al. 1997). The tibialis posterior, originating from the proximal posterior tibia, fibula, and interosseous membrane, functions as the primary invertor (Fig. 13.15). Its tendon courses posterior to the medial malleolus through a specialized retinacular system before inserting on multiple medial and plantar foot bones via complex anatomical expansions that stabilize the medial longitudinal arch.

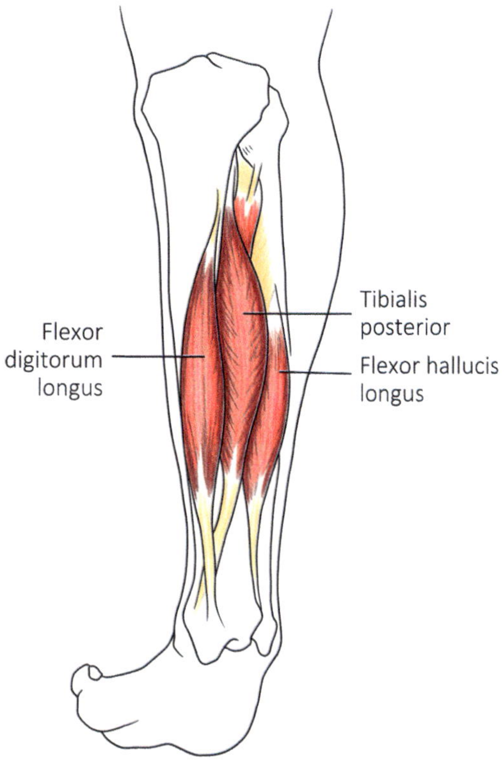

Fig. 13.15 Tibialis posterior, flexor digitorum longus, and flexor hallucis longus. (Adapted with permission from Truong et al. (2009))

The flexor digitorum longus and flexor hallucis longus, positioned in the deep posterior compartment, contribute additional inversion force through their anatomical orientation and mechanical advantage relative to the subtalar joint axis. Electromyographic studies demonstrate sequential activation patterns among these muscles during normal gait, with altered timing and magnitude in spastic conditions (Maharaj et al. 2016).

Eversion

Eversion patterns, though less common in spasticity (5–10% of cases), involve the coordinated action of the peroneal muscle group (Lamontagne et al. 2000). The peroneus longus and brevis, originating along the fibula, provide eversion force through specialized tendon pathways around the lateral malleolus. The longus exhibits a unique course beneath the foot, contributing to first ray plantar flexion and arch stability, while the brevis provides direct lateral foot control through its fifth metatarsal insertion.

Dorsiflexion

Dorsiflexion control involves complex neuromuscular coordination frequently impaired in spastic conditions. The tibialis anterior (Fig. 13.16), with its origin from the lateral tibial condyle and anteromedial shaft, provides approximately 60% of dorsiflexion force. Its anatomical arrangement creates a moment arm that varies with ankle position, maximizing mechanical advantage in plantar flexed positions. The extensor digitorum longus and extensor hallucis longus contribute additional dorsiflexion moments through their oblique force vectors relative to the talocrural axis.

Great Toe Extension

Great toe extension, occurring in 40–50% of spastic cases, primarily involves the extensor hallucis longus (Gracies et al. 2017) (see Fig. 13.16).

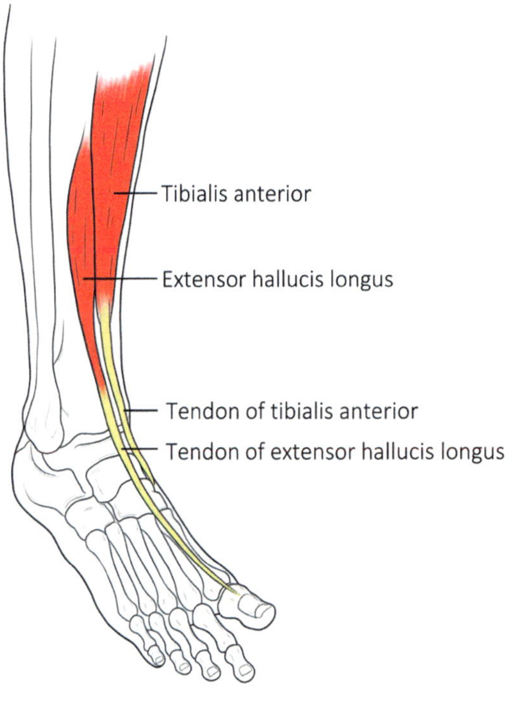

Fig. 13.16 Tibialis anterior and extensor hallucis longus. (Adapted with permission from Truong et al. (2009))

This muscle's distinctive anatomical alignment, extending from its anteromedial fibular origin to the distal hallux insertion, forms a complicated force vector that affects toe movement in both the sagittal and transverse planes. This muscle contains specific mechanoreceptors in the muscle-tendon unit that contribute to proprioceptive feedback and motor control.

13.10.2.2 Spastic Posture Patterns of the Knee

The knee's complexity results from its dual-joint arrangement, wherein the tibiofemoral and patellofemoral joints operate within the same capsular space. The convex femoral condyles glide against the tibial plateau, while menisci enhance the joint's stability and force transmission capabilities. At the front of the knee, the patellofemoral joint creates a pulley-like mechanism that optimizes extensor function throughout movement ranges (Wilson et al. 2003).

Although the knee permits various movement patterns, including limited rotation, spastic conditions primarily affect the flexion–extension cycle. Patients typically develop either flexion or extension patterns, each bringing unique functional implications and therapeutic challenges (Flandry and Hommel 2011).

Extension

Knee extension emerges from the coordinated work of the quadriceps femoris muscle group. Three muscles—vastus lateralis, medialis, and intermedius—attach directly to the femoral shaft. The fourth component, rectus femoris, begins at the anterior inferior iliac spine, crossing both the hip and knee joints. These muscles combine their forces through a common tendon that incorporates the patella before anchoring at the tibial tuberosity. Studies reveal that the rectus femoris contributes about 20% of hip flexion force due to its two-joint design.

The extension pattern in spasticity often becomes a functional adaptation. Many patients use this pattern to stabilize their knee during standing and walking activities, with research showing this occurs in 60–70% of cases. This adaptation raises important treatment considerations, as reducing extensor tone too aggressively might compromise a patient's ability to stand and move safely.

Flexion

Knee flexion relies primarily on the hamstring muscles, with three members—semitendinosus, semimembranosus, and biceps femoris long head—spanning from the ischium to the proximal tibia and fibula. This arrangement allows these muscles to flex the knee while extending the hip. The biceps femoris short head adds focused knee flexion through its connections between the femur and fibula. Together, these muscles create a coordinated flexion system that becomes altered in spastic conditions (Kellis 2016).

The flexion pattern typically creates more functional difficulties than extension (Li et al. 2018). This pattern affects a person's ability to stand upright, limits their walking step length, and changes how their foot contacts the ground during walking. These changes often lead to increased energy requirements during movement, making daily activities more difficult. Treatment approaches must carefully balance the need to reduce abnormal muscle activity while preserving functional abilities.

13.10.2.3 Spastic Posture Patterns of the Hip

The hip joint represents a synovial ball-and-socket articulation where the femoral head, covered with hyaline cartilage, articulates within the acetabulum. This architectural configuration, enhanced by the acetabular labrum, which increases joint surface area by 28%, provides remarkable stability while allowing triaxial motion (Neumann 2010). The joint capsule, reinforced by three major ligaments (iliofemoral, pubofemoral, and ischiofemoral), creates a strong envelope that can withstand forces up to 400% body weight during normal gait (Powers 2010).

Movement occurs in three cardinal planes, with normal ranges including flexion to 120° (knee flexed), extension to 30° (knee extended), abduction to 45°, adduction to 20–30°, and rotations of approximately 45°. These motions result from complex force couples generated by muscles arranged in specific functional groups, whose actions vary based on joint position and load conditions (Kapandji 2007).

Flexion

In spastic conditions, muscle activation patterns demonstrate characteristic alterations. The flexion pattern emerges primarily from three muscle groups working synergistically. The iliopsoas complex, comprising the psoas major (originating from T12–L5 vertebral bodies, discs, and transverse processes) and iliacus (from the iliac fossa), converges into a common tendon inserting on the lesser trochanter. EMG studies reveal altered timing and magnitude of activation in spastic conditions, with co-contraction patterns affecting 60–70% of patients.

Internal Rotation

The hip's rotational control involves complex muscular arrangements. Internal rotation stems from the coordinated action of gluteus minimus and medius, arranged in fan-like patterns on the external iliac fossa, and the tensor fascia latae. These muscles demonstrate specific fiber orientations optimized for their functions: anterior fibers for flexion/internal rotation, middle fibers for abduction, and posterior fibers for extension/external rotation.

External Rotation

External rotation involves six deep muscles collectively known as the lateral rotator group. The piriformis, passing through the greater sciatic foramen, demonstrates variable anatomical relationships with the sciatic nerve, affecting clinical presentations in 15–20% of cases. The obturator internus and externus provide stability through their perpendicular fiber arrangements relative to the femoral neck axis.

Adduction

Adduction patterns represent a significant clinical challenge, involving five primary muscles with distinct anatomical and functional characteristics. The adductor magnus, the largest of the group, shows a unique dual innervation pattern:

anterior fibers (obturator nerve) primarily control adduction, while posterior fibers (sciatic nerve) contribute to hip extension. Biomechanical studies demonstrate that this muscle generates approximately 70% of total adduction torque.

Abduction

Less commonly observed abduction patterns involve muscles arranged in three layers. The gluteus medius, with its three fiber orientations, provides the primary abduction force (40% of total). The gluteus minimus and tensor fasciae latae contribute additional force vectors, while their fascial connections create a dynamic tension system affecting both hip and knee mechanics (Ward et al. 2010).

Extension

Extension patterns, though less prevalent, involve complex biomechanical relationships. The gluteus maximus, the body's largest muscle by volume, demonstrates a multipennate fiber arrangement that optimizes force production. Its superficial fibers, oriented at approximately 35° to the femoral axis, contribute to extension and external rotation, while deeper fibers primarily stabilize the sacroiliac joint. Research indicates that spastic conditions can alter these functional relationships, affecting gait patterns in up to 85% of cases (Arnold et al. 2000).

References

Ada L, O'Dwyer N, O'Neill E. Relation between spasticity, weakness and contracture of the elbow flexors and upper limb activity after stroke: an observational study. Disabil Rehabil. 2006;28(13–14):891–7.

Adams MM, Hicks AL. Spasticity after spinal cord injury. Spinal Cord. 2005;43(10):577–86.

Adams MM, Ginis KAM, Hicks AL. The spinal cord injury spasticity evaluation tool: development and evaluation. Arch Phys Med Rehabil. 2007;88(9):1185–92.

Alstermark B, Isa T. Circuits for skilled reaching and grasping. Annu Rev Neurosci. 2012;35:559–78.

Alvarez FJ, Fyffe RE. The continuing case for the Renshaw cell. J Physiol. 2007;584(1):31–45.

Arnold AS, Asakawa DJ, Delp SL. Do the hamstrings and adductors contribute to excessive internal rotation of the hip in persons with cerebral palsy? Gait Posture. 2000;11(3):181–90.

Baehr M, Frotscher M. Duus' topical diagnosis in neurology. Anatomy, physiology, signs, symptoms. 5th ed. New York: Thieme; 2012.

Baker SN. The primate reticulospinal tract, hand function and functional recovery. J Physiol. 2011;589(23):5603–12.

Banks RW. The motor innervation of mammalian muscle spindles. Prog Neurobiol. 1994;43(4–5):323–62.

Banks RW. An allometric analysis of the number of muscle spindles in mammalian muscles. J Anat. 2006;208(6):753–68.

Banks RW, Hulliger M, Scheepstra KA, Otten E. Pacemaker activity in a sensory ending with multiple encoding sites: the cat muscle spindle primary ending. J Physiol. 1997;498(1):177–99.

Banks RW, Hulliger M, Saed HH, Stacey MJ. A comparative analysis of the encapsulated end-organs of mammalian skeletal muscles and of their sensory nerve endings. J Anat. 2009;214(6):859–87.

Barber L, Carty C, Modenese L, Walsh J, Boyd R. Medial gastrocnemius and soleus muscle-tendon unit, fascicle, and tendon interaction during walking in children with cerebral palsy. Dev Med Child Neurol. 2017;59(8):843–51.

Bareyre FM, Kerschensteiner M, Raineteau O, Mettenleiter TC, Weinmann O, Schwab ME. The injured spinal cord spontaneously forms a new intraspinal circuit in adult rats. Nat Neurosci. 2004;7(3):269–77.

Barnes MP, Johnson GR. Upper motor neurone syndrome and spasticity: clinical management and neurophysiology. 2nd ed. Cambridge: Cambridge University Press; 2008a.

Berardelli A, Hallett M, Kaufman C, Fine E, Berenberg W, Simon SR. Stretch reflexes of triceps surae in normal man. J Neurol Neurosurg Psychiatry. 1982;83(6):513–25.

Bhadra N, Kilgore KL, Peckham PH. Implanted stimulators for restoration of function in spinal cord injury. Med Eng Phys. 2001;23(1):19–28.

Boron WF, Boulpaep EL. Medical physiology. 3rd ed. Philadelphia: Elsevier; 2017.

Bosco G, Poppele RE. Proprioception from a spinocerebellar perspective. Physiol Rev. 2001;81(2):539–68.

Brockett CL, Chapman GJ. Biomechanics of the ankle. Orthop Trauma. 2016;30(3):232–8.

Brown P. Pathophysiology of spasticity. J Neurol Neurosurg Psychiatry. 1994;57(7):773–7.

Brownstone RM, Bui TV. Spinal interneurons providing input to the final common path during locomotion. Prog Brain Res. 2010;187:81–95.

Bucy PC. Studies on the human neuromuscular mechanism: II. Effect of ventromedial chordotomy on muscular spasticity in man. Arch Neurol Psychiatr. 1938;40(4):639–62.

Burke D. Spasticity as an adaptation to pyramidal tract injury. Adv Neurol. 1988;47:401–23.

Burke RE. Sir Charles Sherrington's the integrative action of the nervous system: a centennial appreciation. Brain. 2007;130(4):887–94.

Burke D, Gillies JD, Lance JW. The quadriceps stretch reflex in human spasticity. J Neurol Neurosurg Psychiatry. 1970;33(2):216–23.
Burke D, Gandevia SC, McKeon B. The afferent volleys responsible for spinal proprioceptive reflexes in man. J Physiol. 1983;339:535–52.
Burke D, Wissel J, Donnan GA. Pathophysiology of spasticity in stroke. Neurology. 2013;80(3 Suppl 2):S20–6.
Carpenter RHS, Reddi B. Neurophysiology: conceptual approach. 5th ed. London: Hodder Arnold; 2012.
Cleland CL, Rymer WZ. Neural mechanisms underlying the clasp-knife reflex in the cat. I. Characteristics of the reflex. J Neurophysiol. 1990;64(4):1303–18.
D'Amico JM, Butler AA, Héroux ME, Cotel F, Perrier JM, Butler JE, et al. Human motoneurone excitability is depressed by activation of serotonin 1A receptors with buspirone. J Physiol. 2017;595(5):1763–73.
Datta S, Sinclair E, Nair K, Poage C, Lohman E. Shoulder pain and dysfunction in hemiplegia: effects of functional electrical stimulation. Arch Phys Med Rehabil. 1999;80(3):328–31.
de Bruin M, Smeulders MJ, Kreulen M. Flexor carpi ulnaris tenotomy alone does not eliminate its contribution to wrist torque. Clin Biomech. 2011;26(7):725–8.
Dietz V. Proprioception and locomotor disorders. Nat Rev Neurosci. 2002;3:781–90.
Dietz V, Duysens J. Significance of load receptor input during locomotion: a review. Gait Posture. 2000;11(2):102–10.
Dietz V, Sinkjaer T. Spastic movement disorder: impaired reflex function and altered muscle mechanics. Lancet Neurol. 2007;6(8):725–33.
Dimitrijevic MR. Neurophysiology in spinal cord injury. Paraplegia. 1987;25(3):205–8.
Dimitrijevic MR, Nathan PW. Studies of spasticity in man. II. Analysis of stretch reflexes in spasticity. Brain. 1967;90(2):333–58.
Downey JA, Myers SJ, Gonzalez EG, Lieberman JS. The physiological basis of rehabilitation medicine. Stoneham: Boston: Butterworth-Heinemann; 1994.
Drew T, Dubuc R, Rossignol S. Discharge patterns of reticulospinal and other reticular neurons in chronic, unrestrained cats walking on a treadmill. J Neurophysiol. 1986;55(2):375–401.
Drew T, Prentice S, Schepens B. Cortical and brainstem control of locomotion. Prog Brain Res. 2004;143:251–61.
Dum RP, Strick PL. Frontal lobe inputs to the digit representations of the motor areas. J Neurosci. 2005;25(6):1375–86.
Eccles RM, Lundberg A. Supraspinal control of interneurones mediating spinal reflexes. J Physiol. 1959;147(3):565–84.
Eccles JC, Eccles RM, Lundberg A. The convergence of monosynaptic excitatory afferents on to many different species of alpha motoneurones. J Physiol. 1957;137(1):22–50.
Edgley SA. Organisation of inputs to spinal interneurone populations. J Physiol. 2001;533(1):51–6.
Engberg I, Lundberg A, Ryall RW. Reticulospinal inhibition of transmission in reflex pathways. J Physiol. 1968;194(1):201–23.
Enoka RM. Neuromechanics of human movement. 4th ed. Champaign: Human Kinetics; 2008.
Eriksson J, Olausson B, Jankowska E. Antispastic effects of L-dopa. Exp Brain Res. 1996;111(2):296–304.
Eyre JA. Corticospinal tract development and its plasticity after perinatal injury. Neurosci Biobehav Rev. 2007;31(8):1136–49.
Eyre JA, Taylor JP, Villagra F, Smith M, Miller S. Evidence of activity-dependent withdrawal of corticospinal projections during human development. Neurology. 2001;57(9):1543–54.
FitzGerald MJT, Gruener G, Mtui E. Clinical neuroanatomy and neuroscience. 6th ed. Philadelphia: Elsevier; 2012.
Flandry F, Hommel G. Normal anatomy and biomechanics of the knee. Sports Med Arthrosc Rev. 2011;19(2):82–92.
Flynn JR, Graham BA, Galea MP, Callister RJ. The role of propriospinal interneurons in recovery from spinal cord injury. Neuropharmacology. 2011;60(5):809–22.
Fouad K, Tetzlaff W. Rehabilitative training and plasticity following spinal cord injury. Exp Neurol. 2012;235(1):91–9.
Garcia-Rill E, Hyde J, Kezunovic N, Urbano FJ, Petersen E. The physiology of the pedunculopontine nucleus: implications for deep brain stimulation. J Neural Transm. 2015;122(2):225–35.
Gorassini MA, Knash ME, Harvey PJ. Changes in locomotor muscle activity after treadmill training in subjects with incomplete spinal cord injury. J Neurophysiol. 2009;101(2):969–79.
Gracies JM. Pathophysiology of spastic paresis. I: paresis and soft tissue changes. Muscle Nerve. 2005;31(5):535–51.
Gracies JM. Coefficients of impairment in deforming spastic paresis. Ann Phys Rehabil Med. 2015;58(3):173–8.
Gracies JM, Esquenazi A, Brashear A, Banach M, Kocer S, Jech R, et al. Efficacy and safety of abobotulinumtoxin A in spastic lower limb. Neurology. 2017;89(22):2245–53.
Granit R. The functional role of the muscle spindles: facts and hypotheses. Brain. 1975;98(4):531–56.
Grillner S, Hongo T, Lund S. The vestibulospinal tract. Effects on alpha-motoneurones in the lumbosacral spinal cord in the cat. Exp Brain Res. 1970;10(1):94–120.
Haines DE, Mihailoff GA. Fundamental neuroscience for basic and clinical applications. 5th ed. Philadelphia: Elsevier; 2018.
Heckman CJ, Hyngstrom AS, Johnson MD. Active properties of motoneurone dendrites: diffuse descending neuromodulation, focused local inhibition. J Physiol. 2008;586(5):1225–31.
Heckman CJ, Mottram C, Quinlan K, Theiss R, Schuster J. Motoneuron excitability: the importance of neuromodulatory inputs. Clin Neurophysiol. 2009;120(12):2040–54.

Herman R. The myotatic reflex: clinicophysiological aspects of spasticity and contracture. Brain. 1970;93(2):273–312.

Highstein SM, Holstein GR. The anatomy of the vestibular nuclei. Prog Brain Res. 2006;151:157–203.

Hodges PW, Butler JE, McKenzie DK, Gandevia SC. Contraction of the human diaphragm during rapid postural adjustments. J Physiol. 1997;505(2):539–48.

Holstege G, Kuypers HG. The anatomy of brain stem pathways to the spinal cord in cat. A labeled amino acid tracing study. Prog Brain Res. 1982;57:145–75.

Houk J, Henneman E. Responses of Golgi tendon organs to active contractions of the soleus muscle of the cat. J Neurophysiol. 1967;30(3):466–81.

Hultborn H. Spinal reflexes, mechanisms and concepts: from Eccles to Lundberg and beyond. Prog Neurobiol. 2006;78(3-5):215–32.

Hultborn H, Nielsen JB. Spinal control of locomotion—from cat to man. Acta Physiol. 2007;189(2):111–21.

Hultborn H, Brownstone RM, Toth TI, Gossard JP. Key mechanisms for setting the input-output gain across the motoneuron pool. Prog Brain Res. 2004;143:77–95.

Hunt CC. The reflex activity of mammalian small-nerve fibres. J Physiol. 1951;115(4):456–69.

Hunt CC. Mammalian muscle spindle: peripheral mechanisms. Physiol Rev. 1990;70(3):643–63.

Ivanhoe CB, Reistetter TA. Spasticity: the misunderstood part of the upper motor neuron syndrome. Am J Phys Med Rehabil. 2004;83(10 Suppl):S3–9.

Jami L. Golgi tendon organs in mammalian skeletal muscle: functional properties and central actions. Physiol Rev. 1992;72(3):623–66.

Jang SH. The corticospinal tract from the viewpoint of brain rehabilitation. J Rehabil Med. 2014;46(3):193–9.

Jankowska E. Interneuronal relay in spinal pathways from proprioceptors. Prog Neurobiol. 1992;38(4):335–78.

Jankowska E. Spinal interneuronal systems: identification, multifunctional character and reconfigurations in mammals. J Physiol. 2001;533(1):31–40.

Jankowska E. Spinal interneuronal networks in the cat: elementary components. Brain Res Rev. 2008;57(1):46–55.

Jankowska E, McCrea DA. Shared reflex pathways from Ib tendon organ afferents and Ia muscle spindle afferents in the cat. J Physiol. 1983;338:99–111.

Kagamihara Y, Masakado Y. Excitability of spinal inhibitory circuits in patients with spasticity. J Clin Neurophysiol. 2005;22(2):136–47.

Kamper DG, Rymer WZ. Quantitative features of the stretch response of extrinsic finger muscles in hemiparetic stroke. Muscle Nerve. 2014;49(5):721–33.

Kandel ER, Schwartz JH, Jessell TM, Siegelbaum SA, Hudspeth AJ. Principles of neural science. 5th ed. New York: McGraw-Hill; 2013.

Kapandji IA. The physiology of the joints: lower limb. 6th ed. Edinburgh: Churchill Livingstone; 2007.

Katz R, Pierrot-Deseilligny E. Recurrent inhibition of alpha-motoneurons in patients with upper motor neuron lesions. Brain. 1982;105(1):103–24.

Kay AD, Blazevich AJ. Isometric contractions reduce plantar flexor moment, Achilles tendon stiffness, and neuromuscular activity but remove the subsequent effects of stretch. J Appl Physiol. 2009;107(4):1181–9.

Kelikian AS. Sarrafian's anatomy of the foot and ankle: descriptive, topographic, functional. 3rd ed. Philadelphia: Wolters Kluwer; 2011.

Kellis E. Biceps femoris and semitendinosus tendon/aponeurosis strain during passive and active (isometric) conditions. J Electromyogr Kinesiol. 2016;26:111–9.

Kennedy PR. Corticospinal, rubrospinal and rubro-olivary projections: a unifying hypothesis. Trends Neurosci. 1990;13(12):474–9.

Kholinne E, Zulkarnain RF, Sun YC, Lim S, Chun JM, Jeon IH. The different role of each head of the triceps brachii muscle in elbow extension. Acta Orthop Traumatol Turc. 2018;52(3):201–5.

Ko HY. Management and rehabilitation of spinal cord injuries. 2nd ed. Singapore: Springer; 2022a.

Kollitz KM, Iorio ML, Huang JI. Assessment and treatment of extensor carpi ulnaris tendon pathology: a critical analysis review. JBJS Rev. 2015;3(6):e3.

Kucera J, Walro JM. Formation of muscle spindles in the absence of motor innervation. Neurosci Lett. 1992;145(1):47–50.

Kucera J, Walro JM, Reichler J. Neural organization of spindles in three hindlimb muscles of the rat. Am J Anat. 1991;190(1):74–88.

Kuypers HGJM. The organization of the motor system. Int J Neurol. 1963;4:78–91.

Kuypers HGJM. Anatomy of the descending pathways. In: Brooks VB, editor. Handbook of physiology, section 1: the nervous system, Vol. II, motor control, part 1. Bethesda: American Physiological Society; 1981. p. 597–666.

Lamontagne A, Malouin F, Richards CL. Contribution of passive stiffness to ankle plantarflexor moment during gait after stroke. Arch Phys Med Rehabil. 2000;81(3):351–8.

Lance JW. Pathophysiology of spasticity and clinical experience with Baclofen. In: Lance JW, Feldman RG, Young RR, editors. Spasticity: disordered motor control. Chicago: Year Book; 1980a. p. 185–204.

Lance JW. The control of muscle tone, reflexes, and movement: Robert Wartenberg Lecture. Neurology. 1980b;30(12):1303–13.

Lance JW, De Gail P. Spread of phasic muscle reflexes in normal and spastic subjects. J Neurol Neurosurg Psychiatry. 1965;28(4):328–34.

Lemon RN. Descending pathways in motor control. Annu Rev Neurosci. 2008;31:195–218.

Lemon RN, Griffiths J. Comparing the function of the corticospinal system in different species: organizational differences for motor specialization? Muscle Nerve. 2005;32(3):261–79.

Li S, Francisco GE. New insights into the pathophysiology of post-stroke spasticity. Front Hum Neurosci. 2015;9:192.

Li S, Kamper DG, Rymer WZ. Effects of muscle length on cross-modal excitability modulation in human upper limb muscles. Exp Brain Res. 2006;33(2):183–90.

Li S, Francisco GE, Zhou P. Post-stroke hemiplegic gait: new perspective and insights. Front Physiol. 2018;9:1021.

Li S, Chen YT, Francisco GE, Zhou P, Rymer WZ. A unifying pathophysiological account for post-stroke spasticity and disordered motor control. Front Neurol. 2019;10:468.

Lieber RL, Fridén J. Muscle contracture and passive mechanics in cerebral palsy. J Appl Physiol. 2019;126(5):1492–501.

Lindsley DB, Schreiner LH, Magoun HW. An electromyographic study of spasticity. J Neurophysiol. 1949;12(3):197–205.

Lo SF, Chen SY, Lin HC, Jim YF, Meng NH, Kao MJ. Arthrographic and clinical findings in patients with hemiplegic shoulder pain. Arch Phys Med Rehabil. 2003;84(12):1786–91.

Lundberg A, Malmgren K, Schomburg ED. Reflex pathways from group II muscle afferents. Exp Brain Res. 1987;65(2):282–93.

Lundström E, Terént A, Borg J. Prevalence of disabling spasticity 1 year after first-ever stroke. Eur J Neurol. 2008;15(6):533–9.

Maharaj JN, Cresswell AG, Lichtwark GA. The mechanical function of the tibialis posterior muscle and its tendon during locomotion. J Biomech. 2016;49(14):3238–43.

Matsushita M, Hosoya Y. Spinocerebellar projections from the upper lumbar segments in the cat: an anterograde WGA-HRP study. J Comp Neurol. 1989;281(2):298–319.

Matsuyama K, Mori F, Nakajima K, Drew T, Aoki M, Mori S. Locomotor role of the corticoreticular-reticulospinal-spinal interneuronal system. Prog Brain Res. 2004;143:239–49.

Matthews PB. The differentiation of two types of fusimotor fibre by their effects on the dynamic response of muscle spindle primary endings. Q J Exp Physiol. 1962;47:324–33.

Matthews PB. Muscle spindles and their motor control. Physiol Rev. 1964;44(2):219–88.

Matthews PBC. Mammalian muscle receptors and their central actions. London: Edward Arnold; 1972.

Matthews PB. Historical analysis of the neural control of movement from the bedrock of animal experimentation to human studies. J Appl Physiol. 2004;96(4):1478–85.

Mazzocchio R, Rossi A. Recurrent inhibition in human spinal spasticity. Ital J Neurol Sci. 1989;10(3):337–47.

McPherson JG, Chen A, Ellis MD, Yao J, Heckman CJ, Dewald JPA. Progressive recruitment of contralesional cortico-reticulospinal pathways drives motor impairment post stroke. J Physiol. 2018;596(7):1211–25.

Meinders M, Price R, Lehmann JF, Questad KA. The stretch reflex response in the normal and spastic ankle: effect of ankle position. Arch Phys Med Rehabil. 1996;77(5):487–92.

Moore KL, Dalley AF, Agur AMR. Clinically oriented anatomy. 8th ed. Philadelphia: Wolters Kluwer; 2018.

Mukherjee A, Chakravarty A. Spasticity mechanisms—for the clinician. Front Neurol. 2010;1:149.

Murray KC, Nakae A, Stephens MJ. Recovery of motoneuron and locomotor function after spinal cord injury depends on constitutive activity in 5-HT2C receptors. Nat Med. 2010;16(6):694–700.

Nathan PW, Smith MC. Long descending tracts in man. I. Review of present knowledge. Brain. 1955;78(2):248–303.

Needle AR, Baumeister J, Kaminski TW, Higginson JS, Farquhar WB, Swanik CB. Neuromechanical coupling in the regulation of muscle tone and joint stiffness. Scand J Med Sci Sports. 2014a;24(5):737–48.

Needle AR, Lepley AS, Grooms DR. Central nervous system adaptation after ligamentous injury: a summary of theories, evidence, and clinical interpretation. Sports Med. 2014b;45(4):605–15.

Neumann DA. Kinesiology of the hip: a focus on muscular actions. J Orthop Sports Phys Ther. 2010;40(2):82–94.

Nielsen JB, Crone C, Hultborn H. The spinal pathophysiology of spasticity--from a basic science point of view. Acta Physiol. 2007;189(2):171–80.

Nuyens GE, De Weerdt WJ, Spaepen AJ Jr, Kiekens C, Feys HM. Reduction of spastic hypertonia during repeated passive knee movements in stroke patients. Arch Phys Med Rehabil. 2002;83(7):930–5.

Onodera S, Hicks TP. A comparative neuroanatomical study of the red nucleus of the cat, macaque and human. PLoS One. 2009;4(8):e6623.

O'Sullivan LW, Gallwey TJ. Forearm torque strengths and discomfort profiles in pronation and supination. Ergonomics. 2005;48(6):703–21.

Pandyan AD, Gregoric M, Barnes MP, Wood D, Van Wijck F, Burridge J, et al. Spasticity: clinical perceptions, neurological realities and meaningful measurement. Disabil Rehabil. 2005;27(1–2):2–6.

Peterson BW, Abzug C. Properties of projections from vestibular nuclei to medial reticular formation in the cat. J Neurophysiol. 1975;38(6):1421–35.

Pfaff DW, Volkow ND. Neuroscience in the 21st Century: from basic to clinic. 2nd ed. New York: Springer; 2016.

Phadke CP, Balasubramanian CK, Ismail F, Boulias C. Revisiting physiologic and psychologic triggers that increase spasticity. Am J Phys Med Rehabil. 2013;92(4):357–69.

Pierrot-Deseilligny E. Transmission of the cortical command for human voluntary movement through cervical propriospinal premotoneurons. Prog Neurobiol. 1996;48(4–5):489–517.

Pierrot-Deseilligny E. Propriospinal transmission of part of the corticospinal excitation in humans. Muscle Nerve. 2002;26(2):155–72.

Pierrot-Deseilligny E, Burke D. The circuitry of the human spinal cord: its role in motor control and movement disorders. Cambridge: Cambridge University Press; 2005.

Pompeiano O. Vestibulospinal relations: vestibular influences on gamma motoneurons and primary afferents. Prog Brain Res. 1972;37:197–232.
Porter R, Lemon RN. Corticospinal function and voluntary movement. Oxford: Clarendon Press; 1995.
Powers CM. The influence of abnormal hip mechanics on knee injury: a biomechanical perspective. J Orthop Sports Phys Ther. 2010;40(2):42–51.
Powers RK, Binder MD. Input-output functions of mammalian motoneurons. Rev Physiol Biochem Pharmacol. 2001;143:137–263.
Powers RK, Campbell DL, Rymer WZ. Stretch reflex dynamics in spastic elbow flexor muscles. Ann Neurol. 1989;25(1):32–42.
Prochazka A. Proprioception during voluntary movement. Can J Physiol Pharmacol. 1986;64(4):499–504.
Proske U. What is the role of muscle receptors in proprioception? Muscle Nerve. 2005;31(6):780–7.
Proske U, Gandevia SC. The proprioceptive senses: their roles in signaling body shape, body position and movement, and muscle force. Physiol Rev. 2012;92(4):1651–97.
Proske U, Gandevia SC. Kinesthetic senses. Compr Physiol. 2018;8(3):1157–83.
Proske U, Gregory JE. The discharge rate: tension relation of Golgi tendon organs. Neurosci Lett. 1980;16(3):287–90.
Proske U, Stuart GJ. The initial burst of human primary muscle spindle afferents. J Physiol. 1985;368:1–17.
Purves D, Augustine GJ, Fitzpatrick D, Hall WC, LaMantia AS, Mooney RD, et al. Neuroscience. 6th ed. New York: Oxford University Press; 2018.
Rack PM, Ross HF, Thilmann AF. The ankle stretch reflexes in normal and spastic subjects. Brain. 1984;107(2):637–54.
Raineteau O, Schwab ME. Plasticity of motor systems after incomplete spinal cord injury. Nat Rev Neurosci. 2001;2(4):263–73.
Ralston HJ 3rd. The fine structure of laminae I, II and III of the macaque spinal cord. J Comp Neurol. 1979;184(4):619–42.
Romanes GJ. The motor pools of the spinal cord. Prog Brain Res. 1964;11:93–119.
Rosenzweig ES, Courtine G, Jindrich DL, Brock JH, Ferguson AR, Strand SC, et al. Extensive spontaneous plasticity of corticospinal projections after primate spinal cord injury. Nat Neurosci. 2010;13(12):1505–10.
Rossignol S, Frigon A. Recovery of locomotion after spinal cord injury: some facts and mechanisms. Annu Rev Neurosci. 2011;34:413–40.
Rothwell J. Control of human voluntary movement. 2nd ed. London: Chapman & Hall; 1994.
Rothwell JC. The motor functions of the basal ganglia. J Integr Neurosci. 2011;10(3):303–15.
Rudomin P. In search of lost presynaptic inhibition. Exp Brain Res. 2009;196(1):139–51.
Rudomin P, Schmidt RF. Presynaptic inhibition in the vertebrate spinal cord revisited. Exp Brain Res. 1999;129(1):1–37.
Sakai ST, Davidson AG, Buford JA. Reticulospinal neurons in the pontomedullary reticular formation of the monkey (Macaca fascicularis). Neuroscience. 2009;163(4):1158–70.
Santello M, Baud-Bovy G, Jörntell H. Neural bases of hand synergies. Front Comput Neurosci. 2019;13:32.
Schmit BD, Dhaher Y, Dewald JP, Rymer WZ. Reflex torque response to movement of the spastic elbow: theoretical analyses and implications for quantification of spasticity. Ann Biomed Eng. 1999;27(6):815–29.
Schoultz TW, Swett JE. The fine structure of the Golgi tendon organ. J Neurocytol. 1972;1(1):1–26.
Schreiner LH, Lindsley DB, Magoun HW. Role of brain stem facilitatory systems in maintenance of spasticity. J Neurophysiol. 1949;12(3):207–16.
Sheean G. The pathophysiology of spasticity. Eur J Neurol. 2002;9(Suppl 1):3–9.
Siegel A, Sapru HN. Essential neuroscience. 4th ed. Philadelphia: Wolters Kluwer; 2019.
Sindou M, Georgoulis G, Mertens P. Neurosurgery for spasticity: a practical guide for treating children and adults. Wien: Springer; 2014.
Skoog B. A comparison of the effects of two antispastic drugs, tizanidine and baclofen, on synaptic transmission from muscle spindle afferents to spinal interneurones in cats. Acta Physiol Scand. 1996;156(1):81–90.
Sommerfeld DK, Gripenstedt U, Welmer AK. Spasticity after stroke: an overview of prevalence, test instruments, and treatments. Am J Phys Med Rehabil. 2012;91(9):814–20.
Spanne A, Jörntell H. Processing of multi-dimensional sensorimotor information in the spinal and cerebellar neuronal circuitry: a new perspective. PLoS Comput Biol. 2013;9(3):e1002979.
Standring S. Gray's anatomy: the anatomical basis of clinical practice. 42nd ed. London: Elsevier; 2020.
Stroyan M, Wilk KE. The functional anatomy of the elbow complex. J Orthop Sports Phys Ther. 1993;17(6):279–88.
Thilmann AF, Fellows SJ, Garms E. The mechanism of spastic muscle hypertonus. Variation in reflex gain over the time course of spasticity. Brain. 1991;114(1A):233–44.
Todd AJ. Neuronal circuitry for pain processing in the dorsal horn. Nat Rev Neurosci. 2010;11(12):823–36.
Todd AJ, Spike RC. The localization of classical transmitters and neuropeptides within neurons in laminae I-III of the mammalian spinal dorsal horn. Prog Neurobiol. 1993;41(5):609–45.
Trompetto C, Marinelli L, Mori L, Pelosin E, Currà A, Molfetta L, et al. Pathophysiology of spasticity: implications for neurorehabilitation. Biomed Res Int. 2014;2014:354906.
Truong D, Dressler D, Hallett M. Manual of Botulinum toxin therapy. Cambridge: Cambridge University Press; 2009.
Urban PP, Wolf T, Uebele M, Marx JJ, Vogt T, Stoeter P, et al. Occurrence and clinical predictors of spasticity after ischemic stroke. Stroke. 2010;41(9):2016–20.

van Oudenaarde E, Brandsma JW, Oostendorp RA. The influence of forearm, hand and thumb positions on extensor carpi ulnaris and abductor pollicis longus activity. Acta Anat (Basel). 1997;158(4):296–302.
Vanderah TW, Gould DJ. Nolte's the human brain: an introduction to its functional anatomy. 7th ed. Philadelphia: Elsevier; 2016.
Walsh EG. Clonus: beats provoked by the application of a rhythmic force. J Neurol Neurosurg Psychiatry. 1976;39(3):266–74.
Ward SR, Winters TM, Blemker SS. The architectural design of the gluteal muscle group: implications for movement and rehabilitation. J Orthop Sports Phys Ther. 2010;40(2):95–102.
Watson C, Paxinos G, Kayalioglu G. The spinal cord: a Christopher and Dana Reeve Foundation text and atlas. New York: Academic Press; 2009.
Watson C, Kirkcaldie M, Paxinos G. The brain: an introduction to functional neuroanatomy. 1st ed. Amsterdam: Elsevier; 2010.
Weidner N, Ner A, Salimi N, Tuszynski MH. Spontaneous corticospinal axonal plasticity and functional recovery after adult central nervous system injury. Proc Natl Acad Sci USA. 2001;98(6):3513–8.
Werner BC, Holzgrefe RE, Griffin JW, Lyons ML, Cosgrove CT, Hart JM, et al. Validation of an innovative method of shoulder range-of-motion measurement using a smartphone clinometer application. J Shoulder Elb Surg. 2014;23(11):e275–82.
Whitlock DG. Neuroanatomy of the spinal cord. In: Davidoff RA, editor. Handbook of the spinal cord. New York: Marcel Dekker; 1990. p. 1–40.
Willis WD, Coggeshall RE. Sensory mechanisms of the spinal cord: volume 1 primary afferent neurons and the spinal dorsal horn. 3rd ed. New York: Springer; 2004.
Wilson DR, Apreleva MV, Eichler MJ, Harrold FR. Accuracy and repeatability of a pressure measurement system in the patellofemoral joint. J Biomech. 2003;36(12):1909–15.
Windhorst U. Muscle proprioceptive feedback and spinal networks. Brain Res Bull. 2007;73(4–6):155–202.
Wissel J, Ward AB, Erztgaard P, Bensmail D, Hecht MJ, Lejeune TM, et al. European consensus table on the use of botulinum toxin type A in adult spasticity. J Rehabil Med. 2009;41(1):13–25.
Young RR, Wiegner AW. Spasticity. Clin Orthop Relat Res. 1987;219:50–62.
Zeilhofer HU, Wildner H, Yévenes GE. Fast synaptic inhibition in spinal sensory processing and pain control. Physiol Rev. 2012;92(1):193–235.

Recommended Additional Reading

Barnes MP, Johnson GR. Upper motor neurone syndrome and spasticity. 2nd ed. Cambridge: Cambridge University Press; 2008b.
Brashear A. Spasticity: diagnosis and management. 2nd ed. New York: Demos; 2016.
Gelber DA, Jeffery DR. Clinical evaluation and management of spasticity. Totowa: Humana Press; 2002.
Ko HY. Management and rehabilitation of spinal cord injuries. 2nd ed. Singapore: Springer; 2022b.
Thilmann AF, Burke DJ, Rymer WZ. Spasticity: mechanism and management. New York: Springer; 1993.

Index

H.-Y. Ko, *Practical Functional Anatomy for Spinal Cord Injury Rehabilitation*,
https://doi.org/10.1007/978-981-95-6457-6

B

C

D

E

G

H

J

K

L

M

N

Q

R

U

V

GPSR Compliance

The European Union's (EU) General Product Safety Regulation (GPSR) is a set of rules that requires consumer products to be safe and our obligations to ensure this.

If you have any concerns about our products, you can contact us on ProductSafety@springernature.com

In case Publisher is established outside the EU, the EU authorized representative is:

Springer Nature Customer Service Center GmbH
Europaplatz 3
69115 Heidelberg, Germany

Batch number: 10732805

Printed by Printforce, the Netherlands